NICOLAS BOURBAKI

ELEMENTS OF MATHEMATICS

Commutative Algebra

Chapters 1–7

Springer-Verlag
Berlin Heidelberg New York
London Paris Tokyo

Originally published as
ÉLÉMENTS DE MATHÉMATIQUE
ALGÈBRE COMMUTATIVE 1–4 et 5-7

Mathematics Subject Classification (1991): 13XX

Distribution rights worldwide:
Springer-Verlag Berlin Heidelberg New York London Paris Tokyo

ISBN 978-3-540-64239-8 Springer-Verlag Berlin Heidelberg New York
ISBN 978-3-540-19371-5 2nd printing Springer-Verlag Berlin Heidelberg New York

Softcover edition of the 2nd printing 1989

Library of Congress Cataloging-in-Publication Data
Bourbaki, Nicolas. [Algèbre commutative. English] Commutative algebra / Nicolas Bourbaki. p. cm.-
(Elements of mathematics) Translation of: Algèbre commutative.Bibliography: p. Includes index.
ISBN 0-387-19371-5 (U.S.)
1. Commutative algebra. I. Title. II. Series: Bourbaki, Nicolas. Éléments de mathématique. English.
QA251.3.B6813 1988 512'.24-dc 19 88-31138

SPIN 10670205
41/3143-5 4 3 2 1 0
Printed on acid-free paper

TO THE READER

1. This series of volumes, a list of which is given on pages ix and x, takes up mathematics at the beginning, and gives complete proofs. In principle, it requires no particular knowledge of mathematics on the reader's part, but only a certain familiarity with mathematical reasoning and a certain capacity for abstract thought. Nevertheless, it is directed especially to those who have a good knowledge of at least the content of the first year or two of a university mathematics course.

2. The method of exposition we have chosen is axiomatic and abstract, and normally proceeds from the general to the particular. This choice has been dictated by the main purpose of the treatise, which is to provide a solid foundation for the whole body of modern mathematics. For this it is indispensable to become familiar with a rather large number of very general ideas and principles. Moreover, the demands of proof impose a rigorously fixed order on the subject matter. It follows that the utility of certain considerations will not be immediately apparent to the reader unless he has already a fairly extended knowledge of mathematics; otherwise he must have the patience to suspend judgment until the occasion arises.

3. In order to mitigate this disadvantage we have frequently inserted examples in the text which refer to facts the reader may already know but which have not yet been discussed in the series. Such examples are always placed between two asterisks: * ... *. Most readers will undoubtedly find that these examples will help them to understand the text, and will prefer not to leave them out, even at a first reading. Their omission would of course have no disadvantage, from a purely logical point of view.

4. This series is divided into volumes (here called "Books"). The first six Books are numbered and, in general, every statement in the text assumes as known only those results which have already been discussed in the preceding

volumes. This rule holds good within each Book, but for convenience of exposition these Books are no longer arranged in a consecutive order. At the beginning of each of these Books (or of these chapters), the reader will find a precise indication of its logical relationship to the other Books and he will thus be able to satisfy himself of the absence of any vicious circle.

5. The logical framework of each chapter consists of the *definitions*, the *axioms*, and the *theorems* of the chapter. These are the parts that have mainly to be borne in mind for subsequent use. Less important results and those which can easily be deduced from the theorems are labelled as "propositions," "lemmas," "corollaries," "remarks," etc. Those which may be omitted at a first reading are printed in small type. A commentary on a particularly important theorem appears occasionally under the name of "scholium."

To avoid tedious repetitions it is sometimes convenient to introduce notations or abbreviations which are in force only within a certain chapter or a certain section of a chapter (for example, in a chapter which is concerned only with commutative rings, the word "ring" would always signify "commutative ring"). Such conventions are always explicitly mentioned, generally at the beginning of the chapter in which they occur.

6. Some passages in the text are designed to forewarn the reader against serious errors. These passages are signposted in the margin with the sign Z ("dangerous bend").

7. The Exercises are designed both to enable the reader to satisfy himself that he has digested the text and to bring to his notice results which have no place in the text but which are nonetheless of interest. The most difficult exercises bear the sign ¶.

8. In general, we have adhered to the commonly accepted terminology, except where there appeared to be good reasons for deviating from it.

9. We have made a particular effort always to use rigorously correct language, without sacrificing simplicity. As far as possible we have drawn attention in the text to *abuses of language*, without which any mathematical text runs the risk of pedantry, not to say unreadability.

10. Since in principle the text consists of the dogmatic exposition of a theory, it contains in general no references to the literature. Bibliographical references are gathered together in *Historical Notes*, usually at the end of each chapter. These notes also contain indications, where appropriate, of the unsolved problems of the theory.

The bibliography which follows each historical note contains in general only those books and original memoirs which have been of the greatest importance in the evolution of the theory under discussion. It makes no sort of pretence to completeness; in particular, references which serve only to determine questions of priority are almost always omitted.

As to the exercises, we have not thought it worthwhile in general to indicate their origins, since they have been taken from many different sources (original papers, textbooks, collections of exercises).

11. References to a part of this series are given as follows:

a) If reference is made to theorems, axioms, or definitions presented *in the same section*, they are quoted by their number.
b) If they occur *in another section of the same chapter*, this section is also quoted in the reference.
c) If they occur *in another chapter in the same Book*, the chapter and section are quoted.
d) If they occur *in another Book*, this Book is first quoted by its title.

The *Summaries of Results* are quoted by the letter R: thus *Set Theory*, R signifies "*Summary of Results of the Theory of Sets.*"

CONTENTS
OF
THE ELEMENTS OF MATHEMATICS SERIES

I. Theory of sets

1. Description of formal mathematics. 2. Theory of sets. 3. Ordered sets; cardinals; natural numbers. 4. Structures.

II. Algebra

1. Algebraic structures. 2. Linear algebra. 3. Tensor algebras, exterior algebras, symmetric algebras. 4. Polynomials and rational fractions. 5. Fields. 6. Ordered groups and fields. 7. Modules over principal ideal rings. 8. Semi-simple modules and rings. 9. Sesquilinear and quadratic forms.

III. General topology

1. Topological structures. 2. Uniform structures. 3. Topological groups. 4. Real numbers. 5. One-parameter groups. 6. Real number spaces, affine and projective spaces. 7. The additive groups $\mathbf{R}^n$. 8. Complex numbers. 9. Use of real numbers in general topology. 10. Function spaces.

IV. Functions of a real variable

1. Derivatives. 2. Primitives and integrals. 3. Elementary functions. 4. Differential equations. 5. Local study of functions. 6. Generalized Taylor expansions. The Euler-Maclaurin summation formula. 7. The gamma function. Dictionary.

V. Topological vector spaces

1. Topological vector spaces over a valued field. 2. Convex sets and locally convex spaces. 3. Spaces of continuous linear mappings. 4. Duality in topological vector spaces. 5. Hilbert spaces: elementary theory. Dictionary.

VI. Integration

1. Convexity inequalities. 2. Riesz spaces. 3. Measures on locally compact spaces. 4. Extension of a measure. L^p spaces. 5. Integration of measures. 6. Vectorial integration. 7. Haar measure. 8. Convolution and representation. 9. Integration on Hausdorff topological spaces.

CONTENTS OF THE ELEMENTS OF MATHEMATICS SERIES

Lie groups and lie algebras

1. Lie algebras. 2. Free Lie algebras. 3. Lie groups. 4. Coxeter groups and Tits systems. 5. Groups generated by reflections. 6. Root systems.

Commutative algebra

1. Flat modules. 2. Localization. 3. Graduations, filtrations, and topologies. 4. Associated prime ideals and primary decomposition. 5. Integers. 6. Valuations. 7. Divisors.

Spectral theories

1. Normed algebras. 2. Locally compact groups.

Differentiable and analytic manifolds

Summary of results.

CONTENTS

To the Reader v
Contents of the Elements of Mathematics Series ix

Introduction xix

Chapter I. Flat Modules 1

§ 1. Diagrams and exact sequences 1
1. Diagrams 1
2. Commutative diagrams 2
3. Exact sequences 3
4. The snake diagram 4

§ 2. Flat modules 9
1. Revision of tensor products 9
2. M-flat modules 10
3. Flat modules 12
4. Examples of flat modules 14
5. Flatness of quotient modules 15
6. Intersection properties 17
7. Tensor products of flat modules 19
8. Finitely presented modules 20
9. Extension of scalars in homomorphism modules 22
10. Extension of scalars: case of commutative rings 22
11. Interpretation of flatness in terms of relations 25

§ 3. Faithfully flat modules 27
1. Definition of faithfully flat modules 27
2. Tensor products of faithfully flat modules 30
3. Change of ring 31

4. Restriction of scalars 31
5. Faithfully flat rings 32
6. Faithfully flat rings and finiteness conditions 34
7. Linear equations over a faithfully flat ring 35

§ 4. Flat modules and "Tor" functors 37

Exercises for § 1 39

Exercises for § 2 41

Exercises for § 3 49

Exercises for § 4 50

CHAPTER II. LOCALIZATION 51
§ 1. Prime ideals 51
1. Definition of prime ideals 51
2. Relatively prime ideals 53

§ 2. Rings and modules of fractions 55
1. Definition of rings of fractions 55
2. Modules of fractions 60
3. Change of multiplicative subset 64
4. Properties of modules of fractions 67
5. Ideals in a ring of fractions 70
6. Nilradical and minimal prime ideals 73
7. Modules of fractions of tensor products and homomorphism modules 75
8. Application to algebras 77
9. Modules of fractions of graded modules 78

§ 3. Local rings. Passage from the local to the global 80
1. Local rings 80
2. Modules over a local ring 82
3. Passage from the local to the global 87
4. Localization of flatness 91
5. Semi-local rings 92

§ 4. Spectra of rings and supports of modules 94
1. Irreducible spaces 94
2. Noetherian topological spaces 97
3. The prime spectrum of a ring 98
4. The support of a module 104

§ 5. Finitely generated projective modules. Invertible fractional ideals 108
1. Localization with respect to an element 108
2. Local characterization of finitely generated projective modules 109
3. Ranks of projective modules 111
4. Projective modules of rank 1 114
5. Non-degenerate submodules 116
6. Invertible submodules 117
7. The group of classes of invertible modules 119

Exercises for § 1 121

Exercises for § 2 123

Exercises for § 3 136

Exercises for § 4 140

Exercises for § 5 146

CHAPTER III. GRADUATIONS, FILTRATIONS AND TOPOLOGIES 155
§ 1. Finitely generated graded algebras 155
1. Systems of generators of a commutative algebra 155
2. Criteria of finiteness for graded rings 156
3. Properties of the ring $A^{(d)}$ 157
4. Graded prime ideals 160

§ 2. General results on filtered rings and modules 162
1. Filtered rings and modules 162
2. The order function 165
3. The graded module associated with a filtered module 165
4. Homomorphisms compatible with filtrations 169
5. The topology defined by a filtration 170
6. Complete filtered modules 173
7. Linear compactness properties of complete filtered modules 176
8. The lift of homomorphisms of associated graded modules 177
9. The lift of families of elements of an associated graded module 179
10. Application: examples of Noetherian rings 183
11. Complete $\mathfrak{m}$-adic rings and inverse limits 185
12. The Hausdorff completion of a filtered module 187
13. The Hausdorff completion of a semi-local ring 192

§ 3. $\mathfrak{m}$-adic topologies on Noetherian rings 195
1. Good filtrations 195
2. $\mathfrak{m}$-adic topologies on Noetherian rings 199
3. Zariski rings 201
4. The Hausdorff completion of a Noetherian ring 202
5. The completion of a Zariski ring 206

§ 4. Lifting in complete rings 209
1. Strongly relatively prime polynomials 209
2. Restricted formal power series 212
3. Hensel's Lemma 215
4. Composition of systems of formal power series 218
5. Systems of equations in complete rings 220
6. Application to decompositions of rings 225

§ 5. Flatness properties of filtered modules 226
1. Ideally Hausdorff modules 226
2. Statement of the flatness criterion 227
3. Proof of the flatness criterion 228
4. Applications 230

Exercises for § 1 232

Exercises for § 2 233

Exercises for § 3 245

Exercises for § 4 255

Exercises for § 5 259

CHAPTER IV. ASSOCIATED PRIME IDEALS AND PRIMARY DECOMPOSITION 261

§ 1. Prime ideals associated with a module 261
1. Definition of associated prime ideals 261
2. Localization of associated prime ideals 263
3. Relations with the support 265
4. The case of finitely generated modules over a Noetherian ring 265

§ 2. Primary decomposition 267
1. Primary submodules 267
2. The existence of a primary decomposition 270
3. Uniqueness properties in the primary decomposition 270
4. The localization of a primary decomposition 272

5. Rings and modules of finite length 274
6. Primary decomposition and extension of scalars 279

§ 3. Primary decomposition in graded modules 283
1. Prime ideals associated with a graded module 283
2. Primary submodules corresponding to graded prime ideals 284
3. Primary decomposition in graded modules 285

Exercises for § 1 286

Exercises for § 2 290

Exercises for § 3 301

CHAPTER V. INTEGERS 303
§ 1. Notion of an integral element 303
1. Integral elements over a ring 303
2. The integral closure of a ring. Integrally closed domains 308
3. Examples of integrally closed domains 309
4. Completely integrally closed domains 312
5. The integral closure of a ring of fractions 314
6. Norms and traces of integers 316
7. Extension of scalars in an integrally closed algebra 318
8. Integers over a graded ring 320
9. Application: invariants of a group of automorphisms of an algebra 323

§ 2. The lift of prime ideals 325
1. The first existence theorem 325
2. Decomposition group and inertia group 330
3. Decomposition and inertia for integrally closed domains.. 337
4. The second existence theorem 343

§ 3. Finitely generated algebras over a field 344
1. The normalization lemma 344
2. The integral closure of a finitely generated algebra over a field 348
3. The Nullstellensatz 349
4. Jacobson rings 351

Exercises for § 1 355

Exercises for § 2 362

Exercises for § 3 370

CHAPTER VI. VALUATIONS 375
§ 1. Valuation rings 375
1. The relation of domination between local rings 375
2. Valuation rings 376
3. Characterization of integral elements 378
4. Examples of valuation rings 379
§ 2. Places 381
1. The notion of morphism for laws of composition not everywhere defined 381
2. Places 381
3. Places and valuation rings 383
4. Extension of places 384
5. Characterization of integral elements by means of places 385
§ 3. Valuations 385
1. Valuations on a ring 385
2. Valuations on a field 387
3. Translations 389
4. Examples of valuations 389
5. Ideals of a valuation ring 391
6. Discrete valuations 392
§ 4. The height of a valuation 393
1. Inclusion of valuation rings of the same field 393
2. Isolated subgroups of an ordered group 394
3. Comparison of valuations 395
4. The height of a valuation 396
5. Valuations of height 1 397
§ 5. The topology defined by a valuation 399
1. The topology defined by a valuation 399
2. Topological vector spaces over a field with a valuation 401
3. The completion of a field with a valuation 402
§ 6. Absolute values 403
1. Preliminaries on absolute values 403
2. Ultrametric absolute values 405
3. Absolute values on **Q** 406
4. Structure of fields with a non-ultrametric absolute value 407
§ 7. Approximation theorem 412
1. The intersection of a finite number of valuation rings 412
2. Independent valuations 413
3. The case of absolute values 415

§ 8. Extensions of a valuation to an algebraic extension 416
1. Ramification index. Residue class degree 416
2. Extension of a valuation and completion 418
3. The relation $\sum_i e_i f_i \leqslant n$ 420
4. Initial ramification index 422
5. The relation $\sum_i e_i f_i = n$ 423
6. Valuation rings in an algebraic extension 427
7. The extension of absolute values 428

§ 9. Application: locally compact fields 431
1. The modulus function on a locally compact field 431
2. Existence of representatives 432
3. Structure of locally compact fields 433

§ 10. Extensions of a valuation to a transcendental extension 434
1. The case of a monogenous transcendental extension 434
2. The rational rank of commutative group 437
3. The case of any transcendental extension 438

Exercises for § 1 441

Exercises for § 2 444

Exercises for § 3 446

Exercises for § 4 449

Exercises for § 5 454

Exercises for § 6 459

Exercises for § 7 460

Exercises for § 8 461

Exercises for § 9 470

Exercises for § 10 471

CHAPTER VII. DIVISORS 475

§ 1. Krull domains 475
1. Divisorial ideals of an integral domain 475
2. The monoid structure on D(A) 478
3. Krull domains 480

4. Essential valuations of a Krull domain 482
5. Approximation for essential valuations 484
6. Prime ideals of height 1 in a Krull domain 485
7. Application: new characterizations of discrete valuation rings 487
8. The integral closure of a Krull domain in a finite extension of its field of fractions 487
9. Polynomial rings over a Krull domain 488
10. Divisor classes in Krull domains 489

§ 2. Dedekind domains 493

1. Definition of Dedekind domains 493
2. Characterizations of Dedekind domains 494
3. Decomposition of ideals into products of prime ideals 496
4. The approximation theorem for Dedekind domains 497
5. The Krull-Akizuki Theorem 499

§ 3. Factorial domains 502

1. Definition of factorial domains 502
2. Characterizations of factorial domains 502
3. Decomposition into extremal elements 504
4. Rings of fractions of a factorial domain 505
5. Polynomial rings over a factorial domain 505
6. Factorial domains and Zariski rings 506
7. Preliminaries on automorphisms of rings of formal power series 506
8. The preparation theorem 507
9. Factoriality of rings of formal power series 511

§ 4. Modules over integrally closed Noetherian domains 512

1. Lattices 512
2. Duality; reflexive modules 517
3. Local construction of reflexive modules 521
4. Pseudo-isomorphisms 523
5. Divisors attached to torsion modules 527
6. Relative invariant of two lattices 529
7. Divisor classes attached to finitely generated modules 531
8. Properties relative to finite extensions of the ring of scalars 535
9. A reduction theorem 540
10. Modules over Dedekind domains 543

Exercises for § 1 545

Exercises for § 2 556

Exercises for § 3 563
Exercises for § 4 571
Historical note (Chapters I to VII) 579
Bibliography 603
Index of notation 607
Index of terminology 610
Table of implications 621
Table of invariances — I 622
Table of invariances — II 624
Invariances under completion 625

INTRODUCTION

The questions treated in this Book arose during the development of the theory of algebraic numbers and (later) algebraic geometry (cf. the Historical Note). From the 19th century onwards these two theories began to show remarkable analogies; the attempt to solve the problems they posed led to the isolation of a number of general ideas whose field of application is not limited to rings of algebraic numbers or algebraic functions; and, as always, it is advantageous to consider these in their most general form in order to see their true significance and the repercussions of one study on another. The concepts treated in this Book can be applied in principle to all commutative rings and modules over such rings; it must however be pointed out that substantial results are often obtained only under certain hypotheses of *finiteness* (which always hold in the classical cases), for example by assuming the modules to be finitely generated or the rings to be Noetherian.

The chief notions central to the first chapters are the following:

I. *Localization and globalization.* Let us begin for example with a system of Diophantine equations:

$$(*) \qquad P_i(x_1, \ldots, x_m) = 0 \qquad (1 \leqslant i \leqslant n)$$

where the P_i are polynomials with integer coefficients and solutions (x_i) are sought consisting of rational *integers*. It is possible to start approaching the problem by looking for solutions consisting of *rational numbers*, which involves looking at the same problem with the coefficients of the P_i considered as elements of the *field of fractions* $\mathbf{Q}$ of $\mathbf{Z}$ and the solutions sought with values in $\mathbf{Q}$. A second step consists of seeing whether, given a prime number p, there exist rational solutions whose denominators are not divisible by p (*integer* solutions clearly satisfy this condition); this amounts, in this case, to lying in the subring $\mathbf{Z}_{(p)}$ of $\mathbf{Q}$ consisting of the rational numbers of this form, called the *local ring* of $\mathbf{Z}$ corresponding to the prime number p. Clearly the passage from $\mathbf{Z}$ to $\mathbf{Q}$ and that from $\mathbf{Z}$ to $\mathbf{Z}_{(p)}$ are of the same form: in the two cases, the only denominators allowed do not belong to a certain *prime ideal* (the ideal (0) and the ideal (p) respectively). The same name "local ring" arises in algebraic geometry, where this notion appears in a more natural way: for example for the ring $\mathbf{C}[X]$ of polynomials in one variable with complex coefficients, the local ring corresponding to the prime ideal $(X - \alpha)$ is the ring of rational fractions "regular" at the point α (that is, without a pole at that point).

Every Diophantine problem and more generally every problem on A-modules (A a commutative ring) can be decomposed into two subsidiary problems: its solution is sought in the local rings $A_\mathfrak{p}$ corresponding to the different prime ideals $\mathfrak{p}$ of A ("localization"), then the question is asked whether it is possible to conclude from the existence for all $\mathfrak{p}$ of a solution to the "localized" problem that a solution exists to the problem posed initially ("passage from the local to the global"). Chapter II is devoted to the study of this double process and it is also seen that "localization" is not related only to prime ideals, but has a wider range.

II. *Completion of local rings.* A local ring A shares with fields the property of having *only one* maximal ideal $\mathfrak{m}$. This fact is used to transform, to a certain extent, a problem on A-modules into an analogous problem on *vector spaces* by passing to the quotient ring $A/\mathfrak{m}$, as this latter is a field. If we return for example to the Diophantine system (*) this idea is none other than the principle of "reduction modulo p", transforming the equations into congruences mod. p, which occurred naturally beginning with the very first works in the theory of numbers.

This being so, we clearly cannot hope in this way to obtain complete results for the original problem and it was quickly realized that to obtain more precise information it was necessary to consider, not only congruences modulo $\mathfrak{m}$, but also "higher" congruences modulo $\mathfrak{m}^n$, for arbitrary integers $n > 0$. It is thus found that, the larger n, the closer in some way the original problem is "approached" (in the case $A = \mathbf{Z}$ for example, the reason is that an integer $\neq 0$ cannot be divisible by *all* the powers p^n of a given prime number p; this number will therefore make its presence felt in the reduction mod. p^n provided n is taken large enough). The mathematical translation of this idea consists of considering on A a ring *topology* (cf. *General Topology*, Chapter III, § 6) in which the $\mathfrak{m}^n$ form a fundamental system of neighbourhoods of 0. But when we have for example, solved the system of congruences

$$(**) \qquad P_i(x_1, \ldots, x_m) \equiv 0 \pmod{p^k} \qquad (1 \leqslant i \leqslant n)$$

for *every integer* $k > 0$, it still does not follow that the system (*) has a solution in the local ring $\mathbf{Z}_{(p)}$; the above hypothesis can be interpreted as saying that (*) admits a solution in the *completion* $\hat{\mathbf{Z}}_{(p)}$ of the topological ring $\mathbf{Z}_{(p)}$.

The original problem, thus weakened, is finally transformed into the analogous problem for local rings of the type $A/\mathfrak{m}^n$, which are also nearer to fields than general rings, since they have a nilpotent radical; in classical algebraic geometry this corresponds to a "differential" study of the problem in the neighbourhood of a given point.

Chapter III deals in a general way with these applications of topological notions to the theory of local rings. In Chapter VI a special aspect of this is studied, adapted on the one hand to more detailed studies of algebraic geometry, and above all to the arithmetic of algebraic number fields, where the local rings encountered (such as $\mathbf{Z}_{(p)}$) belong to a particularly simple class, that of "valuation rings", where divisibility is a *total* ordering (cf. *Algebra*, Chapter VI, § 1) of the set of principal ideals.

The study of the passage from a ring A to a local ring $A_{\mathfrak{p}}$ or to a completion $\hat{A}$ brings to light a feature common to these two operations, the property of *flatness* of the A-modules $A_{\mathfrak{p}}$ and $\hat{A}$, which allows amongst other things the use of tensor products of such A-modules with arbitrary A-modules somewhat similar to that of tensor products of vector spaces, that is, without all the precautions surrounding their use in the general case. The properties associated with this notion, which are also applicable to modules over non-commutative rings, are the object of study in Chapter I.

III. *Integers and decomposition of ideals.* The study of divisibility in algebraic number fields necessitated from the start the introduction of the notion of *integer* in such a field K, generalizing the notion of rational integer in the field $\mathbf{Q}$. The general theory of this notion of "algebraic integer", linked, as will be seen, to very strict conditions of finiteness, is developed in Chapter V; it can be applied to *all* commutative rings and is of great interest not only in arithmetic, but in algebraic geometry and even in the modern theory of "analytic spaces" over the field $\mathbf{C}$.

One of the major obstacles to the extension of classical arithmetic to rings of algebraic integers has long been that the classical decomposition of a rational integer into prime factors does not extend in general to these rings. The creation of the theory of ideals was necessary to surmount this difficulty: the unique decomposition sought is then established for ideals, the notion of prime ideal being substituted for that of prime number. Moreover this result can be considered as a typical case where the "passage from the local to the global" is performed satisfactorily: the knowledge, for $x \in K$, of the values at x of *all* the "valuations" on K determines x up to multiplication by an invertible integer.

In less simple rings than rings of algebraic integers (and even for example in rings of polynomials in several indeterminates) this result is no longer valid. However it is possible to associate in a canonical way with every ideal a well-determined set of prime ideals: in algebraic geometry, if we consider for example in K^n (K any commutative field) a subvariety defined by a system of polynomial equations $P_\alpha = 0$, the *irreducible* components of this subvariety correspond bijectively with the minimal elements of the set of prime ideals thus associated with the ideal generated by the P_α. It is moreover possible (if we restrict ourselves to Noetherian rings) to give for every ideal a "decomposition" less precise than a decomposition as a product of prime ideals: here the product is in fact replaced by the intersection and the powers of prime ideals by "primary" ideals connected with the prime ideals associated with the ideal in question (but which are not direct generalizations of powers of prime ideals). The introduction of prime ideals associated with an ideal and the study of their properties is the subject of Chapter IV; here also the existence and certain uniqueness properties of the "primary decompositions" to which we have just alluded are proved; but it seems at present that these decompositions usually only play an accessory role in applications, the essential notion being that of prime ideal associated with an ideal.

In Chapter VII we examine in more detail rings whose properties most nearly approach those of rings of algebraic integers as far as decomposition as a product of prime ideals is concerned; amongst other things it is possible to introduce into these rings the notion of "*divisor*", which is the geometric aspect of this decomposition and plays an important role in algebraic geometry.

Finally Chapters VIII *et seq.* will deal with notions of more interest in algebraic geometry than in arithmetic (where they become trivial) and notably the concept of *dimension.*

With these notions we come to the frontier of algebraic geometry proper, a frontier which is ever moving and difficult to trace. For, if commutative algebra is an essential tool for the development of algebraic geometry in all its generality, conversely (as has already been seen above) the language of geometry proves very convenient for expressing the theorems of commutative algebra and suggesting a certain intuition naturally enough absent from abstract algebra; with the tendency to enlarge more and more the limits of algebraic geometry, algebraic and geometric language tend more than ever to merge.

CHAPTER I(*)

Flat Modules

Unless otherwise stated, all the rings considered in this chapter are assumed to have a unit element; all the ring homomorphisms are assumed to map unit element to unit element. A subring of a ring A *means a subring containing the unit element of* A.

If A *is a ring,* M *a left* A*-module and* U (resp. V) *an additive subgroup of* A (resp. M), *recall that* UV *or* U . V *is used to denote the additive subgroup of* M *generated by the products* uv, *where* $u \in U$, $v \in V$ (*Algebra*, Chapter VIII, § 6, no. 1). *If* $\mathfrak{a}$ *is an ideal of* A, *we write* $\mathfrak{a}^0 = A$. *For any set* E, 1_E (*or* 1 *if no confusion arises*) *is used to denote the identity mapping of* E *to itself.*

Recall that the axioms for a module imply that if E *is a left* (resp. *right*) *module over a ring* A *and* 1 *denotes the unit element of* A, *then* $1 . x = x$ (resp. $x . 1 = x$) *for all* $x \in E$ (*Algebra*, Chapter II, § 1, no. 1). *If* E *and* F *are two left* (resp. *right*) A*-modules, recall that* $\mathrm{Hom}_A(E, F)$ (or simply Hom(E, F)) *is used to denote the additive group of homomorphisms of* E *to* F (*loc. cit.*, § 1, no. 2). *By an abuse of notation* 0 *will often be used to denote a module reduced to its identity element.*

1. DIAGRAMS AND EXACT SEQUENCES

1. DIAGRAMS

Let, for example, A, B, C, D, E be five sets and let f be a mapping from A to B, g a mapping from B to C, h a mapping from D to E, u a mapping from B to D and v a mapping from C to E. To summarize such a situation we often make use of diagrams; for example, the above situation is summarized by the following

(*) With the exception of § 4, the results of this chapter depend only on Books I to VI.

diagram (*Set Theory*, Chapter II, § 3, no. 4):

(1)
$$\begin{array}{ccccc} A & \xrightarrow{f} & B & \xrightarrow{g} & C \\ & & {\scriptstyle u}\downarrow & & \downarrow{\scriptstyle v} \\ & & D & \xrightarrow[h]{} & E \end{array}$$

In such a diagram the group of symbols $A \xrightarrow{f} B$ expresses schematically the fact that f is a mapping from A to B. Where there is no ambiguity about f, we suppress the letter f and write simply $A \rightarrow B$.

When A, B, C, D, E are groups (resp. commutative groups) and f, g, h, u, v group homomorphisms, diagram (1) is called, by way of abbreviation, a *diagram of groups* (resp. *commutative groups*).

In principle, a diagram is not a mathematical object, but only a *figure* designed to facilitate reading an argument. In practice, diagrams are often used as *abbreviatory symbols* to avoid naming all the sets and mappings under consideration; we therefore say "Consider the diagram (1)" instead of saying "Let A, B, C, D, E be five sets . . . and v a mapping from C to E"; see for example the statement of Proposition 2 in no. 4.

2. COMMUTATIVE DIAGRAMS

Consider, for example, the following diagram:

(2)
$$\begin{array}{ccccccc} A & \xrightarrow{f} & B & \xrightarrow{g} & C & \xrightarrow{h} & D \\ {\scriptstyle a}\downarrow & & {\scriptstyle b}\downarrow & & {\scriptstyle c}\downarrow & & {\scriptstyle d}\downarrow \\ A' & \xrightarrow[f']{} & B' & \xrightarrow[g']{} & C' & \xrightarrow[h']{} & D' \end{array}$$

To every path composed of a certain number of segments of the diagram traversed in the direction shown by the arrows corresponds a mapping of the set represented by the beginning of the first segment to the set represented by the end of the last segment, namely the composition of the mappings represented by the various segments traversed. For each vertex of the diagram, for example B, by way of convention there is a path reduced to B, to which corresponds the identity mapping 1_B.

In (2) there are for example three paths beginning at A and ending at C′; the corresponding mappings are $c \circ g \circ f$, $g' \circ b \circ f$ and $g' \circ f' \circ a$. A diagram is said to be *commutative* if, for every pair of paths in the diagram with the same beginning and end, the two corresponding mappings are equal; in particular if the beginning and end of a path coincide the corresponding mapping must be the identity.

For the diagram (2) to be commutative it is necessary and sufficient that the relations

$$(3) \qquad f' \circ a = b \circ f, \qquad g' \circ b = c \circ g, \qquad h' \circ f = d \circ h;$$

hold; in other words, it is necessary and sufficient that the three square diagrams contained in (2) be commutative. For the relations (3) imply $c \circ g \circ f = g' \circ b \circ f$ since $c \circ g = g' \circ b$, and $g' \circ b \circ f = g' \circ f' \circ a$ since $b \circ f = f' \circ a$; thus the three paths beginning at A and ending at C′ give the same mapping. It can be similarly verified that the four paths beginning at A and ending at D′ (resp. the three paths beginning at B and ending at D′) give the same mapping. The relations (3) signify that the two paths beginning at A (resp. B, C) and ending at B′ (resp. C′, D′) give the same mapping. None of the other pairs of vertices of (2) can be joined by more than one path and the diagram (2) is therefore commutative.

In what follows we shall leave the reader to verify analogous results for other types of diagrams.

3. EXACT SEQUENCES

Recall the following definition (*Algebra*, Chapter II, § 1, no. 4):

DEFINITION 1. *Let* A *be a ring*, F, G, H *three right* (resp. *left*) A-*modules*, f *a homomorphism of* F *to* G *and* g *a homomorphism of* G *to* H. *The ordered pair* (f, g) *is called an exact sequence if* $\overset{-1}{g}(0) = f(\mathrm{F})$, *that is if the kernel of* g *is equal to the image of* f.

The diagram

$$(4) \qquad \mathrm{F} \xrightarrow{f} \mathrm{G} \xrightarrow{g} \mathrm{H}$$

is also called an *exact sequence.*

Consider similarly a diagram consisting of four modules and three homomorphisms:

$$(5) \qquad \mathrm{E} \xrightarrow{f} \mathrm{F} \xrightarrow{g} \mathrm{G} \xrightarrow{h} \mathrm{H}.$$

This diagram is said to be *exact at* F if the diagram $\mathrm{E} \xrightarrow{f} \mathrm{F} \xrightarrow{g} \mathrm{G}$ is an exact sequence; it is said to be *exact at* G if $\mathrm{F} \xrightarrow{g} \mathrm{G} \xrightarrow{h} \mathrm{H}$ is an exact sequence. If (5) is exact at F *and* G, it is said to be *exact* or also an *exact sequence.* Exact sequences with any number of terms are similarly defined.

Recall the following results (*loc. cit.*), where E, F, G denote right (resp. left)

A-modules, the arrows represent homomorphisms and 0 denotes a module reduced to its identity element:

(a) To say that $0 \to E \xrightarrow{f} F$ is an exact sequence is equivalent to saying that f is *injective.*

(b) To say that $E \xrightarrow{f} F \to 0$ is an exact sequence is equivalent to saying that f is *surjective.*

(c) To say that $0 \to E \xrightarrow{f} F \to 0$ is an exact sequence is equivalent to saying that f is *bijective,* that is that f is an *isomorphism* of E onto F.

(d) If F is a submodule of E and i denotes the canonical injection of F into E and p the canonical surjection of E onto E/F, the diagram

$$0 \longrightarrow F \xrightarrow{i} E \xrightarrow{p} E/F \longrightarrow 0 \tag{6}$$

is an exact sequence.

(e) If f: $E \to F$ is a homomorphism, the diagram

$$0 \longrightarrow \overset{-1}{f}(0) \xrightarrow{i} E \xrightarrow{f} F \xrightarrow{p} F/f(E) \longrightarrow 0 \tag{7}$$

(where i is the canonical injection of $\overset{-1}{f}(0)$ into E and p the canonical projection of $F/f(E)$) is an exact sequence.

(f) For the diagram

$$E \xrightarrow{f} F \xrightarrow{g} G \tag{8}$$

to be an exact sequence, it is necessary and sufficient that there exist modules S, T and homomorphisms a: $E \to S$, b: $S \to F$, c: $F \to T$ and d: $T \to G$ such that $f = b \circ a$, $g = d \circ c$ and the three sequences

$$\begin{gathered} E \xrightarrow{a} S \longrightarrow 0 \\ 0 \longrightarrow S \xrightarrow{b} F \xrightarrow{c} T \longrightarrow 0 \\ 0 \longrightarrow T \xrightarrow{d} G \end{gathered} \tag{9}$$

be *exact.*

Recall finally that if f: $E \to F$ is an A-module homomorphism, we set $\mathrm{Ker}(f) = \overset{-1}{f}(0)$, $\mathrm{Im}(f) = f(E)$, $\mathrm{Coim}(f) = E/\overset{-1}{f}(0)$ and $\mathrm{Coker}(f) = F/f(E)$. With this notation it is possible to take, in (9), $S = \mathrm{Im}(f) = \mathrm{Ker}(g)$ and $T = \mathrm{Im}(g)$ (canonically isomorphic to $\mathrm{Coker}(f)$).

4. THE SNAKE DIAGRAM

PROPOSITION 1. *Consider a commutative diagram of commutative groups:*

$$\begin{array}{ccccc} A & \xrightarrow{u} & B & \xrightarrow{v} & C \\ {\scriptstyle a}\downarrow & & {\scriptstyle b}\downarrow & & {\scriptstyle c}\downarrow \\ A' & \xrightarrow[u']{} & B' & \xrightarrow[v']{} & C' \end{array} \tag{10}$$

Suppose that the two rows of (10) *are exact. Then:*

(i) *If c is injective, we have*

(11) $$\mathrm{Im}(b) \cap \mathrm{Im}(u') = \mathrm{Im}(u' \circ a) = \mathrm{Im}(b \circ u).$$

(ii) *If a is surjective, we have*

(12) $$\mathrm{Ker}(b) + \mathrm{Im}(u) = \mathrm{Ker}(v' \circ b) = \mathrm{Ker}(c \circ v).$$

Let us prove (i). Clearly

$$\mathrm{Im}(u' \circ a) = \mathrm{Im}(b \circ u) \subset \mathrm{Im}(b) \cap \mathrm{Im}(u').$$

Conversely, let $x \in \mathrm{Im}(b) \cap \mathrm{Im}(u')$. There exists $y \in \mathrm{B}$ such that $x = b(y)$. As $v' \circ u' = 0$, we have $0 = v'(x) = v'(b(y)) = c(v(y))$, whence $v(y) = 0$ since c is injective. As (u, v) is an exact sequence, there exists $z \in \mathrm{A}$ such that $y = u(z)$, whence $x = b(u(z))$.

Let us prove (ii). As $v \circ u = 0$ and $v' \circ u' = 0$, it is clear that

$$\mathrm{Ker}(b) + \mathrm{Im}(u) \subset \mathrm{Ker}(v' \circ b) = \mathrm{Ker}(c \circ v).$$

Conversely, let $x \in \mathrm{Ker}(v' \circ b)$. Then $b(x) \in \mathrm{Ker}(v')$ and there exists $y' \in \mathrm{A}'$ such that $u'(y') = b(x)$, since the sequence (u', v') is exact. As a is surjective, there exists $y \in \mathrm{A}$ such that $a(y) = y'$, whence $b(x) = u'(a(y)) = b(u(y))$; it follows that $x - u(y) \in \mathrm{Ker}(b)$, which completes the proof.

Lemma 1. *Consider a commutative diagram of commutative groups:*

(13) $$\begin{array}{ccc} \mathrm{A} & \xrightarrow{u} & \mathrm{B} \\ {\scriptstyle a}\downarrow & & \downarrow{\scriptstyle b} \\ \mathrm{A}' & \xrightarrow[u']{} & \mathrm{B}' \end{array}$$

Then there exists one and only one homomorphism u_1: $\mathrm{Ker}(a) \to \mathrm{Ker}(b)$ *and one and only one homomorphism u_2:* $\mathrm{Coker}(a) \to \mathrm{Coker(b)}$ *such that the diagrams*

(14) $$\begin{array}{ccc} \mathrm{Ker}(a) & \xrightarrow{u_1} & \mathrm{Ker}(b) \\ {\scriptstyle i}\downarrow & & \downarrow{\scriptstyle j} \\ \mathrm{A} & \xrightarrow[u]{} & \mathrm{B} \end{array}$$

and

(15) $$\begin{array}{ccc} \mathrm{A}' & \xrightarrow{u'} & \mathrm{B}' \\ {\scriptstyle p}\downarrow & & \downarrow{\scriptstyle q} \\ \mathrm{Coker}(a) & \xrightarrow[u_2]{} & \mathrm{Coker}(b) \end{array}$$

are commutative, i and j denoting the canonical injections and p and q the canonical surjections.

If $x \in \mathrm{Ker}(a)$, then $a(x) = 0$ and $b(u(x)) = u'(a(x)) = 0$, hence $u(x) \in \mathrm{Ker}(b)$, and the existence and uniqueness of u_1 are then immediate. Similarly, we have $u'(a(A)) = b(u(A)) \subset b(B)$, then by taking quotients u' gives a homomorphism u_2: $\mathrm{Coker}(a) \to \mathrm{Coker}(b)$, which is the only homomorphism for which (15) is commutative.

Let us now start with the *commutative* diagram (10) of commutative groups; there corresponds to it by Lemma 1 a diagram

(16)

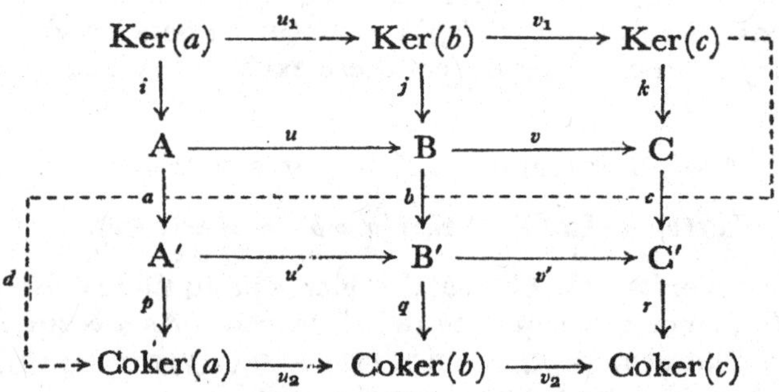

where i, j, k are the canonical injections, p, q, r the canonical surjections and u_1, u_2 (resp. v_1, v_2) the homomorphisms canonically associated with u, u' (resp. v, v') by Lemma 1. It is immediately verified that this diagram is commutative.

PROPOSITION 2. *Suppose that in the commutative diagram* (10) *the sequences* (u, v) *and* (u', v') *are exact. Then:*

(i) $v_1 \circ u_1 = 0$; *if u' is injective, the sequence* (u_1, v_1) *is exact.*
(ii) $v_2 \circ u_2 = 0$; *if v is surjective, the sequence* (u_2, v_2) *is exact.*
(iii) *Suppose that u' is injective and v is surjective. Then there exists one and only one homomorphism* d: $\mathrm{Ker}(c) \to \mathrm{Coker}(a)$ *with the following property: if* $x \in \mathrm{Ker}(c)$, $y \in B$ *and* $t' \in A'$ *satisfy the relations* $v(y) = k(x)$ *and* $u'(t') = b(y)$, *then* $d(x) = p(t')$. *Moreover the sequence*

$$(*) \quad \mathrm{Ker}(a) \xrightarrow{u_1} \mathrm{Ker}(b) \xrightarrow{v_1} \mathrm{Ker}(c) \xrightarrow{d} \mathrm{Coker}(a) \xrightarrow{u_2} \mathrm{Coker}(b) \xrightarrow{v_2} \mathrm{Coker}(c)$$

is exact.

Let us prove (i). As u_1 and v_1 have the same graphs as the restrictions of u and v to Ker(a) and Ker(b) respectively, we have $v_1 \circ u_1 = 0$. Then

$$\mathrm{Ker}(v_1) = \mathrm{Ker}(b) \cap \mathrm{Ker}(v) = \mathrm{Ker}(b) \cap \mathrm{Im}(u) = \mathrm{Im}(j) \cap \mathrm{Im}(u).$$

But, by Proposition 1, (i), $\mathrm{Ker}(v_1) = \mathrm{Im}(j \circ u_1) = \mathrm{Im}(u_1)$ if u' is injective.

Let us prove (ii). As u_2 and v_2 are obtained from u and v by taking quotients, it is clear that $v_2 \circ u_2 = 0$. Suppose that v is surjective; as q and p are surjective, it follows from the hypotheses and Proposition 1, (ii) that

$$\begin{aligned}\mathrm{Ker}(v_2) &= q(\mathrm{Ker}(v_2 \circ q)) = q(\mathrm{Ker}(v') + \mathrm{Im}(b)) = q(\mathrm{Ker}(v')) = q(\mathrm{Im}(u')) \\ &= \mathrm{Im}(q \circ u') = \mathrm{Im}(u_2 \circ p) = \mathrm{Im}(u_2).\end{aligned}$$

Finally let us prove (iii). For $x \in \mathrm{Ker}(c)$ there exists $y \in \mathrm{B}$ such that $v(y) = k(x)$ since v is surjective; moreover $v'(b(y)) = c(k(x)) = 0$ and consequently there exists a *unique* $t' \in \mathrm{A}'$ such that $u'(t') = b(y)$, since u' is injective. We now show that the element $p(t') \in \mathrm{Coker}(a)$ is *independent* of the element $y \in \mathrm{B}$ such that $v(y) = k(x)$. For if $y' \in \mathrm{B}$ is another element such that $v(y') = k(x)$, then $y' = y + u(z)$, where $z \in \mathrm{A}$; we show that if $t'' \in \mathrm{A}'$ is such that $u'(t'') = b(y')$ then $t'' = t' + a(z)$; for

$$u'(t' + a(z)) = u'(t') + u'(a(z)) = b(y) + b(u(z)) = b(y + u(z)) = b(y').$$

Finally, it follows that $p(t'') = p(t') + p(a(z)) = p(t')$. Then it is possible to set $d(x) = p(t')$ and the mapping $d\colon \mathrm{Ker}(c) \to \mathrm{Coker}(a)$ has thus been defined.

If now x_1, x_2 are elements of Ker(c) and $x = x_1 + x_2$, we take elements y_1, y_2 of B such that $v(y_1) = k(x_1)$ and $v(y_2) = k(x_2)$ and choose for $y \in \mathrm{B}$ the element $y_1 + y_2$; it is then immediate that $d(x) = d(x_1) + d(x_2)$ and hence d is a *homomorphism.*

Suppose that $x = v_1(x')$ for some $x' \in \mathrm{Ker}(b)$; then take $y \in \mathrm{B}$ to be the element $j(x')$. As $b(j(x')) = 0$, it follows that $d(x) = 0$, hence $d \circ v_1 = 0$. Conversely, suppose that $d(x) = 0$. In the above notation we then have $t' = a(s)$, where $s \in \mathrm{A}$. In this case we have $b(y) = u'(t') = u'(a(s)) = b(u(s))$, or $b(y - u(s)) = 0$. The element $y - u(s)$ is therefore of the form $j(n)$, where $n \in \mathrm{Ker}(b)$, and we have $k(x) = v(y) = v(u(s) + j(n)) = v(j(n)) = k(v_1(n))$; as k is injective, $x = v_1(n)$, which proves that the sequence (*) is exact at Ker(c).

Finally, we have (always in the same notation)

$$u_2(d(x)) = u_2(p(t')) = q(u'(t')) = q(b(y)) = 0$$

and hence $u_2 \circ d = 0$. Conversely, suppose that an element $w = p(t')$ of Coker(a) is such that $u_2(w) = u_2(p(t')) = 0$ (where $t' \in \mathrm{A}'$). Then $q(u'(t')) = 0$ and consequently $u'(t') = b(y)$ for some $y \in \mathrm{B}$; as $v'(u'(t')) = 0$, we have $v'(b(y)) = 0$, hence $c(v(y)) = 0$, in other words $v(y) = k(x)$ for some $x \in \mathrm{Ker}(c)$, and by definition $w = d(x)$, which shows that the sequence (*) is exact at

Coker(a). It has been seen in (i) that it is exact at Ker(b) and in (ii) it is exact at Coker(b), which completes the proof of (iii).

Remark. If the groups of the diagram (10) are all (for example, right) modules over a ring Λ and the homomorphisms are Λ-module homomorphisms, it is soon verified that the homomorphism d defined in Proposition 2, (iii) is also a Λ-module homomorphism: if $x \in \text{Ker}(c)$ and $\alpha \in \Lambda$, and $y \in B$ is such that $v(y) = k(x)$, it is sufficient to note that $v(y\alpha) = k(x\alpha)$.

COROLLARY 1. *Suppose that the diagram* (10) *is commutative and the two rows are exact. Then:*

(i) *If u', a and c are injective, b is injective.*
(ii) *If v, a and c are surjective, b is surjective.*

Assertion (i) is a consequence of assertion (i) of Proposition 2: for $\text{Ker}(a) = 0$ and $\text{Ker}(c) = 0$, hence $\text{Ker}(b) = 0$.

Assertion (ii) is a consequence of assertion (ii) of Proposition 2: for $\text{Coker}(a) = 0$ and $\text{Coker}(c) = 0$, hence $\text{Coker}(b) = 0$.

COROLLARY 2. *Suppose that the diagram* (10) *is commutative and the two rows are exact. Under these conditions:*

(i) *If b is injective and if a and v are surjective, then c is injective.*
(ii) *If b is surjective and if c and u' are injective, then a is surjective.*

To prove (i) consider the diagram

$$\begin{array}{ccccc} u(A) & \xrightarrow{w} & B & \xrightarrow{v} & C \\ \downarrow{\scriptstyle a'} & & \downarrow{\scriptstyle b} & & \downarrow{\scriptstyle c} \\ u'(A') & \xrightarrow[w']{} & B' & \xrightarrow[v']{} & C' \end{array}$$

where a' is the mapping with the same graph as the restriction of b to $u(A)$ and w and w' are the canonical injections; clearly this diagram is commutative and its rows exact. Moreover w' is injective and by hypothesis v is surjective; then by Proposition 2, (iii) we have an exact sequence

$$0 = \text{Ker}(b) \longrightarrow \text{Ker}(c) \xrightarrow{d} \text{Coker}(a') = 0$$

since b is injective and a' is surjective; whence $\text{Ker}(c) = 0$.

To prove (ii) consider the diagram

$$\begin{array}{ccccc} A & \xrightarrow{u} & B & \xrightarrow{w} & v(B) \\ \downarrow{\scriptstyle a} & & \downarrow{\scriptstyle b} & & \downarrow{\scriptstyle c'} \\ A' & \xrightarrow[w']{} & B' & \xrightarrow[v']{} & v'(B') \end{array}$$

where this time c' is the mapping with the same graph as the restriction of c to

$v(B)$ and w and w' have respectively the same graphs as v and v'; this diagram is commutative and its rows exact. Moreover w is surjective and by hypothesis u' is injective; then by Proposition 2, (iii) we have an exact sequence

$$0 = \mathrm{Ker}(c') \xrightarrow{d} \mathrm{Coker}(a) \longrightarrow \mathrm{Coker}(b) = 0$$

since b is surjective and c' is injective; whence Coker $(a) = 0$.

2. FLAT MODULES(*)

1. REVISION OF TENSOR PRODUCTS

Let A be a ring, E a right A-module and M a left A-module. In *Algebra*, Chapter II, § 3, no. 1 we defined the *tensor product* $E \otimes_A M$, which is a **Z**-module. If E′ (resp. M′) is a right (resp. left) A-module and $u: E \to E'$ (resp. $v: M \to M'$) a homomorphism, we also defined (*loc. cit, no. 2*) a **Z**-*homomorphism*

$$u \otimes v: E \otimes_A M \to E' \otimes_A M'.$$

LEMMA 1. *Let* $M' \xrightarrow{v} M \xrightarrow{w} M'' \to 0$ *be an exact sequence of left* A-*modules and* E *a right* A-*module. Then the sequence*

$$E \otimes_A M' \xrightarrow{1 \otimes v} E \otimes_A M \xrightarrow{1 \otimes w} E \otimes_A M'' \longrightarrow 0$$

is an exact sequence of commutative groups.

This is the Corollary to Proposition 5 of *Algebra*, Chapter II, § 3, no. 6.

It follows that, for any left A-module homomorphism $u: M \to N$,

$$E \otimes_A (\mathrm{Coker}\ u)$$

is canonically identified with $\mathrm{Coker}(1_E \otimes u)$, as Lemma 1 shows when applied to the exact sequence

$$M \xrightarrow{u} N \longrightarrow \mathrm{Coker}\ u \longrightarrow 0.$$

In the notation of Lemma 1, we know (*loc. cit.*) that if v is *injective*, that is if the sequence $0 \to M' \xrightarrow{v} M \xrightarrow{w} M'' \to 0$ is exact, it does not necessarily follow that $1_E \otimes v$ is injective and so $E \otimes_A M'$ cannot in general be identified with a

(*) We inform readers already familiar with Homological Algebra that they will find other characterizations of flat modules in § 4.

subgroup of $E \otimes_A M$. Recall however (*Algebra*, Chapter II, § 3, no. 7, Corollary 5 to Proposition 7) the following result:

LEMMA 2. *If* $v: M' \to M$ *is injective and* $v(M')$ *is a direct factor of* M, *the homomorphism* $1_E \otimes v$ *is injective and its image is a direct factor of* $E \otimes_A M$.

2. M-FLAT MODULES

DEFINITION 1. *Let* A *be a ring,* E *a right* A-*module and* M *a left* A-*module.* E *is said to be flat for* M (*or* M-*flat*) *if, for every left* A-*module* M' *and every injective homomorphism* $v: M' \to M$, *the homomorphism* $1_E \otimes v: E \otimes_A M' \to E \otimes_A M$ *is injective.*

For any right A-module N, the notion of an N-*flat left module* is defined similarly. To say that a right A-module E is flat for a left A-module M is equivalent to saying that E, considered as a left A^0-module (recall that A^0 denotes the opposite ring to A), is flat for the right A^0-module M.

LEMMA 3. *For a right* A-*module* E *to be* M-*flat, it is necessary and sufficient that for every finitely generated submodule* M' *of* M *the canonical homomorphism*

$$1_E \otimes j: E \otimes_A M' \to E \otimes_A M$$

(j *being the canonical injection* $M' \to M$) *be injective.*

Suppose that this condition holds and let N be any submodule of M. Suppose that the canonical image in $E \otimes_A M$ of an element

$$z = \sum_i x_i \otimes y_i \in E \otimes_A N \quad (x_i \in E, y_i \in N)$$

is zero and let M' be the finitely generated submodule of N generated by the y_i; as by hypothesis the composite mapping $E \otimes_A M' \to E \otimes_A N \to E \otimes_A M$ is injective, the sum $z' = \sum_i x_i \otimes y_i$, considered as an element of $E \otimes_A M'$, is zero. As z is the image of z', we have also $z = 0$, whence the lemma.

LEMMA 4. Let E *be a right* A-*module and* M *a left* A-*module such that* E *is* M-*flat. If* N *is either a submodule or a quotient module of* M, *then* E *is* N-*flat.*

The case in which N is a submodule is easy, as, if N' is a submodule of N, the composite homomorphism

$$E \otimes_A N' \to E \otimes_A N \to E \otimes_A M$$

is injective, hence so is $E \otimes_A N' \to E \otimes_A N$. Suppose then that N is a quotient module of M, that is there exists an exact sequence $0 \to R \xrightarrow{i} M \xrightarrow{p} N \to 0$. Let N' be a submodule of N and $M' = \overset{-1}{p}(N')$. Let i' denote the mapping of R to M' with the same graph as i, p' the surjection $M' \to N'$ with the same graph

as the restriction of p to M', r the identity mapping of R to R, m the canonical injection $M' \to M$ and n the canonical injection $N' \to N$. The diagram

$$\begin{array}{ccccccccc} 0 & \longrightarrow & R & \xrightarrow{i'} & M' & \xrightarrow{p'} & N' & \longrightarrow & 0 \\ & & \downarrow{\scriptstyle r} & & \downarrow{\scriptstyle m} & & \downarrow{\scriptstyle n} & & \\ 0 & \longrightarrow & R & \xrightarrow[i]{} & M & \xrightarrow[p]{} & N & \longrightarrow & 0 \end{array}$$

is commutative and its rows are exact.

To simplify the writing, we set $T(Q) = E \otimes_A Q$ for every left A-module Q and $T(v) = 1_E \otimes v$ for every left A-module homomorphism v. The diagram

$$\begin{array}{ccccccc} T(R) & \xrightarrow{T(i')} & T(M') & \xrightarrow{T(p')} & T(N') & \longrightarrow & 0 \\ \downarrow{\scriptstyle T(r)} & & \downarrow{\scriptstyle T(m)} & & \downarrow{\scriptstyle T(n)} & & \\ T(R) & \xrightarrow[T(i)]{} & T(M) & \xrightarrow[T(p)]{} & T(N) & \longrightarrow & 0 \end{array}$$

is commutative and its rows are exact by Lemma 1 of no. 1. Moreover, since E is M-flat, the homomorphism $T(m)$ is injective. As $T(r)$ and $T(p')$ are surjective, it follows from § 1, no. 4, Corollary 2 to Proposition 2 that $T(n)$ is injective, which proves the lemma.

Lemma 5. *Let $(M_\iota)_{\iota \in I}$ be a family of left A-modules, $M = \bigoplus_{\iota \in I} M_\iota$ their direct sum and E a right A-module. If, for all $\iota \in I$, E is flat for M_ι, then E is flat for M.*

(a) Suppose first that $I = \{1, 2\}$, and let M' be a submodule of $M = M_1 \oplus M_2$, M_1 and M_2 being canonically identified with submodules of M. Denote by M'_1 the intersection $M' \cap M_1$ and by M'_2 the image of M' in M_2 under the canonical projection p of M onto M_2. We have a diagram

$$\begin{array}{ccccccccc} 0 & \longrightarrow & M'_1 & \xrightarrow{i'} & M' & \xrightarrow{p'} & M'_2 & \longrightarrow & 0 \\ & & \downarrow{\scriptstyle v_1} & & \downarrow{\scriptstyle v} & & \downarrow{\scriptstyle v_2} & & \\ 0 & \longrightarrow & M_1 & \xrightarrow[i]{} & M & \xrightarrow[p]{} & M_2 & \longrightarrow & 0 \end{array}$$

where v_1, v, v_2, i, i' are the canonical injections and p' is the mapping with the same graph as the restriction of p to M', which is surjective. It is immediately verified that this diagram is commutative and that its rows are exact. With $T(Q)$ and $T(v)$ used in the same sense as in the proof of Lemma 4, we have a commutative diagram

$$\begin{array}{ccccc} T(M'_1) & \xrightarrow{T(i')} & T(M') & \xrightarrow{T(p')} & T(M'_2) \\ \downarrow{\scriptstyle T(v_1)} & & \downarrow{\scriptstyle T(v)} & & \downarrow{\scriptstyle T(v_2)} \\ T(M_1) & \xrightarrow[T(i)]{} & T(M) & \xrightarrow[T(p)]{} & T(M_2) \end{array}$$

By Lemma 1 of no. 1 the two rows of this diagram are exact; as E is flat for M_1 and M_2, $T(v_1)$ and $T(v_2)$ are injective; moreover, by Lemma 2 of no. 1, $T(i)$ is injective. Corollary 2 to Proposition 2 of § 1, no. 4 then shows that $T(v)$ is injective and consequently E is M-flat.

(b) If I is a finite set with *n* elements, the lemma follows by induction on *n* using (a).

(c) In the general case let M′ be a *finitely generated* submodule of M. Then there exists a finite subset J of the indexing set I such that M′ is contained in the direct sum $M_J = \bigoplus_{\iota \in J} M_\iota$. By (b) E is flat for M_J; the canonical homomorphism $E \otimes_A M' \to E \otimes_A M_J$ is therefore injective. On the other hand, as M_J is a direct factor of M, the canonical homomorphism $E \otimes_A M_J \to E \otimes_A M$ is injective (no. 1, Lemma 2). Taking the composition, it follows that $E \otimes_A M_J \to E \otimes_A M$ is injective and E is flat for M by Lemma 3.

3. FLAT MODULES

PROPOSITION 1. *Let* E *be a right* A-*module. The following three properties are equivalent:*

(a) E *is flat for* A_s (*in other words, for every left ideal* $\mathfrak{a}$ *of* A, *the canonical homomorphism* $E \otimes_A \mathfrak{a} \to E \otimes_A A_s = E$ *is injective*).

(b) E *is* M-*flat for every left* A-*module* M.

(c) *For every exact sequence of left* A-*modules and homomorphisms*

$$M' \xrightarrow{v} M \xrightarrow{w} M''$$

the sequence

$$E \otimes_A M' \xrightarrow{1 \otimes v} E \otimes_A M \xrightarrow{1 \otimes w} E \otimes_A M''$$

is exact.

It is immediate that (b) implies (a). Conversely, suppose that (a) holds; by Lemma 5 of no. 2, E is flat for every *free* left A-module; as every left A-module is isomorphic to a quotient of a free module (*Algebra*, Chapter II, § 1, no. 11, Proposition 20), it follows from Lemma 4 of no. 2 that E is flat for M.

We show that (c) implies (b). If v: M′ → M is an injective homomorphism, the sequence $0 \to M' \xrightarrow{v} M$ is exact; by (c) the sequence

$$0 \to E \otimes_A M' \xrightarrow{1 \otimes v} E \otimes_A M$$

is exact; this means that $1 \otimes v$ is injective, in other words, E is M-flat.

Finally, the implication (b) ⇒ (c) is a consequence of the following more precise lemma:

LEMMA 6. *If* $M' \xrightarrow{v} M \xrightarrow{w} M''$ *is an exact sequence of left* A*-modules and if* E *is an* M''*-flat right* A*-module, the sequence*

$$E \otimes_A M' \xrightarrow{1 \otimes v} E \otimes_A M \xrightarrow{1 \otimes w} E \otimes_A M''$$

is exact.

We use the notation T(Q) and T(v) in the same sense as in the proof of Lemma 4 of no. 2. We write $M''_1 = w(M)$ and let $i: M''_1 \to M''$ be the canonical injection and p the mapping of M to M''_1 with the same graph as w. The sequence $M' \xrightarrow{v} M \xrightarrow{p} M''_1 \to 0$ is exact and it follows from Lemma 1 of no. 1 that the sequence

$$T(M') \xrightarrow{T(v)} T(M) \xrightarrow{T(p)} T(M''_1) \longrightarrow 0$$

is exact. Moreover, as E is M''-flat, the mapping $T(i): T(M''_1) \to T(M'')$ is injective, and as $T(i) \circ T(p) = T(w)$, the sequence

$$T(M') \xrightarrow{T(v)} T(M) \xrightarrow{T(w)} T(M'')$$

is exact. (§ 1, no. 3.)

DEFINITION 2. *A right* A*-module* E *is called flat if it has the equivalent properties of Proposition* 1.

Flat left A-modules are defined similarly. To say that a right A-module E is flat equivalent to saying that E, considered as a left A^0-module, is flat.

Remarks

(1) By Lemma 3 of no. 2, for a right A-module E to be flat, it is necessary and sufficient that, for every *finitely generated* left ideal $\mathfrak{a}$ of A, the canonical mapping $E \otimes_A \mathfrak{a} \to E$ (Proposition 1) with image $E\mathfrak{a}$ be injective.

(2) Let E be a flat right A-module. If M′ is a submodule of a left A-module M, the canonical injection $E \otimes_A M' \to E \otimes_A M$ allows us to identify $E \otimes_A M'$ with a subgroup of $E \otimes_A M$. This being so, let N be a left A-module, $u: M \to N$ a homomorphism, $K = \operatorname{Ker} u$, and $I = \operatorname{Im} u$. By considering the exact sequence

$$0 \longrightarrow K \longrightarrow M \xrightarrow{u} N$$

it is easily seen (Proposition 1) that $E \otimes_A (\operatorname{Ker} u)$ *is identified with* $\operatorname{Ker}(1_E \otimes u)$. On the other hand, writing u' for the surjective homomorphism $M \to I$ with the same graph as u, and i for the canonical injection $I \to N$, $1_E \otimes u'$ is surjective (no. 1, Lemma 1) and $1_E \otimes i$ is injective since E is flat. As

$$1_E \otimes u = (1_E \otimes i) \circ (1_E \otimes u'),$$

$E \otimes_A (\operatorname{Im} u)$ *is identified with* $\operatorname{Im}(1_E \otimes u)$.

PROPOSITION 2. (i) *Let* $(E_\iota)_{\iota \in I}$ *be a family of right* A-*modules. For* $E = \bigoplus_{\iota \in I} E_\iota$ *to be flat, it is necessary and sufficient that each of the* E_ι *be flat.*

(ii) *Let* I *be an ordered set and* $(E_\alpha, f_{\beta\alpha})$ *a direct system of right* A-*modules* (*Algebra*, Chapter II, § 6, no. 2). *If each of the* E_α *is flat, then* $E = \varinjlim E_\alpha$ *is flat.*

Let $M' \to M$ be an injective left A-module homomorphism.

(i) For the direct sum homomorphism

$$\bigoplus_{\iota \in I} (E_\iota \otimes_A M') \to \bigoplus_{\iota \in I} (E_\iota \otimes_A M)$$

to be exact, it is necessary and sufficient that each of the homomorphisms $E_\iota \otimes_A M' \to E_\iota \otimes_A M$ be so (*Algebra*, Chapter II, § 1, no. 6, Corollary 1 of Proposition 7), which proves (i) since $\bigoplus_{\iota \in I} (E_\iota \otimes_A M)$ is canonically identified with $E \otimes_A M$ (*Algebra*, Chapter II, § 3, no. 7, Proposition 7).

(ii) By hypothesis each of the sequences

$$0 \to E_\alpha \otimes_A M' \to E_\alpha \otimes_A M$$

is exact; so then is the sequence

$$0 \to E \otimes_A M' \to E \otimes_A M$$

since taking the direct limit commutes with tensor products (*Algebra*, Chapter II, § 6, no. 3, Proposition 7) and preserves exactness (*ibid.*, § 6, no. 2, Proposition 3).

4. EXAMPLES OF FLAT MODULES

(1) For any ring A, A_d is clearly a flat A-module (*Algebra*, Chapter II, § 3, no. 4, Proposition 4). Then it follows from Proposition 2, (i) of no. 3 that every free right A-module, and more generally every *projective* right A-module (*Algebra*, Chapter II, § 2, no. 2), is a flat A-module.

(2) If A is a *semisimple* ring (*Algebra*, Chapter VIII, § 5, no. 1, Definition 1) every right A-module E is semisimple and hence a direct sum of simple modules; as each of these latter is isomorphic to a direct factor of A_d (*ibid.*, § 5, no. 1, Proposition 6), E is projective and therefore flat by (1) (cf. Exercise 16).

(3) In Chapters II and III we shall study in detail two important examples of flat A-modules: rings of fractions $S^{-1}A$ and Hausdorff completions $\hat{A}$ of A for $\mathfrak{I}$-adic topologies.

PROPOSITION 3. *Let* A *be a ring and* E *a right* A-*module.*

(i) *Suppose that* E *is flat. For every element* a *of* A *which is not a right divisor of* 0(*), *the relations* $x \in E$, $xa = 0$ *imply* $x = 0$.

(*) Recall that a *right* (resp. *left*) *divisor* of 0 in a ring A is an element $b \in A$ such that the mapping $x \mapsto xb$ (resp. $x \mapsto bx$) is not injective.

(ii) *Suppose that* A *is an integral domain in which every finitely generated ideal is principal* (for example *a principal ideal domain* (*Algebra*, Chapter VII, § 1, no. 1)). *Then for* E *to be flat it is necessary and sufficient that* E *be torsion-free.*

We prove (i). Let $v: A_s \to A_s$ be the left A-module homomorphism $t \mapsto ta$; the hypothesis implies that v is injective. As E is flat, the homomorphism $1_E \otimes v: E \otimes_A A_s \to E \otimes_A A_s$ is also injective. When $E \otimes_A A_s$ is canonically identified with E, $1_E \otimes v$ becomes the endomorphism $x \mapsto xa$ of E. Thus the relation $xa = 0$ implies $x = 0$.

We prove (ii). By (i), if E is flat, E is torsion-free. Conversely, let E be a torsion-free A-module; we verify that, for every finitely generated ideal $\mathfrak{a}$ of A, the canonical homomorphism $E \otimes_A \mathfrak{a} \to E$ is injective (no. 3, *Remark* 1). This assertion is obvious if $\mathfrak{a} = (0)$; otherwise, by hypothesis $\mathfrak{a} = Aa$ for some $a \in A$, $a \neq 0$, and $t \mapsto ta$ is then an isomorphism v of A onto $\mathfrak{a}$; using i to denote the canonical injection $\mathfrak{a} \to A$, $i \circ v$ is the homothety with ratio a on E and is injective since E is assumed to be torsion-free. Then $1_E \otimes (i \circ v) = (1_E \otimes i) \circ (1_E \otimes v)$; as $1_E \otimes v$ is an isomorphism, $1_E \otimes i$ is injective, which completes the proof.

Example. Applying Proposition 3 to the ring **Z**, it is seen that **Q** is a flat **Z**-module, but that $\mathbf{Z}/n\mathbf{Z}$ (for $n \geqslant 2$) is not a flat **Z**-module.

5. FLATNESS OF QUOTIENT MODULES

PROPOSITION 4. *Let* E *be a right* A*-module. The three following properties are equivalent:*

(a) E *is flat.*

(b) *For every exact sequence of right* A*-modules of the form*

$$0 \longrightarrow G \xrightarrow{v} H \xrightarrow{w} E \longrightarrow 0 \tag{1}$$

and every left A*-module* F, *the sequence*

$$0 \longrightarrow G \otimes_A F \xrightarrow{v \otimes 1} H \otimes_A F \xrightarrow{w \otimes 1} E \otimes_A F \longrightarrow 0 \tag{2}$$

is exact.

(c) *There exists an exact sequence* (1), *where* H *is flat, such that the sequence* (2) *is exact for every left* A*-module* F *of the form* $A_s/\mathfrak{a}$, *where* $\mathfrak{a}$ *is a finitely generated left ideal of* A.

We show first that (a) implies (b). The left A-module F is isomorphic to a quotient of a free module (*Algebra*, Chapter II, § 1, no. 11, Proposition 20); in other words, we have an exact sequence

$$0 \longrightarrow R \xrightarrow{i} L \xrightarrow{p} F \longrightarrow 0$$

where L is free. Consider the diagram

$$(3)\qquad \begin{array}{ccccc}
G \otimes R & \xrightarrow{v \otimes 1_R} & H \otimes R & \xrightarrow{w \otimes 1_R} & E \otimes R \\
\downarrow{\scriptstyle 1_G \otimes i} & & \downarrow{\scriptstyle 1_H \otimes i} & & \downarrow{\scriptstyle 1_E \otimes i} \\
G \otimes L & \xrightarrow[v \otimes 1_L]{} & H \otimes L & \xrightarrow[w \otimes 1_L]{} & E \otimes L \\
\downarrow{\scriptstyle 1_G \otimes p} & & \downarrow{\scriptstyle 1_H \otimes p} & & \\
G \otimes F & \xrightarrow[v \otimes 1_F]{} & H \otimes F & &
\end{array}$$

It follows immediately that this diagram is commutative, and its rows and columns are exact by Lemma 1 of no. 1; moreover, as $1_G \otimes p$ and $1_H \otimes p$ are surjective (no. 1, Lemma 1), we have $G \otimes F = \operatorname{Coker}(1_G \otimes i)$,

$$H \otimes F = \operatorname{Coker}(1_H \otimes i);$$

$w \otimes 1_R$ is surjective (no. 1, Lemma 1); finally, as L is free and hence flat, $v \otimes 1_L$ is injective. Thus the snake diagram (§ 1, no. 4, Proposition 2, (iii)) can be applied to prove the existence of an exact sequence

$$(4)\qquad \operatorname{Ker}(1_H \otimes i) \longrightarrow \operatorname{Ker}(1_E \otimes i) \xrightarrow{d} G \otimes F \xrightarrow{v \otimes 1_F} H \otimes F.$$

This being so, if E is flat, $1_E \otimes i$ is injective, in other words $\operatorname{Ker}(1_E \otimes i) = 0$, and the exact sequence (4) shows that $v \otimes 1_F$ is injective, hence the sequence (2) is exact (taking account of Lemma 1 of no. 1).

As (b) obviously implies (c), it remains to prove that (c) implies (a). Consider the diagram (3) in the case $R = \mathfrak{a}$, $L = A_s$, $F = A_s/\mathfrak{a}$ and apply the exact sequence (4). By hypothesis $v \otimes 1_F$ is injective, hence $\operatorname{Im}(d) = 0$; moreover, as H is flat, we have $\operatorname{Ker}(1_H \otimes i) = 0$; the exactness of the sequence (4) then implies $\operatorname{Ker}(1_E \otimes i) = 0$, in other words, $1_E \otimes i$ is injective and this proves that E is flat (no. 3, *Remark* 1).

PROPOSITION 5. *Let* $0 \rightarrow E' \xrightarrow{v} E \xrightarrow{w} E'' \rightarrow 0$ *be an exact sequence of right* A-*modules. Suppose* E″ *is flat. Then, for* E *to be flat it is necessary and sufficient that* E′ *be flat.*

Let $u\colon F' \rightarrow F$ be an injective homomorphism of left A-modules. Consider the diagram

$$\begin{array}{ccccc}
E' \otimes F' & \xrightarrow{v \otimes 1_{F'}} & E \otimes F' & \xrightarrow{w \otimes 1_{F'}} & E'' \otimes F' \\
\downarrow{\scriptstyle 1_{E'} \otimes u} & & \downarrow{\scriptstyle 1_E \otimes u} & & \downarrow{\scriptstyle 1_{E''} \otimes u} \\
E' \otimes F & \xrightarrow[v \otimes 1_F]{} & E \otimes F & \xrightarrow[w \otimes 1_F]{} & E'' \otimes F
\end{array}$$

It is commutative and its rows are exact (no. 1, Lemma 1). Since E″ is flat,

$1_{E''} \otimes u$ is injective; moreover, Proposition 4 proves that $v \otimes 1_{F'}$ and $v \otimes 1_F$ are injective. This being so, if E is flat, $1_E \otimes u$ is injective, hence also

$$(1_E \otimes u) \circ (v \otimes 1_{F'}) = (v \otimes 1_F) \circ (1_{E'} \otimes u);$$

it follows that $1_{E'} \otimes u$ is injective and consequently E′ is flat. Conversely, if E′ is flat, $1_{E'} \otimes u$ is injective; then it follows from § 1, no. 4, Corollary 1 to Proposition 2 that $1_E \otimes u$ is injective and so E is flat.

Remarks

(1) It is possible for E and E′ to be flat without E″ being so, as the example of the **Z**-modules $E = \mathbf{Z}$, $E' = n\mathbf{Z}$, $E'' = \mathbf{Z}/n\mathbf{Z}$ ($n \geqslant 2$), shows.
(2) A submodule of a flat module is not necessarily a flat module (Exercise 3).

6. INTERSECTION PROPERTIES

LEMMA 7. *Let* E *be a right* A*-module,* F *a left* A*-module and* F′, F″ *two submodules of* F *such that* $F = F' + F''$. *Then the intersection of the canonical images of* $E \otimes F'$ *and* $E \otimes F''$ *in* $E \otimes F$ *is equal to the canonical image of* $E \otimes (F' \cap F'')$.

Consider the diagram

$$\begin{array}{ccccccccc} 0 & \longrightarrow & F' \cap F'' & \longrightarrow & F' & \longrightarrow & F'/(F' \cap F'') & \longrightarrow & 0 \\ & & \downarrow & & \downarrow & & \downarrow{\scriptstyle j} & & \\ 0 & \longrightarrow & F'' & \longrightarrow & F' + F'' & \longrightarrow & (F' + F'')/F'' & \longrightarrow & 0 \end{array}$$

where the unspecified arrows are the canonical injections and surjections and j is the canonical isomorphism defined in *Algebra*, Chapter I, § 6, no. 13, Theorem 6. This diagram is commutative and its rows are exact. We derive (since $F = F' + F''$) a commutative diagram

$$\begin{array}{ccccc} E \otimes (F' \cap F'') & \longrightarrow & E \otimes F' & \longrightarrow & E \otimes (F'/(F' \cap F'')) \\ \downarrow & & \downarrow & & \downarrow{\scriptstyle 1_E \otimes j} \\ E \otimes F'' & \longrightarrow & E \otimes F & \longrightarrow & E \otimes (F/F'') \end{array}$$

The rows of this diagram are exact (no. 1, Lemma 1) and $1_E \otimes j$ is an isomorphism. Our assertion is then a special case of § 1, no. 4, Proposition 1, (i) (cf. Exercise 5).

PROPOSITION 6. *Let* E *be a right* A*-module and* F *a left* A*-module such that* E *is* F*-flat. For every submodule* F′ *of* F, *we denote by* $\phi(F')$ *the image of* $E \otimes F'$ *under the canonical mapping of* $E \otimes F'$ *to* $E \otimes F$ (*which is injective by Definition* 1 *of no.* 2). *Then, if* F′, F″ *are two submodules of* F, *we have*

$$\phi(F' \cap F'') = \phi(F') \cap \phi(F'').$$

As E is F-flat, $\phi(F' + F'')$ is identified with $E \otimes (F' + F'')$, and the submodules $\phi(F')$, $\phi(F'')$ and $\phi(F' \cap F'')$ are identified with the canonical images of $E \otimes F'$, $E \otimes F''$ and $E \otimes (F' \cap F'')$ in $E \otimes (F' + F'')$ respectively. Proposition 6 then follows from Lemma 7.

Remark 1. With the hypotheses of Proposition 6, $E \otimes F'$ is usually identified with $\phi(F')$ for every submodule F' of F, which gives the formula

$$E \otimes_A (F' \cap F'') = (E \otimes_A F') \cap (E \otimes_A F'').$$

PROPOSITION 7. *Let* E *be a right* A*-module,* E' *a submodule of* E, F *a left* A*-module and* F' *a submodule of* F. *Suppose that* E/E' *or* F/F' *is flat. Then the canonical image of* $E' \otimes F'$ *in* $E \otimes F$ *is the intersection of the canonical images of* $E' \otimes F$ *and* $E \otimes F'$ *in* $E \otimes F$.

Suppose for example that E/E' is flat and consider the diagram

$$\begin{array}{ccccc} E' \otimes F' & \longrightarrow & E \otimes F' & \longrightarrow & (E/E') \otimes F' \\ \downarrow & & \downarrow & & \downarrow u \\ E' \otimes F & \longrightarrow & E \otimes F & \longrightarrow & (E/E') \otimes F \end{array}$$

where the arrows are the canonical homomorphisms. This diagram is commutative and its rows are exact (no. 1, Lemma 1). As E/E' is flat, u is injective. Then our assertion is a special case of § 1, no. 4, Proposition 1, (i).

COROLLARY. *Let* E *be a right* A*-module and* E' *a submodule of* E.

(i) *Suppose that* E/E' *is flat. Then, for every left ideal* $\mathfrak{a}$ *of* A,

(5) $$E'\mathfrak{a} = E' \cap E\mathfrak{a}.$$

(ii) *Conversely, suppose that* E *is flat and, for every finitely generated left ideal* $\mathfrak{a}$ *of* A, *relation* (5) *holds. Then* E/E' *is flat.*

(i) It is sufficient to apply Proposition to the case $F = A_s$, $F' = \mathfrak{a}$.

(ii) To prove that E/E' is flat, apply criterion (c) of Proposition 4 of no. 5; it is then necessary to establish that the sequence

$$0 \to E'/E'\mathfrak{a} \to E/E\mathfrak{a} \to E/(E' + E\mathfrak{a}) \to 0$$

is exact at $E'/E'\mathfrak{a}$ for every finitely generated left ideal $\mathfrak{a}$ of A. Now this is precisely what relation (5) expresses.

Remark 2. The conclusion of Proposition 7 remains true if we assume only that E/E' is F-flat or that F/F' is E-flat.

7. TENSOR PRODUCTS OF FLAT MODULES

Let A, B be two rings, E a right A-module and F an (A, B)-bimodule (*Algebra*, Chapter II, § 1, no. 14). Recall (*Algebra*, Chapter II, § 3, no. 4) that $E \otimes_A F$ has a canonical right B-module structure, for which

$$(x \otimes y)b = x \otimes (yb) \quad \text{for} \quad x \in E, \ y \in F, \ b \in B.$$

PROPOSITION 8. *Let* A, B *be two rings,* E *a right A-module and* F *an* (A, B)-*bimodule. Suppose that* E *is flat and that* F *is flat as a* B-*module. Then the* B-*module* $E \otimes_A F$ *is flat.*

Let G be a left B-module and G′ a submodule of G. Since F is flat as a right B-module, the canonical homomorphism $F \otimes_B G' \to F \otimes_B G$ is injective. Since E is flat, the canonical homomorphism

$$E \otimes_A (F \otimes_B G') \to E \otimes_A (F \otimes_B G)$$

is injective. As $E \otimes_A (F \otimes_B G')$ and $E \otimes_A (F \otimes_B G)$ are canonically identified with $(E \otimes_A F) \otimes_B G'$ and $(E \otimes_A F) \otimes_B G$ respectively (*Algebra*, Chapter II, § 3, no. 8, Proposition 8), the canonical homomorphism

$$(E \otimes_A F) \otimes_B G' \to (E \otimes_A F) \otimes_B G$$

is injective, which proves that $E \otimes_A F$ is a flat B-module.

COROLLARY 1. *Let* C *be a commutative ring,* E, F *two flat* C-*modules. Then* $E \otimes_C F$ *is a flat* C-*module.*

F is a (C, C)-bimodule and it is sufficient to apply Proposition 8 with B = A = C.

COROLLARY 2. *Let* ρ *be a homomorphism of a ring* A *to a ring* B. *If* E *is a flat right* A-*module, the right* B-*module* $\rho^*(E) = E_{(B)}$ *obtained by extending the ring of scalars to* B (*Algebra*, Chapter II, § 5, no. 1) *is flat.*

By definition $E_{(B)} = E \otimes_A B$, where B is considered as an (A, B)-bimodule by means of ρ. As the right B-module B_d is flat, it is sufficient to apply Proposition 8.

COROLLARY 3. *Let* R, S *be two rings and* $\phi: R \to S$ *a ring homomorphism. If* M *is a flat right* S-*module and* $\phi_*(S_d)$ *is a flat right* R-*module, then* $\phi_*(M)$ *is a flat right* R-*module.*

Recall that $\phi_*(M)$ is the right R-module defined by $x.r = x.\phi(r)$ for all $x \in M$ and all $r \in R$ (*Algebra*, Chapter II, § 1, no. 13). Then apply Proposition 8 with A = S, B = R, E = M and F = S, S having the structure of a (S, R)-bimodule defined by ϕ; the right R-module $M \otimes_S S$ is then precisely $\phi_*(M)$.

PROPOSITION 9. *Let* $(A_\alpha, f_{\beta\alpha})$ *be a direct system of rings,* $A = \varinjlim A_\alpha$ *its direct limit,* $(E_\alpha, f_{\beta\alpha})$ *a direct system of right* A_α*-modules with the same indexing set and* $E = \varinjlim E_\alpha$ *its direct limit, which is a right* A*-module* (*Algebra*, Chapter II, § 6, no. 2). *If each of the* E_α *is a flat* A_α*-module,* E *is a flat* A*-module.*

Let $E'_\alpha = E_\alpha \otimes_{A_\alpha} A$, where A is considered as a left A_α-module by means of the canonical homomorphism $A_\alpha \to A$; we know that the right A-module E is canonically isomorphic to $\varinjlim E'_\alpha$ (*loc. cit.*, Corollary 2 to Proposition 7). It follows from Corollary 2 to Proposition 8 that E'_α is a flat right A-module for all α, hence E is a flat A-module by Proposition 2 of no. 3.

8. FINITELY PRESENTED MODULES

Let A be a ring. An exact sequence

$$L_1 \to L_0 \to E \to 0 \tag{6}$$

of left (resp. right) A-modules, where L_0 and L_1 are free, is called a *presentation* (or *presentation of length* 1) of a left (resp. right) A-module E.

Every A-module E admits a presentation. We know in fact (*Algebra*, Chapter II, § 1, no. 11, Proposition 20) that there exists a surjective homomorphism $u: L_0 \to E$, where L_0 is free; if R is the kernel of u, there exists similarly a surjective homomorphism $v: L_1 \to R$, where L_1 is free. If v is considered as a homomorphism of L_1 to L_0, the sequence $L_1 \xrightarrow{v} L_0 \xrightarrow{u} 0$ is exact by definition, whence our assertion.

If $\rho: A \to B$ is a ring homomorphism, every presentation (6) of E induces a presentation of $E_{(B)} = E \otimes_A B$:

$$L_1 \otimes_A B \to L_0 \otimes_A B \to E \otimes_A B \to 0 \tag{7}$$

by Lemma 1 of no. 1 and the fact that $L \otimes_A B$ is a free B-module if L is free.

A presentation (6) of a module E is called *finite* if the free modules L_0 and L_1 have finite bases. Clearly, if the presentation (6) is finite, so is the presentation (7). E is called *a finitely presented* A*-module* if it admits a finite presentation.

LEMMA 8. (i) *Every module admitting a finite presentation is finitely generated.*

(ii) *If* A *is a left Noetherian ring, every finitely generated left* A*-module admits a finite presentation.*

(iii) *Every finitely generated projective module admits a finite presentation.*

Assertion (i) follows trivially from the definitions. If A is left Noetherian and there exists a surjective homomorphism $u: L_0 \rightarrow E$, where L_0 is a free left A-module with a finite basis, the kernel R of u is finitely generated (*Algebra*, Chapter VIII, § 2, no. 1, Proposition 1 and no. 3, Proposition 7), hence there is a surjective homomorphism $v: L_1 \rightarrow R$, where L_1 is free and has a finite basis, and the exact sequence $L_1 \xrightarrow{v} L_0 \xrightarrow{u} E \rightarrow 0$ is a finite presentation of E; whence (ii).

Finally, suppose that E is a finitely generated projective module; then it is a direct factor of a finitely generated free module L_0 (*Algebra*, Chapter II, § 2, no. 2, Corollary to Proposition 4); the kernel R of the surjective homomorphism $L_0 \rightarrow E$ is then isomorphic to a quotient of L_0 and hence finitely generated and the proof is completed as for (ii).

LEMMA 9. *Let* A *be a ring and* E *a finitely presented* A-*module. For every exact sequence*

$$0 \longrightarrow F \xrightarrow{j} G \xrightarrow{p} E \longrightarrow 0$$

where G *is finitely generated, the module* F *is finitely generated.*

Let $L_1 \xrightarrow{r} L_0 \xrightarrow{s} E \rightarrow 0$ be a finite presentation; if (e_i) is a basis of L_0, there exists for each i an element $g_i \in G$ such that $p(g_i) = s(e_i)$; the homomorphism $u: L_0 \rightarrow G$ such that $u(e_i) = g_i$ for all i then satisfies $s = p \circ u$. As $s \circ r = 0$, we have $u(r(L_1)) \subset \operatorname{Ker} p$, and as $\operatorname{Ker} p$ is isomorphic to F, it can be seen that there is a homomorphism $v: L_1 \rightarrow F$ such that the diagram

$$\begin{array}{ccccccc} L_1 & \xrightarrow{r} & L_0 & \xrightarrow{s} & E & \longrightarrow & 0 \\ \downarrow{\scriptstyle v} & & \downarrow{\scriptstyle u} & & \downarrow{\scriptstyle 1_E} & & \\ F & \xrightarrow[j]{} & G & \xrightarrow[p]{} & E & \longrightarrow & 0 \end{array}$$

is commutative. As j is injective and s surjective, we can apply the snake diagram (§ 1, no. 4, Proposition 4), in other words there is an exact sequence

$$0 = \operatorname{Ker} 1_E \xrightarrow{d} \operatorname{Coker} v \longrightarrow \operatorname{Coker} u \longrightarrow \operatorname{Coker} 1_E = 0.$$

This shows that $\operatorname{Coker} v$ is isomorphic to $G/u(L_0)$, which is finitely generated by hypothesis. Moreover we have the exact sequence

$$0 \rightarrow v(L_1) \rightarrow F \rightarrow \operatorname{Coker} v \rightarrow 0$$

and as $v(L_1)$ and $\operatorname{Coker} v$ are finitely generated, so is F (*Algebra*, Chapter II, § 1, no. 7, Corollary 5 to Proposition 9).

9. EXTENSION OF SCALARS IN HOMOMORPHISM MODULES

Let A and B be two rings, E a right A-module, F a right B-module and G a (B, A)-bimodule. Recall that we have defined (*Algebra*, Chapter II, § 4, no. 2) a canonical homomorphism of **Z**-modules

$$\nu : F \otimes_B \mathrm{Hom}_A(E, G) \to \mathrm{Hom}_A(E, F \otimes_B G) \tag{8}$$

such that, for all $y \in F$ and $u \in \mathrm{Hom}_A(E, G)$, $\nu(y \otimes u)$ is the A-linear mapping $x \mapsto y \otimes u(x)$.

Proposition 10. *Let* A, B *be two rings*, E *a right* A-*module*, F *a right* B-*module and* G *a* (B, A)-*bimodule. Suppose that* F *is flat. Then, if* E *is finitely generated* (resp. *finitely presented*), *the canonical homomorphism* (8) *is injective* (resp. *bijective*).

Consider A, B, F, G as fixed and for each right A-module E set

$$T(E) = F \otimes_B \mathrm{Hom}_A(E, G), \qquad T'(E) = \mathrm{Hom}_A(E, F \otimes_B G)$$

and denote the homomorphism (8) by ν_E; for every right A-module homomorphism $v : E \to E'$, set $T(v) = 1_F \otimes \mathrm{Hom}(v, 1_G)$ and $T'(v) = \mathrm{Hom}(v, 1_F \otimes 1_G)$. Let $L_1 \xrightarrow{v} L_0 \xrightarrow{w} E \to 0$ be a presentation of E; we suppose the free module L_0 (resp. the free modules L_0 and L_1) to be *finitely generated*. The diagram

$$\begin{array}{ccccccc} 0 \longrightarrow & T(E) & \xrightarrow{T(w)} & T(L_0) & \xrightarrow{T(v)} & T(L_1) \\ & \downarrow{\scriptstyle \nu_E} & & \downarrow{\scriptstyle \nu_{L_0}} & & \downarrow{\scriptstyle \nu_{L_1}} \\ 0 \longrightarrow & T'(E) & \xrightarrow[T'(w)]{} & T'(L_0 & \xrightarrow[T'(v)]{} & T'(L_1) \end{array} \tag{9}$$

is commutative and the second row is exact (*Algebra*, Chapter II, § 2, no. 1, Theorem 1); moreover, the sequence

$$0 \to \mathrm{Hom}_A(E, G) \to \mathrm{Hom}_A(L_0, G) \to \mathrm{Hom}_A(L_1, G)$$

is exact (*loc. cit.*), and as F is *flat*, the first row of (9) is also an exact sequence (no. 3, Proposition 1). Then we know that ν_{L_0} (resp. ν_{L_0} and ν_{L_1}) is *bijective* (resp. are *bijective*) (*Algebra*, Chapter II, § 4, no. 2, Proposition 2). If we assume only that ν_{L_0} is bijective, it follows from (9) that

$$\nu_{L_0} \circ T(w) = T'(w) \circ \nu_E$$

is injective and hence so is ν_E. If we assume that both ν_{L_0} and ν_{L_1} are bijective, it follows from § 1, no. 4, Corollary 2, (ii) to Proposition 2 that ν_E is surjective, and as we have just seen that ν_E is injective, it is bijective.

10. EXTENSION OF SCALARS: CASE OF COMMUTATIVE RINGS

Now let A be a *commutative* ring, B a ring and $\rho : A \to B$ a ring homomorphism such that $\rho(A)$ is contained in the *centre* of B; in other words, ρ defines on B an

A-*algebra* structure. For every A-module E, the right B-module $E_{(B)} = E \otimes_A B$ is then identified with $B \otimes_A E$, the A-module structures of $\rho_*(B_s)$ and $\rho_*(B_d)$ being identical by hypothesis. Recall that for every ordered pair (E, F) of A-modules, we have defined a *canonical* B-*homomorphism*

$$(10) \qquad \omega: (\mathrm{Hom}_A(E, F))_{(B)} \to \mathrm{Hom}_B(E_{(B)}, F_{(B)})$$

such that for all $u \in \mathrm{Hom}_A(E, F)$, $\omega(u \otimes 1) = u \otimes 1_B$ (*Algebra*, Chapter II, § 5, no. 3).

PROPOSITION 11. *Let A be a commutative ring, B a ring, ρ a homomorphism of A to the centre of B, and E and F two A-modules. Suppose that B is a flat A-module and E is finitely generated* (resp. *finitely presented*). *Then the canonical homomorphism* (10) *is injective* (resp. *bijective*).

As ω is composed of the canonical isomorphism

$$\mathrm{Hom}_A(E, B \otimes_A F) \to \mathrm{Hom}_B(E_{(B)}, F_{(B)})$$

and the canonical homomorphism (8)

$$\nu: B \otimes_A \mathrm{Hom}_A(E, F) \to \mathrm{Hom}_A(E, B \otimes_A F)$$

(*loc. cit.*), the proposition is a consequence of Proposition 10 of no. 9.

Suppose now that A and B are commutative and consider three A-modules E_1, E_2, E_3 and an A-bilinear mapping $f: E_1 \times E_2 \to E_3$. Then there exists one and only one B-bilinear mapping $f_B: E_{1(B)} \times E_{2(B)} \to E_{3(B)}$ such that

$$f_B(1 \otimes x_1, 1 \otimes x_2) = 1 \otimes f(x_1, x_2)$$

for all $x_1 \in E_1$, $x_2 \in E_2$ (*Algebra*, Chapter IX, § 1, no. 4, Proposition 1).

In the statement which follows we shall suppose that B is a flat A-module and, for every submodule E′ of E_i ($i = 1, 2, 3$), we shall canonically identify $E'_{(B)}$ with its image in $E_{i(B)}$ (no. 3, *Remark* 2).

PROPOSITION 12. *Let A, B be commutative rings, ρ a homomorphism of A to B,* E_1, E_2, E_3 *three A-modules,* $f: E_1 \times E_2 \to E_3$ *an A-bilinear mapping and*

$$f_B: E_{1(B)} \times E_{2(B)} \to E_{3(B)}$$

its extension. Consider a submodule F_2 *of* E_2, *a submodule* F_3 *of* E_3, *and denote by T the submodule of* E_1 *consisting of those* $x_1 \in E_1$ *such that* $f(x_1, x_2) \in F_3$ *for all* $x_2 \in F_2$. *Suppose that B is a flat A-module and that* F_2 *is finitely generated. Then* $T_{(B)}$ *is the set of those* $x'_1 \in E_{1(B)}$ *such that* $f_B(x'_1, x'_2) \in F_{3(B)}$ *for all* $x'_2 \in F_{2(B)}$.

Let p be the canonical surjection $E_3 \to E_3/F_3$; with each $x_1 \in E_1$ we associate the A-linear mapping $x_2 \mapsto p(f(x_1, x_2))$ of F_2 to E_3/F_3, which we denote by $g(x_1)$; then g is an A-homomorphism of E_1 to $\mathrm{Hom}_A(F_2, E_3/F_3)$ and the kernel of g is precisely T. Since B is a flat A-module, we have the exact sequence

$$0 \longrightarrow T_{(B)} \longrightarrow E_{1(B)} \xrightarrow{1\otimes g} (\mathrm{Hom}_A(F_2, E_3/F_3))_{(B)}$$

(no. 3, Proposition 1). By Proposition 11 the canonical homomorphism

$$\omega: (\mathrm{Hom}_A(F_2, E_3/F_3))_{(B)} \to \mathrm{Hom}_B(F_{2(B)}, (E_3/F_3)_{(B)})$$

is *injective*. On the other hand, as B is a flat A-module, $(E_3/F_3)_{(B)}$ is canonically identified with $E_{3(B)}/F_{3(B)}$; taking the composition of ω with $1 \otimes g$, we obtain a homomorphism u for which the sequence

$$0 \longrightarrow T_{(B)} \longrightarrow E_{1(B)} \xrightarrow{u} \mathrm{Hom}_B(F_{2(B)}, E_{3(B)}/F_{3(B)})$$

is exact. It follows immediately from the definitions that $u(x'_1)$, where

$$x'_1 = 1 \otimes x_1 \in E_{1(B)},$$

is the linear mapping which maps each $x'_2 \in F_{2(B)}$ to the class mod. $F_{3(B)}$ of $f_B(x'_1, x'_2)$; by linearity this is also true for all $x'_1 \in E_{1(B)}$; since the kernel of u is $T_{(B)}$, the proposition is proved.

COROLLARY 1. *Let* A, B *be two commutative rings,* $\rho: A \to B$ *a homomorphism such that* B *is a flat* A*-module and* E *a finitely presented* A*-module. For every finitely generated submodule* F *of* E*, the submodule of the dual of* $E_{(B)}$ *orthogonal to* $F_{(B)}$ *is equal to* $(F')_{(B)}$*, where* F′ *is the submodule of the dual* E* *of* E *orthogonal to* F.

By Proposition 11 $(E^*)_{(B)}$ is canonically isomorphic to the dual $(E_{(B)})^*$ of $E_{(B)}$. Then it suffices to apply Proposition 12 with $E_1 = E^*$, $E_2 = E$, $E_3 = A$, $F_2 = F$, $F_3 = \{0\}$, and f the canonical bilinear form on $E^* \times E$.

COROLLARY 2. *Let* A, B *be two commutative rings,* $\rho: A \to B$ *a homomorphism such that* B *is a flat* A*-module and* E *an* A*-module. Then, for every finitely generated submodule* F *of* E*, the annihilator of* $F_{(B)}$ *is the ideal* $\mathfrak{a}B$ *of* B*, where* $\mathfrak{a}$ *is the annihilator of* F *in* A.

It suffices to apply Proposition 12 with $E_1 = A$, $E_2 = E_3 = E$, $F_2 = F$, $F_3 = \{0\}$.

Remark. If there is no ambiguity over the modules E_i nor the bilinear mapping f, $F_3 : F_2$ is sometimes used to denote the module denoted by T in Proposition 12 and it is called the *transporter* of F_2 to F_3. The conclusion of Proposition 12 then reads

$$F_{3(B)} : F_{2(B)} = (F_3 : F_2)_{(B)}. \tag{11}$$

In the particular case when the E_i are equal to the ring A, f is multiplication and the F_i ideals $\mathfrak{a}_i$, we obtain the *transporter formula*

$$\mathrm{B}(\mathfrak{a}_3 : \mathfrak{a}_2) = \mathrm{B}\mathfrak{a}_3 : \mathrm{B}\mathfrak{a}_2 \tag{12}$$

valid when B is a flat A-module and $\mathfrak{a}_2$ is a finitely generated ideal.

11. INTERPRETATION OF FLATNESS IN TERMS OF RELATIONS (*)

Throughout this no. A denotes a ring, E a right A-module and F a left A-module.

Every element of $\mathrm{E} \otimes_{\mathrm{A}} \mathrm{F}$ can be written in at least one way in the form $z = \sum_{i=1}^{n} e_i \otimes f_i$ where $e_i \in \mathrm{E}$ and $f_i \in \mathrm{F}$. The following lemma gives a condition under which this sum is zero:

LEMMA 10. *Let* $(f_\lambda)_{\lambda \in \mathrm{L}}$ *be a* family of generators *of* F *and* $(e_\lambda)_{\lambda \in \mathrm{L}}$ *a family of elements of* E *of finite support. For* $\sum_{\lambda \in \mathrm{L}} e_\lambda \otimes f_\lambda = 0$, *it is necessary and sufficient that there exist a finite set* J, *a family* $(x_j)_{j \in \mathrm{J}}$ *of elements of* E *and a family* $(a_{j\lambda})$ $(j \in \mathrm{J}, \lambda \in \mathrm{L})$ *of elements of* A *with the following properties:*

(1) *the family* $(a_{j\lambda})$ *has finite support;*

(2) $\sum_{\lambda \in \mathrm{L}} a_{j\lambda} f_\lambda = 0$ *for all* $j \in \mathrm{J}$;

(3) $e_\lambda = \sum_{j \in \mathrm{J}} x_j a_{j\lambda}$ *for all* $\lambda \in \mathrm{L}$.

Loosely speaking, the system of e_λ must be a linear combination with coefficients in E of systems of elements of A which are "relations between the f_λ".

Consider the free A-module $\mathrm{A}_s^{(\mathrm{L})}$, its canonical basis (u_λ) and the homomorphism $g: \mathrm{A}_s^{(\mathrm{L})} \to \mathrm{F}$ such that $g(u_\lambda) = f_\lambda$ for all $\lambda \in \mathrm{L}$; denoting by R the kernel of g, we have (since the f_λ generate F) an exact sequence

$$\mathrm{R} \xrightarrow{i} \mathrm{A}_s^{(\mathrm{L})} \xrightarrow{g} \mathrm{F} \longrightarrow 0$$

where i denotes the canonical injection. By Lemma 1 of no. 1 we derive the exact sequence

$$\mathrm{E} \otimes_{\mathrm{A}} \mathrm{R} \xrightarrow{1 \otimes i} \mathrm{E} \otimes_{\mathrm{A}} \mathrm{A}_s^{(\mathrm{L})} \xrightarrow{1 \otimes g} \mathrm{E} \otimes_{\mathrm{A}} \mathrm{F} \longrightarrow 0. \tag{13}$$

(*) The results of this no. will not be used in the rest of this chapter, except in § 3, no. 7.

Now, $E \otimes_A A_s^{(L)}$ is canonically identified with $E^{(L)}$, a family $e = (e_\lambda) \in E^{(L)}$ being identified with $\sum_{\lambda \in L} e_\lambda \otimes u_\lambda$ (*Algebra*, Chapter II, § 3, no. 7, Corollary 1 to Proposition 7). For such a family to belong to the kernel of $1_E \otimes g$, it is necessary and sufficient that $\sum_{\lambda \in L} e_\lambda \otimes f_\lambda = 0$ in $E \otimes_A F$; taking account of the exact sequence (13), this is equivalent to saying that e belongs to the image of $1_E \otimes i$, in other words there is a relation of the form

$$\sum_{\lambda \in L} e_\lambda \otimes u_\lambda = \sum_{j \in J} x_j \otimes i(r_j) \tag{14}$$

where $x_j \in E$, $r_j \in R$ and J is finite. Writing $i(r_j) = \sum_{\lambda \in L} a_{j\lambda} u_\lambda$, the hypothesis $r_j \in R$ implies the relation $\sum_{\lambda \in L} a_{j\lambda} f_\lambda = 0$ for all $j \in J$; on the other hand the relation (14) implies $e_\lambda = \sum_{j \in J} x_j a_{j\lambda}$ for all $\lambda \in L$ (*Algebra*, Chapter II, § 3, no. 7, Corollary 1 to Proposition 7), which completes the proof.

PROPOSITION 13. *For* E *to be* F*-flat* (no. 2, *Definition* 1), *it is necessary and sufficient that the following condition be satisfied:*

(R) *If* $(e_\iota)_{\iota \in I}$ *and* $(f_\iota)_{\iota \in I}$ *are two finite families of elements of* E *and* F *respectively such that* $\sum_{\iota \in I} e_\iota \otimes f_\iota = 0$ *in* $E \otimes_A F$, *there exists a finite set* J, *a family* $(x_j)_{j \in J}$ *of elements of* E *and a family* $(a_{j\iota})$ $(j \in J,\ \iota \in I)$ *of elements of* A, *with the following properties:*

(1) $\sum_{\iota \in I} a_{j\iota} f_\iota = 0$ *for all* $j \in J$;

(2) $e_\iota = \sum_{j \in J} x_j a_{j\iota}$ *for all* $\iota \in I$.

Suppose that E is F-flat. Let (e_ι) and (f_ι) be finite families of elements such that $\sum_{\iota \in I} e_\iota \otimes f_\iota = 0$ in $E \otimes_A F$, and let F′ be the submodule of F *generated by the* f_ι. Since the canonical mapping $E \otimes_A F' \to E \otimes_A F$ is injective, we have also $\sum_{\iota \in I} e_\iota \otimes f_\iota = 0$ *in* $E \otimes_A F'$ and Lemma 10 can then be applied to E and F′; thus families (x_j) and $(a_{j\iota})$ are obtained satisfying the conditions of (R).

Conversely, suppose that condition (R) holds. Let F′ be a submodule of F and let $y = \sum_{\iota \in I} e_\iota \otimes f_\iota$ be an element of the kernel of the canonical mapping $E \otimes_A F' \to E \otimes_A F$. Since (R) holds, there exist families (x_j) and $(a_{j\iota})$ satisfying conditions (1) and (2). We conclude that, in $E \otimes_A F'$,

$$y = \sum_{\iota, j} x_j a_{j\iota} \otimes f_\iota = \sum_{j \in J} (x_j \otimes \sum_{\iota \in I} a_{j\iota} f_\iota) = 0.$$

Hence $E \otimes_A F' \to E \otimes_A F$ is injective.

Corollary 1. *For a right A-module E to be flat, it is necessary and sufficient that the following condition hold:*

(RP) *If $(e_i)_{i \in I}$ and $(b_i)_{i \in I}$ are two finite families of elements of E and A respectively such that $\sum_{i \in I} e_i b_i = 0$, there exists a finite set J, a family $(x_j)_{j \in J}$ of elements of E and a family (a_{ji}) $(j \in J, i \in I)$ of elements of A such that $\sum_{i \in I} a_{ji} b_i = 0$ for all $j \in J$ and $e_i = \sum_{j \in J} x_j a_{ji}$ for all $i \in I$.*

Condition (RP) is just condition (R) of Proposition 13 applied to the module $F = A_s$.

Loosely speaking, (RP) states: every "relation" between the b_i, with coefficients in E, is a linear combination (with coefficients in E) of "relations" between the b_i with coefficients in A.

Let us consider in particular a homomorphism of A to a ring B, which makes B into a right A-module. We know (no. 3, Proposition 1) that this is equivalent to saying that this A-module is flat, or that it is flat for every left A-module A_s^m $(m \geqslant 1)$. Applying condition (R) of Proposition 13 with $E = B$ and $F = A_s^m$ we obtain the following condition:

Corollary 2. *For the ring B to be a flat right A-module, it is necessary and sufficient that it satisfy the following condition:*

(RP') *Every solution $(y_k)_{1 \leqslant k \leqslant n}$, consisting of elements of B, of a system of homogeneous linear equations*

$$\sum_{k=1}^{n} y_k c_{ki} = 0 \qquad (1 \leqslant i \leqslant m) \tag{15}$$

with coefficients c_{ki} in A, is a linear combination

$$y_k = \sum_{j=1}^{q} b_j z_{jk} \qquad (1 \leqslant k \leqslant n) \tag{16}$$

with coefficients $b_j \in B$, of solutions $(z_{jk})_{1 \leqslant k \leqslant n}$ of the system (15), consisting of elements z_{jk} of A.

3. FAITHFULLY FLAT MODULES

1. DEFINITION OF FAITHFULLY FLAT MODULES

Proposition 1. *Let E be a right A-module. The following four properties are equivalent:*

(a) *For a sequence* $N' \xrightarrow{v} N \xrightarrow{w} N''$ *of left* A*-modules to be exact, it is necessary and sufficient that the sequence*

$$E \otimes_A N' \xrightarrow{1 \otimes v} E \otimes_A N \xrightarrow{1 \otimes w} E \otimes_A N''$$

be exact.

(b) E *is flat and, for every left* A*-module* N, *the relation* $E \otimes_A N = 0$ *implies* $N = 0$.

(c) E *is flat and, for every homomorphism* $v: N' \to N$ *of left* A*-modules, the relation* $1_E \otimes v = 0$ *implies* $v = 0$.

(d) E *is flat and, for every maximal left ideal* $\mathfrak{m}$ *of* A, $E \neq E\mathfrak{m}$.

To simplify the writing we set $T(Q) = E \otimes_A Q$ for every left A-module Q and $T(v) = 1_E \otimes v$ for every homomorphism v of left A-modules.

We prove first the equivalence of (a), (b) and (c).

We prove that (a) implies (b). If (a) holds, clearly E is flat (§ 2, no. 3, Proposition 1). On the other hand, let N be a left A-module such that $T(N) = 0$ and consider the sequence $0 \to N \to 0$; the hypothesis $T(N) = 0$ means that the sequence $0 \to T(N) \to 0$ is exact. By (a) the sequence $0 \to N \to 0$ is exact, whence $N = 0$.

We show that (b) implies (c). Suppose that (b) holds and let $v: N' \to N$ be a homomorphism and I its image. As the image of $T(v)$ is identified with $T(I)$ (§ 2, no. 3, *Remark* 2), the hypothesis $T(v) = 0$ implies $T(I) = 0$, hence $I = 0$ by (b) and consequently $v = 0$.

We show that (c) implies (a). Suppose then that (c) holds and consider a sequence

$$N' \xrightarrow{v} N \xrightarrow{w} N'' \tag{1}$$

of homomorphisms of left A-modules and the corresponding sequence

$$T(N') \xrightarrow{T(v)} T(N) \xrightarrow{T(w)} T(N''). \tag{2}$$

If the sequence (1) is exact, so is (2), since E is flat (§ 2, no. 3, Proposition 1). Conversely, if (2) is exact, we have first $T(w \circ v) = T(w) \circ T(v) = 0$, hence $w \circ v = 0$ by hypothesis. Set $I = v(N')$ and $K = \overset{-1}{w}(0)$; then I is contained in K by the above. Consider the exact sequence

$$0 \longrightarrow I \xrightarrow{i} K \xrightarrow{p} K/I \longrightarrow 0$$

i and p being the canonical mappings. As E is flat, the sequence

$$0 \longrightarrow T(I) \xrightarrow{T(i)} T(K) \xrightarrow{T(p)} T(K/I) \longrightarrow 0$$

is exact, in other words, $T(K/I)$ is isomorphic to $T(K)/T(I)$, which is 0 by hypothesis, since $T(I)$ (resp. $T(K)$) is identified with the image of $T(v)$ (resp. the kernel of $T(w)$) (§ 2, no. 3, *Remark* 2). But the relation $T(p) = 0$ implies $p = 0$ by hypothesis, hence $K = I$, which proves that the sequence (1) is exact.

Finally we show the equivalence of (b) and (d). If (b) holds, then

$$E/E\mathfrak{m} = E \otimes_A (A_s/\mathfrak{m}) \neq 0$$

since $A_s/\mathfrak{m} \neq 0$; whence (d). Conversely, suppose that (d) holds; every left ideal $\mathfrak{a} \neq A$ of A is contained in a maximal left ideal $\mathfrak{m}$ (*Algebra*, Chapter I, § 8, no. 7, Theorem 2), then the hypothesis $E \neq E\mathfrak{m}$ implies $E \neq E\mathfrak{a}$, in other words, $E \otimes_A (A_s/\mathfrak{a}) \neq 0$. That is to say, for every *monogenous* left A-module $N \neq 0$, $T(N) \neq 0$. If now N is any left A-module $\neq 0$, it contains a monogenous submodule $N' \neq 0$; since E is flat, $T(N')$ is identified with a subgroup of $T(N)$; we have just seen that $T(N') \neq 0$, hence $T(N) \neq 0$.

DEFINITION 1. *A right* A*-module* E *is called faithfully flat if it satisfies the four equivalent conditions of Proposition* 1.

Faithfully flat left A-modules are defined similarly; clearly, for a left A-module E to be faithfully flat, it is necessary and sufficient that E, considered as a right A^0-module, be faithfully flat.

Remark. If E is a faithfully flat A-module, E is a *faithful* A-module: for, if an element $a \in A$ is such that $xa = 0$ for all $x \in E$, the homothety $h: b \mapsto ba$ in A is such that $1_E \otimes h = 0$; whence $h = 0$ by property (c) of Proposition 1, that is $a = 0$ since A has a unit element.

Examples

(1) The direct sum of a flat module and faithfully flat module is a faithfully flat module by virtue of property (d) of Proposition 1 and § 2, no. 3, Proposition 2.

(2) As A_s is faithfully flat by virtue of criterion (d) of Proposition 1 and § 2, no. 4, *Example* 1, it follows from (1) that every free module *not reduced to* 0 is faithfully flat. On the other hand, there exist non-zero direct factors of free modules (in other words, non-zero projective modules) which are faithful but not faithfully flat (Exercise 2).

(3) Let A be a *principal ideal domain.* For an A-module E to be faithfully flat, it is necessary and sufficient that it be *torsion-free* and $E \neq Ep$ for every irreducible element (*Algebra*, Chapter VII, § 1, no. 3) p of A; this follows immediately from § 2, no. 4, Proposition 3 and criterion (d) of Proposition 1.

(4) Example (3) shows that the **Z**-module **Q** is a flat and faithful module, but *not faithfully flat.*

PROPOSITION 2. *Let* E *be a faithfully flat right* A*-module and* $u: N' \to N$ *a left* A*-module homomorphism. For* u *to be injective* (resp. *surjective, bijective*), *it is necessary and sufficient that* $1_E \otimes u: E \otimes_A N' \to E \otimes_A N$ *be so.*

This is an immediate consequence of criterion (a) of Proposition 1.

PROPOSITION 3. *Let* $0 \to E' \to E \to E'' \to 0$ *be an exact sequence of right* A-*modules. Suppose that* E' *and* E" *are flat and that one of them is faithfully flat. Then* E *is faithfully flat.*

We know already that E is flat (§ 2, no. 5, Proposition 5). We verify that E has property (b) of Proposition 1. Let N be a left A-module. As E" is flat, there is an exact sequence

$$0 \to E' \otimes_A N \to E \otimes_A N \to E'' \otimes_A N \to 0$$

(§ 2, no. 5, Proposition 4). If $E \otimes_A N = 0$, it follows that $E' \otimes_A N$ and $E'' \otimes_A N$ are zero; as one of the modules E', E" is faithfully flat, this implies that $N = 0$.

2. TENSOR PRODUCTS OF FAITHFULLY FLAT MODULES

PROPOSITION 4. *Let* R, S *be two rings,* E *a right* R-*module and* F *an* (R, S)-*bimodule. Suppose that* E *is faithfully flat. Then, for* F *to be a flat* (resp. *faithfully flat*) S-*module, it is necessary and sufficient that* $E \otimes_R F$ *be so.*

(1) If F is flat, $E \otimes_R F$ is flat (§ 2, no. 7, Proposition 8).

(2) Suppose that $E \otimes_R F$ is flat and let $v: N' \to N$ be an injective left S-module homomorphism. The homomorphism

$$1_E \otimes 1_F \otimes v: E \otimes_R F \otimes_S N' \to E \otimes_R F \otimes_S N$$

is then injective (§ 2, no. 3, Proposition 1). It follows from Proposition 2 of no. 1 that $1_F \otimes v: F \otimes_S N' \to F \otimes_S N$ is injective; then F is a flat S-module (§ 2, no. 3, Proposition 1).

(3) Suppose that F is faithfully flat and let N be a left S-module such that $E \otimes_R F \otimes_S N = 0$. Since E is faithfully flat, this implies that $F \otimes_S N = 0$, whence $N = 0$ since F is faithfully flat; this proves that $E \otimes_R F$ is faithfully flat.

(4) Suppose that $E \otimes_R F$ is faithfully flat and let N be a left S-module such that $F \otimes_S N = 0$. Then $E \otimes_R F \otimes_S N = 0$, whence $N = 0$, which shows that F is faithfully flat.

COROLLARY. *Let* C *be a commutative ring and* E *and* F *two faithfully flat* C-*modules. Then the* C-*module* $E \otimes_C F$ *is faithfully flat.*

Apply Proposition 4 with $R = S = C$.

3. CHANGE OF RING

PROPOSITION 5. *Let* ρ *be a homomorphism from a ring* A *to a ring* B. *If* E *is a faithfully flat right* A-*module, the right* B-*module* $\rho^*(E) = E_{(B)} = E \otimes_A B$ *is faithfully flat.*

Apply Proposition 4 of no. 2 with R = A, S = F = B, noting that the B-module B_d is faithfully flat.

COROLLARY. *If* E *is a faithfully flat right* A-*module and if* $\mathfrak{a}$ *is a two-sided ideal of* A, *the right* $(A/\mathfrak{a})$-*module* $E/E\mathfrak{a}$ *is faithfully flat.*

Apply Proposition 5 with $B = A/\mathfrak{a}$, ρ being the canonical homomorphism.

PROPOSITION 6. *Let* A *be a commutative ring,* B *an algebra over* A *and* $\rho: a \mapsto a.1$ *the canonical homomorphism of* A *to* B. *Suppose that* B *is a faithfully flat* A-*module. Then, for an* A-*module* E *to be flat* (resp. *faithfully flat*), *it is necessary and sufficient that the right* B-*module* $E_{(B)} = E \otimes_A B$ *be flat* (resp. *faithfully flat*).

(1) If E is flat (resp. faithfully flat), $E_{(B)}$ is flat (resp. faithfully flat) by § 2, no. 7, Corollary 2 to Proposition 8 (resp. by Proposition 5).

(2) Suppose that $E_{(B)}$ is flat and let $v: N' \to N$ be an injective A-module homomorphism. By § 2, no. 7, Corollary 3, the A-module $E \otimes_A B$ is flat, hence the homomorphism $1_E \otimes 1_B \otimes v: E \otimes_A B \otimes_A N' \to E \otimes_A B \otimes_A N$ is injective. As the right and left A-module structures on B coincide, this homomorphism is identified with

$$1_E \otimes v \otimes 1_B : E \otimes_A N' \otimes_A B \to E \otimes_A N \otimes_A B.$$

As B is a faithfully flat A-module, it follows that $1_E \otimes v: E \otimes_A N' \to E \otimes_A N$ is injective (no. 1, Proposition 2), which shows that E is flat.

(3) Suppose finally that $E_{(B)}$ is faithfully flat. First of all E is flat by (2). Also let N be an A-module such that $E \otimes_A N = 0$. Then $E \otimes_A N \otimes_A B = 0$, whence, since the right and left A-module structures on B coincide, $E \otimes_A B \otimes_A N = 0$, which may also be written $(E \otimes_A B) \otimes_B (B \otimes_A N) = 0$. As $E_{(B)}$ is a faithfully flat B-module, this implies that $B \otimes_A N = 0$ (no. 1, Proposition 1), whence N = 0 since B is a faithfully flat A-module (no. 1, Proposition 1).

4. RESTRICTION OF SCALARS

PROPOSITION 7. *Let* A, B *be two rings and* ρ *a homomorphism of* A *to* B. *Let* E *be a faithfully flat right* B-*module. For* $\rho^*(E)$ *to be a flat* (resp. *faithfully flat*) *right* A-*module, it is necessary and sufficient that* B *be a flat* (resp. *faithfully flat*) *right* A-*module.*

Applying Proposition 4 of no. 2 with B, A, E, B in place of R, S, E, F respectively, and the right A-module structure on B being defined by ρ, it is seen that

B is a flat (resp. faithfully flat) A-module if and only if $E \otimes_B B = \rho^*(E)$ is a flat (resp. faithfully flat) A-module.

Remarks

(1) Proposition 7 shows that, for B to be a faithfully flat A-module, it is *sufficient* that there exist *one* faithfully flat B-module which is also a faithfully flat A-module.

(2) Let A, B, C be three rings and $\rho: A \to B$, $\sigma: B \to C$ two ring homomorphisms. Proposition 7 shows that if C is a faithfully flat B-module and B a faithfully flat A-module, then C is a faithfully flat A-module. If C is a faithfully flat B-module and a faithfully flat A-module, then B is a faithfully flat A-module (taking the modules as right modules, to fix the ideas). On the other hand B and C may be faithfully flat A-modules without C being a faithfully flat B-module (Exercise 7).

5. FAITHFULLY FLAT RINGS

PROPOSITION 8. *Let* A, B *be two rings and* ρ *a homomorphism from* A *to* B. *Suppose that there exists a right* B*-module* E *such that* $\rho_*(E)$ *is a faithfully flat* A*-module. Then:*

(i) *For every left* A*-module* F, *the canonical homomorphism* $j: F \to F_{(B)} = B \otimes_A F$ (*such that* $j(x) = 1 \otimes x$ *for all* $x \in F$) *is injective.*

(ii) *For every left ideal* $\mathfrak{a}$ *of* A, $\overset{-1}{\rho}(B\mathfrak{a}) = \mathfrak{a}$.

(iii) *The homomorphisms* ρ *is injective.*

(iv) *For every maximal left ideal* $\mathfrak{m}$ *of* A, *there exists a maximal left ideal* $\mathfrak{n}$ *of* B *such that* $\overset{-1}{\rho}(\mathfrak{n}) = \mathfrak{m}$.

We prove (i). We know (*Algebra,* Chapter II, § 5, no. 2, Corollary to Proposition 5) that for every right B-module M, the canonical A-homomorphism $i: M \to \rho_*(M) \otimes_A B = \rho^*(\rho_*(M))$ defined by $i(y) = y \otimes 1$ is *injective* and that the A-module $i(M)$ is a *direct factor* of $\rho_*(M) \otimes_A B$. Hence, for every left A-module F,

$$i \otimes 1_F : \rho_*(M) \otimes_A F \to \rho_*(M) \otimes_A B \otimes_A F$$

is injective (§ 2, no. 1, Lemma 2). Taking M = E, it follows (since $i \otimes 1_F = 1_M \otimes j$) that j is injective (no. 1, Proposition 2).

Assertion (ii) follows from (i) by taking $F = A_s/\mathfrak{a}$ and (iii) from (ii) by taking $\mathfrak{a} = \{0\}$.

Finally, if $\mathfrak{m}$ is a maximal left ideal of A, then $\overset{-1}{\rho}(B\mathfrak{m}) = \mathfrak{m}$ by (ii), and consequently $B\mathfrak{m} \neq B$. Then there exists a maximal left ideal $\mathfrak{n}$ of B containing $B\mathfrak{m}$ (*Algebra,* Chapter I, § 8, no. 7, Theorem 2); then $\mathfrak{m} \subset \overset{-1}{\rho}(\mathfrak{n})$ and as $\rho(1) \notin \mathfrak{n}$, $1 \notin \overset{-1}{\rho}(\mathfrak{n})$. Consequently $\overset{-1}{\rho}(\mathfrak{n}) = \mathfrak{m}$.

If A and B satisfy the conditions of Proposition 8, A is usually identified with a *subring* of B by means of ρ.

COROLLARY. *Under the hypotheses of Proposition* 8, *if* B *is left Noetherian* (resp. *Artinian*), *so is* A.

If $(\mathfrak{a}_n)$ is a non-stationary increasing (resp. decreasing) sequence of left ideals of A, the sequence $(\mathrm{B}\mathfrak{a}_n)$ of ideals of B is non-stationary and increasing (resp. decreasing) since $\overset{-1}{\rho}(\mathrm{B}\mathfrak{a}_n) = \mathfrak{a}_n$, contrary to the hypothesis.

**Remark* (1). If A and B are commutative, we shall see in Chapter II, § 2, no. 5, Corollary 4 to Proposition 11, that the hypothesis of Proposition 8 implies that for every *prime* ideal $\mathfrak{p}$ of A, there exists a *prime* ideal $\mathfrak{q}$ of B such that $\overset{-1}{\rho}(\mathfrak{q}) = \mathfrak{p}$ (where $\mathfrak{p} = \mathrm{A} \cap \mathfrak{q}$ when A is identified with a subring of B).*

An important application of Proposition 8 is when B is itself a *faithfully flat* A-module. But in this case we have the following more precise proposition:

PROPOSITION 9. *Let* A, B *be two rings and* ρ *a homomorphism of* A *to* B. *The following five properties are equivalent:*

(a) *The right* A*-module* B *is faithfully flat.*

(b) *The homomorphism* ρ *is injective and the right* A*-module* $\mathrm{B}/\rho(\mathrm{A})$ *is flat.*

(c) *The right* A*-module* B *is flat and, for every left* A*-module* F, *the canonical homomorphism* $x \mapsto 1 \otimes x$ *of* F *to* $\mathrm{B} \otimes_{\mathrm{A}} \mathrm{F}$ *is injective.*

(d) *The right* A*-module* B *is flat and, for every left ideal* $\mathfrak{a}$ *of* A, $\overset{-1}{\rho}(\mathrm{B}\mathfrak{a}) = \mathfrak{a}$.

(e) *The right* A*-module* B *is flat and, for every maximal left ideal* $\mathfrak{m}$ *of* A, *there exists a maximal left ideal* $\mathfrak{n}$ *of* B *such that* $\overset{-1}{\rho}(\mathfrak{n}) = \mathfrak{m}$.

By Proposition 8, (a) implies each of the properties (c), (d), (e). On the other hand, if (e) holds, then $\mathrm{B}\mathfrak{m} \neq \mathrm{B}$ for every maximal left ideal $\mathfrak{m}$ of A (since there exists a maximal left ideal $\mathfrak{n}$ of B such that $\mathrm{B}\mathfrak{m} \subset \mathfrak{n}$), and B is a faithfully flat A-module by criterion (d) of Proposition 1 of no. 1; hence (e) implies (a).

We shall now prove that (c) $\Rightarrow$ (d) $\Rightarrow$ (b) $\Rightarrow$ (a), which will complete the proof. In the first place, (c) implies (d) by taking $\mathrm{F} = \mathrm{A}_s/\mathfrak{a}$ in (c). If (d) holds, by taking $\mathfrak{a} = \{0\}$ it follows that ρ is injective; (d) and § 2, no. 6, Corollary to Proposition 7 imply that $\mathrm{B}/\rho(\mathrm{A})$ is a flat right A-module, that is (d) implies (b). Finally, if (b) holds, Proposition 3 of no. 1 applied to the exact sequence

$$0 \longrightarrow \mathrm{A}_d \xrightarrow{\rho} \mathrm{B} \longrightarrow \mathrm{B}/\rho(\mathrm{A}) \longrightarrow 0$$

shows that B is a faithfully flat right A-module, since A_d is faithfully flat.

**Remark* (2). If A and B are commutative, we shall see in Chapter II, § 2 no. 5, Corollary 4 to Proposition 11 that the conditions of Proposition 9 are equivalent to the following:

(f) B *is a flat* A-*module and, for every prime ideal* $\mathfrak{p}$ *of* A, *there exists an ideal* $\mathfrak{q}$ *of* B *such that* $\overset{-1}{\rho}(\mathfrak{q}) = \mathfrak{p}$.

Under the conditions of Proposition 9, let us identify A with a *subring* of B by means of ρ. The relation $\overset{-1}{\rho}(B\mathfrak{a}) = \mathfrak{a}$ then reads $A \cap B\mathfrak{a} = \mathfrak{a}$. On the other hand if F is a left A-module, F is identified with its image in $B \otimes_A F$ under the canonical mapping $x \mapsto 1 \otimes x$; if X is an additive subgroup of F, then we denote by BX the left sub-B-module of $B \otimes_A F$ generated by X. With this notation, we have:

Proposition 10. *Let* B *be a ring and* A *a subring of* B *such that* B *is a faithfully flat right* A-*module. Let* F *be a left* A-*module,* F′, F″ *two submodules of* F. *Then:*

(i) *The canonical mapping* $B \otimes_A F' \to B \otimes_A F$ *is an isomorphism of* $B \otimes_A F'$ *onto* BF′.

(ii) $F \cap BF' = F'$.

(iii) $B(F' + F'') = BF' + BF''$.

(iv) $B(F' \cap F'') = BF' \cap BF''$.

As B is a flat right A-module, the canonical mapping

$$B \otimes_A F' \to B \otimes_A F$$

is injective; taking account of the identifications made, its image is BF′, which proves (i). Assertion (ii) follows from § 2, no. 6, Proposition 7, applied with $E = B$, $E' = A$, and using the formulae $A \otimes_A F = F$ and $A \otimes_A F' = F'$. Assertion (iii) is trivial and (iv) follows from § 2, no. 6, Proposition 6.

6. FAITHFULLY FLAT RINGS AND FINITENESS CONDITIONS

Proposition 11. *Let* B *be a ring and* A *a subring of* B *such that* B *is a faithfully flat right* A-*module. For a left* A-*module* F *to be finitely generated* (resp. *finitely presented*), *it is necessary and sufficient that the* B-*module* $B \otimes_A F$ *be finitely generated* (resp. *finitely presented*).

(1) Without any hypothesis on B, clearly, if F is a finitely generated left A-module, $B \otimes_A F$ is a finitely generated left B-module. Conversely, if $B \otimes_A F$ is a finitely generated B-module, it is generated by a finite number of elements of the form $1 \otimes x_i$ with $x_i \in F$; if M is a sub-A-module of F generated by the x_i and j the canonical injection $M \to F$, $1_B \otimes j: B \otimes_A M \to B \otimes_A F$ is a surjective homomorphism, hence j is surjective (no. 1, Proposition 2), which proves that F is finitely generated.

(2) If F admits a finite presentation, so does $B \otimes_A F$ without any hypothesis

on B (§ 2, no. 8). It remains to prove that, if $B \otimes_A F$ admits a finite presentation, so does F. We already know from (1) that F is finitely generated, hence there exists a surjective homomorphism $u: L \to F$, where L is a finitely generated free A-module. Let R be the kernel of u, so that $B \otimes_A R$ is identified with the kernel of the surjective homomorphism $1_B \otimes u: B \otimes_A L \to B \otimes_A F$ (§ 2, no. 3, *Remark* 2). As $B \otimes_A F$ admits a finite presentation by hypothesis, we conclude (§ 2, no. 8, Lemma 9) that $B \otimes_A R$ is finitely generated; then it follows from (1) that R is a finitely generated A-module and consequently F admits a finite presentation.

PROPOSITION 12. *Let* B *be a ring and* A *a commutative subring of the centre of* B *such that* B *is a faithfully flat* A-*module. For an* A-*module* F *to be projective and finitely generated, it is necessary and sufficient that* $B \otimes_A F$ *be a finitely generated projective left* B-*module.*

The condition is obviously necessary without any hypothesis on A or B (*Algebra*, Chapter II, § 5, no. 1, Corollary to Proposition 4); we prove that it is sufficient. If a finitely generated projective module admits a finite presentation (§ 2, no. 8, Lemma 8), the hypothesis implies that F admits a finite presentation by virtue of Proposition 11, hence, for every A-module M, there is a canonical isomorphism

$$\omega: B \otimes_A \mathrm{Hom}_A(F, M) \to \mathrm{Hom}_B(B \otimes_A F, B \otimes_A M)$$

(§ 2, no. 10, Proposition 11). Then let $v: M \to M''$ be a *surjective* A-module homomorphism and consider the commutative diagram

$$\begin{array}{ccc} B \otimes_A \mathrm{Hom}_A(F, M) & \xrightarrow{\omega} & \mathrm{Hom}_B(B \otimes_A F, B \otimes_A M) \\ {\scriptstyle 1_B \otimes \mathrm{Hom}(1_F, v)} \downarrow & & \downarrow {\scriptstyle \mathrm{Hom}(1_{B\otimes F}, 1_B \otimes v)} \\ B \otimes_A \mathrm{Hom}_A(F, M'') & \xrightarrow[\omega]{} & \mathrm{Hom}_B(B \otimes_A F, B \otimes_A M'') \end{array}$$

As $1_B \otimes v$ is surjective and $B \otimes_A F$ is assumed projective, $\mathrm{Hom}(1_{B\otimes F}, 1_B \otimes v)$ is *surjective* (*Algebra*, Chapter II, § 2, no. 2, Proposition 4) and so then is $1_B \otimes \mathrm{Hom}(1_F, v)$. But as B is a faithfully flat A-module, $\mathrm{Hom}(1_F, v)$ is itself surjective (no. 1, Proposition 2), hence F is a projective A-module (*Algebra*, Chapter II, § 2, no. 2, Proposition 4).

7. LINEAR EQUATIONS OVER A FAITHFULLY FLAT RING

Let B be a ring and A a subring of B. We shall say that the ordered pair (A, B) has the *linear extension property* if it satisfies the following condition:

(E) *Every solution* $(y_k)_{1 \leqslant k \leqslant n}$, *consisting of elements of* B, *of a system of linear equations*

$$\sum_{k=1}^{n} y_k c_{ki} = d_i \qquad (1 \leqslant i \leqslant m) \tag{3}$$

whose coefficients c_{ki} *and right-hand sides* d_i *belong to* A, *is of the form*

$$y_k = x_k + \sum_{j=1}^{p} b_j z_{jk} \qquad (1 \leqslant k \leqslant n) \tag{4}$$

where (x_k) *is a solution of* (3) *consisting of elements of* A, *the* b_j *belong to* B *and each of the* $(z_{jk})_{1 \leqslant k \leqslant n}$ *is a solution of the homogeneous linear system associated with* (3), *consisting of elements of* A.

PROPOSITION 13. *Let* A *be a subring of a ring* B. *For the ordered pair* (A, B) *to satisfy the linear extension property, it is necessary and sufficient that* B *be a faithfully flat* A*-module.*

The condition is *sufficient*. For, as B is a flat A-module, every solution with elements in B of the *homogeneous* linear system associated with (3) is a linear combination with coefficients in B of solutions consisting of elements of A (§ 2, no. 11, Corollary 2, to Proposition 13). The problem then reduces to proving that the existence of a solution of (3) with elements in B implies the existence of *one* solution with elements in A. Now if we set

$$c_k = (c_{ki})_{1 \leqslant i \leqslant m} \in \mathrm{A}_s^m, \qquad d = (d_i) \in \mathrm{A}_s^m,$$

system (3) is equivalent to the equation $\sum_{k=1}^{n} y_k \otimes c_k = 1 \otimes d$ in $\mathrm{B} \otimes_{\mathrm{A}} \mathrm{A}_s^m = \mathrm{B}_s^m$. In other words, if M is the sub-A-module of A_s^m generated by the c_k $(1 \leqslant k \leqslant n)$, the existence of the solution (y_k) of (3) is equivalent (with the identifications made in no. 5) to the relation $d \in \mathrm{BM} \cap \mathrm{A}_s^m$; but as $\mathrm{BM} \cap \mathrm{A}_s^m = \mathrm{M}$ (no. 5, Proposition 10, (ii)), it implies $d \in \mathrm{M}$, that is, the existence of a solution (x_k) of system (3) with elements in A.

The condition is *necessary*. For suppose that (A, B) satisfies the linear extension property; we know already that B is a flat right A-module (§ 2, no. 11, Corollary 2 to Proposition 13); we prove that, for every left ideal $\mathfrak{a}$ of A, $\mathrm{B}\mathfrak{a} \cap \mathrm{A} = \mathfrak{a}$, which shows that B is a faithfully flat right A-module (no. 5, Proposition 9, (d)). Now, let $x \in \mathrm{B}\mathfrak{a} \cap \mathrm{A}$; there exists by hypothesis $y_i \in \mathrm{B}$ and $a_i \in \mathfrak{a}$ such that $\sum_i y_i a_i = x$; property (E) applied to this linear equation with coefficients and right hand side in A shows that there exist $x_i \in \mathrm{A}$ such that $x = \sum_i x_i a_i$, hence $x \in \mathfrak{a}$.

4. FLAT MODULES AND "TOR" FUNCTORS

For the benefit of readers conversant with Homological Algebra (*), we shall indicate quickly how the theory of flat modules is related to that of Tor functors.

PROPOSITION 1. *Let* E *be a right* A*-module. The following four properties are equivalent:*

(a) E *is flat.*
(b) *For every left* A*-module* F *and every integer* $n \geqslant 1$, $\mathrm{Tor}_n^A(E, F) = 0$.
(c) *For every left* A*-module* F, $\mathrm{Tor}_1^A(E, F) = 0$.
(d) *For every finitely generated left ideal* $\mathfrak{a}$ *of* A,

$$\mathrm{Tor}_1^A(E, A_s/\mathfrak{a}) = 0.$$

We show that (a) implies (b). Let

$$\dots \to L_n \to L_{n-1} \to \dots \to L_0 \to F \to 0$$

be a free resolution of F. As E is flat, the sequence

$$(1) \qquad \dots \to E \otimes L_n \to E \otimes L_{n-1} \to \dots \to E \otimes L_0 \to E \otimes F \to 0$$

is exact. As the $\mathrm{Tor}_n^A(E, F)$ are isomorphic to homology groups of the complex (1), they are zero for $n \geqslant 1$. It is trivial that (b) implies (c) and (c) implies (d). We show finally that (d) implies (a). The exact sequence

$$0 \to \mathfrak{a} \to A_s \to A_s/\mathfrak{a} \to 0$$

gives the exact sequence

$$\mathrm{Tor}_1^A(E, A_s/\mathfrak{a}) \to E \otimes_A \mathfrak{a} \to E \otimes_A A.$$

As (d) holds, the canonical homomorphism

$$E \otimes_A \mathfrak{a} \to E \otimes_A A = E$$

is injective, which means that E is flat (§ 2, no. 3, Proposition 1).

Proposition 1 provides a characterization of flat modules which is often useful in applications. We shall restrict ourselves, by way of an example, to giving a

(*) See the part of this Treatise devoted to categories and, in particular, Abelian categories (in preparation). Until this is published, the reader can consult H. CARTAN-S. EILENBERG, *Homological Algebra*, Princeton, 1956, or R. GODEMENT, *Théorie des Faisceaux*, Paris (Hermann), 1958.

new proof of Proposition 5 of § 2, no. 5. If E′ and E″ are flat, the exact sequence

$$\mathrm{Tor}_1^A(E', F) \to \mathrm{Tor}_1^A(E, F) \to \mathrm{Tor}_1^A(E'', F)$$

shows that $\mathrm{Tor}_1^A(E, F) = 0$ for every left A-module F, hence E is flat. If E and E″ are flat, the exact sequence

$$\mathrm{Tor}_2^A(E'', F) \to \mathrm{Tor}_1^A(E', F) \to \mathrm{Tor}_1^A(E, F)$$

shows that $\mathrm{Tor}_1^A(E', F) = 0$, hence E′ is flat.

PROPOSITION 2. *Let* R, S *be two rings,* $\rho: R \to S$ *a homomorphism and* F *a left* R-*module. The following two properties are equivalent:*

(a) $\mathrm{Tor}_1^R(\rho_*(E), F) = 0$ *for every right* S-*module* E.

(b) *The left* S-*module* $\rho^*(F) = F_{(S)} = S \otimes_R F$ *is flat and* $\mathrm{Tor}_1^R(\rho_*(S_d), F) = 0$.

Suppose that (a) holds. Taking $E = S_d$, we see that $\mathrm{Tor}_1^R(\rho_*(S_d), F) = 0$. We show also that $F_{(S)}$ is a flat S-module. For that, we note that, if E is a right S-module, the additive group $E \otimes_S F_{(S)}$ is identified with $\rho_*(E) \otimes_R F$. Then if there is an exact sequence of right S-modules

$$0 \to E' \to E \to E'' \to 0$$

we obtain, using (a), an exact sequence

$$0 \to \rho_*(E') \otimes_R F \to \rho_*(E) \otimes_R F \to \rho_*(E'') \otimes_R F \to 0$$

or also

$$0 \to E' \otimes_S F_{(S)} \to E \otimes_S F_{(S)} \to E'' \otimes_S F_{(S)} \to 0$$

which proves that $F_{(S)}$ is flat.

Conversely, if (b) holds, we have first of all, for every *free* right S-module $L = S_d^{(I)}$, $\mathrm{Tor}_1^R(\rho_*(L), F) = (\mathrm{Tor}_1^R(\rho_*(S_d), F))^{(I)} = 0$. Every right S-module E can be written in the form $E = L/H$ for a suitable free S-module L; then we have the exact sequence

$$(2)\quad 0 = \mathrm{Tor}_1^R(\rho_*(L), F) \to \mathrm{Tor}_1^R(\rho_*(E), F) \to \rho_*(H) \otimes_R F \to \rho_*(L) \otimes_R F.$$

But as $F_{(S)}$ is flat, the homomorphism $H \otimes_S F_{(S)} \to L \otimes_S F_{(S)}$ is injective and is identified with the homomorphism

$$\rho_*(H) \otimes_R F \to \rho_*(L) \otimes_R F.$$

Then it follows from (2) that $\mathrm{Tor}_1^R(\rho_*(E), F) = 0$.

Remark. Proposition 2 also follows from the existence of the exact sequence

$$E \otimes_S \mathrm{Tor}_1^R(\rho_*(S_d), F) \to \mathrm{Tor}_1^R(\rho_*(E), F) \to \mathrm{Tor}_1^S(E, S_d \otimes_R F) \to 0$$

arising from the spectral sequence of the "associativity" of the Tor functors.

EXERCISES

§ 1

1. In the commutative diagram (10) suppose that the ordered pair (u', v') is an exact sequence and that $v \circ u = 0$. Show that

$$\operatorname{Im}(b) \cap \operatorname{Im}(u') = b(\operatorname{Ker}(c \circ v)) \quad \text{and} \quad \operatorname{Ker}(b) + \operatorname{Ker}(v) = \overset{-1}{b}(\operatorname{Im}(u' \circ a)).$$

2. Consider a commutative diagram of commutative groups

$$\begin{array}{ccccc} A & \xrightarrow{u} & B & \xrightarrow{v} & C \\ {\scriptstyle a}\downarrow & & {\scriptstyle b}\downarrow & & \downarrow{\scriptstyle c} \\ A' & \xrightarrow[u']{} & B' & \xrightarrow[v']{} & C' \\ {\scriptstyle a'}\downarrow & & {\scriptstyle b'}\downarrow & & \\ A'' & \xrightarrow[u'']{} & B'' & & \end{array}$$

Suppose that: (1) (u, v) and (b, b') are exact sequences; (2) $v' \circ u' = 0$ and $a' \circ a = 0$; (3) c and u' are injective and a' is surjective. Show that under these conditions u'' is injective.

3. Consider a commutative diagram of commutative groups

$$\begin{array}{ccccccc} & & B & \longrightarrow & C & \longrightarrow & D \\ & & \downarrow & & \downarrow & & \downarrow{\scriptstyle d} \\ A' & \longrightarrow & B' & \longrightarrow & C' & \longrightarrow & D' \\ \downarrow & & \downarrow & & \downarrow & & \\ A'' & \xrightarrow{u''} & B'' & \longrightarrow & C'' & & \\ {\scriptstyle a''}\downarrow & & \downarrow & & & & \\ A''' & \xrightarrow[u''']{} & B''' & & & & \end{array}$$

in which the rows and columns are assumed to be exact, d and u'' injective and a'' surjective. Show that under these conditions u''' is injective. Generalize this result.

4. Consider a commutative diagram of commutative groups

$$\begin{array}{ccccccc} A & \longrightarrow & B & \longrightarrow & C & \longrightarrow & D \\ {\scriptstyle a}\downarrow & & {\scriptstyle b}\downarrow & & {\scriptstyle c}\downarrow & & {\scriptstyle d}\downarrow \\ A' & \longrightarrow & B' & \longrightarrow & C' & \longrightarrow & D' \end{array}$$

where the two rows are assumed to be exact.

(a) Show that, if a is surjective and b and d injective, then c is injective.
(b) Show that, if d is injective and a and c surjective, then b is surjective.

5. Suppose that an exact sequence $A' \xrightarrow{u} A \xrightarrow{v} A'' \to 0$ and two surjective homomorphisms $B' \xrightarrow{a'} A'$, $B'' \xrightarrow{a''} A''$ are given, where A, A′, A″, B′, B″ are modules over the same ring. Show that, if B″ is a *projective* module, there exists a surjective homomorphism $a\colon B' \oplus B'' \to A$ such that the diagram

$$\begin{array}{ccccc} B' & \xrightarrow{i} & B' \oplus B'' & \xrightarrow{p} & B'' \\ {\scriptstyle a'}\downarrow & & {\scriptstyle a}\downarrow & & {\scriptstyle a''}\downarrow \\ A' & \xrightarrow{u} & A & \xrightarrow{v} & A'' \end{array}$$

is commutative (i and p being the canonical mappings).

6. Suppose that an exact sequence $0 \to A' \xrightarrow{u} A \xrightarrow{v} A''$ and two injective homomorphisms $A' \xrightarrow{a'} C'$, $A'' \xrightarrow{a''} C''$ are given, where A, A′, A″, C′, C″ are modules over the same ring. Show that, if C′ is an *injective* module (*Algebra*, Chapter II, § 2, Exercise 11), there exists an injective homomorphism $a\colon A \to C' \oplus C''$ such that the diagram

$$\begin{array}{ccccc} A' & \xrightarrow{u} & A & \xrightarrow{v} & A'' \\ {\scriptstyle a'}\downarrow & & {\scriptstyle a}\downarrow & & {\scriptstyle a''}\downarrow \\ C' & \xrightarrow{i} & C' \oplus C'' & \xrightarrow{p} & C'' \end{array}$$

are commutative (i and p being the canonical mappings).

7. Let U, V, W be three commutative groups and $f\colon U \to V$, $g\colon V \to W$ homomorphisms.

(a) Consider the diagram

$$\begin{array}{ccccccccc} 0 & \longrightarrow & U & \xrightarrow{\alpha} & U \times V & \xrightarrow{\beta} & V & \longrightarrow & 0 \\ & & {\scriptstyle f}\downarrow & & {\scriptstyle h}\downarrow & & {\scriptstyle -g}\downarrow & & \\ 0 & \longrightarrow & V & \xrightarrow{\gamma} & W \times V & \xrightarrow{\delta} & W & \longrightarrow & 0 \end{array}$$

where $\alpha(u) = (u, f(u))$, $\beta(u, v) = v - f(u)$, $\gamma(v) = (g(v), v)$, $\delta(w, v) = w - g(v)$, $h(u, v) = (g(f(u)), v)$. Show that this diagram is commutative and that its rows are exact.

(b) Deduce from (a) and Proposition 2 of no. 4 an exact sequence

$$0 \to \operatorname{Ker}(f) \to \operatorname{Ker}(g \circ f) \to \operatorname{Ker}(g) \to \operatorname{Coker}(f) \to \operatorname{Coker}(g \circ f) \to \operatorname{Coker}(g) \to 0.$$

Give a direct definition of this exact sequence.

§ 2

1. Give an example of an exact sequence $0 \to N' \to N \to N'' \to 0$ of left A-modules and a right A-module E such that E is N'-flat and N''-flat, but not N-flat (take for example $N' = N'' = \mathbf{Z}/2\mathbf{Z}$).

2. Let M, N be two submodules of an A-module E such that $M + N$ is flat. For M and N to be flat, it is necessary and sufficient that $M \cap N$ be flat.

3. Let A be the ring K[X, Y] of polynomials in two indeterminates over a field K.

(a) Consider in A the principal ideals $\mathfrak{b} = (X)$, $\mathfrak{c} = (Y)$, which are free A-modules and whose intersection $\mathfrak{b} \cap \mathfrak{c} = (XY)$ is also free. Show that $\mathfrak{a} = \mathfrak{b} + \mathfrak{c}$ is not a flat A-module, although $\mathfrak{a}$ is torsion-free (cf. *Algebra*, Chapter III, § 2, Exercise 4).

(b) In the A-module A^2, let R be the submodule consisting of the elements $(x, -x)$ where $x \in \mathfrak{a}$. In the A-module A^2/R, let M, N be the submodules which are the images of the factor submodules of A^2; show that M and N are isomorphic to A, but that $M \cap N$ is not a flat A-module.

4. (a) Give an example of an exact sequence

$$0 \to E' \to E \to E'' \to 0$$

which does not split and all of whose terms are flat modules (cf. *Algebra*, Chapter VII, § 3, Exercise 8 (b)).

(b) From (a) deduce an example of an exact sequence

$$0 \to E' \to E \to E'' \to 0$$

which does not split and whose terms are right A-modules which are *not flat*, such that for every left A-module F the sequence

$$0 \to E' \otimes F \to E \otimes F \to E'' \otimes F \to 0$$

is exact (use Lemma 2 of no. 1).

5. Give an example of a right A-module E, a left A-module F and two submodules F', F'' of F such that the canonical image of $E \otimes (F' \cap F'')$ in $E \otimes F$ is not the intersection of the canonical images of $E \otimes F'$ and $E \otimes F''$ (cf. Exercise 3(a)).

¶ 6. Let A be a ring and M a left A-module. An exact sequence

$$L_n \to L_{n-1} \to \cdots \to L_1 \to L_0 \to M \to 0$$

where L_i is a free left A-module ($0 \leqslant i \leqslant n$), is called a *presentation* of M *of length n* or an *n-presentation* of M. The presentation is called finite if all the L_i are finitely generated free modules.

If M is a finitely generated left A-module, we denote by $\lambda(M)$ the least upper bound (finite or $+\infty$) of the integers $n \geqslant 0$ such that M has a finite n-presentation. If M is not finitely generated, we set $\lambda(M) = -1$.

(a) Let $0 \to P \to N \to M \to 0$ be an exact sequence of left A-modules. Then $\lambda(N) \geqslant \inf(\lambda(P), \lambda(M))$. (Starting with two n-presentations of P and M respectively, derive one for N using Exercise 5 of § 1.)

(b) Let $M_n \xrightarrow{u_n} M_{n-1} \to \cdots \to M_0 \xrightarrow{u_0} M \to 0$ be a finite n-presentation of M; show that, if $\lambda(M) > n$, $\mathrm{Ker}(u_n)$ is a finitely generated A-module. (Let

$$L_{n+1} \xrightarrow{v_{n+1}} L_n \longrightarrow \cdots \longrightarrow L_2 \xrightarrow{v_2} L_1 \xrightarrow{v_1} L_0 \xrightarrow{v_0} M \longrightarrow 0$$

be a finite $(n+1)$-presentation of M and let $P = \mathrm{Ker}(v_0)$ be such that there is an n-presentation of P:

$$L_{n+1} \xrightarrow{v_{n+1}} L_n \longrightarrow \cdots \longrightarrow L_2 \xrightarrow{v_2} L_1 \xrightarrow{v_1} P \longrightarrow 0.$$

Applying the method of (a) to the exact sequence $0 \to P \to L_0 \to M \to 0$ we obtain an exact sequence

$$M_n \oplus L_{n+1} \xrightarrow{w_n} M_{n-1} \oplus L_n \xrightarrow{w_{n-1}} \cdots \xrightarrow{w_1} M_0 \oplus L_1 \xrightarrow{w_0} L_0 \longrightarrow 0$$

and exact sequences $0 \to \mathrm{Ker}(v_{i+1}) \to \mathrm{Ker}(w_i) \to \mathrm{Ker}(u_i) \to 0$ (§ 1, no. 4, Proposition 2). Observe finally that $\mathrm{Ker}(w_i)$ is a direct factor of $M_i \oplus L_{i+1}$.

(c) Show that with the hypotheses of (a)

$$\lambda(M) \geqslant \inf(\lambda(N), \lambda(P) + 1).$$

(If $n \leqslant \inf(\lambda(N), \lambda(P) + 1)$, show by induction on n that $\lambda(M) \geqslant n$, arguing as in (a) and using (b).)

(d) Show that with the hypotheses of (a)

$$\lambda(P) \geqslant \inf(\lambda(N), \lambda(M) - 1).$$

(Same method as in (c).) Deduce that, if $\lambda(N) = +\infty$, then $\lambda(M) = \lambda(P) + 1$.

(e) Deduce from (a), (c) and (d) that, if $N = M \oplus P$, then

$$\lambda(N) = \inf(\lambda(M), \lambda(P)).$$

In particular, for N to admit a finite presentation it is necessary and sufficient that M and P do so too.

(f) Let N_1, N_2 be two submodules of an A-module M. Suppose that N_1 and N_2 admit finite presentations. For $N_1 + N_2$ to admit a finite presentation it is necessary and sufficient that $N_1 \cap N_2$ be finitely generated.

7. (a) With the notation of Exercise 6, show that, if M is a projective module, then $\lambda(M) = -1$ or $\lambda(M) = +\infty$. If A is a left Noetherian ring, then, for every A-module M, $\lambda(M) = -1$ or $\lambda(M) = +\infty$.

(b) If $\mathfrak{a}$ is a left ideal of a ring A, which is not finitely generated, $A_s/\mathfrak{a}$ is a monogenous A-module which does not admit a finite presentation, in other words $\lambda(A_s/\mathfrak{a}) = 0$ (no. 8, Lemma 9).

(c) Give an example of a monogenous left ideal $\mathfrak{a}$ of a ring A such that $A_s/\mathfrak{a}$ (which has a finite presentation) admits a dual which is not a finitely generated right A-module.

(d) Let K be a commutative field, E the vector space $K^{(\mathbf{N})}$, (e_n) the canonical basis of E and T the tensor algebra of E, which therefore has a basis consisting of the finite products $e_{i_1}e_{i_2}\ldots e_{i_k}$ ($k \geqslant 0$, $i_j \in \mathbf{N}$ for all j). For a given integer n, let $\mathfrak{b}$ be the two-sided ideal of T generated by the products

$$e_1e_0,\ e_2e_1,\ \ldots,\ e_ne_{n-1}$$

and $e_{n+k}e_n$ for all $k \geqslant 1$; let A be the quotient ring $T/\mathfrak{b}$, and for every integer m, let a_m be the canonical image of e_m in A. Show that, if $M = A_s/Aa_0$, then $\lambda(M) = n$ (observe that, for $m \leqslant n-1$, the left annihilator of a_m is Aa_{m+1} and use Exercise 6(b)).

8. Let C be a commutative ring and E, F two C-modules. Show that $\lambda(E \otimes_C F) \geqslant \inf(\lambda(E), \lambda(F))$.

9. Let E be a finitely presented left A-module.

(a) Show that for every family $(F_\iota)_{\iota \in I}$ of right A-modules, the canonical homomorphism $E \otimes_A \left(\prod_{\iota \in I} F_\iota\right) \to \prod_{\iota \in I} (E \otimes_A F_\iota)$ (*Algebra*, Chapter II, § 3, no. 7) is bijective.

(b) Let $(G_\alpha, \phi_{\beta\alpha})$ be a direct system of left A-modules; show that the canonical homomorphism

$$\varinjlim \operatorname{Hom}_A(E, G_\alpha) \to \operatorname{Hom}_A(E, \varinjlim G_\alpha)$$

is bijective.

10. (a) Let A be a ring, I a set and R a submodule of $L = A_d^{(I)}$. Let $\mathfrak{S}$ be the set of ordered pairs (J, S), where J is a finite subset of I and S a finitely generated submodule of $A_d^J \cap R$. $\mathfrak{S}$ is ordered by the relation "$J \subset J'$ and $S \subset S'$"; show that $\mathfrak{S}$ is directed with respect to this order relation, that the family (A_d^J/S) is a direct system of right A-modules with $\mathfrak{S}$ as indexing set and that there exists an isomorphism of L/R onto $\varinjlim_{(J,S)\in\mathfrak{S}} (A_d^J/S)$.

(b) Deduce from (a) that every A-module is a direct limit of finitely presented A-modules.

11. Let E be a right A-module. E is called *pseudo-coherent* if every finitely generated submodule of E is finitely presented; every submodule of a pseudo-coherent module is pseudo-coherent. E is called *coherent* if it is pseudo-coherent and finitely generated (and therefore finitely presented).

(a) Let $0 \to E' \to E \to E'' \to 0$ be an exact sequence of right A-modules. Show that, if E is pseudo-coherent (resp. coherent) and E′ is finitely generated E″ is pseudo-coherent (resp. coherent). Show that, if E′ and E″ are pseudo-coherent (resp. coherent), so is E. Show that, if E and E″ are coherent, so is E′ (use Exercise 6 and Lemma 9 of no. 8).

(b) Let E be a coherent A-module and E′ a pseudo-coherent (resp. coherent) A-module. Show that, for every homomorphism $u: E \to E'$, $\mathrm{Im}(u)$ and $\mathrm{Ker}(u)$ are coherent and that $\mathrm{Coker}(u)$ is pseudo-coherent (resp. coherent) (use (a)).

(c) Show that every direct sum (resp. every finite direct sum) of pseudo-coherent (resp. coherent) modules is a pseudo-coherent (resp. coherent) module.

(d) If E is a pseudo-coherent module and M, N are coherent submodules of E, show that $M + N$ and $M \cap N$ are coherent (use (a) and (c)).

(e) Suppose that A is commutative. Show that, if E is a coherent A-module and F a coherent (resp. pseudo-coherent) A-module, $\mathrm{Hom}_A(E, F)$ is a coherent (resp. pseudo-coherent) A-module. (Reduce it to the case where F is coherent and consider a finite presentation of E, then use (b).)

¶ 12. (a) Let A be a ring. Show that the following four properties are equivalent:

(α) The right A-module A_d is coherent (Exercise 11).
(β) Every finitely presented right A-module is coherent.
(γ) Every A-module A_s^I (I an arbitrary set) is flat.
(δ) Every product of flat left A-modules is flat.

(To prove that (α) implies (β), use Exercise 11(b). To see that (γ) implies (α), use Proposition 13 of no. 11 and argue by *reductio ad absurdum*. To show that (α) implies (δ), use Exercises 9.)

Such a ring is called *right coherent* and the concept of a *left coherent* ring is defined similarly.

(b) Show that every right Noetherian ring is right coherent. Give an example of a right Artinian ring which is not left coherent (cf. *Algebra*, Chapter VIII, § 2, Exercise 4).

(c) The ring of a non-discrete valuation of height 1 (Chapter VI) is coherent but contains ideals which are not coherent and admits (monogenous) quotient modules which are not pseudo-coherent.

(d) Show that, if A is a right coherent ring, then, for every right A-module E, $\lambda(E) = -1$ or $\lambda(E) = 0$ or $\lambda(E) = +\infty$ (Exercise 6).

(e) Let $(A_\alpha, \phi_{\beta\alpha})$ be a direct system of rings whose indexing set is directed and let $A = \varinjlim A_\alpha$. Suppose that, for $\alpha \leqslant \beta$, A_β is a flat right A_α-module. Show that, if the A_α are right coherent, so is A. (Observe that A is a flat A_α-module for all α and that, if E is a finitely generated submodule of A_d, there exists an index α and a finitely generated submodule E_α of $(A_\alpha)_d$ such that $E_\alpha \otimes_{A_\alpha} A$ is isomorphic to E.)

* (f) Deduce from (e) that every polynomial ring (in any finite or infinite set of indeterminates) over a Noetherian commutative ring is coherent. Deduce from this that a quotient ring of a coherent ring is not necessarily coherent. *

(g) For A to be left coherent, it is necessary and sufficient that the left annihilator of every element of A be finitely generated and that the intersection of two finitely generated left ideals of A be finitely generated (use Exercise 6(f)).

¶ 13. Let A, B be two rings, F an (A, B)-bimodule and G a right B-module. Show that, if G is injective (*Algebra* Chapter II, § 2, Exercise 11) and F is a flat left A-module, the right A-module $\mathrm{Hom}_B(F, G)$ is injective. (Use the isomorphism

$$\mathrm{Hom}_A(E, \mathrm{Hom}_B(F, G)) \to \mathrm{Hom}_B(E \otimes_A F, G)$$

for a right A-module E (*Algebra*, Chapter II, § 4, no. 1).)

¶ 14. Let A, B be two rings, E a left A-module, F an (A, B)-bimodule and G a right B-module; consider the canonical homomorphism (*Algebra*, Chapter II, § 4, Exercise 5)

$$\sigma: \mathrm{Hom}_B(F, G) \otimes_A E \to \mathrm{Hom}_B(\mathrm{Hom}_A(E, F), G)$$

such that $(\sigma(u \otimes x))(v) = u(v(x))$ for all $x \in E$, $u \in \mathrm{Hom}_B(F, G)$,

$$v \in \mathrm{Hom}_A(E, F).$$

Show that, if G is an injective B-module (*Algebra*, Chapter II, § 2, Exercise 11) and E finitely presented, σ is bijective. (Consider first the case where E is free and finitely generated.)

¶ 15. Let A be a ring. Show that every left A-module E which is flat and finitely presented is projective. (Given a surjective left A-module homomorphism $u: F \to F''$, let u' be the homomorphism

$$\mathrm{Hom}(1_E, u): \mathrm{Hom}_A(E, F) \to \mathrm{Hom}_A(E, F'')$$

and $\bar{u}$ the homomorphism

$$\mathrm{Hom}(u', 1_G): \mathrm{Hom}_{\mathbf{Z}}(\mathrm{Hom}_A(E, F''), G) \to \mathrm{Hom}_{\mathbf{Z}}(\mathrm{Hom}_A(E, F,) G),$$

where G is a divisible **Z**-module. Using Exercise 14 first, prove that $\bar{u}$ is injective; then, with a suitable choice for G (*Algebra*, Chapter II, § 2, Exercise 14), show that u' is surjective.)

16. Let A be a ring and a an element of A. Show that the following properties are equivalent.

(α) $a \in aAa$.

(β) aA is a direct factor of the module A_d.

(γ) A_d/aA is a flat right A-module.

(δ) For every left ideal $\mathfrak{b}$ of A, $aA \cap \mathfrak{b} = a\mathfrak{b}$.

(Use the Corollary to Proposition 7 of no. 6 to prove the equivalence of (γ) and (δ) and show directly that (δ) implies (α) and (α) implies (β), by proving the existence of an idempotent element $e \in aA$ such that $eA = aA$.)

17. Let A be a ring. Show that the following properties are equivalent:

(α) Every element $a \in A$ satisfies the equivalent properties of Exercise 16.

(β) Every finitely generated right ideal of A is a direct factor of A_d.

(γ) Every left A-module is flat.

(δ) Every right A-module is flat.

Then A is called an *absolutely flat* ring (*).

(To see that (α) implies (β), use Exercise 15(b) of *Algebra*, Chapter VIII, § 6.)

¶ 18. Let A be an absolutely flat ring (Exercise 17).

(a) Let P be a projective right A-module. Show that every finitely generated submodule E of P is a direct factor of P. (Reduce it to the case where P is free and finitely generated. Then note that P/E is finitely presented and use Exercises 15.)

(b) Show that every projective right A-module P is the direct sum of monogenous submodules isomorphic to monogenous right ideals of A. (Use Kaplansky's Theorem (*Algebra*, Chapter II, § 2, Exercises 3) to reduce the problem to the case where P is generated by a countable family of elements, then use (a).)

(c) Give an example of an absolutely flat ring A and a non-projective finitely generated A-module (consider a quotient of A by an ideal which is not finitely generated; cf. *Algebra*, Chapter VIII, § 6, Exercise 15(f) and *Commutative Algebra*, Chapter II, § 4, Exercise 17).

19. Let A be a ring. Show that the following properties are equivalent:

(α) A is semisimple.

(β) Every right ideal of A is an injective A-module.

(γ) Every right A-module is projective.

(δ) Every right A-module is injective.

20. Let A be an integral domain, B an A-algebra which is a flat A-module and M a torsion-free A-module. Show that, if $t \in B$ is not a divisor of zero, the relation $t.z = 0$ for $z \in B \otimes_A M$ implies $z = 0$. (Reduce this to the case where M is finitely generated and, by embedding M in a finitely generated free A-module, to the case where M = A.)

(*) This is a modification of the terminology "regular ring" introduced in *Algebra*, Chapter VIII, § 6, Exercise 15, as this has a completely different meaning in Commutative Algebra.

21. Let S be a commutative ring, R a commutative S-algebra, B an S-algebra (commutative or otherwise) and $B_{(R)}$ the R-algebra obtained from B by extension of scalars. Suppose that R is a flat S-module and that B is a finitely generated S-module. If Z is the centre of B, show that the canonical homomorphism of $Z_{(R)} = Z \otimes_S R$ to $B_{(R)}$ is an isomorphism of $Z_{(R)}$ onto the centre of $B_{(R)}$. (Use the exact sequence

$$0 \longrightarrow Z \longrightarrow B \xrightarrow{\theta} \mathrm{Hom}_S(B, B)$$

where $\theta(x)(y) = xy - yx$, and Proposition 11 of no. 10.)

¶ 22. Let E be a left A-module. For every right ideal $\mathfrak{a}$ of A and every element $a \in A$, denote by $\mathfrak{a}: a$ the set of $x \in A$ such that $ax \in \mathfrak{a}$ and by $\mathfrak{a}E: a$ the set of $y \in E$ such that $ay \in \mathfrak{a}E$. Then clearly $(\mathfrak{a}: a)E \subset \mathfrak{a}E: a$. Show that, for E to be flat, it is necessary and sufficient that, for every right ideal $\mathfrak{a}$ of A and every element $a \in A$, $(\mathfrak{a}: a)E = \mathfrak{a}E: a$. (To see that the condition is necessary, consider the exact sequence of right A-modules $0 \rightarrow (\mathfrak{a}: a)/\mathfrak{a} \xrightarrow{\psi} A_d/\mathfrak{a} \xrightarrow{\varphi} A_d/\mathfrak{a}$, where ψ is the canonical injection and ϕ the mapping obtained by taking quotients under left multiplication by a. To see that the condition is sufficient apply the criterion of Corollary 1 to Proposition 13 of no. 11: starting with a relation $\sum_{i=1}^{n} a_i x_i = 0$, where $a_i \in A$, $x_i \in E$, apply the hypothesis to the ideal $\mathfrak{a}_2 = \sum_{i=2}^{n} a_i A$ and the element a_1 and argue by induction on n.)

23. (a) Let $0 \rightarrow R \rightarrow L \rightarrow E \rightarrow 0$ be an exact sequence of left A-modules, where L is a free A-module; let (e_α) be a basis of L. Show that the following conditions are equivalent:

(α) E is flat.

(β) For all $x \in R$, if $\mathfrak{a}_x$ is the right ideal generated by the components of x with respect to the basis (e_α), then $x \in R\mathfrak{a}_x$.

(γ) For all $x \in R$, there exists a homomorphism $u_x: L \rightarrow R$ such that $u_x(x) = x$.

(δ) For every finite sequence $(x_i)_{1 \leqslant i \leqslant n}$ of elements of R, there exists a homomorphism $u: L \rightarrow R$ such that $u(x_i) = x_i$ for $1 \leqslant i \leqslant n$. (Use the Corollary to Proposition 7 of no. 6.)

(b) Let $\mathfrak{a}$ be a left ideal of A such that $A/\mathfrak{a}$ is a flat A-module. Show that, for every finitely generated left ideal $\mathfrak{b} \subset \mathfrak{a}$, there exists $x \in A$ such that

$$\mathfrak{b} \subset Ax \subset \mathfrak{a}$$

(use condition (δ) of (a)).

(c) Derive from (a) a new proof of the result of Exercise 15.

(d) Let $\mathfrak{r}$ be the Jacobson radical of A and $0 \to R \to L \to E \to 0$ an exact sequence of left A-modules such that L is free. Suppose that E is flat and R is contained in $\mathfrak{r}L$. Show that $R = 0$ (in the notation of (a) observe that $\mathfrak{a}_x$ is a finitely generated ideal and that $\mathfrak{a}_x = \mathfrak{a}_x\mathfrak{r}$).

(e) Let E be a finitely generated flat A-module; suppose that there exists a two-sided ideal $\mathfrak{b}$ of A contained in the Jacobson radical of A such that $E/\mathfrak{b}E$ is a free $(A/\mathfrak{b})$-module. Show that E is then a free A-module (observe that there exists a finitely generated free A-module L such that $L/\mathfrak{b}L$ is isomorphic to $E/\mathfrak{b}E$ and use *Algebra*, Chapter VIII, § 6, no. 3, Corollary 4 to Proposition 6; then apply (d)).

24. A submodule M′ of a right A-module M is called *pure* if, denoting by $j: M' \to M$ the canonical injection, the homomorphism

$$j \otimes 1_N : M' \otimes_A N \to M \otimes_A N$$

is injective for every left A-module N. This is so if M′ is a direct factor of M or if M/M′ is flat, but these two conditions are not necessary (Exercise 4).

(a) Show that, for M′ to be a pure submodule of M, it is necessary and sufficient that, if $(m'_i)_{i \in I}$ is a finite family of elements of M′, $(x_j)_{j \in J}$ a family of elements of M such that $m'_i = \sum_{j \in J} x_j a_{ji}$ for all $i \in I$ and a family (a_{ji}) of elements of A, then there exists a family $(x'_j)_{j \in J}$ of elements of M′ such that $m'_i = \sum_{j \in J} x'_j a_{ji}$ for all $i \in I$. (To see that the condition is sufficient, use Lemma 10 of no. 11 to show that $M' \otimes_A N \to M \otimes_A N$ is injective for every finitely generated left A-module N; to see that the condition is necessary, consider a finitely generated left A-module $N = L/R$, where L is a finitely generated free A-module and R is a finitely generated submodule of L.) Deduce from this criterion that, if A is a *principal ideal domain*, the notion of pure submodule of an A-module coincides with that of *Algebra*, Chapter VII, § 2, Exercise 7.

(b) Let M be a right A-module, M′ a submodule of M and M″ a submodule of M′. Show that, if M′ is a pure submodule of M and M″ a pure submodule of M′, then M″ is a pure submodule of M and M′/M″ is a pure submodule of M/M″. If M″ is a pure submodule of M, M″ is a pure submodule of M′.

(c) Show that, if N and P are two submodules of M such that $N \cap P$ and $N + P$ are pure in M, then N and P are pure submodules of M. Give an example of two submodules N, P of $\mathbf{Z}^2$ which are pure in $\mathbf{Z}^2$ but where $N + P$ is not a pure submodule of $\mathbf{Z}^2$.

(d) Let C be a commutative ring and E, F two C-modules; show that, if E′ (resp. F′) is a pure submodule of E (resp. F), the canonical mapping $E' \otimes_C F' \to E \otimes_C F$ is injective and identifies $E' \otimes_C F'$ with a pure submodule of $E \otimes_C F$.

(e) Let $\rho: A \to B$ be a ring homomorphism, M a right A-module and M′ a pure submodule of M. Show that $M'_{(B)} = M' \otimes_A B$ is canonically identified with a pure submodule of $M_{(B)} = M \otimes_A B$.

§ 3

1. (a) For the direct sum of a family (E_λ) of A-modules to be faithfully flat, it is sufficient that each of the E_λ be flat and that at least one of them be faithfully flat.

(b) Deduce from (a) that, if A is a simple ring, every non-empty A-module is faithfully flat. Is the result true for semisimple rings?

2. Let (p_n) be the strictly increasing sequence of prime numbers and A the product ring $\prod_n \mathbf{Z}/p_n\mathbf{Z}$. Show that the direct sum E of the $\mathbf{Z}/p_n\mathbf{Z}$ is a faithful projective A-module which is not faithfully flat (observe that E is an ideal of A such that $E^2 = E$).

3. Let A be a right coherent ring (§ 2, Exercise 12). For a product of left A-modules to be faithfully flat, it is sufficient that each of them be flat and that at least one of them be faithfully flat.

Deduce that, if A is a coherent commutative ring, the ring of formal power series $A[[X_1, \ldots, X_n]]$ is a faithfully flat A-module.

4. Let A be a simple algebra over a commutative field K and B a subalgebra of A which is semisimple but not simple. Show that A is a faithfully flat (right or left) B-module, but that there exist right B-modules E which are not faithfully flat, whilst $E \otimes_B A$ is always faithfully flat (Exercise 1).

5. Let A be a commutative ring and M a flat A-module containing a submodule N which is not a flat module (cf. § 2, Exercise 3). Let B (resp. C) be the A-module $A \oplus N$ (resp. $A \oplus M$) in which multiplication is defined by $(a, x)(a', x') = (aa', ax' + a'x)$; then B is not a flat A-module, but the B-module C is a faithfully flat A-module and consequently B satisfies the conditions of Proposition 8 of no. 5.

6. Give an example of an integral domain A and a ring B of which A is a subring, such that B is a flat A-module but there exists an A-module E which is neither projective nor finitely generated, for which $B \otimes_A E$ is a finitely generated free B-module.

7. If K is a field, the ring K[X] and the field K(X) are faithfully flat K-modules, but K(X) is not a faithfully flat K[X]-module.

8. Let p be a prime number and A the subring of $\mathbf{Q}$ consisting of the fractions k/p^n, where $k \in \mathbf{Z}$, $n \geqslant 0$. Show that A is a flat $\mathbf{Z}$-module and that

there exists a $\mathbf{Z}$-module E which is not flat but where $A \otimes_{\mathbf{Z}} E$ is a flat A-module.

9. Let A be a commutative ring, B an A-algebra, $(C_\lambda)_{\lambda \in L}$ a family of A-algebras and $B_\lambda = C_\lambda \otimes_A B$ the tensor product algebra of C_λ and B for all $\lambda \in L$. Let E be a left B-module. Set $E_\lambda = B_\lambda \otimes_B E = C_\lambda \otimes_A E$; this is a (B_λ, C_λ)-bimodule. Similarly, if F is a right B-module, set

$$F_\lambda = F \otimes_B B_\lambda = F \otimes_A C_\lambda,$$

which is a (C_λ, B_λ)-bimodule.

(a) Show that the (C_λ, C_λ)-bimodule $F_\lambda \otimes_{B_\lambda} E_\lambda$ is isomorphic to $(F \otimes_B E) \otimes_A C_\lambda$.

(b) Show that, if E is a flat (resp. faithfully flat) B-module, each of the E_λ is a flat (resp. faithfully flat) B_λ-module. The converse is true if we assume further that $\bigoplus_{\lambda \in L} C_\lambda$ is a faithfully flat A-module.

(c) Show that, if L is finite, each of the E_λ is a finitely generated projective B_λ-module and $\bigoplus_{\lambda \in L} C_\lambda$ is a faithfully flat A-module, then E is a finitely generated projective B-module (use Proposition 12 of no. 6).

10. (a) Let $\rho: A \to B$ be a ring homomorphism. Show that, for every left ideal $\mathfrak{a}$ of A which is a left annihilator of a subset M of A, $\overset{-1}{\rho}(B\mathfrak{a}) = \mathfrak{a}$.

(b) Deduce from (a) an example of a homomorphism $\rho: A \to B$ such that right A-module B is not flat but $\overset{-1}{\rho}(B\mathfrak{a}) = \mathfrak{a}$ holds for every left ideal $\mathfrak{a}$ of A (cf. § 2, Exercise 17 and *Algebra*, Chapter VIII, § 3, Exercise 11 and § 2, Exercise 6 and Chapter IX, § 2, Exercise 4).

§ 4

1. Show that in the statement of Proposition 2, condition (a) can be replaced by:

(a′) $\mathrm{Tor}_1^B(\rho_*(E), F) = 0$ for every monogenous right S-module E.

(To prove that (a′) implies (a), consider first the case when E is generated by n elements and argue by induction on n.)

CHAPTER II(*)

Localization

The conventions of Chapter I remain in force in this chapter. Also, unless otherwise stated, all rings are assumed to be commutative.

Let A, B *be two rings,* ρ *a homomorphism from* A *to* B *and* M *a* B*-module. When we speak of* M *as an* A*-module, we mean, unless otherwise stated, with the* A*-module structure* $\rho_*(\mathrm{M})$ *(defined by the external law* $(a, m) \mapsto \rho(a)m$*).*

1. PRIME IDEALS

1. DEFINITION OF PRIME IDEALS

DEFINITION 1. *An ideal* $\mathfrak{p}$ *of a ring* A *is called prime if the ring* $\mathrm{A}/\mathfrak{p}$ *is an integral domain.*

By this definition, an ideal $\mathfrak{p}$ of a ring A is prime if the following two conditions hold:

(1) $\mathfrak{p} \neq \mathrm{A}$;

(2) if x, y are two elements of A such that $x \notin \mathfrak{p}$ and $y \notin \mathfrak{p}$, then $xy \notin \mathfrak{p}$.

These conditions can also be expressed by saying that the product of any *finite family* of elements of $\complement\mathfrak{p}$ belongs to $\complement\mathfrak{p}$, as applying this condition to the empty set yields $1 \notin \mathfrak{p}$.

A *maximal* ideal $\mathfrak{m}$ of A is prime since $\mathrm{A}/\mathfrak{m}$ is a field; then it follows from Krull's theorem (*Algebra*, Chapter I, § 8, no. 7, Theorem 2) that every ideal of

(*) With the exception of the statements placed between two asterisks: $* \cdots *$, the results of this chapter depend only on Books I to VI and Chapter I, §§ 1–3 of this Book.

A *other than* A is contained in at least one prime ideal. In particular, for prime ideals to exist in a ring A, it is necessary and sufficient that A be not reduced to 0.

Let f: A $\to$ B be a ring homomorphism and $\mathfrak{q}$ an ideal of B. Set $\mathfrak{p} = \overset{-1}{f}(\mathfrak{q})$; the homomorphism $\bar{f}$: A/$\mathfrak{p}$ $\to$ B/$\mathfrak{q}$ derived from f by taking quotients is injective. Suppose that $\mathfrak{q}$ is prime; as the ring B/$\mathfrak{q}$ is an integral domain, so is A/$\mathfrak{p}$, being isomorphic to a subring of B/$\mathfrak{q}$; consequently the ideal $\mathfrak{p} = \overset{-1}{f}(\mathfrak{q})$ is prime. In particular, let A be a subring of B; for every ideal $\mathfrak{q}$ of B, $\mathfrak{q} \cap$ A is a prime ideal A. If f is surjective, $\bar{f}$ is an isomorphism; the conditions "$\mathfrak{p}$ is prime" and "$\mathfrak{q}$ is prime" are then equivalent. Hence, if $\mathfrak{p}$ and $\mathfrak{a}$ are ideals of A such that $\mathfrak{a} \subset \mathfrak{p}$, a necessary and sufficient condition for $\mathfrak{p}$ to be prime is that $\mathfrak{p}/\mathfrak{a}$ be prime in A/$\mathfrak{a}$.

PROPOSITION 1. *Let* A *be a ring,* $\mathfrak{a}_1, \mathfrak{a}_2, \ldots, \mathfrak{a}_n$ *ideals of* A *and* $\mathfrak{p}$ *a prime ideal of* A. *If* $\mathfrak{p}$ *contains the product* $\mathfrak{a}_1\mathfrak{a}_2\ldots\mathfrak{a}_n$, *it contains at least one of the* $\mathfrak{a}_i$.

Suppose in fact that $\mathfrak{p}$ contains none of the $\mathfrak{a}_i$. For $1 \leqslant i \leqslant n$ there exists then an element $s_i \in \mathfrak{a}_i \cap \complement\mathfrak{p}$; then $s = s_1s_2\ldots s_n$ is contained in $\mathfrak{a}_1\mathfrak{a}_2\ldots\mathfrak{a}_n$ and is not contained in $\mathfrak{p}$, which is absurd.

COROLLARY. *Let* $\mathfrak{m}$ *be a maximal ideal of* A; *for every integer* $n > 0$, *the only prime ideal containing* $\mathfrak{m}^n$ *is* $\mathfrak{m}$.

Such an ideal $\mathfrak{p}$ must contain $\mathfrak{m}$ by Proposition 1 applied to $\mathfrak{a}_i = \mathfrak{m}$ for $1 \leqslant i \leqslant n$; as $\mathfrak{m}$ is maximal, $\mathfrak{p} = \mathfrak{m}$.

PROPOSITION 2. *Let* A *be a ring,* $\mathfrak{a}$ *a non-empty set of* A *which is closed under addition and multiplication and* $(\mathfrak{p}_i)_{i\in I}$ *a non-empty finite family of ideals of* A. *Suppose that* $\mathfrak{a}$ *is contained in the union of the* $\mathfrak{p}_i$ *and that at most two of the* $\mathfrak{p}_i$ *are not prime. Then* $\mathfrak{a}$ *is contained in one of the* $\mathfrak{p}_i$.

We argue by induction on $n = \mathrm{Card}(I)$; the proposition is trivial if $n = 1$. Suppose that $n \geqslant 2$; if there exists an index j such that $\mathfrak{a} \cap \mathfrak{p}_j \subset \bigcup_{i \neq j} \mathfrak{p}_i$, the set $\mathfrak{a}$, which is the union of the $\mathfrak{a} \cap \mathfrak{p}_i$ where $i \in I$, is contained in $\bigcup_{i \neq j} \mathfrak{p}_i$ and hence in one of the $\mathfrak{p}_i$ by the induction hypothesis. Suppose then that such an index does not exist; for every $j \in I$ let y_j be an element of $\mathfrak{a} \cap \mathfrak{p}_j$ not belonging to any $\mathfrak{p}_i$ for $i \neq j$. Let k be an element of I chosen in such a way that $\mathfrak{p}_k$ is prime if $n > 2$ and chosen arbitrarily if $n = 2$; let $z = y_k + \prod_{i \neq k} y_i$. Then $z \in \mathfrak{a}$, since $\mathfrak{a}$ is closed under addition and multiplication; if $j \neq k$, $\prod_{i \neq k} y_i$

belongs to $\mathfrak{p}_j$, but $y_k \notin \mathfrak{p}_j$, whence $z \notin \mathfrak{p}_j$. On the other hand, $\prod_{i \neq k} y_i$ does not belong to $\mathfrak{p}_k$, as none of the factors y_i $(i \neq k)$ belongs to it and $\mathfrak{p}_k$ is prime if $n - 1 > 1$; as $y_k \in \mathfrak{p}_k$, z does not belong to $\mathfrak{p}_k$ and the proposition is established.

2. RELATIVELY PRIME IDEALS

Let A be a ring; two ideals $\mathfrak{a}$, $\mathfrak{b}$ of A are called *relatively prime* if $\mathfrak{a} + \mathfrak{b} = A$. For this to be true, it is necessary and sufficient that $\mathfrak{a} + \mathfrak{b}$ be contained in no prime ideal (*Algebra*, Chapter I, § 8, no. 7, Theorem 2), in other words, that no prime ideal contain both $\mathfrak{a}$ and $\mathfrak{b}$. Two distinct maximal ideals are relatively prime.

If A is a *principal ideal domain* (*Algebra*, Chapter VII, § 1), for two elements a, b of A to be relatively prime, it is necessary and sufficient, by Bezout's identity (*loc. cit.*, no. 2, Theorem 1), that the ideals Aa and Ab be relatively prime.

PROPOSITION 3. *Let* $\mathfrak{a}$ *and* $\mathfrak{b}$ *be two relatively prime ideals of a ring* A. *Let* $\mathfrak{a}'$ *and* $\mathfrak{b}'$ *be two ideals of* A *such that every element of* $\mathfrak{a}$ (resp. $\mathfrak{b}$) *has a power in* $\mathfrak{a}'$ (resp. $\mathfrak{b}'$). *Then* $\mathfrak{a}'$ *and* $\mathfrak{b}'$ *are relatively prime.*

Under the given hypothesis, every prime ideal which contains $\mathfrak{a}'$ contains $\mathfrak{a}$ and every prime ideal which contains $\mathfrak{b}'$ contains $\mathfrak{b}$. If a prime ideal contains $\mathfrak{a}'$ and $\mathfrak{b}'$, then it contains $\mathfrak{a}$ and $\mathfrak{b}$, which is absurd, since $\mathfrak{a}$ and $\mathfrak{b}$ are relatively prime; hence $\mathfrak{a}'$ and $\mathfrak{b}'$ are relatively prime.

PROPOSITION 4. *Let* $\mathfrak{a}, \mathfrak{b}_1, \ldots, \mathfrak{b}_n$ *be ideals of a ring* A. *If* $\mathfrak{a}$ *is relatively prime to each of the* $\mathfrak{b}_i$ $(1 \leqslant i \leqslant n)$, *it is relatively prime to* $\mathfrak{b}_1\mathfrak{b}_2\ldots\mathfrak{b}_n$.

Let $\mathfrak{p}$ be a prime ideal of A. If $\mathfrak{p}$ contains $\mathfrak{a}$ and $\mathfrak{b}_1\mathfrak{b}_2\ldots\mathfrak{b}_n$, it contains one of the $\mathfrak{b}_i$ (no. 1, Proposition 1), which is absurd since $\mathfrak{a}$ and $\mathfrak{b}_i$ are relatively prime.

PROPOSITION 5. *Let* $(\mathfrak{a}_i)_{i \in I}$ *be a non-empty finite family of ideals of a ring* A. *The following properties are equivalent:*

(a) *For* $i \neq j$, $\mathfrak{a}_i$ *and* $\mathfrak{a}_j$ *are relatively prime.*

(b) *The canonical homomorphism* $\phi: A \rightarrow \prod_{i \neq I} (A/\mathfrak{a}_i)$ (*Algebra*, Chapter II, § 1, no. 7) *is surjective.*

If these hold, the intersection $\mathfrak{a}$ *of the* $\mathfrak{a}_i$ *is equal to their product and the canonical homomorphism* $\psi: A/\mathfrak{a} \rightarrow \prod_{i \in I} (A/\mathfrak{a}_i)$ (*Algebra*, Chapter II, § 1, no. 7) *is bijective.*

We argue by induction on n the number of elements in I, the case $n = 1$ being trivial. Consider first the case $n = 2$. Then the equivalence of (a) and (b) follows from the exactness of the sequence

$$0 \longrightarrow A/(\mathfrak{a}_1 \cap \mathfrak{a}_2) \xrightarrow{\psi} (A/\mathfrak{a}_1) \oplus (A/\mathfrak{a}_2) \longrightarrow A/(\mathfrak{a}_1 + \mathfrak{a}_2) \longrightarrow 0$$

(*Algebra*, Chapter II, § 1, no. 7, formula (30)). Moreover, there exist $e_1 \in \mathfrak{a}_1$ and $e_2 \in \mathfrak{a}_2$ such that $1 = e_1 + e_2$; then, for all $x \in \mathfrak{a} = \mathfrak{a}_1 \cap \mathfrak{a}_2$, $x = xe_1 + xe_2$; but by definition $xe_1 \in \mathfrak{a}_1\mathfrak{a}_2$ and $xe_2 \in \mathfrak{a}_1\mathfrak{a}_2$, hence $x \in \mathfrak{a}_1\mathfrak{a}_2$; whence $\mathfrak{a} \subset \mathfrak{a}_1\mathfrak{a}_2$ and the converse inclusion is obvious.

In the general case, suppose that condition (a) holds and let k be an element of I and $\mathfrak{b}_k = \bigcap_{i \neq k} \mathfrak{a}_i$; the induction hypothesis implies that $\mathfrak{b}_k = \prod_{i \neq k} \mathfrak{a}_i$ and it follows from Proposition 4 that $\mathfrak{a}_k$ and $\mathfrak{b}_k$ are relatively prime; then

$$\mathfrak{a} = \bigcap_{i \in I} \mathfrak{a}_i = \mathfrak{a}_k \cap \mathfrak{b}_k = \mathfrak{a}_k\mathfrak{b}_k = \prod_{i \in I} \mathfrak{a}_i$$

by the first part of the argument and for the same reason the canonical homomorphism $A/\mathfrak{a} \to (A/\mathfrak{a}_k) \times (A/\mathfrak{b}_k)$ is bijective; by the induction hypothesis the canonical homomorphism $A/\mathfrak{b}_k \to \prod_{i \neq k} (A/\mathfrak{a}_i)$ is bijective and so then is the composite homomorphism

$$A/\mathfrak{a} \to (A/\mathfrak{a}_k) \times (A/\mathfrak{b}_k) \to (A/\mathfrak{a}_k) \times \prod_{i \neq k} (A/\mathfrak{a}_i) = \prod_{i \in I} (A/\mathfrak{a}_i)$$

which is precisely ψ; that is, (b) holds. Conversely, suppose that (b) holds. We show that the $\mathfrak{a}_i$ are necessarily relatively prime in pairs. In the contrary case, there would exist an ideal $\mathfrak{c} \neq A$ containing $\mathfrak{a}_i$ and $\mathfrak{a}_j$ for $i \neq j$. We set $\mathfrak{a}'_h = \mathfrak{a}_h$ for h not equal to i or j and $\mathfrak{a}'_i = \mathfrak{a}'_j = \mathfrak{c}$; the canonical homomorphism $\phi' : A \to \prod_{i \in I} (A/\mathfrak{a}'_i)$ can be written as the composite mapping

$$A \xrightarrow{\phi} \prod_{i \in I} (A/\mathfrak{a}_i) \xrightarrow{f} \prod_{i \in I} (A/\mathfrak{a}'_i)$$

f being the product of the canonical homomorphisms $A/\mathfrak{a}_i \to A/\mathfrak{a}'_i$; clearly ϕ' is not surjective, the projection of $\phi'(A)$ onto $(A/\mathfrak{a}'_i) \times (A/\mathfrak{a}'_i)$ being the diagonal of the product $(A/\mathfrak{c}) \times (A/\mathfrak{c})$, which is distinct from this product since $\mathfrak{c} \neq A$. As f is surjective, this shows that ϕ is not surjective.

PROPOSITION 6. *Let $(\mathfrak{a}_i)_{i \in I}$ be a non-empty finite family of ideals of a ring* A *which are relatively prime in pairs; let $\mathfrak{a}$ be the intersection of the $\mathfrak{a}_i$. For every* A*-module* M, *the canonical mapping* $M \to \prod_{i \in I} (M/\mathfrak{a}_iM)$ *is surjective and its kernel is* $\mathfrak{a}M$.

Clearly the canonical mapping of M to $\prod_{i\in I}(M/\mathfrak{a}_i M)$ is zero on $\mathfrak{a}M$; then, by taking quotients, it defines a homomorphism $\lambda: M/\mathfrak{a}M \to \prod_{i\in I}(M/\mathfrak{a}_i M)$. On the other hand, by Proposition 5, the canonical homomorphism

$$\psi: A/\mathfrak{a} \to \prod_{i\in I}(A/\mathfrak{a}_i)$$

is bijective. Then so is $1_M \otimes \psi: M \otimes (A/\mathfrak{a}) \to M \otimes \prod_{i\in I}(A/\mathfrak{a}_i)$. Now $M \otimes (A/\mathfrak{a})$ is identified with $M/\mathfrak{a}M$ and $M \otimes \prod_{i\in I}(A/\mathfrak{a}_i)$ with $\prod_{i\in I} M \otimes (A/\mathfrak{a}_i)$, which is itself identified with $\prod_{i\in I}(M/\mathfrak{a}_i M)$. It is immediately verified that the above identifications transforms $1_M \otimes \psi$ into λ, whence the proposition.

Example. Let K be a field, a_i ($1 \leqslant i \leqslant m$) distinct elements of K and, for each i, let g_i be a polynomial in K[X]; the principal ideal $(X - a_i) = \mathfrak{m}_i$ is maximal in K[X], hence, for every system $(n_i)_{1\leqslant i\leqslant m}$ of m integers $\geqslant 1$, the ideals $\mathfrak{m}_i^{n_i}$ are relatively prime in pairs. Then it follows from Proposition 5 that there exists a polynomial $f \in K[X]$ such that $f(X) \equiv g_i(X)$ (mod. $(X - a_i)^{n_i}$) for $1 \leqslant i \leqslant m$, the difference of two such polynomials being divisible by $\omega(X) = \prod_{i=1}^{m}(X - a_i)^{n_i}$. If all the n_i are taken equal to 1, we find the problem is solved explicitly by Lagrange's interpolation formula (*Algebra*, Chapter IV, § 2, no. 4).

2. RINGS AND MODULES OF FRACTIONS

1. DEFINITION OF RINGS OF FRACTIONS

DEFINITION 1. *Let* A *be a ring. A subset* S *of* A *is called multiplicative if every finite product of elements of* S *belongs to* S.

This is the same as saying that $1 \in S$ and that the product of two elements of S belong to S.

Examples

(1) For every $a \in A$, the set of a^n, where $n \in \mathbf{N}$, is a multiplicative subset of A.

(2) Let $\mathfrak{p}$ be an ideal of A. For $A - \mathfrak{p}$ to be a multiplicative subset of A, it is necessary and sufficient that $\mathfrak{p}$ be *prime*.

(3) The set of elements of A which are not divisors of zero is a multiplicative subset of A.

(4) If S and T are multiplicative subsets of A, the set ST of products st, where $s \in S$ and $t \in T$, is a multiplicative subset.

(5) Let $\mathfrak{S}$ be a *directed* set (with respect to the relation $\subset$) of multiplicative subsets of A. Then $T = \bigcup_{S \in \mathfrak{S}} S$ is a multiplicative subset of A, as any two elements of T belong to some subset $S \in \mathfrak{S}$, hence their product belongs to T.

(6) Every intersection of multiplicative subsets of A is a multiplicative subset.

For every subset S of a ring A, there exist multiplicative subsets of A containing S, for example A itself. The intersection of all these subsets is the smallest multiplicative subset of A containing S; it is said to be *generated* by S. It follows immediately that it is the set consisting of all the finite products of elements of S.

PROPOSITION 1. *Let* A *be a ring and* S *a subset of* A. *There exists a ring* A′ *and a homomorphism* h *of* A *to* A′ *with the following properties:*

(1) *the elements of* $h(S)$ *are invertible in* A′;

(2) *for every homomorphism* u *of* A *to a ring* B *such that the elements of* $u(S)$ *are invertible in* B, *there exists a unique homomorphism* u' *of* A′ *to* B *such that* $u = u' \circ h$.

In other words, (A', h) is a solution of the universal mapping problem (*Set Theory*, Chapter IV, § 3, no. 1) with the following conditions: the species of structure Σ considered is that of a ring, the morphisms are ring homomorphisms and the α-mappings are homomorphisms of A to a ring such that the image of S under such a homomorphism consists of invertible elements. Recall (*loc. cit.*) that, if (A', h) and (A'_1, h_1) are both solutions of this problem, there exists a *unique isomorphism* $j: A' \to A'_1$ such that $h_1 = j \circ h$.

Let $\overline{S}$ be the multiplicative subset of A generated by S. Clearly every solution of the above universal mapping problem is also a solution of the universal mapping problem obtained by replacing S by $\overline{S}$ and conversely.

Consider, in the set $A \times \overline{S}$, the following relation between elements (a, s), (a', s'):

(1) "There exists $t \in \overline{S}$ such that $t(sa' - s'a) = 0$".

This relation is an equivalence relation: it is clearly reflexive and symmetric; it is transitive, for if $t(sa' - s'a) = 0$ and $t'(s'a'' - s''a') = 0$, then $tt's'(sa' - s'a) = 0$ and $tt's' \in \overline{S}$. Let A′ be the quotient set of $A \times \overline{S}$ under this equivalence relation; for every ordered pair $(a, s) \in A \times \overline{S}$ we denote by a/s the canonical image of (a, s) in A′ and we set $h(a) = a/1$ for all $a \in A$. We shall see that A′ can be given a ring structure such that the ordered pair (A', h) solves the problem.

Let $x = a/s$ and $y = b/t$ be two elements of A′. The elements $(ta + sb)/st$

and ab/st depend only on x and y; for if $x = a'/s'$, there exists by hypothesis $r \in \bar{S}$ such that $r(s'a - sa') = 0$, whence

$$r(s't(ta + sb) - st(ta' + s'b)) = 0$$

and $r(s'tab - sta'b) = 0$. It is easily verified that the laws of composition $(x, y) \mapsto x + y = (ta + sb)/st$ and $(x, y) \mapsto xy = ab/st$ define a commutative ring structure on A', under which $0/1$ is the identity element under addition and $1/1$ is the unit element. Moreover, it is immediate that h is a ring homomorphism and that, for all $s \in S$, $s/1$ is invertible in A', its inverse being $1/s$. Finally let B be a ring and $u: A \to B$ a homomorphism such that the elements $u(S)$ are invertible in B; there exists a unique mapping $u': A' \to B$ such that

$$u'(a/s) = u(a)(u(s))^{-1} \qquad (a \in A, s \in \bar{S}). \tag{2}$$

If $a/s = a'/s'$, there exists $t \in \bar{S}$ such that $t(sa' - s'a) = 0$, whence $u(t)(u(s)u(a') - u(s')u(a)) = 0$ and, as $u(t)$, $u(s)$ and $u(s')$ are invertible, $u(a)(u(s))^{-1} = u(a')(u(s'))^{-1}$. It is easily verified that u' is a homomorphism with respect to addition and multiplication; finally, clearly $u' \circ h = u$ and u' is the only homomorphism satisfying this relation, as it implies

$$u'(a/s) = u'((a/1)(1/s)) = u'(1/s)u'(a/1) = u'(1/s)u(a)$$

and $1 = u'(1/1) = u'(s/1)u'(1/s) = u(s)u'(1/s)$, whence formula (2).

DEFINITION 2. *Let A be a ring, S a subset of A and $\bar{S}$ the multiplicative subset generated by S. The ring of fractions of A defined by S and denoted by $A[S^{-1}]$ is the quotient set of $A \times \bar{S}$ under the equivalence relation* (1) *with the ring structure defined by*

$$(a/s) + (b/t) = (ta + sb)/st, \qquad (a/s)(b/t) = (ab)/(st)$$

for a, b in A, s, t in $\bar{S}$. The canonical mapping of A to $A[S^{-1}]$ is the homomorphism $a \mapsto a/1$, which makes $A[S^{-1}]$ into an A-algebra.

In this chapter we usually denote this canonical mapping by i_A^S; the proof of Proposition 1 shows that the ordered pair $(A[S^{-1}], i_A^S)$ satisfies the conditions of the statement of this proposition.

Remarks

(1) Clearly $A[\bar{S}^{-1}] = A[S^{-1}]$.

(2) Two elements of $A[S^{-1}]$ can always be written in the form a/s and a'/s (a, a' in A, $s \in \bar{S}$) with the *same* "denominator" s, for if b/t and b'/t' are two elements of $A[S^{-1}]$, then $b/t = bt'/tt'$ and $b'/t' = b't/tt'$.

(3) The kernel of i_A^S is the set of $a \in A$ such that there exists $s \in \bar{S}$ satisfying $sa = 0$; for i_A^S to be *injective*, it is necessary and sufficient that S *contain no divisor of zero* in A.
(4) If S contains a *nilpotent* element, then $0 \in \bar{S}$ and the ring $A[S^{-1}]$ is *reduced to* 0; this follows easily from Definition 2.
(5) For i_A^S to be a *bijection*, it is necessary and sufficient that every element $s \in S$ be *invertible* in A: the condition is obviously necessary, since $s/1$ is invertible in $A[S^{-1}]$; it is sufficient, since for all $t \in \bar{S}$, t is therefore invertible in A and $a/t = at^{-1}/1$ in $A[S^{-1}]$; hence i_A^S is surjective and it has already been seen in *Remark* 3 that it is injective. Then A and $A[S^{-1}]$ are identified by means of i_A^S.

Example (7). If R is the set of elements in A which are not divisors of 0, the ring $A[R^{-1}]$ is precisely what we have called the *ring of fractions* of A (*Algebra*, Chapter I, § 9, no. 4); to avoid any confusion we shall often call it the *total ring of fractions* of A. In particular, if A is an integral domain, $A[R^{-1}]$ is the *field of fractions* of A (*loc. cit.*).

Proposition 2. *Let* A, B *be two rings,* S *a subset of* A, T *a subset of* B *and* f *a homomorphism from* A *to* B *such that* $f(S) \subset T$. *There exists a unique homomorphism* f' *from* $A[S^{-1}]$ *to* $B[T^{-1}]$ *such that* $f'(a/1) = f(a)/1$ *for all* $a \in A$.

Suppose further that T *is contained in the multiplicative subset of* B *generated by* $f(S)$. *Then, if* f *is surjective* (resp. *injective*) *so is* f'.

The first assertion amounts to saying that there exists a unique homomorphism $f' : A[S^{-1}] \to B[T^{-1}]$ giving a commutative diagram:

$$\begin{array}{ccc} A & \xrightarrow{f} & B \\ {\scriptstyle i_A^S}\downarrow & & \downarrow{\scriptstyle i_B^T} \\ A[S^{-1}] & \xrightarrow[f']{} & B[T^{-1}] \end{array}$$

Now the relation $f(S) \subset T$ implies that $i_B^T(f(s))$ is invertible in $B[T^{-1}]$ for all $s \in S$ and it is sufficient to apply Proposition 1 to $i_B^T \circ f$. It follows easily from (2) that, for all $a \in A$ and $s \in \bar{S}$ (multiplicative subset of A generated by S),

$$f'(a/s) = f(a)/f(s). \tag{3}$$

Suppose that T is contained in the multiplicative subset generated by $f(S)$, which is precisely $f(\bar{S})$. Then it follows from (3) that, if f is surjective, so is f'.

Suppose now that f is injective. Let a/s be an element of the kernel of f'. As the multiplicative subset generated by T is $f(S)$, there is an element $s_1 \in S$ such that $f(s_1)f(a) = 0$, whence $f(s_1 a) = 0$ and consequently $s_1 a = 0$ since f is injective; then $a/s = 0$, which proves that f' is injective.

Remark (6). If the elements of T are invertible in B, $B[T^{-1}]$ is identified with B by means of the isomorphism i_B^T and f' then becomes identical with the unique homomorphism u' of $A[S^{-1}]$ to B such that $u' \circ i_A^S = f$.

COROLLARY 1. *Let* A *be a ring,* S *a subset of* A *and* u *an injective homomorphism of* A *to a ring* B *such that the elements of* $u(S)$ *are invertible in* B. *The unique homomorphism* u' *of* $A[S^{-1}]$ *to* B *such that* $u' \circ i_A^S = u$ *is then injective.*

This is an immediate consequence of Proposition 2 and *Remark* 6.

COROLLARY 2. *Let* A *be a ring and* S *and* T *two subsets of* A *such that* $S \subset T$. *There exists a unique homomorphism* $i_A^{T,S}$ *from* $A[S^{-1}]$ *to* $A[T^{-1}]$ *such that* $i_A^T = i_A^{T,S} \circ i_A^S$.

For all $a \in A$, $i_A^{T,S}$ then maps the element a/s in $A[S^{-1}]$ to the element a/s in $A[T^{-1}]$.

Remark (7). Note that, if i_A^T is injective, so is $i_A^{T,S}$ (Corollary 1). This is what happens if T is the set R of elements of A which are not divisors of 0; then it is possible to identify $A[S^{-1}]$ with the subring of the total ring of fractions $A[R^{-1}]$ generated by A and the inverses in $A[R^{-1}]$ of the elements of S.

COROLLARY 3. *Let* A, B, C *be three rings,* S (resp. T, U) *a multiplicative subset of* A (resp. B, C), $f: A \to B$, $g: B \to C$ *two homomorphisms and* $h: A \to C$ *the composite homomorphism* $g \circ f$; *suppose that* $f(S) \subset T$, $g(T) \subset U$. *Let* $f': A[S^{-1}] \to B[T^{-1}]$, $g': B[T^{-1}] \to C[U^{-1}]$, $h': A[S^{-1}] \to C[U^{-1}]$ *the homomorphisms corresponding to* f, g, h; *then* $h' = g' \circ f'$.

This follows easily from the definitions.

In particular, if S, T, U are three multiplicative subsets of A such that $S \subset T \subset U$, then $i_A^{U,S} = i_A^{U,T} \circ i_A^{T,S}$.

COROLLARY 4. *Let* S *be a subset of a ring* A, B *a subring of* $A[S^{-1}]$ *containing* $i_A^S(A)$ *and* S' *the set* $i_A^S(A)$. *Let* j *be the canonical injection of* B *into* $A[S^{-1}]$; *the unique homomorphism* g *from* $B[S'^{-1}]$ *to* $A[S^{-1}]$ *such that* $g \circ i_A^S = j$ *is an isomorphism.*

The mapping g is injective by Corollary 1; the ring $g(B[S'^{-1}])$ contains $i_A^S(A)$ and the inverse of the elements of S'; hence it is equal to $A[S^{-1}]$.

If A is an *integral domain* and $0 \notin S$, the notation $A[S^{-1}]$ agrees with that of *Algebra*, Chapter IV, § 2, no. 1; also, if S is *multiplicative*, $A[S^{-1}]$ coincides in this case with the set denoted by $S^{-1}A$ in *Algebra*, Chapter I, § 1, no. 1.

As an extension of notation, for any multiplicative *subset* S *of a ring* A, *we henceforth denote by* $S^{-1}A$ *the ring of fractions* $A[S^{-1}]$. *If* S *is the complement of a prime ideal* $\mathfrak{p}$ *of* A, *we write* $A_{\mathfrak{p}}$ *instead of* $S^{-1}A$.

If A is an integral domain and $0 \notin S$, $S^{-1}A$ is always identified with a *subring of the field of fractions* of A, containing A (*Remark* 7).

2. MODULES OF FRACTIONS

The canonical homomorphism $i_A^S : A \to A[S^{-1}]$ defined in no. 1 allows us to consider every $A[S^{-1}]$-module as an A-module.

PROPOSITION 3. *Let* A *be a ring,* S *a subset of* A, M *an* A*-module,* M' *the* A*-module* $M \otimes_A A[S^{-1}]$ *and* f *the canonical* A*-homomorphism* $x \mapsto x \otimes 1$ *of* M *to* M'. *Then:*

(1) *For all* $s \in S$, *the homothety* $z \mapsto sz$ *of* M' *is bijective.*

(2) *For every* A*-module* N *such that, for all* $s \in S$, *the homothety* $y \mapsto sy$ *of* N *is bijective, and every homomorphism* u *of* M *to* N, *there exists a unique homomorphism* u' *of* M' *to* N *such that* $u = u' \circ f$.

In other words, (M', f) is a solution of the universal mapping problem (*Set Theory*, Chapter IV, § 3, no. 1) with the following conditions: the species of structure Σ is that of an A-module in which the homotheties induced by the elements of S are bijective, the morphisms are A-module homomorphisms and the α-mappings are also A-module homomorphisms.

For every A-module N and all $a \in A$, denote by h_a the homothety $y \mapsto ay$ in N; $a \mapsto h_a$ is then a ring homomorphism from A to $\mathrm{End}_A(N)$. To say that h_a is bijective means that h_a is an *invertible* element of $\mathrm{End}_A(N)$. Suppose that, for all $s \in S$, h_s is invertible in $\mathrm{End}_A(N)$; the elements h_a, where $a \in A$, and the inverses of the elements h_s, where $s \in S$, then generate in $\mathrm{End}_A(N)$ a *commutative subring* B and the homomorphism $a \mapsto h_a$ from A to B is such that the images of the elements of S are invertible. Then it follows (no. 1, Proposition 1) that there exists a unique homomorphism h' of $A[S^{-1}]$ to B such that

$$h'(a/s) = h_a(h_s)^{-1};$$

we know (*Algebra*, Chapter II, § 1, no. 14) that such a homomorphism defines on N an $A[S^{-1}]$-*module* structure such that $(a/s).y = h_s^{-1}(a.y)$; the A-module structure derived from this $A[S^{-1}]$-module structure by means of the homomorphism i_A^S is precisely the structure given initially.

Conversely, if N is an $A[S^{-1}]$-module and it is considered as an A-module by means of i_A^S, the homotheties $y \mapsto sy$, for $s \in S$, are bijective, for $y \mapsto (1/s)y$ is the inverse mapping of $y \mapsto sy$; and the $A[S^{-1}]$-module structure on N derived from its A-module structure by the process described above is the $A[S^{-1}]$-module structure given initially. Thus there is a *canonical one-to-one correspondence* between $A[S^{-1}]$-*modules* and A-*modules in which the homotheties induced by the elements of* S *are bijective*; moreover, if N, N′ are two A-modules with this property every A-module homomorphism $u: N \to N'$ is also a homomorphism of the $A[S^{-1}]$-module structures of N and N′, as, for all $y \in N$ and all $s \in S$, we may write $u(y) = u(s.((1/s)y)) = s.u((1/s)y)$, whence $u((1/s)y) = (1/s)u(y)$; the converse is obvious.

This being so, the statement of Proposition 3 is just the characterization of the module obtained from M by *extending the scalars to* $A[S^{-1}]$, taking account of the above interpretation (*Algebra*, Chapter II, § 5, no. 1, *Remark* 1).

DEFINITION 3. *Let* A *be a ring,* S *a subset of* A, $\bar{S}$ *the multiplicative subset of* A *generated by* S *and* M *an A-module. Then the module of fractions of* M *defined by* S *and denoted by* $M[S^{-1}]$ *or* $\bar{S}^{-1}M$ *is the* $A[S^{-1}]$-*module* $M \otimes_A A[S^{-1}]$.

In this chapter we shall usually denote by i_M^S the canonical mapping $m \mapsto m \otimes 1$ of M to $M[S^{-1}]$.

Remarks

(1) Clearly $M[\bar{S}^{-1}] = M[S^{-1}]$.

(2) For $m \in M$ and $s \in \bar{S}$ we also write m/s for the element $m \otimes (1/s)$ of $M[S^{-1}]$. Every element of $M[S^{-1}]$ is of the form, for such an element is of the form $\sum_i m_i \otimes (a_i/s)$, where $m_i \in M$, $a_i \in A$, $s \in S$ (no. 1, *Remark* 2), and

$$m_i \otimes (a_i/s) = (a_i m_i) \otimes (1/s),$$

hence $\sum_i m_i \otimes (a_i/s) = m \otimes (1/s)$, where $m = \sum_i a_i m_i$. Then

$$(m/s) + (m'/s') = (s'm + sm')/ss' \tag{4}$$

$$(a/s)(m/s') = (am)/(ss') \tag{5}$$

where $m, m' \in M$, $a \in A$ and $s, s' \in S$.

(3) If S is the complement of a prime ideal $\mathfrak{p}$ of A, we write $M_{\mathfrak{p}}$ instead of $S^{-1}M$.

(4) Let M be an $A[S^{-1}]$-module; if M is considered canonically as an A-module, i_M^S is a *bijection*, for the ordered pair consisting of M and the identity mapping 1_M is also trivially a solution of the universal mapping problem solved by $M[S^{-1}]$ and i_M^S. Then M is identified with $M[S^{-1}]$.

PROPOSITION 4. *Let* S *be a multiplicative subset of* A *and* M *an* A*-module. For* $m/s = 0$ ($m \in M$, $s \in S$), *it is necessary and sufficient that there exist* $s' \in S$ *such that* $s'm = 0$.

If $s' \in S$ is such that $s'm = 0$, clearly $m/s = (s'm)/(ss') = 0$. Conversely, suppose that $m/s = 0$. As $1/s$ is invertible in $S^{-1}A$, $m/1 = 0$. For every sub-A-module P of $S^{-1}A$ containing 1, we denote by $\beta(P, m)$ the image of $(m, 1)$ under the canonical mapping of $M \times P$ to $M \otimes_A P$; then $\beta(S^{-1}A, m) = 0$. We know (*Algebra*, Chapter II, § 6, no. 3, Corollary 4 to Proposition 7) that there exists a *finitely generated* submodule P of $S^{-1}A$ containing 1 and such that $\beta(P, m) = 0$. For all $t \in S$ we denote by A_t the set of a/t, where $a \in A$; as P is finitely generated, there exists $t \in S$ such that $P \subset A_t$ (no. 1, *Remark* 2), whence $\beta(A_t, m) = 0$. The mapping $a \mapsto a/t$ from A to A_t is surjective; let B be its kernel. It defines a surjective mapping $h: M \otimes_A A \to M \otimes_A A_t$, whose kernel is BM (M being identified with $M \otimes A$); then $\beta(A_t, m) = h(tm)$ and consequently tm can be expressed in the form $\sum_{i=1}^{r} b_i m_i$, where $b_i \in B$, $m_i \in M$ $(1 \leqslant i \leqslant r)$. As $b_i/t = 0$ for $1 \leqslant i \leqslant r$, there exists $t' \in S$ such that $t'b_i = 0$ for $1 \leqslant i \leqslant r$, whence $t'tm = 0$, which proves Proposition 4.

COROLLARY 1. *For* $m/s = m'/s'$ *in* $S^{-1}M$, *it is necessary and sufficient that there exist* $t \in S$ *such that* $t(s'm - sm') = 0$.

$(m/s) - (m'/s') = (s'm - sm')/ss'$.

COROLLARY 2. *Let* M *be a finitely generated* A*-module. For* $S^{-1}M = 0$, *it is necessary and sufficient that there exists* $s \in S$ *such that* $sM = 0$.

Without any conditions on M, clearly the relation $sM = 0$ for some $s \in S$ implies $S^{-1}M = 0$. Conversely, suppose that $S^{-1}M = 0$ and let $(m_i)_{1 \leqslant i \leqslant n}$ be a system of generators of M; the $m_i/1$ generate the $S^{-1}A$-module $S^{-1}M$, hence to say that $S^{-1}M = 0$ amounts to saying that $m_i/1 = 0$ for $1 \leqslant i \leqslant n$; by

Proposition 4 there exist $s_i \in S$ such that $s_i m_i = 0$ and, taking $s = s_1 s_2 \ldots s_n$, $sm_i = 0$ for all i and hence $sM = 0$.

COROLLARY 3. *Let* M *be a finitely generated* A*-module. For an ideal* $\mathfrak{a}$ *of* A *to be such that* $\mathfrak{a}M = M$, *it is necessary and sufficient that there exist* $a \in \mathfrak{a}$ *such that* $(1 + a)M = 0$.

Clearly the relation $(1 + a)M = 0$ implies $M = aM$. To prove the converse we use the following lemma:

LEMMA 1. *For every ideal* $\mathfrak{a}$ *of* A, *the set* S *of elements* $1 + a$, *where* $a \in \mathfrak{a}$, *is a multiplicative subset of* A *and the set* $\mathfrak{a}'$ *of elements of* $S^{-1}A$ *of the form* a/s, *where* $a \in \mathfrak{a}$ *and* $s \in S$, *is an ideal contained in the Jacobson radical of* $S^{-1}A$.

The first assertion is obvious, as well as the fact that $\mathfrak{a}'$ is an ideal of $S^{-1}A$. On the other hand, $(1/1) + (a/s) = (s + a)/s$ and $s + a \in S$ for all $s \in S$ and $a \in \mathfrak{a}$ by definition of S; hence $(1/1) + (a/s)$ is invertible in $S^{-1}A$ for all $a/s \in \mathfrak{a}'$, which completes the proof of the lemma (*Algebra*, Chapter VIII, § 6, no. 3, Theorem 1).

This being so, if we set $N = S^{-1}M$, clearly N is a finitely generated $S^{-1}A$-module; if $\mathfrak{a}M = M$, then $\mathfrak{a}'N = N$ and it follows that $N = 0$ by Nakayama's Lemma (*Algebra*, Chapter VIII, § 6, no. 3, Corollary to Proposition 6); the corollary then follows from Corollary 2.

PROPOSITION 5. *Let* A, B *be two rings,* S *a multiplicative subset of* A, T *a multiplicative subset of* B *and* f *a homomorphism from* A *to* B *such that* $f(S) \subset T$. *Let* M *be an* A*-module,* N *a* B*-module and* u *an* A*-linear mapping from* M *to* N. *Then there exists a unique* $S^{-1}A$*-linear mapping* u' *from* $S^{-1}M$ *to* $T^{-1}N$ *such that* $u'(m/1) = u(m)/1$ *for all* $m \in M$.

The mapping $i_N^T \circ u$ from M to $T^{-1}N$ is A-linear. Moreover, if $s \in S$, then $f(s) \in T$, hence the homothety induced by s on $T^{-1}N$ is bijective. The existence and uniqueness of u' then follow from Proposition 3. Then, for $m \in M$ and $s \in S$,

$$u'(m/s) = u(m)/f(s). \tag{6}$$

With the same notation, let C be a third ring, U a multiplicative subset of C, g a homomorphism from B to C such that $g(T) \subset U$, P a C-module, v a B-linear mapping from N to P and v' the $T^{-1}B$-linear mapping from B to $U^{-1}P$ associated with v. Then

$$(v \circ u)' = v' \circ u' \tag{7}$$

where the left hand side is the A-linear mapping $S^{-1}M \to U^{-1}P$ associated with $v \circ u$. Similarly, if u_1 is a second A-linear mapping from M to N, then

$$(u + u_1)' = u' + u_1', \tag{8}$$

the left-hand side being the A-linear mapping $S^{-1}M \to T^{-1}N$ associated with $u + u_1$.

Remark (5). If, in Proposition 5, we take $B = A$, $T = S$ and $f = 1_A$, it is easily seen that u' is just the mapping $u \otimes 1 : M \otimes S^{-1}A \to N \otimes S^{-1}A$. *We shall henceforth denote it by* $S^{-1}u$; if S is the complement of a prime ideal $\mathfrak{p}$ of A, we write $u_\mathfrak{p}$ instead of $S^{-1}u$.

PROPOSITION 6. *Let f be a homomorphism from a ring A to a ring B and S a multiplicative subset of A. There exists a unique mapping j from $(f(S))^{-1}B$ to $S^{-1}B$ (where B is considered as an A-module by means of f) such that $j(b/f(s)) = b/s$ for all $b \in B$, $s \in S$. If $f' : S^{-1}A \to (f(S))^{-1}B$ is the ring homomorphism associated with f* (no. 1, Proposition 2), *then $j \circ f' = S^{-1}f$. The mapping j is an isomorphism of the $S^{-1}A$-module structure on $(f(S))^{-1}B$ defined by f' onto that on $S^{-1}B$ and also of the B-module structure on $(f(S))^{-1}B$ onto that on $S^{-1}B$ (resulting from the definition $S^{-1}B = (S^{-1}A) \otimes_A B$).*

If b, b' are in B, s, s' in S, the conditions $b/s = b'/s'$ and $b/f(s) = b'/f(s')$ are equivalent, as follows from Corollary 1 to Proposition 4, which establishes the existence of j and shows that j is bijective; the uniqueness of j is obvious. Clearly j is an additive group isomorphism. If $a \in A$, $b \in B$, $s \in S$, $t \in S$, then

$$(a/s).(b/f(t)) = f'(a/s)(b/f(t)) = f(a)/f(s)(b/f(t)) = (f(a)b)/f(st),$$

from which its follows that j is $(S^{-1}A)$-linear. Clearly $j \circ f^1 = S^{-1}f$. Finally, if $b \in B$, $b' \in B$, $s \in S$, then $j(b.(b'/f(s)) = j(bb'/f(s)) = bb'/s = b.(b'/s)$, which proves the last assertion.

The mapping j of Proposition 6 is called the *canonical isomorphism* of $(f(S))^{-1}B$ onto $S^{-1}B$. These two sets are in general identified by means of f; then $f' = S^{-1}f$, $i_B^S = i_B^{f(S)}$.

3. CHANGE OF MULTIPLICATIVE SUBSET

Let A be a ring, S a multiplicative subset of A and M an A-module. If T is a multiplicative subset of A containing S, it follows from Proposition 5 of no. 2 that there exists a unique $S^{-1}A$-linear mapping $i_M^{T,S} : S^{-1}M \to T^{-1}M$ such that

$i_M^T = i_M^{T,S} \circ i_M^S$; the mapping $i_M^{T,S}$ maps the element m/s of $S^{-1}M$ to the element m/s of $T^{-1}M$. It is easily verified that $i_M^{T,S} = i_A^{T,S} \otimes 1_M$. If U is a third multiplicative subset of A such that $T \subset U$, then $i_M^{U,S} = i_M^{U,T} \circ i_M^{T,S}$; moreover, if u: $M \to N$ is an A-module homomorphism, the diagram

$$\begin{array}{ccc} S^{-1}M & \xrightarrow{S^{-1}u} & S^{-1}N \\ {\scriptstyle i_M^{T,S}}\downarrow & & \downarrow{\scriptstyle i_N^{T,S}} \\ T^{-1}M & \xrightarrow[T^{-1}u]{} & T^{-1}N \end{array}$$

is commutative.

Proposition 7. *Let* A *be a ring and* S, T *two multiplicative subsets of* A. *Set* $T' = i(T_A^S)$.

(i) *There exists a unique isomorphism* j *from the ring* $(ST)^{-1}A$ *onto the ring* $T'^{-1}(S^{-1}A)$ *such that the diagram*

$$\begin{array}{ccc} A & \xrightarrow{i_A^S} & S^{-1}A \\ {\scriptstyle i_M^{ST}}\downarrow & & \downarrow{\scriptstyle i_{S^{-1}A}^{T'}} \\ (ST)^{-1}A & \xrightarrow[j]{} & T'^{-1}(S^{-1}A) \end{array}$$

is commutative.

(ii) *Let* M *be an* A*-module. There exists an* $(ST)^{-1}A$*-isomorphism* k *from the* $(ST)^{-1}A$*-module* $(ST)^{-1}M$ *onto the* $T'^{-1}(S^{-1}A)$*-module* $T'^{-1}(S^{-1}M)$ *such that the diagram*

$$\begin{array}{ccc} M & \xrightarrow{i_M^S} & S^{-1}M \\ {\scriptstyle i_A^{ST}}\downarrow & & \downarrow{\scriptstyle i_{S^{-1}M}^{T'}} \\ (ST)^{-1}M & \xrightarrow[k]{} & T'^{-1}(S^{-1}M) \end{array}$$

is commutative.

(i) We use the definition of $(ST)^{-1}A$ as the solution of a universal mapping problem. Let B be a ring and f a homomorphism from A to B such that $f(ST)$ consists of invertible elements. As $f(S)$ consequently consists of invertible elements, there exists a unique homomorphism $f' : S^{-1}A \to B$ such that $f = f' \circ i_A^S$ (no. 1, Proposition 1). For all $t \in T$, $f'(i_A^S(t)) = f(t)$ is invertible in B by hypothesis, hence $f'(T')$ consists of invertible elements; then there exists, by no. 1, Proposition 1, a unique homomorphism f'' from $T'^{-1}(S^{-1}A)$ to B such that $f' = f'' \circ i_{S^{-1}A}^{T'}$, whence $f = f'' \circ u$, setting $u = i_{S^{-1}A}^{T'} \circ i_A^S$.

Moreover, if f''_1: $T'^{-1}(S^{-1}A) \to B$ is a second homomorphism such that $f''_1 \circ u = f$, then $(f''_1 \circ i^{T'}_{S^{-1}A}) \circ i^S_A = (f'' \circ i^{T'}_{S^{-1}A}) \circ i^S_A$, whence $f''_1 \circ i^{T'}_{S^{-1}A} = f'' \circ i^{T'}_{S^{-1}A}$ and consequently $f''_1 = f''$.

As the images under u of the elements of ST in $T'^{-1}(S^{-1}A)$ are invertible, the ordered pair $(T'^{-1}(S^{-1}A), u)$ is a solution of the universal mapping problem (relative to A and ST) considered in no. 1. This shows the existence and uniqueness of j.

(ii) The proof is completely analogous with that of (i), using in this case no. 2, Proposition 3, and is left to the reader.

Proposition 8. *Let* A *be a ring and* S, T *two multiplicative subsets of* A *such that* $S \subset T$. *The following properties are equivalent:*

(a) *The homomorphism* $i^{T,S}_A: S^{-1}A \to T^{-1}A$ *is bijective.*

(b) *For every* A*-module* M, *the homomorphism* $i^{T,S}_A: S^{-1}M \to T^{-1}M$ *is bijective.*

(c) *For all* $t \in T$, *there exists* $a \in A$ *such that* $at \in S$ (in other words, every element of T *divides* an element of S).

(d) *Every prime ideal which meets* T *meets* S.

It has been seen above that $i^{T,S}_A = 1_M \otimes i^{T,S}_A$, which immediately proves the equivalence of (a) and (b). Set $T' = i^S_A(T)$; then (Proposition 7) $T^{-1}A$ is identified with $T'^{-1}(S^{-1}A)$ and (a) is equivalent to saying that the elements of T' are *invertible* in $S^{-1}A$ (no. 1, *Remark* 5). Now, to say that $(t/1)(a/s) = 1/1$ ($t \in T$, $a \in A$, $s \in S$) means that there exists $s' \in S$ such that $tas' = ss'$, which shows the equivalence of (a) and (c). We show that (d) implies (c). Let t be an element of T and suppose that $t/1$ is not invertible in $S^{-1}A$; then there exists a maximal ideal $\mathfrak{m}'$ of $S^{-1}A$ containing $t/1$ (*Algebra*, Chapter I, § 8, no. 7, Theorem 2) and $\mathfrak{p} = (i^S_A)^{-1}(\mathfrak{m}')$ is a prime ideal of A containing t and not meeting S (since the image under i^S_A of an element of S is invertible). Conversely, if there exists a prime ideal $\mathfrak{p}$ which meets T without meeting S, then no element of $\mathfrak{p} \cap T$ can divide an element of S; this proves that (c) implies (d) and completes the proof.

It follows from Proposition 8 that, amongst the multiplicative subsets T of A containing S and satisfying the equivalent conditions of Proposition 8, there exists a *greatest*, consisting of *all* the elements of A which divide an element of S (cf. Exercise 1).

Proposition 9. *Let* I *be a right directed preordered set,* $(S_\alpha)_{\alpha \in I}$ *an increasing family of multiplicative subsets of a ring* A *and* $S = \bigcup_{\alpha \in I} S_\alpha$. *We write* $\rho_{\beta\alpha} = i^{S_\beta, S_\alpha}_A$ *for* $\alpha \leqslant \beta$, $\rho_\alpha = i^{S,S_\alpha}_A$. *Then* $(S_\alpha^{-1}A, \rho_{\beta\alpha})$ *is a direct system of rings and, if, for all* $\alpha \in I$, ρ'_α *is the* *canonical mapping of* $S_\alpha^{-1}A$ *to* $\varinjlim S_\alpha^{-1}A$, *there exists a unique isomorphism j from* $\varinjlim S_\alpha^{-1}A$ *to* $S^{-1}A$ *such that* $j \circ \rho'_\alpha = \rho_\alpha$ *for all* $\alpha \in I$.

For $\alpha \leqslant \beta \leqslant \gamma$, $\rho_{\gamma\alpha} = \rho_{\gamma\beta} \circ \rho_{\beta\alpha}$ (no. 1, Corollary 3 to Proposition 2), hence $(S_\alpha^{-1}A, \rho_{\beta\alpha})$ is a direct system. We write $A' = \varinjlim S_\alpha^{-1}A$; as $\rho_\alpha = \rho_\beta \circ \rho_{\beta\alpha}$ for $\alpha \leqslant \beta$ (no. 1, Corollary 3 to Proposition 2), (ρ_α) is a direct system of homomorphisms and $j = \varinjlim \rho_\alpha$ is the unique homomorphism from A' to $S^{-1}A$ such that $j \circ \rho'_\alpha = \rho_\alpha$ for all $\alpha \in I$. The homomorphisms $\rho'_\alpha \circ i_A^{S_\alpha}: A \to A'$ are all equal, for $\rho_{\beta\alpha} \circ i_A^{S_\alpha} = i_A^{S_\beta}$ for $\alpha \leqslant \beta$; let u be their common value. Clearly the elements of $u(S)$ are invertible in A', which shows that there exists a homomorphism $h: S^{-1}A \to A'$ such that $h \circ i_A^S = u$ (no. 1, Proposition 1). Then

$$j \circ h \circ i_S^A = j \circ u = j \circ \rho'_\alpha \circ i_S^{A_\alpha} = \rho_\alpha \circ i_A^{S_\alpha} = i_A^S$$

for all $\alpha \in I$ and consequently $j \circ h$ is the identity automorphism of $S^{-1}A$. On the other hand, for all $\alpha \in I$,

$$h \circ j \circ \rho'_\alpha \circ i_A^{S_\alpha} = h \circ \rho_\alpha \circ i_A^{S_\alpha} = h \circ i_A^S = u = \rho'_\alpha \circ i_A^{S_\alpha},$$

whence $h \circ j \circ \rho'_\alpha = \rho'_\alpha$ for all $\alpha \in I$; it follows that $h \circ j$ is the identity automorphism of A' and consequently j is an isomorphism.

COROLLARY. *Under the hypotheses of Proposition* 9, *let* M *be an* A*-module. We write* $f_{\beta\alpha} = i_M^{S_\beta, S_\alpha}$ *for* $\alpha \leqslant \beta$, $f_\alpha = i_M^{S, S_\alpha}$ *for all* $\alpha \in I$ *and let* f'_α *be the canonical mapping from* $S_\alpha^{-1}M$ *to* $\varinjlim S_\alpha^{-1}M$; *then there exists an* $S^{-1}A$*-isomorphism g of* $S^{-1}M$ *onto* $\varinjlim S_\alpha^{-1}M$ *such that* $g \circ f_\alpha = f'_\alpha$ *for all* $\alpha \in I$.

The corollary follows immediately from the definitions $S_\alpha^{-1}M = M \otimes_A S_\alpha^{-1}A$ and $S^{-1}M = M \otimes_A S^{-1}A$ and the fact that taking direct limits commutes with tensor products (*Algebra*, Chapter II, § 6, no. 3, Proposition 7).

4. PROPERTIES OF MODULES OF FRACTIONS

Throughout this no., A *denotes a ring and* S *a multiplicative subset of* A.

Let $(M_\alpha, \phi_{\beta\alpha})$ be a direct system of A-modules; then $(S^{-1}M_\alpha, S^{-1}\phi_{\beta\alpha})$ is a direct system of $S^{-1}A$-modules and the fact that taking direct limits commutes with tensor products (*Algebra*, Chapter II, § 6, no. 3, Proposition 7) allows us to define a canonical isomorphism

$$\varinjlim (S^{-1}M_\alpha) \to S^{-1} \varinjlim M_\alpha.$$

Similarly, the fact that taking direct sums commutes with tensor products (*Algebra*, Chapter II, § 3, no. 7, Proposition 7) allows us to define for every family $(M_\iota)_{\iota \in I}$ of A-modules a canonical isomorphism

$$\bigoplus_{\iota \in I} S^{-1}M_\iota \to S^{-1} \bigoplus_{\iota \in I} M_\iota.$$

Finally we note that, if an A-module M is the sum of a family $(N_\iota)_{\iota \in I}$ of submodules, $S^{-1}M$ is the sum of the family of sub-$S^{-1}A$-modules generated by the $i_M^S(N_\iota)$. Then it follows that, if M is a finitely generated A-module (resp. finitely presented A-module), $S^{-1}M = S^{-1}A \otimes_A M$ is a finitely generated $S^{-1}A$-module (resp. finitely presented $S^{-1}A$-module).

THEOREM 1. *The ring* $S^{-1}A$ *is a flat* A-*module* (Chapter I, § 2, no. 1, Definition 2).

If $u: M' \to M$ is an injective homomorphism of A-modules, it is necessary to establish that $S^{-1}u: S^{-1}M' \to S^{-1}M$ is injective. Now, if m'/s ($m' \in M'$, $s \in S$) is such that $u(m')/s = 0$, this implies the existence of an $s' \in S$ such that $s'u(m') = 0$ (no. 2, Proposition 4) or $u(s'm') = 0$; as u is injective, it follows that $s'm' = 0$, whence $m'/s = 0$.

The fact that $S^{-1}A$ is a flat A-module allows us to apply to it the results of Chapter I, § 2. In particular:

(1) If M is an A-module and N a submodule of M, $S^{-1}N$ is canonically identified with a *submodule* of $S^{-1}M$ generated by $i_M^S(N)$ (Chapter I, § 2, no. 3, *Remark* 2); with this identification, $S^{-1}(M/N)$ is identified with $(S^{-1}M)/(S^{-1}N)$ and, if P is a second submodule of M, then

$$S^{-1}(N + P) = S^{-1}N + S^{-1}P, \qquad S^{-1}(N \cap P) = S^{-1}N \cap S^{-1}P$$

(Chapter I, § 2, no. 6, Proposition 2).

(2) If M is a *finitely generated* A-module, then

$$S^{-1}\,\mathrm{Ann}(M) = \mathrm{Ann}(S^{-1}M) \tag{9}$$

(Chapter I, § 2, no. 10, Corollary 2 to Proposition 12).

PROPOSITION 10. *Let* M *be an* A-*module. For every submodule* N′ *of the* $S^{-1}A$-*module* $S^{-1}M$, *let* $\phi(N')$ *be the inverse image of* N′ *under* i_M^S. *Then:*

(i) $S\phi(N') = N'$.

(ii) *For every submodule* N *of* M, *the submodule* $\phi(S^{-1}N)$ *of* M *consists of those* $m \in M$ *for which there exists* $s \in S$ *such that* $sm \in N$.

(iii) ϕ *is an isomorphism (for the orderings defined by inclusion) of the set of sub-*$S^{-1}A$*-modules of* $S^{-1}M$ *onto the set of submodules* Q *of* M *which satisfy the following condition:*

(MS) *If* $sm \in Q$, *where* $s \in S$, $m \in M$, *then* $m \in Q$.

Obviously $S^{-1}\phi(N') \subset N'$; conversely, if $n' = m/s \in N'$, then $m/1 \in N'$, hence $m \in \phi(N')$ and consequently $n' \in S^{-1}(\phi(N'))$; whence (i). For an element $m \in M$ to be such that $m \in \phi(S^{-1}N)$, it is necessary and sufficient that $m/1 \in S^{-1}N$, that is that there exist $s \in S$ and $n \in N$ such that $m/1 = n/s$; this means that there exists $s' \in S$ such that $s'sm = s'n \in N$, whence (ii). Finally, the relation $sm \in \phi(N')$ is equivalent by definition to $sm/1 \in N'$ and as $s/1$ is invertible in $S^{-1}A$, this implies $m/1 \in N'$, or $m \in \phi(N')$, hence $\phi(N')$ satisfies condition (MS); on the other hand, it follows from (ii) that, if N satisfies (MS), then $\phi(S^{-1}N) = N$, which completes the proof of (iii).

The submodule $\phi(S^{-1}N)$ is called the *saturation of* N *in* M *with respect to* S, and the submodules satifying condition (MS) (and hence equal to their saturations) are said to be *saturated with respect to* S. The submodule $\phi(S^{-1}N)$ is the kernel of the composite homomorphism

$$M \xrightarrow{h} M/N \xrightarrow{i^S_{M/N}} S^{-1}M/S^{-1}N$$

where h is the canonical homomorphism, as follows from the commutativity of the diagram

$$\begin{array}{ccc} M & \xrightarrow{h} & M/N \\ {\scriptstyle i^S_M}\downarrow & & \downarrow{\scriptstyle i^S_{M/N}} \\ S^{-1}M & \xrightarrow[S^{-1}h]{} & S^{-1}M/S^{-1}N \end{array}$$

If S is the complement in A of a prime ideal $\mathfrak{p}$, $\phi(S^{-1}N)$ is also called the *saturation of* N *in* M *with respect to* $\mathfrak{p}$.

COROLLARY 1. *Let* N_1, N_2 *be two submodules of an* A*-module* M. *For* $S^{-1}N_1 \subset S^{-1}N_2$, *it is necessary and sufficient that the saturation of* N_1 *with respect to* S *be contained in that of* N_2.

Corollary 2. *If* M *is a Noetherian* (resp. *Artinian*) A-*module,* $S^{-1}M$ *is a Noetherian* (resp. *Artinian*) $S^{-1}A$-*module. In particular, if the ring* A *is Noetherian* (resp. *Artinian*), *so is the ring* $S^{-1}A$.

5. IDEALS IN A RING OF FRACTIONS

Proposition 11. *Let* A *be a ring and* S *a multiplicative subset of* A. *For every ideal* $\mathfrak{b}'$ *of* $S^{-1}A$, *let* $\mathfrak{b} = (i_A^S)^{-1}(\mathfrak{b}')$ *be such that* $\mathfrak{b}' = S^{-1}\mathfrak{b}$.

(i) *Let f be the canonical homomorphism* $A \to A/\mathfrak{b}$. *The homomorphism from* $S^{-1}A$ *to* $(f(S))^{-1}(A/\mathfrak{b})$ *canonically associated with f* (no. 1, Proposition 2) *is surjective and its kernel is* $\mathfrak{b}'$, *which defines, by taking quotients, a canonical isomorphism of* $(S^{-1}A)/\mathfrak{b}'$ *onto* $(f(S))^{-1}(A/\mathfrak{b})$. *Moreover, the canonical homomorphism from* $A/\mathfrak{b}$ *to* $(f(S))^{-1}(A/\mathfrak{b})$ *is injective.*

(ii) *The mapping* $\mathfrak{b}' \mapsto \mathfrak{b} = (i_M^S)^{-1}(\mathfrak{b}')$, *restricted to the set of maximal* (resp. *prime*) *ideals of* $S^{-1}A$, *is an isomorphism* (*with respect to inclusion*) *of this set onto the set of ideals of* A *which are maximal among those which do not meet* S (resp. *the set of prime ideals of* A *not meeting* S).

(iii) *If* $\mathfrak{q}'$ *is a prime ideal of* $S^{-1}A$ *and* $\mathfrak{q} = (i_A^S)^{-1}(\mathfrak{q}')$, *there exists an isomorphism of the ring of fractions* $A_{\mathfrak{q}}$ *onto the ring* $(S^{-1}A)_{\mathfrak{q}'}$, *which maps* a/b *to* $(a/1)/(b/1)$, *where* $a \in A$, $b \in A - \mathfrak{q}$.

(i) $(f(S))^{-1}(A/\mathfrak{b})$ can be identified with $S^{-1}(A/\mathfrak{b})$ by means of the canonical isomorphism between these two modules (no. 2, Proposition 6). The exact sequence $0 \to \mathfrak{b} \to A \to A/\mathfrak{b} \to 0$ then induces an exact sequence

$$0 \to S^{-1}\mathfrak{b} \to S^{-1}A \to S^{-1}(A/\mathfrak{b}) \to 0$$

(no. 4, Theorem 1) whose existence proves the first assertion of (i), taking account of the fact that $\mathfrak{b}' = S^{-1}\mathfrak{b}$. Since $\mathfrak{b}$ is saturated with respect S, the conditions $a \in A$, $s \in S$, $as \in \mathfrak{b}$ imply $a \in \mathfrak{b}$; the homothety of ratio s on $A/\mathfrak{b}$ is then injective, which proves the second assertion of (i).

(ii) We note first that the relation $\mathfrak{b}' = S^{-1}A$ is equivalent to the relation $\mathfrak{b} \cap S \neq \varnothing$, the latter expressing the fact that $\mathfrak{b}'$ contains invertible elements of $S^{-1}A$. It follows from no. 4, Proposition 10 (iii) that $\mathfrak{b}' \mapsto \mathfrak{b} = (i_A^S)^{-1}(\mathfrak{b}')$ is an isomorphism (with respect to inclusion) of the set of ideals of $S^{-1}A$ distinct from $S^{-1}A$ onto the set $\mathfrak{F}$ of ideals of A not meeting S and satisfying condition (MS) of Proposition 10. If $\mathfrak{b}'$ is maximal (resp. prime), clearly $\mathfrak{b}'$ is maximal in $\mathfrak{F}$ (resp. prime) and conversely (by (i)). On the other hand, if $\mathfrak{r}$ is an ideal of A disjoint from S, its saturation $\mathfrak{r}_1$ with respect to S is an ideal of A containing $\mathfrak{r}$ and disjoint from S: for no element $a \in S$ can satisfy $sa \in \mathfrak{r}$ for some $s \in S$, since it would follow that $sa \in \mathfrak{r} \cap S$. We conclude that, if $\mathfrak{r}$ is maximal among the ideals of A meeting S, it is maximal in $\mathfrak{F}$. Similarly, if $\mathfrak{r}$ is a prime ideal not meeting S, it satisfies condition (MS) of no. 4, Proposition 10 by definition of prime ideals and hence belongs to $\mathfrak{F}$ This completes the proof of (ii).

(iii) Suppose that $\mathfrak{q}'$ is prime and such that $\mathfrak{q}$ is also prime. The set $T = A - \mathfrak{q}$ is a multiplicative subset of A which contains S, whence $ST = T$. We write $T' = i_A^S(T)$; it follows from no. 3, Proposition 7 (i) that there exists a unique isomorphism j of $T^{-1}A = A_\mathfrak{q}$ onto $T'^{-1}(S^{-1}A)$ such that

$$j(a/b) = (a/1)/(b/1),$$

where $a \in A$ and $b \in T$. On the other hand T' obviously does not meet $\mathfrak{q}'$; conversely, let $a/s \in S^{-1}A$; since $1/s$ is invertible in $S^{-1}A$, the condition $a/s \notin \mathfrak{q}'$ is equivalent to $i_A^S(a) = a/1 \notin \mathfrak{q}'$ and hence to $a \notin \mathfrak{q}$; it follows that $S^{-1}A - \mathfrak{q}' = S^{-1}T'$ and hence, by Proposition 8 of no. 3, $T'^{-1}(S^{-1}A) = (S^{-1}A)_{\mathfrak{q}'}$.

The isomorphism defined in (iii) is called *canonical.* If A is an *integral domain,* the canonical isomorphisms of $A_\mathfrak{q}$ and $(S^{-1}A)_{\mathfrak{q}'}$ onto subrings of the field of fractions K of A have the same image.

Remark. For an ideal $\mathfrak{a}$ of A to satisfy $S^{-1}\mathfrak{a} = S^{-1}A$ (or, what amounts to the same thing by no. 4, Theorem 1, $S^{-1}(A/\mathfrak{a}) = 0$), it is necessary and sufficient that $\mathfrak{a} \cap S \neq \varnothing$, as follows immediately from the definitions.

Corollary 1. *Let* A *be a ring and* S *a multiplicative subset of* A. *Every ideal* $\mathfrak{p}$ *of* A *which is maximal among those which do not meet* S *is prime.*

By Proposition 11, the hypothesis on $\mathfrak{p}$ means that $\mathfrak{p} = (i_A^S)^{-1}(\mathfrak{m}')$, where $\mathfrak{m}'$ is a maximal ideal of $S^{-1}A$; as $\mathfrak{m}'$ is prime, so is $\mathfrak{p}$.

Corollary 2. *Let* A *be a ring and* S *a multiplicative subset of* A. *For every ideal* $\mathfrak{a}$ *of* A *not meeting* S *there exists a prime ideal containing* $\mathfrak{a}$ *and not meeting* S.

$S^{-1}\mathfrak{a} \neq S^{-1}A$ (*Remark*) and hence there exists a maximal ideal of $S^{-1}A$ containing $S^{-1}\mathfrak{a}$ (*Algebra*, Chapter I, § 8, no. 7, Theorem 2) and the corollary follows from Proposition 11 (ii).

Corollary 3. *Let* A, B *be two rings,* ρ *a homomorphism from* A *to* B *and* $\mathfrak{p}$ *a prime ideal of* A. *For there to exist a prime ideal* $\mathfrak{p}'$ *of* B *such that* $\overset{-1}{\rho}(\mathfrak{p}') = \mathfrak{p}$, *it is necessary and sufficient that* $\overset{-1}{\rho}(B\rho(\mathfrak{p})) = \mathfrak{p}$.

If there exists an ideal $\mathfrak{p}'$ of B such that $\overset{-1}{\rho}(\mathfrak{p}') = \mathfrak{p}$, then $\rho(\mathfrak{p}) \subset \mathfrak{p}'$, whence $B\rho(\mathfrak{p}) \subset \mathfrak{p}'$ and $\overset{-1}{\rho}(B\rho(\mathfrak{p})) \subset \overset{-1}{\rho}(\mathfrak{p}') \subset \mathfrak{p}$; as the converse inclusion is obvious, $\overset{-1}{\sigma}(B\rho(\mathfrak{p})) = \mathfrak{p}$. Conversely, suppose that $\overset{-1}{\rho}(B\rho(\mathfrak{p})) = \mathfrak{p}$ and consider the multiplicative subset $S = \rho(A - \mathfrak{p})$ of B; the hypothesis shows that

$$S \cap B\rho(\mathfrak{p}) = \varnothing;$$

by Corollary 2 there exists a prime ideal $\mathfrak{p}'$ of B containing $B\rho(\mathfrak{p})$ and not meeting S; then $\overset{-1}{\rho}(\mathfrak{p}')$ contains $\mathfrak{p}$ and cannot contain any element of $A - \mathfrak{p}$ and hence is equal to $\mathfrak{p}$.

COROLLARY 4. *Let* A, B *be two rings and* ρ *a homomorphism from* A *to* B.

(i) *Suppose that there exists a* B*-module* E *such that* $\rho_*(E)$ *is a faithfully flat* A*-module. Then, for every prime ideal* $\mathfrak{p}$ *of* A, *there exists a prime ideal* $\mathfrak{p}'$ *of* B *such that* $\overset{-1}{\rho}(\mathfrak{p}') = \mathfrak{p}$.

(ii) *Conversely, suppose that* B *is a flat* A*-module. Then, if, for every prime ideal* $\mathfrak{p}$ *of* A, *there exists an ideal* $\mathfrak{p}'$ *of* B *such that* $\overset{-1}{\rho}(\mathfrak{p}') = \mathfrak{p}$, B *is a faithfully flat* A*-module.*

(i) The hypothesis implies that, for every ideal $\mathfrak{a}$ of A, $\overset{-1}{\rho}(B\rho(\mathfrak{a})) = \mathfrak{a}$ (Chapter I, § 3, no. 5, Proposition 8 (ii)) and it is sufficient to apply Corollary 3.

(ii) It is sufficient to show that, for every maximal ideal $\mathfrak{m}$ of A, there exists a maximal ideal $\mathfrak{m}'$ of B such that $\overset{-1}{\rho}(\mathfrak{m}') = \mathfrak{m}$ (Chapter I, § 2, no. 5, Proposition 9 (e)). Now there exists by hypothesis an ideal $\mathfrak{q}$ of B such that $\overset{-1}{\rho}(\mathfrak{q}) = \mathfrak{m}$; as $\mathfrak{q} \neq B$, there exists a maxiinal ideal $\mathfrak{m}'$ of B containing $\mathfrak{q}$ and consequently $\overset{-1}{\rho}(\mathfrak{m}') \supset \mathfrak{m}$; but, as $\overset{-1}{\rho}(\mathfrak{m}')$ cannot contain 1, $\overset{-1}{\rho}(\mathfrak{m}') = \mathfrak{m}$.

COROLLARY 5. *Let* A *be a ring,* S *a multiplicative subset of* A *and* B *a ring such that* $i_A^S(A) \subset B \subset S^{-1}A$. *Let* $\mathfrak{q}$ *be a prime ideal of* B *such that the prime ideal* $\mathfrak{p} = (i_A^S)^{-1}(\mathfrak{q})$ *of* A *does not meet* S *and let* $\mathfrak{p}'$ *be the prime ideal* $S^{-1}\mathfrak{p}$ *of* $S^{-1}A$. *Then* $\mathfrak{p}' \cap B = \mathfrak{q}$.

Let $S' = i_A^S(S)$; a canonical isomorphism has been defined from $S'^{-1}B$ to $S^{-1}A$ (no. 1, Corollary 4 to Proposition 2); we identify these two rings by means of this isomorphism. As $\mathfrak{q} \cap S' = \varnothing$, $\mathfrak{q}' = S'^{-1}\mathfrak{q}$ is the unique prime ideal of $S^{-1}A = S'^{-1}B$ such that $\mathfrak{q}' \cap B = (i_A^S)^{-1}(\mathfrak{q}') = \mathfrak{q}$ (Proposition 11 (ii)), whence $(i_A^S)^{-1}(\mathfrak{q}') = \mathfrak{p}$; consequently $\mathfrak{q}' = \mathfrak{p}'$ (Proposition 11 (ii)).

In the notation of Corollary 5, there are canonical isomorphisms of $A_{\mathfrak{p}}$ and $B_{\mathfrak{q}}$ onto $(S^{-1}A)_{\mathfrak{p}'}$ (Proposition 11 (iii)) and hence a *canonical isomorphism* $A_{\mathfrak{p}} \to B_{\mathfrak{q}}$.

6. NILRADICAL AND MINIMAL PRIME IDEALS

In a (commutative) ring A the set of nilpotent elements is an *ideal*, for if x, y are elements of A such that $x^m = y^n = 0$, then $(x + y)^{m+n} = 0$ by the binomial theorem.

DEFINITION 4. *The ideal of nilpotent elements of a (commutative) ring* A *is called the nilradical of* A. *If* $\mathfrak{a}$ *is an ideal of* A, *the inverse image, under the canonical mapping* $A \to A/\mathfrak{a}$, *of the nilradical of* $A/\mathfrak{a}$ *is called the radical of* $\mathfrak{a}$.

We often denote by $\mathfrak{r}(\mathfrak{a})$ the radical of an ideal $\mathfrak{a}$ of A.

To say that an element $x \in A$ belongs to the radical of $\mathfrak{a}$ means therefore that there exists an integer $n > 0$ such that $x^n \in \mathfrak{a}$. If f is a homomorphism from A to a ring B and $\mathfrak{b}$ is an ideal of B, the radical of $\overset{-1}{f}(\mathfrak{b})$ is the *inverse image* under f of the radical of $\mathfrak{b}$, for to say that $x^n \in \overset{-1}{f}(\mathfrak{b})$ means that $(f(x))^n \in \mathfrak{b}$.

The nilradical of a ring A is contained in its Jacobson radical (*Algebra*, Chapter VIII, § 6, no. 3, Corollary 3 to Theorem 1) but may be distinct from it; it is always equal to it if A is *Artinian* (*Algebra*, Chapter VIII, § 6, no. 4, Theorem 3).

We say that a prime ideal $\mathfrak{p}$ of a ring A is a *minimal prime ideal* if it is minimal in the set of prime ideals of A ordered by inclusion.

PROPOSITION 12. *Let* $\mathfrak{p}$ *be a minimal prime ideal of a ring* A. *For all* $x \in \mathfrak{p}$, *there exists* $s \in A - \mathfrak{p}$ *and an integer* $k > 0$ *such that* $sx^k = 0$.

The set S of elements of the form sx^k (k an integer >0, $s \in A - \mathfrak{p}$) is a multiplicative subset of A. If $0 \notin S$, there would exist a prime ideal $\mathfrak{p}'$ not meeting S (no. 5, Corollary 2 to Proposition 11). Then $\mathfrak{p}' \subset \mathfrak{p}$ and $\mathfrak{p}' \neq \mathfrak{p}$ since $x \notin \mathfrak{p}'$, contrary to the hypothesis that $\mathfrak{p}$ is minimal.

PROPOSITION 13. *The nilradical of a ring* A *is the intersection of all the prime ideals of* A *and it is also the intersection of the minimal prime ideals of* A.

Clearly, if $x \in A$ is nilpotent, it is contained in every prime ideal of A (§ 1, no. 1, Definition 1). Conversely, let x be a non-nilpotent element of A; the set S of x^k (k an integer $\geqslant 0$) is then a multiplicative subset of A not containing 0 and hence there exists a prime ideal $\mathfrak{p}$ of A not meeting S (no. 5, Corollary 2 to Proposition 11) and *a fortiori* $x \notin \mathfrak{p}$; this establishes the first assertion. To prove the second, it is sufficient to prove

LEMMA 2. *Every prime ideal of a ring* A *contains a minimal prime ideal of* A.

It is sufficient, by Zorn's Lemma, to show that the set P of prime ideals, ordered by the relation $\supset$, is *inductive*. Now, if G is a non-empty totally ordered subset of P, the intersection $\mathfrak{p}_0$ of the ideals $\mathfrak{p} \in G$ is also a prime ideal: for if $x \notin \mathfrak{p}_0$ and $y \notin \mathfrak{p}_0$, there exists an ideal $\mathfrak{p} \in G$ such that $x \notin \mathfrak{p}$ and $y \notin \mathfrak{p}$, whence $xy \notin \mathfrak{p}$ and *a fortiori* $xy \notin \mathfrak{p}_0$.

Remark. In § 4, no. 3, Corollary 3 to Proposition 14, we shall show that in a *Noetherian* ring the set of minimal prime ideals is *finite*; moreover we shall see later that every decreasing sequence of prime ideals in a Noetherian ring is *stationary*.

COROLLARY 1. *The radical of an ideal* $\mathfrak{a}$ *in a ring* A *is the intersection of the prime ideals containing* $\mathfrak{a}$ *and it is also the intersection of the minimal elements of this set of prime ideals.*

COROLLARY 2. *For an ideal* $\mathfrak{a}$ *of a ring* A *we denote by* $\mathfrak{r}(\mathfrak{a})$ *the radical of* $\mathfrak{a}$. *Then, for two ideals of* $\mathfrak{a}$, $\mathfrak{b}$ *of* A,

$$\mathfrak{r}(\mathfrak{a} \cap \mathfrak{b}) = \mathfrak{r}(\mathfrak{a}\mathfrak{b}) = \mathfrak{r}(\mathfrak{a}) \cap \mathfrak{r}(\mathfrak{b});$$

in particular, if $\mathfrak{a} \subset \mathfrak{b}$, *then* $\mathfrak{r}(\mathfrak{a}) \subset \mathfrak{r}(\mathfrak{b})$.

For a prime ideal to contain $\mathfrak{a} \cap \mathfrak{b}$ (or $\mathfrak{a}\mathfrak{b}$), it is necessary and sufficient that it contain one of the ideals $\mathfrak{a}$, $\mathfrak{b}$ (§ 1, no. 1, Proposition 1).

PROPOSITION 14. *For two ideals* $\mathfrak{a}$, $\mathfrak{b}$ *of a ring to be relatively prime, it is necessary and sufficient that their radicals* $\mathfrak{r}(\mathfrak{a})$ *and* $\mathfrak{r}(\mathfrak{b})$ *be so.*

The necessity of the condition is obvious since $\mathfrak{a} \subset \mathfrak{r}(\mathfrak{a})$ and $\mathfrak{b} \subset \mathfrak{r}(\mathfrak{b})$; the condition is sufficient by § 1, no. 2, Proposition 3.

PROPOSITION 15. *In a ring* A *let* $\mathfrak{a}$ *be an ideal and* $\mathfrak{b}$ *a finitely generated ideal contained in the radical of* $\mathfrak{a}$. *Then there exists an integer* $k > 0$ *such that* $\mathfrak{b}^k \subset \mathfrak{a}$.

Let $(b_i)_{1 \leqslant i \leqslant n}$ be a system of generators of $\mathfrak{b}$. By hypothesis there exists an integer h such that $b_i^h \in \mathfrak{a}$ for $1 \leqslant i \leqslant n$. If a product of nh elements, each of which is a linear combination of the b_i with coefficients in A, is expanded, each term is a multiple of a product of nh factors, each of which is equal to a b_i; among these factors, at least h have the same index i for some i, hence, the product belongs to $\mathfrak{a}$ and nh is the required integer k.

COROLLARY. *In a Noetherian ring the nilradical is a nilpotent ideal.*

PROPOSITION 16. *Let* B *be a ring and* A *a subring of* B. *For every minimal prime ideal* $\mathfrak{p}$ *of* A *there exists a minimal prime ideal* $\mathfrak{q}$ *of* B *such that* $\mathfrak{q} \cap A = \mathfrak{p}$.

We write $S = A - \mathfrak{p}$; then the ring $A_\mathfrak{p} = S^{-1}A$ is identified with a subring of $S^{-1}B$ (no. 1, Proposition 2) and on the other hand $A_\mathfrak{p}$ has only *a single* prime ideal $\mathfrak{p}'$ since $\mathfrak{p}$ is minimal (no. 5, Proposition 11). As $S^{-1}B$ is not reduced to 0 (since it contains $A_\mathfrak{p}$), it has at least one prime ideal $\mathfrak{r}'$ and therefore $\mathfrak{r}' \cap A_\mathfrak{p} = \mathfrak{p}'$; if $j = i_B^S$ and we write $\mathfrak{r} = \overset{-1}{j}(\mathfrak{r}')$, then

$$i_A^S(\mathfrak{r} \cap A) \subset \mathfrak{r}' \cap A_\mathfrak{p} = \mathfrak{p}'$$

hence $\mathfrak{r} \cap A \subset \mathfrak{p}$ and, as $\mathfrak{p}$ is minimal, $\mathfrak{r} \cap A = \mathfrak{p}$; moreover, $\mathfrak{r}$ is prime in B; if $\mathfrak{q}$ is a minimal prime ideal of B contained in $\mathfrak{r}$ (Lemma 1), then *a fortiori* $\mathfrak{q} \cap A = \mathfrak{p}$ since $\mathfrak{p}$ is minimal.

DEFINITION 5. *A ring A is called reduced if its nilradical is reduced to* 0, *in other words if no element* $\neq 0$ *of* A *is nilpotent.*

If $\mathfrak{N}$ is the nilradical of a ring A, $A/\mathfrak{N}$ is *reduced*, for if the class mod. $\mathfrak{N}$ of an element $x \in A$ is nilpotent in $A/\mathfrak{N}$, this means that $x^h \in \mathfrak{N}$ for some integer h, hence $x^{hk} = 0$ for some integer k and $x \in \mathfrak{N}$.

PROPOSITION 17. *Let* A *be a ring and* $\mathfrak{N}$ *its nilradical. For every multiplicative subset* S *of* A, $S^{-1}\mathfrak{N}$ *is the nilradical of* $S^{-1}A$. *In particular, if* A *is reduced, then* $S^{-1}A$ *is reduced.*

If $x \in A$, $s \in S$ satisfy $(x/s)^n = x^n/s^n = 0$, there exists $s' \in S$ such that $s'x^n = 0$ (no. 1, *Remark* 3) and *a fortiori* $(s'x)^n = 0$, hence $s'x \in \mathfrak{N}$ and $x/s = s'x/s's \in S^{-1}\mathfrak{N}$; the converse is immediate.

7. MODULES OF FRACTIONS OF TENSOR PRODUCTS AND HOMOMORPHISM MODULES

PROPOSITION 18. *Let* A *be a ring and* S *a multiplicative subset of* A.

(i) *If* M *and* N *are two* A*-modules, the* $S^{-1}A$*-modules* $(S^{-1}M) \otimes_A N$, $M \otimes_A (S^{-1}N)$, $(S^{-1}M) \otimes_{S^{-1}A} (S^{-1}N)$ *and* $S^{-1}(M \otimes_A N)$ *are canonically isomorphic.*

(ii) If M′ *and* N′ *are two* $S^{-1}A$*-modules, the canonical homomorphism*

$$M' \otimes_A N' \to M' \otimes_{S^{-1}A} N'$$

derived from the A*-bilinear mapping* $(x', y') \mapsto x' \otimes y'$ *from* $M' \times N'$ *to* $M' \otimes_{S^{-1}A} N'$ *is bijective.*

Assertion (i) is an immediate consequence of the definition $S^{-1}M =$

$M \otimes_A S^{-1}A$ and the associativity of tensor products, which gives to within canonical isomorphisms

$$(S^{-1}M \otimes_{S^{-1}A} (S^{-1}N) = (S^{-1}M) \otimes_{S^{-1}A} (S^{-1}A \otimes N) = (S^{-1}M) \otimes_A N$$
$$= (S^{-1}A) \otimes_A M \otimes_A N = S^{-1}(M \otimes_A N).$$

To prove (ii), we note first that in M′ and N′, considered as A-modules, the homotheties induced by the elements $s \in S$ are bijective, hence $M' = S^{-1}M'$ and $N' = S^{-1}N'$ (no. 2, *Remark* 4) and similarly $S^{-1}(M' \otimes_A N') = M' \otimes_A N'$; (ii) is then a special case of (i).

COROLLARY. *Let* M *be an* A-*module and* $\mathfrak{a}$ *an ideal of* A. *The sub*-$S^{-1}A$-*modules* $(S^{-1}\mathfrak{a})(S^{-1}M)$, $\mathfrak{a}(S^{-1}M)$, $(S^{-1}\mathfrak{a})j(M)$ (*where* $j: M \to S^{-1}M$ *is the canonical mapping*) *and* $S^{-1}(\mathfrak{a}M)$ *of* $S^{-1}M$ *are identical. In particular, if* $\mathfrak{a}$ *and* $\mathfrak{b}$ *are two ideals of* A, *then*

$$(S^{-1}\mathfrak{a})(S^{-1}\mathfrak{b}) = \mathfrak{a}(S^{-1}\mathfrak{b}) = (S^{-1}\mathfrak{a})\mathfrak{b} = S^{-1}(\mathfrak{a}\mathfrak{b}).$$

Remark. Let M, N, P be three A-modules, $f: M \times N \to P$ an A-bilinear mapping and $S^{-1}f: (S^{-1}M) \times (S^{-1}N) \to (S^{-1}P)$ the $S^{-1}A$-bilinear mapping obtained from f by extending the ring of scalars to $S^{-1}A$ (*Algebra*, Chapter IX, § 1, no. 4, Proposition 1). Let $i: M \to S^{-1}M$, $j: N \to S^{-1}N$ be the canonical homomorphisms; it follows immediately that, if Q is the sub-A-module of P generated by $f(M \times N)$, $S^{-1}Q$ is the sub-$S^{-1}A$-module of $S^{-1}P$ generated by $(S^{-1}f)(i(M) \times j(N))$.

PROPOSITION 19. *Let* A *be a ring and* S *a multiplicative subset of* A.

(i) *If* M *and* N *are two* A-*modules and* M *is finitely generated* (resp. *finitely presented*), *the canonical homomorphism* (Chapter I, § 2, no. 10, formula (10))

$$S^{-1} \operatorname{Hom}_A(M, N) \to \operatorname{Hom}_{S^{-1}A}(S^{-1}M, S^{-1}N)$$

is injective (resp. *bijective*).

(ii) *If* M′, N′ *are two* $S^{-1}A$-*modules, the canonical bijection*

$$\operatorname{Hom}_{S^{-1}A}(M', N') \to \operatorname{Hom}_A(M', N')$$

is bijective.

As $S^{-1}A$ is a flat A-module, (i) is a particular case of Chapter I, § 2, no. 10, Proposition 11. On the other hand, we have already remarked that every A-homomorphism of $S^{-1}A$-modules is necessarily an $S^{-1}A$-homomorphism, in the course of proving Proposition 3 of no. 2; whence (ii).

Proposition 20. *Let* A, A′ *be two rings,* $\rho: A \to A'$ *a homomorphism,* S *a multiplicative subset of* A, $S' = \rho(S)$ *and* $\rho': S^{-1}A \to S'^{-1}A'$ *the homomorphism corresponding to* ρ (no. 1, Proposition 2).

(i) *For every* A′*-module* M′ *there exists a unique* $S^{-1}A$*-isomorphism*

$$j: S^{-1}\rho_*(M') \to \rho'_*(S'^{-1}M')$$

such that $j(m'/s) = m'/\rho(s)$ *for all* $m' \in M'$, $s \in S$.

(ii) *For every* A*-module* M, *there exists a unique isomorphism*

$$j': (S^{-1}M) \otimes_{S^{-1}A} (S'^{-1}A') \to S'^{-1}(M \otimes_A A')$$

of $S'^{-1}A'$*-modules such that* $j'((m/s) \otimes (a'/s')) = (m \otimes a')/(\rho(s)s')$.

(i) If we consider $S'^{-1}M'$ as an A-module by means of the composite homomorphism $i_{M'}^{S'} \circ \rho$, the homotheties induced by the elements of S are bijective, hence there exists a unique homomorphism j with the stated property (no. 2, Proposition 3). As $\rho(S) = S'$, j is surjective; moreover, if $m' \in M'$, $s \in S$, $m'/\rho(s) = 0$, there exists $t' \in S'$ such that $t'm' = 0$; as there exists $t \in S$ such that $\rho(t) = t'$, $t'.m' = 0$ in $\rho_*(M')$, hence $m'/s = 0$ in $S^{-1}\rho_*(M')$.

(ii) As $(S^{-1}M) \otimes_{S^{-1}A} (S'^{-1}A') = (M \otimes_A S^{-1}A) \otimes_{S^{-1}A} (S'^{-1}A')$ and

$$S'^{-1}(M \otimes_A A') = (M \otimes_A A') \otimes_{A'} (S'^{-1}A'),$$

the existence of j' follows from the associativity of tensor products.

8. APPLICATION TO ALGEBRAS

Let A be a ring, B an A-algebra (not necessarily associative or commutative and not necessarily possessing a unit element) and S a multiplicative subset of A. We know that it is possible to define on the $S^{-1}A$-module $S^{-1}B = B \otimes_A S^{-1}A$ a canonical $S^{-1}A$-*algebra* structure, said to be obtained by *extension* of the ring of scalars to $S^{-1}A$ (*Algebra*, Chapter III, § 3) and under which the product $(x/s)(y/t)$ is equal to $(xy)/(st)$. If e is the unit element of B, $e/1$ is the unit element of $S^{-1}B$ and if B is associative (resp. commutative), so is $S^{-1}B$.

Let A be a ring and M an A-module; we denote by T(M) (resp. $\bigwedge$(M), S(M)) the *tensor algebra* (resp. *exterior algebra*, *symmetric algebra*) of M (*Algebra*, Chapter III). It is known that for every commutative A-algebra C there exists a unique isomorphism j of $T(M) \otimes_A C$ onto $T(M \otimes_A C)$ (resp. of $\bigwedge(M) \otimes_A C$ onto $\bigwedge(M \otimes_A C)$, of $S(M) \otimes_A C$ onto $S(M \otimes_A C)$) such that

$$j(x \otimes 1) = x \otimes 1$$

for all $x \in M$, M being canonically identified with a submodule of $T(M)$ (resp. $\bigwedge(M)$, $S(M)$) (*loc. cit.*). Then it is seen in particular that for every multiplicative subset R of A there are *canonical isomorphisms*

$$R^{-1}T(M) \to T(R^{-1}M), \qquad R^{-1}\bigwedge(M) \to \bigwedge(R^{-1}M), \qquad R^{-1}S(M) \to S(R^{-1}M)$$

which reduce to the identity on $R^{-1}M$.

9. MODULES OF FRACTIONS OF GRADED MODULES

Let A be a graded ring, M a graded A-module and Δ the degree monoid; we shall assume in this no. that Δ is a *group*. Recall (*Algebra*, Chapter II, § 11) that A and M are respectively direct sums of additive groups

$$A = \bigoplus_{i \in \Delta} A_i, \qquad M = \bigoplus_{i \in \Delta} M_i$$

with $A_iA_j \subset A_{i+j}$ and $A_iM_j \subset M_{i+j}$ for all $i, j \in \Delta$. Let S be a multiplicative subset of A *all of whose elements are homogeneous*. For all $i \in \Delta$, we write $S_i = S \cap A_i$ and denote by $(S^{-1}M)_i$ the set of elements m' of $S^{-1}M$ for which there exist elements j, k of Δ, an element $m \in M_j$ and an element $s \in S_k$ such that $j - k = i$ and $m' = m/s$. If $(m'_q)_{1 \leqslant q \leqslant r}$ is a finite family of elements of $S^{-1}M$ such that

$$m'_q \in (S^{-1}M)_{i(q)},$$

there exist elements $j(q) \in \Delta$ and $k \in \Delta$, elements $m_q \in M_{j(q)}$ $(1 \leqslant q \leqslant r)$ and $s \in S_k$ such that $m'_q = m_q/s$ for $1 \leqslant q \leqslant r$ (no. 1, *Remark* 2).

PROPOSITION 21. *The ring* $S^{-1}A$ *with the family* $((S^{-1}A)_i)$ *is a graded ring and* $S^{-1}M$ *with the family* $((S^{-1}M)_i)$ *is a graded module over the graded ring* $S^{-1}A$. *The canonical mappings* i_A^S *and* i_M^S *are homogeneous of degree* 0.

Let $m \in M_j$, $s \in S_k$, $m' \in M_{j'}$, $s' \in S_{k'}$ and suppose that $j - k = j' - k' = i$; then $(m/s) - (m'/s') = (s'm - sm')/ss'$ and $s'm - sm' \in M_{j+k'} = M_{j'+k}$ and $ss' \in S_{k+k'}$, hence $(m/s) - (m'/s') \in (S^{-1}M)_i$ by definition; this shows that the $(S^{-1}M)_i$ are additive subgroups of $S^{-1}M$. The sum of these groups is the whole of $S^{-1}M$: for every $x \in S^{-1}M$ may be written as m/s where $m \in M$, $s \in S$; s is *homogeneous* by hypothesis and m is a sum of homogeneous elements m_j; hence x is the sum of the m_j/s, each of which belongs to a subgroup $(S^{-1}M)_i$. Finally the sum of the $(S^{-1}M)_i$ is direct; for let us consider a finite family of elements x_q $(1 \leqslant q \leqslant n)$ such that $x_q \in (S^{-1}M)_{i(q)}$, where the indices $i(q)$ are distinct, and suppose that $\sum_{q=1}^{n} x_q = 0$. Each x_q may be written as $x_q = m_q/s$ where $s \in S_k$ and $m_q \in M_{i(q)+k}$; the hypothesis implies that there exists $s' \in S$ such that $s'\left(\sum_{q=1}^{n} m_q\right) = 0$; if the $s'm_q$ were not all zero, we would have a contradiction since, if $s' \in S_d$, then $s'm_q \in M_{i(q)+d}$ and the $i(q) + d$ are all distinct. It follows that $x_q = 0$ for every index q.

It is immediately verified that, if $a \in (S^{-1}A)_i$ and $x \in (S^{-1}M)_j$, then $ax \in (S^{-1}M)_{i+j}$. Applying this result to the case $M = A$, we see first that $S^{-1}A$ is a ring graded by the $(S^{-1}A)_i$; then we see that $S^{-1}M$ is graded module over $S^{-1}A$. Finally, as $1 \in A_0$, i_A^S and i_M^S are homogeneous of degree 0.

PROPOSITION 22. *Let* A (resp. B) *be a graded ring of type* Δ, M (resp. N) *a graded module over the graded ring* A (resp. B), S (resp. T) *a multiplicative subset of* A (resp. B) *all of whose elements are homogeneous,* $f: A \to B$ *a homogeneous homomorphism of degree* 0 *such that* $f(S) \subset T$ *and* $u: M \to N$ *an* A*-linear mapping which is homogeneous of degree k. Then the homomorphism* $f': S^{-1}A \to T^{-1}B$ *derived from f* (no. 1, Proposition 2) *is homogeneous of degree* 0 *and the* $(S^{-1}A)$*-linear mapping*

$$u': S^{-1}M \to T^{-1}N$$

derived from f and u (no. 2, Proposition 5) *is homogeneous of degree k.*

This follows immediately from the equations $f'(a/s) = f(a)/f(s)$ and $u'(m/s) = u(m)/f(s)$.

Finally we note that, if E is a *graded* A-algebra and S a multiplicative subset of A consisting of homogeneous elements, $S^{-1}E$ with its $(S^{-1}A)$-algebra structure (no. 8) and the graduation $(S^{-1}E)_i$ is a *graded* $S^{-1}A$*-algebra,* as follows immediately from the definitions.

3. LOCAL RINGS. PASSAGE FROM THE LOCAL TO THE GLOBAL

1. LOCAL RINGS

PROPOSITION 1. *Let* A *be a ring and* I *the set of non-invertible elements of* A. *The set* I *is the union of the ideals of* A *which are distinct from* A. *Moreover, the following conditions are equivalent:*

(a) I *is an ideal.*

(b) *The set of ideals of* A *distinct from* A *has a greatest element.*

(c) A *has a unique maximal ideal.*

The relation $x \in \mathrm{I}$ is equivalent to $1 \notin x\mathrm{A}$ and hence $x\mathrm{A} \neq \mathrm{A}$. If $\mathfrak{a}$ is an ideal of A distinct from A and $x \in \mathfrak{a}$, then $x\mathrm{A} \subset \mathfrak{a}$, hence $x\mathrm{A} \neq \mathrm{A}$ and $x \in \mathrm{I}$. Hence every ideal of A distinct from A is contained in I and every element $x \in \mathrm{I}$ belongs to a principal ideal $x\mathrm{A} \neq \mathrm{A}$. This proves the first assertion, which immediately implies the equivalence of (a), (b) and (c).

Remark (1). Note that, if (c) holds, I is the *Jacobson radical* of the ring A (*Algebra*, Chapter VIII, § 6, no. 3, Definition 3).

DEFINITION 1. *A ring* A *is called a local ring if it satisfies the equivalent conditions* (a), (b) *and* (c) *of Proposition* 1. *The quotient of* A *by its Jacobson radical* (*which is then the unique maximal ideal of* A) *is called the residue field of* A.

DEFINITION 2. *Let* A, B *be two local rings and* $\mathfrak{m}$, $\mathfrak{n}$ *their respective maximal ideals. A homomorphism* $u \colon \mathrm{A} \to \mathrm{B}$ *is called local if* $u(\mathfrak{m}) \subset \mathfrak{n}$.

This amounts to saying that $\overset{-1}{u}(\mathfrak{n}) = \mathfrak{m}$, for $\overset{-1}{u}(\mathfrak{n})$ is then an ideal containing $\mathfrak{m}$ and not containing 1 and hence equal to $\mathfrak{m}$. Taking quotients, we then derive canonically from u an *injective homomorphism* $\mathrm{A}/\mathfrak{m} \to \mathrm{B}/\mathfrak{n}$ from the residue field of A to that of B.

Examples

(1) A field is a local ring. A ring which is reduced to 0 is not a local ring.

(2) Let A be a local ring and k its residue field. The ring of formal power series $\mathrm{B} = \mathrm{A}[[\mathrm{X}_1, \ldots, \mathrm{X}_n]]$ is a local ring, for the non-invertible elements of B are the formal power series whose constant terms are not invertible in A (*Algebra*, Chapter IV, § 5, no. 6, Proposition 4). The canonical injection of A into B is a local homomorphism and the corresponding injection of residue fields is an isomorphism.

(3) Let $\mathfrak{b}$ be an ideal of a ring A which is only contained in *a single* maximal

ideal $\mathfrak{m}$; then $A/\mathfrak{b}$ is a *local ring* with maximal ideal $\mathfrak{m}/\mathfrak{b}$ and residue field canonically isomorphic to $A/\mathfrak{m}$. This applies in particular to the case $\mathfrak{b} = \mathfrak{m}^k$, $\mathfrak{m}$ being any maximal ideal of A (§ 1, no. 1, Corollary to Proposition 1). If A itself is a local ring with maximal ideal $\mathfrak{m}$, then for every ideal $\mathfrak{b} \neq A$ of A, $A/\mathfrak{b}$ is a local ring, the canonical homomorphism $A \to A/\mathfrak{b}$ a local homomorphism and the corresponding homomorphism of residue fields an *isomorphism.*

(4) Let X be a topological space, x_0 a point of X and A the ring of germs at the point x_0 of real-valued functions continuous in a neighbourhood of x_0 (*General Topology*, Chapter I, § 6, no. 10). Clearly, for the germ at x_0 of a continuous function f to be invertible in A, it is necessary and sufficient that $f(x_0) \neq 0$, as that implies that $f(x) \neq 0$ in a neighbourhood of x_0. The ring A is therefore a local ring whose maximal ideal $\mathfrak{m}$ is the set of germs of functions which are *zero at* x_0; taking quotients, the mapping $g \mapsto g(x_0)$ of A to $\mathbf{R}$ gives an *isomorphism* of the residue field $A/\mathfrak{m}$ onto $\mathbf{R}$.

PROPOSITION 2. *Let A be a ring and $\mathfrak{p}$ a prime ideal of A. The ring $A_\mathfrak{p}$ is local; its maximal ideal is the ideal $\mathfrak{p}A_\mathfrak{p} = \mathfrak{p}_\mathfrak{p}$, generated by the canonical image of $\mathfrak{p}$ in $A_\mathfrak{p}$; its residue field is canonically isomorphic to the field of fractions of $A/\mathfrak{p}$.*

Let $S = A - \mathfrak{p}$ and $j: A \to A_\mathfrak{p}$ be the canonical homomorphism; the hypothesis that $\mathfrak{p}$ is prime implies that $\mathfrak{p}$ is saturated with respect to S, hence $\overset{-1}{j}(\mathfrak{p}A_\mathfrak{p}) = \mathfrak{p}$ (§ 2, no. 4, Proposition 10) and, as the ideals of A not meeting S are those contained in $\mathfrak{p}$, the first two assertions are special cases of § 2, no. 5, Proposition 11 (ii). Moreover, if f is the canonical homomorphism $A \to A/\mathfrak{p}$, $f(S)$ is the set of elements $\neq 0$ of the integral domain $A/\mathfrak{p}$ and hence the last assertion is a special case of § 2, no. 5, Proposition 11 (i).

DEFINITION 3. *Let A be a ring and $\mathfrak{p}$ a prime ideal of A. The ring $A_\mathfrak{p}$ is called the local ring of A at $\mathfrak{p}$, or the local ring of $\mathfrak{p}$, when there is no ambiguity.*

Remark (2). If A is a local ring and $\mathfrak{m}$ its maximal ideal, the elements of $A - \mathfrak{m}$ are invertible (Proposition 1) and hence $A_\mathfrak{m}$ is canonically identified with A (§ 2, no. 1, *Remark* 5).

Examples

(5) Let p be a prime number. The local ring $\mathbf{Z}_{(p)}$ is the set of rational numbers a/b, where a, b are rational integers with b prime to p; the residue field of $\mathbf{Z}_{(p)}$ is isomorphic to the prime field $\mathbf{F}_p = \mathbf{Z}/(p)$.

* (6) Let V be an affine algebraic variety, A the ring of regular functions on V, W an irreducible subvariety of V and $\mathfrak{p}$ the (necessarily prime) ideal of A consisting of the functions which are zero at every point of W. The ring $A_\mathfrak{p}$ is called the local ring of W on V. *

PROPOSITION 3. *Let* A *be a ring,* $\mathfrak{p}$ *a prime ideal of* A *and* $S = A - \mathfrak{p}$. *For every ideal* $\mathfrak{b}'$ *of* $A_{\mathfrak{p}}$ *distinct from* $A_{\mathfrak{p}}$, *let* $\mathfrak{b}$ *be the ideal* $(i_A^S)^{-1}(\mathfrak{b}')$ *of* A *so that* $\mathfrak{b}' = \mathfrak{b}A_{\mathfrak{p}}$.

(i) *Let* f *be the canonical homomorphism* $A \to A/\mathfrak{b}$. *The homomorphism from* $A_{\mathfrak{p}}$ *to* $(A/\mathfrak{b})_{\mathfrak{p}/\mathfrak{b}}$ *canonically associated with* f (§ 2, no. 1, Proposition 2) *is surjective and its kernel is* $\mathfrak{b}'$, *which defines, by taking quotients, a canonical isomorphism of* $A_{\mathfrak{p}}/\mathfrak{b}'$ *onto* $(A/\mathfrak{b})_{\mathfrak{p}/\mathfrak{b}}$.

(ii) *The mapping* $\mathfrak{b}' \to \mathfrak{b} = (i_A^S)^{-1}(\mathfrak{b}')$, *restricted to the set of prime ideals of* $A_{\mathfrak{p}}$, *is an isomorphism (with respect to inclusion) of this set onto the set of prime ideals of* A *contained in* $\mathfrak{p}$. *If* $\mathfrak{b}'$ *is prime in* $A_{\mathfrak{p}}$, *there exists an isomorphism of the ring* $A_{\mathfrak{b}}$ *onto the ring* $(A_{\mathfrak{p}})_{\mathfrak{b}'}$, *which maps* a/s *to* $(a/1)/(s/1)$ *for all* $a \in A$, $s \in A - \mathfrak{b}$).

This is just a particular case of § 2, no. 5, Proposition 11.

Remarks

(3) If $\mathfrak{a}$ is an ideal of A not contained in $\mathfrak{p}$, then $\mathfrak{a}A_{\mathfrak{p}} = A_{\mathfrak{p}}$ and $(A/\mathfrak{a})_{\mathfrak{p}} = 0$ (§ 2, no. 5, *Remark*).

(4) Let A, B be two rings, $\rho: A \to B$ a homomorphism, $\mathfrak{q}$ a prime ideal of B and $\mathfrak{p}$ the prime ideal $\overset{-1}{\rho}(\mathfrak{q})$ of A. As $\rho(A - \mathfrak{p}) \subset B - \mathfrak{q}$, a canonical homomorphism $\rho_{\mathfrak{q}}: A_{\mathfrak{p}} \to B_{\mathfrak{q}}$ is derived from ρ (§ 2, no. 1, Proposition 2) and it is immediate that $\rho_{\mathfrak{q}}(\mathfrak{p}A_{\mathfrak{p}}) \subset \mathfrak{q}B_{\mathfrak{q}}$, hence $\rho_{\mathfrak{q}}$ is a *local* homomorphism.

2. MODULES OVER A LOCAL RING

PROPOSITION 4. *Let* A *be a ring which is not necessarily commutative,* $\mathfrak{m}$ *a right ideal of* A *contained in the Jacobson radical of* A *and* M *a left* A*-module. Suppose that one of the following conditions holds:*

(i) M *is finitely generated;*

(ii) $\mathfrak{m}$ *is nilpotent.*

Then the relation $(A_d/\mathfrak{m}) \otimes_A M = 0$ *implies* $M = 0$.

The assertion with respect to hypothesis (i) is precisely Corollary 3 to Proposition 6 of *Algebra*, Chapter VIII, § 6, no. 3. On the other hand, the relation $(A_d/\mathfrak{m}) \otimes_A M = 0$ is equivalent to $M = \mathfrak{m}M$ and hence implies $M = \mathfrak{m}^n M$ for every integer $n > 0$; whence the assertion with respect to hypothesis (ii).

COROLLARY 1. *Let* A *be a ring which is not necessarily commutative,* $\mathfrak{m}$ *a right ideal of* A *contained in the Jacobson radical of* A, M *and* N *two left* A*-modules and* $u: M \to N$ *an* A*-linear mapping. If* N *is finitely generated or* $\mathfrak{m}$ *is nilpotent and*

$$1 \otimes u: (A_d/\mathfrak{m}) \otimes_A M \to (A_d/\mathfrak{m}) \otimes_A N$$

is surjective, then u *is surjective.*

$(A_d/\mathfrak{m}) \otimes_A (N/u(M))$ is canonically isomorphic to

$$((A_d/\mathfrak{m}) \otimes_A N)/\mathrm{Im}(1 \otimes u)$$

(*Algebra*, Chapter II, § 3, no. 6, Corollary 1 to Proposition 6); then the hypothesis implies $(A_d/\mathfrak{m}) \otimes_A (N/u(M)) = 0$, hence $N/u(M) = 0$ by Proposition 4.

Corollary 2. *Let* A *be a ring which is not necessarily commutative,* $\mathfrak{m}$ *a two-sided ideal of* A *contained in the Jacobson radical of* A, M *a left* A*-module and* $(x_\iota)_{\iota \in I}$ *a family of elements of* M. *If* M *is finitely generated or* $\mathfrak{m}$ *is nilpotent and the elements* $1 \otimes x_\iota$ $(\iota \in I)$ *generate the left* $(A/\mathfrak{m})$*-module* $(A/\mathfrak{m}) \otimes_A M$, *the* x_ι *generate* M.

Let $(e_\iota)_{\iota \in I}$ be the canonical basis of the left A-module $A_s^{(I)}$: it is sufficient to apply Corollary 1 to the A-linear mapping $u: A_s^{(I)} \to M$ such that $u(e_\iota) = x_\iota$ for all $\iota \in I$.

Proposition 5. *Let* A *be a ring which is not necessarily commutative,* $\mathfrak{m}$ *a two-sided ideal of* A *contained in the Jacobson radical of* A *and* M *a left* A*-module. Suppose that one of the following conditions holds:*

(i) M *is finitely presented;*

(ii) $\mathfrak{m}$ *is nilpotent.*

Then, if $(A/\mathfrak{m}) \otimes_A M = M/\mathfrak{m}M$ *is a free left* $(A/\mathfrak{m})$*-module and the canonical homomorphism from* $\mathfrak{m} \otimes_A M$ *to* M *is injective,* M *is a free* A*-module. To be precise, if* $(x_\iota)_{\iota \in I}$ *is a family of elements of* M *such that* $(1 \otimes x_\iota)$ *is a basis of the* $(A/\mathfrak{m})$*-module* $M/\mathfrak{m}M$, (x_ι) *is a basis of* M.

If $a \in A$, $x \in M$ and $\bar{a}$ is the class of a in $A/\mathfrak{m}$, then $\bar{a} \otimes x = 1 \otimes (ax)$ and hence the hypothesis implies that there exists a family $(x_\iota)_{\iota \in I}$ of elements of M such that $(1 \otimes x_\iota)$ is a basis of the $(A/\mathfrak{m})$-module $(A/\mathfrak{m}) \otimes_A M$. We already know that the x_ι generate M (Corollary 2 to Proposition 4); we shall see that they are linearly independent over A. To this end, let us consider the free A-module $L = A_s^{(I)}$; let (e_ι) be its canonical basis and $u: A_s^{(I)} \to M$ the A-linear mapping such that $u(e_\iota) = x_\iota$ for all $\iota \in I$; if R is the kernel of u, we shall prove that $R = 0$. Under hypothesis (i), $(A/\mathfrak{m}) \otimes_A M$ is a finitely generated $(A/\mathfrak{m})$-module, hence I is necessarily finite and R is a finitely generated A-module by Chapter I, § 2, no. 8, Lemma 9. Then by Proposition 4 it will be sufficient to prove (under either hypothesis) that $R = \mathfrak{m}R$.

Let j be the canonical injection $R \to L$; then there is a commutative diagram

$$\begin{array}{ccccc} \mathfrak{m} \otimes R & \xrightarrow{1 \otimes j} & \mathfrak{m} \otimes L & \xrightarrow{1 \otimes u} & \mathfrak{m} \otimes M \\ {\scriptstyle a}\downarrow & & {\scriptstyle b}\downarrow & & {\scriptstyle c}\downarrow \\ R & \xrightarrow[j]{} & L & \xrightarrow[u]{} & M \end{array}$$

in which the two rows are exact, j is injective and $1 \otimes u$ is surjective (Chapter

I, § 2, no. 1, Lemma 1); as, by hypothesis, $\mathrm{Ker}(c) = 0$, there is an exact sequence

$$0 \xrightarrow{d} \mathrm{Coker}(a) \longrightarrow \mathrm{Coker}(b) \xrightarrow{v} \mathrm{Coker}(c)$$

(Chapter I, § 1, no. 4, Proposition 2); it is sufficient to verify that v is bijective, for then we deduce that $\mathrm{Coker}(a) = 0$, in other words that a is surjective and consequently $\mathrm{R} = \mathfrak{m}\mathrm{R}$. Now, $\mathrm{Coker}(b) = (\mathrm{A}/\mathfrak{m}) \otimes_{\mathrm{A}} \mathrm{L}$ and

$$\mathrm{Coker}(c) = (\mathrm{A}/\mathfrak{m}) \otimes_{\mathrm{A}} \mathrm{M}$$

and by definition $v(1 \otimes e_\iota) = 1 \otimes x_\iota$; as $(1 \otimes e_\iota)$ is a basis of $(\mathrm{A}/\mathfrak{m}) \otimes_{\mathrm{A}} \mathrm{L}$, the definition of the x_ι shows that v is bijective.

COROLLARY 1. *Let* A *be a not necessarily commutative ring,* $\mathfrak{m}$ *the Jacobson radical of* A *and* M *a left* A*-module. Suppose that* $\mathrm{A}/\mathfrak{m}$ *is a field, that the canonical homomorphism from* $\mathfrak{m} \otimes_{\mathrm{A}} \mathrm{M}$ *to* M *is injective and that one of conditions* (i), (ii) *of Proposition* 5 *is satisfied. For a family* (y_λ) *of elements of* M *to be a basis of a direct factor of* M, *it is necessary and sufficient that the family* $(1 \otimes y_\lambda)$ *be free in* $\mathrm{M}/\mathfrak{m}\mathrm{M}$.

If this condition holds, it can be assumed that (y_λ) is a subfamily of a family (x_ι) of elements of M such that $(1 \otimes x_\iota)$ is a basis of $\mathrm{M}/\mathfrak{m}\mathrm{M}$ (*Algebra*, Chapter II, § 7, no. 1, Theorem 2) and Proposition 5 then proves that (x_ι) is a basis of M.

COROLLARY 2. *Let* A *be a not necessarily commutative ring,* $\mathfrak{m}$ *the Jacobson radical of* A *and* M *a left* A*-module. Suppose that* $\mathrm{A}/\mathfrak{m}$ *is a field and that one of the following conditions holds:*

(i) M *is finitely presented;*
(ii) $\mathfrak{m}$ *is nilpotent.*

Then the following properties are equivalent:

(a) M *is free;*
(b) M *is projective;*
(c) M *is flat;*
(d) *the canonical homomorphism* $\mathfrak{m} \otimes_{\mathrm{A}} \mathrm{M} \to \mathrm{M}$ *is injective;*
* (e) $\mathrm{Tor}_1^{\mathrm{A}}(\mathrm{A}/\mathfrak{m}, \mathrm{M}) = 0$. *

The implications (a) ⇒ (b) ⇒ (c) ⇒ (d) are immediate. As $\mathrm{A}/\mathfrak{m}$ is a field, $(\mathrm{A}/\mathfrak{m}) \otimes_{\mathrm{A}} \mathrm{M}$ is a free $(\mathrm{A}/\mathfrak{m})$-module and Proposition 5 shows that (d) implies (a).

* Finally, we know that $\mathrm{Tor}_1^{\mathrm{A}}(\mathrm{A}, \mathrm{M}) = 0$ and from the exact sequence $0 \to \mathfrak{m} \to \mathrm{A} \to \mathrm{A}/\mathfrak{m} \to 0$ we therefore derive the exact sequence

$$0 \to \mathrm{Tor}_1^{\mathrm{A}}(\mathrm{A}/\mathfrak{m}, \mathrm{M}) \to \mathfrak{m} \otimes_{\mathrm{A}} \mathrm{M} \to \mathrm{M};$$

this proves that $\mathrm{Tor}_1^A(A/\mathfrak{m}, M)$ is isomorphic to the kernel of the canonical homomorphism $\mathfrak{m} \otimes_A M \to M$; whence the equivalence of (d) and (e). *

It can be shown that, for every ring A with Jacobson radical $\mathfrak{m}$ such that $A/\mathfrak{m}$ is a field, *every* projective A-module is free (Exercise 3).

Proposition 6. *Let A be a ring which is not necessarily commutative and $\mathfrak{m}$ its Jacobson radical; suppose that $A/\mathfrak{m}$ is a field. Let M and N be two finitely generated free A-modules and u: $M \to N$ a homomorphism. The following properties are equivalent:*

(a) *u is an isomorphism of M onto a direct factor of N;*
(b) *$1 \otimes u$: $(A/\mathfrak{m}) \otimes_A M \to (A/\mathfrak{m}) \otimes_A N$ is injective;*
(c) *u is injective and Coker(u) is a free A-module;*
(d) *the transpose homomorphism ${}^t u$: $N^* \to M^*$ is surjective.*

We know (*Algebra*, Chapter II, § 1, no. 11, Proposition 21) that, if $N/u(M)$ is free, $u(M)$ is a direct factor of N, hence (c) implies (a); conversely, (a) implies that Coker(u), isomorphic to a complement of $u(M)$ in N, is a finitely generated projective A-module and *a fortiori* finitely presented (Chapter I, § 2, no. 8, Lemma 8); hence this module is free by Corollary 2 to Proposition 5 and (a) implies (c). On the other hand, (a) obviously implies (b). For simplicity we write $M' = (A/\mathfrak{m}) \otimes_A M$, $N' = (A/\mathfrak{m}) \otimes_A N$; as M and N are finitely generated, the duals M'^* and N'^* of the $(A/\mathfrak{m})$-modules M' and N' are canonically identified with $M^* \otimes_A (A/\mathfrak{m})$ and $N^* \otimes_A (A/\mathfrak{m})$ and ${}^t(1 \otimes u)$ with $({}^t u) \otimes 1$ (*Algebra*, Chapter II, § 5, no. 4, Proposition 8); as M' and N' are vector spaces over the field $A/\mathfrak{m}$, the hypothesis that $1 \otimes u$ is injective implies that ${}^t(1 \otimes u)$ is surjective (*Algebra*, Chapter II, § 7, no. 5, Proposition 10); Corollary 1 to Proposition 4 then shows that ${}^t u$ is surjective and we have thus proved that (b) implies (d). Finally we show that (d) implies (a). Suppose that ${}^t u$ is surjective; as M^* is free, there exists a homomorphism f from M^* to N^* such that $1_{M^*} = {}^t u \circ f$ (*Algebra*, Chapter II, § 1, no. 11, Proposition 21); as M and N are free and finitely generated, there exists a homomorphism g from N to M such that $f = {}^t g$; hence ${}^t 1_M = 1_{M^*} = {}^t u \circ {}^t g = {}^t(g \circ u)$, whence $1_M = g \circ u$; this proves that u is an isomorphism of M onto a submodule which is a direct factor of N (*Algebra*, Chapter II, § 1, no. 9, Corollary 2 to Proposition 15).

Corollary. *Under the hypotheses of Proposition 6 the following properties are equivalent:*

(a) *u is an isomorphism of M onto N;*
(b) *M and N have the same rank (Algebra, Chapter II, § 7, no. 2) and u is surjective;*
(c) *$1 \otimes u$: $M/\mathfrak{m}M \to N/\mathfrak{m}N$ is bijective.*

Clearly (a) implies (b); (b) implies that $1 \otimes u$ is surjective; moreover the hypothesis that M and N have the same rank implies that so do the vector spaces $(\mathrm{A}/\mathfrak{m}) \otimes_{\mathrm{A}} \mathrm{M}$ and $(\mathrm{A}/\mathfrak{m}) \otimes_{\mathrm{A}} \mathrm{N}$ over $\mathrm{A}/\mathfrak{m}$, hence $1 \otimes u$ is bijective (*Algebra*, Chapter II, § 7, no. 4, Corollary to Proposition 9) and (b) implies (c). Finally, condition (c) implies, by Proposition 6, that N is the direct sum of $u(\mathrm{M})$ and a free submodule P and u is an isomorphism of M onto $u(\mathrm{M})$; if $\mathrm{P} \neq 0$, then $(\mathrm{A}/\mathfrak{m}) \otimes_{\mathrm{A}} \mathrm{P} \neq 0$ and $1 \otimes u$ would not be surjective; hence (c) implies (a).

The propositions proved above in this no. will usually be applied when A is a *local ring* and $\mathfrak{m}$ its *maximal ideal*. Corollary 2 to Proposition 5 is then completed by

PROPOSITION 7. *Let* A *be a reduced local ring,* $\mathfrak{m}$ *its maximal ideal,* $(\mathfrak{p}_\iota)_{\iota \in \mathrm{I}}$ *the family of minimal prime ideals of* A, K_ι *the field of fractions of* $\mathrm{A}/\mathfrak{p}_\iota$ *and* M *a finitely generated* A-*module. For* M *to be free it is necessary and sufficient that*

$$(1) \qquad [(\mathrm{A}/\mathfrak{m}) \otimes_{\mathrm{A}} \mathrm{M} : (\mathrm{A}/\mathfrak{m})] = [\mathrm{K}_\iota \otimes_{\mathrm{A}} \mathrm{M} : \mathrm{K}] \quad \textit{for all } \iota \in \mathrm{I}.$$

If M is free, clearly the two sides of (1) are equal to the rank of M for all $\iota \in \mathrm{I}$. Suppose now that the condition is satisfied and denote by n the common value of the two sides of (1); by Corollary 2 to Proposition 4 M has a system of n generators x_j $(1 \leqslant j \leqslant n)$. Suppose first that A is an *integral domain*, in which case $\mathfrak{p}_\iota = 0$ for all $\iota \in \mathrm{I}$. The elements $1 \otimes x_j$ $(1 \leqslant j \leqslant n)$ generate the vector space $\mathrm{K} \otimes \mathrm{M}$ over the field of fractions K of A; but as by hypothesis this space is of dimension n over K, the elements $1 \otimes x_j$ are linearly independent over K. It follows (*Algebra*, Chapter II, § 1, no. 13, *Remark* 1) that the x_j are linearly independent over A and hence form a basis of M.

Passing to the general case, there exists a surjective homomorphism v from $\mathrm{L} = \mathrm{A}^n$ onto M. Consider the commutative diagram

$$\begin{array}{ccc}
\mathrm{L} & \xrightarrow{\quad v \quad} & \mathrm{M} \\
{\scriptstyle u}\downarrow & & \downarrow{\scriptstyle u'} \\
\prod_\iota ((\mathrm{A}/\mathfrak{p}_\iota) \otimes \mathrm{L}) & \xrightarrow[v']{} & \prod_\iota ((\mathrm{A}/\mathfrak{p}_\iota) \otimes \mathrm{M})
\end{array}$$

where u (resp. u') is the mapping $x \mapsto (\phi_\iota(x))$ (resp. $y \mapsto (\psi_\iota(y))$),

$$\phi_\iota : \mathrm{L} \to (\mathrm{A}/\mathfrak{p}_\iota) \otimes \mathrm{L}$$

(resp. $\psi_\iota : \mathrm{M} \to (\mathrm{A}/\mathfrak{p}_\iota) \otimes \mathrm{M}$) being the canonical mapping, and v' is the product of the $1_{\mathrm{A}/\mathfrak{p}_\iota} \otimes v$. Then $(\mathrm{A}/\mathfrak{p})/(\mathfrak{m}/\mathfrak{p}_\iota) \otimes_{\mathrm{A}/\mathfrak{p}_\iota} ((\mathrm{A}/\mathfrak{p}_\iota) \otimes_{\mathrm{A}} \mathrm{M}) = (\mathrm{A}/\mathfrak{m}) \otimes_{\mathrm{A}} \mathrm{M}$ and, as $\mathrm{A}/\mathfrak{p}_\iota$ is a local integral domain, it follows from the first part of the argument that each of the $1_{\mathrm{A}/\mathfrak{p}_\iota} \otimes v$ is an isomorphism; then so is v'. On the other hand, as A is reduced, $\bigcap_\iota \mathfrak{p}_\iota = (0)$ (§ 2, no. 6, Proposition 13), whence $\bigcap_\iota (\mathfrak{p}_\iota L) = 0$ since L is free (*Algebra*, Chapter II, § 3, no. 7, *Remark*); as $\mathfrak{p}_\iota L$ is the kernel of ϕ_ι, this shows that u is injective. It follows that $v' \circ u = u' \circ v$ is injective, hence v is injective and, as v is surjective by definition, this shows that M is free.

3. PASSAGE FROM THE LOCAL TO THE GLOBAL

PROPOSITION 8. *Let* A *be a ring,* $\mathfrak{m}$ *a maximal ideal of* A *and* M *an* A*-module. If there exists an ideal* $\mathfrak{a}$ *of* A *such that* $\mathfrak{m}$ *is the only maximal ideal of* A *containing* $\mathfrak{a}$ *and* $\mathfrak{a}M = 0$, *then the canonical homomorphism* $M \to M_\mathfrak{m}$ *is bijective.*

$A/\mathfrak{a}$ is then a local ring with maximal ideal $\mathfrak{m}/\mathfrak{a}$; M can be considered as an $(A/\mathfrak{a})$-module; for all $s \in A - \mathfrak{m}$ the canonical image of s in $A/\mathfrak{a}$ is invertible, hence the homothety $x \mapsto sx$ of M is bijective from the definition of $M_\mathfrak{m}$ as the solution of a universal problem (§ 2, no. 2); whence the proposition.

In particular, if there exists $k \geqslant 0$ such that $\mathfrak{m}^k M = 0$, the homomorphism $M \to M_\mathfrak{m}$ is bijective (§ 1, no. 1, Corollary to Proposition 1).

PROPOSITION 9. *Let* A *be a ring,* $\mathfrak{m}$ *a maximal ideal of* A, M *an* A*-module and* k *an integer* $\geqslant 0$. *The canonical homomorphism* $M \to M_\mathfrak{m}/\mathfrak{m}^k M_\mathfrak{m}$ *is surjective, has kernel* $\mathfrak{m}^k M$ *and defines an isomorphism of* $M/\mathfrak{m}^k M$ *onto* $M_\mathfrak{m}/\mathfrak{m}^k M_\mathfrak{m}$.

Since the case $k = 0$ is trivial, suppose that $k \geqslant 1$. It follows from Proposition 8 that the canonical homomorphism $M/\mathfrak{m}^k M \to (M/\mathfrak{m}^k M)_\mathfrak{m}$ is bijective. On the other hand $(M/\mathfrak{m}^k M)_\mathfrak{m}$ is canonically identified with $M_\mathfrak{m}/(\mathfrak{m}^k M)_\mathfrak{m}$ (§ 2, no. 4, Theorem 1) and hence $(\mathfrak{m}^k M)_\mathfrak{m} = \mathfrak{m}^k M_\mathfrak{m}$ (§ 2, no. 7, Corollary to Proposition 18), whence there is an isomorphism of $M/\mathfrak{m}^k M$ onto $M_\mathfrak{m}/\mathfrak{m}^k M_\mathfrak{m}$ which maps the class of an element $x \in M$ to the class of $x/1$.

COROLLARY. *Let* A *be a ring,* $\mathfrak{m}_1, \mathfrak{m}_2, \ldots, \mathfrak{m}_n$ *distinct maximal ideals of* A, M *an* A*-module and* $k_1, k_2, \ldots, k_n$ *integers* $\geqslant 0$. *The canonical homomorphism from* M *to* $\prod_{i=1}^{n} M_\mathfrak{m}/\mathfrak{m}_i^{k_i} M_\mathfrak{m}$ *is surjective and its kernel is* $\left(\bigcap_{i=1}^{n} \mathfrak{m}_i^{k_i}\right)M$.

This follows easily from Proposition 9 and § 1, no. 2, Proposition 6, the $\mathfrak{m}_i^{k_i}$ being relatively prime in pairs (§ 1, no. 2, Proposition 3).

In the rest of this no. A *will denote a ring and* $\Omega(A)$ *(or* Ω*) the set of maximal ideals of* A.

PROPOSITION 10. *The* A-*module* $\bigoplus_{\mathfrak{m} \in \Omega} A_{\mathfrak{m}}$, *the direct sum of the* $A_{\mathfrak{m}}$, *where* $\mathfrak{m} \in \Omega$, *is faithfully flat.*

Each of the $A_{\mathfrak{m}}$ is a flat A-module (§ 2, no. 4, Theorem 1), hence $E = \bigoplus_{\mathfrak{m} \in \Omega} A_{\mathfrak{m}}$ is flat (Chapter I, § 2, no. 3, Proposition 2). Moreover, for every maximal ideal $\mathfrak{m}$ of A, $\mathfrak{m}A_{\mathfrak{m}}$ is the unique maximal ideal of $A_{\mathfrak{m}}$, hence $\mathfrak{m}A_{\mathfrak{m}} \neq A_{\mathfrak{m}}$, whence it follows that $\mathfrak{m}E \neq E$ and consequently E is faithfully flat (Chapter I, § 3, no. 1, Proposition 1 (d)).

THEOREM 1. *Let* M, N *be two* A-*modules*, $u: M \to N$ *an* A-*homomorphism and, for all* $\mathfrak{m} \in \Omega$, *let* $u_{\mathfrak{m}}: M_{\mathfrak{m}} \to N_{\mathfrak{m}}$ *be the corresponding* $A_{\mathfrak{m}}$-*homomorphism* (§ 2, no. 2, *Remark* 5). *For* u *to be injective* (resp. *surjective, bijective, zero*), *it is necessary and sufficient that, for all* $\mathfrak{m} \in \Omega$, $u_{\mathfrak{m}}$ *be injective* (resp. *surjective, bijective, zero*).

To say that, for all $\mathfrak{m} \in \Omega$, $u_{\mathfrak{m}}$ is injective (resp. surjective, bijective, zero) is equivalent to saying that the homomorphism $\bigoplus_{\mathfrak{m}} u_{\mathfrak{m}}: \bigoplus_{\mathfrak{m}} M_{\mathfrak{m}} \to \bigoplus_{\mathfrak{m}} N_{\mathfrak{m}}$ has the same property. But $\bigoplus_{\mathfrak{m}} M_{\mathfrak{m}} = M \otimes_A E$, $\bigoplus_{\mathfrak{m}} N_{\mathfrak{m}} = N \otimes_A E$ and $\bigoplus_{\mathfrak{m}} u_{\mathfrak{m}} = u \otimes 1$, where $E = \bigoplus_{\mathfrak{m}} A_{\mathfrak{m}}$; as E is faithfully flat (Proposition 10), the theorem follows from Chapter I, § 3, no. 1, Proposition 1 (c) and Proposition 2.

COROLLARY 1. *Let* M *be an* A-*module*, N *a submodule of* M *and* x *an element of* M. *For* $x \in N$, *it is necessary and sufficient that, for all* $\mathfrak{m} \in \Omega$, *the canonical image of* x *in* $M_{\mathfrak{m}}$ *belong to* $N_{\mathfrak{m}}$.

Let $\bar{x}$ be the class of x in M/N; to say that $x \in N$ means that the A-linear mapping $u: \alpha \mapsto \alpha\bar{x}$ from A to M/N is zero. Now, $(M/N)_{\mathfrak{m}}$ is identified with $M_{\mathfrak{m}}/N_{\mathfrak{m}}$ (§ 2, no. 4, Theorem 1) and $u_{\mathfrak{m}}: A_{\mathfrak{m}} \to M_{\mathfrak{m}}/N_{\mathfrak{m}}$ with the mapping $\lambda \mapsto \lambda\bar{x}_{\mathfrak{m}}$, where $\bar{x}_{\mathfrak{m}}$ is the class mod. $N_{\mathfrak{m}}$ of the canonical image of x in $M_{\mathfrak{m}}$. As the relation $u = 0$ is equivalent to $u_{\mathfrak{m}} = 0$ for all $\mathfrak{m}$ by Theorem 1, this proves the corollary.

COROLLARY 2. *Let* M *be an* A-*module and, for all* $\mathfrak{m} \in \Omega$, *let* $f_{\mathfrak{m}}$ *be the canonical mapping* $M \to M_{\mathfrak{m}}$. *The homomorphism* $x \mapsto (f_{\mathfrak{m}}(x))$ *of* M *to* $\prod_{\mathfrak{m} \in \Omega} M_{\mathfrak{m}}$ *is injective.*

Applying Corollary 1 in the case $N = 0$, it is seen that the relation $x = 0$ is equivalent to $f_{\mathfrak{m}}(x) = 0$ for all $\mathfrak{m} \in \Omega$.

COROLLARY 3. (i) *Let* $\mathfrak{b}$ *be an ideal of* A *and* a *an element of* A. *For* $a \in \mathfrak{b}$, *it is necessary and sufficient that, for all* $\mathfrak{m} \in \Omega$, *the canonical image of* a *in* $A_{\mathfrak{m}}$ *belong to* $\mathfrak{b}A_{\mathfrak{m}}$.

(ii) *In particular, let b and c be two elements of* A. *For c to be a multiple of b, it is necessary and sufficient that, for all* $\mathfrak{m} \in \Omega$, *the canonical image of c in* $A_\mathfrak{m}$ *be a multiple of that of b.*

As $\mathfrak{b}A_\mathfrak{m} = \mathfrak{b}_\mathfrak{m}$ (§ 2, no. 7, Corollary to Proposition 18), (i) is a special case of Corollary 1; (ii) follows from (i) applied to the ideal Ab.

COROLLARY 4. *Let* A *be an integral domain,* K *its field of fractions and* M *a torsion-free* A*-module such that* M *is canonically identified with a sub-*A*-module of* $K \otimes_A M$. *Then, for all* $\mathfrak{m} \in \Omega$, $M_\mathfrak{m}$ *is canonically identified with a sub-*A*-module of* $K \otimes_A M$ *and* $M = \bigcap_{\mathfrak{m} \in \Omega} M_\mathfrak{m}$.

As M is identified with a submodule of $K \otimes_A M$, $M_\mathfrak{m}$ is identified with a sub-$A_\mathfrak{m}$-module of $(K \otimes_A M)_\mathfrak{m} = K_\mathfrak{m} \otimes_A M$ (§ 2, no. 4, Theorem 1); as $K_\mathfrak{m} = K$, we see straightway that $M_\mathfrak{m}$ is torsion-free; moreover, the commutativity of the diagram

$$\begin{array}{ccc} M & \longrightarrow & K \otimes_A M \\ \downarrow & & \downarrow \\ M_\mathfrak{m} & \longrightarrow & (K \otimes_A M)_\mathfrak{m} \end{array}$$

proves that the canonical mapping $M \to M_\mathfrak{m}$ is injective. The corollary then follows from Corollary 1 applied to the A-module $K \otimes_A M$ and its submodule M.

In particular, for every *integral domain* A,

$$A = \bigcap_{\mathfrak{m} \in \Omega} A_\mathfrak{m}. \tag{2}$$

COROLLARY 5. *Let* A *be a ring. Every system of generators of the* A*-module* A^n *with n elements is a basis of* A^n.

Let $(e_i)_{1 \leqslant i \leqslant n}$ be the canonical basis of A^n, $(x_i)_{1 \leqslant i \leqslant n}$ a system of generators of A^n with n elements and $u: A^n \to A^n$ the A-linear mapping such that $u(e_i) = x_i$ for $1 \leqslant i \leqslant n$. By hypothesis u is surjective and it is necessary to show that u is injective. By virtue of Theorem 1 this can immediately be reduced to the case where A is a *local* ring; if $\mathfrak{m}$ is the maximal ideal of A, the elements $1 \otimes x_i$ $(1 \leqslant i \leqslant n)$ in $(A/\mathfrak{m})^n$ then form a system of generators of the free $(A/\mathfrak{m})$-module $(A/\mathfrak{m})^n$; as $A/\mathfrak{m}$ is a field, this system is a basis of $(A/\mathfrak{m})^n$; as A^n is a free A-module, we deduce from Proposition 5 that (x_i) is a basis of A^n.

PROPOSITION 11. *Let* M *be an* A*-module,* N *a finitely generated* A*-module and* $u: M \to N$ *a homomorphism. For u to be surjective, it is necessary and sufficient that, for* *all* $\mathfrak{m} \in \Omega$, *the homomorphism* $M/\mathfrak{m}M \to N/\mathfrak{m}N$ *derived from* u *by taking quotients be surjective.*

It follows from Theorem 1 that, for u to be surjective, it is necessary and sufficient that $u_\mathfrak{m}: M_\mathfrak{m} \to N_\mathfrak{m}$ be surjective for all $\mathfrak{m} \in \Omega$. As $A_\mathfrak{m}$ is a local ring and $N_\mathfrak{m}$ is a finitely generated $A_\mathfrak{m}$-module, this amounts to saying that the homomorphism $u'_\mathfrak{m}: M_\mathfrak{m}/\mathfrak{m}M_\mathfrak{m} \to N_\mathfrak{m}/\mathfrak{m}N_\mathfrak{m}$, obtained by taking quotients, is surjective (no. 2, Corollary 1 to Proposition 4); but $M_\mathfrak{m}/\mathfrak{m}M_\mathfrak{m}$ (resp. $N_\mathfrak{m}/\mathfrak{m}N_\mathfrak{m}$) is identified with $M/\mathfrak{m}M$ (resp. $N/\mathfrak{m}N$) (Proposition 9), whence the proposition.

PROPOSITION 12. *Let* E, F, G *be three* A-*modules and* $v: G \to F$, $u: E \to F$ *homomorphisms. Suppose that* E *is finitely presented. For there to exist a homomorphism* $w: E \to G$ *such that* u *factors into* $E \xrightarrow{w} G \xrightarrow{v} F$, *it is necessary and sufficient that, for all* $\mathfrak{m} \in \Omega$, *there exist a homomorphism* $w^\mathfrak{m}: E_\mathfrak{m} \to G_\mathfrak{m}$ *such that* $u_\mathfrak{m}: E_\mathfrak{m} \to F_\mathfrak{m}$ *factors into* $E_\mathfrak{m} \xrightarrow{w^\mathfrak{m}} G_\mathfrak{m} \xrightarrow{v_\mathfrak{m}} F_\mathfrak{m}$.

The existence of w satisfying the above statement is equivalent to the following property: u belongs to the image P of the mapping

$$r = \mathrm{Hom}(1_E, v): \mathrm{Hom}_A(E, G) \to \mathrm{Hom}_A(E, F).$$

Now, $(\mathrm{Hom}_A(E, F))_\mathfrak{m}$ (resp. $(\mathrm{Hom}_A(E, G))_\mathfrak{m}$) is canonically identified with $\mathrm{Hom}_{A_\mathfrak{m}}(E_\mathfrak{m}, F_\mathfrak{m})$ (resp. $\mathrm{Hom}_{A_\mathfrak{m}}(E_\mathfrak{m}, G_\mathfrak{m})$) (§ 2, no. 7, Proposition 19 (i)), the canonical image of u in $(\mathrm{Hom}_A(E, F))_\mathfrak{m}$ is identified with $u_\mathfrak{m}$, $r_\mathfrak{m}$ is identified with $\mathrm{Hom}_{A_\mathfrak{m}}(1_{E_\mathfrak{m}}, v_\mathfrak{m})$ and $P_\mathfrak{m}$ with the image of $r_\mathfrak{m}$. The proposition then follows from Corollary 1 to Theorem 1 applied to $\mathrm{Hom}_A(E, F)$ and its submodule P.

COROLLARY 1. *Let* M *be an* A-*module and* N *a submodule of* M *such that* M/N *is finitely presented. For* N *to be a direct factor of* M, *it is necessary and sufficient that, for all* $\mathfrak{m} \in \Omega$, $N_\mathfrak{m}$ *be a direct factor of* $M_\mathfrak{m}$.

To say that N is a direct factor of M means that the identity homomorphism on M/N factors into $M/N \xrightarrow{w} M \xrightarrow{\phi} M/N$ where ϕ is the canonical homomorphism and w a homomorphism (*Algebra*, Chapter II, § 1, no. 9, Proposition 14); as $(M/N)_\mathfrak{m} = M_\mathfrak{m}/N_\mathfrak{m}$ and $\phi_\mathfrak{m}$ is the canonical homomorphism $M_\mathfrak{m} \to M_\mathfrak{m}/N_\mathfrak{m}$, the corollary follows easily from Proposition 12.

COROLLARY 2. *Let* M *be a finitely generated free* A-*module and* N *a submodule of* M *which is a finitely generated free* A-*module. For* N *to be a direct factor of* M, *it is necessary and sufficient that, for all* $\mathfrak{m} \in \Omega$, $\mathfrak{m}N = N \cap (\mathfrak{m}M)$.

By definition M/N is finitely presented; on the other hand, $N_\mathfrak{m}$ and $M_\mathfrak{m}$

are finitely generated free A_m-modules. For N_m to be a direct factor of M_m, it is necessary and sufficient that the canonical mapping $N_m/mN_m \to M_m/mM_m$ be injective (no. 2, Proposition 6); this is the same as saying that the canonical mapping $N/mN \to M/mM$ must be injective (Proposition 9), and as its kernel is $(N \cap mM)/mN$, this proves the corollary.

Proposition 12 (resp. its Corollary 1) will be applied in particular when A is *Noetherian* and E (resp. M/N) a *finitely generated* A-module (Chapter I, § 2, no. 8, Lemma 8).

4. LOCALIZATION OF FLATNESS

PROPOSITION 13. *Let* S *be a multiplicative subset of a ring* A *and* M *an A-module. If* M *is flat* (resp. *faithfully flat*), $S^{-1}M$ *is a flat* (resp. *faithfully flat*) $S^{-1}A$*-module and a flat A-module.*

As $S^{-1}M = M \otimes_A S^{-1}A$, the first assertion follows from Chapter I, § 2, no. 7, Corollary 2 to Proposition 8 (resp. Chapter I, § 3, no. 3, Proposition 5); moreover, $S^{-1}A$ is a flat A-module (§ 2, no. 4, Theorem 1); hence if M is a flat A-module, so is $S^{-1}M$ by virtue of Chapter I, § 2, no. 7, Corollary 3 to Proposition 8.

Remark. If N is an $S^{-1}A$-module, $S^{-1}N$ is identified with N and this is consequently *equivalent* to saying that N is a flat $S^{-1}A$-module or a flat A-module.

PROPOSITION 14. *Let* A *be a ring,* B *a commutative A-algebra and* T *a multiplicative subset of* B. *If* N *is a B-module which is flat as an A-module,* $T^{-1}N$ *is a flat A-module.*

$T^{-1}N = T^{-1}B \otimes_B N$; the proposition then follows from Chapter I, § 2, no. 7, Proposition 8, applied with A replaced by B, B by A, E by $T^{-1}B$ and F by N.

PROPOSITION 15. *Let* A, B *be two rings,* $\phi: A \to B$ *a homomorphism and* N *a B-module. The following properties are equivalent:*

(a) N *is a flat A-module.*

(b) *For every maximal ideal* $\mathfrak{n}$ *of* B, $N_\mathfrak{n}$ *is a flat A-module.*

(c) *For every maximal ideal* $\mathfrak{n}$ *of* B, *if we write* $\mathfrak{m} = \overset{-1}{\phi}(\mathfrak{n})$, $N_\mathfrak{n}$ *is a flat* $A_\mathfrak{m}$*-module.*

For all $a \notin \mathfrak{m}$, the homothety of $N_\mathfrak{n}$ induced by a is bijective, hence $N_\mathfrak{n}$ is canonically identified with $(N_\mathfrak{n})_\mathfrak{m}$ and the equivalence of (b) and (c) follows from the

Remark following Proposition 13; the fact that (a) implies (b) is a special case of Proposition 14. It remains to prove that (b) implies (a), that is, that, if (b) holds, for every injective A-module homomorphism $u: M \to M'$, the homomorphism $v = 1 \otimes u: N \otimes_A M \to N \otimes_A M'$ is injective. Now, v is also a B-module homomorphism and, for it to be injective, it is necessary and sufficient that $v_n: (N \otimes_A M)_n \to (N \otimes_A M')_n$ be so for every maximal ideal $\mathfrak{n}$ of B (no. 3, Theorem 1). As

$$(N \otimes_A M)_\mathfrak{n} = B_\mathfrak{n} \otimes_B (N \otimes_A M) = N_\mathfrak{n} \otimes_A M,$$

$v_\mathfrak{n}$ is just the homomorphism $1 \otimes u: N_\mathfrak{n} \otimes_A M \to N_\mathfrak{n} \otimes_A M'$, which is injective since $N_\mathfrak{n}$ is a flat A-module by hypothesis.

COROLLARY. *For an* A-*module* M *to be flat* (resp. *faithfully flat*), *it is necessary and sufficient that, for every maximal ideal* $\mathfrak{m}$ *of* A, $M_\mathfrak{m}$ *be a flat* (resp. *faithfully flat*) $A_\mathfrak{m}$-*module.*

The necessity of the conditions follows from Proposition 13. Conversely, if $M_\mathfrak{m}$ is a flat $A_\mathfrak{m}$-module for every maximal ideal $\mathfrak{m}$ of A, M is a flat A-module by virtue of Proposition 15 applied to the case where ϕ is the identity. Finally, if $M_\mathfrak{m}$ is a faithfully flat $A_\mathfrak{m}$-module for all $\mathfrak{m}$, then $\mathfrak{m}M_\mathfrak{m} = \mathfrak{m}A_\mathfrak{m}M_\mathfrak{m} \neq M_\mathfrak{m}$, hence $\mathfrak{m}M \neq M$ for all $\mathfrak{m}$ (no. 3, Proposition 9), which proves that M is a faithfully flat A-module (Chapter I, § 3, no. 1, Proposition 1 (d)).

5. SEMI-LOCAL RINGS

PROPOSITION 16. *Let* A *be a ring. The following properties are equivalent:*

(a) *the set of maximal ideals of* A *is finite;*

(b) *the quotient of* A *by its Jacobson radical is the direct composition of a finite number of fields.*

Suppose that the quotient of A by its Jacobson radical $\mathfrak{R}$ is a direct composition of a finite number of fields. Then $A/\mathfrak{R}$ possesses only a finite number of ideals and *a fortiori* only a finite number of maximal ideals. As every maximal ideal contains $\mathfrak{R}$ (*Algebra*, Chapter VIII, § 6, no. 2, Definition 2), the maximal ideals of A are the inverse images of the maximal ideals of $A/\mathfrak{R}$ under the canonical homomorphism $A \to A/\mathfrak{R}$; hence they are finite in number.

Conversely, suppose that A has only a finite number of distinct maximal ideals $\mathfrak{m}_1, \ldots, \mathfrak{m}_n$. The $A/\mathfrak{m}_i$ are fields and it follows from § 1, no. 2, Proposition 5 that the canonical mapping $A \to \prod_{i=1}^{n} A/\mathfrak{m}_i$ is surjective; as its kernel $\bigcap_{i=1}^{n} \mathfrak{m}_i$ is the Jacobson radical $\mathfrak{R}$ (*Algebra*, Chapter VIII, § 6, no. 2, Definition 2), $A/\mathfrak{R}$ is isomorphic to $\prod_{i=1}^{n} A/\mathfrak{m}_i$.

DEFINITION 4. *A ring is called semi-local if it satisfies the equivalent conditions* (a), (b) *of Proposition* 16.

Examples. Every local ring is semi-local. Every quotient of a semi-local ring is semi-local. Every finite product of semi-local rings is semi-local.

* If A is a Noetherian semi-local ring and B is an A-algebra which is a finitely generated A-module, then B is semi-local (Chapter IV, § 2, no. 5, Corollary 3 to Proposition 9). *

Another example, generalizing the construction of the local rings $A_{\mathfrak{p}}$, is provided by the following proposition:

PROPOSITION 17. *Let* A *be a ring and* $\mathfrak{p}_1, \dots, \mathfrak{p}_n$ *prime ideals of* A. *We write* $S = \bigcap_{i=1}^{n} (A - \mathfrak{p}_i) = A - \bigcup_{i=1}^{n} \mathfrak{p}_i$.

(a) *The ring* $S^{-1}A$ *is semi-local; if* $\mathfrak{q}_1, \dots, \mathfrak{q}_r$ *are the distinct maximal elements (with respect to inclusion) of the set of* $\mathfrak{p}_i$, *the maximal ideals of* $S^{-1}A$ *are the* $S^{-1}\mathfrak{q}_j$ $(1 \leqslant j \leqslant r)$ *and these ideals are distinct.*

(b) *The ring* $A_{\mathfrak{p}_i}$ *is canonically isomorphic to* $(S^{-1}A)_{S^{-1}\mathfrak{p}_i}$ *for* $1 \leqslant i \leqslant n$.

(c) *If* A *is an integral domain, then* $S^{-1}A = \bigcap_{i=1}^{n} A_{\mathfrak{p}_i}$ *in the field of fractions of* A.

(a) The ideals of A not meeting S are the ideals contained in the union of the $\mathfrak{p}_i$ and hence in at least one of the $\mathfrak{p}_i$ (§ 1, no. 1, Proposition 2); the $\mathfrak{q}_j$ are therefore the maximal elements of the set of ideals not meeting S; consequently, the $S^{-1}\mathfrak{q}_j$ are the maximal ideals of $S^{-1}A$ by § 2, no. 5, Proposition 11 (ii).

(b) is a special case of § 2, no. 5, Proposition 11 (iii).

(c) Suppose that A is an integral domain. If $\mathfrak{p}_i \subset \mathfrak{p}_k$, then $A_{\mathfrak{p}_i} \supset A_{\mathfrak{p}_k}$; to prove (c), we may therefore suppose that no two $\mathfrak{p}_i$ are comparable. Then it follows from (a) and no. 3, Corollary 4 to Theorem 1 that $S^{-1}A = \bigcap_{i=1}^{n} (S^{-1}A)_{S^{-1}\mathfrak{p}_i}$; whence (c) by virtue of (b).

If A is an integral domain, so is $S^{-1}A$ and Proposition 17 then provides an example of a semi-local ring which is not a direct composition of local rings (cf. Chapter III, § 2, no. 13).

COROLLARY. *Let* A *be an integral domain and* $\mathfrak{p}_1, \dots, \mathfrak{p}_n$ *prime ideals of* A, *no two of which are comparable with respect to inclusion. If* $A = \bigcap_{i=1}^{n} A_{\mathfrak{p}_i}$ *in the field of fractions of* A, *the maximal ideals of* A *are* $\mathfrak{p}_1, \dots, \mathfrak{p}_n$.

Setting $S = \bigcap_{i=1}^{n} (A - \mathfrak{p}_i)$, $S^{-1}A = A$ by Proposition 17 (c); hence the elements of S are invertible in A and $S^{-1}\mathfrak{p}_i = \mathfrak{p}_i$ for all i. Our assertion then follows by virtue of Proposition 17 (a).

4. SPECTRA OF RINGS AND SUPPORTS OF MODULES

1. IRREDUCIBLE SPACES

DEFINITION 1. *A topological space X is called irreducible if every finite intersection of non-empty open sets of X is non-empty.*

By considering the empty family of open sets of X it is seen that an irreducible space is *non-empty*; for a topological space X to be irreducible, it is necessary and sufficient that it be non-empty and that the intersection of two non-empty open sets of X be always non-empty (or, what amounts to the same thing, that the union of two closed sets distinct from X be always distinct from X).

PROPOSITION 1. *Let X be a non-empty topological space. The following conditions are equivalent:*

(a) X *is irreducible;*
(b) *every non-empty open set of* X *is dense in* X*;*
(c) *every open set of* X *is connected.*

By definition, a set which is dense in X is a set which meets every non-empty open set, hence (a) and (b) are equivalent. It is immediate that (c) implies (a), for if U_1 and U_2 are disjoint non-empty open sets, $U_1 \cup U_2$ is a disconnected open set. Finally let us show that (a) implies (c): if U is a disconnected open set, it is the union of two disjoint non-empty sets U′, U″ which are open in U and hence also open in X, which implies that X is not irreducible.

A *Hausdorff* space is irreducible only if it consists of a single point.

A subset E of a topological space X is called an *irreducible set* if the subspace E of X is irreducible. For this to be so, it is necessary and sufficient that, for every pair of sets U, V which are open in X and meet E, $U \cap V$ also meet E, or (what amounts to the same thing) that, for every pair of sets F, G which are closed in X and satisfy $E \subset F \cup G$, $E \subset F$ or $E \subset G$. By induction on n, we deduce that, if $(F_i)_{1 \leqslant i \leqslant n}$ is a finite family of closed sets of X such that $E \subset \bigcup_{i=1}^{n} F_i$, there exists an index i such that $E \subset F_i$.

Proposition 2. *In a topological space* X, *for a set* E *to be irreducible, it is necessary and sufficient that its closure* $\overline{E}$ *be so.*

For an open set of X to meet E, it is necessary and sufficient that it meet $\overline{E}$; then the proposition follows from the above remarks.

Proposition 3. (i) *If* X *is an irreducible space, every non-empty open subset of* X *is irreducible.*

(ii) *Let* $(U_\alpha)_{\alpha \in A}$ *be a non-empty covering of a topological space* X *consisting of open sets such that* $U_\alpha \cap U_\beta \neq \varnothing$ *for every pair of indices* (α, β). *If the sets* U_α *are irreducible, the space* X *is irreducible.*

(i) If X is irreducible, $U \subset X$ non-empty and open in X and $V \subset U$ non-empty and open in U, V is also open in X, therefore dense in X and *a fortiori* dense in U. Then U is irreducible (Proposition 1).

(ii) Let us show that, for every non-empty open set V in X, $V \cap U_\alpha \neq \varnothing$ for all $\alpha \in A$: it follows that $V \cap U_\alpha$ is dense in U_α by hypothesis, hence that V is dense in X and this proves that X is irreducible (Proposition 1). Now there exists at least one index γ such that $V \cap U_\gamma \neq \varnothing$; as $U_\alpha \cap U_\gamma \neq \varnothing$ for all α and $V \cap U_\gamma$ is dense in U_γ, $U_\alpha \cap U_\gamma \cap V \neq \varnothing$ and *a fortiori* $U_\alpha \cap V \neq \varnothing$, which completes the proof of (ii).

Proposition 4. *Let* X *and* Y *be two topological spaces and* f *a continuous mapping from* X *to* Y. *For every irreducible subset* E *of* X, $f(E)$ *is an irreducible subset of* Y.

If U, V are two open sets of Y which meet $f(E)$, $\overset{-1}{f}(U)$ and $\overset{-1}{f}(V)$ are open sets of X which meet E. Consequently, $\overset{-1}{f}(U) \cap \overset{-1}{f}(V) = \overset{-1}{f}(U \cap V)$ meets E, which implies that $U \cap V$ meets $f(E)$ and proves the proposition.

Definition 2. *Every maximal irreducible subset of a topological space* X *is called an irreducible component of* X.

It follows from Proposition 2 that every irreducible component of X is *closed* in X.

Proposition 5. *Let* X *be a topological space. Every irreducible subset of* X *is contained in an irreducible component of* X *and* X *is the union of its irreducible components.*

To prove the first assertion, it is sufficient, by virtue of Zorn's Lemma, to prove that the set $\mathfrak{F}$ of irreducible subsets of X is *inductive*. Let $\mathfrak{G}$ be a subset of $\mathfrak{F}$ totally ordered by inclusion; we show that the union E of the sets $F \in \mathfrak{G}$ is irreducible. Let U, V be two open sets of X which meet E; as $\mathfrak{G}$ is totally ordered, there exists a set $F \in \mathfrak{G}$ meeting U and V; as F is irreducible, $U \cap V$ meets F and hence also E, which proves that E is irreducible and hence that $\mathfrak{F}$ is inductive. The second assertion follows from the first, for every subset of X consisting of a single point is irreducible.

Corollary. *Every connected component of a topological space* X *is a union of irreducible components of* X.

Every irreducible subspace of X is connected by Proposition 1 and hence is contained in a connected component of X.

Note that two distinct irreducible components of X may have points in common (Exercise 11).

Proposition 6. *Let* X *be a topological space and* $(P_i)_{1 \leqslant i \leqslant n}$ *a finite covering of* X *consisting of irreducible closed sets. Then the irreducible components of* X *are the maximal elements* (*with respect to inclusion*) *of the set of* P_i.

We may restrict ourselves to the case where no two P_i are comparable. If E is an irreducible subset of X, then $E \subset \bigcup_{i=1}^{n} P_i$ and hence E is contained in one of the closed sets P_i; this proves that the P_i are the only maximal irreducible subsets of X.

Corollary. *Let* X *be a topological space and* E *a subspace with only a finite number of distinct irreducible components* Q_i $(1 \leqslant i \leqslant n)$; *then the irreducible components of the closure* $\overline{E}$ *in* X *are the closures* $\overline{Q}_i$ *of the* Q_i $(1 \leqslant i \leqslant n)$ *and* $\overline{Q}_i \neq \overline{Q}_j$ *for* $i \neq j$.

$\overline{E}$ is the union of the $\overline{Q}_i$, which are irreducible (Proposition 2); as Q_i is closed in E, $\overline{Q}_i \cap E = Q_i$; as $Q_i \not\subset Q_j$ for $i \neq j$, $\overline{Q}_i \not\subset \overline{Q}_j$, whence the corollary by virtue of Proposition 6.

Remark. Suppose that X has only a *finite* number of distinct irreducible components X_i $(1 \leqslant i \leqslant n)$; then $U_i = \complement\left(\bigcup_{j \neq i} X_j\right)$ is *open in* X and *dense in* X_i since $X_i \not\subset \bigcup_{j \neq i} X_j$; the U_i $(1 \leqslant i \leqslant n)$ are therefore non-empty open sets of X which are irreducible (Proposition 2), pairwise disjoint and with their union dense in X.

Proposition 7. *Let* U *be an open subset of a topological space* X. *The mapping* $V \mapsto \overline{V}$ (*closure in* X) *is a bijection of the set of irreducible subsets of* U *which are closed in* U *onto the set of irreducible subsets of* X *which are closed in* X *and meet* U; *the inverse bijection is* $Z \mapsto Z \cap U$. *In particular, this bijection maps the set of irreducible components of* U *onto the set of irreducible components of* X *which meet* U.

If V is closed in U and irreducible, $\overline{V}$ is irreducible (Proposition 2) and $V = \overline{V} \cap U$. Conversely, if Z is irreducible and closed in X and meets U, $Z \cap U$ is a non-empty open subset of Z, hence irreducible (Proposition 3) and dense in Z and, as Z is closed, $Z = \overline{Z \cap U}$. This proves the proposition.

2. NOETHERIAN TOPOLOGICAL SPACES

DEFINITION 3. *A topological space X is called Noetherian if every non-empty set of closed subsets of X, ordered by inclusion, has a minimal element.*

It amounts to the same to say that every non-empty set of open subsets of X, ordered by inclusion, has a maximal element, or that every decreasing (resp. increasing) sequence of closed (resp. open) sets is stationary (*Set Theory*, Chapter III, § 6, no. 5, Proposition 6).

PROPOSITION 8. (i) *Every subspace of a Noetherian space is Noetherian.*

(ii) *Let* $(A_\iota)_{\iota \in I}$ *be a finite covering of a topological space X. If the subspaces* A_ι *of X are Noetherian, X is Noetherian.*

(i) Let X be a Noetherian space, A a subspace of X and (F_n) a decreasing sequence of subsets of A which are closed *in* A; then $F_n = \overline{F}_n \cap A$ and the closures $\overline{F}_n$ of the F_n in X form a decreasing sequence of closed subsets of X. As this sequence is stationary, so is the sequence (F_n).

(ii) Let $(G_n)_{n \geqslant 0}$ be a decreasing sequence of closed subsets of X; by hypothesis, each of the sequences $(G_n \cap A_\iota)_{n \geqslant 0}$ is stationary. As I is finite, there is an integer n_0 such that, for $n \geqslant n_0$, $G_n \cap A_\iota = G_{n_0} \cap A_\iota$ for all $\iota \in I$. But

$$G_n = \bigcup_{\iota \in I} (G_n \cap A_\iota)$$

and hence the sequence (G_n) is stationary and X is Noetherian.

PROPOSITION 9. *For a topological space X to be Noetherian, it is necessary and sufficient that every open set in X be quasi-compact.*

To show that the condition is necessary, it is sufficient, by virtue of Proposition 8, to prove that every Noetherian space X is quasi-compact. Let $(U_\iota)_{\iota \in I}$ be an open covering of X; the set of finite unions of sets U_ι is non-empty and hence admits a maximal element $V = \bigcup_{\iota \in H} U_\iota$, where H is a finite subset of I. By definition, $V \cup U_\iota = V$ for all $\iota \in I$ and hence $V = X$.

Conversely, suppose that every open set of X is quasi-compact and let (U_n) be an increasing sequence of open subsets of X. The union V of the U_n is open and hence quasi-compact; as (U_n) is an open covering of V, there is a finite subfamily of (U_n) which is a covering of V and hence $V = U_n$ for some index n, which proves that the sequence (U_n) is stationary.

LEMMA 1 ("Principle of Noetherian induction"). *Let* E *be an ordered set every non-empty subset of which admits a minimal element. Let* F *be a subset of* E *with the following property: if $a \in E$ is such that the relation $x < a$ implies $x \in F$, then $a \in F$. Then* $F = E$.

Suppose $F \neq E$; then $\complement F$ would have a minimal element b. By definition, $x \in F$ for all $x < b$, which implies that $b \in F$, which is a contradiction.

PROPOSITION 10. *If* X *is a Noetherian space, the set of irreducible components of* X *(and a fortiori the set of connected components of* X*) is finite.*

It is sufficient to prove that X is a finite union of irreducible closed subsets (no. 1, Proposition 6). Let us show that the principle of Noetherian induction can be applied taking E to be the set of closed subsets of X, ordered by inclusion, and F to be the set of finite unions of irreducible closed subsets. Let Y be a closed subset of X such that every closed subset $\neq Y$ of Y belongs to F. If Y is irreducible, then $Y \in F$ by definition; otherwise, Y is the union of two closed subsets Y_1, Y_2 which are distinct from Y. Then $Y_1 \in F$ and $Y_2 \in F$ by hypothesis, whence $Y \in F$ by definition of F.

In particular it follows that a *Hausdorff* Noetherian space is necessarily *finite.*

3. THE PRIME SPECTRUM OF A RING

Let A be a ring and X the set of prime ideals of A. For every subset M of A, we denote by V(M) the set of prime ideals of A containing M; clearly, if $\mathfrak{a}$ is the ideal of A generated by M, $V(M) = V(\mathfrak{a})$; if M consists of a single element f, we write $V(f)$ instead of $V(\{f\})$ and we have $V(f) = V(Af)$. The mapping $M \mapsto V(M)$ is *decreasing* with respect to inclusion in A and X. Moreover, the following formulae hold:

$$V(0) = X, \qquad V(1) = \varnothing; \tag{1}$$

$$V\Big(\bigcup_{\iota \in I} M_\iota\Big) = V\Big(\sum_{\iota \in I} M_\iota\Big) = \bigcap_{\iota \in I} V(M_\iota) \tag{2}$$

for every family $(M_\iota)_{\iota \in I}$ of subsets of A;

$$V(\mathfrak{a} \cap \mathfrak{a}') = V(\mathfrak{a}\mathfrak{a}') = V(\mathfrak{a}) \cup V(\mathfrak{a}') \tag{3}$$

for every pair of ideals $\mathfrak{a}$, $\mathfrak{a}'$ in A. Formulae (1) and (2) are obvious; on the other hand, formula (3) means that, for a prime ideal $\mathfrak{p}$ of A to contain one of the ideals $\mathfrak{a}$ or $\mathfrak{a}'$, it is necessary and sufficient that it contain $\mathfrak{a}\mathfrak{a}'$ or that it contain $\mathfrak{a} \cap \mathfrak{a}'$; then it is a consequence of § 1, no. 1, Proposition 1. The second formula (1) has the following converse: if $\mathfrak{a}$ is an ideal of A such that $V(\mathfrak{a}) = \varnothing$, then $\mathfrak{a} = A$, for there is no maximal ideal of A containing $\mathfrak{a}$. Finally, if $\mathfrak{a}$ is an ideal of A and $\mathfrak{r}(\mathfrak{a})$ is its *radical* (§ 2, no. 6, Definition 4), then

$$V(\mathfrak{a}) = V(\mathfrak{r}(\mathfrak{a})) \tag{4}$$

as follows from § 2, no. 6, Corollary 1 to Proposition 13.

Formulae (1) to (3) show that the subsets V(M) of X satisfy the *closed set* axioms of a topology (*General Topology*, Chapter I, § 1, no. 4).

Definition 4. *Let* A *be a ring. The set* X *of prime ideals of* A, *with the topology whose closed sets are the sets* V(M), *where* M *runs through* $\mathfrak{P}(A)$, *is called the prime spectrum of* A *and denoted by* Spec(A). *The topology thus defined is called the spectral or Zariski topology on* X.

Clearly the relation $\operatorname{Spec}(A) = \varnothing$ is equivalent to $A = \{0\}$.

Let X be the prime spectrum of a ring A; for all $f \in A$, let us denote by X_f the set of prime ideals of A *not containing* f; then $X_f = X - V(f)$ and X_f is therefore an *open* set. By (2), every closed subset of X is an intersection of closed sets of the form $V(f)$ and hence the X_f form a *base* of the spectral topology on X. Moreover, it follows immediately from the definitions that

$$X_0 = \varnothing, \qquad X_1 = X, \tag{5}$$

and more generally $X_f = X$ for every invertible element f of A;

$$X_{fg} = X_f \cap X_g \qquad \text{for all } f, g \text{ in A}. \tag{6}$$

For every subset Y of X, let us denote by $\mathfrak{J}(Y)$ the intersection of the prime ideals of A which belong to Y. Clearly $\mathfrak{J}(Y)$ is an ideal of A and the mapping $Y \mapsto \mathfrak{J}(Y)$ is *decreasing* with respect to inclusion in X and A. Clearly the relations

$$\mathfrak{J}(\varnothing) = A \tag{7}$$

$$\mathfrak{J}\Big(\bigcup_{\lambda \in L} Y_\lambda\Big) = \bigcap_{\lambda \in L} \mathfrak{J}(Y_\lambda) \tag{8}$$

hold for every family $(Y_\lambda)_{\lambda \in L}$ of subsets of X. Moreover:

Proposition 11. *Let* A *be a ring,* $\mathfrak{a}$ *an ideal of* A *and* Y *a subset of* $X = \operatorname{Spec}(A)$.

(i) $V(\mathfrak{a})$ *is closed in* X *and* $\mathfrak{J}(Y)$ *is an ideal of* A *which is equal to its radical.*

(ii) $\mathfrak{J}(V(\mathfrak{a}))$ *is the radical of* $\mathfrak{a}$ *and* $V(\mathfrak{J}(Y))$ *is the closure of* Y *in* X.

(iii) *The mappings* $\mathfrak{I}$ *and* V *define decreasing bijections, one of which is the inverse of the other, between the set of closed subsets of* X *and the set of ideals of* A *which are equal to their radicals.*

Assertion (i) and the first assertion of (ii) follow from the definitions and § 2, no. 6, Corollary 1 to Proposition 13. If a closed set $V(M)$ (for some $M \subset A$) contains Y, then $M \subset \mathfrak{p}$ for every prime ideal $\mathfrak{p} \in Y$, whence $M \subset \mathfrak{I}(Y)$ and consequently $V(M) \supset V(\mathfrak{I}(Y))$; as $Y \subset V(\mathfrak{I}(Y))$, $V(\mathfrak{I}(Y))$ is the smallest closed set of X containing Y, which completes the proof of (ii). Finally, it follows from (ii) that, if $\mathfrak{a}$ is a prime ideal equal to its radical, then $\mathfrak{I}(V(\mathfrak{a})) = \mathfrak{a}$ and that, if Y is closed in X, then $V(\mathfrak{I}(Y)) = Y$; this proves (iii).

It follows immediately from Proposition 11 that, if M is any subset of A and Y is any subset of X, then $V(M) = V(\mathfrak{I}(V(M)))$ and $\mathfrak{I}(Y) = \mathfrak{I}(V(\mathfrak{I}(Y)))$.

COROLLARY 1. *For every family* $(Y_\lambda)_{\lambda \in L}$ *of closed subsets of* X, $\mathfrak{I}\left(\bigcap_{\lambda \in L} Y_\lambda\right)$ *is the radical of the sum of the ideals* $\mathfrak{I}(Y_\lambda)$.

It follows from Proposition 11 (iii) that $\mathfrak{I}\left(\bigcap_{\lambda \in L} Y_\lambda\right)$ is the smallest ideal which is equal to its radical and contains all the $\mathfrak{I}(Y_\lambda)$; this ideal then contains $\sum_{\lambda \in L} \mathfrak{I}(Y_\lambda)$ and therefore also the radical of $\sum_{\lambda \in L} \mathfrak{I}(Y_\lambda)$ (§ 2, no. 6, Corollary 2 to Proposition 13), whence the corollary.

COROLLARY 2. *Let* $\mathfrak{r}(\mathfrak{a})$ *denote the radical of an ideal* $\mathfrak{a}$ *of* A; *if* $\mathfrak{a}$ *and* $\mathfrak{b}$ *are two ideals of* A, *the relation* $V(\mathfrak{a}) \subset V(\mathfrak{b})$ *is equivalent to* $\mathfrak{b} \subset \mathfrak{r}(\mathfrak{a})$ *and* $\mathfrak{r}(\mathfrak{b}) \subset \mathfrak{r}(\mathfrak{a})$.

It is immediate that the relations $\mathfrak{b} \subset \mathfrak{r}(\mathfrak{a})$ and $r(\mathfrak{b}) \subset \mathfrak{r}(\mathfrak{a})$ are equivalent and, as $V(\mathfrak{a}) = V(\mathfrak{r}(\mathfrak{a}))$, the corollary follows immediately from Proposition 11, (iii).

COROLLARY 3. *Let* $(f_\lambda)_{\lambda \in L}$ *be a family of elements of* A. *For an element* $g \in A$ *to satisfy* $X_g \subset \bigcup_{\lambda \in L} X_{f_\lambda}$, *it is necessary and sufficient that there exist an integer* $n > 0$ *such that* g^n *belongs to the ideal generated by the* f_λ.

The relation $X_g \subset \bigcup_{\lambda \in L} X_{f_\lambda}$ is equivalent to $V(g) \supset \bigcap_{\lambda \in L} V(f_\lambda)$ and it is sufficient to apply Corollary 2.

COROLLARY 4. *For two elements* f, g *of* A *to satisfy* $X_f = X_g$, *it is necessary and sufficient that there exist two integers* $m > 0$, $n > 0$ *such that* $f^m \in Ag$ *and* $g^n \in Af$.

COROLLARY 5. *For $f \in A$ to satisfy $X_f = \varnothing$, it is necessary and sufficient that f be nilpotent.*

This follows immediately from Corollary 4.

COROLLARY 6. *The closure of a set consisting of a point $\mathfrak{p} \in X = \mathrm{Spec}(A)$ is the set $V(\mathfrak{p})$ of prime ideals containing $\mathfrak{p}$. For the set $\{\mathfrak{p}\}$ to be closed in X* (or, as we shall also say by an abuse of language, for $\mathfrak{p}$ to be a *closed point* of X), *it is necessary and sufficient that $\mathfrak{p}$ be maximal.*

COROLLARY 7. *If A is a Noetherian ring, $X = \mathrm{Spec}(A)$ is a Noetherian space.*

PROPOSITION 12. *For all $f \in A$, the open set X_f in $X = \mathrm{Spec}(A)$ is quasi-compact; in particular, the space X is quasi-compact.*

As the X_g form a base of the topology, it is sufficient to prove that, if $(g_\lambda)_{\lambda \in L}$ is a family of elements of A such that $X_f \subset \bigcup_{\lambda \in L} X_{g_\lambda}$, then there exists a finite subfamily $(g_\lambda)_{\lambda \in H}$ such that $X_f \subset \bigcup_{\lambda \in H} X_{g_\lambda}$. But the relation $X_f \subset \bigcup_{\lambda \in L} X_{g_\lambda}$ means that there exists an integer $n > 0$ and a finite subfamily $(g_\lambda)_{\lambda \in L}$ such that f^n belongs to the ideal generated by that subfamily (Corollary 3 to Proposition 11); whence the proposition.

PROPOSITION 13. *Let A, A' be two rings, $X = \mathrm{Spec}(A)$, $X' = \mathrm{Spec}(A')$ and h a homomorphism from A to A'; the mapping ${}^a h\colon \mathfrak{p}' \mapsto \overset{-1}{h}(\mathfrak{p}')$ from X' to X is continuous.*

For $M \subset A$, the set $({}^a h)^{-1}(V(M))$ is the set of prime ideals $\mathfrak{p}'$ of A' such that $M \subset \overset{-1}{h}(\mathfrak{p}')$, which is equivalent to $h(M) \subset \mathfrak{p}'$; this set is then equal to $V(h(M))$ and is therefore closed.

We call ${}^a h$ the mapping *associated* with the homomorphism h.

Remark. If h is surjective and $\mathfrak{a}$ is its kernel, it follows from the definition of the spectral topology that ${}^a h$: is a homeomorphism of X' onto the closed subspace $V(\mathfrak{a})$ of X; for a prime ideal $\mathfrak{p}'$ of A' to contain an ideal $\mathfrak{b}'$ of A', it is necessary and sufficient that $\overset{-1}{h}(\mathfrak{p}')$ contain $\overset{-1}{h}(\mathfrak{b}')$; we see first that ${}^a h$ is injective by taking $\mathfrak{b}'$ to be prime; moreover, for every ideal $\mathfrak{b}'$ of A', the image under ${}^a h$ of $V(\mathfrak{b}')$ is $V(\overset{-1}{h}(\mathfrak{b}'))$, whence our assertion, the ideals of the form $\overset{-1}{h}(\mathfrak{b}')$ all being ideals of A which contain $\mathfrak{a}$.

COROLLARY. *Let S be a multiplicative subset of A, $A' = S^{-1}A$ and h the canonical homomorphism i_A^S; then ${}^a h$ is a homeomorphism of $X' = \mathrm{Spec}(A')$ onto the subspace of $X = \mathrm{Spec}(A)$ consisting of the prime ideals of A which do not meet S.*

Let $f' = f/s$, where $f \in A$, $s \in S$; then $X'_{f'} = X'_{f/1}$ since $s/1$ is invertible in A'. We know already that ${}^a h$ is injective and that, for all $\mathfrak{p}' \in X'$, the relations $f/1 \in \mathfrak{p}'$ and $f \in \overset{-1}{h}(\mathfrak{p}') = {}^a h(\mathfrak{p}')$ are equivalent and hence the conditions $\mathfrak{p}' \in X'_{f/1}$ and ${}^a h(\mathfrak{p}') \in X_f$ are equivalent; this shows that ${}^a h(X'_{f'})$ is equal to $X_f \cap {}^a h(X')$, whence the first assertion, since the X_f (resp. $X'_{f'}$) form a base of the topology of X (resp. X'). The second assertion follows from § 2, no. 5, Proposition 11 (ii).

PROPOSITION 14. *Let* A *be a ring. For a subset* Y *of* X = Spec(A) *to be irreducible, it is necessary and sufficient that the ideal* $\mathfrak{J}(Y)$ *be prime.*

Writing $\mathfrak{p} = \mathfrak{J}(Y)$, we note that, for an element $f \in A$, the relation $f \in \mathfrak{p}$ is equivalent to $Y \subset V(f)$. Suppose that Y is irreducible and let f, g be elements of A such that $fg \in \mathfrak{p}$. Then

$$Y \subset V(fg) = V(f) \cup V(g);$$

as Y is irreducible and $V(f)$ and $V(g)$ are closed, $Y \subset V(f)$ or $Y \subset V(g)$, hence $f \in \mathfrak{p}$ or $g \in \mathfrak{p}$, which proves that $\mathfrak{p}$ is prime.

Suppose now that $\mathfrak{p}$ is prime; then $\overline{Y} = V(\mathfrak{p})$ (Proposition 11 (ii)) and, as $\mathfrak{p}$ is prime, $\mathfrak{p} = \mathfrak{J}(\{\mathfrak{p}\})$, whence $\overline{Y} = V(\mathfrak{J}(\{\mathfrak{p}\})) = \overline{\{\mathfrak{p}\}}$ (Proposition 11 (ii)). As a set consisting of a single point is irreducible, Y is irreducible (no. 1, Proposition 2).

COROLLARY 1. *For a ring* A *to be such that* X = Spec(A) *is irreducible, it is necessary and sufficient that the quotient of* A *by its nilradical* $\mathfrak{N}$ *be an integral domain.*

By Proposition 11 (i), $\mathfrak{J}(X)$ is the radical of the ideal (0), that is $\mathfrak{N}$.

COROLLARY 2. *The mapping* $\mathfrak{p} \mapsto V(\mathfrak{p})$ *is a bijection of* X = Spec(A) *onto the set of irreducible closed subsets of* X*; in particular the irreducible components of a closed subset* Y *of* X *are the sets* $V(\mathfrak{p})$, *where* $\mathfrak{p}$ *runs through the set of minimal elements of the set of prime ideals of* A *which contain* $\mathfrak{J}(Y)$.

As $\mathfrak{J}(V(\mathfrak{p})) = \mathfrak{p}$ for every prime ideal $\mathfrak{p}$ of A and $Y = V(\mathfrak{J}(Y))$ for every closed subset Y of X, the first assertion follows from Proposition 14; on the other hand, for $Y \supset V(\mathfrak{p})$, it is necessary and sufficient that

$$\mathfrak{p} = \mathfrak{J}(V(\mathfrak{p})) \supset \mathfrak{J}(Y)$$

(Proposition 11), whence the second assertion.

COROLLARY 3. *The set of minimal prime ideals of a Noetherian ring* A *is finite.*

X = Spec(A) has only a finite number of irreducible components (Corollary 7 to Proposition 11 and no. 2, Proposition 10) and the corollary follows from Corollary 2 above.

Proposition 15. *Let A be a ring, I a finite set and E the set of orthogonal families $(e_\iota)_{\iota\in I}$ of idempotents $e_\iota \neq 0$ of A such that $\sum_{\iota\in I} e_\iota = 1$. For all $(e_\iota)_{\iota\in I} \in E$, we set $\varpi((e_\iota)_{\iota\in I}) = (V(A(1-e_\iota)))_{\iota\in I}$, $\sigma((e_\iota)_{\iota\in I}) = (Ae_\iota)_{\iota\in I}$. Then ϖ is a bijection of E onto the set P of partitions $(U_\iota)_{\iota\in I}$ of $X = \operatorname{Spec}(A)$ into open sets and σ is a bijection of E onto the set S of families $(\mathfrak{a}_\iota)_{\iota\in I}$ of ideals $\neq 0$ of A such that A is the direct sum of the $\mathfrak{a}_\iota$.*

Let $(e_\iota)_{\iota\in I}$ be an element of E and set $Y_\iota = V(A(1-e_\iota))$; if $\iota \neq \jmath$, then $1 = 1 - e_\iota + e_\iota(1-e_\jmath) \in A(1-e_\iota) + A(1-e_\jmath)$, whence $Y_\iota \cap Y_\jmath = \varnothing$ (formulae (1) and (2)). On the other hand,

$$\bigcup_{\iota\in I} Y_\iota = V\Big(\prod_{\iota\in I} A(1-e_\iota)\Big) \qquad \text{(formula (3))};$$

by hypothesis $\prod_{\iota\in I}(1-e_\iota) = 1 - \sum_{\iota\in I} e_\iota = 0$, whence $\bigcup_{\iota\in I} Y_\iota = X$ (formula (1)). As the Y_ι are closed, they are also open, whence $\varpi(E) \subset P$. Also, obviously $A = \sum_{\iota\in I} Ae_\iota$; if $0 = \sum_{\iota\in I} a_\iota e_\iota$ where $a_\iota \in A$, we obtain, by multiplying by e_ι, $0 = a_\iota e_\iota^2 = a_\iota e_\iota$ for all ι; this proves that $\sigma(E) \subset S$.

Lemma 2. *If e, f are two idempotents of A such that Ae and Af have the same radical then $e = f$.*

There exists by hypothesis integers $m \geqslant 0$, $n \geqslant 0$ such that $e = e^m \in Af$ and $f = f^n \in Ae$; let x, y be elements of A such that $e = xf$, $f = ye$; then $ef = xf^2 = xf = e$ and similarly $ef = ye^2 = ye = f$, whence $e = f$.

Lemma 2 and Corollary 2 to Proposition 11 show that the mappings ϖ and σ are *injective*.

Let us show that σ is surjective. If $(\mathfrak{a}_\iota)_{\iota\in I}$ is an element of S, there are elements $e_\iota \in \mathfrak{a}_\iota$ such that $1 = \sum_{\iota\in I} e_\iota$; if $\iota \neq \jmath$, then $e_\iota e_\jmath \in \mathfrak{a}_\iota \cap \mathfrak{a}_\jmath = \{0\}$, whence

$$e_\iota = \sum_{\jmath\in I} e_\iota e_\jmath = e_\iota^2;$$

finally, $Ae_\iota \subset \mathfrak{a}_\iota$ for all $\iota \in I$ and $\sum_{\iota\in I} Ae_\iota = A$, whence $Ae_\iota = \mathfrak{a}_\iota$.

It remains to prove that ϖ is surjective. Let $(U_\iota)_{\iota\in I}$ be an element of P and set $Z_\iota = \complement U_\iota = \bigcup_{\jmath\neq\iota} U_\jmath$; as U_ι and Z_ι are closed, there exist ideals $\mathfrak{a}_\iota$, $\mathfrak{b}_\iota$ of A such that $U_\iota = V(\mathfrak{a}_\iota)$, $Z_\iota = V(\mathfrak{b}_\iota)$. We now show that it is possible to suppose further that $\mathfrak{a}_\iota \cap \mathfrak{b}_\iota = 0$. Now $U_\iota \cap Z_\iota = \varnothing$, whence $\mathfrak{a}_\iota + \mathfrak{b}_\iota = A$; let $a_\iota \in \mathfrak{a}_\iota$, $b_\iota \in \mathfrak{b}_\iota$ be such that $a_\iota + b_\iota = 1$. Then $X = U_\iota \cup Z_\iota = V(\mathfrak{a}_\iota\mathfrak{b}_\iota)$ (formula (3)); every element of $\mathfrak{a}_\iota\mathfrak{b}_\iota$ is therefore nilpotent (Corollary 2 to Proposition 11); let p be an integer such that $a_\iota^p b_\iota^p = 0$. Then $U_\iota \subset V(Aa_\iota) = V(Aa_\iota^p)$,

$$Z_\iota \subset V(Ab_\iota) = V(Ab_\iota^p)$$

and $V(A\mathfrak{a}_i) \cap V(A\mathfrak{b}_i) = V(A\mathfrak{a}_i + A\mathfrak{b}_i) = \varnothing$, hence $U_i = V(A\mathfrak{a}_i^p)$ and $Z_i = V(A\mathfrak{b}_i^p)$, which establishes our assertion by replacing $\mathfrak{a}_i$ by $A\mathfrak{a}_i^p$ and $\mathfrak{b}_i$ by $A\mathfrak{b}_i^p$. The ideals $\mathfrak{a}_i$ and $\mathfrak{b}_i$ thus chosen, it follows from the fact that σ is bijective that there exist two idempotents $f_i \in \mathfrak{a}_i$, $e_i \in \mathfrak{b}_i$ such that $1 = e_i + f_i$, $e_i f_i = 0$, $\mathfrak{a}_i = Af_i$, $\mathfrak{b}_i = Ae_i$. If $i \neq j$, then $X = Z_i \cup Z_j = V(Ae_ie_j)$, and as e_ie_j is idempotent, Lemma 2 shows that $e_ie_j = 0$. Finally $e = \sum_{i\in I} e_i$ is idempotent and $e_i \in Ae$ for all $i \in I$, whence $V(Ae) \subset Z_i$ for all i; it follows that

$$V(Ae) = \varnothing = V(A.1)$$

and Lemma 2 shows also that $e = 1$.

COROLLARY 1. *Let A be a ring, $\mathfrak{r}$ a nil ideal of A and $h: A \to A/\mathfrak{r}$ the canonical homomorphism. For every finite orthogonal family $(e'_i)_{i\in I}$ of idempotents of $A/\mathfrak{r}$ such that $\sum_{i\in I} e'_i = 1$, there exists a finite orthogonal family $(e_i)_{i\in I}$ of idempotents of A such that $\sum_{i\in I} e_i = 1$ and $h(e_i) = e'_i$ for all $i \in I$.*

We write $A' = A/\mathfrak{r}$. We know (*Remark* following Proposition 13) that

$${}^a h: \mathrm{Spec}(A') \to \mathrm{Spec}(A)$$

is a homeomorphism, every prime ideal of A containing $\mathfrak{r}$ by hypothesis. Proposition 15 shows that there exists in A a finite orthogonal family $(e_i)_{i\in I}$ of idempotents such that $\sum_{i\in I} e_i = 1$ and that the image under ${}^a h$ of $V(A'(1 - e'_i))$ is $V(A(1 - e_i))$. But clearly $V(A(1 - e_i))$ is also the image under ${}^a h$ of

$$V(A'(1 - h(e^i)));$$

as $1 - e'_i$ and $1 - h(e_i)$ are idempotent, Lemma 2 shows that $e'_i = h(e_i)$, whence the corollary.

COROLLARY 2. *For the prime spectrum $X = \mathrm{Spec}(A)$ of a ring A to be connected, it is necessary and sufficient that A contain no idempotents other than 0 and 1.*

To say that X is not connected means that there exists in X a set which is open and closed and distinct from $\varnothing$ and X.

4. THE SUPPORT OF A MODULE

DEFINITION 5. *Let A be a ring and M an A-module. The set of prime ideals $\mathfrak{p}$ of A such that $M_{\mathfrak{p}} \neq 0$ is called the support of M and is denoted by* Supp(M).

As every maximal ideal of A is prime, it follows immediately from § 3, no. 3, Corollary 2 to Theorem 1, that for A-module M to be equal to 0, *it is necessary and sufficient that* $\mathrm{Supp}(M) = \varnothing$.

Example. Let $\mathfrak{a}$ be an ideal of A; in the notation of no. 3, we have

$$V(\mathfrak{a}) = \mathrm{Supp}(A/\mathfrak{a}). \tag{9}$$

If $\mathfrak{p}$ is a prime of A such that $\mathfrak{a} \not\subset \mathfrak{p}$, then $(A/\mathfrak{a})_\mathfrak{p} = 0$ (§ 3, no. 1, *Remark* 3); if on the other hand $\mathfrak{a} \subset \mathfrak{p}$, $\mathfrak{a}A_\mathfrak{p}$ is contained in the maximal ideal $\mathfrak{p}A_\mathfrak{p}$ of $A_\mathfrak{p}$ and $(A/\mathfrak{a})_\mathfrak{p}$ is isomorphic to $A_\mathfrak{p}/\mathfrak{a}A_\mathfrak{p}$ and hence is non-zero (§ 3, no. 1, Proposition 3); whence our assertion.

In particular, Supp(A) = Spec(A).

PROPOSITION 16. *Let* A *be a ring and* M *an* A*-module.*

(i) *If* N *is a submodule of* M, *then*

$$\mathrm{Supp}(M) = \mathrm{Supp}(N) \cup \mathrm{Supp}(M/N).$$

(ii) *If* M *is the sum of a family* $(N_\iota)_{\iota \in I}$ *of submodules, then*

$$\mathrm{Supp}(M) = \bigcup_{\iota \in I} \mathrm{Supp}(N_\iota).$$

(i) From the exact sequence $0 \to N \to M \to M/N \to 0$, we derive, for every prime ideal $\mathfrak{p}$ of A, the exact sequence

$$0 \to N_\mathfrak{p} \to M_\mathfrak{p} \to (M/N)_\mathfrak{p} \to 0$$

(§ 2, no. 4, Theorem 1). For $M_\mathfrak{p}$ to be reduced to 0, it is necessary and sufficient that $N_\mathfrak{p}$ and $(M/N)_\mathfrak{p}$ be so. In other words, the relation $\mathfrak{p} \notin \mathrm{Supp}(M)$ is equivalent to "$\mathfrak{p} \notin \mathrm{Supp}(N)$ and $\mathfrak{p} \notin \mathrm{Supp}(M/N)$", which proves (i).

(ii) For every prime ideal $\mathfrak{p}$ of A, $M_\mathfrak{p}$ is the sum of the family of submodules $(N_\iota)_\mathfrak{p}$ (§ 2, no. 4). To say that $M_\mathfrak{p} \neq 0$ means that there exists $\iota \in I$ such that $(N_\iota)_\mathfrak{p} \neq 0$, whence (ii).

COROLLARY. *Let* A *be a ring,* M *an* A*-module,* $(m_\iota)_{\iota \in I}$ *a system of generators of* M *and* $\mathfrak{a}_\iota$ *the annihilator of* m_ι. *Then* $\mathrm{Supp}(M) = \bigcup_{\iota \in I} V(\mathfrak{a}_\iota)$.

$\mathrm{Supp}(M) = \bigcup_{\iota \in I} \mathrm{Supp}(Am_\iota)$ by Proposition 16 (ii). On the other hand, Am_ι is isomorphic to the A-module $A/\mathfrak{a}_\iota$ and we have seen that

$$\mathrm{Supp}(A/\mathfrak{a}_\iota) = V(\mathfrak{a}_\iota)$$

(*Example* above).

PROPOSITION 17. *Let* A *be a ring,* M *an* A*-module and* $\mathfrak{a}$ *its annihilator; if* M *is finitely generated, then* $\mathrm{Supp}(M) = V(\mathfrak{a})$ *and* Supp(M) *is therefore closed in* Spec(A).

Let $(m_i)_{1 \leqslant i \leqslant n}$ be a system of generators of M and let $\mathfrak{a}_i$ be the annihilator of m_i; then $\mathfrak{a} = \bigcap_{i=1}^{n} \mathfrak{a}_i$, hence $V(\mathfrak{a}) = \bigcup_{i=1}^{n} V(\mathfrak{a}_i)$ (no. 3, equation (3)) and the proposition follows from the Corollary to Proposition 16.

COROLLARY 1. *Let* A *be a ring,* M *a finitely generated* A*-module and* a *an element of* A. *For* a *to belong to every prime ideal of the support of* M, *it is necessary and sufficient that the homothety of* M *with ratio* a *be nilpotent.*

It follows from Proposition 17 that the intersection of the prime ideals belonging to Supp(M) is the radical of the annihilator $\mathfrak{a}$ of M (no. 3, Proposition 11 (ii)). To say that a belongs to this radical is equivalent to saying that there exist a power $a^k \in \mathfrak{a}$ and hence that $a^k M = 0$.

COROLLARY 2. *Let* A *be a Noetherian ring,* M *a finitely generated* A*-module and* $\mathfrak{a}$ *an ideal of* A. *For* $\operatorname{Supp}(M) \subset V(\mathfrak{a})$, *it is necessary and sufficient that there exist an integer* k *such that* $\mathfrak{a}^k M = 0$.

If $\mathfrak{b}$ is the annihilator of M, the relation $\operatorname{Supp}(M) \subset V(\mathfrak{a})$ is equivalent to $V(\mathfrak{b}) \subset V(\mathfrak{a})$ by Proposition 17 and hence to $\mathfrak{a} \subset \mathfrak{r}(\mathfrak{b})$, where $\mathfrak{r}(\mathfrak{b})$ is the radical of $\mathfrak{b}$ (no. 3, Corollary 2 to Proposition 11). Since A is Noetherian, this condition is also equivalent to the existence of an integer $k > 0$ such that $\mathfrak{a}^k \subset \mathfrak{b}$ (§ 2, no. 6, Proposition 15).

PROPOSITION 18. *Let* M, M′ *be two finitely generated modules over a ring* A; *then*

$$\operatorname{Supp}(M \otimes_A M') = \operatorname{Supp}(M) \cap \operatorname{Supp}(M'). \tag{10}$$

We need to prove that, if $\mathfrak{p}$ is a prime ideal of A, the relations $(M \otimes_A M')_{\mathfrak{p}} \neq 0$ and "$M_{\mathfrak{p}} \neq 0$ and $M'_{\mathfrak{p}} \neq 0$" are equivalent. As the $A_{\mathfrak{p}}$-modules $M_{\mathfrak{p}} \otimes_{A_{\mathfrak{p}}} M'_{\mathfrak{p}}$ and $(M \otimes_A M')_{\mathfrak{p}}$ are isomorphic (§ 2, no. 7, Proposition 18), our assertion follows from the following lemma:

LEMMA 3. *Let* B *be a local ring and* E *and* E′ *two finitely generated* B*-modules. If* $E \neq 0$ *and* $E' \neq 0$, *then* $E \otimes_B E' \neq 0$.

Let k be the residue field of B. By virtue of § 3, no. 2, Proposition 4, $k \otimes_B E \neq 0$ and $k \otimes_B E' \neq 0$; then we deduce that

$$(k \otimes_B E) \otimes_k (k \otimes_B E') \neq 0$$

(*Algebra*, Chapter II, § 3, no. 7). But, since the tensor product is associative (*loc. cit.*, § 3, no. 8), this tensor product is isomorphic to

$$E \otimes_B ((k \otimes_k k) \otimes_B E') = E \otimes_B (k \otimes_B E')$$

and therefore to $k \otimes_B (E \otimes_B E')$, whence the lemma.

COROLLARY. *Let* M *be a finitely generated* A*-module and* $\mathfrak{n}$ *its annihilator. For every ideal* $\mathfrak{a}$ *of* A, $\operatorname{Supp}(M/\mathfrak{a}M) = V(\mathfrak{a}) \cap V(\mathfrak{n}) = V(\mathfrak{a} + \mathfrak{n})$.

$M/\mathfrak{a}M = M \otimes_A (A/\mathfrak{a})$ and $A/\mathfrak{a}$ is finitely generated.

Proposition 19. *Let* A, B *be two rings*, $\phi: A \to B$ *a homomorphism and*

$$^a\phi: \mathrm{Spec}(B) \to \mathrm{Spec}(A)$$

the continuous mapping associated with ϕ (Proposition 13). *For every* A-*module* M, $\mathrm{Supp}(M_{(B)}) \subset {}^a\overset{-1}{\phi}(\mathrm{Supp}(M))$; *if also* M *is finitely generated, then*

$$\mathrm{Supp}(M_{(B)}) = {}^a\overset{-1}{\phi}(\mathrm{Supp}(M)).$$

Let $\mathfrak{q}$ be a prime ideal of B and $\mathfrak{p} = \overset{-1}{\phi}(\mathfrak{q})$. Suppose that $\mathfrak{q}$ belongs to $\mathrm{Supp}(M_{(B)})$; then $M_{(B)} \otimes_B B_\mathfrak{q} = (M \otimes_A B) \otimes_B B_\mathfrak{q} = M \otimes_A B_\mathfrak{q} = (M \otimes_A A_\mathfrak{p}) \otimes_A B_\mathfrak{q}$, since the homomorphism $A \to B \to B_\mathfrak{q}$ factors into $A \to A_\mathfrak{p} \to B_\mathfrak{q}$ (§ 2, no. 1, Proposition 2); the hypothesis $M_{(B)} \otimes_B B_\mathfrak{q} \neq 0$ implies therefore $M \otimes_A A_\mathfrak{p} \neq 0$, whence the first assertion. As the homomorphism $\phi_\mathfrak{q}: A_\mathfrak{p} \to B_\mathfrak{q}$ is local, the second assertion follows from the following lemma:

Lemma 4. *Let* A, B *be two local rings*, $\rho: A \to B$ *a local homomorphism and* E *a finitely generated* A-*module. If* $E \neq 0$, *then* $E_{(B)} \neq 0$.

Let $\mathfrak{m}$ be the maximal ideal of A and $k = A/\mathfrak{m}$ the residue field; the hypothesis implies that $B \otimes_A k = B/\mathfrak{m}B \neq 0$; since the tensor product is associative, $(E \otimes_A B) \otimes_A k$ is isomorphic to $E \otimes_A (B \otimes_A k)$, hence also to $E \otimes_A (k \otimes_k (B \otimes_A k))$ and finally to $(E \otimes_A k) \otimes_k (B \otimes_A k)$; by § 3, no. 2, Proposition 4, $E \otimes_A k \neq 0$, hence $(E \otimes_A B) \otimes_A k \neq 0$ (*Algebra*, Chapter II, § 3, no. 7) and *a fortiori* $E \otimes_A B \neq 0$.

Proposition 20. *Let* A *be a ring and* M *a finitely generated* A-*module. For every prime ideal* $\mathfrak{p} \in \mathrm{Supp}(M)$, *there exists a non-zero* A-*homomorphism* $w: M \to A/\mathfrak{p}$.

Let $\mathfrak{p} \in \mathrm{Supp}(M)$. As M is finitely generated and $M_\mathfrak{p} \neq 0$,

$$M_\mathfrak{p}/\mathfrak{p}M_\mathfrak{p} = M_\mathfrak{p} \otimes_A (A_\mathfrak{p}/\mathfrak{p}A_\mathfrak{p}) \neq 0$$

(§ 3, no. 2, Proposition 4). Let $K = A_\mathfrak{p}/\mathfrak{p}A_\mathfrak{p}$ be the field of fractions of the integral domain $A/\mathfrak{p}$; since $M_\mathfrak{p}/\mathfrak{p}M_\mathfrak{p}$ is a vector space over K, which is not reduced to 0, there exists a non-zero linear form $u: M_\mathfrak{p}/\mathfrak{p}M_\mathfrak{p} \to K$. If $(x_i)_{1 \leqslant i \leqslant n}$ is a system of generators of M, $\bar{x}_i$ the image of x_i in the $(A/\mathfrak{p})$-module $M_\mathfrak{p}/\mathfrak{p}M_\mathfrak{p}$, there exists an element $\alpha \neq 0$ of $A/\mathfrak{p}$ such that the $\alpha u(\bar{x}_i)$ belong to $A/\mathfrak{p}$ for $1 \leqslant i \leqslant n$; hence $v = \alpha u$ is a non-zero $(A/\mathfrak{p})$-linear mapping from $M_\mathfrak{p}/\mathfrak{p}M_\mathfrak{p}$ to $A/\mathfrak{p}$. The composition

$$w: M \longrightarrow M_\mathfrak{p} \longrightarrow M_\mathfrak{p}/\mathfrak{p}M_\mathfrak{p} \xrightarrow{v} A/\mathfrak{p}$$

is therefore the required homomorphism.

5. FINITELY GENERATED PROJECTIVE MODULES. INVERTIBLE FRACTIONAL IDEALS

1. LOCALIZATION WITH RESPECT TO AN ELEMENT

Let A be a ring and M an A-module. For every element $f \in A$, we shall write $A_f = A[f^{-1}]$, $M_f = M[f^{-1}] = M \otimes_A A[f^{-1}]$ (§ 2, nos. 1 and 2); if S_f is the set of f^n for $n \geqslant 0$, then $A_f = S_f^{-1}A$, $M_f = S_f^{-1}M$. If f is invertible in A, A_f (resp. M_f) is canonically identified with A (resp. M); if f is nilpotent, then $A_f = 0$ and $M_f = 0$. For every A-module homomorphism $u: M \to N$, we write $u_f = u \otimes 1: M_f \to N_f$.

Let g be another element of A; A_{fg} (resp. M_{fg}) is canonically identified with $(A_f)_{g/1}$ (resp. $(M_f)_{g/1}$), where $g/1$ is the image of g in A_f, and u_{fg} with $(u_f)_{g/1}$ (§ 2, no. 3, Proposition 7).

PROPOSITION 1. *Let f be an element of a ring A and $\phi: A \to A_f$ the canonical mapping. The mapping ${}^a\phi: \mathrm{Spec}(A_f) \to \mathrm{Spec}(A)$ is a homeomorphism of $\mathrm{Spec}(A_f)$ onto the open subspace X_f of $X = \mathrm{Spec}(A)$ (§ 4, no. 3).*

This a particular case of § 4, no. 3, Corollary to Proposition 13.

PROPOSITION 2. *Let A be a ring, $u: M \to N$ an A-module homomorphism and $\mathfrak{p}$ a prime ideal of A.*

(i) *Suppose that $u_\mathfrak{p}: M_\mathfrak{p} \to N_\mathfrak{p}$ is surjective and that N is finitely generated. Then there exists $f \in A - \mathfrak{p}$ such that $u_f: M_f \to N_f$ is surjective.*

(ii) *Suppose that $u_\mathfrak{p}$ is bijective, that M is finitely generated and that N is finitely presented. Then there exists $f \in A - \mathfrak{p}$ such that u_f is bijective.*

Let R and Q be the kernel and cokernel of u; if $g \in A$, the kernel and cokernel of u_g (resp. $u_\mathfrak{p}$) are R_g and Q_g (resp. $R_\mathfrak{p}$ and $Q_\mathfrak{p}$) (§ 2, no. 4, Theorem 1). Then $Q_\mathfrak{p} = 0$; as N is finitely generated, so is Q and there exists $g' \in A - \mathfrak{p}$ such that $g'Q = 0$ (§ 2, no. 2, Corollary 2 to Proposition 4), whence $Q_{g'} = 0$. Under the hypotheses of (ii), the sequence $0 \to R_{g'} \to M_{g'} \to N_{g'} \to 0$ is exact, hence $R_{g'}$ is finitely generated (Chapter I, § 2, no. 8, Lemma 9). Now,

$$(R_{g'})_{\mathfrak{p}R_{g'}} = R_\mathfrak{p} = 0;$$

hence there exists $g_1 \in A_{g'} - \mathfrak{p}A_{g'}$ such that $g_1 R_{g'} = 0$ (§ 2, no. 2, Corollary 2 to Proposition 4). Then $g_1 = g''/g'^h$, where $g'' \in A - \mathfrak{p}$; as $g'/1$ is invertible in $R_{g'}$, $(g''/1)R_{g'} = 0$, whence $R_{g'g''} = (R_{g'})_{g''/1} = 0$. If $f = g'g''$, $f \in A - \mathfrak{p}$, $Q_f = 0$ and $R_f = 0$, so that u_f is bijective.

COROLLARY. *If N is finitely presented and $N_\mathfrak{p}$ is a free $A_\mathfrak{p}$-module of rank p, there exists $f \in A - \mathfrak{p}$ such that N_f is a free A_f-module of rank p.*

There exist by hypothesis p elements $x_i \in N$ $(1 \leqslant i \leqslant p)$ such that the $x_i/1$ form a basis of the free $A_\mathfrak{p}$-module $N_\mathfrak{p}$. Consider the homomorphism $u: A^p \rightarrow N$ such that $u(e_i) = x_i$ for $1 \leqslant i \leqslant p$, $(e_i)_{1 \leqslant i \leqslant p}$ being the canonical basis of A^p. As $u_\mathfrak{p}$ is bijective by hypothesis, there exists $f \in A - \mathfrak{p}$ such that u_f is bijective, by virtue of Proposition 2.

PROPOSITION 3. *Let $(f_i)_{i \in I}$ be a finite family of elements of a ring* A, *generating the ideal* A *of* A. *The ring* $B = \prod_{i \in I} A_{f_i}$ *is then a faithfully flat* A*-module.*

By § 2, no. 4, Theorem 1, each of the A_{f_i} is a flat A-module, hence so is B (Chapter I, § 2, no. 3, Proposition 2). On the other hand, if $\mathfrak{p}$ is a prime ideal of A, there exists an index i such that $f_i \notin \mathfrak{p}$ and $\mathfrak{p}_f = \mathfrak{p}A_f$ is therefore a prime ideal of A_{f_i}. Then $\mathfrak{p}B \subset \mathfrak{p}A_{f_i} \times \prod_{j \neq i} A_{f_j} \neq B$ since $\mathfrak{p}A_{f_i} \neq A_{f_i}$; this suffices to imply that B is a faithfully flat A-module (Chapter I, § 3, no. 1, Proposition 1).

COROLLARY. *Under the hypotheses of Proposition* 3, *for an* A*-module* M *to be finitely generated* (resp. *finitely presented*), *it is necessary and sufficient that, for every index i, the* A_{f_i}*-module* M_{f_i} *be finitely generated* (resp. *finitely presented*).

The condition is obviously necessary (§ 2, no. 4). Conversely, if all the M_{f_i} are finitely generated (resp. finitely presented), $M' = \prod_{i \in I} M_{f_i}$ is a finitely generated (resp. finitely presented) B-module, for we can obviously suppose that for each i there is an exact sequence $A_f^m \rightarrow A_{f_i}^n \rightarrow M_{f_i} \rightarrow 0$, where m and n are *independent of i*). Now, $M' = M \otimes_A B$. The corollary then follows from Proposition 3 and Chapter I, § 3, no. 6, Proposition 11.

Note that the condition on the f_i means that the open sets X_{f_i} form a *covering* of Spec(A) (§ 4, no. 3, Corollary 3 to Proposition 11).

2. LOCAL CHARACTERIZATION OF FINITELY GENERATED PROJECTIVE MODULES

THEOREM 1. *Let* A *be a ring and* P *an* A*-module. The following properties are equivalent:*

(a) P *is a finitely generated projective module.*

(b) P *is a finitely presented module and, for every maximal ideal* $\mathfrak{m}$ *of* A, $P_\mathfrak{m}$ *is a free* $A_\mathfrak{m}$*-module.*

(c) P *is a finitely generated module, for all* $\mathfrak{p} \in \mathrm{Spec}(A)$, *the* $A_\mathfrak{p}$*-module* $P_\mathfrak{p}$ *is free and, if we denote its rank by* $r_\mathfrak{p}$, *the function* $\mathfrak{p} \mapsto r_\mathfrak{p}$ *is locally constant in the topological space* Spec(A) (that is, every point of Spec(A) admits a neighbourhood in which this function is constant).

(d) *There exists a finite family* $(f_i)_{i \in I}$ *of elements of* A, *generating the ideal* A, *such that, for all* $i \in I$, *the* A_{f_i}*-module* P_{f_i} *is free with finite rank.*

(e) *For every maximal ideal* $\mathfrak{m}$ *of* A, *there exists* $f \in A - \mathfrak{m}$ *such that* P_f *is a free* A_f*-module of finite rank.*

We show the theorem by proving the following scheme of implications

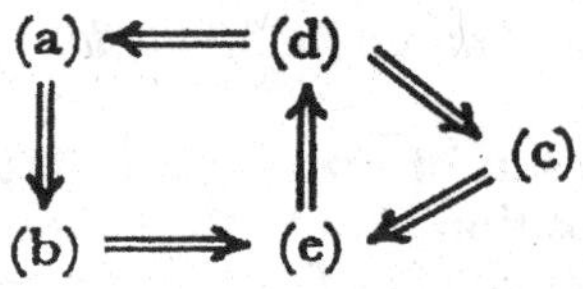

(a) ⇒ (b): We know that a finitely generated projective module is finitely presented (Chapter I, § 2, no. 8, Lemma 8 (iii)); if P is a projective A-module, $P_{\mathfrak{m}} = P \otimes_A A_{\mathfrak{m}}$ is a projective $A_{\mathfrak{m}}$-module (*Algebra*, Chapter II, § 5, no. 1, Corollary to Proposition 4); finally, as $A_{\mathfrak{m}}$ is a local ring, every finitely presented projective $A_{\mathfrak{m}}$-module is free (§ 3, no. 2, Corollary to Proposition 5).

(b) ⇒ (e): This follows from the Corollary to Proposition 2 of no. 1.

(c) ⇒ (e): Let $\mathfrak{m}$ be a maximal ideal of A; write $r_{\mathfrak{m}} = n$ and let $(x_i)_{1 \leqslant i \leqslant n}$ be a basis of $P_{\mathfrak{m}}$. We can assume that the x_i are canonical images of elements $p_i \in P$ $(1 \leqslant i \leqslant n)$ to within multiplication by an invertible element of $A_{\mathfrak{m}}$. Let $(e_i)_{1 \leqslant i \leqslant n}$ be the canonical basis of A^n and $u: A^n \to P$ the homomorphism such that $u(e_i) = p_i$ for $1 \leqslant i \leqslant n$. As P is finitely generated, it follows from Proposition 2 of no. 1 that there exists $f \in A - \mathfrak{m}$ such that u_f is surjective. We conclude that u_{fg} is also surjective for all $g \in A - \mathfrak{m}$ and by hypothesis there exists $g \in A - \mathfrak{m}$ such that $r_{\mathfrak{p}} = n$ for $\mathfrak{p} \in X_g$. Then, replacing f by fg, we may assume that $r_{\mathfrak{p}} = n$ for all $\mathfrak{p} \in X_f$. Then $u_{\mathfrak{p}}: A_{\mathfrak{p}}^n \to P_{\mathfrak{p}}$ is a surjective homomorphism and $P_{\mathfrak{p}}$ and $A_{\mathfrak{p}}$ are both free $A_{\mathfrak{p}}$-modules of the same rank; hence (§ 3, no. 2, Corollary to Proposition 6) $u_{\mathfrak{p}}$ is bijective for all $\mathfrak{p} \in X_f$. Let $\mathfrak{p}'$ be a prime ideal of A_f and let $\mathfrak{p}$ be its inverse image in A under the canonical mapping; if $(A_f^n)_{\mathfrak{p}'}$ and $(P_f)_{\mathfrak{p}'}$ are identified with $A_{\mathfrak{p}}^n$ and $P_{\mathfrak{p}}$ under the canonical isomorphisms, $(u_f)_{\mathfrak{p}'}$ is identified with $u_{\mathfrak{p}}$ and is consequently bijective. We conclude that u_f is bijective (§ 3, no. 3, Theorem 1), which establishes (e).

(e) ⇒ (d): Let E be the set of $f \in A$ such that P_f is a finitely generated free A_f-module. The hypothesis implies that E is contained in no maximal ideal of A, hence E generates the ideal A and there therefore exist a finite family $(f_i)_{1 \leqslant i \leqslant n}$ of elements of E and $a_i \in A$ $(1 \leqslant i \leqslant n)$ such that $1 = \sum_{i=1}^{n} a_i f_i$; whence (d).

(d) ⇒ (c): It follows from no. 1, Corollary to Proposition 3 that P is finitely generated. On the other hand, for every prime ideal $\mathfrak{p}$ of A, there exists an index i such that $\mathfrak{p} \in X_{f_i}$; if $\mathfrak{p}' = \mathfrak{p}_{f_i}$, then $P_{\mathfrak{p}} = (P_{f_i})_{\mathfrak{p}'}$ (§ 2, no. 5, Proposition

10) and hence by hypothesis $P_\mathfrak{p}$ is free and of the same rank as P_{f_i}, which proves (c).

(d) ⇒ (a): Consider the ring $B = \prod_{i \in I} A_{f_i}$ and the B-module

$$M = \prod_{i \in I} P_{f_i} = P \otimes_A B.$$

For every index i, there exists a free A_{f_i}-module L_i such that P_{f_i} is a direct factor of L_i and it may be assumed that all the L_i have the same rank; then $L = \prod_{i \in I} L_i$ is a free B-module of which M is a direct factor, in other words M is a finitely generated projective B-module. As B is a faithfully flat A-module (no. 1, Proposition 3), we conclude that P is a finitely generated projective A-module (Chapter I, § 3, no. 6, Proposition 12).

COROLLARY 1. *Suppose that the equivalent properties of the statement of Theorem 1 hold. Let m be an integer > 0 such that, for every family $(x_i)_{1 \leqslant i \leqslant m}$ of elements of* P, *there exists a family $(a_i)_{1 \leqslant i \leqslant m}$ of elements of* A, *which are not all divisors of zero and for which $\sum_{i=1}^{m} a_i x_i = 0$. Then, for all $\mathfrak{p} \in \operatorname{Spec}(A)$, $r_\mathfrak{p} \leqslant m$.*

Let $\mathfrak{p}$ be a prime ideal of A; set $r = r_\mathfrak{p}$ and let $(y_j)_{1 \leqslant j \leqslant r}$ be a basis of the free $A_\mathfrak{p}$-module $P_\mathfrak{p}$. There exist elements x_j $(1 \leqslant j \leqslant r)$ of P and $s \in A - \mathfrak{p}$ such that $y_j = x_j/s$ for all j. Then for every family $(a_j)_{1 \leqslant j \leqslant r}$ of elements of A such that $\sum_{j=1}^{r} a_j x_j = 0$, $\sum_{j=1}^{r} (a_j/1) y_j = 0$ in $P_\mathfrak{p}$, whence $a_j/1 = 0$ for $1 \leqslant j \leqslant r$. As $A - \mathfrak{p}$ does not contain 0, this shows that the a_j are all divisors of zero in A (§ 2, no. 1, *Remark* 3), hence of necessity $r \leqslant m$.

COROLLARY 2. *Every finitely presented flat module is projective.*

If P is a finitely presented flat A-module and $\mathfrak{m}$ a maximal ideal of A, the $A_\mathfrak{m}$-module $P_\mathfrak{m}$ is flat (§ 3, no. 4, Proposition 13) and finitely presented (§ 2, no. 4) and hence free (§3, no. 2, Corollary 2 to Proposition 5), Condition (b) of Theorem 1 therefore holds.

Remarks

(1) There exist finitely generated flat modules which are not projective (Exercise 7).

(2) Corollary 2 to Theorem 1 extends to modules over non-commutative rings (Chapter I, § 2, Exercise 15).

3. RANKS OF PROJECTIVE MODULES

DEFINITION 1. *Let* P *be a finitely generated projective* A*-module. For every prime ideal $\mathfrak{p}$ of* A, *the rank of the free $A_\mathfrak{p}$-module $P_\mathfrak{p}$ is called the rank of* P *at $\mathfrak{p}$ and is denoted by* $\operatorname{rg}_\mathfrak{p}(P)$.

By Theorem 1 the integer-valued function $\mathfrak{p} \mapsto \mathrm{rg}_{\mathfrak{p}}(\mathrm{P})$ is *locally constant* on $\mathrm{X} = \mathrm{Spec}(\mathrm{A})$; it is therefore constant if X is *connected* and in particular if the ring A is an *integral domain* (§ 4, no. 3, Corollary 2 to Proposition 15).

Definition 2. *Let n be an integer* $\geqslant 0$. *A projective* A-*module* P *is said to be of rank n if it is finitely generated and* $\mathrm{rg}_{\mathfrak{p}}(\mathrm{P}) = n$ *for every prime ideal* $\mathfrak{p}$ *of* A.

Clearly every finitely generated *free* A-module L is of rank n in the sense of Definition 2, n being equal to the *dimension* (or *rank*) of L defined in *Algebra*, Chapter II, § 7, no. 2.

A projective module of rank 0 is zero (§ 3, no. 3, Corollary 2 to Theorem 1). If A is not reduced to 0 and a projective A-module P is of rank n, the integer n is determined uniquely; it is then denoted by $\mathrm{rg}(\mathrm{P})$.

Theorem 2. *Let* P *be an* A-*module and n an integer* $\geqslant 0$. *The following properties are equivalent:*

(a) P *is projective of rank n.*

(b) P *is finitely generated and, for every maximal ideal* $\mathfrak{m}$ *of* A, *the* $\mathrm{A}_{\mathfrak{m}}$-*module* $\mathrm{P}_{\mathfrak{m}}$ *is free of rank n.*

(c) P *is finitely generated and, for every prime ideal* $\mathfrak{p}$ *of* A, *the* $\mathrm{A}_{\mathfrak{p}}$-*module* $\mathrm{P}_{\mathfrak{p}}$ *is free of rank n.*

(d) *For every maximal ideal* $\mathfrak{m}$ *of* A, *there exists* $f \in \mathrm{A} - \mathfrak{m}$ *such that the* A_f-*module* P_f *is free of rank n.*

By Definition 2 and Theorem 1, (a) and (c) are equivalent; (b) implies (c), as, for every prime ideal $\mathfrak{p}$ of A, there exists a maximal ideal $\mathfrak{m}$ containing $\mathfrak{p}$ and, writing $\mathfrak{p}' = \mathfrak{p}_{\mathfrak{m}}$, $\mathrm{P}_{\mathfrak{p}}$ is isomorphic to $(\mathrm{P}_{\mathfrak{m}})_{\mathfrak{p}'}$ (§ 2, no. 5, Proposition 11); if $\mathrm{P}_{\mathfrak{m}}$ is free of rank n, so then is $\mathrm{P}_{\mathfrak{p}}$. Property (c) implies (d) by virtue of Theorem 1 and the fact that, if $f \in \mathrm{A} - \mathfrak{m}$ and $\mathfrak{m}' = \mathfrak{m}_f$, $\mathrm{P}_{\mathfrak{m}}$ is isomorphic to $(\mathrm{P}_f)_{\mathfrak{m}'}$ and therefore the ranks of P_f and $\mathrm{P}_{\mathfrak{m}}$ are equal. Finally, this last argument and Theorem 1 show that (d) implies (b).

Remark. If A is an *integral domain*, a projective A-module admits a well-defined rank (in the sense of Definition 2), as has been observed above; moreover, this rank coincides with the rank defined in *Algebra*, Chapter II, § 7, no. 2; it is sufficient to apply Theorem 2 (c) with $\mathfrak{p} = (0)$.

Let E and F be two finitely generated projective A-modules. We know (*Algebra*, Chapter II, §§ 2 and 3) that $\mathrm{E} \times \mathrm{F}$, $\mathrm{E} \otimes_{\mathrm{A}} \mathrm{F}$, $\mathrm{Hom}_{\mathrm{A}}(\mathrm{E}, \mathrm{F})$ and the dual E^* of E are projective and finitely generated; so is the exterior power $\overset{k}{\wedge} \mathrm{E}$ for every integer $k > 0$ (*Algebra*, Chapter III). Also, it follows immediately from Definition 1 and § 2, no. 7, Propositions 18 and 19 and no. 8, that, for every prime ideal $\mathfrak{p}$ of A:

(1) $$\mathrm{rg}_{\mathfrak{p}}(\mathrm{E} \times \mathrm{F}) = \mathrm{rg}_{\mathfrak{p}}(\mathrm{E}) + \mathrm{rg}_{\mathfrak{p}}(\mathrm{F})$$

(2) $$\mathrm{rg}_{\mathfrak{p}}(\mathrm{E} \otimes_{\mathrm{A}} \mathrm{F}) = \mathrm{rg}_{\mathfrak{p}}(\mathrm{E}).\mathrm{rg}_{\mathfrak{p}}(\mathrm{F})$$

(3) $$\mathrm{rg}_{\mathfrak{p}}(\mathrm{Hom}_{\mathrm{A}}(\mathrm{E}, \mathrm{F})) = \mathrm{rg}_{\mathfrak{p}}(\mathrm{E}).\mathrm{rg}_{\mathfrak{p}}(\mathrm{F})$$

(4) $$\mathrm{rg}_{\mathfrak{p}}(\mathrm{E}^*) = \mathrm{rg}_{\mathfrak{p}}(\mathrm{E})$$

(5) $$\mathrm{rg}_{\mathfrak{p}}(\bigwedge^{k} \mathrm{E}) = \binom{\mathrm{rg}_{\mathfrak{p}}(\mathrm{E})}{k}.$$

If the ranks of E and F are defined, so are those of $\mathrm{E} \times \mathrm{F}$, $\mathrm{E} \otimes_{\mathrm{A}} \mathrm{F}$, $\mathrm{Hom}_{\mathrm{A}}(\mathrm{E}, \mathrm{F})$, E^* and $\bigwedge^{k} \mathrm{E}$ and the above equations also hold with the index $\mathfrak{p}$ omitted. Moreover:

Corollary. *For a finitely generated projective A-module P to be of rank n, it is necessary and sufficient that* $\bigwedge^{n} \mathrm{P}$ *be of rank* 1.

Proposition 4. *Let B be a commutative A-algebra and P a projective A-module of rank n. The B-module* $\mathrm{P}_{(\mathrm{B})} = \mathrm{B} \otimes_{\mathrm{A}} \mathrm{P}$ *is then projective of rank n.*

We know that $\mathrm{P}_{(\mathrm{B})}$ is projective and finitely generated (*Algebra*, Chapter II, § 5, no. 1, Corollary to Proposition 4). If $\mathfrak{q}$ is a prime ideal of B and $\mathfrak{p}$ its inverse image in A, then

$$(\mathrm{P}_{(\mathrm{B})})_{\mathfrak{q}} = (\mathrm{P} \otimes_{\mathrm{A}} \mathrm{B}) \otimes_{\mathrm{B}} \mathrm{B}_{\mathfrak{q}} = \mathrm{P} \otimes_{\mathrm{A}} \mathrm{B}_{\mathfrak{q}} = (\mathrm{P} \otimes_{\mathrm{A}} \mathrm{A}_{\mathfrak{p}}) \otimes_{\mathrm{A}} \mathrm{B}_{\mathfrak{q}}$$

and, as, by hypothesis, $\mathrm{P} \otimes_{\mathrm{A}} \mathrm{A}_{\mathfrak{p}}$ is a free $\mathrm{A}_{\mathfrak{p}}$-module of rank n, $(\mathrm{P}_{(\mathrm{B})})_{\mathfrak{q}}$ is a free $\mathrm{B}_{\mathfrak{q}}$-module of rank n.

Proposition 5. *Let A be a semi-local ring and P a finitely generated projective A-module. If the rank of P is defined, P is a free A-module.*

Suppose first that A is isomorphic to a product of fields K_i ($1 \leqslant i \leqslant n$). The K_i are then identified with the minimal ideals (*Algebra*, Chapter VIII, § 3, no. 1) of A and, for all i, the sum $\mathfrak{p}_i$ of the K_j of index $j \neq i$ is a maximal ideal of A, the $\mathfrak{p}_i$ ($1 \leqslant i \leqslant n$) being the only prime ideals of A. Every finitely generated A-module P is therefore the direct sum of its isotypical components P_i ($1 \leqslant i \leqslant n$), P_i being isomorphic to a direct sum of a finite number r_i of A-modules isomorphic to K_i (*Algebra*, Chapter VIII, § 5, no. 1, Proposition 1 and no. 3, Proposition 11); the ring $\mathrm{A}_{\mathfrak{p}_i}$ is identified with K_i and annihilates the P_j of index $j \neq i$, hence $r_i = \mathrm{rg}_{\mathfrak{p}_i}(\mathrm{P})$; if all the r_i are equal to the same number r, P is isomorphic to A^r, whence the proposition in this case. In the general case, let $\mathfrak{R}$ be the Jacobson radical of A and $\mathrm{B} = \mathrm{A}/\mathfrak{R}$; as B is a product of fields, the projective B-module $\mathrm{P}_{(\mathrm{B})}$ is free by the remarks preceding Proposition 4. Also P is a flat A-module and the proposition then follows from § 3, no. 2, Proposition 5.

4. PROJECTIVE MODULES OF RANK 1

THEOREM 3. *Let* A *be a ring and* M *a finitely generated* A*-module.*

(i) *If there exists an* A*-module* N *such that* $M \otimes_A N$ *is isomorphic to* A, *the module* M *is projective of rank* 1.

(ii) *Conversely, if* M *is projective of rank* 1 *and* M^* *is the dual of* M, *the canonical homomorphism* $u: M \otimes_A M^* \to A$ *corresponding to the canonical bilinear form* $(x, x^*) \to \langle x, x^* \rangle$ *on* $M \times M^*$ (*Algebra*, Chapter II, § 2, no. 3) *is bijective.*

(i) It is required to prove that, for every maximal ideal $\mathfrak{m}$ of A, the $A_\mathfrak{m}$-module $M_\mathfrak{m}$ is free of rank 1 (Theorem 2 (b)); we are free to replace A by $A_\mathfrak{m}$ and hence may assume that A is a *local* ring (§ 2, no. 7, Proposition 18). Let $k = A/\mathfrak{m}$. The isomorphism $v: M \otimes_A N \to A$ defines an isomorphism

$$v \otimes 1_k: (M/\mathfrak{m}M) \otimes_k (N/\mathfrak{m}N) \to k$$

as the rank over k of $(M/\mathfrak{m}M) \otimes_k (N/\mathfrak{m}N)$ is the product of the ranks of $M/\mathfrak{m}M$ and $N/\mathfrak{m}N$, these latter are necessarily equal to 1, in other words $M/\mathfrak{m}M$ is monogenous. It follows that M is monogenous (§ 3, no. 2, Corollary 2 to Proposition 4); on the other hand, the annihilator of M also annihilates $M \otimes_A N$ and hence is zero, which proves that M is isomorphic to A.

(ii) It is sufficient to prove that, for every maximal ideal $\mathfrak{m}$ of A, $u_\mathfrak{m}$ is an isomorphism (§ 3, no. 3, Theorem 1). As M is finitely presented (Chapter I, § 2, no. 8, Lemma 8), $(M^*)_\mathfrak{m}$ is canonically identified with the dual $(M_\mathfrak{m})^*$ (§ 2, no. 7, Proposition 19) and, as $M_\mathfrak{m}$ is free of rank 1 like its dual $(M_\mathfrak{m})^*$, clearly the canonical homomorphism $u_\mathfrak{m}: (M_\mathfrak{m}) \otimes_{A_\mathfrak{m}} (M_\mathfrak{m})^* \to A_\mathfrak{m}$ is bijective, which completes the proof.

Remark (1). If *M* is projective of rank 1 and N is such that $M \otimes_A N$ is isomorphic to A, then N is isomorphic to M^*: there are isomorphisms

$$N \to N \otimes A \to N \otimes M \otimes M^* \to A \otimes M^* \to M^*.$$

PROPOSITION 6. *Let* M *and* N *be projective* A*-modules of rank* 1. *Then* $M \otimes_A N$, $\mathrm{Hom}_A(M, N)$ *and the dual* M^* *of* M *are projective of rank* 1.

This follows immediately from formulae (2), (3) and (4).

Let us now note that every finitely generated A-module is isomorphic to a quotient module of $L = A^{(\mathbf{N})}$; we may therefore speak of the *set* $\boldsymbol{F}(A)$ *of classes of finitely generated* A*-modules* with respect to isomorphism (*Set Theory*, Chapter I, § 6, no. 9); we denote by $\boldsymbol{P}(A)$ the subset of $\boldsymbol{F}(A)$ consisting of the classes of projective A-modules of rank 1 and by cl(M) the image in $\boldsymbol{P}(A)$ of a projective A-module M of rank 1. It is immediate that, for two projective A-modules M,

N of rank 1, $\mathrm{cl}(M \otimes_A N)$ depends only on cl(M) and cl(N); as definition we set

$$\mathrm{cl}(M) + \mathrm{cl}(N) = \mathrm{cl}(M \otimes_A N) \tag{6}$$

and an internal law of composition is thus defined on $\boldsymbol{P}(A)$.

PROPOSITION 7. *The set* $\boldsymbol{P}(A)$ *of classes of projective* A*-modules of rank* 1, *with the law of composition* (6), *is a commutative group. If* M *is a projective* A*-module of rank* 1 *and* M^* *is its dual, then*

$$\mathrm{cl}(M^*) = -\mathrm{cl}(M) \quad \textit{and} \quad \mathrm{cl}(A) = 0. \tag{7}$$

The associativity and commutativity of the tensor product show that the law of composition (6) is associative and commutative; the isomorphism between $A \otimes_A M$ and M prove that cl(A) is the identity element under this law and, by virtue of Theorem 3, $\mathrm{cl}(M) + \mathrm{cl}(M^*) = \mathrm{cl}(A)$, whence the proposition.

Let B be a commutative A-algebra and M a projective A-module of rank 1; then $M_{(B)} = B \otimes_A M$ is a projective B-module of rank 1 (no. 3, Proposition 4). Then there exists a mapping called canonical $\phi: \boldsymbol{P}(A) \to \boldsymbol{P}(B)$ such that

$$\phi(\mathrm{cl}(M)) = \mathrm{cl}(M_{(B)}). \tag{8}$$

The equation $M_{(B)} \otimes_B N_{(B)} = (M \otimes_A N)_{(B)}$ for two A-modules M, N proves that the mapping ϕ is a commutative group *homomorphism*.

**Remark* (2). Condition (e) of Theorem 1 (equivalent to the fact that P is projective and finitely generated) may also be expressed by saying that *the sheaf of modules* $\tilde{P}$ *over* X = Spec(A) *associated* (*) *with* P *is locally free and of finite type* and may therefore be interpreted as the sheaf of sections of a vector bundle over X. Conversely, every vector bundle over X arises from a finitely generated projective module, which is determined to within a unique isomorphism; the projective modules of rank n thus correspond to the vector bundles all of whose fibres have dimension n. In particular, the vector bundles of rank 1 correspond to the projective modules of rank 1. If we denote by $\mathcal{O}_X$ the structure sheaf $\tilde{A}$ and by $\mathcal{O}_X^*$ the *sheaf of units* of $\mathcal{O}_X$ (whose sections over an open set U of X are the invertible elements of the ring of sections of $\mathcal{O}_X$ over U), it follows that the group $\boldsymbol{P}(A)$ is isomorphic to the first cohomology group $H^1(X, \mathcal{O}_X^*)$.*

(*) See A. GROTHENDIECK, *Éléments de géométrie algébrique*, I (§ 1) (*Publ. Math. I.H.E.S.*, no. 4, 1960).

5. NON-DEGENERATE SUBMODULES

In this no. and the two following, A *denotes a ring,* S *a multiplicative subset of* A *consisting of elements* which are not divisors of zero in A, *and* B *the ring* $S^{-1}A$; A *is canonically identified with a subring of* B (§ 2, no. 1, *Remark* 3). *The elements of* S *are therefore invertible in* B.

One of the most important special cases for applications is that where A is an *integral domain* and S is the set of elements $\neq 0$ of A; B is then the *field of fractions* of A.

DEFINITION 3. *Let* M *be a sub-*A*-module of* B. M *is called non-degenerate if* $B.M = B$.

If B is a field, this condition simply means that M is not reduced to 0.

PROPOSITION 8. *Let* M *be a sub-*A*-module of* B. *The following conditions are equivalent:*

(a) M *is non-degenerate.*

(b) M *meets* S.

(c) *If* $j: M \to B$ *is the canonical injection, the homomorphism* $u = S^{-1}j: S^{-1}M \to B$ *is bijective.*

(a) implies (b), for if $B.M = B$, there exists $a \in A$, $s \in S$ and $x \in M$ such that $(a/s)x = 1$, hence $ax = s$ belongs to $S \cap M$. To see that (b) implies (c), note that u is already injective (§ 2, no. 4, Theorem 1); moreover, if $x \in M \cap S$, the image under u of $x/x \in S^{-1}M$ in B is equal to 1 and u is therefore surjective. Finally, (c) clearly implies (a).

COROLLARY. *If* M *and* N *are two non-degenerate sub-*A*-modules of* B, *the* A*-modules* $M + N$, $M.N$ *and* $M \cap N$ *are non-degenerate.*

The assertion is trivial for $M + N$; on the other hand if $s \in S \cap M$ and $t \in S \cap N$, then $st \in S \cap (M.N)$ and $st \in S \cap (M \cap N)$.

Given two sub-A-modules M and N of B, let us denote by $N:M$ the sub-A-module of B consisting of those $b \in B$ such that $bM \subset N$ (Chapter I, § 2, no. 10, *Remark*). If every $b \in N:M$ is mapped to the homomorphism $h_b: x \mapsto bx$ of M to N, a *canonical homomorphism* $b \mapsto h_b$ is obtained from $N:M$ to $\mathrm{Hom}_A(M, N)$.

PROPOSITION 9. *Let* M, N *be two sub-*A*-modules of* B. *If* M *is non-degenerate, the canonical homomorphism from* $N:M$ *to* $\mathrm{Hom}_A(M, N)$ *is bijective.*

Let $s \in S \cap M$. If $b \in N:M$ is such that $bx = 0$ for all $x \in M$, then $bs = 0$, whence $b = 0$ since s is not a divisor of 0 in B. On the other hand, let

$f \in \mathrm{Hom}_A(M, N)$ and set $b = f(s)/s$; for all $x \in M$, there exists $t \in S$ such that $tx \in A$. Then

$$f(x) = s^{-1}t^{-1}f(stx) = s^{-1}t^{-1}txf(s) = bx,$$

whence $b \in N : M$ and $f = h_b$, which proves the proposition.

Remark. In particular, A: M is canonically identified with the dual M* of M, the canonical bilinear form on M × M* being identified with the restriction to M × (A: M) of the multiplication B × B → B.

6. INVERTIBLE SUBMODULES

(*We preserve the notation of no.* 5.)

DEFINITION 4. *A sub-A-module* M *of* B *is called invertible if there exists a sub-A-module* N *of* B *such that* M.N = A.

Example. If b is invertible element of B, the A-module Ab is invertible, as is seen by taking $N = Ab^{-1}$.

PROPOSITION 10. *Let* M *be an invertible sub-A-module of* B. *Then:*

(i) *There exists* $s \in S$ *such that* $As \subset M \subset As^{-1}$ (*and in particular* M *is non-degenerate*).

(ii) A: M *is the only sub-A-module* N *of* B *such that* M.N = A.

If M.N = A, then B.M = B.(B.M) ⊃ B.(M.N) = B.A = B, hence M is non-degenerate. Similarly N is non-degenerate. If $t \in S \cap M$ and $u \in S \cap N$ (no. 5, Proposition 8), the element $s = tu$ belongs to $S \cap M \cap N$, whence $Ms \subset M.N = A$ and therefore $As \subset M \subset As^{-1}$.

On the other hand, obviously N ⊂ A: M, whence

$$A = M.N \subset M.(A: M) \subset A$$

and M.(A: M) = A; multiplying the two sides by N, we deduce A: M = N, which completes the proof.

THEOREM 4. *Let* M *be a non-degenerate sub-A-module of* B. *The following properties are equivalent:*

(a) M *is invertible.*

(b) M *is projective.*

(c) M *is projective of rank* 1.

(d) M *is a finitely generated* A*-module and, for every maximal ideal* $\mathfrak{m}$ *of* A, *the* $A_{\mathfrak{m}}$*-module* $M_{\mathfrak{m}}$ *is monogenous.*

Let us show first the equivalence of properties (a), (b) and (c). If (a) holds and N is sub-A-module of B such that M.N = A, then there is a relation

$$\sum_{i=1}^{p} m_i n_i = 1 \qquad (m_i \in M,\ n_i \in N \text{ for all } i). \tag{9}$$

For all $x \in M$, set $v_i(x) = n_i x$; the v_i are linear forms on M and by (9) $x = \sum_{i=1}^{n} m_i v_i(x)$ for all $x \in M$; this proves (*Algebra*, Chapter II, § 2, no. 6, Proposition 12) that M is projective and generated by the m_i; hence M is a finitely generated projective module.

Let $\mathfrak{m}$ be a maximal ideal of A; we show that the integer $r = \mathrm{rg}_{\mathfrak{m}}(M)$ is equal to 1. Let S′ be the image of S in $A_{\mathfrak{m}}$; as the elements of S are not divisors of 0 in A, those of S′ are not divisors of 0 in $A_{\mathfrak{m}}$, since $A_{\mathfrak{m}}$ is a flat A-module (§ 2, no. 4, Theorem 1 and Chapter I, § 2, no. 4, Proposition 3); then $S'^{-1}A_{\mathfrak{m}} \neq 0$ and, as $M_{\mathfrak{m}}$ is a free $A_{\mathfrak{m}}$-module of rank r, $S'^{-1}M_{\mathfrak{m}}$ is a free $S'^{-1}A_{\mathfrak{m}}$-module of rank r. But if T′ is the image of $A - \mathfrak{m}$ in $S^{-1}A$, $S'^{-1}A_{\mathfrak{m}}$ (resp. $S'^{-1}M_{\mathfrak{m}}$) is canonically identified with $T'^{-1}(S^{-1}A)$ (resp. $T'^{-1}(S^{-1}M)$) (§ 2, no. 3, Proposition 7). Now $S^{-1}M = B$ (Proposition 8 (c)) and hence $T'^{-1}(S^{-1}M)$ is a free A-module of rank 1 over $T'^{-1}(S^{-1}A)$, which proves that $r = 1$ and shows the implication (a) ⇒ (c).

The implication (c) ⇒ (b) is trivial. Let us show that (b) ⇒ (a). There exists by hypothesis a family (not necessarily finite) $(f_\lambda)_{\lambda \in L}$ of linear forms on M and a family $(m_\lambda)_{\lambda \in L}$ of elements of M such that, for all $x \in M$, the family $(f_\lambda(x))$ has finite support and $x = \sum_{\lambda \in L} m_\lambda f_\lambda(x)$ (*Algebra*, Chapter II, § 2, no. 6, Proposition 12). Since M is non-degenerate, $f_\lambda(x) = n_\lambda x$ for some $n_\lambda \in A: M$ by virtue of Proposition 9 of no. 5. Taking x as an element of $M \cap S$ (no. 5, Proposition 8), it is seen that of necessity $n_\lambda = 0$ except for a finite number of indices and $\sum_{\lambda \in L} m_\lambda n_\lambda = 1$. This obviously implies M.(A: M) = A, whence (a).

By virtue of Definition 2 of no. 3, (c) implies (d). Let us show the converse. As M is non-degenerate, its annihilator is zero (Proposition 8 (b)), then so is the annihilator of $M_{\mathfrak{m}}$ (§ 2, no. 4, formula (9)). As $M_{\mathfrak{m}}$ is assumed to be a monogenous $A_{\mathfrak{m}}$-module, it is therefore free of rank 1 and it then follows from no. 3, Theorem 2 that M is projective of rank 1.

Corollary. *Every invertible sub-A-module of* B *is flat and finitely presented.*
This follows from Theorem 4 (c).

Proposition 11. *Let* M, N *be two sub-A-modules of* B. *Suppose that* M *is invertible. Then:*

(i) *The canonical homomorphism* $M \otimes_A N \to M.N$ *is bijective.*

(ii) N: M = N.(A: M) *and* N = (N: M).M.

Let j be the canonical injection $N \to B$. Since M is a flat A-module (Corollary to Theorem 4), $1 \otimes j: M \otimes_A N \to M \otimes_A B$ is injective. But, as $B = S^{-1}A$, the B-module $M \otimes_A B$ is equal to $S^{-1}M$ and hence is identified with B since M is non-degenerate (no. 5, Proposition 8). If this identification is made, the image of $1 \otimes j$ is $M.N$, whence (i).

Let us set $M' = A : M$. Then obviously $M'.N \subset N : M$ and $M.(N : M) \subset N$. On the other hand, since $M.M' = A$ (Proposition 10),

$$N : M = M'.M.(N : M) \subset M'.N$$

and $N = M.M'.N \subset M.(N : M)$, whence (ii).

Remark. The proof of (i) in Proposition 11 uses only the fact that M is flat and non-degenerate.

7. THE GROUP OF CLASSES OF INVERTIBLE MODULES

(*We preserve the notation of nos.* 5 *and* 6.)

Under multiplication, the sub-A-modules of B form a commutative monoid $\mathfrak{M}$, with A as identity element. Then the invertible modules are the invertible elements of $\mathfrak{M}$ and therefore form a commutative group $\mathfrak{I}$. We have seen (no. 6, Proposition 10) that the inverse of $M \in \mathfrak{I}$ is $A : M$.

Let A^* (resp. B^*) be the multiplicative group of invertible elements of A (resp. B) and let u denote the canonical injection $A \to B$. For all $b \in B^*$, $\theta(b) = bA$ is an invertible sub-A-module. The mapping $\theta : B^* \to \mathfrak{I}$ is a homomorphism whose kernel is $u(A^*)$; its cokernel will be denoted by $\mathfrak{C}$ or $\mathfrak{C}(A)$. The group $\mathfrak{C}$ is called the *group of classes of invertible sub-A-modules of* B. The following exact sequence has been constructed

$$(10) \qquad (1) \longrightarrow A^* \xrightarrow{u} B^* \xrightarrow{\theta} \mathfrak{I} \xrightarrow{\rho} \mathfrak{C} \longrightarrow (1)$$

where (1) denotes the group consisting only of the identity element and ρ is the canonical mapping $\mathfrak{I} \to \mathfrak{C} = \mathfrak{I}/\theta(B^*)$.

As every invertible sub-A-module M of B is projective of rank 1 (no. 6, Theorem 4), the element $\mathrm{cl}(M) \in \boldsymbol{P}(A)$ is defined (no. 4).

PROPOSITION 12. *The mapping* $\mathrm{cl} : \mathfrak{I} \to \boldsymbol{P}(A)$ *defines, by taking quotients, an isomorphism from* $\mathfrak{C} = \mathfrak{I}/\theta(B^*)$ *onto the kernel of the canonical homomorphism*

$$\phi : \boldsymbol{P}(A) \to \boldsymbol{P}(B)$$

(no. 4).

In other words, there is an exact sequence

$$(11) \qquad (1) \longrightarrow A^* \xrightarrow{u} B^* \xrightarrow{\theta} \mathfrak{I} \xrightarrow{\mathrm{cl}} \boldsymbol{P}(A) \xrightarrow{\phi} \boldsymbol{P}(B).$$

It follows from Proposition 11 of no. 6 and the definition of addition in P(A) that cl(M.N) = cl(M) + cl(N) for M, N in $\mathfrak{F}$, which shows that cl is a homomorphism. If $M \in \mathfrak{F}$ is isomorphic to A, there exists $b \in B$ such that $M = Ab$ and, as M is invertible, there exists $b' \in B$ such that $b'b = 1$, in other words b is invertible in B; the converse is immediate. Hence the kernel of cl in $\mathfrak{F}$ is $\theta(B^*)$.

Let us now determine the image of cl. If $M \in \mathfrak{F}$, then $M \otimes_A B = S^{-1}M = B$ (no. 5, Proposition 8 (c)), whence $\text{cl}(M) \in \text{Ker}(\phi)$. Conversely, let P be a projective A-module of rank 1 such that $P_{(B)} = P \otimes_A B$ is B-isomorphic to B. As P is a flat A-module, the injection $u: A \rightarrow B$ defines an injection $u \otimes 1: P \rightarrow P_{(B)} = B$ and P is thus identified with a sub-A-module of B; by virtue of Proposition 8 (c) of no. 5, P is non-degenerate and Theorem 4 of no. 6 shows that P is invertible. The kernel of ϕ is therefore equal to the image of $\text{cl}: \mathfrak{F} \rightarrow \boldsymbol{P}(A)$.

Corollary 1. *For two invertible sub-A-modules of* B *to have the same image in* $\mathfrak{C}$, *it is necessary and sufficient that they be isomorphic.*

Corollary 2. *If the ring* B *is semi-local, the group* $\mathfrak{C}$ *of classes of invertible sub-*A-*modules of* B *is canonically identified with the group* $\boldsymbol{P}(A)$ *of classes of projective* A-*modules of rank* 1.

In this case P(B) = 0 (no. 3, Proposition 5).

Remark. The hypothesis of Corollary 2 is fulfilled in the two following cases:

(1) A is an integral domain and S is the set of elements $\neq 0$ of A, B then being the field of fractions of A. The invertible sub-A-modules of B are also called in this case *invertible fractional ideals*; those which are monogenous free A-modules Ab ($b \neq 0$ in B) are just the *fractional principal ideals* defined in *Algebra*, Chapter VI, § 1, no. 5.

(2) The ring A is Noetherian and S is the set of elements of A which are not divisors of 0 such that B is the total ring of fractions of A. In this case $S = A - \bigcup_i \mathfrak{p}_i$, where the $\mathfrak{p}_i$ are the elements (finite in number) of Ass(A) (Chapter IV, § 1), hence B is semi-local (§ 3, no. 5, Proposition 17).

EXERCISES

§1

¶ 1. (a) Show that a group G cannot be the union of two subgroups distinct from G. Show that, for every set I with at least two elements, the commutative group $G = \mathbf{F}_2^{(I)}$ is the union of three subgroups distinct from G.

(b) Let $(H_\iota)_{\iota \in I}$ be a finite family of subgroups of a group G such that each of the H_ι is a subgroup of G of infinite index. Show that G cannot be the union of a finite number of left cosets of the H_ι. (Argue by induction on the number of elements in I; if there exist two distinct indices i, j such that the index $(H_i : (H_i \cap H_j))$ is finite, H_i may be suppressed; if on the other hand $H_i \cap H_j$ is of infinite index in H_i for every ordered pair (i, j) of distinct indices, consider an index k such that H_k is maximal in the set of H_i and show that H_k is the union of a finite number of left cosets of the $H_k \cap H_i$ where $i \neq k$.)

(c) Give an example of a commutative ring A and four ideals $\mathfrak{a}, \mathfrak{b}_1, \mathfrak{b}_2, \mathfrak{b}_3$ of A such that $\mathfrak{a} \not\subset \mathfrak{b}_i$ $(i = 1, 2, 3)$, but $\mathfrak{a} = \bigcup_i \mathfrak{b}_i$ (use (a)).

2. Let A be a ring which is not necessarily commutative and $\mathfrak{a}, \mathfrak{p}_1, \ldots, \mathfrak{p}_n$ two-sided ideals of A. Suppose that $\mathfrak{a}$ is contained in the union of the $\mathfrak{p}_i$ and that all the $\mathfrak{p}_i$ except at most two are prime ideals (*Algebra*, Chapter VIII, § 8, Exercise 6). Show that $\mathfrak{a}$ is contained in one of the $\mathfrak{p}_i$.

3. In the product ring $A = \mathbf{R}^{\mathbf{N}}$, let (for each $n \in \mathbf{N}$) $\mathfrak{m}_n$ be the maximal ideal consisting of the $f : \mathbf{N} \to \mathbf{R}$ such that $f(n) = 0$. Show that $\mathfrak{a} = \mathbf{R}^{(\mathbf{N})}$ is an ideal of A contained in the union of the $\mathfrak{m}_n$, but in none of these maximal ideals.

4. In a ring A let $\mathfrak{p}$ be a prime ideal and a an element such that $\mathfrak{p} \subset Aa$ but $a \notin \mathfrak{p}$. Show that $\mathfrak{p} = a\mathfrak{p}$.

5. (a) Let A be a Noetherian ring and $\mathfrak{a}$ an ideal of A. Show that there exists a finite number of prime ideals $\mathfrak{p}_i$ $(1 \leqslant i \leqslant r)$ such that $\mathfrak{p}_1\mathfrak{p}_2 \ldots \mathfrak{p}_r \subset \mathfrak{a}$. (Argue by *reductio ad absurdum* considering among the ideals containing $\mathfrak{a}$, distinct from A and containing no finite product of prime ideals, a maximal element;

observe on the other hand that, if the ideal $\mathfrak{b}$ is not prime, there exist two ideals $\mathfrak{c}$, $\mathfrak{d}$ containing $\mathfrak{b}$, distinct from $\mathfrak{b}$ and satisfying $\mathfrak{c}\mathfrak{d} \subset \mathfrak{b}$.) (Cf. Chapter IV.)

If A is an integral domain and $\mathfrak{a} \neq 0$, we may assume that the $\mathfrak{p}_i \neq 0$.

(b) Give an example of a non-Noetherian ring A for which the result of (a) is false. (Take for example A to be the ring of continuous real-valued functions on [0, 1].)

¶ 6. (a) Let A be a ring and $\mathfrak{a}$, $\mathfrak{b}$ two ideals of A such that $\mathfrak{b}$ is finitely generated; show that, if the quotient rings $A/\mathfrak{a}$ and $A/\mathfrak{b}$ are Noetherian, so is $A/\mathfrak{a}\mathfrak{b}$ (observe that $\mathfrak{b}/\mathfrak{a}\mathfrak{b}$ is a finitely generated $(A/\mathfrak{a})$-module).

(b) Show that, if a ring A is such that every prime ideal of A is finitely generated, A is Noetherian. (Argue by *reductio ad absurdum* by showing that in the set of ideals of A which are not finitely generated there would be a maximal element $\mathfrak{c}$ which is not prime by hypothesis; there would therefore be two ideals $\mathfrak{a} \supset \mathfrak{c}$, $\mathfrak{b} \supset \mathfrak{c}$ distinct from $\mathfrak{c}$ and satisfying $\mathfrak{a}\mathfrak{b} \subset \mathfrak{c}$; then use (a).)

7. Let $A = \prod_{i=1}^{n} A_i$ be the product of a finite family of rings, the A_i being canonically identified with ideals of A. Let B be a subring of A such that $\mathrm{pr}_i B = A_i$ for $1 \leqslant i \leqslant n$.

(a) Show that, if the A_i are Noetherian (resp. Artinian), B is Noetherian (resp. Artinian) (cf. *Algebra*, Chapter VIII, § 2, Exercise 12).

(b) Let $\mathfrak{n}_i$ be the ideal of B which is the kernel of the restriction of pr_i to B. Show that every prime ideal $\mathfrak{p}$ of B is contained in one of the $\mathfrak{n}_i$ (use Proposition 1); deduce that, for this index i, $\mathrm{pr}_i\,\mathfrak{p} \neq A_i$. Show that, if each of the A_i contains only a finite number k_i of maximal ideals, B contains at most $\sum_{i=1}^{n} k_i$ maximal ideals.

(c) Let $\mathfrak{R}(A)$, $\mathfrak{R}(B)$ be the Jacobson radicals of A and B respectively. Show that $\mathfrak{R}(B) = B \cap \mathfrak{R}(A)$. If each of the A_i contains only a finite number of maximal ideals, show that, for every integer $k > 1$, $\mathrm{pr}_i((\mathfrak{R}(B))^k) = (\mathfrak{R}(A_i))^k$ for all i and $(\mathfrak{R}(B))^k = (\mathfrak{R}(A))^k \cap B$ (write $\mathfrak{R}(B)$ as a product of distinct maximal ideals and note that, if $\mathfrak{m}$ is a maximal ideal of B such that $\mathrm{pr}_i(\mathfrak{m}) = \mathfrak{m}_i$ is a maximal ideal of A_i, then $\mathrm{pr}_i(\mathfrak{m}^k) = \mathfrak{m}_i^k$).

8. (a) Let $\mathfrak{a}$, $\mathfrak{b}$ be two relatively prime ideals of a ring A. Show that $\mathfrak{a} : \mathfrak{b} = \mathfrak{a}$, $\mathfrak{b} : \mathfrak{a} = \mathfrak{b}$. If $\mathfrak{c}$ is an ideal of A such that $\mathfrak{a}\mathfrak{c} \subset \mathfrak{b}$, then $\mathfrak{c} \subset \mathfrak{b}$.

(b) Let $\mathfrak{p}$, $\mathfrak{q}$ be two prime ideals neither of which is contained in the other; then $\mathfrak{p} : \mathfrak{q} = \mathfrak{p}$ and $\mathfrak{q} : \mathfrak{p} = \mathfrak{q}$. Give an example of two principal prime ideals $\mathfrak{p}$, $\mathfrak{q}$ in the polynomial ring $A = K[X, Y]$ (where K is a field), neither of which is contained in the other and which are not relatively prime.

(c) Let a be an element which is not a divisor of 0 in A. Show that, if the principal ideal $\mathfrak{p} = Aa$ is prime, the relation $\mathfrak{p} = \mathfrak{b}\mathfrak{c}$ for two ideals $\mathfrak{b}$, $\mathfrak{c}$ of A implies $\mathfrak{b} = A$ or $\mathfrak{c} = A$.

9. (a) Let $(\mathfrak{a}_i)_{1 \leqslant i \leqslant n}$ be a finite family of ideals of a ring A which are relatively prime in pairs and such that no $\mathfrak{a}_i$ is of the form $\mathfrak{bc} = \mathfrak{b} \cap \mathfrak{c}$, where $\mathfrak{b}$ and $\mathfrak{c}$ are two relatively prime ideals. If $(\mathfrak{a}'_j)_{1 \leqslant j \leqslant m}$ is another family of ideals of A which are relatively prime in pairs, such that $\mathfrak{a}_1\mathfrak{a}_2\ldots\mathfrak{a}_n = \mathfrak{a}'_1\mathfrak{a}'_2\ldots\mathfrak{a}'_m$, show that $m \leqslant n$; if $m = n$, there exists a permutation π on $[1, n]$ such that $\mathfrak{a}'_i = \mathfrak{a}_{\pi(i)}$ for $1 \leqslant i \leqslant n$ (use Proposition 5 and *Algebra*, Chapter VIII, § 1, Exercise 1 (d)).

(b) Let A be a Noetherian ring. Show that every ideal $\mathfrak{a}$ of A is equal to the product of a finite number of ideals which are relatively prime in pairs, none of which is the product of two relatively prime ideals. (Argue by *reductio ad absurdum* considering a maximal element of the set of ideals not having this property.)

10. Give an example of a ring A and an infinite family of distinct maximal ideals $(\mathfrak{m}_n)_{n \in \mathbf{N}}$ of A such that $\bigcap_n \mathfrak{m}_n = 0$ but the canonical mapping

$$A \rightarrow \prod_n (A/\mathfrak{m}_n)$$

is not surjective.

11. Let A be a ring which is not necessarily commutative and let $\mathfrak{a}$, $\mathfrak{b}$ be two two-sided ideals of A such that $\mathfrak{a} + \mathfrak{b} = A$. Show that $\mathfrak{a} \cap \mathfrak{b} = \mathfrak{ab} + \mathfrak{ba}$. Give an example where $\mathfrak{a} \cap \mathfrak{b} \neq \mathfrak{ab}$ (consider the ring of lower triangular matrices of order 2 over a field).

§ 2

1. A multiplicative subset S of a ring A is called *saturated* if the relation $xy \in S$ implies $x \in S$ and $y \in S$.

(a) For a subset S of A to be multiplicative and saturated, it is necessary and sufficient that A — S be a union of prime ideals of A.

(b) Let S be a multiplicative subset of A and $\tilde{S}$ the set of $x \in A$ for which there exists $y \in A$ such that $xy \in S$. Show that $\tilde{S}$ is the smallest saturated multiplicative subset containing S, A — $\tilde{S}$ is the union of the prime ideals of A not meeting S and, for every A-module M, the canonical mapping from $S^{-1}M$ to $\tilde{S}^{-1}M$ is bijective.

(c) Let S and T be two multiplicative subsets of A. Show that the two following properties are equivalent: (α) $S \subset \tilde{T}$; (β) for every A-module M such that $S^{-1}M = 0$, $T^{-1}M = 0$. (To see that (β) implies (α), consider a quotient A-module A/As, where $s \in S$.)

2. Let A be a ring and S a multiplicative subset of A. Show that the set $\mathfrak{n}$ of $x \in A$ such that $sx = 0$ for some $s \in S$ is an ideal of A. Set $A_1 = A/\mathfrak{n}$ and denote by S_1 the canonical image of S in A_1. Show that no element of S_1 is a

divisor of 0 in A_1 and that the canonical homomorphism $S^{-1}A \to S_1^{-1}A_1$ is bijective.

Deduce that, for $S^{-1}A$ to be a finitely generated A-module, it is necessary and sufficient that all the elements of S_1 be invertible in A_1, in which case $S^{-1}A$ is identified with A_1.

3. Let $S = \mathbf{Z}^*$ (the complement of $\{0\}$ in $\mathbf{Z}$), p an integer > 1 and M the $\mathbf{Z}$-module the direct sum of the modules $\mathbf{Z}/p^n\mathbf{Z}$ for $n \in \mathbf{N}$. Show that $S^{-1}M = 0$ although M is a faithful $\mathbf{Z}$-module and that the canonical mapping

$$S^{-1} \operatorname{End}_{\mathbf{Z}}(M) \to \operatorname{End}_{S^{-1}\mathbf{Z}}(S^{-1}M)$$

is not injective.

4. Let $S = \mathbf{Z}^*$ and M be a free $\mathbf{Z}$-module with infinite basis. Show that the canonical mapping $S^{-1} \operatorname{End}_{\mathbf{Z}}(M) \to \operatorname{End}_{S^{-1}\mathbf{Z}}(S^{-1}M)$ is not surjective. Deduce from this and Exercise 3 an example of a $\mathbf{Z}$-module M such that the canonical mapping $S^{-1}\operatorname{End}_{\mathbf{Z}}(M) \to \operatorname{End}_{S^{-1}\mathbf{Z}}(S^{-1}M)$ is neither injective nor surjective.

5. Give an example of a sequence (P_n) of sub-$\mathbf{Z}$-modules of $\mathbf{Z}$ such that, for $S = \mathbf{Z}^*$, the submodule $S^{-1}\left(\bigcap_n P_n\right)$ is distinct from $\bigcap_n (S^{-1}P_n)$.

6. Give an example of a $\mathbf{Z}$-module M such that, for $S = \mathbf{Z}^*$, $\operatorname{Ann}(M) = 0$, but $\operatorname{Ann}(S^{-1}M) = \mathbf{Q}$ (cf. Exercise 3).

7. Let S be a multiplicative subset of a ring A; for every family $(M_\iota)_{\iota \in I}$ of A-modules, define a canonical homomorphism $S^{-1} \prod_{\iota \in I} M_\iota \to \prod_{\iota \in I} S^{-1}M_\iota$ of $S^{-1}A$-modules. Give an example where this homomorphism is neither injective nor surjective (cf. Exercise 3).

8. Let A be a ring, $(S_\lambda)_{\lambda \in L}$ a family of multiplicative subsets of A and $M = \bigoplus_{\lambda \in L} S_\lambda^{-1}A$. Show that the two following properties are equivalent: (α) M is a faithfully flat A-module; (β) for every maximal ideal $\mathfrak{m}$ of A, there exists $\lambda \in L$ such that $\mathfrak{m} \cap S_\lambda = \varnothing$. In particular, if S is a multiplicative subset of A, the A-module $S^{-1}A$ is faithfully flat only if the elements of S are invertible, in which case $S^{-1}A$ is identified with A.

9. Let A be a ring, S a multiplicative subset of A and $\mathfrak{q}$ a prime ideal of A. Show that, if T is the image of $A - \mathfrak{q}$ in $S^{-1}A$, the ring $T^{-1}(S^{-1}A)$ is isomorphic to $S'^{-1}A$, where S′ is the complement of the union of the prime ideals of A contained in $\mathfrak{q}$ and not meeting S.

10. (a) Let K be a commutative field, A the polynomial ring K[X, Y], $\mathfrak{p} = (X)$, $\mathfrak{q} = (Y)$ and S the complement of $\mathfrak{p} \cup \mathfrak{q}$ in A. Show that the saturation of $\mathfrak{p} + \mathfrak{q}$ with respect to S is distinct from the sum of the saturations of $\mathfrak{p}$ and $\mathfrak{q}$.

(b) Let K be a commutative field, A the polynomial ring K[X, Y, Z], $\mathfrak{p} = (X) + (Z)$, $\mathfrak{q} = (Y) + (Z)$ and S the complement of $\mathfrak{p} \cup \mathfrak{q}$ in A. Show that the saturation of $\mathfrak{p}\mathfrak{q}$ with respect to S is distinct from the product of the saturations of $\mathfrak{p}$ and $\mathfrak{q}$.

11. Let p be a prime number; which ideals of **Z** are saturated with respect to the ideal (p)?

12. In a ring A let $\mathfrak{r}(\mathfrak{a})$ denote the radical of an ideal $\mathfrak{a}$.

(a) Show that, for three ideals $\mathfrak{a}$, $\mathfrak{b}$, $\mathfrak{c}$ of A,

$$\mathfrak{r}(\mathfrak{a} + \mathfrak{b}\mathfrak{c}) = \mathfrak{r}(\mathfrak{a} + (\mathfrak{b} \cap \mathfrak{c})) = \mathfrak{r}(\mathfrak{a} + \mathfrak{b}) \cap \mathfrak{r}(\mathfrak{a} + \mathfrak{c}).$$

(b) Give an example of an infinite sequence of ideals $(\mathfrak{a}_n)$ of A such that $\mathfrak{r}\left(\bigcap_n \mathfrak{a}_n\right) \neq \bigcap_n \mathfrak{r}(\mathfrak{a}_n)$.

(c) Give an example of a principal ideal $\mathfrak{a}$ such that $\mathfrak{r}(\mathfrak{a})$ is not finitely generated and there exists no integer $n > 0$ such that $(\mathfrak{r}(\mathfrak{a}))^n \subset \mathfrak{a}$. (Consider the ring of a non-discrete valuation of height 1.)

13. Let A be a ring.

(a) Give another proof of Exercise 6 (b) of *Algebra*, Chapter VIII, § 6, by considering the image of the polynomial f in the rings $(A/\mathfrak{p})[X]$, where $\mathfrak{p}$ runs through the set of prime ideals of A.

(b) Let N be a square matrix of order r over A; for N to be nilpotent, it is necessary and sufficient that each of the coefficients (other than the dominant coefficient) of its characteristic polynomial be so. (To see that the condition is necessary, use the same method as in (a); to see that it is sufficient, use the Cayley-Hamilton Theorem.)

(c) Show that, for every ordered pair of positive integers k, r, there exists a least positive integer $m(k, r)$ independent of the ring A such that the relation $N^k = 0$ implies $(\mathrm{Tr}(N))^{m(k,r)} = 0$. (Consider the ring $\Lambda = \mathbf{Z}[T_{ij}]$, where the T_{ij} are r^2 indeterminates, the matrix $X = (T_{ij})$ over Λ and the element $t = \sum_{i=1}^{r} T_{ii} = \mathrm{Tr}(X)$ of Λ; if $\mathfrak{a}_k$ is the ideal of Λ generated by the elements of X^k, show, with the aid of (b), that there exists an integer m such that $t^m \in \mathfrak{a}_k$.)

(d) Show that $m(2, 2) = 4$.

¶ 14. (a) Let A be a left Noetherian ring (not necessarily commutative) and $\mathfrak{a}$ a left ideal of A. Suppose that there exists a left ideal $\mathfrak{b} \neq 0$ such that $\mathfrak{a} \cap \mathfrak{b} = 0$.

Show that every element $a \in \mathfrak{a}$ is a right divisor of 0. (Let $b \neq 0$ be an element of $\mathfrak{b}$ and let $\mathfrak{a}_n = Ab + Aba + \cdots + Aba^n$; consider the smallest integer n such that $\mathfrak{a}_n = \mathfrak{a}_{n+1}$.)

(b) Let E be an infinite-dimensional vector space over a field K and A the ring $\mathrm{End}_K(E)$. Give an example of two elements u, v of A such that neither u nor v is a right divisor of 0 and $Au \cap Av = 0$.

¶ 15. Let X be a completely regular topological space and βX its Stone-Čech compactification (*General Topology*, Chapter IX, § 1, Exercise 7). For all $x \in X$, let $\mathfrak{J}(x)$ denote the maximal ideal of the ring $\mathscr{C}(X; \mathbf{R})$ consisting of those $f \in \mathscr{C}(X; \mathbf{R})$ such that x belongs to the closure of the sub set $\overset{-1}{f}(0)$ of X (*General Topology*, Chapter X, § 4, Exercise 15); let $\mathfrak{U}(x)$ denote the ideal consisting of those $f \in \mathscr{C}(X; \mathbf{R})$ such that $\overset{-1}{f}(0)$ is the trace on X of a neighbourhood of x in X; $\mathfrak{U}(x)$ is equal to its radical.

An ideal $\mathfrak{a}$ of $\mathfrak{C}(X; \mathbf{R})$ is called *isolated* (resp. *absolutely isolated*) if, for all $f \geqslant 0$ on $\mathfrak{a}$, every g such that $0 \leqslant g \leqslant f$ (resp. $|g| \leqslant f$) belongs to $\mathfrak{a}$ (cf. *Algebra*, Chapter IV, § 1, Exercise 4); then $\mathscr{C}(X; \mathbf{R})/\mathfrak{a}$ possesses an order structure (resp. a lattice-order structure) compatible with its ring structure and for which the elements $\geqslant 0$ are the classes mod. $\mathfrak{a}$ of the elements $\geqslant 0$ to $\mathscr{C}(X; \mathbf{R})$ (*loc. cit.*).

(a) Show that an ideal $\mathfrak{a}$ of $\mathscr{C}(X; \mathbf{R})$ which is equal to its radical is absolutely isolated (if $f \in \mathfrak{a}$ and $|g| \leqslant |f|$, consider the function equal to 0 when $f(x) = 0$ and to $(g(x))^2/f(x)$ when $f(x) \neq 0$). In this case for $\mathscr{C}(X; \mathbf{R})/\mathfrak{a}$ to be totally ordered, it is necessary and sufficient that $\mathfrak{a}$ be prime; the set of prime ideals of $\mathscr{C}(X; \mathbf{R})/\mathfrak{a}$ is then totally ordered by inclusion.

(b) For an isolated ideal $\mathfrak{a}$ of $\mathscr{C}(X; \mathbf{R})$ to be such that $\mathscr{C}(X; \mathbf{R})/\mathfrak{a}$ is totally ordered, it is necessary that $\mathfrak{a}$ be contained in a single maximal ideal $\mathfrak{J}(x)$ and then $\mathfrak{a}$ necessarily contains $\mathfrak{U}(x)$; it is sufficient that $\mathfrak{a}$ contain a prime ideal and then $\mathfrak{a}$ is absolutely isolated (note that the relation $fg = 0$ then implies $f \in \mathfrak{a}$ or $g \in \mathfrak{a}$).

(c) Take $X = \mathbf{R}$, $x = 0$; show that the principal ideal generated by the function $t \mapsto |t|$ is contained in $\mathfrak{J}(0)$ and contains $\mathfrak{U}(0)$ but is not isolated; the principal ideal generated by the identity mapping $t \mapsto t$ is contained in $\mathfrak{J}(0)$, contains $\mathfrak{U}(0)$ and is isolated but not absolutely isolated.

(d) Show that, if $\mathfrak{p}$ and $\mathfrak{q}$ are two prime ideals in $\mathscr{C}(X; \mathbf{R})$, then $\mathfrak{p}\mathfrak{q} = \mathfrak{p} \cap \mathfrak{q}$ (and in particular $\mathfrak{p}^2 = \mathfrak{p}$) and $\mathfrak{p} + \mathfrak{q}$ is prime or equal to $\mathscr{C}(X; \mathbf{R})$. (If $\mathfrak{p} + \mathfrak{q} \neq \mathscr{C}(X; \mathbf{R})$, observe that $\mathfrak{p} + \mathfrak{q}$ is absolutely isolated and equal to its radical and use (a).)

(e) For $\mathfrak{U}(x)$ to be prime, it is necessary and sufficient that, for every function $f \in \mathscr{C}(X; \mathbf{R})$, the closure in βX of one of the two sets $\overset{-1}{f}([0, +\infty[)$, $\overset{-1}{f}(]-\infty, 0])$ is a neighbourhood of x. Deduce an example where $\mathfrak{U}(x)$ is prime and not maximal (consider the topological space associated with a non-trivial ultrafilter; cf. *General Topology*, Chapter I, § 6, no. 5).

(f) Suppose that $x \in X$. For $\mathfrak{U}(x) = \mathfrak{J}(x)$, it is necessary and sufficient that every countable intersection of neighbourhoods of x in X be a neighbourhood of x in X. (Cf. *General Topology*, Chapter I, § 7, Exercise 7.)

(g) Show that $\mathscr{C}(X; \mathbf{R})$ is canonically identified with the total ring of fractions of $\mathscr{C}^{\infty}(X; \mathbf{R})$.

16. Let A be a topological ring (not necessarily commutative). The topology on A is called (left) *linear* if the open left ideals of A form a fundamental system of neighbourhoods of 0. If this is so, the set $\mathscr{F}$ of open left ideals of A satisfies the following conditions:

(1) Every left ideal of A containing an ideal $\mathfrak{m} \in \mathscr{F}$ belongs to $\mathscr{F}$.

(2) Every finite intersection of ideals of $\mathscr{F}$ belongs to $\mathscr{F}$.

(3) If $\mathfrak{m} \in \mathscr{F}$ and $a \in A$, the left ideal $\mathfrak{n}$ consists of those $x \in A$ such that $xa \in \mathfrak{m}$ belongs to $\mathscr{F}$.

Conversely, if a set $\mathscr{F}$ of left ideals of a ring A satisfies these three conditions, there exists a unique topology compatible with the ring structure on A and for which $\mathscr{F}$ is the set of open left ideals of A. Such a set $\mathscr{F}$ of left ideals is called (left) *topologizing*.

17. (*) Let A be a ring (not necessarily commutative) and $\mathscr{F}$ a topologizing set (Exercise 16) of left ideals of A.

(a) A (left) A-module M is called *$\mathscr{F}$-negligible* if the annihilator of every element of M belongs to $\mathscr{F}$. If M is $\mathscr{F}$-negligible, so is every submodule and every quotient module of M. If M and N are two $\mathscr{F}$ negligible A-modules, $M \oplus N$ is $\mathscr{F}$-negligible. For $A_s/\mathfrak{m}$ to be $\mathscr{F}$-negligible, it is necessary and sufficient that $\mathfrak{m} \in \mathscr{F}$. If $(0) \in \mathscr{F}$, every A-module is $\mathscr{F}$-negligible; if consists only of the single left ideal A_s, every $\mathscr{F}$-negligible A-module is reduced to 0.

(b) For every A-module M there exists a greatest $\mathscr{F}$-negligible submodule of M which is denoted by $\mathscr{F}M$. For every A-module homomorphism $u: M \to N$, $u(\mathscr{F}M) \subset \mathscr{F}N$; let $\mathscr{F}u$ be the mapping from $\mathscr{F}M$ to $\mathscr{F}N$ whose graph coincides with that of $u|\mathscr{F}M$. If

$$0 \longrightarrow M \xrightarrow{u} N \xrightarrow{v} P$$

is an exact sequence of A-modules, the sequence

$$0 \longrightarrow \mathscr{F}M \xrightarrow{\mathscr{F}u} \mathscr{F}N \xrightarrow{\mathscr{F}v} \mathscr{F}P$$

is exact. Show that $\mathscr{F}A_s$ is a two-sided ideal of A.

(c) Let $\mathfrak{m}$, $\mathfrak{n}$ be two elements of $\mathscr{F}$ such that $\mathfrak{m} \subset \mathfrak{n}$; for every left A-module M the canonical injection $j: \mathfrak{m} \to \mathfrak{n}$ defines a commutative group homomorphism

$$u_{\mathfrak{m},\mathfrak{n}} = \mathrm{Hom}_A(j, 1_M) : \mathrm{Hom}_A(\mathfrak{n}, M) \to \mathrm{Hom}_A(\mathfrak{m}, M).$$

(*) Exercises 17 to 25 (unedited) were communicated to us by P. Gabriel.

If $\mathscr{F}$ is ordered by the relation $\supset$, the $u_{\mathfrak{m},\mathfrak{n}}$ define a direct system of commutative groups, whose direct limit will be denoted by $M_{(\mathscr{F})}$. The mappings $u_{\mathfrak{m},A_s}$ form a direct system, whose direct limit is a mapping (called canonical) from $M = \mathrm{Hom}_A(A_s, M)$ to $M_{(\mathscr{F})}$. Similarly, if $v: M \to N$ is a left A-module homomorphism, the $v_\mathfrak{m} = \mathrm{Hom}_A(1, v): \mathrm{Hom}_A(\mathfrak{m}, M) \to \mathrm{Hom}_A(\mathfrak{m}, N)$ form a direct system, whose direct limit is a commutative group homomorphism

$$v_{(\mathscr{F})}: M_{(\mathscr{F})} \to N_{(\mathscr{F})}.$$

If $0 \to M \xrightarrow{v} N \xrightarrow{w} P$ is an exact sequence of A-modules, the sequence

$$0 \longrightarrow M_{(\mathscr{F})} \xrightarrow{v(\mathscr{F})} N_{(\mathscr{F})} \xrightarrow{w(\mathscr{F})} P_{(\mathscr{F})}$$

is exact.

18. (a) Let A be a ring and $\mathscr{F}$ and $\mathscr{G}$ two topologizing sets (Exercise 16) of left ideals of A. Let $\mathscr{F}.\mathscr{G}$ denote the set of left ideals $\mathfrak{l}$ of A for which there exists a left ideal $\mathfrak{n} \in \mathscr{G}$ such that $\mathfrak{n}/\mathfrak{l}$ is $\mathscr{F}$-negligible. Show that $\mathscr{F}.\mathscr{G}$ is topologizing; for an A-module M to be $\mathscr{F}.\mathscr{G}$-negligible, it is necessary and sufficient that there exist an $\mathscr{F}$-negligible submodule M′ of M such that M/M′ is $\mathscr{G}$-negligible. Show that the composition law $(\mathscr{F}, \mathscr{G}) \mapsto \mathscr{F}.\mathscr{G}$ on the set of topologizing sets of left ideals of A is associative.

$\mathscr{F}$ is called *idempotent* if $\mathscr{F}.\mathscr{F} = \mathscr{F}$.

(b) Show that $\mathscr{F}.\mathscr{G}$ contains all the ideals of the form $\mathfrak{m}.\mathfrak{n}$ where $\mathfrak{m} \in \mathscr{F}$ and $\mathfrak{n} \in \mathscr{G}$.

(c) Let K be a commutative field and B the ring of polynomials with coefficients in K in an infinite system of indeterminates (X_i) $(i \geqslant 0)$. Let $\mathfrak{b}$ be the ideal of B generated by the elements X_iX_j $(i \neq j)$; let A be the ring $B/\mathfrak{b}$, ξ_i the class mod. $\mathfrak{b}$ of X_i and $\mathfrak{m}$ the ideal of A generated by the ξ_i; take $\mathscr{F}$ to be the set of ideals containing a power of $\mathfrak{m}$. Show that $\mathscr{F}$ is topologizing and contains the product of any two ideals of $\mathscr{F}$ but is not idempotent.

(d) Suppose that A is *commutative* and $\mathscr{F}$ is a topologizing set of ideals of A such that every ideal $\mathfrak{m} \in \mathscr{F}$ contains a *finitely generated* ideal $\mathfrak{n} \in \mathscr{F}$. Show that, if the product of two ideals of $\mathscr{F}$ belongs to $\mathscr{F}$, $\mathscr{F}$ is idempotent.

¶ 19. Let $\mathscr{F}$ be a set of left ideals of a ring A; suppose that $\mathscr{F}$ is topologizing and idempotent (Exercises 16 and 18).

(a) Show that, if $\mathfrak{m} \in \mathscr{F}$, $\mathfrak{n} \in \mathscr{F}$ and $u \in \mathrm{Hom}_A(\mathfrak{n}, A_s)$, then $\overset{-1}{u}(\mathfrak{m}) \in \mathscr{F}$ (consider the exact sequence $0 \to \mathfrak{n}/\overset{-1}{u}(\mathfrak{m}) \to A_s/\overset{-1}{u}(\mathfrak{m}) \to A_s/\mathfrak{n} \to 0$). For every A-module M and all $v \in \mathrm{Hom}_A(\mathfrak{m}, M)$, let $u.v$ denote the canonical image in $M_{(\mathscr{F})}$ (Exercise 17 (c)) of the composite homomorphism $\overset{-1}{u}(\mathfrak{m}) \xrightarrow{u} \mathfrak{m} \xrightarrow{v} M$. Show that the mapping $(u, v) \mapsto u.v$ is $\mathbf{Z}$-bilinear; there corresponds to it a $\mathbf{Z}$-linear mapping

$$\psi_{\mathfrak{m},\mathfrak{n}}: \mathrm{Hom}_A(\mathfrak{n}, A_s) \otimes_{\mathbf{Z}} \mathrm{Hom}_A(\mathfrak{m}, M) \to M_{(\mathscr{F})}.$$

These mappings form a direct system and their direct limit therefore corresponds canonically to a **Z**-bilinear mapping

$$\psi: A_{(\mathscr{F})} \times M_{(\mathscr{F})} \to M_{(\mathscr{F})}.$$

Show that, if $M = A_s$, ψ is an internal law of composition which makes $A_{(\mathscr{F})}$ into a ring and the canonical mapping $A \to A_{(\mathscr{F})}$ is a ring homomorphism for this structure. For any left A-module M, ψ is an external law defining on $M_{(\mathscr{F})}$ a left $A_{(\mathscr{F})}$-module structure; with this structure (and the canonical mapping $A \to A_{(\mathscr{F})}$) the canonical mapping $M \to M_{(\mathscr{F})}$ is an A-module homomorphism.

(b) If $i: M \to M_{(\mathscr{F})}$ is the canonical homomorphism, show that

$$\operatorname{Ker}(i) = \mathscr{F}M$$

and that the A-module Coker $(i) = M_{(\mathscr{F})}/i(M)$ is $\mathscr{F}$-negligible.

(c) For every left A-module M let $M_{\mathscr{F}}$ denote the left A-module $(M/\mathscr{F}M)_{(\mathscr{F})}$ and j_M the composite canonical mapping

$$M \to M/\mathscr{F}M \to M_{\mathscr{F}};$$

then $\operatorname{Ker}(j_M) = \mathscr{F}M$ and the A-module $\operatorname{Coker}(j_M) = M_{\mathscr{F}}/j_M(M)$ is $\mathscr{F}$-negligible.

Let $u: P \to M$, $h: P \to Q$ be A-module homomorphisms such that $\operatorname{Ker}(h)$ and $\operatorname{Coker}(h)$ are $\mathscr{F}$-negligible; show that there exists a unique A-homomorphism $v: Q \to M_{\mathscr{F}}$ which makes the following diagram commutative

$$\begin{array}{ccc} P & \xrightarrow{u} & M \\ {\scriptstyle h}\downarrow & & \downarrow{\scriptstyle j_M} \\ Q & \xrightarrow[v]{} & M_{\mathscr{F}} \end{array}$$

(First reduce it to the case where h is injective; then, identifying P with a submodule of Q, note that, for $x \in Q$, there exists $m \in \mathscr{F}$ such that $mx \in P$ and map x to the canonical image in $M_{\mathscr{F}}$ of the composite homomorphism

$$m \xrightarrow{x} mx \to P \xrightarrow{u} M \to M/\mathscr{F}M.)$$

(d) For every A-module homomorphism $u: M \to N$ show that there exists a unique homomorphism $u_{\mathscr{F}}: M_{\mathscr{F}} \to N_{\mathscr{F}}$ which makes the following diagram commutative

$$\begin{array}{ccc} M & \xrightarrow{u} & N \\ {\scriptstyle j_M}\downarrow & & \downarrow{\scriptstyle j_N} \\ M_{\mathscr{F}} & \xrightarrow[u_{\mathscr{F}}]{} & N_{\mathscr{F}} \end{array}$$

(use (a), (b) and (c)). Show that, if

$$0 \longrightarrow M \xrightarrow{u} N \xrightarrow{v} P$$

is an exact sequence of A-modules, the sequence

$$0 \longrightarrow M_{\mathscr{F}} \xrightarrow{u_{\mathscr{F}}} N_{\mathscr{F}} \xrightarrow{v_{\mathscr{F}}} P_{\mathscr{F}}$$

is exact (note that the mapping $M/\mathscr{F}M \to N/\mathscr{F}N$ derived from u by taking quotients is injective).

Show that the mappings $j_{M_{\mathscr{F}}}$ and $(j_M)_{\mathscr{F}}$ coincide and are isomorphisms (and therefore $\mathscr{F}M_{\mathscr{F}} = 0$). If N′ is a sub-A-module of $M_{\mathscr{F}}$, $(\overset{-1}{j_M}(N'))_{\mathscr{F}}$ is canonically identified with $N'_{\mathscr{F}}$ (observe that, if we write $N = \overset{-1}{j_M}(N')$, $N'/j_M(N)$ is $\mathscr{F}$-negligible).

(e) Let M be a left A-module and $\phi: A \times M \to M$ the **Z**-bilinear mapping $(a, m) \mapsto am$. Show that there exists a unique **Z**-bilinear mapping $\phi_{\mathscr{F}}: A_{\mathscr{F}} \times M_{\mathscr{F}} \to M_{\mathscr{F}}$ which makes the following diagram commutative

$$\begin{array}{ccc} A \times M & \xrightarrow{\phi} & M \\ {\scriptstyle j_A \times j_M}\downarrow & & \downarrow{\scriptstyle j_M} \\ A_{\mathscr{F}} \times M_{\mathscr{F}} & \xrightarrow[\phi_{\mathscr{F}}]{} & M_{\mathscr{F}} \end{array}$$

We also define on $A_{\mathscr{F}}$ a ring structure and on $M_{\mathscr{F}}$ a left $A_{\mathscr{F}}$-module structure. If $u: M \to N$ is an A-module homomorphism, $u_{\mathscr{F}}: M_{\mathscr{F}} \to N_{\mathscr{F}}$ is an $A_{\mathscr{F}}$-module homomorphism. In order that, for every A-module M, the mapping $u \mapsto u_{\mathscr{F}}$ of $\mathrm{Hom}_A(M, N)$ to $\mathrm{Hom}_{A_{\mathscr{F}}}(M_{\mathscr{F}}, N_{\mathscr{F}})$ be bijective, it is necessary and sufficient that j_N be bijective.

(f) Take A to be the polynomial ring K[X, Y] over a commutative field K and $\mathscr{F}$ to be the (topologizing and idempotent) set of ideals of A containing a power of the ideal $\mathfrak{m} = (X) + (Y)$. If $M = A/AX$ and $u: A \to M$ is the canonical mapping, show that $u_{\mathscr{F}}: A_{\mathscr{F}} \to M_{\mathscr{F}}$ is not surjective. (Show that $A_{\mathscr{F}}$ is identified with A and $M_{\mathscr{F}}$ with $K[\overline{Y}, 1/\overline{Y}]$, where $\overline{Y}$ is the class of Y in M, so that $X\overline{Y} = 0$ and $X(1/\overline{Y}) = 0$.)

(g) If we set $\mathscr{F}^1M = M_{\mathscr{F}}/j_M(M)$, show that, for every exact sequence $0 \to M \to N \to P \to 0$ of A-modules, there is an exact sequence

$$0 \to \mathscr{F}M \to \mathscr{F}N \to \mathscr{F}P \to \mathscr{F}^1M \to \mathscr{F}^1N \to \mathscr{F}^1P$$

(use the snake diagram).

(h) Let $\mathscr{F}'$ be the set of left ideals $\mathfrak{l}'$ of $A_{\mathscr{F}}$ such that $(A_{\mathscr{F}})_s/\mathfrak{l}'$ is an $\mathscr{F}$-negligible A-module: show that $\mathscr{F}'$ is the set of left ideals of $A_{\mathscr{F}}$ containing the $j_A(\mathfrak{m})$, where $\mathfrak{m} \in \mathscr{F}$ (observe that $\mathfrak{m}_{\mathscr{F}} = A_{\mathscr{F}}$ for all $\mathfrak{m} \in \mathscr{F}$, using the exact sequence of (d); consequently $A_{\mathscr{F}}/j_A(\mathfrak{m})$ is $\mathscr{F}$-negligible). For every $A_{\mathscr{F}}$-module M', $\mathscr{F}M' = \mathscr{F}'M'$ (M' being considered as an A-module by means of j_A), $\mathscr{F}'$ is topologizing and idempotent and $M'_{\mathscr{F}'}$ is canonically identified with $M'_{\mathscr{F}}$.

20. Let $\mathscr{F}$ be a topologizing and idempotent set of left ideals of a ring A.

(a) Suppose that every ideal of $\mathscr{F}$ contains a *finitely generated* ideal belonging to $\mathscr{F}$. Let M be an A-module and $(N_\iota)_{\iota \in I}$ a right directed family of submodules of M, whose union is M. Show that $M_{\mathscr{F}}$ is the union of its sub-$A_{\mathscr{F}}$-modules $(N_\iota)_{\mathscr{F}}$. In particular, for every family $(M_\lambda)_{\lambda \in L}$ of A-modules, $\left(\bigoplus_{\lambda \in L} M_\lambda\right)_{\mathscr{F}}$ is canonically isomorphic to $\bigoplus_{\lambda \in L} (M_\lambda)_{\mathscr{F}}$.

(b) Suppose that condition (a) holds and also that, for every surjective A-module homomorphism $u: M \to N$, $u_{\mathscr{F}}: M_{\mathscr{F}} \to N_{\mathscr{F}}$ is surjective. Then, the canonical mapping $A_{\mathscr{F}} \otimes_A M \to M_{\mathscr{F}}$ defined by the external law on $M_{\mathscr{F}}$ as an $A_{\mathscr{F}}$-module is an $A_{\mathscr{F}}$-module isomorphism (reduce it to the case where M is free); the ring $A_{\mathscr{F}}$ is a flat left A-module and the left $A_{\mathscr{F}}$-modules coincide with the left A-modules M such that the mapping $j_M: M \to M_{\mathscr{F}}$ is bijective.

(c) Suppose that A is an infinite product $\prod_{\iota \in I} A_\iota$ of rings; let e_ι denote the unit element of A_ι and take $\mathscr{F}$ to be the set of left ideals of A containing the two-sided ideal $\mathfrak{a} = \bigoplus_{\iota \in I} A_\iota$ (the A_ι being canonically identified with ideals of A). Show that, for every left A-module M, $M_{\mathscr{F}}$ is identified with the product $\prod_{\iota \in I} (e_\iota M)$ and in particular $A_{\mathscr{F}} = A$. Deduce that, for every surjective A-homomorphism $u: M \to N$, $u_{\mathscr{F}}: M_{\mathscr{F}} \to N_{\mathscr{F}}$ is surjective, but give examples of A-modules M such that j_M is not bijective.

(d) Suppose that condition (a) holds and also that A is *commutative*. Show that, if M is $\mathscr{F}$-negligible, $M_{(\mathscr{F})} = 0$.

21. Let A be a *commutative* ring and $\mathscr{F}$ a topologizing and idempotent set of ideals of A.

(a) Then the ring $A_{\mathscr{F}}$ is commutative (note that, if $\mathfrak{m} \in \mathscr{F}$ and

$$u \in \mathrm{Hom}_A(\mathfrak{m}, A),$$

then $u(\mathfrak{m}^2) \subset \mathfrak{m}$; if v is another element of $\mathrm{Hom}_A(\mathfrak{m}, A)$, show that

$$v(u(xy)) = u(v(xy))$$

for $x \in \mathfrak{m}$, $y \in \mathfrak{m}$).

(b) Let $\mathscr{F}'$ be the family of ideals $\mathfrak{l}'$ of $A_{\mathscr{F}}$ such that $A_{\mathscr{F}}/\mathfrak{l}'$ is $\mathscr{F}$-negligible (exercise 19 (h)). Show that the mapping $\mathfrak{p} \mapsto \mathfrak{p}_{\mathscr{F}}$ is a bijection of the set of ideals $\mathfrak{p}$ of A not belonging to $\mathscr{F}$ onto the set of prime ideals of $A_{\mathscr{F}}$ not belonging to $\mathscr{F}'$. (Show first that, if A is an integral domain, so is $A_{\mathscr{F}}$, making the same note as in (a); then use the exact sequence of Exercise 19 (d) to show that $\mathfrak{p}_{\mathscr{F}}$ is prime. If $\mathfrak{p}'$ is a prime ideal of $A_{\mathscr{F}}$ not belonging to $\mathscr{F}'$ and $\mathfrak{p} = \overset{-1}{j_A}(\mathfrak{p}')$, recall that $\mathfrak{p}' = \mathfrak{p}_{\mathscr{F}}$.)

(c) Let B be an A-algebra and $\mathscr{G}$ the set of left ideals $\mathfrak{l}$ of B such that $B_s/\mathfrak{l}$ is $\mathscr{F}$-negligible. Show that $\mathscr{G}$ is the set of left ideals of B containing an ideal of the form $B.\mathfrak{m}$, where $\mathfrak{m} \in \mathscr{F}$; deduce that $\mathscr{G}$ is topologizing and idempotent and that, for every B-module N, $\mathscr{F}N = \mathscr{G}N$ and $N_{\mathscr{F}}$ is canonically identified with $N_{\mathscr{G}}$.

(d) Let S be a multiplicative subset of A and $\mathscr{F}$ the set of ideals of A meeting S. Show that $\mathscr{F}$ is topologizing and idempotent and that, for every A-module M, $M_{(\mathscr{F})}$ and $M_{\mathscr{F}}$ are canonically identified with $S^{-1}M$ (note that every ideal of $\mathscr{F}$ contains a principal ideal As, where $s \in S$).

¶ 22. Let A be a ring (not necessarily commutative) and S a multiplicative subset of A.

(a) Let $\mathscr{F}$ be the set of left ideals $\mathfrak{l}$ of A such that, for all $a \in A$, there exists $s \in S$ such that $sa \in \mathfrak{l}$; this implies in particular that $\mathfrak{l} \cap S \neq \varnothing$. Show that $\mathscr{F}$ is topologizing and idempotent.

(b) Let B be a ring and $\phi: A \to B$ a ring homomorphism. The ordered pair (B, ϕ) is called a *ring of left fractions* of A with denominators in S if it satisfies the following conditions:

(I) If $\phi(a) = 0$, there exists $s \in S$ such that $sa = 0$.

(II) If $s \in S$, $\phi(s)$ is invertible in B.

(III) Every element of B is of the form $(\phi(s))^{-1}\phi(a)$, where $a \in A$.

Show that the following properties are equivalent:

(α) The ring A possesses a ring of left fractions with denominators in S.

(β) The following conditions hold:

(β_1) For all $s \in S$, $a \in A$, there exists $t \in S$ and $b \in A$ such that $ta = bs$.

(β_2) If $a \in A$ and $s \in S$ satisfy $as = 0$, there exists $t \in S$ such that $ta = 0$.

(γ) The canonical images of the elements of S in $A_{\mathscr{F}}$ are invertible.

Moreover these properties imply the following:

(δ) The principal ideals As, where $s \in S$, belong to $\mathscr{F}$ and every ideal $\mathfrak{m} \in \mathscr{F}$ contains one of these principal ideals.

(ε) The left annihilator of every $s \in S$ is contained in $\mathscr{F}A_s$.

(ζ) For every exact sequence $0 \to M \xrightarrow{h} N \xrightarrow{p} P \to 0$ of A-modules, the sequence $0 \longrightarrow M_{\mathscr{F}} \xrightarrow{h_{\mathscr{F}}} N_{\mathscr{F}} \xrightarrow{p_{\mathscr{F}}} P_{\mathscr{F}} \longrightarrow 0$ is exact.

(Show that (α) $\Rightarrow$ (β) $\Rightarrow$ (γ) $\Rightarrow$ (α). To see that (γ) implies (α), note that, for every A-module M and every $x \in \mathscr{F}M$, there exists $s \in S$ such that $sx = 0$. To see that (β) implies (γ), prove first that (β) implies (δ), (ε) and (ζ). To show that (β) implies (ζ), establish, using (δ), that, if $\mathfrak{m} \in \mathscr{F}$ and

$$u \in \operatorname{Hom}_A(\mathfrak{m}, P/\mathscr{F}P),$$

then there exists $s \in S$ such that $As \subset \mathfrak{m}$ and $v \in \operatorname{Hom}_A(As, N/\mathscr{F}N)$ such that the diagram

$$\begin{array}{ccc} As & \xrightarrow{v} & N/\mathscr{F}N \\ {\scriptstyle i}\downarrow & & \downarrow{\scriptstyle p'} \\ \mathfrak{m} & \xrightarrow[u]{} & P/\mathscr{F}P \end{array}$$

is commutative, i being the canonical injection and p' the mapping derived from p by taking quotients. Consider finally, for all $s \in S$, the exact sequence

$$0 \longrightarrow N \longrightarrow A \xrightarrow{\mu_s} A \longrightarrow A/As \longrightarrow 0$$

where μ_s is the homothety $\lambda \mapsto \lambda s$; use (ε), (ζ) and Exercise 4 of *Algebra*, Chapter I, § 8.)

(c) Deduce from (b) that, if A possesses a ring of left fractions (B, ϕ) with denominators in S, this ring has the following universal property: for every ring homomorphism $f: A \to C$ such that $f(s)$ is invertible in C for all $s \in S$, there exists a unique homomorphism $g: B \to C$ such that $f = g \circ \phi$. In particular B is canonically isomorphic to $A_{\mathscr{F}}$.

(d) The notion of a *ring of right fractions* of A with denominators in S is similarly defined. Deduce from (c) that, if there exist both a ring of left fractions and a ring of right fractions of A, these two rings are isomorphic.

(e) Let E be a vector space over a field K, $(e_\iota)_{\iota \in I}$ a basis of E and T the tensor algebra of E. Let J be a non-empty subset of I distinct from I and S the multiplicative subset of T generated by the e_ι such that $\iota \in J$. Show that T admits no ring of left or right fractions with denominators in S.

(f) With the notation of (e) suppose that $I = \mathbf{N}$ and consider the quotient ring A of T by the two-sided ideal generated by the elements e_0^2 and $e_0 e_i - e_{i+1} e_0$, for all $i > 0$, and $e_i e_j - e_j e_i$, for $i > 0, j > 0$. Let S′ be the multiplicative set generated by the canonical images of the e_i in A, for $i > 0$; show that A admits a ring of left fractions, but not a ring of right fractions, with denominators in S′.

(g) Suppose that A admits a ring of left fractions with denominators in S. In the set $S \times A$, let R be the relation between (s, x) and (t, y): "there exists $r \in S$, $u \in A$, $v \in A$ such that $r = us = vt$ and $ux = vy$". Show that R is an equivalence relation and define on the set $B = (S \times A)/R$ a ring structure such that the ordered pair consisting of S and the mapping $\phi: A \to B$, which maps $x \in A$ to the class of $(1, x)$, is a ring of left fractions of A with denominators in S (use the fact that, for $s \in S$ and $t \in S$, there exists $u \in A$ and $v \in A$ such that $r = us = tv$ belongs to S). Consider the case $S = A - \{0\}$ (cf. *Algebra*, Chapter I, § 9, Exercise 8).

23. (a) Let A be a left Noetherian ring with no divisor of 0. Show that A possesses a field of left fractions (use Exercise 14 (a) to show that condition (β_1) of Exercise 22 (b) is fulfilled for $S = A - \{0\}$).

(b) Let A be a left and right Noetherian ring with no divisor of 0. Show that every finitely generated left A-module M, each of whose elements $\neq 0$ is free, is a submodule of a free A-module. (Embed M canonically in $K \otimes_A M$, where K is the field of fractions (left or right) of A considered as a right A-module; if (x_ι) is a basis of the left vector K-space $K \otimes_A M$, show that there exists $a \neq 0$ in A such that M is contained in the sub-A-module of $K \otimes_A M$ generated by the $a^{-1}x_\iota$.)

¶ 24. Let A be a ring, commutative or not, and $\mathscr{F}$ the set of left ideals $\mathfrak{l}$ of A such that $\mathfrak{l} \cap \mathfrak{m} \neq 0$ for *every* left ideal $\mathfrak{m} \neq 0$ of A.

In order that $\mathfrak{l} \in \mathscr{F}$, it is necessary and sufficient that, for every element $a \neq 0$ of A, there exist $b \in A$ such that $ba \neq 0$ and $ba \in \mathfrak{l}$. Every ideal $\mathfrak{l}$ of $\mathscr{F}$ contains the *left socle* of A (*Algebra*, Chapter VIII, § 5, Exercise 9) and, if A is Artinian, $\mathscr{F}$ consists of the ideals containing the left socle of A.

(a) Show that $\mathscr{F}$ is a topologizing set (Exercise 16). Deduce that the set $\mathfrak{a}$ of $a \in A$ such that $\mathfrak{l}a = 0$ for some $\mathfrak{l} \in \mathscr{F}$ (union of the right annihilators of the $\mathfrak{l} \in \mathscr{F}$) is a *two-sided* ideal of A, which contains no idempotent. If A is left Noetherian, show that $\mathfrak{a}$ is nilpotent (prove first that every element $a \in \mathfrak{a}$ is nilpotent, by considering the left annihilators of the a^n; then use Exercise 26 (b) of *Algebra*, Chapter VIII, § 6).

Let K be a commutative field and B the ring of polynomials with coefficients in K in indeterminates Y and X_i $(i \geqslant 1)$. Let $\mathfrak{b}$ be the ideal of B generated by the X_iX_j $(i \neq j)$ and the $Y^{i-1}X_i$ and let A be the quotient ring $B/\mathfrak{b}$; show that in A the ideal $\mathfrak{a}$ defined above contains the class of Y, which is not nilpotent.

(b) Show that, for every left ideal $\mathfrak{m}$ of A, there exists a left ideal $\mathfrak{n}$ of A such that $\mathfrak{m} \cap \mathfrak{n} = 0$ and $\mathfrak{m} + \mathfrak{n} \in \mathscr{F}$. (Take $\mathfrak{n}$ to be an ideal which is maximal amongst the left ideals $\mathfrak{l}$ such that $\mathfrak{m} \cap \mathfrak{l} = 0$.)

(c) A ring A is called (left) *neat* if the two-sided ideal $\mathfrak{a}$ defined in (a) is reduced to 0. If this is so, the set $\mathscr{F}$ is idempotent (let $\mathfrak{l}$, $\mathfrak{m}$, $\mathfrak{n}$ be left ideals of A such that $\mathfrak{n} \in \mathscr{F}$, $\mathfrak{l} \subset \mathfrak{n}$, $\mathfrak{n}/\mathfrak{l}$ being $\mathscr{F}$-negligible, and $\mathfrak{m} \neq 0$. If $x \neq 0$ belongs to $\mathfrak{m} \cap \mathfrak{n}$, there exists $\mathfrak{r} \in \mathscr{F}$ such that $\mathfrak{r}.x \subset \mathfrak{l} \cap \mathfrak{m}$).

¶ 25. With the notation of Exercise 24, suppose that A is a left neat ring.

(a) Show that the canonical mapping $j_A : A \to A_{\mathscr{F}}$ is injective, which allows us to identify A with a subgroup of $A_{\mathscr{F}}$. Let $\mathscr{F}'$ be the set of left ideals $\mathfrak{l}'$ of $A_{\mathscr{F}}$ such that $\mathfrak{l}'_{\mathscr{F}} = A_{\mathscr{F}}$; show that $\mathscr{F}'$ is the set of left ideals $\mathfrak{l}'$ of $A_{\mathscr{F}}$ such that $\mathfrak{l}' \cap \mathfrak{m}' \neq 0$ for every left ideal $\mathfrak{m}' \neq 0$ of $A_{\mathscr{F}}$. (Show first that, for every left ideal $\mathfrak{m}' \neq 0$ of $A_{\mathscr{F}}$, $A \cap \mathfrak{m}' \neq 0$; note on the other hand that the ideals $\mathfrak{l}' \in \mathscr{F}'$ are precisely the left ideals of $A_{\mathscr{F}}$ containing an ideal $\mathfrak{m} \in \mathscr{F}$.) Show that the ring $A_{\mathscr{F}}$ is left neat.

(b) Show that $A_{\mathscr{F}}$ is an *injective* A-module, that is (*Algebra*, Chapter II, § 2, Exercise 11) that, for every left ideal $\mathfrak{l}$ of A and all

$$u \in \operatorname{Hom}_A(\mathfrak{l}, (A_{\mathscr{F}})_s),$$

there exists $a' \in A_{\mathscr{F}}$ such that $u(x) = xa'$ for all $x \in \mathfrak{l}$. (Use Exercise 24 (b) and Exercise 19 (c).) Deduce that every endomorphism of the A-module $A_{\mathscr{F}}$ is of form $u: x \mapsto xa'$ where $a' \in A_{\mathscr{F}}$ (note that $\mathrm{Hom}_A(A, A_{\mathscr{F}}) = \mathrm{Hom}_{A_{\mathscr{F}}}(A_{\mathscr{F}}, A_{\mathscr{F}})$ by Exercise 19 (e)); also $\mathrm{Ker}(u) = (\mathrm{Ker}(u))_{\mathscr{F}}$.

(c) Show that, for every left ideal $\mathfrak{l}$ of A, $\mathfrak{l}_{\mathscr{F}}$ is a direct factor of $A_{\mathscr{F}}$. (Let $\mathfrak{m}$ be a left ideal of A such that $\mathfrak{m} \cap \mathfrak{l} = 0$ and $\mathfrak{m} + \mathfrak{l} \in \mathscr{F}$; extend the projection of $\mathfrak{m} + \mathfrak{l}$ onto $\mathfrak{m}$, which is zero on $\mathfrak{l}$, to an endomorphism p of $A_{\mathscr{F}}$; show that $p^2 = p$ and that $\mathrm{Ker}(p) = \mathfrak{l}_{\mathscr{F}}$.)

(d) Show that $A_{\mathscr{F}}$ is an absolutely flat ring (Chapter I, § 2, Exercise 17). (Reduce it to the case where $A = A_{\mathscr{F}}$. If u is the endomorphism $x \mapsto xa$ of the left A-module A_s, then $\mathrm{Ker}(u) = (\mathrm{Ker}(u))_{\mathscr{F}}$, and $\mathrm{Ker}(u)$ therefore admits a complementary ideal $\mathfrak{m}$ in A_s; note that $\mathrm{Im}(u)$ is isomorphic to $\mathfrak{m}$, hence *injective* (*Algebra*, Chapter II, § 2, Exercise 11) and therefore a direct factor of A_s (*loc. cit.*). Finally, use Exercise 13 of *Algebra*, Chapter II, § 1.)

¶ 26. (a) Show that every Zorn ring A (*Algebra*, Chapter VIII, § 6, Exercise 13) with no nil ideal $\neq 0$ is left neat; in particular, every absolutely flat ring and every primitive ring whose socle is not reduced to 0 is a left neat ring.

(b) Let A be a left primitive ring whose socle is not reduced to 0; show that, if A is represented as a dense subring of a ring of endomorphisms $\mathrm{End}_D(V)$ of a vector space (*Algebra*, Chapter VIII, § 5, Exercise 10), the ring $A_{\mathscr{F}}$ corresponding to A (Exercise 25) is isomorphic to $\mathrm{End}_D(V)$.

(c) Show that every quasi-simple ring A (*Algebra*, Chapter VIII, § 5, Exercise 5) is left and right neat and that the corresponding ring $A_{\mathscr{F}}$ is quasi-simple. Deduce that, for every ring B not reduced to 0, there exists a homomorphism $\phi: B \to R$ to a ring R which is quasi-simple, absolutely flat and such that R_s is an injective R-module.

(d) Let A be a left Noetherian ring with no nilpotent ideals. Show that A is left neat and that the corresponding ring $A_{\mathscr{F}}$ is a *semi-simple* ring ("*Goldie's Theorem*"; use Exercise 24 (a) and, on the other hand, use *Algebra*, Chapter VIII, § 2, Exercise 7, noting that in the absolutely flat ring $A_{\mathscr{F}}$ there cannot exist any family of idempotents which are pairwise orthogonal, unless A contained an infinite direct sum of left ideals $\neq 0$). Show that, if $s \in A$ is not a right divisor of 0 in A, it is invertible in $A_{\mathscr{F}}$ and conversely; the set S of these elements is a multiplicative subset of A. Show that $\mathscr{F}$ is the set of left ideals of A which meet S (to see that the ideals of A meeting S belong to $\mathscr{F}$, note that the elements of S are not left divisors of 0 in A; to see that every $\mathfrak{l} \in \mathscr{F}$ contains an element of S, represent each simple component of $A_{\mathscr{F}}$ as an endomorphism ring of an n-dimensional vector space E and define by induction n elements $u_1, \ldots, u_n$ such that $u_i u_j = 0$ for $j < i$ and $u_i^2 \neq 0$, the u_i being of rank 1; finally show that the kernel of $s = u_1 + \cdots + u_n$ is zero). Deduce that $(A_{\mathscr{F}}, j_A)$ is a ring of left fractions of A with denominators in S (Exercise 22).

27. (a) Let A be a ring and M a finitely generated faithful A-module. Show that, if $\mathfrak{a}$, $\mathfrak{b}$ are two ideals of A such that $\mathfrak{a}M = \mathfrak{b}M$, the radicals of $\mathfrak{a}$ and $\mathfrak{b}$ are equal (use no. 2, Corollary 2 to Proposition 4).

(b) Deduce from (a) that, if in an integral domain A $\mathfrak{p}$ is a finitely generated prime ideal, then $\mathfrak{p}^m \neq \mathfrak{p}^n$ for $m \neq n$.

(c) Let K be a field and A the polynomial ring K[X, Y] in two indeterminates. Let $\mathfrak{a}$ be the ideal $AX^4 + AX^3Y + AXY^3 + AY^4$, $\mathfrak{b}$ the ideal $\mathfrak{a} + AX^2Y^2$ of A; show that $\mathfrak{b} \neq \mathfrak{a}$ and $\mathfrak{b}^2 = \mathfrak{a}^2$.

§ 3

1. (a) Let A be a ring which is not necessarily commutative and I_s (resp. I_d) the set of elements of A which have no left (resp. right) inverse. Show that the following conditions are equivalent: (1) the sum of two elements of I_s belongs to I_s; (2) I_s is a left ideal; (3) there is a greatest left ideal $\mathfrak{J}$ in the set of left ideals $\neq$A. If these conditions hold, the opposite ring A^0 also satisfies these conditions, $I_s = I_d = \mathfrak{J}$, $\mathfrak{J}$ is the unique maximal (left or right) ideal of A and hence the Jacobson radical of A, every element $x \notin \mathfrak{J}$ is invertible and $A/\mathfrak{J}$ is a field (cf *Algebra*, Chapter VIII, § 6, no. 3). Then we also say that A is a *local ring*.

(b) Show that in a local ring there exists no idempotent other than 0 and 1.

2. (a) Let A be a (commutative) local ring whose maximal ideal $\mathfrak{m}$ is principal and such that $\bigcap_{n \geqslant 1} \mathfrak{m}^n = 0$ (cf. Chapter III, § 3, no. 2, Corollary to Proposition 4). Show that the only ideals $\neq$0 and $\neq$A of A are the powers $\mathfrak{m}^n$; deduce that A is Noetherian and is *either a discrete valuation ring (Chapter VI)$_*$ or a *quasi-principal* ring in which 1 is an indecomposable idempotent (*Algebra*, Chapter VII, § 1, Exercise 6).

(b) Let A be the ring of germs at the point $t = 0$ of real-valued functions defined and continuous in a neighbourhood of 0 and differentiable at the point 0 (*General Topology*, Chapter I, 3rd ed., § 6, no. 10). Show that A is a local ring whose maximal ideal $\mathfrak{m}$ is generated by the germ of the function $j: t \mapsto t$; $\mathfrak{m}^n \neq 0$ for all n but A is not a principal ideal domain. If c is the germ of the function $t \mapsto \exp(-1/t^2)$ and $\mathfrak{p}$ a prime ideal of A not containing c, show that the quotient ring $B = A/\mathfrak{p}$ is a local integral domain which is not a principal ideal domain and in which the maximal ideal is principal.

¶ 3. Let A be a local ring (not necessarily commutative, cf. Exercise 1) and $\mathfrak{m}$ its Jacobson radical.

(a) Let L be a free left A-module and $x \neq 0$ an element of L. Let $(e_\iota)_{\iota \in I}$ be a basis of L for which the number of components $\neq$0 of x is the least possible; if $x = \sum_{\iota \in I} \xi_\iota e_\iota$, let J be the set of $\iota \in I$ such that $\xi_\iota \neq 0$. Show that none of the ξ_ι ($\iota \in J$) belongs to the right ideal of A generated by the others.

(b) With the hypotheses and notation of (a), let P, Q be two complementary submodules of L such that $x \in P$; for all $\iota \in I$, let $e_\iota = y_\iota + z_\iota$, where $y_\iota \in P$, $z_\iota \in Q$. Show that the family of y_ι for $\iota \in J$ and e_ι for $\iota \notin J$ is a basis of L. (Prove, using (a), that, if $z_\kappa = \sum_{\iota \in I} \zeta_{\kappa\iota} e_\iota$, the components $\zeta_{\kappa\iota}$ such that $\kappa \in J$ and $\iota \in J$ necessarily belong to $\mathfrak{m}$ and use Corollary 2 to Proposition 4.) Deduce that there exists a free submodule of P which contains x and is a direct factor of P.

(c) Show that every projective A-module P is free (using Kaplansky's Theorem (*Algebra*, Chapter II, § 2, Exercise 4), reduce it to the case where P is generated by a countable family of elements and apply (b) to this case).

(d) Give an example of a local ring A and a flat A-module M which is not free and satisfies $\mathfrak{m}M = M$ (take $A = \mathbf{Z}_{(p)}$). *(In Chapter III, § 3, we shall give examples of faithfully flat A-modules which are not free over a Noetherian local ring A.)*

(e) Let A be a local ring and M a finitely generated flat A-module. Show that M is a free A-module (use Exercise 23 (e) of Chapter I, § 2).

4. Let A be a local ring (not necessarily commutative) and $\mathfrak{m}$ its Jacobson radical. Let M and N be two free A-modules (finitely generated or not) and $u: M \to N$ a homomorphism such that $1 \otimes u: M/\mathfrak{m}M \to N/\mathfrak{m}N$ is bijective. Show that u is bijective. (Prove first that u is surjective and then that u is injective, reducing it to the case where M and N are finitely generated and using respectively Corollary 1 to Proposition 4 and Proposition 6.)

5. Show by an example that Proposition 7 does not extend to the case where the local ring A is not reduced.

6. (a) Let M be the **Z**-module defined in *Algebra*, Chapter VII, § 3, Exercise 22 (h); let T be the torsion submodule of M. Show that, for every prime number p, the submodule $T_{(p)}$ of $M_{(p)}$ is a direct factor of $M_{(p)}$, although T is not a direct factor of M.

(b) Let M be the **Z**-module defined in *Algebra*, Chapter VII, § 3, Exercise 5; show that, for every prime number p, $M_{(p)}$ is a free $\mathbf{Z}_{(p)}$-module, although M is not a free **Z**-module.

7. Let A be a ring and $(S_\lambda)_{\lambda \in L}$ a family of multiplicative subsets of A such that, for every maximal ideal $\mathfrak{m}$ of A, there exists $\lambda \in L$ such that $\mathfrak{m} \cap S_\lambda = \varnothing$ (cf. § 2, Exercise 8). Let M, N be two A-modules and $u: M \to N$ a homomorphism. For u to be surjective (resp. injective, bijective, zero), it is necessary and sufficient that, for all $\lambda \in L$, $S_\lambda^{-1}u: S_\lambda^{-1}M \to S_\lambda^{-1}N$ be so. How can the Corollaries to Theorem 1 of no. 3 be generalized?

8. Let A be a ring, B an A-algebra (not necessarily commutative) and E a finitely generated left B-module.

(a) Let S be a multiplicative subset of A and let $S^{-1}B = S^{-1}A \otimes_A B$ be given its $S^{-1}A$-algebra structure; $S^{-1}E$ can then be considered as a left $S^{-1}B$-module (*Algebra*, Chapter VIII, § 7, no. 1) isomorphic to $S^{-1}B \otimes_B E$. Show that, if E is a flat B-module, $S^{-1}E$ is a flat $S^{-1}B$-module.

(b) Let $(S_\lambda)_{\lambda \in L}$ be a family of multiplicative subsets of A such that, for every maximal ideal $\mathfrak{m}$ of A, there exists $\lambda \in L$ such that $\mathfrak{m} \cap S_\lambda = \varnothing$. Show that $\bigoplus_{\lambda \in L} S_\lambda^{-1}B$ is a faithfully flat B-module (left or right; cf. § 2, Exercise 8). Show that, if, for all $\lambda \in L$, $S_\lambda^{-1}E$ is a flat (resp. faithfully flat) $S_\lambda^{-1}B$-module, then E is a flat (resp. faithfully flat) B-module (use Chapter I, § 3, Exercise 9, with $C_\lambda = S_\lambda^{-1}A$).

(c) Suppose that the S_λ satisfy condition (b) and that L is finite. Show that, if, for every $\lambda \in L$, $S_\lambda^{-1}E$ is a finitely generated projective $S_\lambda^{-1}B$-module, then E is a finitely generated projective B-module.

9. Show that, for a (commutative) ring A to be absolutely flat (Chapter I, § 2, Exercise 17), it is necessary and sufficient that, for every maximal ideal $\mathfrak{m}$ of A, $A_\mathfrak{m}$ be a field (note that an absolutely flat local ring is necessarily a field and use the Corollary to Proposition 15).

10. Let A, B be two rings, $\rho: A \to B$ a homomorphism, $\mathfrak{q}$ a prime ideal of B and $\mathfrak{p} = \overset{-1}{\rho}(\mathfrak{q})$. Suppose that there exists a B-module N such that $N_\mathfrak{q}$ is a flat A-module which is not reduced to 0 and $N_\mathfrak{q}$ is a finitely generated $A_\mathfrak{p}$-module; then the prime ideal $\mathfrak{p}$ is minimal in A. (Observe that $N_\mathfrak{p}$ is then a faithfully flat $A_\mathfrak{p}$-module and use Corollary 4 to Proposition 11 of § 2, no. 5.)

11. Let A be a ring, $\mathfrak{a}$ an ideal of A and S the multiplicative subset of A consisting of the elements whose canonical images in $A/\mathfrak{a}$ are not divisors of 0; let Φ be the set of maximal elements of the set of ideals of A not meeting S (such that the ideals $\mathfrak{p} \in \Phi$ are prime, cf. § 2, no. 5, Proposition 11). Show that $\mathfrak{a}$ is the intersection of its saturations with respect to the ideals $\mathfrak{p} \in \Phi$ (reduce it to the case where $\mathfrak{a} = 0$ and use Corollary 1 to Theorem 1).

¶ 12. Let $A = \prod_{i=1}^{n} A_i$ be the product of a finite family of local rings A_i, canonically identified with ideals of A. Let B be a subring of A such that $\mathrm{pr}_i B = A_i$ for $1 \leqslant i \leqslant n$. Show that B is the direct composition of at most n local rings and can only be the direct composition of n local rings if B = A. (Proceed by induction on n, considering the ideals $\mathfrak{a}_i = B \cap A_i$ of B. Examine successively the case where $\mathfrak{a}_i = A_i$ for at least one index i and the case where $\mathfrak{a}_i \neq A_i$ for all i. Note that, if $\mathfrak{a}_i \neq A_i$, every maximal ideal of B contains $\mathfrak{a}_i$, and considering the ring $B/\mathfrak{a}_i$, conclude that there are at most $n - 1$ distinct maximal ideals in B; if then B is not a local ring, reduce it to the case where $\mathfrak{a}_j = A_j$ for some $j \neq i$ using the induction hypothesis and Exercise 1 (b).) Give an example where $n > 1$ and B is a local ring but not isomorphic to any A_i (take the A_i to be algebras over the same field K, whose maximal ideals $\mathfrak{m}_i$ are of zero square).

¶ 13. Let G be a finite group whose order n is a power p^f of a prime number p and K a field of characteristic p.

(a) Let E be a finite set on which G operates (*Algebra*, Chapter I, § 7, no. 2); if E′ is the set of elements of E which are invariant for all $g \in G$, show that $\mathrm{Card}(E) \equiv \mathrm{Card}(E') \pmod{p}$. Deduce that, if G is not reduced to its identity element e, the centre Z of G is not reduced to e (make G operate on itself by inner automorphisms). Conclude that G is solvable (*Algebra*, Chapter I, § 6, Exercise 14).

(b) Let V be a vector space over K and ρ a homomorphism of G to $\mathbf{GL}(V)$. Let V′ be the set of $x \in V$ which are invariant under the linear mappings $\rho(g)$, where $g \in G$. Show that, if V′ is reduced to 0, V is necessarily reduced to 0. (If $x \in V$ and $x \neq 0$, apply (a) to the additive subgroup E of V generated by the elements $\rho(g).x$, where $g \in G$.)

(c) Let $A = K^{(G)}$ be the algebra of the group G over K (*Algebra*, Chapter III, § 1) and let I be the vector subspace of A (over K) generated by the elements $1 - g$ where $g \in G$. Show that I is the Jacobson radical of A and that A/I is isomorphic to K (use (b) and the definition of Jacobson radical); deduce that I is nilpotent.

(d) Let V be a finitely generated A-module and V′ the set of elements x of V such that $g.x = x$ for all $g \in G$. Show that the following inequalities hold

$$(*)\ \dim_K(V) \leqslant n.\dim_K(V/IV); \qquad (**)\ \dim_K(V) \leqslant n.\dim_K(V').$$

(To establish (*), use Corollary 2 to Proposition 4; deduce (**) by considering the dual of the vector space V over K.) For the two sides of the inequality (*) (resp. (**)) to be equal, it is necessary and sufficient that V be a free A-module (cf. Corollary 1 to Proposition 5).

14. Let A be a ring, which is not necessarily commutative and whose Jacobson radical $\mathfrak{m}$ is such that $A/\mathfrak{m}$ is a field, and M a finitely generated free A-module.

(a) Let M′ be another finitely generated free A-module and $u: M \to M'$ a homomorphism such that $u(M)$ is a direct factor of M′; the number dim(M) is called the *rank* of u and denoted by $\mathrm{rg}(u)$. For such a homomorphism $\mathrm{Ker}(u)$ is a direct factor of M and $\mathrm{rg}(u) = \dim(M) - \dim(\mathrm{Ker}(u))$; for u to be injective (resp. surjective), it is necessary and sufficient that $\mathrm{rg}(u) = \dim(M)$ (resp. $\mathrm{rg}(u) = \dim(M')$). The transpose ${}^t u$ is such that ${}^t u(M'^*)$ is a direct factor of M^* and $\mathrm{rg}({}^t u) = \mathrm{rg}(u)$.

(b) Let $n = \dim(M)$. The direct factors of M of dimension 1 (resp. $n - 1$) are called *lines* (resp. *hyperplanes*). An automorphism $u \in \mathbf{GL}(M)$ distinct from the identity is called a *transvection* if there exists a hyperplane H of M all of whose elements are invariant under u; then $u(x) = x + a\phi(x)$, where ϕ is a linear form on M such that $H = \operatorname{Ker}(\phi)$ and $a \in H$; obtain the converse. If A is commutative, show that every automorphism $u \in \mathbf{GL}(M)$ of determinant 1 is a product of transvections (observe that, in the matrix of u with respect to any basis of M, every column contains at least one invertible element of A and a matrix of the form $I + E_{ij}$ $(i \neq j)$ is the matrix of a transvection).

(c) Give an example of an automorphism $u \in \mathbf{GL}(M)$ such that the kernel of $1 - u$ is not a direct factor of M (take A to be the local ring K[[X]] of formal power series in one indeterminate over a field K and $M = A$).

(d) Give an example of direct factors N, P of M (necessarily free) such that $N + P$ and $N \cap P$ are not direct factors of M. (Take $A = K[X]/(X^2)$, where K is a field and $M = A^2$.)

¶ 15. Let A be a (commutative) ring and M an A-module.

(a) For a submodule M′ of M to be pure (Chapter I, § 2, Exercise 24), it is necessary and sufficient that, for every maximal ideal $\mathfrak{m}$ of A, $M'_{\mathfrak{m}}$ be a pure sub-$A_{\mathfrak{m}}$-module of $M_{\mathfrak{m}}$ (use Theorem 1 of no. 3).

(b) Suppose that A is a local ring with maximal ideal $\mathfrak{m}$ and M a finitely generated free A-module. For a finitely generated submodule M′ of M to be pure, it is necessary and sufficient that it be a direct factor of M. (Using Corollary 1 to Proposition 5 of no. 2, reduce it to proving that, if $M' \subset \mathfrak{m}M$, M′ can only be a pure submodule of M if $M' = 0$.)

16. Let $(A_\lambda, f_{\mu\lambda})$ be a direct system of local rings, such that the $f_{\mu\lambda}$ are local homomorphisms; let $\mathfrak{m}_\lambda$ be the maximal ideal of A_λ and $K_\lambda = A_\lambda/\mathfrak{m}_\lambda$. Then $A = \varinjlim A_\lambda$ is a local ring whose maximal ideal is $\mathfrak{m} = \varinjlim \mathfrak{m}_\lambda$ and residue field is $K = \varinjlim K_\lambda$. Moreover, if $\mathfrak{m}_\mu = A_\mu \mathfrak{m}_\lambda$ for $\lambda < \mu$, then $\mathfrak{m} = A\mathfrak{m}_\lambda$ for all λ.

§ 4

1. Let $(X_\alpha, f_{\alpha\beta})$ be an inverse system of topological spaces, whose indexing set is directed, $X = \varprojlim X_\alpha$ and f_α be the canonical mapping from X to X_α. Suppose that, for all α, f_α is surjective; show that, if the X_α are irreducible, X is irreducible (cf. *General Topology*, Chapter I, § 4, no. 4, Corollary to Proposition 9). In particular, every product of irreducible spaces is irreducible.

2. A point x in an irreducible space X is called *generic* if $\{x\}$ is everywhere dense in X.

(a) If X is a Kolmogoroff space (*General Topology*, Chapter I, § 1, Exercise 2), it admits at least one generic point; if X is an accessible space (*General*

Topology, Chapter I, § 8, Exercise 1), it cannot admit a generic point unless it consists of a single point.

(b) Give an example of an infinite irreducible accessible space (cf. *General Topology*, Chapter I, § 8, Exercise 5).

(c) Let X, Y be two irreducible spaces each admitting a generic point; suppose also that Y admits only one generic point y. Let $f: X \to Y$ be a continuous mapping; for $f(X)$ to be dense in Y, it is necessary and sufficient that, for every generic point x of X, $f(x) = y$.

(d) Let $(X_\alpha, f_{\alpha\beta})$ be an inverse system of irreducible spaces whose indexing set is directed; suppose that each of the X_α admits a single generic point x_α. Show that, if, for $\alpha \leqslant \beta$, $f_{\alpha\beta}(X_\beta)$ is dense in X_α, then $X = \varprojlim X_\alpha$ is irreducible and admits a single generic point (same method as in Exercise 1).

3. Let X be a topological space and (Y_α) a right directed family of subspaces of X; show that if each of the subspaces Y_α is irreducible and X is the union of the family (Y_α), X is irreducible. Deduce an example where X is a Kolmogoroff space, each of the Y_α admits a generic point, but X does not admit a generic point.

4. Let Y be an irreducible space admitting a single generic point y, X a topological space and $f: X \to Y$ a continuous mapping.

(a) For every irreducible component Z of X meeting $\overset{-1}{f}(y)$, $f(Z)$ is dense in Y.

(b) Give an example where $f(X)$ is dense in Y but $\overset{-1}{f}(y)$ is empty (take X to be a subspace of Y).

(c) Show that, if Z admits a generic point z and $f(Z)$ is dense in Y, then $f(z) = y$ and z is a generic point of $Z \cap \overset{-1}{f}(y)$.

5. Let G be a connected semi-topological group (*General Topology*, Chapter III, § 1, Exercise 2). Show that, if G only admits a finite number of irreducible components, it is necessarily irreducible (observe that the irreducible components are derived from the irreducible component containing the identity element e by left or right translation).

6. Let X be a topological space.

(a) Let x be a point of X and U an open neighbourhood of x with only a finite number of irreducible components. Show that there exists a neighbourhood V of x such that every neighbourhood contained in V is connected.

(b) Suppose that every point of X possesses an open neighbourhood with only a finite number of irreducible components. Show that the following properties are equivalent:

(α) The irreducible components of X are open.

(β) The irreducible components of X are identical with its connected components.

(γ) The connected components of X are irreducible.

(δ) Two distinct irreducible components of X do not meet.

7. Let X be a compact metrizable space and R an open equivalence relation on X such that the quotient space X/R is irreducible. Show that there exists a point $x \in X$ whose class mod. R is everywhere dense. (Note first that every open subset of X which is saturated with respect to R is everywhere dense. Then use Baire's Theorem.)

8. Show that the product of two Noetherian spaces is Noetherian. (If X and Y are Noetherian spaces and A is an open subset of X × Y, show that, for all $x \in \mathrm{pr}_1 A$, there exists an open neighbourhood V of x such that $V \times A(x) \subset A$ and deduce that A is quasi-compact.)

9. (a) Show that the prime spectrum of a commutative ring is a Kolmogoroff space (*General Topology*, Chapter I, § 1, Exercise 2), in which every irreducible component admits a (unique) generic point.

(b) Let A be an integral domain and $\xi = (0)$ the generic point of $X = \mathrm{Spec}(A)$. For the point ξ to be isolated in X, it is necessary and sufficient that the intersection of the prime ideals $\neq 0$ of A be an ideal $\neq 0$. Give an example of a local ring with this property.

10. Let A be a ring and $\mathfrak{N}$ its nilradical. For the spectrum of A to be discrete, it is necessary and sufficient that $A/\mathfrak{N}$ be the direct composition of a finite number of fields.

11. Let K be a field, A the polynomial ring K[X, Y] and B the quotient ring A/(XY); show that Spec(B) is connected but has two distinct irreducible components.

12. Let Y be a completely regular space, $A = \mathscr{C}^\infty(Y; \mathbf{R})$ and $B = \mathscr{C}(Y; \mathbf{R})$. Show that the Stone-Čech compactification βY of Y is canonically identified with an everywhere dense subspace of Spec(A) and with an everywhere dense subspace of Spec(B).

13. Let A be a ring and X = Spec(A) its spectrum.

(a) If a subset F of X is closed, it has the following two properties: (α) for all $\mathfrak{p} \in F$, $V(\mathfrak{p}) \subset F$; (β) for all $\mathfrak{p} \notin F$, there exists a closed subset of $V(\mathfrak{p})$ which contains $F \cap V(\mathfrak{p})$ and does not contain $\mathfrak{p}$.

(b) Suppose X is Noetherian. Show that every subset F of X with properties (α) and (β) of (a) is closed in X (consider the irreducible components of $\overline{F}$ and use Exercise 9 (a)).

(c) With the notation of Exercise 12, take Y to be the interval [0, 1] of **R**. Show that Y is not closed in X = Spec(A) (cf. § 2, Exercise 15) but satisfies conditions (α) and (β) of (a).

14. Let A be a ring, X = Spec(A) its spectrum and P the set of minimal prime ideals of A.

(a) Let R{$\mathfrak{p}$, $\mathfrak{p}'$} be the symmetric and reflexive relation: "there exists a prime ideal $\mathfrak{p}''$ of A containing $\mathfrak{p} + \mathfrak{p}'$" between the elements $\mathfrak{p}$, $\mathfrak{p}'$ of P; let S be the equivalence relation whose graph is the smallest of the graphs of equivalences containing the graph of R (*Set Theory*, Chapter II, § 6, Exercise 10). Show that, if I is an equivalence class with respect to S, the set $V_I = \bigcup_{\mathfrak{p} \in I} V(\mathfrak{p})$ is connected.

(b) Show that, if P is finite (and in particular if A is Noetherian), the V_I are the connected components of X. Does this result extend to the case where P is infinite (cf. Exercise 12)?

¶ 15. Let A be a ring and $\mathfrak{a}$ a finitely generated ideal of A. Show that the following properties are equivalent: (α) $\mathfrak{a}^2 = \mathfrak{a}$; (β) A is generated by an idempotent; (γ) V($\mathfrak{a}$) is both open and closed in X = Spec(A) and $\mathfrak{a}$ is minimal among the ideals $\mathfrak{b}$ such that $\mathfrak{r}(\mathfrak{b}) = \mathfrak{r}(\mathfrak{a})$ (use § 2, Corollary 3 to Proposition 4). Give an example where $\mathfrak{a}^2 = \mathfrak{a}$, $\mathfrak{a}$ is not finitely generated and does not contain an idempotent (cf. § 2, Exercise 15).

¶ 16. (a) Let A be an *absolutely flat* (commutative) ring (Chapter I, § 2, Exercise 17). Show that X = Spec(A) is a totally disconnected compact space, that every prime ideal $\mathfrak{p}$ of A is maximal and that $A_{\mathfrak{p}}$ is canonically isomorphic to the field $A/\mathfrak{p}$ (show first that, for all $f \in A$, V(f) is both open and closed in X; on the other hand use § 3, Exercise 9).

(b) Show that the mapping $\mathfrak{a} \mapsto V(\mathfrak{a})$ is a bijection of the set of ideals of A onto the set of closed subsets of X. Moreover, the following conditions are equivalent: (α) V($\mathfrak{a}$) is open; (β) $\mathfrak{a}$ is finitely generated; (γ) $\mathfrak{a}$ is generated by an idempotent; (δ) $A/\mathfrak{a}$ is a projective A-module.

* (c) Let $\tilde{A}$ be the structure sheaf of rings over X. Show that every $\tilde{A}$-Module $\mathscr{F}$ is of the form $\tilde{E}$, where E is an A-module defined up to a unique isomorphism (take $E = \Gamma(X, \mathscr{F})$ and use the fact that every point of X admits a fundamental system of neighbourhoods which are both open and closed). For every ideal $\mathfrak{a}$ of A, show that $\mathrm{Hom}_A(\mathfrak{a}, E)$ is identified with $\Gamma(X - V(\mathfrak{a}), \tilde{E})$ (note that $X - V(\mathfrak{a})$ is the union of the X_e, where e runs through the set of idempotents $e \in \mathfrak{a}$). Deduce that, for E to be an injective A-module, it is necessary and sufficient that $\tilde{E}$ be flabby. *

(d) Let A be a ring and $\mathfrak{N}$ its nilradical. The following properties are equivalent:

(α) $A/\mathfrak{N}$ is absolutely flat.

(β) X = Spec(A) is Hausdorff.

(γ) Every point of X is closed (in other words, every prime ideal of A is maximal).

(Use Exercises 9 of § 3.)

¶ 17. * (a) Let X be a totally disconnected compact space and $\mathcal{O}$ a sheaf of rings over X such that, for all $x \in X$, $\mathcal{O}_x$ is a commutative field. Show that, if $A = \Gamma(X, \mathcal{O})$, the ring A is absolutely flat and its spectrum (with the sheaf of rings $\tilde{A}$) is canonically identified with X (with the sheaf $\mathcal{O}$). (Observe that for all $f \in A$ the set of $x \in X$ such that $f_x = 0$ is both open and closed in X and deduce the existence of $g \in A$ such that $f = gf^2$.) *

(b) Let X be a totally disconnected compact space, K a (commutative) field and A the ring of locally constant mappings from X to K. Show that A is absolutely flat (argue directly or use (a)).

(c) Let A be the ring of real-valued step functions defined on $I = [0, 1[$, whose points of discontinuity are of the form $k/2^n$ ($n \in \mathbf{N}$, $0 \leqslant k < 2^n$) and which are right continuous. Let $\mathfrak{Z}$ be the set of finite unions of half-open intervals $[x, y[\subset I$ whose extremities are of the form $k/2^n$. For every $f \in A$, the set $Z(f) = \overset{-1}{f}(0)$ belongs to $\mathfrak{Z}$. If $\mathfrak{a}$ is an ideal of A, the set $\mathfrak{B}$ of $Z(f)$, where f runs through $\mathfrak{a}$, is such that: (1) the intersection of two sets of $\mathfrak{B}$ belongs to $\mathfrak{B}$; (2) every set of $\mathfrak{Z}$ containing a set of $\mathfrak{B}$ belongs to $\mathfrak{B}$. Obtain the converse. For $\mathfrak{a}$ to be prime, it is necessary and sufficient that $\mathfrak{B}$ be the base of a filter which converges to a point of $[0, 1]$. Show that A is an absolutely flat reduced ring and that, for every maximal ideal $\mathfrak{m}$ of A, $A_\mathfrak{m}$ is isomorphic to the field $\mathbf{R}$. Describe the space Spec(A).

¶ 18. (a) Let Y be a completely regular space. Show that the following conditions are equivalent:

(α) Every prime ideal of $\mathscr{C}(Y; \mathbf{R})$ is maximal.

(β) Every countable intersection of open subsets of Y is open.

(γ) Every continuous real-valued function on Y is locally constant.

(δ) The ring $\mathscr{C}(Y; \mathbf{R})$ is absolutely flat.

(Use Exercise 15 (f) of § 2.)

If these hold, the space $X = \operatorname{Spec}(\mathscr{C}(Y; \mathbf{R}))$ is canonically identified with the Stone-Čech compactification βY of Y. Such a space Y is called *flat*.

(b) Every subspace of a flat space is flat. Every completely regular quotient space of a flat space is flat. Every finite product of flat spaces is flat. Every sum of flat spaces is a flat space.

(c) In a flat space, every countable subspace is closed and discrete. In particular every countable flat space is discrete.

(d) Every Weierstrassian completely regular space (*General Topology*, Chapter IX, § 1, Exercise 22) (and *a fortiori* every compact space), which is flat, is finite.

(e) The space associated with a non-trivial ultrafilter is extremely disconnected (*General Topology*, Chapter I, § 11, Exercise 21) but is not flat.

(f) The discrete space $\mathbf{N}$ is flat but its Stone-Čech compactification is not flat and the spectra of the rings $\mathscr{C}(\mathbf{N}; \mathbf{R})$ and $\mathscr{C}^\infty(\mathbf{N}; \mathbf{R})$ are therefore not isomorphic.

19. (a) Let $\phi: A \to B$ be a ring homomorphism. Show that, for every ideal $\mathfrak{b}$ of B, $\overline{{}^a\phi(V(\mathfrak{b}))} = V(\overset{-1}{\phi}(\mathfrak{b}))$.

(b) Deduce from (a) that, for ${}^a\phi(\mathrm{Spec}(B))$ to be dense in Spec(A), it is necessary and sufficient that $\mathrm{Ker}(\phi)$ be a nil ideal of A.

(c) Show that, if every $b \in B$ may be written as $b = h\phi(a)$, where h is invertible in B and $a \in A$, ${}^a\phi$ is a homeomorphism of Spec(B) onto a subspace of Spec(A).

20. Give an example of a ring homomorphism $\phi: A \to B$ such that every maximal ideal of A is of the form $\overset{-1}{\phi}(\mathfrak{m})$, where $\mathfrak{m}$ is a maximal ideal of B, but there exist prime ideals of A which are not inverse images under ϕ of ideals of B (take B to be a field).

21. Let $(A_\alpha, \phi_{\beta\alpha})$ be a direct system of rings whose indexing set is directed, $A = \varinjlim A_\alpha$ and $\phi_\alpha: A_\alpha \to A$ the canonical homomorphism. If we write $X_\alpha = \mathrm{Spec}(A_\alpha)$, $(X_\alpha, {}^a\phi_{\beta\alpha})$ is an inverse system of topological spaces and, if $X = \mathrm{Spec}(A)$, the ${}^a\phi_\alpha: X \to X_\alpha$ form an inverse system of continuous mappings. Show that $u = \varprojlim {}^a\phi_\alpha$ is a homeomorphism of X onto $\varprojlim X_\alpha$. (Prove first that, if, for all α, $\mathfrak{p}_\alpha$ is a prime ideal of A_α such that $\mathfrak{p}_\alpha = \overset{-1}{\phi}_{\beta\alpha}(\mathfrak{p}_\beta)$ for all $\alpha \leqslant \beta$, the union $\mathfrak{p}$ of the $\phi_\alpha(\mathfrak{p}_\alpha)$ is a prime ideal of A; conversely, every prime ideal $\mathfrak{p}$ of A is the union of the $\phi_\alpha(\mathfrak{p}_\alpha)$, where $\mathfrak{p}_\alpha = \overset{-1}{\phi}_\alpha(\mathfrak{p})$. Finally, if $f_\alpha \in A_\alpha$ and $f = \phi_\alpha(f_\alpha)$, note that the relation $\mathfrak{p} \in X_f$ is equivalent to

$$u(\mathfrak{p}) \in \overset{-1}{\mathrm{pr}}_\alpha((X_\alpha)_{f_\alpha}).$$

22. (a) Let M be an A-module and (N_ι) a finite family of submodules of M. Show that $\mathrm{Supp}\big(M/\bigcap_\iota N_\iota\big) = \bigcup_\iota \mathrm{Supp}(M/N_\iota)$ (note that $M/\bigcap_\iota N_\iota$ is isomorphic to a submodule of the direct sum of the M/N_ι).

(b) Let p be a prime number; the support of the **Z**-module **Z** is distinct from the closure of the union of the supports of the **Z**-modules $\mathbf{Z}/p^k\mathbf{Z}$ ($k \in \mathbf{N}$).

(c) Let M be the direct sum of the **Z**-modules $\mathbf{Z}/p^k\mathbf{Z}$ ($k \in \mathbf{N}$); show that Supp(M) is closed but distinct from $V(\mathfrak{a})$, where $\mathfrak{a}$ is the annihilator of M.

(d) Let N be the direct sum of the **Z**-modules $\mathbf{Z}/n\mathbf{Z}$ ($n \geqslant 1$); show that Supp(N) is not closed in Spec(**Z**).

(e) Deduce from (c) and (d) an example of an A-module M such that, if $\mathfrak{a}$ is its annihilator, Supp(M) is not closed and its closure is distinct from $V(\mathfrak{a})$.

(f) Show that the support of the **Z**-module $\prod_{k=1}^{\infty} (\mathbf{Z}/p^k\mathbf{Z})$ is distinct from the closure of the union of the supports of the factor modules $\mathbf{Z}/p^k\mathbf{Z}$.

23. Let p be a prime number; give an example of **Z**-modules M, N, where N

is finitely generated, such that $M_{(p)} \neq 0$ and $N_{(p)} \neq 0$, but $(M \otimes_{\mathbf{Z}} N)_{(p)} = 0$ (take $M = \mathbf{Q}$).

24. (a) Let A be a ring and M, N two A-modules such that M is finitely generated. Show that $\mathrm{Supp}(\mathrm{Hom}_A(M, N))$ is contained in the intersection $\mathrm{Supp}(M) \cap \mathrm{Supp}(N)$.

(b) Give an example where M and N are both finitely presented and $\mathrm{Supp}(\mathrm{Hom}_A(M, N))$ is strictly contained in $\mathrm{Supp}(M) \cap \mathrm{Supp}(N)$ (take $A = \mathbf{Z}_{(p)}$ and $N = A$).

(c) Let M be the **Z**-module the direct sum of the $\mathbf{Z}/p^k\mathbf{Z}$ (p prime, $k \in \mathbf{N}$). Show that $\mathrm{Supp}(\mathrm{Hom}_{\mathbf{Z}}(M, M))$ is not contained in $\mathrm{Supp}(M)$.

25. Let A be a Noetherian ring, $X = \mathrm{Spec}(A)$ its spectrum, M a finitely generated A-module and $Y = \mathrm{Supp}(M)$. Let Y_k ($1 \leqslant k \leqslant n$) be the distinct connected components of Y.

(a) Show that there exists a unique decomposition of M as a direct sum $M = M_1 \oplus M_2 \oplus \cdots \oplus M_n$ such that $\mathrm{Supp}(M_k) = Y_k$ for all k.

(b) If $\mathfrak{a} = \mathrm{Ann}(M)$ and $\mathfrak{a}_k = \mathrm{Ann}(M_k)$, show that the $\mathfrak{a}_k$ are relatively prime in pairs and that $\prod_{k=1}^{n} \mathfrak{a}_k = \mathfrak{a}$.

(Using Proposition 17 of no. 4, reduce it to the case where $\mathfrak{a} = 0$ and $Y = X$ and apply Proposition 15 of no. 3.)

26. Let $\phi: A \to B$ be a ring homomorphism such that the mapping

$$^a\phi: \mathrm{Spec}(B) \to \mathrm{Spec}(A)$$

is surjective. Let N be a finitely generated A-module; show that, if $u: M \to N$ is an A-module homomorphism such that $u \otimes 1: M_{(B)} \to N_{(B)}$ is surjective, then u is surjective (apply Proposition 19 of no. 4 to $\mathrm{Coker}(u)$).

27. Give an example of a local homomorphism $\rho: A \to B$ and an A-module M (not finitely generated) such that $\mathrm{Supp}(M_{(B)})$ is not equal to $^a\rho^{-1}(\mathrm{Supp}(M))$ (take $A = \mathbf{Z}_{(p)}$ and $B = \mathbf{Z}/p\mathbf{Z}$ for some prime number p).

28. Give an example of a **Z**-module M such that, for some prime number p such that $(p) \in \mathrm{Supp}(M)$, there exists no **Z**-homomorphism $\neq 0$ from M to $\mathbf{Z}/p\mathbf{Z}$.

§ 5

1. Define an endomorphism u of the **Z**-module $M = \mathbf{Z}^{(\mathbf{N})}$ such that, for the prime ideal $\mathfrak{p} = (0)$ of **Z**, $u_{\mathfrak{p}}$ is an automorphism of $M_{\mathfrak{p}}$ but that there exists no $f \neq 0$ in **Z** such that u_f is a surjective endomorphism of M_f.

2. Let K be a field and B the ring of polynomials with coefficients in K in an infinite system of indeterminates X_i ($i \in \mathbf{N}$); let $\mathfrak{b}$ be the ideal of B generated

by the products X_iX_j $(i \neq j)$, A the quotient ring $B/\mathfrak{b}$, ξ_i the class of X_i mod. $\mathfrak{b}$ $(i \in N)$, $\mathfrak{p}$ the ideal of A generated by the ξ_i of index $i \geqslant 1$, which is prime, and u the canonical mapping $A \to A/\mathfrak{p}$. Show that $u_\mathfrak{p}: A_\mathfrak{p} \to (A/\mathfrak{p})_\mathfrak{p}$ is bijective but that there exists no $f \in A - \mathfrak{p}$ such that u_f is bijective.

3. Let A be a ring and P a finitely generated projective A-module. Show that A is isomorphic to a finite product of rings $\prod_{i \in I} A_i$ and P is isomorphic to a product $\prod_{i \in I} P_i$, where P_i is a projective A_i-module of rank n_i, the n_i being distinct (use Theorem 1 and the fact that Spec(A) is quasi-compact).

4. Let A be a (commutative) ring and B a (not necessarily commutative) ring containing A. Suppose that the left A-module B is projective and finitely generated. Show that A is a direct factor of B. (Reduce it to the case where A is a local ring and use § 3, no. 2, Corollary 1 to Proposition 5.)

5. Let A be a reduced ring and M a finitely generated A-module. Suppose that there exists an integer $n > 0$ such that, for every homomorphism ϕ of A to a field K_ϕ, $[M \otimes_A K_\phi : K_\phi] = n$. Show that M is a projective module of rank n. (Reduce it to the case where A is a local ring and use § 2, no. 2, Proposition 7.)

¶ 6. Let A be an *integral domain*, K its field of fractions and M an A-module. Show that the following properties are equivalent:

(α) M is a projective A-module such that $[M \otimes_A K : K]$ is finite.

(β) M is finitely generated projective A-module.

(γ) M is finitely generated and, for every maximal ideal $\mathfrak{m}$ of A, the $A_\mathfrak{m}$-module $M_\mathfrak{m}$ is free.

(To show that (α) implies (β), use *Algebra*, Chapter II, § 5, no. 5, Proposition 9. To show that (γ) implies (α), observe first that M is torsion-free and conclude that, for every prime ideal $\mathfrak{p}$ of A, $M_\mathfrak{p}$ is a free $A_\mathfrak{p}$-module of constant rank.)

7. Let A be an absolutely flat (commutative) ring (cf. § 4, Exercise 16) and let X be its spectrum.

(a) Let F be a closed subset of X and $\mathfrak{a}$ the corresponding ideal of A (*loc. cit.*). Show that $M = A/\mathfrak{a}$ is a monogenous flat A-module such that $M_\mathfrak{m}$ is a free $A_\mathfrak{m}$-module for all $\mathfrak{m} \in X$, but that M is projective only if F is open.

(b) Let E be the module the direct sum of the $A/\mathfrak{m}$, for $\mathfrak{m} \in X$. Show that E is a flat A-module such that $E_\mathfrak{m}$ is a free $A_\mathfrak{m}$-module of rank 1 for all $\mathfrak{m} \in X$; for E to be a projective A-module, it is necessary and sufficient that X be finite (note that, if $A/\mathfrak{m}$ is a projective A-module, $\mathfrak{m}$ is finitely generated).

8. Let A be a ring and M a projective A-module of rank n. Show that there exists a finitely generated (commutative) A-algebra B such that the A-module

B is faithfully flat and $M_{(B)} = M \otimes_A B$ is a free B-module of rank n (use Theorem 1 (d) and Proposition 3). Obtain the converse.

¶ 9. (a) Let A be a ring, $(f_i)_{i \in I}$ a finite family of elements of A generating the ideal A and M an A-module. Suppose given in each of the A_{f_i}-modules M_{f_i} an element z_i such that, for $i \neq j$, the canonical images of z_i and z_j in $M_{f_i f_j}$ are equal. Show that there exists a unique $z \in M$ such that, for all i, the canonical image of z in M_{f_i} is equal to z_i. (Note that, for every integer $k > 0$, there exists a family $(g_i)_{i \in I}$ of elements of A such that $\sum_i g_i f_i^k = 1$.)

(b) Let A be a ring and P a projective A-module of rank 1. Show that the ring $\text{End}_A(P)$ is canonically isomorphic to A (use Theorem 3).

(c) Let A be a ring and P a projective A-module of rank n. For every endomorphism u of P, there corresponds canonically to $\overset{n}{\wedge} u$, by (b), an element $\det(u)$ of A, called the *determinant* of u, such that $\det(u \circ v) = (\det u)(\det v)$ for two endomorphisms u, v of P. The element $\chi_u(T) = \det(T.1 - u)$ of A[T] is called the *characteristic polynomial* of u (cf. no. 3, Proposition 4); show that χ_u is a monic polynomial of degree n and that $\chi_u(u) = 0$ (use (a) and § 3, no. 3, Theorem 1). Deduce that, for u to be bijective, it is necessary and sufficient that $\det(u)$ be invertible in A. If A is an integral domain, $\chi_u(T) = \prod_{i=1}^{n} (T - \alpha_i)$ the decomposition into the linear factors of χ_u in an algebraically closed extension of the field of fractions of A, and $\chi_{q(u)}(T) = \prod_{i=1}^{n} (T - q(\alpha_i))$ for every polynomial $q \in A[T]$. Generalize similarly Proposition 14 of *Algebra*, Chapter VII, § 5, no. 6.

10. Let A be a ring and $\boldsymbol{E}(A)$ the set of classes of finitely generated projective A-modules; let G be the **Z**-module of formal linear combinations of elements of $\boldsymbol{E}(A)$ and N the submodule of G consisting of the elements $\xi - \xi' - \xi''$, where ξ, ξ', ξ'' are the classes of three projective A-modules M, M', M'' such that there exists an exact sequence

$$0 \to M' \to M \to M'' \to 0$$

(it amounts to the same to say that M is isomorphic to the direct sum $M' \oplus M''$ by virtue of *Algebra*, Chapter II, § 2, no. 2). Let $\boldsymbol{K}(A)$ be the quotient **Z**-module G/N and, for every finitely generated projective A-module, let $\kappa(M)$ (or $\kappa_A(M)$) be its class in $\boldsymbol{K}(A)$.

(a) Show that there exists on $\boldsymbol{K}(A)$ a unique commutative ring structure whose addition is that the **Z**-module $\boldsymbol{K}(A)$ and multiplication is such that $\kappa(M)\kappa(N) = \kappa(M \otimes_A N)$ for two finitely generated projective A-modules M, N; $\kappa(A)$ is the unit element of $\boldsymbol{K}(A)$.

(b) For every finitely generated projective A-module M and every integer $p \geqslant 0$ let λ^p denote the element $\kappa(\bigwedge^p M)$ of $\boldsymbol{K}(A)$. Show that

$$\lambda^p(M \oplus N) = \sum_{r+s=p} \lambda^r(M) . \lambda^s(N).$$

Let $\lambda_T(M)$ denote the element $\sum_{p=0}^{\infty} \lambda^p(M)T^p$ in the ring $\boldsymbol{K}(A)[[T]]$ of formal power series in one indeterminate over the ring $\boldsymbol{K}(A)$; show that

$$\lambda_T(M \oplus N) = \lambda_T(M)\lambda_T(N);$$

derive a homomorphism also written $x \mapsto \lambda_T(x)$ of $\boldsymbol{K}(A)$ to the multiplicative group of formal power series of $\boldsymbol{K}(A)[[T]]$ whose constant term is equal to 1. Let $\lambda^p(x)$ denote the coefficient of T^p in $\lambda_T(x)$; then

$$\lambda^p(x + y) = \sum_{r+s=p} \lambda^r(x)\lambda^s(y)$$

for all x, y in $\boldsymbol{K}(A)$.

(c) If A is a local ring, the ring $\boldsymbol{K}(A)$ is canonically identified with $\mathbf{Z}$ and $\lambda^p(n) = \binom{n}{p}$. Deduce that, if A is the direct composition of m local rings, $\boldsymbol{K}(A)$ is isomorphic to $\mathbf{Z}^m$.

(d) Let $\phi: A \to B$ be a ring homomorphism. Show that there exists a unique ring homomorphism $\phi^!: \boldsymbol{K}(A) \to \boldsymbol{K}(B)$ such that $\phi^!(\kappa_A(M)) = \kappa_B(M_{(B)})$ for every finitely generated projective A-module M; then $\phi^!(\lambda^p(x)) = \lambda^p(\phi^!(x))$ for all $x \in \boldsymbol{K}(A)$; if $\psi: B \to C$ is another ring homomorphism, $(\psi \circ \phi)^! = \psi^! \circ \phi^!$.

(e) Let $\phi: A \to B$ be a ring homomorphism for which B is a finitely generated projective A-module; for every finitely generated projective B-module N, $\phi_*(N)$ is then a finitely generated projective A-module; deduce that there exists a $\mathbf{Z}$-module homomorphism $\phi_!: \boldsymbol{K}(B) \to \boldsymbol{K}(A)$ such that

$$\phi_!(\kappa_B(N)) = \kappa_A(\phi_*(N)).$$

Show that, for all $x \in \boldsymbol{K}(A)$ and $y \in \boldsymbol{K}(B)$, $\phi_!(y.\phi^!(x)) = \phi_!(y).x$. If $\psi: B \to C$ is another ring homomorphism such that C is a finitely generated projective B-module, then $(\psi \circ \phi)_! = \phi_! \circ \psi_!$.

11. Let K be a field of characteristic $\neq 2$; consider a polynomial $g(X) \in K[X]$ of even degree $\geqslant 2$ with all its roots distinct (in an algebraically closed extension of K) and satisfying $g(0) \neq 0$. Set $B = K[X]$,

$$A = B[Y]/(Y^2 - f(X))$$

where $f(X) = Xg(X)$.

(a) Let y be the class of Y in A. Show that every $a \in A$ may be written uniquely in the form $a = P + yQ$, where P and Q are in B. Let $\bar{a} = P - yQ$

and $N(a) = a\bar{a} = P^2 - fQ^2$. Show that the relation $N(a) = 0$ implies $a = 0$; deduce that A is an integral domain. Show that the invertible elements of A are the elements $\neq 0$ of K.

(b) Let $\mathfrak{m}$ be the ideal $AX + Ay$ of A; show that $\mathfrak{m}$ is maximal and that the ideal $\mathfrak{m}_{\mathfrak{m}}$ of $A_{\mathfrak{m}}$ is generated by y (note that $X = y^2/g(X)$). Deduce that $\mathfrak{m}$ is an invertible A-module (use Theorem 1). Show that $\mathfrak{m}^2 = AX$; deduce that $\mathfrak{m}$ is not a principal ideal of A and that the group of classes of invertible A-modules in the field of fractions of A is not reduced to the identity element. (If $\mathfrak{m} = At$, then $X = \lambda t^2$ where $\lambda \in K$, whence $\lambda^{-1}X = P^2 + fQ^2$ where $P \in B$ and $Q \in B$; prove that such a relation is impossible.)

¶ 12. Let A be a ring, I an A-module and B the unique ring whose underlying additive group is $A \oplus I$, A being a subring of B and I an ideal of zero square, whose A-module structure obtained by restriction of the scalars is the given structure (*Algebra*, Chapter II, § 1, Exercise 7).

(a) Suppose that every non-invertible element of A belongs to the annihilator of some element $\neq 0$ of I. Show that every element in B which is not a divisor of 0 is invertible, in other words that B is equal to its total ring of fractions.

(b) Let $(\mathfrak{m}_\lambda)_{\lambda \in L}$ be a family of maximal ideals of A such that every non-invertible ideal of A belongs to at least one $\mathfrak{m}_\lambda$. Show that, if I is taken to be the direct sum of the A-modules $A/\mathfrak{m}_\lambda$, condition (a) is satisfied.

(c) Deduce from (b) that, if there exists a non-free projective A-module of rank 1, I can be chosen such that B is equal to its total ring of fractions but there exists a non-free projective B-module of rank 1 (this module is necessarily not invertible since B is the only non-degenerate sub-B-module of B).

(d) Suppose that there exists in A a maximal ideal $\mathfrak{m}$ such that, for all $x \in \mathfrak{m}$, there exists a maximal ideal $\mathfrak{m}' \neq \mathfrak{m}$ of A containing x. If I is taken to be the A-module the direct sum of the $A/\mathfrak{m}'$, where $\mathfrak{m}'$ runs through the set of maximal ideals of A distinct from $\mathfrak{m}$, the corresponding ring $B = A \oplus I$ is equal to its total ring of fractions; if we write $\mathfrak{n} = \mathfrak{m} \oplus I$, which is a maximal ideal of B, show that $B_{\mathfrak{n}}$ is isomorphic to $A_{\mathfrak{m}}$; if A is an integral domain (necessarily not a field by hypothesis), $B_{\mathfrak{n}}$ is then an integral domain distinct from its field of fractions. If further $\mathfrak{m}A_{\mathfrak{m}}$ is a principal ideal of the integral domain $A_{\mathfrak{m}}$, show that the ideal $\mathfrak{m}B = \mathfrak{m} \otimes_A B$ of B is a non-free degenerate projective B-module of rank 1. (We shall see in Chapter VII that there are Dedekind domains A in which there exist maximal ideals $\mathfrak{m}$ none of whose powers is a principal ideal; for such a ring A, all the preceding hypotheses are satisfied.)

¶ 13. Let A be a (commutative) ring and B an A-algebra (not necessarily commutative). Suppose that the A-module B is faithful and generated by a finite number of elements b_i $(1 \leqslant i \leqslant n)$; recall that B^0 denotes the opposite algebra of B and that a canonical homomorphism is defined of the A-algebra

$B \otimes_A B^0$ to the A-algebra $\mathrm{End}_A(B)$ mapping $b \otimes b'$ to the endomorphism $z \mapsto bzb'$.

(a) Show that the two following conditions are equivalent:

(α) The b_i form a basis of the A-module B and the canonical homomorphism $B \otimes_A B^0 \to \mathrm{End}_A(B)$ is bijective.

(β) The matrix $U = (u_{ij})$ such that $u_{ij} = b_j b_i$ is invertible in $\mathbf{M}_n(B)$.

(b) Suppose further that A is a local ring with maximal ideal $\mathfrak{m}$; let β_i denote the canonical image of b_i in $B/\mathfrak{m}B$. Show that conditions (α) and (β) are then equivalent to each of the conditions:

(γ) The β_i form a basis of the $A/\mathfrak{m}$-module $B/\mathfrak{m}B$ and $B/\mathfrak{m}B$ is a simple central $A/\mathfrak{m}$-algebra (*Algebra*, Chapter VIII, § 5, no. 4).

(δ) The matrix $(\beta_j\beta_i)$ is invertible in $\mathbf{M}_n(B/\mathfrak{m}B)$.

(To prove that (δ) implies (β), use the fact that $\mathfrak{m}B$ is contained in the Jacobson radical of B and Exercise 5 of *Algebra*, Chapter VIII, § 6).

14. Let A be a (commutative) ring and B an A-algebra such that the A-module B is faithfully flat and finitely presented. Show that the following conditions are equivalent:

(α) For every maximal ideal $\mathfrak{m}$ of A, the $A/\mathfrak{m}$-algebra $B/\mathfrak{m}B$ is simple and central.

(β) B is a projective A-module and, if B^0 denotes the opposite algebra to B, the canonical homomorphism $B \otimes_A B^0 \to \mathrm{End}_A(B)$ is bijective.

(Reduce it to the case where A is a local ring and apply Exercise 13.) B is then called an *Azumaya algebra* over A.

¶ 15. Let B be an Azumaya A-algebra (Exercise 14).

(a) Show that the centre of B is identical with the subring A of B and is a direct factor of the A-module B (cf. Chapter I, § 2, Exercise 21 and Chapter II, § 5, Exercise 4).

(b) Show that, for every two-sided ideal $\mathfrak{b}$ of B, $\mathfrak{b} = (\mathfrak{b} \cap A)B$. (Note first that B is stable under every homomorphism of the A-module B and use the existence of a projector of B onto A.)

16. (a) Let A be a (commutative) ring, A′ a commutative A-algebra which is a flat A-module and B an A-algebra. Show that, if B is an Azumaya algebra, $B' = A' \otimes_A B$ is an Azumaya A′-algebra; the converse is true if A′ is assumed to be a faithfully flat A-module (cf. *Algebra*, Chapter II, § 5, no. 1, Corollary to Proposition 4 and *Commutative Algebra*, Chapter I, § 2, no. 9, Proposition 10 and § 3, no. 6, Proposition 12).

(b) Show that the tensor product $B \otimes_A C$ of two Azumaya A-algebras B, C is an Azumaya A-algebra (cf. *Algebra*, Chapter VIII, § 7, no. 4, Corollary 2 to Theorem 2).

¶ 17. Let A be a ring and P a faithful finitely generated projective A-module.

(a) Show that the A-algebra $B = \mathrm{End}_A(P)$ is an Azumaya algebra. (Observe first that the A-module B is isomorphic to $P \otimes_A P^*$ and hence finitely presented, then use § 2, no. 7, Proposition 19.)

(b) Show that P is a left projective B-module (use Theorem 1 (d) and § 3, Exercise 8 (c)).

(c) Let P′ be another faithful finitely generated projective A-module. Show that $P \otimes_A P'$ is a faithful finitely generated projective A-module and that $\mathrm{End}_A(P \otimes_A P')$ is canonically identified with $\mathrm{End}_A(P) \otimes_A \mathrm{Eng}_A(P')$ (use Theorem 1 (e), § 2, no. 7, Propositions 18 and 19 and § 3, no. 3, Theorem 1). Consider the case where P′ is of rank 1.

¶ 18. Let A be a ring, P a faithful finitely generated projective A-module, B the A-algebra $\mathrm{End}_A(P)$, E a left B-module and F the A-module $\mathrm{Hom}_B(P, E)$. Show that the canonical homomorphism

$$\beta : P \otimes_A F \to E$$

defined in *Algebra*, Chapter VIII, § 1, no. 4, is bijective. (Reduce it to the case where A is a local ring as in Exercise 17 (c); in this case apply Exercise 9 of *Algebra*, Chapter VIII, § 1, observing that F^n and $\mathrm{Hom}_B(P, E^n)$ are canonically isomorphic A-modules.)

Show that there exists a strictly increasing bijection of the set of sub-A-modules of F onto the set of sub-B-modules of E (same method).

¶ 19. Let A be a ring, B and C two A-algebra and $\phi : B \to C$ an A-algebra homomorphism. Suppose that B is an Azumaya algebra (Exercise 14). Let C′ be the sub-A-algebra of C consisting of the elements which commute with the elements of $\phi(B)$. Show that C is isomorphic to the algebra $B \otimes_A C'$. (Consider C as a $B \otimes_A B^0$-module and apply Exercise 18, taking account of the fact that $B \otimes_A B^0$ is isomorphic to $\mathrm{End}_A(B)$.)

¶ 20. Let A be a ring and P, P′ two faithful finitely generated projective A-modules. Show that, if $\theta : \mathrm{End}_A(P) \to \mathrm{End}_A(P')$ is an A-algebra isomorphism, there exists a projective A-module F of rank 1 and an A-isomorphism $\phi : P' \to P \otimes_A F$ such that $\theta(u) = \overset{-1}{\phi} \circ (u \otimes 1) \circ \phi$ for all $u \in \mathrm{End}_A(P)$. (Apply Exercise 18 to E = P′ considered as an $(\mathrm{End}_A(P))$-module by means of θ; to prove that $\mathrm{Hom}_B(P, P')$ is a projective A-module of rank 1, use Theorem 2 of no. 3 and *Algebra*, Chapter VIII, § 5, Exercise 7 (c).)

¶ 21. Let A be a ring and n an integer $\geqslant 0$.

(a) Show that A-modules F such that F^n is isomorphic to A^n are projective of rank 1 and that their classes in $\boldsymbol{P}(A)$ form a subgroup $\boldsymbol{P}_n(A)$ every element of which is annihilated by n. (To show the first assertion, use Proposition 5 of § 2, no. 2; for the second, observe that $\overset{n}{\wedge} F^n$ and $\overset{n}{\otimes} F$ are canonically isomorphic.)

(b) Let $\mathrm{Aut}_n(\mathrm{A})$ be the group of automorphisms of the A-algebra of matrices $\mathbf{M}_n(\mathrm{A})$ and let $\mathrm{Int}_n(\mathrm{A})$ be the subgroup of inner automorphisms of $\mathbf{M}_n(\mathrm{A})$. Show that the quotient group $\mathrm{Aut}_n(\mathrm{A})/\mathrm{Int}_n(\mathrm{A})$ is isomorphic to $\boldsymbol{P}_n(\mathrm{A})$ (use Exercise 20 with $\mathrm{P} = \mathrm{P}' = \mathrm{A}^n$ or Exercise 9 of *Algebra*, Chapter VIII, § 1). Deduce that, if A is a semi-local ring or a principal ideal domain, then

$$\mathrm{Aut}_n(\mathrm{A}) = \mathrm{Int}_n(\mathrm{A}).$$

* (c) If A is a Dedekind domain, show that $\boldsymbol{P}_n(\mathrm{A})$ is identical with the subgroup of $\boldsymbol{P}(\mathrm{A})$ consisting of the elements annihilated by n. *

¶ 22. Let A be a ring and M a finitely generated projective A-module. For a finitely presented submodule M′ of M to be a direct factor of M, it is necessary and sufficient that it be a pure submodule of M (use Theorem 1 and § 3, Exercise 15).

23. Let A be a ring and $(f_i)_{i \in \mathrm{I}}$ a finite family of elements of A generating the ideal A.

(a) Let M be an A-module. Show that, if, for all $i \in \mathrm{I}$, M_{f_i} is a finitely generated A_{f_i}-module, M is a finitely generated A-module.

(b) Let B be a (commutative) A-algebra. Show that, if, for all $i \in \mathrm{I}$, B_{f_i} is a finitely generated A_{f_i}-algebra, B is a finitely generated A-algebra.

24. Let A be a ring such that Spec(A) is connected.

(a) If M is a projective A-module of rank n, every direct factor N of M not reduced to 0 admits a rank $\leqslant n$.

(b) Let u be a homomorphism of a projective A-module of finite rank M to a projective A-module of finite rank M′ such that $u(\mathrm{M})$ is a direct factor of M′. Show that Ker(u) is then a direct factor of M (cf. Exercise 23) and

$$\mathrm{rg}(\mathrm{Ker}(u)) + \mathrm{rg}(\mathrm{Im}(u)) = \mathrm{rg}(\mathrm{M}).$$

Moreover the transpose ${}^t u$ is such that ${}^t u(\mathrm{M}'^*)$ is a direct factor of M^* and $\mathrm{rg}({}^t u(\mathrm{M}'^*)) = \mathrm{rg}(u(\mathrm{M}))$ (cf. § 3, Exercise 14 (a)).

25. Let A be a ring and S a multiplicative subset of A. An ideal $\mathfrak{a}$ of A is called S-invertible if there exists $s \in \mathrm{S}$ and an ideal $\mathfrak{b}$ of A such that $\mathfrak{a}\mathfrak{b} = \mathrm{A}s$; two ideals $\mathfrak{a}$, $\mathfrak{b}$ of A are called S-equivalent if there exists s, t in S such that $s\mathfrak{a} = t\mathfrak{b}$. Show that the S-equivalence classes of the S-invertible ideals of A form a multiplicative group $\mathfrak{C}_{\mathrm{S}}$. If S is saturated (§ 2, Exercise 1) and no element of S is a divisor of 0 in A, show that the group $\mathfrak{C}_{\mathrm{S}}$ is isomorphic to the classes of invertible sub-$\mathrm{S}^{-1}\mathrm{A}$-modules of the total ring of fractions B of A (and $\mathrm{S}^{-1}\mathrm{A}$).

CHAPTER III(*)

Graduations, Filtrations and Topologies

All the rings considered in this chapter are assumed to have a unit element; all the ring homomorphisms are assumed to map unit element to unit element. By a subring of a ring A *we mean a subring containing the unit element of* A. *Unless otherwise stated, all the modules are left modules.*

1. FINITELY GENERATED GRADED ALGEBRAS

1. SYSTEMS OF GENERATORS OF A COMMUTATIVE ALGEBRA

Let A be a commutative ring and B a commutative A-algebra. Let us recall (*Algebra*, Chapter IV, § 2, no. 1) that if $x = (x_\iota)_{\iota \in I}$ is a family of elements of B, the mapping $f \mapsto f(x)$ from the polynomial algebra $A[X_\iota]_{\iota \in I}$ to B is homomorphism of $A[X_\iota]_{\iota \in I}$ onto the subalgebra of B generated by the x_ι, whose kernel $\mathfrak{a}$ is the ideal of polynomials f such that $f(x) = 0$, called the *ideal of algebraic relations* (*with coefficients in* A) *between the* x_ι.

DEFINITION 1. *In a commutative algebra* B *over a commutative ring* A, *a family* $(x_\iota)_{\iota \in I}$ *of elements of* B *is called algebraically free over* A (or *the* x_ι *are called algebraically independent over* A) *if the ideal of algebraic relations between the* x_ι, *with coefficients in* A, *is reduced to* 0. *A family* (x_ι) *which is not algebraically free over* A *is also called algebraically related* (or its elements are also called *algebraically dependent* over A).

This definition generalizes that given in *Algebra*, Chapter V, § 5, no. 1, Definition 1 for families of elements of a commutative field.

(*) Except in § 5, which uses the results of Chapter I, § 4 and therefore homological algebra, no use is made in this chapter of any books other than Books I to VI and Chapters I and II of this book.

To say that a family $(x_\iota)_{\iota \in I}$ is algebraically free over A amounts to the same as saying that the monomials $\prod_\iota x_\iota^{n_\iota}$ in the x_ι are *linearly independent* over A; in particular the x_ι are then linearly independent over A.

DEFINITION 2. *A commutative algebra* B *over a commutative ring* A *is called finitely generated if it is generated by a finite family of elements.*

It amounts to the same to say that B is isomorphic to an A-algebra of the form $A[X_1, \ldots, X_n]/\mathfrak{a}$ (where the X_i are indeterminates and $\mathfrak{a}$ is an ideal of the polynomial ring $A[X_1, \ldots, X_n]$).

If the A-algebra B is a *finitely generated* A-*module*, it is obviously a finitely generated A-algebra; the converse is false as the example of polynomial algebras shows (cf. Chapter V).

If B is a finitely generated A-algebra and A′ is any commutative A-algebra, $B_{(A')} = B \otimes_A A'$ is a finitely generated A′-algebra, for if $(x_\iota)_{\iota \in I}$ is a system of generators of the A-algebra B, clearly the $x_\iota \otimes 1$ form a system of generators of the A′-algebra $B_{(A')}$.

If B is a finitely generated A-algebra and C a finitely generated B-algebra, then C is a finitely generated A-algebra; for it follows immediately from the definitions that if $(b_\lambda)_{\lambda \in L}$ is a system of generators of the A-algebra B and $(c_\mu)_{\mu \in M}$ is a system of generators of the B-algebra C, every element of C is equal to a polynomial, with coefficients in A, in the b_λ and the c_μ.

2. CRITERIA OF FINITENESS FOR GRADED RINGS

In this no. and the following, all the graduations considered (*Algebra*, Chapter II, § 11) *are assumed to be of type* **Z**. *If* A (resp. M) *is a graded ring* (resp. *graded module*), A_i (resp. M_i) *will denote the set of homogeneous elements of degree i in* A (resp. M).

If $A_i = \{0\}$ (resp. $M_i = \{0\}$) for $i < 0$, A (resp. M) will, to abbreviate, be called a *graded ring* (resp. *module*) *with positive degrees.*

PROPOSITION 1. *Let* $A = \bigoplus_{i \in \mathbf{Z}} A_i$ *be a graded commutative ring with positive degrees,* $\mathfrak{m}$ *the graded ideal* $\bigoplus_{i \geqslant 1} A_i$ *and* $(x_\lambda)_{\lambda \in L}$ *a family of homogeneous elements of* A *of degrees* $\geqslant 1$. *The following conditions are equivalent:*

(a) *The ideal of* A *generated by the family* (x_λ) *is equal to* $\mathfrak{m}$.

(b) *The family* (x_λ) *is a system of generators of the* A_0*-algebra* A.

(c) *For all* $i \geqslant 0$, *the* A_0*-module* A_i *is generated by the elements of the form* $\prod_\lambda x_\lambda^{n_\lambda}$ *which are of degree i in* A.

Clearly conditions (b) and (c) are equivalent. If they hold, every element of $\mathfrak{m}$ is of the form $f((x_\lambda))$ where f is a polynomial of $A_0[X_\lambda]_{\lambda \in L}$ with no constant term; then $\mathfrak{m} = \sum_{\lambda \in L} Ax_\lambda$, which proves that (c) implies (a). Conversely, suppose that condition (a) holds. Let $A' = A_0[x_\lambda]_{\lambda \in L}$ be the sub-A_0-algebra of A generated by the family (x_λ) and let us show that $A' = A$. For this, it is sufficient to show that $A_i \subset A'$ for all $i \geqslant 0$. We proceed by induction on i, the property being obvious for $i = 0$. Then let $y \in A_i$ with $i \geqslant 1$. Since $y \in \mathfrak{m}$, there exists a family $(a_\lambda)_{\lambda \in L}$ of elements of A of finite support such that $y = \sum_\lambda a_\lambda x_\lambda$ and we may assume that each of the a_λ is homogeneous of degree $i - \deg(x_\lambda)$ (by replacing it if need be by its homogeneous component of that degree); as $\deg(x_\lambda) > 0$, the induction hypothesis shows that $a_\lambda \in A'$ for all $\lambda \in L$, whence $y \in A'$ and $A_i \subset A'$, which proves that (a) implies (b).

Corollary. *Let* $A = \bigoplus_{i \in \mathbf{Z}} A_i$ *be a graded commutative ring with positive degrees and* $\mathfrak{m}$ *the graded ideal* $\bigoplus_{i \geqslant 1} A_i$.

(i) *The following conditions are equivalent:*

(a) *The ideal* $\mathfrak{m}$ *is a finitely generated A-module.*

(b) *The ring A is a finitely generated* A_0*-module.*

(ii) *Suppose that the conditions in* (i) *hold and let* $M = \bigoplus_{i \in \mathbf{Z}} M_i$ *be a finitely generated graded A-module. Then, for all* $i \in \mathbf{Z}$, M_i *is a finitely* A_0*-module and there exists* i_0 *such that* $M_i = \{0\}$ *for* $i < i_0$.

(i) If a family (y_μ) of elements of A is a system of generators of the A-module $\mathfrak{m}$ (resp. of the A_0-module A), so is the family consisting of the homogeneous components of the y_μ; the equivalence of conditions (a) and (b) then follows from Proposition 1.

(ii) We may suppose that A is generated (as an A_0-algebra) by homogeneous elements a_i $(1 \leqslant i \leqslant r)$ of degree $\geqslant 1$ and M is generated (as an A-module) by homogeneous elements x_j $(1 \leqslant j \leqslant s)$; let $h_i = \deg(a_i)$, $k_j = \deg(x_j)$. Clearly M_n consists of the linear combinations with coefficients in A_0 of the elements $a_1^{\alpha_1} a_2^{\alpha_2} \ldots a_r^{\alpha_r} x_j$ such that the α_i are integers $\geqslant 0$ satisfying the relation $k_j + \sum_{i=1} \alpha_i h_i = n$; for each n there is only a finite number of families $(\alpha_i)_{1 \leqslant i \leqslant r}$ satisfying these conditions, since $h_i \geqslant 1$ for all i; we conclude that M_n is a finitely generated A_0-module and moreover clearly $M_n = \{0\}$ when $n < \inf_j(k_j)$.

3. PROPERTIES OF THE RING $A^{(d)}$

Let $A = \bigoplus_{i \in \mathbf{Z}} A_i$ be a graded ring and $M = \bigoplus_{i \in \mathbf{Z}} M_i$ a graded A-module; for every ordered pair of integers (d, k) such that $d \geqslant 1$ and $0 \leqslant k \leqslant d-1$, set

$$A^{(d)} = \bigoplus_{i \in \mathbf{Z}} A_{id}, \qquad M^{(d,k)} = \bigoplus_{i \in \mathbf{Z}} M_{id+k}.$$

Clearly $A^{(d)}$ is a graded subring of A and $M^{(d,k)}$ a graded $A^{(d)}$-module; moreover, if N is a graded submodule of M, $N^{(d,k)}$ is a graded sub-$A^{(d)}$-module of $M^{(d,k)}$. We shall write $M^{(d)}$ instead of $M^{(d,0)}$; for each $d \geqslant 1$, M is the direct sum of the $A^{(d)}$-modules $M^{(d,k)}$ $(0 \leqslant k \leqslant d-1)$.

Proposition 2. *Let* $A = \bigoplus_{i \in \mathbf{Z}} A_i$ *be a graded commutative ring with positive degrees and* $M = \bigoplus_{i \in \mathbf{Z}} M_i$ *a graded A-module. Suppose that A is a finitely generated* A_0-*algebra and M a finitely generated A-module. Then, for every ordered pair* (d, k) *of integers such that* $d \geqslant 1$, $0 \leqslant k \leqslant d-1$:

(i) $A^{(d)}$ *is a finitely generated* A_0-*module.*

(ii) $M^{(d,k)}$ *is a finitely generated* $A^{(d)}$-*module.*

Let us show that A is a *finitely generated* $A^{(d)}$-*module*. Let $(a_i)_{1 \leqslant i \leqslant s}$ be a system of generators of the A_0-algebra A consisting of homogeneous elements. The elements of A (finite in number) of the form $a_1^{\alpha_1} a_2^{\alpha_2} \ldots a_s^{\alpha_s}$ such that $0 \leqslant \alpha_i \leqslant d$ for $1 \leqslant i \leqslant s$ constitute a system of generators of the $A^{(d)}$-module A; for every system of integers $n_i \geqslant 0$ $(1 \leqslant i \leqslant s)$, there are positive integers q_i, r_i such that $n_i = q_i d + r_i$ where $r_i < d$ $(1 \leqslant i \leqslant s)$; then

$$a_1^{n_1} a_2^{n_2} \ldots a_s^{n_s} = (a_1^{q_1} \ldots a_s^{q_s})^d (a_1^{r_1} \ldots a_s^{r_s})$$

which proves our assertion, for every homogeneous element $x \in A$ satisfies $x^d \in A^{(d)}$. Then, if M is a finitely generated A-module, it is also a finitely generated $A^{(d)}$-module; as M is the *direct* sum of the $M^{(d,k)}$ $(0 \leqslant k \leqslant d-1)$, each of the $M^{(d,k)}$ is a finitely generated $A^{(d)}$-module, which proves (ii).

Let us apply the above to the graded A-module $\mathfrak{m} = \bigoplus_{i \geqslant 1} A_i$, which is finitely generated by the Corollary to Proposition 1 of no. 2; it is seen that $\mathfrak{m}^{(d)}$ is a finitely generated $A^{(d)}$-module; therefore (no. 2, Corollary to Proposition 1) $A^{(d)}$ is a finitely generated A_0-algebra.

Lemma 1. *Let A be a graded commutative ring such that* $A = A_0[A_1]$, *M a graded A-module and* $(y_\lambda)_{\lambda \in L}$ *a system of homogeneous generators of M such that* $\deg(y_\lambda) \leqslant n_0$ *for all* $\lambda \in L$. *Then, for all* $n \geqslant n_0$ *and all* $k \geqslant 0$, $M_{n+k} = A_k . M_n$.

Let $n \geqslant n_0$ and $x \in M_{n+1}$. Since the y_λ generate M, there exists a family $(a_\lambda)_{\lambda \in L}$ of elements of A of finite support such that $x = \sum_\lambda a_\lambda y_\lambda$; we may further suppose that each a_λ is homogeneous and of degree $n + 1 - \deg(y_\lambda)$ (replacing it if need be by its homogeneous component of that degree). As

$A = A_0[A_1]$ and $\deg(a_\lambda) > 0$, each a_λ is a sum of elements of the form bb' where $b \in A_1$, $b' \in A$, whence $x \in A_1M_n$. Then $M_{n+1} = A_1M_n$, whence $M_{n+k} = A_kM_n$ by induction on k.

Lemma 2. *Let A be a graded commutative ring such that $A = A_0[A_1]$ and let $S = \bigoplus_{i \geq 0} S_i$ be a graded commutative A-algebra with positive degrees, which is a finitely generated A-module. Then there exists an integer $n_0 \geqslant 0$ such that:*

(i) *For $n \geqslant n_0$ and $k \geqslant 0$, $S_{n+k} = S_k . S_n$.*

(ii) *For $d \geqslant n_0$, $S^{(d)} = S_0[S_d]$.*

By Lemma 1 there exists an integer $n_0 \geqslant 0$ such that, for $n \geqslant n_0$ and $k \geqslant 0$, $S_{n+k} = A_kS_n$, whence *a fortiori* $S_{n+k} = S_kS_n$, which establishes (i). Then, for $d \geqslant n_0$ and $m > 0$, $S_{md} = (S_d)^m$ as follows by induction on m applying (i); this establishes (ii).

Proposition 3. *Let $R = \bigoplus_{i \geq 0} R_i$ be a graded commutative ring with positive degrees which is a finitely generated R_0-algebra. There exists an integer $e \geqslant 1$ such that $R^{(me)} = R_0[R_{me}]$ for all $m \geqslant 1$.*

Let $(x_j)_{1 \leqslant j \leqslant s}$ be a system of homogeneous generators of the R_0-algebra R, whose degrees are $\geqslant 1$. Let $h_j = \deg(x_j)$, let q be a common multiple of the h_j and let us write $q_j = q/h_j$ for $1 \leqslant j \leqslant s$; the elements $x_j^{q_j}$ are then all of degree q. Let B be the sub-R_0-algebra of R generated by the $x_j^{q_j}$; it is a graded subalgebra of R and $B_i = 0$ if i is not a multiple of q. Let A (resp. S) be the graded ring whose underlying ring is B (resp. $R^{(q)}$) and whose graduation consists of the $A_i = B_{iq}$ (resp. $S_i = R_{iq}$). Then $A = A_0[A_1]$ by definition of B. Consider the elements of R (finite in number) of the form $x_1^{\alpha_1}x_2^{\alpha_2}\ldots x_s^{\alpha_s}$, where $0 \leqslant \alpha_j \leqslant q_j$ and $\alpha_1h_1 + \cdots + \alpha_sh_s \equiv 0 \pmod{q}$; let us show that they generate the B-module $R^{(q)}$. It is sufficient to prove that every element of $R^{(q)}$ of the form $x_1^{n_1}x_2^{n_2}\ldots x_s^{n_s}$ is a B-linear combination of the above elements. Now, there exist positive integers k_j, r_j such that $n_j = k_jq_j + r_j$ where $r_j < q_j$ $(1 \leqslant j \leqslant s)$; then

$$x_1^{n_1}x_2^{n_2}\ldots x_s^{n_s} = (x_1^{q_1})^{k_1}\ldots(x_s^{q_s})^{k_s}.(x_1^{r_1}\ldots x_s^{r_s})$$

and by hypothesis $\sum_{j=1}^{s} n_jh_j \equiv 0 \pmod{q}$, hence $\sum_{j=1}^{s} r_jh_j \equiv 0 \pmod{q}$; as the $x_j^{q_j}$ belong to B by definition, this proves our assertion. As S is a finitely generated A-module, Lemma 2 can be applied: there exists n_0 such that for $d \geqslant n_0$, $S^{(d)} = S_0[S_d]$ and hence $R^{(qd)} = R_0[R_{qd}]$ for $d \geqslant n_0$. The proposition follows with $e = qn_0$.

4. GRADED PRIME IDEALS

Let $A = \bigoplus_{i \geqslant 0} A_i$ be a graded commutative ring with *positive* degrees and $\mathfrak{m}$ the graded ideal $\bigoplus_{i \geqslant 1} A_i$; we shall call the two graded ideals $\mathfrak{a} = \bigoplus_{i \geqslant 0} \mathfrak{a}_i$, $\mathfrak{b} = \bigoplus_{i \geqslant 0} \mathfrak{b}_i$ of A *equivalent* if there exists an integer n_0 such that $\mathfrak{a}_n = \mathfrak{b}_n$ for $n \geqslant n_0$ (clearly it is an equivalence relation). A graded ideal is called *essential* if it is not equivalent to $\mathfrak{m}$.

PROPOSITION 4. *Let $\mathfrak{p} = \bigoplus_{i \geqslant 0} \mathfrak{p}_i$ be a graded ideal of* A; *for $\mathfrak{p}$ to be prime, it is necessary and sufficient that, if $x \in A_m$, $y \in A_n$ satisfy $x \notin \mathfrak{p}$ and $y \notin \mathfrak{p}$, then $xy \notin \mathfrak{p}$.*

The condition is obviously necessary. Conversely, if it is fulfilled, then in the graded ring $A/\mathfrak{p} = \bigoplus_{i \geqslant 0} A_i/\mathfrak{p}_i$ the product of two homogeneous elements $\neq 0$ is $\neq 0$ and hence $A/\mathfrak{p}$ is an integral domain (*Algebra*, Chapter II, § 11, no. 4, Proposition 7).

PROPOSITION 5. *Let $\mathfrak{a} = \bigoplus_{i \geqslant 0} \mathfrak{a}_i$ be a graded ideal of* A *and n_0 an integer* >0. *For there to exist a graded prime ideal $\mathfrak{p} = \bigoplus_{i \geqslant 0} \mathfrak{p}_i$ such that $\mathfrak{p}_n = \mathfrak{a}_n$ for $n \geqslant n_0$, it is necessary and sufficient that, for pair of homogeneous elements x, y of degrees $\geqslant n_0$, the relation $xy \in \mathfrak{a}$ implies "$x \in \mathfrak{a}$ or $y \in \mathfrak{a}$". If there exists $n \geqslant n_0$ such that $\mathfrak{a}_n \neq A_n$, the prime ideal satisfying the above conditions is unique.*

The condition of the statement is obviously necessary. If $\mathfrak{a}_n = A_n$ for all $n \geqslant n_0$, clearly every prime ideal containing $\mathfrak{m}$ is a solution to the problem; there may therefore be several prime ideals which solve the problem; however, any two of these ideals are obviously equivalent. Suppose then that there exists a homogeneous element $a \in A_d$ (with $d \geqslant n_0$) not belonging to $\mathfrak{a}_d$. Let $\mathfrak{p}$ be the set of $x \in A$ such that $ax \in \mathfrak{a}$. Clearly $\mathfrak{p}$ is an ideal of A; as the homogeneous components of ax are the products by a of those of x and $\mathfrak{a}$ is a graded ideal, $\mathfrak{p}$ is a graded ideal; moreover, $1 \notin \mathfrak{p}$ and hence $\mathfrak{p} \neq A$. To prove that $\mathfrak{p}$ is prime, it is sufficient to show that, if $x \in A_m$, $y \in A_n$ satisfy $x \notin \mathfrak{p}$ and $y \notin \mathfrak{p}$, then $xy \notin \mathfrak{p}$ (Proposition 4). Then $ax \notin \mathfrak{a}_{m+d}$, $ay \notin \mathfrak{a}_{n+d}$, whence by hypothesis

$$a^2xy \notin \mathfrak{a}_{m+n+2d};$$

we conclude that $axy \notin \mathfrak{a}_{m+n+d}$ since $xy \notin \mathfrak{p}$. Finally, if $n \geqslant n_0$ and $x \in A_n$, the conditions $x \in \mathfrak{a}_n$ and $ax \in \mathfrak{a}_{n+d}$ are equivalent by hypothesis and hence $\mathfrak{p} \cap A_n = \mathfrak{a}_n$, which completes the proof of the existence of the graded prime ideal $\mathfrak{p}$ which solves the problem. If also $\mathfrak{p}'$ is another graded prime ideal of A such that $\mathfrak{p}' \cap A_n = \mathfrak{a}_n$ for $n \geqslant n_0$, then $a \notin \mathfrak{p}'$ and $ax \in \mathfrak{p}'$ for all $x \in \mathfrak{p}$, whence $\mathfrak{p} \subset \mathfrak{p}'$ since $\mathfrak{p}'$ is prime. On the other hand, if x is a homogeneous element of degree $n \geqslant 0$ of $\mathfrak{p}'$, ax is homogeneous of degree $n + d \geqslant n_0$ and therefore belongs to $\mathfrak{p}' \cap A_{n+d} = \mathfrak{a}_{n+d}$, whence by definition $x \in \mathfrak{p}$, which shows that $\mathfrak{p}' \subset \mathfrak{p}$ and finally $\mathfrak{p}' = \mathfrak{p}$.

PROPOSITION 6. *Let d be an integer $\geqslant 1$.*

(i) *For every essential graded ideal $\mathfrak{p}$ of A, $\mathfrak{p} \cap A^{(d)}$ is an essential graded prime ideal of $A^{(d)}$.*

(ii) *Conversely, for every essential graded prime ideal $\mathfrak{p}'$ of $A^{(d)}$, there exists a unique (necessarily essential) graded prime ideal $\mathfrak{p}$ of A such that $\mathfrak{p} \cap A^{(d)} = \mathfrak{p}'$.*

(i) If $a \in A_k$ does not belong to $\mathfrak{p}_k$, a^{kd} does not belong to $\mathfrak{p}_{kd}$ and hence $\mathfrak{p} \cap A^{(d)}$ is essential.

(ii) For all $n \geqslant 0$, the set $\mathfrak{p} \cap A_n$ must be equal to the set $\mathfrak{a}_n$ of $x \in A_n$ such that $x^d \in \mathfrak{p}'$. Let us show that $\mathfrak{a} = \bigoplus_{n \geqslant 0} \mathfrak{a}_n$ is a graded prime ideal; as $\mathfrak{a}_n = \mathfrak{p}'_n$ when n is a multiple of d, since $\mathfrak{p}'$ is prime, this will prove the uniqueness of $\mathfrak{p}$. Now, if $x \in \mathfrak{a}_n$, $y \in \mathfrak{a}_n$, $(x - y)^{2d}$ is the sum of terms each of which is a product of x^d or y^d by a homogeneous element of degree nd and hence $(x - y)^{2d} \in \mathfrak{p}'$ and, since $\mathfrak{p}'$ is prime, $(x - y)^d \in \mathfrak{p}'$ and therefore $\mathfrak{a}_n$ is a subgroup of A. As $\mathfrak{p}'$ is an ideal of $A^{(d)}$, $\mathfrak{a}$ is a graded ideal of A; finally, the relation $(xy)^d \in \mathfrak{p}'$ implies $x^d \in \mathfrak{p}'$ or $y^d \in \mathfrak{p}'$, which completes the proof by virtue of Proposition 4.

Let A be a graded commutative ring with positive degrees and $\mathfrak{p}$ an essential graded prime ideal of A. The set S of *homogeneous* elements of A not belonging to $\mathfrak{p}$ is multiplicative and the ring of fractions $S^{-1}A$ is therefore graded canonically (Chapter II, § 2, no. 9) (note that there will in general be homogeneous elements $\neq 0$ of negative degree in this graduation). We shall denote by $A_{(\mathfrak{p})}$ the subring of $S^{-1}A$ consisting of the homogeneous elements of degree 0, in other words the set of fractions x/s, where x and s are homogeneous of the same degree in A and $s \notin \mathfrak{p}$. Similarly, for every graded A-module M, $S^{-1}M$ is graded canonically (*loc. cit.*) and we shall denote by $M_{(\mathfrak{p})}$ the subgroup of homogeneous elements of degree 0, which is obviously an $A_{(\mathfrak{p})}$-module.

PROPOSITION 7. *Let $\mathfrak{p}$ be a graded prime ideal of A, d an integer $\geqslant 1$ and $\mathfrak{p}'$ the graded prime ideal $\mathfrak{p} \cap A^{(d)}$ of $A^{(d)}$; for every graded A-module M, the homomorphism $(M^{(d)})_{(\mathfrak{p}')} \to M_{(\mathfrak{p})}$ derived from the canonical injection $M^{(d)} \to M$ is bijective.*

If S is the set of homogeneous elements of A not belonging to $\mathfrak{p}$ and $S' = S \cap A^{(d)}$, the canonical homomorphism $\phi: S'^{-1}M^{(d)} \to S^{-1}M$ is a homogeneous homomorphism of degree 0 and it is injective, for, if $x \in M_{nd}$ satisfies $sx = 0$ for $s \in A_m$, $s \notin \mathfrak{p}$, then also $s^d x = 0$ and $s^d \in A_{md}$, $s^d \notin \mathfrak{p}'$. It remains to show that the image under ϕ of $(M^{(d)})_{(\mathfrak{p}')}$ is the whole of $M_{(\mathfrak{p})}$; but if $x \in M_n$, $s \in A_n$ and $s \notin \mathfrak{p}$, then also $x/s = (xs^{d-1})/s^d$ where $xs^{d-1} \in A_{nd}$, $s^d \in A_{nd}$ and $s^d \notin \mathfrak{p}'$, whence our assertion.

PROPOSITION 8. *Let* $\mathfrak{m} = \bigoplus_{i \geqslant 1} A_i$; *let* $(\mathfrak{p}^{(k)})_{1 \leqslant k \leqslant n}$ *be a finite family of graded prime ideals of* A *and* $\mathfrak{a}$ *a graded ideal of* A *such that* $\mathfrak{a} \cap \mathfrak{m} \not\subset \mathfrak{p}^{(k)}$ *for all* k; *then there exists a homogeneous element* $z \in \mathfrak{a} \cap \mathfrak{m}$ *not belonging to any of the* $\mathfrak{p}^{(k)}$.

We argue by induction on n, the proposition being trivial for $n = 1$. If there exists an index j such that $\mathfrak{a} \cap \mathfrak{m} \cap \mathfrak{p}^{(j)}$ is contained in one of the $\mathfrak{p}^{(k)}$ of index $k \neq j$, it follows from the induction hypothesis that there is a homogeneous element $z' \in \mathfrak{a} \cap \mathfrak{m}$ not belonging to any of the $\mathfrak{p}^{(k)}$ for $k \neq j$ and therefore not belonging to $\mathfrak{p}^{(j)}$ either and this element solves the problem. Suppose then that for every index j, $\mathfrak{a} \cap \mathfrak{m} \cap \mathfrak{p}^{(j)}$ is not contained in any of the $\mathfrak{p}^{(k)}$ of index $k \neq j$; the induction hypothesis therefore implies the existence of a homogeneous element $y_j \in \mathfrak{a} \cap \mathfrak{m} \cap \mathfrak{p}^{(j)}$ belonging to none of the $\mathfrak{p}^{(k)}$ for $k \neq j$; as the y_j are all of degrees $\geqslant 1$, we may assume by replacing them by suitable powers (since the $\mathfrak{p}^{(k)}$ are prime) that y_1 and $\prod_{j=2}^{n} y_j$ are of the same degree. Then $z = y_1 + \prod_{j=2}^{n} y_j$ is homogeneous of degree $\geqslant 1$ and the same argument as in Chapter II, § 1, no. 1, Proposition 2 shows that z solves the problem.

§2. GENERAL RESULTS ON FILTERED RINGS AND MODULES

1. FILTERED RINGS AND MODULES

DEFINITION 1. *An increasing* (resp. *decreasing*) *sequence* $(G_n)_{n \in \mathbf{Z}}$ *of subgroups of a group* G *is called an increasing* (resp. *decreasing*) *filtration on* G.

A group with a filtration is called a filtered group.

If $(G_n)_{n \in \mathbf{Z}}$ is an increasing (resp. decreasing) filtration on a group G and we write $G'_n = G_{-n}$, clearly $(G'_n)_{n \in \mathbf{Z}}$ is a decreasing (resp. increasing) filtration on G. We may therefore restrict our study to *decreasing* filtrations and hence forth when we speak of a filtration, we shall mean a decreasing filtration, unless otherwise stated.

Given a decreasing filtration $(G_n)_{n \in \mathbf{Z}}$ on a group G, clearly $\bigcap_{n \in \mathbf{Z}} G_n$ and $\bigcup_{n \in \mathbf{Z}} G_n$ are subgroups of G; the filtration is called *separated* if $\bigcup_{n \in \mathbf{Z}} G_n$ is reduced to the identity element and *exhaustive* if $\bigcup_{n \in \mathbf{Z}} G_n = G$.

DEFINITION 2. *Given a ring* A, *a filtration* $(A_n)_{n \in \mathbf{Z}}$ *over the additive group* A *is called compatible with the ring structure on* A *if*

$$A_m A_n \subset A_{m+n} \quad \textit{for} \quad m \in \mathbf{Z}, \quad n \in \mathbf{Z} \tag{1}$$

(2) $$1 \in A_0.$$

The ring A with this filtration is then called a filtered ring.

Conditions (1) and (2) show that A_0 is a *subring* of A and the A_n (left and right) A_0-modules. The set $B = \bigcup_{n \in \mathbf{Z}} A_n$ is a *subring* of A and the set $\mathfrak{n} = \bigcap_{n \in \mathbf{Z}} A_n$ a *two-sided ideal* of B; for if $x \in \mathfrak{n}$ and $a \in A_p$, for all $k \in \mathbf{Z}$, $x \in A_{k-p}$, whence $ax \in A_k$ and $xa \in A_k$ by (1); therefore $ax \in \mathfrak{n}$ and $xa \in \mathfrak{n}$.

An important particular case is that in which $A_0 = A$; then $A_n = A$ for $n \leqslant 0$ and all the A_n are *two-sided ideals* of A.

DEFINITION 3. *Let A be a filtered ring, $(A_n)_{n \in \mathbf{Z}}$ its filtration and E an A-module. A filtration $(E_n)_{n \in \mathbf{Z}}$ on E is called compatible with its module structure over the filtered ring A if*

(3) $$A_m E_n \subset E_{m+n} \quad \textit{for} \quad m \in \mathbf{Z}, \quad n \in \mathbf{Z}.$$

The A-module E with this filtration is called a filtered module.

The E_n are all A_0-modules; if $B = \bigcup_{n \in \mathbf{Z}} A_n$, clearly $\bigcup_{n \in \mathbf{Z}} E_n$ is a B-module and so is $\bigcap_{n \in \mathbf{Z}} E_n$ by the same argument as above for $\bigcap_{n \in \mathbf{Z}} A_n$. If $A_0 = A$, *all* the E_n are *submodules* of E.

Examples

(1) Let A be a graded ring of type **Z**; for all $i \in \mathbf{Z}$, let $A_{(i)}$ be the subgroup of homogeneous elements of degree i in A. Let us write $A_n = \sum_{i \geqslant n} A_{(i)}$; then it is immediate that (A_n) is an *exhaustive* and *separated* decreasing filtration which is compatible with the ring structure on A; this filtration is said to be *associated* with the graduation $(A_{(i)})_{i \in \mathbf{Z}}$ and the filtered ring A is said to be *associated* with the given graded ring A.

Now let E be a graded module of type **Z** over the graded ring A and for all $i \in \mathbf{Z}$ let $E_{(i)}$ be the subgroup of homogeneous elements of degree i of E. Let us write $E_n = \sum_{i \geqslant n} E_{(i)}$; then (E_n) is an *exhaustive* and *separated* decreasing filtration which is compatible with the module structure on E over the filtered ring A; this filtration is said to be *associated* with the graduation $(E_{(i)})_{i \in \mathbf{Z}}$ and the filtered module E is said to be *associated* with the given graded module E.

(2) Let A be a filtered ring, $(A_n)_{n \in \mathbf{Z}}$ its filtration and E an A-module. Let us write $E_n = A_n E$; it follows from (1) that

$$A_m E_n = A_m A_n E \subset A_{m+n} E = E_{m+n},$$

and from (2) that $E_0 = E$; then (E_n) is an *exhaustive* filtration which is compatible with the A-module structure on E. This filtration is said to be *derived* from the given filtration (A_n) on A; note that it is not necessarily separated, even if (A_n) is separated and E and the A_n are finitely generated A-modules (cf. § 3, Exercise 2; see however § 3, no. 3, Proposition 5 and no. 2, Corollary to Proposition 4).

(3) Let A be a ring and $\mathfrak{m}$ a two-sided ideal of A. Let us write $A_n = \mathfrak{m}^n$ for $n \geqslant 0$, $A_n = A$ for $n < 0$; it is immediate that (A_n) is an exhaustive filtration on A, called the $\mathfrak{m}$-*adic filtration*. Let E be an A-module; the filtration (E_n) derived from the $\mathfrak{m}$-adic filtration on A is called the $\mathfrak{m}$-*adic filtration* on E; in other words, $E_n = \mathfrak{m}^n E$ for $n \geqslant 0$ and $E_n = E$ for $n < 0$.

If A is commutative and B is an A-*algebra*, $\mathfrak{n} = \mathfrak{m}B$ is a two-sided ideal of B and, for every B-module F, $\mathfrak{n}^k F = \mathfrak{m}^k F$ and hence the $\mathfrak{n}$-adic filtration on F *coincides* with the $\mathfrak{m}$-adic filtration (if F is considered as an A-module).

(4) If A is a filtered ring and (A_n) its filtration, the left A-module A_s is a filtered A-module with the filtration (A_n). On the other hand, clearly (A_n) is a filtration which is compatible with the ring structure on the opposite ring A^0 and A_d is a filtered (left) A^0-module with the filtration (A_n).

(5) On a ring A the sets A_n such that $A_n = 0$ for $n > 0$, $A_n = A$ for $n \leqslant 0$ form what is called a *trivial* filtration associated (*Example* 1) with the trivial graduation on A; on an A-module E, every filtration (E_n) consisting of sub-A-modules is then compatible with the module structure on E over the filtered ring A. Then it is possible to say that every filtered commutative group G is a filtered $\mathbf{Z}$-module, if $\mathbf{Z}$ is given the trivial filtration.

Let G be a filtered group and $(G_n)_{n \in \mathbf{Z}}$ its filtration; clearly, for every subgroup H of G, $(H \cap G_n)_{n \in \mathbf{Z}}$ is a filtration said to be *induced* by that on G; it is exhaustive (resp. separated) if that on G is. Similarly, if H is a *normal* subgroup of G, the family $((H.G_n)/H)_{n \in \mathbf{Z}}$ is a filtration on the group G/H, called the *quotient* under H of the filtration on G; it is exhaustive if (G_n) is. If G′ is another filtered group and $(G'_n)_{n \in \mathbf{Z}}$ its filtration, $(G_n \times G'_n)$ is a filtration on $G \times G'$ called the *product* of the filtrations on G and G′, which is exhaustive (resp. separated) if (G_n) and (G'_n) are.

Now let A be a filtered ring and (A_n) its filtration; on every subring B of A, clearly the filtration induced by that on A is compatible with the ring structure on B. If $\mathfrak{b}$ is a two-sided ideal of A, the quotient filtration on $A/\mathfrak{b}$ of that on A is compatible with the structure of this ring, for

$$(A_n + \mathfrak{b})(A_m + \mathfrak{b}) \subset A_{n+m} + \mathfrak{b}.$$

If A′ is another filtered ring, the product filtration on $A \times A'$ is compatible with the structure of this ring.

Finally let E be a filtered A-module and (E_n) its filtration; on every submodule F of E, the filtration induced by that on E is compatible with the A-module structure on F and, on the quotient module E/F, the quotient filtration of that on E is compatible with the A-module structure, as

$$A_n(F + E_m) \subset F + A_n E_m \subset F + E_{m+n}.$$

Note that if the filtration on E is *derived* from that on A (*Example* 2), so is the quotient filtration on E/F, *but not in general the filtration induced on* F (Exercise 1; see however § 3, no. 2, Theorem 2).

If E' is another filtered A-module, the product filtration on E × E' is compatible with its A-module structure. If the filtrations on E and E' are derived from that on A (*Example* 2), so is their product filtration.

2. THE ORDER FUNCTION

Let A be a filtered ring, E a filtered A-module and (E_n) the filtration of E. For all $x \in E$ let $v(x)$ denote the *least upper bound* in $\overline{\mathbf{R}}$ of the set on integers $n \in \mathbf{Z}$ such that $x \in E_n$. Then the following equivalences hold:

$$(4)\qquad \begin{cases} v(x) = -\infty \Leftrightarrow x \notin \bigcup_{n\in\mathbf{Z}} E_n \\ v(x) = p \quad \Leftrightarrow x \in E_p \text{ and } x \notin E_{p+1} \\ v(x) = +\infty \Leftrightarrow x \in \bigcap_{n\in\mathbf{Z}} E_n \end{cases}$$

The mapping $v\colon E \to \overline{\mathbf{R}}$ is called the *order function* of the filtered module E. If v is known then so are the E_n, for E_n is the set of $x \in E$ such that $v(x) \geqslant n$; the fact that the E_n are additive subgroups of E implies the relation

$$(5)\qquad v(x-y) \geqslant \inf(v(x), v(y)).$$

The above definition applies in particular to the filtered A-module A_s; let w be its order function. It follows from equation (3) of no. 1 that for $a \in A$ and $x \in E$,

$$(6)\qquad v(ax) \geqslant w(a) + v(x)$$

whenever the right-hand side is defined; in particular, for $a \in A$ and $b \in A$,

$$(7)\qquad w(ab) \geqslant w(a) + w(b)$$

whenever the right hand side is defined.

The *order function* is defined similarly on a filtered group G which is not necessarily commutative; the corresponding relation to (5) is then written

$$(5')\qquad v(yx^{-1}) = v(xy^{-1}) \geqslant \inf(v(x), v(y)).$$

3. THE GRADED MODULE ASSOCIATED WITH A FILTERED MODULE

Let G be a commutative group (written additively) and (G_n) a filtration on G. Let us write

$$(8)\qquad \mathrm{gr}_n(G) = G_n/G_{n+1} \quad \text{for } n \in \mathbf{Z}$$
$$\mathrm{gr}(G) = \bigoplus_{n\in\mathbf{Z}} \mathrm{gr}_n(G).$$

The commutative group gr(G) is then a graded group of type **Z**, called the *graded group associated* with the filtered group G, the homogeneous elements of degree n of gr(G) being those of $\mathrm{gr}_n(\mathrm{G})$.

Now let A be a filtered ring, (A_n) its filtration, E a filtered A-module and (E_n) its filtration. For all $p \in \mathbf{Z}$, $q \in \mathbf{Z}$, a mapping

$$\mathrm{gr}_p(\mathrm{A}) \times \mathrm{gr}_q(\mathrm{E}) \to \mathrm{gr}_{p+q}(\mathrm{E}) \tag{9}$$

is defined as follows: given $\alpha \in \mathrm{gr}_p(\mathrm{A})$, $\xi \in \mathrm{gr}_q(\mathrm{E})$, two representatives a, a' of α and two representatives x, x' of ξ, $ax \in \mathrm{E}_{p+q}$, $a'x' \in \mathrm{E}_{p+q}$ and $ax \equiv a'x'$ (mod. E_{p+q+1}), for

$$ax - a'x' = (a - a')x + a'(x - x')$$

and $a - a' \in \mathrm{A}_{p+1}$ and $x - x' \in \mathrm{E}_{q+1}$ and hence our assertion follows from formula (3) of no. 1. We may therefore denote by $\alpha\xi$ the class in

$$\mathrm{E}_{p+1}/\mathrm{E}_{p+q+1} = \mathrm{gr}_{p+q}(\mathrm{E})$$

of the product ax of any representative $a \in \alpha$ and any representative $x \in \xi$. It is immediate that the mapping (9) is **Z**-*bilinear*; by linearity, we derive a **Z**-bilinear mapping

$$\mathrm{gr}(\mathrm{A}) \times \mathrm{gr}(\mathrm{E}) \to \mathrm{gr}(\mathrm{E}). \tag{10}$$

If this definition is first applied to the case $\mathrm{E} = \mathrm{A}_s$, the mapping (10) is an internal law of composition on gr(A), which it is immediately verified is *associative* and has an identity element which is the canonical image in $\mathrm{gr}_0(\mathrm{A})$ of the unit element of A; it therefore defines on gr(A) a ring structure and the graduation $(\mathrm{gr}_n(\mathrm{A}))_{n \in \mathbf{Z}}$ is by definition compatible with this structure. The graded ring gr(A) (of type **Z**) thus defined is called the *graded ring associated* with the filtered ring A; it is obviously commutative if A is commutative; $\mathrm{gr}_0(\mathrm{A})$ is a subring of gr(A). The mapping (10) is on the other hand a gr(A)-module external law on gr(E), the module axioms being trivially satisfied, and the graduation $(\mathrm{gr}_n(\mathrm{E}))_{n \in \mathbf{Z}}$ on gr(E) is obviously compatible with this module structure. The graded gr(A)-module gr(E) (of type **Z**) thus defined is called the *graded module associated* with the filtered A-module E.

Examples

(1) Let A be a commutative ring and t an element of A which is not a divisor of 0. Let us give A the (t)-*adic filtration* (no. 1, *Example* 3). Then the associated graded ring gr(A) is canonically isomorphic to the *polynomial ring* $(\mathrm{A}/(t))[\mathrm{X}]$. For $\mathrm{gr}_n(\mathrm{A}) = 0$ for $n < 0$ and by definition the ring $\mathrm{gr}_0(\mathrm{A})$ is the ring $\mathrm{A}/(t)$. We now note that by virtue of the hypothesis on t the relation $at^n \equiv 0$ (mod. t^{n+1}) is equivalent to $a \equiv 0$ (mod. t); if τ is the canonical image of t in $\mathrm{gr}_1(\mathrm{A})$, every element of $\mathrm{gr}_n(\mathrm{A})$ may then be written uniquely in the form $\alpha\tau^n$ where $\alpha \in \mathrm{gr}_0(\mathrm{A})$; whence our assertion.

(2) Let K be a commutative ring, A the ring of formal power series

$$K[[X_1, \ldots, X_r]]$$

(*Algebra*, Chapter IV, § 5) and $\mathfrak{m}$ the ideal of A whose elements are the formal power series with no constant term. Let us give A the $\mathfrak{m}$-*adic filtration* (no. 1, *Example* 3); if $M_1, \ldots, M_s$ are the distinct monomials in $X_1, \ldots, X_r$ of total degree $n - 1$, clearly every formal power series u of total order $\omega(u) \geqslant n$ (*loc. cit.*, no. 2) may be written as $\sum_{k=1}^{s} u_k M_k$, where the u_k belong to $\mathfrak{m}$; it is seen that $\mathfrak{m}^n$ is the set of formal power series u such that $\omega(u) \geqslant n$, which shows that ω is the *order function* for the $\mathfrak{m}$-adic filtration. Then clearly, for every formal power series $u \in \mathfrak{m}^n$, there exists a unique *homogeneous polynomial* of degree n in the X_i which is congruent to u mod. $\mathfrak{m}^{n+1}$, namely the sum of terms of degree n of u; we conclude that gr(A) is canonically isomorphic to the *polynomial ring* $K[X_1, \ldots, X_r]$.

(3) More generally, let A be a commutative ring, $\mathfrak{b}$ an ideal of A and A be given the $\mathfrak{b}$-adic filtration. If we write $B = \operatorname{gr}_0(A)$, $F = \operatorname{gr}_1(A) = \mathfrak{b}/\mathfrak{b}^2$, we know (*Algebra*, Chapter III,) that the identity mapping of the B-module F onto itself can be extended uniquely to a homomorphism u from the *symmetric algebra* S(F) of F to the B-algebra gr(A); it follows from the definition of gr(A) that u is a *surjective homomorphism of graded algebras*; for $n \geqslant 1$, every element of $\operatorname{gr}_n(A)$ is a sum of classes mod $\mathfrak{b}^{n+1}$ of elements of the form $y = x_1 x_2 \ldots x_n$, where $x_i \in \mathfrak{b}$ $(1 \leqslant i \leqslant n)$; if ξ_i is the class of x_i mod. $\mathfrak{b}^2$, clearly the class of y mod. $\mathfrak{b}^{n+1}$ is the element $u(\xi_1) \ldots u(\xi_n)$, whence our assertion. In particular, every system of generators of the B-module F is a system of generators of the B-algebra gr(A).

If now E is an A-module and E is given the $\mathfrak{b}$-adic filtration, it is seen similarly that the graded gr(A)-module gr(E) is *generated* by $\operatorname{gr}_0(E) = E/\mathfrak{b}E$. To be precise, the restriction ϕ to $\operatorname{gr}(A) \times \operatorname{gr}_0(E)$ of the external law on the gr(A)-module gr(E) is a **Z**-bilinear mapping of $\operatorname{gr}(A) \times \operatorname{gr}_0(E)$ to gr(E); moreover gr(A) is a $(\operatorname{gr}_0(A), \operatorname{gr}_0(A))$-bimodule and $\operatorname{gr}_0(E)$ a $\operatorname{gr}_0(A)$-module; it is immediately verified that, for $\alpha \in \operatorname{gr}(A)$, $\alpha_0 \in \operatorname{gr}_0(A)$, $\xi \in \operatorname{gr}_0(E)$,

$$\phi(\alpha\alpha_0, \xi) = \phi(\alpha, \alpha_0\xi)$$

and hence ϕ defines a *surjective* $\operatorname{gr}_0(A)$-*linear* mapping

$$\gamma_E \colon \operatorname{gr}(A) \otimes_{\operatorname{gr}_0(A)} \operatorname{gr}_0(E) \to \operatorname{gr}(E) \qquad (11)$$

which is called *canonical*.

* (4) Let K be a commutative ring, $\mathfrak{g}$ a Lie algebra over K and U the enveloping algebra of $\mathfrak{g}$. An *increasing* filtration $(U_n)_{n \in \mathbf{Z}}$ is defined on U by taking $U_n = \{0\}$ for $n < 0$ and denoting by U_n for $n \geqslant 0$ the set of elements of U which can be expressed as a sum of products of at most n elements of $\mathfrak{g}$; then

$U_0 = K$ and gr(U) is a *commutative* K-algebra (*Lie Groups and Lie Algebras*, Chapter I, § 2, no. 6). The canonical mapping of $\mathfrak{g}$ to $gr_1(U) = U_1/U_0$ can be extended uniquely to a homomorphism h of the symmetric algebra $S(\mathfrak{g})$ of the K-module $\mathfrak{g}$ to the K-algebra gr(U); the homomorphism h is surjective and, if the K-module $\mathfrak{g}$ is free, h is bijective (*loc. cit.*, no. 7, Theorem 1).

(5) Let A be a graded ring of type **Z** and E a graded A-module of type **Z**; let $A_{(i)}$ (resp. $E_{(i)}$) be the subgroup of homogeneous elements of degree i of A (resp. E). Let A and E be given the filtrations associated with their graduations (no. 1, *Example* 1) and let A′ and E′ denote the filtered ring and filtered A′-module thus obtained. Then it is immediate that the **Z**-linear mapping $A \to gr(A')$ which maps an element of $A_{(n)}$ to its canonical image in

$$gr_n(A) = \Big(\bigoplus_{i \geqslant n} A_{(i)}\Big)\Big/\Big(\bigoplus_{i \geqslant n+1} A_{(i)}\Big)$$

is a graded ring isomorphism. A canonical graded A-module isomorphism $E \to gr(E')$ is defined similarly.

PROPOSITION 1. *Let* A *be a filtered ring,* $(A_n)_{n \in \mathbf{Z}}$ *its filtration and v its order function. Suppose that* gr(A) *is a ring with no divisor of zero. Then, for every ordered pair of elements a, b of the ring* $B = \bigcup_{n \in \mathbf{Z}} A_n$, $v(ab) = v(a) + v(b)$.

As $\mathfrak{n} = \bigcap_{n \in \mathbf{Z}} A_n$ is a two-sided ideal of the ring B, the formula holds if $v(a)$ or $v(b)$ is equal to $+\infty$. If not, $v(a) = r$ and $v(b) = s$ are integers; the classes α of a mod. A_{r+1} and β of b mod. A_{s+1} are $\neq 0$ by definition, whence by hypothesis $\alpha\beta \neq 0$ in gr(A) and therefore $ab \notin A_{r+s+1}$; as $ab \in A_{r+s}$,

$$v(ab) = v(a) + v(b).$$

COROLLARY. *Let* A *be a filtered ring and* $(A_n)_{n \in \mathbf{Z}}$ *its filtration; let us set* $B = \bigcup_{n \in \mathbf{Z}} A_n$, $\mathfrak{n} = \bigcap_{n \in \mathbf{Z}} A_n$. *If the ring* gr(A) *has no divisors of zero, neither has the ring* $B/\mathfrak{n}$.

If a and b are elements of B not belonging to $\mathfrak{n}$, then $v(a) \neq +\infty$ and $v(b) \neq +\infty$, whence $v(ab) \neq +\infty$ and therefore $ab \notin \mathfrak{n}$.

Note that the ring A can be an integral domain and the filtration (A_n) exhaustive and separated without gr(A) being an integral domain (Exercise 2).

Remark. Let G be a group which is not necessarily commutative with a filtration $(G_n)_{n \in \mathbf{Z}}$ such that G_{n+1} is *normal in* G_n for all $n \in \mathbf{Z}$; again let $gr_n(G) = G_n/G_{n+1}$. The *restricted product* of the family $(gr_n(G))_{n \in \mathbf{Z}}$, that is the subgroup of the product $\prod_{n \in \mathbf{Z}} gr_n(G)$ consisting of the elements (ξ_n) all of whose components, except at most a finite number, are equal to the identity element, is also called the *graded group associated* with G and denoted by gr(G).

4. HOMOMORPHISMS COMPATIBLE WITH FILTRATIONS

Let G, G′ be two commutative groups (written additively), (G_n) a filtration on G and (G'_n) a filtration on G′; a homomorphism $h: G \to G'$ is called *compatible with the filtrations* on G and G′ if $h(G_n) \subset G'_n$ for all $n \in \mathbf{Z}$. The composite homomorphism $G_n \xrightarrow{h|G_n} G'_n \longrightarrow G'_n/G'_{n+1}$ is zero on G_{n+1} and hence defines by taking quotients a homomorphism $h_n: G_n/G_{n+1} \to G'_n/G'_{n+1}$; there is therefore a unique additive group homomorphism $\mathrm{gr}(h): \mathrm{gr}(G) \to \mathrm{gr}(G')$ such that, for all $n \in \mathbf{Z}$, $\mathrm{gr}(h)$ coincides with h_n on $\mathrm{gr}_n(G) = G_n/G_{n+1}$. $\mathrm{gr}(h)$ is called the *graded group homomorphism associated with h*. If G″ is a third filtered group and $h': G' \to G''$ a homomorphism which is compatible with the filtrations, $h' \circ h$ is a homomorphism which is compatible with the filtrations and

$$\mathrm{gr}(h' \circ h) = \mathrm{gr}(h') \circ \mathrm{gr}(h). \tag{12}$$

PROPOSITION 2. *Let* G *be a filtered commutative group and* H *a subgroup of* G; *let* H *be given the induced filtration and* G/H *the quotient filtration. If* $j: H \to G$ *is the canonical injection and* $p: G \to G/H$ *the canonical surjection, j and p are compatible with the filtrations and the sequence*

$$0 \longrightarrow \mathrm{gr}(H) \xrightarrow{\mathrm{gr}(j)} \mathrm{gr}(G) \xrightarrow{\mathrm{gr}(p)} \mathrm{gr}(G/H) \longrightarrow 0 \tag{13}$$

is exact.

The first assertion is obvious; if (G_n) is the filtration on G, then

$$(H \cap G_n) \cap G_{n+1} = H \cap G_{n+1}$$

and hence $\mathrm{gr}(j)$ is injective; moreover the canonical mapping

$$G_n \to (H + G_n)/H$$

is surjective, hence so is $\mathrm{gr}(p)$ and $\mathrm{gr}(p) \circ \mathrm{gr}(j) = 0$ by (12). Finally, let $\xi \in G_n/G_{n+1}$ belong to the kernel of $\mathrm{gr}(p)$; then there exists $x \in \xi$ such that $x \in H + G_{n+1}$; but as $G_{n+1} \subset G_n$,

$$G_n \cap (H + G_{n+1}) = (H \cap G_n) + G_{n+1}$$

and hence $x = y + z$ where $y \in H \cap G_n$ and $z \in G_{n+1}$; this proves that ξ is the class mod. G_{n+1} of $j(y)$, in other words it belongs to the image of $\mathrm{gr}(H)$ under $\mathrm{gr}(j)$.

Note that, given an exact sequence $0 \to G' \xrightarrow{u} G \xrightarrow{v} G'' \to 0$ of filtered commutative groups, where u and v are compatible with the filtrations, the sequence $0 \longrightarrow \mathrm{gr}(G') \xrightarrow{\mathrm{gr}(u)} \mathrm{gr}(G) \xrightarrow{\mathrm{gr}(v)} \mathrm{gr}(G'') \longrightarrow 0$ is not necessarily exact (Exercise 4).

If now A and B are two filtered rings and $h: A \to B$ a ring homomorphism which is compatible with the filtrations, it is immediately verified that the graded group homomorphism $gr(h): gr(A) \to gr(B)$ is also a ring homomorphism. In particular, if A′ is a subring of A with the induced filtration, gr(A′) is canonically identified with a graded subring of gr(A) (Proposition 2); if $\mathfrak{b}$ is a two-sided ideal of A and $A/\mathfrak{b}$ is given the quotient filtration, $gr(A/\mathfrak{b})$ is canonically identified with the quotient graded ring $gr(A)/gr(\mathfrak{b})$ (Proposition 2).

Finally, let A be a filtered ring, E, F two filtered A-modules and $u: E \to F$ an homomorphism compatible with the filtrations. Then it is immediate that $gr(u): gr(E) \to gr(F)$ is a gr(A)-*linear* mapping and hence a homogeneous homomorphism of degree 0 of graded gr(A)-modules. Moreover, if $u': E \to F$ is another A-homomorphism compatible with the filtrations, so is $u + u'$ and

$$gr(u + u') = gr(u) + gr(u'). \tag{14}$$

Remarks

(1) Clearly filtered ring homomorphisms (resp. homomorphisms of filtered modules over a given filtered ring A) compatible with the filtrations can be taken as *morphisms* for the filtered ring structure (resp. filtered A-module structure) (*Set Theory*, Chapter IV, § 2, no. 1).

(2) Let E and F be two modules over a filtered ring A and let them have the filtrations *derived* from the filtration (A_n) on A (no. 1, *Example* 2). Then *every* A-linear mapping $u: E \to F$ is compatible with the filtrations, since

$$u(A_nE) = A_n u(E) \subset A_nF.$$

(3) Note that a filtered A-module homomorphism $u: E \to F$ which is compatible with the filtrations may satisfy $gr(u) = 0$ without being zero; this is so for example of the endomorphism $x \mapsto nx$ of the additive group $\mathbf{Z}$ with the (n)-adic filtration (for any integer $n > 1$). The relation $gr(u) = gr(v)$ for two homomorphisms u, v of E to F, compatible with the filtrations, does not therefore imply necessarily $u = v$.

(4) The definitions at the beginning of this no. extend immediately to two groups G, G′, which are not necessarily commutative and are filtered by subgroups G_n, G'_n such that G_{n+1} (resp. G'_{n+1}) is normal in G_n (resp. G'_n). Proposition 2 is also valid with the same hypotheses on the G_n and assuming that H is invariant in G, the proof remaining unchanged except for notation.

5. THE TOPOLOGY DEFINED BY A FILTRATION

Let G be a group filtered by a family $(G_n)_{n \in \mathbf{Z}}$ of *normal* subgroups of G. There exists a unique topology on G which is compatible with the group structure and for which the G_n constitute a fundamental system of neighbourhoods of the identity element e of G (*General Topology*, Chapter III, § 1, no. 2, *Example*); it is called *the topology on* G *defined by the filtration* (G_n). When we use topological notions concerning a filtered group, we shall mean, unless otherwise stated, with the topology defined by the filtration. Note that the G_n, being subgroups of G, are *both open and closed* (*General Topology*, Chapter III, § 2, no. 1, Corollary to Proposition 4).

As each G_n is normal in G, the entourages of the left and right uniformities on G coincide; we deduce that G admits a *Hausdorff completion* group $\hat{G}$ (*General Topology*, Chapter III, § 3, no. 4, Theorem 1 and no. 1, Proposition 2).

For every subset M of G, the *closure* of M in G is equal to

$$\bigcap_{n \in \mathbf{Z}} (M.G_n) = \bigcap_{n \in \mathbf{Z}} (G_n.M)$$

(*General Topology*, Chapter III, § 3, no. 1, formula (1)); in particular $\bigcap_{n \in \mathbf{Z}} G_n$ is the closure of $\{e\}$; thus it is seen that for the topology on G to be *Hausdorff* it is necessary and sufficient that the filtration (G_n) be separated. For the topology on G to be *discrete*, it is necessary and sufficient that there exist $n \in \mathbf{Z}$ such that $G_n = \{e\}$ (in which case $G_m = \{e\}$ for $m \geqslant n$); then the filtration (G_n) is called *discrete*.

Since the Hausdorff group associated with G is $H = G/\left(\bigcap_{n \in \mathbf{Z}} G_n\right)$, the associated graded groups gr(G) and gr(H) (if H is given the quotient filtration) are canonically identified.

Not let G′ be another filtered group and $u: G \to G'$ a homomorphism compatible with the filtrations; the definition of the topologies on G and G′ shows immediately that u is *continuous* (*). If H is a subgroup (resp. normal subgroup) of G, the topology induced on H by that on G (resp. the quotient topology with respect to H of that on G) is the topology on H (resp. G/H) defined by the filtration induced by that on G (resp. quotient topology of that on G). The product topology of those on G and G′ is the topology defined by the product of filtrations on G and G′.

Let v be the order function (no. 2) on G. The hypothesis on the G_n implies that $v(xyx^{-1}) = v(y)$ and hence $v(xy^{-1}) = v(yx^{-1}) = v(x^{-1}y) = v(y^{-1}x)$ for all x, y in G. Let ρ be a real number such that $0 < \rho < 1$ (for example take

(*) Throughout this chapter we shall use the words "continuous homomorphism" in the sense of what is called "continuous representation" in *General Topology*, Chapter III, § 2, no. 8; the word "homomorphism" will *never* be used in the sense of *General Topology*, Chapter III, § 2, no. 8, Definition 1; for this notion we shall always use the term "strict morphism" in order to avoid any confusion.

$\rho = 1/e$) and let $d(x, y) = \rho^{v(xy^{-1})}$ for all x, y in G. Then $d(x, x) = 0$, $d(x, y) = d(y, x)$ and inequality (5') of no. 2 gives

$$(15) \qquad d(x, y) \leqslant \sup(d(x, z), d(y, z))$$

for all x, y, z in G, which implies the triangle inequality

$$d(x, y) \leqslant d(x, z) + d(y, z).$$

Thus d is a *pseudometric* on G which is invariant under left and right translations and G_n is the set of $x \in G$ such that $d(e, x) \leqslant \rho^n$; the uniform structure defined by d is then the uniform structure on the topological group G. If G is Hausdorff, G is a *zero-dimensional* metrizable topological space (*General Topology*, Chapter IX, § 6, no. 4); d is a *distance* on G if also the filtration (G_n) is exhaustive.

Given a topological ring A, recall that a *left topological* A-*module* is an A-module E with a topology compatible with its additive group structure and such that the mapping $(a, x) \mapsto ax$ from $A \times E$ to E is continuous (*General Topology*, Chapter III, § 6, no. 6).

PROPOSITION 3. *Let* A *be a filtered ring,* (A_n) *its filtration,* B *the subring* $\bigcup_{n \in \mathbf{Z}} A_n$ *of* A, E *a filtered* B-*module,* (E_n) *its filtration and* F *the sub*-B-*module* $\bigcup_{n \in \mathbf{Z}} E_n$ *of* E. *Then the mapping* $(a, x) \mapsto ax$ *from* $B \times F$ *to* F *is continuous.*

Let $a_0 \in B$, $x_0 \in F$; there exists by hypothesis integers r, s such that $a_0 \in A_r$ and $x_0 \in E_s$. The relation

$$ax - a_0x_0 = (a - a_0)x_0 + a_0(x - x_0) + (a - a_0)(x - x_0)$$

shows that if $a - a_0 \in A_i$ and $x - x_0 \in E_j$, $ax - a_0x_0$ belongs to

$$E_{i+s} + E_{j+r} + E_{i+j}.$$

Then, given an integer n, $ax - a_0x_0 \in E_n$ provided $i \geqslant n - s$, $j \geqslant n - r$ and $i + j \geqslant n$, that is so long as i and j are sufficiently large.

COROLLARY. *The ring* B *is a topological ring and the* B-*module* F *a topological* B-*module.*

The first assertion is obtained by applying Proposition 3 to $F = B_s$.

It is seen in particular that a filtered ring A whose filtration is *exhaustive* is a topological ring; if this is so every filtered A-module whose filtration is *exhaustive* is a topological A-module.

PROPOSITION 4. *Let* A *be a commutative ring filtered by an exhaustive filtration* (A_n) *and* $\mathfrak{p}$ *an ideal of* A. *Suppose that the ideal* $\mathrm{gr}(\mathfrak{p}) = \bigoplus_{n \in \mathbf{Z}} (\mathfrak{p} \cap A_n)/(\mathfrak{p} \cap A_{n+1})$ *of the ring* gr(A) *is prime. Then the closure of* $\mathfrak{p}$ *in* A *is a prime ideal.*

We know that gr(A/$\mathfrak{p}$) is isomorphic to gr(A)/gr($\mathfrak{p}$) (no. 4, Proposition 2) and hence an integral domain; we conclude that $A/\bigcap_{n \in \mathbf{Z}} (\mathfrak{p} + A_n)$ is an integral domain (no. 3, Corollary to Proposition 1). Then the closure $\bigcap_{n \in \mathbf{Z}} (\mathfrak{p} + A_n)$ of $\mathfrak{p}$ is a prime ideal.

Let A be a ring and $\mathfrak{m}$ a two-sided ideal of A; the topology defined on A by the $\mathfrak{m}$-adic filtration (no. 1, *Example* 3) is called the $\mathfrak{m}$-*adic topology*; as the $\mathfrak{m}$-adic filtration is exhaustive, A is a topological ring with this topology (Corollary to Proposition 3). Similarly, for every A-module E, the topology defined by the $\mathfrak{m}$-adic filtration is called the $\mathfrak{m}$-*adic topology* on E; E is a topological A-module under this topology.

Let $\mathfrak{m}'$ be another two-sided ideal of A; for the $\mathfrak{m}'$-adic topology on A to be *finer* than the $\mathfrak{m}$-adic topology, it is necessary and sufficient that there exist an integer $n > 0$ such that $\mathfrak{m}'^n \subset \mathfrak{m}$; the condition is necessary and, if it is fulfilled, $\mathfrak{m}'^{hn} \subset \mathfrak{m}^h$ for all $h > 0$ and hence the condition is sufficient. If A is a *commutative Noetherian ring*, it amounts to the same to say that $V(\mathfrak{m}) \subset V(\mathfrak{m}')$ in the prime spectrum of A (Chapter II, § 4, no. 3, Corollary 2 to Proposition 11 and § 2, no. 6, Proposition 15).

6. COMPLETE FILTERED MODULES

PROPOSITION 5. *Let* G *be a filtered group whose filtration* (G_n) *consists of invariant subgroups of* G. *The following conditions are equivalent:*

(a) G *is a complete topological group.*

(b) *The associated Hausdorff group* $G' = G/\left(\bigcap_{n \in \mathbf{Z}} G_n\right)$ *is complete.*

(c) *Every Cauchy sequence in* G *is convergent.*

If G *is commutative and written additively, these conditions are also equivalent to the following:*

(d) *Every family* $(x_\lambda)_{\lambda \in L}$ *of elements of* G′ *which converges to* 0 *with respect to the filter* $\mathfrak{F}$ *of complements of finite subsets of* L *is summable in* G′.

For a filter on G to be a Cauchy filter (resp. a convergent filter), it is necessary and sufficient that its image under the canonical mapping $G \to G'$ be a Cauchy (resp. convergent) filter (*General Topology*, Chapter II, § 3, no. 1, Proposition 4); whence first of all the equivalence of (a) and (b); on the other hand, as G′ is metrizable, the equivalence of (b) and (c) follows from Proposition 9 of *General Topology*, Chapter IX, § 2, no. 6.

Suppose now that G is commutative. Suppose that G′ is complete and let $(x_\lambda)_{\lambda \in L}$ be a family of elements of G′ which converge to 0 with respect to $\mathfrak{F}$. For every neighbourhood V′ of 0 in G′ which is a subgroup of G′, there exists a finite subset J of L such that the condition $\lambda \in L - J$ implies $x_\lambda \in V'$; then $\sum_{\lambda \in H} x_\lambda \in V'$ for every finite subset H of L not meeting J, which shows that the family $(x_\lambda)_{\lambda \in L}$ is summable (*General Topology*, Chapter III, § 5, no. 2, Theorem 1).

Conversely, suppose that condition (d) holds and let (x_n) be a Cauchy sequence on G′; the family $(x_{n+1} - x_n)$ is then summable and in particular the series with general term $x_{n+1} - x_n$ is convergent and hence the sequence (x_n) is convergent.

Let G be a filtered group whose filtration (G_n) consists of normal subgroups of G; the quotient groups G/G_n are *discrete* and hence complete, since the G_n are open in G. Let f_n be the canonical mapping $G \to G/G_n$ and for $m \leqslant n$ let f_{mn} be the canonical mapping $G/G_n \to G/G_m$; $(G/G_n, f_{mn})$ is an inverse system of discrete groups with **Z** as indexing set (*General Topology*, Chapter III, § 7, no. 3). Let $\tilde{G}$ be the topological group the inverse limit of this inverse system and for all n let $g_n: \tilde{G} \to G/G_n$ be the canonical mapping; let $f: G \to \tilde{G}$ be the inverse limit of the inverse system of mappings (f_n) such that $f_n = g_n \circ f$ for all n; finally, let j be the canonical mapping of G to its Hausdorff completion $\hat{G}$; as the G/G_n are complete, there exists a unique topological group isomorphism $i: \hat{G} \to \tilde{G}$ such that $f = i \circ j$ (*loc. cit.*, Corollary 1 to Proposition 2); we shall call it the *canonical* isomorphism of $\hat{G}$ onto $\tilde{G}$.

Let H be another filtered group whose filtration (H_n) consists of normal subgroups of H and let $u: G \to H$ be a homomorphism compatible with the filtrations (no. 4). Set $\tilde{H} = \varprojlim H/H_n$; for all n, u defines by taking quotients a homomorphism $u_n: G/G_n \to H/H_n$ and the u_n obviously form an inverse system of mappings; set $\tilde{u} = \varprojlim u_n$. Moreover let $\hat{H}$ be the Hausdorff completion of H and $\hat{u}: \hat{G} \to \hat{H}$ the homomorphism derived from u by passing to the Hausdorff completions (*General Topology*, Chapter II, § 3, no. 7, Proposition 15). It follows immediately from the definitions that if $\hat{G}$ is identified with $\tilde{G}$ and $\hat{H}$ with $\tilde{H}$ by means of the canonical isomorphisms, $\hat{u}$ is identified with $\tilde{u}$. We conclude in particular that, if, for all n, u_n is an isomorphism, then $\hat{u}$ is an isomorphism of topological groups.

Examples of complete filtered groups and rings

(1) Let G be a complete filtered group. Every *closed* subgroup of G with the induced filtration is complete (*General Topology*, Chapter II, § 3, no. 4, Proposition 8). Every quotient group of G with the quotient filtration is complete (*General Topology*, Chapter IX, § 3, no. 1, *Remark* 1).

(2) Let A be a filtered commutative ring whose filtration we denote by $(\mathfrak{a}_n)_{n \in \mathbf{Z}}$; let A′ be the ring of formal power series $A[[X_1, \ldots, X_s]]$. For all $e = (e_1, \ldots, e_s) \in \mathbf{N}^s$, we write $|e| = \sum_{i \in I}^{s} e_i$, $X^e = \prod_{i \in I}^{s} X_i^{e_i}$ so that every element $P \in A'$ can be written uniquely $P = \sum_{e \in \mathbf{N}^s} \alpha_{e,P} X^e$ where $\alpha_{e,P} \in A$. For all $n \in \mathbf{Z}$,

let $\mathfrak{a}'_n$ denote the set of $P \in A'$ such that $\alpha_{e,P} \in \mathfrak{a}_{n-|e|}$ for all $e \in \mathbf{N}^s$; we show that $\mathfrak{a}'_n$ is an *ideal* of A'. Clearly $\mathfrak{a}'_n$ is an additive subgroup of A'; on the other hand, if $P \in \mathfrak{a}'_n$ and $Q \in A'$, then, for all $e \in \mathbf{N}^s$, $\alpha_{e,PQ} = \sum_{e'+e''=e} \alpha_{e',Q}\alpha_{e'',P}$; as the relation $e' + e'' = e$ implies $|e''| \leqslant |e|$, $PQ \in \mathfrak{a}'_n$. Moreover, if $Q \in \mathfrak{a}'_m$, then, for $e' + e'' = e$, $\alpha_{e',Q}\alpha_{e'',P} \in \mathfrak{a}_{m-|e'|}\mathfrak{a}_{n-|e''|} \subset \mathfrak{a}_{m+n-|e|}$, which proves that $(\mathfrak{a}'_n)_{n\in\mathbf{Z}}$ is a *filtration* compatible with the ring structure on A' (for obviously $1 \in \mathfrak{a}'_0$). When in future we speak of A' as a filtered ring, we shall mean, unless otherwise stated, with the filtration $(\mathfrak{a}'_n)$. Clearly $\bigcap_{n\in\mathbf{Z}} \mathfrak{a}'_n$ is the set of formal power series all of whose coefficients belong to $\bigcap_{n\in\mathbf{Z}} \mathfrak{a}_n$; then, if A is Hausdorff, so is A'. If $\mathfrak{a}_0 = A$, then $\mathfrak{a}'_0 = A'$.

PROPOSITION 6. *With the above notation, suppose that $\mathfrak{a}_0 = A$ and let h denote the mapping $P \mapsto (\alpha_{e,P})_{e\in\mathbf{N}^s}$. Then h is an isomorphism of the additive topological group A' onto the additive topological group $A^{\mathbf{N}^s}$. The polynomial ring $A[X_1, \ldots, X_s]$ is dense in A'; if A is complete, so is A'.*

Clearly h is bijective; $V_n = h(\mathfrak{a}'_n)$ is the set of $(a_e) \in A^{\mathbf{N}^s}$ such that $a_e \in \mathfrak{a}_{n-|e|}$ for all $e \in \mathbf{N}^s$ such that $|e| \leqslant n$; as these elements e are finite in number, V_n is a neighbourhood of 0 in $A^{\mathbf{N}^s}$. Conversely, if V is a neighbourhood of 0 in $A^{\mathbf{N}^s}$, there is a finite subset E of $\mathbf{N}^s$ and an integer ν such that the conditions $a_e \in \mathfrak{a}_\nu$ for all $e \in E$ imply $(a_e) \in V$; if then n is the greatest of the integers $\nu + |e|$ for $e \in E$, then $h(\mathfrak{a}'_n) \subset V$, which proves the first assertion of Proposition 6. Moreover, with n and E defined as above, $h\left(P - \sum_{e\in E} \alpha_{e,P}X^e\right) \in V$ for all $P \in A'$, which shows that $A[X_1, \ldots, X_s]$ is dense in A'. The last assertion follows from the first and the fact that a product of complete spaces is complete.

Let $\mathfrak{m}$ be an ideal of A and suppose that $(\mathfrak{a}_n)$ is the $\mathfrak{m}$-*adic* filtration; then, if $\mathfrak{n}$ is the ideal of A' generated by $\mathfrak{m}$ and the X_i $(1 \leqslant i \leqslant s)$, the filtration $(\mathfrak{a}'_n)$ is the $\mathfrak{n}$-*adic* filtration. For clearly, for all $k \geqslant 0$, $\mathfrak{n}^k$ is generated by the elements aX^e such that $a \in \mathfrak{m}^{k-|e|}$ for all $e \in \mathbf{N}^s$ such that $|e| \leqslant k$, whence $\mathfrak{n}^k \subset \mathfrak{a}'_k$. Let us prove conversely that $\mathfrak{a}'_k \subset \mathfrak{n}^k$. For all $P \in \mathfrak{a}'_k$, $P = P' + P''$, where $P' = \sum_{|e|<k} \alpha_{e,P}X^e$, $P'' = \sum_{|e|\geqslant k} \alpha_{e,P}X^e$. Clearly it is possible to write $P'' = \sum_{|e|=k} X^eQ_e$, where the Q_e are elements of A', whence $P'' \in \mathfrak{n}^k$; on the other hand, clearly $\alpha_{e,P}X^e \in \mathfrak{n}^k$ for all $e \in \mathbf{N}^s$, whence $P' \in \mathfrak{n}^k$. Then $\mathfrak{n}^k = \mathfrak{a}'_k$.

COROLLARY. *Let A be a commutative ring,*

$$A' = A[[X_1, \ldots, X_s]]$$

the ring of formal power series in s indeterminates over A and $\mathfrak{n}$ the ideal of A' consisting of the formal power series with no constant term. The ring A' is Hausdorff and complete *with the* $\mathfrak{n}$*-adic topology and the polynomial ring* $A[X_1, \ldots, X_s]$ *is everywhere dense in* A'.

It is sufficient to apply what has just been said to the case $\mathfrak{m} = \{0\}$.

7. LINEAR COMPACTNESS PROPERTIES OF COMPLETE FILTERED MODULES

Recall that, if E is an A-module, an *affine subset* (or an *affine linear variety*) of E is any subset F which is empty or of the form $a + M$, where $a \in E$ and M is a submodule of E called the *direction* of F (*Algebra*, Chapter II, § 9, nos. 1 and 3).

PROPOSITION 7. *Let* A *be a filtered ring,* E *a filtered* A*-module and* (E_n) *the filtration of* E; *suppose that* $E_0 = E$, *that the* E_n *are submodules of* E, *that the* A*-modules* E/E_n *are Artinian and finally that the topological group* E *is Hausdorff and complete. Then the intersection of a decreasing sequence of non-empty closed affine subsets of* E *is non-empty.*

We have seen in no. 6 that, since E is Hausdorff and complete, it is identified with $\hat{E} = \varprojlim E/E_n$. Let (W_p) be a decreasing sequence of non-empty closed affine subsets of E and, for all $n \geqslant 0$, let $W_{p,n}$ be the canonical image of W_p in E/E_n; we are going to construct a sequence $x = (x_n) \in \hat{E}$ such that $x_n \in W_{p,n}$ for all p and all n; hence $x \in W_p + E_n$ for all p and all n and, as the W_p are closed, $x \in W_p$ for all p (no. 5), which will prove the proposition.

As $E/E_0 = 0$, we shall take $x_0 = 0$. Suppose that the x_i are defined for $0 \leqslant i \leqslant n - 1$ and let $W'_{p,n}$ be the set of elements of $W_{p,n}$ whose canonical image in E/E_{n-1} is x_{n-1}; as $x_{n-1} \in W_{p,n-1}$ and $W_{p,n-1}$ is the canonical image of $W_{p,n}$, $W'_{p,n}$ is non-empty and is obviously an affine subset of E/E_n; moreover the sequence $(W'_{p,n})_{p \in \mathbf{N}}$ is decreasing. As E/E_n is Artinian, this sequence is stationary (otherwise the sequence of submodules of E/E_n which are the directions of the $W'_{p,n}$ would be strictly decreasing, which is absurd). It is sufficient then to take x_n to be an element of $\bigcap_{p \in \mathbf{N}} W'_{p,n}$ and the construction of (x_n) can then be performed by induction.

PROPOSITION 8. *Suppose that* A *and* E *satisfy the hypotheses of Proposition* 7. *Let* (F_p) *be a decreasing sequence of closed submodules of* E *such that* $\bigcap_p F_p = 0$. *Then, for every neighbourhood* V *of* 0 *in* E, *there exists* p *such that* $F_p \subset V$ (in other words, the base of the filter (F_p) converges to 0).

We may assume that V is one of the E_n, in which case E/V is Artinian. Let us write $F'_p = (F_p + V)/V$; as the F'_p form a decreasing sequence of submodules of E/V, there exists an integer j such that $F'_p = F'_j$ for all $p \geqslant j$. We shall see that $F'_j = \{0\}$, which will complete the proof. Let $x \in F'_j$ and let W_p be the set of elements of F_p whose image in E/V is x ($p \geqslant j$); by definition of j, the W_p are *non-empty* closed affine subsets of E and obviously $W_{p+1} \subset W_p$;

then it follows from Proposition 7 that there exists an element y belonging to all the W_p. As $W_p \subset F_p$ and $\bigcap_{p \in N} F_p = \{0\}$, $y = 0$; since x is the canonical image of y in E/V, $x = 0$ (cf. Exercises 15 to 21).

8. THE LIFT OF HOMOMORPHISMS OF ASSOCIATED GRADED MODULES

THEOREM 1. *Let X, Y be two filtered groups whose filtrations (X_n), (Y_n) consist of normal subgroups; let $u: X \to Y$ be a homomorphism compatible with the filtrations.*

(i) *Suppose that the filtration (X_n) is exhaustive. For* $\mathrm{gr}(u)$ *to be injective, it is necessary and sufficient that* $\overset{-1}{u}(Y_n) = X_n$ *for all* $n \in \mathbf{Z}$.

(ii) *Suppose that one of the following hypotheses holds:* (α) *X is complete and Y Hausdorff;* (β) *Y is discrete. Then, for* $\mathrm{gr}(u)$ *to be surjective, it is necessary and sufficient that* $Y_n = u(X_n)$ *for all* $n \in \mathbf{Z}$.

(i) To say that the mapping $\mathrm{gr}_n(u)$ is injective means that

$$X_n \cap \overset{-1}{u}(Y_{n+1}) \subset X_{n+1}.$$

This is obviously the case of $\overset{-1}{u}(Y_{n+1}) = X_{n+1}$. Conversely, if

$$X_n \cap \overset{-1}{u}(Y_{n+1}) \subset X_{n+1}$$

for all n, we deduce by induction on k that $X_{n-k} \cap \overset{-1}{u}(Y_{n+1}) \subset X_{n+1}$ for all $n \in \mathbf{Z}$ and all $k \geqslant 0$. As the filtration (X_n) is exhaustive, we see that, for all n, X is the union of the X_{n-k} $(k \geqslant 0)$, hence $\overset{-1}{u}(Y_{n+1}) \subset X_{n+1}$ for all n and therefore $X_{n+1} \subset \overset{-1}{u}(Y_{n+1})$, which completes the proof.

(ii) To say that the mapping $\mathrm{gr}_n(u)$ is surjective means that

$$Y_n = u(X_n)Y_{n+1}.$$

This is obviously the case if $Y_n = u(X_n)$. Conversely, suppose that $Y_n = u(X_n)Y_{n+1}$ for all $n \in \mathbf{Z}$. Let n be an integer and y an element of Y_n; we shall define a sequence $(x_k)_{k \geqslant 0}$ of elements of X_n such that $x_k \in X_n$, $x_{k+1} \equiv x_k \pmod{X_{n+k}}$ and $u(x_k) \equiv y \pmod{Y_{n+k}}$ for all $k \geqslant 0$. We shall take x_0 equal to the identity element of X, which certainly gives $u(x_0) \equiv y$ (mod. Y_n). Suppose that an $x_k \in X_n$ has been constructed such that $u(x_k) \equiv y$ (mod. Y_{n+k}); then $(u(x_k))^{-1}y \in Y_{n+k}$; the hypothesis implies that there exists $t \in X_{n+k}$ such that $u(t) \equiv (u(x_k))^{-1}y$ (mod. Y_{n+k+1}) and hence $u(x_k t) \equiv y$ (mod. Y_{n+k+1}); it is sufficient to take $x_{k+1} = x_k t$ to carry out the induction. This being so, if Y is discrete, there exists $k \geqslant 0$ such that $Y_{n+k} = \{e'\}$ (the identity element of Y), whence $u(x_k) = y$ and hence in this case it has been proved that $u(X_n) = Y_n$ for all n. Suppose now that X is complete and Y Hausdorff. As $x_h^{-1}x_k \in X_{n+k}$ for $h \geqslant k \geqslant 0$, (x_k) is a Cauchy sequence in X_n; as

X_n is closed in X and hence complete, this sequence has at least one limit x in X_n. By virtue of the continuity of u, $u(x)$ is the unique limit of the sequence $(u(x_k))$ in Y, Y being Hausdorff. But the relations $u(x_k) \equiv y \pmod{Y_{n+k+1}}$ show that y is also a limit of this sequence, whence $u(x) = y$ and it has also been proved that $u(X_n) = Y_n$.

COROLLARY 1. *Suppose that* X *is Hausdorff and its filtration exhaustive. Then, if* gr(u) *is injective,* u *is injective.*

Let e, e' be the identity elements of X and Y respectively. Then

$$\overset{-1}{u}(e') \subset \bigcap_n \overset{-1}{u}(Y_n) = \bigcap_n X_n = \{e\}$$

by hypothesis, whence the corollary.

COROLLARY 2. *Suppose that one of the following hypotheses holds:*

(α) X *is complete,* Y *is Hausdorff and its filtration is exhaustive;*

(β) Y *is discrete and its filtration is exhaustive.*

Then, if gr(u) *is surjective,* u *is surjective.*

In this case $Y = \bigcup_n Y_n = \bigcup_n u(X_n) \subset u(X)$.

COROLLARY 3. *Suppose that* X *and* Y *are Hausdorff, the filtrations of* X *and* Y *exhaustive and* X *complete. Then, if* gr(u) *is bijective,* u *is bijective.*

Let A be a local ring, $\mathfrak{m}$ its maximal ideal and M an A-module; let A and M be given the $\mathfrak{m}$-*adic* filtrations and let gr(A) and gr(M) be the graded ring and the graded gr(A)-module associated with A and M. We have seen (no. 3, *Example* 3) that the canonical mapping (11) is always *surjective*; we are going to consider the following property of M:

(GR) *The canonical mapping*

$$\gamma_M : \mathrm{gr}(A) \otimes_{\mathrm{gr}_0(A)} \mathrm{gr}_0(M) \to \mathrm{gr}(M)$$

is bijective.

PROPOSITION 9. *Let* A *be a local ring,* $\mathfrak{m}$ *its maximal ideal,* M, N *two* A*-modules and* $u: N \to M$ *an* A*-homomorphism.* M *and* N *are given the* $\mathfrak{m}$*-adic filtrations and suppose that:* (1) M *satisfies property* (GR); (2) $\mathrm{gr}_0(u): \mathrm{gr}_0(N) \to \mathrm{gr}_0(M)$ *is injective. Then* $\mathrm{gr}(u): \mathrm{gr}(N) \to \mathrm{gr}(M)$ *is injective,* N *and* $P = \mathrm{Coker}(u)$ *satisfy property* (GR) *and* $\mathfrak{m}^n N = \overset{-1}{u}(\mathfrak{m}^n M)$ *for every integer* $n > 0$.

It is immediately verified that the diagram

$$\begin{array}{ccc} \mathrm{gr}(A) \otimes_{\mathrm{gr}_0(A)} \mathrm{gr}_0(N) & \xrightarrow{1 \otimes \mathrm{gr}_0(u)} & \mathrm{gr}(A) \otimes_{\mathrm{gr}_0(A)} \mathrm{gr}_0(M) \\ \downarrow{\scriptstyle \gamma_N} & & \downarrow{\scriptstyle \gamma_M} \\ \mathrm{gr}(N) & \xrightarrow[\mathrm{gr}(u)]{} & \mathrm{gr}(M) \end{array}$$

is commutative. As $\mathrm{gr}_0(A) = A/\mathfrak{m}$ is a field, the hypothesis implies that $1 \otimes \mathrm{gr}_0(u)$ is injective; as by hypothesis γ_M is injective, so is $\gamma_M \circ (1 \otimes \mathrm{gr}_0(u))$. This implies first that γ_N is injective and hence bijective and therefore that $\mathrm{gr}(u)$ is injective. The formula $\overset{-1}{u}(\mathfrak{m}^n M) = \mathfrak{m}^n N$ is then a consequence of Theorem 1 (i).

Also, let us write $N' = u(N)$ and let $j: N' \to M$ be the canonical injection. If $p: M \to P = M/N'$ is the canonical homomorphism, then in the commutative diagram

$$\begin{array}{ccccccc}
\mathrm{gr}(A) \otimes \mathrm{gr}_0(N') & \xrightarrow{1 \otimes \mathrm{gr}_0(j)} & \mathrm{gr}(A) \otimes \mathrm{gr}_0(M) & \xrightarrow{1 \otimes \mathrm{gr}_0(p)} & \mathrm{gr}(A) \otimes \mathrm{gr}_0(P) & \to & 0 \\
\downarrow{\scriptstyle \gamma_{N'}} & & \downarrow{\scriptstyle \gamma_M} & & \downarrow{\scriptstyle \gamma_P} & & \\
\mathrm{gr}(N') & \xrightarrow[\mathrm{gr}(j)]{} & \mathrm{gr}(M) & \xrightarrow[\mathrm{gr}(p)]{} & \mathrm{gr}(P) & \longrightarrow & 0
\end{array}$$

the lower row is exact (no. 4, Proposition 2) and so is the upper row by virtue of Proposition 2 of no. 4 and the fact that $\mathrm{gr}_0(A)$ is a field. Moreover, $\mathrm{gr}(j)$ is injective (no. 4, Proposition 2) and hence $\mathrm{gr}_0(j)$ is injective. The first part of the argument applied to j shows that $\gamma_{N'}$ is bijective; as γ_M is also bijective by hypothesis, we conclude that γ_P is bijective (Chapter I, § 1, no. 4, Corollary 2 to Proposition 2).

COROLLARY. *Under the hypotheses of Proposition* 9, *if we assume also that* N *is Hausdorff with the* $\mathfrak{m}$*-adic filtration, then u is injective.*

This follows from the fact that $\mathrm{gr}(u)$ is injective (Corollary 1 to Theorem 1).

* *Remark.* Suppose that the hypotheses of Proposition 9 hold and also *one* of the following conditions:

(1) $\mathfrak{m}$ is nilpotent;

(2) A is Noetherian and P is ideally Hausdorff (cf. § 5, no. 1);

then P is a *flat* A-module. This follows from the fact that γ_P is bijective and § 5, no. 2, Theorem 1 (iv), since $A/\mathfrak{m}$ is a field. *

9. THE LIFT OF FAMILIES OF ELEMENTS OF AN ASSOCIATED GRADED MODULE

Let A be a filtered commutative ring, $(A_n)_{n \in \mathbf{Z}}$ its filtration and C a subring of A_0 such that $C \cap A_1 = \{0\}$. The restriction to C of the canonical mapping $A_0 \to A_0/A_1 = \mathrm{gr}_0(A)$ is then injective, which allows us to identify C with a submodule of $\mathrm{gr}_0(A)$; this is what we shall usually do in similar cases. If $A_1 \neq A_0$ and K is any *subfield* of A_0, then $K \cap A_1 = \{0\}$ since $K \cap A_1$ is an ideal of K not containing 1; then K may be identified with a subfield of $\mathrm{gr}_0(A)$.

PROPOSITION 10. *Let* A *be a filtered commutative ring and* (A_n) *its filtration; suppose that there exists a subring* C *of* A_0 *such that* $C \cap A_1 = \{0\}$ *and* C *is identified with a subring of* $\mathrm{gr}_0(A)$. *Let* $(x_i)_{1 \leqslant i \leqslant q}$ *be a finite family of elements of* A; *suppose that* $x_i \in A_{n_i}$ *for* $1 \leqslant i \leqslant q$ *and let* ξ_i *be the class of* x_i *in* $\mathrm{gr}_{n_i}(A)$ *for* $1 \leqslant i \leqslant q$.

(i) *If the family* (ξ_i) *of elements of* gr(A) *is algebraically free over* C, *the family* (x_i) *is algebraically free over* C.

(ii) *If the filtration on* A *is exhaustive and discrete and* (ξ_i) *is a system of generators of the* C*-algebra* gr(A), *then* (x_i) *is a system of generators of the* C*-algebra* A.

Let A′ be the polynomial algebra $C[X_1, \ldots, X_q]$ over C; let A′ be given the graduation (A'_n) of type **Z** where A'_n is the set of C-linear combinations of the monomials $X_1^{s(1)} \ldots X_q^{s(q)}$ such that $\sum_{i=1}^{q} n_i s(i) = n$. Let u be the homomorphism $f \mapsto f(x_1, \ldots, x_q)$ from the C-algebra A′ to the C-algebra A; by definition, $u(A'_n) \subset A_n$ for all $n \in \mathbf{Z}$ and hence u is compatible with the filtrations (A′ being given its filtered ring structure associated with its graded ring structure, cf. no. 1, *Example* 1). This being so, the hypothesis of (i) means that

$$\mathrm{gr}(u) : A' = \mathrm{gr}(A') \to \mathrm{gr}(A)$$

is injective; as the filtration on A′ is exhaustive and separated, Corollary 1 to Theorem 1 of no. 8 may be applied and u is injective, which proves the conclusion of (i). Similarly, the hypothesis (ii) on the (ξ_i) means that $\mathrm{gr}(u)$ is surjective; as A is discrete and its filtration is exhaustive, Corollary 2 to Theorem 1 of no. 8 may be applied and u is surjective, which proves the conclusion of (ii).

PROPOSITION 11. *Let* A *be a complete Hausdorff filtered commutative ring,* C *a subring of* A_0 *such that* $C \cap A_1 = 0$ *and* $(x_i)_{1 \leqslant i \leqslant q}$ *a finite family of elements of* A *such that* $x_i \in A_{n_i}$ *where* $n_i > 0$ *for* $1 \leqslant i \leqslant q$; *let* ξ_i *be the class of* x_i *in* $\mathrm{gr}_{n_i}(A)$ *for* $1 \leqslant i \leqslant q$.

(i) *There exists a unique* C*-homomorphism* v *from the algebra of formal power series* $A'' = C[[X_1, \ldots, X_q]]$ *to* A *such that* $v(X_i) = x_i$ *for* $1 \leqslant i \leqslant q$.

(ii) *If the family* (ξ_i) *is algebraically free over* C, *the homomorphism* v *is injective.*

(iii) *If the filtration on* A *is exhaustive and the family* (ξ_i) *is a system of generators of the* C*-algebra* gr(A), *the homomorphism* v *is surjective.*

As $n_i \geqslant 1$ for all i, $\sum_{i=1}^{q} n_i s(i) \geqslant \sum_{i=1}^{q} s(i)$ for every monomial $X_1^{s(1)} \ldots X_q^{s(q)}$ and on the other hand $\sum_{i=1}^{q} n_i s(i) \leqslant r . \sum_{i=1}^{q} s(i)$ if r is the greatest of the n_i. If A''_n denotes the set of formal power series whose non-zero terms $a_s X_1^{s(1)} \ldots X_q^{s(q)}$ satisfy $\sum_{i=1}^{q} n_i s(i) \geqslant n$, it follows from no. 6, Corollary to Proposition 6 that A″ is Hausdorff and complete with the exhaustive filtration (A''_n) and that

$$A' = C[X_1, \ldots, X_q]$$

is dense in A″; moreover the homomorphism u defined in the proof of Proposition 10 is *continuous* on A′ and can be extended uniquely to a continuous homomorphism $v: A'' \to A$, since A is Hausdorff and complete (*General Topology*, Chapter III, § 3, no. 3, Proposition 5), which proves (i); also, $\mathrm{gr}(A'') = \mathrm{gr}(A')$ and $\mathrm{gr}(v) = \mathrm{gr}(u)$; (ii) and (iii) then follow respectively from Corollaries 1 and 2 to Theorem 1 of no. 8 in view of the hypotheses on A.

The conclusion of (ii) (resp. (iii)) of the proposition is sometimes expressed by saying that the family (x_i) is *formally free* over C (resp. a *formal system of generators* of A).

PROPOSITION 12. *Let* A *be a filtered ring,* E *a filtered* A*-module and* (A_n) *and* (E_n) *the respective filtrations on* A *and* E. *Suppose that* A *is complete and the filtration* (E_n) *is exhaustive and separated. Let* $(x_i)_{i \in I}$ *be a finite family of elements of* E *and for* $i \in I$ *let* $n(i)$ *be an integer such that* $x_i \in E_{n(i)}$; *finally let* ξ_i *be the class of* x_i *in* $\mathrm{gr}_{n(i)}(E)$. *Then, if* (ξ_i) *is a system of generators of the* gr(A)*-module* gr(E), (x_i) *is a system of generators of the* A*-module* E.

In the A-module $L = A_s^I$ let L_n denote the set (a_i) such that $a_i \in A_{n-n(i)}$ for all $i \in I$; if p and q are the least and greatest of the $n(i)$, then $A_{n-q}^I \supset L_n \supset A_{n-p}^I$ and the topology defined on L by the definition (L_n) is the same as the product topology; hence L is a *complete* filtered A-module. As L is free, there exists an A-linear mapping $u: L \to E$ such that $u((a_i)) = \sum_{i \in I} a_i x_i$ and it is obviously compatible with the filtrations; we must prove that u is surjective and for this it is sufficient, by virtue of Corollary 2 to Theorem 1, no. 8, to show that

$$\mathrm{gr}(u): \mathrm{gr}(L) \to \mathrm{gr}(E)$$

is surjective or also that, for all $x \in E_n$, there exist a family (a_i) such that $a_i \in A_{n-n(i)}$ for all $i \in I$ and $x \equiv \sum_{i \in I} a_i x_i \pmod{E_{n+1}}$. Let ξ be the class of x in $\mathrm{gr}_n(E)$; since the ξ_i generate the gr(A)-module gr(E), there exist $\alpha_i \in \mathrm{gr}(A)$ such that $\xi = \sum_{i \in I} \alpha_i \xi_i$ and we may assume that $\alpha_i \in \mathrm{gr}_{n-n(i)}(A)$ by replacing if need be α_i by its homogeneous component of degree $n - n(i)$. Then α_i is the image of an element $a_i \in A_{n-n(i)}$ and the family (a_i) has the required property.

COROLLARY 1. *Let* A *be a complete filtered ring and* E *a filtered* A*-module whose filtration is exhaustive and separated. If* gr(E) *is a finitely generated* (resp. *Noetherian*) gr(A)*-module, then* E *is a finitely generated* (resp. *Noetherian*) A*-module.*

If gr(E) is finitely generated, there is a finite system of homogeneous generators and Proposition 12 shows that E is finitely generated. Suppose now that gr(E) is Noetherian and let F be a submodule of E; the filtration induced on F by that on E is exhaustive and separated and gr(F) is identified with a sub-gr(A)-module of gr(E) (no. 4, Proposition 2) and hence is finitely generated by hypothesis; we conclude that F is a finitely generated A-module and hence E is Noetherian.

COROLLARY 2. *Let* A *be a complete Hausdorff filtered ring whose filtration is exhaustive. If* gr(A) *is a left Noetherian ring, so is* A.

It is sufficient to apply Corollary 1 with $E = A_s$.

COROLLARY 3. *Let* A *be a complete filtered ring,* (A_n) *its filtration,* E *a Hausdorff filtered* A*-module,* (E_n) *its filtration and* F *a finitely generated submodule of* E; *suppose that* $A_0 = A$ *and* $E_0 = E$.

(i) *If, for all* $k \geqslant 0$, $E_k = E_{k+1} + A_k F$, *then* $F = E$.

(ii) *If it is further supposed that the filtration on* E *is derived from that on* A (no. 1, *Example* 2), *the relation* $E = E_1 + F$ *implies* $F = E$.

Let ξ_i $(1 \leqslant i \leqslant n)$ be the classes mod. E_1 of a finite system of generators of F. It follows from the given hypothesis that for all $k \geqslant 0$ every element of $\mathrm{gr}_k(E)$ can be expressed in the form $\sum_{i=1}^{n} \alpha_i \xi_i$ where $\alpha_i \in \mathrm{gr}(A)$; the ξ_i therefore generate the gr(A)-module gr(E), which proves (i) by virtue of Proposition 12. If the filtration on E is derived from that on A, the relation $E = E_1 + F$ implies

$$E_k = A_k E = A_k E_1 + A_k F = A_k A_1 E + A_k F \subset A_{k+1} E + A_k F$$
$$= E_{k+1} + A_k F \subset E_k,$$

whence (ii).

PROPOSITION 13. *Let* A *be a ring,* $\mathfrak{m}$ *a two-sided ideal of* A *contained in the Jacobson radical of* A *and* E *an* A*-module. Let* A *and* E *be given the* $\mathfrak{m}$*-adic filtrations* (no. 1, Example 3). *Suppose that one of the following conditions holds:*

(a) E *is a finitely generated* A*-module and* A *is Hausdorff;*

(b) $\mathfrak{m}$ *is nilpotent.*

For E *to be a free* A*-module, it is necessary and sufficient that* $E/\mathfrak{m}E$ *be a free* $(A/\mathfrak{m})$*-module and that* E *satisfy property* (GR) (no. 8).

If E is a free A-module and (e_λ) a basis of E, $\mathfrak{m}^k E$ is the direct sum of the submodules $\mathfrak{m}^k e_\lambda$ of E for all $k \geqslant 0$ (*Algebra*, Chapter II, § 3, no. 7, *Remark*); then $\mathfrak{m}^k E/\mathfrak{m}^{k+1} E$ is identified with the direct sum of the $\mathfrak{m}^k e_\lambda/\mathfrak{m}^{k+1} e_\lambda$ (*Algebra*, Chapter II, § 1, no. 6, Proposition 7). We deduce first (for $k = 0$) that the classes $1 \otimes e_\lambda$ of the e_λ in $E/\mathfrak{m}E = (A/\mathfrak{m}) \otimes_A E$ form a basis of the $(A/\mathfrak{m})$-module $E/\mathfrak{m}E$, since the canonical mapping

$$(\mathfrak{m}^k/\mathfrak{m}^{k+1}) \otimes_A (E/\mathfrak{m}E) \to \mathfrak{m}^k E/\mathfrak{m}^{k+1} E$$

is bijective for all $k \geqslant 0$; hence γ_E is bijective. Note that this part of the proof uses neither condition (a) nor condition (b).

Suppose conversely that the conditions of the statement hold and let $(x_\iota)_{\iota \in I}$ be a family of elements of E whose classes mod. $\mathfrak{m}E$ form a basis of the $(A/\mathfrak{m})$-module $E/\mathfrak{m}E$; let L be the free A-module $A_s^{(I)}$, $(f_\iota)_{\iota \in I}$ its canonical basis and $u: L \to E$ the A-linear mapping such that $u(f_\iota) = x_\iota$ for all $\iota \in I$. The hypotheses already imply that u is surjective (Chapter II, § 3, no. 2, Corollary 1 to Proposition 4) and it remains to prove that u is injective. Now, each of the hypotheses (a) and (b) implies that A is Hausdorff and hence so is L with the $\mathfrak{m}$-adic filtration, since $\mathfrak{m}^k L = (\mathfrak{m}^k)^{(I)}$ (*Algebra*, Chapter II, § 3, no. 7, *Remark*) and gr(L) is identified with $\operatorname{gr}(A) \otimes_{A/\mathfrak{m}} (L/\mathfrak{m}L)$ from the first part of the proof; the homomorphism u is compatible with the filtrations and it is possible to write $\operatorname{gr}(u) = \gamma_E \circ v$ where v is the bijection of gr(L) onto

$$\operatorname{gr}(A) \otimes_{A/\mathfrak{m}} (E/\mathfrak{m}E)$$

mapping the class of f_ι mod. $\mathfrak{m}M$ onto $1 \otimes \bar{x}_\iota$, where $\bar{x}_\iota$ is the class of x_ι mod. $\mathfrak{m}E$. The hypothesis then implies that $\operatorname{gr}(u)$ is injective and the conclusion follows with the aid of Corollary 1 to Theorem 1, no. 8.

10. APPLICATION: EXAMPLES OF NOETHERIAN RINGS

LEMMA 1. *Let A be a graded ring of type* **Z**, *whose graduation* (A_n) *is such that* $A_n = 0$ *for all* $n < 0$ *or* $A_n = 0$ *for all* $n > 0$. *Let M be a graded A-module of type* **Z**. *For M to be a Noetherian A-module, it is necessary and sufficient that every graded submodule of M be finitely generated.*

As $n \mapsto -n$ is an automorphism of the group **Z**, we may restrict our attention to the case $A_n = 0$ for all $n > 0$. Let A′ and M′ denote the ring A and the module M with the filtrations associated with their respective graduations (no. 1, *Example* 1), which are exhaustive and separated; the hypothesis on A implies that A′ is *discrete* and hence complete. If E is a sub-A-module of M, the filtered A′-module E′ obtained by giving E the induced filtration is Hausdorff and its filtration is exhaustive; moreover gr(E′) is identified with a graded sub-A-module of M = gr(M′) and hence is finitely generated by hypothesis. The conclusion then follows from Corollary 1 to Proposition 12 of no. 9.

THEOREM 2. *Let A be a graded ring of type* **N**, *M a graded A-module of type* **N** *and* (A_n) *and* (M_n) *their respective graduations. Suppose that there exists an element* $a \in A_1$ *such that* $A_n = A_0 a^n$ *and* $M_n = a^n M_0$ *for all* $n > 0$. *Then, if* M_0 *is a Noetherian* A_0*-module, M is a Noetherian A-module.*

By virtue of Lemma 1 it is sufficient to prove that every *graded* submodule N of M is finitely generated. For all $r \geqslant 0$, let $N_r = N \cap M_r$ and let L_r be the set of $m \in M_0$ such that $a^r m \in N_r$. As

$$a^r A_0 \subset A_r = A_0 a^r, \qquad a^r A_0 L_r \subset A_0 a^r L_r \subset A_0 N_r \subset N_r$$

and hence the L_r are sub-A_0-modules of M_0; moreover,

$$aN_r \subset N \cap aM_r = N \cap M_{r+1} = N_{r+1}$$

and hence the sequence $(L_r)_{r \geqslant 0}$ is increasing. The hypothesis implies that there exists an integer $n \geqslant 0$ such that $L_r = L_n$ for $r \geqslant n$. For each $r \leqslant n$, let $(m_{r,s})_{1 \leqslant s \leqslant k_r}$ be a system of generators of the A_0-module L_r. We shall show that the elements $a^r m_{r,s}$ for $1 \leqslant s \leqslant k_r$, $0 \leqslant r \leqslant n$ form a system of generators of the A-module N. As $M_r = a^r M_0$ for all r, $N_r = a^r L_r$ for all r by definition of L_r. Then, for $r \leqslant n$,

$$N_r = a^r L_r = \sum_{s=1}^{k_r} a^r A_0 m_{r,s} \subset \sum_{s=1}^{k_r} A_0 a^r m_{rs},$$

and, for $r > n$,

$$N_r = a^r L_n = \sum_{s=1}^{k_n} a^r A_0 m_{n,s} \subset \sum_{s=1}^{k_n} A_0 a^r m_{n,s} \subset \sum_{s=1}^{k_n} A_0 a^{r-n}.(a^n m_{n,s})$$

which completes the proof (cf. Exercise 10).

COROLLARY 1 (Hilbert's Theorem). *For every commutative Noetherian ring* C, *the polynomial ring* C[X] *is Noetherian* (cf. Exercise 10).

COROLLARY 2. *For every commutative Noetherian ring* C *and every integer* $n > 0$, *the polynomial ring* $C[X_1, \ldots, X_n]$ *is Noetherian.*

This follows from Corollary 1 by induction on n.

COROLLARY 3. *If* C *is a commutative Noetherian ring, every finitely generated commutative* C-*algebra is a Noetherian ring.*

Such an algebra is isomorphic to a quotient of a polynomial algebra $C[X_1, \ldots, X_n]$ (§ 1, no. 1).

COROLLARY 4. *Let* A *be a graded commutative ring of type* **N** *and let* (A_n) *be its graduation. For* A *to be Noetherian, it is necessary and sufficient that* A_0 *be Noetherian and that* A *be a finitely generated* A_0-*algebra.*

The condition is sufficient by Corollary 3. Conversely, suppose A is Noetherian; $\mathfrak{m} = \sum_{n \geqslant 1} A_n$, which is an ideal of A, is then finitely generated; then A is a finitely generated A_0-algebra (§ 1, no. 2, Corollary to Proposition 1); on the other hand A_0, which is isomorphic to $A/\mathfrak{m}$, is a Noetherian ring.

COROLLARY 5. *Let* A *be a commutative ring and* $\mathfrak{m}$ *an ideal of* A *such that* $A/\mathfrak{m}$ *is Noetherian,* $\mathfrak{m}/\mathfrak{m}^2$ *is a finitely generated* $(A/\mathfrak{m})$-*module and* A *is Hausdorff and complete with the* $\mathfrak{m}$-*adic topology. Then* gr(A) *and* A *are Noetherian.*

gr(A) is an $(A/\mathfrak{m})$-module generated by $\mathfrak{m}/\mathfrak{m}^2$ (no. 3, *Example* 3) and hence the ring gr(A) is Noetherian by Corollary 3. From this we deduce that A itself is Noetherian (no. 9, Corollary 2 to Proposition 12).

COROLLARY 6. *For every commutative Noetherian ring* C *and every integer* $n > 0$, *the ring of formal power series* $C[[X_1, \dots, X_n]]$ *is Noetherian.*

This follows from Corollary 5 and no. 6, Corollary to Proposition 6, for if $\mathfrak{m}$ is the ideal of $A = C[[X_1, \dots, X_n]]$ consisting of the formal power series with no constant term, $A/\mathfrak{m}$ is isomorphic to C and $\mathfrak{m}/\mathfrak{m}^2$ to the C-module C^n.

Remarks

(1) Corollaries 2, 3 and 6 apply in particular if C is a commutative *field.*

* (2) Let $\mathfrak{g}$ be a Lie algebra over a commutative Noetherian ring C and suppose that $\mathfrak{g}$ is a finitely generated C-module. Let the enveloping algebra U of $\mathfrak{g}$ be given the *increasing* filtration (U_n) defined in no. 3, *Example* 4. With the corresponding topology, U is discrete and hence Hausdorff and complete; the associated graded ring gr(U) is a finitely generated C-algebra, being a quotient of the symmetric algebra $S(\mathfrak{g})$, hence gr(U) is a Noetherian ring (Corollary 3) and we deduce that U is a left and right Noetherian ring (no. 9, Corollary 2 to Proposition 12). *

11. COMPLETE $\mathfrak{m}$-ADIC RINGS AND INVERSE LIMITS

We have seen in no. 6 that, if A is a commutative ring and $\mathfrak{m}$ an ideal of A such that A is *Hausdorff* and *complete* with the $\mathfrak{m}$-adic topology, then the topological ring A is canonically identified with the inverse limit of the discrete rings $A_i = A/\mathfrak{m}^{i+1}$ ($i \in \mathbf{N}$) with respect to the canonical mappings

$$h_{ij}: A/\mathfrak{m}^{j+1} \to A/\mathfrak{m}^{i+1} \quad (i \leqslant j);$$

note that h_{ij} is surjective and that, if $\mathfrak{n}_{ij}$ is its kernel, then

$$\mathfrak{n}_{ij} = \mathfrak{m}^{i+1}/\mathfrak{m}^{j+1} = (\mathfrak{m}/\mathfrak{m}^{j+1})^{i+1} = (\mathfrak{n}_{0j})^{i+1};$$

in particular $(\mathfrak{n}_{0j})^{j+1} = 0$. Conversely:

PROPOSITION 14. *Let* (A_i, h_{ij}) *be an inverse system of discrete commutative rings, whose indexing set is* $\mathbf{N}$ *and let* (M_i, u_{ij}) *be an inverse system of modules over the inverse system of rings* (A_i, h_{ij}). *Let* $\mathfrak{n}_j$ *denote the kernel of* $h_{0j}: A_j \to A_0$ *and set* $A = \varprojlim A_i$, $M = \varprojlim M_i$. *Suppose that*

(a) *for all* $i \in \mathbf{N}$, h_{ii} *is the identity mapping on* A_i *and, for* $i \leqslant j$, h_{ij} *and* u_{ij} *are surjective;*

(b) *for* $i \leqslant j$, *the kernels of* h_{ij} *and* u_{ij} *are* $\mathfrak{n}_j^{i+1}$ *and* $\mathfrak{n}_j^{i+1}M_j$ *respectively.*

Then:

(i) A *is a complete Hausdorff topological ring,* M *is a complete Hausdorff topological* A*-module and the canonical mappings* $h_i: A \to A_i$, $u_i: M \to M_i$ *are surjective.*

(ii) *If* M_0 *is a finitely generated* A_0*-module,* M *is a finitely generated* A*-module;*

to be precise, every finite subset S *of* M *such that* $u_0(S)$ *generates* M_0 *is a system of generators of* M.

The assertions in (i) follow from *General Topology*, Chapter II, § 3, no. 5, Corollary to Proposition 10 and Corollary 1 to Theorem 1.

For all $i \in \mathbf{N}$, let us write $\mathfrak{m}_{i+1} = \operatorname{Ker}(h_i)$, $N_{i+1} = \operatorname{Ker}(u_i)$; then

$$\mathfrak{m}_{i+1} = \varprojlim_{k \geqslant 0} \overset{-1}{h}_{i,i+k}(0) = \varprojlim_{k} \mathfrak{n}_{i+1}^{i+k}$$

and $N_{i+1} = \varprojlim \mathfrak{n}_{i+k}^{i+1} M_{i+k}$; as h_{i+k} and u_{i+k} are surjective,

$$(16) \qquad h_{i+k}(\mathfrak{m}_{i+1}) = \mathfrak{n}_{i+k}^{i+1}, \qquad u_{i+k}(N_{i+1}) = \mathfrak{n}_{i+k}^{i+1} M_{i+k}.$$

Let us show that $\mathfrak{m}_i N_j \subset N_{i+j}$ for $i \geqslant 1$ and $j \geqslant 1$, which amounts to proving that $u_{i+j-1}(\mathfrak{m}_i N_j) = 0$; now

$$u_{i+j-1}(\mathfrak{m}_i N_j) = h_{i+j-1}(\mathfrak{m}_i) u_{i+j-1}(N_j)$$

is equal to $\mathfrak{n}_{i+j-1}^{i}(\mathfrak{n}_{i+j-1}^{j} M_{i+j-1}) = 0$, as, for all $k \geqslant 0$, $\mathfrak{n}_k^{k+1}$, which is the kernel of h_{kk}, is equal to 0. We see similarly that $\mathfrak{m}_i \mathfrak{m}_j \subset \mathfrak{m}_{i+j}$. If for $i \leqslant 0$ we set $\mathfrak{m}_i = A$ and $N_i = M$, $(\mathfrak{m}_i)_{i \in \mathbf{Z}}$ is a filtration of A and $(N_i)_{i \in \mathbf{Z}}$ a filtration of M compatible with the filtration on A; the topologies on A and M are obviously those defined by these filtrations. This being so, let $\mathfrak{a}$ be an ideal of A such that $h_1(\mathfrak{a}) = \mathfrak{n}_1$ and M′ the submodule of M generated by S; we are going to prove that

$$(17) \qquad N_i = \mathfrak{a}^i M' + N_{i+1} \quad \text{for} \quad i \geqslant 0.$$

Let us write $\mathfrak{a}_i = h_i(\mathfrak{a})$, $M'_i = u_i(M')$; it is sufficient to show that

$$u_i(N_i) = \mathfrak{a}_i^i M'_i.$$

This is true if $i = 0$, for $N_0 = M$ and $M'_0 = M_0$ by hypothesis. If $i \geqslant 1$, then $u_i(N_i) = \mathfrak{n}_i^i M_i$ by (16). As h_{1i} is surjective and $h_{0i} = h_{01} \circ h_{1i}$, h_{1i} maps the kernel $\mathfrak{n}_i$ of h_{0i} *onto* the kernel $\mathfrak{n}_1$ of h_{01} and $\mathfrak{n}_i = \overset{-1}{h}_{1i}(\mathfrak{n}_1)$; then

$$h_{1i}(\mathfrak{a}_i) = h_1(\mathfrak{a}) = \mathfrak{n}_1 = h_{1i}(\mathfrak{n}_i)$$

and, as the kernel of h_{1i} is $\mathfrak{n}_i^2$, $\mathfrak{n}_i \subset \mathfrak{a}_i + \mathfrak{n}_i^2$ and $\mathfrak{a}_i \subset \mathfrak{n}_i$, whence $\mathfrak{n}_i = \mathfrak{a}_i + \mathfrak{n}_i^2$. Moreover $u_{0i}(M'_i) = u_0(M') = M_0 = u_{0i}(M_i)$ and, as $\operatorname{Ker}(u_{0i}) = \mathfrak{n}_i M_i$, $M_i = M'_i + \mathfrak{n}_i M_i$; whence

$$\mathfrak{n}_i^i M_i = (\mathfrak{a}_i + \mathfrak{n}_i^2)^i (M'_i + \mathfrak{n}_i M_i).$$

Now, $\mathfrak{a}_i^k \mathfrak{n}_i^{i+1-k} \subset \mathfrak{n}_i^{i+1} = 0$ for $0 \leqslant k \leqslant i$; then it certainly follows that $u_i(N_i) = \mathfrak{n}_i M_i = \mathfrak{a}_i^i M'_i$, which proves (17).

Moreover $\mathfrak{m}_1 = \overset{-1}{h}_1(\mathfrak{n}_1)$, whence $\mathfrak{a} \subset \mathfrak{m}_1$ and therefore $\mathfrak{a}^i \subset \mathfrak{m}_1^i \subset \mathfrak{m}_i$, whence

$N_i \subset \mathfrak{m}_i M' + N_{i+1}$; on the other hand obviously $\mathfrak{m}_i M \subset N_i$ and hence $N_i = \mathfrak{m}_i M' + N_{i+1}$ for all $i \geqslant 0$; then it follows from Corollary 3 to Proposition 12 of no. 9 that $M' = M$, which completes the proof.

COROLLARY 1. *With the notation and hypotheses of Proposition 14 suppose further that* M_0 *is a finitely generated* A_0*-module and that the ideal* $\mathfrak{n}_1$ *of* A_1 *is finitely generated. Let* $\mathfrak{m}_1$ *be the kernel of* h_0*; the topologies on* A *and* M *are then the* $\mathfrak{m}_1$*-adic topologies on this ring and this module respectively; to be precise, for all* $i \geqslant 0$*, the kernels of* h_i *and* u_i *are* $\mathfrak{m}_1^{i+1}$ *and* $\mathfrak{m}_1^{i+1}M$ *respectively; further* $\mathfrak{m}_1/\mathfrak{m}_1^2$ *is a finitely generated* A*-module.*

We preserve the notation of the proof of Proposition 14; the hypotheses here allow us to assume that the ideal $\mathfrak{a}$ is *finitely generated.* Let $i \geqslant 0$ be any integer; for all $j \geqslant 0$, by (17), $N_{i+j} = \mathfrak{a}^j(\mathfrak{a}^i M) + N_{i+j+1} \subset \mathfrak{m}_j(\mathfrak{a}^i M) + N_{i+j+1}$; conversely, $\mathfrak{m}_j(\mathfrak{a}^i M) \subset \mathfrak{m}_j \mathfrak{m}_i M \subset \mathfrak{m}_{i+j} M \subset N_{i+j}$, whence

$$N_{i+j} = \mathfrak{m}_j(\mathfrak{a}^i M) + N_{i+j+1}.$$

As $\mathfrak{a}$ and M are finitely generated A-modules, so is $\mathfrak{a}^i M$. Applying Corollary 3 to Proposition 12 of no. 9 to the module N_i with the filtration $(N_{ij})_{j \in \mathbf{Z}}$ defined by $N_{ij} = N_i$ if $j < 0$, $N_{ij} = N_{i+j}$ if $j \geqslant 0$, we obtain $N_i = \mathfrak{a}^i M$, whence $N_i \subset \mathfrak{m}_1^i M$. But also $\mathfrak{m}_1^i M \subset \mathfrak{m}_i M \subset N_i$, whence $N_i = \mathfrak{m}_1^i M$. Applying this to the case where $M_i = A_i$, $u_{ij} = h_{ij}$, we obtain $\mathfrak{m}_i = \mathfrak{m}_1^i$. Moreover, $\mathfrak{m}_1 = \mathfrak{a} + \mathfrak{m}_1^2$ by (17), which proves the last assertion of the corollary.

COROLLARY 2. *Under the hypotheses of Corollary* 1, *for* A *to be Noetherian, it is necessary and sufficient that* A_0 *be so.*

The condition is necessary since A_0 is isomorphic to a quotient of A; it is sufficient by virtue of no. 10, Corollary 5 to Theorem 2.

12. THE HAUSDORFF COMPLETION OF A FILTERED MODULE

Let G be a filtered group whose filtration (G_n) consists of *normal* subgroups of G; we have already recalled (no. 6) that the *Hausdorff completion* $\hat{G}$ of the topological group G is canonically identified with the inverse limit $\varprojlim G/G_n$ of the *discrete* groups G/G_n, the canonical homomorphism $i: G \to \hat{G}$ having image the Hausdorff group associated with G (everywhere dense in $\hat{G}$) and kernel the closure $\bigcap G_n$ of $\{0\}$ in G. The Hausdorff completion $\hat{G}_n$ of the subgroup G_n of G is identified with the closure of $i(G_n)$ in $\hat{G}$ (*General Topology*, Chapter II, § 3, no. 9, Corollary 1 to Proposition 18) and, since G_n is closed in G,

$$G_n = \overset{-1}{i}(\hat{G}_n) = \overset{-1}{i}(\hat{G}_n \cap i(G)). \tag{18}$$

Moreover, the $\hat{G}_n$ form a fundamental system of neighbourhoods of 0 in $\hat{G}$ (*General Topology*, Chapter III, § 3, no. 4, Proposition 7) and are therefore normal *open* subgroups of $\hat{G}$ (*General Topology*, Chapter III, § 2, no. 3, Proposition 8); the topology on $\hat{G}$ is defined by the filtration $(\hat{G}_n)$, which is always separated by definition. As $i(G)$ is dense in $\hat{G}$ and $\hat{G}_n$ is open,

$$\hat{G} = i(G).\hat{G}_n \tag{19}$$

and similarly

$$\hat{G}_{n-1} = i(G_{n-1}).\hat{G}_n. \tag{20}$$

We deduce from (18) and (19) that the filtration $(\hat{G}_n)$ is exhaustive if and only if (G_n) is.

The second isomorphism theorem (*Algebra*, Chapter I, § 6, no. 13, Theorem 6 (d)) and equations (18), (19) and (20) show that the canonical homomorphisms

$$G_{n-1}/G_n \to \hat{G}_{n-1}/\hat{G}_n, \qquad G/G_n \to \hat{G}/\hat{G}_n, \tag{21}$$

are *bijective* and hence so is the canonical homomorphism

$$\mathrm{gr}(G) \to \mathrm{gr}(\hat{G}). \tag{22}$$

Now let A be a filtered ring, E a filtered A-module and (A_n) and (E_n) the respective filtrations of A and E; we shall assume that these filtrations are *exhaustive* so that for the corresponding topologies A is a topological ring and E a topological A-module (no. 5, Proposition 3). Then we have defined (*General Topology*, Chapter III, § 6, nos. 5 and 6) $\hat{A}$ as a topological ring and $\hat{E}$ as a topological $\hat{A}$-module. If $i: A \to \hat{A}$ is the canonical homomorphism, then $i(A_m)i(A_n) \subset i(A_{m+n})$, whence by the continuity of multiplication in $\hat{A}$,

$$\hat{A}_m\hat{A}_n \subset \hat{A}_{m+n} \tag{23}$$

since $\hat{A}_n$ is the closure of $i(A_n)$ in $\hat{A}$. It can be similarly shown that

$$\hat{A}_m\hat{E}_n \subset \hat{E}_{m+n}; \tag{24}$$

in other words:

Proposition 15. *Let* A *be a filtered ring and* E *a filtered* A*-module, the respective filtrations* (A_n), (E_n) *of* A *and* E *being exhaustive. Then* $(\hat{A}_n)$ *is a filtration compatible with the ring structure on* $\hat{A}$ *and* $(\hat{E}_n)$ *a filtration compatible with the module structure on* $\hat{E}$ *over the filtered ring* $\hat{A}$; *moreover these filtrations are exhaustive and define respectively the topologies on* $\hat{A}$ *and* $\hat{E}$. *Finally, the canonical mappings* $\mathrm{gr}(A) \to \mathrm{gr}(\hat{A})$ *and* $\mathrm{gr}(E) \to \mathrm{gr}(\hat{E})$ *of graded* **Z***-modules are respectively a graded ring isomorphism and a graded* $\mathrm{gr}(A)$*-module isomorphism.*

In what follows, for every uniform space X, j_X will denote the canonical mapping from X to its Hausdorff completion $\hat{X}$ and $X_0 = j_X(X)$ the uniform subspace of $\hat{X}$, which is the Hausdorff space *associated* with X. Recall that the topology on X is the inverse image under j_X of that on X_0 (*General Topology*,

Chapter II, § 3, no. 7, Proposition 12). Recall also that, for every uniformly continuous mapping $f: X \to Y$, $\hat{f}$ denotes the uniformly continuous mapping from $\hat{X}$ to $\hat{Y}$ such that $\hat{f} \circ j_X = j_Y \circ f$ (*loc. cit.*, Proposition 15); if X is a uniform subspace of Y and f the canonical injection, $\hat{X}$ is identified with a uniform subspace of $\hat{Y}$ and $\hat{f}$ is the canonical injection of $\hat{X}$ into $\hat{Y}$ (*loc. cit.*, no. 9. Corollary 1 to Proposition 18).

LEMMA 2. *Let* $X \xrightarrow{f} Y \xrightarrow{g} Z$ *be an exact sequence of strict morphisms of topological groups* (*Algebra*, Chapter II, § 1, no. 4, *Remark*). *Suppose that* X, Y, Z *admit Hausdorff completion groups and that the identity elements of* X, Y, Z *admit countable fundamental systems of neighbourhoods. Then* $\hat{X} \xrightarrow{\hat{f}} \hat{Y} \xrightarrow{\hat{g}} \hat{Z}$ *is an exact sequence of strict morphisms.*

Let N_f, N_g be the respective kernels of f and g; let us write

$$f = f_3 \circ f_2 \circ f_1$$

where f_1 is the canonical mapping $X \to X/N_f$, f_2 is an isomorphism of X/N_f onto N_g and f_3 is the canonical injection $N_g \to Y$. We already know that $\hat{f}_2$ is an isomorphism of $(X/N_f)^\wedge$ onto $\hat{N}_g$ and we have just recalled that $\hat{f}_3$ is an injective strict morphism of $\hat{N}_g$ to $\hat{Y}$; if we show that $\hat{f}_1$ is a surjective strict morphism, it will follow that $\hat{f}$ is a strict morphism (*General Topology*, Chapter III, § 2, no. 8, *Remark* 2). Let g_1 be the canonical mapping $Y \to Y/N_g$; if we show that $\hat{g}_1$ is a surjective strict morphism of the kernel $\hat{N}_g$, we shall see as above that $\hat{g}$ is a strict morphism and the sequence $\hat{X} \xrightarrow{\hat{f}} \hat{Y} \xrightarrow{\hat{g}} \hat{Z}$ will be exact. Thus we have reduced the problem to proving that, if $Y = X/N$ (where N is a normal subgroup of X) and $f: X \to Y$ is the canonical mapping, $\hat{f}$ is a *surjective strict morphism with kernel* $\hat{N}$.

Let $f_0: X_0 \to Y_0$ be the mapping which coincides with $\hat{f}$ on X_0; as j_X (resp. j_Y) is a surjective strict morphism of X onto X_0 (resp. Y onto Y_0), f_0 is a surjective strict morphism (*General Topology*, Chapter III, § 2, no. 8, *Remark* 3). Now X_0 and Y_0 are metrizable (*General Topology*, Chapter IX, § 3, no. 1, Proposition 1); then it follows from *General Topology*, Chapter IX, § 3, no. 1, Corollary 1 to Proposition 4 and Lemma 1, that $\hat{f}_0 = \hat{f}$ is a surjective strict morphism and has kernel the closure $\hat{N}'_0$ *in* $\hat{X}$ of the kernel N'_0 of f_0. Then it will be sufficient for us to prove that $\hat{N}'_0 = \hat{N}$. Now N'_0 obviously contains $N_0 = j_X(N)$; it will be sufficient to show that N'_0 is contained in the closure $\overline{N}_0$ of N_0 *in* X. Now,

$$U = \overset{-1}{j_X}(X_0 - \overline{N}_0) = X - \overset{-1}{j_X}(\overline{N}_0)$$

is an open set in X which does not meet N; as f is a surjective strict morphism, $V = f(U)$ is an open set in Y not containing the identity element e' of Y and hence not meeting the closure of e'; then $j_Y(V)$ does not contain the identity element of Y_0. But $j_Y(V) = f_0(X_0 - \overline{N}_0)$ and hence $N'_0 \subset \overline{N}_0$, which complete the proof of Lemma 2.

PROPOSITION 16. *Let* A *be a filtered ring,* (A_n) *its filtration,* E *an* A*-module and* (E_n) *the filtration on* E *derived from that on* A *consisting of the* $E_n = A_nE$. *Suppose that the filtration* (A_n) *is exhaustive and the module* E *is finitely generated. If* $i: E \to \hat{E}$ *is the canonical mapping, then, for all* $n \in \mathbf{Z}$,

$$\hat{E}_n = \hat{A}_n\hat{E} = \hat{A}_n i(E) \quad \textit{and} \quad \hat{E} = \hat{A}.i(E). \tag{25}$$

In particular $\hat{E}$ *is a finitely generated* $\hat{A}$*-module.*

The equation $A_nE = E_n$ implies, by virtue of the continuity of the external law on the $\hat{A}$-module $\hat{E}$, $\hat{A}_n\hat{E} \subset \hat{E}_n$ and obviously $\hat{A}_n\hat{E} \supset \hat{A}_n i(E)$. By hypothesis there exists a surjective homomorphism $u: L \to E$, where $L = A_s^I$, I being a finite set; let L be given the product filtration, consisting of the $L_n = A_n^I$, which define on L the product topology; then $\hat{L} = \hat{A}_s^I$ and $\hat{L}_n = \hat{A}_n^I$ (*General Topology*, Chapter II, § 3, no. 9, Corollary 2 to Proposition 18). Let $j: L \to \hat{L}$ be the canonical mapping and $(e_\iota)_{\iota \in I}$ the canonical basis of L; for an element $\sum_{\iota \in I} a_\iota j(e_\iota)$ (where $a_\iota \in \hat{A}$ for all $i \in I$) to belong to $\hat{L}_n$, it is necessary and sufficient that $a_\iota \in \hat{A}_n$ for all i; then $\hat{L}_n = \hat{A}_n.j(L)$. This being so, by definition $u(L_n) = A_nE = E_n$ and hence u is a *strict morphism* of L onto E (*General Topology*, Chapter III, § 2, no. 8, Proposition 24). Lemma 2 then shows that $\hat{u}: \hat{L} \to \hat{E}$ is a *surjective strict morphism.* As $\hat{L}_n$ is an open subgroup of $\hat{L}$, $\hat{u}(\hat{L}_n)$ is an open (and therefore closed) subgroup of $\hat{E}$; but $\hat{u}(\hat{L}_n) = \hat{A}_n\hat{u}(j(L)) = \hat{A}_n i(E)$ and, as $i(E_n) \subset A_n i(E) \subset \hat{A}_n i(E)$, finally $\hat{E}_n \subset \hat{A}_n i(E) \subset \hat{A}_n\hat{E} \subset \hat{E}_n$ and therefore $\hat{E}_n = \hat{A}_n\hat{E} = \hat{A}_n i(E)$; setting $n = 0$, we obtain the second formula of (25).

COROLLARY 1. *Under the conditions of Proposition* 16, *if* A *is complete, so is* E.

As the canonical mapping $A \to \hat{A}$ is then surjective (no. 6, Proposition 5), $\hat{E} = i(E)$ by (25) and the conclusion follows by Proposition 5 of no. 6.

COROLLARY 2. *Let* A *be a commutative ring,* $\mathfrak{m}$ *a finitely generated ideal of* A *and* $\hat{A}$ *the Hausdorff completion of* A *with respect to the* $\mathfrak{m}$*-adic topology. Then* $\widehat{\mathfrak{m}^n} = (\hat{\mathfrak{m}})^n = \mathfrak{m}^n.\hat{A}$ *for every integer* $n > 0$ *and the topology on* $\hat{A}$ *is the* $\hat{\mathfrak{m}}$*-adic topology.*

Let us write $A_n = \mathfrak{m}^n$, which is a finitely generated ideal of A. The formula $\mathfrak{m}^p A_n = \mathfrak{m}^{n+p}$ shows that the topology induced on A_n by the $\mathfrak{m}$-adic topology coincides with the $\mathfrak{m}$-adic topology on the A-module A_n (no. 1, *Example* 3). By Proposition 16 applied to $E = A_n$, $\hat{A}_n = \hat{A}.A_n$, in other words $\widehat{\mathfrak{m}^n} = \mathfrak{m}^n.\hat{A}$. In particular $\hat{\mathfrak{m}} = \mathfrak{m}.\hat{A}$, whence

$$(\hat{\mathfrak{m}})^n = \mathfrak{m}^n.\hat{A} = \hat{A}.A_n$$

(cf. Exercise 12).

Examples of Hausdorff completions of filtered rings

(1) Let A be a graded ring of type **N** and let $(A_n)_{n \geqslant 0}$ be its graduation; let it be given the associated filtration which is separated and exhaustive (no. 1, *Example* 1). The additive group A is canonically identified with a subgroup of $B = \prod_{n \in \mathbf{N}} A_n$; if B is given the topology the product of the discrete topologies, the topology induced on A is the topology defined by the filtration on A; also B is a complete topological group and A is *dense* in B (*General Topology*, Chapter III, § 2, no. 9, Proposition 25). The additive topological group B is then identified with the *completion* $\hat{A}$ of the Hausdorff additive group A and it follows from Proposition 15 that it has a unique ring structure which makes it the completion of the topological ring A. To define multiplication in this ring, note that, if we write $A'_n = \sum_{i > n} A_i$, the closure in B of the two-sided ideal A'_n is the set B_n of $x = (x_i) \in B$ such that $x_i = 0$ for $i \leqslant n$. Then let $x = (x_i), y = (y_i)$ be two elements of B and $z = (z_i)$ their product. Then, for all $n > 0$, $x \equiv x'_n \pmod{B_n}$, $y \equiv y'_n \pmod{B_n}$, where $x'_n = (x_i)_{0 \leqslant i \leqslant n}$, $y'_n = (y_i)_{0 \leqslant i \leqslant n}$, whence $z \equiv x'_n y'_n \pmod{B_n}$. But x'_n and y'_n belong to A and it is therefore seen that, for all $n \in \mathbf{N}$,

$$z_n = \sum_{j=0}^{n} x_j y_{n-j}. \tag{26}$$

In particular, we again obtain the Corollary to Proposition 6 of no. 6: if C is a commutative ring, the completion of the polynomial ring $C[X_1, \ldots, X_r]$, with the filtration associated with its usual graduation (by total degree) is canonically identified with the ring of formal power series $C[[X_1, \ldots, X_r]]$ (cf. *Algebra*, Chapter IV, § 5, no. 10).

* (2) Let K be a complete commutative field with a valuation. The completion of the ring of convergent series in r variables over K is canonically identified with the ring of formal power series $K[[X_1, \ldots, X_r]]$. *

(3) Let α be a non-zero non-invertible element of a *principal ideal domain*; the (α)-adic topology on A is also called the α-*adic* topology; it is *Hausdorff*, for the intersection of the ideals (α^n) reduces to 0 (*Algebra*, Chapter VII, § 1, no. 3). Note that the completion of A with respect to this topology is not necessarily an integral domain (cf. no. 13, *Remark* 3). The associated graded ring $\operatorname{gr}(A) = \operatorname{gr}(\hat{A})$ is canonically isomorphic to $(A/\mathfrak{a})[X]$ (no. 3, *Example* 1). If $A = \mathbf{Z}$, the completion of $\mathbf{Z}$ with respect to the n-adic topology ($n > 1$) is denoted by $\mathbf{Z}_n$ and its elements are called *n-adic integers*.

Every element of $\mathbf{Z}/n^k\mathbf{Z}$ admits a unique representative of the form $\sum_{i=0}^{k-1} a_i n^i$ where $0 \leqslant a_i \leqslant n-1$ for all i; moreover, its canonical image in $\mathbf{Z}/n^{k-1}\mathbf{Z}$

is the class of $\sum_{i=0}^{k-2} a_i n^i$. These remarks and the fact that $\mathbf{Z}_n$ is canonically identified with the inverse limit $\varprojlim_k \mathbf{Z}/n^k\mathbf{Z}$ show immediately that every element of $\mathbf{Z}_n$ can be written uniquely in the form $\sum_{i=0}^{\infty} a_i n^i$ where $0 \leqslant a_i < n$ and conversely that such a series is convergent in $\mathbf{Z}_n$.

13. THE HAUSDORFF COMPLETION OF A SEMI-LOCAL RING

PROPOSITION 17. *Let* A *be a commutative ring and* $(\mathfrak{m}_\lambda)_{\lambda\in L}$ *a family of ideals of* A, *distinct from* A, *such that* $\mathfrak{m}_\lambda$ *and* $\mathfrak{m}_\mu$ *are relatively prime for* $\lambda \neq \mu$. *For every family* $s = (s(\lambda))_{\lambda\in L}$ *of integers* $\geqslant 0$, *of finite support, set* $\mathfrak{a}_s = \bigcap_{\lambda\in L} \mathfrak{m}_\lambda^{s(\lambda)}$ (*equal to the product of the* $\mathfrak{m}^{s(\lambda)}$ *for the* λ *such that* $s(\lambda) \neq 0$; cf. Chapter II, § 1, no. 2, Propositions 3 and 5); *the* $\mathfrak{a}_s$ *form a fundamental system of neighbourhoods of* 0 *with respect to a topology* $\mathscr{T}$ *compatible with the ring structure on* A; *let* $\hat{A}$ *be the Hausdorff completion of* A *with respect to this topology. On the other hand, for all* $\lambda \in L$, *let* A_λ *be the ring* A *with the* $\mathfrak{m}_\lambda$*-adic topology and let* $\hat{A}_\lambda$ *be its Hausdorff completion. If* $u: A \to \prod_{\lambda\in L} A_\lambda$ *denotes the diagonal homomorphism, u is continuous and the corresponding homomorphism* $\hat{u}$:

$$\hat{A} \to \Big(\prod_{\lambda\in L} A_\lambda\Big)^{\wedge} = \prod_{\lambda\in L} \hat{A}_\lambda$$

(*General Topology*, Chapter III, § 6, no. 5 and Chapter II, § 3, no. 9, Corollary 2 to Proposition 18) *is a topological ring isomorphism.*

The first assertion follows from *General Topology*, Chapter III, § 6, no. 3, *Example* 3. Let us set $B = \prod_{\lambda\in L} A_\lambda$; as the topology on A is finer than each of the $\mathfrak{m}_\lambda$-adic topologies, the mappings $\mathrm{pr}_\lambda \circ u$ are continuous and hence u is continuous. Also, $u(\mathfrak{a}_s)$ is the intersection of the diagonal Δ of B and the open set $\bigcap_{\lambda\in L} \overset{-1}{\mathrm{pr}}_\lambda(\mathfrak{m}_\lambda^{s(\lambda)})$ of B; it follows that u is a strict morphism from the additive group A to B with image Δ. Now Δ is *dense* in B. For let $b = (a_\lambda)_{\lambda\in L}$ be an element of B; every neighbourhood of b in B contains a set of the form $b + V$, where $V = \bigcap_{\lambda\in L} \overset{-1}{\mathrm{pr}}_\lambda(\mathfrak{m}_\lambda^{s(\lambda)})$ for a family $s = (s(\lambda))_{\lambda\in L}$ with finite support of integers $\geqslant 0$. As the $\mathfrak{m}_\lambda^{s(\lambda)}$ are relatively prime in pairs (Chapter II, § 1, no. 2, Proposition 3), there exists $x \in A$ such that $x \equiv a_\lambda \pmod{\mathfrak{m}_\lambda^{s(\lambda)}}$ for all λ (*loc. cit.*, Proposition 5) and hence $(b + V) \cap \Delta \neq \varnothing$. The Hausdorff completion of the group B/Δ is then $\{0\}$; applying Lemma 2 of no. 12 to the exact sequences $0 \to A \xrightarrow{u} B$, $A \xrightarrow{u} B \to B/\Delta$, we see that $\hat{u}$ is an isomorphism of $\hat{A}$ onto $\hat{B}$.

COROLLARY. *Let* A *be a principal ideal domain and* P *a representative system of extremal elements of* A (*Algebra*, Chapter VII, § 1, no. 3). *The topology on* A *with respect to which the ideals* $\neq 0$ *of* A *form a fundamental system of neighbourhoods of* 0, *which is compatible with the ring structure on* A, *is Hausdorff and the completion of* A *with this topology is canonically isomorphic to the product of the completions of* A *with respect to the* π*-adic topologies, where* π *runs through* P.

The principal ideals (π) where $\pi \in P$ are maximal and distinct and hence relatively prime, we have already seen (no. 12, *Example* 3) that the π-adic topologies are Hausdorff and hence so is the topology defined in the statement of Proposition 17, which is finer than each of the π-adic topologies.

If the Corollary to Proposition 17 is applied when $A = \mathbf{Z}$, we denote by $\hat{\mathbf{Z}}$ the completion of $\mathbf{Z}$ with respect to the topology for which all the ideals $\neq 0$ of $\mathbf{Z}$ form a fundamental system of neighbourhoods of 0, the ring isomorphic to the product $\prod_{p \in P} \mathbf{Z}_p$ of the rings of p-adic integers (P being the set of prime numbers).

Remarks

(1) Clearly, under the conditions of Proposition 17, the topology $\mathcal{T}$ is the *least upper bound* of the $\mathfrak{m}_\lambda$-adic topologies on A.

(2) Every *closed* ideal $\mathfrak{a}$ of $\prod_{\lambda \in L} \hat{A}_\lambda$ is identical with the *product* of its projections $\mathfrak{a}_\lambda = \mathrm{pr}_\lambda(\mathfrak{a})$, which are *closed* ideals in the $\hat{A}_\lambda$; for $\hat{A}_\lambda$ is canonically identified with a closed ideal A'_λ of $\prod_\lambda \hat{A}_\lambda$ and $\mathfrak{a}_\lambda$ with $\mathfrak{a} \cap A'_\lambda$ (*Algebra*, Chapter I, § 8, no. 10, Proposition 6), the sum of the $\mathfrak{a}_\lambda$ is *dense* in the product $\prod_\lambda \mathfrak{a}_\lambda$ (*General Topology*, Chapter III, § 2, no. 9, Proposition 25) and the latter is closed in $\prod_\lambda \hat{A}_\lambda$, whence our assertion.

PROPOSITION 18. *Let* A *be a commutative ring,* $(\mathfrak{m}_i)_{1 \leqslant i \leqslant q}$ *a finite family of distinct maximal ideals of* A, $\mathfrak{r}$ *the product ideal* $\mathfrak{m}_1\mathfrak{m}_2 \ldots \mathfrak{m}_q = \mathfrak{m}_1 \cap \mathfrak{m}_2 \cap \cdots \cap \mathfrak{m}_q$ *and* S *the multiplicative subset* $\bigcap_{i=1}^{q} (A - \mathfrak{m}_i)$. *Let* A *be given the* $\mathfrak{r}$*-adic topology, the ring* $B = S^{-1}A$ *the* $\mathfrak{r}B$*-adic topology and each of the local rings* $A_{\mathfrak{m}_i}$ *the* $(\mathfrak{m}_i A_{\mathfrak{m}_i})$*-adic topology. Let* $u: A \to B$, $v_i: B \to A_{\mathfrak{m}_i}$ *be the canonical homomorphisms* (Chapter II, § 2, no. 1, Corollary 2 to Proposition 2) *and* v *the homomorphism* $(v_i): B \to \prod_{i=1}^{q} A_{\mathfrak{m}_i}$. *The homomorphisms* u *and* v *are continuous and the corresponding homomorphisms* $\hat{u}: \hat{A} \to \hat{B}$ *and* $\hat{v}: \hat{B} \to \prod_{i=1}^{q} (A_{\mathfrak{m}_i})^\wedge$ *are topological ring isomorphisms.*

$\mathfrak{m}_i \cap S = \varnothing$ for $1 \leqslant i \leqslant q$, hence the ideal $\mathfrak{m}'_i = \mathfrak{m}_i B$ of B is maximal (Chapter II, § 2, no. 5, Proposition 11) and

$$\mathfrak{r}B = \mathfrak{m}'_1 \cap \mathfrak{m}'_2 \cap \cdots \cap \mathfrak{m}'_q$$

(Chapter II, § 2, no. 4); finally, $B_{\mathfrak{m}'_i} = A_{\mathfrak{m}_i}$ up to a canonical isomorphism (Chapter II, § 2, no. 5, Proposition 11). As $\overset{-1}{u}(\mathfrak{r}B) = \mathfrak{r}$ and $\overset{-1}{v_i}(\mathfrak{m}_i A_{\mathfrak{m}_i}) \supset \mathfrak{r}B$, u and v are continuous. Then it is sufficient to prove that,

$$w = v \circ u : A \to \prod_{i=1}^{q} A_{\mathfrak{m}_i},$$

$\hat{w}$ is an isomorphism of $\hat{A}$ onto $\prod_{i=1}^{q} \hat{A}_{\mathfrak{m}_i}$, for this result applied to B and the $\mathfrak{m}'_i$ will show that $\hat{v}$ is an isomorphism and therefore also $\hat{u}$. Note that every product of powers of the $\mathfrak{m}_i$ contains a power of $\mathfrak{r}$ and hence the $\mathfrak{r}$-adic topology is the least upper bound of the $\mathfrak{m}_i$-adic topologies; moreover, if A_i denotes the ring A with the $\mathfrak{m}_i$-adic topology and $\phi: A \to \prod_{i=1}^{q} A_i$ the diagonal mapping, $\hat{\phi}: \hat{A} \to \prod_{i=1}^{q} \hat{A}_i$ is an isomorphism (Proposition 17). Then it all amounts to proving that, if $u_i: A_i \to A_{\mathfrak{m}_i}$ is the canonical mapping, $\hat{u}_i: \hat{A}_i \to \hat{A}_{\mathfrak{m}_i}$ is an isomorphism. Now, for all n, the mapping

$$u_{i,n}: A/\mathfrak{m}_i^n \to A_{\mathfrak{m}_i}/\mathfrak{m}_i^n A_{\mathfrak{m}_i}$$

derived from u_i by taking quotients is an isomorphism (Chapter II, § 3, no. 3, Proposition 9); our assertion follows from the fact that $\hat{A}_i$ (resp. $\hat{A}_{\mathfrak{m}_i}$) is the inverse limit of the discrete rings $A/\mathfrak{m}_i^n$ (resp. $A_{\mathfrak{m}_i}/\mathfrak{m}_i^n A_{\mathfrak{m}_i}$) (cf. no. 6).

Remark (3). We see that an *integral domain* A can be such that its Hausdorff completion $\hat{A}$ admits non-zero divisors of zero.

PROPOSITION 19. *Let* A *be a commutative ring and* $\mathfrak{m}$ *a maximal ideal of* A. *The Hausdorff completion* $\hat{A}$ *of* A *with respect to the* $\mathfrak{m}$*-adic topology is a local ring whose maximal ideal is* $\hat{\mathfrak{m}}$.

If $\mathfrak{a} = \bigcap_{k \geqslant 1} \mathfrak{m}^k$, $\hat{A}$ is the completion of the Hausdorff ring $A/\mathfrak{a}$ associated with A and, as $\mathfrak{m}/\mathfrak{a}$ is maximal in $A/\mathfrak{a}$, we may assume that A is Hausdorff with respect to the $\mathfrak{m}$-adic topology. As $A/\mathfrak{m}$ and $\hat{A}/\hat{\mathfrak{m}}$ are isomorphic rings (no. 12, formula (21)), $\hat{\mathfrak{m}}$ is maximal in $\hat{A}$. As the topology on $\hat{A}$ is defined by the filtration $(\mathfrak{m}^n)^\wedge$ (no. 12), the proposition will be a consequence of the following lemma:

LEMMA 3. *Let* A *be a complete Hausdorff topological ring, in which there exists a fundamental system* $\mathfrak{S}$ *of neighbourhoods of* 0 *consisting of additive subgroups of* A.

(i) *For all* $x \in A$ *such that* $\lim_{n\to\infty} x^n = 0$, $1 - x$ *is invertible in* A *and its inverse is equal to* $\sum_{n=0}^{\infty} x^n$.

(ii) *Let* $\mathfrak{a}$ *be a two-sided ideal of* A *such that* $\lim_{n\to\infty} x^n = 0$ *for all* $x \in \mathfrak{a}$. *For an* *element y of* A *to be invertible, it is necessary and sufficient that its class* mod. $\mathfrak{a}$ *be invertible in* $A/\mathfrak{a}$; *in particular* $\mathfrak{a}$ *is contained in the Jacobson radical of* A.

(i) As

$$(1 - x)(1 + x + \cdots + x^n) = (1 + x + \cdots + x^n)(1 - x) = 1 - x^{n+1},$$

it all amounts to proving that the series with general term x^n is convergent in A; now, by hypothesis, for every neighbourhood $V \in \mathfrak{S}$ of 0 in A, there exists an integer $p > 0$ such that $x^n \in V$ for all $n \geqslant p$. We conclude that

$$x^p + x^{p+1} + \cdots + x^q \in V$$

for all $q \geqslant p$ and our assertion then follows from Cauchy's criterion (*General Topology*, Chapter III, § 5, no. 2, Theorem 1).

(ii) Suppose that there exists $y' \in A$ such that $yy' \equiv 1 \pmod{\mathfrak{a}}$ and $y'y \equiv 1 \pmod{\mathfrak{a}}$. The hypothesis on $\mathfrak{a}$ implies, by (i), that yy' and $y'y$ are invertible in A and hence y is invertible in A. In particular, every $x \in \mathfrak{a}$ is such that $1 - x$ is invertible in A and, as $\mathfrak{a}$ is a two-sided ideal of A, it is contained in the Jacobson radical of A (*Algebra*, Chapter VIII, § 6, no. 3, Theorem 1).

Having established this lemma, it is sufficient to apply it to the topological ring $\hat{A}$ and the ideal $\hat{\mathfrak{m}}$, as, for all $x \in \hat{\mathfrak{m}}$, $x^n \in (\hat{\mathfrak{m}})^n \subset (\mathfrak{m}^n)^\wedge$ and the sequence (x^n) therefore tends to 0.

If we take $A = \mathbf{Z}$, every maximal ideal of $\mathbf{Z}$ is of the form $p\mathbf{Z}$ where p is prime. The ring of p-adic numbers $\mathbf{Z}_p$ is then a local ring of which $p\mathbf{Z}_p$ is the maximal ideal (Corollary 2 to Proposition 16) and whose residue field is isomorphic to $\mathbf{Z}/p\mathbf{Z} = \mathbf{F}_p$, and $\mathbf{Z}_{(p)}$ with the $p\mathbf{Z}_{(p)}$-adic topology is identified with a topological subring of $\mathbf{Z}_p$ containing $\mathbf{Z}$.

Corollary. *Let* A *be a semi-local ring* (Chapter II, § 3, no. 5), $\mathfrak{m}_i$ *its distinct maximal ideals* $(1 \leqslant i \leqslant q)$ *and*

$$\mathfrak{r} = \mathfrak{m}_1 \cap \mathfrak{m}_2 \cap \cdots \cap \mathfrak{m}_q$$

its Jacobson radical. The Hausdorff completion $\hat{A}$ *of* A *with respect to the* $\mathfrak{r}$*-adic topology is a semi-local ring, canonically isomorphic to the product* $\prod_{i=1}^{q} \hat{A}_{\mathfrak{m}_i}$, *where* $\hat{A}_{\mathfrak{m}_i}$ *is the Hausdorff completion ring of the local ring* $A_{\mathfrak{m}_i}$ *with respect to the* $(\mathfrak{m}_i A_{\mathfrak{m}_i})$*-adic topology.*

3. $\mathfrak{m}$-ADIC TOPOLOGIES ON NOETHERIAN RINGS

All the filtrations considered in this paragraph are assumed to be exhaustive.

1. GOOD FILTRATIONS

Let A be a filtered commutative ring, E a filtered A-module and (A_n) and (E_n)

the respective filtrations of A and E; suppose that $A_0 = A$. In the polynomial ring A[X] the set $A' = \sum_{n \geqslant 0} A_n X^n$ is a *graded sub-A-algebra of type* **N**; the subgroup $E' = \sum_{n \geqslant 0} E_n \otimes_A AX^n$ of $E \otimes_A A[X]$ is a *graded* A'*-module* of type **N**, since

$$A_m X^m (E_n \otimes_A AX^n) \subset (A_m E_n \otimes_A AX^{m+n}) \subset E_{m+n} \otimes_A AX^{m+n}.$$

DEFINITION 1. *Let* A *be a commutative ring,* $\mathfrak{m}$ *an ideal of* A, E *an* A*-module and* (E_n) *a filtration on the additive group* E *consisting of submodules of* E. *The filtrations* (E_n) *is called* $\mathfrak{m}$*-good if:*

(1) $\mathfrak{m}E_n \subset E_{n+1}$ *for all* $n \in \mathbf{Z}$;

(2) *there exists an integer* n_0 *such that* $\mathfrak{m}E_n = E_{n+1}$ *for* $n \geqslant n_0$.

Then by induction on q, $\mathfrak{m}^q E_n = E_{n+q}$ for $n \geqslant n_0$, $q \geqslant 1$. Note that condition (1) means that the filtration (E_n) is compatible with the A-module structure on E if A is given the $\mathfrak{m}$-adic filtration. Clearly, on every A-module E, the $\mathfrak{m}$-adic filtration is $\mathfrak{m}$-good. If a filtration on an A-module E is $\mathfrak{m}$-good, the quotient filtration on any quotient module of E is $\mathfrak{m}$-good.

THEOREM 1. *Let* A *be a commutative ring,* $\mathfrak{m}$ *an ideal of* A, E *an* A*-module and* (E_n) *a filtration of the additive group* E *consisting of finitely generated sub-*A*-modules. Suppose that* $\mathfrak{m}E_n \subset E_{n+1}$ *for all* n. *Let* A' *be the graded subalgebra* $\sum_{n \geqslant 0} \mathfrak{m}^n X^n$ *of* A[X] *and* E' *the graded* A'*-module* $\sum_{n \geqslant 0} E_n \otimes_A AX^n$. *The two following conditions are equivalent:*

(a) *The filtration* (E_n) *is* $\mathfrak{m}$*-good.*

(b) E' *is a finitely generated* A'*-module.*

Suppose that $\mathfrak{m}E_{n-1} = E_n$ for $n > n_0 \geqslant 0$. For $i \leqslant n_0$, let $(e_{ij})_{1 \leqslant j \leqslant r_i}$ be a finite system of generators of the A-module E_i. As the A-module $E_n \otimes_A AX^n$ is generated by the elements $e_{nj} \otimes X^n$ for $0 \leqslant n \leqslant n_0$ and is equal to

$$\mathfrak{m}^{n-n_0} E_{n_0} \otimes_A AX^n$$

for $n > n_0$, the A'-module E' is generated by the elements $e_{nj} \otimes X^n$ for $0 \leqslant n \leqslant n_0$ and $1 \leqslant j \leqslant r_n$; then it is certainly finitely generated.

Conversely, if E' is a finitely generated A'-module, it is generated by a finite family of elements of the form $e_k \otimes X^{n(k)}$, where $e_k \in E_{n(k)}$. Let n_0 be the greatest of the integers $n(k)$. Then for $n \geqslant n_0$ and $f \in E_n$,

$$f \otimes X^n = \sum_k t_k (e_k \otimes X^{n(k)})$$

where $t_k \in A'$; replacing if need be t_k by its homogeneous component of degree $n - n(k)$, we may assume that $t_k = a_k X^{n-n(k)}$, where $a_k \in \mathfrak{m}^{n-n(k)}$.

As the unique element X^n forms a basis of the A-module AX^n, the equation $f \otimes X^n = \left(\sum_k a_k e_k\right) \otimes X^n$ implies $f = \sum_k a_k e_k$. Then $E_n \subset \mathfrak{m}^{n-n_0} E_{n_0}$; since the opposite inclusion is obvious,

$$E_n = \mathfrak{m}^{n-n_0} E_{n_0},$$

whence $E_n = \mathfrak{m} E_{n-1}$ for $n > n_0$.

Lemma 1. *Let* A *be a commutative Noetherian ring and* $\mathfrak{m}$ *an ideal of* A. *Then the subring* $A' = \sum_{n \geqslant 0} \mathfrak{m}^n X^n$ *of* A[X] *is Noetherian.*

A′ is an A-algebra generated by $\mathfrak{m}X$; as A is Noetherian, $\mathfrak{m}X$ is a finitely generated A-module and the conclusion then follows from § 2, no. 10, Corollary 3 to Theorem 2.

Proposition 1. *Let* A *be a commutative Noetherian ring and* $\mathfrak{m}$ *an ideal of* A; *let* A *be given the* $\mathfrak{m}$*-adic filtration. Let* E, F *be two filtered* A*-modules and* $j: F \to E$ *an injective homomorphism compatible with the filtrations. If* E *is finitely generated and its filtration is* $\mathfrak{m}$*-good, then* F *is finitely generated and its filtration is* $\mathfrak{m}$*-good.*

As F is isomorphic to a submodule of E, it is finitely generated since A is Noetherian and E is finitely generated. Let (E_n), (F_n) be the respective filtrations on E and F, which consist of finitely generated submodules; preserving the notation of Lemma 1, we set $E' = \sum_{n \geqslant 0} E_n \otimes_A AX^n$, $F' = \sum_{n \geqslant 0} F_n \otimes_A AX^n$; as by hypothesis F_n is isomorphic to a submodule of E_n, we see that F′ is isomorphic to a submodule of E′. By Theorem 1, E′ is a finitely generated A′-module and hence so is F′ since A′ is Noetherian (Lemma 1). Hence the conclusion by virtue of Theorem 1.

Corollary 1 (the Artin-Rees Lemma). *Let* A *be a commutative Noetherian ring,* $\mathfrak{m}$ *an ideal of* A, E *a finitely generated* A*-module and* F *a submodule of* E. *The filtration induced on* F *by the* $\mathfrak{m}$*-adic filtration on* E *is* $\mathfrak{m}$*-good.*

In other words, there exists an integer n_0 such that

$$\mathfrak{m}((\mathfrak{m}^n E) \cap F) = (\mathfrak{m}^{n+1} E) \cap F \tag{1}$$

for all $n \geqslant n_0$.

Corollary 2. *Let* A *be a commutative Noetherian ring and* $\mathfrak{a}$, $\mathfrak{b}$ *two ideals of* A. *There exists an integer* $h > 0$ *such that* $\mathfrak{a}^h \cap \mathfrak{b} \subset \mathfrak{a}\mathfrak{b}$.

There exists n such that $\mathfrak{a}^{n+1} \cap \mathfrak{b} = \mathfrak{a}(\mathfrak{a}^n \cap \mathfrak{b}) \subset \mathfrak{a}\mathfrak{b}$ by Corollary 1 applied to $E = A$, $F = \mathfrak{b}$.

Corollary 3. *Let* A *be a commutative Noetherian ring,* $\mathfrak{m}$ *an ideal of* A *and* x *an element of* A *which is not a divisor of* 0. *There exists an integer* $k > 0$ *such that, for all* $n \geqslant k$, *the relation* $xy \in \mathfrak{m}^n$ *implies* $y \in \mathfrak{m}^{n-k}$.

Corollary 1 applied to $E = A$, $F = Ax$ shows that there exists k such that, for all $n \geqslant k$, $\mathfrak{m}^n \cap Ax = \mathfrak{m}^{n-k}(\mathfrak{m}^k \cap Ax)$. Then, if $xy \in \mathfrak{m}^n$,

$$xy \in \mathfrak{m}^n \cap Ax \subset \mathfrak{m}^{n-k}x$$

and, as x is not a divisor of 0, we deduce that $y \in \mathfrak{m}^{n-k}$.

In the notation of transporters (Chapter I, § 2, no. 10), the conclusion of Corollary 3 reads

$$\mathfrak{m}^n : Ax \subset \mathfrak{m}^{n-k}. \tag{2}$$

COROLLARY 4. *Let* A *be a commutative Noetherian ring,* $\mathfrak{m}$ *an ideal of* A, E *a finitely generated* A-*module and* (E_n) *and* (E'_n) *two filtrations consisting of submodules of* E. *Suppose that the filtrations* (E_n) *and* (E'_n) *are compatible with the* A-*module structure on* E *when* A *is given the* $\mathfrak{m}$-*adic filtration. If the filtration* (E_n) *is* $\mathfrak{m}$-*good and* $E'_n \subset E_n$ *for all* $n \in \mathbf{Z}$, *the filtration* (E'_n) *is* $\mathfrak{m}$-*good.*

This a special case of Proposition 1.

LEMMA 2. *Let* A, B *be two commutative Noetherian rings,* $\phi: A \to B$ *a ring homomorphism,* E *a finitely generated* A-*module and* F *a finitely generated* B-*module. Then* $\mathrm{Hom}_A(E, \phi_*(F))$ *is a finitely generated* B-*module.*

There exists by hypothesis a surjective A-homomorphism $v: A^n \to E$; the mapping $u \mapsto u \circ v$ of $\mathrm{Hom}_A(E, \phi_*(F))$ to $\mathrm{Hom}_A(A^n, \phi_*(F))$ is therefore injective and, as B is Noetherian, it is sufficient to prove that $\mathrm{Hom}_A(A^n, \phi_*(F))$ is a finitely generated B-module; which is immediate since it is isomorphic to F^n.

PROPOSITION 2. *Let* A *be a commutative Noetherian ring,* $\mathfrak{m}$ *an ideal of* A *and* E, F *two finitely generated* A-*modules. If* (F_n) *is an* $\mathfrak{m}$-*good filtration on* F, *the submodules* $\mathrm{Hom}_A(E, F_n)$ *form an* $\mathfrak{m}$-*good filtration on the* A-*module* $\mathrm{Hom}_A(E, F)$.

As $\mathfrak{m}^k F_n \subset F_{n+k}$ for $n \in \mathbf{Z}$, $k \geqslant 0$, it is also true that

$$\mathfrak{m}^k \,\mathrm{Hom}_A(E, F_n) \subset \mathrm{Hom}_A(E, F_{n+k});$$

the family $(\mathrm{Hom}_A(E, F_n))_{n \in \mathbf{Z}}$ is then a filtration on $\mathrm{Hom}_A(E, F)$ compatible with its module structure over the ring A filtered by the $\mathfrak{m}$-adic filtration. Since E is finitely generated, there exists an integer $r > 0$ and a surjective A-homomorphism $u: A^r \to E$ which defines an injective A-homomorphism

$$v = \mathrm{Hom}(u, 1_F): \mathrm{Hom}_A(E, F) \to \mathrm{Hom}_A(A^r, F);$$

clearly v is compatible with the filtrations $(\mathrm{Hom}_A(E, F_n))$ and $(\mathrm{Hom}_A(A^r, F_n))$. As $\mathrm{Hom}_A(E, F)$ and $\mathrm{Hom}_A(A^r, F)$ are finitely generated (Lemma 2), it is sufficient by virtue of Proposition 1 to show that the filtration $(\mathrm{Hom}_A(A^r, F_n))$ is $\mathfrak{m}$-good; but this is immediate by virtue of the existence of the canonical isomorphism $\mathrm{Hom}_A(A^r, F_n) \to F_n^r$ and the fact that the relation $\mathfrak{m}F_n = F_{n+1}$ implies $\mathfrak{m}(F_n^r) = (\mathfrak{m}F_n)^r = F_{n+1}^r$ (*Algebra*, Chapter II, § 3, no. 7, *Remark*).

PROPOSITION 3. *Let* A *be a Noetherian ring and* $\mathfrak{m}$ *an ideal of* A *such that* A *is Hausdorff and complete with respect to the* $\mathfrak{m}$*-adic topology. Let* E *be a filtered* A*-module over the filtered ring* A, *the filtration* (E_n) *of* A *being such that* $E_0 = E$ *and* E *is Hausdorff with respect to the topology defined by* (E_n). *Then the following conditions are equivalent:*

(a) E *is a finitely generated* A*-module and* (E_n) *is an* $\mathfrak{m}$*-good filtration.*

(b) gr(E) *is a finitely generated* gr(A)*-module.*

(c) *For all* $n \geqslant 0$, $\mathrm{gr}_n(E)$ *is a finitely generated* A*-module and there exists* n_0 *such that for* $n \geqslant n_0$ *the canonical homomorphism*

$$\mathrm{gr}_1(A) \otimes_A \mathrm{gr}_n(E) \to \mathrm{gr}_{n+1}(E) \tag{3}$$

is surjective.

It follows immediately from the definitions that (a) implies (c). The fact that (b) implies (c) is a consequence of § 1, no. 3, Lemma 1; conversely, if (c) holds, clearly gr(E) is generated as a gr(A)-module by the sum of the $\mathrm{gr}_p(E)$ for $p \leqslant n_0$ and hence by hypothesis admits a finite system of generators. It remains to prove that (c) implies (a); as the $\mathrm{gr}_n(E)$ are finitely generated and $E_0 = E$, clearly first, by induction on n, E/E_n is a finitely generated A-module for all n; it will therefore be sufficient to prove that, for $n > n_0$, E_n is a finitely generated A-module and that $\mathfrak{m}E_n = E_{n+1}$. Now, consider the A-module E_{n+1} with the exhaustive and separated filtration $(E_{n+k})(k \geqslant 1)$; $\mathfrak{m}E_n \subset E_{n+1}$; hypothesis (c) implies that the image of $\mathfrak{m}E_n$ in $\mathrm{gr}_{n+1}(E) = E_{n+1}/E_{n+2}$ is equal to $\mathrm{gr}_{n+1}(E)$ and *generates* the graded gr(A)-module $\mathrm{gr}(E_{n+1})$. As $\mathrm{gr}_{n+1}(E)$ is by hypothesis a finitely generated A-module, it follows from § 2, no. 9, Proposition 12 that $\mathfrak{m}E_n = E_{n+1}$ and that E_{n+1} is a finitely generated A-module.

2. m-ADIC TOPOLOGIES ON NOETHERIAN RINGS

PROPOSITION 4. *Let* A *be a commutative Noetherian ring,* $\mathfrak{m}$ *an ideal of* A *and* E *a finitely generated* A*-module. All the* $\mathfrak{m}$*-good filtrations on* E *define the same topology (namely the* $\mathfrak{m}$*-adic topology).*

Let (E_n) be an $\mathfrak{m}$-good filtration on E. As this filtration is exhaustive, every element of E belongs to one of the E_n and, as E is finitely generated and the E_n are A-modules, there exists an integer n_1 such that $E_{n_1} = E$. On the other hand let n_0 be such that $\mathfrak{m}E_n = E_{n+1}$ for $n \geqslant n_0$; then, for $n > n_0 - n_1$, $\mathfrak{m}^n E \subset E_{n+n_1} = \mathfrak{m}^{n+n_1-n_0}E_{n_0} \subset \mathfrak{m}^{n+n_1-n_0}E$, which proves the proposition.

THEOREM 2 (Krull). *Let* A *be a commutative Noetherian ring,* $\mathfrak{m}$ *an ideal of* A, E *a finitely generated* A-*module and* F *a submodule of* E. *Then the* $\mathfrak{m}$-*adic topology on* F *is induced by the* $\mathfrak{m}$-*adic topology on* E.

It follows from no. 1, Proposition 1 that the filtration induced on F by the $\mathfrak{m}$-adic filtration on E is $\mathfrak{m}$-good and the conclusion then follows from Proposition 4.

COROLLARY. *Let* A *be a commutative Noetherian ring,* $\mathfrak{m}$ *an ideal of* A, E *an* A-*module and* F *a finitely generated* A-*module. Every* A-*linear mapping* $u\colon E \rightarrow F$ *is a strict morphism* (*General Topology*, Chapter III, § 2, no. 8) *for the* $\mathfrak{m}$-*adic topologies.*

As $u(\mathfrak{m}^n E) = \mathfrak{m}^n u(E)$, u is a strict morphism of E onto $u(E)$ for the $\mathfrak{m}$-adic topologies on these two modules and the $\mathfrak{m}$-adic topology on $u(E)$ is induced by the $\mathfrak{m}$-adic topology on F by Theorem 2.

PROPOSITION 5. *Let* A *be a commutative Noetherian ring,* $\mathfrak{m}$ *an ideal of* A *and* E *a finitely generated* A-*module. The closure* $\bigcap_{n=1}^{\infty} \mathfrak{m}^n E$ *of* $\{0\}$ *in* E *with respect to the* $\mathfrak{m}$-*adic topology is the set of* $x \in E$ *for which there exists an element* $m \in \mathfrak{m}$ *such that* $(1 - m)x = 0$.

If $x = mx$ where $m \in \mathfrak{m}$, $x = m^n x \in \mathfrak{m}^n E$ for every integer $n \geqslant 0$ and hence $x \in F = \bigcap_{n=0}^{\infty} \mathfrak{m}^n E$. Conversely, if $x \in F$, Ax is contained in the intersection of the neighbourhoods of 0 in E; it then follows from Theorem 2 that the $\mathfrak{m}$-adic topology on Ax, which is induced by that on E, is the coarsest topology; as $\mathfrak{m}x$ is by definition a neighbourhood of 0 with this topology, $\mathfrak{m}x = Ax$ and hence there exists $m \in \mathfrak{m}$ such that $x = mx$.

We may also say that $\mathfrak{m}F = F$ since

$$F = \bigcap_{n=1}^{\infty} \mathfrak{m}^{n+1}E \subset \mathfrak{m} . \bigcap_{n=1}^{\infty} \mathfrak{m}^n E = \mathfrak{m}F;$$

as A is Noetherian, F is a finitely generated A-module; it is then sufficient to apply Chapter II, § 2, no. 2, Corollary 3 to Proposition 4.

COROLLARY (Krull). *Let* A *be a commutative Noetherian ring and* $\mathfrak{m}$ *an ideal of* A. *The ideal* $\bigcap_{n=1}^{\infty} \mathfrak{m}^n$ *is the set of elements* $x \in A$ *for which there exists* $m \in \mathfrak{m}$ *such that* $(1 - m)x = 0$. *In particular, for* $\bigcap_{n=1}^{\infty} \mathfrak{m}^n = \{0\}$, *it is necessary and sufficient that no element of* $1 + \mathfrak{m}$ *be a divisor of* 0 *in* A.

It is sufficient to apply Proposition 5 to $E = A_s$.

Remark. The hypothesis that A is *Noetherian* is essential in this corollary. For example, let A be the ring of infinitely differentiable mappings from **R** to itself and let $\mathfrak{m}$ be the (maximal) ideal of A consisting of the functions f such that $f(0) = 0$. It is immediate that $\bigcap_{n=0}^{\infty} \mathfrak{m}^n$ is the set of functions f such that

$f^{(n)}(0) = 0$ for all $n \geqslant 0$ and there exist such functions with $f(x) \neq 0$, for all $x \neq 0$, for example the function f defined by $f(x) = e^{-1/x^2}$ for $x \neq 0$ and $f(0) = 0$.

DEFINITION 1. *Let* A *be a topological ring. If a two-sided ideal* $\mathfrak{m}$ *of* A *is such that the given topology on* A *is the* $\mathfrak{m}$*-adic topology,* $\mathfrak{m}$ *is called a defining ideal of the topology on* A.

Let A be a commutative Noetherian ring, $\mathfrak{m}$ an ideal of A and $\mathfrak{t}$ its radical (Chapter II, § 2, no. 6). If $\mathfrak{m}'$ is a defining ideal of the $\mathfrak{m}$-adic topology, there exists an integer $n > 0$ such that $\mathfrak{m}'^n \subset \mathfrak{m}$ (§ 2, no. 5) and hence $\mathfrak{m}' \subset \mathfrak{t}$; conversely, since A is Noetherian, there exists an integer $k > 0$ such that $\mathfrak{t}^k \subset \mathfrak{m}$ (Chapter II, § 2, no. 6, Proposition 15) and hence $\mathfrak{t}$ is *the largest defining ideal* of the $\mathfrak{m}$-adic topology.

3. ZARISKI RINGS

PROPOSITION 6. *Let* A *be a commutative Noetherian ring and* $\mathfrak{m}$ *an ideal of* A. *The following properties are equivalent:*

(a) $\mathfrak{m}$ *is contained in the Jacobson radical of* A.

(b) *Every finitely generated* A*-module is Hausdorff with the* $\mathfrak{m}$*-adic topology.*

(c) *For every finitely generated* A*-module* E, *every submodule of* E *is closed with respect to the* $\mathfrak{m}$*-adic topology on* E.

(d) *Every maximal ideal of* A *is closed with respect to the* $\mathfrak{m}$*-adic topology.*

Let us show that (a) implies (b). Suppose that $\mathfrak{m}$ is contained in the Jacobson radical of A and let E be a finitely generated A-module. If $x \in$ E and $m \in \mathfrak{m}$ are such that $(1 - m)x = 0$, then $x = 0$, for $1 - m$ is invertible in A. Then (no. 2, Proposition 5) E is Hausdorff with the $\mathfrak{m}$-adic topology.

Let us prove that (b) implies (c). Suppose (b) holds. Let E be a finitely generated A-module and F a submodule of E. Then E/F is Hausdorff with the $\mathfrak{m}$-adic topology, which is the quotient topology of the $\mathfrak{m}$-adic topology on E; then F is closed in E.

Clearly (c) implies (d). Let us show finally that (d) implies (a). It follows from (d) that, for every maximal ideal $\mathfrak{a}$ of A, the A-module $A/\mathfrak{a}$ is Hausdorff with the $\mathfrak{m}$-adic topology. This implies $\mathfrak{m}(A/\mathfrak{a}) \neq A/\mathfrak{a}$, unless the $\mathfrak{m}$-adic topology on $A/\mathfrak{a}$ were the coarsest topology and $A/\mathfrak{a}$ were reduced to 0, which is absurd since $A/\mathfrak{a}$ is a *field*. The canonical image of $\mathfrak{m}$ in $A/\mathfrak{a}$ is therefore an ideal of $A/\mathfrak{a}$ distinct from $A/\mathfrak{a}$ and hence reduced to 0; then $\mathfrak{m} \subset \mathfrak{a}$, which proves that $\mathfrak{m}$ is contained in the Jacobson radical of A.

DEFINITION 2. *A topological ring* A *is called a Zariski ring if it is commutative and Noetherian and there exists a defining ideal* $\mathfrak{m}$ *for the topology on* A *satisfying the equivalent conditions of Proposition* 6.

A Zariski ring A is necessarily *Hausdorff* (Proposition 6) and *every* defining ideal of its topology is contained in the Jacobson radical of A.

Examples of Zariski rings

(1) Let A be a commutative Noetherian ring and $\mathfrak{m}$ an ideal of A. If A is *Hausdorff and complete with the* $\mathfrak{m}$*-adic topology*, A is a Zariski ring with this topology, by virtue of § 2, no. 13, Lemma 3.

(2) Every *quotient ring* $A/\mathfrak{b}$ of a Zariski ring is a Zariski ring, for it is Noetherian and, if $\mathfrak{m}$ is a defining ideal of A, $\mathfrak{m}(A/\mathfrak{b}) = (\mathfrak{m} + \mathfrak{b})/\mathfrak{b}$ is contained in the Jacobson radical of $A/\mathfrak{b}$ (*Algebra*, Chapter VIII, § 6, no. 3, Proposition 7).

(3) Let A be a *Noetherian semi-local* ring and $\mathfrak{r}$ its Jacobson radical. Then A with the $\mathfrak{r}$-adic topology is a Zariski ring. This will always be the topology in question (unless otherwise stated) when we consider a Noetherian semi-local ring as a topological ring.

PROPOSITION 7. *Let* A, A′ *be two commutative rings and* $h: A \to A'$ *a ring homomorphism. Suppose that* A *is Noetherian and that* A′ *is a finitely generated* A*-module (with the structure defined by* h*). Let* $\mathfrak{m}$ *be an ideal of* A *and let* $\mathfrak{m}' = \mathfrak{m}A'$*. Then:*

(i) *For the* $\mathfrak{m}'$*-adic topology on* A′ *to be Hausdorff, it is necessary and sufficient that the elements of* $1 + h(\mathfrak{m})$ *be not divisors of* 0 *in* A′.

(ii) *If* A *with the* $\mathfrak{m}$*-adic topology is a Zariski ring, then* A′ *with the* $\mathfrak{m}'$*-adic topology is a Zariski ring.*

(iii) *If* h *is injective (thus identifying* A *with a subring of* A′*), the* $\mathfrak{m}'$*-adic topology on* A′ *induces on* A *the* $\mathfrak{m}$*-adic topology.*

Recall that the $\mathfrak{m}'$-adic filtration on A′ coincides with the $\mathfrak{m}$-adic filtration on the A-*module* A′ (§ 2, no. 1, *Example* 3). Assertion (i) is thus a special case of Proposition 5 of no. 2 and assertion (iii) a special case of Theorem 2 of no. 2, Finally let us show (ii). Suppose that A is a Zariski ring with the $\mathfrak{m}$-adic topology and let E′ be a finitely generated A′-module; it is also a finitely generated A-module and the $\mathfrak{m}$-adic and $\mathfrak{m}'$-adic filtrations on E′ coincide; then E′ is Hausdorff with the $\mathfrak{m}'$-adic topology. Finally the A-module A′ is Noetherian and hence the ring A′ is Noetherian, which completes the proof that A′ is a Zariski ring.

4. THE HAUSDORFF COMPLETION OF A NOETHERIAN RING

Let A be a commutative ring, $\mathfrak{m}$ an ideal of A and E an A-module; let $\hat{A}$ and $\hat{E}$ denote the respective Hausdorff completions of A and E with respect to the $\mathfrak{m}$-adic topology and j_E the canonical mapping $E \to \hat{E}$. The A-bilinear mapping $(a, x) \mapsto a j_E(x)$ of $\hat{A} \times E$ to $\hat{E}$ defines an A-linear mapping

$$\alpha_E: \hat{A} \otimes_A E \to \hat{E},$$

called *canonical*. Let $u: E \to F$ be an A-module homomorphism and let

$\hat{u}: \hat{E} \to \hat{F}$ be the mapping obtained by passing to the Hausdorff completions; for $a \in \hat{A}$, $x \in E$,

$$\alpha_F(a \otimes u(x)) = a j_F(u(x)) = a\hat{u}(j_E(x)) = \hat{u}(\alpha_E(a \otimes x)),$$

in other words, the diagram

$$\begin{array}{ccc} \hat{A} \otimes_A E & \xrightarrow{1 \otimes u} & \hat{A} \otimes_A F \\ \downarrow{\scriptstyle \alpha_E} & & \downarrow{\scriptstyle \alpha_F} \\ \hat{E} & \xrightarrow{\hat{u}} & \hat{F} \end{array}$$

is commutative. Finally, it follows from § 2, no. 12, Proposition 16 that, if E is *finitely generated*, the homomorphism α_E is *surjective*.

THEOREM 3. *Let A be a commutative Noetherian ring, $\mathfrak{m}$ an ideal of A and E, F, G three finitely generated A-modules. Then:*

(i) *If $E \xrightarrow{u} F \xrightarrow{v} G$ is an exact sequence of A-linear mappings, the sequence $\hat{E} \xrightarrow{\hat{u}} \hat{F} \xrightarrow{\hat{v}} \hat{G}$ obtained by passing to the Hausdorff completions (with respect to the $\mathfrak{m}$-adic topologies) is exact.*

(ii) *The canonical $\hat{A}$-linear mapping $\alpha_E: \hat{A} \otimes_A E \to \hat{E}$ is bijective.*

(iii) *The A-module $\hat{A}$ is flat.*

We have seen that u and v are strict morphisms of topological groups (no. 2, Corollary to Theorem 2). Assertion (i) then follows from § 2, no. 12, Lemma 2. Assertion (ii) is obvious when E = A and the case where E is a free finitely generated A-module can be immediately reduced to that. In the general case, E admits a finite presentation

$$L \xrightarrow{u} L' \xrightarrow{v} E \to 0$$

(Chapter I, § 2, no. 8, Lemma 8). We derive a commutative diagram

$$\begin{array}{ccccccc} \hat{A} \otimes_A L & \xrightarrow{1 \otimes u} & \hat{A} \otimes_A L' & \xrightarrow{1 \otimes v} & \hat{A} \otimes_A E & \longrightarrow & 0 \\ \downarrow{\scriptstyle \alpha_L} & & \downarrow{\scriptstyle \alpha_{L'}} & & \downarrow{\scriptstyle \alpha_E} & & \\ \hat{L} & \xrightarrow{\hat{u}} & \hat{L}' & \xrightarrow{\hat{v}} & \hat{E} & \longrightarrow & 0 \end{array}$$

The first row is exact (Chapter I, § 2, no. 1, Lemma 1) and so is the second by (i). We already know that α_E is surjective (§ 2, no. 12, Proposition 16); on the other hand, as α_L and $\alpha_{L'}$ are bijective and $1 \otimes v$ is surjective, α_E is injective by virtue of Chapter I, § 1, no. 4, Corollary 2 to Proposition 2; this shows (ii).

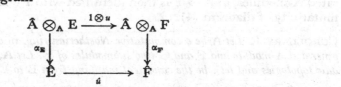

Then it follows from (i) and (ii) that, if $\mathfrak{a}$ is an ideal of A (necessarily finitely generated), the canonical mapping $\hat{A} \otimes_A \mathfrak{a} \to \hat{A}$ is injective, being the composition of $\hat{\mathfrak{a}} \to \hat{A}$ and α_E, which proves that $\hat{A}$ is a flat A-module (Chapter I, § 2, no. 3, Proposition 1).

Under the conditions of Theorem 3 $\hat{A} \otimes_A E$ is often identified with $\hat{E}$ by means of the canonical mapping α_E. If $u: E \to F$ is a homomorphism of finitely generated A-modules, $\hat{u}: \hat{E} \to \hat{F}$ is then identified with $1 \otimes u$ by virtue of the commutativity of diagram (4).

COROLLARY 1. *Let* A *be a commutative Noetherian ring,* $\mathfrak{m}$ *an ideal of* A, E *a finitely generated* A*-module and* F *and* G *two submodules of* E. *Let* A, E, F, G *be given the* $\mathfrak{m}$*-adic topologies and let* i *be the canonical mapping from* E *to* $\hat{E}$. *Then:*

$$\hat{F} = \hat{A}.i(F), \qquad (F + G)^\wedge = \hat{F} + \hat{G}, \qquad (F \cap G)^\wedge = \hat{F} \cap \hat{G},$$
$$(F : G)^\wedge = \hat{F} : \hat{G}.$$

Moreover, if $\mathfrak{a}$ *and* $\mathfrak{b}$ *are two ideals of* A *and* $\mathfrak{c} = \mathfrak{a}\mathfrak{b}$, $\hat{\mathfrak{c}} = \hat{\mathfrak{a}}\hat{\mathfrak{b}}$.

By Theorem 3, $\hat{E}$, $\hat{F}$, $\hat{G}$ are canonically identified with $\hat{A} \otimes_A E$, $\hat{A} \otimes_A F$, $\hat{A} \otimes_A G$, which establishes the first two formulae. The third and fourth follow respectively from Chapter I, § 2, no. 6, Proposition 6 and no. 10, Proposition 12. Finally, as $\hat{\mathfrak{a}} = \hat{A}i(\mathfrak{a})$, $\hat{\mathfrak{b}} = \hat{A}i(\mathfrak{b})$, $\hat{\mathfrak{c}} = \hat{A}i(\mathfrak{c})$,

$$\hat{\mathfrak{c}} = \hat{A}i(\mathfrak{a}\mathfrak{b}) = \hat{A}i(\mathfrak{a})i(\mathfrak{b}) = \hat{\mathfrak{a}}\hat{\mathfrak{b}}.$$

COROLLARY 2. *Let* A *be a commutative Noetherian ring,* $\mathfrak{m}$ *an ideal of* A *and* $\hat{A}$ *the Hausdorff completion of* A *with respect to the* $\mathfrak{m}$*-adic topology. If an element* $a \in A$ *is not a divisor of* 0 *in* A, *its canonical image* a' *in* $\hat{A}$ *is not a divisor of* 0 *in* $\hat{A}$.

As $\hat{A}$ is a flat A-module, the corollary is a special case of Chapter I, § 2, no. 4, Proposition 3 (i).

COROLLARY 3. *If* A *is a commutative Noetherian ring, the ring of formal power series* $A[[X_1, \ldots, X_n]]$ *is a flat* A*-module.*

It is the completion of the polynomial ring

$$B = A[X_1, \ldots, X_n]$$

with respect to the $\mathfrak{m}$-adic topology, where $\mathfrak{m}$ is the set of polynomials with no constant term (§ 2, no. 12, *Example* 1); as B is Noetherian (§ 2, no. 10, Corollary 2 to Theorem 2), $A[[X_1, \ldots, X_n]]$ is a flat B-module by Theorem 3 and, as B is free A-module, $A[[X_1, \ldots, X_n]]$ is a flat A-module (Chapter I, § 2, no. 7, Corollary 3 to Proposition 8).

PROPOSITION 8. *Let* A *be a commutative Noetherian ring,* $\mathfrak{m}$ *an ideal of* A, $\hat{A}$ *the Hausdorff completion of* A *with respect to the* $\mathfrak{m}$*-adic topology and* j *the canonical mapping from* A *to* $\hat{A}$. *The :*

(i) $\hat{A}$ *is a Zariski ring and* $\hat{\mathfrak{m}} = \hat{A}.j(\mathfrak{m})$ *is a defining ideal of* $\hat{A}$.

(ii) *The mapping* $\mathfrak{n} \mapsto \hat{\mathfrak{n}} = \hat{A}.j(\mathfrak{n})$ *is a bijection of the set of maximal ideals of* A *containing* $\mathfrak{m}$ *onto the set of maximal ideals of* $\hat{A}$ *and* $\mathfrak{q} \mapsto \overset{-1}{j}(\mathfrak{q})$ *is the inverse bijection.*

(iii) *Let* $\mathfrak{n}$ *be a maximal ideal of* A *containing* $\mathfrak{m}$. *The homomorphism* $j' : A_{\mathfrak{n}} \to \hat{A}_{\hat{\mathfrak{n}}}$ *derived from* j *is injection; if* $A_{\mathfrak{n}}$ *is identified by means of* j *with a subring of* $\hat{A}_{\hat{\mathfrak{n}}}$, *the* $(\mathfrak{n}A_{\mathfrak{n}})$*-adic topology on* $A_{\mathfrak{n}}$ *is induced by the* $\hat{\mathfrak{n}}$*-adic topology on* $\hat{A}_{\hat{\mathfrak{n}}}$ *and* $A_{\mathfrak{n}}$ *is dense in* $\hat{A}_{\hat{\mathfrak{n}}}$ *with the* $\hat{\mathfrak{n}}$*-adic topology.*

Let us show (i). As $\mathfrak{m}$ is a finitely generated ideal, $(\mathfrak{m}^n)^{\wedge} = (\hat{\mathfrak{m}})^n = \mathfrak{m}^n\hat{A}$ (§ 2, no. 12, Corollary 2 to Proposition 16) and the topology on $\hat{A}$ is the $\hat{\mathfrak{m}}$-adic topology. As $\hat{A}/\hat{\mathfrak{m}}$ is isomorphic to $A/\mathfrak{m}$, it is a Noetherian ring and $\hat{\mathfrak{m}} = \mathfrak{m}\hat{A}$ is a finitely generated $\hat{A}$-module and therefore $\hat{A}$ is Noetherian (§ 2, no. 10, Corollary 5 to Theorem 2); finally, as $\hat{A}$ is Hausdorff and complete with respect to the $\hat{\mathfrak{m}}$-adic topology, $\hat{A}$ is a Zariski ring (no. 3, *Example* 1).

Assertion (ii) follows immediately from the fact that the canonical homomorphism $A/\mathfrak{m} \to \hat{A}/\hat{\mathfrak{m}}$ derived from j is bijective and the fact that every maximal ideal of $\hat{A}$ contains $\hat{\mathfrak{m}}$, since $\hat{A}$ is a Zariski ring and the Jacobson radical of $\hat{A}$ then contains $\hat{\mathfrak{m}}$ (no. 3, Proposition 6).

Finally let us prove (iii). As $\mathfrak{n} = \overset{-1}{j}(\hat{\mathfrak{n}})$, $j(A - \mathfrak{n}) \subset \hat{A} - \hat{\mathfrak{n}}$ and j certainly defines a homomorphism $j' : A_{\hat{\mathfrak{n}}} \to \hat{A}_{\hat{\mathfrak{n}}}$ (Chapter II, § 2, no. 1, Proposition 2). Let us show that j' is injective; let $a \in A$, $s \in A - \mathfrak{n}$ be such that

$$j'(a/s) = j(a)/j(s) = 0;$$

then there exists $s' \in \hat{A} - \hat{\mathfrak{n}}$ such that $s'j(a) = 0$ (Chapter II, § 2, no. 1, *Remark* 3) and the annihilator of $j(a)$ in $\hat{A}$ is therefore not contained in $\hat{\mathfrak{n}}$. Now, if $\mathfrak{b}$ is the annihilator of a in A, the annihilator of $j(a)$ in $\hat{A}$ is $\hat{\mathfrak{b}}$ (Corollary 1 to Theorem 3); hence $\mathfrak{b} \not\subset \mathfrak{n}$, which shows that $a/s = 0$.

Moreover, there is a commutative diagram

$$(5) \qquad \begin{array}{ccc} A/\mathfrak{n}^k & \longrightarrow & A_{\mathfrak{n}}/(\mathfrak{n}A_{\mathfrak{n}})^k \\ {\scriptstyle h}\downarrow & & \downarrow{\scriptstyle h'} \\ \hat{A}/\hat{\mathfrak{n}}^k & \longrightarrow & \hat{A}_{\hat{\mathfrak{n}}}/(\hat{\mathfrak{n}}\hat{A}_{\hat{\mathfrak{n}}})^k \end{array}$$

where h and h' are derived from j and j' respectively and the horizontal arrows are the canonical isomorphisms of Chapter II, § 3, no. 3, Proposition 9. As $\mathfrak{n}^k$ is an open ideal of A (since it contains $\mathfrak{m}^k$), h is bijective and hence so is h'. This shows first that $(\mathfrak{n}A_{\mathfrak{n}})^k = j'((\hat{\mathfrak{n}}\hat{A}_{\hat{\mathfrak{n}}})^k)$ and hence the topology on $A_{\mathfrak{n}}$ is induced by that on $\hat{A}_{\hat{\mathfrak{n}}}$; moreover, $\hat{A}_{\hat{\mathfrak{n}}} = A_{\mathfrak{n}} + (\hat{\mathfrak{n}}\hat{A}_{\hat{\mathfrak{n}}})^k$ for all $k > 0$ and hence $A_{\mathfrak{n}}$ is everywhere dense in $\hat{A}_{\hat{\mathfrak{n}}}$.

COROLLARY. *Let* A *be a Noetherian local* (resp. *semi-local*) *ring and* $\mathfrak{m}$ *its Jacobson radical. Then* $\hat{A}$ *is a Noetherian local* (resp. *semi-local*) *ring whose Jacobson radical is* $\hat{\mathfrak{m}}$.

$\hat{A}$ is Noetherian by Proposition 8 (i) and the rest follows from Proposition 8 (ii) and the third formula in Corollary 1 to Theorem 3.

5. THE COMPLETION OF A ZARISKI RING

PROPOSITION 9. *Let* A *be a commutative Noetherian ring and* $\mathfrak{m}$ *an ideal of* A; *let* A *be given the* $\mathfrak{m}$*-adic topology. For* $\hat{A}$ *to be a faithfully flat* A*-module, it is necessary and sufficient that* A *be a Zariski ring.*

For every finitely generated A-module M, the canonical mapping $M \to M \otimes_A \hat{A}$ is identified with the canonical mapping $M \to \hat{M}$ from M to its Hausdorff completion with respect to the $\mathfrak{m}$-adic topology (no. 4, Theorem 3) and the kernel of this mapping is then the closure of $\{0\}$ in M with respect to this topology. As we already know that $\hat{A}$ is a flat A-module (no. 4, Theorem 3), the proposition follows from the characterization of faithfully flat modules (Chapter I, § 3, no. 1, Proposition 1 (b)) and the characterization of Zariski rings (no. 3, Proposition 6).

If A is a Zariski ring and E is a finitely generated A-module, we may (by virtue of Proposition 9) identify E with a subset of $\hat{E}$ by means of the canonical mapping $j_E: E \to \hat{E}$. With this identification:

COROLLARY 1. *Let* A *be a Zariski ring,* E *a finitely generated* A*-module and* F *a submodule of* E. *Then* $F = \hat{F} \cap E = (\hat{A}F) \cap E$.

This is a special case of Chapter I, § 3, no. 5, Proposition 10 (ii) and it also follows from no. 3, Proposition 6.

COROLLARY 2. *Let* A *be a Zariski ring and* E *a finitely generated* A*-module. If* $\hat{E}$ *is a free* $\hat{A}$*-module,* E *is a free* A*-module.*

Let $\mathfrak{m}$ be a defining ideal of A, which is therefore contained in the Jacobson radical of A. We apply the criterion of Chapter II, § 3, no. 5, Proposition 5: the canonical mapping $j_E: E \to \hat{E}$ defines a bijection $i_E: E/\mathfrak{m}E \to \hat{E}/(\mathfrak{m}E)^\wedge$; similarly the canonical mapping $j_A: A \to \hat{A}$ defines a bijection $i_A: A/\mathfrak{m} \to \hat{A}/\hat{\mathfrak{m}}$, which is a ring isomorphism. Then $(\mathfrak{m}E)^\wedge = \hat{A}.\mathfrak{m}E = \hat{\mathfrak{m}}\hat{E}$ (no. 4, Theorem 3), so that $\hat{E}/(\mathfrak{m}E)^\wedge$ is given an $(\hat{A}/\hat{\mathfrak{m}})$-module structure and hence (by means of i_A) an $(A/\mathfrak{m})$-module structure. It is immediate that i_E is $(A/\mathfrak{m})$-linear, so that it is an $(A/\mathfrak{m})$-module isomorphism. As $\hat{E}/\hat{\mathfrak{m}}\hat{E}$ is a free $(\hat{A}/\hat{\mathfrak{m}})$-module, $E/\mathfrak{m}E$ is a free $(A/\mathfrak{m})$-module.

On the other hand, let $v: \mathfrak{m} \otimes_A E \to E$ be the canonical homomorphism; as $(\mathfrak{m} \otimes_A E) \otimes_A \hat{A}$ is canonically identified with $\hat{\mathfrak{m}} \otimes_{\hat{A}} \hat{E}$ and $E \otimes_A \hat{A}$ with $\hat{E}$ (no. 4, Theorem 3), the hypothesis that $\hat{E}$ is a free $\hat{A}$-module implies that the homomorphism $v \otimes 1: \hat{\mathfrak{m}} \otimes_{\hat{A}} \hat{E} \to \hat{E}$ is injective. As $\hat{A}$ is a faithfully flat A-module (Proposition 9), we conclude that v is injective (Chapter I, § 3, no. 1, Proposition 2) and the conditions for applying the above mentioned criterion are indeed fulfilled.

COROLLARY 3. *Let* A *be a Zariski ring such that* $\hat{A}$ *is an integral domain and let* $\mathfrak{a}$ *be an ideal of* A. *If the ideal* $\mathfrak{a}\hat{A}$ *of* $\hat{A}$ *is principal,* $\mathfrak{a}$ *is principal.*

This is a special case of Corollary 2.

COROLLARY 4. *Let* A *be a Zariski ring such that* $\hat{A}$ *is an integral domain,* L *the field of fractions of* $\hat{A}$ *and* K ⊂ L *the field of fractions of* A; *then* $\hat{A} \cap K = A$.

Clearly $A \subset \hat{A} \cap K$; on the other hand, if $x \in \hat{A} \cap K$, then $\hat{A}x \subset \hat{A}$ and hence, as $\hat{A}x = \hat{A} \otimes_A (Ax)$ (no. 4, Theorem 3), $\hat{A} \otimes_A ((Ax + A)/A) = 0$. As $\hat{A}$ is a faithfully flat A-module (Proposition 9), we deduce that $Ax \subset A$, whence $x \in A$.

COROLLARY 5. *Let* A *be a commutative Noetherian ring,* E, F *two finitely generated* A-*modules and* $u: E \to F$ *an* A-*homomorphism. For every maximal ideal* $\mathfrak{m}$ *of* A, *let* $A(\mathfrak{m})$ (resp. $E(\mathfrak{m})$, $F(\mathfrak{m})$) *denote the Hausdorff completion of* A (resp. E, F) *with respect to the* $\mathfrak{m}$-*adic topology and* $u(\mathfrak{m}): E(\mathfrak{m}) \to F(\mathfrak{m})$ *the corresponding homomorphism to* u. *For* u *to be injective* (resp. *surjective, bijective, zero*), *it is necessary and sufficient that* $u(\mathfrak{m})$ *be so for every maximal ideal* $\mathfrak{m}$ *of* A.

We know that for u to be injective (resp. surjective, bijective, zero), it is necessary and sufficient that $u_{\mathfrak{m}}: E_{\mathfrak{m}} \to F_{\mathfrak{m}}$ be so for every maximal ideal $\mathfrak{m}$ of A (Chapter II, § 3, no. 3, Theorem 1). We now note that $A_{\mathfrak{m}}$ is a Noetherian local ring (Chapter II, § 2, no. 4, Corollary 2 to Proposition 10) and hence a Zariski ring and there is a canonical A-algebra isomorphism $\hat{A}_{\mathfrak{m}} \to A(\mathfrak{m})$ (§ 2, no. 13, Proposition 18). On the other hand (beginning of no. 4), there is a commutative diagram

$$\begin{array}{ccccc} E_{\mathfrak{m}} \otimes_{A_{\mathfrak{m}}} A(\mathfrak{m}) & \longrightarrow & E \otimes_A A(\mathfrak{m}) & \longrightarrow & E(\mathfrak{m}) \\ \downarrow{\scriptstyle u_{\mathfrak{m}} \otimes 1} & & \downarrow{\scriptstyle u \otimes 1} & & \downarrow{\scriptstyle u(\mathfrak{m})} \\ F_{\mathfrak{m}} \otimes_{A_{\mathfrak{m}}} A(\mathfrak{m}) & \longrightarrow & F \otimes_A A(\mathfrak{m}) & \longrightarrow & F(\mathfrak{m}) \end{array}$$

where the horizontal arrows on the left arise from the associativity of the tensor product and the isomorphisms $E_{\mathfrak{m}} \to E \otimes_A A_{\mathfrak{m}}$, $F_{\mathfrak{m}} \to F \otimes_A A_{\mathfrak{m}}$; as E and F are finitely generated A-modules, it follows from no. 4, Theorem 3 that the rows of this diagram consist of isomorphisms; thus we are reduced to proving that $u_{\mathfrak{m}}$ being injective (resp. surjective, bijective, zero) is equivalent to $u_{\mathfrak{m}} \otimes 1$ being so. But this follows from the fact that $\hat{A}_{\mathfrak{m}}$ (and hence also $A(\mathfrak{m})$) is a faithfully flat $A_{\mathfrak{m}}$-module by Proposition 9 (Chapter I, § 3, no. 1, Propositions 1 and 2).

PROPOSITION 10. *Let* A, B *be two Zariski rings,* $\hat{A}$, $\hat{B}$ *their completions,* $f: A \to B$ *a continuous ring homomorphism and* $\hat{f}: \hat{A} \to \hat{B}$ *the homomorphism obtained from* f *by passing to the completions; if* $\hat{f}$ *is bijective, the* A-*module* B *is faithfully flat.*

As A and B are Hausdorff, the hypothesis that $\hat{f}$ is bijective implies first that f is injective. Identifying (algebraically) A with $f(A)$ by means of f and $\hat{A}$ with $\hat{B}$ by means of $\hat{f}$, we then obtain the inclusions $A \subset B \subset \hat{A} = \hat{B}$; we know that $\hat{A}$ is a faithfully flat A-module and a faithfully flat B-module (Proposition 9); we conclude that B is a faithfully flat A-module (Chapter I, § 3, no. 4, *Remark* 2).

Proposition 11. *Let* A *be a Noetherian local ring,* $\mathfrak{m}$ *its maximal ideal,* $\hat{A}$ *its* $\mathfrak{m}$*-adic completion and* B *a ring such that* $A \subset B \subset \hat{A}$. *Suppose that* B *is a Noetherian local ring whose maximal ideal* $\mathfrak{n}$ *satisfies the relation* $\mathfrak{n} = \mathfrak{m}B$. *Then*

$$\mathfrak{n}^k = \mathfrak{m}^k B = \hat{\mathfrak{m}}^k \cap B$$

for all $k \geqslant 1$, *the* $\mathfrak{n}$*-adic topology on* B *is induced by the* $\hat{\mathfrak{m}}$*-adic topology on* $\hat{A}$, B *is a faithfully flat* A*-module and there is an isomorphism of* $\hat{A}$ *onto the* $\mathfrak{n}$*-adic completion* $\hat{B}$ *of* B, *which extends the canonical injection* $A \rightarrow B$.

It is sufficient to verify the relation $\mathfrak{n}^k = \hat{\mathfrak{m}}^k \cap B$, for, as B is dense in $\hat{A}$ and the $\mathfrak{n}$-adic topology is induced by the $\hat{\mathfrak{m}}$-adic topology, the last assertion will follow from *General Topology*, Chapter II, § 3, no. 9, Corollary 1 to Proposition 18 and the last but one from Proposition 10. The injection $j_A: A \rightarrow \hat{A}$ (resp. $j_B: B \rightarrow \hat{A}$) defines by taking quotients an injective homomorphism

$$i_A: A/(\hat{\mathfrak{m}} \cap A) \rightarrow \hat{A}/\hat{\mathfrak{m}}$$

(resp. $i_B: B/(\hat{\mathfrak{m}} \cap B) \rightarrow \hat{A}/\hat{\mathfrak{m}}$). We know that $\hat{\mathfrak{m}} \cap A = \mathfrak{m}$ and that i_A is bijective, hence i_B is bijective, which shows that $B/(\hat{\mathfrak{m}} \cap B)$ is a field, hence that $\hat{\mathfrak{m}} \cap B$ is a maximal ideal of B and therefore $\hat{\mathfrak{m}} \cap B = \mathfrak{n}$. As $\hat{A} = A + \hat{\mathfrak{m}}$, $B = A + \mathfrak{n} = A + \mathfrak{m}B$; by induction on k we deduce that

$$B = A + \mathfrak{m}^k B = A + \mathfrak{n}^k$$

for all $k > 1$. As $\mathfrak{n}^k \subset \hat{\mathfrak{m}}^k \cap B$, it is sufficient to show that $\hat{\mathfrak{m}}^k \cap B \subset \mathfrak{n}^k$; if $b \in \hat{\mathfrak{m}}^k \cap B$, we may write $b = a + z$ where $a \in A$, $z \in \mathfrak{n}^k$; whence

$$a = b - z \in \hat{\mathfrak{m}}^k \cap A = \mathfrak{m}^k \subset \mathfrak{n}^k$$

and $b \in \mathfrak{n}^k$.

* An important case where this applies is the following: B is the ring of integral series in n variables over a complete valued field, which *converge* in the neighbourhood of 0, A is the local ring

$$K[X_1, \ldots, X_n]_{\mathfrak{p}}$$

where $\mathfrak{p}$ is the maximal ideal consisting of the polynomials with no constant term and $\hat{A}$ is the ring of formal power series $K[[X_1, \ldots, X_n]]$. *

Remark. A local ring B such that $A \subset B \subset \hat{A}$, whose maximal ideal $\mathfrak{n}$ is equal to $\mathfrak{m}B$ and whose $\mathfrak{n}$-adic topology is induced by the $\mathfrak{m}$-adic topology on $\hat{A}$, is not necessarily Noetherian (Exercise 14).

Proposition 12. *Let* A *be a commutative Noetherian ring,* $\mathfrak{m}$ *an ideal of* A, S *the multiplicative subset* $1 + \mathfrak{m}$ *of* A *and* E *a finitely generated* A*-module. Under these conditions:*

(i) $S^{-1}A$ *is a Zariski ring with the* $(S^{-1}\mathfrak{m})$*-adic topology.*

(ii) *The canonical mapping* f: $\mathrm{E} \to \mathrm{S}^{-1}\mathrm{E}$ *is continuous if* E *is given the* $\mathfrak{m}$*-adic topology and* $\mathrm{S}^{-1}\mathrm{E}$ *the* $(\mathrm{S}^{-1}\mathfrak{m})$*-adic topology and* $\hat{f}$: $\hat{\mathrm{E}} \to (\mathrm{S}^{-1}\mathrm{E})^\wedge$ *is an isomorphism.*

Every element of $1 + (\mathrm{S}^{-1}\mathfrak{m})$ is of the form

$$1 + (m/(1 + m')) = (1 + m + m')/(1 + m)$$

where $m \in \mathfrak{m}$ and $m' \in \mathfrak{m}$; it is therefore invertible in $\mathrm{S}^{-1}\mathrm{A}$, which proves that $\mathrm{S}^{-1}\mathfrak{m}$ is contained in the Jacobson radical of $\mathrm{S}^{-1}\mathrm{A}$; as $\mathrm{S}^{-1}\mathrm{A}$ is Noetherian (Chapter II, § 2, no. 4, Corollary 2 to Proposition 10), $\mathrm{S}^{-1}\mathrm{A}$ is a Zariski ring with the $(\mathrm{S}^{-1}\mathfrak{m})$-adic topology, which proves (i). Let us show (ii). For all $n > 0$,

$$\overset{-1}{f}((\mathrm{S}^{-1}\mathfrak{m})^n\mathrm{E}) = \overset{-1}{f}(\mathrm{S}^{-1}(\mathfrak{m}^n\mathrm{E})) = \mathfrak{m}^n\mathrm{E}:$$

for clearly first of all $f(\mathfrak{m}^n\mathrm{E}) \subset \mathrm{S}^{-1}\mathfrak{m}^n\mathrm{E}$; conversely, let x be an element of $\overset{-1}{f}(\mathrm{S}^{-1}\mathfrak{m}^n\mathrm{E})$; then there exist elements m', m'' of $\mathfrak{m}$ and $x'' \in \mathfrak{m}^n\mathrm{E}$ such that $(1 + m')((1 + m'')x - x'') = 0$, whence $(1 - m)x = x'$, where

$$m = -(m' + m'' + m'm'') \in \mathfrak{m}$$

and $x' = (1 + m')x'' \in \mathfrak{m}^n\mathrm{E}$; we conclude that

$$x = (1 + m + \cdots + m^{n-1})x' + m^n x \in \mathfrak{m}^n\mathrm{E}.$$

This proves that f is a strict morphism. Moreover, the kernel of f, which is the set of $x \in \mathrm{E}$ for which there exists some $s \in \mathrm{S}$ such that $sx = 0$, is identical with the kernel of the canonical mapping j: $\mathrm{E} \to \hat{\mathrm{E}}$ (no. 2, Proposition 5). Then there is a topological isomorphism $f_0 : j(\mathrm{E}) \to f(\mathrm{E})$ such that $f = f_0 \circ j$; as $\hat{j}$ is a topological isomorphism, the problem reduces to verifying that $f(\mathrm{E})$ is dense in $\mathrm{S}^{-1}\mathrm{E}$. Now every element of $\mathrm{S}^{-1}\mathrm{E}$ may be written as $x/(1 - m)$, where $m \in \mathfrak{m}$, and it is immediately verified that

$$x/(1 - m) \equiv ((1 + m + \cdots + m^{n-1})x)/1 \ (\mathrm{mod.}\ \mathrm{S}^{-1}\mathfrak{m}^n\mathrm{E})$$

which completes the proof.

4. LIFTING IN COMPLETE RINGS

1. STRONGLY RELATIVELY PRIME POLYNOMIALS

Let R be a commutative ring. Two elements x, y of R are called *strongly relatively prime* if the principal ideals Rx and Ry are relatively prime, in other words (Chapter II, § 1, no. 2) if $Rx + Ry = R$; it amounts to the same to say that there exist two elements a, b of R such that $ax + by = 1$.

LEMMA 1 ("Euclid's Lemma"). *Let x, y be two strongly relatively prime elements of* R; *if $z \in R$ is such that x divides yz, then x divides z.*

If $1 = ax + by$, then $z = x(az) + (yz)b$.

If x and y are strongly relatively prime in R, then

$$Rxy = (Rx) \cap (Ry)$$

(Chapter II, § 1, no. 2, Proposition 5); if R is an *integral domain*, two strongly relatively prime elements then have an *lcm* equal to their product (*Algebra*, Chapter VI, § 1, no. 8) and are therefore *relatively prime* in the sense of *Algebra*, Chapter VI, § 1, no. 12. Conversely, if R is a *principal ideal domain*, two relatively prime elements are also strongly relatively prime, as follows from Bezout's identity (*Algebra*, Chapter VII, § 1, no. 2, Theorem 1).

For polynomial rings there is the following result:

PROPOSITION 1. *Let* A *be a commutative ring and* P *and* P′ *two strongly relatively prime polynomials in* A[X]. *Suppose that* P *is monic and of degree s. Then every polynomial* T *in* A[X] *may be written uniquely in the form*

$$T = PQ + P'Q' \tag{1}$$

where $Q \in A[X]$, $Q' \in A[X]$ *and* $\deg(Q') < s$.

If further $\deg(T) \leqslant t$ *and* $\deg(P') \leqslant t - s$, *then* $\deg(Q) \leqslant t - s$.

As P is monic, $PR \neq 0$ for every polynomial $R \neq 0$ of A[X] and in this case $\deg(PR) = s + \deg(R)$.

Let T be any polynomial in A[X]. As the ideal generated by P and P′ is the whole of A[X], there exist polynomials Q_1 and Q'_1 such that

$$T = PQ_1 + P'_1Q'_1;$$

as P is monic of degree s, Euclidean division (*Algebra*, Chapter IV, § 1, no. 5) shows that there exist two polynomials Q', Q'' such that $Q'_1 = PQ'' + Q'$ where $\deg(Q') < s$; then we deduce that

$$T = PQ_1 + P'(PQ'' + Q') = PQ + P'Q'$$

where $Q = Q_1 + P'Q''$. To show the uniqueness of formula (1), it is sufficient to prove that the relations

$$0 = PQ + P'Q', \qquad \deg(Q') < s \tag{2}$$

imply $Q = Q' = 0$. Now, if (2) holds, P divides $-PQ = P'Q'$ and, as P and

P′ are strongly relatively prime, P divides Q′ by Lemma 1; if $Q' \neq 0$, there would exist a polynomial $S \neq 0$ such that $Q' = PS$, whence

$$\deg(Q') = s + \deg(S) \geqslant s,$$

which is a contradiction. We conclude that $Q' = 0$, whence $PQ = 0$ and finally $Q = 0$ by the remark at the beginning.

Finally, suppose that $\deg(T) \leqslant t$ and $\deg(P') \leqslant t - s$; with the polynomial T in the form (1),

$$\deg(P'Q') \leqslant \deg(P') + \deg(Q') < s + \deg(P') \leqslant t$$

and therefore

$$s + \deg(Q) = \deg(PQ) = \deg(T - P'Q') \leqslant t$$

whence $\deg(Q) \leqslant t - s$.

Example. For a polynomial $P \in A[X]$ to be strongly relatively prime to $X - a$ (where $a \in A$), it is necessary and sufficient that $P(a)$ be *invertible* in A. For if P and $X - a$ are strongly relatively prime, it follows from Proposition 1 that there exist $c \in A$ and a polynomial $Q \in A[X]$ such that $cP + (X - a)Q = 1$, whence $cP(a) = 1$ and $P(a)$ is invertible. Conversely, by Euclidean division

$$P = (X - a)R + P(a)$$

and, if $P(a) = b^{-1}$, where $b \in A$, we deduce that $1 = bP - b(X - a)R$, which shows that P and $X - a$ are strongly relatively prime.

Let A and B be two commutative rings and $f\colon A \to B$ a ring homomorphism. If $P = \sum_{i \geqslant 0} a_i X^i$ is a formal power series in $A[[X]]$, let $\bar{f}(P)$ denote the formal power series $\sum_{i \geqslant 0} f(a_i) X^i$ in $B[[X]]$. If P is a polynomial, so is $\bar{f}(P)$ and, if further P is monic, then $\bar{f}(P)$ is monic of the same degree as P. Finally, $P \mapsto \bar{f}(P)$ is clearly a homomorphism of $A[[X]]$ to $B[[X]]$ which extends f and maps X to X. *The notation $\bar{f}$ will be constantly used in this sense in the rest of this paragraph.*

PROPOSITION 2. *Let* A *and* B *be two commutative rings, f a homomorphism from* A *to* B *and* P, P′ *two polynomials in* A[X]. *If* P *and* P′ *are strongly relatively prime in* A[X], *then $\bar{f}(P)$ and $\bar{f}(P')$ are strongly relatively prime in* B[X]. *The converse is true if f is surjective, if its kernel is contained in the Jacobson radical of* A *and if* P *is monic.*

Suppose that P and P′ are strongly relatively prime; then there exist polynomials Q, Q′ in A[X] such that $PQ + P'Q' = 1$; we deduce that

$$\bar{f}(P)\bar{f}(Q) + \bar{f}(P')\bar{f}(Q') = 1,$$

whence the first assertion. To show the second, let $\mathfrak{a}$ denote the kernel of f;

let $E = A[X]$ and F the ideal of E generated by P and P'; as $\bar{f}$ is surjective and $\bar{f}(P)$ monic, Proposition 1 shows that for every polynomial $T \in A[X]$ there exist two polynomials Q, Q' in A[X] such that

$$\bar{f}(T) = \bar{f}(P)\bar{f}(Q) + \bar{f}(P')\bar{f}(Q'),$$

whence the relation $E = F + \mathfrak{a}E$. Now, E/F is a finitely generated A-module, for every polynomial is congruent mod. P to a polynomial of degree $< \deg(P)$, P being monic. As $E/F = \mathfrak{a}(E/F)$ and $\mathfrak{a}$ is contained in the Jacobson radical of A, Nakayama's Lemma shows that $E/F = 0$ (*Algebra*, Chapter VIII, § 6, no. 3, Corollary 2 to Proposition 6), which means that P and P' are strongly relatively prime.

2. RESTRICTED FORMAL POWER SERIES

DEFINITION 1. *A commutative topological ring* A *is said to be linearly topologized* (and its topology is said to be *linear*) *if there exists a fundamental system* $\mathscr{B}$ *of neighbourhoods of* 0 *consisting of ideals of* A.

Note that in such a ring, the ideals $\mathfrak{I} \in \mathscr{B}$ are *open* and *closed* (*General Topology*, Chapter III, § 2, no. 1, Corollary to Proposition 4). For all $\mathfrak{I} \in \mathscr{B}$, the quotient topological ring $A/\mathfrak{I}$ is then discrete; for $\mathfrak{I} \in \mathscr{B}$, $\mathfrak{I}' \in \mathscr{B}$, $\mathfrak{I}' \subset \mathfrak{I}$, let

$$h_{\mathfrak{I}\mathfrak{I}'} : A/\mathfrak{I}' \to A/\mathfrak{I}$$

be the canonical mapping. We know (*General Topology*, Chapter III, § 7, no. 3) that $(A/\mathfrak{I}, h_{\mathfrak{I}\mathfrak{I}'})$ is an *inverse system* of discrete rings (relative to the indexing set $\mathscr{B}$ which is ordered by $\supset$ and directed), whose inverse limit is a complete Hausdorff linearly topologized ring $\tilde{A}$; further (*loc. cit.*, Proposition 2), a strict morphism $i : A \to \tilde{A}$ is defined, whose kernel is the closure of $\{0\}$ in A and whose image is everywhere dense in $\tilde{A}$, so that $\tilde{A}$ is canonically identified with the *Hausdorff completion* of A.

DEFINITION 2. *Given a commutative topological ring* A, *a formal power series*

$$T = \sum_{(n_i)} c_{n_1 n_2 \ldots n_p} X_1^{n_1} X_2^{n_2} \ldots X_p^{n_p}$$

in the ring $A[[X_1, \ldots, X_p]]$ *is called restricted if, for every neighbourhood* V *of* 0 *in* A, *there is only a finite number of coefficients* $c_{n_1 n_2 \ldots n_p}$ *not belonging to* V (in other words, the family $(c_{n_1 n_2 \ldots n_p})$ *tends to* 0 *in* A with respect to the filter of complements of finite subsets of $\mathbf{N}^p$).

If A is *linearly topologized*, the restricted formal power series in $A[[X_1, \ldots, X_p]]$ form a *subring* of $A[[X_1, \ldots, X_p]]$, denoted by $A\{X_1, \ldots, X_p\}$: for if $T = \sum_{(n_i)} c_{n_1 \ldots n_p} X_1^{n_1} \ldots X_p^{n_p}$, $T' = \sum_{(n_i)} c'_{n_1 \ldots n_p} X_1^{n_1} \ldots X_p^{n_p}$ are two restricted formal power series and $\mathfrak{I}$ a neighbourhood of 0 in A which is an ideal of A, there exists an integer m such that $c_{n_1\ldots n_p} \in \mathfrak{I}$ and $c'_{n_1\ldots n_p} \in \mathfrak{I}$ for every system $(n_1, \ldots, n_p)$ such that $n_k \geqslant m$ for at least one index k; now, if

$$T'' = TT' = \sum_{(n_i)} c''_{n_1\ldots n_p} X_1^{n_1} \ldots X_p^{n_p},$$

then $c''_{n_1\ldots n_p} = \sum c_{r_1\ldots r_p} c'_{s_1\ldots s_p}$ for all systems (r_k), (s_k) such that $r_k + s_k = n_k$ for $1 \leqslant k \leqslant p$; we conclude that if $n_k \geqslant 2m$, then $r_k \geqslant m$ or $s_k \geqslant m$ and hence, since $\mathfrak{I}$ is an ideal, $c''_{n_1\ldots n_p} \in \mathfrak{I}$ so long as $n_k \geqslant 2m$ for at least one k, which establishes our assertion. Moreover, every *derivative* $\partial T/\partial X_i$ $(1 \leqslant i \leqslant p)$ of a restricted formal power series is restricted, as follows immediately from the definition and the fact that the neighbourhoods $\mathfrak{I} \in \mathscr{B}$ are additive subgroups of A.

If A is *discrete*, the ring of restricted formal power series is just the polynomial ring $A[X_1, \ldots, X_p]$.

Let us always assume that A is *linearly topologized* and let $\mathscr{B}$ be a fundamental system of neighbourhoods of 0 in A consisting of ideals of A; for all $\mathfrak{I} \in \mathscr{B}$, let $p_{\mathfrak{I}}: A \to A/\mathfrak{I}$ be the canonical homomorphism. By definition, for every restricted formal power series $T \in A\{X_1, \ldots, X_p\}$,

$$\bar{p}_{\mathfrak{I}}(T) \in (A/\mathfrak{I})[X_1, \ldots, X_p].$$

Clearly

$$((A/\mathfrak{I})[X_1, \ldots, X_p], \bar{h}_{\mathfrak{I}\mathfrak{I}'})$$

is an inverse system of rings (relative to the directed indexing set $\mathscr{B}$) and $(\bar{p}_{\mathfrak{I}})$ is an inverse system of homomorphisms $A\{X_1, \ldots, X_p\} \to (A/\mathfrak{I})[X_1, \ldots, X_p]$; as every polynomial is a restricted formal power series, $\bar{p}_{\mathfrak{I}}$ is surjective; its kernel $N_{\mathfrak{I}}$ is the ideal of $A\{X_1, \ldots, X_p\}$ consisting of the restricted formal power series all of whose coefficients belong to $\mathfrak{I}$; we shall give $A\{X_1, \ldots, X_n\}$ the (linear) topology for which the $N_{\mathfrak{I}}$ (for $\mathfrak{I} \in \mathscr{B}$) form a fundamental system of neighbourhoods of 0 (a topology which obviously depends only on that on A). Then it follows from *General Topology*, Chapter III, § 7, no. 3, Proposition 2 that

$$(3) \qquad \pi = \varprojlim_{\mathfrak{I}} \bar{p}_{\mathfrak{I}}: A\{X_1, \ldots, X_p\} \to \varprojlim_{\mathfrak{I}} (A/\mathfrak{I})[X_1, \ldots, X_p]$$

is a *strict morphism* whose kernel is the closure of $\{0\}$ in $A\{X_1, \ldots, X_p\}$ and whose image is dense in

$$A' = \varprojlim_{\mathfrak{I}} (A/\mathfrak{I})[X_1, \ldots, X_p].$$

PROPOSITION 3. *If the linearly topologized commutative ring* A *is Hausdorff and complete, the canonical homomorphism* π *is a topological ring isomorphism.*

For all $(n_1, \ldots, n_p) \in \mathbf{N}^p$ and all $\mathfrak{J} \in \mathscr{B}$, let $\phi^{\mathfrak{J}}_{n_1 \ldots n_p}$ be the mapping $(A/\mathfrak{J})[X_1, \ldots, X_p] \to A/\mathfrak{J}$ which maps every polynomial to the coefficient of $X_1^{n_1} \ldots X_p^{n_p}$ in this polynomial; clearly the $\phi^{\mathfrak{J}}_{n_1 \ldots n_p}$ form an inverse system of $(A/\mathfrak{J})$-module homomorphisms (relative to the ordered set $\mathscr{B}$) and, as A is canonically identified with $\varprojlim_{\mathfrak{J}} (A/\mathfrak{J})$ by hypothesis, $\phi_{n_1 \ldots n_p} = \varprojlim_{\mathfrak{J}} \phi^{\mathfrak{J}}_{n_1 \ldots n_p}$ is a continuous A-homomorphism from A' to A. For every element $S = (S_{\mathfrak{J}})_{\mathfrak{J} \in \mathscr{B}}$ of A', we shall see that the formal power series $T = \sum_{(n_i)} \phi_{n_1 \ldots n_p}(S) X_1^{n_1} \ldots X_p^{n_p}$ is restricted and satisfies $\pi(T) = S$. For all $\mathfrak{J} \in \mathscr{B}$ and all $\mathfrak{J}' \in \mathscr{B}$ such that $\mathfrak{J}' \subset \mathfrak{J}$, the relation $\phi^{\mathfrak{J}}_{n_1 \ldots n_p}(S_{\mathfrak{J}}) = 0$ implies

$$\phi^{\mathfrak{J}'}_{n_1 \ldots n_p}(S_{\mathfrak{J}'}) \in \mathfrak{J}/\mathfrak{J}';$$

as $S_{\mathfrak{J}}$ is a polynomial, we see that $\phi_{n_1 \ldots n_p}(S) \in \mathfrak{J}$ except for those $(n_1, \ldots, n_p)$, finite in number, such that $\phi^{\mathfrak{J}}_{n_1 \ldots n_p}(S_{\mathfrak{J}}) \neq 0$, which proves our first assertion; the second follows from the definitions. As A is Hausdorff, the intersection of the $N_{\mathfrak{J}}$ reduces to 0 and hence π is *bijective*, which completes the proof, since π is a strict morphism.

PROPOSITION 4. *Let* A, B *be two linearly topologized commutative rings,* B *being Hausdorff and complete, and* $u: A \to B$ *a continuous homomorphism. For every family* $\mathbf{b} = (b_i)_{1 \leqslant i \leqslant p}$ *of elements of* B, *there exists a unique continuous homomorphism*

$$\tilde{u}: A\{X_1, \ldots, X_p\} \to B$$

such that $\tilde{u}(a) = u(a)$ *for all* $a \in A$ *and* $\tilde{u}(X_i) = b_i$ *for* $1 \leqslant i \leqslant p$.

There exists a unique homomorphism $v: A[X_1, \ldots, X_p] \to B$ such that $v(a) = u(a)$ for $a \in A$ and $v(X_i) = b_i$ for $1 \leqslant i \leqslant p$. Moreover, if $\mathfrak{H}$ is a neighbourhood of 0 in B which is an ideal, $\overset{-1}{u}(\mathfrak{H}) = \mathfrak{J}$ is an ideal of A which is a neighbourhood of 0 and, for every polynomial $P \in N_{\mathfrak{J}}$, clearly $v(P) \in \mathfrak{H}$ and hence v is continuous. As $A[X_1, \ldots, X_p]$ is dense in $A\{X_1, \ldots, X_p\}$, the existence and uniqueness of $\tilde{u}$ follow from *General Topology*, Chapter III, § 3, no. 3, Proposition 5 and the principle of extension of identities.

In the special case where $A = B$ and u is the identity mapping we shall write $f(b_1, \ldots, b_p)$ or $f(\mathbf{b})$ for the value of $\tilde{u}(f)$ for every restricted formal power series $f \in A\{X_1, \ldots, X_p\}$.

Remarks

(1) Proposition 4 proves that for every *closed* ideal $\mathfrak{a}$ in a ring A which is assumed to be Hausdorff and complete, the relations $b_i \in \mathfrak{a}$ for $1 \leqslant i \leqslant p$ imply $f(b_1, \ldots, b_p) \in \mathfrak{a}$ for every restricted formal power series $f \in A\{X_1, \ldots, X_p\}$.

(2) Suppose that A is linearly topologized; let r be an integer such that

$1 \leqslant r \leqslant p$ and let the ring $A\{X_1, \ldots, X_r\}$ be given the topology defined above. Then the topological ring $A\{X_1, \ldots, X_p\}$ is identified with the ring of restricted formal power series

$$(A\{X_1, \ldots, X_r\})\{X_{r+1}, \ldots, X_p\}$$

as follows immediately from the definitions.

(3) With the notation of *Remark* 2, suppose further that A is Hausdorff and complete and let us write every restricted formal power series $f \in A\{X_1, \ldots, X_p\}$ in the form

$$f = \sum_{(n_i)} c_{n_{r+1}\ldots n_p}(X_1, \ldots, X_r) X_{r+1}^{n_{r+1}} \ldots X_p^{n_p}$$

where the $c_{n_{r+1}\ldots n_p}$ are restricted formal power series. For every system $\mathbf{x} = (x_1, \ldots, x_r)$ of elements of A, let

$$b_{n_{r+1}\ldots n_p} = c_{n_{r+1}\ldots n_p}(x_1, \ldots, x_p).$$

It follows immediately from *Remark* 1 that $\sum_{(n_i)} b_{n_{r+1}\ldots n_p} X_{r+1}^{n_{r+1}} \ldots X_p^{n_p}$ is a *restricted* formal power series denoted by $f(x_1, \ldots, x_r, X_{r+1}, \ldots, X_p)$; it is said to be obtained by *substituting* the x_i for the X_i for $1 \leqslant i \leqslant r$ in f.

3. HENSEL'S LEMMA

In a topological ring A, an element x is called *topologically nilpotent* if 0 is a limit of the sequence $(x^n)_{n \geqslant 0}$. If A is a *linearly topologized* commutative ring, to say that x is topologically nilpotent means that for every open ideal $\mathfrak{I}$ of A the canonical image of x in $A/\mathfrak{I}$ is a *nilpotent* element of that ring. If $\mathfrak{r}_{\mathfrak{I}}$ is the nilradical of $A/\mathfrak{I}$, clearly $(\mathfrak{r}_{\mathfrak{I}})$ is an inverse system of subsets and the set $\mathfrak{t}$ of topological nilpotent elements of A is the inverse image of $\mathfrak{r} = \varprojlim_{\mathfrak{I}} \mathfrak{r}_{\mathfrak{I}}$ under the canonical homomorphism $A \to \varprojlim A/\mathfrak{I}$; it is therefore a *closed ideal* of A. If also A is *Hausdorff* and *complete*, this ideal is contained in the Jacobson radical of A and, for an element $x \in A$ to be invertible, it is necessary and sufficient that its class mod. $\mathfrak{t}$ be invertible in $A/\mathfrak{t}$ (§ 2, no. 13, Lemma 3).

Note that if A is a ring and $\mathfrak{m}$ a two-sided ideal of A, the elements of $\mathfrak{m}$ are topologically nilpotent with respect to the $\mathfrak{m}$-*adic* topology.

THEOREM 1 (Hensel). *Let* A *be a complete Hausdorff linearly topologized commutative ring. Let* $\mathfrak{m}$ *be a closed ideal of* A *whose elements are topologically nilpotent. Let* $B = A/\mathfrak{m}$ *be the quotient topological ring and* $\varphi: A \to B$ *the canonical mapping. Let* R *be a restricted formal power series in* A{X}, $\bar{P}$ *a monic polynomial in* B[X] *and* $\bar{Q}$ *a restricted formal power series in* B{X}. *Suppose that* $\bar{\varphi}(R) = \bar{P}.\bar{Q}$ *and that* $\bar{P}$ *and* $\bar{Q}$ *are strongly relatively prime in* B{X}. *Then there exists a unique ordered pair* (P, Q) *consisting of a monic polynomial* $P \in A[X]$ *and a restricted formal power series* $Q \in A\{X\}$ *such that*

(4) $$R = P.Q, \quad \bar{\varphi}(P) = \bar{P}, \quad \bar{\varphi}(Q) = \bar{Q}.$$

Moreover, P *and* Q *are strongly relatively prime in* $A\{X\}$ *and, if* R *is a polynomial, so is* Q.

The proof is divided into four steps. In the first three we assume that A is *discrete*, in which case R and $\bar{Q}$ are polynomials.

(1) $\mathfrak{m}^2 = 0$

Let S, T be two polynomials of A[X] such that S is monic and $\bar{\varphi}(S) = \bar{P}$, $\bar{\varphi}(T) = \bar{Q}$; Proposition 2 of no. 1 shows that S and T are strongly relatively prime; hence (no. 1, Proposition 1) there exists a unique ordered pair of polynomials (S′, T′) of A[X] such that

(5) $$R - ST = ST' + TS' \quad \text{and} \quad \deg(S') < \deg(S) = \deg(\bar{P}).$$

The polynomials $P = S + TS'$, $Q = T + T'$ are then solutions to the problem; then in fact

(6) $$\bar{P}.\bar{\varphi}(T') + \bar{Q}.\bar{\varphi}(S') = \bar{\varphi}(ST' + TS') = \bar{\varphi}(R - ST) = 0.$$

As $\bar{P}$ is monic, $\bar{P}$ and $\bar{Q}$ strongly relatively prime and

$$\deg(\bar{\varphi}(S')) < \deg(\bar{P}),$$

Proposition 1 of no. 1 shows that $\bar{\varphi}(S') = \bar{\varphi}(T') = 0$, in other words the coefficients of S′ and T′ belong to $\mathfrak{m}$ and the relation $\mathfrak{m}^2 = 0$ gives

$$PQ = ST + ST' + TS' = R,$$

which satisfies relation (4). Since $\bar{\varphi}(P) = \bar{P}$ and $\bar{\varphi}(Q) = \bar{Q}$, P and Q are strongly relatively prime (no. 1, Proposition 2); finally, if P_1 and Q_1 are two other polynomials in A[X] satisfying (4) and such that P_1 is monic, then necessarily, setting $S'_1 = P_1 - S$, $T'_1 = Q_1 - T$, $\deg(S'_1) < \deg(S)$ and $R - ST = ST'_1 + TS'_1$ since S'_1 and T'_1 have their coefficients in $\mathfrak{m}$; but Proposition 1 then proves that $S' = S'_1$ and $T' = T'_1$, which proves the uniqueness of the ordered pair (P, Q).

(2) $\mathfrak{m}$ *is nilpotent*

Let *n* be the smallest integer such that $\mathfrak{m}^n = 0$ and let us argue by induction on $n > 2$, the theorem having been shown for $n = 2$. Let $A' = A/\mathfrak{m}^{n-1}$, $\mathfrak{m}' = \mathfrak{m}/\mathfrak{m}^{n-1}$; as $\mathfrak{m}'^{n-1} = 0$, there exists a unique ordered pair (P′, Q′) of polynomials in A′[X] such that P′ is monic, $R' = P'Q'$, $\bar{\psi}(P') = \bar{P}$ and $\bar{\psi}(Q') = \bar{Q}$, where ψ denotes the canonical homomorphism $A' \to A'/\mathfrak{m}' = B$, θ the canonical homomorphism $A \to A'$ and $R' = \bar{\theta}(R)$. On the other hand, as $(\mathfrak{m}^{n-1})^2 = 0$, there exists a unique ordered pair (P, Q) of polynomials in A[X] such that P is monic and $R = PQ$, $\bar{\theta}(P) = \bar{P}$, $\bar{\theta}(Q) = \bar{Q}$; as $\phi = \psi \circ \theta$, this shows the existence and uniqueness of P and Q satisfying (4); moreover P′ and Q′ are strongly relatively prime by the induction hypothesis and hence so are P and Q.

(3) A *is discrete*

Note that in this case $\mathfrak{m}$ is no longer necessarily nilpotent, but it is always a *nilideal* by hypothesis. Let P_0, Q_0 be two polynomials of A[X] such that $\bar{\varphi}(P_0) = \bar{P}$, $\bar{\varphi}(Q_0) = \bar{Q}$ and P_0 is monic. Let us consider the ideal $\mathfrak{n}$ of A generated by the coefficients of $R - P_0Q_0$; it is finitely generated and contained in $\mathfrak{m}$, hence it is *nilpotent* (Chapter II, § 2, no. 6, Proposition 15) and by definition, if $\psi: A \to A/\mathfrak{n}$ is the canonical mapping, then $\bar{\psi}(R) = \bar{\psi}(P_0)\bar{\psi}(Q_0)$. Moreover, $\bar{\psi}(P_0)$ and $\bar{\psi}(Q_0)$ are strongly relatively prime, as follows from the hypothesis on $\bar{P}$ and $\bar{Q}$ and Proposition 2 of no. 1 applied to the canonical homomorphism $A/\mathfrak{n} \to A/\mathfrak{m}$. By virtue of case (2), there therefore exists an ordered pair (P, Q) of polynomials in A[X] such that P is monic and relations (4) hold. The fact that $\bar{P}$ and $\bar{Q}$ are strongly relatively prime implies also here that P and Q are strongly relatively prime in A[X] by virtue of no. 1, Proposition 2, for $\mathfrak{m}$ is contained in the Jacobson radical of A. Suppose finally that P_1, Q_1 are two polynomials in A[X] satisfying (4) and such that P_1 is monic and let $\mathfrak{n}_1$ be the finitely generated ideal of A generated by the coefficients of $P - P_1$ and the coefficients of $Q - Q_1$; as $\mathfrak{n}_1$ is contained in $\mathfrak{m}$, it is nilpotent and, if $\psi_1; A \to A/\mathfrak{n}_1$ is the canonical mapping, $\bar{\psi}_1(P) = \bar{\psi}_1(P_1)$ and

$$\bar{\psi}_1(Q) = \bar{\psi}_1(Q_1);$$

the uniqueness property for case (2) therefore implies $P = P_1$, $Q = Q_1$.

(4) *General case*

Let $\mathscr{B}$ be a fundamental system of neighbourhoods of 0 in A consisting of ideals of A. For all $\mathfrak{J} \in \mathscr{B}$, let $f_\mathfrak{J}$ be the canonical mapping $A \to A/\mathfrak{J}$, $\phi_\mathfrak{J}$ the canonical mapping

$$A/\mathfrak{J} \mapsto (A/\mathfrak{J})/((\mathfrak{m} + \mathfrak{J})/\mathfrak{J}) = A/(\mathfrak{m} + \mathfrak{J}),$$

$g_\mathfrak{J}$ the canonical mapping $B = A/\mathfrak{m} \to A(\mathfrak{m} + \mathfrak{J})$ and write $R_\mathfrak{J} = \bar{f}_\mathfrak{J}(R)$, $\bar{P}_\mathfrak{J} = \bar{g}_\mathfrak{J}(\bar{P})$, $\bar{Q}_\mathfrak{J} = \bar{g}_\mathfrak{J}(\bar{Q})$. As each ring $A/\mathfrak{J}$ is discrete, case (3) can be applied to it and we see that there exists a unique ordered pair $(P_\mathfrak{J}, Q_\mathfrak{J})$ of polynomials in $(A/\mathfrak{J})[X]$ such that $P_\mathfrak{J}$ is monic and $R_\mathfrak{J} = P_\mathfrak{J}Q_\mathfrak{J}$, $\bar{\varphi}_\mathfrak{J}(P_\mathfrak{J}) = \bar{P}_\mathfrak{J}$, $\bar{\varphi}_\mathfrak{J}(Q_\mathfrak{J}) = \bar{Q}_\mathfrak{J}$. The uniqueness of this ordered pair implies that, if $\mathfrak{J}' \subset \mathfrak{J}$, $\mathfrak{J}' \in \mathscr{B}$ and $f_{\mathfrak{J}\mathfrak{J}'}: A/\mathfrak{J}' \to A/\mathfrak{J}$ is the canonical mapping, then $P_\mathfrak{J} = \bar{f}_{\mathfrak{J}\mathfrak{J}'}(P_{\mathfrak{J}'})$, $Q_\mathfrak{J} = \bar{f}_{\mathfrak{J}\mathfrak{J}'}(Q_{\mathfrak{J}'})$. Then it follows from the canonical identification of A{X} with $\varprojlim_\mathfrak{J}(A/\mathfrak{J})[X]$

(no. 2, Proposition 3) that there exists $P \in A\{X\}$ and $Q \in A\{X\}$ such that $R = PQ$ and $\bar{f}_{\mathfrak{I}}(P) = P_{\mathfrak{I}}$, $\bar{f}_{\mathfrak{I}}(Q) = Q_{\mathfrak{I}}$ for all $\mathfrak{I} \in \mathscr{B}$. Moreover,

$$\bar{g}_{\mathfrak{I}}(\overline{P} - \bar{\varphi}(P)) = 0, \qquad \bar{g}_{\mathfrak{I}}(\overline{Q} - \bar{\varphi}(Q)) = 0$$

for all $\mathfrak{I} \in \mathscr{B}$, which means that for all $\mathfrak{I} \in \mathscr{B}$ the coefficients of $\overline{P} - \bar{\varphi}(P)$ and $\overline{Q} - \bar{\varphi}(Q)$ all belong to $(\mathfrak{m} + \mathfrak{I})/\mathfrak{m}$. But, as $\mathfrak{m}$ is closed in A, $\bigcap_{\mathfrak{I}} (\mathfrak{m} + \mathfrak{I}) = \mathfrak{m}$, whence $\overline{P} = \bar{\varphi}(P)$, $\overline{Q} = \bar{\varphi}(Q)$ and P and Q then certainly satisfy (4); moreover, as the $P_{\mathfrak{I}}$ are monic and of the same degree, the restricted formal power series P is a monic polynomial. If (P', Q') were another ordered pair satisfying (4) and such that P' is a monic polynomial, we would deduce that $R_{\mathfrak{I}} = \bar{f}_{\mathfrak{I}}(P')\bar{f}_{\mathfrak{I}}(Q')$, $\bar{\varphi}_{\mathfrak{I}}(\bar{f}_{\mathfrak{I}}(P')) = \overline{P}_{\mathfrak{I}}$ and $\bar{\varphi}_{\mathfrak{I}}(\bar{f}_{\mathfrak{I}}(Q')) = \overline{Q}_{\mathfrak{I}}$ and by the uniqueness in case (3) $\bar{f}_{\mathfrak{I}}(P') = P_{\mathfrak{I}}$, $\bar{f}_{\mathfrak{I}}(Q') = Q_{\mathfrak{I}}$ for all $\mathfrak{I} \in \mathscr{B}$, which implies that $P = P'$ and $Q = Q'$. Let us show finally that P and Q are strongly relatively prime; by virtue of case (3) and Proposition 1 of no. 1, for all $\mathfrak{I} \in \mathscr{B}$, there exists a unique ordered pair $(S_{\mathfrak{I}}, T_{\mathfrak{I}})$ of polynomials in $(A/\mathfrak{I})[X]$ such that

(7) $$1 = P_{\mathfrak{I}}S_{\mathfrak{I}} + Q_{\mathfrak{I}}T_{\mathfrak{I}} \quad \text{and} \quad \deg(T_{\mathfrak{I}}) < \deg(P_{\mathfrak{I}}) = \deg(\overline{P}).$$

The uniqueness of this ordered pair shows immediately that, for $\mathfrak{I}' \in \mathscr{B}$, $\mathfrak{I}' \subset \mathfrak{I}$, $S_{\mathfrak{I}} = \bar{f}_{\mathfrak{I}\mathfrak{I}'}(S_{\mathfrak{I}'})$, $T_{\mathfrak{I}} = \bar{f}_{\mathfrak{I}\mathfrak{I}'}(T_{\mathfrak{I}'})$; taking account of no. 2, Proposition 3, we conclude that there exist two restricted formal power series S, T of $A\{X\}$ such that $S_{\mathfrak{I}} = \bar{f}_{\mathfrak{I}}(S)$, $T_{\mathfrak{I}} = \bar{f}_{\mathfrak{I}}(T)$ and $1 = PS + QT$.

It remains to verify that, if R is a polynomial, so is Q. Now, the $Q_{\mathfrak{I}}$ are polynomials by construction and, as $P_{\mathfrak{I}}$ is monic, the relation $R_{\mathfrak{I}} = P_{\mathfrak{I}}Q_{\mathfrak{I}}$ implies

$$\deg(Q_{\mathfrak{I}}) \leqslant \deg(R_{\mathfrak{I}}) \leqslant \deg(R)$$

for all $\mathfrak{I} \in \mathscr{B}$; whence immediately the required result by definition of Q.

4. COMPOSITION OF SYSTEMS OF FORMAL POWER SERIES

Let A be a commutative ring; we shall say that a system

(8) $$\mathbf{f} = (f_1, \dots, f_p) \in (A[[X_1, \dots, X_q]])^p$$

of formal power series in the X_j $(1 \leqslant j \leqslant q)$, with coefficients in A, is *without constant term* if this is true of all the f_j. For every system (8) of formal power series and every system

(9) $$\mathbf{g} = (g_1, \dots, g_q) \in (A[[X_1, \dots, X_r]])^q$$

of q formal power series without constant term, we shall denote by $\mathbf{f} \circ \mathbf{g}$ (or $\mathbf{f}(\mathbf{g})$) the system of formal power series $f_j(g_1, \dots, g_q)$ $(1 \leqslant j \leqslant p)$ in

$$(A[[X_1, \dots, X_r]])^p$$

(*Algebra*, Chapter IV, § 5, no. 5). If

$$\mathbf{h} = (h_1, \ldots, h_r) \in (A[[X_1, \ldots, X_s]])^r$$

is a third system without constant term, then

$$(\mathbf{f} \circ \mathbf{g}) \circ \mathbf{h} = \mathbf{f} \circ (\mathbf{g} \circ \mathbf{h}). \tag{10}$$

For every integer m,

$$(\mathbf{f}^{(m)} \circ \mathbf{g}^{(m)}) \circ \mathbf{h}^{(m)} = \mathbf{f}^{(m)} \circ (\mathbf{g}^{(m)} \circ \mathbf{h}^{(m)})$$

where $\mathbf{f}^{(m)}$, $\mathbf{g}^{(m)}$, $\mathbf{h}^{(m)}$ denote the systems of polynomials consisting of terms of total degree $\leqslant m$ in the systems of formal power series $\mathbf{f}$, $\mathbf{g}$, $\mathbf{h}$. But clearly the terms of total degree $\leqslant m$ in the series of $(\mathbf{f} \circ \mathbf{g}) \circ \mathbf{h}$ (resp. $\mathbf{f} \circ (\mathbf{g} \circ \mathbf{h})$) are the same as in $(\mathbf{f}^{(m)} \circ \mathbf{g}^{(m)}) \circ \mathbf{h}^{(m)}$ (resp. $\mathbf{f}^{(m)} \circ (\mathbf{g}^{(m)} \circ \mathbf{h}^{(m)})$), whence our assertion.

For every system (8), we shall denote by $M_{\mathbf{f}}$ or $M_{\mathbf{f}}(\mathbf{X})$, the *Jacobian matrix* $(\partial f_i/\partial X_j)$ ($1 \leqslant i \leqslant p$, $1 \leqslant j \leqslant q$) where i is the index of the rows and j that of the columns; for two systems (8) and (9), where $\mathbf{g}$ is without constant term,

$$M_{\mathbf{f} \circ \mathbf{g}} = (M_{\mathbf{f}}(\mathbf{g})) . M_{\mathbf{g}} \tag{11}$$

where $M_{\mathbf{f}}(\mathbf{g})$ is the matrix whose elements are obtained by substituting g_j for X_j ($1 \leqslant j \leqslant q$) in each series element of $M_{\mathbf{f}}$; this formula is just a reformulation of formula (9) of *Algebra*, Chapter IV, § 5, no. 8. We shall denote by $M_{\mathbf{f}}(0)$ the matrix of constant terms of the elements of $M_{\mathbf{f}}$; then we deduce from (11) that

$$M_{\mathbf{f} \circ \mathbf{g}}(0) = M_{\mathbf{f}}(0) . M_{\mathbf{g}}(0). \tag{12}$$

Given an integer $n > 0$, we shall write

$$\mathbf{1}_n = \mathbf{X} = (X_1, \ldots, X_n) \in (A[[X_1, \ldots, X_n]])^n, \tag{13}$$

which will be considered as a matrix with a single column.

For every system $\mathbf{f} = (f_1, \ldots, f_n) \in (A[[X_1, \ldots, X_n]])^n$, $M_{\mathbf{f}}$ is a square matrix of order n; we shall denote by $J_{\mathbf{f}}$ or $J_{\mathbf{f}}(\mathbf{X})$ its determinant and by $J_{\mathbf{f}}(0)$ the constant term of $J_{\mathbf{f}}$, equal to $\det(M_{\mathbf{f}}(0))$; if $\mathbf{g} = (g_1, \ldots, g_n)$ is a system without constant term in $(A[[X_1, \ldots, X_n]])^n$, then, by (11) and (12),

$$J_{\mathbf{f} \circ \mathbf{g}} = J_{\mathbf{f}}(\mathbf{g}) \cdot J_{\mathbf{g}} \tag{14}$$

$$J_{\mathbf{f} \circ \mathbf{g}}(0) = J_{\mathbf{f}}(\mathbf{g}) J_{\mathbf{g}}(0). \tag{15}$$

PROPOSITION 5. *Let* A *be a commutative ring and* $\mathbf{f} = (f_1, \ldots, f_n)$ *a system without constant term of* n *series in* $A[[X_1, \ldots, X_n]]$. *Suppose that* $J_{\mathbf{f}}(0)$ *is invertible in* A. *Then there exists a system without constant term* $\mathbf{g} = (g_1, \ldots, g_n)$ *of* n *series in* $A[[X_1, \ldots, X_n]]$ *such that*

$$\mathbf{f} \circ \mathbf{g} = \mathbf{1}_n. \tag{16}$$

This system is unique and

(17) $$\mathbf{g} \circ \mathbf{f} = \mathbf{1}_n.$$

The existence and uniqueness of $\mathbf{g}$ follows from *Algebra*, Chapter IV, § 5, no. 9, Proposition 10, applied to the n formal power series

$$f_i(Y_1, \ldots, Y_n) - X_i \qquad (1 \leqslant i \leqslant n).$$

It follows from (15) and (16) that $J_{\mathbf{f}}(0)J_{\mathbf{g}}(0) = 1$ and hence $J_{\mathbf{g}}(0)$ is also invertible. We conclude that there exists a system $\mathbf{h} = (h_1, \ldots, h_n)$ of n series without constant term in $A[[X_1, \ldots, X_n]]$ such that $\mathbf{g} \circ \mathbf{h} = \mathbf{1}_n$; from this relation and (16) it then follows, with the aid of (10), that

$$\mathbf{h} = \mathbf{1}_n \circ \mathbf{h} = (\mathbf{f} \circ \mathbf{g}) \circ \mathbf{h} = \mathbf{f} \circ (\mathbf{g} \circ \mathbf{h}) = \mathbf{f} \circ \mathbf{1}_n = \mathbf{f}.$$

Proposition 5 and formulae (10) and (15) show that under the law of composition $(\mathbf{f}, \mathbf{g}) \mapsto \mathbf{f} \circ \mathbf{g}$ the set of systems $\mathbf{f} = (f_1, \ldots, f_n)$ of n series without constant term in $A[[X_1, \ldots, X_n]]$ for which $J_{\mathbf{f}}(0)$ is invertible in A, is a *group*.

5. SYSTEMS OF EQUATIONS IN COMPLETE RINGS

To abbreviate, we shall say in what follows that a ring *satisfies Hensel's conditions* if it is commutative, linearly topologized, Hausdorff and complete; given an ideal $\mathfrak{m}$ in such a ring, $\mathfrak{m}$ (or the ordered pair $(A, \mathfrak{m})$) will be said to *satisfy Hensel's conditions* if $\mathfrak{m}$ is closed in A and its elements are topologically nilpotent. The ideal $\mathfrak{t}$ of A consisting of all the topologically nilpotent elements satisfies Hensel's conditions (no. 3).

In particular, if A is a commutative ring and $\mathfrak{m}$ an ideal of A and A is Hausdorff and complete with respect to the $\mathfrak{m}$-*adic* topology, the ordered pair $(A, \mathfrak{m})$ satisfies Hensel's conditions.

PROPOSITION 6. *Let* A *be a commutative ring,* B *a ring satisfying Hensel's conditions and* $u: A \to B$ *a homomorphism. For every family* $\mathbf{x} = (x_1, \ldots, x_n)$ *of topologically nilpotent elements of* B, *there exists a unique homomorphism* $\tilde{u}$ *from* $A[[X_1, \ldots, X_n]]$ *to* B *such that* $\tilde{u}(a) = u(a)$ *for all* $a \in A$ *and* $\tilde{u}(X_i) = x_i$ *for* $1 \leqslant i \leqslant n$. *Moreover, if* $\mathfrak{m}$ *denotes the ideal of series without constant term in* $A[[X_1, \ldots, X_n]]$, $\tilde{u}$ *is continuous for the* $\mathfrak{m}$-*adic topology.*

Let $\mathfrak{a}$ be the finitely generated ideal generated in B by the x_i $(1 \leqslant i \leqslant n)$; for every open ideal $\mathfrak{H}$ of B, the images of the x_i in $B/\mathfrak{H}$ are nilpotent, hence the ideal $(\mathfrak{a} + \mathfrak{H})/\mathfrak{H}$ is nilpotent in $B/\mathfrak{H}$ and there exists an integer k such that, for $\sum_{i=1}^{n} p_i \geqslant k$, $x_1^{p_1} \ldots x_n^{p_n} \in \mathfrak{H}$. As every element of $\mathfrak{m}^k$ is a finite sum of formal power series of the form $X_1^{p_1} \ldots X_n^{p_n} g(X_1, \ldots, X_n)$, where $\sum_{i=1}^{n} p_i \geqslant k$, it is seen that, if $\tilde{u}$ solves the problem, then $\tilde{u}(\mathfrak{m}^k) \subset \mathfrak{H}$, which proves the continuity of $\tilde{u}$. There obviously exists a unique homomorphism

$$v: A[X_1, \ldots, X_n] \to B$$

such that $v(a) = u(a)$ for $a \in A$ and $v(X_i) = x_i$ for $1 \leqslant i \leqslant n$ and the above argument shows that v is continuous with respect to the topoloogy induced on $A[X_1, \ldots, X_n]$ by the $\mathfrak{m}$-adic topology. As $A[X_1, \ldots, X_n]$ is dense in $A[[X_1, \ldots, X_n]]$ with the $\mathfrak{m}$-adic topology and B is Hausdorff and complete, this completes the proof of the existence and uniqueness of $\tilde{u}$.

Note that this proposition gives us again as a special case (i) of Proposition 11 of § 2, no. 9.

If A itself is linearly topologized, the restriction of $\tilde{u}$ to $A\{X_1, \ldots, X_n\}$ *coincides* with the homomorphism derived from u in Proposition 4 of no. 2. This follows immediately from the fact that $A[X_1, \ldots, X_n]$ is *dense* in $A\{X_1, \ldots, X_n\}$ if this ring is given the topology with fundamental system of neighbourhoods of 0 the ideals $\mathfrak{m}^k \cap N_{\mathfrak{J}}$ (in the notation of no. 2, this topology is the least upper bound of the topology induced on $A\{X_1, \ldots, X_n\}$ by the $\mathfrak{m}$-adic topology on $A[[X_1, \ldots, X_n]]$ and the topology defined in no. 2).

If $B = A$ and u is the identity mapping, we shall write $f(x_1, \ldots, x_n)$ or $f(\mathbf{x})$ for the element $\tilde{u}(f)$ for every formal power series $f \in A[[X_1, \ldots, X_n]]$; for every system $\mathbf{f} = (f_1, \ldots, f_r)$ of formal power series of $A[[X_1, \ldots, X_n]]$, let $\mathbf{f}(\mathbf{x})$ denote the element $(f_1(\mathbf{x}), \ldots, f_r(\mathbf{x}))$ of A^r, then it is said to be obtained by *substituting* the x_i for the X_i in $\mathbf{f}$. If $n \leqslant m$ and F is a formal power series of $A[[X_1, \ldots, X_m]]$, it is possible to consider F as a formal power series in $X_{n+1}, \ldots, X_m$ with coefficients in $A[[X_1, \ldots, X_n]]$; let

$$F(x_1, \ldots, x_n, X_{n+1}, \ldots, X_m)$$

denote the formal power series in $A[[X_{n+1}, \ldots, X_m]]$ obtained by substituting the x_i for the X_i in the coefficients of F, for $1 \leqslant i \leqslant n$.

Let us take B to be the ring of formal power series $A[[X_1, \ldots, X_r]]$ and let $\mathfrak{n}$ be the ideal of series in B without constant term, so that $(B, \mathfrak{n})$ satisfies Hensel's conditions (§ 2, no. 6, Corollary to Proposition 6). Proposition 6 may be applied by taking the $x_i \in B$ to be series without constant term; then, for every series $f \in A[[X_1, \ldots, X_n]]$, $\tilde{u}(f)$ is just the formal power series $f(x_1, \ldots, x_n)$ defined in *Algebra*, Chapter IV, § 5, no. 5. This is obvious if f is a polynomial and it follows from the proposition in the general case by remarking that $f \mapsto f(x_1, \ldots, x_n)$ is continuous on $A[[X_1, \ldots, X_n]]$ with respect to the $\mathfrak{m}$-adic topology.

COROLLARY. *Let* A *be a ring satisfying Hensel's condition and* $\mathbf{x} = (x_1, \ldots, x_n)$ *a family of topologically nilpotent elements of* A. *Let* $\mathbf{g} = (g_1, \ldots, g_q)$ *be a system without constant term of series in* $A[[X_1, \ldots, X_n]]$ *and* $\mathbf{f} = (f_1, \ldots, f_p)$ *a system of formal power series in* $A[[X_1, \ldots, X_q]]$. *Then* $\mathbf{g}(\mathbf{x}) = (g_1(\mathbf{x}), \ldots, g_q(\mathbf{x}))$ *is a family of topologically nilpotent elements of* A *and*

$$(\mathbf{f} \circ \mathbf{g})(\mathbf{x}) = \mathbf{f}(\mathbf{g}(\mathbf{x})). \tag{18}$$

The fact that the $g_i(\mathbf{x})$ are topologically nilpotent follows immediately from Proposition 6 and the fact that in A the ideal of topologically nilpotent elements is closed. Relation (18) is obvious when the f_j are polynomials; on the other hand, if $\mathfrak{m}$ and $\mathfrak{m}'$ are the ideals of series without constant term in

$$A[[X_1, \ldots, X_q]] \quad \text{and} \quad A[[X_1, \ldots, X_n]]$$

respectively, clearly the relation $f \in \mathfrak{m}^k$ implies $f(g_1, \ldots, g_q) \in \mathfrak{m}'^k$. The two sides of (18) are therefore continuous functions of $\mathbf{f}$ to $(A[[X_1, \ldots, X_q]])^p$ if $A[[X_1, \ldots, X_q]]$ is given the $\mathfrak{m}$-adic topology, by virtue of the above remark and Proposition 6; whence relation (18).

In what follows, for a ring A and an ideal $\mathfrak{m}$ of A we shall denote by $\mathfrak{m}^{\times n}$ the product set $\prod_{i=1}^{n} \mathfrak{m}_i$ in A^n, where $\mathfrak{m}_i = \mathfrak{m}$ for $1 \leqslant i \leqslant n$, to avoid ambiguity.

PROPOSITION 7. *Let* A *be a ring and* $\mathfrak{m}$ *an ideal of* A *such that the ordered pair* (A, $\mathfrak{m}$) *satisfies Hensel's conditions. Let* $\mathbf{f} = (f_1, \ldots, f_n)$ *be a system without constant term of series in* $A[[X_1, \ldots, X_n]]$ *such that* $J_{\mathbf{f}}(0)$ *is invertible in* A. *Then, for all* $\mathbf{x} \in \mathfrak{m}^{\times n}$, $\mathbf{f}(\mathbf{x}) \in \mathfrak{m}^{\times n}$ *and* $\mathbf{x} \mapsto \mathbf{f}(\mathbf{x})$ *is a bijection of* $\mathfrak{m}^{\times n}$ *onto itself, the inverse bijection being* $\mathbf{x} \mapsto \mathbf{g}(\mathbf{x})$, *where* $\mathbf{g}$ *is given by relation* (16) *in* no. 4.

The fact that $\mathbf{f}(\mathbf{x}) \in \mathfrak{m}^{\times n}$ is obvious when the f_i are polynomials and follows in the general case from Proposition 6 and the fact that $\mathfrak{m}$ is closed in A. The other assertions of the proposition are then immediate consequences of (16), (17) and (18).

COROLLARY. *Let* $\mathfrak{q}$ *be a closed ideal of* A *contained in* $\mathfrak{m}$. *Then the relation* $\mathbf{x} \equiv \mathbf{x}'$ (mod. $\mathfrak{q}^{\times n}$) *is equivalent to* $\mathbf{f}(\mathbf{x}) \equiv \mathbf{f}(\mathbf{x}')$ (mod. $\mathfrak{q}^{\times n}$).

For every formal power series $f \in A[[X_1, \ldots, X_n]]$,

$$f(X_1, \ldots, X_n) - f(Y_1, \ldots, Y_n) = \sum_{i=1}^{n} (X_i - Y_i) h_i(X_1, \ldots, X_n, Y_1, \ldots, Y_n)$$

where the h_i belong to $A[[X_1, \ldots, X_n, Y_1, \ldots, Y_n]]$ (*Algebra*, Chapter IV, § 5, no. 8, Proposition 9); it follows immediately that the relation $\mathbf{x} \equiv \mathbf{x}'$ (mod. $\mathfrak{q}^{\times n}$) implies $\mathbf{f}(\mathbf{x}) \equiv \mathbf{f}(\mathbf{x}')$ (mod. $\mathfrak{q}^{\times n}$). The converse is obtained by replacing $\mathbf{f}$ by its "inverse" $\mathbf{g}$.

THEOREM 2. *Let* A *be a ring and* $\mathfrak{m}$ *an ideal of* A *such that the ordered pair* (A, $\mathfrak{m}$) *satisfies Hensel's conditions. Let* $\mathbf{f} = (f_1, \ldots, f_n)$ *be a system of* n *elements of*

$A\{X_1, \ldots, X_n\}$ *and let* $\mathbf{a} \in A^n$; *let us write* $J_{\mathbf{f}}(\mathbf{a}) = e$. *There exists a system* $\mathbf{g} = (g_1, \ldots, g_n)$ *of restricted formal power series without constant term in* $A\{X_1, \ldots, X_n\}$ *such that*

(i) $M_{\mathbf{g}}(0) = I_n$ *(unit matrix).*

(ii) *For all* $\mathbf{x} \in A^n$,

(19) $$\mathbf{f}(\mathbf{a} + e\mathbf{x}) = \mathbf{f}(\mathbf{a}) + M_{\mathbf{f}}(\mathbf{a}).e\mathbf{g}(\mathbf{x}).$$

(iii) *Let* $\mathbf{h} = (h_1, \ldots, h_n)$ *be the system of formal power series without constant term* (not necessarily restricted) *such that* $\mathbf{g} \circ \mathbf{h} = \mathbf{1}_n$ (Proposition 5). *For all* $y \in \mathfrak{m}^{\times n}$,

(20) $$\mathbf{f}(\mathbf{a} + e\mathbf{h}(\mathbf{y})) = \mathbf{f}(\mathbf{a}) + M_{\mathbf{f}}(\mathbf{a}).e\mathbf{y}.$$

For every formal power series $f \in A[[X_1, \ldots, X_n]]$,

(21) $$f(\mathbf{X} + \mathbf{Y}) = f(\mathbf{X}) + M_f(\mathbf{X}).\mathbf{Y} + \sum_{1 \leqslant i \leqslant j \leqslant n} G_{ij}(\mathbf{X}, \mathbf{Y})Y_iY_j$$

where the G_{ij} are well determined formal power series in

$$A[[X_1, \ldots, X_n, Y_1, \ldots, Y_n]].$$

If f is *restricted,* so are the elements of M_f and the G_{ij}, for these formal power series are polynomials if f is a polynomial and it follows from their uniqueness that for every open ideal $\mathfrak{I}$ of A, denoting by $p_{\mathfrak{I}}: A \to A/\mathfrak{I}$ the canonical mapping, the image of G_{ij} under $\bar{p}_{\mathfrak{I}}$ is the coefficient of Y_iY_j in $\bar{p}_{\mathfrak{I}}(F)$ where F is the formal power series $f(\mathbf{X} + \mathbf{Y})$ in $A[[X_1, \ldots, X_n, Y_1, \ldots, Y_n]]$; whence our assertion.

This being so, writing formula (21) for each series f_i $(1 \leqslant i \leqslant n)$, we obtain for all $\mathbf{x} \in A^n$ (no. 2, Proposition 4),

(22) $$\mathbf{f}(\mathbf{a} + e\mathbf{x}) = \mathbf{f}(\mathbf{a}) + M_{\mathbf{f}}(\mathbf{a}).e\mathbf{x} + e^2\mathbf{r}(\mathbf{x})$$

where $\mathbf{r} = (r_1, \ldots, r_n)$ is a system of restricted formal power series each of which is of total order $\geqslant 2$. It follows from formulae (18) of *Algebra*, Chapter III, § 6, no. 5 that there exists a square matrix $M' \in \mathbf{M}_n(A)$ such that

(23) $$M_{\mathbf{f}}(\mathbf{a}).M' = eI_n,$$

whence using this in (22)

(24) $$\mathbf{f}(\mathbf{a} + e\mathbf{x}) = \mathbf{f}(\mathbf{a}) + M_{\mathbf{f}}(\mathbf{a}).e\mathbf{x} + M_{\mathbf{f}}(\mathbf{a})M'.e\mathbf{r}(\mathbf{x}).$$

Writing $\mathbf{g} = \mathbf{1}_n + M'.\mathbf{r}$, we see that $\mathbf{g}$ satisfies conditions (i) and (ii); then it is sufficient to replace $\mathbf{x}$ by $\mathbf{h}(\mathbf{y})$ to obtain (iii).

COROLLARY 1. *Let* A *be a ring and* $\mathfrak{m}$ *an ideal of* A *such that the ordered pair* (A, $\mathfrak{m}$) *satisfies Hensel's conditions. Let* $f \in A\{X\}$, $a \in A$ *and write* $e = f'(a)$. *If* $f(a) \equiv 0$ (mod. $e^2\mathfrak{m}$), *then there exists* $b \in A$ *such that* $f(b) = 0$ *and* $b \equiv a$ (mod. $e\mathfrak{m}$). *If*

b' is another element of A *such that* $f(b') = 0$ *and* $b' \equiv a \pmod{e\mathfrak{m}}$, *then* $e(b - b') = 0$. *In particular, b is unique if e is not a divisor of zero in* A.

Let $f(a) = e^2c$ where $c \in \mathfrak{m}$; formula (20) for $n = 1$ gives

$$f(a + eh(y)) = e^2(c + y)$$

and it is therefore sufficient to take $y = -c$, $b = a + eh(-c)$. Moreover if $b = a + ex$, $b' = a + ex'$, $x \in \mathfrak{m}$, $x' \in \mathfrak{m}$, $f(b) = f(b') = 0$, we deduce from (19) that $e^2(g(x) - g(x')) = 0$. As $g(X) - g(Y) = (X - Y)u(X, Y)$, where u is restricted and $u(0, 0) = 1$, $g(x) - g(x') = (x - x')v$, where $v \in A$ is invertible, for, since $\mathfrak{m}$ is closed, $v - 1 = u(x, x') - 1 \in \mathfrak{m}$ and $\mathfrak{m}$ is contained in the Jacobson radical of A; whence the relation $e(b - b') = 0$.

Remark. The corollary applies notably when e is invertible in A; we can then also deduce the existence of b from Hensel's Theorem, for the canonical image of $f(X)$ in $(A/\mathfrak{m})\{X\}$ is of the form $(X - \alpha)f_1(X)$, $X - \alpha$ and $f_1(X)$ being strongly relatively prime, for $f_1(\alpha) = f'(\alpha)$ is the image of e (no. 1, *Example*).

Examples

(1) Let p be a prime number $\neq 2$ and n an integer whose class mod. p is a square $\neq 0$ in the prime field $\mathbf{F}_p$. If $\mathbf{Z}_p$ is the ring of p-adic integers (§ 2, no. 12, Example 3), the application of Corollary 1 to the polynomial $X^2 - n$ shows that n is a square in $\mathbf{Z}_p$; for example 7 is a square in $\mathbf{Z}_3$.

(2) Let $A = K[[Y]]$ be the ring of formal power series in one indeterminate with coefficients in a commutative field K; with the (Y)-adic topology, the ring A is Hausdorff and complete (§ 2, no. 6, Corollary to Proposition 6) and the mapping $f(Y) \mapsto f(0)$ defines by passing to the quotient ring an isomorphism of A/(Y) onto the field K. By Corollary 1, if $F(Y, X)$ is a polynomial in X with coefficients in A and a is a simple root of $F(0, X)$ in K, there exists a unique formal power series $f(Y)$ such that $f(0) = a$ and $F(Y, f(Y)) = 0$.

COROLLARY 2. *Let* A *be a ring and* $\mathfrak{m}$ *an ideal of* A *such that the ordered pair* (A, $\mathfrak{m}$) *satisfies Hensel's conditions. Let r, n be integers such that $0 \leqslant r < n$ and* $\mathbf{f} = (f_{r+1}, \dots, f_n)$ *is a system of $n - r$ elements of* $A\{X_1, \dots, X_n\}$; *let* $J_{\mathbf{f}}^{(n-r)}(\mathbf{X})$ *denote the minor of* $M_{\mathbf{f}}(\mathbf{X})$ *consisting of the columns of index j such that* $r + 1 \leqslant j \leqslant n$. *Let* $\mathbf{a} \in A^n$ *be such that* $J_{\mathbf{f}}^{(n-r)}(\mathbf{a})$ *is invertible in* A *and* $\mathbf{f}(\mathbf{a}) \equiv 0 \pmod{\mathfrak{m}^{\times(n-r)}}$. *Then there exists a unique* $\mathbf{x} = (x_1, \dots, x_n) \in A^n$ *such that* $x_k = a_k$ *for* $1 \leqslant k \leqslant r$, $\mathbf{x} \equiv \mathbf{a} \pmod{\mathfrak{m}^{\times n}}$ *and* $\mathbf{f}(\mathbf{x}) = 0$.

Substituting a_k for X_k for $1 \leqslant k \leqslant r$ in the f_i (no. 2, *Remark* 3), we see immediately that we may restrict our attention to the case where $r = 0$ to prove the corollary. Theorem 1 and Proposition 7 show then that $\mathbf{f}$ defines a bijection of $\mathbf{a} + \mathfrak{m}^{\times n}$ onto $\mathbf{f}(\mathbf{a}) + \mathfrak{m}^{\times n} = \mathfrak{m}^{\times n}$; the corollary follows from the fact that $0 \in \mathfrak{m}^{\times n}$.

COROLLARY 3. *With the notation of Corollary 2, let* $\mathbf{a} \in A^n$; *let us write* $e = J_{\mathbf{f}}^{(n-r)}(\mathbf{a})$ *(not necessarily invertible in* A) *and suppose that* $\mathbf{f}(\mathbf{a}) \equiv 0$ (mod. $e^2\mathfrak{m}^{\times(n-r)}$). *Then there exist* $n - r$ *formal power series without constant term* ϕ_i $(r + 1 \leqslant i \leqslant n)$ *in* $A[[X_1, \ldots, X_r]]$ *such that, for all* $\mathbf{t} = (t_1, \ldots, t_r) \in \mathfrak{m}^{\times r}$,

(25) $f_i(a_1 + e^2t_1, \ldots, a_r + e^2t_r, a_{r+1} + e\phi_{r+1}(\mathbf{t}), \ldots, a_n + e\phi_n(\mathbf{t})) = 0$

for $r + 1 \leqslant i \leqslant n$.

For $1 \leqslant i \leqslant r$, let $f_i(X) = X_i - a_i$ and let $\mathbf{u} = (f_1, \ldots, f_n)$; then $J_{\mathbf{u}}(\mathbf{a}) = e$ and Theorem 2 may be applied to the system $\mathbf{u}$. With the notation of Theorem 2 it follows from the above definitions that $g_i(X) = X_i$ for $1 \leqslant i \leqslant r$, whence $h_i(X) = X_i$ for $1 \leqslant i \leqslant r$; moreover, if $M' \in \mathbf{M}_n(A)$ is such that $M_{\mathbf{u}}(\mathbf{a}) . M' = eI_n$, M' is of the form

$$\begin{pmatrix} eI_r & 0 \\ * & * \end{pmatrix}.$$

Replacing $\mathbf{y}$ by $M'.\mathbf{z}$ (where $\mathbf{z} = (z_1, \ldots, z_n) \in \mathfrak{m}^{\times n}$) in formula (20), we obtain

(26) $$f_i(a_1 + e^2z_1, \ldots, a_r + e^2z_r, a_{r+1} + eh_{r+1}(M'.\mathbf{z}), \ldots, a_n + eh_n(M'.\mathbf{z})) = f_i(a) + e^2z_i \quad \text{for} \quad 1 \leqslant i \leqslant n.$$

By hypothesis, $f_j(a) = e^2b_j$ where $b_j \in \mathfrak{m}$ for $r + 1 \leqslant j \leqslant n$. Let us write $\psi_j(X_1, \ldots, X_n) = h_j(M'.\mathbf{X})$ and

$$\phi_j(X_1, \ldots, X_r) = \psi_j(X_1, \ldots, X_r, -b_{r+1}, \ldots, -b_n)$$

for $r + 1 \leqslant j \leqslant n$. For $r + 1 \leqslant i \leqslant n$, substituting t_j for z_j for $1 \leqslant j \leqslant r$ and b_j for z_j for $r + 1 \leqslant j \leqslant n$ in (26), we obtain relations (25) for all $\mathbf{t} \in \mathfrak{m}^{\times r}$.

6. APPLICATION TO DECOMPOSITIONS OF RINGS

LEMMA 2. *Let* A *be a ring and* $\mathfrak{m}$ *an ideal of* A *such that the ordered pair* (A, $\mathfrak{m}$) *satisfies Hensel's conditions. Let* B *be the quotient ring* A/$\mathfrak{m}$ *and* $\pi: A \to B$ *be the canonical homomorphism. For every idempotent* c *of* B *there exists a unique idempotent* e *of* A *such that* $\pi(e) = c$.

Let a be such that $\pi(a) = c$; Corollary 1 to Theorem 2 of no. 5 may be applied to the polynomial $f(X) = X^2 - X$ in A[X] and the element $a \in A$. Then $f'(a) = 2a - 1$ and, as $\pi(2a - 1) = 2c - 1$ and $(2c - 1)^2 = 1$ in B, $2c - 1$ is invertible in B and hence $2a - 1$ is invertible in A (§ 2, no. 13, Lemma 3). As $f(a) \in \mathfrak{m}$, Corollary 1 to Theorem 2 of no. 5 immediately gives the existence and uniqueness of e.

PROPOSITION 8. *Let* A *be a ring and* $\mathfrak{m}$ *an ideal of* A *such that the ordered pair* (A, $\mathfrak{m}$) *satisfies Hensel's conditions. Let* B *be the quotient ring* A/$\mathfrak{m}$ *and* $\pi: A \to B$ *the canonical homomorphism. If* B *is the direct composition of a finite family* $(\mathfrak{b}_i)_{i \in I}$ *of ideals, there exists a unique family* $(\mathfrak{a}_i)_{i \in I}$ *of ideals of* A *such that* $\pi(\mathfrak{a}_i) = \mathfrak{b}_i$ *for all* $i \in I$ *and* A *is the direct composition of the family* $(\mathfrak{a}_i)$.

Let $1 = \sum_i c_i$ where $c_i \in \mathfrak{b}_i$ for all i; the c_i are idempotents of B such that $c_i c_j = 0$ for $i \neq j$. By Lemma 2 there therefore exist idempotents e_i of A $(i \in I)$ such that $\pi(e_i) = c_i$ for all i; as $e_i e_j$ is an idempotent such that

$$\pi(e_i e_j) = c_i c_j = 0$$

for $i \neq j$, $e_i e_j = 0$ for $i \neq j$ (Lemma 2); as $1 - \sum_i e_i$ is an idempotent such that $\pi\left(1 - \sum_i e_i\right) = 1 - \sum_i c_i = 0$, similarly $1 = \sum_i e_i$. It follows that A is the direct composition of the ideals $\mathfrak{a}_i = e_i$A and that $\pi(\mathfrak{a}_i) = \pi(e_i)$B $= \mathfrak{b}_i$.

It remains to show the uniqueness of such a decomposition. Now, suppose that A is the direct composition of another family $(\mathfrak{a}'_i)_{i \in I}$ of ideals such that $\pi(\mathfrak{a}'_i) = \mathfrak{b}_i$ for all i; then $1 = \sum_i e'_i$ where $e'_i \in \mathfrak{a}'_i$, whence, in B, $1 = \sum_i \pi(e'_i)$ where $\pi(e'_i) \in \mathfrak{b}_i$, which implies $\pi(e'_i) = c_i$; as e'_i and e_i are idempotents, necessarily $e'_i = e_i$ (Lemma 2), which completes the proof.

Remark. Proposition 8 again gives the structure of a semi-local ring on A which is Hausdorff and complete with the $\mathfrak{r}$-adic topology ($\mathfrak{r}$ the Jacobson radical of A), which has already been obtained as a consequence of § 2, no. 13, Corollary to Proposition 19.

5. FLATNESS PROPERTIES OF FILTERED MODULES

1. IDEALLY HAUSDORFF MODULES

DEFINITION 1. *Let* A *be a commutative ring and* $\mathfrak{I}$ *an ideal of* A. *An* A*-module* M *is called ideally Hausdorff with respect to* $\mathfrak{I}$ (or simply *ideally Hausdorff* if there is no ambiguity) *if, for every finitely generated ideal* $\mathfrak{a}$ *of* A, *the* A*-module* $\mathfrak{a} \otimes_A$ M *is Hausdorff with the* $\mathfrak{I}$*-adic topology.*

Putting $\mathfrak{a} = $ A in this definition, we have already seen that M is necessarily *Hausdorff* with the $\mathfrak{I}$-adic topology.

Examples

(1) If A is Noetherian and $\mathfrak{I}$ is contained in the Jacobson radical of A (in other words if A is a Zariski ring with the $\mathfrak{I}$-adic topology), every *finitely generated* A-module is ideally Hausdorff (§ 3, no. 3, Proposition 6).

(2) Every direct sum of ideally Hausdorff modules is an ideally Hausdorff module, by virtue of the relations

$$\mathfrak{I}^n\Big(\mathfrak{a} \otimes_A \bigoplus_{\lambda\in L} M_\lambda\Big) = \mathfrak{I}^n \bigoplus_{\lambda\in L} (\mathfrak{a} \otimes_A M_\lambda) = \bigoplus_{\lambda\in L} \mathfrak{I}^n(\mathfrak{a} \otimes_A M_\lambda).$$

(3) If an A-module M is *flat* and *Hausdorff* with the $\mathfrak{I}$-adic topology it is ideally Hausdorff, for $\mathfrak{a} \otimes_A M$ is then identified with a submodule of M and the $\mathfrak{I}$-adic topology on $\mathfrak{a} \otimes_A M$ is *finer* than the topology induced on $\mathfrak{a} \otimes_A M$ by the $\mathfrak{I}$-adic topology on M, which is Hausdorff by hypothesis.

2. STATEMENT OF THE FLATNESS CRITERION

Let A be a ring, $\mathfrak{I}$ a two-sided ideal of A, M a left-module and gr(A) and gr(M) the graded ring and graded gr(A)-module associated respectively with the ring A and with the module M with the $\mathfrak{I}$-adic filtrations (§ 2, no. 3). We have seen (*loc. cit.*) that for every integer $n \geqslant 0$ there is a *surjective* **Z**-module homomorphism

$$\gamma_n : (\mathfrak{I}^n/\mathfrak{I}^{n+1}) \otimes_{A/\mathfrak{I}} (M/\mathfrak{I}M) \to \mathfrak{I}^n M/\mathfrak{I}^{n+1}M$$

and a graded homomorphism of degree 0 of graded gr(A)-modules

$$\gamma_M : \mathrm{gr}(A) \otimes_{\mathrm{gr}_0(A)} \mathrm{gr}_0(M) \to \mathrm{gr}(M)$$

whose restriction to $\mathrm{gr}_n(A) \otimes_{\mathrm{gr}_0(A)} \mathrm{gr}_0(M)$ is γ_n for all n and which is therefore *surjective*.

THEOREM 1. *Let* A *be a commutative ring,* $\mathfrak{I}$ *an ideal of* A *and* M *an* A*-module. Consider the following properties:*

(i) M *is a flat* A*-module.*

(ii) $\mathrm{Tor}_1^A(N, M) = 0$ *for every* A*-module* N *annihilated by* $\mathfrak{I}$.

(iii) $M/\mathfrak{I}M$ *is a flat* $(A/\mathfrak{I})$*-module and the canonical mapping* $\mathfrak{I} \otimes_A M \to \mathfrak{I}M$ *is bijective (the latter condition being equivalent to* $\mathrm{Tor}_1^A(A/\mathfrak{I}, M) = 0$ *by virtue of the relation* $\mathrm{Tor}_1^A(A, M) = 0$ *and the exact sequence*

$$\mathrm{Tor}_1^A(A, M) \to \mathrm{Tor}_1^A(A/\mathfrak{I}, M) \to \mathfrak{I} \otimes_A M \to M).$$

(iv) $M/\mathfrak{I}M$ *is a flat* $(A/\mathfrak{I})$*-module and the canonical homomorphism*

$$\gamma_M : \mathrm{gr}(A) \otimes_{\mathrm{gr}_0(A)} \mathrm{gr}_0(M) \to \mathrm{gr}(M)$$

is bijective (property (GR) of § 2, no. 8).

(v) *For all* $n \geqslant 1$, $M/\mathfrak{I}^n M$ *is a flat* $(A/\mathfrak{I}^n)$*-module.*

Then (i) ⇒ (ii) ⇔ (iii) ⇒ (iv) ⇔ (v).

If further $\mathfrak{I}$ *is nilpotent or if* A *is Noetherian and* M *is ideally Hausdorff, properties* (i), (ii), (iii), (iv) *and* (v) *are equivalent.*

Remark. If $A/\mathfrak{I}$ is a field (as often happens in applications) the condition "$M/\mathfrak{I}M$ is a flat $(A/\mathfrak{I})$-module" holds automatically for every A-module M, which simplifies the statement of properties (iii) and (iv); moreover, in this case, property (v) is equivalent to saying that $M/\mathfrak{I}^n M$ is a *free* $(A/\mathfrak{I}^n)$-module for every integer $n \geqslant 1$ (Chapter II, § 3, no. 2, Corollary 2 to Proposition 5).

3. PROOF OF THE FLATNESS CRITERION

(A) *The implications* (i) ⇒ (ii) ⇔ (iii)

The implication (i) ⇒ (ii) is immediate (Chapter I, § 4). The equivalence (ii) ⇔ (iii) is a special case of Chapter I, § 4, Proposition 2 applied to $R = A$, $S = A/\mathfrak{J}$, $F = M$, $E = N$, taking account of the fact that being given an $(A/\mathfrak{J})$-module structure on N is equivalent to being given an A-module structure under which N is annihilated by $\mathfrak{J}$.

Remark (1). Condition (ii) is also equivalent to the following:

(ii′) $\mathrm{Tor}_1^A(N, M) = 0$ *for every* A*-module* N *annihilated by a power of* $\mathfrak{J}$.

Clearly (ii′) implies (ii). Conversely, if (ii) holds, then in particular $\mathrm{Tor}_1^A(\mathfrak{J}^nN/\mathfrak{J}^{n+1}N, M) = 0$ for all n; from the exact sequence

$$0 \to \mathfrak{J}^{n+1}N \to \mathfrak{J}^nN \to \mathfrak{J}^nN/\mathfrak{J}^{n+1}N \to 0$$

we derive the exact sequence

$$\mathrm{Tor}_1^A(\mathfrak{J}^{n+1}N, M) \to \mathrm{Tor}_1^A(\mathfrak{J}^nN, M) \to \mathrm{Tor}_1^A(\mathfrak{J}^nN/\mathfrak{J}^{n+1}N, M)$$

and, as there exists an integer m such that $\mathfrak{J}^mN = 0$, we deduce by descending induction on n that $\mathrm{Tor}_1^A(\mathfrak{J}^nN, M) = 0$ for all $n \leqslant m$ and in particular for $n = 0$.

It follows from this that if $\mathfrak{J}$ is *nilpotent*, (ii) implies (i), for (ii′) then means that $\mathrm{Tor}_1^A(N, M) = 0$ for *every* A-module N and hence that M is flat (Chapter I, § 4).

(B) Let us prove the following proposition:

PROPOSITION 1. *Let* A *be a commutative ring,* $\mathfrak{J}$ *an ideal of* A *and* M *an* A*-module. The following conditions are equivalent:*

(a) *For all* $n \geqslant 1$, $\mathrm{Tor}_1^A(A/\mathfrak{J}^n, M) = 0$.

(b) *For all* $n \geqslant 1$, *the canonical homomorphism*

$$\theta_n : \mathfrak{J}^n \otimes_A M \to \mathfrak{J}^nM$$

is bijective.

Moreover these conditions imply:

(c) *The canonical homomorphism* $\gamma_M : \mathrm{gr}(A) \otimes_{\mathrm{gr}_0(A)} \mathrm{gr}_0(M) \to \mathrm{gr}(M)$ *is bijective. Conversely, if* $\mathfrak{J}$ *is nilpotent,* (c) *implies* (a) *and* (b).

The equivalence of (a) and (b) follows from the exact sequence

$$0 = \mathrm{Tor}_1^A(A, M) \to \mathrm{Tor}_1^A(A/\mathfrak{J}^n, M) \to \mathfrak{J}^n \otimes_A M \to M.$$

Consider next the diagram

$$(1)\qquad \begin{array}{ccccccccc} & & \mathfrak{J}^{n+1} \otimes_A M & \longrightarrow & \mathfrak{J}^n \otimes_A M & \longrightarrow & (\mathfrak{J}^n/\mathfrak{J}^{n+1}) \otimes_A (M/\mathfrak{J}M) & \longrightarrow & 0 \\ & & \downarrow{\scriptstyle\theta_{n+1}} & & \downarrow{\scriptstyle\theta_n} & & \downarrow{\scriptstyle\gamma_n} & & \\ 0 & \longrightarrow & \mathfrak{J}^{n+1}M & \longrightarrow & \mathfrak{J}^nM & \longrightarrow & \mathrm{gr}_n(M) & \longrightarrow & 0 \end{array}$$

where we note that $(\mathfrak{J}^n/\mathfrak{J}^{n+1}) \otimes_A (M/\mathfrak{J}M)$ is canonically identified with $(\mathfrak{J}^n/\mathfrak{J}^{n+1}) \otimes_{A/\mathfrak{J}} (M/\mathfrak{J}M)$. This diagram is commutative by definition of γ_n and its rows are exact. If (b) holds, θ_n and θ_{n+1} are bijective and so therefore is γ_n by definition of cokernel, hence (b) implies (c). Conversely, assuming that $\mathfrak{J}$ is nilpotent, let us show that (c) implies (b); we shall argue by descending induction on n, since $\mathfrak{J}^n \otimes_A M = \mathfrak{J}^n M = 0$ for n sufficiently large. Suppose then that in diagram (1), γ_n and θ_{n+1} are bijective; then so is θ_n by virtue of Chapter I, § 1, no. 4, Corollary 1 to Proposition 2.

(C) *The implication* (ii) ⇒ (iv)

If (ii) holds, so does (ii') by *Remark* 1; Proposition 1 then shows that γ_M is an isomorphism. On the other hand, we already know that (ii) implies (iii) and hence $M/\mathfrak{J}M$ is a flat $(A/\mathfrak{J})$-module, which completes the proof that (ii) implies (iv).

Remark (2). Proposition 1 shows that, if $\mathfrak{J}$ is nilpotent, (iv) implies (iii); taking account of *Remark* 1, we have therefore proved in this case that (i), (ii), (iii) and (iv) are equivalent.

(D) *The equivalence* (iv) ⇔ (v)

For all $n \geqslant 1$, $M/\mathfrak{J}^n M$ has a canonical $(A/\mathfrak{J}^n)$-module structure. If it is filtered by the $(\mathfrak{J}/\mathfrak{J}^n)$-adic filtration, it is immediate that $\mathrm{gr}_m(M/\mathfrak{J}^n M) = \mathrm{gr}_m(M)$ if $m < n$ and $\mathrm{gr}_m(M/\mathfrak{J}^n M) = 0$ if $m \geqslant n$. For all $k \geqslant 1$, let $A_k = A/\mathfrak{J}^k$, $\mathfrak{J}_k = \mathfrak{J}/\mathfrak{J}^k$ $M_k = M/\mathfrak{J}^k M$; let $(\mathrm{iv})_k$ (resp. $(\mathrm{v})_k$) denote the assertion derived from (iv) (resp. (v)) by replacing A, $\mathfrak{J}$, M by A_k, $\mathfrak{J}_k$, M_k. It follows from what has just been said that (iv) is equivalent to "for all $k \geqslant 1$, $(\mathrm{iv})_k$" and obviously (v) is equivalent to "for all $k \geqslant 1$, $(\mathrm{v})_k$". Then it will suffice to establish the equivalence $(\mathrm{iv})_k \Leftrightarrow (\mathrm{v})_k$ for all k or also to show that (iv) ⇔ (v) when $\mathfrak{J}$ is *nilpotent*. Now (*Remark* 2) we have seen that in that case (iv) is equivalent to (i). As $M/\mathfrak{J}^n M$ is isomorphic to $M \otimes_A (A/\mathfrak{J}^n)$, (i) implies (v) (Chapter I, § 2, no. 7, Corollary 2 to Proposition 8); moreover clearly (v) then implies (i). We have therefore shown the equivalence (iv) ⇔ (v) in all cases and also that of all the properties of the theorem in the case where $\mathfrak{J}$ is nilpotent.

(E) *The implication* (v) ⇒ (i) *when* A *is Noetherian and* M *ideally Hausdorff*

It is sufficient to prove that for every ideal $\mathfrak{a}$ of A the canonical mapping $j: \mathfrak{a} \otimes_A M \to M$ is *injective* (Chapter I, § 2, no. 3, Proposition 1). Let $x \in \mathrm{Ker}\, j$; as $\mathfrak{a} \otimes_A M$ is Hausdorff with the $\mathfrak{J}$-adic topology, it suffices to verify that, for every integer $n > 0$, $x \in \mathfrak{J}^n(\mathfrak{a} \otimes_A M)$. Let $f: \mathfrak{J}^n\mathfrak{a} \to \mathfrak{a}$ be the canonical injection; it suffices to show that $x \in \mathrm{Im}(f \otimes 1_M)$; for if $b \in \mathfrak{J}^n$, $a \in \mathfrak{a}$ and $m \in M$, the image under $f \otimes 1_M$ of the element $(ba) \otimes m$ of $(\mathfrak{J}^n\mathfrak{a}) \otimes_A M$ is the element $(ba) \otimes m = b(a \otimes m)$ of $\mathfrak{a} \otimes_A M$ and hence $\mathrm{Im}(f \otimes 1_M) \subset \mathfrak{J}^n(\mathfrak{a} \otimes_A M)$. By virtue of Krull's Theorem (§ 3, no. 2, Theorem 2), there exists an integer k

such that $\mathfrak{a}_k = \mathfrak{a} \cap \mathfrak{J}^k \subset \mathfrak{J}^n\mathfrak{a}$; if $i: \mathfrak{a}_k \to \mathfrak{a}$ is the canonical injection, it will then be sufficient to show that $x \in \mathrm{Im}(i \otimes 1_M)$. Now, denoting by $p: \mathfrak{a} \to \mathfrak{a}/\mathfrak{a}_k$ and $h: \mathfrak{a}/\mathfrak{a}_k \to A/\mathfrak{J}^k$ the canonical mappings, there is a commutative diagram

$$\begin{array}{ccccccc} \mathfrak{a}_k \otimes_A M & \xrightarrow{i\otimes 1_M} & \mathfrak{a} \otimes_A M & \xrightarrow{p\otimes 1_M} & (\mathfrak{a}/\mathfrak{a}_k) \otimes_A M & \longrightarrow & 0 \\ & & \downarrow{\scriptstyle j} & & \downarrow{\scriptstyle h\otimes 1_M} & & \\ & & M & \longrightarrow & (A/\mathfrak{J}^k) \otimes_A M & & \end{array}$$

in which the first row is exact. It suffices to prove that $x \in \mathrm{Ker}(p \otimes 1_M)$ and, as $x \in \mathrm{Ker}\, j$ by hypothesis, it will suffice to verify that the mapping $h \otimes 1_M$ is *injective*. Now, it may also be written (*Algebra*, Chapter II, § 3, no. 6, Corollary 3 to Proposition 6)

$$h \otimes 1_{M/\mathfrak{J}^kM}: (\mathfrak{a}/\mathfrak{a}_k) \otimes_{A/\mathfrak{J}^k} (M/\mathfrak{J}^kM) \to M/\mathfrak{J}^kM$$

and, as h is injective and, by (v), $M/\mathfrak{J}^kM$ is a *flat* $(A/\mathfrak{J}^k)$-module, this completes the proof.

4. APPLICATIONS

PROPOSITION 2. *Let* A *be a commutative ring,* $\mathfrak{J}$ *an ideal of* A *and* B *a commutative Noetherian* A*-algebra such that* B *is contained in the Jacobson radical of* B. *Then every finitely generated* B*-module* M *is an ideally Hausdorff* A*-module with respect to* $\mathfrak{J}$.

We shall see more generally that for every finitely generated A-module N, $N \otimes_A M$ is Hausdorff with the $\mathfrak{J}$-adic topology. For $N_{(B)} = N \otimes_A B$ is a finitely generated B-module and the B-module $N \otimes_A M$ is canonically identified with $N_{(B)} \otimes_B M$ by virtue of the associativity of the tensor product. Let $\mathfrak{R}$ be the Jacobson radical of B; as $\mathfrak{J}B$ is contained in $\mathfrak{R}$, the $\mathfrak{J}$-adic topology on $N \otimes_A M$ is therefore identified with a finer topology than the $\mathfrak{R}$-adic topology on $N_{(B)} \otimes_B M$; but this latter topology is Hausdorff since $N_{(B)} \otimes_B M$ is a finitely generated B-module (no. 1, *Example* 1), whence the conclusion.

PROPOSITION 3. *Let* A *be a commutative ring,* B *a commutative* A*-algebra,* $\mathfrak{J}$ *an ideal of* A *and* M *a* B*-module. Suppose that* B *is a Noetherian ring and a flat* A*-module and that* M *is ideally Hausdorff with respect to* $\mathfrak{J}B$. *The following conditions are equivalent:*

(a) M *is a flat* B*-module.*

(b) M *is a flat* A*-module and* $M/\mathfrak{J}M = M/(\mathfrak{J}B)M$ *is a flat* $(B/\mathfrak{J}B)$*-module.*

If further the canonical homomorphism $A/\mathfrak{J} \to B/\mathfrak{J}B$ *is bijective, conditions* (a) *and* (b) *are also equivalent to:*

(c) M *is a flat* A*-nodule.*

Condition (a) implies (b) by Chapter I, § 2, no. 7, Corollaries 2 and 3 to Proposition 8 and the fact that $M/\mathfrak{J}M$ is isomorphic to $M \otimes_B (B/\mathfrak{J}B)$. Suppose condition (b) holds; to show that M is a flat B-module, we shall apply Theorem 1 of no. 2 with A replaced by B and $\mathfrak{J}$ by $\mathfrak{J}B$. It will therefore be sufficient to show that the canonical mapping $f: \mathfrak{J}B \otimes_B M \to \mathfrak{J}M$ is injective. Let f_1 be the canonical mapping $\mathfrak{J} \otimes_A B \to \mathfrak{J}B$ and f_2 the canonical isomorphism $\mathfrak{J} \otimes_A M \to (\mathfrak{J} \otimes_A B) \otimes_B M$; $f \circ (f_1 \circ 1_M) \circ f_2$ is the canonical mapping $f': \mathfrak{J} \otimes_A M \to \mathfrak{J}M$, as is easily verified. Now f' is an isomorphism since M is a flat A-module, whilst f_1 is an isomorphism because B is flat over A; f is then an isomorphism.

Let $\rho: A/\mathfrak{J} \to B/\mathfrak{J}B$ be the canonical homomorphism; the $(A/\mathfrak{J})$-module structure on $M/\mathfrak{J}M$ derived by means of ρ from its $(B/\mathfrak{J}B)$-module structure is isomorphic to that on $M \otimes_A (A/\mathfrak{J})$. Then it follows that, if M is a flat A-module, $M/\mathfrak{J}M$ is a flat $(A/\mathfrak{J})$-module and hence also a flat $(B/\mathfrak{J}B)$-module if ρ is an isomorphism; we have thus proved that (c) $\Rightarrow$ (b) in that case.

Corollary. *Let* A *be a commutative Noetherian ring,* $\mathfrak{J}$ *an ideal of* A, $\hat{A}$ *the Hausdorff completion of* A *with respect to the* $\mathfrak{J}$*-adic topology and* M *an ideally Hausdorff* $\hat{A}$*-module with respect to* $\mathfrak{J}\hat{A}$. *For* M *to be a flat* A*-module, it is necessary and sufficient that* M *be a flat* $\hat{A}$*-module.*

We know in fact that $\hat{A}$ is a Noetherian ring (§ 3, no. 4, Proposition 8) and a flat A-module (§ 3, no. 4, Theorem 3), that $\mathfrak{J}\hat{A} = \hat{\mathfrak{J}}$ (§ 2, no. 12, Proposition 16) and that the canonical homomorphism $A/\mathfrak{J} \to \hat{A}/\hat{\mathfrak{J}}$ is bijective (§ 2, no. 12, Proposition 15); Proposition 3 can therefore be applied.

Proposition 4. *Let* A *and* B *be two commutative Noetherian rings,* $h: A \to B$ *a ring homomorphism,* $\mathfrak{J}$ *an ideal of* A *and* $\mathfrak{L}$ *an ideal of* B *containing* $\mathfrak{J}B$ *and contained in the Jacobson radical of* B. *Let* $\hat{A}$ *be the Hausdorff completion of* A *with respect to the* $\mathfrak{J}$*-adic topology and* $\hat{B}$ *the Hausdorff completion of* B *with respect to the* $\mathfrak{L}$*-adic topology;* h *is continuous with these topologies and* $\hat{h}: \hat{A} \to \hat{B}$ *therefore makes* $\hat{B}$ *into an* $\hat{A}$*-algebra. Let* M *be a finitely generated* B*-module and* $\hat{M}$ *its Hausdorff completion with respect to the* $\mathfrak{L}$*-adic topology; the following properties are equivalent:*

(a) M *is a flat* A*-module.*
(b) $\hat{M}$ *is a flat* A*-module.*
(c) $\hat{M}$ *is a flat* $\hat{A}$*-module.*

As B with the $\mathfrak{L}$-adic topology is a Zariski ring, $\hat{B}$ is a faithfully flat B-module (§ 3, no. 5, Proposition 9) and $\hat{M}$ is canonically isomorphic to $M \otimes_B \hat{B}$ (§ 3, no. 4, Theorem 3); it is immediately verified that this canonical isomorphism is an isomorphism of the A-module structure on $\hat{M}$ onto the A-module structure on $M \otimes_B \hat{B}$ derived from that on M. Applying Proposition 4 of Chapter I, § 3, no. 2 with R replaced by B, S by A, E by $\hat{B}$, F by M, we see that for M to be a flat A-module, it is necessary and sufficient that $\hat{M}$ be a flat A-module. Moreover, $\hat{M}$ is a finitely generated $\hat{B}$-module and $\mathfrak{J}\hat{B}$ is contained in $\hat{\mathfrak{L}} = \mathfrak{L}\hat{B}$ and hence in the Jacobson radical of $\hat{B}$ (§ 3, no. 4, Proposition 8); therefore $\hat{M}$ is an ideally Hausdorff $\hat{A}$-module with respect to $\mathfrak{J}\hat{A}$ (Proposition 2). Conditions (b) and (c) are therefore equivalent by the Corollary to Proposition 3.

EXERCISES

§ 1

1. Let A be a graded commutative ring of type **Z** and (A_n) its graduation; we set $A^{\geqslant} = \bigoplus_{n \geqslant 0} A_n$, $A^{\leqslant} = \bigoplus_{n \geqslant 0} A_n$; these are graded subrings of A.

(a) For every homogeneous element f of A of degree d, the ring of fractions A_f (corresponding to the multiplicative subset consisting of the f^n, where $n \geqslant 0$) has a canonical graded ring structure (*Algebra*, Chapter II, § 11); we denote by $A_{(f)}$ the subring of A_f consisting of elements of degree 0. Show that, if $d > 0$, then $(A^{\geqslant})_f = A_f$, $A_{(f)}$ is isomorphic to $A^{(d)}/(f-1)A^{(d)}$ and $((A_f)^{\geqslant})_{f/1}$ is a graded ring isomorphic to A_f.

(b) Let the polynomial ring $B = A[X] = A \otimes_{\mathbf{Z}} \mathbf{Z}[X]$ be given the graduation the tensor product of those on A and $\mathbf{Z}[X]$, a graduation compatible with the ring structure on B. Show that, if $d > 0$, $B_{(f)}$ is isomorphic to $(A_f)^{\leqslant}$ and $(A^{(d)})_f$ is a graded ring isomorphic to $A_{(f)} \otimes_{\mathbf{Z}} \mathbf{Z}[X, X^{-1}]$.

(c) Let g be another homogeneous element of A of degree e. Show that, if $d > 0$ and $e > 0$, $A_{(fg)}$ is isomorphic to $(A_{(f)})_{g^d/f^e}$.

* (d) Suppose $d > 0$ and $A_n = \{0\}$ for $n < 0$. Show that, if A is Noetherian, so is $A_{(f)}$ (use § 2, no. 10, Corollary 4 to Theorem 2). *

2. Let A be a graded commutative ring of type **Z** such that $A_n = 0$ for $n < 0$. For all $n \geqslant 0$, let $A_{[n]}$ denote the graded ideal $\bigoplus_{m \geqslant n} A_m$ of A; the A-algebra $A^{\natural} = \bigoplus_{n \geqslant 0} A_{[n]}$ is a ring graded by the $A_{[n]} = A^{\natural}_n$; for all $f \in A_d$ $(d > 0)$, let $f^{\natural}$ denote the element of $A^{\natural}$ whose component are zero except for that of degree d, which is equal to f.

(a) Show that, if $f \in A_d$, where $d > 0$, the ring $A^{\natural}_{(f^{\natural})}$ is isomorphic to $(A_f)^{\geqslant}$ (notation of Exercise 1).

(b) For $A^{\natural}$ to be a finitely generated A-algebra, it is necessary and sufficient that A be a finitely generated A_0-algebra.

(c) In order that $A^{\natural}_{n+1} = A^{\natural}_1 A^{\natural}_n$ (resp. $A^{\natural}_n = (A^{\natural}_1)^n$) for $n \geqslant n_0$, it is necessary and sufficient that $A_{n+1} = A_1A_n$ (resp. $A_n = (A_1)^n$) for $n \geqslant n_0$.

3. Let K be a field, B the polynomial ring K[X, Y] with the graduation defined by the total degree, C the graded subring of B generated by X and Y^2 and $\mathfrak{a}$ the graded ideal of C generated by Y^4. Show that, in the graded ring $A = C/\mathfrak{a}$, $A_{n+1} = A_1A_n$ for $n \geqslant 2$ but $A_n \neq (A_1)^n$ for $n \geqslant 6$.

§ 2

1. Let K be a commutative field, A the polynomial ring K[X, Y] in two indeterminates and $\mathfrak{a}_n$ the principal ideal (XY^n) in A; the sequence $(\mathfrak{a}_n)_{n \geqslant 1}$ forms with $\mathfrak{a}_0 = A$ an exhaustive and separated filtration on A. Let $\mathfrak{b}$ be the principal ideal (X) of A; show that the topology on $\mathfrak{b}$ induced by that on A is strictly coarser than the topology defined by the filtration on $\mathfrak{b}$ associated with that on A (and *a fortiori* this latter filtration is distinct from the filtration induced by that on A).

2. Let K be a commutative field of characteristic $\neq 2$ and A the ring K[[X, Y]] of formal power series in two indeterminates.

(a) Show that in A the principal ideal $\mathfrak{p} = (X^2 - Y^3)$ is prime. (If a product $f(X, Y)\,g(X, Y)$ of two formal power series is divisible by $X^2 - Y^3$, note first that $f(T^3, T^2) = 0$ or $g(T^3, T^2) = 0$ in the ring of formal power series K[[T]]; assuming for example $f(T^3, T^2) = 0$, show first that $f(X, Y^2)$ is divisible by $X - Y^3$ and by $X + Y^3$ and, by considering $f(-Y^3 + X, Y^2)$, prove finally that $f(X, Y^2)$ is divisible by $X^2 - Y^6$.)

(b) Let $\mathfrak{m}$ be the maximal ideal AX + AY of A. Show that $\mathfrak{p}$ is closed with respect to the $\mathfrak{m}$-adic topology on A (with which A is Hausdorff and complete) but that in the ring gr(A), gr($\mathfrak{p}$) is not a prime ideal ($\mathfrak{p}$ being given the filtration induced by that on A).

3. Let A be a filtered ring and E a finitely generated A-module. Show that, if E is given the filtration induced by that on A, gr(E) is a finitely generated gr(A)-module (cf. Exercise 5 (c)).

4. Give an example of a bijective A-linear mapping u: E → F, where E and F are two filtered A-modules, such that u is compatible with the filtrations but gr(u) is neither injective nor surjective. (Take E = A, F = A with another filtration and u the identity mapping.)

5. Let A be a filtered commutative ring with filtration $(\mathfrak{a}_n)_{n \geqslant 0}$ such that $\mathfrak{a}_0 = A$.

(a) For the ring gr(A) to be generated by a family of elements whose degrees are bounded, it is necessary and sufficient that there exist an integer q such that, for all n, $\mathfrak{a}_n = \mathfrak{a}_{n+1} + \mathfrak{b}_n$, where $\mathfrak{b}_n$ is the sum of the ideals $\mathfrak{a}_1^{\alpha_1}\mathfrak{a}_2^{\alpha_2}\ldots\mathfrak{a}_q^{\alpha_q}$ for all systems of integers $\alpha_i \geqslant 0$ $(1 \leqslant i \leqslant q)$ such that $\sum_{i=1}^{q} i\alpha_i = n$. Then $\mathfrak{a}_n = \mathfrak{a}_{n+k} + \mathfrak{b}_n$ for all $k > 0$.

(b) For $\mathrm{gr}(A)$ to be Noetherian, it is necessary and sufficient that $A/\mathfrak{a}_1$ be Noetherian, that the condition of (a) hold and that for $i \leqslant q$ the $\mathfrak{a}_i/\mathfrak{a}_{i+1}$ be finitely generated $(A/\mathfrak{a}_1)$-modules (use Corollary 4 to Theorem 2 of no. 10).

(c) Let K be a commutative field and A the polynomial ring in two indeterminates K[X, Y]. A total ordering is defined on the set of monomials X^mY^n by setting $X^mY^n \leqslant X^pY^q$ if $m + n \leqslant p + q$ or if $m + n = p + q$ and $m \leqslant p$. Let (M_k) be the sequence of monomials in X, Y thus arranged in ascending order and let $\mathfrak{a}_n$ be the ideal of A generated by the M_k of index $k \geqslant n$. Show that $(\mathfrak{a}_n)$ is a filtration compatible with the ring structure on A; with this filtration $\mathrm{gr}(A)$ admits divisors of 0 and is not Noetherian, although A is a Noetherian integral domain (use the criterion in (b)); in particular, $\mathrm{gr}(\mathfrak{a}_1)$ (with the filtration induced on $\mathfrak{a}_1$ by that on A) is not a finitely generated $\mathrm{gr}(A)$-module, although $\mathfrak{a}_1$ is a finitely generated A-module (cf. § 1, no. 2, Corollary to Proposition 1).

¶ 6. Let A be a commutative ring, E and F two A-modules and (E_n) (resp. (F_n)) an exhaustive filtration on E (resp. F) consisting of sub-A-modules. On the tensor product $G = E \otimes_A F$, consider the exhaustive filtration consisting of the $G_n = \sum_{i+j=n} \mathrm{Im}(E_i \otimes_A F_j)$.

(a) Show that the composite canonical homomorphisms

$$(E_i/E_{i+1}) \otimes_A (F_j/F_{j+1}) \to (E_i \otimes_A F_j)/(\mathrm{Im}(E_i \otimes_A F_{j+1}) + \mathrm{Im}(E_{i+1} \otimes_A F_j)) \to G_{i+j}/G_{i+j+1}$$

are the restrictions of a graded homomorphism of degree 0 (called canonical) $\mathrm{gr}(E) \otimes_A \mathrm{gr}(F) \to \mathrm{gr}(E \otimes_A F)$, which is surjective.

(b) Show that, if $\mathrm{gr}(E)$ is a flat A-module, the A-modules E_m/E_n for $m \leqslant n$ and E/E_n for $n \in \mathbf{Z}$ are flat.

(c) Suppose in the following that $\mathrm{gr}(E)$ is a flat A-module and that E is a flat A-module (the second hypothesis being a consequence of the first if (E_n) is a discrete filtration). Then the $H_{ij} = E_i \otimes_A F_j$ are canonically identified with submodules of $G = E \otimes_A F$ (Chapter I, § 2, no. 5, Proposition 4). Show that, for any two finite subsets R, S of $\mathbf{Z} \times \mathbf{Z}$,

$$(*) \qquad \Big(\sum_{(i,j)\in R} H_{ij}\Big) \cap \Big(\sum_{(h,k)\in S} H_{hk}\Big) = \sum H_{\sup(i,h),\,\sup(j,k)}$$

where in the second sum (i, j, h, k) run through $R \times S$. (If R and S are each reduced to a single element, use Chapter I, § 2, no. 6, Proposition 7. If p is the greatest of the indices i and h appearing in the elements of R or S, q the least of the indices j and k appearing in the elements of R or S, consider the images of the two sides of (*) under the canonical homomorphism

$$E \otimes_A F_q \rightarrow (E/E_p) \otimes_A F_q$$

and argue by induction on Card(R) + Card(S).)

(d) Deduce from (c) that the canonical homomorphism defined in (a) is then bijective.

7. Let A be a commutative ring, E an A-module, F a submodule of E, B the exterior algebra $\wedge(E)$ and $\mathfrak{J}$ the two-sided ideal of B generated by the canonical images in B of the elements of F. Let $\mathrm{gr}^{\mathfrak{J}}(B)$ be the graded ring associated with the ring B filtered by the $\mathfrak{J}$-adic filtration.

(a) Define a surjective canonical (non-graded) A-algebra homomorphism $(\wedge(F)) \otimes_A^g (\wedge(E/F)) \rightarrow \mathrm{gr}^{\mathfrak{J}}(B)$ (where we mean the skew tensor product of graded algebras (*Algebra*, Chapter III).

(b) Show that, if F admits a complement in E, the homomorphism defined in (a) is an isomorphism.

(c) Take $A = \mathbf{Z}$, $E = \mathbf{Z}/4\mathbf{Z}$, $F = 2\mathbf{Z}/4\mathbf{Z}$; show that the homomorphism defined in A is not injective in this case.

8. Let A be a filtered ring, (A_n) its filtration, E a filtered A-module and (E_n) its filtration. Show that if $A_0 = A$ and $E_0 = E$, the mapping $(a, x) \mapsto ax$ of $A \times E$ to E is uniformly continuous with the topologies defined by the filtrations.

9. Give an example of two filtered rings A, B whose filtrations are exhaustive and separated and a non-surjective homomorphism $u: A \rightarrow B$ compatible with the filtrations and such that $\mathrm{gr}(u)$ is bijective. From this deduce a counter-example to Corollary 1 to Proposition 12 when the ring A is not assumed to be complete and a counter-example to Proposition 13 when E is no longer assumed to be finitely generated (use *Algebra*, Chapter VIII, § 7, Exercise 3 (b)).

10. Let A be a Noetherian ring and σ an automorphism of A. Show that the ring E defined in *Algebra*, Chapter IV, § 5, Exercise 10 (b) is a (left or right) Noetherian ring.

11. Show that, if E is a vector space of dimension $\geqslant 2$ over a commutative field, the tensor algebra of E is a ring which is neither left nor right Noetherian (if a, b are two linearly independent vectors in E, consider the left ideal (or right ideal) generated by the elements $a^n b^n$ for $n \geqslant 1$).

¶ **12.** Let K be a commutative field, A the polynomial ring $K[X_\iota]_{\iota \in I}$ in an arbitrary infinite family of indeterminates and $\mathfrak{m}$ the (maximal) ideal of A generated by the X_ι. If we set $A_i = A/\mathfrak{m}^{i+1}$, the A_i and the canonical homomorphisms $h_{ij}: A/\mathfrak{m}^{j+1} \rightarrow A/\mathfrak{m}^{i+1}$ for $i \leqslant j$ satisfy the conditions of Proposition

14; the ring $\varprojlim A_\iota$ is the completion $\hat{A}$ of A with respect to the $\mathfrak{m}$-adic topology and the kernel of the canonical homomorphism $\hat{A} \to A_\iota$ is equal to $(\mathfrak{m}^\iota)^\wedge$, the closure of $\mathfrak{m}^\iota$ in $\hat{A}$.

(a) Show that $\hat{A}$ is canonically identified with the ring of formal power series in the X_ι, each of which has only a *finite* number of terms of given degree (*Algebra*, Chapter IV, § 5, Exercise 1).

(b) From now on take $I = \mathbf{N}$. Show that $\hat{\mathfrak{m}} \neq \hat{A}.\mathfrak{m}$ (consider the formal power series $\sum_{n=1}^{\infty} X_n^n$).

(c) Suppose that K is a finite field. Show that $(\hat{\mathfrak{m}})^2 \neq (\mathfrak{m}^2)^\wedge$. (Show first the following result: for every integer $k > 0$, there exists an integer n_k and for every integer $n \geqslant n_k$ a homogeneous polynomial F_n of degree n in n^2 indeterminates with coefficients in K such that F_n cannot be the sum of the terms of degree n in *any* polynomial of the form $P_1Q_1 + \cdots + P_kQ_k$, where the P_i and Q_i are polynomials *without constant term* in the same n^2 indeterminates).

Deduce that $\hat{A}$ is not complete with respect to the $\hat{\mathfrak{m}}$-adic topology.

13. Let K be a field, $A = K[[X]]$ the ring of formal power series and $\mathfrak{m}$ its maximal ideal, so that A is Hausdorff and complete with the $\mathfrak{m}$-adic topology (no. 6, Corollary to Proposition 6). On the additive group A consider the filtration (E_n) such that $E_0 = A$ and E_n is the intersection of $\mathfrak{m}^n$ and the ring $K[X]$; this filtration is exhaustive and separated, the topology $\mathcal{T}$ which it defines on A is compatible with the additive group structure on A and is finer than the $\mathfrak{m}$-adic topology but A is not a complete group with the topology $\mathcal{T}$ (consider the sequence of polynomials $(1 - X^n)/(1 - X)$).

14 Let A be a ring (not necessarily commutative) and E a left A-module. A topology on E is called *linear* if it is invariant under translations and 0 admits a fundamental system of neighbourhoods which are submodules of E; E is then said to be *linearly topologized.* A linear topology on E is compatible with its additive group structure and defines on E a topological A-module structure if A is given the discrete topology. On any A-module the discrete topology and the coarsest topology are linear topologies.

(a) If E is a linearly topologized module and F a submodule of E, the induce topology on F and the quotient topology of that on E by F are linear topologies. If $(E_\alpha, f_{\alpha\beta})$ is an inverse system of linearly topologized A-modules, where the $f_{\alpha\beta}$ are continuous linear mappings, the topological A-module $E = \varprojlim E_\alpha$ is linearly topologized.

(b) Let E be a linearly topologized A-module. There exists a fundamental system $(V_\lambda)_{\lambda \in L}$ of neighbourhoods of 0 in E consisting of open (and closed) submodules; if $\phi_{\lambda\mu}: E/V_\mu \to E/V_\lambda$ is the canonical mapping when $V_\lambda \supset V_\mu$, the family $(E/V_\lambda, \phi_{\lambda\mu})$ is an inverse system of *discrete* A-modules; the topological A-module $\hat{E} = \varprojlim E/V_\lambda$ is identified with the Hausdorff completion of E, which is therefore linearly topologized (*General Topology*, Chapter III, § 7, no. 3).

(c) A Hausdorff linearly topologized A-module E is discrete if it is Artinian or if there exists a least element in the set of submodules $\neq 0$ of E.

15. (a) Let E be a linearly topologized A-module (Exercise 14). For a filter base $\mathfrak{B}$ on E consisting of linear varieties to admit a cluster point, it is necessary and sufficient that there exists a convergent filter base $\mathfrak{B}' \supset \mathfrak{B}$ consisting of linear affine varieties. E is said to be *linearly compact* if it is Hausdorff and every filter base on E consisting of affine linear varieties admits at least one cluster point. Every Artinian module is linearly compact with the discrete topology. Every linearly compact submodule of a Hausdorff linearly topologized module F is closed in F.

(b) If E is a linearly compact A-module and u a continuous linear mapping of E to a Hausdorff linearly topologized A-module F, $u(\mathrm{F})$ is a linearly compact submodule of F.

(c) Let E be a Hausdorff linearly topologized A-module and F a closed submodule of E. For E to be linearly compact, it is necessary and sufficient that F and E/F be so.

(d) Every product of linearly compact modules is linearly compact (consider a filter base which is maximal among the filter bases consisting of affine linear varieties.

(e) Let $(\mathrm{E}_\alpha, f_{\alpha\beta})$ be an inverse system of linearly topologized modules relative to a directed indexing set; suppose that the $f_{\alpha\beta}$ are continuous linear mappings and that, for $\alpha \leqslant \beta$, $\overset{-1}{f_{\alpha\beta}}(0)$ is a linearly compact submodule of E_β. Let $\mathrm{E} = \varprojlim \mathrm{E}_\alpha$ and let f_α be the canonical mapping $\mathrm{E} \rightarrow \mathrm{E}_\alpha$; show that, for all α, $f_\alpha(\mathrm{E}) = \bigcap_{\alpha \leqslant \beta} f_{\alpha\beta}(\mathrm{E}_\beta)$ (in particular, if the $f_{\alpha\beta}$ are surjective, so are the f_α) and $\overset{-1}{f_\alpha}(0)$ is linearly compact (use (d) and *General Topology*, Chapter I, Appendix, no. 2, Theorem 1).

¶ 16. (a) Let E be a Hausdorff linearly topologized A-module. Show that the following properties are equivalent:

(α) E is linearly compact (Exercise 15).

(β) For every continuous linear mapping u from E to a Hausdorff linearly topologized A-module F, $u(\mathrm{E})$ is a closed submodule of F.

(γ) With every linear topology (Hausdorff or not) on E coarser than the given topology, E is complete.

(δ) E is complete and there is a fundamental system (U_λ) of open neighbourhoods of 0 in E, consisting of submodules and such that the $\mathrm{E}/\mathrm{U}_\lambda$ are linearly compact discrete A-modules (cf. § 3, Exercise 5).

(To see that (γ) implies (δ) consider an open submodule F of E and a filter base $\mathfrak{B}$ on E/F consisting of affine linear varieties. For all $\mathrm{V} \in \mathfrak{B}$, let M_V be the inverse image in E of the direction submodule of V in E/F; consider on E the linear topology in which the M_V form a fundamental system of neighbourhoods of 0.)

(b) Let E be a Hausdorff linearly topologized A-module and M a linearly compact submodule of E. Show that, for every closed submodule F of E, M + F is closed in E (consider the image of M in E/F).

(c) Let E be a linearly compact A-module and F a Hausdorff linearly topologized A-module. Show that for ever closed submodule M of E × F, the projection of M onto F is closed. Obtain the converse (cf. *General Topology*, Chapter I, § 10, no. 2).

(d) Let E be a linearly compact A-module and u a continuous linear mapping of E to a Hausdorff linearly topologized A-module F. Show that, for every filter base $\mathfrak{B}$ on E consisting of affine linear varieties, the image under u of the set of cluster points of $\mathfrak{B}$ is the set of cluster points of $u(\mathfrak{B})$. In particular, for every closed submodule M of E, $\bigcap_{N \in \mathfrak{B}} (M + N) = M + \bigcap_{N \in \mathfrak{B}} N$.

17. A linear topology $\mathscr{T}$ on an A-module E is called *minimal* if it is Hausdorff and there exists no Hausdorff linear topology strictly coarser than $\mathscr{T}$.

(a) For a Hausdorff linear topology $\mathscr{T}$ on E to be minimal, it is necessary and sufficient that every filter base $\mathfrak{B}$ on E, consisting of affine linear varieties and having a single cluster point, be convergent to this point. (To see that the condition is necessary, observe that, when M runs through $\mathfrak{B}$ and V a fundamental system of neighbourhoods of 0 consisting of submodules, the M + V form a filter base with the same cluster points as $\mathfrak{B}$. To see that the condition is sufficient, note that a filter base consisting of open submodules, whose intersection is reduced to 0, is a fundamental system of neighbourhoods of 0 in a Hausdorff linear topology coarser than $\mathscr{T}$.)

(b) For a discrete topology on an A-module E to be minimal, it is necessary and sufficient that E be Artinian.

(c) If $\mathscr{T}$ is minimal, the topology induced by $\mathscr{T}$ on any closed submodule of E is minimal.

¶ 18. In an A-module E, a submodule $M \neq E$ is called *sheltered* if there exists a least element in the set of submodules $\neq 0$ in E/M.

(a) Show that every submodule $N \neq E$ of E is an intersection of sheltered submodules (for all $x \notin N$, consider a maximal element in the set of submodules of E containing N and not containing x).

(b) Let $\mathscr{T}$ be a Hausdorff linear topology on E and let $\mathscr{T}^*$ be the linear topology with fundamental system of neighbourhoods of 0 the filter base generated by the submodules of E which are open under $\mathscr{T}$ and sheltered. Show that $\mathscr{T}^*$ is Hausdorff and that every submodule which is closed under $\mathscr{T}$ is also closed under $\mathscr{T}^*$ (note that every submodule which is closed under $\mathscr{T}$ is an intersection of submodules which are open under $\mathscr{T}$). Deduce that, if E is complete with respect to $\mathcal{T}^*$, it is complete with respect to $\mathcal{T}$ (*General Topology*, Chapter III, § 3, no. 5, Proposition 9).

(c) Suppose that $\mathcal{T}$ is linearly compact. Then show that $\mathcal{T}^*$ is linearly compact and is the coarsest Hausdorff linear topology coarser than $\mathcal{T}$; in particular $\mathcal{T}^*$ is a minimal linear topology (Exercise 17). (Let $\mathfrak{B}$ be a filter base consisting of submodules closed under $\mathcal{T}$ whose intersection is 0; show that, for every submodule U which is open under $\mathcal{T}$ and sheltered, there exists $M \in \mathfrak{B}$ such that $M \subset U$, using Exercise 16 (d)).

(d) Let F be a Hausdorff linearly topologized A-module and $\mathcal{T}_1$ its topology; show that, if $u: E \to F$ is a continuous linear mapping with respect to the topologies $\mathcal{T}$ and $\mathcal{T}_1$, u is also continuous with respect to the topologies $\mathcal{T}^*$ and $\mathcal{T}_1^*$.

¶ 19. (a) Let E be a linearly compact A-module. Show that the following conditions are equivalent:

(α) For every continuous linear mapping u from E to a Hausdorff linearly topologized A-module F, u is a strict morphism (*General Topology*, Chapter III, § 2, no. 8) from E to F.

(β) For every closed submodule F of E, the quotient topology on E/F is a minimal topology (Exercise 17).

(γ) E is complete and there is a fundamental system (U_λ) of open neighbourhoods of 0 in E, consisting of submodules and such that the E/U_λ are Artinian modules.

(To see that (γ) implies (β), reduce it to the case where $F = 0$ and use Exercises 17 and 18.)

If E satisfies the equivalent conditions (α), (β) and (γ), E is called *strictly linearly compact*.

(b) Let E be a Hausdorff linearly topologized A-module and F a closed submodule of E. For E to be strictly linearly compact, it is necessary and sufficient that F and E/F be so.

(c) Every inverse limit of strictly linearly compact modules is strictly linearly compact.

(d) Let E be a Hausdorff linearly topologized A-module and u a linear mapping from E to a strictly linearly compact A-module F. Show that if the graph of u is closed in $E \times F$, u is continuous (use Exercise 16 (c) and the fact that, if M is closed in E and E/M is Artinian, M is open in E).

¶ 20. (a) Show that a discrete linearly compact module cannot be the direct sum of an infinity of submodules which are not reduced to 0.

(b) Give an example of a minimal linear topology (Exercise 17) which is not linearly compact (consider an infinite direct sum of simple modules no two of which are isomorphic).

(c) Let E be a linearly compact module such that there exists a family of simple submodules whose sum is dense in E. Show that: (1) E is strictly linearly compact (apply (a) to the quotient of E by an open submodule); (2) E is isomorphic to the (topological) product of a family of discrete simple modules. (Consider the sets $\mathfrak{O}$ of maximal open submodules of E such that for every finite sequence $(G_k)_{1 \leqslant k \leqslant n}$ of n distinct elements of $\mathfrak{O}$, $E\Big/\Big(\bigcap_{k=1}^{n} G_k\Big)$ is the direct sum of n simple submodules. Show that there exists a maximal set $\mathfrak{O}$ (with respect to inclusion) and that the intersection of all the submodules G belonging to such a set is reduced to 0; use the fact that $\{0\}$ is an intersection of maximal open submodules and that for every maximal open submodules G of E there is at least one simple submodule of E not contained in G. Conclude by using the fact that E is strictly linearly compact.)

(d) Deduce from (c) that every linearly compact vector space over a field K is strictly linearly compact and isomorphic to a product K_s^I.

¶ 21. Recall that on a ring A a topology is called (left) *linear* if it is a linear topology on the A-module A_s and it is compatible with the ring structure on A (Chapter II, § 2, Exercise 16). A topological ring A is called (left) *linearly compact* (resp. (left) *strictly linearly compact*) if A_s is a linearly compact (resp. strictly linearly compact) A-module.

(a) Show that, if A is left linearly compact with the topology $\mathcal{T}$, the topology $\mathcal{T}^*$ defined in Exercise 18 (b) is also compatible with the ring structure on A (use Exercise 18 (d)).

(b) Suppose that A is commutative; let $u \neq 0$ be an idempotent of $A/\mathfrak{R}$, where $\mathfrak{R}$ is the Jacobson radical of A. Show that if A is linearly compact with the discrete topology, there exists a unique idempotent $e \in A$ whose image in $A/\mathfrak{R}$ is equal to u. (Show that among the linear affine varieties $x + \mathfrak{b}$, where $\mathfrak{b} \subset \mathfrak{R}$ and $x^2 - x \in \mathfrak{b}$, which are contained in the class u, there is a minimal element $e + \mathfrak{a}$. Deduce that $e^2 = e$ and $\mathfrak{a} = 0$, by considering the element $e^2 - e = r$ and showing that $r \in Ar^2$, using *Algebra*, Chapter VIII, § 6, Exercise 10 (a). Prove the uniqueness of e by showing that if e_1, e_2 are two idempotents of A such that $e_1e_2 \in \mathfrak{R}$, then $e_1e_2 = 0$.)

(c) Show that every commutative ring A which is linearly compact with the discrete topology is a direct composition of a finite number of local rings (which are linearly compact with the discrete topology). (Consider first the case, where $\mathfrak{R} = 0$, using Exercise 15 (b) and Chapter II, § 1, no. 2, Proposition 5, then apply (b) to the idempotents of $A/\mathfrak{R}$.)

(d) Show that every linearly compact (resp. strictly linearly compact) commutative ring A is the product of a family of linearly compact (resp. strictly linearly compact) local rings. (Let $(\mathfrak{m}_\lambda)$ be the family of open maximal ideals of A; for every open ideal $\mathfrak{a}$ not containing $\mathfrak{m}_\lambda$, let $A'_\lambda \times A''_\lambda$ be the decomposition of $A/\mathfrak{a}$ as a direct composition of two rings such that A'_λ is the local ring with maximal ideal $(\mathfrak{m}_\lambda + \mathfrak{a})/\mathfrak{a}$ defined in (c); let $e'_\lambda(\mathfrak{a})$ be the unit element of A'_λ considered as a class mod. $\mathfrak{a}$ in A; the $e'_\lambda(\mathfrak{a})$ form a filter base which converges in A to an idempotent e_λ and Ae_λ is a closed ideal of A of which e_λ is the unit element. Show that the ring Ae_λ is a local ring and that A is isomorphic to the product of the Ae_λ.)

¶ 22. (a) Let A be a local ring and $\mathfrak{m}$ its maximal ideal. Show that, if, for a linear topology $\mathscr{T}$ on A, A is strictly linearly compact, $\mathscr{T}$ is coarser than the $\mathfrak{m}$-adic topology.

(b) Let A be a commutative ring and $\mathfrak{m}$ an ideal of A. For A to be strictly linearly compact with the $\mathfrak{m}$-adic topology, it is necessary and sufficient that it be Hausdorff and complete with this topology, that $A/\mathfrak{m}$ be an Artinian ring and $\mathfrak{m}/\mathfrak{m}^2$ and $(A/\mathfrak{m})$-module of finite length * (in other words, A is a complete Noetherian semi-local ring) *.

(c) Let I be an infinite indexing set, K a finite field and $A = K[[X_\iota]]_{\iota \in I}$ the algebra of formal power series in the family of indeterminates $(X_\iota)_{\iota \in I}$ (*Algebra*, Chapter IV, § 5, Exercise 1); A is a local ring. As a vector space over K, A may be identified with the product space $K^{\mathbf{N}^{(I)}}$; if K is given the discrete topology and A the product topology $\mathscr{T}$, show that $\mathscr{T}$ is a linear topology on the ring A with which A is strictly linearly compact; if $\mathfrak{m}$ is the maximal ideal of A, show, by arguing as in Exercise 12 (c), that A is not complete with the $\mathfrak{m}$-adic topology.

¶ 23. Let A be a commutative ring with a linear topology $\mathscr{T}$.

(a) Show that the ideals $\mathfrak{a}$ of A, which are closed under $\mathscr{T}$ and such that $A/\mathfrak{a}$ is Artinian, form a fundamental system of neighbourhoods of 0 under a linear topology $\mathscr{T}^{(c)}$ on A, which is coarser than $\mathscr{T}$ (use *Algebra*, Chapter VIII, § 2, Exercise 12); then $(\mathscr{T}^{(c)})^{(c)} = \mathscr{T}^{(c)}$. Let B be the Hausdorff completion ring of A with respect to the topology $\mathscr{T}^{(c)}$; show that B is strictly linearly compact; B is said to be the strictly linearly compact ring *associated* with A.

(b) Let $\mathscr{T}_\omega(A)$ be the discrete topology on A, $\mathscr{T}_c(A)$ the topology $(\mathscr{T}_\omega(A))^{(c)}$. For the topology $\mathscr{T}^{(c)}$ on A to be Hausdorff, it is necessary and sufficient that $\mathscr{T}$ be Hausdorff and that, for every ideal $\mathfrak{b}$ of A open under $\mathscr{T}$, the topology $\mathscr{T}_c(A/\mathfrak{b})$ be Hausdorff. * If A is the ring of a non-discrete valuation of height 1 (Chapter VI), $\mathscr{T}_c(A)$ is not Hausdorff. *

(c) Suppose that A is linearly compact under $\mathscr{T}$; then A is complete (but in general not Hausdorff) with $\mathscr{T}^{(c)}$. Then for $\mathscr{T}^{(c)}$ to be Hausdorff, it is necessary and sufficient that there exist on A a linear topology which is strictly linearly compact and coarser than $\mathscr{T}$; $\mathscr{T}^{(c)}$ is then the unique topology with these properties.

(d) Let E be a linearly topologized and strictly linearly compact A-module. Show that, if A is given the topology $\mathscr{T}_c(A)$, E is a topological ring (note that if F is an Artinian A-module, then, for all $x \in F$, the annihilator $\mathfrak{a}$ of x is such that $A/\mathfrak{a}$ is Artinian). Deduce that, if B is the Hausdorff completion of A with respect to the topology $\mathscr{T}_c(A)$, E can be considered as a topological B-module; the ring B being isomorphic to a product of strictly linearly compact local rings

B_λ (Exercise 21 (d)), show that E is isomorphic to a product of strictly linearly compact submodules E_λ, where E_λ is annihilated by the B_μ of index $\mu \neq \lambda$ and can therefore be considered as a topological B_λ-module (if e_λ is the unit element of B, take $E = e_\lambda . E$).

24. On a commutative ring A let $\mathscr{T}_m(A)$ denote the linear topology which has a fundamental system of neighbourhoods of 0 the products (equal to the intersections) of a finite number of powers of maximal ideals (cf. no. 13, Proposition17); if every finite intersection of ideals $\neq 0$ of A is $\neq 0$, let $\mathscr{T}_u(A)$ denote the linear topology for which a fundamental system of neighbourhoods of 0 consists of all the ideals $\neq \{0\}$ of A.

(a) Show that $\mathscr{T}_c(A)$ (Exercise 23) is coarser than $\mathscr{T}_m(A)$ (observe that the Jacobson radical of an Artinian ring is nilpotent); give an example where $\mathscr{T}_c(A) \neq \mathscr{T}_m(A)$ (cf. Exercise 22 (c)). * If A is the ring of a non-discrete valuation of height 1 (Chapter VI), then $\mathscr{T}_c(A) = \mathscr{T}_m(A)$ and $\mathscr{T}_u(A) \neq \mathscr{T}_m(A)$. *

(b) The topology $\mathscr{T}_m(A)$ is the coarsest linear topology on A under which the maximal ideals of A are closed and every power of an open ideal is an open ideal (note that under a linear topology on A, under which a maximal ideal $\mathfrak{m}$ is closed, $\mathfrak{m}$ is necessarily open).

(c) If A is a Noetherian ring, then $\mathscr{T}_m(A) = \mathscr{T}_c(A)$ (observe that, if $\mathfrak{m}$ is a maximal ideal of A, $A/\mathfrak{m}^k$ is an Artinian ring for every integer $k \geqslant 1$, noting that $\mathfrak{m}^h/\mathfrak{m}^{h+1}$ is an $(A/\mathfrak{m})$-module which is necessarily of finite length). Give an example of a Noetherian local ring A for which $\mathscr{T}_m(A) \neq \mathscr{T}_u(A)$ (consider a local ring of a polynomial ring over a field).

25. Let A be a topological ring and E a finitely generated left A-module. Show that there exists on E a topology (called *canonical*) compatible with its A-module structure and finer than all the others (write E in the form A_s^n/R, give A_s^n the product topology and E the quotient topology). If E′ is a topological A-module and $u: E \to E'$ an A-linear mapping, show that u is continuous with the canonical topology on E. If further E′ is finitely generated and has the canonical topology, u is a strict morphism. If the topology on A is defined by a filtration (A_n), show that the canonical topology on E is defined by the filtration (A_nE).

26. Let A be a commutative ring, $\mathfrak{m}$ an ideal of A, $(a_\lambda)_{\lambda \in L}$ a system of generators of $\mathfrak{m}$ and $\hat{A}$ the Hausdorff completion of A with respect to the $\mathfrak{m}$-adic topology.

Show that, if A′ denotes the ring of formal power series in a family $(T_\lambda)_{\lambda \in L}$ of indeterminates with only a finite number of terms of given degree (Exercise 12), $\hat{A}$ is isomorphic to the quotient of A′ by the closure $\mathfrak{b}$ of the ideal of A′

generated by the $T_\lambda - a_\lambda$ (use Theorem 1 of no. 8). In particular, the ring $\mathbf{Z}_p$ of p-adic integers is isomorphic to the quotient $\mathbf{Z}[[T]]/(T - p)$.

¶ 27. (a) Let A be a commutative ring with a linear topology. For every multiplicative subset S of A, let $A\{S^{-1}\}$ denote the Hausdorff completion of the ring $S^{-1}A$ with the topology for which a fundamental system of neighbourhoods of 0 consists of the ideals $S^{-1}U_\lambda$, where (U_λ) is a fundamental system of neighbourhoods of 0 in A consisting of ideals of A. Show that $A\{S^{-1}\}$ is canonically isomorphic to the inverse limit of the rings $S_\lambda^{-1}(A/U_\lambda)$ where S_λ is the canonical image of S in A/U_λ. If $\hat{A}$ is the Hausdorff completion of A, $A\{S^{-1}\}$ is canonically isomorphic to $\hat{A}\{S'^{-1}\}$, where S' is the canonical image of S in $\hat{A}$.

(b) For $A\{S^{-1}\}$ to be reduced to 0, it is necessary and sufficient that, in A, 0 belong to the closure of S. Deduce an example where A is Hausdorff and complete but $S^{-1}A$ is not so with the topology defined in (a).

(c) For every continuous homomorphism u of A to a complete Hausdorff linearly topologized ring B such that $u(S)$ consists of invertible elements of B, show that $u = u' \circ j$, where $j: A \to A\{S^{-1}\}$ is the canonical mapping and u' is continuous; moreover u' is determined uniquely.

(d) Let S_1, S_2 be two multiplicative subsets of A and let S_2' be the canonical image of S_2 in $A\{S^{-1}\}$; define a canonical isomorphism of $A\{(S_1S_2)^{-1}\}$ onto $A\{S_1^{-1}\}\{S_2'^{-1}\}$.

(e) Let $\mathfrak{a}$ be an *open* ideal of A and let $\mathfrak{a}\{S^{-1}\}$ be the Hausdorff completion of $S^{-1}\mathfrak{a}$ with respect to the topology induced by that on $S^{-1}A$; show that $\mathfrak{a}\{S^{-1}\}$ is canonically identified with an open ideal of $A\{S^{-1}\}$ and that the discrete ring $A\{S^{-1}\}/\mathfrak{a}\{S^{-1}\}$ is isomorphic to $S^{-1}(A/\mathfrak{a})$. Conversely, if $\mathfrak{a}'$ is an open ideal of $A\{S^{-1}\}$, its inverse image $\mathfrak{a}$ in A is an open ideal such that $\mathfrak{a}' = \mathfrak{a}\{S^{-1}\}$. In particular, the mapping $\mathfrak{p} \mapsto \mathfrak{p}\{S^{-1}\}$ is an increasing bijection of the set of open prime ideals of A not meeting S onto the set of open prime ideals of $A\{S^{-1}\}$.

(f) Let $\mathfrak{p}$ be an *open* prime ideal of A and let $S = A - \mathfrak{p}$. Show that $A\{S^{-1}\}$ is a local ring whose residue field is isomorphic to the field of fractions of $A/\mathfrak{p}$.

(g) For every element $f \in A$, let $A_{\{f\}}$ denote the ring $A\{S_f^{-1}\}$, where S_f is the multiplicative set of the f^n ($n \geqslant 0$). If f runs through a multiplicative subset S of A, the $A_{\{f\}}$ form a direct system of rings whose direct limit is denoted by $A_{\{S\}}$ (without any topology); define a canonical homomorphism

$$A_{\{S\}} \to A\{S^{-1}\}.$$

If $S = A - \mathfrak{p}$, where $\mathfrak{p}$ is an *open* prime ideal of A, show that $A_{\{S\}}$ is a local ring, the canonical homomorphism $A_{\{S\}} \to A\{S^{-1}\}$ is local and the residue fields of $A_{\{S\}}$ and $A\{S^{-1}\}$ are canonically isomorphic (cf. Chapter II, § 3, Exercise 16).

(h) Suppose that the topology on A is the $\mathfrak{m}$-adic topology for some ideal $\mathfrak{m}$ of A, that A is Hausdorff and complete and that $\mathfrak{m}/\mathfrak{m}^2$ is a finitely generated $(A/\mathfrak{m})$-module. Show that, if $\mathfrak{m}' = \mathfrak{m}\{S^{-1}\}$, the topology on $A' = A\{S^{-1}\}$ is the $\mathfrak{m}'$-adic topology, $\mathfrak{m}' = \mathfrak{m}A'$ and $\mathfrak{m}'/\mathfrak{m}'^2$ is a finitely generated $(A'/\mathfrak{m}')$-module (use Proposition 14 of no. 10). If A is Noetherian, so is A'.

28. Let A be a commutative ring with a linear topology and E, F two topological A-modules which are linearly topologized. If V (resp. W) runs through the set of open submodules of E (resp. F), the submodules

$$\operatorname{Im}(V \otimes_A F) + \operatorname{Im}(E \otimes_A W)$$

of $E \otimes_A F$ form a fundamental system of neighbourhoods of 0 in $E \otimes_A F$ for a topology which is compatible with its module structure over the topological ring A and is called the *tensor product* of the given topologies on E and F. The Hausdorff completion $(E \otimes_A F)^\wedge$ of this A-module is an $\hat{A}$-module called the *completed tensor product* of E and F.

(a) Show that, if (V_λ) (resp. (W_μ)) is a fundamental system of neighbourhoods of 0 in E (resp. F) consisting of submodules, $(E \otimes_A F)^\wedge$ is canonically isomorphic to the inverse limit of the inverse system of A-modules

$$(E/V_\lambda) \otimes_A (F/W_\mu);$$

deduce that $(E \otimes_A F)^\wedge$ is an $\hat{A}$-module canonically isomorphic to $(\hat{E} \otimes_{\hat{A}} \hat{F})^\wedge$; it is also denoted by $\hat{E} \hat{\otimes}_{\hat{A}} \hat{F}$.

(b) Let E', F' be two topological A-modules which are linearly topologized and $u: E \to E'$ $v: F \to F'$ two continuous A-linear mappings; show that $u \otimes v: E \otimes F \to E' \otimes F'$ is continuous with the tensor product topologies on $E \otimes F$ and $E' \otimes F'$; we denote by $u \hat{\otimes} v$ the continuous linear mapping $(E \otimes F)^\wedge \to (E' \otimes F')^\wedge$ corresponding to $u \otimes v$.

(c) Let B, C be two commutative A-algebras with linear topologies such that the canonical mappings $A \to B$, $A \to C$ are continuous (to abbreviate, B, C are called *topological A-algebras*). Show that $(B \otimes_A C)^\wedge$ has a canonical topological A-algebra structure called the *completed tensor product* of the algebras B and C. Define canonical continuous representations

$$\rho: B \to (B \otimes_A C)^\wedge, \qquad \sigma: C \to (B \otimes_A C)^\wedge$$

with the following property: for every complete Hausdorfff commutative topological A-algebra D and every ordered pair of continuous A-homomorphisms $u: B \to D$, $v: C \to D$, there exists a unique continuous A-homomorphism $w: (B \otimes_A C)^\wedge \to D$ such that $u = w \circ \rho$ and $v = w \circ \sigma$.

(d) Let $\mathfrak{m}$ be an ideal of A such that the topology on A is the $\mathfrak{m}$-adic topology. If E and F are each given the $\mathfrak{m}$-adic topology, show that on $E \otimes_A F$ the tensor product of the topologies on E and F is the $\mathfrak{m}$-adic topology. Deduce that $(E \otimes_A F)^\wedge$ is isomorphic to the inverse limits $\varprojlim((E/\mathfrak{m}^{n+1}) \otimes_A F)$ and $\varprojlim(E \otimes_A (F/\mathfrak{m}^{n+1}F))$.

29. Let A be a commutative ring, $\mathfrak{m}$ an ideal of A such that $\bigcap_{n \geqslant 0} \mathfrak{m}^n = 0$ and c is an element of A which is not a divisor of 0. Show that the transporters $\mathfrak{q}_n = \mathfrak{m}^n : Ac$ are open under the $\mathfrak{m}$-adic topology and have zero intersection. If A is strictly linearly compact with the $\mathfrak{m}$-adic topology (Exercise 22(b)), $(\mathfrak{q}_n)$ is a fundamental system of neighbourhoods of 0 in A for this topology.

30. Let K be an infinite commutative field and A the ring of formal power series K[[X, Y]] in two indeterminates, which is a complete Hausdorff Noetherian local ring. Define a multiplicative subset S of A such that $S^{-1}A$ is not a semi-local ring (if (λ_n) is an infinite sequence of distinct elements of K, consider the prime ideals $\mathfrak{p}_n = A(X + \lambda_n Y)$ of A).

31. Show that if A is a semi-local ring, so is the ring of formal power series A[[X]] (consider the Jacobson radical of this ring).

§ 3

1. (a) Let K be a commutative field of characteristic $p > 0$ and B the ring of formal power series with coefficients in K in two infinite systems of indeterminates (X_n), (Y_n) with only a finite number of terms of given degree (§ 2, Exercise 12); B is a Hausdorff local ring with the $\mathfrak{m}$-adic topology, $\mathfrak{m}$ being its maximal ideal. Let $\mathfrak{b}$ be the closed ideal of B generated by the monomials $Y_i X_j$ for $i \geqslant 0$, $0 \leqslant j \leqslant i$; let A be the local ring $B/\mathfrak{b}$, which is Hausdorff with the $\mathfrak{n}$-adic topology, where $\mathfrak{n} = \mathfrak{m}/\mathfrak{b}$ is its maximal ideal. Let c be the element of A equal to the class mod. $\mathfrak{b}$ of $\sum_{n=1}^{\infty} X_n^{p^n}$; show that, for $k \geqslant 3$, $\mathfrak{n}^k \cap (Ac) \not\subset \mathfrak{n}^2 c$ (and *a fortiori* the relation

$$\mathfrak{n}^2(\mathfrak{n}^{2k} \cap (Ac)) = \mathfrak{n}^{2k+2} \cap (Ac)$$

cannot hold).

(b) Let C′ be the commutative ring with underlying set A × A, with multiplication $(a, x)(a', x') = (aa', ax' + a'x)$; let C be the subring A × (Ac) of C′; C and C′ are local rings, whose maximal ideals we denote by $\mathfrak{r}$ and $\mathfrak{r}'$, and C′ is a finitely generated C-module. Show that $\mathfrak{r} = \mathfrak{r}' \cap C$ but that the topology induced on C by the $\mathfrak{r}'$-adic topology is not the $\mathfrak{r}$-adic topology.

2. Let A be a Noetherian integral domain which is not a field, K its field of fractions and $\mathfrak{m}$ an ideal $\neq$A of A. The $\mathfrak{m}$-adic topology on A is not induced by the $\mathfrak{m}$-adic topology on K and the latter is not Hausdorff.

* 3. If A is the ring of a non-discrete valuation of height 1 (Chapter VI) and $\mathfrak{m}$ is its maximal ideal, A is not Hausdorff with the $\mathfrak{m}$-adic topology, the closure of $\{0\}$ being $\mathfrak{m}$. *

¶ 4. Let A be a semi-local ring and $\mathfrak{r}$ its Jacobson radical. For A to be Noetherian, it is necessary and sufficient that every ideal of A be closed under the $\mathfrak{r}$-adic topology and that every maximal ideal of A be finitely generated. (If these conditions hold, show first that the Hausdorff completion $\hat{A}$ of A is Noetherian, using § 2, no. 13, Corollary to Proposition 19 and no. 10, Corollary 5 to Theorem 1. Then observe that the $\mathfrak{r}$-adic topology is Hausdorff on A and that there exists an increasing injection of the set of ideals of A into the set of ideals of $\hat{A}$.)

¶ 5. Let A be a commutative Noetherian ring and $\mathfrak{m}$ its Jacobson radical.

(a) For A to be strictly linearly compact with a linear topology $\mathscr{T}$ (§ 2, Exercise 19), it is necessary and sufficient that A be semi-local and complete with the $\mathfrak{m}$-adic topology; $\mathscr{T}$ is then necessarily identical with this latter topology. (Use § 2, Exercise 21 (d) and 22 (a) and also § 3, no. 3, Proposition 6.)

(b) For A to be linearly compact with a linear topology $\mathscr{T}$, it is necessary and sufficient that A be semi-local and complete with the $\mathfrak{m}$-adic topology and that $\mathscr{T}$ be finer than this latter topology. (To see that it is necessary, reduce it first to the case where A is a local ring, using Exercise 21 (d) of § 2. Consider first the case where $\mathscr{T}$ is the discrete topology and note that A is then linearly compact with the $\mathfrak{m}$-adic topology. In the general case, reduce it to the case where $\mathscr{T}$ is a minimal topology (§ 2, Exercise 18 (c)); show that A is then strictly linearly compact and use criterion (γ) of § 2, Exercise 19 (a) and Exercise 18 (b) of § 2.)

6. Let A be a Noetherian local integral domain, $\mathfrak{m}$ its maximal ideal and K its field of fractions; suppose that A is complete with the $\mathfrak{m}$-adic topology. Consider on A a linear topology $\mathscr{T}$ which is finer than the $\mathfrak{m}$-adic topology. Show that, if $\mathscr{T}'$ is the linear topology on K for which a fundamental system of neighbourhoods of 0 consists of the neighbourhoods of 0 in A with the topology $\mathscr{T}$, $\mathscr{T}'$ is compatible with the field structure on K and K is complete with the topology $\mathscr{T}'$.

7. Let A be a commutative Noetherian ring and E an A-module. Let A be given the topology $\mathscr{T}_m(A)$ (§ 2, Exercise 24) and let $\mathscr{T}_m(E)$ denote the linear topology on E for which a fundamental system of neighbourhoods of 0 consists of the submodules $\mathfrak{a}.E$, where $\mathfrak{a}$ runs through a fundamental system of neighbourhoods of 0 for $\mathscr{T}_m(A)$, consisting of ideals of A. Show that if E is finitely generated, E is a Hausdorff topological A-module (with $\mathscr{T}_m(A)$ and $\mathscr{T}_m(E)$) where every submodule is closed. A fundamental system of neighbourhoods of 0 for $\mathscr{T}_m(E)$ then consists of the submodules F of E such that E/F is a

module of finite length and, for every submodule M of E, the topology $\mathcal{T}_{\mathfrak{m}}(M)$ is induced by $\mathcal{T}_{\mathfrak{m}}(E)$.

¶ 8. Let A be a Noetherian semi-local ring, $\mathfrak{n}$ its nilradical (which is the largest nilpotent ideal of A) and $\mathfrak{m}$ its Jacobson radical. Show that, if A (with the $\mathfrak{m}$-adic topology) is such that $A/\mathfrak{n}$ is complete, then A is complete. (Reduce it to the case where $\mathfrak{n}$ is generated by a single element c such that $c^2 = 0$; using Exercise 9 of *General Topology*, Chapter III, § 3, reduce it to showing that $\mathfrak{n} = Ac$ is complete and use the fact that $\mathfrak{n}$ is an $(A/\mathfrak{n})$-module.)

¶ 9. (a) Let A be a Zariski ring, $\mathfrak{m}$ a defining ideal of the topology on A and B a ring such that $A \subset B \subset \hat{A}$, which is a finitely generated A-module; show that, if $\mathfrak{m}B$ is open in B under the topology induced by the topology on $\hat{A}$, of necessity $B = A$ (use Nakayama's Lemma).

* (b) Let A be a commutative Noetherian ring and $\mathfrak{m}$ an ideal of A. Show that, if the Hausdorff completion $\hat{A}$ of A with respect to the $\mathfrak{m}$-adic topology is a finitely generated A-module, A is complete with the $\mathfrak{m}$-adic topology (reduce it to the case where A is Hausdorff; use Chapter V, § 2, no. 1, Proposition 1 and Theorem 1 to show that A is a Zariski ring with the $\mathfrak{m}$-adic topology and conclude with the aid of (a)). *

10. Let A be a Zariski ring, $\mathfrak{m}$ a defining ideal of the topology on A and E a finitely generated A-module. Show that, if $\hat{E}$ is an $\hat{A}$-module admitting a system of generators with r elements, then the A-module E admits a system of generators with r elements (note that $E/\mathfrak{m}E$ and $\hat{E}/\mathfrak{m}\hat{E}$ are isomorphic and use Chapter II, § 3, Corollary 2 to Proposition 4). Is the result valid if E is not assumed to be finitely generated (cf. Exercise 9)? Is it valid if we only assume that A is Noetherian (take $A = \mathbf{Z}$)?

11. Let A be the ring $\mathbf{Z}_p$ of p-adic integers (p prime) with the p-adic topology with which it is a complete Zariski ring. Let E be the A-module $A^{(\mathbf{N})}$ with the p-adic topology.

(a) Show that the completion $\hat{E}$ of E is identified with the submodule of $A^{\mathbf{N}}$ consisting of the sequences $(a_n)_{n \in \mathbf{N}}$ of elements of A such that $\lim_{n \to \infty} a_n = 0$.

(b) Let e_n be the n-th vector of the canonical basis of E. Consider in E the submodule F generated by the vectors e_{2n-1} and the submodule G generated by the vectors $p^{2n}e_{2n} - e_{2n-1}$ ($n \geqslant 1$). Show that the topologies induced on F and G by the p-adic topology on E are the p-adic topologies on F and G but that, in $\hat{E}$,

$$\overline{F + G} \neq \overline{F} + \overline{G}.$$

(c) In $\hat{E}$, let $a_r = \sum_{n=0}^{\infty} p^n e_{r+2}$ ($r \geqslant 0$); let H be the submodule of $\hat{E}$ generated by the a_r ($r \geqslant 0$). Show that on H the topology induced by the p-adic topology on $\hat{E}$ is the p-adic topology on H but that $\overline{E \cap H} \neq E \cap \bar{H}$.

(d) Let L be the submodule of E generated by the $p^n e_n$; show that on L the topology induced by the p-adic topology on E is distinct from the p-adic topology on L.

12. (a) Let A be a commutative Noetherian ring and $\mathfrak{m}_1$, $\mathfrak{m}_2$ two ideals of A contained in the Jacobson radical of A; let A_i be the completion of A with respect to the $\mathfrak{m}_i$-adic topology. If $\mathfrak{m}_1 \subset \mathfrak{m}_2$, show that the identity mapping on A can be extended by continuity to an injective representation of A_1 into A_2 (cf. *General Topology*, Chapter III, § 3, no. 5, Proposition 9).

(b) Let $\mathfrak{n} = A_1\mathfrak{m}_2$. Show that, if A_1 is canonically identified with a subring of A_2, A_2 is identified with the completion of A_1 with respect to the $\mathfrak{n}$-adic topology.

13. Let A be a Noetherian local ring with maximal ideal $\mathfrak{m}$ and B a ring such that $A \subset B \subset \hat{A}$. Suppose that B is a Noetherian local ring with maximal ideal $\mathfrak{n}$ (Exercise 14).

(a) Show that $\mathfrak{n} = B \cap \hat{\mathfrak{m}}$ and $B = A + \mathfrak{n}$. If further $\mathfrak{n}^2 = B \cap \hat{\mathfrak{m}}^2$, then $\mathfrak{n} = B\mathfrak{m}$ (note that then $\mathfrak{n} = B\mathfrak{m} + \mathfrak{n}^2$).

(b) Let K be a commutative field, C the polynomial ring K[X], $\mathfrak{p}$ the prime ideal CX, A the (Noetherian) local ring $C_{\mathfrak{p}}$ and $\mathfrak{m}$ its maximal ideal; the completion $\hat{A}$ is identified with the ring of formal power series K[[X]] (no. 4, Proposition 8). Let $u = u(X)$ be an element of $\hat{A}$ which is transcendental over the field of rational fractions K(X) (cf. *Algebra*, Chapter V, § 5, Exercise 13 or *Functions of a Real Variable*, Chapter III, § 1, Exercise 14 (a)) with no constant term; let B be the subring of $\hat{A}$ consisting of the quotients

$$P(X, u(X))/Q(X, u(X)),$$

where P and Q are polynomials in K[X, Y] such that $Q(0, 0) \neq 0$. Show that B is a Noetherian local ring containing A, whose maximal ideal $\mathfrak{n}$ is generated by X and $u(X)$; but on B the $\mathfrak{n}$-adic topology is strictly coarser than the $\mathfrak{m}$-adic topology.

(c) With the same definitions as in (b), let B′ be the subring of B generated by A and $u(X)$; show that B′ is not a local ring (note that $B' \cap \mathfrak{n}$ is a maximal ideal of B′ but that there are elements of B′ not invertible in B′ and not belonging to $B' \cap \mathfrak{n}$).

¶ 14. Let K be a commutative field, C the ring K[X, Y], $\mathfrak{p}$ the maximal ideal CX + CY of C, A the local ring $C_{\mathfrak{p}}$, $\mathfrak{m}$ its maximal ideal and $\hat{A}$ the completion of A, which is identified with the ring of formal power series K[[X, Y]] (no. 4, Proposition 8). Let B be the subring of $\hat{A}$ consisting of the formal power series of the type $Xf(X, Y) + (P(Y)/Q(Y))$, where $f \in \hat{A}$ and P and Q are two polynomials of K[Y] such that $Q(0) \neq 0$.

(a) Show that $A \subset B$, that B is a local ring whose maximal ideal $\mathfrak{n}$ is equal to $B\mathfrak{m}$ and that $\mathfrak{n}^k = B \cap \hat{\mathfrak{m}}^k$ for every integer $k \geqslant 1$; B is therefore everywhere dense in $\hat{A}$ and the topology induced on B by the $\hat{\mathfrak{m}}$-adic topology on $\hat{A}$ is the $\mathfrak{n}$-adic topology.

(b) Show that in B the ideal $\mathfrak{b}$ generated by all the elements of the form $Xf(Y)$, where $f(Y) \in K[[Y]]$, is not finitely generated (cf. *Algebra*, Chapter V, § 5, Exercise 13) and therefore B is not Noetherian, although $B/\mathfrak{n}$ and $\operatorname{gr}(B) = \operatorname{gr}(\hat{A})$ are Noetherian and the ideal $\mathfrak{n}$ is finitely generated; $\mathfrak{b} = B \cap \hat{A}X$ and $\mathfrak{b}$ is the closure in B of the principal ideal (not closed) BX; finally B is not a flat A-module and $\hat{B} = \hat{A}$ is not a flat B-module (use Chapter I, § 3, no. 5, Proposition 9).

(c) Let $f(Y)$ be an invertible formal power series in $K[[Y]]$, which is not an element of $K(Y)$. If $\mathfrak{c}$ is the ideal of B generated by X and $Xf(Y)$, show that on $\mathfrak{c}$ the $\mathfrak{n}$-adic topology is strictly finer than the topology induced on $\mathfrak{c}$ by the $\mathfrak{n}$-adic topology on B. Show that the canonical mapping $\hat{B} \otimes_B \mathfrak{c} \to \hat{\mathfrak{c}}$ (where $\hat{\mathfrak{c}}$ is the completion of $\mathfrak{c}$ with respect to the $\mathfrak{n}$-adic topology) is not injective (consider the images of $f(Y) \otimes X$ and $1 \otimes Xf(Y)$; to show that these two elements are distinct in $\hat{B} \otimes_B \mathfrak{c}$, consider this tensor product as a quotient of the tensor product $\hat{B} \otimes_{K(Y)} \mathfrak{c}$).

(d) Let $f_1(Y), f_2(Y)$ be two invertible formal power series in $K[[Y]]$ such that $1, f_1$ and f_2 are linearly independent over $K(Y)$. Let $\mathfrak{c}_1, \mathfrak{c}_2$ be the principal ideals of B generated respectively by $Xf_1(Y)$ and $Xf_2(Y)$; the $\mathfrak{n}$-adic topologies on $\mathfrak{c}_1$ and $\mathfrak{c}_2$ are respectively identical with the topologies induced by that on B and the closures of these ideals in $\hat{A} = \hat{B}$ are both identical with the principal ideal $\hat{A}X$ of $\hat{A}$. Deduce that the closure of $\mathfrak{c}_1 \cap \mathfrak{c}_2$ in $\hat{A}$ is not equal to $\bar{\mathfrak{c}}_1 \cap \bar{\mathfrak{c}}_2$ and that the closure of $\mathfrak{c}_1 : \mathfrak{c}_2$ is not equal to $\bar{\mathfrak{c}}_1 : \bar{\mathfrak{c}}_2$.

15. (a) Let A be a Noetherian integral domain, $\mathfrak{m}$ an ideal of A and $\hat{A}$ the Hausdorff completion of A with respect to the $\mathfrak{m}$-adic topology. Show that, if M is a torsion-free A-module, then, for every element b which is not a divisor of 0 in $\hat{A}$, the homothety with ratio b in the $\hat{A}$-module $\hat{A} \otimes_A M$ is injective. (Reduce it to the case where M is finitely generated; then there exists a maximal free system $(m_j)_{1 \leqslant j \leqslant n}$ in M and $a \in A$ such that for all $m \in M$, $am = \sum_j a_j m_j$ where $a_j \in A$; every $x \in \hat{A} \otimes_A M$ therefore satisfies $ax = \sum_j b_j \otimes m_j$ where $b_j \in \hat{A}$; then use the flatness of the A-module $\hat{A}$.)

(b) Let K be an algebraically closed field of characteristic 0, B the polynomial ring $K[X, Y]$ and $P(X, Y) = X(X^2 + Y^2) + (X^2 - Y^2)$. Show that the ideal BP is prime in B; consider the quotient ring $A = B/BP$ which is a Noetherian integral domain. Let $\mathfrak{m}$ be the maximal ideal of A, the canonical image of the maximal ideal $\mathfrak{n} = BX + BY$ of B. Show that the Hausdorff completion $\hat{A}$ of A with respect to the $\mathfrak{m}$-adic topology is not an integral domain (observe that in the ring of formal power series $K[[X, Y]]$, P decomposes into a product of two formal power series ("double point of an irreducible cubic")).

16. Let A be a commutative Noetherian ring, $\mathfrak{m}$ an ideal of A and E a finitely generated A-module with the $\mathfrak{m}$-adic topology. For every infinite set I, let E_I denote the set of families $(x_\iota)_{\iota \in I}$ of elements of E such that $\lim x_\iota = 0$ with respect to the filter of complements of finite subsets of I; it is a submodule of the product A-module E^I.

(a) Show that, if $0 \to E' \to E \to E'' \to 0$ is an exact sequence of finitely generated A-modules, the corresponding sequence $0 \to E'_I \to E_I \to E''_I \to 0$ is exact.

(b) Define a canonical A-homomorphism $A_I \otimes_A E \to E_I$ for every A-module E and show that it is an isomorphism (first verify this when E is free and finitely generated and then use (a)).

(c) Deduce from (b) that A_I is a faithfully flat A-module.

¶ 17. (a) Let K be a commutative field, $A = K[[X_1, \ldots, X_n]]$ the ring of formal power series in n indeterminates with coefficients in K and V a vector space over K. With the notation of § 2, no. 6, *Example* 1, let $V[[X_1, \ldots, X_n]]$ denote the vector space $V^{\mathbf{N}^n}$ with the A-module structure defined by

$$\left(\sum_\alpha c_\alpha X^\alpha\right)(v_\alpha) = (w_\alpha),$$

where $w_\alpha = \sum_{\beta+\gamma=\alpha} c_\beta v_\gamma$. Show that, if V admits a basis with I as indexing set, the A-module $V[[X_1, \ldots, X_n]]$ is isomorphic to A^I if I is finite and to the A-module A_I defined in Exercise 16 if I is infinite. Deduce that $V[[X_1, \ldots, X_n]]$ is a flat A-module and is faithfully flat if V is not reduced to 0.

(b) Let L be an extension of the field K. Deduce from (a) that the ring of formal power series $L[[X_1, \ldots, X_n]]$ is a faithfully flat module over the ring $K[[X_1, \ldots, X_n]]$. If L has finite rank over K, $L[[X_1, \ldots, X_n]]$ is isomorphic to $L \otimes_K K[[X_1, \ldots, X_n]]$.

(c) If L is an *algebraic* extension of K, show that the ring $L[[X_1, \ldots, X_n]]$ is a faithfully flat module over the ring

$$L \otimes_K K[[X_1, \ldots, X_n]]$$

(consider L as the direct limit of its sub-extensions of finite rank over K and use Chapter I, § 2, no. 7, Proposition 9). Deduce that the ring

$$B = L \otimes_K K[[X_1, \ldots, X_n]]$$

is a Noetherian local ring whose completion is identified with $L[[X_1, \ldots, X_n]]$. In order that $\hat{B} = B$, it is necessary and sufficient that $n = 0$ or $[L:K] < +\infty$.

¶ 18. Let A be a local ring with maximal ideal $\mathfrak{m}$; an A-module M is called *quasi-finite* if $M/\mathfrak{m}M$ is a vector space of finite rank over the residue field $k = A/\mathfrak{m}$. In particular, if A is an integral domain, the field of fractions K of A is a quasi-finite A-module.

(a) Show that if A is Noetherian and M is a quasi-finite A-module, its Hausdorff completion $\hat{M}$ with respect to the $\mathfrak{m}$-adic topology is a finitely generated $\hat{A}$-module (use § 2, no. 11, Corollary 2 to Proposition 14). In particular, if A is complete and M Hausdorff with the $\mathfrak{m}$-adic topology, M is a finitely generated A-module.

(b) Let B be another local ring, $\mathfrak{n}$ its maximal ideal, $\phi: A \to B$ a local homomorphism and M a finitely generated B-module. Show that, if B is Noetherian and M is a quasi-finite A-module, then the $\mathfrak{m}$-adic and $\mathfrak{n}$-adic topologies on M are identical. (Note that $M/\mathfrak{m}M$ is a B-module of finite length and deduce that, if $\mathfrak{b}$ is the annihilator of the B-module $M/\mathfrak{m}M$, then $V(\mathfrak{b}) = \{\mathfrak{n}\}$ in Spec(B) and therefore $V(\mathfrak{m}B + \mathfrak{b}) = \{\mathfrak{n}\}$. Conclude using Chapter II, § 4, no. 3, Corollary 2 to Proposition 11.)

(c) Under the hypotheses of (b), show that, if $M \neq 0$, $B/\mathfrak{b}$ is a quasi-finite A-module (note that $M \neq \mathfrak{n}M$ and $M/\mathfrak{n}M$ is a vector space of finite rank over k; deduce that $B/\mathfrak{n}$ is of finite rank over k).

¶ 19. Let A be a commutative Noetherian ring, $\mathfrak{m}$ an ideal of A and S a multiplicative subset of A. Let A be a given the $\mathfrak{m}$-adic topology.

(a) Show that the ring $A\{S^{-1}\}$ (§ 2, Exercise 27) is a flat A-module.

(b) Let S′ be another multiplicative subset of A contained in S. Show that $A\{S^{-1}\}$ is a flat $(A\{S'^{-1}\})$-module (use Exercise 27 (d) of § 2).

(c) Show that $A\{S^{-1}\}$ is a flat $A_{(S)}$-module (§ 2, Exercise 27 (g)).

(d) Suppose that $S = A - \mathfrak{p}$, where $\mathfrak{p}$ is an *open* prime ideal of A. Show that $A\{S^{-1}\}$ is a faithfully flat $A_{(S)}$-module and deduce that the ring $A_{(S)}$ is Noetherian (cf. § 2, Exercise 27 (g)).

20. Let A be a commutative ring and $\mathfrak{m}$ an ideal of A; let A be given the $\mathfrak{m}$-adic topology. Let B be a commutative topological A-algebra (§ 2, Exercise 28); suppose that B is a Zariski ring; let $\mathfrak{n}$ be a defining ideal of B. Show that, if M is a finitely generated A-module with the $\mathfrak{m}$-adic topology, then on the B-module $B \otimes_A M$ the tensor product of the topology on B and the $\mathfrak{m}$-adic topology on M is the $\mathfrak{n}$-adic topology; then the completion tensor product $(B \otimes_A M)^\wedge$ is isomorphic to $\hat{B} \otimes_A M$.

21. Let A be a commutative Notherian ring, $\mathfrak{m}$ an ideal of A and M and N two finitely generated A-modules.

(a) Suppose that M is Hausdorff with the $\mathfrak{m}$-adic topology. Show that in $\mathrm{Hom}_A(M, N)$ with the $\mathfrak{m}$-adic topology the set of injective homomorphisms is open (use no. 1, Proposition 2 and the Artin-Rees Lemma).

(b) Suppose that A is a complete Zariski ring and $\mathfrak{m}$ a defining ideal of

A. For every integer i, let $A_i = A/\mathfrak{m}^{i+1}$, $M_i = M/\mathfrak{m}^{i+1}M$, $N_i = N/\mathfrak{m}^{i+1}N$; show that the topological A-module $\mathrm{Hom}_A(M, N)$ is isomorphic to

$$\varprojlim \mathrm{Hom}_A(M_i, N_i).$$

Deduce that in $\mathrm{Hom}_A(M, N)$ the set of surjective homomorphisms is open.

¶ 22. (*) Let A be a ring, commutative or not; all the A-modules to be considered are left A-modules. Let $\boldsymbol{P}$ be a property such that: (α) if f: $M \to N$ is an injective A-module homomorphism and N has property $\boldsymbol{P}$, then M has property $\boldsymbol{P}$; (β) the direct sum of two A-modules with property $\boldsymbol{P}$ has property $\boldsymbol{P}$.

(a) Let M be an A-module. Show that the submodules M′ of M such that M/M′ has property $\boldsymbol{P}$ form a fundamental system of neighbourhoods of 0 for a linear topology $\mathscr{T}_P(M)$ on M. Show that $\mathscr{T}_P(A_s)$ is compatible with the ring structure on A and that M with the topology $\mathscr{T}_P(M)$ is a topological module over A with $\mathscr{T}_P(A_s)$. Every A-module homomorphism f: $M \to N$ is continuous with the topologies $\mathscr{T}_P(M)$ and $\mathscr{T}_P(N)$.

(b) Suppose that the following condition holds: (γ) if N is a submodule of M with property $\boldsymbol{P}$ and, for every submodule $L \neq 0$ of M, $N \cap L \neq 0$, then M has property $\boldsymbol{P}$. Show that with these conditions, if F is a submodule of an A-module E, the topology $\mathscr{T}_P(F)$ is induced by $\mathscr{T}_P(E)$. (Let F′ be a submodule of F such that F/F′ has property $\boldsymbol{P}$; consider a maximal element G among the submodules of E such that $G \cap F = F'$ and show that E/G has property $\boldsymbol{P}$.)

¶ 23. (a) Let A be a commutative Noetherian ring and $\mathfrak{m}$ an ideal of A. Let $\boldsymbol{P}\{M\}$ denote the following property: M is an A-module and every finitely generated submodule of M is annihilated by a power of $\mathfrak{m}$.

Show that conditions (α), (β) and (γ) of Exercise 22 are fulfilled. (To prove (γ), reduce it to the case where M is finitely generated; for all $a \in \mathfrak{m}$, there exists by hypothesis $k > 0$ such that $a^k N = 0$; use the fact that there exists $r > 0$ such that $\mathrm{Ker}(a_M^r) \cap \mathrm{Im}(a_M^r) = 0$ (*Algebra*, Chapter VIII, § 2, no. 2, Lemma 2).)

(b) Show that, if M is a finitely generated A-module, the topology $\mathscr{T}_P(M)$ is identical with the $\mathfrak{m}$-adic topology. Give an example of an A-module M for which $\mathscr{T}_P(M)$ is strictly finer that the $\mathfrak{m}$-adic topology (cf. Exercise 11).

(c) Show that the conclusion of (a) does not extend to the case where A is a non-commutative left Noetherian ring and $\mathfrak{m}$ is a two-sided ideal of A (consider the ring of lower triangular matrices of order 2 over a commutative field).

¶ 24. A ring (commutative or not) A, not reduced to 0, is called a *principal ideal ring* if it has no divisors of zero and every left or right ideal of A is monogenous. Such a ring is left and right Noetherian.

---

* Exercises 22 to 25 were communicated to us by P. Gabriel.

(a) Show that, for every element $c \neq 0$ of A, A/Ac is an A-module of finite length. (Observe that, if a decreasing sequence of ideals Aa_n of A contains Ac, then $c = b_n a_n$ for all n and consider the right ideals b_nA.)

(b) Show that every submodule of a (left or right) free A-module is free (same argument as in *Algebra*, Chapter VII, § 3, Theorem 1).

(c) In every A-module M, the set of elements of M which are not free is a submodule T of M called the *torsion submodule* of M (use Chapter II, § 2, Exercise 14 (a)); M is called a *torsion module* if T = M; M is called *torsion-free* if T = 0.

(d) Show that every finitely generated torsion-free A-module is free (use Chapter II, § 2, Exercise 23 (b)).

(e) Show that every finitely generated A-module is the direct sum of a free module and a torsion module.

(f) Let $\mathfrak{a}$ be a two-sided ideal $\neq 0$ of A. Show that there exists an element $a \in \mathfrak{a}$ and an automorphism σ of A such that $\mathfrak{a} = Aa = a$A and $ax = \sigma(x)a$ for all $x \in$ A. (If b is a generator of the left ideal $\mathfrak{a}$, there exists an endomorphism τ of A such that $bx = \tau(x)b$ for all $x \in$ A; show that, if $a = ub$ is a generator of the right ideal $\mathfrak{a}$, u is invertible, using *Algebra*, Chapter VIII, § 2, Exercise 8 (b).)

¶ 25. Let A be a principal ideal ring (Exercise 24), $\mathfrak{a}$ a two-sided ideal $\neq 0$ of A and a an element of $\mathfrak{a}$ with the properties stated in Exercise 24 (f). Let $\boldsymbol{P}\{M\}$ denote the following property: M is a left A-module and every finitely generated submodule of M is annihilated by a power of a.

(a) Show that the conditions (α), (β) and (γ) of Exercise 22 are fulfilled (consider the homothety a_M and argue as in Exercise 23 (a), observing that $\mathrm{Ker}(a_M^r)$ and $\mathrm{Im}(a_M^r)$ are submodules of M).

(b) Show that every torsion A-module M (Exercise 24 (c)) is the direct sum of a submodule M_a which has property $\boldsymbol{P}$ and a submodule M'_a such that the restriction to M'_a of the homothety a_M is bijective (observe that, if N is a torsion A-module and a_N is injective, then a_N is bijective; for this, reduce it to the case where N is monogenous and use Exercise 24 (a) and also *Algebra*, Chapter VIII, § 2, no. 2, Lemma 2).

(c) Let S be the set of elements $s \in$ A whose canonical image in the ring A/$\mathfrak{a}$ is invertible. Show that S is a multiplicative subset of A and that the following conditions, for any left ideal $\mathfrak{l}$ of A, are equivalent:

(α) $\mathfrak{l} \cap S \neq \varnothing$; (β) $\mathfrak{l} + \mathfrak{a} = A$; (γ) $(A/\mathfrak{l})_{\mathfrak{a}} = 0$ (in the notation of (b)); (δ) for all $x \in$ A, there exists $s \in$ S such that $sx \in \mathfrak{l}$.

Deduce that A has a (left or right) ring of fractions with respect to S (Chapter II, § 2, Exercise 22), which is a principal ideal ring and whose only non-zero two-sided ideals are generated by the canonical images of the ideals $\mathfrak{a}^n$; moreover the canonical mapping of A to this ring of fractions is injective.

(d) Suppose now that the two-sided ideal $\mathfrak{a}$ is *maximal*. Show that, for every integer $n > 0$, the ring $A/\mathfrak{a}^n$ is isomorphic to a matrix ring $\mathbf{M}_r(B_n)$ over a completely primary ring (*Algebra*, Chapter VIII, § 6, Exercise 20). If $\mathfrak{b}_n$ is the maximal ideal of B_n, show that $\mathfrak{b}_n^n = 0$, that every (left or right) ideal of B_n is of the form $\mathfrak{b}_n^k$ and is monogenous. (Note on the one hand that

$$\mathfrak{a}/\mathfrak{a}^n = \mathbf{M}_r(\mathfrak{b}_n)$$

(*Algebra*, Chapter VIII, § 6, Exercise 5) and on the other that, for $k < n$, $\mathfrak{b}_n^k/\mathfrak{b}_n^{k+1}$ is necessarily a simple B_n-module, without which $\mathbf{M}_r(\mathfrak{b}_n^k)$ could not be a monogenous $(A/\mathfrak{a}^n)$-module.)

Deduce that the completion $\hat{A}$ of A with respect to the topology $\mathcal{T}_P(A_s)$ is a matrix ring $\mathbf{M}_r(B)$ over a ring B with no divisors of zero and each of whose ideals (left or right) is a power of the same maximal two-sided ideal.

26. Let B be a commutative ring and A a complete Noetherian semi-local subring of B. Let $\mathfrak{n}$ be an ideal of B containing a power of the Jacobson radical of A and such that the $\mathfrak{n}$-adic topology on B is Hausdorff. Then show that the topology on A is induced by the $\mathfrak{n}$-adic topology on B (use Proposition 8 of § 2, no. 7).

¶ 27. Let A be a ring, B a commutative A-algebra which is a Zariski ring and N a finitely generated B-module.

(a) Suppose that, for an ideal $\mathfrak{I}$ of B contained in the Jacobson radical of B, the A-modules $N/\mathfrak{I}^{n+1}N$ are flat for $n \geqslant 0$. Show that N is a flat A-module. (If $v: M \to M'$ is an injective homomorphism of *finitely generated* A-modules, it is necessary to prove that $u = v \otimes 1: M \otimes_A N \to M' \otimes_A N$ is injective. Reduce it to proving that, if we write $N_n = N/\mathfrak{I}^{n+1}N$ and $u_n = v \otimes 1_{N_n}$, the homomorphism $\varprojlim u_n$ is injective, using the fact that the $\mathfrak{I}$-adic topologies on $M \otimes_A N$ and $M' \otimes_A N$ are Hausdorff.)

(b) Let b be an element contained in the Jacobson radical of B and such that the homothety with ratio b on N is injective. Show that, if N/bN is a flat A-module, N is a flat A-module (reduce it to the case in (a)).

(c) Suppose further that A is a local ring with maximal ideal $\mathfrak{m}$ and residue field $k = A/\mathfrak{m}$ and that $\mathfrak{m}B$ is contained in the Jacobson radical of B. Let P be a B-module which is a flat A-module and $u: N \to P$ a homomorphism such that $u \otimes 1_k: N \otimes_A k \to P \otimes_A k$ is injective. Then show that N is a flat A-module and that u is injective. (Reduce it to showing that, for every finitely generated A-module M, the homomorphism $u \otimes 1_M: N \otimes_A M \to P \otimes_A M$ is injective; observe that the $\mathfrak{m}$-adic topology on $N \otimes_A M$ is Hausdorff; then use § 2, no. 8, Corollary 1 to Theorem 1, considering the commutative diagram

$$\begin{array}{ccc}
\mathrm{gr}_0(N) \otimes_k \mathrm{gr}(M) & \xrightarrow{\mathrm{gr}_0(u) \otimes 1} & \mathrm{gr}_0(P) \otimes_k \mathrm{gr}(M) \\
\downarrow & & \downarrow \\
\mathrm{gr}(N \otimes_A M) & \xrightarrow[\mathrm{gr}(u \otimes 1)]{} & \mathrm{gr}(P \otimes_A M)
\end{array}$$

and using the flatness of P, the vertical arrows being the canonical homomorphisms of § 2, Exercise 6.)

§ 4

1. Let A be a commutative ring which is Hausdorff and complete with a filtration $(\mathfrak{a}_n)_{n\geqslant 0}$ such that $\mathfrak{a}_0 = A$. Let M, M′, N be three A-modules, each with the filtration induced by that on A and the topology defined by that filtration; we write $\overline{M} = M/\mathfrak{a}_1 M$, $\overline{M}' = M'/\mathfrak{a}_1 M'$, $\overline{N} = N/\mathfrak{a}_1 N$. Let $f: M \times M' \to N$ be a bilinear mapping and $\bar{f}: \overline{M} \times \overline{M}' \to \overline{N}$ the $(A/\mathfrak{a}_1)$-bilinear mapping derived from f by taking quotients. Let $y \in N$, $\bar{x} \in \overline{M}$, $\bar{x}' \in \overline{M}'$ be such that: (1) $\bar{f}(\bar{x}, \bar{x}')$ is the class $\bar{y}$ of y in $\overline{N}$; (2) every element of $\overline{N}$ can be written as

$$\bar{f}(\bar{x}, \bar{z}') + \bar{f}(\bar{z}, \bar{x}'),$$

where $\bar{z} \in \overline{M}$ and $\bar{z}' \in \overline{M}'$. Show that, if N is Hausdorff and M and M′ are complete, there exists $x \in \bar{x}$ and $x' \in \bar{x}'$ such that $f(x, x') = y$ (argue by induction as in the proof of Hensel's lemma). Under what conditions are x and x' determined uniquely?

¶ 2. Let A be a local ring, $\mathfrak{m}$ its maximal ideal, $k = A/\mathfrak{m}$ its residue field and $f: A \to k$ the canonical homomorphism. Let $P \in A[X]$ be a monic polynomial of degree n. We write $B = A[X]/P.A[X]$ and denote by x the canonical image of X in B.

(a) Let Q, Q′ be two strongly relatively prime monic polynomials in A[X] such that $P = QQ'$. Show that the ring B is the direct sum of the ideals $B.Q(x)$ and $B.Q'(x)$.

(b) Conversely, let $B = \mathfrak{b} \oplus \mathfrak{b}'$ be a decomposition of B as a direct sum of two ideals. Show that there exist polynomials Q, Q′ in A[X] satisfying the hypotheses of (a) and such that $\mathfrak{b} = B.Q(x)$, $\mathfrak{b}' = B.Q'(x)$. (Show first that $\mathfrak{b}/\mathfrak{m}\mathfrak{b}$ and $\mathfrak{b}'/\mathfrak{m}\mathfrak{b}'$ are generated by the images in $B/\mathfrak{m}B$ of monic polynomials $Q_0 \in \overset{-1}{\bar{f}}(\mathfrak{b})$, $Q'_0 \in \overset{-1}{\bar{f}}(\mathfrak{b}')$ such that

$$\bar{f}(P) = \bar{f}(Q_0)\bar{f}(Q'_0).$$

Let $r = \deg(Q_0)$, $s = \deg(Q'_0)$. Show that $\mathfrak{b}$ is a free A-module with basis $Q_0(x), xQ_0(x, \ldots, x^{s-1}Q_0(x)$ (apply Chapter II, § 3, no. 2, Proposition 5); we may then write $x^s Q_0(x) = a_0 Q_0(x) + a_1 x Q_0(x) + \cdots + a_{s-1}x^{s-1}Q_0(x)$, where $a_i \in A$ $(0 \leqslant i \leqslant s-1)$; show that, if we write

$$Q'(X) = X^s - (a_0 + a_1 X + \cdots + a_{s-1}X^{s-1}),$$

then $\bar{f}(P) = \bar{f}(Q_0)\bar{f}(Q')$ and $Q' \in \overset{-1}{\bar{f}}(\mathfrak{b}')$. Similarly define Q starting with Q′ and Q_0 and show that Q and Q′ solve the problem, using Proposition 2 of no. 1.)

¶ 3. Let A be a local ring, $\mathfrak{m}$ its maximal ideal, $k = A/\mathfrak{m}$ its residue field and $f: A \to k$ the canonical homomorphism. Show that the two following conditions are equivalent:

(H) For every monic polynomial $P \in A[X]$ and every decomposition of $\bar{f}(P) \in k[X]$ as a product $\bar{f}(P) = \bar{Q}.\bar{Q}'$ of relatively prime monic polynomials, there exist two monic polynomials Q, Q′ in A[X] such that $\bar{f}(Q) = \bar{Q}$, $\bar{f}(Q') = \bar{Q}'$ and $P = QQ'$.

(C) Every commutative algebra over A which is a finitely generated A-module is a direct composition of A-algebras which are *local* rings.

(To prove that (H) implies (C), argue as in Proposition 8 of no. 6, using Exercise 2 (a). To see that (C) implies (H), show first that, for every commutative A-algebra B which is a finitely generated A-module, every decomposition of $B/\mathfrak{m}B$ as a direct sum of two ideals is necessarily of the form $\mathfrak{b}/\mathfrak{m}\mathfrak{b} \oplus \mathfrak{b}'/\mathfrak{m}\mathfrak{b}'$, where $B = \mathfrak{b} \oplus \mathfrak{b}'$ is a decomposition of B as a direct sum of two ideals; then use Exercise 2 (b).)

A local ring satisfying conditions (H) and (C) is called *Henselian*. Every *complete Hausdorff* local ring is Henselian. If A is Henselian and B is a commutative A-algebra which is a local ring and a finitely generated A-module, then B is Henselian.

4. (a) Let $(A_\alpha, \phi_{\alpha\beta})$ be a direct system of Henselian local rings, the homomorphisms $\phi_{\alpha\beta}$ being *local*. Show that the local ring $A = \varinjlim A_\alpha$ (Chapter II, § 3, Exercise 16) is Henselian (use criterion (H) of Exercise 3).

(b) Let K be a commutative field and L an algebraic extension of K. Deduce from (a) that the ring $L \otimes_K K[[X_1, \ldots, X_n]]$ is Henselian.

* (c) Let A be a Henselian local ring and B a commutative A-algebra which is *integral* over A (Chapter V) and is a local ring. Show that B is a Henselian ring (use (a)). *

¶ 5. Let A be a Henselian local ring, B a (not necessarily commutative) A-algebra which is finitely generated A-module, $\mathfrak{b}$ a two-sided ideal of B and let $\bar{B} = B/\mathfrak{b}$.

(a) Show that every idempotent ε in $\bar{B}$ is the canonical image of an idempotent in B (reduce it to the commutative case, considering the subalgebra of B generated by a single element).

(b) Let $(\varepsilon_n)_{n \geq 1}$ be an infinite sequence of elements of $\bar{B}$ such that

$$\varepsilon_i \varepsilon_j = \delta_{ij} \varepsilon_j$$

for every ordered pair of indices (i, j). Show that there exists in B an orthogonal sequence $(e_n)_{n \geq 1}$ of idempotents such that ε_n is the canonical image of e_n for all n. (Argue by induction as in *Algebra*, Chapter VIII, § 6, Exercise 10; observe that, if e, e' are two idempotents of B such that $ee' = 0$, $e' - e'e = e''$ is an idempotent such that $ee'' = e''e = 0$.)

(c) Suppose now that $\mathfrak{b}$ is the *Jacobson radical* of B. Let n be an integer and (ε_{ij}) $(1 \leqslant i \leqslant n, 1 \leqslant j \leqslant n)$ a family of elements of $\overline{\mathrm{B}}$ such that $\varepsilon_{ij}\varepsilon_{hk} = \delta_{jh}\varepsilon_{ik}$ and $1 = \sum_{i=1}^{n} \varepsilon_{ii}$. Show that there exists in B a family (e_{ij}) $(1 \leqslant i \leqslant n, 1 \leqslant j \leqslant n)$ such that $e_{ij}e_{hk} = \delta_{jh}e_{ik}$ and $1 = \sum_{i=1}^{n} e_{ii}$ and that ε_{ij} is the canonical image of e_{ij} for every ordered pair of indices. (Use (b), Exercise 11 of *Algebra*, Chapter VIII, § 6, and Exercise 9 of *Algebra*, Chapter VIII, § 1.) Deduce that, if $\overline{\mathrm{B}}$ is isomorphic to a matrix ring $\mathbf{M}_r(\overline{\mathrm{D}})$, where $\overline{\mathrm{D}}$ is a not necessarily commutative field, then B is isomorphic to a matrix ring $\mathbf{M}_r(\mathrm{D})$, where D is an A-algebra which is a finitely generated A-module and whose Jacobson radical $\mathfrak{d}$ is such that $\mathrm{D}/\mathfrak{d}$ is isomorphic to $\overline{\mathrm{D}}$ (cf. *Algebra*, Chapter VIII, § 1, Exercises 9 and 3). Show that, if further B is an Azumaya algebra over A (Chapter II, § 5, Exercise 14), so is D.

6. Give an example of a non-commutative Artinian ring A, which is filtered by a sequence $(\mathfrak{a}_n)$ of two-sided ideals such that $\mathfrak{a}_0 = \mathrm{A}$ and $\mathfrak{a}_2 = 0$ and for which Proposition 8 of no. 6 does not hold (cf. *Algebra*, Chapter VIII, § 2, Exercise 6).

¶ 7. (a) Let A be a commutative ring and $\mathfrak{m}$ an ideal of A such that A is Hausdorff and complete with the $\mathfrak{m}$-adic topology and $\mathfrak{m}/\mathfrak{m}^2$ is a finitely generated A-module. Show that the topology on $\mathrm{A}' = \mathrm{A}\{\mathrm{X}_1, \ldots, \mathrm{X}_r\}$ is the $\mathfrak{m}'$-adic topology, where $\mathfrak{m}' = \mathfrak{m}\mathrm{A}'$ and that $\mathfrak{m}'/\mathfrak{m}'^2$ is a finitely generated $(\mathrm{A}'/\mathfrak{m}')$-module (use Proposition 14 of § 2, no. 11). In particular, if A is Noetherian, so is A'.

(b) Let A be a commutative Noetherian ring and $\mathfrak{m}$ an ideal of A such that A is Hausdorff and complete with the $\mathfrak{m}$-adic topology. Let $u\colon \mathrm{A} \to \mathrm{B}$ be a continuous homomorphism from A to a Hausdorff commutative topological ring B making B into an A-algebra. Show that the following conditions are equivalent:

(α) B is Noetherian, its topology is the $\mathfrak{m}\mathrm{B}$-adic topology, B is complete and $\mathrm{B}/\mathfrak{m}\mathrm{B}$ is a finitely generated algebra over $\mathrm{A}/\mathfrak{m}$.

(β) B is topologically A-isomorphic to $\varprojlim \mathrm{B}_n$, where $(\mathrm{B}_n)_{n \geqslant 1}$ is an inverse system of discrete A-algebras such that the mappings $\phi_{nm}\colon \mathrm{B}_m \to \mathrm{B}_n$ for $m \geqslant n$ are surjective, the kernel of ϕ_{nm} is $\mathfrak{m}^{n+1}\mathrm{B}_m$ and B_1 is a finitely generated algebra over $\mathrm{A}/\mathfrak{m}$.

(γ) B is topologically A-isomorphic to a quotient of an algebra of the form $\mathrm{A}\{\mathrm{X}_1, \ldots, \mathrm{X}_r\}$ by a closed ideal.

(To prove that (β) implies (γ), use Proposition 14 of § 2, no. 11 and Theorem 1 of § 2, no. 8.)

8. Let A be a linearly topologized commutative ring. The additive group $A[[X_1, \dots, X_p]]$ is identified with the product group $A^{\mathbf{N}^p}$ and is given the product topology $\mathscr{T}$.

(a) Show that $\mathscr{T}$ is compatible with the ring structure on $A[[X_1, \dots, X_p]]$ and is a linear topology for which the mappings $f \mapsto \partial f/\partial X_i$ are continuous.

(b) For every element $P = \sum_e \alpha_{e,P} X^e$ of $A[[X_1, \dots, X_p]]$ (notation of § 1, no. 6) and every complete Hausdorff topological A-algebra B (§ 2, Exercise 28), an element $\mathbf{x} = (x_1, \dots, x_p) \in B^p$ is said to be *substitutable* in P if the family $(\alpha_{e,P}\mathbf{x}^e)$ (where we have written $\mathbf{x}^e = x_1^{e_1} \dots x_p^{e_p}$ for $e = (e_1, \dots, e_p)$) converges to 0 in B with respect to the filter of complements of finite subsets of $\mathbf{N}^p$. This family is then summable and its sum is denoted by $P(\mathbf{x})$. Show that the set of formal power series $P \in A[[X_1, \dots, X_p]]$ such that $\mathbf{x}$ is substitutable in P is a subring S_x of $A[[X_1, \dots, X_p]]$ and the mapping $P \mapsto P(\mathbf{x})$ is a homomorphism from S_x to B.

(c) Show that, if $\mathbf{x}$ is substitutable in P and $\mathbf{y} = (y_1, \dots, y_p)$ is an element of B^p such that the y_i are topologically nilpotent for $1 \leqslant i \leqslant p$, then $\mathbf{x} + \mathbf{y}$ is substitutable in P.

In particular, if there exists in B a neighbourhood of 0 consisting of topologically nilpotent elements, then, for all $P \in A[[X_1, \dots, X_p]]$, the set D of $\mathbf{x} \in B^p$ which are substitutable in P is open and the mapping $\mathbf{x} \mapsto P(\mathbf{x})$ from D to B is continuous.

(d) Suppose from now on that A is *Hausdorff and complete* and let $A[[X_1, \dots, X_p]]$ be given the topology $\mathscr{T}$. For a system of q formal power series $P_1, \dots, P_q$ of $A[[X_1, \dots, X_p]]$ to be substitutable in a formal power series

$$Q \in A[[X_1, \dots, X_q]],$$

it is necessary and sufficient that the system $(P_1(0), \dots, P_q(0))$ be substitutable in Q.

(e) Suppose that $\mathbf{x}$ is substitutable in each of the P_k $(1 \leqslant k \leqslant q)$ and that the system $(P_1, \dots, P_q)$ is substitutable in Q. Show that $\mathbf{x}$ is substitutable in $Q(P_1, \dots, P_q)$, that the system $(P_1(\mathbf{x}), \dots, P_q(\mathbf{x}))$ is substitutable in Q and that $Q(P_1(\mathbf{x}), \dots, P_q(\mathbf{x})) = ((Q(P_1, \dots, P_q))(\mathbf{x})$ if further *one* of the following hypotheses is assumed to be satisfied: (α) there is in B an ideal $\mathfrak{n}$ such that the ordered pair $(B, \mathfrak{n})$ satisfies Hensel's conditions and the x_i $(1 \leqslant i \leqslant p)$ belong to $\mathfrak{n}$; (β) the formal power series Q is restricted.

(f) Take $B = A = \mathbf{F}_2$ with the discrete topology, $P = X + X^2$, $Q = \sum_{n=0}^{\infty} X^{2^n}$; then P is substitutable in Q, $x = 1$ is substitutable in P and in $Q(P)$ and $P(x)$ is substitutable in Q, but $Q(P(x)) = 0$ and $(Q(P))(x) = 1$.

§ 5

1. Let A be a commutative ring and $\mathfrak{I}$ an ideal of A. An A-module M is called *absolutely Hausdorff* with respect to $\mathfrak{I}$ if, for every finitely generated A-module N, the A-module $N \otimes_A M$ is Hausdorff with the $\mathfrak{I}$-adic topology; an absolutely Hausdorff A-module is ideally Hausdorff.

(a) For M to be absolutely Hausdorff with respect to $\mathfrak{I}$, it is necessary and sufficient that, for every finitely generated A-module N and every submodule N' of N, $\operatorname{Im}(N' \otimes_A M)$ be closed in $N \otimes_A M$ with the $\mathfrak{I}$-adic topology.

(b) Let B be a commutative A-algebra and $\mathfrak{L}$ an ideal of B containing $\mathfrak{I}B$. If M is an absolutely Hausdorff B-module with respect to $\mathfrak{L}$, M is an absolutely Hausdorff A-module with respect to $\mathfrak{I}$.

2. Show that every **Z**-module which is Hausdorff with the p-adic topology (p a prime number) is ideally Hausdorff, but give an example of a finitely generated **Z**-module which is Hausdorff but not absolutely Hausdorff with respect to p (use Exercise 1 (a)).

3. With the notation of Exercise 11 of § 3, let N be the submodule of $\hat{E}$ which is the closure in $\hat{E}$ of the submodule of $\hat{E}$ generated by the vectors $pe_{2n-1} - p^n e_{2n}$ $(n \geqslant 1)$ and let $M = \hat{E}/N$. Show that under the p-adic topology the submodule pM of M is not closed in M and deduce that M is a $\mathbf{Z}_p$-module which is ideally Hausdorff but not absolutely Hausdorff with respect to p.

¶ 4. Let A be a commutative ring, $\mathfrak{I}$ an ideal of A and $S = 1 + \mathfrak{I}$. Show that, if M is an absolutely Hausdorff A-module with respect to $\mathfrak{I}$, then $S^{-1}M = M$, in other words, for all $a \in S$, $x \mapsto ax$ is a bijection of M onto itself. (Show first that the hypothesis that M is Hausdorff with the $\mathfrak{I}$-adic topology implies that $x \mapsto ax$ is injective; then prove that the submodule aM of M is dense in M with the $\mathfrak{I}$-adic topology and use Exercise 1 (a).)

5. Take $A = \mathbf{Z}$ and $\mathfrak{I} = p\mathbf{Z}$, where p is prime.

(a) Show that, if q is a prime number distinct from p, the **Z**-module $\mathbf{Z}/q\mathbf{Z}$ satisfies property (ii) of Theorem 1, but not property (i).

(b) Show that the **Z**-module $\mathbf{Q}/\mathbf{Z}$ satisfies property (iv) of Theorem 1 but not property (ii).

6. Let A be a commutative Noetherian ring, $\mathfrak{I}$ an ideal of A and M an A-module. Show that condition (v) of Theorem 1 is equivalent to the following:

(v') For every finitely generated A-module N and every submodule N' of N, the canonical mapping $(M \otimes_A N')^\wedge \to (M \otimes_A N)^\wedge$ (where the two sides are the Hausdorff completions with respect to the $\mathfrak{I}$-adic topologies on $M \otimes_A N'$ and $M \otimes_A N$ respectively) is injective. (To prove that (v) implies (v'), argue as in part (E) of the proof of Theorem 1. To see that (v') implies (v), consider A-modules N annihilated by a power of $\mathfrak{I}$.)

¶ 7. Let $(A_\lambda, f_{\mu\lambda})$ be a directed direct system of Noetherian local rings; if $\mathfrak{m}_\lambda$ is the maximal ideal of A_λ, suppose that, for $\lambda \leqslant \mu$, $\mathfrak{m}_\mu = \mathfrak{m}_\lambda A_\mu$ and that A_μ is a flat A_λ-module. Show then that $A = \varinjlim A_\lambda$ is Noetherian and a flat A_λ-module for all λ. (If $\mathfrak{m} = \mathfrak{m}_\lambda A$ is the maximal ideal of A (Chapter II, § 3, Exercise 16), show that on A the $\mathfrak{m}$-adic topology is Hausdorff, observing that A is a faithfully flat A_λ-module for all λ. Then prove that, if $\hat{A}$ is the completion of A with respect to the $\mathfrak{m}$-adic topology, $\hat{A}$ is Noetherian, using § 2, no. 10, Corollary 5 to Theorem 2; finally, prove that, for all λ, $\hat{A}$ is a flat A_λ-module using Theorem 1 of no. 2 and Proposition 2 of no. 3.)

¶ 8. Let A be a Noetherian local ring, $\mathfrak{m}$ its maximal ideal and $k = A/\mathfrak{m}$ its residue field. Let K be an extension of k; show that there exists a local homomorphism from A to a Noetherian local ring B such that $B/\mathfrak{m}B$ is isomorphic to K and B is a flat A-module. (Consider first the case where $K = k(t)$, distinguishing two cases according to whether t is algebraic or transcendental over k; consider next a family (K_λ) of subfields of K containing k, which is well ordered by inclusion and such that, if K_λ has a predecessor K_μ, $K_\lambda = K_\mu(t_\mu)$ for some $t_\mu \in K_\mu$. Finally, apply Exercise 7.)

CHAPTER IV(*)

Associated Prime Ideals and Primary Decomposition

All the rings considered in this chapter are assumed to be commutative and to possess a unit element; all the ring homomorphisms are assumed to map unit element to unit element. By a subring of a ring A *we shall mean a subring containing the unit element of* A.

Recall that for every A*-module* E *and all* $x \in E$, $\mathrm{Ann}(x)$ *denotes the annihilator of* x, *the set of* $a \in A$ *such that* $ax = 0$.

1. PRIME IDEALS ASSOCIATED WITH A MODULE

1. DEFINITION OF ASSOCIATED PRIME IDEALS

DEFINITION 1. *Let* M *be a module over a ring* A. *A prime ideal* $\mathfrak{p}$ *is said to be associated with* M *if there exists* $x \in M$ *such that* $\mathfrak{p}$ *is equal to the annihilator of* x. *The set of prime ideals associated with* M *is denoted by* $\mathrm{Ass}_A(M)$, *or simply* $\mathrm{Ass}(M)$.

* *Example.* Let $\mathfrak{a}$ be an ideal in the polynomial ring $A = \mathbf{C}[X_1, \dots, X_r]$, V the corresponding affine algebraic variety and $V_1, \dots, V_p$ the irreducible components of V. If M is taken to be the ring $A/\mathfrak{a}$ of functions which are regular on V, the set of prime ideals associated with M consists of the ideals of $V_1, \dots, V_p$ and in general other prime ideals each of which contains one of the ideals of the V_i. *

As the annihilator of 0 is A, an element $x \in M$ whose annihilator is a prime ideal is necessarily $\neq 0$. To say that a prime idea $\mathfrak{p}$ is associated with M

(*) The results of this chapter depend only on Books I to VI and Chapters I to III of this Book, excluding Chapter I, § 4 and Chapter III, § 5.

amounts to saying that M contains a submodule *isomorphic to* $A/\mathfrak{p}$ (namely Ax, for all $x \in M$ whose annihilator is $\mathfrak{p}$).

If an A-module M is the union of a family $(M_\iota)_{\iota \in I}$ of submodules, then clearly

$$\text{(1)} \qquad \operatorname{Ass}(M) = \bigcup_{\iota \in I} \operatorname{Ass}(M_\iota).$$

PROPOSITION 1. *For every prime ideal* $\mathfrak{p}$ *of a ring* A *and every submodule* $M \neq 0$ *of* $A/\mathfrak{p}$, $\operatorname{Ass}(M) = \{\mathfrak{p}\}$.

As the ring $A/\mathfrak{p}$ is an integral domain, the annihilator of an element $\neq 0$ of $A/\mathfrak{p}$ is $\mathfrak{p}$.

PROPOSITION 2. *Let* M *be a module over a ring* A. *Every maximal element of the set of ideals* $\operatorname{Ann}(x)$ *of* A, *where* x *runs through the set of elements* $\neq 0$ *of* M, *belongs to* $\operatorname{Ass}(M)$.

Let $\mathfrak{a} = \operatorname{Ann}(x)$ ($x \in M$, $x \neq 0$) be such a maximal element; it is sufficient to show that $\mathfrak{a}$ is prime. As $x \neq 0$, $\mathfrak{a} \neq A$. Let b, c be elements of A such that $bc \in \mathfrak{a}$ and $c \notin \mathfrak{a}$. Then $cx \neq 0$, $b \in \operatorname{Ann}(cx)$ and $\mathfrak{a} \subset \operatorname{Ann}(cx)$. As $\mathfrak{a}$ is maximal, $\operatorname{Ann}(cx) = \mathfrak{a}$, whence $b \in \mathfrak{a}$, so that $\mathfrak{a}$ is prime.

COROLLARY 1. *Let* M *be a module over a Noetherian ring* A. *Then the condition* $M \neq \{0\}$ *is equivalent to* $\operatorname{Ass}(M) \neq \varnothing$.

If $M = \{0\}$, clearly $\operatorname{Ass}(M)$ is empty (without any hypothesis on A). If $M \neq \{0\}$, the set of ideals of the form $\operatorname{Ann}(x)$, where $x \in M$ and $x \neq 0$, is non-empty and consists of ideals $\neq A$; as A is Noetherian, this set has a maximal element; then it suffices to apply Proposition 2.

COROLLARY 2. *Let* A *be a Noetherian ring,* M *an* A*-module and* a *an element of* A. *For the homothety on* M *with ratio* a *to be injective, it is necessary and sufficient that* a *belong to no prime ideal associated with* M.

If a belongs to a prime ideal $\mathfrak{p} \in \operatorname{Ass}(M)$, then $\mathfrak{p} = \operatorname{Ann}(x)$, where $x \in M$, $x \neq 0$; whence $ax = 0$ and the homothety with ratio a is not injective. Conversely, if $ax = 0$ for some $x \in M$ such that $x \neq 0$, then $Ax \neq \{0\}$, whence $\operatorname{Ass}(x) \neq \varnothing$ (Corollary 1). Let $\mathfrak{p} \in \operatorname{Ass}(Ax)$; then obviously $\mathfrak{p} \in \operatorname{Ass}(M)$ and $\mathfrak{p} = \operatorname{Ann}(bx)$, where $b \in A$; whence $a \in \mathfrak{p}$, since $abx = 0$.

COROLLARY 3. *The set of divisors of zero in a Noetherian ring* A *is the union of the ideals* $\mathfrak{p} \in \operatorname{Ass}(A)$.

PROPOSITION 3. *Let* A *be a ring,* M *an* A*-module and* N *a submodule of* N. *Then*

$$\text{(2)} \qquad \operatorname{Ass}(N) \subset \operatorname{Ass}(M) \subset \operatorname{Ass}(N) \cup \operatorname{Ass}(M/N).$$

The inclusion $\operatorname{Ass}(N) \subset \operatorname{Ass}(M)$ is obvious. Let $\mathfrak{p} \in \operatorname{Ass}(M)$, E be a submodule of M isomorphic to $A/\mathfrak{p}$ and $F = E \cap N$. If $F = \{0\}$, E is isomorphic to a submodule of M/N, whence $\mathfrak{p} \in \mathrm{Ass}(M/N)$. If $F \neq \{0\}$, the annihilator of every element $\neq 0$ of F is $\mathfrak{p}$ (Proposition 1) and hence $\mathfrak{p} \in \mathrm{Ass}(F) \subset \mathrm{Ass}(N)$.

COROLLARY 1. *If an A-module M is the direct sum of a family* $(M_\iota)_{\iota \in I}$ *of submodules, then* $\mathrm{Ass}(M) = \bigcup_{\iota \in I} \mathrm{Ass}(M_\iota)$.

It may be reduced to the case where I is finite by means of (1), then to the case where $\mathrm{Card}(I) = 2$ by induction on $\mathrm{Card}(I)$. Then let $I = \{i, j\}$, where $i \neq j$; as M/M_ι is isomorphic to M_j, $\mathrm{Ass}(M) \subset \mathrm{Ass}(M_i) \cup \mathrm{Ass}(M_j)$ (Proposition 3); moreover, $\mathrm{Ass}(M_\iota)$ and $\mathrm{Ass}(M_j)$ are contained in $\mathrm{Ass}(M)$ (Proposition 3), whence the result.

COROLLARY 2. *Let M be an A-module and* $(Q_\iota)_{\iota \in I}$ *a finite family of submodules of M. If* $\bigcap_{\iota \in I} Q_\iota = \{0\}$, *then*

$$\mathrm{Ass}(M) \subset \bigcup_{\iota \in I} \mathrm{Ass}(M/Q_\iota).$$

The canonical mapping $M \to \bigoplus_{\iota \in I} (M/Q_\iota)$ is injective; then it suffices to apply Proposition 3 and its Corollary 1.

PROPOSITION 4. *Let M be an A-module and* Φ *a subset of* $\mathrm{Ass}(M)$. *Then there exists a submodule N of M such that* $\mathrm{Ass}(N) = \mathrm{Ass}(M) - \Phi$ *and* $\mathrm{Ass}(M/N) = \Phi$.

Let $\mathfrak{E}$ be the set of submodules P of M such that $\mathrm{Ass}(P) \subset \mathrm{Ass}(M) - \Phi$. Formula (1) shows that the set $\mathfrak{E}$, ordered by inclusion, is *inductive*; moreover, $\{0\} \in \mathfrak{E}$ and hence $\mathfrak{E} \neq \varnothing$. Let N be a maximal element of $\mathfrak{E}$. Then $\mathrm{Ass}(N) \subset \mathrm{Ass}(M) - \Phi$. We shall see that $\mathrm{Ass}(M/N) \subset \Phi$, which, by Proposition 3, will complete the proof. Let $\mathfrak{p} \in \mathrm{Ass}(M/N)$; then M/N contains a submodule F/N isomorphic to A/$\mathfrak{p}$. By Propositions 1 and 3, $\mathrm{Ass}(F) \subset \mathrm{Ass}(N) \cup \{\mathfrak{p}\}$. Since N is maximal in $\mathfrak{E}$, $F \notin \mathfrak{E}$ and hence $\mathfrak{p} \in \Phi$.

2. LOCALIZATION OF ASSOCIATED PRIME IDEALS

PROPOSITION 5. *Let A be a ring, S a multiplicative subset of A,* Φ *the set of prime ideals of A which do not meet S and M an A-module. Then:*

(i) *The mapping* $\mathfrak{p} \mapsto S^{-1}\mathfrak{p}$ *is a bijection of* $\mathrm{Ass}_A(M) \cap \Phi$ *onto a subset of* $\mathrm{Ass}_{S^{-1}A}(S^{-1}M)$.

(ii) *If* $\mathfrak{p} \in \Phi$ *is a finitely generated ideal and* $S^{-1}\mathfrak{p} \in \mathrm{Ass}_{S^{-1}A}(S^{-1}M)$, *then* $\mathfrak{p} \in \mathrm{Ass}_A(M)$.

Recall (Chapter II, § 2, no. 5, Proposition 11) that the mapping $\mathfrak{p} \mapsto S^{-1}\mathfrak{p}$ is a bijection of Φ onto the set of prime ideals of $S^{-1}A$. If $\mathfrak{p} \in \mathrm{Ass}_A(M) \cap \Phi$, $\mathfrak{p}$ is the annihilator of a monogenous submodule N of M; then $S^{-1}\mathfrak{p}$ is the annihilator of the monogenous submodule $S^{-1}N$ of $S^{-1}M$ (Chapter II, § 2, no. 4, formula

(9)) and hence $S^{-1}\mathfrak{p} \in \mathrm{Ass}_{S^{-1}A}(S^{-1}M)$. Conversely, suppose that $\mathfrak{p} \in \Phi$ is finitely generated and such that $S^{-1}\mathfrak{p}$ is associated with $S^{-1}M$; then there exists $x \in M$ and $t \in S$ such that $S^{-1}\mathfrak{p}$ is the annihilator of x/t. Let $(a_i)_{1 \leqslant i \leqslant n}$ be a system of generators of $\mathfrak{p}$; then $(a_i/1)(x/t) = 0$ and hence there exists $s_i \in S$ such that $s_i a_i x = 0$ $(1 \leqslant i \leqslant n)$. Let us write $s = s_1 s_2 \ldots s_n$; for all $a \in \mathfrak{p}$, $sax = 0$, whence $\mathfrak{p} \subset \mathrm{Ann}(sx)$; on the other hand, if $b \in A$ satisfies $bsx = 0$, then $b/1 \in S^{-1}\mathfrak{p}$ by definition, whence $b \in \mathfrak{p}$. Then $\mathfrak{p} = \mathrm{Ann}(sx)$ and $\mathfrak{p} \in \mathrm{Ass}_A(M)$.

COROLLARY. *If the ring* A *is Noetherian, the mapping* $\mathfrak{p} \mapsto S^{-1}\mathfrak{p}$ *is a bijection of* $\mathrm{Ass}_A(M) \cap \Phi$ *onto* $\mathrm{Ass}_{S^{-1}A}(S^{-1}M)$.

If A is not Noetherian, the mapping $\mathfrak{p} \mapsto S^{-1}\mathfrak{p}$ of $\mathrm{Ass}_A(M) \cap \Phi$ to $\mathrm{Ass}_{S^{-1}A}(S^{-1}M)$ is not necessarily surjective (Exercise 1).

PROPOSITION 6. *Let* A *be a Noetherian ring,* M *an* A*-module,* S *a multiplicative subset of* A *and* Ψ *the set of elements of* $\mathrm{Ass}_A(M)$ *which do not meet* S. *Then the kernel* N *of the canonical mapping* $M \to S^{-1}M$ *is the unique submodule of* M *which satisfies the relations*

$$\mathrm{Ass}(N) = \mathrm{Ass}(M) - \Psi, \qquad \mathrm{Ass}(M/N) = \Psi. \tag{3}$$

By Proposition 4 of no. 1, there exists a submodule N′ of M which satisfies the relations $\mathrm{Ass}(N') = \mathrm{Ass}(M) - \Psi$ and $\mathrm{Ass}(M/N') = \Psi$. We need to prove $N' = N$. Consider the commutative diagram

$$\begin{array}{ccc} M & \xrightarrow{p} & M/N' \\ {\scriptstyle u}\downarrow & & \downarrow{\scriptstyle v} \\ S^{-1}M & \xrightarrow[S^{-1}p]{} & S^{-1}(M/N') \end{array}$$

where p, u, v are the canonical homomorphisms. We shall show that $S^{-1}p$ and v are injective, which will prove that u and p have the same kernel and hence $N' = N$.

As $\mathrm{Ass}(N') \cap \Psi = \varnothing$, every element of $\mathrm{Ass}(N')$ meets S. Then $\mathrm{Ass}_{S^{-1}A}(S^{-1}N') = \varnothing$ (Corollary to Proposition 5), whence $S^{-1}N' = \{0\}$ (no. 1, Corollary 1 to Proposition 2), which proves that $S^{-1}p$ is injective (Chapter II, § 2, no. 4, Theorem 1). On the other hand, if x belongs to the kernel K of v, then $\mathrm{Ann}(x) \cap S \neq \varnothing$ (Chapter II, § 2, no. 2, Proposition 4); hence $\mathrm{Ass}(K) = \varnothing$ since $\mathrm{Ass}(K) \subset \mathrm{Ass}(M/N') = \Psi$; we deduce that $K = \{0\}$ (no. 1, Corollary 1 to Proposition 2) and v is injective.

3. RELATIONS WITH THE SUPPORT

Let M be a module over a ring A. Recall that the set of prime ideals $\mathfrak{p}$ of A such that $M_{\mathfrak{p}} \neq 0$ is called the *support* of M and is denoted by Supp(M) (Chapter II, § 4, no. 4, Definition 5).

PROPOSITION 7. *Let* A *be a ring and* M *an* A-module.

(i) *Every prime ideal* $\mathfrak{p}$ *of* A *containing an element of* Ass(M) *belongs to* Supp(M).

(ii) *Conversely, if* A *is Noetherian, every ideal* $\mathfrak{p} \in$ Supp(M) *contains an element of* Ass(M).

If $\mathfrak{p}$ contains an element $\mathfrak{q}$ of Ass(M), then $\mathfrak{q} \cap (A - \mathfrak{p}) = \varnothing$ and hence, if we write $S = A - \mathfrak{p}$, $S^{-1}\mathfrak{p}$ is a prime ideal associated with $S^{-1}M = M_{\mathfrak{p}}$ (no. 2, Proposition 5) and *a fortiori* $M_{\mathfrak{p}} \neq 0$, hence $\mathfrak{p} \in$ Supp(M). Conversely, if A is Noetherian, so is $A_{\mathfrak{p}}$ (Chapter II, § 2, no. 4, Corollary 2 to Proposition 10). If $M_{\mathfrak{p}} \neq 0$, then $\text{Ass}_{A_{\mathfrak{p}}}(M_{\mathfrak{p}}) \neq 0$ (no. 1, Corollary 1 to Proposition 2) and hence there exists $\mathfrak{q} \in \text{Ass}_A(M)$ such that $\mathfrak{q} \cap (A - \mathfrak{p}) = \varnothing$ (no. 2, Corollary to Proposition 5).

COROLLARY 1. *If* M *is a module over a Noetherian ring, then* Ass(M) ⊂ Supp(M) *and these two sets have the same minimal elements.*

COROLLARY 2. *The nilradical of a Noetherian ring* A *is the intersection of the ideals* $\mathfrak{p} \in$ Ass(A).

We know that the nilradical of A is the intersection of the minimal elements of Spec(A) = Supp(A) (Chapter II, § 2, no. 6, Proposition 13).

4. THE CASE OF FINITELY GENERATED MODULES OVER A NOETHERIAN RING

THEOREM 1. *Let* A *be a Noetherian ring and* M *a finitely generated* A-module. *There exists a composition series* $(M_i)_{0 \leqslant i \leqslant n}$ *of* M *such that, for* $0 \leqslant i \leqslant n - 1$, M_i/M_{i+1} *is isomorphic to* $A/\mathfrak{p}_i$, *where* $\mathfrak{p}_i$ *is a prime ideal of* A.

Let $\mathfrak{E}$ be the set of submodules of M which have a composition series with the property of the statement. As $\mathfrak{E}$ is non-empty (for {0} belongs to $\mathfrak{E}$) and M is Noetherian, $\mathfrak{E}$ has a maximal element N. If $M \neq N$, then $M/N \neq 0$ and hence Ass(M/N) $\neq \varnothing$ (no. 1, Corollary 1 to Proposition 2); M/N therefore contains a submodule N'/N isomorphic to an A-module of the form $A/\mathfrak{p}$, where $\mathfrak{p}$ is prime; then by definition $N' \in \mathfrak{E}$, which contradicts the maximal character of N. Then necessarily $N = M$.

THEOREM 2. *Let* M *be a finitely generated module over a Noetherian ring* A *and* $(M_i)_{0 \leqslant i \leqslant n}$ *a composition series of* M *such that, for* $0 \leqslant i \leqslant n - 1$, M_i/M_{i+1} *is isomorphic to* $A/\mathfrak{p}_i$ *where* $\mathfrak{p}_i$ *is a prime ideal of* A. *Then*

(4) $$\text{Ass}(M) \subset \{\mathfrak{p}_0, \ldots, \mathfrak{p}_{n-1}\} \subset \text{Supp}(M);$$

the minimal elements of these three sets are the same and coincide with the minimal elements of the set of prime ideals containing Ann(M).

The inclusion $\mathrm{Ass}(M) \subset \{\mathfrak{p}_0, \ldots, \mathfrak{p}_{n-1}\}$ follows immediately from Propositions 1 and 3 of no. 1. For $0 \leqslant i \leqslant n-1$,

$$\mathfrak{p}_i \in \mathrm{Supp}(A/\mathfrak{p}_i) = \mathrm{Supp}(M_i/M_{i+1})$$

(Chapter II, § 4, no. 4, *Example*), whence $\mathfrak{p}_i \in \mathrm{Supp}(M_i) \subset \mathrm{Supp}(M)$ (Chapter II, § 4, no. 4, Proposition 16), which shows the inclusion

$$\{\mathfrak{p}_0, \ldots, \mathfrak{p}_{n-1}\} \subset \mathrm{Supp}(M).$$

Corollary 1 to Proposition 7 of no. 3 shows that Ass(M) and Supp(M) have the same minimal elements and (4) shows that these are just the minimal elements of $\{\mathfrak{p}_0, \ldots, \mathfrak{p}_{n-1}\}$. The last assertion then follows from Chapter II, § 4, no. 4, Proposition 17.

Corollary. *If* M *is a finitely generated module over a Noetherian ring* A, Ass(M) *is finite.*

Under the conditions of Theorem 2, the set $\{\mathfrak{p}_0, \ldots, \mathfrak{p}_{n-1}\}$ is not necessarily determined uniquely by M; in particular it may be distinct from Ass(M) (Exercise 6).

Proposition 8. *Let* A *be a Noetherian ring,* $\mathfrak{a}$ *an ideal of* A *and* M *a finitely generated* A*-module. The following conditions are equivalent:*

(a) *there exists an element* $x \neq 0$ *of* M *such that* $\mathfrak{a}x = 0$.

(b) *for all* $a \in \mathfrak{a}$, *there exists an element* $x \neq 0$ *of* M *such that* $ax = 0$;

(c) *there exists* $\mathfrak{p} \in \mathrm{Ass}(M)$ *such that* $\mathfrak{a} \subset \mathfrak{p}$.

Clearly (a) implies (b). By virtue of no. 1, Corollary 2 to Proposition 2, condition (b) means that the ideal $\mathfrak{a}$ is contained in the union of the prime ideals associated with M and hence in one of them since Ass(M) is finite (Chapter II, § 1, no. 1, Proposition 2); thus (b) implies (c). Finally, if there exists $\mathfrak{p} \in \mathrm{Ass}(M)$ such that $\mathfrak{a} \subset \mathfrak{p}$, $\mathfrak{p}$ is the annihilator of an element $x \neq 0$ of M (no. 1, Definition 1) and $\mathfrak{a}x = 0$; thus (c) implies (a).

Proposition 9. *Let* A *be a Noetherian ring,* $\mathfrak{a}$ *an ideal of* A *and* M *a finitely generated* A*-module. For there to exist an integer* $n > 0$ *such that* $\mathfrak{a}^n M = 0$, *it is necessary and sufficient that* $\mathfrak{a}$ *be contained in the intersection of the prime ideals associated with* M.

This intersection is also that of the minimal elements of Supp(M) (no. 3, Corollary 1 to Proposition 7) and to say that $\mathfrak{a}$ is contained in this intersection is equivalent to saying that $V(\mathfrak{a}) \supset \mathrm{Supp}(M)$ in the notation of Chapter II, § 4; the conclusion then follows from Chapter II, § 4, no. 4, Corollary 2 to Proposition 17.

DEFINITION 2. *Given an A-module M, an endomorphism u of M is called almost nilpotent if, for all $x \in M$, there exists an integer $n(x) > 0$ such that $u^{n(x)}(x) = 0$.*

If M is finitely generated, every almost nilpotent endomorphism is nilpotent.

COROLLARY. *Let A be a Noetherian ring. M an A-module and a an element of A. For the homomorphism $a_M : x \mapsto ax$ of M to be almost nilpotent, it is necessary and sufficient that a belong to every ideal of* Ass(M).

The condition of the statement is equivalent to saying that for all $x \in M$ there exists $n(x) > 0$ such that $(Aa)^{n(x)}(Ax) = 0$; by Proposition 9 this means also that a belongs to all the prime ideals associated with the submodule Ax of M; the corollary then follows from the fact that Ass(M) is the union of the Ass(Ax) where x runs through M (no. 1, formula (1)).

PROPOSITION 10. *Let A be a Noetherian ring, E a finitely generated A-module and F an A-module. Then*

$$\text{Ass}(\text{Hom}_A(E, F)) = \text{Ass}(F) \cap \text{Supp}(E). \tag{5}$$

By hypothesis, E is isomorphic to an A-module of the form A^n/R, hence $\text{Hom}_A(E, F)$ is isomorphic to a submodule of $\text{Hom}_A(A^n, F)$ and the latter is isomorphic to F^n; now, $\text{Ass}(F^n) = \text{Ass}(F)$ (no. 1, Corollary 1 to Proposition 3); thus $\text{Ass}(\text{Hom}_A(E, F)) \subset \text{Ass}(F)$. On the other hand,

$$\text{Supp}(\text{Hom}_A(E, F)) \subset \text{Supp}(E):$$

for every prime ideal $\mathfrak{p}$ of A, $\text{Hom}_{A_\mathfrak{p}}(E_\mathfrak{p}, F_\mathfrak{p})$ is isomorphic to $(\text{Hom}_A(E, F))_\mathfrak{p}$ (Chapter II, § 2, no. 7, Proposition 19), whence our assertion immediately; then we conclude from Theorem 2 that

$$\text{Ass}(\text{Hom}_A(E, F)) \subset \text{Supp}(E).$$

Conversely, let $\mathfrak{p}$ be a prime ideal of A belonging to $\text{Ass}(F) \cap \text{Supp}(E)$. By definition, F contains a submodule isomorphic to $A/\mathfrak{p}$. On the other hand, since E is finitely generated and $E_\mathfrak{p} \neq 0$, we know that there exists a homomorphism $w \neq 0$ from E to $A/\mathfrak{p}$ (Chapter II, § 4, no. 4, Proposition 20). As there exists an injective homomorphism j from $A/\mathfrak{p}$ to F, $j \circ w \in \text{Hom}(E, F)$ and $j \circ w \neq 0$. On the other hand, the relation $aw = 0$ for some $a \in A$ is equivalent to $a \in \mathfrak{p}$, the annihilator of every element $\neq 0$ of $A/\mathfrak{p}$ being $\mathfrak{p}$; then certainly $\mathfrak{p} \in \text{Ass}(\text{Hom}_A(E, F))$.

2. PRIMARY DECOMPOSITION

1. PRIMARY S BMODULES

PROPOSITION 1. *Let A be a Noetherian ring and M an A-module.*

The following conditions are equivalent:

(a) Ass(M) *is reduced to a single element.*

(b) $M \neq 0$ *and every homothety of* M *is either injective or almost nilpotent* (§ 1, no. 4).

If these conditions are fulfilled and $\mathfrak{p}$ *is the set of* $a \in A$ *such that the homothety* a_M *is almost nilpotent, then* $\mathrm{Ass}(M) = \{\mathfrak{p}\}$.

This follows immediately from § 1, no. 4, Corollary to Proposition 9 and no. 1, Corollary 2 to Proposition 2.

DEFINITION 1. *Let* A *be a Noetherian ring,* N *an* A*-module and* Q *a submodule of* N. *If the module* $M = N/Q$ *satisfies the conditions of Proposition* 1, Q *is called* $\mathfrak{p}$*-primary with respect to* N (or *in* N).

When there is no ambiguity, we shall simply say that Q is "$\mathfrak{p}$-primary" or "primary"; clearly for every submodule $N' \neq Q$ of N containing Q, Q is $\mathfrak{p}$-primary in N'.

Definition 1 applies in particular to the case $N = A$; the submodules of N are then the *ideals* of A and hence an ideal $\mathfrak{q}$ of A is called *primary* if $\mathrm{Ass}(A/\mathfrak{q})$ has a single element or, what amounts to the same, if $A \neq \mathfrak{q}$ and every divisor of zero in the ring $A/\mathfrak{q}$ is *nilpotent*. If $\mathfrak{q}$ is $\mathfrak{p}$-primary, it follows from Definition 1 that $\mathfrak{p}$ is the *radical* (Chapter II, § 2, no. 6) of the ideal $\mathfrak{q}$.

Remark. Let $\mathfrak{q}$ be a $\mathfrak{p}$-primary submodule of an A-module N. If N/Q is *finitely generated*, there exists an integer $k > 0$ such that $\mathfrak{p}^k N \subset Q$ by § 1, no. 4, Proposition 9.

Examples

(1) If $\mathfrak{p}$ is a prime ideal of A, $\mathfrak{p}$ is $\mathfrak{p}$-primary (§ 1, no. 1, Proposition 1).

(2) Let $\mathfrak{q}$ be an ideal of A such that there exists *a single* prime ideal $\mathfrak{m}$ (necessarily maximal) containing $\mathfrak{q}$; then, if M is an A-module such that $\mathfrak{q}M \neq M$, $\mathfrak{q}M$ is $\mathfrak{m}$-*primary* with respect to M. For every element of $\mathrm{Ass}(M/\mathfrak{q}M)$ contains $\mathfrak{q}$, hence is equal to $\mathfrak{m}$ and $\mathrm{Ass}(M/\mathfrak{q}M) \neq \varnothing$ (§ 1, no. 1, Corollary 1 to Proposition 2). In particular $\mathfrak{q}$ is an $\mathfrak{m}$-*primary* ideal in A.

(3) Let $\mathfrak{m}$ be a *maximal* ideal of A; the $\mathfrak{m}$-primary ideals are then the ideals $\mathfrak{q}$ of A for which there exists an integer $n \geqslant 1$ such that $\mathfrak{m}^n \subset \mathfrak{q} \subset \mathfrak{m}$. For if $\mathfrak{m}^n \subset \mathfrak{q} \subset \mathfrak{m}$, $\mathfrak{m}$ is the only prime ideal containing $\mathfrak{q}$ (Chapter II, § 1, no. 1, Corollary to Proposition 1) and the conclusion follows from *Example* 2; conversely, if $\mathfrak{q}$ is $\mathfrak{m}$-primary, $\mathfrak{m}$ is the radical of $\mathfrak{q}$ and there therefore exists $n \geqslant 1$ such that $\mathfrak{m}^n \subset \mathfrak{q}$ (Chapter II, § 2, no. 6, Proposition 15).

(4) In a *principal ideal domain* A, the primary ideals are (0) and the ideals of the form Ap^n, where p is an extremal element and $n \geqslant 1$; this follows immediately from *Example* 3.

(5) The powers of any prime ideal are not necessarily primary ideals (Exercise 1). On the other hand, there exist primary ideals which are not powers of prime ideals (Exercise 1).

PROPOSITION 2. *Let M be a module over a Noetherian ring A, $\mathfrak{p}$ a prime ideal of A and $(Q_\iota)_{\iota\in I}$ a non-empty finite family of submodules of M which are $\mathfrak{p}$-primary with respect to M. Then $\bigcap_{\iota\in I} Q_\iota$ is $\mathfrak{p}$-primary with respect to M.*

$M/(\bigcap_{\iota\in I} Q_\iota)$ is isomorphic to a submodule $\neq 0$ of the direct sum $\bigoplus_{\iota\in I}(M/Q_\iota)$. Now

$$\mathrm{Ass}\Big(\bigoplus_{\iota\in I}(M/Q_\iota)\Big) = \bigcup_{\iota\in I}\mathrm{Ass}(M/Q_\iota) = \{\mathfrak{p}\}$$

(§ 1, no. 1, Corollary 1 to Proposition 3). Hence $\mathrm{Ass}\big(M/(\bigcap_{\iota\in I} Q_\iota)\big) = \{\mathfrak{p}\}$ (§ 1, no. 1, Proposition 3 and Corollary 1 to Proposition 2).

PROPOSITION 3. *Let A be a Noetherian ring, S a multiplicative subset of A, $\mathfrak{p}$ a prime ideal of A, M an A-module, N a submodule of M and $i = i_A^S$ the canonical mapping of M to $S^{-1}M$.*

(i) *Suppose that $\mathfrak{p} \cap S \neq \varnothing$. If N is $\mathfrak{p}$-primary with respect to M, then $S^{-1}N = S^{-1}M$.*

(ii) *Suppose that $\mathfrak{p} \cap S = \varnothing$. For N to be $\mathfrak{p}$-primary with respect to M, it is necessary and sufficient that N be of the form $\overset{-1}{i}(N')$, where N′ is a sub-$S^{-1}A$-module of $S^{-1}M$ which is $(S^{-1}\mathfrak{p})$-primary with respect to $S^{-1}M$; then $N' = S^{-1}N$.*

(i) If $\mathfrak{p} \cap S \neq \varnothing$ and N is $\mathfrak{p}$-primary with respect to M, then

$$\mathrm{Ass}_{S^{-1}A}(S^{-1}(M/N)) = \varnothing$$

(§ 1, no. 2, Corollary to Proposition 5) and hence $S^{-1}(M/N) = 0$ (§ 1, no. 1, Corollary 1 to Proposition 2), whence $S^{-1}M/S^{-1}N = 0$.

(ii) Suppose that $\mathfrak{p} \cap S = \varnothing$. If N is $\mathfrak{p}$-primary with respect to M, then $\mathrm{Ass}_{S^{-1}A}(S^{-1}(M/N)) = \{S^{-1}\mathfrak{p}\}$ (§ 1, no. 2, Corollary to Proposition 5) and hence the submodule $N' = S^{-1}N$ of $S^{-1}M$ is $(S^{-1}\mathfrak{p})$-primary; moreover, if $s \in S$ and $m \in M$ are such that $sm \in N$, then $m \in N$, for the homothety with ratio s in M/N is injective, whence $N = \overset{-1}{i}(N')$ (Chapter II, § 2, no. 4, Proposition 10). Conversely, let N′ be a submodule of $S^{-1}M$ which is $(S^{-1}\mathfrak{p})$-primary with respect to $S^{-1}M$; let us write $N = \overset{-1}{i}(N')$; then $N' = S^{-1}N$ (Chapter II, § 2, no. 4, Proposition 10) and $\mathrm{Ass}_{S^{-1}A}(S^{-1}(M/N)) = \mathrm{Ass}_{S^{-1}A}((S^{-1}M)/N') = \{S^{-1}\mathfrak{p}\}$. As the canonical mapping $M/N \to S(M/N)$ is injective, no prime ideal of A associated with M/N meets S (§ 1, no. 2, Proposition 6); it follows that $\mathrm{Ass}(M/N) = \{\mathfrak{p}\}$ (§ 1, no. 2, Corollary to Proposition 5), so that N is $\mathfrak{p}$-primary with respect to M.

2. THE EXISTENCE OF A PRIMARY DECOMPOSITION

DEFINITION 2. *Let A be a Noetherian ring, M an A-module and N a submodule of M. A finite family $(Q_\iota)_{\iota \in I}$ of submodules of M which are primary with respect to M and such that $N = \bigcap_{\iota \in I} Q_\iota$ is called a primary decomposition of N in M.*

Example. Let us take $A = \mathbf{Z}$, $M = \mathbf{Z}$, $N = n\mathbf{Z}$ for some integer $n > 0$. If $n = p_1^{\alpha_1} \ldots p_k^{\alpha_k}$ is the decomposition of n into prime factors,

$$n\mathbf{Z} = (p_1^{\alpha_1}\mathbf{Z}) \cap \cdots \cap (p_k^{\alpha_k}\mathbf{Z})$$

is a primary decomposition of $n\mathbf{Z}$ in $\mathbf{Z}$ by *Example* 4 of no. 1.

By an abuse of language, the relation $N = \bigcap_{\iota \in I} Q_\iota$ is called a primary decomposition of N in M. It amounts to the same to say that $\{0\} = \bigcap_{\iota \in I} (Q_\iota/N)$ is a primary decomposition of $\{0\}$ in M/N. If $(Q_\iota)_{\iota \in I}$ is a primary decomposition of N in M, the canonical mapping from M/N to $\bigoplus_{\iota \in I} (M/Q_\iota)$ is injective. Conversely let N be a submodule of M and f an injective homomorphism from M/N to a finite direct sum $P = \bigoplus_{\iota \in I} P_\iota$, where each set $\mathrm{Ass}(P_\iota)$ is reduced to a single element $\mathfrak{p}_\iota$; let f_ι be the homomorphism $M/N \rightarrow P_\iota$ obtained by taking the composition of f with the projection $P \rightarrow P_\iota$ and let Q_ι/N be the kernel of f_ι; then the Q_ι distinct from M are primary with respect to M (no. 1, Definition 1) and $N = \bigcap_{\iota \in I} Q_\iota$. Moreover, $\mathrm{Ass}(M/N) \subset \bigcup_{\iota \in I} \{\mathfrak{p}_\iota\}$ by virtue of § 1, no. 1, Proposition 3.

THEOREM 1. *Let M be a finitely generated module over a Noetherian ring and let N be a submodule of M. There exists a primary decomposition of N of the form*

$$N = \bigcap_{\mathfrak{p} \in \mathrm{Ass}(M/N)} Q(\mathfrak{p}) \tag{1}$$

where, for all $\mathfrak{p} \in \mathrm{Ass}(M/N)$, $Q(\mathfrak{p})$ is $\mathfrak{p}$-primary with respect to M.

We may replace M by M/N and therefore suppose that $N = 0$. By § 1, no. 4, Corollary to Theorem 2, Ass(M) is finite; by § 1, no. 1, Proposition 4, there exists, for each $\mathfrak{p} \in \mathrm{Ass}(M)$, a submodule $Q(\mathfrak{p})$ of M such that $\mathrm{Ass}(M/Q(\mathfrak{p})) = \{\mathfrak{p}\}$ and $\mathrm{Ass}(Q(\mathfrak{p})) = \mathrm{Ass}(M) - \{\mathfrak{p}\}$. Let us write $P = \bigcap_{\mathfrak{p} \in \mathrm{Ass}(M)} Q(\mathfrak{p})$; for all $\mathfrak{p} \in \mathrm{Ass}(M)$, $\mathrm{Ass}(P) \subset \mathrm{Ass}(Q(\mathfrak{p}))$ and hence $\mathrm{Ass}(P) = \varnothing$, which implies $P = 0$ (§ 1, no. 1, Corollary 1 to Proposition 2) and therefore proves the theorem.

3. UNIQUENESS PROPERTIES IN THE PRIMARY DECOMPOSITION

DEFINITION 3. *Let M be a module over a Noetherian ring and N a submodule of M. A primary decomposition $N = \bigcap_{\iota \in I} Q_\iota$ of N in M is called reduced if the following conditions are fulfilled:*

(a) *there exists no index* $i \in I$ *such that* $\bigcap_{j \neq i} Q_j \subset Q_i$;
(b) *if* $\mathrm{Ass}(M/Q_i) = \{\mathfrak{p}_i\}$, *the* $\mathfrak{p}_i$ $(i \in I)$ *are distinct.*

From every primary decomposition $N = \bigcap_{i \in I} Q_i$ of N in M a *reduced* primary decomposition of M in N can be deduced as follows: let J be a minimal element of the set of subsets I′ of I such that $N = \bigcap_{i \in I'} Q_i$. Clearly $(Q_i)_{i \in J}$ satisfies condition (a). Then let Φ be the set of $\mathfrak{p}_i$ for $i \in J$; for all $\mathfrak{p} \in \Phi$, let $H(\mathfrak{p})$ be the set of $i \in J$ such that $\mathfrak{p}_i = \mathfrak{p}$ and let $Q(\mathfrak{p}) = \bigcap_{i \in H(\mathfrak{p})} Q_i$; it follows from Proposition 2 of no. 1 that $Q(\mathfrak{p})$ is $\mathfrak{p}$-primary with respect to M; further $N = \bigcap_{\mathfrak{p} \in \Phi} Q(\mathfrak{p})$ and the family $Q((\mathfrak{p}))_{\mathfrak{p} \in \Phi}$ is therefore a reduced primary decomposition of N in M.

We shall see that the primary decomposition defined in the proof of Theorem 1 of no. 2 is *reduced*; this follows from the following proposition:

PROPOSITION 4. *Let* M *be a module over a Noetherian ring,* N *a submodule of* M, $N = \bigcap_{i \in I} Q_i$ *a primary decomposition of* N *in* M *and, for all* $i \in I$, *let* $\{\mathfrak{p}_i\} = \mathrm{Ass}(M/Q_i)$. *For this decomposition to be reduced, it is necessary and sufficient that the* $\mathfrak{p}_i$ *be distinct and belong to* $\mathrm{Ass}(M/N)$; *then*

$$\mathrm{Ass}(M/N) = \bigcup_{i \in I} \{\mathfrak{p}_i\} \tag{2}$$

$$\mathrm{Ass}(Q_i/N) = \bigcup_{j \neq i} \{\mathfrak{p}_j\} \quad \textit{for all } i \in I. \tag{3}$$

If the condition of the statement is fulfilled, $N = \bigcap_{j \neq i} Q_j$ cannot hold, for we would deduce that $\mathrm{Ass}(M/N) \subset \bigcup_{j \neq i} \{\mathfrak{p}_j\}$ (§ 1, no. 1, Corollary 2 to Proposition 3) contrary to the hypothesis; the primary decomposition $(Q_i)_{i \in I}$ of N is then certainly reduced. Conversely, $\mathrm{Ass}(M/N) \subset \bigcup_{i \in I} \{\mathfrak{p}_i\}$ always holds (§ 1, no. 1, Corollary 2 to Proposition 3); on the other hand, for all $i \in I$, let us write $P_i = \bigcap_{j \neq i} Q_j$; then $P_i \cap Q_i = N$ and $P_i \neq N$ if $(Q_i)_{i \in I}$ is reduced, hence P_i/N is non-zero and is isomorphic to the submodule $(P_i + Q_i)/Q_i$ of M/Q_i, whence $\{\mathfrak{p}_i\} = \mathrm{Ass}(P_i/N)$ (§ 1, no. 1, Proposition 3 and Corollary 1 to Proposition 2); as $P_i/N \subset M/N$, $\mathfrak{p}_i \in \mathrm{Ass}(M/N)$, which completes the proof of the necessity of the condition in the statement and formula (2). Finally, as $N = \bigcap_{j \neq i} (Q_j \cap Q_i)$, $\mathrm{Ass}(Q_i/N) \subset \bigcup_{j \neq i} \mathrm{Ass}(Q_i/(Q_j \cap Q_i))$ (§ 1, no. 1, Corollary 2 to Proposition 3); but $Q_i/(Q_j \cap Q_i)$ is isomorphic to the submodule $(Q_i + Q_j)/Q_j$ of M/Q_j, hence $\mathrm{Ass}(Q_i/(Q_j \cap Q_i)) \subset \{\mathfrak{p}_j\}$ and

$$\mathrm{Ass}(Q_i/N) \subset \bigcup_{j \neq i} \{\mathfrak{p}_j\};$$

taking account of (2) and Proposition 3 of § 1, no. 1, formula (3) follows easily.

COROLLARY. *Let* A *be a Noetherian ring,* M *an* A*-module,* N *a submodule of* M *and* $(Q_\iota)_{\iota\in I}$ *a primary decomposition of* N *in* M. *Then* $\text{Card}(I) \geqslant \text{Card}(\text{Ass}(M/N))$; *for* $(Q_\iota)_{\iota\in I}$ *to be a reduced primary decomposition, it is necessary and sufficient that* $\text{Card}(I) = \text{Card}(\text{Ass}(M/N))$.

It follows from the remarks preceding Proposition 4 that there exists a reduced primary decomposition $(R_j)_{j\in J}$ of N in M such that $\text{Card}(J) \leqslant \text{Card}(I)$; the first assertion then follows from the second and the latter is a consequence of Proposition 4.

PROPOSITION 5. *Let* A *be a Noetherian ring,* M *an* A*-module,* N *a submodule of* M, $N = \bigcap_{\iota\in I} Q_\iota$ *a reduced primary decomposition of* N *in* M *and, for all* $\iota \in I$, *let* $\{\mathfrak{p}_\iota\} = \text{Ass}(M/Q_\iota)$. *If* $\mathfrak{p}_\iota$ *is a minimal element of* Ass(M/N), Q_ι *is equal to the saturation on* N *with respect to* $\mathfrak{p}_\iota$ (Chapter II, § 2, no. 4) (cf. Exercise 2).

We can obviously restrict our attention to the case where $N = 0$, replacing if need be M by M/N. If $\mathfrak{p}_\iota$ is minimal in Ass(M), the set of elements of Ass(M) which do not meet $A - \mathfrak{p}_\iota$ reduces to $\mathfrak{p}_\iota$; the proposition then follows from formula (3) above and § 1, no. 2, Proposition 6, the kernel of the canonical mapping $M \to M_{\mathfrak{p}_\iota}$ being equal to the saturation of 0 with respect to $\mathfrak{p}_\iota$ (Chapter II, § 2, no. 4).

Remark. The prime ideals $\mathfrak{p}_\iota \in \text{Ass}(M/N)$ which are not minimal elements of this set are sometimes called the *immersed* prime ideals associated with M/N; if M/N is finitely generated, for $\mathfrak{p}_0 \in \text{Ass}(M/N)$ to be immersed, it is necessary and sufficient that $V(\mathfrak{p}_0)$ *be not* an irreducible component of Supp(M/N) (Chapter II, § 4, no. 3, Corollary 2 to Proposition 14); if $(Q(\mathfrak{p}))_{\mathfrak{p}\in \text{Ass}(M/N)}$ and $(Q'(\mathfrak{p}))_{\mathfrak{p}\in \text{Ass}(M/N)}$ are two reduced primary decompositions of N in M, it may be that $Q'(\mathfrak{p}_0) \neq Q(\mathfrak{p}_0)$ (Exercise 24 (c)); a *canonical* reduced primary decomposition of N in M may always be defined by imposing supplementary conditions on the primary submodules which appear in it (Exercise 4).

4. THE LOCALIZATION OF A PRIMARY DECOMPOSITION

Given a submodule N of a module M over a Noetherian ring A, to simplify we shall denote by $D_I(M/N)$, in this no., the set of reduced primary decompositions of N in M whose indexing set is I (equipotent to Ass(M/N)).

PROPOSITION 6. *Let* A *be a Noetherian ring,* M *an* A*-module,* N *a submodule of* M *and* $I = \text{Ass}(M/N)$. *Let* S *be a multiplicative subset of* A *and* J *the subset of* I *consisting* *of the indices i such that* $S \cap \mathfrak{p}_i = \varnothing$. *Let* N' *be the saturation of* N *with respect to* S *in* M. *Then:*

(i) *If* $(Q_i)_{i \in I}$ *is an element of* $D_I(M/N)$, *the family* $(Q_i)_{i \in J}$ *is an element of* $D_J(M/N')$ *and the family* $(S^{-1}Q_i)_{i \in J}$ *is an element of* $D_J(S^{-1}M/S^{-1}N)$.

(ii) *The mapping* $(Q_i)_{i \in J} \to (S^{-1}Q_i)_{i \in J}$ *is a bijection of* $D_J(M/N')$ *onto* $D_J(S^{-1}M/S^{-1}N)$.

(iii) *If* $(Q_i)_{i \in J}$ *is an element of* $D_J(M/N')$ *and* $(R_i)_{i \in I}$ *an element of* $D_I(M/N)$, *the family* $(T_i)_{i \in I}$ *such that* $T_i = Q_i$ *for* $i \in J$ *and* $T_i = R_i$ *for* $i \in I - J$ *is an element of* $D_I(M/N)$.

(i) We know (no. 1, Proposition 3) that for $i \in J$, $S^{-1}Q_i$ is primary for $S^{-1}\mathfrak{p}_i$ and that for $i \in I - J$, $S^{-1}Q_i = S^{-1}M$; as $S^{-1}N = \bigcap_{i \in I} S^{-1}Q_i$ (Chapter II, § 2, no. 4), then also $S^{-1}N = \bigcap_{i \in J} S^{-1}Q_i$. The $S^{-1}\mathfrak{p}_i$ for $i \in J$ are distinct and their set is $\mathrm{Ass}(S^{-1}M/S^{-1}N)$ (§ 1, no. 2, Corollary to Proposition 5); then (Proposition 4) $(S^{-1}Q_i)_{i \in J}$ is a reduced primary decomposition of $S^{-1}N$. Moreover, $Q_i = (i_M^S)^{-1}(S^{-1}Q_i)$ (no. 1 Proposition 3), hence $N' = (i_M^S)^{-1}(S^{-1}N) = \bigcap_{i \in J} Q_i$ and $(Q_i)_{i \in J}$ is obviously a reduced primary decomposition of N' in M.

(ii) As $S^{-1}N' = S^{-1}N$, we may replace N by N', that is suppose that $J = I$. Let $(P_i)_{i \in I}$ be a reduced primary decomposition of $S^{-1}N$ in $S^{-1}M$ and let us write $Q_i = (i_M^S)^{-1}(P_i)$; it follows from no. 1, Proposition 3 that Q_i is primary for $\mathfrak{p}_i$ $(i \in I)$ and $(Q_i)_{i \in I}$ is then a reduced primary decomposition of N in M by virtue of no. 3, Corollary to Proposition 4. Finally, as, for all $i \in I$ and every submodule Q'_i of which M is $\mathfrak{p}_i$-primary with respect to M, $Q'_i = (i_M^S)^{-1}(S^{-1}Q'_i)$ by virtue of no. 1, Proposition 3 and the hypothesis $J = I$, we see that two mappings $D_I(M/N) \to D_I(S^{-1}M/S^{-1}N)$ and $D_I(S^{-1}M/S^{-1}N) \to D_I(M/N)$ have been defined whose compositions are the identities on $D_I(M/N)$ and $D_I(S^{-1}M/S^{-1}N)$, which proves (ii).

(iii) By virtue of (i), $N' = \bigcap_{i \in J} R_i$, whence

$$N = N' \cap \bigcap_{i \in I - J} R_i = \Big(\bigcap_{i \in J} Q_i\Big) \cap \Big(\bigcap_{i \in I - J} R_i\Big)$$

and it follows immediately from no. 3, Corollary to Proposition 4 that this primary decomposition is reduced.

COROLLARY. *The mappings*

$$D_I(M/N) \to D_J(M/N') \quad \textit{and} \quad D_I(M/N) \to D_J(S^{-1}M/S^{-1}N)$$

defined in Proposition 6 (i) *are surjective.*

Proposition 6 (iii) shows that the mapping $D_I(M/N) \to D_J(M/N')$ is surjective and Proposition 6 (ii) then shows that the mapping

$$D_I(M/N) \to D_J(S^{-1}M/S^{-1}N)$$

is surjective.

5. RINGS AND MODULES OF FINITE LENGTH

If an A-module M is of finite length, we shall denote this length by $\mathrm{long}_A(M)$ or long(M). Recall that every Artinian ring is Noetherian (*Algebra*, Chapter VIII, § 6, no. 5, Corollary 3 to Proposition 12) and that every finitely generated module over an Artinian ring is of finite length (*loc. cit.*, Corollary 1 to Proposition 12). Moreover, every Artinian integral domain is a field (*Algebra*, Chapter VIII, § 6, no. 4, Proposition 9).

PROPOSITION 7. *Let* M *be a finitely generated module over a Noetherian ring* A. *The following properties are equivalent:*

(a) M *is of finite length.*
(b) *Every ideal* $\mathfrak{p} \in \mathrm{Ass}(M)$ *is a maximal ideal of* A.
(c) *Every ideal* $\mathfrak{p} \in \mathrm{Supp}(M)$ *is a maximal ideal of* A.

Let $(M_i)_{0 \leqslant i \leqslant n}$ be a composition series of M such that, for $0 \leqslant i \leqslant n-1$, M_i/M_{i+1} is isomorphic to $A/\mathfrak{p}_i$, where $\mathfrak{p}_i$ is prime (§ 1, no. 4, Theorem 1). If M is of finite length, so is each of the A-modules $A/\mathfrak{p}_i$, which implies that each of the rings $A/\mathfrak{p}_i$ is Artinian; but as $A/\mathfrak{p}_i$ is an integral domain, it is therefore a field, in other words $\mathfrak{p}_i$ is maximal; we conclude that (a) implies (b) (§ 1, no. 4, Theorem 2). Condition (b) implies (c) by § 1, no. 3, Proposition 7. Finally, if all the ideals of Supp(M) are maximal, so are the $\mathfrak{p}_i$ (§ 1, no. 4, Theorem 2), hence the $A/\mathfrak{p}_i$ are simple A-modules and M is of finite length, which completes the proof.

COROLLARY 1. *For every module of finite length* M *over a Noetherian ring* A, Ass(M) = Supp(M).

Every element of Supp(M) is then minimal in Supp(M) and the conclusion follows from § 1, no. 3, Corollary 1 to Proposition 7.

COROLLARY 2. *Let* M *be a finitely generated module over a Noetherian ring* A *and* $\mathfrak{p}$ *a prime ideal of* A. *For* $M_\mathfrak{p}$ *to be a non-zero* $A_\mathfrak{p}$*-module of finite length, it is necessary and sufficient that* $\mathfrak{p}$ *be a minimal element of* Ass(M).

By § 1, no. 2, Corollary to Proposition 5, $\mathrm{Ass}_{A_\mathfrak{p}}(M_\mathfrak{p})$ is the set of ideals $\mathfrak{q}_\mathfrak{p}$, where $\mathfrak{q}$ runs through the set of ideals of Ass(M) which are contained in $\mathfrak{p}$. On the other hand, $\mathfrak{p}_\mathfrak{p}$ is the unique maximal ideal of $A_\mathfrak{p}$; by Proposition 7, for $M_\mathfrak{p}$ to be an $A_\mathfrak{p}$-module of finite length, it is necessary and sufficient that no element of Ass(M) be strictly contained in $\mathfrak{p}$. On the other hand, for $M_\mathfrak{p} \neq 0$, it is necessary and sufficient by definition that $\mathfrak{p} \in \mathrm{Supp}(M)$ (Chapter II, § 4, no. 4), that is that $\mathfrak{p}$ contain an element of Ass(M) (§ 1, no. 3, Proposition 7). This proves the corollary.

Remark (1). Let M be a finitely generated module over a Noetherian ring A; let $(M_i)_{0 \leqslant i \leqslant n}$ be a composition series of M such that, for $0 \leqslant i \leqslant n-1$, M_i/M_{i+1} is isomorphic to $A/\mathfrak{p}_i$, where $\mathfrak{p}_i$ is a prime ideal of A (§ 1, no. 4, Theorem 1). If $\mathfrak{p}$ is a minimal element of Ass(M), the length $\mathrm{long}_{A_\mathfrak{p}}(M_\mathfrak{p})$ is equal to the *number of indices i such that* $\mathfrak{p}_i = \mathfrak{p}$. For the $(M_i)_\mathfrak{p}$ form a composition series of $M_\mathfrak{p}$ and $(M_i)_\mathfrak{p}/(M_{i+1})_\mathfrak{p}$ is isomorphic to $(A/\mathfrak{p}_i)_\mathfrak{p}$ and hence to $\{0\}$ if $\mathfrak{p}_i \neq \mathfrak{p}$ (since $\mathfrak{p}$ is minimal in the set of $\mathfrak{p}_i$ by § 1, no. 4, Theorem 2) and to $(A/\mathfrak{p})_\mathfrak{p}$ which is a field, if $\mathfrak{p}_i = \mathfrak{p}$.

PROPOSITION 8. *Let* M *be a module of finite length over a Noetherian ring* A.

(i) *There only exists a single primary decomposition of* {0} *with respect to* M *indexed by* Ass(M) *(necessarily reduced); let* $\{0\} = \bigcap_{\mathfrak{p} \in \mathrm{Ass}(M)} Q(\mathfrak{p})$ *be this decomposition, where* $Q(\mathfrak{p})$ *is* $\mathfrak{p}$*-primary with respect to* M.

(ii) *There exists an integer* n_0 *such that, for all* $n \geqslant n_0$ *and all* $\mathfrak{p} \in \mathrm{Ass}(M)$, $Q(\mathfrak{p}) = \mathfrak{p}^n M$.

(iii) *For all* $\mathfrak{p} \in \mathrm{Ass}(M)$, *the canonical mapping of* M *to* $M_\mathfrak{p}$ *is surjective and its kernel is* $Q(\mathfrak{p})$.

(iv) *The canonical injection of* M *into* $\bigoplus_{\mathfrak{p} \in \mathrm{Ass}(M)} (M/Q(\mathfrak{p}))$ *is bijective.*

As every element $\mathfrak{p} \in \mathrm{Ass}(M)$ is minimal in Ass(M) (Proposition 7), assertion (i) follows from no. 3, Proposition 5. As M is finitely generated, there exists n_0 such that $\mathfrak{p}^n M \subset Q(\mathfrak{p})$ for all $\mathfrak{p} \in \mathrm{Ass}(M)$ and all $n \geqslant n_0$ (no. 1, *Remark*); but as $\mathfrak{p}$ is a maximal ideal, $\mathfrak{p}^n M$ is $\mathfrak{p}$-primary with respect to M (no. 1, *Examples* 2 and 3) and, as $\bigcap_{\mathfrak{p} \in \mathrm{Ass}(M)} \mathfrak{p}^n M = \{0\}$, it follows from (i) that necessarily $\mathfrak{p}^n M = Q(\mathfrak{p})$ for all $\mathfrak{p} \in \mathrm{Ass}(M)$; whence (ii). As the $\mathfrak{p}^n$, for $\mathfrak{p} \in \mathrm{Ass}(M)$, are relatively prime in pairs (Chapter II, § 1, no. 2, Proposition 3), the canonical mapping $M \to \bigoplus_{\mathfrak{p} \in \mathrm{Ass}(M)} (M/\mathfrak{p}^n M)$ is surjective (Chapter II, § 1, no. 2, Proposition 6), whence (iv). Then $\mathrm{Ass}(Q(\mathfrak{p})) = \mathrm{Ass}(M) - \{\mathfrak{p}\}$ and $\mathrm{Ass}(M/Q(\mathfrak{p})) = \{\mathfrak{p}\}$ (no. 3, Proposition 4); as the elements of Ass(M) are maximal ideals, $\mathfrak{p}$ is the only element of Ass(M) which does not meet $A - \mathfrak{p}$; $Q(\mathfrak{p})$ is therefore the kernel of the canonical mapping $j: M \to M_\mathfrak{p}$ (§ 1, no. 2, Proposition 6). If $s \in A - \mathfrak{p}$, the homothety of $M/Q(\mathfrak{p})$ with ratio s is injective by virtue of the relation $\mathrm{Ass}(M/Q(\mathfrak{p})) = \{\mathfrak{p}\}$ (no. 1, Proposition 1); since $M/Q(\mathfrak{p})$ is Artinian, this homothety is bijective (*Algebra*, Chapter VIII, § 1, no. 2, Lemma 3). The canonical mapping $M \to M/Q(\mathfrak{p})$ is then written $f \circ j$, where $f: M_\mathfrak{p} \to M/Q(\mathfrak{p})$ is an A-homomorphism (Chapter II, § 2, no. 2, Proposition 3); as $\mathrm{Ker}(j) = \mathrm{Ker}(f \circ j) = Q(\mathfrak{p})$, f is injective; we conclude that j is surjective and f bijective.

COROLLARY. *If* M *is a module of finite length over a Noetherian ring* A, *then*

$$\text{long}_A(M) = \sum_{\mathfrak{p} \in \text{Ass}(M)} \text{long}_{A_\mathfrak{p}}(M_\mathfrak{p}). \tag{4}$$

This will follow from Proposition 8 (iv) if we prove that

$$\text{long}_A(M/Q(\mathfrak{p})) = \text{long}_{A_\mathfrak{p}}(M_\mathfrak{p}).$$

Now, it follows from Proposition 1 of no. 1 that for all $s \in A - \mathfrak{p}$ the homothety with ratio s on $M/Q(\mathfrak{p})$ is injective; the homothety with ratio s on every submodule R of $M/Q(\mathfrak{p})$ is therefore injective and, as R is Artinian, it is bijective (*Algebra*, Chapter VIII, § 2, no. 2, Lemma 3); we conclude that the sub-A-modules of $M/Q(\mathfrak{p})$ are the images under the bijection $f: M_\mathfrak{p} \to M/Q(\mathfrak{p})$ of the sub-$A_\mathfrak{p}$-modules of $M_\mathfrak{p}$ (Chapter II, § 2, no. 3), whence our assertion.

PROPOSITION 9. *Let* A *be a Noetherian ring. The following conditions are equivalent:*

(a) A *is Artinian.*

(b) *All the prime ideals of* A *are maximal ideals.*

(c) *All the elements of* Ass(A) *are maximal ideals.*

If these conditions are fulfilled, A *has only a finite number of prime ideals, which are all maximal and associated with the* A*-module* A; *further,* A *is a semi-local ring and its Jacobson radical is nilpotent.*

To say that A is Artinian is equivalent to saying that A is an A-module of finite length; hence (a) and (c) are equivalent by Proposition 7. Clearly (b) implies (c). Finally, (a) implies (b) since every Artinian integral domain is a field. The properties (a), (b) and (c) are therefore equivalent.

Suppose they hold. As every prime ideal of A belongs to Supp(A) and every element of Supp(A) contains an element of Ass(A) (§ 1, no. 3, Proposition 7), it follows from (c) that Ass(A) is the set of all prime ideals of A; then A has only a finite number of prime ideals, all of them maximal and associated with the A-module A. This obviously implies that A is semi-local; finally, we know that the Jacobson radical of an Artinian ring is nilpotent (*Algebra*, Chapter VIII, § 6, no. 4, Theorem 3).

Remark (2) The conditions of Proposition 9 for a *Noetherian* ring A imply that the spectrum of A is *finite* and *discrete*, every point of Spec(A) being therefore closed (Chapter II, § 4, no. 3, Corollary 6 to Proposition 11). Conversely, for a *Noetherian* ring A, to say that *every point of* Spec(A) *is closed* means that every prime ideal of A is maximal (*loc. cit.*) and hence this condition is equivalent to that of Proposition 9.

COROLLARY 1. *Every Artinian ring* A *is isomorphic to the direct composition of a finite family of Artinian local rings.*

It follows from Proposition 9 and Proposition 8 (iii) and (iv) that, if $(\mathfrak{m}_i)_{1 \leqslant i \leqslant n}$ is the family of maximal ideals of A, the canonical mapping $A \to \prod_i A_{\mathfrak{m}_i}$ is bijective.

Remark (3). This corollary can also be deduced from the fact that Spec(A) is finite and discrete and Chapter II, § 4, no. 3, Proposition 15.

COROLLARY 2. *Let* A *be a Noetherian ring and* $\mathfrak{m}$ *an ideal of* A. *The following conditions are equivalent:*

(a) A *is a semi-local ring and* $\mathfrak{m}$ *is a defining ideal of* A.
(b) A *is a Zariski ring with the* $\mathfrak{m}$*-adic topology and* $A/\mathfrak{m}$ *is Artinian.*

If (a) holds, A is a Zariski ring with the $\mathfrak{m}$-adic topology (Chapter III, § 3, no. 3, *Example* 3); further, as by hypothesis $\mathfrak{m}$ contains a power of the Jacobson radical $\mathfrak{r}$ of A, every prime ideal of A which contains $\mathfrak{m}$ also contains $\mathfrak{r}$ (Chapter II, § 1, no. 1, Proposition 1); it is therefore maximal, since $\mathfrak{r}$ is a finite intersection of maximal ideals (*loc. cit.*); Proposition 9 then shows that $A/\mathfrak{m}$ is Artinian. Conversely, if (b) holds, every maximal ideal $\mathfrak{p}$ of A contains the Jacobson radical of A and hence contains $\mathfrak{m}$ (Chapter III, § 3, no. 3, Proposition 6); as $A/\mathfrak{m}$ is Artinian, the ideals $\mathfrak{p}/\mathfrak{m}$ are finite in number (Proposition 9) and hence A has only a finite number of maximal ideals, which implies that it is semi-local.

COROLLARY 3. *Let* A, A′ *be two rings and* h *a homomorphism from* A *to* A′. *Suppose that* A *is semi-local and Noetherian and that* A′ *is a finitely generated* A*-module. Then the ring* A′ *is semi-local and Noetherian; if* $\mathfrak{m}$ *is a defining ideal of* A, $\mathfrak{m}A'$ *is a defining ideal of* A′.

We know that A′ is a Zariski ring with the $\mathfrak{m}A'$-adic topology (Chapter III, § 3, no. 3, Proposition 7). As $A/\mathfrak{m}$ is Artinian (Corollary 2) and $A'/\mathfrak{m}A'$ is a finitely generated $(A/\mathfrak{m})$-module, $A'/\mathfrak{m}A'$ is an Artinian ring, hence A′ is semi-local and $\mathfrak{m}A'$ is a defining ideal of A′ (Corollary 2).

COROLLARY 4. *Let* A *be a complete semi-local Noetherian ring,* $\mathfrak{m}$ *a defining ideal of* A, E *a finitely generated* A*-module and* (F_n) *a decreasing sequence of submodules of* E *such that* $\bigcap_n F_n = 0$. *Then, for all* $p > 0$, *there exists* $n > 0$ *such that* $F_n \subset \mathfrak{m}^p E$.

As A is a Zariski ring, E is Hausdorff and the F_n are closed under the $\mathfrak{m}$-adic topology. On the other hand, E is complete (Chapter III, § 2, no. 12, Corollary 1 to Proposition 16). Finally, $E/\mathfrak{m}^p E$ is a finitely generated module over the ring $A/\mathfrak{m}^p$, which is Artinian (Corollary 2); then $E/\mathfrak{m}^p E$ is an Artinian $(A/\mathfrak{m}^p)$-module and hence an Artinian A-module. The corollary then follows from Chapter III, § 2, no. 7, Proposition 8.

COROLLARY 5. *In a complete semi-local Noetherian ring every decreasing sequence of ideals whose intersection is 0 is a filter base which converges to 0.*

It is sufficient to apply Corollary 4 to the A-module A.

PROPOSITION 10. *Let A be a Noetherian ring and* $\mathfrak{p}_1, \ldots, \mathfrak{p}_n$ *the prime ideals associated with the A-module A, where* $\mathfrak{p}_i \neq \mathfrak{p}_j$ *for* $i \neq j$.

(i) *The set* $S = \bigcap_{i=1}^{n} (A - \mathfrak{p}_i)$ *is the set of elements which are not divisors of* 0 *in* A.

(ii) *If all the* $\mathfrak{p}_i$ *are minimal elements of* Ass(A), *the total ring of fractions* $S^{-1}A$ *of A is Artinian.*

(iii) *If the ring A is reduced, all the* $\mathfrak{p}_i$ *are minimal elements of* Ass(A) (*and therefore are the minimal elements of* Spec(A)) *and each of the* $A_{\mathfrak{p}_i}$ *is a field; for each index i, the canonical homomorphism* $S^{-1}A \rightarrow A_{\mathfrak{p}_i}$ (Chapter II, § 2, no. 1, Corollary 1 to Proposition 2) *is surjective and its kernel is* $S^{-1}\mathfrak{p}_i$; *finally the canonical homomorphism from* $S^{-1}A$ *to* $\prod_{i=1}^{n} (S^{-1}A/S^{-1}\mathfrak{p}_i)$ *is bijective.*

The fact that S is the set of elements which are not divisors of 0 in A has already been seen (§ 1, no. 1, Corollary 3 to Proposition 2). The prime ideals of $S^{-1}A$ are of the form $S^{-1}\mathfrak{p}$, where $\mathfrak{p}$ is a prime ideal of A contained in $\bigcup_{i=1}^{n} \mathfrak{p}_i$ (Chapter II, § 2, no. 5, Proposition 10) that is contained in one of the $\mathfrak{p}_i$ (Chapter II, § 1, no. 1, Proposition 2). If $\mathfrak{p}_i$ is a minimal element of Ass(A), it is a minimal element of Spec(A) (§ 1, no. 3, Corollary to Proposition 7); if each of the $\mathfrak{p}_i$ is a minimal element of Ass(A), we then see that the prime ideals of $S^{-1}A$ are the $S^{-1}\mathfrak{p}_i$ and they are therefore all maximal, which proves that $S^{-1}A$ is Artinian (Proposition 9).

Suppose finally that the ring A is reduced. Then $\bigcap_{i=1}^{n} \mathfrak{p}_i = \{0\}$ (§ 1, no. 3, Corollary 2 to Proposition 7). We deduce that $\{0\} = \bigcap_{i=1}^{n} \mathfrak{p}_i$ is a *reduced* primary decomposition of the ideal {0} (no. 3, Corollary to Proposition 4); in particular, none of the $\mathfrak{p}_i$ can contain a $\mathfrak{p}_j$ of index $j \neq i$ and therefore the $\mathfrak{p}_i$ are all minimal elements of Ass(A). The ring $S^{-1}A$ is then Artinian by (ii). The $S^{-1}\mathfrak{p}_i$ are prime ideals associated with the $S^{-1}A$-module $S^{-1}A$ (§ 1, no. 2, Corollary to Proposition 5) and $\{0\} = S^{-1}\left(\bigcap_{i=1}^{n} \mathfrak{p}_i\right) = \bigcap_{i=1}^{n} S^{-1}\mathfrak{p}_i$ (Chapter II, § 2, no. 4); as the $S^{-1}\mathfrak{p}_i$ are distinct, $(S^{-1}\mathfrak{p}_i)_{1 \leqslant i \leqslant n}$ is a reduced primary decomposition of {0} in $S^{-1}A$ (no. 3, Corollary to Proposition 4). Proposition 8 then shows that the canonical homomorphism $g_i: S^{-1}A \to (S^{-1}A)_{\mathfrak{p}_i}$ is surjective and has kernel $S^{-1}\mathfrak{p}_i$ and the canonical homomorphism $S^{-1}A \to \prod_{i=1}^{n} (S^{-1}A/S^{-1}\mathfrak{p}_i)$ is bijective. We know moreover that the canonical homomorphism $S^{-1}A \to A_{\mathfrak{p}_i}$ is composed of g_i and an isomorphism $(S^{-1}A)_{S^{-1}\mathfrak{p}_i} \to A_{\mathfrak{p}_i}$ (Chapter II, § 2, no. 3, Proposition 7). Finally, it follows from Proposition 8 that $(S^{-1}A)_{S^{-1}\mathfrak{p}_i}$ is isomorphic to $S^{-1}A/S^{-1}\mathfrak{p}_i$ and hence is a field since $S^{-1}\mathfrak{p}_i$ is a maximal ideal.

6. PRIMARY DECOMPOSITION AND EXTENSION OF SCALARS

In this no., A and B will denote two rings and we shall consider a ring homomorphism $\rho: A \to B$ which makes B into an A-algebra; recall that, for every B-module F, $\rho_*(F)$ is the commutative group F with the A-module structure defined by $a.y = \rho(a)y$ for all $a \in A$, $y \in F$.

LEMMA 1. *Let A be a Noetherian ring, $\mathfrak{p}$ a prime ideal of A, E an A-module whose annihilator contains a power of $\mathfrak{p}$ and such that $\mathrm{Ass}(E) = \{\mathfrak{p}\}$ and F a B-module such that $\rho_*(F)$ is a flat A-module. The condition $\mathfrak{P} \in \mathrm{Ass}_B(E \otimes_A F)$ then implies $\overset{-1}{\rho}(\mathfrak{P}) = \mathfrak{p}$.*

If n is such that $\mathfrak{p}^n E = 0$, then $\mathfrak{p}^n B \subset \mathrm{Ann}(E \otimes_A F)$, whence $\mathfrak{p}^n B \subset \mathfrak{P}$, which implies $\mathfrak{p}^n \subset \overset{-1}{\rho}(\mathfrak{P})$ and therefore $\mathfrak{p} \subset \overset{-1}{\rho}(\mathfrak{P})$ since $\overset{-1}{\rho}(\mathfrak{P})$ is prime. Moreover, if $a \in A - \mathfrak{p}$, the homothety h with ratio a on E is injective (§ 1, no. 1, Corollary 2 to Proposition 2); as $h \otimes 1_F$ is the homothety h' with ratio $\rho(a)$ on $E \otimes_A F$ and $\rho_*(F)$ is flat, h' is injective (Chapter I, § 2, no. 2, Definition 1); this proves that $\rho(a) \notin \mathfrak{P}$, whence $\overset{-1}{\rho}(\mathfrak{P}) = \mathfrak{p}$.

THEOREM 2. *Let $\rho: A \to B$ be a ring homomorphism, E an A-module and F a B-module such that $\rho_*(F)$ is a flat A-module. Then*

$$\mathrm{Ass}_B(E \otimes_A F) \supset \bigcup_{\mathfrak{p} \in \mathrm{Ass}_A(E)} \mathrm{Ass}_B(F/\mathfrak{p}F). \tag{5}$$

When A is Noetherian, the two sides of (5) *are equal.*

Let $\mathfrak{p} \in \mathrm{Ass}_A(E)$; by definition there exists an exact sequence

$$0 \to A/\mathfrak{p} \to E.$$

Since F is a flat A-module we derive an exact sequence

$$0 \to F/\mathfrak{p}F \to E \otimes_A F$$

whence $\mathrm{Ass}_B(F/\mathfrak{p}F) \subset \mathrm{Ass}_B(E \otimes_A F)$, which proves the inclusion (5).

Suppose now that A is *Noetherian* and let us prove the opposite inclusion.

We proceed in stages:

(i) Suppose first that E is a *finitely generated* A-module and that $\mathrm{Ass}_A(E)$ is reduced to *a single element* $\mathfrak{p}$. By § 1, no. 4, Theorem 1 there exists a composition series $(E_i)_{0 \leqslant i \leqslant n}$ of E such that E_i/E_{i+1} is isomorphic to $A/\mathfrak{p}_i$, where $\mathfrak{p}_i$ is a prime ideal of A; moreover (§ 1, no. 4, Theorem 2 and no. 3, Proposition 7) all the $\mathfrak{p}_i$ contain $\mathfrak{p}$. As F is a flat A-module, the $E_i \otimes_A F$ form a composition series of $E \otimes_A F$ and $(E_i \otimes_A F)/(E_{i+1} \otimes_A F)$ is identified with

$$(A/\mathfrak{p}_i) \otimes_A F = F/\mathfrak{p}_i F.$$

Then by virtue of § 1, no. 1, Proposition 3

$$\mathrm{Ass}_B(E \otimes_A F) \subset \bigcup_{i=0}^{n-1} \mathrm{Ass}_B(F/\mathfrak{p}_i F).$$

We know that E is annihilated by a power of $\mathfrak{p}$ (no. 1, *Remark*); Lemma 1 then shows that, for all $\mathfrak{P} \in \mathrm{Ass}_B(E \otimes_A F)$, $\overset{-1}{\rho}(\mathfrak{P}) = \mathfrak{p}$. As $F/\mathfrak{p}_i F$ is isomorphic to $(A/\mathfrak{p}_i) \otimes_A F$, $\overset{-1}{\rho}(\mathfrak{P}') = \mathfrak{p}_i$ for all $\mathfrak{P}' \in \mathrm{Ass}_B(F/\mathfrak{p}_i F)$ by Lemma 1, whence $\mathrm{Ass}_B(E \otimes_A F) \cap \mathrm{Ass}(F/\mathfrak{p}_i F) = \varnothing$ if $\mathfrak{p}_i \neq \mathfrak{p}$, which proves the theorem in the case considered.

(ii) Suppose only that E is a *finitely generated* A-module. Let $\mathfrak{p}_i$ $(1 \leqslant i \leqslant n)$ be the elements of $\mathrm{Ass}_A(E)$ and let $\{0\} = \bigcap_{i=1}^{n} Q_i$ be a corresponding reduced primary decomposition (no. 3); E is then isomorphic to a submodule of the direct sum of the $E_i = E/Q_i$ and, as F is a flat A-module, $E \otimes_A F$ is isomorphic to a submodule of the direct sum of B-modules $E_i \otimes_A F$. We deduce (§ 1, no. 1, Proposition 3 and Corollary 1 to Proposition 3) that

$$\mathrm{Ass}_B(E \otimes_A F) \subset \bigcup_{i=1}^{n} \mathrm{Ass}_B(E_i \otimes_A F).$$

But E_i is a finitely generated A-module such that $\mathrm{Ass}_A(E_i)$ is reduced to a single element $\mathfrak{p}_i$ (no. 1, Definition 1). By (i), $\mathrm{Ass}_B(E_i \otimes_A F) = \mathrm{Ass}_B(F/\mathfrak{p}_i F)$, whence the theorem in this case.

(iii) *General case.* The B-module $E \otimes_A F$ is the union of the submodules $E' \otimes_A F$, where E′ runs through the set of finitely generated submodules of the A-module E. If $\mathfrak{P}$ belongs to $\mathrm{Ass}_B(E \otimes_A F)$, then there exists a finitely generated submodule E′ of E such that $\mathfrak{P} \in \mathrm{Ass}_B(E' \otimes_A F)$. By (ii), there exists $\mathfrak{p} \in \mathrm{Ass}_A(E')$ such that $\mathfrak{P} \in \mathrm{Ass}_B(F/\mathfrak{p}F)$; as $\mathrm{Ass}_A(E') \subset \mathrm{Ass}_A(E)$, this completes the proof of Theorem 2.

COROLLARY 1. *If* A *is Noetherian and* $\mathfrak{P} \in \mathrm{Ass}_B(E \otimes_A F)$, *then* $\overset{-1}{\rho}(\mathfrak{P}) \in \mathrm{Ass}_A(E)$ *and* $\overset{-1}{\rho}(\mathfrak{P})$ *is the only prime ideal* $\mathfrak{p}$ *of* A *such that* $\mathfrak{P} \in \mathrm{Ass}_B(F/\mathfrak{p}F)$.

This follows from Theorem 2 and Lemma 1 applied to the case where $E = A/\mathfrak{p}$.

COROLLARY 2. *Suppose that* A *and* B *are Noetherian and that* B *is a flat* A*-module. Let* $\mathfrak{p}$ *be a prime ideal of* A, $Q \subset E$ *a* $\mathfrak{p}$*-primary submodule and* $\mathfrak{P}$ *a prime ideal of* B. *For* $Q \otimes_A B$ *to be a* $\mathfrak{P}$*-primary submodule of* $E \otimes_A B$, *it is necessary and sufficient that* $\mathfrak{p}B$ *be a* $\mathfrak{P}$*-primary ideal of* B.

Let us apply Theorem 2 to the A-module E/Q and the B-module B; then $\mathrm{Ass}_A(E/Q) = \{\mathfrak{p}\}$ and $(E/Q) \otimes_A B$ is isomorphic to $(E \otimes_A B)/(Q \otimes_A B)$ and hence $\mathrm{Ass}_B((E \otimes_A B)/(Q \otimes_A B)) = \mathrm{Ass}_B(B/\mathfrak{p}B)$. To say that $Q \otimes_A B$ is $\mathfrak{P}$-primary in $E \otimes_A B$ therefore means that $\mathrm{Ass}_B(B/\mathfrak{p}B)$ is reduced to $\mathfrak{P}$, whence the corollary.

Remark. Suppose that A *and* B are *Noetherian.* Let $\mathfrak{P}$ be a prime ideal of B and let $\mathfrak{p} = \overset{-1}{\rho}(\mathfrak{P})$; let us write $S = A - \mathfrak{p}$ and let $k(\mathfrak{p}) = S^{-1}(A/\mathfrak{p})$ be the field of fractions of $A/\mathfrak{p}$. Since $\mathfrak{P}$ contains $\mathfrak{p}B$, $\mathfrak{P}/\mathfrak{p}B$ is a prime ideal of $B/\mathfrak{p}B$. If ρ' is the composite homomorphism $A \xrightarrow{\rho} B \to B/\mathfrak{p}B$, we know that $S^{-1}(B/\mathfrak{p}B)$ is identified with the ring $(\rho'(S))^{-1}(B/\mathfrak{p}B)$ and $\mathfrak{P}' = S^{-1}(\mathfrak{P}/\mathfrak{p}B)$ with an ideal of this ring (Chapter II, § 2, no. 2, Proposition 6); as $\mathfrak{P}/\mathfrak{p}B$ does not meet $\rho'(S)$, $\mathfrak{P}'$ is a prime ideal of $S^{-1}(B/\mathfrak{p}B)$ (Chapter II, § 2, no. 5, Proposition 11); moreover there are canonical isomorphisms between $S^{-1}(B/\mathfrak{p}B)$, $S^{-1}((A/\mathfrak{p}) \otimes_A B)$ and $(S^{-1}(A/\mathfrak{p})) \otimes_A B = k(\mathfrak{p}) \otimes_A B$; similarly $S^{-1}(F/\mathfrak{p}F)$ is canonically identified with $k(\mathfrak{p}) \otimes_A F$. This being so, under the hypotheses of Theorem 2, *in order that* $\mathfrak{P} \in \mathrm{Ass}_B(E \otimes_A F)$, *it is necessary and sufficient that* $\mathfrak{p} \in \mathrm{Ass}_A(E)$ *and*

$$\mathfrak{P}' \in \mathrm{Ass}_{k(\mathfrak{p})\otimes_A B}(k(\mathfrak{p}) \otimes_A F).$$

For by Theorem 2 and its Corollary 1, it amounts to verifying that the conditions

$$\text{“}\mathfrak{P} \in \mathrm{Ass}_B(F/\mathfrak{p}F)\text{”} \quad \text{and} \quad \text{“}\mathfrak{P}' \in \mathrm{Ass}_{k(\mathfrak{p})\otimes_A B}(k(\mathfrak{p}) \otimes_A F)\text{”}$$

are equivalent; but, as B is Noetherian, this follows from § 1, no. 2, Corollary to Proposition 5 and the above identifications.

PROPOSITION 11. *Suppose that* A *and* B *are Noetherian and that* B *is a flat* A*-module. Let* E *be an* A*-module and* E′ *a submodule of* E *such that, for every ideal* $\mathfrak{p} \in \mathrm{Ass}_A(E/E')$, $\mathfrak{p}B$ *is a prime ideal of* B *or equal to* B. *Let* $E' = \bigcap_{\mathfrak{p} \in \mathrm{Ass}(E/E')} Q(\mathfrak{p})$ *be a reduced primary decomposition of* E′ *in* E, $Q(\mathfrak{p})$ *being* $\mathfrak{p}$*-primary for all* $\mathfrak{p} \in \mathrm{Ass}(E/E')$.

(i) *If* $\mathfrak{p} \in \mathrm{Ass}(E/E')$ *and* $\mathfrak{p}B = B$, *then* $Q(\mathfrak{p}) \otimes_A B = E \otimes_A B$.

(ii) *If* $\mathfrak{p} \in \mathrm{Ass}(E/E')$ *and* $\mathfrak{p}B$ *is prime,* $Q(\mathfrak{p})$ *is* $\mathfrak{p}B$*-primary in* $E \otimes_A B$.

(iii) *If* Φ *is the set of* $\mathfrak{p} \in \mathrm{Ass}(E/E')$ *such that* $\mathfrak{p}B$ *is prime, then*

$$E' \otimes_A B = \bigcap_{\mathfrak{p} \in \Phi} (Q(\mathfrak{p}) \otimes_A B)$$

and this relation is a reduced primary decomposition of $E' \otimes_A B$ *in* $E \otimes_A B$.

If $\mathfrak{p}B = B$, Theorem 2 applied to $E/Q(\mathfrak{p})$ and B shows that

$$\mathrm{Ass}_B((E/Q(\mathfrak{p})) \otimes_A B) = \varnothing$$

and, as B is Noetherian and is a flat A-module, we conclude (§ 1, no. 1, Corollary 1 to Proposition 2) that $Q(\mathfrak{p}) \otimes_A B = E \otimes_A B$. Assertion (ii) follows from Corollary 2 to Theorem 2, taking $\mathfrak{P} = \mathfrak{p}B$. Finally the relation $E' \otimes_A B = \bigcap_{\mathfrak{p} \in \Phi} (Q(\mathfrak{p}) \otimes_A B)$ follows from the fact that B is a flat A-module (Chapter I, § 2, no. 6, Proposition 6); as $\mathfrak{p} = \overset{-1}{\rho}(\mathfrak{p}B)$ for $\mathfrak{p} \in \Phi$ (Lemma 1), $\mathfrak{p}B \neq \mathfrak{p}'B$ for two distinct ideals $\mathfrak{p}, \mathfrak{p}'$ of the set Φ; on the other hand,

$$\mathrm{Ass}((E \otimes_A B)/(E' \otimes_A B)) = \Phi$$

by Theorem 2; we conclude from no. 3, Proposition 4 that

$$E' \otimes_A B = \bigcap_{\mathfrak{p} \in \Phi} (Q(\mathfrak{p}) \otimes_A B)$$

is a reduced primary decomposition.

COROLLARY. *Suppose that* $\mathfrak{p}B$ *is prime for all* $\mathfrak{p} \in \mathrm{Ass}_A(E/E')$. *Then, if* $\mathfrak{p}_1, \ldots, \mathfrak{p}_n$ *are the minimal elements of* $\mathrm{Ass}_A(E/E')$, *the* $\mathfrak{p}_i B$ *are minimal elements of*

$$\mathrm{Ass}_A((E \otimes_A B)/(E' \otimes_A B)).$$

It follows from Proposition 11 that in this case $\mathfrak{p}_i B \neq \mathfrak{p}_j B$ for $i \neq j$.

Examples

(1) Let us take $B = S^{-1}A$, where S is a multiplicative subset of A; if A is Noetherian, the hypotheses of Proposition 11 are satisfied and we recover a part of Proposition 6 of no. 4.

(2) Let A be a Noetherian ring, $\mathfrak{m}$ an ideal of A and B the Hausdorff completion of A with respect to the $\mathfrak{m}$-adic topology; then B is a flat A-module and Theorem 2 may be applied with $F = B$; but in general the hypotheses of Proposition 11 are not satisfied for the prime ideals of A (Chapter III, § 2, Exercise 15 (b)).

(3) Let A be a Noetherian ring and B the polynomial algebra $A[X_1, \ldots, X_n]$; B is Noetherian and is a free A-module and therefore flat. Also, if $\mathfrak{p}$ is a prime ideal of A, $B/\mathfrak{p}B$ is isomorphic to $(A/\mathfrak{p})[X_1, \ldots, X_n]$, which is an integral domain, and hence $\mathfrak{p}B$ is prime; the hypotheses of Proposition 11 are therefore satisfied for every A-module E and every submodule E' of E.

(4) Let A be a finitely generated algebra over a field k, K an extension of k and $B = A \otimes_k K$ the algebra over K obtained by extension of scalars; A and B are Noetherian and B is a free A-module and hence Theorem 2 may be applied to $F = B$. In certain cases (for example if k is algebraically closed) it can be shown that for every prime ideal $\mathfrak{p}$ of A, $\mathfrak{p}B$ is prime or equal to B; we shall return later to this example.

3. PRIMARY DECOMPOSITION IN GRADED MODULES

1. PRIME IDEALS ASSOCIATED WITH A GRADED MODULE

PROPOSITION 1. *Let Δ be a torsion-free commutative group, A a graded ring of type Δ and M a graded A-module of type Δ. Every prime ideal associated with M is graded and is the annihilator of a homogeneous element of M.*

We know that Δ can be given a total order structure compatible with its group structure (*Algebra*, Chapter II, § 11, no. 4, Lemma 2). Let $\mathfrak{p}$ be a prime ideal associated with M, the annihilator of an element $x \in M$, and let $(x_\iota)_{\iota \in \Delta}$ be the family of homogeneous components of x; let $i(1) < i(2) < \cdots < i(r)$ be the values of i for which $x_\iota \neq 0$. Consider an element $a \in \mathfrak{p}$ and let $(a_\iota)_{\iota \in \Delta}$ be the family of its homogeneous components; we shall prove that $a_\iota \in \mathfrak{p}$ for all $i \in \Delta$, which will show that $\mathfrak{p}$ is a *graded* ideal.

We argue by induction on the number of indices i such that $a_\iota \neq 0$. Our assertion is obvious if this number is 0; if not, let m be the greatest of the indices i for which $a_\iota \neq 0$; if we prove that $a_m \in \mathfrak{p}$, the induction hypothesis applied to $a - a_m$ will give the conclusion. Now, $ax = 0$; for all $j \in \Delta$, using the fact that the homogeneous component of degree $m + j$ of ax is 0, we obtain $\sum_{\iota \in \Delta} a_{m-\iota} x_{j+\iota} = 0$; we conclude that $a_m x_j$ is a linear combination of the x_ι of indices $i > j$. In particular therefore $a_m x_{\iota(r)} = 0$, whence, by descending induction on $n < r$, $a_m^{r-n+1} x_{\iota(n)} = 0$. Then $a_m^r x = 0$, whence $a_m^r \in \mathfrak{p}$ and, as $\mathfrak{p}$ is prime, $a_m \in \mathfrak{p}$.

We now show that $\mathfrak{p}$ is the annihilator of a homogeneous element of M. Let us write $\mathfrak{b}_n = \operatorname{Ann}(x_{\iota(n)})$ for $1 \leqslant n \leqslant r$. For every homogeneous element b of $\mathfrak{p}$ and all n the homogeneous component of bx of degree $i(n) + \deg(b)$ is $bx_{\iota(n)}$, hence $bx_{\iota(n)} = 0$ and therefore $b \in \mathfrak{b}_n$; as $\mathfrak{p}$ is generated by its homogeneous elements, $\mathfrak{p} \subset \mathfrak{b}_n$. On the other hand, clearly $\bigcap_{n=1}^{r} \mathfrak{b}_n \subset \operatorname{Ann}(x) = \mathfrak{p}$; as $\mathfrak{p}$ is prime, there exists an n such that $\mathfrak{b}_n \subset \mathfrak{p}$ (Chapter II, § 1, no. 1, Proposition 1), whence $\mathfrak{b}_n = \mathfrak{p} = \operatorname{Ann}(x_{\iota(n)})$, which completes the proof.

COROLLARY. *For every (necessarily graded) prime ideal $\mathfrak{p}$ associated with a graded A-module M, there exists an index $k \in \Delta$ such that the graded A-module $(A/\mathfrak{p})(k)$ obtained from the graded A-module $A/\mathfrak{p}$ by diminishing the degrees by k (Algebra, Chapter II, § 11, no. 2) is isomorphic to a graded submodule of* M.

With the notation of the proof of Proposition 1, consider the homomorphism obtained, by taking quotients, from the homomorphism $a \mapsto ax_{\iota(n)}$ of A to M; the latter is a graded homomorphism of degree $i(n)$ and hence it gives on taking quotients a graded bijective homomorphism of degree $i(n)$ of $A/\mathfrak{p}$ onto a graded submodule of M.

Proposition 2. *Let Δ be a torsion-free commutative group, A a graded Noetherian ring of type Δ and M a graded finitely generated A-module of type Δ. There exists a composition series $(M_i)_{0 \leqslant i \leqslant n}$ consisting of graded submodules of M such that for $0 \leqslant i \leqslant n - 1$ the graded module M_i/M_{i+1} is isomorphic to a shifted graded module $(A/\mathfrak{p}_i)(k_i)$, where $\mathfrak{p}_i$ is a graded prime ideal of A and $k_i \in \Delta$.*

It is sufficient to retrace the argument of § 1, no. 4, Theorem 1 taking on this occasion $\mathfrak{C}$ to be the set of *graded* submodules of M with a composition series with the properties of the statement; we conclude using the Corollary to Proposition 1.

2. PRIMARY SUBMODULES CORRESPONDING TO GRADED PRIME IDEALS

Proposition 3. *Let Δ be a torsion-free commutative group, A a graded Noetherian ring of type Δ, $\mathfrak{p}$ a graded ideal of A and M a graded A-module of type Δ not reduced to 0. Suppose that for every homogeneous element a of $\mathfrak{p}$ the homothety of ratio a on M is almost nilpotent and that for every homogeneous element b of $A - \mathfrak{p}$ the homothety of ratio b on M is injective. Then $\mathfrak{p}$ is prime and the submodule $\{0\}$ of M is $\mathfrak{p}$-primary.*

It suffices to show that $\mathrm{Ass}(M) = \{\mathfrak{p}\}$ (§ 2, no. 1, Proposition 1). Let $\mathfrak{q}$ be a prime ideal associated with M; it is a graded ideal and it is the annihilator of a homogeneous element $x \neq 0$ of M (no. 1, Proposition 1). For every homogeneous element a of $\mathfrak{q}$, $ax = 0$ and hence the homothety of ratio a on M is not injective, whence $a \in \mathfrak{p}$. Conversely, let b be a homogeneous element of $\mathfrak{p}$; there exists an integer $n > 0$ such that $b^n x = 0$, whence $b^n \in \mathrm{Ann}(x) = \mathfrak{q}$ and, as $\mathfrak{q}$ is prime, $b \in \mathfrak{q}$. As $\mathfrak{p}$ and $\mathfrak{q}$ are generated by their respective homogeneous element, $\mathfrak{p} = \mathfrak{q}$, which proves that $\mathrm{Ass}(M) \subset \{\mathfrak{p}\}$. As $M \neq \{0\}$, $\mathrm{Ass}(M) \neq \varnothing$ (§ 1, no. 1, Corollary 1 to Proposition 2), whence $\mathrm{Ass}(M) = \{\mathfrak{p}\}$.

Proposition 4. *Let Δ be a torsion-free commutative group, A a graded Noetherian ring of type Δ and M a graded A-module of type Δ. Let $\mathfrak{p}$ be a prime ideal of A and N, a submodule of M which is $\mathfrak{p}$-primary with respect to M.*

(i) *The largest graded ideal $\mathfrak{p}'$ of A contained in $\mathfrak{p}$ (Algebra, Chapter II, § 11, no. 3) is prime.*

(ii) *The largest graded submodule N' of N is $\mathfrak{p}'$-primary with respect to M.*

We know (*loc. cit.*) that the homogeneous elements of $\mathfrak{p}'$ (resp. N') are just the homogeneous elements of $\mathfrak{p}$ (resp. N). Let a be a homogeneous element of $\mathfrak{p}$; if x is a homogeneous element of M, there exists an integer $n > 0$ such that $a^n x \in N$; as $a^n x$ is homogeneous, $a^n x \in N'$; as every $y \in M$ is the direct sum of a finite number of homogeneous elements, we conclude that there exists an integer $q > 0$ such that $a^q y \in N'$, so that the homothety with ratio a in M/N' is almost nilpotent.

Consider now a homogeneous element b of $A - \mathfrak{p}'$; then $b \notin \mathfrak{p}$ since b is homogeneous. Let x be an element of M such that $bx \in N'$ and let $(x_i)_{i \in \Delta}$ be the family of homogeneous components of x. As N' is graded, $bx_i \in N'$ for all i, hence $bx_i \in N$ and, as $b \notin \mathfrak{p}$, we conclude that $x_i \in N$; as x_i is homogeneous, $x_i \in N'$, whence $x \in N'$ and the homothety with ratio b on M/N' is injective. Proposition 4 then follows from Proposition 3 applied to $\mathfrak{p}'$ and M/N'.

3. PRIMARY DECOMPOSITION IN GRADED MODULES

PROPOSITION 5. *Let Δ be a torsion-free commutative group, A a graded Noetherian ring of type Δ, M a graded A-module of type* M, *N a graded submodule of* M *and* $N = \bigcap_{i \in I} Q_i$ *a primary decomposition of* N *in* M.

(i) *Let Q'_i be the largest graded submodule of* M *contained in Q_i. Then the Q'_i are primary and* $N = \bigcap_{i \in I} Q'_i$.

(ii) *If the primary decomposition* $N = \bigcap_{i \in I} Q_i$ *is reduced, so is the primary decomposition* $N = \bigcap_{i \in I} Q'_i$ *and for all $i \in I$ the prime ideals corresponding to Q_i and Q'_i are equal.*

(iii) *If Q_i corresponds to a prime ideal $\mathfrak{p}_i$ which is a minimal element of* Ass(M/N), *Q_i is a graded submodule of* M.

We have seen (no. 2, Proposition 4) that the Q'_i are primary with respect to M and $N \subset Q'_i \subset Q_i$, which proves (i). Proposition 4 of no. 2 also shows that the prime ideal $\mathfrak{p}'_i$ corresponding to Q'_i is the largest graded ideal contained in the prime ideal $\mathfrak{p}_i$ corresponding to Q_i. If the decomposition $N = \bigcap_{i \in I} Q_i$ is reduced, $\mathfrak{p}_i \in \mathrm{Ass}(M/N)$ for all i (§ 2, no. 3, Proposition 4), hence $\mathfrak{p}_i$ is a graded ideal (no. 1, Proposition 1) and therefore $\mathfrak{p}'_i = \mathfrak{p}_i$; then $\mathrm{Ass}(M/N) = \bigcup_{i \in I} \{\mathfrak{p}'_i\}$ (§ 2, no. 3, Proposition 4), which proves that the decomposition $N = \bigcap_{i \in I} Q'_i$ is reduced (§ 2, no. 3, Proposition 4). Finally, if $\mathfrak{p}_i$ is a minimal element of Ass(M/N), then $\mathfrak{p}'_i = \mathfrak{p}_i$ since $\mathfrak{p}_i$ is graded (no. 1, Proposition 1), whence $Q'_i = Q_i$ by virtue of § 2, no. 3, Proposition 5.

EXERCISES

§ 1

1. (a) Let A be an absolutely flat ring (Chapter I, § 2, Exercise 17). Show that Ass(A) is the set of *isolated* points of Spec(A) (note that Ass(A) is the set of prime ideals which are annihilators of an idempotent of A).

(b) Deduce from (a) an example of a ring A such that $A \neq 0$ and $\text{Ass}(A) = \varnothing$ (cf. Chapter II, § 4, Exercise 17) and for which the conclusion of no. 1, Corollary 2 to Proposition 2 does not hold.

(c) Deduce from (b) an example of a ring A and a multiplicative subset S of A such that the mapping $\mathfrak{p} \mapsto S^{-1}\mathfrak{p}$ of $\text{Ass}(A) \cap \Phi$ to $\text{Ass}(S^{-1}A)$ (no. 2, Proposition 5) is not surjective.

* 2. Let K be a commutative field and A the valuation ring, whose order group is $\mathbf{Q}$, consisting of the "formal power series" $\sum_r c_r T^r$, where $c_r \in K$, $r \in \mathbf{Q}_+$ and the set of r such that $c_r \neq 0$ is well ordered. Let α be an irrational number >0, $\mathfrak{a}$ the ideal of A consisting of the elements with valuation $>\alpha$ and C the ring $A/\mathfrak{a}$. Then Spec(C) reduces to a point $\mathfrak{p}$; for all $x \neq 0$ in C, $\mathfrak{p}x \neq 0$, but for all $\lambda \in \mathfrak{p}$, there exists $y \neq 0$ in C such that $\lambda y = 0$. In particular, $\text{Ass}(C) = \varnothing$, although $\text{Supp}(C) = \text{Spec}(C) = \{\mathfrak{p}\}$. *

3. Let M be an A-module and N a submodule of M. Show that every prime ideal $\mathfrak{p} \in \text{Ass}(M/N)$ which does not contain Ann(N) is associated with M. Give an example of an ideal $\mathfrak{p}$ belonging to Ass(M/N) but not to Ass(M) (take A to be an integral domain, M = A).

4. Give an example of a ring A such that M = A does not satisfy the conclusion of Theorem 1 of no. 4 (cf. Exercise 1 (b)).

5. Let A = K[X, Y] be the polynomial algebra in two indeterminates over a commutative field K and $\mathfrak{a}$ the maximal ideal AX + AY of A. Show that $\text{Supp}(\mathfrak{a}) = \text{Spec}(A)$ is infinite and $\text{Ass}(\mathfrak{a}) = \{0\}$. Prove that there is no

composition series of $\mathfrak{a}$ all of whose factor modules are isomorphic to A. (The existence of such a series would imply that $\mathfrak{a}$ was isomorphic to a module A^n; then necessarily $n = 1$, which is absurd.)

¶ 6. Let A be a ring and E an A-module.

(a) Show that for a prime ideal $\mathfrak{p}$ of A to belong to Supp(E), it is necessary and sufficient that there exist a submodule F of E such that $\mathfrak{p} \in \mathrm{Ass}(E/F)$ (to see that the condition is necessary, consider a submodule F of E of the form $\mathfrak{p}x$, where $x \in E$; to see that the condition is sufficient, use Proposition 7 of no. 3).

(b) Suppose that A is Noetherian and E finitely generated. Show that for every prime ideal $\mathfrak{p} \in \mathrm{Supp}(E)$, there exists a composition series $(E_i)_{0 \leqslant i \leqslant n}$ of E such that, for $0 \leqslant i \leqslant n - 1$, E_i/E_{i+1} is isomorphic to $A/\mathfrak{p}_i$, where $\mathfrak{p}_i$ is a prime ideal and one of the $\mathfrak{p}_i$ is equal to $\mathfrak{p}$.

¶ 7. Let A be a Noetherian ring, M a finitely generated A-module and $\mathfrak{a} = \mathrm{Ann}(M)$.

(a) Show that, if $\mathfrak{a}$ is prime, $\mathfrak{a}$ is the least element of Ass(M). (Note that $\mathfrak{a} = \bigcap_{\mathfrak{p} \in \mathrm{Ass}(M)} \mathfrak{p}$.)

(b) Show that every prime ideal associated with $A/\mathfrak{a}$ is associated with M. (Note that if $\mathfrak{p}$ is a prime ideal of A the annihilator of the class in $A/\mathfrak{a}$ of an element $\alpha \in A$, then $\mathfrak{p} = \mathrm{Ann}(\alpha M)$ and use (a).)

(c) Let $\mathfrak{p}$, $\mathfrak{q}$ be two distinct prime ideals of A such that $\mathfrak{p} \subset \mathfrak{q}$. Show that, if $M = (A/\mathfrak{p}) \oplus (A/\mathfrak{q})$, then $\mathrm{Ass}(A/\mathfrak{a}) \neq \mathrm{Ass}(M)$.

8. Let A be a Noetherian ring, $\mathfrak{a}$ an ideal of A, M a finitely generated A-module and P the submodule of M consisting of the $x \in M$ such that $\mathfrak{a}x = 0$. Show that $\mathrm{Ass}(M/P) \subset \mathrm{Ass}(M)$ (note that for all $x \in M$ the annihilator of $(Ax + P)/P$ is also that of $\mathfrak{a}x$ and use Exercise 7 (a)).

9. Let A be a Noetherian ring and $\mathfrak{a}$ an ideal of A.

(a) For an ideal $\mathfrak{b}$ of A to satisfy $\mathfrak{a} : \mathfrak{b} \neq \mathfrak{a}$, it is necessary and sufficient that $\mathfrak{b}$ be contained in a prime ideal $\mathfrak{p} \in \mathrm{Ass}(A/\mathfrak{a})$.

(b) Let A be the polynomial algebra K[X, Y, Z] in 3 indeterminates over a field K. Let $\mathfrak{n} = AX$, $\mathfrak{m} = AX + AY + AZ$, $\mathfrak{a} = \mathfrak{n} \cap \mathfrak{m}^2$. Show that there is a prime ideal $\mathfrak{p}$ containing $\mathfrak{a}$ such that $\mathfrak{a} : \mathfrak{p} \neq \mathfrak{a}$ but which is not a prime ideal associated with A.

10. (a) Give an example of a **Z**-module M such that $\mathrm{Ass}(\mathrm{Hom}_{\mathbf{Z}}(M, M))$ is not contained in Supp(M) (cf. Chapter II, § 4, Exercise 24 (c)).

(b) Give an example of **Z**-modules E, F such that

$$\mathrm{Ass}(\mathrm{Hom}_{\mathbf{Z}}(E, F)) = \varnothing,$$

but $\mathrm{Ass}(F) \cap \mathrm{Supp}(E) \neq \varnothing$ (take $F = \mathbf{Z}$).

11. (a) Let A be a Noetherian ring and E an A-module. Show that the

canonical homomorphism of E to the product $\prod_{\mathfrak{p} \in \mathrm{Ass}(E)} E_{\mathfrak{p}}$ is injective (if N is the kernel of this homomorphism, show that $\mathrm{Ass}(N) = \varnothing$).

(b) Take A to be the polynomial ring K[X, Y, Z] over a field K; let $\mathfrak{p}_1 = AX + AY$, $\mathfrak{p}_2 = AX + AZ$, which are prime ideals and $\mathfrak{a}$ the ideal $\mathfrak{p}_1\mathfrak{p}_2$. The set $\mathrm{Ass}(A/\mathfrak{a})$ consists of $\mathfrak{p}_1$, $\mathfrak{p}_2$ and the maximal ideal $\mathfrak{m} = \mathfrak{p}_1 + \mathfrak{p}_2$; show that the canonical homomorphism from $E = A/\mathfrak{a}$ to $E_{\mathfrak{p}_1} \times E_{\mathfrak{p}_2}$ is not injective.

12. Let A be a Noetherian ring and P a projective A-module. Show that, if, for all $\mathfrak{p} \in \mathrm{Ass}(A)$, $P_{\mathfrak{p}}$ is a finitely generated $A_{\mathfrak{p}}$-module, then P is a finitely generated A-module (embed A in the product $\prod_{\mathfrak{p} \in \mathrm{Ass}(A)} A_{\mathfrak{p}}$ (Exercise 11) and use *Algebra*, Chapter II, § 5, no. 5, Proposition 9).

¶ 13. Let A be a Noetherian ring, E a finitely generated A-module, $\mathfrak{p}$, $\mathfrak{q}$ two prime ideals of A such that $\mathfrak{p} \subset \mathfrak{q}$ and a an element of $\mathfrak{q}$. Suppose that $\mathfrak{p} \in \mathrm{Ass}(E)$ and that the homothety a_E is injective. Show that there exists a prime ideal $\mathfrak{n} \in \mathrm{Ass}(E/aE)$ such that $\mathfrak{p} + Aa \subset \mathfrak{n} \subset \mathfrak{q}$. (Replacing A by $A_{\mathfrak{q}}$, we may assume that A is local with maximal ideal $\mathfrak{q}$. Let $F \neq 0$ be the submodule of E consisting of the x such that $\mathfrak{p}x = 0$. Show that the relation $F \subset aE$ would imply $F = aF$ and obtain a contradiction using Nakayama's Lemma. Then use Proposition 8 of no. 4.)

14. Let A be an integral domain, $a \neq 0$ an element of A and $\mathfrak{p}$ an element of $\mathrm{Ass}(A/aA)$. Show that, for every element $b \neq 0$ of $\mathfrak{p}$, $\mathfrak{p} \in \mathrm{Ass}(A/bA)$ (if $c \in A$ is such that the relation $xc \in aA$ is equivalent to $x \in \mathfrak{p}$, show that there exists $d \in A$ such that $xd \in bA$ is equivalent to $xc \in aA$).

15. Let A be a Noetherian ring, E a finitely generated A-module, F a submodule of E and $\mathfrak{m}$ an ideal of A. For F to be closed in E under the $\mathfrak{m}$-adic topology, it is necessary and sufficient that $\mathfrak{p} + \mathfrak{m} \neq A$ for every ideal $\mathfrak{p} \in \mathrm{Ass}(E/F)$. (Reduce it to the case where $F = 0$, use Chapter III, § 3, no. 2, Corollary to Proposition 5 and apply Corollary 2 to Proposition 2 of no. 1 to an element $1 + m$, where $m \in \mathfrak{m}$.)

16. Let A be a Noetherian ring. For A to be isomorphic to a finite product of integral domains, it is necessary and sufficient that for every prime ideal $\mathfrak{p}$ of A the local ring $A_{\mathfrak{p}}$ be an integral domain. (To see that the condition is sufficient, note first that it implies that that ring A is reduced; deduce that $\{0\} = \bigcap_i \mathfrak{p}_i$, where the $\mathfrak{p}_i$ $(1 \leqslant i \leqslant n)$ are the minimal prime ideals of A. Show that, for $i \neq j$, of necessity $\mathfrak{p}_i + \mathfrak{p}_j = A$; for this, note that, if there existed a maximal ideal $\mathfrak{m}$ containing $\mathfrak{p}_i + \mathfrak{p}_j$, the ring $A_{\mathfrak{m}}$ would not be an integral domain, using Corollary 3 to Proposition 2 of no. 1, and the Corollary to Proposition 5 of no. 2.)

¶ 17. Let A be a ring and M an A-module. A prime ideal $\mathfrak{p}$ of A is said to be *weakly associated* with M if there exists $x \in M$ such that $\mathfrak{p}$ is a minimal element of the set of prime ideals containing $\mathrm{Ann}(x)$; we denote by $\mathrm{Ass}_f(M)$ the set of ideals weakly associated with M. Then $\mathrm{Ass}(M) \subset \mathrm{Ass}_f(M)$.

(a) Show that the relation $M \neq 0$ is equivalent to $\mathrm{Ass}_f(M) \neq \varnothing$.

(b) For $a \in A$ to be such that a_M is injective, it is necessary and sufficient that a belong to no element of $\mathrm{Ass}_f(M)$ (note that, if a belongs to the radical of an ideal $\mathrm{Ann}(x)$ where $x \neq 0$ in M, there exists $y \neq 0$ in M such that $a \in \mathrm{Ann}(y)$. To show that the condition is necessary, reduce it, by considering the ring $A/\mathrm{Ann}(x)$, to proving that in a ring A every element belonging to a prime ideal $\mathfrak{p}$ which is minimal in the set of prime ideals of A is necessarily a divisor of 0 (Chapter II, § 2, no. 6, Proposition 12).)

For $a \in A$ to be such that a_M is almost nilpotent, it is necessary and sufficient that a belongs to all the elements of $\mathrm{Ass}_f(M)$.

(c) If N is a submodule of M, then

$$\mathrm{Ass}_f(N) \subset \mathrm{Ass}_f(M) \subset \mathrm{Ass}_f(N) \cup \mathrm{Ass}_f(M/N).$$

(Note that, if $\mathfrak{p}$ is a prime ideal, $a \notin \mathfrak{p}$, $x \in M$ such that $ax \in N$ and $\mathrm{Ann}(x) \subset \mathfrak{p}$, then $\mathrm{Ann}(ax) \subset \mathfrak{p}$.)

(d) Let S be a multiplicative subset of A and Φ the set of prime ideals of A not meeting S. Show that $\mathfrak{p} \mapsto S^{-1}\mathfrak{p}$ is a bijection of $\mathrm{Ass}_f(M) \cap \Phi$ onto $\mathrm{Ass}_f(S^{-1}M)$. (Observe that the inverse image under the canonical mapping $A \to S^{-1}A$ of the annihilator of an element x/s, where $x \in M$, $s \in S$, is the saturation with respect to S of $\mathrm{Ann}(x)$.)

(e) Let S be a multiplicative subset of A; show that, if N is the kernel of the canonical homomorphism $M \to S^{-1}M$, $\mathrm{Ass}_f(N)$ is the set of $\mathfrak{p} \in \mathrm{Ass}_f(M)$ which meet S and $\mathrm{Ass}_f(M/N)$ the set of $\mathfrak{p} \in \mathrm{Ass}_f(M)$ which do not meet S. (To prove the latter point, consider a prime ideal $\mathfrak{q}$ which is minimal in the set of those containing $\mathrm{Ann}(\bar{y})$, where $\bar{y} \in M/N$; note that, if $y \in \bar{y}$ and $t \in \mathfrak{q}$, there exists $c \notin \mathfrak{q}$ and an integer $n > 0$ such that $ct^n y \in N$ (Chapter II, § 2, no. 6, Proposition 12); deduce first that $\mathfrak{q} \cap S = \varnothing$. There exists $s \in S$ such that $sct^n y = 0$; conclude that there can be no prime ideal $\mathfrak{q}' \neq \mathfrak{q}$ such that $\mathrm{Ann}(y) \subset \mathfrak{q}' \subset \mathfrak{q}$.)

(f) For a prime ideal of A to belong to $\mathrm{Supp}(M)$, it is necessary and sufficient that it contain an element of $\mathrm{Ass}_f(M)$.

(g) Show that, if A is Noetherian, $\mathrm{Ass}_f(M) = \mathrm{Ass}(M)$.

(h) If A is an absolutely flat ring, then $\mathrm{Ass}_f(A) = \mathrm{Spec}(A)$.

(i) For every A-module E the canonical homomorphism from E to the product $\prod_{\mathfrak{p} \in \mathrm{Ass}_f(E)} E_{\mathfrak{p}}$ is injective; in other words, the intersection of the *saturations* of $\{0\}$ in E with respect to the ideals $\mathfrak{p} \in \mathrm{Ass}_f(E)$ is reduced to 0. Generalize Exercise 12 similarly.

(j) Let M be a finitely generated A-module. Show that the minimal elements of the set of prime ideals containing $\mathfrak{a} = \text{Ann}(M)$ belong to $\text{Ass}_f(M)$ (if $(x_i)_{1 \leqslant i \leqslant n}$ is a system of generators of M, show that such an ideal contains one of the $\text{Ann}(x_i)$). Deduce that $\text{Ass}_f(A/\mathfrak{a})$ is contained in $\text{Ass}_f(M)$ (if a prime ideal $\mathfrak{p}$ contains the annihilator of the class mod. $\mathfrak{a}$ of an element $\alpha \in A$, note that $\mathfrak{p}$ contains the annihilator of the submodule αM of M); show that $\text{Ass}_f(A/\mathfrak{a})$, Ass(M) and the set of prime ideals containing $\mathfrak{a}$ have the same minimal elements.

18. Let A be a ring and M an A-module. For an element $c \in A$ to be such that, for all $a \in A$ such that the homothety a_M is injective, b_M is an injective homothety for all b of the form $a + \lambda c$, where $\lambda \in A$, it is sufficient that c belong to the intersection of the maximal elements of $\text{Ass}_f(M)$ (use Exercise 17 (b)). Is this condition necessary?

* 19. Let A be the ring of a valuation of height 2, Γ its order group, Γ_1 the unique isolated subgroup of Γ distinct from $\{0\}$ and Γ and $\gamma > 0$ an element of Γ not belonging to Γ_1. Let $\mathfrak{a}$ be the ideal of those $x \in A$ such that $v(x) \geqslant \gamma$, $\mathfrak{b}$ the ideal of those $x \in A$ such that $v(x) > \gamma$, $E = A/\mathfrak{a}$ and $F = A/\mathfrak{b}$. Show that $\text{Ass}_f(\text{Hom}_A(E, F))$ is distinct from $\text{Ass}_f(F) \cap \text{Supp}(E)$. *

§ 2

1. (a) Show that in the Noetherian polynomial ring $A = \mathbf{Z}[X]$ in one indeterminate over $\mathbf{Z}$, the ideal $\mathfrak{m} = 2A + AX$ is maximal and that the ideal $\mathfrak{q} = 4A + AX$ is $\mathfrak{m}$-primary, but is not equal to a power of $\mathfrak{m}$.

(b) In the ring $B = \mathbf{Z}[2X, X^2, X^3] \subset A$ which is Noetherian, show that the ideal $\mathfrak{p} = 2BX + BX^2$ is prime, but that $\mathfrak{p}^2$ is not $\mathfrak{p}$-primary, although its radical is equal to $\mathfrak{p}$.

(c) Let K be a commutative field and A the quotient ring of K[X, Y, Z] by the ideal generated by $Z^2 - XY$; let x, y, z be the canonical images of X, Y, Z in A. Show that $\mathfrak{p} = Ax + Az$ is prime, that $\mathfrak{p}^2$ is not primary and that $\mathfrak{p}^2 = \mathfrak{a} \cap \mathfrak{b}^2$ is a primary decomposition of $\mathfrak{p}^2$, where $\mathfrak{a} = Ax$ and $\mathfrak{b} = Ax + Ay + Az$.

2. In the example of Exercise 11 (b) of § 1, show that $\mathfrak{a} = \mathfrak{m}^2 \cap \mathfrak{p}_1 \cap \mathfrak{p}_2$ is a reduced primary decomposition of $\mathfrak{a}$ in A and that the saturation of $\mathfrak{a}$ with respect to $\mathfrak{m}$ is equal to $\mathfrak{a}$ (and is therefore not primary).

3. Let A be a Noetherian ring and M a finitely generated A-module. Let Q be a $\mathfrak{p}$-primary submodule in M. The greatest lower bound of the integers $n \geqslant 1$ such that $\mathfrak{p}^n M \subset Q$ is called the *exponent* of Q in M and denoted by $e(M/Q)$.

Let $(Q_\lambda)_{\lambda \in L}$ be a family of $\mathfrak{p}$-primary submodules in M. For $Q = \bigcap_{\lambda \in L} Q_\lambda$

to be $\mathfrak{p}$-primary in M, it is necessary and sufficient that the family $(e(M/Q_\lambda))_{\lambda \in L}$ be bounded above; then $e(M/Q) \geqslant e(M/Q_\lambda)$ for all $\lambda \in L$.

¶ 4. Let A be a Noetherian ring and M a finitely generated A-module. Let $\mathfrak{p}$ be an element of Ass(M) and let $\mathfrak{E}\mathfrak{p}$ be the set of submodules Q of M such that $\mathrm{Ass}(M/Q) = \{\mathfrak{p}\}$ and $\mathrm{Ass}(Q) = \mathrm{Ass}(M) - \{\mathfrak{p}\}$ (a set which is not empty by § 1, no. 1, Proposition 4). We write $e_\mathfrak{p}(M) = \inf_{Q \in \mathfrak{E}_\mathfrak{p}} e(M/Q)$ (Exercise 3).

(a) Let $n_\mathfrak{p}$ be an integer $\geqslant e_\mathfrak{p}(M)$ and let $\mathfrak{F}(n_\mathfrak{p})$ be the subset of $\mathfrak{E}_\mathfrak{p}$ consisting of the submodules Q such that $e(M/Q) \leqslant n_\mathfrak{p}$. Show that $\mathfrak{F}(n_\mathfrak{p})$ has a least element $Q(\mathfrak{p}, n_\mathfrak{p})$.

(b) Let $(n_\mathfrak{p})_{\mathfrak{p} \in \mathrm{Ass}(M)}$ be a family of integers such that $n_\mathfrak{p} \geqslant e_\mathfrak{p}(M)$ for all $\mathfrak{p} \in \mathrm{Ass}(M)$. Show that the submodules $Q(\mathfrak{p}, n_\mathfrak{p})$ corresponding to this family form a reduced primary decomposition of $\{0\}$ in M said to be *canonically determined by the family* $(n_\mathfrak{p})$. If we take $n_\mathfrak{p} = e_\mathfrak{p}(M)$ for all $\mathfrak{p} \in \mathrm{Ass}(M)$, the primary decomposition consisting of the $Q(\mathfrak{p}) = Q(\mathfrak{p}, e_\mathfrak{p}(M))$ is called the *canonical primary decomposition* of $\{0\}$ in M. Let $\{0\} = \bigcap_{\mathfrak{p} \in \mathrm{Ass}(M)} Q'(\mathfrak{p})$ be any reduced primary decomposition of $\{0\}$ in M. Show that $e(M/Q(\mathfrak{p})) \leqslant e(M/Q'(\mathfrak{p}))$ for all $\mathfrak{p} \in \mathrm{Ass}(M)$ and that, if $e(M/Q(\mathfrak{p})) = e(M/Q'(\mathfrak{p}))$ for some $\mathfrak{p}$, then $Q(\mathfrak{p}) \subset Q'(\mathfrak{p})$ ("*Theorem of Ortiz*").

(c) Show that $Q(\mathfrak{p}, n_\mathfrak{p})$ is the saturation of $\mathfrak{p}^{n_\mathfrak{p}} M$ with respect to $\mathfrak{p}$ (use § 1, no. 2, Proposition 6); $\mathfrak{p}$ is the least element of $\mathrm{Supp}(M/\mathfrak{p}^n M)$.

(d) Let S be a multiplicative subset of A not meeting a prime ideal $\mathfrak{p} \in \mathrm{Ass}(M)$ and let Q be a $\mathfrak{p}$-primary submodule of M. Show that $e(M/Q) = e(S^{-1}M/S^{-1}Q)$. Show that, if S is a multiplicative subset of A, Φ the set of $\mathfrak{p} \in \mathrm{Ass}(M)$ such that $S \cap \mathfrak{p} = \varnothing$ and N the saturation of $\{0\}$ in M with respect to S, the $Q(\mathfrak{p}, n_\mathfrak{p})$ where $\mathfrak{p} \in \Phi$ form the reduced primary decomposition of N in M canonically determined by the family $(n_\mathfrak{p})_{\mathfrak{p} \in \Phi}$ and the $S^{-1}Q(\mathfrak{p}, n_\mathfrak{p})$ where $\mathfrak{p} \in \Phi$ the reduced primary decomposition of $\{0\}$ in $S^{-1}M$ canonically determined by the family $(n_\mathfrak{p})_{\mathfrak{p} \in \Phi}$. Consider the particular case of canonical primary decompositions.

5. Let L be a finitely generated free **Z**-module and T a finite commutative group whose order is a power of a prime number p. Show that the canonical primary decomposition (Exercise 4) of $\{0\}$ in $M = L \oplus T$ is $\{0\} = p^n L \cap T$, where n is the least integer $\geqslant 0$ such that $p^n T = 0$ (note that $p^n M = p^n L$).

6. In Exercise 5 of § 1, $\{0\}$ is primary in the A-module $\mathfrak{a}$ relative to the prime ideal $\{0\}$ of A. Show that the submodule AX of $\mathfrak{a}$ is primary with respect to $\mathfrak{a}$ relative to a prime ideal $\neq \{0\}$ and that the submodule AXY of $\mathfrak{a}$ is not primary with respect to $\mathfrak{a}$.

7. Let A be a Noetherian ring, M a finitely generated A-module and $(\mathfrak{p}_i)_{1 \leqslant i \leqslant n}$ a sequence obtained by arranging the elements of Ass(M) in any

order. Show that there exists a composition series $(M_i)_{0 \leqslant i \leqslant n}$ of M such that, for $0 \leqslant i \leqslant n-1$, M_{i+1} is $\mathfrak{p}_i$-primary in M_i (use Proposition 4 of no. 3).

¶ 8. Let A be a Noetherian ring, E a finitely generated A-module, F a submodule of E and $\mathfrak{m}$ an ideal of A. Let $F = \bigcap_{\mathfrak{p} \in \mathrm{Ass}(E/F)} Q(\mathfrak{p})$ be a reduced primary decomposition of F in E. Show that the closure of F in E under the $\mathfrak{m}$-adic topology on E is equal to $\bigcap_{\mathfrak{p} \in \Phi} Q(\mathfrak{p})$, where Φ is the set of $\mathfrak{p} \in \mathrm{Ass}(E/F)$ such that $\mathfrak{p} + \mathfrak{m} \neq A$. (Consider first the case where F is $\mathfrak{p}$-primary in E and show that, if $\mathfrak{p} + \mathfrak{m} = A$, F is dense in E; proceed to the general case using Exercise 15 of § 1 and Chapter III, § 3, no. 4, Corollary 1 to Theorem 3).

9. Let A be a Noetherian ring, $\mathfrak{m}$ a maximal ideal of A and M an A-module such that {0} is $\mathfrak{m}$-primary in M. If $\hat{A}$ is the Hausdorff completion of A with respect to the $\mathfrak{m}$-adic topology, show that the canonical mapping $M \to M \otimes_A \hat{A}$ is bijective (consider first the case where M is finitely generated, using Proposition 7 of no. 5; in the general case, consider M as the direct limit of its finitely generated submodules).

10. Let A be a Noetherian ring and M an A-module. Show that the following properties are equivalent:

(α) Every finitely generated submodule of M is of finite length.
(β) M is a direct limit of A-modules of finite length.
(γ) Every element $\mathfrak{p} \in \mathrm{Ass}(M)$ is a maximal ideal of A.
(δ) Every element $\mathfrak{p} \in \mathrm{Supp}(M)$ is a maximal ideal of A.

11. Let A be a ring, M an A-module and N a submodule of M. The set of $a \in A$ such that the homothety of ratio a on M/N is almost nilpotent is called the *radical* of N in M and denoted by $\mathfrak{r}_M(N)$. Show that $\mathfrak{r}_M(N)$ is an ideal of A and that, for all $\mathfrak{p} \in \mathrm{Ass}_f(M/N)$, $\mathfrak{r}_M(N) \subset \mathfrak{p}$. If N_1, N_2 are two submodules of M, then $\mathfrak{r}_M(N_1 \cap N_2) = \mathfrak{r}_M(N_1) \cap \mathfrak{r}_M(N_2)$; if $\mathfrak{a}$ is an ideal of A, $\mathfrak{r}_M(\mathfrak{a}N) \supset \mathfrak{r}(\mathfrak{a}) \cap \mathfrak{r}_M(N)$ and $\mathfrak{r}_A(\mathfrak{a}) = \mathfrak{r}(\mathfrak{a})$.

¶ 12. Let A be a ring, M an A-module and N a submodule of M.

(a) For $\mathrm{Ass}_f(M/N)$ (§ 1, Exercise 17) to be reduced to a single element $\mathfrak{p}$, it is necessary and sufficient that $N \neq M$ and that, for all $a \in A$, the homothety with ratio a on M/N be injective or almost nilpotent; then $\mathfrak{p} = \mathfrak{r}_M(N)$ (Exercise 11) and we also say that N is *primary* (or $\mathfrak{p}$-*primary*) in M. For every submodule M′ of M such that $N \subset M'$ and $N \neq M'$, N is then $\mathfrak{p}$-primary in M′.

(b) If $\mathfrak{r}_M(N)$ is a maximal ideal of A, show that N is primary in M. In particular, every ideal $\mathfrak{q}$ of A which is only contained in a single prime ideal $\mathfrak{m}$ (necessarily maximal) is $\mathfrak{m}$-primary.

(c) Let A be an integral domain and x an element of A such that $\mathfrak{p} = Ax$ is prime; then, for every integer $n \geqslant 1$, Ax^n is $\mathfrak{p}$-primary. Show that this result does not extend to the case where A is not an integral domain (take $A = B/\mathfrak{b}$, where $B = K[X, Y]$ (K being a field), $\mathfrak{b} = BX^2 + BXY$).

(d) If A is an absolutely flat ring (Chapter I, § 2, Exercise 17), the primary ideals of A are identical with the prime ideals of A.

(e) Give an example of a **Z**-module M such that {0} is p-primary (for a prime number p) but that for no $a \neq 0$ in **Z** is the homothety a_M nilpotent (cf. Chapter II, § 2, Exercise 3).

13. (a) Let $u: M \to M'$ be a surjective A-module homomorphism. Show that, if N′ is a primary submodule in M′, $N = \overset{-1}{u}(N')$ is primary in M and $\mathfrak{r}_M(N) = \mathfrak{r}_{M'}(N')$.

(b) Every intersection of a non-empty finite family of $\mathfrak{p}$-primary submodules in an A-module M is $\mathfrak{p}$-primary in M. Does the proposition extend to arbitrary intersections?

(c) For an A-module M to be such that {0} is $\mathfrak{p}$-primary in M, it is necessary and sufficient that, for every submodule $N \neq 0$ of M, $\text{Supp}(N) = V(\mathfrak{p})$. (Note that, for all $x \in M$, M contains a submodule isomorphic to $A/\text{Ann}(x)$.)

14. (a) Generalize Proposition 3 of no. 1 to arbitrary rings.

(b) Let M be an A-module such that, for every multiplicative subset S of A, the kernel of the canonical mapping $M \to S^{-1}M$ is {0} or M. Show that {0} is primary in M (consider for all $a \neq 0$ in A the multiplicative subset of the a^n $(n \geqslant 0)$.

¶ 15. Let $\mathfrak{q}$ be a primary ideal in a ring A and $\mathfrak{p}$ its radical. Show that in the polynomial ring $B = A[X]$ the ideal $B\mathfrak{q}$ is primary and has radical the ideal $B\mathfrak{p}$. (Let $f(X) = \sum_j a_j X^j$ and $g(X) = \sum_j b_j X^j$ be two non-constant polynomials such that $fg \in B\mathfrak{q}$ and $f \notin B\mathfrak{p}$. Let a_m be the coefficient of smallest index not belonging to $\mathfrak{p}$; let $\mathfrak{a}$ be the ideal of A generated by $a_0, a_1, \ldots, a_{m-1}$; there exists an integer k such that $\mathfrak{a}^k \subset \mathfrak{q}$. Let $\mathfrak{q}_i$ be the transporter of $\mathfrak{a}^{k-i}$ in $\mathfrak{q}$ for $i \leqslant k$. Show by induction on i that $g \in B\mathfrak{q}_i$, arguing by *reductio ad absurdum.*

16. In a ring A let $\mathfrak{p}$ be a non-maximal prime ideal and $\mathfrak{q}$ a $\mathfrak{p}$-primary ideal distinct from $\mathfrak{p}$. Let x be an element of $\mathfrak{p} - \mathfrak{q}$ and y an element of $A - \mathfrak{p}$ such that $\mathfrak{p} + Ay \neq A$. Show that the ideal $\mathfrak{a} = \mathfrak{q} + Axy$, such that $\mathfrak{p} \supset \mathfrak{a} \supset \mathfrak{q}$, is not primary (note that $x \in \mathfrak{a}$ is impossible).

17. Let M be an A-module and N a $\mathfrak{p}$-primary submodule in M.

(a) Let F be a submodule of M not contained in N. Show that the transporter $N:F$ is an ideal of A contained in $\mathfrak{p}$.

(b) Let $\mathfrak{a}$ be an ideal of A. Show that if $\mathfrak{a} \subset N:M$, then $N: a = M$. If $\mathfrak{a} \not\subset N:M$, $N: \mathfrak{a}$ is a $\mathfrak{p}$-primary submodule in M. If $\mathfrak{a} \not\subset \mathfrak{p}$, then $N: \mathfrak{a} = N$.

18. (a) Let A be a ring and M an A-module. Show that, if $\mathfrak{p}$ is a minimal element of $\mathrm{Ass}_f(\mathrm{M})$, the saturation of $\{0\}$ with respect to $\mathfrak{p}$ in M is $\mathfrak{p}$-primary in M (cf. § 1, Exercise 17 (b)).

(b) Let $\mathfrak{p}$ be a prime ideal of A. Show that for every integer $n > 0$ the saturation in A of $\mathfrak{p}^n$ with respect to $\mathfrak{p}$ is $\mathfrak{p}$-primary in A (use Exercise 14 (a)); this saturation is denoted by $\mathfrak{p}^{(n)}$ and is called the *n-th symbolic power* of $\mathfrak{p}$.

(c) Show that, if every prime ideal of A is maximal, then, for every A-module M and every submodule N of M, N is the intersection of a family (finite or otherwise) of primary submodules in M (use (a) and Exercise 17 (i) of § 1).

* 19. Determine the primary ideals in the ring of a valuation of height 2 (cf. Chapter VI); deduce an example of a ring where there are ideals which are not intersections of a family (finite or otherwise) of primary ideals, although there exist in the ring only two distinct prime ideals. *

¶ 20. The notion of *primary decomposition* and that of *reduced primary decomposition* may be defined for any ring A as in nos. 2 and 3 (using the notion of primary submodule defined in Exercise 12).

(a) Generalize Propositions 4 and 6 by replacing Ass by Ass_f throughout.

(b) Let E be a finitely generated A-module and F a submodule of E. For every multiplicative subset S of A let $\mathrm{sat}_{\mathrm{S}}(\mathrm{F})$ denote the saturation of F in E with respect to S. Show that, if F admits a primary decomposition in E, every saturation of F in E is of the form $\mathrm{F} : (a)$ for some $a \in \mathrm{A}$. (Let $\mathfrak{p}_i$ $(1 \leqslant i \leqslant r)$ be the elements of $\mathrm{Ass}_f(\mathrm{E/F})$. Show first that, using (a), that for every multiplicative subset S of A there exists $b \in \mathrm{A}$ such that, if T is the multiplicative subset of the b^n $(n \geqslant 0)$, then $\mathrm{sat}_{\mathrm{T}}(\mathrm{F}) = \mathrm{sat}_{\mathrm{S}}(\mathrm{F})$, taking b such that $b \in \mathfrak{p}_i$ if $\mathfrak{p}_i \cap \mathrm{S} \neq 0$ and $b \notin \mathfrak{p}_i$ otherwise. Then prove that we may take $a = b^m$ for m sufficiently large.)

(c) Give an example of a **Z**-module E such that $\{0\}$ is primary with respect to E but there exist saturations $\mathrm{sat}_{\mathrm{S}}(0)$ which are not of the form $0 : (a)$ (cf. *Algebra*, Chapter VII, § 2, Exercise 3).

21. Let A be a ring, F a monic polynomial of A[X], B the ring $\mathrm{A[X]/F.A[X]}$, $\mathfrak{m}$ a maximal ideal of A, $k = \mathrm{A}/\mathfrak{m}$ the quotient field and $\overline{\mathrm{F}}$ the canonical image of F in $k[\mathrm{X}]$.

(a) Let $\overline{\mathrm{F}} = \prod_{i=1}^{n} f_i^{e_i}$, where the f_i are distinct irreducible polynomials of $k[\mathrm{X}]$. Let $\mathrm{F}_i \in \mathrm{A[X]}$ be a monic polynomial such that its canonical image in $k[\mathrm{X}]$ is f_i. Let $\mathfrak{M}_i$ denote the ideal $\mathfrak{m}\mathrm{B} + \mathrm{F}_i.\mathrm{B}$ of B; show that the $\mathfrak{M}_i$ are distinct and are the only maximal ideals of B containing $\mathfrak{m}\mathrm{B}$ (note that $\mathrm{B}/\mathfrak{m}\mathrm{B}$ is an algebra of finite rank over $\mathrm{A}/\mathfrak{m} = k$ generated by the roots of $\overline{\mathrm{F}}$).

(b) For all i let us write $\mathfrak{Q}_i = \mathfrak{m}\mathrm{B} + \mathrm{F}_i^{e_i}.\mathrm{B}$. Show that the $\mathfrak{Q}_i$ are $\mathfrak{M}_i$-primary in B and that $\mathfrak{m}\mathrm{B} = \mathfrak{Q}_1 \cap \mathfrak{Q}_2 \cap \cdots \cap \mathfrak{Q}_n = \mathfrak{Q}_1\mathfrak{Q}_2 \cdots \mathfrak{Q}_n$.

(c) For B to be a local ring, it is necessary and sufficient that A be a local ring and that $\overline{F}$ be a power of an irreducible polynomial of $k[X]$.

¶ 22. Let E be an A-module and F a submodule of E.

(a) Let $\mathfrak{p}$ be a prime ideal of A and a an element of A such that $a \notin \mathfrak{p}$ and $Q = F:(a)$ is $\mathfrak{p}$-primary in E. Show that $F = Q \cap (F + aE)$.

(b) Suppose that there exist $b \in A$ and $x \in E$ such that $b^n x \notin F$ for all $n > 0$. Show that, if there exists an integer $n > 0$ such that $F:b^{n+1} = F:b^n$, then $F = (F + Ab^n x) \cap (F:(b^n))$.

(c) Suppose that F is *irreducible* with respect to E (*Algebra*, Chapter II, § 2, Exercise 16) and consider the three following properties:

(α) F is primary in E.

(β) For every prime ideal $\mathfrak{p}$ of A, the saturation of F with respect to $\mathfrak{p}$ in E is of the form $F:(a)$ for some $a \notin \mathfrak{p}$.

(γ) For all $b \in A$, the sequence $(F:(b^n))_{n \geqslant 1}$ is stationary.

Show that each of conditions (β), (γ) implies (α) (use (a) and (b) and Exercise 18 (a)). If E is finitely generated, the three conditions (α), (β) and (γ) are equivalent (cf. Exercise 20 (b) and 20 (c)).

(d) Let K be a field, A the ring $K[X, Y]$ and $\mathfrak{m}$ the maximal ideal $AX + AY$ of A; show that the ideal $\mathfrak{m}^2$ is primary but not irreducible.

(e) Let A be a not necessarily commutative ring and E a left Noetherian A-module. Show that every submodule F of E is the intersection of a finite family of irreducible submodules in E (consider a submodule of E which is maximal among those which are not finite intersections of irreducible submodules).

(f) Deduce from (e) and (c) a new proof of Theorem 1 of no. 2.

¶ 23. Let A be a ring. An A-module E is called *Laskerian* if it finitely generated and if every submodule of E has a primary decomposition in E.

(a) For a finitely generated A-module E to be Laskerian, it is necessary and sufficient that it satisfy the two following axioms:

(LA_I) For every submodule F of E and every prime ideal $\mathfrak{p}$ of A, the saturation of F with respect to $\mathfrak{p}$ in E is of the form $F:(a)$ for some $a \notin \mathfrak{p}$.

(LA_{II}) For every submodule F of E, every decreasing sequence $(\text{sat}_{S_n}(F))$ (where (S_n) is any decreasing sequence of multiplicative subsets of A) is stationary.

(To show that the conditions are sufficient, prove first, using (LA_I) and Exercises 18 (a) and 22 (a), that, for every submodule F of E, there exists a submodule Q of E which is primary for some ideal $\mathfrak{p} \supset F:E$ and a submodule $G = F + aE$, where $a \notin \mathfrak{p}$, such that $F = Q \cap G$. Then argue by *reductio ad absurdum:* show that there would exist an infinite sequence (Q_n) of $\mathfrak{p}_n$-primary submodules of E and a strictly increasing sequence (G_n) of submodules of E such that, if we write $H_n = Q_1 \cap Q_2 \cap \cdots \cap Q_n$: (1) $F = H_n \cap G_n$; (2) G_n is maximal among the submodules G containing G_{n-1} and such that

$F = H_n \cap G$; (3) there exists $a_n \notin \mathfrak{p}_n$ such that $a_n E \subset G_n$. Show then that, if $S_n = \bigcap_{1 \leqslant i \leqslant n} (A - \mathfrak{p}_i)$, S_n meets $G_n : E$ and deduce that $\text{sat}_{S_n}(F) = H_n$. Conclude with the aid of (LA_{II}).)

* (b) Give an example of a ring A such that the A-module A does not satisfy (LA_I) but where every ideal $\mathfrak{a} \subset A$ is irreducible and the set of saturations $\text{sat}_S(\mathfrak{a})$ for all the multiplicative subsets S of A is finite (cf. Exercise 19). *

(c) Let K be a field, $B = K[[X]]$ the algebra of formal power series over K and $\mathfrak{m}$ the maximal ideal of B consisting of the formal power series without constant term. In the product algebra $B^{\mathbf{N}}$, consider the subalgebra A generated by 1 and the ideal $\mathfrak{n} = \mathfrak{m}^{(\mathbf{N})}$. Show that in A the only prime ideals distinct from $\mathfrak{n}$ are the ideals the direct sums in $\mathfrak{n}$ of all the component ideals except one; every strictly increasing sequence of prime ideals of A then has at most two elements. Let $\mathfrak{a} \neq A$ be an ideal of A and let S be a multiplicative subset of A not meeting $\mathfrak{a}$; if $\mathfrak{a}_n$ (resp. S_n) is the projection of $\mathfrak{a}$ (resp. S) onto the n-th factor B_n of $B^{\mathbf{N}}$, show that $\text{sat}_S(\mathfrak{a})$ is the direct sum of the ideals $\text{sat}_{S_n}(\mathfrak{a}_n)$ (observe that, if $s \in S$, $s_n \in S_n$ is the n-th projection of s and $x_n \in \text{sat}_{S_n}(\mathfrak{a}_n)$ is such that $s_n x_n \in \mathfrak{a}_n$, then $s^2 x_n \in \mathfrak{a}$). Deduce that A satisfies axiom (LA_I) but not (LA_{II}) (use Exercise 20 (b)).

(d) Show that, if an A-module E satisfies axiom (LA_I), every submodule of E is the intersection of a family (finite or otherwise) of primary submodules in E (argue as in (a), by transfinite induction).

(e) Let E be a finitely generated A-module and $E' \subset E$ a finitely generated submodule. For E to be Laskerian, it is necessary and sufficient that E' and E/E' be so. In particular, if A is a Laskerian ring, every finitely generated A-module is Laskerian. If E is a Laskerian A-module, $S^{-1}E$ is a Laskerian $S^{-1}A$-module for every multiplicative subset S of A.

¶ 24. In a Laskerian ring A (Exercise 23), let $\mathfrak{p}_1, \mathfrak{p}_2, \mathfrak{p}_3$ be three distinct prime ideals such that $\mathfrak{p}_1 \subset \mathfrak{p}_2 \subset \mathfrak{p}_3$; let $x_2 \in \mathfrak{p}_2 - \mathfrak{p}_1$, $x_3 \in \mathfrak{p}_3 - \mathfrak{p}_2$. Replacing if need be $\mathfrak{p}_2$ and $\mathfrak{p}_3$ by prime ideals contained in $\mathfrak{p}_2, \mathfrak{p}_3$ respectively and containing respectively x_2 and x_3, we may suppose that $\mathfrak{p}_2$ is minimal in $\text{Ass}_f(A/(\mathfrak{p}_1 + Ax_2))$ and $\mathfrak{p}_3$ is minimal in $\text{Ass}_f(A/\mathfrak{p}_2 + Ax_3))$. Consider the ideal $\mathfrak{a} = \mathfrak{p}_1 + x_2\mathfrak{p}_3$.

(a) Show that $x_2 \notin \mathfrak{a}$ and $x_3^k \notin \mathfrak{a}$ for every integer $k > 0$; deduce that $\mathfrak{a}$ is not primary in A. Show that $\mathfrak{p}_2$ is a minimal element of $\text{Ass}_f(A/\mathfrak{a})$.

(b) Let $\mathfrak{a} = \bigcap_{i=1}^{n} \mathfrak{q}_i$ be a reduced primary decomposition of $\mathfrak{a}$ in A and let $\mathfrak{p}'_i$ be the radical of $\mathfrak{q}_i$; suppose that $\mathfrak{p}_3 \not\subset \mathfrak{p}'_i$ for $1 \leqslant i \leqslant s$, $\mathfrak{p}_3 \subset \mathfrak{p}'_i$ for $s + 1 \leqslant i \leqslant n$; show that of necessity $s < n$ and $x_2 \in \mathfrak{q}_i$ for $1 \leqslant i \leqslant s$. Show that there exists an index $i \geqslant s + 1$ such that $\mathfrak{p}_3 = \mathfrak{p}'_i$. (Argue by *reductio ad absurdum:* in the opposite case, there would exist $y \notin \mathfrak{p}_3$ such that $y \in \mathfrak{q}_i$ for all $i \geqslant s + 1$; then $x_2 y \in \mathfrak{a}$ and prove that this relation contradicts $y \notin \mathfrak{p}_3$.)

Conclude that $\mathfrak{p}_3$ is a non-minimal element of $\text{Ass}_f(A/\mathfrak{a})$; we denote by $\mathfrak{q}'_3$ the primary component of $\mathfrak{a}$ in A which corresponds to it in the above decomposition.

(c) Let $\mathfrak{b} = \mathfrak{p}_2 \cap \mathfrak{q}'_3$; on the other hand let $m > 0$ be such that $x_3^m \in \mathfrak{q}'_3$ and let $\mathfrak{c} = \mathfrak{b} + Ax_3^{2m}$. Show that the saturation of $\mathfrak{c}$ with respect to $\mathfrak{p}_3$ is a $\mathfrak{p}_3$-primary ideal $\mathfrak{q}''_3$ (show that every element of $\mathfrak{p}_3$ is a power in $\mathfrak{q}''_3$ by using the fact that the saturation of $\mathfrak{p}_2 + Ax_3$ with respect to $\mathfrak{p}_3$ is primary). Prove on the other hand that $x_3^m \notin \mathfrak{q}''_3$ and that $\mathfrak{b} = \mathfrak{p}_2 \cap \mathfrak{q}''_3$; then there are two distinct reduced primary decompositions of the ideal $\mathfrak{b}$.

¶ 25. Let A be a Laskerian ring. Let $\mathfrak{a}$ be an ideal of A and $\mathfrak{a} = \bigcap_{1 \leqslant i \leqslant n} \mathfrak{q}_i$ a reduced primary decomposition of $\mathfrak{a}$ in A; let $\mathfrak{p}_i$ denote the radical of $\mathfrak{q}_i$; suppose that the minimal elements of $\text{Ass}_f(A/\mathfrak{a})$ are the $\mathfrak{p}_i$ such that $1 \leqslant i \leqslant s$ and that $s < n$. We write $\mathfrak{b} = \bigcap_{1 \leqslant i \leqslant s} \mathfrak{q}_i$, $\mathfrak{c} = \bigcap_{s+1 \leqslant i \leqslant n} \mathfrak{q}_i$.

(a) Show that there exists $x \in \mathfrak{c}$ such that $x \notin \mathfrak{p}_i$ for $1 \leqslant i \leqslant s$ and that $\mathfrak{a} = \mathfrak{b} \cap (\mathfrak{a} + Ax)$.

(b) Suppose for example that $\mathfrak{p}_{s+1}$ is minimal in $\text{Ass}_f(A/\mathfrak{c})$. Show that $x\mathfrak{p}_{s+1} + \mathfrak{a} \neq Ax + \mathfrak{a}$ and $\mathfrak{a} = \mathfrak{b} \cap (x\mathfrak{p}_{s+1} + \mathfrak{a})$; moreover, $\mathfrak{p}_{s+1}$ is minimal in $\text{Ass}_f(A/\mathfrak{d})$, where $\mathfrak{d} = x\mathfrak{p}_{s+1} + \mathfrak{a}$. Show that the saturation of $\mathfrak{d}$ with respect to $\mathfrak{p}_{s+1}$ is different from $\mathfrak{q}_{s+1}$. (Note that this saturation cannot contain x.)

¶ 26. Let A be a Laskerian ring in which every ideal $\neq$A has a *unique* reduced primary decomposition in A.

(a) Show that for every ideal $\mathfrak{a} \neq A$ all the elements of $\text{Ass}_f(A/\mathfrak{a})$ are minimal in this set (use Exercise 25 (b)). Let $\{0\} = \bigcap_{1 \leqslant i \leqslant n} \mathfrak{q}_i$ be a reduced primary decomposition in A and let $\mathfrak{p}_i$ be the radical of $\mathfrak{q}_i$ ($1 \leqslant i \leqslant n$). Show that every prime ideal of A distinct from the $\mathfrak{p}_i$ is maximal (use Exercise 24 (c)). If $\mathfrak{p}$ is a prime ideal of A, $\mathfrak{q}$ an ideal contained in $\mathfrak{p}$ and such that $\mathfrak{p}/\mathfrak{q}$ is a nil ideal in $A/\mathfrak{q}$, show that every ideal $\mathfrak{a}$ such that $\mathfrak{q} \subset \mathfrak{a} \subset \mathfrak{p}$ is $\mathfrak{p}$-primary. Deduce that if one of the $\mathfrak{p}_i$ is not maximal, $\mathfrak{p}_i$ is the only $\mathfrak{p}_i$-primary ideal and $\mathfrak{p}_i^2 = \mathfrak{p}_i$ (use Exercise 16).

(b) Show that in A every increasing sequence of ideals equal to their radicals is stationary. Deduce that every strictly increasing sequence $(\mathfrak{a}_n)$ of ideals such that every quotient $\mathfrak{a}_n/\mathfrak{a}_{n-1}$ contains non-nilpotent elements in $A/\mathfrak{a}_{n-1}$ is finite. Conclude from this that for every prime ideal $\mathfrak{p}$ of A there exists a primary ideal with radical $\mathfrak{p}$ which is finitely generated. In particular, each of the $\mathfrak{p}_i$ which is not maximal is finitely generated and therefore contains an idempotent e_i which generates it (Chapter II, § 4, Exercise 15).

(c) Show that, for $i \neq j$ $\mathfrak{p}_i + \mathfrak{p}_j = A$ (consider $e_i + e_j - e_ie_j$ if $\mathfrak{p}_i$ and $\mathfrak{p}_j$ are not maximal). Conclude that A is isomorphic to the product of a finite number of rings $A_i = A/\mathfrak{q}_i$ such that: either A_i is a local ring whose maximal ideal is a nil ideal; or A is an integral domain in which every prime ideal $\neq\{0\}$ is maximal and in which every ideal $\neq\{0\}$ is only contained in a finite number of maximal ideals * (cf. § 1, Exercise 2). * Obtain the converse.

¶ 27. Let M be an A-module and Q a $\mathfrak{p}$-primary submodule of M. Q is said to be of *exponent m* if m is the smallest integer k such that $\mathfrak{p}^k M \subset Q$ (cf. Exercise 3); if there exists no integer k with this property, Q is said to be of *infinite* exponent. If Q is of finite exponent, Q is also said to be *strongly primary*.

(a) Show that, if $Q : \mathfrak{p} = Q$, Q is of infinite exponent and that there exists $\alpha \in \mathfrak{p}$ such that $Q : (\alpha)$ is distinct from Q and is a $\mathfrak{p}$-primary submodule of M of infinite exponent.

* (b) Give an example of a ring A and a $\mathfrak{p}$-primary ideal $\mathfrak{q}$ of infinite exponent and such that $\mathfrak{q} : \mathfrak{p} = \mathfrak{q}$ (cf. § 1, Exercise 2). *

(c) Let K be a field, A the polynomial ring $K[X_n]_{n \in \mathbf{N}}$, $\mathfrak{p}$ the prime ideal of A generated by the X_n and $\mathfrak{q}$ the $\mathfrak{p}$-primary ideal generated by the products $X_i X_j$ $(i \neq j)$ and the elements X_n^{n+1}. Show that for all $\alpha \in \mathfrak{p} - \mathfrak{q}$, $\mathfrak{q} : (\alpha)$ is of finite exponent, that $\mathfrak{q} : \mathfrak{p} \neq \mathfrak{q}$ and that $\mathfrak{q} : \mathfrak{p}$ is a $\mathfrak{p}$-primary ideal of infinite exponent.

(d) Let Q be a $\mathfrak{p}$-primary submodule of M of finite exponent e. Show that the $\mathfrak{p}^k M$ for $k < e$ are all distinct and that the submodules $Q : \mathfrak{p}^k$ for $k < e$ are all distinct. For every ideal $\mathfrak{a}$ of A show that either $Q : \mathfrak{a} = Q$ or $Q : \mathfrak{a}$ is a $\mathfrak{p}$-primary submodule in M distinct from M and of exponent $\leqslant e - 1$ (observe that if $\mathfrak{a} \subset \mathfrak{p}$ then $\mathfrak{p}^{e-1} M \subset Q : \mathfrak{a}$).

¶ 28. A finitely generated A-module E is called *strongly Laskerian* if every submodule of E admits a primary decomposition in E all of whose elements are strongly primary (Exercise 27).

(a) Show that for a finitely generated A-module E to be strongly Laskerian, it is necessary and sufficient that it satisfy axiom (LA_{II}) of Exercise 23 and the axiom:

(LA_{III}) For every sequence $(\mathfrak{b}_k)$ of ideals of A and every submodule F of E, the increasing sequence of submodules $F : (\mathfrak{b}_1 \mathfrak{b}_2 \ldots \mathfrak{b}_k)$ is stationary.

(To show that the conditions are necessary, use Exercise 27 (d). To see that they are sufficient, prove first that (LA_{III}) implies (LA_{I}) and then that it implies that every primary ideal is of finite exponent, using Exercise 27 (a).)

(b) Show that the ring A defined in Exercise 23 (c) satisfies (LA_{III}).

(c) Let K be a field, V an infinite dimensional vector K-space and $A = K \oplus V$ the ring whose multiplication is defined by

$$(a, x)(a', x') = (aa', ax' + a'x).$$

Show that in A every ideal $\neq A$ is strongly primary, although A is not Noetherian.

(d) Let E be a finitely generated A-module and $E' \subset E$ a finitely generated submodule. For E to be strongly Laskerian, it is necessary and sufficient that E' and E/E' be so. In particular, if A is a strongly Laskerian ring, every finitely generated A-module is strongly Laskerian. If E is a strongly Laskerian A-module, $S^{-1}E$ is a strongly Laskerian $S^{-1}A$-module for every multiplicative subset S of A.

(e) Generalize Exercises 3 and 4 to strongly Laskerian modules.

¶ 29. Let A be a strongly Laskerian ring (Exercise 28).

(a) Let $\mathfrak{a}$, $\mathfrak{b}$ be two ideals of A, $\mathfrak{ab} = \bigcap_{i=1}^{n} \mathfrak{q}_i$ a reduced primary decomposition of $\mathfrak{ab}$ in A and $\mathfrak{c}$ the intersection of those $\mathfrak{q}_i$ whose radicals do not contain $\mathfrak{b}$; show that $\mathfrak{a} \subset \mathfrak{c}$ (if $\mathfrak{q}_i$ is such that its radical $\mathfrak{p}_i$ does not contain $\mathfrak{b}$ and $x \in \mathfrak{b}$ and $x \notin \mathfrak{p}_i$, consider the product $\mathfrak{a}x$).

(b) Deduce from (a) that there exist three ideals $\mathfrak{a}'$, $\mathfrak{b}'$, $\mathfrak{c}$ and an integer $s > 0$ such that $\mathfrak{a}^s \subset \mathfrak{a}'$, $\mathfrak{b}^s \subset \mathfrak{b}'$, $(\mathfrak{a} + \mathfrak{b})^s \subset \mathfrak{c}$ and

$$\mathfrak{ab} = \mathfrak{a}' \cap \mathfrak{b} = \mathfrak{a} \cap \mathfrak{b}' = \mathfrak{a} \cap \mathfrak{b} \cap \mathfrak{c}.$$

(c) Show that for every finite family $(\mathfrak{a}_k)$ of ideals of A, there exists an integer $m > 0$ such that $\bigcap_k \mathfrak{a}_k^m \subset \prod_k \mathfrak{a}_k$ (apply (b) arguing by induction on the number of ideals $\mathfrak{a}_k$).

(d) For every ideal $\mathfrak{a}$ of A show that the intersection $\mathfrak{b} = \bigcap_{n=1}^{\infty} \mathfrak{a}^n$ is the ideal of the $x \in A$ such that $x \in x\mathfrak{a}$ (apply (b) to the product $(Ax)\mathfrak{a}$). Deduce that $\mathfrak{b}$ is the intersection of the primary components of $\{0\}$ (for a reduced primary decomposition of $\{0\}$) whose radicals meet $1 + \mathfrak{a}$.

(e) Deduce from (d) that, for every prime ideal $\mathfrak{p}$ of A, the intersection of the symbolic powers $\mathfrak{p}^{(n)}$ (Exercise 18 (b)) for $n \in \mathbf{N}$ is the ideal of those $x \in A$ such that $x \in x\mathfrak{p}$ (consider the ring $A_{\mathfrak{p}}$).

30. Let M be an A-module and let N be a submodule of M admitting a reduced primary decomposition $N = \bigcap_{i=1}^{n} Q_i$; let $\mathfrak{p}_i$ be the prime ideal associated with Q_i.

(a) Show that, if $\mathfrak{b}$ is an ideal of A which is not contained in any of the $\mathfrak{p}_i$, then $N : \mathfrak{b} = N$ (cf. Exercise 17).

(b) If each of the Q_i is strongly primary in M, show conversely that, if $N : \mathfrak{b} = N$, $\mathfrak{b}$ is not contained in any of the $\mathfrak{p}_i$. (Note that, if $\mathfrak{b} \subset \mathfrak{p}_i$ and $P = \bigcap_{j \neq i} Q_j$, then $\mathfrak{b}^r P \subset N$ for some suitable integer r and deduce that, if $N : \mathfrak{b} = N$, then $P = N$.) Deduce that there then exists $\beta \in \mathfrak{b}$ such that $N : (\beta) = N$.

¶ 31. (a) Let A be a Noetherian ring, $\mathfrak{m}$ a maximal ideal of A and $\mathfrak{p}$ a prime ideal contained in $\mathfrak{m}$; show that, if the Hausdorff completion $\hat{A}$ of A with respect to the $\mathfrak{m}$-adic topology is an integral domain, the filter base of symbolic power $\mathfrak{p}^{(n)}$ of $\mathfrak{p}$ tends to 0 with respect to the $\mathfrak{m}$-adic topology on A. (Reduce it to the case where A is local; let $\mathfrak{q}$ be a prime ideal of $\hat{A}$ whose trace on A is $\mathfrak{p}$ and for all $n > 0$ let $c_n \notin \mathfrak{p}$ be such that $c_n\mathfrak{p}^{(n)} \subset \mathfrak{p}^n$; show that $c_n\mathfrak{p}^{(n)}\hat{A} \subset \mathfrak{q}^{(n)}$. Use Exercise 29 (e) and Chapter III, § 2, no. 7, Proposition 8.)

(b) Let A be a Noetherian ring, $\mathfrak{n}$ an ideal of A, $\mathfrak{p}$ a minimal prime ideal in $\mathrm{Ass}(A/\mathfrak{n})$ and $\mathfrak{m}_0$, $\mathfrak{m}$ two maximal ideals of A containing $\mathfrak{p}$; let $A(\mathfrak{n})$, $A(\mathfrak{m}_0)$, $A(\mathfrak{m})$ denote the Hausdorff completions of A with respect to the $\mathfrak{n}$-adic, $\mathfrak{m}_0$-adic, $\mathfrak{m}$-adic topologies respectively. Let z be an element of $A(\mathfrak{n})$ whose canonical image in $A(\mathfrak{m}_0)$ is zero. Show that if $A(\mathfrak{m})$ is an integral domain, the canonical image of z in $A(\mathfrak{m})$ is also zero. (Take z as the limit of a sequence (x_n) of elements of A such that $x_n - x_m \in \mathfrak{n}^m$ for $n \geqslant m$ and $x_n \in \mathfrak{m}_0^n$; deduce that x_m belongs to the closure of $\mathfrak{n}^m$ with respect to the $\mathfrak{m}_0$-adic topology; using Exercise 8, conclude that $x_m \in \mathfrak{p}^{(m)}$ for all m and complete with the aid of (a).)

(c) Let A be a Zariski ring and $\mathfrak{r}$ a defining ideal of A; suppose that Spec(A) is connected and that for every maximal ideal $\mathfrak{m}$ of A the Hausdorff completion $A(\mathfrak{m})$ of A with respect to the $\mathfrak{m}$-adic topology is an integral domain. Show then that the completion $\hat{A}$ of A with respect to the $\mathfrak{r}$-adic topology is an integral domain. (Show that for every maximal ideal $\mathfrak{m}$ of A, the canonical homomorphism $\hat{A} \to A(\mathfrak{m})$ is injective. For this, we may assume that $\mathfrak{r}$ is the intersection of a finite number of prime ideals which are minimal in $\mathrm{Ass}(A/\mathfrak{r})$, let $\mathfrak{r} = \mathfrak{p}_1 \cap \mathfrak{p}_2 \cap \cdots \cap \mathfrak{p}_s$; show that the hypothesis on Spec(A) implies for each i the existence of a maximal ideal $\mathfrak{m}_i$ containing $(\mathfrak{p}_1 \cap \cdots \cap \mathfrak{p}_{i-1}) + \mathfrak{p}_i$. Using (b), show that, if the canonical image of $z \in \hat{A}$ in $A(\mathfrak{m})$ is zero and for example $\mathfrak{p}_1 \subset \mathfrak{m}$, then the canonical image of z in each of the $A(\mathfrak{m}_i)$ is zero, then conclude that the canonical image of z in $A(\mathfrak{m}')$ is zero for every maximal ideal $\mathfrak{m}'$ of A.)

32. A ring A is called *primary* if it is a local ring and its maximal ideal $\mathfrak{m}$ is a nil ideal (*). Every ideal of A contained in $\mathfrak{m}$ is then $\mathfrak{m}$-primary in A.

(a) Let A be a strongly Laskerian primary ring, so that in particular the maximal ideal $\mathfrak{m}$ is nilpotent. Show that, if $\mathfrak{a}$, $\mathfrak{b}$ are two ideals contained in $\mathfrak{m}$, necessarily $\mathfrak{a} : \mathfrak{b} \neq \mathfrak{a}$ (use Exercise 30 (b)).

(b) A primary Noetherian ring is Artinian. Deduce that, for every finitely generated module M over a Noetherian ring A and every $\mathfrak{p}$-primary submodule N of M, there exists an integer m such that every strictly decreasing sequence $(M_i/N)_{0 \leqslant i \leqslant k}$ of submodules of M/N for which the M_i are $\mathfrak{p}$-primary has length $k \leqslant m$ (consider the $A_\mathfrak{p}$-module $M_\mathfrak{p}$ and note that $M_\mathfrak{p}/N_\mathfrak{p}$ is annihilated by a power of $\mathfrak{p}A_\mathfrak{p}$).

(c) Show that an Artinian ring A in which every prime ideal $\neq \{0\}$ is maximal is the direct composition of a finite number of primary rings.

33. Let A be a commutative ring. An ideal $\mathfrak{a}$ of A is called primal if, in $A/\mathfrak{a}$, the set of divisors of 0 is an ideal $\mathfrak{p}/\mathfrak{a}$; the ideal $\mathfrak{p}$ is then prime.

(*) For (commutative) Artinian rings this definition coincides with that of *Algebra*, Chapter VIII, § 6, Exercise 20.

(a) Show that every primary ideal and every irreducible ideal is primal.

(b) Let $\mathfrak{a}$ be an ideal of A such that the set of ideals $\mathfrak{b}$ for which $\mathfrak{a} : \mathfrak{b} \neq \mathfrak{a}$ has a greatest element $\mathfrak{p}$; show that $\mathfrak{p}$ is prime and that $\mathfrak{a}$ is primal. * Give an example of a primal ideal $\mathfrak{a}$ for which the above condition is not satisfied (cf. § 1, Exercise 2). *

(c) In a Noetherian ring A characterize the primal ideals and give an example of a non-primary primal ideal (cf. § 1, Exercise 11 (b)).

34. In a commutative ring A an ideal $\mathfrak{a}$ is called quasi-prime if for every ordered pair of ideals $\mathfrak{b}$, $\mathfrak{c}$ of A, the relation $\mathfrak{b} \cap \mathfrak{c} \subset \mathfrak{a}$ implies $\mathfrak{b} \subset \mathfrak{a}$ or $\mathfrak{c} \subset \mathfrak{a}$; every prime ideal is quasi-prime and every quasi-prime ideal is irreducible. If the Chinese remainder theorem is valid in A, show that every irreducible ideal is quasi-prime (*Algebra*, Chapter VI, § 1, Exercise 25).

* 35. Let K be a commutative field, B the valuation ring, whose order group is $\mathbf{R}$, consisting of the "formal power series" $\sum_{x} c_x T^x$, where $c_x \in K$, $x \in \mathbf{R}_+$ and the set of x such that $c_x \neq 0$ is well ordered; the ring A defined in § 1, Exercise 2 is a subring of B and B is a flat A-module. If C is the A-module defined in § 1, Exercise 2, show that $\mathrm{Ass}_B(C \otimes_A B) \neq \varnothing$ although $\mathrm{Ass}_A(C) = \varnothing$ (consider an element of $C \otimes_A B = B/\mathfrak{a}B$ the canonical image of an element of B of valuation α). *

§ 3

1. Under the hypotheses of Proposition 1 of no. 1, show that every prime ideal $\mathfrak{p} \in \mathrm{Ass}_f(M)$ (§ 2, Exercise 17) is graded and is a minimal element of the set of prime ideals containing the annihilator of a homogeneous element of M. (Show first that the largest graded ideal $\mathfrak{p}'$ contained in a prime ideal $\mathfrak{p}$ of A is prime, observing that, if the product of two elements x, y belongs to $\mathfrak{p}'$, so does the product of their homogeneous components $\neq 0$ of highest degree. Note then that, if a prime ideal contains the annihilator of an element $z \in M$, it also contains the annihilator of at least one of the homogeneous components $\neq 0$ of z.)

2. In a graded ring A of type Δ (where Δ is torsion-free) let $\mathfrak{p}$ be a minimal element of the set of prime ideals of A, which is necessarily graded (Exercise 1). Show that for every homogeneous element $t \in \mathfrak{p}$ there exists a homogeneous element $s \notin \mathfrak{p}$ and an integer $n > 0$ such that $st^n = 0$. Deduce generalizations of Propositions 3, 4 and 5 to graded rings which are not necessarily Noetherian (with the definitions of § 2, Exercises 12 and 20).

3. Let A be a graded ring of type Δ (where Δ is torsion-free) and M a strongly Laskerian finitely generated graded A-module (§ 2, Exercise 28). Show that the submodules $Q(\mathfrak{p}) = Q(\mathfrak{p}, e_{\mathfrak{p}}(M))$ (where $\mathfrak{p}$ runs through $\mathrm{Ass}_f(M)$) defined in Exercise 4 of § 2 are graded submodules.

CHAPTER V(*)

Integers

Unless otherwise stated, all the rings and all the algebras considered in this chapter are assumed to be commutative and to possess a unit element; all the ring homomorphisms are assumed to map unit element to unit element. By a subring of a ring A *we mean a subring containing the unit element of* A.

1. NOTION OF AN INTEGRAL ELEMENT

1. INTEGRAL ELEMENTS OVER A RING

THEOREM 1. *Let* A *be a (commutative) ring,* R *an algebra over* A (not necessarily commutative) *and* x *an element of* R. *The following properties are equivalent:*

(E_I) *x is a root of a monic polynomial in the polynomial ring* A[X].

(E_{II}) *The subalgebra* A[x] *of* R *is a finitely generated* A-*module.*

(E_{III}) *There exists a faithful module over the ring* A[x] *which is a finitely generated* A-*module.*

Let us show first that (E_I) implies (E_{II}). Let

$$X^n + a_1X^{n-1} + \cdots + a_n$$

be a monic polynomial in A[X] with x as a root; for every integer $q \geqslant 0$ let M_q be the sub-A-module of R generated by $1, x, \ldots, x^{n+q}$. Then

$$x^{n+q} = -a_1x^{n+q-1} - \cdots - a_nx^q \in M_{q-1}$$

for all $q \geqslant 1$, whence, by induction on q,

$$M_q = M_{q-1} = \cdots = M_0.$$

(*) The results of this chapter and the next one depend on no Book other than Books I to VI, nor on Chapter I, § 4, nor on Chapter III, § 5.

We conclude that A[x] is equal to M_0 and is therefore a finitely generated A-module.

As the commutative ring A[x] is a faithful module over itself, (E_{II}) implies (E_{III}).

Finally, the fact that (E_{III}) implies (E_I) will follow from the following more precise lemma:

Lemma 1. *Let* A *be a ring,* R *an algebra* (not necessarily commutative) *over* A *and* x *an element of* R. *Let* M *be a faithful module over* A[x] *which is a finitely generated* A*-module. If* $\mathfrak{q}$ *is an ideal of* A *such that* $x\mathrm{M} \subset \mathfrak{q}\mathrm{M}$, *then* x *is a root of a monic polynomial with coefficients in* A *all of whose coefficients other than the dominant coefficient belong to* $\mathfrak{q}$.

Let $(u_i)_{1 \leqslant i \leqslant n}$ be a finite family of elements of M such that $\mathrm{M} = \sum_{i=1}^{n} \mathrm{A}u_i$. For all i, there exists by hypothesis a finite family $(q_{ij})_{1 \leqslant j \leqslant n}$ of elements of $\mathfrak{q}$ such that

$$xu_i = \sum_{j=1}^{n} q_{ij}u_j \qquad \text{for} \quad 1 \leqslant i \leqslant n.$$

Therefore (*Algebra*, Chapter III, § 8) if d is the determinant of the matrix $(q_{ij} - \delta_{ij}x)$ with elements in A[x] (δ_{ij} denoting the Kronecker index), then $du_i = 0$ for all i and hence $d\mathrm{M} = 0$; as M is assumed to be a faithful A[x]-module, necessarily $d = 0$. This means that x is a root of the polynomial $\det(q_{ij} - \delta_{ij}\mathrm{X})$ in A[X] which, up to a sign, is a monic polynomial whose coefficients other than the dominant coefficient belong to $\mathfrak{q}$.

Definition 1. *Let* A *be a ring and* R *an* A*-algebra* (not necessarily commutative). *An element* $x \in \mathrm{R}$ *is called integral over A if it satisfies the equivalent properties* (E_I), (E_{II}), (E_{III}) *of Theorem* 1.

A relation of the form $\mathrm{P}(x) = 0$, where P is a *monic* polynomial in A[X] is also called an *equation of integral dependence* with coefficients in A.

Examples

(1) Let K be a (commutative) field and R a K-algebra; to say that an element $x \in \mathrm{R}$ is integral over K is equivalent to saying that x is a root of a *non-constant* polynomial in the ring K[X]; generalizing the terminology introduced when R is an *extension* of K (*Algebra*, Chapter V, § 3, no. 3), the elements $x \in \mathrm{R}$ which are integral over K are also called the *algebraic* elements of R over K.

* (2) The elements of $\mathbf{Q}(i)$ which are integral over $\mathbf{Z}$ are the elements of the form $a + ib$ where $a \in \mathbf{Z}$ and $b \in \mathbf{Z}$ ("*Gaussian integers*"); the elements of $\mathbf{Q}(\sqrt{5})$ which are integral over $\mathbf{Z}$ are the elements of the form $(a + b\sqrt{5})/2$ where a and b belong to $\mathbf{Z}$ and are both even or both odd (for these two examples see Exercise 1). *

(3) The *complex numbers* integral over **Z** are also called *algebraic integers*.

Remarks

(1) Let A′ be the subring of R (contained in the centre of R) the image of A under the ring homomorphism A → R which defines the A-algebra structure on R. Clearly it is equivalent to say that an element of R is integral over A or that it is integral over A′.

(2) Let R′ be a sub-A-algebra of R; the elements of R′ which are integral over A are just the elements of R which are integral over A and belong to R′; this often allows us not to specify the algebra to which an integral element over A belongs, when no confusion can arise.

PROPOSITION 1. *Let* A *be a ring,* R *an algebra over* A (not necessarily commutative) *and x an element of* R. *For x to be integral over* A, *it is necessary and sufficient that* A[x] *be contained in a subalgebra* R′ *of* R *which is a finitely generated* A*-module.*

The condition is obviously necessary by virtue of property (E_{II}); it is also sufficient by virtue of (E_{III}), for R′ is a faithful A[x]-module (since it contains the unit element of R).

COROLLARY. *Let* A *be a* Noetherian *ring,* R *an* A*-algebra* (not necessarily commutative) *and x an element of* R. *For x to be integral over* A, *it is necessary and sufficient that there exist a finitely generated submodule of* R *containing* A[x].

The condition is necessary by virtue of (E_{II}); it is sufficient for if A[x] is a sub-A-module of a finitely generated A-module, it is itself a finitely generated A-module (*Algebra*, Chapter VIII, § 2, no. 3, Proposition 7).

The hypothesis that A is Noetherian cannot be omitted from the statement (Exercise 2).

DEFINITION 2. *Let* A *be a ring. An* A*-algebra* R (not necessarily commutative) *is called integral over* A *if every element of* R *is integral over* A. R *is called finite over* A *if* R *is a finitely generated* A*-module.*

It follows from Proposition 1 that every finite A-algebra is *integral*; if R is commutative and a finite A-algebra, R is obviously a *finitely generated* A-algebra; the converse is false.

Example (4) If M is a finitely generated A-module, the algebra $\mathrm{End}_A(M)$ of endomorphisms of M is integral over A by virtue of (E_{III}); in particular, for every integer n, the matrix algebra $M_n(A) = \mathrm{End}_A(A^n)$ is integral (and even finite) over A.

PROPOSITION 2. *Let* A, A′ *be two rings,* R *an* A*-algebra,* R′ *an* A′*-algebra* (not necessarily commutative) *and* $f: A \rightarrow A'$ *and* $g: R \rightarrow R'$ *two ring homomorphisms such that the diagram*

$$\begin{array}{ccc} A & \xrightarrow{f} & A' \\ \downarrow & & \downarrow \\ R & \xrightarrow{g} & R' \end{array}$$

is commutative. If an element $x \in R$ *is integral over* A, *then* $g(x)$ *is integral over* A'.

If $x^n + a_1 x^{n-1} + \cdots + a_n = 0$ where $a_i \in A$ for $1 \leqslant i \leqslant n$, we deduce that

$$(g(x))^n + f(a_1)(g(x))^{n-1} + \cdots + f(a_n) = 0.$$

Corollary 1. *Let* A *be a ring,* B *a* (*commutative*) A-*algebra and* C *a* (not necessarily commutative) B-*algebra. Then every element* $x \in C$ *which is integral over* A *is integral over* B.

Corllary 2. *Let* K *be a field,* L *an extension of* K *and* x, x' *two elements of* L *which are conjugate over* K (*Algebra*, Chapter V, § 6, no. 2). *If* A *is a subring of* K *and* x *is integral over* A, x' *is also integral over* A.

There exists a K-isomorphism f of $K(x)$ onto $K(x')$ such that $f(x) = x'$ and the elements of A are invariant under f.

Corollary 3. *Let* A *be a ring,* B *a* (*commutative*) A-*algebra and* C *a* (not necessarily commutative) B-*algebra. If* C *is integral over* A, C *is integral over* B.

Proposition 3. *Let* $(R_i)_{1 \leqslant i \leqslant n}$ *be a finite family of* (not necessarily commutative) A-*algebras and let* $R = \prod_{i=1}^{n} R_i$ *be their product. For an element* $x = (x_i)_{1 \leqslant i \leqslant n}$ *of* R *to be integral over* A, *it is necessary and sufficient that each of the* x_i *be integral over* A. *For* R *to be integral over* A, *it is necessary and sufficient that each of the* R_i *be integral over* A.

It is obviously sufficient to prove the first assertion. The condition is necessary by Proposition 2. Conversely, if each of the x_i is integral over A, the subalgebra $A[x_i]$ of R_i is a finitely generated A-module and hence so is the subalgebra $\prod_{i=1}^{n} A[x_i]$ of R; as $A[x]$ is contained in this subalgebra, x is integral over A by Proposition 1.

Proposition 4. *Let* A *be a ring,* R *an* A-*algebra* (not necessarily commutative) *and* $(x_i)_{1 \leqslant i \leqslant n}$ *a finite family of elements of* R *which are pairwise permutable. If, for all* i, x_i *is integral over* $A[x_1, \cdots, x_{i-1}]$ (*and in particular if all the* x_i *are integral over* A), *then the subalgebra* $A[x_1, \ldots, x_n]$ *of* R *is a finitely generated* A-*module.*

We argue by induction on n, the proposition being just (E_{II}) for $n = 1$. The induction hypothesis implies that $B = A[x_1, \cdots, x_{n-1}]$ is a finitely generated A-module; as x_n is integral over B, $B[x_n] = A[x_1, \cdots, x_n]$ is a finitely generated B-module and therefore also a finitely generated A-module (*Algebra*, Chapter II, § 1, no. 13, Proposition 25).

COROLLARY 1. *Let* A *be a ring and* R *a (commutative)* A*-algebra. The set of elements of* R *integral over* A *is a subalgebra of* R.

In fact, if x, y are two elements of R integral over A, it follows from Proposition 4 that $A[x, y]$ is a finitely generated A-module; as it contains $x + y$ and xy, the corollary follows from Proposition 1.

In a non-commutative algebra the sum and product of two integral elements over A are not necessarily integral over A (Exercise 4).

COROLLARY 2. *Let* A *be a ring,* R *an* A*-algebra* (not necessarily commutative) *and* E *a set of elements of* R *which are pairwise permutable and integral over* A. *Then the sub-*A*-algebra* B *of* R *generated by* E *is integral over* A.

Every element of B belongs to a sub-A-algebra of B generated by a finite subset of E.

Remark (3) It follows from Proposition 4 that every *commutative* A-algebra integral over A is the union of a right directed family of *finite* subalgebras over A.

PROPOSITION 5. *Let* A *be a ring and* A' *and* R *two (commutative)* A*-algebras. If* R *is integral over* A, $R \otimes_A A'$ *is integral over* A'.

Consider any element $x' = \sum_{i=1}^{n} x_i \otimes a'_i$ of $R \otimes_A A'$, where the x_i belong to R and the a'_i to A'; as $x_i \otimes a'_i = (x_i \otimes 1)a'_i$ and the $x_i \otimes 1$ are integral over A' (Proposition 2), so is x.

COROLLARY. *Let* R *be a ring and* A, B, C *subrings of* R *such that* $A \subset B$. *If* B *is integral over* A, C[B] *is integral over* C[A].

$B \otimes_A C[A]$ is integral over C[A] by Proposition 5 and hence so is the canonical image C[B] of $B \otimes_A C[A]$ in R (considered as an A-algebra) by Proposition 2.

PROPOSITION 6. *Let* A *be a ring,* B *a (commutative)* A*-algebra and* C *a* (not necessarily commutative) B*-algebra. If* B *is integral over* A *and* C *is integral over* B, *then* C *is integral over* A.

It is sufficient to verify that every $x \in C$ is integral over A. By hypothesis there exists a monic polynomial $X^n + b_1X^{n-1} + \cdots + b_n$ with coefficients in B with x as a root; then x is integral over $B' = A[b_1, \ldots, b_n]$ and $B'[x]$ is therefore a finitely generated B'-module. But as B is integral over A, B' is a finitely generated A-module (Proposition 4); we conclude that $B'[x]$ is also a finitely generated A-module (*Algebra*, Chapter II, § 1, no. 13, Proposition 25) and therefore x is integral over A.

COROLLARY. *Let* A *be a ring and* R, R′ *two* (*commutative*) *A-algebras integral over* A. *Then* $R \otimes_A R'$ *is integral over* A.

$R \otimes_A R'$ is integral over R′ (Proposition 5) and hence the conclusion follows from Proposition 6.

2. THE INTEGRAL CLOSURE OF A RING. INTEGRALLY CLOSED DOMAINS

DEFINITION 3. *Let* A *be a ring and* R *a* (*commutative*) *A-algebra. The sub-A-algebra* A′ *of* R *consisting of the elements of* R *integral over* A (no. 1, Corollary 1 to Proposition 4) *is called the integral closure of* A *in* R. *If* A′ *is equal to the canonical image of* A *in* R, A *is called integrally closed in* R.

Remarks

(1) If $h: A \to R$ is the ring homomorphism defining the A-algebra structure on R, the integral closure of A in R is also that of $h(A)$ in R. On the other hand, if R′ is a subalgebra of R, the integral closure of A in R′ is $A' \cap R'$.

(2) If A is a *field*, the integral closure A′ of A in R consists of the elements of R which are *algebraic* over A (no. 1, *Example* 1); generalizing the terminology used for field extensions (*Algebra*, Chapter V, § 3, no. 3), A′ is then also called the *algebraic closure* of the field A in the algebra R and A is called *algebraically closed in* R if $A' = A$.

DEFINITION 4. *If* A *is an integral domain, the integral closure of* A *in its field of fractions is called the integral closure of* A. *An integral domain is called integrally closed if it is equal to its integral closure.*

Note that an integrally closed domain is not necessarily closed in any ring containing it, as the example of a field which is not algebraically closed shows.

PROPOSITION 7. *Let* A *be a ring and* R *an A-algebra. The integral closure* A′ *of* A *in* R *is a subring integrally closed in* R.

The integral closure of A′ in R is integral over A by no. 1, Proposition 6; it is therefore equal to A′.

COROLLARY. *The integral closure of an integral domain* A *is an integrally closed domain.*

Let K be the field of fractions of A and B the integral closure of A. Clearly

K is the field of fractions of B and it is sufficient to apply Proposition 7 to R = K.

PROPOSITION 8. *Let R be a ring, $(B_\lambda)_{\lambda \in L}$ a family of subrings of R and for each $\lambda \in L$ let A_λ be a subring of B_λ. If each A_λ is integrally closed in B_λ, then $A = \bigcap_{\lambda \in L} A_\lambda$ is integrally closed in $B = \bigcap_{\lambda \in L} B_\lambda$.*

This follows immediately from Definition 3 and no. 1, Corollary 1 to Proposition 2.

COROLLARY. *Every intersection of a non-empty family of integrally closed subdomains of an integral domain A is an integrally closed domain.*

It is sufficient to apply Proposition 8 taking R and the B_λ equal to the field of fractions K of A and noting that a subring of K integrally closed in K is *a fortiori* an integrally closed domain since its field of fractions is contained in K.

PROPOSITION 9. *Let A be a ring, $(R_i)_{1 \leqslant i \leqslant n}$ a finite family of A-algebras and A'_i the integral closure of A in R_i $(1 \leqslant i \leqslant n)$. Then the integral closure of A in $R = \prod_{i=1}^{n} R_i$ is equal to $\prod_{i=1}^{n} A'_i$.*

This is an immediate consequence of no. 1, Proposition 3.

COROLLARY 1. *Let A be a reduced Noetherian ring, $\mathfrak{p}_i$ $(1 \leqslant i \leqslant n)$ its distinct minimal prime ideals, K_i the field of fractions of the integral domain $A/\mathfrak{p}_i$ (canonically isomorphic to the local ring $A_{\mathfrak{p}_i}$* (Chapter IV, § 2, no. 5, Proposition 10)) *and A'_i the integral closure of A in K_i $(1 \leqslant i \leqslant n)$. Then the canonical isomorphism of the total ring of fractions B of A onto $\prod_{i=1}^{n} K_i$ (loc. cit.) maps the integral closure of A in B onto the product ring $\prod_{i=1}^{n} A'_i$.*

COROLLARY 2. *For a reduced Noetherian ring to be integrally closed in its ring of fractions it is necessary and sufficient that it be a direct composition of integrally closed (Noetherian) domains.*

3. EXAMPLES OF INTEGRALLY CLOSED DOMAINS

PROPOSITION 10. *Every principal ideal domain is integrally closed.*

Let A be a principal ideal domain, K its field of fractions and x an element of K. There exist two relatively prime elements a, b of A such that $x = ab^{-1}$ (*Algebra*, Chapter VII, § 1, no. 2, Proposition 1 and Chapter VI, § 1, no. 11, Proposition 9 (DIV)). If x is integral over A, it is a root of a polynomial

$X^n + c_1X^{n-1} + \cdots + c_n$ of A[X]. Then $a^n = b(-c_1a^{n-1} - \cdots - c_nb^{n-1})$, which proves that b divides a^n. Since a and b are relatively prime, this implies that b is invertible in A (*Algebra*, Chapter VI, § 1, no. 12, Corollary 1 to Proposition 11 (DIV)); hence $x \in$ A.

LEMMA 2. *Let* R *be a ring and* P *a monic polynomial in* R[X]. *There exists a ring* R′ *containing* R *such that in the polynomial ring* R′[X] *the polynomial* P *is a product of monic polynomials of degree* 1.

We proceed by induction on the degree n of P, the lemma being obvious for $n = 0$ and $n = 1$. Suppose therefore that $n > 1$. Let $\mathfrak{a}$ be the ideal of R[X] generated by P and let f be the canonical homomorphism of R[X] onto $B = R[X]/\mathfrak{a}$. Since P is monic, for every polynomial $Q \in R[X]$, $\deg(PQ) = \deg(P) + \deg(Q)$, whence $\mathfrak{a} \cap R = 0$; the restriction of f to R is therefore injective. Identifying R with the subring $f(R)$ of B by means of f and writing $b = f(X)$, we see that b is a root of P in B, P being considered as a polynomial in B[X]. Then there exists a monic polynomial Q in B[X] of degree $n - 1$ such that $P(X) = (X - b)Q(X)$ (*Algebra*, Chapter IV, § 1, no. 4, Proposition 5). By the induction hypothesis there exists a ring $R' \supset B$ such that in R′[X] the polynomial Q is a product of monic polynomials of degree 1; clearly in R′[X] P is then a product of monic polynomials of degree 1.

PROPOSITION 11. *Let* A *be a ring,* R *an* A*-algebra and* P *and* Q *monic polynomials in* R[X]. *If the coefficients of* PQ *are integral over* A, *the coefficients of* P *and* Q *are integral over* A.

By a double application of Lemma 2, we see that there exists a ring R′ containing R and families of elements $(a_i)_{1 \leqslant i \leqslant m}$, $(b_j)_{1 \leqslant j \leqslant n}$ of R′ such that in R′[X] $P(X) = \prod_{i=1}^{\infty} (X - a_i)$, $Q(X) = \prod_{i=1}^{n} (X - b_j)$; the coefficients of PQ belong to the integral closure A′ of A in R′ and hence (no. 2, Proposition 7) the elements a_i $(1 \leqslant i \leqslant m)$ and b_j $(1 \leqslant j \leqslant n)$ belong to A′. It follows that the coefficients of P and Q are integral over A (no. 1, Corollary 1 to Proposition 4).

Let A be an integral domain, K its field of fractions and K′ a K-algebra (not necessarily commutative). Given an element $x \in K'$ *algebraic* over K, the polynomials $P \in K[X]$ such that $P(x) = 0$ form an ideal $\mathfrak{a} \neq 0$ of K[X], necessarily principal (*Algebra*, Chapter IV, § 1, no. 5, Proposition 7). There exists a unique *monic* polynomial which generates $\mathfrak{a}$; generalizing the terminology introduced in *Algebra*, Chapter V, § 3, no. 1, Definition 3, this monic polynomial will be called the *minimal polynomial* of x over K.

COROLLARY. *Let* A *be an integral domain,* K *its field of fractions and x an element of a* K*-algebra* K′ (not necessarily commutative). *If x is integral over* A, *the coeffi-*

cients of the minimal polynomial P *of* x *over* K *are integral over* A (and they therefore belong to A if A is *integrally closed*).

There exists by hypothesis (no. 1, Theorem 1) a *monic* polynomial $Q \in A[X]$ such that $Q(x) = 0$. As P divides Q in K[X], it follows from Proposition 11 that the coefficients of P are integral over A.

Let A be a ring and R a (commutative) A-algebra; the homomorphism $\phi: A \to R$ defining the A-algebra structure on R can be extended uniquely to a homomorphism $A[X] \to R[X]$ of polynomial rings over A and R, leaving X invariant and hence R[X] is given a canonical A[X]-algebra structure.

PROPOSITION 12. *Let* A *be a ring,* R *an* A*-algebra and* P *a polynomial in* $R[X_1, \dots, X_n]$. *For* P *to be integral over* $A[X_1, \dots, X_n]$, *it is necessary and sufficient that the coefficients of* P *be integral over* A.

By considering the polynomials of $R[X_1, \dots, X_n]$ as polynomials in X_n with coefficients in $R[X_1, \dots, X_{n-1}]$, we see immediately that it is reduced to proving the proposition for $n = 1$. Then let P be a polynomial in R[X]; it follows immediately from no. 1, Proposition 5 that, if the coefficients of P are in the integral closure B of A in R, the element P, which belongs to $B[X] = B \otimes_A A[X]$, is integral over A[X]. Conversely, suppose that P is integral over A[X] and let

$$Q(Y) = Y^m + F_1 Y^{m-1} + \cdots + F_m$$

be a monic polynomial with coefficients $F_i \in A[X]$ with P as a root. Let r be an integer strictly greater than all the degrees of the polynomials P and F_i $(1 \leqslant i \leqslant m)$ and let us write

$$P_1(X) = P(X) - X^r.$$

Then P_1 is a root of the polynomial

$$Q_1(Y) = Q(Y + X^r) = Y^m + G_1 Y^{m-1} + \cdots + G_m$$

with coefficients in A[X]; we may therefore write

$$(1) \qquad -P_1(P_1^{m-1} + G_1 P_1^{m-2} + \cdots + G_{m-1}) = G_m.$$

Now the choice of r implies that $-P_1$ is a *monic* polynomial of R[X] and so is $G_m(X) = Q(X^r)$, the degrees of the polynomials $F_k(X)X^{r(m-k)}$ being all $< rm$ for $k \geqslant 1$. We conclude first of all that the polynomial

$$P_1^{m-1} + G_1 P_1^{m-2} + \cdots + G_{m-1}$$

of R[X] is also monic; moreover, as the coefficients of G_m belong to A, Proposition 11 shows that P_1 has coefficients integral over A and the coefficients of P are therefore certainly integral over A.

Proposition 13. *Let* A *be a ring,* R *an* A*-algebra and* A′ *the integral closure of* A *in* R. *Then the integral closure of* $A[X_1, \ldots, X_n]$ *in* $R[X_1, \ldots, X_n]$ *is equal to* $A'[X_1, \ldots, X_n]$.

This follows from Proposition 12 and Definition 3 of no. 2.

Corollary 1. *Let* A *be an integral domain and* A′ *its integral closure. Then the integral closure of the polynomial ring* $A[X_1, \ldots, X_n]$ *is* $A'[X_1, \ldots, X_n]$.

Arguing by induction on n, the problem is immediately reduced to the case $n = 1$. Let K be the field of fractions of A, which is also that of A′; if an element P of the field of fractions K(X) of A[X] is integral over A[X], it belongs to the polynomial ring K[X], for the latter is a principal ideal domain (*Algebra*, Chapter IV, § 1, no. 5, Proposition 7) and hence integrally closed (Proposition 10); the corollary then follows from Proposition 13 applied to R = K.

Corollary 2. *Let* A *be an integral domain. For the polynomial ring* $A[X_1, \ldots, X_n]$ *to be integrally closed, it is necessary and sufficient that* A *be integrally closed.*

Corollary 3. *If* K *is a field, every polynomial algebra* $K[X_1, \ldots, X_n]$ *is an integrally closed domain.*

4. COMPLETELY INTEGRALLY CLOSED DOMAINS

Definition 5. *An integral domain* A *is called completely integrally closed if the following condition is satisfied: every element x of the field of fractions* K *of* A *such that all the powers* x^n $(n \geqslant 0)$ *are contained in a finitely generated sub-*A*-module of* K, *belongs to* A.

Note that the hypothesis that the x^n are contained in a finitely generated sub-A-module of K can also be expressed by saying that there exists a non-zero element $d \in A$ such that $dx^n \in A$ for all $n \geqslant 0$; for the latter condition means that $x^n \in Ad^{-1}$; and conversely, if $(b_i)_{1 \leqslant i \leqslant m}$ is a finite sequence of elements of K, there exists $d \in A$ such that $db_i \in A$ for $1 \leqslant i \leqslant m$, whence $dM \subset A$ for the sub-A-module M of K generated by the b_i.

Clearly a completely integrally closed domain is integrally closed; conversely, the Corollary to Proposition 1 of no. 1 shows that an integrally closed *Noetherian* domain is completely integrally closed. * On the other hand, the ring of a valuation of height $\geqslant 2$ (Chapter VI, § 4, no. 4) is integrally closed but not completely integrally closed. * If (A_ι) is a family of completely integrally closed domains with the same field of fractions K, $A = \bigcap_\iota A_\iota$ is completely integrally closed. For if $x \in K$ is such that for some non-zero d in A, dx^n belongs to A for all $n > 0$, the hypothesis implies that $x \in A_\iota$ for all ι and hence $x \in A$.

PROPOSITION 14. *Let A be a completely integrally closed domain. Then every polynomial ring* $A[X_1, \ldots, X_n]$ (resp. *every ring of formal power series* $A[[X_1, \ldots, X_n]]$) *is completely integrally closed.*

By induction on n, it is sufficient to prove that $A[X]$ (resp. $A[[X]]$) is completely integrally closed. Then let P be an element of the field of fractions of $A[X]$ (resp. $A[[X]]$) and suppose that there exists a non-zero element $Q \in A[X]$ (resp. $Q \in A[[X]]$) such that $QP^m \in A[X]$ (resp. $QP^m \in A[[X]]$) for every integer $m \geqslant 0$. If K is the field of fractions of A, $A[X]$ (resp. $A[[X]]$) is a subring of $K[X]$ (resp. $K[[X]]$) and $K[X]$ (resp. $K[[X]]$) is a principal ideal domain (*Algebra*, Chapter VII, § 1, no. 1) and hence integrally closed (no. 4, Proposition 10) and Noetherian (*Algebra*, Chapter VIII, § 2, no. 3) and therefore completely integrally closed; then we have already seen that $P \in K[X]$ (resp. $P \in K[[X]]$). Let $P = \sum_{k=0}^{\infty} a_k X^k$ $(a_k \in K)$ and $Q = \sum_{k=0}^{\infty} b_k X^k$ $(b_k \in A)$ and we argue by *reductio ad absurdum* by supposing that the a_k do not all belong to A; then there is a least index i such that $a_i \notin A$; if we write $P_1 = \sum_{k=0}^{i-1} a_k X^k \in A[X]$, it follows immediately from the hypothesis that also $Q(P - P_1)^m \in A[X]$ (resp. $Q(P - P_1)^m \in A[[X]]$) for all $m \geqslant 0$. Let j be the least integer such that $b_j \neq 0$; clearly in $Q(P - P_1)^m$ the term of least degree with a coefficient $\neq 0$ is $b_j a_i^m X^{j+mi}$ and hence $b_j a_i^m \in A$ for all $m \geqslant 0$; but as A is completely integrally closed this implies $a_i \in A$, contrary to the hypothesis.

PROPOSITION 15. *Let A be a filtered ring whose filtration is exhaustive and such that every principal ideal of A is closed under the topology defined by the filtration. If the associated graded ring* gr(A) (Chapter III, § 2, no. 3) *is a completely integrally closed domain, then A is a completely integrally closed domain.*

Let $(A_n)_{n \in \mathbf{Z}}$ be the filtration defined on A; as $\bigcap_{n \in \mathbf{Z}} A_n$ is the closure of the ideal (0) (Chapter III, § 2, no. 5), the hypothesis implies first that the filtration (A_n) is separated and, as gr(A) is an integral domain, so then is A (Chapter III, § 2, no. 3, Corollary to Proposition 1). Let $x = b/a$ be an element of the field of fractions K of A $(a \in A, b \in A)$ for which there exists an element $d \neq 0$ of A such that $dx^n \in A$ for all $n \geqslant 0$. We must prove that $b \in Aa$ and, as by hypothesis the ideal Aa is closed, it is sufficient to show that, for all $n \in \mathbf{Z}$, $b \in Aa + A_n$. As the filtration of A is exhaustive, there exists an integer $q \in \mathbf{Z}$ such that $b \in Aa + A_q$. It will therefore suffice to prove that the relation $b \in Aa + A_m$ implies $b \in Aa + A_{m+1}$.

Suppose then that $b = ay + z$ where $y \in A$, $z \in A_m$. Then by hypothesis $dx^n \in A$ for all $n \geqslant 0$, whence we obtain immediately $d(x - y)^n \in A$ for all $n \geqslant 0$; in other words, $dz^n = a^n t_n$ where $t_n \in A$ for all $n \geqslant 0$. We can obviously limit our attention to the case where $z \neq 0$. Let v denote the order function on A (Chapter III, § 1, no. 2) and let us write $v(d) = n_1$, $v(z) = n_2 \geqslant m$, $v(a) = n_3$; let d', z', a' be the respective images of d, z, a in A_{n_1}/A_{n_1+1}, A_{n_2}/A_{n_2+1}, A_{n_3}/A_{n_3+1}. For all $n \geqslant 0$, $v(dz^n) = n_1 + nn_2$ (Chapter III, § 2, no. 3, Proposition 1) and hence the canonical image in $\mathrm{gr}(A)$ of dz^n is $d'z'^n$; similarly it is seen that the canonical image in $\mathrm{gr}(A)$ of $a^n t_n$ is of the form $a'^n t'_n$ where $t'_n \in \mathrm{gr}(A)$ and, as $a' \neq 0$ we deduce that, for all $n \geqslant 0$, $d'(z'/a')^n \in \mathrm{gr}(A)$. The hypothesis that $\mathrm{gr}(A)$ is completely integrally closed therefore implies the existence of an $s' \in \mathrm{gr}(A)$ such that $z' = a's'$; decomposing s' into a sum of homogeneous elements, it is further seen (since z' and a' are homogeneous) that s' may be assumed to be homogeneous, that is the image of an element $s \in A$; then $v(as) = v(z) = n_2$ and $z \equiv as \pmod{A_{n_2+1}}$; as $n_2 \geqslant m$, *a fortiori* $z \equiv as$ (mod, A_{m+1}), hence $b \equiv a(y + s) \pmod{A_{m+1}}$.

5. THE INTEGRAL CLOSURE OF A RING OF FRACTIONS

Let A be a ring, R an A-algebra and S a multiplicative subset of A. Recall (Chapter II, § 2, no. 8) that $S^{-1}R$ has a canonical $S^{-1}A$-algebra structure.

PROPOSITION 16. *Let* A *be a ring,* R *an* A*-algebra,* A′ *the integral closure of* A *in* R *and* S *a multiplicative subset of* A. *Then the integral closure of* $S^{-1}A$ *in* $S^{-1}R$ *is* $S^{-1}A'$.

Let b/s be an element of $S^{-1}A'$ ($s \in S$, $b \in A'$). Since the diagram

$$\begin{array}{ccc} A & \xrightarrow{i_A^S} & S^{-1}A \\ {\scriptstyle h}\downarrow & & \downarrow{\scriptstyle S^{-1}h} \\ R & \xrightarrow[i_R^S]{} & S^{-1}R \end{array}$$

is commutative, $b/1$ is integral over $S^{-1}A$ (no. 1, Proposition 2). As $1/s \in S^{-1}A$, $b/s = (b/1)(1/s)$ is integral over $S^{-1}A$.

Conversely, let r/t ($r \in R$, $t \in S$) be an element of $S^{-1}R$ which is integral over $S^{-1}A$; then $r/1 = (t/1)(r/t)$ is integral over $S^{-1}A$. Then there is a relation of the form

$$(r/1)^n + (a_1/s)(r/1)^{n-1} + \cdots + (a_n/s) = 0,$$

where $a_i \in A$ ($1 \leqslant i \leqslant n$) and $s \in S$. This relation may also be written

$$(sr^n + a_1 r^{n-1} + \cdots + a_n)/s = 0$$

and therefore there exists $s' \in S$ such that $s'^n(sr^n + a_1 r^{n-1} + \cdots + a_n) = 0$; we deduce that $(s'sr)^n + s'a_1(s'sr)^{n-1} + \cdots + s'^n s^{n-1} a_n = 0$. By definition therefore $s'sr \in A'$, whence $r/1 \in S^{-1}A'$ and $r/t \in S^{-1}A'$.

Corollary 1. *Let A be an integral domain, A' its integral closure and S a multiplicative subset of A such that $0 \notin S$. Then the integral closure of $S^{-1}A$ is $S^{-1}A'$.*

The field of fractions R of A is also the field of fractions of $S^{-1}A$ since $0 \notin S$ (Chapter II, § 1, no. 1, *Remark* 7); Proposition 16 is then applied to R.

Corollary 2. *Let A be an integral domain, K its field of fractions, R an algebra over K and B the integral closure of A in R. The elements of R which are algebraic over K (no. 1, Example 1) are the elements of the form $a^{-1}b$ where $b \in B$ and $a \in A$, $a \neq 0$; if L is the algebraic closure of K in R, there exists a basis of L over K contained in B.*

The first assertion follows from Proposition 16 applied in the case $S = A - \{0\}$. If $(x_\iota)_{\iota \in I}$ is a basis of L over K, then there exists for all $\iota \in I$ an element $a_\iota \neq 0$ of A such that $a_\iota x_\iota \in B$; then $(a_\iota x_\iota)_{\iota \in I}$ is also a basis of L over K.

Corollary 3. *Let A be an integral domain and Ω the set of maximal ideals of A. For A to be integrally closed, it is necessary and sufficient that, for all $\mathfrak{m} \in \Omega$, $A_\mathfrak{m}$ be integrally closed.*

It follows from Corollary 1 that the condition is necessary. The condition is sufficient, for $A = \bigcap_{\mathfrak{m} \in \Omega} A_\mathfrak{m}$ (Chapter II, § 3, no. 3, formula (2)) and it is sufficient to apply the Corollary to Proposition 8 of no. 2.

Corollary 4. *Let A be an integral domain, K its field of fractions and S a multiplicative subset of A such that $0 \notin S$.*

(i) *Let B be a subring of K which is integral over A and let $\mathfrak{f}$ be the annihilator of the A-module B/A. Then $S^{-1}\mathfrak{f}$ is contained in the annihilator of the $(S^{-1}A)$-module $S^{-1}B/S^{-1}A$ and is equal to this annihilator if B is a finitely generated A-module.*

(ii) *Let A' be the integral closure of A. For $S^{-1}A$ to be integrally closed, it is sufficient that the annihilator $\mathfrak{f}$ of the A-module A'/A meet S. This condition is also necessary if A' is a finitely generated A-module.*

(i) As $\mathfrak{f}B \subset A$, $(S^{-1}\mathfrak{f})(S^{-1}B) \subset S^{-1}A$ and hence $S^{-1}\mathfrak{f}$ is contained in Ann $(S^{-1}B/S^{-1}A)$. If B is a finitely generated A-module, the equation $S^{-1}\mathfrak{f} = \mathrm{Ann}(S^{-1}B/S^{-1}A)$ is a special case of formula (9) of Chapter II, § 2, no. 4, $S^{-1}B/S^{-1}A$ being canonically identified with $S^{-1}(B/A)$.

(ii) By Corollary 1 $S^{-1}A'$ is the integral closure of $S^{-1}A$. As the relations $\mathfrak{f} \cap S \neq \varnothing$ and $S^{-1}\mathfrak{f} = S^{-1}A$ are equivalent (Chapter II, § 2, no. 5, *Remark*) (ii) is an immediate consequence of (i).

If B is a subring of K which is integral over A, the annihilator $\mathfrak{f}$ of B/A (equal by definition to the transporter A: B (Chapter I, § 2, no. 10)) is sometimes called the *conductor* of B in A.

COROLLARY 5. *Let* A *be an integral domain,* A' *its integral closure and* $\mathfrak{f}$ *the annihilator of the* A*-module* A'/A. *Suppose that* A' *is a finitely generated* A*-module. The prime ideals* $\mathfrak{p}$ *of* A *such that* $A_{\mathfrak{p}}$ *is not integrally closed are those which contain* $\mathfrak{f}$.

This follows immediately from Corollary 4 (ii) applied to $S = A - \mathfrak{p}$.

Note that under the hypotheses of Corollary 5 $\mathfrak{f} \neq 0$ since A' is a finitely generated A-module and every element of K/A (K the field of fractions of A) has an annihilator $\neq 0$.

* In algebraic geometry, Corollary 5 and the above remark show that the points where an affine variety V is not normal form a closed set distinct from V. *

6. NORMS AND TRACES OF INTEGERS

PROPOSITION 17. *Let* A *be a ring,* B *a (commutative)* A*-algebra and* X *a square matrix of order* n *over* B; *the following properties are equivalent:*

(a) X *is integral over* A.

(b) *There exists a finitely generated sub-*A*-module* M *of* B^n *such that* $X.x \in M$ *for all* $x \in M$ *and* M *is a system of generators of the* B*-module* B^n.

(c) *The coefficients of the characteristic polynomial of* X *are integral over* A.

If $\chi(T) = \det(T.1 - X)$ is the characteristic polynomial of X, the Cayley-Hamilton Theorem shows that $\chi(X) = 0$ (*Algebra*, Chapter VII, § 5, no. 4, *Remark* 1) and, as χ is a monic polynomial, (c) implies (a) by no. 1, Proposition 6.

Suppose in the second place that (a) holds. If $(e_i)_{1 \leqslant i \leqslant n}$ is the canonical basis of B^n, the sub-A-module M of B generated by the $X^k.e_i$ $(1 \leqslant i \leqslant n, k \geqslant 0)$ is a finitely generated A-module, since the A-algebra $A[X]$ is a finitely generated A-module (no. 1, Theorem 1); as M contains the e_i, it is seen that (a) implies (b); the converse is a consequence of no. 1, Theorem 1, condition (E_{III}).

Finally let us prove that (a) implies (c); as X is integral over A and *a fortiori* over the polynomial ring A[T], $T.1 - X$ is also integral over A[T] and by no. 3, Proposition 12 the problem is seen to reduce (by replacing X by $T.1 - X$ and A by A[T]) to proving that, if X is integral over A, $d = \det(X)$ is an element of B which is integral over A. Now, we have seen above that the endomorphism u of B^n defined by X leaves stable a finitely generated sub-A-module M containing the e_i; the n-vectors $x_1 \wedge x_2 \wedge \cdots \wedge x_n$, where $x_i \in M$ for $1 \leqslant i \leqslant n$, therefore generate in $\bigwedge^n(B^n)$ a finitely generated sub-A-module containing $e_1 \wedge e_2 \wedge \cdots \wedge e_n$ and which is stable under $\bigwedge^n u$, in other words under the homothety of ratio d; as the annihilator of

$$e_1 \wedge e_2 \wedge \cdots \wedge e_n$$

in B reduces to 0, condition (E_{III}) of no. 1, Theorem 1 proves that d is integral over A.

COROLLARY 1. *Let* A *be an integral domain*, K *its field of fractions and* K′ *a finite-dimensional* K*-algebra* (not necessarily commutative). *If* $x \in K'$ *is integral over* A, *the coefficients of the characteristic polynomial* $Pc_{K'/K}(x; X)$ (*Algebra*, Chapter VIII, § 12, no. 2) *are integral over* A.

If $z \mapsto M(z)$ is the regular representation of the algebra K′ (considered as a matrix representation; cf. *Algebra*, Chapter VIII, § 13) $Pc_{K'/K}(x; X)$ is by definition the characteristic polynomial of the matrix $M(x)$; if x integral over A, the matrix $M(x)$ is integral over A (no. 1, Proposition 2) and it is sufficient to apply Proposition 17.

COROLLARY 2. *With the same hypotheses and notation as in Corollary* 1, $Tr_{K'/K}(x)$ *and* $N_{K'/K}(x)$ *are integral over* A.

$Tr_{K'/K}(x)$ and $N_{K'/K}(x)$ are, to within a sign, coefficients of $Pc_{K'/K}(x; X)$ (*Algebra*, Chapter VIII, § 12, no. 1, equations (4)) and hence are integral.

Remark (1) If K′ is a simple central algebra over K and $x \in K'$ is integral over A, the coefficients of the *reduced* characteristic polynomial of x (*Algebra*, Chapter VIII, § 12, no. 3) are *integral over* A. For there is a power of this polynomial equal to $Pc_{K'/K}(x; X)$ (*loc. cit.*, Proposition 8) and it is sufficient to apply Proposition 17 and no. 3, Proposition 11.

PROPOSITION 18. *Let* A *be an integrally closed domain*, K *its field of fractions*, K′ *a finite-dimensional* separable K*-algebra* (*Algebra*, Chapter VIII, § 7, no. 5) *and* A′ *the integral closure of* A *in* K′. *Then* A′ *is contained in a finitely generated* A*-module.*

The proposition will follow from the following more precise lemma:

LEMMA 3. *Under the hypotheses of Proposition* 18, *let* $(w_1, \dots, w_n)$ *be a basis of* K′ *over* K *contained in* A′ (no. 5, Corollary 2 to Proposition 16); *then there is a unique basis* $(w_1^*, \dots, w_n^*)$ *of* K′ *over* K *for which* $Tr_{K'/K}(w_i w_j^*) = \delta_{ij}$ (Kronecker index); *if* $d = D_{K'/K}(w_1, \dots, w_n)$ *is the discriminant of the basis* $(w_1, \dots, w_n)$ (*Algebra*, Chapter IX, § 2), *then* $d \neq 0$ *and*

$$(2) \qquad \sum_{i=1}^{n} Aw_i \subset A' \subset \sum_{i=1}^{n} Aw_i^* \subset d^{-1}\left(\sum_{i=1}^{n} Aw_i\right).$$

In particular, if d *is an invertible element of* A, A′ *is a free* A*-module with basis* $(w_1, \dots, w_n)$.

As K′ is a separable K-algebra, $d \neq 0$ (*Algebra*, Chapter IX, § 2, Proposition 5) and the K-bilinear form

$$(x, y) \mapsto Tr_{K'/K}(xy)$$

over K′ is therefore non-degenerate (*loc. cit.*, Proposition 4); this shows the existence and uniqueness of the basis $(w_i^*)_{1 \leqslant i \leqslant n}$ (*Algebra*, Chapter IX, § 1, no. 6, Corollary to Proposition 6). This being so, the first inclusion of (2) is obvious. Let x be an element of A′; let us write $x = \sum_{i=1}^{n} \xi_i w_i^*$ where $\xi_i \in K$; for all i, $\xi_i = \mathrm{Tr}_{K'/K}(xw_i)$, hence ξ_i is integral over A (Corollary 2 to Proposition 17) and, as A is integrally closed, $\xi_i \in A$ for $1 \leqslant i \leqslant n$; this shows the second inclusion (2). Finally, let us write $w_j^* = \sum_{i=1}^{n} \alpha_{ji} w_i$ where $\alpha_{ji} \in K$; then $\sum_{i=1}^{n} \alpha_{ji} \mathrm{Tr}_{K'/K}(w_i w_k) = \delta_{jk}$ for all j and k; Cramer's formulae show that the α_{ji} belong to $d^{-1}A$, whence the third inclusion (2). The last assertion follows immediately from (2), which in this case gives $A' = \sum_{i=1}^{n} Aw_i$.

In the two corollaries which follow the hypotheses and notations are those of Proposition 18.

COROLLARY 1. *If* A *is a Noetherian ring, the* A*-module* A′ *is finitely generated and in particular the ring* A′ *is Noetherian.*

A′ is a submodule of a finitely generated A-module.

COROLLARY 2. *If* A *is a principal ideal domain,* A′ *is a free* A*-module of rank* n.

Every submodule of a free A-module is then free (*Algebra*, Chapter VII, § 3, Theorem 1).

COROLLARY 3. *Let* E *be an extension of degree* n *of the field* $\mathbf{Q}$ *of rational numbers. The additive group of the integral closure in* E *of the ring* $\mathbf{Z}$ *of rational integers is a free commutative group of rank* n.

$\mathbf{Z}$ is integrally closed (no. 3, Proposition 10) and E is separable since $\mathbf{Q}$ is of characteristic 0. Corollary 2 can therefore be applied to the case where $A = \mathbf{Z}$, $K = \mathbf{Q}$ and $K' = E$.

Remark (2). The conclusions of Corollary 1 are not necessarily true if K′ is not assumed to be separate over K, even if K′ is an extension field of K (Exercise 20). On the other hand, if A is a *finitely generated integral* K_0-*algebra*, where K_0 is a *field*, the integral closure of A in *any* extension of finite degree of the field of fractions of A is a finitely generated A-module and a Noetherian ring, as we shall see in § 3, no. 2, Theorem 2.

7. EXTENSION OF SCALARS IN AN INTEGRALLY CLOSED ALGEBRA

PROPOSITION 19. *Let* k *be a field,* L *a separable extension of* k *and* R *an integrally closed* k*-algebra. If the ring* $L \otimes_k R$ *is an integral domain, it is integrally closed.*

Let K be the field of fractions of R; as k is a field, $L \otimes_k R$ is canonically identified with a sub-k-algebra of $L \otimes_k K$ and L and R with sub-k-algebras of $L \otimes_k R$. Moreover, since an element $s \neq 0$ of R is not a divisor of 0 in R, $1 \otimes s$ is not a divisor of zero in $L \otimes_k R$ since L is flat over k (Chapter I, § 2, no. 3); identifying s with $1 \otimes s$, it is therefore seen that, if $S = R - \{0\}$, $L \otimes_k K$ is identified with $S^{-1}(L \otimes_k K)$; as $L \otimes_k R$ is assumed to be an *integral domain*, $L \otimes_k K$ is thus identified with a subring of the field of fractions Ω of $L \otimes_k R$.

(1) Suppose that L is a *finite* extension of k; then $L \otimes_k K$ is an algebra of finite rank over K and by hypothesis has no divisor of 0; hence it is a *field* (*Algebra*, Chapter V, § 2, no. 1, Proposition 1) and therefore it is in this case the *field of fractions* Ω of $L \otimes_k R$. Let $(w_1, \ldots, w_n)$ be a basis of L over k, which is therefore also a basis of $L \otimes_k K$ over K. There exists a basis $(w_1^*, \ldots, w_n^*)$ of L such that $\mathrm{Tr}_{L/k}(w_i w_j^*) = \delta_{ij}$ (no. 6, Lemma 3); every $z \in L \otimes_k K$ may be written uniquely $z = \sum_{i=1}^{n} a_i w_i$ where $a_i \in K$; then

$$\mathrm{Tr}_{(L\otimes K)/K}(zw_j^*) = \sum_{i=1}^{n} a_i \mathrm{Tr}_{(L\otimes K)/K}(w_i w_j^*)$$

and as in L the traces $\mathrm{Tr}_{(L\otimes K)/K}$ and $\mathrm{Tr}_{L/K}$ coincide (*Algebra*, Chapter VIII, § 12, no. 2, formula (13)) finally $\mathrm{Tr}_{(L\otimes K)/K}(zw_j^*) = a_j$ for $1 \leqslant j \leqslant n$. Note on the other hand that the elements of L are integral over k and hence also over R (no. 1, Corollary 1 to Proposition 2); therefore (no. 1, Proposition 5) $L \otimes_k R$ is integral over R. This being so, suppose that $z \in L \otimes_k K$ is integral over $L \otimes_k R$; then z is also integral over R (no. 1, Proposition 6), hence so is zw_j^* and therefore also $a_j = \mathrm{Tr}_{(L\otimes K)/K}(zw_j^*)$ for $1 \leqslant j \leqslant n$ (no. 6, Corollary 2 to Proposition 17). As R is integrally closed, $a_j \in R$ for all j and hence $z \in L \otimes_k R$, which proves the proposition in this case.

(2) Suppose now that L is a *finitely generated* separable extension of K; then there exists a *separating* transcendence basis $(x_1, \ldots, x_d)$ of L over k (*Algebra*, Chapter V, § 9, no. 3, Theorem 2); as L and K are algebraically disjoint over k in the field Ω (*Algebra*, Chapter V, § 5, no. 4), the x_i are algebraically independent over K; hence $R[x_1, \ldots, x_d]$ is integrally closed (no. 3, Corollary 2 to Proposition 13). Let T be the set of elements $\neq 0$ of the ring $A = k[x_1, \ldots, x_d] \subset L$, so that the field $k_1 = k(x_1, \ldots, x_d) \subset L$ is equal to $T^{-1}k[x_1, \ldots, x_d]$; then

$$k_1 \otimes_k R = (T^{-1}A) \otimes_k R = T^{-1}A \otimes_A (A \otimes_k R) = T^{-1}(A \otimes_k R) = T^{-1}R[x_1, \ldots, x_d]$$

by the associativity of the tensor product, hence this domain is integrally closed (no. 5, Corollary 1 to Proposition 16). But $L \otimes_k R$ is identified with

$L \otimes_{k_1} (k_1 \otimes_k R)$ and by definition L is a finite separable extension of k_1; it follows therefore from (1) that $L \otimes_k R$ is integrally closed.

(3) *General case.* If z is an element of Ω which is integral over $L \otimes_k R$, it satisfies a relation of the form $z^m + b_1 z^{m-1} + \cdots + b_m = 0$, where the b_i belong to $L \otimes_k R$; then there exists a finitely generated sub-extension L′ of L over k such that the b_i belong to $L' \otimes_k R$ for $1 \leqslant i \leqslant m$ and z to $L' \otimes_k K$. Then it follows from (2) that $z \in L' \otimes_k R$ and hence $L \otimes_k R$ is integrally closed.

* Let V be an affine irreducible algebraic variety, k a defining field of V and R the ring of functions regular on V defined over k; if R is integrally closed, V is called *normal over k*; Proposition 19 shows that, if V is normal over k, it remains normal over every separable extension L of k. *

COROLLARY. *Let k be a field and* R *and* S *two integrally closed k-algebras. Suppose that the ring* $R \otimes_k S$ *is an integral domain and that the fields of fractions* K *and* L *of* R *and* S *respectively are separable over k. Then the ring* $R \otimes_k S$ *is an integrally closed domain.*

As R and S are identified with subalgebras of $R \otimes_k S$, K and L are identified with subfields of the field of fractions Ω of $R \otimes_k S$ which are linearly disjoint over k (*Algebra*, Chapter V, § 2, no. 3, Proposition 5). It then follows from Proposition 19 that $R \otimes_k L$ and $K \otimes_k S$ are integrally closed domains; as their intersection is $R \otimes_k S$ (Chapter I, § 2, no. 6, Proposition 7), $R \otimes_k S$ is an integrally closed domain (no. 2, Corollary to Proposition 8).

* Given two irreducible affine varieties V, W defined over k, their product $V \times W$ is an affine variety and the ring of functions regular on $V \times W$ is identified with the tensor product over k of the ring of functions regular on V and the ring of functions regular on W. The Corollary to Proposition 19 shows that, if V and W are normal over k, then $V \times W$ is normal over k. *

8. INTEGERS OVER A GRADED RING

All the graduations considered in this no. are of type **Z**; *if* A *is a graded ring and* $i \in \mathbf{Z}$, A_i *denotes the set of homogeneous elements of degree i of the ring* A.

Let A be a graded ring and B a graded A-algebra. Let x be a *homogeneous* element of B which is integral over A; then there is a relation

(3) $$x^n + a_1 x^{n-1} + \cdots + a_n = 0 \quad \text{where} \quad a_i \in A \quad \text{for} \quad 1 \leqslant i \leqslant n.$$

Let $m = \deg(x)$ and let a'_i be the homogeneous component of degree mi of a_i $(1 \leqslant i \leqslant n)$; then obviously

(4) $$x^n + a'_1 x^{n-1} + \cdots + a'_n = 0$$

in other words x satisfies an equation of integral dependence with *homogeneous* coefficients.

Let $A[X, X^{-1}]$ denote the ring of fractions $S^{-1}A[X]$ of the polynomial ring $A[X]$ in one indeterminate, S being the multiplicative subset of $A[X]$ consisting of the powers X^n of X ($n \geqslant 0$); as X is not a divisor of 0 in $A[X]$ it is immediate that the X^i ($i \in \mathbf{Z}$) form a *basis* over A of the A-module $A[X, X^{-1}]$. For every element $a \in A$ with homogeneous components a_i ($i \in \mathbf{Z}$), we write

$$j_A(a) = \sum_{i \in \mathbf{Z}} a_i X^i \in A[X, X^{-1}] \tag{5}$$

it is immediate that $j_A: A \to A[X, X^{-1}]$ is an injective ring homomorphism.

PROPOSITION 20. *Let* $A = \bigoplus_{i \in \mathbf{Z}} A_i$ *be a graded ring and* B *a (commutative) graded* A-*algebra. The set* A′ *of elements of* B *integral over* A *is a graded subalgebra of* B. *If* $A_i = 0$ *for* $i < 0$ *and* B *is a reduced ring, then* $A'_i = 0$ *for* $i < 0$.

The diagram

$$\begin{array}{ccc} A & \xrightarrow{\ \rho\ } & B \\ {\scriptstyle j_A}\downarrow & & \downarrow{\scriptstyle j_B} \\ A[X, X^{-1}] & \xrightarrow{\ \rho'\ } & B[X, X^{-1}] \end{array}$$

(where ρ is the homomorphism defining the A-algebra structure on B and ρ' the homomorphism canonically derived from it) is commutative, as is immediately verified from the definition (5). Let x be an element of B integral over A; then $j_B(x)$ is integral over $A[X, X^{-1}]$ (no. 1, Proposition 2) and it therefore follows from no. 5, Proposition 16 that there exists an integer $m > 0$ such that $X^m j_B(x)$ is an element of $B[X]$ integral over $A[X]$. We then deduce from no. 3, Proposition 12 that the coefficients of the polynomial $X^m j_B(x)$ are integral over A; as these coefficients are by definition the homogeneous components of x, it is seen that these are integral over A, which proves that A′ is a *graded* subalgebra of B.

Suppose now that $x \in A'_i$ where $i < 0$; the remark at the beginning of this no. shows that x satisfies an equation of the form (4) where $a'_k \in A_{ki}$ for $1 \leqslant k \leqslant n$. If $A_j = 0$ for $j < 0$, then $x^n = 0$ and if B is a reduced ring we conclude that $x = 0$ and hence $A'_i = 0$ for all $i < 0$ in this case.

Recall (Chapter II, § 2, no. 9) that, if $A = \bigoplus_{i \in \mathbf{Z}} A_i$ is a graded ring and S is a multiplicative subset of A consisting of *homogeneous* elements, a graded ring structure is defined on $S^{-1}A$ by taking the set $(S^{-1}A)_i$ of homogeneous elements of degree i to be the set of elements of the form a/s, where $a \in A$ and $s \in S$ are homogeneous and such that $\deg(a) - \deg(s) = i$.

Lemma 4. *Let* $A = \bigoplus_{i \in \mathbf{Z}} A_i$ *be a graded integral domain and* S *the set of homogeneous elements* $\neq 0$ *of* A.

(i) *Every homogeneous element* $\neq 0$ *of* $S^{-1}A$ *is invertible, the ring* $K_0 = (S^{-1}A)_0$ *is a field and the set of* $i \in \mathbf{Z}$ *such that* $(S^{-1}A)_i \neq 0$ *is a subgroup* $q\mathbf{Z}$ *of* $\mathbf{Z}$ *(where* $q \geqslant 0$).

(ii) *Suppose that* $q \geqslant 1$ *and let* t *be a non-zero element of* $(S^{-1}A)_q$. *Then the* K_0-*homomorphism* f *of the polynomial ring* $K_0[X]$ *to* $S^{-1}A$ *which maps* X *to* t *extends to an isomorphism of* $K_0[X, X^{-1}]$ *onto* $S^{-1}A$ *and* $S^{-1}A$ *is integrally closed.*

The assertions in (i) follow immediately from the definitions and the hypothesis that A is an integral domain, for if a/s and a'/s' are two homogeneous elements $\neq 0$ of $S^{-1}A$ of degrees i and i', aa'/ss' is a homogeneous element $\neq 0$ and of degree $i + i'$. To show (ii), we note that since t is invertible in $S^{-1}A$ the homomorphism f extends in a unique way to a homomorphism $\bar{f}: K_0[X, X^{-1}] \to S^{-1}A$ and necessarily $\bar{f}(X^{-1}) = t^{-1}$. On the other hand, by definition of q, every homogeneous element $\neq 0$ of $S^{-1}A$ is of degree qn ($n \in \mathbf{Z}$) and hence can be written *uniquely* in the form λt^n where $\lambda \in K_0$ (since $S^{-1}A$ is an integral domain); hence $\bar{f}$ is bijective. Finally, we know that $K_0[X]$ is integrally closed (no. 3, Proposition 10) and hence so is $K_0[X, X^{-1}]$ (no. 5, Corollary 1 to Proposition 16), which completes the proof of the Lemma.

Proposition 21. *Let* $A = \bigoplus_{i \in \mathbf{Z}} A_i$ *be a graded integral domain and* S *the set of homogeneous elements* $\neq 0$ *of* A. *The integral closure* A' *of* A *is then a graded subring of* $S^{-1}A$. *If further* $A_i = 0$ *for* $i < 0$, *then* $A'_i = 0$ *for* $i < 0$.

If $A = A_0$, the proposition is trivial. Otherwise we may apply Lemma 4; the ring $S^{-1}A$ is an integrally closed domain and therefore $A' \subset S^{-1}A$; as $S^{-1}A$ is graded, so is A' by Proposition 20; the latter assertion also follows from Proposition 20.

Corollary 1. *With the hypotheses and notation of Proposition* 21, *if every homogeneous element of* $S^{-1}A$ *which is integral over* A *belongs to* A, *then* A *is integrally closed.*

Then $A'_i \subset A$ for all $i \in \mathbf{Z}$ and hence $A' = A$.

Corollary 2. *If* $A = \bigoplus_{i \in \mathbf{Z}} A_i$ *is an integrally closed graded domain, the domain* A_0 *is integrally closed.*

The field of fractions K_0 of A_0 is identified (in the notation of Proposition 21) with a subring of the ring of homogeneous elements of degree 0 of $S^{-1}A$; every element of K_0 integral over A_0 (and *a fortiori* over A) belongs therefore by hypothesis to A_0.

COROLLARY 3. *Let* $A = \bigoplus_{i \in \mathbf{Z}} A_i$ *be an integrally closed graded domain. Then, for every integer* $d > 0$, *the ring* $A^{(d)}$ (Chapter III, § 1, no. 3) *is an integrally closed domain.*

Let U be the set of homogeneous elements $\neq 0$ of $A^{(d)}$ and let x be a homogeneous element of $U^{-1}A^{(d)}$ integral over $A^{(d)}$ and hence over A; as $x \in S^{-1}A$, x belongs to A by hypothesis; as its degree is divisible by d, it belongs to $A^{(d)}$ and it then follows from Corollary 1 that $A^{(d)}$ is integrally closed.

9. APPLICATION: INVARIANTS OF A GROUP OF AUTOMORPHISMS OF AN ALGEBRA

Given a ring K, a K-algebra A and a group $\mathscr{G}$, we shall say that $\mathscr{G}$ *operates on* A if: (1) the set A has group of operators $\mathscr{G}$ (*Algebra*, Chapter I, § 7, no. 2); (2) for all $\sigma \in \mathscr{G}$, the mapping $x \mapsto \sigma . x$ is an *endomorphism* of the K-algebra A (and therefore an *automorphism* since it is bijective (*loc. cit.*)). We shall denote by $A^{\mathscr{G}}$, the set of elements of A which are *invariant* under $\mathscr{G}$; clearly it is a *sub-K-algebra* of A.

We shall say that $\mathscr{G}$ is a *locally finite* group of operators on A if every orbit of $\mathscr{G}$ in A (*Algebra*, Chapter I, *Corrections to Fascicule* IV) is *finite*.

PROPOSITION 22. *Let* A *be a (commutative)* K-*algebra and* $\mathscr{G}$ *a locally finite group of operators on* A. *Then* A *is integral over the subalgebra* $A^{\mathscr{G}}$.

For all $x \in A$, let x_i ($1 \leqslant i \leqslant n$) be the distinct elements of the orbit of x under $\mathscr{G}$; for all $\sigma \in \mathscr{G}$, there exists a permutation π_σ of the set $\{1, 2, \ldots, n\}$ such that $\sigma . x_i = x_{\pi_\sigma}(i)$ for $1 \leqslant i \leqslant n$; therefore the elementary symmetric functions of the x_i are elements of A which are invariant under $\mathscr{G}$, in other words elements of $A^{\mathscr{G}}$. As x is a root of the monic polynomial $\prod_{i=1}^{n} (X - x_i)$ and the coefficients of this polynomial belong to $A^{\mathscr{G}}$, x is integral over $A^{\mathscr{G}}$.

THEOREM 2. *Let* A *be a finitely generated* K-*algebra and* $\mathscr{G}$ *a locally finite group of operators on* A. *Then* A *is a finitely generated* A-*module; if further* K *is Noetherian,* $A^{\mathscr{G}}$ *is a finitely generated* K-*algebra.*

Let $(a_j)_{1 \leqslant j \leqslant m}$ be a system of generators of the K-algebra A; as *a fortiori* $A = A^{\mathscr{G}}[a_1, \ldots, a_m]$ and the a_j are integral over $A^{\mathscr{G}}$ by Proposition 22, the first assertion follows from no. 1, Proposition 4. The second is a consequence of the following lemma:

LEMMA 5. *Let* K *be a Noetherian ring,* B *a finitely generated* K-*algebra and* C *a sub-K-algebra of* B *such that* B *is integral over* C. *Then* C *is a finitely generated* A-*algebra.*

Let $(x_i)_{1 \leqslant i \leqslant n}$ be a finite system of generators of the K-algebra B. For all i, there exists by hypothesis a monic polynomial $P_i \in C[X]$ such that $P_i(x_i) = 0$. Let C′ be the sub-K-algebra of C generated by the coefficients of the P_i $(1 \leqslant i \leqslant n)$; clearly the x_i are integral over C′ and $B = C'[x_1, \ldots, x_n]$; hence B is a finitely generated C′-module (no. 1, Proposition 4). On the other hand C′ is a Noetherian ring (Chapter III, § 2, no. 10, Corollary 3 to Theorem 2); hence C is a finitely generated C′-module, which proves that C is a finitely generated K-algebra.

Remark. The set of $\sigma \in \mathscr{G}$ such that $\sigma a_j = a_j$ for $1 \leqslant j \leqslant m$ obviously leaves invariant every element of A. The normal subgroup $\mathscr{H}$ of $\mathscr{G}$ leaving invariant every element of A is therefore of *finite* index in $\mathscr{G}$ and A may be considered as having a *finite* group of operators $\mathscr{G}/\mathscr{H}$; obviously $A^{\mathscr{G}/\mathscr{H}} = A^{\mathscr{G}}$.

Let S be a multiplicative subset of a ring A and $\mathscr{G}$ a group operating on A and for which S is *stable*; then, for all $\sigma \in \mathscr{G}$, there exists a unique endomorphism $z \mapsto \sigma . z$ of the ring $S^{-1}A$ such that $\sigma.(a/1) = (\sigma.a)/1$ for all $a \in A$; it is given by the formula $\sigma.(a/s) = (\sigma.a)/(\sigma.s)$ for $a \in A$ and $s \in S$ (Chapter II, § 2, no. 1, Proposition 2); if τ is another element of $\mathscr{G}$, clearly $\sigma.(\tau.z) = (\sigma\tau).z$ for all $z \in S^{-1}A$ and hence the group $\mathscr{G}$ *operates on the ring* $S^{-1}A$.

PROPOSITION 23. *Let* A *be a* K-*algebra,* $\mathscr{G}$ *a locally finite group of operators on* A, S *a multiplicative subset of* A *stable under* $\mathscr{G}$ *and* $S^{\mathscr{G}}$ *the set* $S \cap A^{\mathscr{G}}$. *Then the canonical mapping of* $(S^{\mathscr{G}})^{-1}A$ *to* $S^{-1}A$ (Chapter II, § 2, no. 1, Corollary 2 to Proposition 2) *is an isomorphism which maps* $(S^{\mathscr{G}})^{-1}A^{\mathscr{G}}$ *to* $(S^{-1}A)^{\mathscr{G}}$.

For all $s \in S$, let $s, s_1, \ldots, s_q$ be the distinct elements of the orbit of s under $\mathscr{G}$; as $ss_1 \ldots s_q \in S^{\mathscr{G}}$, the first assertion follows from Chapter II, § 2, no. 3, Proposition 8. Identifying canonically $(S^{\mathscr{G}})^{-1}A$ with $S^{-1}A$, clearly every element of $(S^{\mathscr{G}})^{-1}A^{\mathscr{G}}$ is invariant under $\mathscr{G}$. Conversely, let a/t be an element of $(S^{\mathscr{G}})^{-1}A$ which is invariant under $\mathscr{G}$ $(a \in A,\ t \in S^{\mathscr{G}})$; if a_j $(1 \leqslant j \leqslant m)$ are the distinct elements of the orbit of a under $\mathscr{G}$, then $a_j/t = a/t$ for $1 \leqslant j \leqslant m$ and therefore there exists $s \in S^{\mathscr{G}}$ such that $s(a_j - a) = 0$ for $1 \leqslant j \leqslant m$; in other words, sa is invariant under $\mathscr{G}$ and, as $a/t = (sa)/(st)$, certainly $a/t \in (S^{\mathscr{G}})^{-1}A^{\mathscr{G}}$.

COROLLARY. *Let* A *be an integral domain,* K *its field of fractions and* $\mathscr{G}$ *a locally finite group of operators on* A. *Then* $\mathscr{G}$ *operates on* K *and* $K^{\mathscr{G}}$ *is the field of fractions of* $A^{\mathscr{G}}$.

$A - \{0\}$ is stable under $\mathscr{G}$.

2. THE LIFT OF PRIME IDEALS

1. THE FIRST EXISTENCE THEOREM

DEFINITION 1. *Let* A, A′ *be two rings and* $h\colon A \to A'$ *a ring homomorphism. An ideal* $\mathfrak{a}'$ *of* A′ *is said to lie above an ideal* $\mathfrak{a}$ *of* A *if* $\mathfrak{a} = \overset{-1}{h}(\mathfrak{a}')$.

To say that a prime ideal $\mathfrak{p}'$ of A′ lies above an ideal $\mathfrak{p}$ of A therefore means that $\mathfrak{p}$ is the *image* of $\mathfrak{p}'$ under the continuous mapping ${}^a h\colon \mathrm{Spec}(A') \to \mathrm{Spec}(A)$ associated with h (Chapter II, § 4, no. 3).

Note that for there to exist an ideal of A′ lying above the ideal (0) of A, it is necessary and sufficient that $h\colon A \to A'$ be *injective*.

Let $\mathfrak{a}$ be an ideal of A; by taking quotients the homomorphism h gives a homomorphism $h_1\colon A/\mathfrak{a} \to A'/\mathfrak{a}'$; to say that $\mathfrak{a}'$ is an ideal of A′ lying above $\mathfrak{a}$ is equivalent to saying that $\mathfrak{a}A' \subset \mathfrak{a}'$ and that $\mathfrak{a}'/\mathfrak{a}A'$ is an ideal of $A'/\mathfrak{a}A'$ *lying above* (0).

LEMMA 1. *Let* $h\colon A \to A'$ *be a ring homomorphism,* S *a multiplicative subset of* A, $i = i_A^S\colon A \to S^{-1}A$, $i' = i_{A'}^{h(S)}\colon A' \to S^{-1}A' = (h(S))^{-1}A'$ *the canonical homomorphisms and* $h_1 = S^{-1}h\colon S^{-1}A \to S^{-1}A'$, *so that there is a commutative diagram*

$$\begin{array}{ccc} A & \xrightarrow{\ h\ } & A' \\ {\scriptstyle i}\big\downarrow & & \big\downarrow{\scriptstyle i'} \\ S^{-1}A & \xrightarrow[\ h_1\ ]{} & S^{-1}A' \end{array}$$

Let $\mathfrak{p}$ *be a prime ideal of* A *such that* $\mathfrak{p} \cap S = \varnothing$. *Then* $\mathfrak{a}' \mapsto S^{-1}\mathfrak{a}'$ *is a surjective mapping of the set* $\mathscr{F}$ *of ideals of* A′ *lying above* $\mathfrak{p}$ *onto the set* $\mathscr{F}_1$ *of ideals of* $S^{-1}A'$ *lying above* $S^{-1}\mathfrak{p}$ *and the mapping* $\mathfrak{a}'_1 \mapsto \overset{-1}{i'}(\mathfrak{a}'_1)$ *is a bijection of* $\mathscr{F}_1$ *onto the set of ideals belonging to* $\mathscr{F}$ *and is saturated with respect to* $h(S)$; *in particular* $\mathfrak{p}' \mapsto S^{-1}\mathfrak{p}'$ *is a bijection of the set of prime ideals of* A′ *lying above* $\mathfrak{p}$ *onto the set of prime ideals of* $S^{-1}A'$ *lying above* $S^{-1}\mathfrak{p}$.

We know that $S^{-1}\mathfrak{p}$ is a prime ideal of $S^{-1}A$ and that $\overset{-1}{i}(S^{-1}\mathfrak{p}) = \mathfrak{p}$ (Chapter II, § 2, no. 5, Proposition 11); if there exists an ideal $\mathfrak{b}'$ of $S^{-1}A'$ lying above $S^{-1}\mathfrak{p}$, then $\overset{-1}{h}(\overset{-1}{i'}(\mathfrak{b}')) = \overset{-1}{i}(\overset{-1}{h_1}(\mathfrak{b}')) = \mathfrak{p}$; as $S^{-1}.\overset{-1}{i'}(\mathfrak{b}') = \mathfrak{b}'$ (*loc. cit.*), this already shows that the image of $\mathscr{F}$ under the mapping $\mathfrak{a}' \mapsto S^{-1}\mathfrak{a}'$ *contains* $\mathscr{F}_1$. On the other hand, if $\mathfrak{a}' \in \mathscr{F}$, $a \in A$ and $s \in S$, there are the following equivalences

$$\begin{aligned} h_1(a/s) \in S^{-1}\mathfrak{a}' &\Leftrightarrow h(a)/h(s) \in S^{-1}\mathfrak{a}' \\ &\Leftrightarrow \text{there exists } t \in S \text{ such that } h(t)h(a) \in \mathfrak{a}' \\ &\Leftrightarrow \text{there exists } t \in S \text{ such that } ta \in \mathfrak{p} \\ &\Leftrightarrow a/z \in S^{-1}\mathfrak{p}. \end{aligned}$$

Hence $h_1^{-1}(S^{-1}\mathfrak{a}') = S^{-1}\mathfrak{p}$, which completes the proof that the image of $\mathscr{F}$ under the mapping $\mathfrak{a}' \mapsto S^{-1}\mathfrak{a}'$ is equal to $\mathscr{F}_1$; the other assertions follow from Chapter II, § 2, no. 5, Proposition 11.

PROPOSITION 1. *Let $h: A \to A'$ be a ring homomorphism such that A' is integral over A, $\mathfrak{p}'$ a prime ideal of A' and $\mathfrak{p} = h^{-1}(\mathfrak{p}')$. For $\mathfrak{p}$ to be maximal, it is necessary and sufficient that $\mathfrak{p}'$ be so.*

Let us write $B = A/\mathfrak{p}$, $B' = A'/\mathfrak{p}'$ and let $h_1: B \to B'$ be the homomorphism derived from h by taking quotients; B and B' are integral domains and B' is integral over B (§ 1, no. 1, Proposition 2). To say that $\mathfrak{p}$ (resp. $\mathfrak{p}'$) is maximal means that B (resp. B') is a field. The proposition then follows from the following lemma:

LEMMA 2. *Let B be an integral domain and A a subring of B such that B is integral over A. For B to be a field, it is necessary and sufficient that A be a field.*

If A is a field, then, for all $y \neq 0$ in B, $A[y]$ is by hypothesis (§ 1, Theorem 1) a finitely generated A-module; as $A[y]$ is an integral domain, it is a field (*Algebra*, Chapter V, § 2, no. 1, Proposition 1) and *a fortiori* y is invertible in B and hence B is a field. Conversely, suppose that B is a field and let $z \neq 0$ in A; as $z^{-1} \in B$, z^{-1} is integral over A, in other words there is an equation of integral dependence

$$z^{-n} + a_1 z^{-(n-1)} + \cdots + a_n = 0$$

where the $a_i \in A$; now this relation shows that

$$-z^{-1} = a_1 + a_2 z + \cdots + a_n z^{n-1} \in A,$$

hence A is certainly a field.

COROLLARY 1. *Let $h: A \to A'$ a ring homomorphism such that A' is integral over A, $\mathfrak{p}$ a prime ideal of A and $\mathfrak{p}'$ and $\mathfrak{a}'$ two ideals of A' lying above $\mathfrak{p}$ such that $\mathfrak{p}' \subset \mathfrak{a}'$. If $\mathfrak{p}'$ is prime, then $\mathfrak{a}' = \mathfrak{p}'$.*

Let us write $S = A - \mathfrak{p}$; then $S^{-1}A'$ is integral over $S^{-1}A$ (§ 1, no. 5, Proposition

16), $S^{-1}\mathfrak{p}$ is a maximal ideal of $S^{-1}A$ (Chapter II, § 2, no. 5, Proposition 11), $S^{-1}\mathfrak{a}'$ and $S^{-1}\mathfrak{p}'$ are ideals of $S^{-1}A'$ lying above $S^{-1}\mathfrak{p}$ (Lemma 1) and $S^{-1}\mathfrak{a}' \supset S^{-1}\mathfrak{p}'$. As $S^{-1}\mathfrak{p}'$ is prime, it is maximal by Proposition 1 and hence $S^{-1}\mathfrak{p}' = S^{-1}\mathfrak{a}'$; therefore $\mathfrak{a}'$ is contained in the saturation of $\mathfrak{p}'$ with respect to $h(S)$, which is equal to $\mathfrak{p}'$ (Chapter II, § 2, no. 5, Proposition 11).

COROLLARY 2. *Let* A′ *be an integral domain,* A *a subring of* A′ *such that* A′ *is integral over* A *and* f *a homomorphism from* A′ *to a ring* B. *If the restriction of* f *to* A *is injective,* f *is injective.*

If $\mathfrak{a}'$ is the kernel of f, the hypothesis means that $\mathfrak{a}' \cap A = (0)$; as A′ is an integral domain, Corollary 1 may be applied taking $\mathfrak{p}$ and $\mathfrak{p}'$ to be the ideal (0) of A and the ideal (0) of A′ respectively, whence $\mathfrak{a}' = (0)$.

COROLLARY 3. *Let* $h: A \to A'$ *be a ring homomorphism such that* A′ *is integral over* A *and* $\mathfrak{m}$ *a maximal ideal of* A *and suppose that there are in* A′ *only a finite number of distinct maximal ideals* $\mathfrak{m}'_j$ $(1 \leqslant j \leqslant n)$ *lying above* $\mathfrak{m}$. *Let* $\mathfrak{q}'_j$ *be the saturation of* $\mathfrak{m}A'$ *with respect to* $\mathfrak{m}'_j$ (Chapter II, § 2, no. 4). *Then:*

(i) *In the ring* $A'/\mathfrak{q}'_j$ *the divisors of zero are the elements of* $\mathfrak{m}'_j/\mathfrak{q}'_j$ *and they are nilpotent* $(1 \leqslant j \leqslant n)$.

(ii) $\mathfrak{m}A' = \bigcap_j \mathfrak{q}'_j = \prod_j \mathfrak{q}'_j$.

(iii) *The canonical homomorphism* $A'/\mathfrak{m}A' \to \prod_j (A'/\mathfrak{q}'_j)$ *is bijective.*

For a prime ideal of A′ to contain $\mathfrak{m}A'$, it is necessary and sufficient that its inverse image under h contain $\mathfrak{m}$ and hence that it lie above $\mathfrak{m}$, since $\mathfrak{m}$ is maximal in A; the $\mathfrak{m}'_j$ are therefore the only prime ideals of A′ containing $\mathfrak{m}A'$ (Proposition 1) and therefore $\mathfrak{r}' = \bigcap_j \mathfrak{m}'_j$ is the radical of $\mathfrak{m}A'$ (Chapter II, § 2, no. 6, Corollary 1 to Proposition 13). By definition of $\mathfrak{q}'_j$, the class mod. $\mathfrak{q}'_j$ of an element of $A' - \mathfrak{m}'_j$ is not a divisor of 0 in $A'/\mathfrak{q}'_j$; on the other hand, as the $\mathfrak{m}'_j$ are distinct maximal ideals, for every index j there exists an element a'_j belonging to $\bigcap_{i \neq j} \mathfrak{m}'_i$ and not to $\mathfrak{m}'_j$ (Chapter II, § 1, no. 1, Proposition 4); then, for all $x \in \mathfrak{m}'_j$, $a'_j x \in \mathfrak{r}'$, hence the class mod. $\mathfrak{q}'_j$ of $a'_j x$ is nilpotent and, as that of a'_j is not a divisor of 0, we conclude that the class of x is nilpotent; in other words $\mathfrak{m}'_j$ is the *radical* of $\mathfrak{q}'_j$, which proves (i). It follows that the $\mathfrak{q}'_j$ are relatively prime in pairs (Chapter II, § 1, no. 1, Proposition 3); (iii) will therefore be a consequence of (ii), taking account of Chapter II, § 1, no. 2, Proposition 5. To establish (ii), we note that in the ring $A'/\mathfrak{m}A'$ the $\mathfrak{m}'_j/\mathfrak{m}A'$ are the only maximal ideals and $\mathfrak{q}'_j/\mathfrak{m}A'$ is the saturation of (0) with respect to $\mathfrak{m}'_j/\mathfrak{m}A'$ (Chapter II, § 2, no. 4); we may therefore restrict our attention to the case $\mathfrak{m}A' = (0)$; the assertion of (ii) then follows from Chapter II, § 3, no. 3, Corollary 2 to Theorem 1.

Remark (1) If A′ is Noetherian, it follows from (i) and (ii) that $(\mathfrak{q}'_j)_{1 \leqslant j \leqslant n}$ is the *unique primary decomposition* of $\mathfrak{m}$A′ (Chapter IV, § 2, no. 3).

THEOREM 1. *Let* h: A → A′ *be an* injective *ring homomorphism such that* A′ *is integral over* A *and* $\mathfrak{p}$ *a prime ideal of* A. *There exists a prime ideal* $\mathfrak{p}'$ *of* A′ *lying above* $\mathfrak{p}$.

Suppose first that A is a local ring and $\mathfrak{p}$ the maximal ideal of A; then, for every maximal ideal $\mathfrak{m}'$ of A′, $\overset{-1}{h}(\mathfrak{m}')$ is a maximal ideal of A (Proposition 1) and hence equal to $\mathfrak{p}$, which proves the theorem in this case (since A′ contains A by hypothesis and is therefore not reduced to 0). In the general case, let us write S = A — $\mathfrak{p}$; then $S^{-1}A$ is a local ring whose maximal ideal is $S^{-1}\mathfrak{p}$ (Chapter II, § 2, no. 5, Proposition 11), $S^{-1}h$: $S^{-1}A \to S^{-1}A$, is injective (Chapter II, § 2, no. 4, Theorem 1) and $S^{-1}A'$ is integral over $S^{-1}A$ (§ 1, no. 5, Proposition 16); then there exists a prime ideal $\mathfrak{q}'$ of $S^{-1}A'$ lying above $S^{-1}\mathfrak{p}$ and we know that $\mathfrak{q}' = S^{-1}\mathfrak{p}'$, where $\mathfrak{p}'$ is a prime ideal of A′ lying above $\mathfrak{p}$ (Lemma 1).

If h: A → A′ is not injective, Theorem 1 is no longer necessarily true, as the example of the homomorphism **Z** → **Z**/n**Z** ($n > 1$) shows. However Theorem 1 can be applied to the canonical injection h(A) → A′; in other words, the statement of Theorem 1 is true for prime ideals $\mathfrak{p}$ *containing* Ker(h).

COROLLARY 1. *With the hypotheses and notation of Theorem* 1, $\overset{-1}{h}(\mathfrak{p}A') = \mathfrak{p}$.

$\mathfrak{p}A' \subset \mathfrak{p}'$ and $\overset{-1}{h}(\mathfrak{p}') = \mathfrak{p}$.

COROLLARY 2. *Let* h: A → A′ *be a ring homomorphism such that* A′ *is integral over* A, $\mathfrak{a}$ *and* $\mathfrak{p}$ *two ideals of* A *such that* $\mathfrak{a} \subset \mathfrak{p}$ *and* $\mathfrak{a}'$ *an ideal of* A′ *lying above* $\mathfrak{a}$. *Suppose that* $\mathfrak{p}$ *is prime. Then there exists a prime ideal* $\mathfrak{p}'$ *of* A′ *lying above* $\mathfrak{p}$ *and containing* $\mathfrak{a}'$.

If h_1: A/$\mathfrak{a}$ → A′/$\mathfrak{a}'$ is the homomorphism derived from h by taking quotients, h_1 is injective by hypothesis and A′/$\mathfrak{a}'$ is integral over A/$\mathfrak{a}$ (§ 1, no. 1, Proposition 2); then there exists a prime ideal $\mathfrak{p}'/\mathfrak{a}'$ of A′/$\mathfrak{a}'$ ($\mathfrak{p}'$ prime in A′) lying above $\mathfrak{p}/\mathfrak{a}$ (Theorem 1) and $\mathfrak{p}'$ is the required ideal.

COROLLARY 3. *Let* A *be a ring and* A′ *a ring containing* A *and integral over* A. *If* $\mathfrak{R}'$ *is the Jacobson radical of* A′, $\mathfrak{R}' \cap$ A *is the Jacobson radical of* A.

Let $\mathfrak{R}$ be the Jacobson radical of A. For every maximal ideal $\mathfrak{m}'$ of A′, $\mathfrak{m}' \cap$ A is a maximal ideal of A (Proposition 1), hence $\mathfrak{R} \subset \mathfrak{m}' \cap$ A and therefore $\mathfrak{R} \subset \mathfrak{R}' \cap$ A (*Algebra*, Chapter VIII, § 5, no. 3, Definition 3). Conversely, let $x \in \mathfrak{R}' \cap$ A; for every maximal ideal $\mathfrak{m}$ of A, there exists a prime ideal of A′ lying above $\mathfrak{m}$ (Theorem 1) and this ideal $\mathfrak{m}'$ is necessarily maximal (Proposition 1), hence $x \in \mathfrak{m}' \cap A = \mathfrak{m}$ and therefore $x \in \mathfrak{R}$.

COROLLARY 4. *Let* A *be a ring,* A′ *a ring containing* A *and integral over* A *and* f *a homomorphism from* A *to an algebraically closed field* L. *Then* f *can be extended to a homomorphism from* A′ *to* L.

Let $\mathfrak{p}$ be the kernel of f, which is a prime ideal since $f(A) \subset L$ is an integral domain; let $\mathfrak{p}'$ be a prime ideal of A′ lying above $\mathfrak{p}$ (Theorem 1). Then $A/\mathfrak{p}$ is canonically identified with a subring of $A'/\mathfrak{p}'$ and $A'/\mathfrak{p}'$ is integral over $A/\mathfrak{p}$ (§ 1, no. 1, Proposition 2). The homomorphism f defines, by taking the quotient, an isomorphism of $A/\mathfrak{p}$ onto the subring $f(A)$ of L, which can be extended to an isomorphism g of the field of fractions K of $A/\mathfrak{p}$ onto a subfield of L. As the field of fractions K′ of $A'/\mathfrak{p}'$ is algebraic over K, g can be extended to an isomorphism g' of K′ onto a subfield of L (*Algebra*, Chapter V, § 4, no. 2, Corollary to Theorem 1); if $\pi': A' \to A'/\mathfrak{p}'$ is the canonical homomorphism, $g' \circ \pi'$ is a homomorphism from A′ to L extending f.

Remark (2). Let $h: A \to A'$ be a ring homomorphism such that A′ is integral over A; then the associated continuous mapping ${}^a h: \operatorname{Spec}(A') \to \operatorname{Spec}(A)$ is *closed*. For every ideal $\mathfrak{a}'$ of A′, $A'/\mathfrak{a}'$ is integral over A′, hence also over A (§ 1, no. 1, Proposition 6) and $\operatorname{Spec}(A'/\mathfrak{a}')$ is identified with a closed subspace $V(\mathfrak{a}')$ of $\operatorname{Spec}(A')$; to show that ${}^a h$ is closed, we see then (replacing A′ by $A'/\mathfrak{a}'$) that it is sufficient to prove that the image of $\operatorname{Spec}(A)'$ under ${}^a h$ is a *closed* subset of Spec(A); now it follows from Theorem 1 that this image is just the set of prime ideals of A containing the ideal Ker(h) and this set is closed by definition of the topology on Spec(A).

PROPOSITION 2. *Let* $h: A \to A'$ *be a ring homomorphism such that* A′ *is integral over* A, $\mathfrak{p}$ *a prime ideal of* A, $S = A - \mathfrak{p}$, $(\mathfrak{p}'_\iota)_{\iota \in I}$ *the family of all the prime ideals of* A′ *lying above* $\mathfrak{p}$ *and* $S' = \bigcap_{\iota \in I} (A' - \mathfrak{p}_\iota)$; *then* $S^{-1}A' = S'^{-1}A'$.

In fact, by definition $h(S) \subset S'$ and, as

$$h(S)^{-1}A' = S^{-1}A',$$

it suffices to prove, by virtue of Chapter II, § 2, no. 3, Proposition 8, that, if a prime ideal $\mathfrak{q}'$ of A′ does not meet $h(S)$, it does not meet S′ either. Now, suppose that $\mathfrak{q}' \cap h(S) = \varnothing$ and let $\mathfrak{q} = \overset{-1}{h}(\mathfrak{q}')$; then $\mathfrak{q} \cap S = \varnothing$, in other words $\mathfrak{q} \subset \mathfrak{p}$. As $\mathfrak{q}'$ lies above $\mathfrak{q}$ by definition, it follows from Corollary 2 to Proposition 1 that there is an index ι such that $\mathfrak{q}' \subset \mathfrak{p}'_\iota$ and hence $\mathfrak{q}' \cap S' = \varnothing$, which completes the proof.

PROPOSITION 3. *Let* $h: A \to A'$ *be a ring homomorphism such that* A′ *is a finitely generated* A*-module; then, for every prime ideal* $\mathfrak{p}$ *of* A, *the set of prime ideals of* A′ *lying above* $\mathfrak{p}$ *is finite.*

Let $S = A - \mathfrak{p}$; by Lemma 1 we may replace A by $S^{-1}A$, A′ by $S^{-1}A'$ (which is a finitely generated $S^{-1}A$-module) and $\mathfrak{p}$ by $S^{-1}\mathfrak{p}$; in other words, we may assume that A is a *local* ring and $\mathfrak{p}$ is its maximal ideal. Then (by the remark made at the beginning of this no.) A may be replaced by $A/\mathfrak{p}$, A′ by $A'/\mathfrak{p}A'$ and $\mathfrak{p}$ by (0), for $A'/\mathfrak{p}A' = (A/\mathfrak{p}) \otimes_A A'$ is a finitely generated $(A/\mathfrak{p})$-module. Thus we have finally reduced the problem to proving the proposition when A is a *field* and $\mathfrak{p} = (0)$; A′ is then an A-algebra of finite rank and therefore *Artinian* and we know that in such an algebra there is only a *finite* number of prime ideals (Chapter IV, § 2, no. 5, Proposition 9).

2. DECOMPOSITION GROUP AND INERTIA GROUP

DEFINITION 2. *Let* A′ *be a ring and* $\mathcal{G}$ *a group operating on* A′ (§ 1, no. 9). *Given a prime ideal* $\mathfrak{p}'$ *of* A′ *the subgroup of elements* $\sigma \in \mathcal{G}$ *such that* $\sigma.\mathfrak{p}' = \mathfrak{p}'$ *is called the decomposition group of* $\mathfrak{p}'$ *(with respect to* $\mathcal{G}$*) and is denoted by* $\mathcal{G}^Z(\mathfrak{p}')$. *The ring of elements of* A′ *invariant under* $\mathcal{G}^Z(\mathfrak{p}')$ *is called the decomposition ring of* $\mathfrak{p}'$ *(with respect to* $\mathcal{G}$*) and is denoted by* $A^Z(\mathfrak{p}')$ (*).

We often write $\mathcal{G}^Z$ and A^Z instead of $\mathcal{G}^Z(\mathfrak{p}')$ and $A^Z(\mathfrak{p}')$ respectively, when there is no ambiguity.

For all $\sigma \in \mathcal{G}^Z(\mathfrak{p}')$ we also denote by $z \mapsto \sigma.z$ the endomorphism of the ring $A'/\mathfrak{p}'$ derived from the endomorphism $x \mapsto \sigma.x$ of A′ by taking quotients; clearly the group $\mathcal{G}^Z(\mathfrak{p}')$ operates in this way on the ring $A'/\mathfrak{p}'$.

DEFINITION 3. *With the notation of Definition* 2, *the subgroup of* $\mathcal{G}^Z(\mathfrak{p}')$ *consisting of those* σ *such that the endomorphism* $z \mapsto \sigma.z$ *of* $A'/\mathfrak{p}'$ *is the identity is called the inertia group of* $\mathfrak{p}'$ *(with respect to* $\mathcal{G}$*) and denoted by* $\mathcal{G}^T(\mathfrak{p}')$ *(or* $\mathcal{G}^T$*). The ring of elements of* A′ *invariant under* $\mathcal{G}^T(\mathfrak{p}')$ *is called the inertia ring of* $\mathfrak{p}'$ *(with respect to* $\mathcal{G}$*) and is denoted by* $A^T(\mathfrak{p}')$ *(or* A^T*)* (†).

If A is the subring of A′ consisting of the invariants of $\mathcal{G}$, clearly

$$A \subset A^Z(\mathfrak{p}') \subset A^T(\mathfrak{p}') \subset A'. \tag{1}$$

It follows from Definitions 2 and 3 that, for all $\rho \in \mathcal{G}$,

$$\mathcal{G}^Z(\rho.\mathfrak{p}') = \rho\mathcal{G}^Z(\mathfrak{p}')\,\rho^{-1}, \qquad \mathcal{G}^T(\rho.\mathfrak{p}') = \rho\mathcal{G}^T(\mathfrak{p}')\,\rho^{-1}. \tag{2}$$

If, for all $\sigma \in \mathcal{G}^Z(\mathfrak{p}')$, $\bar{\sigma}$ is the automorphism $z \mapsto \sigma.z$ of $A'/\mathfrak{p}'$, $\sigma \mapsto \bar{\sigma}$ is a homomorphism (called *canonical*) of $\mathcal{G}^Z$ to the group Γ_0 of automorphisms of

---

(*) The letter Z is the initial of the German word "Zerlegung" which means "decomposition."

(†) The letter T is the initial of the German word "Trägheit" which means "inertia."

$A'/\mathfrak{p}'$ leaving invariant the elements of $A^Z/(\mathfrak{p}' \cap A^Z)$ (canonically identified with a subring of $A'/\mathfrak{p}'$) and by definition $\mathscr{G}^T(\mathfrak{p}')$ is the *kernel* of this canonical homomorphism; $\mathscr{G}^T$ is therefore a *normal subgroup* of $\mathscr{G}^Z$. If k' is the field of fractions of $A'/\mathfrak{p}'$, every automorphism of $A'/\mathfrak{p}'$ can be extended uniquely to an automorphism of k', so that $\sigma \mapsto \bar{\sigma}$ can be considered as a homomorphism from $\mathscr{G}^Z(\mathfrak{p}')$ to the group of automorphisms of k'. Note finally that, since $\mathscr{G}^T$ is normal in $\mathscr{G}^Z$, A^T is *stable* under $\mathscr{G}^Z$.

LEMMA 3. *Let* A' *be a ring,* $\mathscr{G}$ *a group operating on* A', A *the ring of invariants of* $\mathscr{G}$, $\mathfrak{p}'$ *a prime ideal of* A' *and* S *a multiplicative subset of* A *not meeting* $\mathfrak{p}'$. *Then* $\mathscr{G}^Z(S^{-1}\mathfrak{p}') = \mathscr{G}^Z(\mathfrak{p}')$, $\mathscr{G}^T(S^{-1}\mathfrak{p}') = \mathscr{G}^T(\mathfrak{p}')$ *and, if* $\mathscr{G}$ *is locally finite,* $S^{-1}A^Z(\mathfrak{p}') = A^Z(S^{-1}\mathfrak{p}')$ *and* $S^{-1}A^T(\mathfrak{p}') = A^T(S^{-1}\mathfrak{p}')$.

As the elements of S are invariant under $\mathscr{G}$, clearly, if $\sigma.\mathfrak{p}' = \mathfrak{p}'$, also $\sigma.(S^{-1}\mathfrak{p}') = S^{-1}\mathfrak{p}'$. Conversely, suppose that $\sigma \in \mathscr{G}$ is such that $\sigma.(S^{-1}\mathfrak{p}') = S^{-1}\mathfrak{p}'$; then, if $x \in \mathfrak{p}'$, $(\sigma.x)/1 \in S^{-1}\mathfrak{p}'$ and there therefore exists $s \in S$ such that $s(\sigma.x) \in \mathfrak{p}'$, whence $\sigma.x \in \mathfrak{p}'$ since $\mathfrak{p}'$ is prime and $s \notin \mathfrak{p}'$; this proves that $\sigma.\mathfrak{p}' \subset \mathfrak{p}'$ and it can be similarly shown that $\sigma^{-1}.\mathfrak{p}' \subset \mathfrak{p}'$, hence $\sigma.\mathfrak{p}' = \mathfrak{p}'$ and $\sigma \in \mathscr{G}^Z(\mathfrak{p}')$. If $\sigma \in \mathscr{G}^T(\mathfrak{p}')$, then $\sigma.x - x \in \mathfrak{p}'$ for all $x \in A'$, hence also, for all $s \in S$,

$$\sigma.(x/s) - (x/s) = (\sigma.x - x)/s \in S^{-1}\mathfrak{p}'$$

and therefore $\sigma \in \mathscr{G}^T(S^{-1}\mathfrak{p}')$. Conversely, suppose that $\sigma \in \mathscr{G}^T(S^{-1}\mathfrak{p}')$; then, for all $x \in A'$, $\sigma.(x/1) - (x/1) \in S^{-1}\mathfrak{p}'$ and therefore there exists $s \in S$ such that $s(\sigma.x - x) \in \mathfrak{p}'$, whence as above $\sigma.x - x \in \mathfrak{p}'$, which proves that $\sigma \in \mathscr{G}^T(\mathfrak{p}')$. The last two assertions follow from § 1, no. 9, Proposition 23.

THEOREM 2. *Let* A' *be a ring,* $\mathscr{G}$ *a finite group operating on* A' *and* A *the ring of invariants of* $\mathscr{G}$, *so that* A' *is integral over* A (§ 1, no. 9, Proposition 22).

(i) *Given two prime ideals* $\mathfrak{p}'$, $\mathfrak{q}'$ *of* A' *lying above the same prime ideal* $\mathfrak{p}$ *of* A, *there exists* $\sigma \in \mathscr{G}$ *such that* $\mathfrak{q}' = \sigma.\mathfrak{p}'$; *in other words,* $\mathscr{G}$ *operates transitively on the set of prime ideals of* A' *lying above* $\mathfrak{p}$.

(ii) *Let* $\mathfrak{p}'$ *be a prime ideal of* A', $\mathfrak{p} = \mathfrak{p}' \cap A$ *and* k (resp. k') *the field of fractions of* $A/\mathfrak{p}$ (resp. $A'/\mathfrak{p}'$). *Then* k' *is a quasi-Galois extension* (*) *of* k *and the canonical homomorphism* $\sigma \mapsto \bar{\sigma}$ *from* $\mathscr{G}^Z(\mathfrak{p}')$ *to the group* Γ *of* k*-automorphisms of* k' *defines, by taking the quotient, an isomorphism of* $\mathscr{G}^Z(\mathfrak{p}')/\mathscr{G}^T(\mathfrak{p}')$ *onto* Γ.

(*) In order to avoid confusion with other meanings of the word "normal" we henceforth use the term "quasi-Galois extension" as synonymous with the term "normal extension" defined in *Algebra*, Chapter V, § 6, no. 2, Definition 2.

(i) If $x \in \mathfrak{q}'$, then $\prod_{\sigma \in \mathscr{G}} \sigma.x \in \mathfrak{q}' \cap A = \mathfrak{p} \subset \mathfrak{p}'$; then there exists $\sigma \in \mathscr{G}$ such that $\sigma.x \in \mathfrak{p}'$, that is $x \in \sigma^{-1}.\mathfrak{q}'$. We conclude that $\mathfrak{q}' \subset \bigcup_{\sigma \in \mathscr{G}} \sigma.\mathfrak{p}'$ and hence (since $\mathscr{G}$ is finite and the $\sigma.\mathfrak{p}'$ prime) there exists $\sigma \in \mathscr{G}$ such that $\mathfrak{q}' \subset \sigma.\mathfrak{p}'$ (Chapter II, § 1, no. 1, Proposition 2); as $\mathfrak{q}'$ and $\sigma.\mathfrak{p}'$ both lie above $\mathfrak{p}$, $\mathfrak{q}' = \sigma.\mathfrak{p}'$ (no. 1, Corollary 1 to Proposition 1).

(ii) To see that k' is a quasi-Galois extension of k, it suffices to prove that every element $\bar{x} \in A'/\mathfrak{p}'$ is a root of a polynomial P in $k[X]$ all of whose roots are in $A'/\mathfrak{p}'$ (*Algebra*, Chapter V, § 6, no. 3, Corollary 3 to Proposition 9). Now, let $x \in A'$ be a representative of the class $\bar{x}$; the polynomial $Q(X) = \prod_{\sigma \in \mathscr{G}} (X - \sigma.x)$ has all its coefficients in A; let P(X) be the polynomial in $(A/\mathfrak{p})[X]$ whose coefficients are the images of those of Q under the canonical homomorphism $\pi: A \to A/\mathfrak{p}$. As π may be considered as the restriction to A of the canonical homomorphism $\pi': A' \to A'/\mathfrak{p}'$, it is seen that P is the product in $(A'/\mathfrak{p}')[X]$ of the linear factors $X - \pi'(\sigma.x)$ and therefore solves the problem since $\bar{x} = \pi'(x)$.

Clearly, for all $\sigma \in \mathscr{G}^Z$, $\bar{\sigma}$ is a k-automorphism of k'; it remains to verify that $\sigma \mapsto \bar{\sigma}$ maps $\mathscr{G}^Z$ *onto* the group of *all* k-automorphisms of k'. Let us write $S = A - \mathfrak{p}$; k and k' are not changed by replacing A' and $\mathfrak{p}'$ by $S^{-1}A'$ and $S^{-1}\mathfrak{p}'$ respectively, by virtue of § 1, no. 9, Proposition 23 and the relation $S^{-1}\mathfrak{p}' \cap S^{-1}A = S^{-1}(A \cap \mathfrak{p}') = S^{-1}\mathfrak{p}$ (Chapter II, § 2, no. 4); it follows from Lemma 3 that neither $\mathscr{G}^Z$ nor its operation on k' is changed; we may therefore restrict our attention to the case where $\mathfrak{p}$ is *maximal*, in which case we know that so is $\mathfrak{p}'$ (no. 1, Proposition 1) and every element of k' is therefore of the form $\pi'(x)$ for some $x \in A'$; it has been seen above that such an element is a root of a polynomial in $k[X]$ of degree $\leqslant \text{Card}(\mathscr{G})$. As every finite separable extension of k admits a primitive element (*Algebra*, Chapter V, § 7, no. 7, Proposition 12 and § 11, no. 4, Proposition 4), it is seen that every finite separable extension of k contained in k' is of degree $\leqslant \text{Card}(\mathscr{G})$, whence it follows that the *greatest* separable extension k'_s of k contained in k' (*Algebra*, Chapter V, § 7, no. 6, Proposition 11) is of degree $\leqslant \text{Card}(\mathscr{G})$ (*Algebra*, Chapter V, § 3, no. 2, *Remark* 2). Let $y \in A'$ be an element such that $\pi'(y)$ is a primitive element of k'_s. The ideals $\sigma.\mathfrak{p}'$ for $\sigma \in \mathscr{G} - \mathscr{G}^Z$ are maximal and distinct from $\mathfrak{p}'$ by definition; there therefore exists $x \in A'$ such that $x \equiv y \pmod{\mathfrak{p}'}$ and $x \in \sigma^{-1}\mathfrak{p}'$ for $\sigma \in \mathscr{G} - \mathscr{G}^Z$ (Chapter II, § 1, no. 2, Proposition 5). This being so, let u be a k-automorphism of k' and let $P(X) = \prod_{\sigma \in \mathscr{G}} (X - \pi'(\sigma.x))$; as $\pi'(x)$ is a root of P and $P \in k[X]$, $u(\pi'(x))$ is also a root of P in k' and hence there exists $\tau \in \mathscr{G}$ such that

$$u(\pi'(x)) = \pi'(\tau.x);$$

but $u(\pi'(x)) \neq 0$ and, for $\sigma \in \mathscr{G} - \mathscr{G}^Z$, $\sigma.x \in \mathfrak{p}'$ and hence $\pi'(\sigma.x) = 0$; we conclude that necessarily $\tau \in \mathscr{G}^Z$. But as u and $\bar{\tau}$ have the same value for the primitive element $\pi'(y) = \pi'(x)$ of k'_s, they coincide on k'_s and, as k' is a radicial extension of k'_s, they coincide on k'.

COROLLARY. *With the hypotheses and notation of Theorem 2, let f_1, f_2 be two homomorphisms of* A′ *to a field* L *with the same restriction to* A′. *Then there exists $\sigma \in \mathscr{G}$ such that*

$$f_2(x) = f_1(\sigma.x)$$

for all $x \in$ A′.

Let $\mathfrak{p}'_i$ be the kernel of f_i $(i = 1, 2)$ which is a prime ideal of A′; by hypothesis $\mathfrak{p}'_1 \cap \mathrm{A} = \mathfrak{p}'_2 \cap \mathrm{A}$ and this intersection is a prime ideal $\mathfrak{p}$ of A; there therefore exists $\tau \in \mathscr{G}$ such that $\tau.\mathfrak{p}'_2 = \mathfrak{p}'_1$ (Theorem 2 (i)); replacing f_1 by the homomorphism $x \mapsto f_1(\tau.x)$ we may then assume that $\mathfrak{p}'_2 = \mathfrak{p}'_1$ (an ideal which we shall denote by $\mathfrak{p}'$). By taking the quotient we then derive from f_1 and f_2 two injective homomorphisms f'_1, f'_2 from $\mathrm{A}'/\mathfrak{p}'$ to L which therefore extend to two injective homomorphisms f''_1, f''_2 from the field of fractions k' of $\mathrm{A}'/\mathfrak{p}'$ to L. As k' is a quasi-Galois extension of k, so is $k''_1 = f''_1(k')$ and $k''_2 = f''_2(k')$ (k being identified with a subfield of L) and, as there is a k-isomorphism of k''_1 onto k''_2, $k''_1 = k''_2$ (*Algebra*, Chapter V, § 6, Proposition 6). Thus $\overset{-1}{f''_1} \circ f''_2$ is a k-automorphism of k'; by Theorem 2 (ii) it is therefore of the form $\bar{\sigma}$, where $\sigma \in \mathscr{G}^Z(\mathfrak{p}')$. In particular, for all $x \in \mathrm{A}'$ the elements $f_2(x)$ and $f_1(\sigma.x)$ are equal.

Remarks

(1) Note that under the hypotheses of Theorem 2 k' may be *infinite* over k if k' is not separable over k (Exercise 9).

(2) Clearly k' is a *Galois* extension *of* k if the field k is *perfect*. It is then finite over k.

PROPOSITION 4. *Let* A′ *be a ring, $\mathscr{G}$ a finite group operating on* A′, *$\mathscr{H}$ a subgroup of $\mathscr{G}$,* A *and* B *the rings of invariants of $\mathscr{G}$ and $\mathscr{H}$ respectively and $\mathfrak{p}'$ a prime ideal of* A′; *let us write $\mathfrak{p} = \mathrm{A} \cap \mathfrak{p}'$ and $\mathfrak{p}(\mathrm{B}) = \mathrm{B} \cap \mathfrak{p}'$.*

(i) *For $\mathscr{H}$ to be contained in the decomposition group $\mathscr{G}^Z(\mathfrak{p}')$, it is necessary and sufficient that $\mathfrak{p}'$ be the only prime ideal of* A *lying above $\mathfrak{p}(\mathrm{B})$.*

(ii) *If $\mathscr{H}$ contains $\mathscr{G}^Z(\mathfrak{p}')$, the following conditions are satisfied:*

(a) *The rings $\mathrm{A}/\mathfrak{p}$ and $\mathrm{B}/\mathfrak{p}(\mathrm{B})$ have the same field of fractions.*

(b) *The maximal ideal of the local ring $\mathrm{B}_{\mathfrak{p}(\mathrm{B})}$ is equal to $\mathfrak{p}\mathrm{B}_{\mathfrak{p}(\mathrm{B})}$.*

(iii) *Suppose further that* A′ *is an integral domain and that $\bigcap_{n>0} \mathfrak{p}^n\mathrm{A}'_{\mathfrak{p}'} = 0$; then conditions* (a) *and* (b) *of* (ii) *imply that $\mathscr{G}^Z(\mathfrak{p}')$ leaves invariant the elements of* B.

(i) It follows from Theorem 2 (i) that the prime ideals of A′ lying above $\mathfrak{p}(\mathrm{B})$ are the ideals of the form $\sigma.\mathfrak{p}'$, where $\sigma \in \mathscr{H}$; whence immediately (i).

(ii) We write $S = A - \mathfrak{p}$; we know that the rings of invariants of $\mathscr{G}$ and $\mathscr{H}$ in $S^{-1}A'$ are respectively $S^{-1}A$ and $S^{-1}B$ (§ 1, no. 9, Proposition 23) and $\mathscr{G}^Z(S^{-1}\mathfrak{p}) = \mathscr{G}^Z(\mathfrak{p}')$ (Lemma 3); finally $S^{-1}\mathfrak{p}(B) = S^{-1}\mathfrak{p}' \cap S^{-1}B$ (Chapter II, § 2, no. 4), the local ring of the prime ideal $S^{-1}\mathfrak{p}(B)$ of the ring $S^{-1}B$ is canonically isomorphic to $B_{\mathfrak{p}(B)}$ and its residue field is isomorphic to the field of fractions of $B/\mathfrak{p}(B)$ (Chapter II, § 2, no. 5, Proposition 11). We can therefore show (ii) restricting our attention to the case where $\mathfrak{p}$ is *maximal*. To establish (a) it will be sufficient to prove that

$$B = A + \mathfrak{p}(B) \tag{3}$$

for this will show that the fields $A/\mathfrak{p}$ and $B/\mathfrak{p}(B)$ are *canonically isomorphic*. By Theorem 2 there is only a finite number of prime ideals of A′ lying above $\mathfrak{p}$ and by Theorem 1 of no. 1 there is at least one prime ideal of A′ lying above every prime ideal of B; this implies that there is only a *finite* number of prime ideals of B lying above $\mathfrak{p}$; let $\mathfrak{n}_j$ $(1 \leqslant j \leqslant r)$ denote those of these ideals which are $\neq \mathfrak{p}(B)$. Let x be an element of B; as the ideals $\mathfrak{p}(B)$ and $\mathfrak{n}_j$ are maximal (no. 1, Proposition 1), there exists $y \in B$ such that $y \equiv x \pmod{\mathfrak{p}(B)}$ and $y \in \mathfrak{n}_j$ for $1 \leqslant j \leqslant r$ (Chapter II, § 1, no. 2, Proposition 5). Let $y_1 = y, y_2, \ldots, y_q$ be distinct elements of the orbit of y under $\mathscr{G}$; clearly

$$z = y_1 + y_2 + \cdots + y_q \in A,$$

and to establish (3) it will be sufficient to show that $y_i \in \mathfrak{p}'$ for $i \geqslant 2$, for then we shall deduce that $z - y \in \mathfrak{p}' \cap B = \mathfrak{p}(B)$, whence $x \in A + \mathfrak{p}(B)$ since $x \equiv y \pmod{\mathfrak{p}(B)}$. Then let $i \geqslant 2$ and $\sigma \in \mathscr{G}$ such that $\sigma.y = y_i$; we show that $\sigma^{-1}.\mathfrak{p}'$ does not lie above $\mathfrak{p}(B)$. For otherwise there would exist $\tau \in \mathscr{H}$ such that $\sigma^{-1}.\mathfrak{p}' = \tau.\mathfrak{p}'$ (Theorem 2 (i)), whence $(\tau^{-1}\sigma^{-1}).\mathfrak{p}' = \mathfrak{p}'$, in other words $\tau^{-1}\sigma^{-1} \in \mathscr{G}^Z \subset \mathscr{H}$ by hypothesis, whence $\sigma \in \mathscr{H}$; but as $y \in B$ and $\sigma.y \neq y$, this is absurd. We conclude that $\sigma^{-1}.\mathfrak{p}'$ lies above one of the ideals $\mathfrak{n}_j$ and, as $y \in \mathfrak{n}_j$ by construction, certainly $y \in \sigma^{-1}.\mathfrak{p}'$ or $y_i = \sigma.y \in \mathfrak{p}'$.

To prove (b) it will suffice to establish that $\mathfrak{p}(B)$ is contained in the *saturation* $\mathfrak{q}$ of the ideal $\mathfrak{p}B$ with respect to $\mathfrak{p}(B)$ (Chapter II, § 2, no. 4, Proposition 10); as $\mathfrak{p}(B)$ is contained in none of the $\mathfrak{n}_j$ $(1 \leqslant j \leqslant r)$, it will suffice even to prove that

$$\mathfrak{p}(B) \subset \mathfrak{q} \cup \mathfrak{n}_1 \cup \cdots \cup \mathfrak{n}_r \tag{4}$$

by Chapter II, § 1, no. 1, Proposition 2. For this, we consider an element $u \in \mathfrak{p}(B)$ belonging to none of the $\mathfrak{n}_j$ $(1 \leqslant j \leqslant r)$ (Chapter II, § 1, no. 1, Proposition 2); let $u_1 = u, u_2, \ldots, u_m$ be the distinct elements of the orbit of u under $\mathscr{G}$; we write $w = u_1u_2\ldots u_m$, $v = u_2\ldots u_m$; clearly $w \in A$; on the other hand, if $\tau \in \mathscr{H}$, $\tau.u = u$ and hence necessarily $\tau.u_i \neq u$ for $i \geqslant 2$, which shows that $\tau.v = v$ and hence $v \in B$. It can be shown as in the proof of (a) that, if $\sigma \in \mathscr{G}$ is such that $\sigma.u = u_i$ where $i \geqslant 2$, $\sigma^{-1}.\mathfrak{p}'$ lies above one of the $\mathfrak{n}_j$ and, as $u \notin \mathfrak{n}_j$, also $u \notin \sigma^{-1}.\mathfrak{p}'$, in other words $u_i \notin \mathfrak{p}'$. We conclude that $v \notin \mathfrak{p}'$ and therefore $v \notin \mathfrak{p}(B)$. On the other hand clearly $w \in \mathfrak{p}' \cap A = \mathfrak{p}$ and the relation $w = uv$ shows that u is in the saturation of $\mathfrak{p}B$ with respect to $\mathfrak{p}(B)$ and hence establishes (4).

(iii) Suppose that A′ is an integral domain, that $\bigcap_{n>0} \mathfrak{p}^n A'_{\mathfrak{p}'} = 0$ and that conditions (a) and (b) of (ii) hold. With the same notation as in (ii), clearly $S^{-1}A'$ is an integral domain and $S^{-1}A'_{\mathfrak{p}'} = A'_{\mathfrak{p}'}$; it is therefore possible to replace A′ and $\mathfrak{p}'$ by $S^{-1}A'$ and $S^{-1}\mathfrak{p}'$, in other words, suppose also that the ideal $\mathfrak{p}$ is *maximal*. The hypotheses (a) and (b) then imply that

$$B_{\mathfrak{p}(B)} = A + \mathfrak{p}B_{\mathfrak{p}(B)} \tag{5}$$

By induction on n we deduce that $B_{\mathfrak{p}(B)} = A + \mathfrak{p}^n B_{\mathfrak{p}(B)}$ for all $n > 0$. Then let σ be an element of $\mathscr{G}^Z$ and x be an element of B. For all $n > 0$, there exists $a_n \in A$ such that $x - a_n \in \mathfrak{p}^n B_{\mathfrak{p}(B)} \subset \mathfrak{p}^n A'_{\mathfrak{p}'}$; as $\sigma.a_n = a_n$ and $\sigma.\mathfrak{p}' = \mathfrak{p}'$, we deduce that $\sigma.x - x \in \mathfrak{p}^n A'_{\mathfrak{p}'}$. Since this relation holds for all n, we conclude from the hypothesis that $\sigma.x = x$.

Remarks

(3) If A′ is an integral domain and Noetherian, the condition $\bigcap_{n \geqslant 1} \mathfrak{p}^n A'_{\mathfrak{p}'} = 0$ always holds (Chapter III, § 3, no. 2, Corollary to Proposition 5). It can be shown that this condition is also satisfied if A′ is assumed to be an integral domain and A to be Noetherian.

(4) If $\mathfrak{p}$ is not a maximal ideal of A relation (3) does not necessarily hold under the hypotheses of (ii) and therefore $A/\mathfrak{p}$ and $B/\mathfrak{p}(B)$ are not necessarily isomorphic even if we take $\mathscr{H} = \mathscr{G}^Z$, whence $B = A^Z$ (Exercise 10).

COROLLARY 1. *Under the hypotheses of Theorem 2 the rings* $A/\mathfrak{p}$ *and* $A^Z/(\mathfrak{p}' \cap A^Z)$ *have the same field of fractions and the maximal ideal of the local ring* $(A^Z)_{\mathfrak{p}' \cap A^Z}$ *is generated by* $\mathfrak{p}$.

COROLLARY 2. *Let* A′ *be an integral domain,* $\mathscr{G}$ *a finite group operating on* A′, A′ *the ring of invariants of* $\mathscr{G}$ *and* $\mathfrak{p}'$ *a prime ideal of* A′; *let* K, K^Z *and* K′ *be the fields of fractions of* A, A^Z *and* A′ *respectively. Then* K′ *is a Galois extension of* K *and the subfields* L *of* K′ *containing* K *and such that* $\mathfrak{p}'$ *is the only prime ideal of* A′ *lying above the ideal* $\mathfrak{p}' \cap L$ *of* $A' \cap L$ *are just those which contain* K^Z.

$\mathscr{G}$ operates on K′ and K is the field of invariants of $\mathscr{G}$ in K′ (§ 1, no. 9, Proposition 23 applied to $S = A - \{0\}$) and similarly K^Z is the field of invariants of $\mathscr{G}^Z$; by definition K′ is therefore a Galois extension of K. If $\mathscr{H}$

is the subgroup of $\mathscr{G}$ consisting of those $\sigma \in \mathscr{G}$ leaving invariant the elements of L, to say that L contains K^Z means that $\mathscr{H}$ is contained in $\mathscr{G}^Z$ (*Algebra*, Chapter V, § 10, no. 5, Theorem 3) and, as L is the field of invariants of $\mathscr{H}$ in K', A' $\cap$ L is the ring of invariants of $\mathscr{H}$ in A'; the second assertion then follows from Proposition 4 (i).

DEFINITION 4. *With the hypotheses and notation of Corollary* 2 *to Proposition* 4, *a prime ideal* $\mathfrak{p}$ *of* A *is said to decompose completely in* K' *if the number of prime ideals of* A' *lying above* $\mathfrak{p}$ *is equal to* [K' : K].

It amounts to the same to say that, for a prime ideal $\mathfrak{p}'$ of A' lying over $\mathfrak{p}$, the subgroup $\mathscr{G}^Z(\mathfrak{p}')$ is equal to the subgroup $\mathscr{M}$ leaving invariant all the elements of A', or that $A^Z(\mathfrak{p}') = A'$, or that $\mathscr{G}/\mathscr{M}$ *operates* faithfully on the set of prime ideals of A' lying above $\mathfrak{p}$.

COROLLARY 3. *Let* A' *be an integral domain,* $\mathscr{G}$ *a finite* commutative *group operating on* A', A *the ring of invariants of* $\mathscr{G}$, $\mathfrak{p}$ *a prime ideal of* A *and* K *and* K' *the fields of fractions of* A *and* A' *respectively. Then the prime ideals of* A' *lying above* $\mathfrak{p}$ *all have the same decomposition ring* A^Z *and the field of fractions* K^Z *of* A^Z *is the greatest intermediate field between* K *and* K' *in which* $\mathfrak{p}$ *decomposes completely.*

If $\mathfrak{p}'$ is a prime ideal of A' lying above $\mathfrak{p}$, then $\mathscr{G}^Z(\sigma.\mathfrak{p}') = \mathscr{G}^Z(\mathfrak{p}')$ since $\mathscr{G}$ is commutative (formula (2)), hence (Theorem 2 (i)) all the prime ideals of A' lying above $\mathfrak{p}$ have the same decomposition group $\mathscr{G}^Z$ and therefore the same decomposition ring A^Z; their number is $(\mathscr{G}:\mathscr{G}^Z)$. Let L be an intermediate field between K and K' and let $\mathscr{H}$ be the subgroup of $\mathscr{G}$ leaving invariant the elements of L; the decomposition group of $\mathfrak{p}'$ with respect to $\mathscr{H}$ is $\mathscr{G}^Z \cap \mathscr{H}$; as A' $\cap$ L is the ring of invariants of $\mathscr{H}$ in A', the number of prime ideals of A' lying above $\mathfrak{p}' \cap$ L is $(\mathscr{H}:(\mathscr{G}^Z \cap \mathscr{H})) = (\mathscr{H}\mathscr{G}^Z:\mathscr{G}^Z)$ (since $\mathscr{G}$ is commutative). The number of prime ideals of A' $\cap$ L lying above $\mathfrak{p}$ is therefore $(\mathscr{G}:\mathscr{H}\mathscr{G}^Z)$. For $\mathfrak{p}$ to decompose completely in L, it is necessary and sufficient therefore that $(\mathscr{G}:\mathscr{H}\mathscr{G}^Z) = [L:K] = (\mathscr{G}:\mathscr{H})$ and, as $\mathscr{H} \subset \mathscr{H}\mathscr{G}^Z$, this is equivalent to $\mathscr{H}\mathscr{G}^Z = \mathscr{H}$ or also to $\mathscr{G}^Z \subset \mathscr{H}$ and finally to $L \subset K^Z$.

PROPOSITION 5. *With the hypotheses and notation of Theorem* 2, *the field of fractions* k^T *of* $A^T/(\mathfrak{p}' \cap A^T)$ *is equal to the greatest separable extension* k'_s *of* k *contained in* k'.

As in Proposition 4 this may be reduced to the case where $\mathfrak{p}$ is a maximal ideal of A, which implies that $\mathfrak{p}'$, $\mathfrak{p}' \cap A^Z$ and $\mathfrak{p}' \cap A^T$ are maximal in A', A^Z and A^T respectively (no. 1, Proposition 1).

For all $x \in A'$, the polynomial $P(X) = \prod_{\sigma \in \mathscr{G}^T} (X - \sigma.x)$ has its coefficients in the inertia ring A^T and, by definition of $\mathscr{G}^T$, all its roots in A' are congruent mod. $\mathfrak{p}'$; the polynomial $\pi'(P)$ over $A^T/(\mathfrak{p}' \cap A^T)$ whose coefficients are the canonical images of those of P under the homomorphism $\pi': A' \to A'/\mathfrak{p}'$ therefore has all its roots in $A'/\mathfrak{p}'$ equal to the image of x, which shows that k' is a *radicial* extension of k^T; whence $k'_s \subset k^T$, since every element of k'_s is separable over k and *a fortiori* over k^T.

We know that k'_s is a Galois extension of k (*Algebra*, Chapter V, § 10, no. 9, Proposition 14) and it follows from Theorem 2 that its Galois group is isomorphic to $\mathscr{G}' = \mathscr{G}^Z/\mathscr{G}^T$. As k^T is a radicial extension of k'_s, k^T is a quasi-Galois extension of k and the separable factor of the degree of k^T over k is $q = (\mathscr{G}^Z:\mathscr{G}^T)$. It remains to see that k^T is a *separable* extension of k. We have seen above that $\mathscr{G}'$ is identified with an automorphism group of A^T and that A^Z is the ring of invariants of $\mathscr{G}'$. If $x \in A^T$, the polynomial $Q(X) = \prod_{\sigma' \in \mathscr{G}'} (X - \sigma'(x))$ therefore has its coefficients in A^Z; the polynomial over $A^Z/(\mathfrak{p}' \cap A^Z)$ whose coefficients are the images of those of Q under π' is of degree q and has a root $\pi'(x) \in A^T/(\mathfrak{p}' \cap A^T)$. As $A^Z/(\mathfrak{p}' \cap A^Z) = k$ by Proposition 4 (ii), we see that every element of k^T is of degree $\leqslant q$ over k.

This being so, let k_1 be the field of invariants of the group of k-automorphisms of the quasi-Galois extension k^T of k; then $[k^T:k_1] = q$ (*Algebra*, Chapter V, § 10, no. 9, Proposition 14). Let u be a primitive element of k^T over k_1; as it is of degree q over k_1 and of degree $\leqslant q$ over k, it is of degree q over k and its minimal polynomial over k_1 has coefficients in k; this shows that u is separable over k. On the other hand, for all $v \in k_1$, there exists a power p^f of the characteristic exponent p such that $v^{p^f} \in k$. We conclude that $k(u - v)$, which contains

$$(u - v)^{p^f} = u^{p^f} - v^{p^f},$$

contains u^{p^f} and consequently $k(u^{p^f})$. But as u is separable over k, $k(u) = k(u^{p^f})$ (*Algebra*, Chapter V, § 8, no. 3, Proposition 4), whence $k(u) \subset k(u - v)$. As u is of degree q over k and $u - v$ of degree $\leqslant q$, it follows that $k(u) = k(u - v)$, whence $v \in k(u)$. This shows that v is separable over k, hence $k_1 = k$ and k^T is separable over k.

COROLLARY. *If the order of the inertia group $\mathscr{G}^T(\mathfrak{p}')$ is relatively prime to the characteristic exponent $\mathfrak{p}$ of k, the field k' is a Galois extension of k.*

With the notation of the proof of Proposition 5, the polynomial $\pi'(P)$ has coefficients in $k^T = k'_s$ and all its roots equal to $\pi'(x)$; we immediately deduce that $\pi'(P)$ is a power of a minimal polynomial of $\pi'(x)$ over k'_s; but the latter has degree equal to a power of $\mathfrak{p}$ and hence, as the degree of $\pi'(P)$ is equal to the order of $\mathscr{G}^T$, the hypothesis implies that $\pi'(x) \in k'_s$, in other words $k'_s = k'$.

3. DECOMPOSITION AND INERTIA FOR INTEGRALLY CLOSED DOMAINS

LEMMA 4. *Let A be an integrally closed domain, K its field of fractions, p the characteristics exponent of K, K′ a radicial extension of K and A′ a subring of K′ containing A and integral over A. For every prime ideal $\mathfrak{p}$ of A, there exists a unique prime ideal*

$\mathfrak{p}'$ of A' lying above $\mathfrak{p}$ and $\mathfrak{p}'$ is the set of $x \in A'$ such that there exists an integer $m \geqslant 0$ for which $x^{p^m} \in \mathfrak{p}$.

The existence of $\mathfrak{p}'$ follows from no. 1, Theorem 1. If $x \in \mathfrak{p}'$, there exists $m \geqslant 0$ such that $x^{p^m} \in K$, whence $x^{p^m} \in A$ since A is integrally closed, hence $x^{p^m} \in \mathfrak{p}' \cap A = \mathfrak{p}$. Conversely, if $x \in A'$ is such that $x^{p^m} \in \mathfrak{p} \subset \mathfrak{p}'$, then $x \in \mathfrak{p}$ since $\mathfrak{p}'$ is prime.

Remark (1). It follows from § 1, no. 3, Corollary to Proposition 11 that the integral closure of A in K' is the set of $x \in K'$ such that there exists $m \geqslant 0$ for which $x^{p^m} \in A$ (*Algebra*, Chapter V, § 8, no. 1, Proposition 1).

PROPOSITION 6. *Let A be an integrally closed domain, K its field of fractions, K' a quasi-Galois extension of K and A' the integral closure of A in K'. Then:*

(i) *For every prime ideal $\mathfrak{p}$ of A, the group $\mathscr{G}$ of K-automorphisms of K' operates transitively on the set of prime ideals of A' lying over $\mathfrak{p}$.*

(i) *For every prime ideal $\mathfrak{p}'$ of A', the field of fractions k' of $A'/\mathfrak{p}'$ is a quasi-Galois extension of the field of fractions k of $A/(A \cap \mathfrak{p}')$ and the canonical homomorphism $\sigma \mapsto \bar{\sigma}$ from $\mathscr{G}^Z(\mathfrak{p}')$ to the group Γ of k-automorphisms of k' defines, by taking the quotient, a bijection of $\mathscr{G}^Z(\mathfrak{p}')/\mathscr{G}^T(\mathfrak{p}')$ onto Γ.*

(A) Suppose first that K' is a *finite Galois* extension of K. Then $A = A' \cap K$ since A is integrally closed and A is therefore the ring of invariants of $\mathscr{G}$ in A'. As $\mathscr{G}$ is finite, the proposition follows in this case from no. 2, Theorem 2.

(B) Suppose secondly that K' is *any Galois* extension of K. Then K' is the union of a right directed family $(K_\alpha)_{\alpha \in I}$ of finite Galois extensions of K. To show (i), consider two prime ideals $\mathfrak{p}'$, $\mathfrak{q}'$ of A' lying above $\mathfrak{p}$. For all $\alpha \in I$, $\mathfrak{p}' \cap K_\alpha$ and $\mathfrak{q}' \cap K_\alpha$ are two prime ideals of $A' \cap K_\alpha$ lying above $\mathfrak{p}$. Since $A' \cap K_\alpha$ is the integral closure of A in K_α and the restrictions to K_α of the elements of $\mathscr{G}$ form the group of K-automorphisms of K, it follows from case (A) that there exists $\sigma \in \mathscr{G}$ such that $\sigma.(\mathfrak{p}' \cap K_\alpha) = \mathfrak{q}' \cap K_\alpha$. Let $\mathscr{E}_\alpha$ be the set of $\sigma \in \mathscr{G}$ which have the above property. Let $\sigma \in \mathscr{G} - \mathscr{E}_\alpha$; then, for all $\tau \in \mathscr{G}$ leaving invariant the elements of K_α, $(\sigma\tau).(\mathfrak{p}' \cap K_\alpha) = \sigma.(\mathfrak{p}' \cap K_\alpha) \neq \mathfrak{q}' \cap K_\alpha$ and hence $\sigma\tau \in \mathscr{G} - \mathscr{E}_\alpha$. It follows that $\mathscr{E}_\alpha$ is *closed* in the topological Galois group $\mathscr{G}$ (*Algebra*, Chapter V, Appendix II, no. 1) and clearly the family $(\mathscr{E}_\alpha)_{\alpha \in I}$ is left directed. As $\mathscr{G}$ is compact (*loc. cit.*, no. 2, Proposition 3) and the $\mathscr{E}_\alpha$ are non-empty, the intersection $\mathscr{E}$ of the family $(\mathscr{E}_\alpha)$ is non-empty and $\sigma.\mathfrak{p}' = \mathfrak{q}'$ for all $\sigma \in \mathscr{E}$, whence (i).

To show (ii), note that k' is the union of the right directed family $(k_\alpha)_{\alpha \in I}$, where k_α is the field of fractions of $(A' \cap K_\alpha)/(\mathfrak{p}' \cap K_\alpha)$. As each k_α is a quasi-Galois extension of k by (A), so is k' (*Algebra*, Chapter V, § 6, no. 3, Proposition 8). On the other hand, let u be a k-automorphism of k' and let $\pi' : A' \to A'/\mathfrak{p}'$ be the canonical homomorphism. By virtue of no. 2, Theorem 2 applied to $A' \cap K_\alpha$, there exists for all α a non-empty set $\mathscr{F}_\alpha$ of elements $\sigma \in \mathscr{G}$ such that $\sigma.(\mathfrak{p}' \cap K_\alpha) = \mathfrak{p}' \cap K_\alpha$ and $u(\pi'(x)) = \pi'(\sigma.x)$ for all $x \in A' \cap K_\alpha$. As above it is seen that $\mathscr{F}_\alpha$ is closed in $\mathscr{G}$ and, as $(\mathscr{F}_\alpha)$ is a left directed set, its intersection $\mathscr{F}$ is non-empty. Clearly for $\sigma \in \mathscr{F}$, $\sigma \in \mathscr{G}^Z(\mathfrak{p}')$ and $\bar{\sigma} = u$, which completes the proof of (ii) in this case.

(C) *General case.* The field of invariants K_1 of $\mathscr{G}$ is a radicial extension of K (*Algebra*, Chapter V, § 10, no. 9, Proposition 14); there therefore exists a single prime ideal $\mathfrak{p}_1$ of $A_1 = A' \cap K_1$ lying above $\mathfrak{p}$ (Lemma 4). If $\mathfrak{p}'$ and $\mathfrak{q}'$ are two prime ideals of A′ lying above $\mathfrak{p}$, then they lie above $\mathfrak{p}_1$; as K′ is a Galois extension of K_1 and $A' \cap K_1$ is integrally closed (§ 1, no. 2, Proposition 7 and Corollary to Proposition 8), it follows from (B) that there exists $\sigma \in \mathscr{G}$ such that $\sigma.\mathfrak{p}' = \mathfrak{q}'$; whence (i). On the other hand, clearly the field of fractions k_1 of $A_1/\mathfrak{p}_1$ is a radicial extension of k (A being integrally closed); as k' is a quasi-Galois extension of k_1 by (B), k' is a quasi-Galois extension of k, every k-isomorphism of k' to an algebraically closed extension of k' being a k_1-isomorphism. This last remark shows also, taking account of (B), that every k-automorphism of k' is of the form $\bar{\sigma}$ where $\sigma \in \mathscr{G}^Z(\mathfrak{p}')$, which completes the proof of (ii).

Remark

(2) Suppose that K′ is a *Galois* extension of K and let us keep the notation of the proof of Proposition 6; for all α let $\mathscr{G}^Z_\alpha$ (resp. $\mathscr{G}^T_\alpha$) be the subgroup of $\mathscr{G}$ consisting of the σ whose restriction to $A' \cap K_\alpha$ belongs to $\mathscr{G}^Z(\mathfrak{p}' \cap K_\alpha)$ (resp. $\mathscr{G}^T(\mathfrak{p}' \cap K_\alpha)$). The proof of Proposition 6 shows that these subgroups are *closed* in $\mathscr{G}$ and that

$$\mathscr{G}^Z(\mathfrak{p}') = \bigcap_\alpha \mathscr{G}^Z_\alpha \quad \text{and} \quad \mathscr{G}^T(\mathfrak{p}') = \bigcap_\alpha \mathscr{G}^T_\alpha.$$

Moreover, the set of restrictions to $A' \cap K_\alpha$ of the elements of $\mathscr{G}$ (resp. $\mathscr{G}^T_\alpha$) is the whole of the group $\mathscr{G}^Z(\mathfrak{p}' \cap K_\alpha)$ (resp. $\mathscr{G}^T(\mathfrak{p}' \cap K_\alpha)$), every K-automorphism of K_α extending to an element of $\mathscr{G}$.

With the same hypotheses, the ring $A^Z(\mathfrak{p}')$ (resp. $A^T(\mathfrak{p}')$) is the *union* of the directed family of the $A^Z(\mathfrak{p}' \cap K_\alpha)$ (resp. $A^T(\mathfrak{p}' \cap K_\alpha)$): in fact, every $x \in A^Z(\mathfrak{p}')$ (resp. every $x \in A^T(\mathfrak{p}')$) belongs to one of the K_α and by the above there exists a β such that $K_\alpha \subset K_\beta$ and the restrictions to $A' \cap K_\alpha$ of the elements of $\mathscr{G}^Z(\mathfrak{p}')$ (resp. $\mathscr{G}^T(\mathfrak{p}')$) are the same as the restrictions to $A' \cap K_\alpha$ of the elements of $\mathscr{G}^Z(\mathfrak{p}' \cap K_\beta)$ (resp. $\mathscr{G}^T(\mathfrak{p}' \cap K_\beta)$), the groups $\mathscr{G}^Z(\mathfrak{p}' \cap K_\alpha)$ and $\mathscr{G}^T(\mathfrak{p}' \cap K_\beta)$ being finite; hence x belongs to $A^Z(\mathfrak{p}' \cap K_\beta)$ (resp. $A^T(\mathfrak{p}' \cap K_\beta)$).

COROLLARY 1. *Under the hypotheses of Proposition* 6, *let f be a homomorphism from* A *to a field* L *and g_1, g_2 two homomorphisms from* A′ *to* L *which extend f. Then there exists a* K*-automorphism σ of* K′ *such that $g_1 = g_2 \circ \sigma$.*

The proof starting from Proposition 6 is the same as that of the Corollary to Theorem 2 starting from the latter.

COROLLARY 2. *Let* A *be an integrally closed domain,* K *its field of fractions,* K′ *a finite algebraic extension of* K *and* A′ *a subring of* K′ *containing* A *and integral over* A.

(i) *For every prime ideal* $\mathfrak{p}$ *of* A *the set of prime ideals of* A′ *lying above* $\mathfrak{p}$ *is finite.*

(ii) *If* $\mathfrak{p}'$ *is a prime ideal of* A′ *lying above* $\mathfrak{p}$, *every element of* $A'/\mathfrak{p}'$ *is of degree* $\leqslant [K' : K]$ *over the field of fractions of* $A/\mathfrak{p}$.

(i) Let K″ be the quasi-Galois extension of K generated by K′ in an algebraic closure of K′ and A″ the integral closure of A in K″. The field K″ is a finite extension of K (*Algebra*, Chapter V, § 6, no. 3, Corollary 1 to Proposition 9) and hence its group of K-automorphisms is finite; it follows that the set of prime ideals of A″ lying above $\mathfrak{p}$ is finite (Proposition 6 (i)). On the other hand, as A″ is integral over A, the mapping $\mathfrak{p}'' \mapsto \mathfrak{p}'' \cap A'$ of the set of prime ideals of A″ lying above $\mathfrak{p}$ to the set of prime ideals of A′ lying above x is surjective (no. 1, Theorem 1).

(ii) The coefficients of the minimal polynomial (over K) of any element $x' \in A'$ belong to A (§ 1, no. 3, Corollary to Proposition 10); applying the canonical homomorphism $\pi' : A' \to A'/\mathfrak{p}'$ to the coefficients of this polynomial, an equation of integral dependence with coefficients in $A/\mathfrak{p}$ and of degree $\leqslant [K' : K]$ is obtained for the class mod. $\mathfrak{p}'$ of x; whence the conclusion.

COROLLARY 3. *With the hypotheses and notation of Corollary* 2, *if* A *is semi-local, so is* A′.

For every maximal ideal $\mathfrak{m}'$ of A′, $\mathfrak{m}' \cap A$ is a maximal ideal of A (no. 1, Proposition 1); the corollary then follows from Corollary 2, since by hypothesis the set of maximal ideals of A is finite.

COROLLARY 4. *Let* A *be an integrally closed domain,* K *its field of fractions,* K′ *a Galois extension of* K, A′ *the integral closure of* A *in* K′, $\mathfrak{p}'$ *a prime ideal of* A′, $\mathfrak{p} = A \cap \mathfrak{p}'$ *and* k *and* k' *the fields of fractions of* $A/\mathfrak{p}$ *and* $A'/\mathfrak{p}'$ *respectively. Then:*

(i) *The field of fractions of* $A^Z/(\mathfrak{p}' \cap A^Z)$ *is equal to* k *and the maximal ideal of the local ring of* A^Z *relative to* $\mathfrak{p}' \cap A^Z$ *is generated by* $\mathfrak{p}$.

(ii) *The field of fractions* k^T *of* $A^T/(\mathfrak{p}' \cap A^T)$ *is the greatest separable extension of* k *contained in* k'.

The ring A is the ring of invariants in A′ of the Galois group of K′ over K; if K′ is of *finite* degree over K, the corollary then follows from Propositions 4 and 5 of no. 2. Consider now the general case, K′ therefore being the union of a right directed set (K_α) of finite Galois extensions of K. Then:

(i) If x, y are two elements of A^Z, where $y \notin \mathfrak{p}'$, there is an index α such that x and y belong to $A^Z(\mathfrak{p}' \cap K_\alpha)$ (*Remark* 2); by Proposition 4 of no. 2, there are x_0, y_0 in A with $y_0 \notin \mathfrak{p}'$ such that $xy_0 - x_0 y \in \mathfrak{p}'$, which proves the first assertion of (i); if further $x \in \mathfrak{p}'$, we may assume that y_0 satisfies

$$xy_0 \in \mathfrak{p}A^Z(\mathfrak{p}' \cap K_\alpha) \subset \mathfrak{p}A^Z(\mathfrak{p}'),$$

which proves the second assertion of (i).

(ii) Suppose now that $x \in A^T$; there exists α such that $x \in A^T(\mathfrak{p}' \cap K_\alpha)$ (*Remark* 2) and Proposition 5 of no. 2 shows that the class $\bar{x}$ of x mod. $(\mathfrak{p}' \cap K_\alpha \cap A^T)$ is algebraic and separable over k; *a fortiori* the class mod. $(\mathfrak{p}' \cap A^T)$ of x is separable over k; to complete the proof of the corollary, it is sufficient to show that k' is a *radicial* extension of k^T. Now, k' is the union of the right directed family of fields of fractions k_α of the rings $(A' \cap K_\alpha)/(\mathfrak{p}' \cap K_\alpha)$. It follows therefore from Proposition 5 that, if an element of k' belongs to k_α, it is radicial over the field of fractions of

$$A^T(\mathfrak{p}' \cap K_\alpha)/(\mathfrak{p}' \cap A^T(\mathfrak{p}' \cap K_\alpha))$$

and *a fortiori* over k^T (by virtue of *Remark* 2).

DEFINITION 5. *With the hypotheses and notation of Proposition* 6, *the field of invariants* $K^Z(\mathfrak{p}')$ (resp. $K^T(\mathfrak{p}')$) *of the group* $\mathscr{G}^Z(\mathfrak{p}')$ (resp. $\mathscr{G}^T(\mathfrak{p}')$) *in the field* K' *is called the decomposition field* (resp. *inertia field*) *of* $\mathfrak{p}'$ *with respect to* K.

We also write K^Z (resp. K^T) in place of $K^Z(\mathfrak{p}')$ (resp. $K^T(\mathfrak{p}')$). It follows from § 1, no. 9, Proposition 23 that K^Z (resp. K^T) is the *field of fractions* of the ring A^Z (resp. A^T); A^Z (resp. A^T) is the integral closure of A in K^Z (resp. K^T).

Remarks

(3) Under the conditions of Corollary 4 of Proposition 6 and assuming that $[K':K]$ is *finite*, the number of distinct prime ideals lying above $\mathfrak{p}$ is $[K^Z:K]$, this degree being equal to the index $(\mathscr{G}:\mathscr{G}^Z)$ by Galois theory; moreover, it follows from Galois theory that

(6) $$[K^T:K^Z] = (\mathscr{G}^Z:\mathscr{G}^T) = [k^T:k].$$

(4) Let A be an integrally closed domain, K its field of fractions, K′ a *finite* algebraic extension of K and A′ the integral closure of A in K′. Then, for every prime ideal $\mathfrak{p}$ of A, *the number of prime ideals of* A′ *lying above* $\mathfrak{p}$ *is at most* $[K':K]_s$ (the separable factor of the degree of K′ over K). We may first restrict our attention to the case where K′ is a separable extension of K, for in general K′ is a radicial extension of the greatest separable extension K_0 of K contained in K′, $[K':K]_s = [K_0:K]$ by definition and, if A_0 is the integral closure of A in K_0, the prime ideals of A_0 and A′ are in one-to-one correspondence (Lemma 4). Suppose therefore that K′ is separable over K and let N be the Galois extension of K generated by K′ in an algebraic closure of K, $\mathscr{G}$ its Galois group, B the integral closure of A in N and $\mathfrak{P}$ a prime ideal of B lying above $\mathfrak{p}$. Let $\mathscr{H}$ be the Galois group of N over K′ and $\mathscr{G}^Z$ the decomposition group of $\mathfrak{P}$; the prime ideals of B lying above $\mathfrak{p}$ are the $s.\mathfrak{P}$ where $s \in \mathscr{G}$ (no. 2, Theorem 2) and the relation $s.\mathfrak{P} = s'.\mathfrak{P}$ means that $s' = sg$ where $g \in \mathscr{G}^Z$. On the other hand in order that $s.\mathfrak{P} \cap K' = s'.\mathfrak{P} \cap K'$, it is necessary and sufficient that $s'.\mathfrak{P} = ts.\mathfrak{P}$, where $t \in \mathscr{H}$ (no. 2, Theorem 2), whence finally $s' = tsg$ where $t \in \mathscr{H}$ and $g \in \mathscr{G}^Z$. The number of prime

ideals of A′ lying above $\mathfrak{p}$ is therefore equal to the *number of classes of* $\mathscr{G}$ *under the equivalence relation "there exists* $t \in \mathscr{H}$ *and* $g \in \mathscr{G}^{\mathrm{Z}}$ *such that* $s' = tsg$*" between* s *and* s'; clearly this number is at most equal to the index $(\mathscr{G}:\mathscr{H})$, the number of right cosets of $\mathscr{H}$ in $\mathscr{G}$, and $(\mathscr{G}:\mathscr{H}) = [\mathrm{K}':\mathrm{K}]$ by Galois theory.

PROPOSITION 7. *Let* A *be an integrally closed domain,* K *its field of fractions,* K′ *a Galois extension of* K, $\mathscr{G}$ *its Galois group,* A′ *the integral closure of* A *in* K′, $\mathfrak{p}'$ *a prime ideal of* A′ *and* $\mathfrak{p} = \mathrm{A} \cap \mathfrak{p}'$. *Finally, let* L *be a subfield of* K′ *containing* K *and let* $\mathfrak{p}(\mathrm{L}) = \mathfrak{p}' \cap \mathrm{L}$.

(i) *The decomposition field* (resp. *inertia field*) *of* $\mathfrak{p}'$ *with respect to* L *is* $\mathrm{L}(\mathrm{K}^{\mathrm{Z}})$ (resp. $\mathrm{L}(\mathrm{K}^{\mathrm{T}})$)*; if further* L *is a Galois extension of* K, *the decomposition field of* $\mathfrak{p}(\mathrm{L})$ *with respect to* K *is* $\mathrm{L} \cap \mathrm{K}^{\mathrm{Z}}$.

(ii) *If* L *is contained in* K^{Z}, $\mathrm{A}/\mathfrak{p}$ *and* $(\mathrm{A}' \cap \mathrm{L})/\mathfrak{p}(\mathrm{L})$ *have the same field of fractions and in the local ring* $\mathrm{A}' \cap \mathrm{L}$ *corresponding to the prime ideal* $\mathfrak{p}(\mathrm{L})$, *the maximal ideal is generated by* $\mathfrak{p}$. *Conversely, if these two conditions hold and further* $\bigcap_{n>0} \mathfrak{p}^n\mathrm{A}'_{\mathfrak{p}'} = 0$, L *is contained in* K^{Z}.

(i) If $\mathscr{H}$ is the subgroup of $\mathscr{G}$ leaving invariant the elements of L, clearly the decomposition group (resp. inertia group) of $\mathfrak{p}'$ with respect to L is $\mathscr{G}^{\mathrm{Z}} \cap \mathscr{H}$ (resp. $\mathscr{G}^{\mathrm{T}} \cap \mathscr{H}$) and the first assertion follows from Galois theory if K′ is a *finite* Galois extension of K (*Algebra*, Chapter V, § 10, no. 6, Corollary 1 to Theorem 3); in the general case it follows from the fact that A^{Z} (resp. A^{T}) is the union of the $\mathrm{A}^{\mathrm{Z}}(\mathfrak{p}' \cap \mathrm{K}_\alpha)$ (resp. $\mathrm{A}^{\mathrm{T}}(\mathfrak{p}' \cap \mathrm{K}_\alpha)$) in the notation of *Remark* 2: every element $x \in \mathrm{K}'$ belongs to some K_α and if it is invariant under $\mathscr{G}^{\mathrm{Z}}(\mathfrak{p}') \cap \mathscr{H}$ (resp. $\mathscr{G}^{\mathrm{T}}(\mathfrak{p}') \cap \mathscr{H}$) it is also invariant under $\mathscr{G}^{\mathrm{Z}}(\mathfrak{p}' \cap \mathrm{K}_\beta) \cap \mathscr{H}$ (resp. $\mathscr{G}^{\mathrm{T}}(\mathfrak{p}' \cap \mathrm{K}_\beta) \cap \mathscr{H}$) for some suitable β; hence it belongs by the beginning of the argument to

$$\mathrm{L}(\mathrm{K}^{\mathrm{Z}}(\mathfrak{p}' \cap \mathrm{K}_\alpha)) \subset \mathrm{L}(\mathrm{K}^{\mathrm{Z}}) \text{ (resp. } \mathrm{L}(\mathrm{K}^{\mathrm{T}}(\mathfrak{p}' \cap \mathrm{K}_\beta)) \subset \mathrm{L}(\mathrm{K}^{\mathrm{T}})).$$

Suppose now that L is a Galois extension of K; the restriction to L of every $\sigma \in \mathscr{G}^{\mathrm{Z}}$ then leaves invariant $\mathfrak{p}(\mathrm{L}) = \mathfrak{p}' \cap \mathrm{L}$ and hence belongs to the decomposition group of $\mathfrak{p}(\mathrm{L})$ with respect to K. Conversely, let τ be an automorphism of L belonging to this group and let σ be an extension of τ to a K-automorphism of K′; we write $\mathfrak{q}' = \sigma.\mathfrak{p}'$. As $\mathfrak{p}'$ and $\mathfrak{q}'$ both lie above $\mathfrak{p}(\mathrm{L})$, there exists an automorphism $\rho \in \mathscr{H}$ such that $\mathfrak{q}' = \rho.\mathfrak{p}'$, whence $\rho^{-1}\sigma \in \mathscr{G}^{\mathrm{Z}}$ and τ is the restriction of $\rho^{-1}\sigma$ to L; in other words, the decomposition group of $\mathfrak{p}(\mathrm{L})$ with respect to K is identical with the group of restrictions to L of the automorphisms $\sigma \in \mathscr{G}^{\mathrm{Z}}$, which proves the second assertion.

(ii) To say that $\mathrm{L} \subset \mathrm{K}^{\mathrm{Z}}$ means that $\mathscr{H} \supset \mathscr{G}^{\mathrm{Z}}$ and the assertions of (ii) are therefore special cases of no. 2, Proposition 4 (ii) and (iii) when $[\mathrm{K}':\mathrm{K}]$ is finite. In the general case the argument is as in the proof of Proposition 6.

4. THE SECOND EXISTENCE THEOREM

THEOREM 3. *Let* A *be an integrally closed domain and* A′ *a ring containing* A *and integral over* A. *Suppose that* 0 *is the only element of* A *which is a divisor of* 0 *in* A′. *Let* $\mathfrak{p}$, $\mathfrak{q}$ *be two prime ideals of* A *such that* $\mathfrak{q} \supset \mathfrak{p}$ *and* $\mathfrak{q}'$ *a prime ideal of* A′ *lying above* $\mathfrak{q}$. *Then there exists a prime ideal* $\mathfrak{p}'$ *of* A′ *lying above* $\mathfrak{p}$ *and such that* $\mathfrak{q}' \supset \mathfrak{p}'$.

Suppose first that A′ is an *integral domain*. Let K, K′ be the fields of fractions of A and A′ respectively; let K″ be the algebraic closure of K′ and A″ the integral closure of A in K″; then $A \subset A' \subset A''$. Let $\mathfrak{p}''$ be a prime ideal of A″ lying above $\mathfrak{p}$ (no. 1, Theorem 1), $\mathfrak{q}''$ a prime ideal of A″ lying above $\mathfrak{q}$ and such that $\mathfrak{p}'' \subset \mathfrak{q}''$ (no. 1, Corollary 2 to Theorem 1) and finally $\mathfrak{q}''_1$ a prime ideal of A″ lying above $\mathfrak{q}'$ (no. 1, Theorem 1). By no. 3, Proposition 6 (i), there exists a K-automorphism σ of K″ such that $\sigma.\mathfrak{q}'' = \mathfrak{q}''_1$. Then $\sigma.\mathfrak{p}''$ is a prime ideal of A″ lying above $\mathfrak{p}$ such that $\sigma.\mathfrak{p}'' = \mathfrak{q}''_1$ and hence $\mathfrak{p}' = A' \cap \sigma.\mathfrak{p}''$ is a prime ideal of A′ lying above $\mathfrak{p}$ and contained in $A' \cap \mathfrak{q}''_1 = \mathfrak{q}'$.

We pass to the general case. As A is an integral domain and $\mathfrak{q}'$ is prime, the subsets $A - \{0\}$ and $A' - \mathfrak{q}'$ of A′ are multiplicative; then their product $S = (A - \{0\}(A' - \mathfrak{q}')$ is a multiplicative subset of A′ which does not contain 0 since the non-zero elements of A are not divisors of 0 in A′. Then there exists (Chapter II, § 2, no. 5, Corollary 2 to Proposition 11) a prime ideal $\mathfrak{m}'$ of A′ disjoint from S, in other words such that $\mathfrak{m}' \subset \mathfrak{q}'$ and $\mathfrak{m}' \cap A = 0$. Let h be the canonical homomorphism $A' \to A'/\mathfrak{m}'$. The restriction of h to A is injective and hence $h(A)$ is integrally closed. As $\mathfrak{m}' \subset \mathfrak{q}'$, $h(\mathfrak{q}')$ is a prime ideal of $A'/\mathfrak{m}'$ lying above $h(\mathfrak{q})$; since $A'/\mathfrak{m}'$ is an integral domain, the first part of the proof proves that there exists a prime ideal $\mathfrak{n}'$ of $A'/\mathfrak{m}'$ such that $\mathfrak{n}' \cap h(A) = h(\mathfrak{p})$ and $h(\mathfrak{q}') \supset \mathfrak{n}'$. The ideal $\mathfrak{p}' = \overset{-1}{h}(\mathfrak{n}')$ is a prime ideal of A′ and $\mathfrak{q}' \supset \mathfrak{p}'$, since $\mathfrak{q}'$ contains the kernel of h. As $\mathfrak{n}' \supset h(\mathfrak{p})$, $\mathfrak{p}' \supset \mathfrak{p}$. Finally, for $x \in \mathfrak{p}' \cap A$, $h(x) \in \mathfrak{n}' \cap h(A) = h(\mathfrak{p})$ and hence $x \in \mathfrak{p}$ since the restriction of h to A is injective; hence $\mathfrak{p}' \cap A = \mathfrak{p}$.

COROLLARY. *Under the hypotheses on* A *and* A′ *of Theorem* 3, *let* $\mathfrak{p}$ *be a prime ideal of* A. *The prime ideals of* A′ *lying above* $\mathfrak{p}$ *are the minimal elements of the set* $\mathscr{E}$ *of prime ideals of* A′ *containing* $\mathfrak{p}A'$.

A prime ideal of A′ lying above $\mathfrak{p}$ is minimal in $\mathscr{E}$ by virtue of no. 1, Corollary 1 to Proposition 1. Conversely, let $\mathfrak{q}'$ be a minimal element of $\mathscr{E}$. As $\mathfrak{q}' \cap A \supset \mathfrak{p}$, Theorem 3 shows that there exists a prime ideal $\mathfrak{p}'$ of A′ lying above $\mathfrak{p}$ such that $\mathfrak{q}' \supset \mathfrak{p}'$. As $\mathfrak{p}'$ contains $\mathfrak{p}A'$, the hypothesis made on $\mathfrak{q}'$ implies that $\mathfrak{q}' = \mathfrak{p}'$ and hence $\mathfrak{q}'$ lies above $\mathfrak{p}$.

* Let V, V′ be two affine algebraic varieties and f a morphism of V′ to V such that $f(V')$ is dense in V. Let A (resp. A′) be the ring of functions regular on V (resp. V′); having been given f we can identify A with a subring of A′; suppose that A′ is *integral* over A. Theorem 1 of no. 1 shows that

for every irreducible subvariety W of V there exists an irreducible subvariety W′ of V′ such that $f(W')$ is a dense subset of W; in particular, every point of V is the image of an irreducible subvariety of V′, which shows that f is *surjective*. Similarly, the restriction of f to every irreducible subvariety W′ of V′ maps W′ *onto* an irreducible subvariety of V. Corollary 2 to Theorem 1, no. 1 shows that, if W and X are two irreducible subvarieties of V such that $W \supset X$ and if W′ is an irreducible subvariety of V′ such that $f(W') = W$, then there exists an irreducible subvariety X′ of V′ *contained in* W′ and such that $f(X') = X$.

If A is integrally closed, V is called *normal*. Theorem 3 shows that, if V is normal, if W and X are irreducible subvarieties of V such that $W \supset X$ and if X′ is an irreducible subvariety of V′ such that $f(X') = X$, then there exists a subvariety W′ of V′ *containing* X′ and such that $f(W') = W$. Finally, the Corollary to Theorem 3 shows that, if V is normal and W is an irreducible subvariety of V, the irreducible subvarieties W′ of V′ such that $f(W') = W$ are just the irreducible components of $\overset{-1}{f}(W)$.

3. FINITELY GENERATED ALGEBRAS OVER A FIELD

1. THE NORMALIZATION LEMMA

In this no. and the following, k denotes a *commutative field*.

THEOREM 1 (Normalization Lemma). *Let* A *be a finitely generated k-algebra and let $\mathfrak{a}_1 \subset \mathfrak{a}_2 \subset \cdots \subset \mathfrak{a}_p$ be an increasing finite sequence of ideals of* A *such that $p \geqslant 1$ and $\mathfrak{a}_p \neq A$. There exists a finite sequence $(x_i)_{1 \leqslant i \leqslant n}$ of elements of* A *which are algebraically independent over k* (Chapter III, § 1, no. 1) *and such that:*

(a) A *is integral over the ring* $B = k[x_1, \ldots, x_n]$.

(b) *For all j such that $1 \leqslant j \leqslant p$, there exists an increasing sequence $(h(j))_{1 \leqslant j \leqslant p}$ of integers such that for all j the ideal $\mathfrak{a}_j \cap B$ of* B *is generated $x_1, \ldots, x_{h(j)}$.*

Note first that it is sufficient to prove the theorem *when* A *is a polynomial algebra* $k[Y_1, \ldots, Y_m]$. For in the general case, A is isomorphic to a quotient of such an algebra A′ by an ideal $\mathfrak{a}'_0$; let $\mathfrak{a}'_j$ denote the inverse image of $\mathfrak{a}_j$ in A′ and let x'_i $(1 \leqslant i \leqslant r)$ be elements of A′ satisfying the conditions of the statement for the ring A′ and the increasing sequence of ideals $\mathfrak{a}'_0 \subset \mathfrak{a}'_1 \subset \cdots \subset \mathfrak{a}'_p$. Then the images x_i of the x'_i in A for $i > h(0)$ satisfy the desired conditions; this is obvious for condition (b) and for condition (a) this follows from § 1, no. 1, Proposition 2; finally, if the $x_i (h(0) + 1 \leqslant i \leqslant r)$ were not algebraically independent over k, there would be a non-zero polynomial

$$Q \in k[X_{h(0)+1}, \ldots, X_r]$$

such that $Q(x'_{h(0)+1}, \ldots, x'_r) \in \mathfrak{a}'_0 \cap B'$, writing $B' = k[x'_1, \ldots, x'_r]$; but by hypothesis, every element of $\mathfrak{a}'_0 \cap B'$ may be written uniquely as a polynomial in the x'_j $(1 \leqslant j \leqslant r)$ with coefficients in k, each monomial of which contains at least one of the x'_j for $1 \leqslant j \leqslant h(0)$; we therefore obtain a contradiction, which proves our assertion.

We shall therefore suppose in the rest of the proof that $A = k[Y_1, \ldots, Y_m]$ and we shall argue by induction on p.

(A) $p = 1$. We distinguish two cases:

(A1) *The ideal* $\mathfrak{a}_1$ *is a principal ideal generated by an element* $x_1 \notin k$.

$x_1 = P(Y_1, \ldots, Y_m)$, where P is a non-constant polynomial. We shall see that, for a suitable choice of integers $r_i > 0$, the ring A is *integral over* $B = k[x_1, x_2, \ldots, x_m]$, where

$$x_i = Y_i - Y_1^{r_i} \quad (2 \leqslant i \leqslant m). \tag{1}$$

For this it is sufficient to choose the r_i such that Y_1 is *integral over* B (§ 1, no. 1, Proposition 4). Now, there is the relation

$$P(Y_1, x_2 + Y_1^{r_2}, \ldots, x_m + Y_1^{r_m}) - x_1 = 0. \tag{2}$$

Let $P = \sum_{\mathbf{p}} a_{\mathbf{p}} Y^{\mathbf{p}}$, where $\mathbf{p} = (p_1, \ldots, p_m)$, the p_i being integers $\geqslant 0$, $Y^{\mathbf{p}} = Y_1^{p_1} \ldots Y_m^{p_m}$, the $a_{\mathbf{p}}$ are elements $\neq$ of k and at least one of the indices $\mathbf{p}$ is distinct from $(0, \ldots, 0)$; relation (2) may be written

$$\sum_{\mathbf{p}} a_{\mathbf{p}} Y_1^{p_1} (x_2 + Y_1^{r_2})^{p_2} \ldots (x_m + Y_1^{r_m})^{p_m} - x_1 = 0. \tag{3}$$

Let us write $f(\mathbf{p}) = p_1 + r_2 p_2 + \cdots + r_m p_m$ and suppose that the r_i are chosen so that the $f(\mathbf{p})$ are distinct (it suffices for example to take $r_i = h^i$, where h is an integer strictly greater than all the p_j (*Set Theory*, Chapter III, § 5, no. 7, Proposition 8)). Then there will be a unique system $\mathbf{p} = (p_1, \ldots, p_m)$ such that $f(\mathbf{p})$ is maximal and relation (3) may be written

$$a_{\mathbf{p}} Y_1^{f(\mathbf{p})} + \sum_{j < f(\mathbf{p})} Q_j(x_1, \ldots, x_m) Y_1^j = 0 \tag{4}$$

where the Q_j are polynomials in $k[Y_1, \ldots, Y_m]$; as $a_{\mathbf{p}} \neq 0$ is invertible in k, (4) is certainly an equation of integral dependence with coefficients in B, whence our assertion.

The field of fractions $k(Y_1, \ldots, Y_m)$ of A is therefore algebraic over the field of fractions $k(x_1, \ldots, x_m)$ of B, which proves (*Algebra*, Chapter V, § 5, no. 3, Theorem 4) that the x_i $(1 \leqslant i \leqslant m)$ are algebraically independent. Moreover, $\mathfrak{a}_1 \cap B = Bx_1$; for every element $z \in \mathfrak{a}_1 \cap B$ may be written $z = x_1 z'$ where $z' \in A \cap k(x_1, \ldots, x_m)$; but $A \cap k(x_1, \ldots, x_m) = k[x_1, \ldots, x_m] = B$

since B is integrally closed (§ 1, no. 3, Corollary 2 to Proposition 13); therefore $z' \in \mathrm{B}$, which completes the proof of properties (a) and (b) in this case.

(A2) *General case* ($p = 1$).

We argue by induction on m, the case $m = 0$ being trivial. We may obviously suppose that $\mathfrak{a}_1 \neq 0$ (otherwise we may take $x_i = \mathrm{Y}_i$ for $1 \leqslant i \leqslant m$ and $h(1) = 0$). Let x_1 be a non-zero element of $\mathfrak{a}_1$; by (A1) there exist $t_2, \ldots, t_m$ such that $x_1, t_2, \ldots, t_m$ are algebraically independent over k, A is integral over $\mathrm{C} = k[x_1, t_2, \ldots, t_m]$ and $x_1\mathrm{A} \cap \mathrm{C} = x_1\mathrm{C}$. By the induction hypothesis there exist elements $x_2, \ldots, x_m$ of $k[t_2, \ldots, t_m]$ and an integer h such that $k[t_2, \ldots, t_m]$ is integral over $\mathrm{B}' = k[x_2, \ldots, x_m]$, $x_2, \ldots, x_m$ are algebraically independent over k and the ideal $\mathfrak{a}_1 \cap \mathrm{B}'$ is generated by $x_2, \ldots, x_h$. Then C is integral over $\mathrm{B} = k[x_1, x_2, \ldots, x_m]$ (§ 1, no. 1, Corollary to Proposition 5) and hence so is A (§ 1, no. 1, Proposition 6); the same argument as in the case (A1) shows that $x_1, \ldots, x_m$ are algebraically independent over k; finally, as $x_1 \in \mathfrak{a}_1$ and $\mathrm{B} = \mathrm{B}'[x_1]$, $\mathfrak{a}_1 \cap \mathrm{B} = \mathrm{B}x_1 + (\mathfrak{a}_1 \cap \mathrm{B}')$ and, as $\mathfrak{a}_1 \cap \mathrm{B}'$ is generated (in B') by $x_2, \ldots, x_h$, $\mathfrak{a}_1 \cap \mathrm{B}$ is generated (in B) by $x_1, x_2, \ldots, x_h$.

(B) *Passage from* $p - 1$ *to* p.

Let $t_1, \ldots, t_m$ be elements of A satisfying the conditions of the theorem for the increasing sequence of ideals $\mathfrak{a}_1 \subset \cdots \subset \mathfrak{a}_{p-1}$ and let us write $r = h(p-1)$. By (A2) there exist elements $x_{r+1}, \ldots, x_m$ of $k[t_{r+1}, \ldots, t_m]$ and an integer s such that

$$\mathrm{C} = k[t_{r+1}, \ldots, t_m]$$

is integral over $\mathrm{B}' = k[x_{r+1}, \ldots, x_m]$, $x_{r+1}, \ldots, x_m$ are algebraically independent over k and the ideal $\mathfrak{a}_p \cap \mathrm{B}'$ is generated by $x_{r+1}, \ldots, x_s$. Writing $x_i = t_i$ for $i \leqslant r$ the family $(x_i)_{1 \leqslant i \leqslant m}$ obtained solves the problem with $h(p) = s$. For A is integral over $\mathrm{C}[t_1, \ldots, t_r] = \mathrm{C}[x_1, \ldots, x_r]$ and hence also over $\mathrm{B} = k[x_1, \ldots, x_m] = \mathrm{B}'[x_1, \ldots, x_r]$ since C is integral over B' (§ 1, no. 1, Corollary to Proposition 5 and Proposition 6); it can be shown as in the case (A1) that the x_i are algebraically independent over k. On the other hand, for $j \leqslant p - 1$, the ideal

$$\mathfrak{a}_j \cap k[x_1, \ldots, x_r, t_{r+1}, \ldots, t_m]$$

is by hypothesis the set of polynomials in $x_1, \ldots, x_r, t_{r+1}, \ldots, t_m$ all of whose monomials contain one of the elements $x_1, \ldots, x_{h(j)}$; as $x_{r+1}, \ldots, x_m$ are polynomials in $t_{r+1}, \ldots, t_m$ with coefficients in k, it is seen immediately that a polynomial in $x_1, \ldots, x_r, x_{r+1}, \ldots, x_m$ (with coefficients in k) can belong to $\mathfrak{a}_j$ only if all its monomials contain one of the elements $x_1, \ldots, x_{h(j)}$. Finally, as $x_1, \ldots, x_r$ belong to $\mathfrak{a}_{p-1}$ and hence also to $\mathfrak{a}_p$, $\mathfrak{a}_p \cap \mathrm{B}'[x_1, \ldots, x_r]$ consists of the polynomials in $x_1, \ldots, x_r$ with coefficients in B' whose constant term belongs to $\mathfrak{a}_p \cap \mathrm{B}'$; this ideal is therefore generated by $x_1, \ldots, x_r, x_{r+1}, \ldots, x_s$.

COROLLARY 1. *Let* A *be an integral domain and* B *a finitely generated* A*-algebra containing* A *as a subring. Then there exist an element* $s \neq 0$ *of* A *and a subalgebra* B′ *of* B *isomorphic to a polynomial algebra* $A[Y_1, \ldots, Y_n]$ *such that* $B[s^{-1}]$ (Chapter II, § 2, no. 1) *is integral over* $B'[s^{-1}]$.

We write $S = A - \{0\}$ and let $k = S^{-1}A$ the field of fractions of A; clearly $S^{-1}B$ is a finitely generated k-algebra and, as it contains k by hypothesis (Chapter II, § 2, no. 4, Theorem 1), it is not reduced to 0. By Theorem 1 (applied to $p = 1$ and $\mathfrak{a}_1 = 0$) there exists therefore a finite sequence $(x_i)_{1 \leqslant i \leqslant n}$ of elements of $S^{-1}B$ which are algebraically independent over k and such that $S^{-1}B$ is integral over $k[x_1, \ldots, x_n]$. Let $(z_j)_{1 \leqslant j \leqslant m}$ be a system of generators of the A-algebra B; in $S^{-1}B$ each of the $z_j/1$ satisfies an equation of integral dependence

$$(z_j/1)^{q_j} + \sum_{h < q_j} P_{hj}(x_1, \ldots, x_n)(z_j/1)^h = 0 \tag{5}$$

where the P_{hj} are polynomials in the x_i with coefficients in k. There exists an element $s \neq 0$ of A such that we may write $x_i = y_i/s$ where $y_i \in B$ for $1 \leqslant i \leqslant n$ and all the coefficients of the P_{hj} are of the form c/s where $c \in A$; finally, replacing if need be, s by a product of elements of S, we may assume that in B

$$s z_j^{q_j} + \sum_{h < q_j} Q_{hj} z_j^h = 0 \tag{6}$$

where the Q_{hj} are polynomials in $y_1, \ldots, y_n$ with coefficients in A; if we write $z'_j = s z_j$ it is seen, multiplying (6) by $s^{q_j - 1}$, that z'_j is integral over $B' = A[y_1, \ldots, y_n]$. We show that the y_i are algebraically independent over A; if there is a relation of the form $\sum_{\mathbf{p}} a_{\mathbf{p}} y_1^{p_1} \ldots y_n^{p_n} = 0$ where $a_{\mathbf{p}} \in A$ for all $\mathbf{p}$, we deduce that $\sum_{\mathbf{p}} a'_{\mathbf{p}} x_1^{p_1} \ldots x_n^{p_n} = 0$ in $S^{-1}B$, where $a'_{\mathbf{p}} = a_{\mathbf{p}} s^{p_1 + \cdots + p_n}$ in k; by hypothesis therefore $a'_{\mathbf{p}} = 0$ for all $\mathbf{p}$, whence $a_{\mathbf{p}} = 0$ for all $\mathbf{p}$. Moreover, in the ring $B[s^{-1}]$ each of the $z'_j/1$ is integral over $B'[s^{-1}]$ (§ 1, no. 1, Proposition 2) and, as $z_j/1 = (z'_j/1)(1/s)$ in $B[s^{-1}]$, it is seen that the $z_j/1$ are integral over $B'[s^{-1}]$, which completes the proof (§ 1, no. 1, Proposition 4).

COROLLARY 2. *Let* K *be a field,* A *a subring of* K *and* L *the field of fractions of* A. *If* K *is a finitely generated* A*-algebra,* [K:L] *is finite and there exists* $a \neq 0$ *in* A *such that* $L = A[a^{-1}]$.

It follows from Corollary 1 that there exist elements $x_1, \ldots, x_n$ of K and an element $a \neq 0$ of A such that $x_1, \ldots, x_n$ are algebraically independent over A (and therefore over L) and that K is integral over the subring $A[x_1, \ldots, x_n, a^{-1}]$. Then it follows from § 2, no. 1, Lemma 2 that $A[x_1, \ldots, x_n, a^{-1}]$ is a field. But the only invertible elements of a polynomial ring $C[Y_1, \ldots, Y_n]$ over an integral domain C are the invertible elements of C; applying this remark to $C = A[a^{-1}]$, it is seen that necessarily $n = 0$ and that $A[a^{-1}]$ is a field equal to L by definition of the latter. As K is integral over L and is a finitely generated L-algebra, the degree $[K:L]$ is finite (§ 1, no. 1, Proposition 4).

Corollary 3. *Let* A *be an integral domain,* B *a finitely generated* A*-algebra and* b *an element of* B *such that* $zb^n \neq 0$ *for all* $z \neq 0$ *in* A *and every integer* $n > 0$. *Let* $\rho: A \rightarrow B$ *be the canonical homomorphism; there exists* $a \neq 0$ *in* A *such that, for every homomorphism* f *from* A *to an algebraically closed field* L *such that* $f(a) \neq 0$, *there exists a homomorphism* g *from* B *to* L *such that* $g(b) \neq 0$ *and* $f = g \circ \rho$.

The hypothesis on b implies that, if h is the canonical homomorphism $x \mapsto x/1$ of B to $B[b^{-1}]$, the homomorphism $h \circ \rho$ of A to $B[b^{-1}]$ is injective. By Corollary 1 there therefore exist an element $a \neq 0$ of A and a subring B′ of $B[b^{-1}]$ such that $B[b^{-1}, a^{-1}]$ is integral over $B'[a^{-1}]$ and B′ is isomorphic to a polynomial algebra $A[Y_1, \ldots, Y_n]$. Let f be a homomorphism from A to an algebraically closed field L such that $f(a) \neq 0$; there exists a homomorphism from $A[Y_1, \ldots, Y_n]$ to L extending f and hence there exists a homomorphism f' from B′ to L extending f. As $f'(a) \neq 0$ in L, there exists a homomorphism f'' from $B'[a^{-1}]$ to L such that

$$f''(x/a^n) = f'(x).(f(a))^{-n}$$

for all $x \in B'$ and all $n > 0$ (Chapter II, § 2, no. 1, Proposition 1). Finally, as $B[b^{-1}, a^{-1}]$ is integral over $B'[a^{-1}]$, there exists a homomorphism f''' from $B[b^{-1}, a^{-1}]$ to L extending f'' (§ 2, no. 1, Corollary 4 to Theorem 1). If $j: x \mapsto x/1$ is the canonical homomorphism from B to $B[b^{-1}, a^{-1}]$, $g = f''' \circ j$ solves the problem for $j(b)$ is invertible in $B[b^{-1}, a^{-1}]$ and hence $f'''(j(b)) \neq 0$ in L.

Note that, if B is assumed to be an integral domain and $A \subset B$ in Corollary 3, the hypothesis on b is equivalent to $b \neq 0$.

2. THE INTEGRAL CLOSURE OF A FINITELY GENERATED ALGEBRA OVER A FIELD

Theorem 2. *Let* A *be a finitely generated integral* k*-algebra,* K *its field of fractions and* A′ *the integral closure of* A *in a field* K′ *which is a finite algebraic extension of* K. *Then* A′ *is a finitely generated* A*-module and a finitely generated* k*-algebra.*

By Theorem 1 there exists a subalgebra C of A isomorphic to a polynomial algebra $k[X_1, \ldots, X_n]$ and such that A is integral over C; A′ is obviously the integral closure of C in K′ (§ 1, no. 1, Proposition 6); we may therefore confine our attention to the case where $A = k[X_1, \dots, X_n]$. Let N be the quasi-Galois extension of K′ (in an algebraic closure of K′) generated by K′, which is a finite algebraic extension of K (*Algebra*, Chapter V, § 6, no. 3, Corollary 1 to Proposition 9). It will suffice to prove that the integral closure B of A in N is a finitely generated A-module, for A′ is a sub-A-module of B and A is a Noetherian ring (Chapter III, § 2, no. 10, Corollary 2 to Theorem 2). We may therefore confine our attention to the case where K′ is a *quasi-Galois* extension of K. Then we know (*Algebra*, Chapter V, § 10, no. 9, Proposition 14) that K′ is a (finite) *Galois* extension of a (finite) *radicial* extension K″ of K. If A″ is the integral closure of A in K″, A′ is the integral closure of A″ in K′ and it will suffice to prove that A″ is a finitely generated A-module and A′ is a finitely generated A″-module. Now, if it has been proved that A″ is a finitely generated A-module, it is a Noetherian domain, integrally closed by definition; the fact that A′ is a finitely generated A″-module will follow from § 1, no. 6, Corollary 1 to Proposition 18.

We see therefore that we may confine our attention to the case where $A = k[X_1, \dots, X_n]$ and where K′ is a finite *radicial* extension of $K = k(X_1, \dots, X_n)$. Then K′ is generated by a finite family of elements $(y_i)_{1 \leqslant i \leqslant m}$ and there exists a power q of the characteristic exponent of k such that $y_i^q \in k(X_1, \dots, X_n)$. Let c_j $(1 \leqslant j \leqslant r)$ be the coefficients of the numerators and denominators of the rational functions in $X_1, \dots, X_n$ equal to y_i^q $(1 \leqslant i \leqslant m)$. Then K′ is contained in the extension $L = k'(X_1^{q^{-1}}, \dots, X_n^{q^{-1}})$, where $k' = k(c_1^{q^{-1}}, \dots, c_n^{q^{-1}})$ (we are in an algebraic closure of K′) and A′ is contained in the algebraic closure B′ of A in L′. Now, k' is algebraic over k and hence $C' = k'[X_1, \dots, X_n]$ is integral over A (§ 1, no. 1, Proposition 5); as $k'[X_1^{q^{-1}}, \dots, X_n^{q^{-1}}]$ is integrally closed (§ 1, no. 3, Corollary 2 to Proposition 13), it is seen that this ring is the integral closure of C′ in L′ and hence also that of A (§ 1, no. 1, Proposition 6), in other words $B' = k'[X_1^{q^{-1}}, \dots, X_n^{q^{-1}}]$. Now clearly B′ is a finitely generated C′-module (§ 1, no. 1, Proposition 4) and, as k' is a finite extension of k, C′ is a finitely generated A-module and hence B′ is a finitely generated A-module; since A is Noetherian and $A' \subset B'$, A′ is a finitely generated A-module.

3. THE NULLSTELLENSATZ

PROPOSITION 1. *Let* A *be a finitely generated algebra over a field* k *and* L *the algebraic closure of* k.

(i) *If* $A \neq \{0\}$, *there exists a* k*-homomorphism from* A *to* L.

(ii) *Let* f_1, f_2 *be two* k*-homomorphisms from* A *to* L. *For* f_1 *and* f_2 *to have the same kernel, it is necessary and sufficient that there exist a* k*-automorphism* s *of* L *such that* $f_2 = s \circ f_1$.

(iii) *Let $\mathfrak{a}$ be an ideal of* A. *For $\mathfrak{a}$ to be maximal, it is necessary and sufficient that it be the kernel of a k-homomorphism from* A *to* L.

(iv) *For an element x of* A *to be such that $f(x) = 0$ for every k-homomorphism f from* A *to* L, *it is necessary and sufficient that x be nilpotent.*

Assertion (i) follows from no. 1, Corollary 3 to Theorem 1 applied replacing A by k, B by A, b by the unit element of B and f by the canonical injection of k into L.

If f is a k-homomorphism from A to L, $f(A)$ is a subring of L containing k; as L is an algebraic extension of k, $f(A)$ is a *field* (*Algebra*, Chapter V, § 3, no. 2, Proposition 3) and, if $\mathfrak{a}$ is the kernel of f, $A/\mathfrak{a}$, isomorphic to $f(A)$, is therefore a field, which proves that $\mathfrak{a}$ is maximal. Conversely, if $\mathfrak{a}$ is a maximal ideal of A, it follows from (i) that there exists a k-homomorphism from $A/\mathfrak{a}$ to L and hence a k-homomorphism of A to L whose kernel $\mathfrak{b}$ contains $\mathfrak{a}$; but as $\mathfrak{a}$ is maximal, $\mathfrak{b} = \mathfrak{a}$; this proves (iii).

We now prove (ii). If s is a k-automorphism of L such that $f_2 = s \circ f_1$, clearly f_1 and f_2 have the same kernel. Conversely, suppose that f_1 and f_2 have the same kernel; then there exists a k-isomorphism s_0 of the field $f_1(A)$ onto the field $f_2(A)$ such that $f_2 = s_0 \circ f_1$; but by *Algebra*, Chapter V, § 6, no. 3, Proposition 7, s_0 extends to a k-automorphism s of L and hence $f_2 = s \circ f_1$.

Finally, if $x \in A$ is such that $x^n = 0$, for every k-homomorphism f from A to L, $(f(x))^n = f(x^n) = 0$ and hence $f(x) = 0$ since L is a field. Conversely, suppose that $x \in A$ is not nilpotent; then $A[x^{-1}]$ is a finitely generated A-algebra (and therefore a finitely generated k-algebra) not reduced to 0 (Chapter II, § 2, no. 1, *Remark* 3) and hence there exists a k-homomorphism g from $A[x^{-1}]$ to L by (i). If $j: A \to A[x^{-1}]$ is the canonical homomorphism, $f = g \circ j$ is a k-homomorphism from A to L and $f(x)g(1/x) = g(x/1)g(1/x) = g(1) = 1$, whence $f(x) \neq 0$.

Let k be a field and L an extension field of k; an element $\mathbf{x} = (x_1, \ldots, x_n)$ of L^n is called a *zero in* L^n *of an ideal* $\mathfrak{r}$ *of the polynomial ring* $k[X_1, \ldots, X_n]$ if

$$P(\mathbf{x}) = P(x_1, \ldots, x_n) = 0$$

for *all* $P \in \mathfrak{r}$.

Lemma 1. *Let* A *be a finitely generated algebra over a field k, $(a_i)_{1 \leqslant i \leqslant n}$ a system of generators of this algebra and $\mathfrak{r}$ the ideal of algebraic relations between the a_i with coefficients in k* (*Algebra*, Chapter IV, § 2, no. 1). *For every extension field* L *of k, the mapping $f \mapsto (f(a_i))_{1 \leqslant i \leqslant n}$ is a bijection of the set of k-homomorphisms from* A *to* L *onto the set of zeros of $\mathfrak{r}$ in L^n.*

There exists a unique k-algebra homomorphism h of $k[X_1, \ldots, X_n]$ onto A such that $h(X_i) = a_i$ for $1 \leqslant i \leqslant n$ and by definition $\mathfrak{r}$ is the kernel of h. The mapping $f \mapsto f \circ h$ is a bijection of the set of k-homomorphisms from A to L onto the set of k-homomorphisms from $k[X_1, \ldots, X_n]$ to L which are zero on $\mathfrak{r}$. For every polynomial $P \in k[X_1, \ldots, X_n]$ and every element $\mathbf{x} = (x_1, \ldots, x_n) \in L^n$ we write $h_{\mathbf{x}}(P) = P(\mathbf{x})$; then the mapping $\mathbf{x} \mapsto h_{\mathbf{x}}$ is a bijection of L^n onto the set of k-homomorphisms from $k[X_1, \ldots, X_n]$ to L (such a homomorphism being determined by its values at the X_i $(1 \leqslant i \leqslant n)$); to say that $h_{\mathbf{x}}$ is zero on $\mathfrak{r}$ means that $\mathbf{x}$ is a zero of $\mathfrak{r}$ in L^n, whence the lemma.

If Proposition 1 is applied to the algebra $A = k[X_1, \ldots, X_n]/\mathfrak{r}$, where $\mathfrak{r}$ is an ideal of $k[X_1, \ldots, X_n]$ distinct from the whole ring, we obtain by Lemma 1 the following statement:

PROPOSITION 2 (Hilbert's Nullstellensatz). *Let k be a field and L an algebraic closure of k.*

(i) *Every ideal $\mathfrak{r}$ of $k[X_1, \ldots, X_n]$ not containing 1 admits at least one zero in L^n.*

(ii) *Let $\mathbf{x} = (x_1, \ldots, x_n)$, $\mathbf{y} = (y_1, \ldots, y_n)$ be two elements of L^n; for the set of polynomials of $k[X_1, \ldots, X_n]$ zero at $\mathbf{x}$ to be identical with the set of polynomials of $k[X_1, \ldots, X_n]$ zero at $\mathbf{y}$, it is necessary and sufficient that there exists a k-automorphism s of L such that $y_i = s(x_i)$ for $1 \leqslant i \leqslant n$.*

(iii) *For an ideal $\mathfrak{a}$ of $k[X_1, \ldots, X_n]$ to be maximal, it is necessary and sufficient that there exist an $\mathbf{x}$ in L^n such that $\mathfrak{a}$ is the set of polynomials of $k[X_1, \ldots, X_n]$ zero at $\mathbf{x}$.*

(iv) *For a polynomial Q of $k[X_1, \ldots, X_n]$ to be zero on the set of zeros in L^n o, an ideal $\mathfrak{r}$ of $k[X_1, \ldots, X_n]$, it is necessary and sufficient that there exist an integer $m > 0$ such that $Q^m \in \mathfrak{r}$.*

4. JACOBSON RINGS

DEFINITION 1. *A ring A is called a Jacobson ring if every prime ideal of A is the intersection of a family of maximal ideals.*

Examples

(1) Every field is a Jacobson ring.

(2) The ring **Z** is a Jacobson ring, the unique prime ideal which is not maximal (0) being the intersection of the maximal ideals (p) of **Z**, where p runs through the set of prime numbers (cf. Proposition 4).

(3) Let A be a Jacobson ring and let $\mathfrak{a}$ be an ideal of A. Then $A/\mathfrak{a}$ is a Jacobson ring, for the ideals of $A/\mathfrak{a}$ are of the form $\mathfrak{b}/\mathfrak{a}$, where $\mathfrak{b}$ is an ideal of A containing $\mathfrak{a}$ and $\mathfrak{b}/\mathfrak{a}$ is prime (resp. maximal) if and only if $\mathfrak{b}$ is.

PROPOSITION 3. *For a ring A to be a Jacobson ring, it is necessary and sufficient that, for every ideal $\mathfrak{a}$ of A, the Jacobson radical of $A/\mathfrak{a}$ be equal to its nilradical* (Chapter II, § 2, no. 6).

The Jacobson radical (resp. nilradical) of $A/\mathfrak{a}$ is the intersection of the maximal (resp. prime) ideals of $A/\mathfrak{a}$ (*Algebra*, Chapter VIII, § 6, no. 3, Definition 3 and *Commutative Algebra*, Chapter II, § 2, no. 6, Proposition 13). The stated condition means therefore that for every ideal $\mathfrak{a}$ of A the intersection of the prime ideals containing $\mathfrak{a}$ is equal to the intersection of the maximal ideals containing $\mathfrak{a}$. This condition obviously holds for every ideal $\mathfrak{a}$ of A if A is a Jacobson ring; conversely, if it holds for every prime ideal of A, A is a Jacobson ring by definition.

COROLLARY. *Let* A *be a Jacobson ring; for every ideal* $\mathfrak{a}$ *of* A, *the radical of* $\mathfrak{a}$ *is the intersection of the maximal ideals of* A *containing* $\mathfrak{a}$.

It is sufficient to note that $A/\mathfrak{a}$ is a Jacobson ring.

PROPOSITION 4. *Let* A *be a principal ideal domain and* $(p_\lambda)_{\lambda \in L}$ *a representative system of extremal elements of* A (*Algebra*, Chapter VII, § 1, no. 3, Definition 2). *For* A *to be a Jacobson ring, it is necessary and sufficient that* L *be infinite.*

The maximal ideals of A are the Ap_λ (*loc. cit.*, no. 2, Proposition 2). If L is finite, their intersection is the ideal Ax, where $x = \prod_{\lambda \in L} p_\lambda$ (*ibid.*) and hence different from (0); on the other hand, if L is infinite, the intersection of the Ap_λ is (0), every element $\neq 0$ of A being divisible by only a finite number of extremal elements (*loc. cit.*, no. 3, Theorem 2). The proposition then follows from the fact that (0) is the only prime ideal which is not maximal in A (*Algebra*, Chapter VI, § 1, no. 13, Proposition 14 (DIV)).

PROPOSITION 5. *Let* A *be a ring and* B *an* A*-algebra integral over* A. *If* A *is a Jacobson ring, so is* B.

Replacing A by its canonical image in B, we may assume that $A \subset B$. Let $\mathfrak{p}'$ be a prime ideal of B and let $\mathfrak{p} = A \cap \mathfrak{p}'$. There exists by hypothesis a family $(\mathfrak{m}_\lambda)_{\lambda \in L}$ of maximal ideals of A whose intersection is equal to $\mathfrak{p}$. For all $\lambda \in L$ there exists a maximal ideal $\mathfrak{m}'_\lambda$ of B lying above $\mathfrak{m}_\lambda$ and containing $\mathfrak{p}'$ (§ 2, no. 1, Proposition 1 and Corollary 2 to Theorem 1). If we write $\mathfrak{q}' = \bigcap_{\lambda \in L} \mathfrak{m}'_\lambda$, then $\mathfrak{q}' \cap A = \bigcap_{\lambda \in L} \mathfrak{m}_\lambda = \mathfrak{p}$ and $\mathfrak{q}' \supset \mathfrak{p}'$, whence $\mathfrak{q}' = \mathfrak{p}'$ (§ 2, no. 1, Corollary 1 to Proposition 1).

THEOREM 3. *Let* A *be a Jacobson ring,* B *a finitely generated* A*-algebra and* $\rho: A \to B$ *the canonical homomorphism. Then:*

(i) B *is a Jacobson ring.*

(ii) *For every maximal ideal* $\mathfrak{m}'$ *of* B, $\mathfrak{m} = \overset{-1}{\rho}(\mathfrak{m}')$ *is a maximal ideal of* A *and* $B/\mathfrak{m}'$ *is a finite algebraic extension of* $A/\mathfrak{m}$.

Let $\mathfrak{p}'$ be a prime ideal of B and $\mathfrak{p} = \overset{-1}{\rho}(\mathfrak{p}')$. Let v be an element $\neq 0$ of $B/\mathfrak{p}'$.

As $B/\mathfrak{p}'$ is a finitely generated integral $(A/\mathfrak{p})$-algebra and the canonical homomorphism $\phi: A/\mathfrak{p} \to B/\mathfrak{p}'$ is injective, there exists an element $u \neq 0$ of $A/\mathfrak{p}$ such that, for every homomorphism f from $A/\mathfrak{p}$ to an algebraically closed field L whose kernel does not contain u, there exists a homomorphism g from $B/\mathfrak{p}'$ to L whose kernel does not contain v and for which $f = g \circ \phi$ (no. 1, Corollary 3 to Theorem 1). Since A is a Jacobson ring, there exists a maximal ideal $\mathfrak{m}$ of A containing $\mathfrak{p}$ and such that $u \notin \mathfrak{m}/\mathfrak{p}$. We take L to be an algebraic closure of $A/\mathfrak{m}$ and f to be the canonical homomorphism $A/\mathfrak{p} \to L$; let

$$g: B/\mathfrak{p}' \to L$$

be a homomorphism such that $f = g \circ \phi$ and $g(v) \neq 0$. Then

$$A/\mathfrak{m} \subset g(B/\mathfrak{p}') \subset L,$$

hence $g(B/\mathfrak{p}')$ is a subfield of K (*Algebra*, Chapter V, § 3, no. 2, Proposition 3) and the kernel of g is therefore a maximal ideal of $B/\mathfrak{p}'$ not containing v. Thus it is seen that the intersection of the maximal ideals of $B/\mathfrak{p}'$ is reduced to 0, which proves that B is a Jacobson ring. Moreover, if $\mathfrak{p}'$ is maximal, g is necessarily injective and hence $\mathfrak{p} = \mathfrak{m}$ is maximal; finally $B/\mathfrak{p}'$ is then a finitely generated algebra over the field $A/\mathfrak{m}$ and hence is a finite extension of $A/\mathfrak{m}$ (no. 1, Corollary 2 to Theorem 1).

COROLLARY 1. *Every finitely generated algebra* A *over* $\mathbf{Z}$ *is a Jacobson ring; for a prime ideal* $\mathfrak{p}$ *of* A *to be maximal, it is necessary and sufficient that the ring* $A/\mathfrak{p}$ *be finite.*

If the integral domain $A/\mathfrak{p}$ is finite, it is a field, as, for every $u \neq 0$ in $A/\mathfrak{p}$, the mapping $v \mapsto uv$ of $A/\mathfrak{p}$ to itself is injective and hence bijective since $A/\mathfrak{p}$ is finite. Conversely, for every maximal ideal $\mathfrak{m}$ of A, the inverse image of $\mathfrak{m}$ in $\mathbf{Z}$ is a maximal ideal (p) and $A/\mathfrak{m}$ is finite over the prime field $\mathbf{Z}/(p) = \mathbf{F}_p$ by Theorem 3.

COROLLARY 2. *Let* $(P_\lambda)_{\lambda \in L}$ *be a family of polynomials in* $\mathbf{Z}[X_1, \dots, X_n]$ *and let* Q *be a polynomial in* $\mathbf{Z}[X_1, \dots, X_n]$ *such that, for every system of elements* $(x_i)_{1 \leqslant i \leqslant n}$ *belonging to a finite field and for which* $P_\lambda(x_1, \dots, x_n) = 0$ *for all* λ, *also* $Q(x_1, \dots, x_n) = 0$. *Then, if* $\mathfrak{a}$ *is the ideal of* $\mathbf{Z}[X_1, \dots, X_n]$ *generated by the* P_λ, *there exists an integer* $m > 0$ *such that* $Q^m \in \mathfrak{a}$. *Moreover, for every reduced ring* R *and every system* $(y_i)_{1 \leqslant i \leqslant n}$ *of elements of* R *such that* $P_\lambda(y_1, \dots, y_n) = 0$ *for all* λ, *also* $Q(y_1, \dots, y_n) = 0$.

The second assertion follows from the first since the ideal of $\mathbf{Z}[X_1, \dots, X_n]$ consisting of the polynomials P such that $P(y_1, \dots, y_n) = 0$ contains $\mathfrak{a}$. To show the first assertion, it suffices to note that, for every maximal ideal $\mathfrak{m}$ of $A = \mathbf{Z}[X_1, \dots, X_n]$ containing $\mathfrak{a}$, $A/\mathfrak{m}$ is a finite field (Corollary 1) and the hypothesis implies that the canonical image of Q in $A/\mathfrak{m}$ is zero; then Q belongs to the intersection of the maximal ideals of A containing $\mathfrak{a}$, which is the radical of $\mathfrak{a}$ (Corollary to Proposition 3).

Corollary 3. *Let* A *be a Jacobson ring. If there exists a finitely generated* A*-algebra* B *containing* A *and which is a field, then* A *is a field and* B *is an algebraic extension of* A.

It suffices to apply Theorem 3 (ii) with $\mathfrak{m}' = (0)$.

EXERCISES

§ 1

1. Let d be a rational integer not divisible by a square in $\mathbf{Z}$. Show that the elements of the field $K = \mathbf{Q}(\sqrt{d})$ which are integral over $\mathbf{Z}$ are of the form $a + b\sqrt{d}$ where $a \in \mathbf{Z}$, $b \in \mathbf{Z}$ if $d - 1 \not\equiv 0 \pmod{4}$ and the elements $(a + b\sqrt{d})/2$ where $a \in \mathbf{Z}$, $b \in \mathbf{Z}$, a and b both even or a and b both odd, if $d - 1 \equiv 0 \pmod{4}$ (cf. *Algebra*, Chapter VII, § 1, Exercise 8). Deduce that $A = \mathbf{Z}[\sqrt{5}]$ is not integrally closed and give an example of an element of $\mathbf{Q}[\sqrt{5}]$ which is integral over A but whose minimal polynomial over the field of fractions of A does not have coefficients in A.

2. Let K be a commutative field and A the sub-K-algebra of the polynomial algebra K[X, Y] generated by the monomials $Y^k X^{k+1}$ ($k \geqslant 0$). Show that XY is such that A[XY] is contained in a finitely generated A-module but XY is not integral over A.

3. Give an example of an infinite sequence (K_n) of extensions of a commutative field K of finite degree over K such that the K-algebra $\prod_n K_n$ is not algebraic over K.

4. In the matrix ring $\mathbf{M}_2(\mathbf{Q})$ give an example of two elements integral over $\mathbf{Z}$ but neither the sum nor the product of which is integral over $\mathbf{Z}$ (consider matrices of the form $I + N$, where N is nilpotent).

5. Let A be a commutative ring, B a commutative ring containing A and x an invertible element of B. Show that every element of $A[x] \cap A[x^{-1}]$ is integral over A. (If $y = a_0 + \cdots + a_p x^{-p} = b_0 + \cdots + b_q x^q$, where the a_i and b_j are in A, show that the sub-A-module of B generated by $1, x, \ldots, x^{p+q+1}$ is a faithful A[y]-module.)

6. Let A be a commutative ring and B a commutative A-algebra; suppose

that the *minimal* prime ideals of B are *finite* in number; let $\mathfrak{p}_i$ $(1 \leqslant i \leqslant n)$ be these ideals (note that this hypothesis holds if B is Noetherian). For an element $x \in B$ to be integral over A, it is necessary and sufficient that each of its canonical images x_i in $B/\mathfrak{p}_i$ be integral over A. (If this condition holds, show first that x is integral over the subalgebra of B generated by the nilradical $\mathfrak{N} = \bigcap_i \mathfrak{p}_i$ of B and use the fact that the elements of $\mathfrak{N}$ are nilpotent.)

7. Give an example of a reduced ring A which is integrally closed in its total ring of fractions and which is not a product of integral domains (take the integral closure of a field K in the infinite product $\prod_n K_n$ of suitable algebraic extensions of K).

8. Let A be a commutative ring and B a finitely generated A-algebra. Show that there exists a finite A-algebra C which is a *free* A-module and a surjective A-homomorphism $C \to B$.

9. Show that the quotient domain $\mathbf{Z}[X]/(X^2 + 4)$ is not integrally closed even though $\mathbf{Z}[X]$ is.

10. Let A be a completely integrally closed domain. Show that for every set I the polynomial domain $A[X_\iota]_{\iota \in I}$ and the domain of formal power series $A[[X_\iota]]_{\iota \in I}$ (*Algebra*, Chapter IV, § 5, Exercise 1) are completely integrally closed. (Use Proposition 4 of no. 4 and the following lemma: for every finite subset J of I and all $P \in A[[X_\iota]]_{\iota \in J}$, let P_J be the formal power series in $A[[X_\iota]]_{\iota \in J}$ derived from P by replacing by 0 in P all the X_ι of index $\iota \notin J$; if P, Q are two formal power series such that P_J is divisible by Q_J in $A[[X_\iota]]_{\iota \in J}$ for all J, then P is divisible by Q in $A[[X_\iota]]_{\iota \in I}$. For this note that, if J, J′ are two finite subsets of I such that $J \subset J'$ and $P_{J'} = L'Q_{J'}$ and $P_J = LQ_J$, then L is obtained from L′ by replacing by 0 in L′ the X_ι of index $\iota \in J' - J$.)

11. (a) Let R be an integral domain, (A_α) a right directed family of subrings of R and A the union of the A_α. Show that, if each of the A_α is integrally closed, so is A.

(b) Let K be a commutative field and $R = K(X, Y)$ the field of rational functions in two indeterminates over K. For every integer $n > 0$ let A_n denote the subring $K[X, Y, YX^{-n}]$ of R. Show that the A_n are completely integrally closed but that their union A is not. (For an element of the form $P(X, Y).X^{-ns}$, where $P \in K[X, Y]$, to belong to A_n show that it is necessary and sufficient that P belong to $\mathfrak{J}^s$, where $\mathfrak{J}$ denotes the ideal of K[X, Y] generated by X^n and Y.)

¶ * 12. Let A be the ring of integral functions (with complex values) of a complex variable. It is an integral domain whose field of fractions is the field of meromorphic functions of a complex variable.

(a) Show that A is completely integrally closed.

(b) For all $x \in A$, let $Z(x)$ be the set of zeros of x (which is a closed subset of $\mathbf{C}$, which is discrete (and therefore countable) if $x \neq 0$). If $\mathfrak{a}$ is an ideal of A distinct from A and (0), show that the $Z(x)$ for $x \in \mathfrak{a}$ form a filter base $\mathfrak{F}_{\mathfrak{a}}$. (If x and y are two integral functions with no zero in common, show, using the Mittag-Leffler Theorem, that it is possible to write $1/xy = u/x + v/y$, where u and v are in A, and deduce that $Ax + Ay = A$). Conversely, for every filter $\mathfrak{F}$ on $\mathbf{C}$ to which a closed discrete subset of $\mathbf{C}$ belongs, the set $\mathfrak{a}(\mathfrak{F})$ of $x \in A$ for which $Z(x) \in \mathfrak{F}$ is an ideal of A. Show that $\mathfrak{F}_{\mathfrak{a}(\mathfrak{F})} = \mathfrak{F}$ for every filter $\mathfrak{F}$ to which a closed discrete subset of $\mathbf{C}$ belongs and $\mathfrak{a}(\mathfrak{F}) \supset \mathfrak{a}$ for every ideal $\mathfrak{a}$ of A distinct from A and (0).

(c) If $\mathfrak{p} \neq (0)$ is a prime ideal of A, show that $\mathfrak{F}_{\mathfrak{p}}$ is an ultrafilter. (Restricting it to the case where all the sets of $\mathfrak{F}_{\mathfrak{p}}$ are infinite, show that, if $M \in \mathfrak{F}_{\mathfrak{p}}$ is the union of two infinite sets M′, M″ with no point in common, one of these sets belongs to $\mathfrak{F}_{\mathfrak{p}}$.) Conversely, for every ultrafilter $\mathfrak{U}$ on $\mathbf{C}$ to which a closed discrete subset of $\mathbf{C}$ belongs, $\mathfrak{a}(\mathfrak{U})$ is a maximal ideal of A; obtain the converse.

(d) Let $\mathfrak{U}$ be an ultrafilter on $\mathbf{C}$, the image under the canonical injection $\mathbf{Z} \to \mathbf{C}$ of a non-trivial ultrafilter on $\mathbf{Z}$. Show that there exist non-maximal prime ideals $\mathfrak{p}$ of A such that $\mathfrak{F}_{\mathfrak{p}} = \mathfrak{U}$. (For all $x \in \mathfrak{a}(\mathfrak{U})$ and all $n \in \mathbf{Z}$, let $\omega_x(n)$ be the order of x at the point n ($\omega_x(n) = 0$ if $x(n) \neq 0$); consider the set $\mathfrak{p}$ of $x \in \mathfrak{a}(\mathfrak{U})$ such that $\lim_{\mathfrak{U}} \omega_x = +\infty$.)

(e) With the notation of (d), let $\mathfrak{m}$ be the maximal ideal $\mathfrak{a}(\mathfrak{U})$. Show that the domain $A_{\mathfrak{m}}$ is not completely integrally closed (consider the meromorphic function $1/(\sin \pi z)$). *

¶ 13. Let A be a local integral domain and K its field of fractions. Consider the two following properties:

(i) A is Henselian (Chapter III, § 4, Exercise 3).

(ii) For every finite extension L of K, the integral closure A′ of A in L is a local ring.

Show that (i) implies (ii) and that the converse is true if A is integrally closed. (To see that (i) implies (ii), note that A′ is the union of a right directed family of finite A-algebras and use the definition of a Henselian ring. To see that (ii) implies (i) if A is integrally closed, show that, if $P \in A[X]$ is an irreducible monic polynomial and $f: A \to k$ is the canonical homomorphism of A onto its residue field k, $\bar{f}(P)$ is irreducible in K[X] and consider the field $L = K[X]/(f)$.)

If condition (ii) is fulfilled, the integral closure A′ of A in any extension of K is a local ring. Consider the case of complete Hausdorff local rings; if A is a complete Hausdorff Noetherian local ring, every A-algebra contained in A′ and which is a finitely generated A-module is a complete Hausdorff local ring.

14. Let A be a completely integrally closed domain and K its field of fractions. Show that for every extension L of K the integral closure A′ of A in

L is a completely integrally closed domain. (Reduce it to the case where L is a finite quasi-Galois extension of K of degree m: if $x \in L$ is such that there exists $d \in A'$ such that $dx^n \in A'$ for all $n \geqslant 0$, show first that it may be assumed that $d \in A$, then deduce that, if c_i $(1 \leqslant i \leqslant m)$ are the coefficients of the minimal polynomial of x over K, then $d^m c_i^n \in A$ for all $n \geqslant 0$.)

15. Let K be a commutative field of characteristic 0 containing all the roots of unity and such that there exist Galois extensions of K whose Galois group is not solvable (cf. for example *Algebra*, Chapter V, Appendix I, no. 1, Proposition 2). Let Ω be an algebraic closure of K and let Ω_0 be the set of elements $z \in \Omega$ such that the Galois extension of K generated by z in Ω admits a solvable Galois group.

(a) Show that Ω_0 is a subgroup of Ω distinct from Ω (cf. *Algebra*, Chapter V, § 10, no. 4, Theorem 1).

(b) Let A be the subring of $\Omega[X]$ consisting of the polynomials P such that $P(0) \in \Omega_0$. Show that A is not integrally closed but that every element of its field of fractions $\Omega(X)$ a power of which belongs to A is itself in A.

16. Let B be a ring, A a subring of B and $\mathfrak{f}$ the ideal of A the annihilator of the A-module B/A. Let U be the complement in $X = \operatorname{Spec}(A)$ of the closed set $V(\mathfrak{f})$ and let $u = {}^a\rho$, where $\rho: A \to B$ is the canonical injection. Show that the restriction of u to $\overset{-1}{u}(U)$ is a homeomorphism of $\overset{-1}{u}(U)$ onto U (for every element $g \in \mathfrak{f}$, consider the spectrum of the ring A_g identified with the open set $X_g \subset U$).

17. Let K be a field, B the polynomial ring $K[X_n]_{n \in \mathbf{N}}$ and A the subring of B generated by the monomials X_n^2 and X_n^3 for all $n \in \mathbf{N}$; show that B is the integral closure of A and that the conductor of B in A is reduced to 0; but, if $S = A - \{0\}$, the conductor of $S^{-1}B$ in $S^{-1}A$ is not reduced to 0 and $S^{-1}A$ is integrally closed.

18. Let A be an integral domain, K its field of fractions, M a finitely generated A-module, u an endomorphism of M and $u \otimes 1$ the corresponding endomorphism of the finite dimensional vector K-space $M_{(K)} = M \otimes_A K$. Show that the coefficients of the characteristic polynomial of $u \otimes 1$ are integral over A (use condition (E_{III}) of Theorem 1 of no. 1).

19. (a) Let A be an integrally closed domain, K its field of fractions, L a separable algebraic extension of K, x an element of L integral over A, K′ the field $K(x) \subset L$ and F the minimal polynomial of x over K. If x is of degree n over K, show that the integral closure of A in K′ is contained in the A-module generated by the elements $1/F'(x), x/F'(x), \dots, x^{n-1}/F'(x)$ of K′. (If $z \in K'$ is integral over A, write

$$zF'(x) = g(x), \quad \text{where} \quad g(X) = b_0 + b_1X + \cdots + b_{n-1}X^{n-1} \in K[X].$$

Show with the aid of Lagrange's interpolation formula that

$$g(X) = \mathrm{Tr}_{K'[X]/K[X]}\left(z.\frac{F(X)}{X - x}\right).)$$

(b) Let A be an integrally closed domain, K its field of fractions and p the characteristic exponent of K. Suppose that the subring $A^{1/p}$ of $K^{1/p}$ consisting of the p-th roots of the elements of A is a finitely generated A-module. Show that for every finite extension E of K the integral closure of A in E is a finitely generated A-module (reduce to the case where E is a quasi-Galois extension).

¶ * 20. Let K_0 be a perfect field of characteristic $p > 0$, K the field of rational functions $K_0(X_n)_{n \in \mathbf{N}}$ and B the ring of formal power series K[[Z]], where Z is an indeterminate; if $\mathfrak{m}$ is the maximal ideal of B, B is Hausdorff and complete with the $\mathfrak{m}$-adic topology and is a discrete valuation ring. Let L = K((Z)) be the field of fractions of B and E_0 the subfield of L generated by L^p and the elements Z and X_n (for $n \in \mathbf{N}$).

(a) Show that the element $c = \sum_{n=0}^{\infty} X_n Z^n$ of L does not belong to E_0.

(b) Let E be a maximal element of the set of subfields of L containing E_0 and not containing c; we write $A = B \cap E$. Show that A is a discrete valuation ring (and therefore Noetherian) whose field of fractions is E; if $\mathfrak{m}'$ is the maximal ideal of A, then $\mathfrak{m}'^k = \mathfrak{m}^k \cap A$ and A is dense in B with the $\mathfrak{m}$-adic topology; $L = K(c)$ is an extension of K of degree p and B is the integral closure of A in L but B is not a finitely generated A-module (use Chapter III, § 3, Exercise 9 (b)). *

¶ 21. (a) Let K_0 be a perfect field of characteristic $p > 0$, L the field of rational functions $K_0(X_n)_{n \in \mathbf{N}}$, A' the ring of formal power series L[[Y, Z]], where Y and Z are two indeterminates, and A the tensor product ring $K[[Y, Z]] \otimes_K L$, where we write $K = L^p \subset L$; A is a Noetherian local ring, which is not complete and whose completion is identified with A' (Chapter III, § 3, Exercise 17) and A' is integral over A. Let c_n denote the element

$$X_n Y + X_{n+1} YZ + X_{n+2} YZ^2 + \cdots$$

of A'. Let B be the subring of A' generated by A and the elements c_n ($n \geqslant 0$). If $C = A[c_0]$, C is Noetherian and the ring B (which is not integrally closed) is contained in the integral closure of C and contains C; but show that B is not Noetherian (observe that c_n does not belong to the ideal of B generated by the c_i of index $i < n$).

(b) Suppose $p = 2$; K_0, K and L being used with the same meaning as in (a), consider this time the ring of formal power series A' = L[[Y, Z, T]] in three indeterminates, the subring

$$A = K[[Y, Z, T]] \otimes_K L,$$

and write

$$b = Y \sum_{n \geqslant 0} X_{2n} T^n + Z \sum_{n \geqslant 0} X_{2n+1} T^n.$$

Show that the ring A$[b]$ is Noetherian but that its integral closure B is not a Noetherian ring. (As $b^2 \in$ A, every element of the field of fractions of A$[b]$ is of the form P + Qb, where P and Q are linear combinations of a finite number of X_k with coefficients in the field of fractions of K[[Y, Z, T]]; for such an element to be integral over A$[b]$, show that it is necessary and sufficient that it belong to A′. Show that B is generated by the elements

$$b_i = Y \sum_{n \geqslant 0} X_{2(i+n)} T^n + Z \sum_{n \geqslant 0} X_{2(i+n)+1} T^n \qquad \text{for} \quad i \geqslant 0$$

and conclude the argument as in (a).)

22. Let A be an integrally closed domain and K its field of fractions; show that for every extension L of K there exists an integrally closed subdomain B of L whose field of fractions is L and such that B ∩ K = A (consider L as an algebraic extension of a pure transcendental extension E = $K(X_\iota)_{\iota \in I}$ of K and thus reduce it to the case where L is an algebraic extension of K, using Exercise 11 (a)).

23. (a) Let K be a commutative field, n an integer prime to the characteristic of K and P a polynomial in K[X] with only simple roots (in an algebraic closure of K). If we write $f(X, Y) = Y^n - P(X)$, show that, if K contains the n-th roots of unity, the domain A = K[X, Y]/(f) is integrally closed and has in its field of fractions the separable extension L of K(X) generated by a root y of f (considered as a polynomial in K(X)[Y]). (Show that, if an element z of L is integral over K[X], it belongs to A; for this, write z in the form

$$z = \sum_{i=0}^{n-1} a_i(X) y^i,$$

where $a_i(X) \in K(X)$. Show that the elements $a_i(X)y^i$ $(0 \leqslant i \leqslant n - 1)$ are integral over K[X] and deduce that the $(a_i(X))^n(P(X))^i$ are elements of K[X]; conclude that so are the $a_i(X)$.)

(b) Suppose that K is an imperfect field of characteristic $p \neq 2$; let $a \in$ K be an element not belonging to K^p, let E be the field $K(a^{1/p})$ and let us write $P(X) = X^p - a$, $f(X, Y) = Y^2 - P(X)$. If A is the integrally closed domain K[X, Y]/(f), show that B = E $\otimes_K$ A is an integral domain but not integrally closed (consider the element $y/(X - a^{1/p})$ of the field of fractions of B).

24. Let Δ be a torsion-free commutative group written additively. Let A be a commutative ring, B a commutative A-algebra and h: A → B the ring homomorphism defining the algebra structure on B. Then $h^{(\Delta)}: A^{(\Delta)} \to B^{(\Delta)}$ (*Algebra*,

Chapter II, § 11, Exercise 1) is a (ring) homomorphism of the algebra $A^{(\Delta)}$ of the group Δ with respect to A to the algebra $B^{(\Delta)}$ of the group Δ with respect to B. Let $P = \sum_{\alpha \in \Delta} b_\alpha \otimes X^\alpha$ be an element of $B^{(\Delta)}$. Show that for P to be integral over $A^{(\Delta)}$, it is necessary and sufficient that each of the elements $b_\alpha \in B$ is integral over A. (Reduce it to the case where Δ is finitely generated and then to the case $\Delta = \mathbf{Z}$.) Show that if A is an integral domain and integrally closed so is $A^{(\Delta)}$.

25. Let Δ be a totally ordered commutative group. Extend Propositions 20 and 21 of no. 8 to graded rings of type Δ (for Proposition 20 use Exercise 24; for Proposition 21 generalize Lemma 4 of no. 8 to the case where Δ is finitely generated and then prove Proposition 21 by taking the direct limit).

¶ 26. (a) Let A be an integral domain such that for all $a \neq 0$ in A, the ideal Aa is the intersection of a family $(\mathfrak{q}_\iota)$ of saturations of Aa with respect to prime ideals $\mathfrak{p}_\iota$ which are minimal among those which contain a, such that the domains $A_{\mathfrak{p}_\iota}$ are integrally closed (resp. completely integrally closed). Then show that A is integrally closed (resp. completely integrally closed).

(b) Let A be a strongly Laskerian local ring (Chapter IV, § 2, Exercise 28) and $\mathfrak{m}$ its maximal ideal. Suppose that, for some $a \in A$, $\mathfrak{m}$ is an *immersed* prime ideal weakly associated with A/Aa (Chapter IV, § 1, Exercise 17). Show that $Aa : \mathfrak{m}$ is contained in every ideal which is primary for every prime ideal $\mathfrak{p} \neq \mathfrak{m}$ weakly associated with A/Aa (consider the transporter $\mathfrak{q} : \mathfrak{m}$ for such a primary ideal $\mathfrak{q}$, cf. Chapter IV, § 2, Exercise 30). If further A is assumed to be an integral domain, show that the relation $\mathfrak{p} . (Aa : \mathfrak{m}) = Aa$, for a prime ideal $\mathfrak{p}$ minimal among those which contain Aa, is impossible (if $\mathfrak{q}$ is the saturation of A with respect to $\mathfrak{p}$, note that the above relation would imply that $\mathfrak{q}A_\mathfrak{p} = \mathfrak{q}\mathfrak{p}A_\mathfrak{p}$ in $A_\mathfrak{p}$ and conclude with the aid of Exercise 29 (d) of Chapter IV, § 2).

(c) Show that, if A is a strongly Laskerian integral domain and there exists an element $a \neq 0$ of A such that there are prime ideals which are weakly associated with A/Aa and *immersed*, A is not completely integrally closed. (Reduce it to the case considered in (b); in the notation of (b), there then exists $b \in Aa : \mathfrak{m}$ such that $b \notin Aa$ and $b\mathfrak{p} \subset a\mathfrak{m}$; deduce by induction on n that $b^n\mathfrak{p} \subset a^n\mathfrak{m}$ for all n.)

(d) Deduce in particular from (a) and (c) that, for a Noetherian integral domain to be integrally closed, it is necessary and sufficient that for all $a \neq 0$ in A the prime ideals $\mathfrak{p}$ associated with A/Aa be not immersed and be such that $A_\mathfrak{p}$ is integrally closed (cf. Chapter VII, § 1, no. 4, Proposition 8).

27. Let A be an integrally closed but not completely integrally closed domain and K its field of fractions; let $a \neq 0$ be a non-invertible element of A for which there exists $d \neq 0$ in A such that $da^{-n} \in A$ for every integer $n \geqslant 0$.

Show that in the field of formal power series K((X)) there exists a formal power series $\sum_{k=1}^{\infty} u_k X^k = f(X)$ satisfying the equation

$$(f(X))^2 - af(X) + X = 0,$$

hence integral over A[[X]], belonging to the field of fractions of A[[X]] but not belonging to A[[X]], so that A[[X]] is not integrally closed.

28. Let k be a field and A_0 the polynomial ring $k[X_n, Y_n]_{n \in \mathbf{Z}}$ in two infinite families of indeterminates. Let G be an infinite monogenous group (therefore isomorphic to $\mathbf{Z}$) and let σ be a generator of $\mathscr{G}$. G is defined as operating on A_0 by the conditions $\sigma(Y_n) = Y_n$ and $\sigma(X_n) = X_{n+1}$ for all $n \in \mathbf{Z}$. Let $\mathfrak{I}$ be the ideal of A_0 generated by the elements $Y_n(X_n - X_{n+1})$ for all $n \in \mathbf{Z}$ and let A be the quotient ring $A_0/\mathfrak{I}$; as $\sigma(\mathfrak{I}) \subset \mathfrak{I}$, the group $\mathscr{G}$ also operates on A by passing to the quotient. Let S be the multiplicative subset of A generated by the canonical images of the Y_n in A, so that S consists of the elements invariant under G. Show that $S^{-1}(A^{\mathscr{G}}) \neq (S^{-1}A)^{\mathscr{G}}$.

29. Let k be a field of characteristic $\neq 2$, A the polynomial ring $k[T]$ in one indeterminate and a, b two elements of k such that $a \neq 0$, $b \neq 0$ and $a \neq b$. Let $K = k(T)$ be the field of fractions of A and L, M the extensions of K adjoining to K respectively a root of $X^2 - T(T-a)$ and a root of $X^2 - T(T-b)$. Let B, C be the integral closures of A in L and M respectively, which are finitely generated free A-modules. Show that $L \otimes_K M$ is a field, a separable finite extension of K, but that $B \otimes_A C$, which is canonically identified with a subring of $L \otimes_K M$, is not the integral closure of A in $L \otimes_K M$.

§ 2

1. Give an example of an injective ring homomorphism $\rho: A \to B$ such that B is finite A-algebra, the mapping ${}^a\rho: \mathrm{Spec}(B) \to \mathrm{Spec}(A)$ is surjective but B is not a flat A-module. Conversely, give an example of a faithfully flat A-algebra which is finitely generated but not integral over A.

2. Let A be a ring such that Spec(A) is Hausdorff (Chapter II, § 4, Exercise 16). Show that, for every A-algebra B integral over A, Spec(B) is Hausdorff.

3. Let A be a ring and A′ a ring containing A and integral over A. For every ideal $\mathfrak{a}$ of A, show that the radical of $\mathfrak{a}$ is the intersection of A and the radical of $\mathfrak{a}A'$. Show that the latter is the set of elements $x \in A'$ satisfying an equation of integral dependence whose coefficients (other than the dominant coefficient) belong to $\mathfrak{a}$ (note that, if y, z are two elements of A′ satisfying such an equation of integral dependence, so does $y - z$).

4. Let A be a Noetherian ring, $\rho: A \to B$ a ring homomorphism making B into a finitely generated A-module and N a finitely generated B-module. Show that the prime ideals $\mathfrak{p} \in \mathrm{Ass}(\rho_*(N))$ are the inverse images under ρ of the prime ideals $\mathfrak{q} \in \mathrm{Ass}(N)$.

5. Let K be a field; in the polynomial domain K[X, Y] in two indeterminates consider the subrings $A = K[X^4, Y^4]$, $A' = K[X^4, X^3Y, XY^3, Y^4]$; A is Noetherian and A′ is a finite A-algebra. Let $\mathfrak{p} = AY^4$, which is a prime ideal of A; show that $\mathfrak{m} = A'X^4 + A'X^3Y + A'XY^3 + A'Y^4$ is an (immersed) prime ideal associated with the ideal $\mathfrak{p}A'$ of A′ but does not lie above $\mathfrak{p}$ (note that $\mathfrak{m}$ is the annihilator of the class of X^2Y^6 in the ring $A'/\mathfrak{p}A'$) (cf. § 1, Exercise 26 (d)).

6. Define an increasing sequence of algebraic number fields $K_0 = \mathbf{Q}, K_1, \ldots, K_n, \ldots$ such that, if A_n denotes the integral closure of $\mathbf{Z}$ in K_n, p a prime number $\neq 2$, K_n is a quadratic extension of K_{n-1} and there exists in A_n a prime ideal $\mathfrak{p}_n$ lying above (p) such that there are in A_{n+1} two distinct prime ideals $\mathfrak{p}_{n+1}, \mathfrak{p}'_{n+1}$ lying above $\mathfrak{p}_n$ (observe that, in a finite field of characteristic $\neq 2$, there are always elements which are not squares). Let K be the union of the K_n and A′ the integral closure of $\mathbf{Z}$ in K; show that there exists an infinity of (maximal) ideals of A′ lying above the maximal ideal (p) of $\mathbf{Z}$.

¶ 7. Let A be a Noetherian integral domain and A′ its integral closure. Show that for every prime ideal $\mathfrak{p}$ of A there exists only a finite number of prime ideals of A′ lying above $\mathfrak{p}$. (Reduce it first to the case where A is a local ring with maximal ideal $\mathfrak{p}$; consider its completion $\hat{A}$ and the total ring of fractions B of $\hat{A}$ and note that the field of fractions K of A is identified with a subring of B (Chapter III, § 3, no. 4, Corollary 2 to Theorem 3). Note that the spectrum of B is finite and therefore that in B there is only a finite number of idempotents which are orthogonal to one another. Then let $C \supset A$ be a subring of K which is a finitely generated A-module and therefore semi-local. Note that the completion $\hat{C}$ of C is identified with a subring of B; conclude by noting that $\hat{C}$ is the product of a finite number of local rings, this number being equal to the number of maximal ideals of C). Generalize to the integral closure of a Noetherian ring in its total ring of fractions (reduce it to the case of a reduced ring).

8. A local ring A is called *unibranch* if its integral closure is a local ring.

(a) Let A be a local integral domain and K its field of fractions. For A to be unibranch, it is necessary and sufficient that every sub-A-algebra of K which is a finitely generated A-module be a local ring.

(b) Show that, if the completion of a Noetherian local ring A is an integral domain, A is unibranch (if B is a finite A-algebra contained in K, show that the subring of the field of fractions of $\hat{A}$ generated by B and $\hat{A}$ is isomorphic to B $\otimes_A \hat{A}$, using the flatness of the A-module $\hat{A}$ and the fact that B is contained in a finitely generated free A-module contained in K; then use (a) and Chapter III, § 2, no. 13, Corollary to Proposition 19).

9. Let k_0 be a field and A′ the polynomial ring $k_0[X_n]_{n \in \mathbf{N}}$ in an infinite sequence of indeterminates; it is an integrally closed domain whose field of fractions is $K' = k_0(X_n)_{n \in \mathbf{N}}$. Let σ denote the k_0-automorphism of A′ such that

$$\sigma(X_{2n}) = -X_{2n} + X_{2n+1}, \qquad \sigma(X_{2n+1}) = X_{2n+1}$$

for all $n \geqslant 0$; σ and the identity automorphism form a group $\mathscr{G}$ of automorphisms of A′. Let A be the ring of invariants of $\mathscr{G}$ and let K be its field of fractions; K′ is a separable extension of K of degree 2 and A′ is the integral closure of A in K′.

(a) Show that A′ is an A-module admitting an *infinite* basis. (If k_0 is of characteristic $\neq 2$, show that, if we write $Y_n = X_{2n} - \frac{1}{2}X_{2n+1}$, A is the subring of A′ generated by the X_{2n+1} and the products Y_nY_m for $n \geqslant 0$, $m \geqslant 0$. If k_0 is of characteristic 2, A′ is generated by the X_{2n+1} and the

$$X_{2n}^2 + X_{2n}X_{2n+1}$$

for $n \geqslant 0$.)

(b) Let $\mathfrak{m}'$ be the maximal ideal of A′ generated by all the X_n and let $\mathfrak{m} = A \cap \mathfrak{m}'$; show that $\mathfrak{m}' = \mathfrak{m}A'$, that $\mathfrak{m}'$ is the only ideal of A′ lying above $\mathfrak{m}$ and that the fields $A/\mathfrak{m}$ and $A'/\mathfrak{m}'$ are canonically isomorphic.

(c) Let $\mathfrak{p}'$ be the prime ideal of A′ generated by the X_{2n+1} for $n \geqslant 0$ and let $\mathfrak{p} = \mathfrak{p}' \cap A$; let k, k' be the fields of fractions of $A/\mathfrak{p}$ and $A'/\mathfrak{p}'$ respectively; $\mathfrak{p}'$ is the only prime ideal of A′ lying above $\mathfrak{p}$. If k_0 is not of characteristic 2, k' is a separable extension of k of degree 2; if k_0 is of degree 2, k' is an infinite radicial extension of k.

10. (a) Let A′ be the polynomial ring $\mathbf{Z}[X, Y]$ in two indeterminates and let σ be the automorphism of A′ leaving Y invariant and such that $\sigma(X) = -X$; σ and the identity automorphism form a group $\mathscr{G}$ of automorphisms of A′ and the ring of invariants of $\mathscr{G}$ is $A = \mathbf{Z}[X^2, Y]$. Let $\mathfrak{p}'$ be the prime ideal $A'(Y - 2X)$, $\mathfrak{p} = A \cap \mathfrak{p}'$; the decomposition group $\mathscr{G}^Z(\mathfrak{p}')$ is reduced to the identity element. Show that $A/\mathfrak{p}$ and $A'/\mathfrak{p}'$ are not isomorphic (consider the tensor products of these rings with $\mathbf{Z}/2\mathbf{Z}$).

(b) Let K be a field of characteristic $\neq 2$, A′ the polynomial ring K[X, Y] and $\mathscr{G}$ the automorphism group of A′ consisting of the identity and the K[Y]-automorphism σ of A′ such that $\sigma(X) = -X$; the ring of invariants A of $\mathscr{G}$ is $K[X^2, Y]$. Let $\mathfrak{p}'$ be the prime ideal $A'(X^3 - Y)$ of A′ and $\mathfrak{p} = A \cap \mathfrak{p}'$; the decomposition group $\mathscr{G}^Z(\mathfrak{p}')$ is reduced to the identity element but $A/\mathfrak{p}$ and $A'/\mathfrak{p}'$ are not isomorphic.

¶ 11. (a) Let A be a local integrally closed domain, $\mathfrak{m}$ its maximal ideal, K its field of fractions, K′ a finite Galois extension of K, A′ the integral closure of A in K′, $\mathfrak{m}'$ a maximal ideal of A′ lying above $\mathfrak{m}$, $B = A^Z(\mathfrak{m}')$ its decomposition ring and $\mathfrak{n} = \mathfrak{m}' \cap B$. For every integer $k > 0$, $B = A + \mathfrak{n}^k$ (argue as in Proposition 4). Moreover there exists a primitive element z of K^Z (field of fractions of B) such that $z \in \mathfrak{n}$; deduce that $\mathfrak{n} \cap A[z] = \mathfrak{m}A[z]$.

(b) Show that, for every integer k, $\mathfrak{m}^k B_\mathfrak{n} \cap A = \mathfrak{m}^k$. (If $x \in \mathfrak{m}^k B_\mathfrak{n} \cap A$, note that there exists $t \in A[z]$ (where z is defined in (a)) such that $t \notin \mathfrak{n}$ and $tx \in \mathfrak{m}^k A[z]$; writing $t = t_0 + t_1 z + \cdots + t_{r-1} z^{r-1}$ (where r is the degree $[K^Z : K]$), deduce that $t_0 x \in \mathfrak{m}^k$ and $t_0 \notin \mathfrak{m}$.)

(c) Suppose now that K′ is *any* Galois extension of K. Keeping the same notation as above, show that $B = A + \mathfrak{n}^k$ and $\mathfrak{m}^k B_\mathfrak{n} \cap A = \mathfrak{m}^k$ still (note that every element of $A^Z(\mathfrak{m}')$ is contained in the decomposition ring of $\mathfrak{m}' \cap C$, where C is the integral closure of A in a finite Galois sub-extension of K′).

(d) Under the hypotheses of (c), show that, if A is Noetherian, so is $B_\mathfrak{n}$. (Observe that the Hausdorff completion of $B_\mathfrak{n}$ with respect to the $\mathfrak{m}$-adic topology is identified with the completion $\hat{A}$ of A by (c); if $\phi : B_\mathfrak{n} \to \hat{A}$ is the canonical homomorphism, show that, for every ideal $\mathfrak{a}$ of $B_\mathfrak{n}$, $\overset{-1}{\phi}(\mathfrak{a}\hat{A}) = \mathfrak{a}$, using the fact that in $\hat{A}$ every ideal is finitely generated and hence that $\mathfrak{a}\hat{A}$ is generated by a finite number of elements of $\mathfrak{a}$ and thus reduce it to the case where K′ is a finite extension of K.)

12. Let K be a field, K′ a finite Galois extension of K, C the polynomial ring K′[X, Y], B the quotient ring C/CXY and x, y the canonical images of X, Y in B. Let A′ be the subring $K[x] + yK'[y]$ of B. The Galois group $\mathscr{G}$ of K′ over K operates faithfully on A′, the ring of invariants A of $\mathscr{G}$ in A′ being $K[x, y]$. Let S be the set of powers of x in A′. Show that $S^{-1}A' = S^{-1}A$ and that $\mathscr{G}$ does not operate faithfully on $S^{-1}A'$.

¶ 13. (a) Let K be a field of characteristic 0 and $\mathfrak{n}$ the ideal of the polynomial ring K[X, Y, Z] generated by $Y^2 - X^2 - X^3$. Show that $\mathfrak{n}$ is prime and that the integral domain $A = K[X, Y, Z]/\mathfrak{n}$ is not integrally closed: if x, y, z are the images of X, Y, Z in A, the element $t = y/x$ of the field of fractions L of A is integral over A but does not belong to A. Let $A' \supset A$ be the subring $K[x, t, z]$ of L, which is integral over A. Consider in A the prime ideal $\mathfrak{p}$ generated by $xz - y$ and $z^2 - 1 - x$ and the maximal ideal $\mathfrak{q} \supset \mathfrak{p}$ generated by $x, y, z - 1$. Consider on the other hand in A′ the maximal ideal $\mathfrak{q}'$ generated by x, $t + 1$ and $z - 1$; show that $\mathfrak{q}'$ lies above $\mathfrak{q}$ but that there exists no prime ideal $\mathfrak{p}' \subset \mathfrak{q}'$ of A′ lying above $\mathfrak{p}$. (Consider a homomorphism from $A'/\mathfrak{p}A'$ to an extension field of K and show that the image of $\mathfrak{q}'$ under such a homomorphism is necessarily 0.)

(b) Let A′ be the quotient ring of the polynomial ring $\mathbf{Z}[X]$ by the ideal $\mathfrak{n}$ generated by 2X and $X^2 - X$; let x be the image of X in A′; A′ is integral over **Z** but the number 2 is a divisor of 0 in A'. Let q' be the prime ideal of A' generated by 2 and $x - 1$; then $\mathfrak{q}' \cap \mathbf{Z} = 2\mathbf{Z}$; if $\mathfrak{p}$ is the prime ideal (0) in **Z**, show that there exists no prime ideal $\mathfrak{p}' \subset \mathfrak{q}'$ contained in A' which lies above $\mathfrak{p}$.

(c) Let K be a field, $\mathfrak{n}$ the ideal of the polynomial ring K[X, Y, Z] generated by X^2, XY, XZ, YZ $-$ Y and $Z^2 - Z$, A' the quotient ring K[X, Y, Z]/$\mathfrak{n}$, x, y, z the images of X, Y, Z in A' and A the subring K[x, y] of A'; A is integrally closed in its total ring of fractions and A' is integral over A. Consider in A the prime ideal $\mathfrak{p}$ generated by x and the maximal ideal $\mathfrak{q}$ generated by x and y. Consider on the other hand in A' the maximal ideal $\mathfrak{q}'$ generated by x, y, z; show that $\mathfrak{q}'$ lies above $\mathfrak{q}$ but that there exists no prime ideal $\mathfrak{p}' \subset \mathfrak{q}'$ lying above $\mathfrak{p}$ (same method as in (a)).
* Interpret the examples in (a) and (c) geometrically. *

¶ 14. (a) Let A be a subring of the field K and x an element $\neq 0$ of K. Show that, if x is not integral over A, there exists a maximal ideal of A[x^{-1}] which contains x^{-1} and that, for every maximal ideal $\mathfrak{m}$ of A[x^{-1}] containing x, the ideal A $\cap$ $\mathfrak{m}$ is maximal in A (observe that x does not belong to the ring A[x^{-1}]).

(b) Let A be a local integral domain, $\mathfrak{m}$ its maximal ideal, K a field containing A and k the residue field of A. Let $x \in$ K be such that $x \notin$ A and $x^{-1} \notin$ A and suppose that A is integrally closed in the ring A[x]. Show that there exists an A-algebra homomorphism A[x] $\rightarrow k$[X] which extends the canonical homomorphism A $\rightarrow k$ and which maps x to X. (It is necessary to prove that, if $\sum_{i=0}^{n} a_i x^i = 0$, where $a_i \in$ A for all i, then all the a_i belong to $\mathfrak{m}$. Otherwise there would be a relation $a_n x^{n-p} + \cdots + a_p = -(a_{p+1}x^{-1} + \cdots + a_0 x^{-p})$, where a_p is invertible in A; using Exercise 5 of § 1, show that the common value b of the two sides belongs to A and, using (a), show that necessarily $b \in \mathfrak{m}$; conclude that x^{-1} would be integral over A, which is absurd.)

(c) Under the hypotheses of (b), show that the ideal $\mathfrak{m}$A[x] of A[x] is prime and not maximal.

15. Let A be a ring and $\mathfrak{p}$ a prime ideal of A. Show that, in the polynomial ring B = A[X], the ideal $\mathfrak{p}$B = $\mathfrak{q}$ is prime; the maximal ideal of the local ring $B_{\mathfrak{q}}$ is equal to $\mathfrak{p}B_{\mathfrak{q}}$; if k is the field of fractions of A/$\mathfrak{p}$, $B_{\mathfrak{q}}/\mathfrak{p}B_{\mathfrak{q}}$ is isomorphic to the field of rational fractions k(X). If A is integrally closed, so is $B_{\mathfrak{q}}$.

16. Let A be an integral domain, K its field of fractions, K' a finite extension of K of degree n and A' a subring of K' containing A, with K' as field of fractions and integral over A. Let $\mathfrak{m}$ be a maximal ideal of A and B the local ring $A[X]_{\mathfrak{m}A[X]}$ (Exercise 15).

(a) The subring $B' = B[A']$ of $K'(X)$ is integral over B; its field of fractions is $K'(X)$ and $[K'(X):K(X)] = n$.

(b) For every maximal ideal $\mathfrak{m}'$ of A' containing $\mathfrak{m}$, $\mathfrak{n}' = \mathfrak{m}'B'$ is a maximal ideal of B'; if we write $\mathfrak{n} = \mathfrak{m}B$, then

$$[(A'/\mathfrak{m}'):(A/\mathfrak{m})] = [(B'/\mathfrak{n}'):(B/\mathfrak{n})]$$

and

$$[(A'/\mathfrak{m}'):(A/\mathfrak{m})]_s = [(B'/\mathfrak{n}'):(B/\mathfrak{n})]_s.$$

(Observe that $(A'/\mathfrak{m}') \otimes_A B$ is isomorphic to the field $(A'/\mathfrak{m}')(X)$ and deduce that $B'/\mathfrak{n}'$ is isomorphic to $(A'/\mathfrak{m}')(X)$.)

(c) The mapping $\mathfrak{m}' \mapsto \mathfrak{m}'B'$ is the bijection of the set of maximal ideals of A' containing $\mathfrak{m}$ onto the set of maximal ideals of B'; the inverse bijection is $\mathfrak{n}' \mapsto \mathfrak{n}' \cap A'$.

¶ 17. (a) Let K be an infinite field, E a finite separable algebraic extension of K and g a polynomial of K[X]. Show that there exists an element $x \in E$ such that $E = K(x)$ and the minimal polynomial of x over K is relatively prime to g (argue as in *Algebra*, Chapter V, § 7, no. 7, Proposition 12).

(b) Let k be a field, A a primary k-algebra (Chapter IV, § 2, Exercise 32) and $\mathfrak{m}$ its unique prime ideal. If $A/\mathfrak{m}$ is a transcendental extension of k, there exists in A elements which are not algebraic over k. If $A/\mathfrak{m}$ is algebraic over k and d is an integer at least equal to the separable factor of the degree of $A/\mathfrak{m}$ over k (and hence any integer if this separable factor is infinite), show that there exists in A elements whose minimal polynomial over k is of degree $\geqslant d$. If further $A/\mathfrak{m}$ is not separable over k or if $\mathfrak{m} \neq 0$, there exists in A an element whose minimal polynomial over k is of degree $> d$. (If $A/\mathfrak{m}$ is not separable over k, use *Algebra*, Chapter V, § 8, Exercise 5. If $A/\mathfrak{m}$ is separable and of degree d over k and $x \in A$ is such that $\xi \in A/\mathfrak{m}$, the class of x, is a primitive element of $A/\mathfrak{m}$, whose minimal polynomial is $f \in k[X]$, observe that $f'(\xi) \neq 0$ and deduce that, if $y \neq 0$ is an element of $\mathfrak{m}$, then $f(x+y) \neq 0$.)

(c) Let k be an infinite field and A a k-algebra the direct composition of r primary algebras A_i $(1 \leqslant i \leqslant r)$; let $\mathfrak{m}_i$ be the maximal ideal of A_i and d_i an integer at least equal to the separable factor of the degree of $A_i/\mathfrak{m}_i$ over k if A_i is algebraic and this factor is finite, or an arbitrary integer otherwise. Show that there exists in A an element which is not algebraic over k or whose minimal polynomial over k is of degree $\geqslant \sum_{i=1}^{r} d_i$; if further one of the $\mathfrak{m}_i$ is $\neq 0$ or if one of the fields $A_i/\mathfrak{m}_i$ is not a separable algebraic extension of k, there exists in A an element which is not algebraic over k or whose minimal polynomial over k is of degree $> \sum_{i=1}^{r} d_i$ (use (a) and (b)).

¶ 18. Let A be a local integrally closed domain, K its field of fractions, $\mathfrak{m}$

its maximal ideal and $k = A/\mathfrak{m}$ its residue field. Let K′ be a finite algebraic extension of K of degree n, A′ a subring of K′ containing A, integral over A and with K′ as field of fractions and $\mathfrak{m}_i$ $(1 \leqslant i \leqslant r)$ the maximal ideals of A′. An ideal $\mathfrak{m}$ is called *unramified* in A′ if there exists a basis $(w_i)_{1 \leqslant i \leqslant n}$ of K′ over K consisting of elements of A′ and whose discriminant $D_{K'/K}(w_1, \ldots, w_n)$ (*Algebra*, Chapter IX, § 2) belongs to $A - \mathfrak{m}$.

(a) Show that, if $\mathfrak{m}$ is unramified in A′, K′ is a separable extension of K, A′ is the integral closure of A in K′ and is a *free* A-module every basis of which over A is a basis of K′ over K; $\mathfrak{m}A'$ is the Jacobson radical of A′ and $A'/\mathfrak{m}A'$ is a semi-simple k-algebra of rank n over k which is separable over k and the direct composition of the fields $A'/\mathfrak{m}'_i$.

(b) Let d_i be the separable factor of the degree of $A'/\mathfrak{m}'_i$ over $A/\mathfrak{m}$. Show that, if k is infinite, $\sum_i d_i \leqslant n$ (use Exercise 17 (c)); moreover, the following conditions are equivalent:

(α) $\mathfrak{m}$ is unramified in A′.

(β) A′ is a finitely generated A-module and $A'/\mathfrak{m}A'$ is a semi-simple k-algebra and separable over k.

(γ) $\sum_{i=1}^{r} d_i \leqslant n$.

(To see that (β) implies (γ), note that there is a system of generators of the A-module A′ whose classes mod. $\mathfrak{m}A'$ form a basis of $A'/\mathfrak{m}A'$ over k, using Chapter II, § 3, no. 2, Corollary 2 to Proposition 4; deduce that

$$[K' : K] \leqslant [(A'/\mathfrak{m}A') : k].$$

To see that (γ) implies (α), use Exercise 17 (c) to show that there exists a basis of $A'/\mathfrak{m}A'$ over k consisting of the powers ξ^i of an element ξ $(0 \leqslant i \leqslant n - 1)$; if $x \in A'$ is an element of the class ξ, deduce that $D_{K'/K}(1, x, \ldots, x^{n-1})$ belongs to $A - \mathfrak{m}$.) Give an example where K′ is separable over K, $A'/\mathfrak{m}A' = A/\mathfrak{m}$ and A′ is not a finitely generated A-module (cf. Exercise 9 (c)).

(c) Extend the results of (b) to the case where k is finite (in the notation of Exercise 16, show that, if $\mathfrak{n}$ is unramified in B′, then $\mathfrak{m}$ is unramified in A′; taking account of the fact that $B/\mathfrak{n}$ is infinite, use Exercise 16 (c), thus obtaining an element $z \in B'$ such that

$$D_{K'(X)/K(X)}(1, z, \ldots, z^{n-1}) \in B - \mathfrak{n};$$

show that z may be assumed to be a polynomial in A′[X] and, by expressing the above discriminant as a polynomial in X, obtain the existence of a system of n elements w_i $(1 \leqslant i \leqslant n)$ of A′ such that $D_{K'/K}(w_1, \ldots, w_n) \in A - \mathfrak{m}$).

¶ 19. Let A be an integrally closed domain, K its field of fractions, K′ a finite algebraic extension of K of degree n and A′ a subring of K′ containing A, integral over A and with K′ as field of fractions; if $\mathfrak{p}$ is a prime ideal of A, let $A'_\mathfrak{p}$ denote the ring of fractions of A′ whose denominators are in A — $\mathfrak{p}$. $\mathfrak{p}$ is called *unramified in* A′ if $\mathfrak{p}A_\mathfrak{p}$ is unramified in $A'_\mathfrak{p}$.

(a) Let $\mathfrak{p}'_i$ $(1 \leqslant i \leqslant r)$ be the prime ideals of A′ lying above $\mathfrak{p}$. Let d_i be the separable factor of the degree of the field of fractions of $A'/\mathfrak{p}'_i$ over the field of fractions of $A/\mathfrak{p}$; show that $\sum_{i=1}^{r} d_i \leqslant n$. (Reduce it to Exercise 18.)

(b) The following conditions are equivalent:

(α) The ideal $\mathfrak{p}$ is unramified in A′.

(β) A′ contains a basis of K′ over K whose discriminant belongs to A — $\mathfrak{p}$.

(γ) $\sum_{i=1}^{r} d_i = n$.

In particular, if $r = n$, $\mathfrak{p}$ is unramified in A′.

Give an example where $\mathfrak{p}$ is unramified in A′ but A′ is not a finitely generated A-module (cf. Exercise 9 (a)).

(c) A is called *unramified in* A′ (or in K′) if every prime ideal of A is unramified in A′. Show that A′ is then a *faithfully flat* A-module (cf. Chapter II, § 3, no. 4, Corollary to Proposition 15). Let K″ be a finite algebraic extension of K′ of degree n'; let A′ be the integral closure of A in K′ and A″ the integral closure of A in K″; for A to be unramified in A″, it is necessary and sufficient that A be unramified in A′ and A′ be unramified in A″.

(d) Let k be a field, A an integrally closed integral k-algebra and K its field of fractions. Let k' be a finite separable algebraic extension of k of degree n over k. Show that, if $k' \otimes_k K$ is a field, A is unramified in $k' \otimes_k K$ (consider a primitive element x of k' over k and the basis of $k' \otimes_k K$ consisting of the x^i for $0 \leqslant i \leqslant n - 1$).

20. Let A be the integral domain defined in Chapter III, § 3, Exercise 15 (b), C the local ring $A_\mathfrak{m}$, $\mathfrak{n} = \mathfrak{m}A_\mathfrak{m}$ the maximal ideal of C and L the field of fractions of A and C. If x, y are the canonical images of X and Y in C, show that $t = y/x$ does not belong to C but that $t^2 \in$ C, so that C is not integrally closed; moreover $C' = C[t]$ is the integral closure of C and is isomorphic to $S^{-1}K[T]$ (T an indeterminate), where S is the multiplicative subset of K[T] consisting of the polynomials which are not divisible by $T - 1$ nor by $T + 1$. Show that C′ is a finitely generated C-module and that the (maximal) ideals of C′ lying above $\mathfrak{n}$ are the principal ideals $\mathfrak{q}_1 = (t - 1)C'$ and $\mathfrak{q}_2 = (t + 1)C'$; but the local rings $C'_{\mathfrak{q}_1}$ and $C'_{\mathfrak{q}_2}$ are not C-algebras integral over C.

21. Let A be an integrally closed domain containing the field $\mathbf{Q}$, K its field of fractions, L an algebraic extension of K and B the integral closure of A in L. Show that, for every ideal $\mathfrak{a}$ of A, $A \cap \mathfrak{a}B = \mathfrak{a}$.

¶ 22. Let A be a Noetherian integral domain, K its field of fractions and A′ a subring of K containing A. Suppose that A is integrally closed in A′ and that for every prime ideal $\mathfrak{p}$ of A there exists a prime ideal $\mathfrak{p}'$ of A' lying above $\mathfrak{p}$. Argue by *reductio ad absurdum*: let $x \in A' \cap \complement A$ and let $\mathfrak{a} = A \cap (x^{-1} A)$, which is an ideal distinct from A. Let $\mathfrak{p}$ be a prime ideal associated with $A/\mathfrak{a}$; the transporter $\mathfrak{b} = \mathfrak{a} : \mathfrak{p}$ is then distinct from $\mathfrak{a}$ (Chapter IV, § 1, no. 4, Proposition 9) and there therefore exists $y \in \mathfrak{b}$ such that $xy \notin A$ and

$$xy\mathfrak{p} \subset \mathfrak{p}A' \cap A = \mathfrak{p};$$

deduce a contradiction.)

§ 3

¶ 1. Let A be a ring, $\mathfrak{p}$ a prime ideal of A, B a finitely generated A-algebra and let $\mathfrak{r}_1 \subset \cdots \subset \mathfrak{r}_m$ be a non-empty increasing sequence of ideals of B lying above $\mathfrak{p}$. Show that there exist an element $a \in A - \mathfrak{p}$ and a finite family $(y_k)_{1 \leqslant k \leqslant q}$ of elements of B with the following properties:

(1) If we write $B' = A[y_1, \ldots, y_q]$ and if S is the set of powers of a, the ring $S^{-1}B$ is integral over $S^{-1}B'$.

(2) For all i such that $1 \leqslant i \leqslant m$ there exists an integer $h(i) \geqslant 0$ such that $\mathfrak{r}_i$ contains $y_1, \ldots, y_{h(i)}$ and for every polynomial $P \in A[X_{h(i)+1}, \ldots, X_q]$ the relation $P(y_{h(i)+1}, \ldots, y_q) \in \mathfrak{r}_i$ implies that all the coefficients of P belong to $\mathfrak{p}$.

The ideal $\mathfrak{r}_i \cap B'$ is then generated by $\mathfrak{p}$ and the y_k such that $k \leqslant h(i)$. (Argue by induction on m, thus reducing it to the case $m = 1$. Then let $\mathfrak{r} = \mathfrak{r}_1\alpha$; argue by induction on the number n of generators of the A-algebra B. Let $(x_i)_{1 \leqslant i \leqslant n}$ be such a system of generators and let $\mathfrak{s}$ be the ideal of $A[X_1, \ldots, X_n]$ consisting of those P such that $P(x_1, \ldots, x_n) \in \mathfrak{r}$; then $\mathfrak{s} \supset \mathfrak{p}A[X_1, \ldots, X_n]$; distinguish two cases according to whether $\mathfrak{s}$ is equal or not to $\mathfrak{p}A[X_1, \ldots, X_n]$. In the second case, there is a polynomial $Q \in A[X_1, \ldots, X_n]$ none of whose coefficients is in $\mathfrak{p}$ and which belongs to $\mathfrak{s}$. Show that it is possible to determine integers m_i $(2 \leqslant i \leqslant n)$ such that, if we write $x'_1 = Q(x_1, \ldots, x_n)$, $x'_i = x_i - x_1^{m_i}$ $(2 \leqslant i \leqslant n)$, there exists an element $b \in A - \mathfrak{p}$ for which bx_1 is integral over $C = A[x'_2, \ldots, x'_n]$; apply the induction hypothesis to the A-algebra

$$B_1 = A[x'_2, \ldots, x'_n]$$

and the ideal $\mathfrak{r} \cap B_1$.)

2. Let K be a field and B the quotient of the polynomial ring K[X, Y] in two indeterminates by the ideal $\mathfrak{a}$ of K[X, Y] generated by X^2 and XY; denote by x and y the images in B of the elements X, Y; let A be the subring $K[x]$ of B. Show that the elements x and y^n $(n \geqslant 0)$ form a basis of B over K and that B is not integral over A. On the other hand, there exists in B no element which is algebraically free over A. Deduce a counter-example to Corollary 1 to Theorem 1 of no. 1 when A is not an integral domain.

3. Let K be an infinite field, L an extension field of K, $(x_i)_{1 \leqslant i \leqslant n}$ a finite family of elements of L and d the transcendence degree of $K(x_1, \ldots, x_n)$ over K. Show that there exist d elements $z_j = \sum_{i=1}^{n} a_{ji}x_i$ of L (where the $a_{ji} \in K$) such that $K[x_1, \ldots, x_n]$ is integral over $K[z_1, \ldots, z_d]$. If we suppose further that $K(x_1, \ldots, x_n)$ is a separable extension of K, show that the a_{ji} may be chosen such that the z_j form a separating transcendence basis of $K(x_1, \ldots, x_n)$ over K (argue by induction on n when $n > d$).

4. Let B be an integral domain and A a subring of B. Suppose that there exist $n \geqslant 1$ elements $t_1, \ldots, t_n$ of B which are algebraically independent over A and such that B is integral over $A[t_1, \ldots, t_n]$. Let $\mathfrak{n}$ be a maximal ideal of B and $\mathfrak{p} = \mathfrak{n} \cap A$. Then there is a prime ideal $\mathfrak{q}$ of B contained in $\mathfrak{n}$, distinct from $\mathfrak{n}$ and containing $\mathfrak{p}$. (Reduce it first to the case where A is integrally closed by considering the integral closure A′ of A (contained in the field of fractions of B) and the ring B′ = B[A′]; in the case where A is integrally closed, use Theorem 3 of § 2, no. 4.)

¶ 5. Let A be an integral domain and B ⊃ A a subring of the field of fractions K of A; suppose that A is integrally closed in B and that B contains an element t such that B is a finitely generated $A[t]$-module.

(a) Let F be a polynomial $\neq 0$ of A[X] such that $F(t)$ belongs to the conductor $\mathfrak{f}$ of B with respect to $A[t]$; show that, if c is the dominant coefficient of F, c belongs to the radical of $\mathfrak{f}$. (By considering the rings $A[c^{-1}]$ and $B[c^{-1}]$, reduce it to showing that, if $c = 1$, necessarily $B = A[t]$. On the other hand we may assume that $F(t) \neq 0$, otherwise $t \in A$. For all $x \in B$, by hypothesis

$$xF(t) = F(t)Q(t) + R(t)$$

where Q and R are polynomials in A[X] and deg(R) < deg(F); attention may be confined to the case where $R(t) \neq 0$; if $y = R(t)(F(t))^{-1} = x - Q(t)$, show that t is integral over $A[y^{-1}]$ and deduce that y is integral over A.)

(b) Suppose further that A is Noetherian; let $\mathfrak{q}$ be a prime ideal of B belonging to $\mathrm{Ass}(B/\mathfrak{f})$. Show that, if ω is the canonical homomorphism from B to the field of fractions of $B/\mathfrak{q}$, $\omega(t)$ is transcendental over the field of fractions of $A/(\mathfrak{q} \cap A)$. (If $\mathfrak{p} = \mathfrak{q} \cap A[t]$, there is an element $y \in B$ such that $\mathfrak{q}$ is the set of $z \in B$ for which $zy \in \mathfrak{f}$; show that $\mathfrak{p}$ is the conductor of $B' = A[t] + By$ with respect to $A[t]$. Then argue by *reductio ad absurdum* using (a).)

(c) Suppose that A is Noetherian. Let $\mathfrak{n}$ be a maximal ideal of B and $\mathfrak{p} = A \cap \mathfrak{n}$. Show that, if $A_{\mathfrak{p}} \neq B_{\mathfrak{n}}$, there exists a prime ideal of B contained in $\mathfrak{n}$, distinct from $\mathfrak{n}$ and containing $\mathfrak{p}$. (Let $\mathfrak{f}$ be the conductor of B with respect to $A[t]$. Suppose first that $\mathfrak{n} \supset \mathfrak{f}$; consider a minimal element $\mathfrak{q}$ in the set of prime ideals of B contained in $\mathfrak{n}$ and containing $\mathfrak{f}$; consider the canonical image of A in $B/\mathfrak{q}$ and use (b) and Exercise 4. If $\mathfrak{n} \not\supset \mathfrak{f}$, the local ring $B_\mathfrak{n}$ is equal to the local ring $A[t]_{\mathfrak{n}_1}$, where $\mathfrak{n}_1 = \mathfrak{n} \cap A[t]$ (§ 1, Exercise 16). If $t \in A_\mathfrak{p}$, show that $A_\mathfrak{p} = B_\mathfrak{n}$. If on the contrary $A_\mathfrak{p} \neq B_\mathfrak{n}$, deduce from § 2, Exercise 14 (c) that $\mathfrak{p}A_\mathfrak{p}[t]$ is a non-maximal prime ideal of $A_\mathfrak{p}[t]$ and conclude that $\mathfrak{p}B_\mathfrak{n}$ is a prime ideal of $B_\mathfrak{n}$ distinct from $\mathfrak{n}B_\mathfrak{n}$.)

6. An integral domain A is called *integrally Noetherian* if it is Noetherian and if, for every finite sequence $(t_i)_{1 \leqslant i \leqslant n}$ of elements of an extension field of the field of fractions of A, the integral closure of $A[t_1, \dots, t_n]$ is a finitely generated $A[t_1, \dots, t_n]$-module. Every finitely generated integral algebra over a field is an integrally Noetherian domain (no. 2, Theorem 2).
(a) Every finitely generated integral algebra over an integrally Noetherian domain is an integrally Noetherian domain. Every ring of fractions of an integrally Noetherian domain not reduced to 0 is an integrally Noetherian domain.
(b) Let A be an integrally Noetherian domain $(t_i)_{1 \leqslant i \leqslant n}$ a finite sequence of elements of an extension field of the field of fractions K of A and B a subring of $K(t_1, \dots, t_n)$ containing A and integral over A. Then B is a finitely generated A-module. (Note that the field of fractions of B is a finite algebraic extension of K, using *Algebra*, Chapter V, § 5, no. 7.)

¶ 7. Let A be an integrally Noetherian domain (Exercise 6), $(t_i)_{1 \leqslant i \leqslant n}$ a finite sequence of elements of an extension field of the field of fractions of A, B the ring $A[t_1, \dots, t_n]$, $\mathfrak{n}$ a maximal ideal of B, $\mathfrak{p} = A \cap \mathfrak{n}$, A′ the integral closure of A in B and $\mathfrak{p}' = \mathfrak{n} \cap A'$. Then, if the local rings $A'_{\mathfrak{p}'}$ and $B_\mathfrak{n}$ are distinct, there exists a prime ideal of B contained in $\mathfrak{n}$, distinct from $\mathfrak{n}$ and containing $\mathfrak{p}$ ("*Zariski's Principal Theorem*"). (Using Exercise 6 (b), reduce it to the case where $A' = A$, then reduce it to the case where A is a local ring with maximal ideal $\mathfrak{p}$. Show that $B/\mathfrak{n}$ is then a finite algebraic extension of $A/\mathfrak{p}$ (cf. no. 4, Theorem 3); deduce that, for every ring C such that $A \subset C \subset B$, $\mathfrak{n} \cap C$ is a maximal ideal of C. Then argue by induction on n, denoting by B_i the integral closure of $A[t_1, \dots, t_i]$ in B; distinguish two cases according to whether t_{i+1} is transcendental or algebraic over the field of fractions of B_i; in the former case, use Exercise 4 and, in the latter, Exercise 5 (c).)

8. Let A be a ring. For A to be a Jacobson ring, it is necessary and sufficient that, for every maximal ideal $\mathfrak{m}'$ of the polynomial ring A[X], $\mathfrak{m}' \cap A$ is a maximal ideal of A. (Note that if an integral domain B admits a Jacobson radical $\mathfrak{R} \neq 0$ and if $b \in \mathfrak{R}$, $b \neq 0$, the intersection of B and a maximal ideal of B[X] containing $1 + bX$ cannot contain b.)

9. Give an example of a Jacobson domain A and a Jacobson ring B containing A and such that there exists a maximal ideal of B whose intersection with A is not a maximal ideal.

¶ 10. Let k be a field, L an infinite set and A the polynomial ring $k[X_\lambda]_{\lambda \in L}$. If k_0 is the prime field contained in k, let b denote the cardinal of a transcendence basis of k over k_0 and let us write $c = \mathrm{Card}(L)$.

(a) For every ideal $\mathfrak{a}$ of A, show that there exists a system of generators S of $\mathfrak{a}$ such that $\mathrm{Card}(S) \leqslant c$. (Consider the intersections of $\mathfrak{a}$ and the subrings $k[X_\lambda]_{\lambda \in J}$ for the finite subsets J of L.)

(b) Suppose that $c < b$. Show that, if $\mathfrak{m}$ is a maximal ideal of A, $A/\mathfrak{m}$ is an algebraic extension of k. (Let K be the subfield of k generated over k_0 by the coefficients of the polynomials forming a system of generators of $\mathfrak{m}$ of cardinal $\leqslant c$; let $\mathfrak{m}_0 = \mathfrak{m} \cap K[X_\lambda]_{\lambda \in L}$, so that $\mathfrak{m} = \mathfrak{m}_0 \otimes_K k$ and

$$A/\mathfrak{m} = (K[X_\lambda]_{\lambda \in L}/\mathfrak{m}_0) \otimes_K k.$$

Note finally that there exists a K-isomorphism of the field of fractions of $K[X_\lambda]_{\lambda \in L}/\mathfrak{m}_0$ onto an algebraic closure of k.) Deduce that A is then a Jacobson ring (use Exercise 8).

(c) Suppose that $b < c$; then $\mathrm{Card}(k) \leqslant c$, so that there exists a bijection $\lambda \mapsto \xi_\lambda$ of a subset L′ of L onto k, L′ having a non-empty complement in L. Let $\lambda_0 \in L - L'$; show that there exists a k-homomorphism u of A onto a subfield of the field of rational functions $k(Y)$ such that $u(X_{\lambda_0}) = Y$,

$$u(X_\lambda) = 1/(Y - \xi_\lambda)$$

for $\lambda \in L'$, $u(X_\lambda) = 0$ for $\lambda \in L - (L' \cup \{\lambda_0\})$; if $\mathfrak{m}$ is the kernel of u, $A/\mathfrak{m}$ is therefore not an algebraic extension of k.

(d) Under the same hypotheses as in (c), show that A is not a Jacobson ring. (Note that there exist k-algebras B which are not Jacobson rings (for example local rings) such that $\mathrm{Card}(B) = b$.)

CHAPTER VI

Valuations

Unless otherwise stated, all the rings considered in this chapter are assumed to be commutative and to possess a unit element. All the ring homomorphisms are assumed to map unit element to unit element. Every subring of a ring A *is assumed to contain the unit element of* A.

If A *is a local ring, its maximal ideal will be denoted by* $\mathfrak{m}(A)$, *its residue field* $A/\mathfrak{m}(A)$ *will be denoted by* $\kappa(A)$ *and the multiplicative group of invertible elements of* A *will be denoted by* $U(A)$; *then* $U(A) = A - \mathfrak{m}(A)$.

1. VALUATION RINGS

1. THE RELATION OF DOMINATION BETWEEN LOCAL RINGS

DEFINITION 1. *Let* A *and* B *be two local rings.* B *is said to dominate* A *if* A *is a subring of* B *and* $\mathfrak{m}(A) = A \cap \mathfrak{m}(B)$.

PROPOSITION 1. *Let* A *and* B *be local rings such that* A *is a subring of* B. *The following conditions are equivalent:*

(a) $\mathfrak{m}(A) \subset \mathfrak{m}(B)$;

(b) B *dominates* A;

(c) *the ideal* $B\mathfrak{m}(A)$ *generated by* $\mathfrak{m}(A)$ *in* B *does not contain* 1.

If $\mathfrak{m}(A) \subset \mathfrak{m}(B)$, $\mathfrak{m}(B) \cap A$ is an ideal of A which does not contain 1 and which contains the maximal ideal $\mathfrak{m}(A)$; it is therefore equal to it and (a) implies (b). If B dominates A, the ideal $B\mathfrak{m}(A)$ is contained in $\mathfrak{m}(B)$ and hence does not contain 1; thus (b) implies (c). If (c) holds, $B\mathfrak{m}(A)$ is contained in the unique maximal ideal $\mathfrak{m}(B)$ of B, whence (a).

Note that, if K is a ring, the relation "B dominates A" is an *order relation* on the set of local subrings of K.

Let A and B be two local rings such that B dominates A. The canonical injection $A \to B$ defines by taking quotients an isomorphism of the field $\kappa(A)$ onto a subfield of $\kappa(B)$; this isomorphism allows us to identify $\kappa(A)$ with a subfield of $\kappa(B)$.

Examples

(1) Let A be a Noetherian local ring and $\hat{A}$ its completion; the local ring $\hat{A}$ then dominates A (Chapter III, § 3, no. 5, Proposition 9).
(2) Let B be an integral domain, A a subring of B, $\mathfrak{p}'$ a prime ideal of B and $\mathfrak{p} = A \cap \mathfrak{p}'$. Then $\mathfrak{p}A_{\mathfrak{p}} \subset \mathfrak{p}'B_{\mathfrak{p}'}$, so that $B_{\mathfrak{p}'}$ dominates $A_{\mathfrak{p}}$.

2. VALUATION RINGS

THEOREM 1. *Let* K *be a field and* V *a subring of* K. *The following conditions are equivalent:*

(a) V *is a maximal element of the set of local subrings of* K, *this set being ordered by the relation* "B *dominates* A" *between* A *and* B.

(b) *There exists an algebraically closed field* L *and a homomorphism* h *from* V *to* L *which is maximal in the set of homomorphisms from subrings of* K *to* L, *ordered by the relation* "g *is an extension of* f" *between* f *and* g.

(c) *If* $x \in K - V$, *then* $x^{-1} \in V$.

(d) *The field of fractions of* V *is* K *and the set of principal ideals of* V *is totally ordered by the relation of inclusion.*

(e) *The field of fractions of* V *is* K *and the set of ideals of* V *is totally ordered by the relation of inclusion.*

We shall show the theorem by proving the following implications

$$(a) \Rightarrow (b) \Rightarrow (c) \Rightarrow (d) \Rightarrow (e) \Rightarrow (a).$$

Suppose that (a) holds. Then V is a local ring. Let L be an algebraic closure of the residue field $\kappa(V)$ and h the canonical homomorphism from V to L. Let V' be a subring of K containing V and h' a homomorphism from V' to L extending h. If $\mathfrak{p}'$ is the kernel of h', then $\mathfrak{p}' \cap V = \mathfrak{m}(V)$; hence (no. 1, *Example* 2) $V'_{\mathfrak{p}'}$ dominates V, which implies $V'_{\mathfrak{p}'} = V$ and $V' = V$. Thus (b) holds.

Suppose (b) holds. Let L be an algebraically closed field and h a homomorphism from V to L; suppose that h is maximal in the set of homomorphisms from subrings of K to L; let $\mathfrak{p}$ be the kernel of h. The elements of $h(V - \mathfrak{p})$ being invertible in L, h can be extended to a homomorphism from $V_{\mathfrak{p}}$ to L (Chapter II, § 2, no. 1, Proposition 1); hence $V = V_{\mathfrak{p}}$, which shows that V is a local ring and that $\mathfrak{p}$ is its maximal ideal. Let x be a non-zero element of K; we must show that one at least of the elements x, x^{-1} belongs to V, that is, by virtue of the maximal character of h, that h can be extended to $V[x]$ or $V[x^{-1}]$. If x is integral over V, this follows from Chapter V, § 2, no. 1, Corollary 4 to Theorem 1. If x is not integral over V, we shall use the following lemma:

Lemma 1. *Let* A *be a subring of a ring* B *and* x *an element of* B *not integral over* A; *then the ring of fractions* B_x (Chapter II, § 5, no. 1) *is not reduced to* 0 *and there exists in the subring* $A[1/x]$ *of* B_x *maximal ideals containing* $1/x$; *moreover, if* $\mathfrak{M}$ *is any of these maximal ideals, the inverse image of* $\mathfrak{M}$ *in* A *is a maximal ideal.*

As x is not integral over A, it is not nilpotent and therefore $B_x \neq 0$; moreover, $x/1 \notin A[1/x]$, otherwise there would be a relation of the form

$$x/1 = a_0/1 + a_1/x + \cdots + a_n/x^n$$

for some $n \geqslant 0$ (where $a_i \in A$ for $0 \leqslant i \leqslant n$), which is equivalent to

$$x^{n+h} - a_0x^{n+h-1} - a_1x^{n+h-2} - \cdots - a_nx^{h-1} = 0$$

for some convenient $h \geqslant 1$; but such a relation would imply that x is integral over A, contrary to the hypothesis. The existence of a maximal ideal of $A[1/x]$ containing $1/x$ therefore follows from the fact that $1/x$ is not invertible in $A[1/x]$ (*Algebra*, Chapter I, § 8, no. 7, Theorem 2).

Then let $\mathfrak{M}$ be a maximal ideal of $A[1/x]$ containing $1/x$; let $\phi: A \to A[1/x]$, $p: A[1/x] \to A[1/x]/\mathfrak{M}$ be the canonical homomorphisms; then

$$p(A[1/x]) = p(\phi(A))[p(1/x)] = p(\phi(A))$$

since $p(1/x) = 0$; this proves that $p(\phi(A))$ is a field and hence the inverse image $\overset{-1}{\phi}(\mathfrak{M})$ is a maximal ideal of A.

We apply this lemma with A = V and B = K; then there is a maximal ideal $\mathfrak{M}$ of $V[x^{-1}]$ containing x^{-1} and $\mathfrak{M} \cap V$ is a maximal ideal of V; then $\mathfrak{M} \cap V = \mathfrak{p}$ since V is local; denoting by f the canonical homomorphism of $V[x^{-1}]$ onto $V[x^{-1}]/\mathfrak{M}$, $f(x^{-1}) = 0$, whence $V/\mathfrak{p} = f(V) = f(V[x^{-1}])$; as h defines by taking the quotient an injective homomorphism $\bar{h}$ from $V/\mathfrak{p}$ to L, $\bar{h} \circ f$ is a homomorphism from $V[x^{-1}]$ to L extending h. Hence (c) holds.

Suppose now that (c) holds. Clearly K is the field of fractions of V. Let a and b be elements of V such that $Va \not\subset Vb$; we show that $Vb \subset Va$. It is true if $b = 0$; otherwise the relation $a \notin Vb$ implies $b^{-1}a \notin V$, whence by (c) $a^{-1}b \in V$ and therefore $Vb \subset Va$. Hence (d) holds.

Suppose that (d) holds. Let $\mathfrak{a}$ and $\mathfrak{b}$ be ideals of V such that $\mathfrak{a} \not\subset \mathfrak{b}$. There exists $a \in \mathfrak{a}$ such that $a \notin \mathfrak{b}$. For all $b \in \mathfrak{b}$, $a \notin Vb$, whence $Va \not\subset Vb$ and therefore $Vb \subset Va \subset \mathfrak{a}$ (by (c)) and $b \in \mathfrak{a}$. Then $\mathfrak{b} \subset \mathfrak{a}$, which shows that condition (e) is satisfied.

Suppose finally that (e) holds. As V has a maximal ideal, it has only one and is therefore a local ring. Let V′ be a local subring of K dominating V and let x be a non-zero element of V′; we write $x = ab^{-1}$ where $a \in V$, $b \in V$. One of the ideals Va, Vb is contained in the other. If $Va \subset Vb$, then $x \in V$. If $Vb \subset Va$, then $x^{-1} \in V$; as the ideal $V'\mathfrak{m}(V)$ does not contain 1 (no. 1, Proposition 1), $x^{-1} \notin \mathfrak{m}(V)$, whence again $x \in V$ since V is local. Every element of V′ therefore belongs to V; we conclude that (a) holds.

Definition 2. *In the notation of Theorem* 1, V *is called a valuation ring for the field* K *if the equivalent conditions* (a), (b), (c), (d), (e) *hold. A ring is called a valuation ring if it is an integral domain and it is a valuation ring for its field of fractions.*

Theorem 2. *Let* K *be a field and h a homomorphism from a subring* A *of* K *to an algebraically closed field* L. *Then there exist a valuation ring* V *for* K *and a homomorphism h′ from* V *to* L *such that* V *contains* A, *h′ extends h and* $\overset{-1}{h'}(0) = \mathfrak{m}(V)$.

Let $\mathfrak{H}$ be the set of homomorphisms of subrings of K to L, ordered by the relation of extension. This set is inductive; for if $(h_\alpha)_{\alpha \in I}$ is a non-empty totally ordered family of elements of $\mathfrak{H}$ and B_α is the defining ring of h_α, the B_α form a totally ordered family of subrings of K and their union B is therefore a subring of K; there therefore exists a unique mapping $\bar{h}$ from B to L which extends the h_α (*Set Theory*, Chapter II, § 4, no. 6, Proposition 7) and it is immediately seen that $\bar{h}$ is a homomorphism from B to L. Zorn's Lemma then shows that there exists a maximal element h' of $\mathfrak{H}$ which extends h. The defining ring V of h' is a valuation ring of K (Theorem 1); if $\mathfrak{p}$ is the kernel of h', h' can be extended to a homomorphism from $V_\mathfrak{p}$ to L (Chapter II, § 2, no. 1, Proposition 1), whence $V_\mathfrak{p} = V$ and $\mathfrak{p} = \mathfrak{m}(V)$.

Corollary. *Every local subring* A *of a field* K *is dominated by at least one valuation ring of* K.

Apply Theorem 2 to the canonical homomorphism h from A to an algebraic closure L of $A/\mathfrak{m}(A)$.

3. CHARACTERIZATION OF INTEGRAL ELEMENTS

Theorem 3. *Let* A *be a subring of a field* K. *The integral closure* A′ *of* A *in* K *is the intersection of the valuation rings of* K *which contain* A; *if* A *is local,* A′ *is the intersection of the valuation rings of* K *which dominate* A.

Let x be an element of A′ and V a valuation ring of K containing A; as x is integral over V, there exists a prime ideal $\mathfrak{p}'$ of $V[x]$ such that $\mathfrak{p}' \cap V = \mathfrak{m}(V)$ (Chapter V, § 2, no. 1, Theorem 1); clearly then the local ring $(V[x])_{\mathfrak{p}'}$ dominates V and hence is equal to it; whence $x \in V$. Conversely let y be an element of K which is not integral over A; then there exists a maximal ideal $\mathfrak{M}$ of $A[y^{-1}]$ which contains y^{-1} (no. 2, Lemma 1); there exists also a valuation ring V of K which dominates $(A[y^{-1}])_\mathfrak{M}$ (no. 2, Corollary to Theorem 2); as $y^{-1} \in \mathfrak{m}(V)$, $y \notin V$. Further $\mathfrak{M} \cap A$ is a maximal ideal of A (no. 2, Lemma 1); hence, if A is local, $\mathfrak{M} \cap A = \mathfrak{m}(A)$ and V dominates A.

COROLLARY 1. *Every valuation ring is integrally closed.*

COROLLARY 2. *For an integral domain to be integrally closed, it is necessary and sufficient that it be the intersection of a family of valuation rings of its field of fractions.*

In the case of a Noetherian ring, Corollary 2 can be made more precise (Chapter VII, § 1, no. 3, Corollary to Theorem 1).

COROLLARY 3. *Let* K *be a field,* K' *an extension of* K *and* A *a valuation ring of* K. *The integral closure of* A *in* K' *is the intersection of the valuation rings* V' *of* K' *such that* $V' \cap K = A$.

Theorem 1 (c) show that, if V' is a valuation ring of K', $V' \cap K$ is a valuation ring of K and V' dominates $V' \cap K$. For V' to dominate A, it is necessary and sufficient that $V' \cap K$ dominate A and therefore be equal to it.

4. EXAMPLES OF VALUATION RINGS

(1) Every field is a valuation ring.

(2) If V' is a valuation ring of a field K' and K is a subfield of K', $V' \cap K$ is, by no. 2, Theorem 1 (c), a valuation ring of K.

(3) The following proposition provides numerous examples of valuation rings:

PROPOSITION 2. *Let* A *be a local ring whose maximal ideal is a principal ideal* Ap. *If* $\bigcap_{n=1}^{\infty} Ap^n = (0)$ (*for example if* A *is Noetherian,* cf. Chapter III, § 3, no. 2, Corollary to Proposition 5), *the only ideals of* A *are* (0) *and the* Ap^n; *then either* p *is nilpotent or* A *is a valuation ring.*

Let A be filtered by the Ap^n and let v denote the corresponding order function (Chapter III, § 2, no. 2). As

$$\bigcap_{n=1}^{\infty} Ap^n = (0),$$

the relation $v(x) = +\infty$ implies $x = 0$. Let $\mathfrak{a}$ be an ideal $\neq (0)$ of A and a an element of $\mathfrak{a}$ at which v takes its least value; let us write $v(a) = s$ ($s \neq +\infty$). Then $\mathfrak{a} \subset Ap^s$. In particular, there exists $u \in A$ such that $a = up^s$; as $a \notin Ap^{s+1}$, $u \notin Ap$; hence u is invertible and $p^s \in Aa \subset \mathfrak{a}$. It follows that $\mathfrak{a} = Ap^s$, whence our first assertion. It is also seen that every element $a \neq 0$ of A may be written in the form $a = up^{v(a)}$ where u is invertible. If $a' = u'p^{v(a')}$ (u' invertible) is another non-zero element of A, then $aa' = uu'p^{v(a)+v(a')}$; hence, if p is not nilpotent, $aa' \neq 0$ and A is an integral domain. Then, as the set of ideals of A is totally ordered by inclusion, we conclude that A is a valuation ring (Theorem 1 (e)).

For example, if p is a prime number, the local ring $\mathbf{Z}_{(p)}$ is a valuation ring. Let $B = K[X_1, \ldots, X_n]$ be the polynomial ring in n indeterminates over a field K; the ideal BX_1 is prime, since B/BX_1 is isomorphic to $K[X_2, \ldots, X_n]$; hence B_{BX_1} is a valuation ring; it is composed of rational functions PQ^{-1}, where P and Q are polynomials and $Q(0, X_2, \ldots, X_n) \neq 0$.

* More generally, we shall see that, if F is an extremal element of

$$B = K[X_1, \ldots, X_n],$$

B_{BF} is a valuation ring (cf. Chapter VII, § 3, no. 5). *

The ring of formal power series $K[[T]]$ in one indeterminate over a field K is a Noetherian local integral domain whose maximal ideal is principal; it is therefore a valuation ring. On the other hand the ring $K[[T_1, T_2]]$ of formal power series in two indeterminates, which is a Noetherian local integral domain is not a valuation ring, for neither of the elements T_1, T_2 is a multiple of the other.

PROPOSITION 3. *Let* A *be a principal ideal domain and* K *its field of fractions. The valuation rings of* K *containing* A *and distinct from* K *are the rings of the form* A_{Ap}, *where* p *is an extremal element of* A.

Clearly A_{Ap} (p extremal) is a valuation ring containing A and distinct from K (Proposition 2). Conversely, let V be a valuation ring distinct from K and containing A. As $V \neq K$, $\mathfrak{m}(V)$ contains an element $x \neq 0$; writing $x = a/b$ where $a \in A$ and $b \in A$, it is seen that $A \cap \mathfrak{m}(V)$ contains the non-zero element a. As $A \cap \mathfrak{m}(V)$ is prime, it is of the form Ap where p is extremal in A. Then $A_{Ap} \subset V$, $pA_{Ap} \subset \mathfrak{m}(V)$, so that V dominates A_{Ap}; as A_{Ap} is a valuation ring of K (Proposition 2), $V = A_{Ap}$.

COROLLARY 1. *Every valuation ring of the field* $\mathbf{Q}$ *and distinct from* $\mathbf{Q}$ *is of the form* $\mathbf{Z}_{(p)}$ *where* p *is a prime number.*

Every subring of $\mathbf{Q}$ contains $\mathbf{Z}$.

COROLLARY 2. *Let* K *be a field,* K(X) *the field of rational functions in one indeterminate over* K *and* V *a valuation ring of* K(X) *containing* K *and distinct from* K(X). *If* $X \in V$, *there exists an irreducible polynomial* $P \in K[X]$ *such that* $V = (K[X])_{(P)}$; *otherwise* V *is the local ring of* $K[X^{-1}]$ *at the prime ideal* $X^{-1}K[X^{-1}]$ *(in other words* V *is composed of the fractions* A/B, *where* $A \in K[X]$ *and* $B \in K[X]$, *such that* $\deg(A) \leqslant \deg(B)$).

If $X \in V$, then $K[X] \subset V$ and the assertion made follows from Proposition 3. If $X \notin V$, then $X^{-1} \in V$, whence $K[X^{-1}] \subset V$; then V is the local ring of $K[X^{-1}]$ at a prime ideal $\mathfrak{p}$ (Proposition 3) and this ideal contains X^{-1} since X^{-1} is not invertible in V; then $\mathfrak{p} = X^{-1}K[X^{-1}]$ since the latter ideal is maximal. Finally let us consider a rational function $A(X)/B(X)$, where A and B are polynomials of respective degrees a and b; then $A(X) = X^a A'(X^{-1})$ and $B(X) = X^b B'(X^{-1})$, where A' and B' are polynomials such that $A'(0) \neq 0$ and $B'(0) \neq 0$; hence, for $A(X)/B(X)$ to belong to the local ring of $K[X^{-1}]$ at $X^{-1}K[X^{-1}]$, it is necessary and sufficient that $a \leqslant b$.

2. PLACES

1. THE NOTION OF MORPHISM FOR LAWS OF COMPOSITION NOT EVERYWHERE DEFINED

DEFINITION 1. *Let* E *and* E' *be two sets each with an internal law of composition denoted by* $(x, y) \mapsto x * y$, *not necessarily everywhere defined. A mapping* $f: E \to E'$ *is a morphism if, for all* x, y *in* E *such that* $f(x) * f(y)$ *is defined, the composition* $x * y$ *is also defined and:*

$$f(x * y) = f(x) * f(y). \tag{1}$$

More briefly, we may say that formula (1) must hold every time the *right* hand side has a meaning.

The notion of morphism is distinct from that of *representation* (*Algebra*, Chapter I, § 1, no. 1), where it is demanded that equation (1) hold whenever the *left* hand side has a meaning. Of course, the two notions coincide for laws of composition everywhere defined.

DEFINITION 2. *Let* E *and* E' *be two sets each with a family of internal laws of composition* $(x, y) \mapsto x *_\alpha y$, $\alpha \in I$. *A mapping* $f: E \to E'$ *is a morphism if it is a morphism for each of the laws of composition* $(x, y) \mapsto x *_\alpha y$.

Just as representations, so morphisms satisfy axioms (MO_I), (MO_{II}), (MO_{III}) of *Set Theory*, Chapter IV, § 2. If $f: E \to E'$ is a morphism, $f(E)$ is a stable subset of R'.

2. PLACES

If K is a field, recall that $\tilde{K}$ denotes the set the sum of K and an element denoted by ∞ (*Algebra*, Chapter II, § 9, no. 9); the laws of composition of K extend to $\tilde{K}$ by setting (*loc. cit.*)

$$a + \infty = \infty \quad \text{for } a \in K, \quad a \neq \infty, \tag{2}$$

$$\infty . a = a . \infty = \infty \quad \text{for } a \in \tilde{K}, \quad a \neq 0. \tag{3}$$

The only compositions not defined are therefore the compositions $\infty + \infty$, $\infty . 0$ and $0 . \infty$. On the other hand, the mappings $x \mapsto -x$ and $x \mapsto x^{-1}$ extend similarly to $\check{K}$ by setting $-\infty = \infty$, $0^{-1} = \infty$, $\infty^{-1} = 0$. We shall also write $x + (-y) = x - y$.

The set $\check{K}$, called the *projective field* associated with K, can be identified with the *projective line* $\mathbf{P}_1(K)$ (*loc. cit.*).

DEFINITION 3. *Let* K *and* L *be two fields. Every morphism f of* $\check{K}$ *to* $\check{L}$ (*for addition and multiplication*) *such that* $f(1) = 1$ *is called a place of* K *with values in* L.

In other words, if x and y are elements of $\check{K}$ and $f(x) + f(y)$ (resp. $f(x)f(y)$) is defined, then $x + y$ (resp. xy) is defined and

$$f(x + y) = f(x) + f(y) \tag{4}$$

$$f(xy) = f(x)f(y). \tag{5}$$

As $\infty + \infty$ is not defined, neither is $f(\infty) + f(\infty)$, which shows that

$$f(\infty) = \infty. \tag{6}$$

Similarly, since $0.\infty$ is not defined, neither is $f(0)f(\infty)$, which, by virtue of (6), implies

$$f(0) = 0. \tag{7}$$

On the other hand

$$f(a^{-1}) = f(a)^{-1} \qquad \text{for all} \quad a \in \check{K}. \tag{8}$$

If $f(a)f(a^{-1})$ is defined, aa^{-1} is defined and hence is equal to 1; then $f(a)f(a^{-1}) = f(1) = 1$, which proves (8) in this case. If $f(a)f(a)^{-1}$ is not defined, then, either $f(a) = 0$ and $f(a^{-1}) = \infty$ or $f(a) = \infty$ and $f(a^{-1}) = 0$ and (8) still holds.

Similarly it can be shown that

$$f(-a) = -f(a) \qquad \text{for all} \quad a \in \check{K}. \tag{9}$$

From formulae (8) and (9) it follows that f is also a morphism for the laws of composition $(x, y) \mapsto x - y$ and $(x, y) \mapsto xy^{-1}$.

For $x \in \check{K}$, f is called finite at x if $f(x) \neq \infty$; this implies $x \in K$ by (6).

If $f: \check{K} \to \check{L}$ is a place, $f(\check{K})$ is a subset of $\check{L}$ which is stable for the laws of composition $(x, y) \mapsto x + y$, $(x, y) \mapsto x - y$, $(x, y) \mapsto xy$ and $(x, y) \mapsto xy^{-1}$ and which contains 1. If E is the set of finite elements of $f(\check{K})$, E is a subfield of L and $f(\check{K}) = \check{E}$. By an abuse of language E is called the *value field* of f.

The *composite* mapping of two places is a place.

Let F be an isomorphism of a field K onto a subfield of a field L; let us extend f to $\check{K}$ by setting $f(\infty) = \infty$. Thus we obtain a place of K with values in L which is called *trivial* and which is often identified with the isomorphism f.

3. PLACES AND VALUATION RINGS

PROPOSITION 1. *Let* K *be a field,* A *a valuation ring of* K *and* $\kappa(A)$ *the residue field of* A. *We extend the canonical mapping of* A *onto* $\kappa(A)$ *to a mapping* $h_A : \tilde{K} \to (\kappa(A))^{\sim}$ *by the equation* $h_A(x) = \infty$ *if* $x \notin A$. *The mapping* h_A *thus defined is a place of* K *whose value field is* $\kappa(A)$.

Clearly $h_A(1) = 1$.

We show that h_A is a morphism for addition. Let x, y be two elements of $\tilde{K}$ such that $h_A(x) + h_A(y)$ is defined. One of the two elements x, y belongs then to A and hence $x + y$ is defined. If $x \in A$ and $y \in A$, clearly

$$h_A(x) + h_A(y) = h_A(x + y)$$

holds. If $x \in A$ and $y \notin A$, then $x + y \notin A$ and the two sides of the above formula equal ∞.

We show finally that h_A is a morphism for multiplication. Let $x \in \tilde{K}$, $y \in \tilde{K}$ be such that $h_A(x)h_A(y)$ is defined. If $x \in A$ and $y \in A$, clearly xy is defined and $h_A(x)h_A(y) = h_A(xy)$. Suppose now that one of the elements x, y, for example y, does not belong to A; as $h_A(y) = \infty$, $h_A(x) \neq 0$, that is $x \notin \mathfrak{m}(A)$, whence $x^{-1} \in A$; it follows that xy is defined and that $xy \notin A$, whence

$$h_A(xy) = \infty = h_A(x)h_A(y).$$

This proves Proposition 1.

If j is an isomorphism of $\kappa(A)$ onto a subfield of a field L, $j \circ h_A : \tilde{K} \to \tilde{L}$ is a place of K with values in L. The above process in fact provides *all* the places on K. To be precise:

PROPOSITION 2. *Let* K *and* L *be two fields and* f *a place of* K *with values in* L. *Then there exists a valuation ring* A *of* K *and an isomorphism* j *of* $\kappa(A)$ *onto a subfield of* L *such that* $f = j \circ h_A$; *these conditions determine* A *and* j *uniquely. The ring* A *is the set of* $x \in K$ *such that* $f(x) \neq \infty$ *and* $\mathfrak{m}(A)$ *is the set of* $x \in K$ *such that* $f(x) = 0$.

If $f = j \circ h_A$, the condition $f(x) \neq \infty$ (resp. $f(x) = 0$) is equivalent to the condition $h_A(x) \neq \infty$ (resp. $h_A(x) = 0$) and hence to the condition $x \in A$ (resp. $x \in \mathfrak{m}(A)$). Hence A is determined uniquely and, as h_A is surjective, j is also unique.

Now let f be any place of K with values in L; let A denote the set of $x \in K$ such that $f(x) \neq \infty$. If $x \in A$ and $y \in A$, the compositions $f(x) - f(y)$ and $f(x)f(y)$ are defined and $\neq \infty$, which shows that $x - y \in A$ and $xy \in A$; hence A is a subring of K. If $x \notin A$, then $f(x) = \infty$, hence $f(x^{-1}) = 0$ and x^{-1} belongs to the kernel $\mathfrak{m}$ of the homomorphism f' obtained by restricting f to A. Conversely if $y \in \mathfrak{m}$, then $y^{-1} \notin A$. This shows that A is a valuation ring of K and that $\mathfrak{m}$ is its maximal ideal. Let j be the injective homomorphism from $\kappa(A)$ to L derived from

f' by passing to the quotient. Then $f(x) = j(h_A(x))$ for all $x \in A$ and this equation remains true for $x \notin A$, the two sides being then equal to ∞.

The decomposition $f = j \circ h_A$ is called the *canonical decomposition* of the place f. A is called the *ring* of f, $\mathfrak{m}(A)$ the *ideal* of f and $\kappa(A)$ the *residue field* of f. For two places $f: \tilde{K} \to \tilde{L}$ and $f': \tilde{K} \to \tilde{L}'$ to have the same ring, it is necessary and sufficient that there exist an isomorphism s of the value field of f onto that of f' such that $f' = s \circ f$; then f and f' are called *equivalent*. It is seen that every result on valuation rings can be translated into a result on places and conversely; this is what we shall do in the following nos.

Examples of places

(1) Let K be a field. The identity mapping on K is a trivial place with ring K and ideal (0).

(2) Let k be a field. For all $u \in k((T))^{\sim}$, let us write $f(u) = \infty$ if $u \notin k[[T]]$ and define $f(u)$ to be the constant term of u if $u \in k[[T]]$. Then f is a place of $k((T))$, with residue field k and ring $k[[T]]$. For $k[[T]]$ is a valuation ring of $k((T))$ (§ 1, no. 4, *Example* 3) and the restriction of f to $k[[T]]$ is identified with the canonical homomorphism of $k[[T]]$ onto its residue field.

(3) Let k be a field, a an element of k and A the set of $u \in k(X)$ such that a is substitutable in u (*Algebra*, Chapter IV, § 3, no. 2). If $\mathfrak{p}$ denotes the prime ideal $(X - a)$ of $k[X]$, then $A = k[X]_{\mathfrak{p}}$, so that A is a valuation ring of $k(X)$ (§ 1, no. 4, Proposition 2). For all $u \in k(X)^{\sim}$, let us write $f(u) = \infty$ if $u \notin A$ and $f(u) = u(a)$ if $u \in A$. Then f is a place of $k(X)$ with residue field k and ring A; for the restriction of f to A is a homomorphism of A onto k (*Algebra*, Chapter IV, § 3, Proposition 2) of kernel $\mathfrak{p}A = \mathfrak{m}(A)$. The element $f(u) \in \tilde{k}$ is said to be obtained by putting $X = a$ in u.

* (4) Let S be a connected complex analytic variety of dimension 1 and K the field of meromorphic functions on S. For all $z_0 \in S$, the mapping $f \mapsto f(z_0)$ from K to $\tilde{\mathbf{C}}$ is a place of K whose ring is the set of $f \in K$ which are holomorphic at z_0 and whose ideal is the set of $f \in K$ which are zero at z_0. It is this example and other analogues which are the origin of the term "place".*

4. EXTENSION OF PLACES

PROPOSITION 3. *Let* K *be a field,* S *a subring of* K *and f a homomorphism from* S *to an algebraically closed field* L. *Then there exists a place of* K *with values in* L *which extends f.*

Taking account of Proposition 1, this is a translation of Theorem 2 of § 1, no. 2.

PROPOSITION 4. *Let* K *be a field, f a place of* K *with values in a field* L *and* K′ *an extension of* K. *Then there exists an extension* L′ *of* L *and a place f' of* K′ *with values in* L′ *which extends f. If $x_1, \ldots, x_n$ are elements of* K′ *which are algebraically independent* *over* K *and* $a_1, \ldots, a_n$ *any elements of* L, f' *may be chosen such that* $f(x_i) = a_i$ *for* $1 \leqslant i \leqslant n$.

Let V be the ring of f, g the restriction of f to V and g' the extension of g to $V[x_1, \ldots, x_n]$ such that $g'(x_i) = a_i$ for $1 \leqslant i \leqslant n$. It is sufficient to take L′ to be an algebraic closure of L and apply Proposition 3 to g' and L′: we obtain a place $f' : \tilde{K}' \to \tilde{L}'$ which extends g'; if $x \in \tilde{K} - V$, then $x^{-1} \in \mathfrak{m}(V)$, whence $f(x^{-1}) = g(x^{-1}) = 0$ and $f'(x) = \infty = f(x)$; hence f' extends f.

5. CHARACTERIZATION OF INTEGRAL ELEMENTS BY MEANS OF PLACES

PROPOSITION 5. *Let* K *be a field,* S *a subring of* K, *h a homomorphism from* S *to a field and* $\mathfrak{p}$ *the kernel of h. For an element x of* K *to be integral over the local ring* $S_\mathfrak{p}$, *it is necessary and sufficient that every place of* K *extending h be finite at x.*

If f is a place of K extending h, f is finite on $S_\mathfrak{p}$ and zero on $\mathfrak{p}S_\mathfrak{p}$ and hence the ring of the place f dominates $S_\mathfrak{p}$. Conversely, if V is a valuation ring of K which dominates $S_\mathfrak{p}$, V is the ring of a place f whose restriction to S is a homomorphism with the same kernel as h; replacing f by an equivalent place, it is seen that V is the ring of a place of K which extends h. To say that every place of K extending h is finite at x is equivalent to saying that x belongs to all the valuation rings of K which dominate $S_\mathfrak{p}$. The proposition then follows from Theorem 3 of § 1, no. 3.

PROPOSITION 6. *Let* K *be a field and* S *a subring of* K. *For an element* $x \in K$ *to be integral over* S, *it is necessary and sufficient that every place of* K *which is finite on* S *be finite at x.*

This is also a consequence of Theorem 3 of § 1, no. 3.

3. VALUATIONS

1. VALUATIONS ON A RING

Let Γ be a *totally ordered* commutative group written additively. In the rest of this chapter, we shall have to consider, for such a group, the set obtained by adjoining to Γ an element denoted by $+\infty$; we shall denote this set by Γ_∞ and we shall give it: (1) a total ordering for which $+\infty$ is the *greatest element*, in other words, such that $\alpha < +\infty$ for all $\alpha \in \Gamma$; (2) a commutative monoid structure whose law induces on Γ the given group law and is defined by the equations

$$(+\infty) + (+\infty) = +\infty, \qquad \alpha + (+\infty) = +\infty$$

for all $\alpha \in \Gamma$; it is immediately verified that this law is associative and commutative and that the relation $\alpha \leqslant \beta$ in Γ_∞ implies $\alpha + \gamma \leqslant \beta + \gamma$ for all $\gamma \in \Gamma_\infty$.

DEFINITION 1. *Let* C *be a (not necessarily commutative) ring and* Γ *a totally ordered commutative group written additively. A valuation on* C *with values in* Γ *is any mapping* $v: \mathrm{C} \to \Gamma_\infty$ *which satisfies the following conditions:*

($\mathrm{VL_I}$) $v(xy) = v(x) + v(y)$ *for* $x \in \mathrm{C}, y \in \mathrm{C}$.

($\mathrm{VL_{II}}$) $v(x + y) \geqslant \inf(v(x), v(y))$ *for* $x \in \mathrm{C}, y \in \mathrm{C}$.

($\mathrm{VL_{III}}$) $v(1) = 0$ *and* $v(0) = +\infty$.

If C has no divisor of zero other than 0, the unique mapping v_0 of C to Γ_∞ such that $v_0(x) = 0$ for $x \neq 0$ and $v_0(0) = +\infty$ is a valuation, called the *improper valuation* on C. If $z \in \mathrm{C}$ is such that $z^n = 1$ for some integer $n \geqslant 1$, then, by ($\mathrm{VL_I}$), $nv(z) = v(z^n) = 0$ and hence $v(z) = 0$ for *every* valuation v on C, since Γ is a totally ordered group. In particular $v(-1) = 0$, whence $v(-x) = v(x)$ for all $x \in \mathrm{C}$. Moreover, it follows from ($\mathrm{VL_I}$) that $v(xy) = v(yx)$ for all x, y in C. If x is invertible in C, then $v(x^{-1}) = -v(x)$.

PROPOSITION 1. *Let* v *be a valuation on a (not necessarily commutative) ring* C. *For any elements* $x_i \in \mathrm{C}$ $(1 \leqslant i \leqslant n)$,

$$(1) \qquad v\left(\sum_{i=1}^{n} x_i\right) \geqslant \inf_{1 \leqslant i \leqslant n} v(x_i)$$

Moreover, if there exists a single index k such that $v(x_k) = \inf_{1 \leqslant i \leqslant n} v(x_i)$, *the two sides of* (1) *are equal. In particular, if* $v(x) \neq v(y)$, *then* $v(x + y) = \inf(v(x), v(y))$.

Relation (1) follows from axiom ($\mathrm{VL_{II}}$) by induction on n. If there exists a single index k such that $v(x_k) = \inf_{1 \leqslant i \leqslant n} v(x_i)$, then, writing $y = \sum_{i \neq k} x_i$ and $z = \sum_{i=1}^{n} x_i$, $v(y) > v(x_k)$ and $v(z) \geqslant v(x_k)$ by (1); if $v(z) > v(x_k)$, the relation $x_k = z - y$ would give $v(x_k) \geqslant \inf(v(z), v(y)) > v(x_k)$, which is absurd; whence $v(z) = v(x_k)$, which proves the second assertion.

COROLLARY. *If a finite sequence of elements* $(x_i)_{1 \leqslant i \leqslant n}$ *of* C *(for* $n \geqslant 2$*) is such that* $\sum_{i=1}^{n} x_i = 0$, *there exist at least two distinct indices j, k such that*

$$v(x_j) = v(x_k) = \inf_{1 \leqslant i \leqslant n} v(x_i).$$

If there were only a single index k such that $v(x_k) = \inf_{1 \leqslant i \leqslant n} v(x_i)$, Proposition 1 would show that $v(x_k) = v(0) = +\infty$, whence $v(x_i) = +\infty$ for all i, contrary to the relation $n \geqslant 2$ and the hypothesis made on k.

Remarks

(1) If $v: C \to \Gamma_\infty$ is a valuation on C and $u: B \to C$ a homomorphism of a ring B to C, it is immediate that the composite mapping $B \xrightarrow{u} C \xrightarrow{v} \Gamma_\infty$ is a *valuation on* B with values in Γ.

(2) Conditions (VL_I) and (VL_{II}) show immediately that the set $\overset{-1}{v}(+\infty)$ is a *two-sided ideal* $\mathfrak{p}$ in C distinct from C by virtue of (VL_{III}); moreover, if x, y are two elements of C such that $v(xy) = +\infty$, it follows from (VL_I) that necessarily $v(x) = +\infty$ or $v(y) = +\infty$; in other words, the quotient ring $C/\mathfrak{p}$ *has no divisor of* 0 *other than* 0; it is immediately verified that the mapping $\bar{v}: C/\mathfrak{p} \to \Gamma_\infty$ derived from v by passing to the quotient is a *valuation on* $C/\mathfrak{p}$, the inverse image of $+\infty$ under this valuation reducing to 0.

2. VALUATIONS ON A FIELD

PROPOSITION 2. *Let* K *be a (not necessarily commutative) field and* v *a valuation on* K *with values in* Γ. *Then:*

(i) *For* $x \neq 0$, $v(x) \neq +\infty$.

(ii) *The set* A *of* $x \in K$ *such that* $v(x) \geqslant 0$ *is a subring of* K.

(iii) *For all* $\alpha \geqslant 0$ *in* Γ, *the set* V_α (resp. V'_α) *of* $x \in A$ *such that* $v(x) > \alpha$ (resp. $v(x) \geqslant \alpha$) *is a two-sided ideal of* A *and every (left or right) ideal* $\neq (0)$ *of* A *contains one of the* V'_α.

(iv) *The set* $\mathfrak{m}(A)$ *of* $x \in A$ *such that* $v(x) > 0$ *is the greatest ideal* $\neq A$ *in* A; $U(A) = A - \mathfrak{m}(A)$ *is the set of invertible elements of* A *and* $\kappa(A) = A/\mathfrak{m}(A)$ *is a (not necessarily commutative) field.*

(v) *For all* $x \in K - A$, $x^{-1} \in \mathfrak{m}(A)$.

Assertion (i) follows from the fact that $\overset{-1}{v}(+\infty)$ is an ideal of K not equal to K. The verification of the fact that A is a ring and the V_α and V'_α two-sided ideals is trivial by virtue of axioms (VL_I), (VL_{II}) and (VL_{III}). If $\mathfrak{a}$ is a (left, for example) ideal of A and $x \neq 0$ belongs to A, every $y \in A$ such that $v(y) \geqslant v(x)$ can be written $y = zx$ where $z = yx^{-1}$, hence $v(z) = v(y) - v(x) \geqslant 0$ and therefore $z \in A$; in other words the left ideal Ax contains the V'_α for $\alpha \geqslant v(x)$. The set $U(A) = A - \mathfrak{m}(A)$ is the set of $x \in K$ such that $v(x) = 0$; if $x \in U(A)$, then

$$v(x^{-1}) = -v(x) = 0,$$

whence $x^{-1} \in U(A)$; conversely, if $y \in A$ is invertible in A, then $v(y) \geqslant 0$, $v(y^{-1}) \geqslant 0$ and $v(y) + v(y^{-1}) = 0$, whence $v(y) = 0$ and $y \in U(A)$; this proves (iv) and (v) follows immediately from the definitions.

A (resp. $\mathfrak{m}(A)$, $\kappa(A)$) is called the *ring* (resp. *ideal*, *residue field*) of the valuation v on K.

Clearly U(A) is the *kernel* of the homomorphism $v: K^* \to \Gamma$ and the *image* $v(K^*)$ under v of the multiplicative group K^* is a *subgroup* of the additive group Γ, called the *order group* or *value group* of v, which is therefore isomorphic to $K^*/U(A)$; for $x \in K$, the element $v(x)$ of Γ_∞ is sometimes called the *valuation* or *order of* x for v. Two valuations v, v' on K are called *equivalent* if they have the *same ring*.

PROPOSITION 3. *For two valuations v, v' over a (not necessarily commutative) field* K *to be equivalent, it is necessary and sufficient that there exist an isomorphism λ of the ordered group $v(K^*)$ onto the ordered group $v'(K^*)$ such that $v' = \lambda \circ v$.*

Suppose v and v' are equivalent; by hypothesis, the ring A of the valuation v being the same as that of v', v and v' (restricted to K^*) factor into homomorphisms $K^* \to K^*/U(A) \xrightarrow{\mu} v(K^*)$, $K^* \to K^*/U(A) \xrightarrow{\nu} v'(K^*)$, where μ and ν are isomorphisms; moreover, the set of positive elements of $v(K^*)$ (resp. $v'(K^*)$) is the image under μ (resp. ν) of the set of classes mod. U(A) of elements $\neq 0$ of $\mathfrak{m}(A)$; we conclude that $\lambda = \nu \circ \overset{-1}{\mu}$ solves the problem, the converse being obvious.

Suppose now that K is a *commutative* field; then, for all valuations v on K, the ring A of the valuation v is a *valuation ring of* K in the sense of § 1, no. 2, Definition 2 (which justifies the terminology); this follows immediately from Proposition 2 (c) and § 1 no. 2, Theorem 1 (c). *Conversely*, recall that for every integral domain B whose field of fractions is K the relation of divisibility $x|y$ (equivalent to $y \in Bx$) makes K^* into a preordered group, whose *associated ordered group* Γ_B is the quotient $K^*/U(B)$ of K^* by the group U(B) of invertible elements of B, the positive elements of this group being those of $B^*/U(B)$ (where $B^* = B - \{0\}$); the mapping $x \mapsto Bx$ defines, by passing to the quotient, an isomorphism of the ordered group $K^*/U(B)$ onto the group (ordered by the relation $\supset$) of non-zero principal fractional ideals of K (*Algebra*, Chapter VI, § 1, no. 5). The rings A with field of fractions K and for which the group $\Gamma_A = K^*/U(A)$ is *totally ordered* are precisely the *valuation rings of* K (§ 1, no. 2, Theorem 1 (d)). If v_A denotes the canonical homomorphism of K^* onto Γ_A, it is immediate that v_A (extended by $v_A(0) = +\infty$) is a *valuation* (called *canonical*) on K whose ring is A; every valuation equivalent to v_A may be written $v = \sigma \circ v_A$, where σ is an isomorphism of Γ_A onto a subgroup of the group where v takes its values (Proposition 3); $\sigma \circ v_A$ is called the *canonical factorization of* v.

PROPOSITION 4. *Let* C *be an integral domain,* K *its field of fractions,* $C^* = C - \{0\}$ *and $v: C \to \Gamma_\infty$ a valuation on* C. *Then there exists a unique valuation w on* K *which extends v and $w(K^*)$ is the subgroup of Γ generated by $v(C^*)$.*

By Theorem 2 of *Algebra*, Chapter I, § 2, no. 7, there exists a unique homomorphism w from K^* to Γ which extends $v \mid C^*$ and $w(K^*)$ is generated by $v(C^*)$. It remains to prove that w satisfies axiom (VL_{II}). Then let $x' \in K^*$, $y \in K^*$ be such that $x + y \in K^*$; there exists $a \in C^*$ such that $ax \in C^*$ and $ay \in C^*$, whence $a(x + y) \in C^*$. Since the restriction of w to C^* satisfies (VL_{II}),

$$w(a(x + y)) \geqslant \inf(w(ax), w(ay)).$$

Eliminating $w(a)$ from both sides, we obtain

$$w(x + y) \geqslant \inf(w(x), w(y)).$$

3. TRANSLATIONS

Let K be a (commutative) field, f a place of K, v a valuation on K and A a valuation ring of K. We shall say that A, f and v are *associated* if A is the ring of f and the ring of v. By virtue of no. 1 and § 2, no. 3, each of the three objects A, f and v then determines the other two (up to an equivalence as far as places and valuations are concerned). In particular there are the following equivalences:

$$\begin{array}{lll} x \in A & \Leftrightarrow f(x) \neq \infty & \Leftrightarrow v(x) \geqslant 0 \\ x \in \mathfrak{m}(A) & \Leftrightarrow f(x) = 0 & \Leftrightarrow v(x) > 0 \\ x \in A - \mathfrak{m}(A) = U(A) & \Leftrightarrow f(x) \neq 0 \text{ and } f(x) \neq \infty & \Leftrightarrow v(x) = 0 \\ x \in K - A & \Leftrightarrow f(x) = \infty & \Leftrightarrow v(x) < 0 \end{array}$$

Every result relating to valuation rings, places or valuations can be translated into a result relating to the other two notions. Thus Proposition 4 of § 2, no. 4 gives:

PROPOSITION 5. *Let* K *be a field, v a valuation on* K *and* K' *an extension of* K. *There exists a valuation v' on* K' *whose restriction to* K *is equivalent to v.*

Let Γ_v and $\Gamma_{v'}$ be the order groups of v and v'. Since the restriction of v' to K is equivalent to v, there exists an isomorphism λ of Γ_v onto a subgroup of $\Gamma_{v'}$, such that $v' = \lambda \circ v$ on K. If Γ_v is identified with $\lambda(\Gamma_v)$ by means of λ, it is seen that v' extends v.

Note that $\Gamma_{v'}$ is in general *distinct from* $\lambda(\Gamma_v)$ and the equivalence class of v' is not necessarily unique. We shall return to this in § 8.

Translating Theorem 3 of § 1, no. 3 (or Proposition 6 of § 2, no. 5), we obtain:

PROPOSITION 6. *Let* K *be a field,* A *a subring of* K *and x an element of* K. *For x to be integral over* A, *it is necessary and sufficient that every valuation on* K *which is positive on* A *be positive at x.*

From now on, we shall in general leave to the reader the trouble of performing translations analogous to the above.

4. EXAMPLES OF VALUATIONS

The examples of valuation rings given in § 1, no. 4 provide us with *Examples* 1 to 4 below:

Example (1) Every valuation on a *finite* field F is improper, since every element of F^* is a root unity.

Example (2) If K is a subfield of a field K', the *restriction* to K of a valuation on K' is a valuation on K.

Example (3) Let k be a field and $K = k((T))$. The mapping v which maps every non-zero formal power series to its *order* (*Algebra*, Chapter IV, § 5, no. 7) is a valuation on K whose order group is **Z** and whose ring is $k[[T]]$. The associated place is the canonical homomorphism $f: k[[T]] \to k$ extended to $k((T))$ by setting $f(u) = \infty$ if $u \notin k[[T]]$.

Example (4) Let A be a principal ideal domain, K its field of fractions and p an extremal element of A. For $x \in K^*$ let $v_p(x)$ denote the exponent of p in the decomposition of x into extremal elements (*Algebra*, Chapter VII, § 1, no. 3, Theorem 2); it is immediately seen that v_p is a valuation whose order group is **Z** and whose ring is A_{Ap}. By Proposition 3 of § 1, no. 4 we thus obtain, up to equivalence, all the valuations on K which are not improper and are positive on A. Taking $A = \mathbf{Z}$ we recover the p-adic valuations on **Q** (*General Topology*, Chapter IX, § 3, no. 2); these valuations are, up to equivalence, the only valuations on **Q** which are not improper (§ 1, no. 4, Corollary 1 to Proposition 3). Taking $A = k[X]$, where k is a field, the non-improper valuations on $k(X)$ whose restrictions to k are improper are (up to equivalence): on the one hand the valuations v_P where P runs through the set of irreducible monic polynomials of $k[X]$ and on the other hand the valuation v defined by

$$v(P/Q) = \deg(Q) - \deg(P)$$

for $P \in k[X]$ and $Q \in k[X]$ (§ 1, no. 4, Corollary 2 to Proposition 3); all these valuations obviously have **Z** as order group and their residue fields are monogenous algebraic extensions of k (*Algebra*, Chapter V, § 3, no. 1).

Example (5) The mapping $P(X, Y) \mapsto P(T, e^T)$ of $\mathbf{C}[X, Y]$ to $\mathbf{C}((T))$ is injective (*Functions of a real variable*, Chapter IV, § 2, Proposition 9) and therefore can be extended to an isomorphism of $\mathbf{C}(X, Y)$ onto a subfield of $\mathbf{C}((T))$. The restriction to this subfield of the valuation on $\mathbf{C}((T))$ defined in Example 3 defines a valuation on $\mathbf{C}(X, Y)$ which is improper on **C**, whose order group is **Z** and whose residue field is **C**.

Proposition 4 of no. 2 allows us to construct a valuation whose order group and residue field are given:

Example (6) Let Γ be a totally ordered group and k a field. Let Γ_+ be the monoid of positive elements of Γ and C the algebra of Γ_+ over k. By definition, C has a basis $(x_\alpha)_{\alpha \in \Gamma_+}$ over k whose multiplication table is $x_\alpha x_\beta = x_{\alpha+\beta}$. If $x = \sum_\alpha a_\alpha x_\alpha$ is a non-zero element of C, we write $v(x) = \inf_{a_\alpha \neq 0} (\alpha)$ and

$v(0) = +\infty$; it is immediately verified that the mapping v of C to Γ_∞ satisfies conditions (VL_I) and (VL_{II}) of no. 1 and that C is an integral domain. Let K be the field of fractions of C and w the valuation on K which extends v (Proposition 4, no. 2). As every element of Γ is the difference of two positive elements, w admits Γ as order group. Let A be the ring of w and $\mathfrak{m}$ its maximal ideal; we shall show that A is the direct sum of $\mathfrak{m}$ and k (identified with $k.1$), which will prove that the residue field of w is isomorphic to k. Clearly $\mathfrak{m} \cap k = (0)$. On the other hand, denoting by $\mathfrak{p}$ the ideal of C generated by the x_α where $\alpha > 0$, every element x of valuation 0 in K can be written in the form $(a + y)/(b + z)$ where $a \in k^*$, $b \in k^*$, $y \in \mathfrak{p}$ and $z \in \mathfrak{p}$; then

$$x = ab^{-1} + (by - az)b^{-1}(b + z)^{-1}$$

whence $w(x - ab^{-1}) > 0$ and $x \equiv ab^{-1}$ (mod. $\mathfrak{m}$); this shows our assertion.

If $\Gamma = \mathbf{Z} \times \mathbf{Z}$, then $K = k(X, Y)$ and the above construction then provides valuations on $k(X, Y)$ which are improper on k, whose order group is $\mathbf{Z} \times \mathbf{Z}$ and whose residue field is k. These valuations depend on the order structure chosen on $\mathbf{Z} \times \mathbf{Z}$. For example, $\mathbf{Z} \times \mathbf{Z}$ can be given the lexicographic ordering. Or indeed, for an irrational number α, $\mathbf{Z} \times \mathbf{Z}$ may be identified with a subgroup of $\mathbf{R}$ under the homomorphism $(m, n) \mapsto m + n\alpha$ (a homomorphism which is injective since α is irrational) and given the ordering induced by that on $\mathbf{R}$.

Other constructions of valuations using Proposition 4 of no. 2 will be described in § 10.

5. IDEALS OF A VALUATION RING

DEFINITION 2. *Let G be an ordered set. A subset of G is called major if the relations $x \in M$ and $y \geqslant x$ imply $y \in M$.*

Let K be a field, v a valuation on K, A the ring of v and G the order group of v. For every major subset $M \subset G$, let $\mathfrak{a}(M)$ be the set of $x \in K$ such that $v(x) \in M \cup \{+\infty\}$. Clarly $\mathfrak{a}(M)$ is a sub-A-module of K.

PROPOSITION 7. *The mapping $M \mapsto \mathfrak{a}(M)$ is an increasing bijection of the set of major subsets of G onto the set of sub-A-modules of K.*

Let $\mathfrak{b}$ be a sub-A-module of K. The set of $v(x)$ for $x \in \mathfrak{b} - (0)$ is a major subset $M(\mathfrak{b})$ of G. Proposition 7 will be shown if the following equations are proved:

(2) $M(\mathfrak{a}(N)) = N$ for every major subset N of G;

(3) $\mathfrak{a}(M(\mathfrak{b})) = \mathfrak{b}$ for every sub-A-module $\mathfrak{b}$ of K.

Formula (2) is easy, since, for all $m \in N$, there exists $x \in K$ such that $v(x) = m$. Then obviously $\mathfrak{b} \subset \mathfrak{a}(M(\mathfrak{b}))$; conversely, let $x \in \mathfrak{a}(M(\mathfrak{b}))$ and suppose $x \neq 0$;

then $v(x) \in M(\mathfrak{b})$ and therefore there exists $y \in \mathfrak{b}$ such that $v(x) = v(y)$; whence $x = uy$ where $v(u) = 0$, which proves that $x \in Ay \subset \mathfrak{b}$ and completes the proof.

COROLLARY. *Let* G_+ *be the set of positive elements in* G. *The mapping* $M \mapsto \mathfrak{a}(M)$ *is a bijection of the set of major subsets of* G_+ *onto the set of ideals of* A.

As $A = \mathfrak{a}(G_+)$, $\mathfrak{a}(M) \subset A$ is equivalent to $M \subset G_+$.

For example the maximal ideal $\mathfrak{m}(A)$ is equal to $\mathfrak{a}(S)$, where S denotes the set of strictly positive elements of G.

6. DISCRETE VALUATIONS

DEFINITION 3. *Let* K *be a (not necessarily commutative) field, v a valuation on* K *and* Γ *the order group of v. v is called discrete if there exists a (necessarily unique) isomorphism of the ordered group* Γ *onto* **Z**. *Let* γ *be the element of* Γ *corresponding to* 1 *under this isomorphism; every element u of* K *such that* $v(u) = \gamma$ *is called a uniformizer of v. A discrete valuation is called normed if its order group is* **Z**.

For example the valuation v_p defined by an extremal element p of a principal ideal *or factorial* domain is a normed discrete valuation which admits p as a uniformizer. In particular, if k is a field, $k[[T]]$ is the ring of a discrete valuation on $k((T))$ which admits T as a uniformizer. * Let S be a connected complex analytic variety of dimension 1, K the field of meromorphic functions on S and z_0 a point of S; the set of $f \in K$ which are holomorphic at z_0 is the ring of a discrete valuation v; for a function $f \in K$ to be uniformizing for v, it is necessary and sufficient that it be holomorphic and zero at z_0 and that there exist a neighbourhood V of z_0 in S such that the restriction of f to V be a homomorphism of V onto a neighbourhood of the origin in **C**. It is this example and other analogues which are the origin of the word "uniformizer".*

PROPOSITION 8. *Let* K *be a (not necessarily commutative) field, v a discrete valuation on* K, A *the ring of v and u a uniformizer for v. The non-zero ideals of* A *are two-sided and of the form* Au^n $(n \geqslant 0)$.

It may be assumed that v is *normed*, so that $v(u) = 1$. For all $x \in K^*$, there is an integer $n \in \mathbf{Z}$ such that $(v(x) = n = v(u^n)$ and hence we may write

$$x = zu^n = u^n z',$$

where z, z' are two *invertible* elements of the ring A; whence the proposition.

PROPOSITION 9. *Let* A *be a local integral domain distinct from its field of fractions. The following conditions are equivalent:*

(a) A *is the ring of a discrete valuation.*
(b) A *is a principal ideal domain.*
(c) *The ideal* $\mathfrak{m}(A)$ *is principal and* $\bigcap_{n=1}^{\infty} \mathfrak{m}(A)^n = (0)$.

(d) A *is a Noetherian ring and* $\mathfrak{m}(A)$ *is principal.*
(e) A *is a Noetherian valuation ring.*

Proposition 8 shows that (a) implies (b), (d) and (e). If A is a principal ideal domain, then $\mathfrak{m}(A) = Au$ and every non-zero ideal of A is of the form Au^n since A is local (*Algebra*, Chapter VII, § 1, no. 3, Theorem 2); therefore $\bigcap_{n=1}^{\infty} \mathfrak{m}(A)^n = 0$; this shows that (b) implies (c). On the other hand (d) implies (c) (Chapter III, § 3, no. 2, Corollary to Proposition 5); by Proposition 2 of § 1, no. 4, (c) implies (a). Thus conditions (a), (b), (c), (d) are equivalent and imply (e). Finally suppose (e) holds and let us show that (b) holds; it will be sufficient to prove the following lemma:

Lemma 1. *Let* A *be a valuation ring. Every finitely generated torsion-free* A*-module is free. Every finitely generated ideal of* A *is principal. Every torsion-free* A*-module is flat.*

Let E be a finitely generated torsion-free A-module and let $x_1, \ldots, x_n$ be generators of E which are minimal in number; we show that they are linearly independent. If $\sum_{i=1}^{n} a_i x_i = 0$ ($a_i \in A$) is a non-trivial relation between the x_i, one of the a_i, say a_1, divides all the others since the set of principal ideals of A is totally ordered by inclusion (§ 1, no. 2, Theorem 1); then $a_1 \neq 0$ since the relation is non-trivial. As E is torsion-free, we can divide by a_1, which amounts to assuming that $a_1 = 1$. But then x_1 is a linear combination of $x_2, \ldots, x_n$, contrary to the minimal character of n. Hence E is free.

In particular every finitely generated ideal $\mathfrak{a}$ of A is principal, all the elements of a system of generators of $\mathfrak{a}$ being multiples of one of them. Proposition 3 of Chapter I, § 2, no. 4 then shows that every torsion-free A-module is flat.

4. THE HEIGHT OF A VALUATION

1. INCLUSION OF VALUATION RINGS OF THE SAME FIELD

Proposition 1. *Let* K *be a field and* A *a valuation ring of* K. *Then:*

(a) *Every ring* B *such that* $A \subset B \subset K$ *is a valuation ring of* K;

(b) *The maximal ideal* $\mathfrak{m}(B)$ *of such a ring is contained in* A *and it is a prime ideal of* A;

(c) *The mapping* $\mathfrak{p} \mapsto A_{\mathfrak{p}}$ *is a decreasing bijection of the set of prime ideals of* A *onto the set of rings* B *such that* $A \subset B \subset K$; *its inverse bijection is the mapping* $B \mapsto \mathfrak{m}(B)$.

If B is a ring such that $A \subset B \subset K$ and $x \in K - B$, then $x \in K - A$, whence $x^{-1} \in \mathfrak{m}(A) \subset B$, which proves both that B is a valuation ring of K and that $\mathfrak{m}(B) \subset \mathfrak{m}(A)$; as $\mathfrak{m}(B) = \mathfrak{m}(B) \cap A$ is a prime ideal of A, we have shown (a) and (b). Moreover, $A_{\mathfrak{m}(B)} \subset B$; conversely, if $x \in B - A$, then $x^{-1} \in A$ and $x^{-1} \notin \mathfrak{m}(B)$ and hence $x \in A_{\mathfrak{m}(B)}$; thus $A_{\mathfrak{m}(B)} = B$. Finally let $\mathfrak{p}$ be a prime ideal of A; we write $B = A_{\mathfrak{p}}$; then $\mathfrak{m}(B) \cap A = \mathfrak{p}$. (Chapter II, § 2, no. 5, Proposition 11) and $\mathfrak{m}(B) \subset A$ by (b); hence $\mathfrak{m}(B) = \mathfrak{p}$, which shows that the mappings $\mathfrak{p} \mapsto A_{\mathfrak{p}}$ and $B \mapsto \mathfrak{m}(B)$ of the statement are inverse bijections.

Corollary. *The set of subrings of* K *containing* A *is totally ordered by inclusion.*

The set of prime ideals of A is totally ordered by inclusion (§ 1, no. 2, Theorem 1 (e)) and the mapping $\mathfrak{p} \mapsto A_{\mathfrak{p}}$ reverses the inclusion relations.

Proposition 2. *Let* K *be a field,* B *a valuation ring of* K *and* h_B *the place of* K *associated with* B *(with values in* $\kappa(B)$*). Then the mapping* $A \mapsto h_B(A)$ *defines a bijection of the set* $\mathfrak{A}$ *of valuation rings of* K *contained in* B *onto the set* $\mathfrak{A}'$ *of valuation rings of* $\kappa(B)$.

If $A \in \mathfrak{A}$, then $h_B(A) \in \mathfrak{A}'$: for if $x' = h_B(x)$ (where $x \in B$) is an element of $\kappa(B) - h_B(A)$, then $x \notin A$, hence $x^{-1} \in A$ and $h_B(x)^{-1} \in h_B(A)$. On the other hand, for $A \in \mathfrak{A}$, $A \supset \mathfrak{m}(B)$ (Proposition 1 (b)) and hence the mapping, $A \mapsto h_B(A)$ is injective. Finally, let $A' \in \mathfrak{A}'$ and $A = \overset{-1}{h_B}(A') \subset B$; we shall show, which will complete the proof, that $A \in \mathfrak{A}$; if $x \in K - A$, then either $x \notin B$, or $x \in B$; if $x \notin B$, then $x^{-1} \in \mathfrak{m}(B) \subset A$; if $x \in B$, then $h_B(x) \in \kappa(B)$ and $h_B(x) \notin A'$, hence $h_B(x^{-1}) \in A'$ and we conclude again that $x^{-1} \in A$; hence $A \in \mathfrak{A}$.

Corollary. *Let* A *and* B *be two valuation rings of* K*, where* $A \subset B$*; let* $A' = h_B(A)$*, which is a valuation ring of* $\kappa(B)$*. The residue field* $\kappa(A')$ *of* A' *is canonically isomorphic to the residue field* $\kappa(A)$ *of* A *and the place* h_A *associated with* A *is the composition* $h_{A'} \circ h_B$ *of the places associated with* A' *and* B.

Since the local ring A' is a quotient of the local ring A, their residue fields are canonically isomorphic and the equation $h_A(x) = h_{A'}(h_B(x))$ holds for $x \in A$. On the other hand, if $x \in B - A$, then $h_B(x) \notin A'$ and the two sides of the equation are equal to ∞; the same is true if $x \in K - B$.

Remark. Conversely, let f be a place of K with values in K' and f' a place of K' with values in K''. Then $f' \circ f$ is a place of K whose ring is contained in the ring of the place f.

2. ISOLATED SUBGROUPS OF AN ORDERED GROUP

To study the situation in no. 1 from the point of view of valuations we shall need Definition 1 and Proposition 3 below.

Definition 1. *A subgroup* H *of an ordered group* G *is called isolated if the relations* $0 \leqslant y \leqslant x$ *and* $x \in H$ *imply* $y \in H$.

Example (1) Let A and B be two ordered groups; let A × B be given the lexicographic order (i.e. "$(a, b) \leqslant (a', b')$" is equivalent to "$(a < a')$ or $(a = a'$ and $b \leqslant b')$"). The second factor B of A × B is then, as is seen immediately, an isolated subgroup of A × B.

PROPOSITION 3. *Let* G *be an ordered group and* P *the set of its positive elements.*

(a) *The kernel of an increasing homomorphism of* G *to an ordered group is an isolated subgroup of* G.

(b) *Conversely, let* H *be an isolated subgroup of* G *and* g *the canonical homomorphism of* G *onto* G/H. *Then* g(P) *is the set of positive elements of an ordered group structure on* G/H. *Moreover, if* G *is totally ordered, so is* G/H.

(a) Let f be an increasing homomorphism from G to an ordered group; let H denote the kernel of f. If $0 \leqslant y \leqslant x$ and $x \in H$, then $0 \leqslant f(y) \leqslant f(x) = 0$, whence $f(y) = 0$, that is $y \in H$. Hence H is isolated.

(b) Let H be an isolated subgroup of G and $g: G \to G/H$. Let $P' = g(P)$. Clearly $P' + P' \subset P'$. Also

$$P' \cap (-P') = \{0\},$$

for, if x and y are elements of P such that $g(x) = -g(y)$, then $x + y \in H$, whence $x \in H$ and $y \in H$ since H is isolated; hence $g(x) = g(y) = 0$. Thus P' is the set of positive elements of an ordered group structure on G/H (*Algebra*, Chapter VI, § 1, no. 3, Proposition 3). Finally, if G is totally ordered, then $P \cap (-P) = G$, whence $P' \cup (-P') = G/H$ and therefore G/H is totally ordered (*loc. cit.*).

Example (2) If we reconsider the example where G is a lexicographic product A × B and H = B, the ordered group G/H is canonically identified with A.

3. COMPARISON OF VALUATIONS

Let K be a field and A a valuation ring of K. For every subring B of K containing A, $U(A) \subset U(B)$. Then there is a canonical homomorphism λ of $\Gamma_A = K^*/U(A)$ *onto* $\Gamma_B = K^*/U(B)$, whose kernel is $U(B)/U(A)$. Then, letting v_A and v_B denote the canonical valuations on K defined by A and B (§ 3, no. 2),

$$v_B = \lambda \circ v_A. \tag{1}$$

As $A \subset B$, λ maps the positive elements of Γ_A to positive elements of Γ_B and hence is increasing. Therefore (Proposition 3) the kernel H_B of λ is an isolated subgroup of Γ_A and λ factors into $\Gamma_A \to \Gamma_A/H_B \xrightarrow{\mu} \Gamma_B$, where μ is an increasing bijective homomorphism and hence an *isomorphism* of totally ordered groups; hence Γ_B is identified with the quotient totally ordered group Γ_A/H_B.

PROPOSITION 4. *The mapping* $B \mapsto H_B$ *is an increasing bijection of the set of subrings of* K *containing* A *onto the set of isolated subgroups of* Γ_A.

Given H_B, v_B is defined up to equivalence and hence B is determined uniquely. On the other hand, let H be an isolated subgroup of Γ_A; considering Γ_A/H as a totally ordered group (Proposition 3), the composite mapping

$$K^* \xrightarrow{v_A} \Gamma_A \longrightarrow \Gamma_A/H$$

is a valuation on K whose ring contains A.

Remark. Under the above hypotheses, let f denote the canonical homomorphism of B onto $\kappa(B)$ and $A' = f(A)$; it is a valuation ring of $\kappa(B)$ (Proposition 2, no. 1). Then $\overset{-1}{f}(\kappa(B)^*) = U(B), \overset{-1}{f}(A') = A, \overset{-1}{f}(\mathfrak{m}(A')) = \mathfrak{m}(A)$, hence

$$\overset{-1}{f}(U(A')) = U(A).$$

Then there is a canonical isomorphism of $U(B)/U(A)$ onto $\kappa(B)^*/U(A') = \Gamma_{A'}$. The exact sequence

$$0 \rightarrow U(B)/U(A) \rightarrow \Gamma_A \rightarrow \Gamma_B \rightarrow 0$$

then gives an exact sequence

$$0 \rightarrow \Gamma_{A'} \rightarrow \Gamma_A \rightarrow \Gamma_B \rightarrow 0.$$

Example. Let k be a field,

$$E = k(X) \quad \text{and} \quad K = k(X, Y) = E(Y)$$

(X, Y indeterminates). Let $B = E[Y]_{(Y)}$ be the valuation ring of K defined by the extremal element Y of the principal ideal domain E[Y] (§ 1, no. 4, Proposition 3). The residue field $\kappa(B)$ is canonically identified with $E[Y]/(Y) = E$. Similarly, let $A' = k[X]_{(X)}$ be the valuation ring of $E = k(X)$ defined by the extremal element X of $k[X]$. Denoting by h_B the place of E associated with B and writing $A = \overset{-1}{h_B}(A')$, a valuation ring A of K is defined which is contained in B and $\kappa(A) = \kappa(A') = k$. The canonical place $h_A: K \rightarrow k$ can be described as follows: if $f(X, Y)$ is an element of K, then we first put $Y = 0$ in f (which gives an element of $\tilde{E} = k(X)^{\sim}$), then $X = 0$ in the result obtained. The groups $\Gamma_{A'}$ and Γ_B are canonically isomorphic to **Z** (§ 3, no. 4, *Example* 4). *It is not difficult to show (cf. § 10, no. 2, Lemma 2) that the group Γ_A is isomorphic to the lexicographic product **Z** × **Z** and that the valuation v_A is equivalent to the valuation defined in § 3, no. 4, end of *Example* 6. *

4. THE HEIGHT OF A VALUATION

Let G be a totally ordered group. Given two isolated subgroups H and H′ of G, one of them is contained in the order: for otherwise there would exist a positive element x of H not belonging to H′ and a positive element x' of H′ not belonging to H; let, for example $x \geqslant x'$; as H is isolated, $x' \in$ H, which is a contradiction.

This also follows from Proposition 4 of no. 3 and the Corollary to Proposition 1 of no. 1, taking account of the fact that every totally ordered group is the order group of a valuation (§ 3, no. 4, *Example* 6).

DEFINITION 2. *Let* G *be a totally ordered group. If the number of isolated subgroups of* G *distinct from* G *is finite and equal to n,* G *is said to be of height n. If this number is infinite,* G *is said to be of infinite height.*

Examples

(1) The height of the group G = {0} is 0.

(2) The groups **Z** and **R** are of height 1.

(3) Let G be a totally ordered group and H an isolated subgroup of G. If h(H) and h(G/H) denote the heights of the totally ordered groups H and G/H, then

$$h(\mathrm{G}) = h(\mathrm{H}) + h(\mathrm{G/H}), \tag{2}$$

since the set of isolated subgroups of G is totally ordered by inclusion. In particular, if G is the lexicographic product of two totally ordered groups H and H′, then

$$h(\mathrm{G}) = h(\mathrm{H}) + h(\mathrm{H'}) \tag{3}$$

(cf. no. 2, *Example* 2); thus the lexicographic product **Z** × **Z** is of height 2.

On the other hand the height of **Z** × **Z** ordered by embedding in **R** (cf. § 3, no. 4, end of *Example* 6) is equal to 1 (cf. Proposition 8 below).

DEFINITION 3. *The height of the order group of a valuation is called the height of that valuation.*

For example a discrete valuation is of height 1. Only improper valuations are of height 0. Propositions 1 and 4 imply:

PROPOSITION 5. *The height of a valuation is equal to the number of non-zero prime ideals in its ring.*

5. VALUATIONS OF HEIGHT 1

PROPOSITION 6. *Let* K *be a field and* A *a subring of* K. *Suppose that* A *is not a field. Then the following conditions are equivalent:*

(a) A *is the ring of a valuation of height* 1 *on* K;

(b) A *is a valuation ring of* K *and has no prime ideals other than* (0) *and* $\mathfrak{m}$(A);

(c) A *is maximal among the subrings of* K *distinct from* K.

Proposition 5 of no. 4 shows that (a) implies (b) and Proposition 1 of no. 1 shows that (b) implies (c). It remains to show that (c) implies (a). Suppose A is maximal among the subrings of K distinct from K. Let $\mathfrak{m}$ be a maximal ideal of A and V a valuation ring of K dominating $A_{\mathfrak{m}}$ (§ 1, no. 2, Corollary to Theorem 2); as $\mathfrak{m}(V) \cap A = \mathfrak{m}$ and $\mathfrak{m} \neq (0)$ (since A is not a field), $V \neq K$, whence $V = A$, which proves that A is not the ring of a valuation v on K. This being so, v is of height 1 by Propositions 1 (no. 1) and 5 (no. 4).

Proposition 7. *For a valuation on a field to be of height* 1, *it is necessary and sufficient that its order group be isomorphic to a non-zero ordered subgroup of* **R**.

This follows in fact from the following proposition:

Proposition 8. *Let* G *be a totally ordered group not reduced to* 0. *The following conditions are equivalent:*

(a) G *is of height* 1;

(b) *for all* $x > 0$ *and* $y \geqslant 0$ *in* G, *there exists an integer* $n \geqslant 0$ *such that* $y \leqslant nx$;

(c) G *is isomorphic to a subgroup of the ordered additive group* **R** *which is not reduced to* 0.

Let x be a positive element of G and let H_x be the set of $y \in G$ such that there exists an integer $n \geqslant 0$ satisfying $|y| \leqslant nx$. It is easily verified that H_x is an isolated subgroup of G and that every isolated subgroup of G containing x contains H_x. Condition (a) is therefore equivalent to "$H_x = G$ for all $x > 0$", that is, to condition (b).

Clearly (c) implies (b). Conversely, suppose condition (b) holds and let Q denote the set of elements >0 of G. Suppose first that Q has a least element x; for all $y \in Q$, let n be the least integer such that $y \leqslant nx$; if $y < nx$, then also $nx - y \geqslant x$, whence $y \leqslant (n-1)x$ contrary to the choice of n; then $y = nx$, which shows that $G = \mathbf{Z}x$ is isomorphic to $\mathbf{Z} \subset \mathbf{R}$. Suppose now that Q has no least element; we apply to the ordered set $P = Q \cup \{0\}$ Proposition 1 of *General Topology*, Chapter V, § 2 (which is possible, since condition (b) is just "Archimedes' axiom"); it is seen that there exists a strictly increasing mapping f of P to $\mathbf{R}_+$ such that

$$f(x+y) = f(x) + f(y)$$

for $x \in P$ and $y \in P$; by linearity f can be extended to an isomorphism of G onto a subgroup of **R**, which proves that (b) implies (c).

Proposition 9. *Let* K *be a field,* v *a non-improper valuation on* K *and* A *the ring of* v. *For* A *to be completely integrally closed* (Chapter V, § 1, no. 4, Definition 5), *it is necessary and sufficient that* v *be of height* 1.

Suppose v is of height 1. Let $x \in K$ be such that the x^n ($n \geqslant 0$) are all contained in a finitely generated sub-A-module of K. There exists $d \in A - \{0\}$ such that $dx^n \in A$ for all $n \geqslant 0$. Then $v(d) + nv(x) \geqslant 0$, that is $n(-v(x)) \leqslant v(d)$ for all

$n \geqslant 0$, whence $-v(x) \leqslant 0$ (Proposition 8 (b)) and $x \in A$. Thus A is completely integrally closed.

Suppose now that v is not of height 1. Then there exist $y \in \mathfrak{m}(A)$ and $t \in A$ such that $nv(y) < v(t)$ for all $n \geqslant 0$ (Proposition 8 (b)). Then $ty^{-n} \in A$ for all $n \geqslant 0$, but $y^{-1} \notin A$. Hence A is not completely integrally closed.

COROLLARY. *Let K be a field, $(v_\alpha)_{\alpha \in I}$ a family of valuations of height 1 on K and A the intersection of the rings of the v_α. Then A is a completely integrally closed domain.*

A completely integrally closed domain is not always an intersection of valuation rings of height 1 (Exercise 6).

5. THE TOPOLOGY DEFINED BY A VALUATION

1. THE TOPOLOGY DEFINED BY A VALUATION

Let K be a not necessarily commutative field, v a valuation on K and G the totally ordered group $v(K^*)$. For all $\alpha \in G$ let V_α be the set of $x \in K$ such that $v(x) > \alpha$; this set is an additive subgroup of K (§ 3, no. 1). There exists a unique topology $\mathscr{T}_v$ on K for which the V_α form a fundamental system of neighbourhoods of 0 (*General Topology*, Chapter III, § 1, no. 2, *Example*). For v to be improper, it is necessary and sufficient that $\mathscr{T}_v$ be the discrete topology.

LEMMA 1. *Let $x \in K^*$, $y \in K^*$ and $\alpha \in G$. If*

$$v(x - y) > \sup(\alpha + 2v(y), v(y)),$$

then $v(x^{-1} - y^{-1}) > \alpha$.

$x^{-1} - y^{-1} = x^{-1}(y - x)y^{-1}$ and hence

$$v(x^{-1} - y^{-1}) = v(x - y) - v(x) - v(y).$$

If $v(x - y) > v(y)$, Proposition 1 of § 3, no. 1 implies that $v(x) = v(y)$, since $x = y + (x - y)$. Moreover, if $v(x - y) > \alpha + 2v(y)$, then

$$v(x^{-1} - y^{-1}) > \alpha + 2v(y) - 2v(y) = \alpha.$$

PROPOSITION 1. *The topology $\mathscr{T}_v$ is Hausdorff and compatible with the field structure on K. The mapping $v\colon K^* \to G$ is continuous if G is given the discrete topology.*

Let $x \in K^*$ and $\alpha = v(x)$; then $x \notin V_\alpha$, which shows that $\mathscr{T}_v$ is Hausdorff. For all $x_0 \in K$ and $\alpha \in G$, there exists $\beta \in G$ such that $x_0 V_\beta \subset V_\alpha$ and $V_\beta x_0 \subset V_\alpha$ (it is sufficient to take $\beta \geqslant \alpha - v(x_0)$). On the other hand, if $\alpha \geqslant 0$, then $V_\alpha V_\alpha \subset V_\alpha$. The axioms ($AV_I$) and ($AV_{II}$) of *General Topology*, Chapter III,

§ 6, no. 3 being thus satisfied, $\mathcal{T}_v$ is compatible with the ring structure on K, Let $x_0 \in K^*$; if $x \in K^*$ satisfies $v(x - x_0) > \sup(\alpha + 2v(x_0), v(x_0))$, then $v(x^{-1} - x_0^{-1}) > \alpha$ (Lemma 1), which shows that $x \mapsto x^{-1}$ is continuous and that $\mathcal{T}_v$ is therefore compatible with the field structure on K. Finally, the single condition $v(x - x_0) > v(x_0)$ implies $v(x) = v(x_0)$ (§ 3, no. 1, Proposition 1) and hence the mapping $v: K^* \to G$ is continuous if G is given the discrete topology.

Let $\alpha \in G$ and V'_α be the set of $x \in K$ such that $v(x) \geqslant \alpha$. If $\beta < \alpha$, then $V_\beta \supset V'_\alpha \supset V_\alpha$. If v is not improper, it is therefore seen that the V'_α form a fundamental system of neighbourhoods of 0 for $\mathcal{T}_v$.

The V_α and the V'_α are open additive subgroups and therefore closed in K and therefore the topological field K is *totally disconnected.* As every non-zero ideal of the ring of v contains a V_α, it is *open and closed* in K. The quotient topology on the residue field of v is therefore *discrete.*

Let A be the ring of v. If v is discrete, Proposition 8 of § 3, no. 6 shows that the topology induced by $\mathcal{T}_v$ on A is the $\mathfrak{m}(A)$-adic topology. This is not so in general (Exercise 4).

PROPOSITION 2. *Let K be a not necessarily commutative field, v a non-improper valuation on K, A the ring of v and $\mathfrak{m}$ the ideal of v. For K with the topology $\mathcal{T}_v$ to be locally compact, it is necessary and sufficient that the following conditions be fulfilled:*

(i) *K is complete;*

(ii) *v is discrete;*

(iii) *the residue field $\kappa(A)$ is finite.*

If so, A is compact.

Suppose K is locally compact. Then it is complete (*General Topology*, Chapter III, § 3, no. 3, Corollary 1 to Proposition 4); further there exists a compact neighbourhood of 0, which contains a neighbourhood V'_α, where α belongs to the value group of v; in other words, there exists $a \neq 0$ in K^* such that $A.a$ is compact and it follows that $A = (A.a)a^{-1}$ is compact. As every ideal $\mathfrak{b} \neq (0)$ of A is open, $A/\mathfrak{b}$ is compact and discrete (*General Topology*, Chapter III, § 2, no. 5, Proposition 14) and therefore finite and in particular $\kappa(A) = A/\mathfrak{m}$ is finite. Moreover, for $y \neq 0$ in $\mathfrak{m}$, the ring A/Ay being finite, there is only a finite number of ideals of A containing Ay and the set P of elements of the form $v(x)$ such that

$$0 < v(x) \leqslant v(y)$$

is finite; as $v(K^*)$ is totally ordered, P has a least element γ. Then for all $x \in A$ such that $v(x) > 0$, either $v(x) > v(y) \geqslant \gamma$, or $v(x) \leqslant v(y)$ and then $v(x) \geqslant \gamma$ by definition, so that γ is the least of the elements >0 of $v(K^*)$. As P is finite, there is a greatest integer $m \geqslant 0$ such that $m\gamma \in P$, whence

$m\gamma \leqslant v(y) < (m+1)\gamma$. We deduce that $0 \leqslant v(y) - m\gamma < \gamma$ and by definition of γ this implies $v(y) = m\gamma$. Therefore $v(\mathrm{K}^*) = \mathbf{Z}.\gamma$ and the valuation v is discrete.

Conversely, suppose conditions (i), (ii), (iii) hold. We may restrict our attention to the case where v is normed; let u be a uniformizer for v. Then $\kappa(\mathrm{A}) = \mathrm{A}/\mathrm{A}u$ and hence $\mathrm{A}/\mathrm{A}u$ is finite. As $x \mapsto xu^n$ defines by taking quotients an isomorphism of the additive group $\mathrm{A}/\mathrm{A}u$ onto $\mathrm{A}u^n/\mathrm{A}u^{n+1}$, $\mathrm{A}/\mathrm{A}u^j$ is finite for all $j \geqslant 0$. As A is closed in K, it is complete and hence isomorphic to the inverse limit of the $\mathrm{A}/\mathrm{A}u^j$ (*General Topology*, Chapter III, § 7, no. 3, Proposition 2) and therefore compact. Since A is open in K, it is therefore seen that K is locally compact.

Remark. Note that it is sufficient in this proof to suppose that A is *complete.*

We shall see in § 9 that a field K fulfilling the conditions of Proposition 2 admits a centre which is either a finite algebraic extension of a p-adic field, or a field $\mathbf{F}_q((\mathrm{T}))$ of formal power series over a finite field; moreover K is of finite rank over its centre.

2. TOPOLOGICAL VECTOR SPACES OVER A FIELD WITH A VALUATION

Throughout let K be a (not necessarily commutative) field, v a valuation on K and G its order group. K is given the topology $\mathscr{T}_v$.

PROPOSITION 3. *Let* E *be a left topological vector space over* K *which is Hausdorff and of dimension* 1. *Suppose that v is not improper. For all $x_0 \neq 0$ in* E, *the mapping $a \mapsto ax_0$ of K_s onto* E *is a topological isomorphism.*

This mapping is a continuous algebraic isomorphism. It is sufficient to show that it is bicontinuous. Let $\alpha \in \mathrm{G}$. We need to show that there exists a neighbourhood V of 0 in E such that the relation $ax_0 \in \mathrm{V}$ implies $v(a) > \alpha$. Let $a_0 \in \mathrm{K}^*$ be such that $v(a_0) = \alpha$. As E is Hausdorff, there exists a neighbourhood W of 0 in E such that $a_0x_0 \notin \mathrm{W}$. As v is not improper, there exist a neighbourhood W′ of 0 in E and an element β of G such that the relations $y \in \mathrm{W}'$, $v(a) \geqslant \beta$ imply $ay \in \mathrm{W}$. Let $a_1 \in \mathrm{K}^*$ be such that $v(a_1) = -\beta$. The relations $ax_0 \in a_1^{-1}\mathrm{W}'$ and $v(a) \leqslant \alpha$ imply $a_1ax_0 \in \mathrm{W}'$ and $v(a_0a^{-1}a_1^{-1}) = \alpha + \beta - v(a) \geqslant \beta$ and hence $a_0x_0 = a_0a^{-1}a_1^{-1}(a_1ax_0) \in \mathrm{W}$, which is absurd; in other words, the relation $ax_0 \in a_1^{-1}\mathrm{W}$ implies $v(a) > \alpha$.

COROLLARY. *Let* E *be a left topological vector space over* K, H *a closed hyperplane of* E *and* D *a* 1*-dimensional vector subspace of* E *an algebraic supplement of* H. *Suppose that v is not improper. Then* D *is a topological supplement of* H.

Taking account of Propositions 1 and 3, the proof is the same as that of *Topological Vector Spaces*, Chapter I, § 2, Corollary 2 to Theorem 1.

Proposition 4. *Suppose that v is not improper and K is complete. Let E be a left topological vector space over K, which is Hausdorff and of finite dimension n. For every basis $(e_i)_{1 \leqslant i \leqslant n}$ of E over K, the mapping $(a_i) \mapsto \sum_{i=1}^{n} a_i e_i$ of K_s^n onto E is a topological vector space isomorphism.*

Taking account of Proposition 3 and its corollary, the proof is the same as that of *Topological Vector Spaces*, Chapter I, § 2, Theorem 2.

Corollary. *Suppose that v is not improper and K is complete. Let E be a Hausdorff topological vector space over K and F a finite-dimensional vector subspace of E. Then F is closed.*

F is complete.

3. THE COMPLETION OF A FIELD WITH A VALUATION

Proposition 5. *Let K be a not necessarily commutative field, v a valuation on K and G the group $v(K^*)$ with the discrete topology.*

(a) *The completion ring $\hat{K}$ of K (with $\mathcal{T}_v$) is a topological field.*

(b) *The mapping $v: K^* \to G$ can be extended uniquely to a continuous mapping $\hat{v}: \hat{K}^* \to G$. The mapping $\hat{v}$ (extended by $\hat{v}(0) = +\infty$) is a valuation on $\hat{K}$ and $\hat{v}(\hat{K}^*) = v(K^*)$.*

(c) *The topology on $\hat{K}$ is the topology defined by the valuation $\hat{v}$.*

(d) *For all $\alpha \in G$ let V_α, V'_α be the subgroups of K defined by the conditions $v(x) > \alpha$, $v(x) \geqslant \alpha$. Then the closures $\overline{V}_\alpha$, $\overline{V}'_\alpha$ of V_α, V'_α in $\hat{K}$ are defined by the conditions $\hat{v}(x) > \alpha$, $\hat{v}(x) \geqslant \alpha$ respectively.*

(e) *The ring of $\hat{v}$ is the completion $\hat{A}$ of the ring A of v; the ideal of $\hat{v}$ is the completion $\hat{\mathfrak{m}}$ of the ideal $\mathfrak{m}$ of v.*

(f) $\hat{A} = A + \hat{\mathfrak{m}}$; *the residue field of $\hat{v}$ is canonically identified with that of v.*

To prove (a) it suffices (*General Topology*, Chapter III, § 6, no. 8, Proposition 7) to show the following: let $\mathfrak{F}$ be a Cauchy filter (with respect to the additive uniform structure) on K^* for which 0 is not a cluster point; then the image of $\mathfrak{F}$ under the bijection $x \mapsto x^{-1}$ is a Cauchy filter (with respect to the additive uniform structure). For since 0 is not a cluster point of $\mathfrak{F}$, there exists $M \in \mathfrak{F}$ and $\beta \in G$ such that β is an upper bound of $v(M)$. Let $\alpha \in G$. If M' is an element of $\mathfrak{F}$ contained in M and such that $v(x - y) > \sup(\alpha + 2\beta, \beta)$ for $x \in M'$ and $y \in M'$, then $v(x^{-1} - y^{-1}) > \alpha$ for $x \in M'$ and $y \in M'$ (no. 1, Lemma 1). Whence (a).

By Proposition 1 of no. 1, $v|K^*$ is a continuous homomorphism from K^* to G and hence can be extended uniquely to a continuous homomorphism $\hat{v}$ from $\hat{K}^*$ to G. the Relation

$$\hat{v}(x + y) \geqslant \inf(\hat{v}(x), \hat{v}(y))$$

holds in K^* and hence also holds in $\hat{K}^*$ by continuity. Thus $\hat{v}$ (extended by $\hat{v}(0) = +\infty$) is a valuation on $\hat{K}$ and (b) is proved.

We now show (d). Let $\alpha \in G$ and $x \in \overline{V}_\alpha - \{0\}$. For y in V_α sufficiently close to x, $\hat{v}(x) = \hat{v}(y) = v(y)$ and hence $\hat{v}(x) > \alpha$. Conversely, let $x \in \hat{K}^*$ be such that $\hat{v}(x) > \alpha$; for y in K^* sufficiently close to x, $v(y) = \hat{v}(y) = \hat{v}(x)$ and therefore $y \in V_\alpha$, whence $x \in \overline{V}_\alpha$. Thus $\overline{V}_\alpha$ is the set of $x \in \hat{K}$ such that $\hat{v}(x) > \alpha$. The argument is analogous for V'_α. This proves (d).

Taking account of Proposition 7 of *General Topology*, Chapter III, § 3, no. 4, assertion (c) is a consequence of (d). Assertion (e) is a special case of (d). Finally let $x \in \hat{A}$; there exists $y \in A$ such that $\hat{v}(x - y) > 0$; then $z = x - y \in \hat{\mathfrak{m}}$ and hence $x = y + z \in A + \hat{\mathfrak{m}}$; thus $\hat{A} = A + \hat{\mathfrak{m}}$, which shows (f).

Remark. For all $x \in \hat{K}$ not belonging to $\hat{A}$, there exists $x_0 \in K$ such that $\hat{v}(x - x_0) > 0$, $\hat{v}(x) = \hat{v}(x_0) = v(x_0) < 0$; then $x_0^{-1}x \in \hat{A}$ and, as $x_0^{-1} \in A$, it is seen that, if we set $S = A - \{0\}$, it is possible to write $\hat{K} = S^{-1}\hat{A}$.

6. ABSOLUTE VALUES

1. PRELIMINARIES ON ABSOLUTE VALUES

Let K be a field (commutative or not). Recall (*General Topology*, Chapter IX, § 3, no. 2, Definition 2) that an *absolute value* on K is any mapping f from K to $\mathbf{R}_+$ satisfying the following axioms:

(VA_I) *The relation* $f(x) = 0$ *is equivalent to* $x = 0$.

(VA_{II}) $f(xy) = f(x)f(y)$ *for all* x, y *in* K.

(VA_{III}) $f(x + y) \leqslant f(x) + f(y)$ *for all* x, y *in* K.

It follows from (VA_I) and (VA_{II}) that $f(1) = 1, f(-1) = 1$ and

$$f(x^{-1}) = \frac{1}{f(x)}$$

for $x \neq 0$.

For a mapping f from K to $\mathbf{R}_+$ and a real number $A > 0$, let (U_A) denote the relation

$$f(x + y) \leqslant A.\sup(f(x), f(y)) \qquad \textit{for all } x, y \textit{ in } K.$$

We shall denote by $\mathscr{V}(K)$ *the set of mappings* f *from* K *to* $\mathbf{R}_+$ *satisfying* (VA_I) *and* (VA_{II}) *and for which there exists an* $A > 0$ (*depending on* f) *such that* (U_A) *holds.*

Note that if $f \in \mathscr{V}(K)$, then, putting $x = 1, y = 0$ in (U_A),

$$1 = f(1) \leqslant A.\sup(f(1), f(0)) = A.$$

PROPOSITION 1. *For a mapping* f *from* K *to* $\mathbf{R}_+$ *satisfying* (VA_I) *and* (VA_{II}) *to belong to* $\mathscr{V}(K)$, *it is necessary and sufficient that* $f(1 + x)$ *be bounded in the set of* $x \in K$ *such that* $f(x) \leqslant 1$.

If f satisfies $(\mathrm{U_A})$, then $f(1 + x) \leqslant \mathrm{A}$ if $f(x) \leqslant 1$. Conversely, suppose that $f(x + 1) \leqslant \mathrm{A}$ for the $x \in \mathrm{K}$ such that $f(x) \leqslant 1$ (which implies that $\mathrm{A} \geqslant f(1) = 1$); then, if $x = 0$ or $y = 0$, condition $(\mathrm{U_A})$ is fulfilled; if on the other hand $x \neq 0$ and $y \neq 0$, we may assume for example that $f(y) \leqslant f(x)$, hence, by $(\mathrm{VA_{II}})$, $f(yx^{-1}) \leqslant 1$ and therefore $f(1 + yx^{-1}) \leqslant \mathrm{A}$, which gives, by virtue of $(\mathrm{VA_{II}})$, $f(x + y)f(x)^{-1} \leqslant \mathrm{A}$; whence

$$f(x + y) \leqslant \mathrm{A}f(x) \leqslant \mathrm{A}.\sup(f(x), f(y)).$$

If f is an absolute value on K, then $f(n.1) \leqslant n$ by induction on the integer $n > 0$ starting from $(\mathrm{VA_{III}})$; conversely:

PROPOSITION 2. *Let f be a mapping of* K *to* $\mathbf{R}_+$ *belonging to* $\mathscr{V}(\mathrm{K})$; *if there exists* $\mathrm{C} > 0$ *such that* $f(n.1) \leqslant \mathrm{C}.n$ *for every integer* $n > 0$, f *is an absolute value on* K.

By induction on $r > 0$ we deduce from $(\mathrm{U_A})$ the relation

$$(1) \qquad f(x_1 + x_2 + \cdots + x_{2^r}) \leqslant \mathrm{A}^r \sup_{1 \leqslant i \leqslant 2^r} f(x_i)$$

for every family (x_i) of 2^r elements of K. We set $n = 2^r - 1$; for all $x \in \mathrm{K}$, we deduce from (1)

$$(f(1 + x))^n = f((1 + x)^n) = f\left(\sum_{i=0}^{n} \binom{n}{i} x^i\right) \leqslant \mathrm{A}^r \sup\left(f\left(\binom{n}{i}\right)(f(x))^i\right)$$

$$\leqslant \mathrm{CA}^r \sum_{i=0}^{n} \binom{n}{i} (f(x))^i = \mathrm{CA}^r(1 + f(x))^n$$

for $f\left(\binom{n}{i}\right) \leqslant \mathrm{C}\binom{n}{i}$; therefore

$$f(1 + x) \leqslant \mathrm{C}^{1/n}\mathrm{A}^{r/n}(1 + f(x)).$$

Letting r tend to $+\infty$, we obtain $f(1 + x) \leqslant 1 + f(x)$ for all $x \in \mathrm{K}$; applying this inequality with x replaced by xy^{-1} (where $y \neq 0$) and taking account of $(\mathrm{VA_{II}})$, we obtain relation $(\mathrm{VA_{III}})$, which proves the proposition.

COROLLARY 1. *For a mapping f from* K *to* $\mathbf{R}_+$ *to be an absolute value, it is necessary and sufficient that it satisfy conditions* $(\mathrm{VA_I})$, $(\mathrm{VA_{II}})$ *and* $(\mathrm{U_2})$.

It is necessary, for $(\mathrm{VA_{II}})$ implies

$$f(x + y) \leqslant f(x) + f(y) \leqslant 2\sup(f(x), f(y)).$$

Conversely, suppose f satisfies $(\mathrm{VA_I})$, $(\mathrm{VA_{II}})$ and $(\mathrm{U_2})$; for every integer $n > 0$, let r be the least integer such that $2^r \geqslant n$; if in (1) A is replaced by 2, the x_i of index $i \leqslant n$ by 1 and the x_i of index $i > n$ by 0, we obtain

$$f(n.1) \leqslant 2^r < 2n;$$

then Proposition 2 may be applied with $\mathrm{C} = 2$ and hence f is an absolute value.

COROLLARY 2. *For a mapping f of K to $\mathbf{R}_+$ to belong to $\mathscr{V}$(K), it is necessary and sufficient that it be of the form g^t, where $t > 0$ and g is an absolute value on* K.

To say that f satisfies (U_A) is equivalent to saying that f^s satisfies (U_{A^s}); as there exists $s > 0$ such that $A^s \leqslant 2$, Corollary 1 shows that for such a value of s, f^s is an absolute value.

2. ULTRAMETRIC ABSOLUTE VALUES

A mapping f of K to $\mathbf{R}_+$ is called an *ultrametric absolute value* if it satisfies conditions (VA_I), (VA_{II}) and (U_1) (which obviously implies that f is an absolute value).

PROPOSITION 3. *Let f be a mapping of* K *to $\mathbf{R}_+$. The following properties are equivalent:*

(a) *f is an ultrametric absolute value.*

(b) *There exists a valuation v on* K *with values in* **R** *and a real number a such that $0 < a < 1$ and $f = a^v$.*

(c) *f belongs to $\mathscr{V}$(K) and $f(n.1) \leqslant 1$ for every integer $n > 0$.*

(d) *For all $s > 0$, f^s is an absolute value.*

For every real number c such that $0 < c < 1$, the mapping $t \mapsto c^t$ is an isomorphism of the ordered group **R** (with the opposite ordering to the usual ordering) on the ordered group $\mathbf{R}_+^*$; this shows the equivalence of (a) and (b). Clearly (a) implies (c); (c) implies (d), for we deduce from (c) that

$$(f(n.1))^s \leqslant 1 \leqslant n$$

for every integer $n > 0$ and Proposition 2 of no. 1 shows that f^s is an absolute value. Finally (d) implies (a): for if f^s is an absolute value, it satisfies (U_2) and hence f satisfies ($U_{2^{1/s}}$) for all $s > 0$ and therefore also (U_1), letting s tend to $+\infty$.

COROLLARY. *If* K *is a (not necessarily commutative) field of characteristic $p > 0$, every function on $\mathscr{V}$(K) is an ultrametric absolute value.*

Every non-zero element $z = n.1$ (n an integer >0) belongs to the prime subfield $\mathbf{F}_p$ of K and hence satisfies the relation $z^{p-1} = 1$, which implies $f(z) = 1$ and we may apply Proposition 3 (c).

Given a real number c such that $0 < c < 1$, the formulae

$$f(x) = c^{v(x)}, \qquad v(x) = \log_c f(x)$$

therefore establish a one-to-one correspondence between ultrametric absolute values on K and valuations on K with real values. The improper valuation corresponds to the improper absolute value (*General Topology*, Chapter IX, § 3, no. 2). Let v_1, v_2 be two valuations on K with real values and f_1, f_2 the corresponding absolute values; for v_1 and v_2 to be equivalent, it is necessary and sufficient that f_1 and f_2 be so: for to say that v_1 and v_2 are equivalent amounts to saying that the relations $v_1(x) \geqslant 0$ and $v_2(x) \geqslant 0$ are equivalent or again that the relations $f_1(x) \leqslant 1$ and $f_2(x) \leqslant 1$ are equivalent; it is therefore sufficient to apply Proposition 5 of *General Topology*, Chapter IX, § 3, no. 2. Moreover (*loc. cit.*) for the topologies defined on K by f_1 and f_2 to be identical, it is necessary and sufficient that f_1 and f_2 be equivalent.

3. ABSOLUTE VALUES ON Q

PROPOSITION 4. *Let f be a mapping of $\mathbf{Q}$ to $\mathbf{R}_+$ belonging to $\mathscr{V}(\mathbf{Q})$. Then:*

(i) *Either f is the improper absolute value on $\mathbf{Q}$.*

(ii) *Or there exists a real number a and a prime number p such that $0 < a < 1$ and $f = a^{v_p}$, where v_p is the p-adic valuation.*

(iii) *Or there exists $s > 0$ such that $f(x) = |x|^s$ for all $x \in \mathbf{Q}$.*

In case (iii) *for f to be an absolute value on $\mathbf{Q}$, it is necessary and sufficient that* $0 < s \leqslant 1$.

Suppose first that $f(n) \leqslant 1$ for every integer $n > 0$. By Proposition 3 of no. 2 there exist a real number b and a valuation v on $\mathbf{Q}$ such that $0 < b < 1$ and $f = b^v$. Now, we know (§ 3, no. 4, *Example* 4) that the only valuations on $\mathbf{Q}$ are (up to equivalence) the improper valuation and the p-adic valuations v_p; we therefore have either case (i) or case (ii).

Suppose from now on that there exists an integer $h > 0$ such that $f(h) > 1$; by no. 1, Corollary 2 to Proposition 2, there exists a number $\rho > 0$ such that f^ρ is an absolute value; let us write

$$g(x) = \rho \log(f(x))/\log|x|$$

for every rational number $x \neq 0$. Let a, b be two integers $\geqslant 2$; for every integer $n \geqslant 2$ let $q(n)$ denote the integral part of $n.\log a/\log b$, in other words the least integer m such that $a^n < b^{m+1}$; the expansion of a^n to base b is therefore

$$a^n = c_0 + c_1 b + \cdots + c_{q(n)} b^{q(n)} \tag{2}$$

where $0 \leqslant c_i < b$ for $0 \leqslant i \leqslant q(n)$. As f^ρ is an absolute value, $f^\rho(c_i) \leqslant c_i \leqslant b$ and we therefore deduce from (2) that

$$\begin{aligned}(f(a))^{n\rho} = (f(a^n))^\rho &\leqslant b(1 + (f(b))^\rho + \cdots + (f(b))^{\rho q(n)}) \\ &\leqslant b(q(n) + 1)(\sup(1, (f(b))^\rho))^{q(n)}.\end{aligned}$$

Taking logarithms on both sides of this inequality and dividing by $n.\log a$, we obtain

$$g(a) \leqslant \frac{\log b}{n.\log a} + \frac{\log(q(n)+1)}{q(n)} \cdot \frac{q(n)}{n.\log a} + \frac{\sup(0, \rho \log f(b))}{\log a} \cdot \frac{q(n)}{n}. \tag{3}$$

Note now that as n tends to $+\infty$, $q(n)/n$ tends to $\log a/\log b$; therefore $q(n)$ tends to $+\infty$ and

$$\log(q(n) + 1)/q(n)$$

tends to 0 (*Functions of a Real Variable*, Chapter III, § 2, no. 1). Taking the limit in (3), we obtain

$$g(a) \leqslant \frac{\sup(0, \rho \log f(b))}{\log b} = \sup(0, g(b)). \tag{4}$$

But $f(h) > 1$, whence $g(h) > 0$; if a is replaced by h in (4), we obtain $\sup(0, g(b)) > 0$ and hence

$$\sup(0, g(b)) = g(b).$$

Then, for any integers a, b at least equal to 2, $g(a) \leqslant g(b)$ and therefore $g(a) = g(b)$, exchanging the roles of a and b. In other words, there exists a constant λ such that $g(a) = \lambda$ for every integer $a \geqslant 2$; if we write $s = \lambda/\rho$, then $f(a) = |a|^s$ for every integer $a \geqslant 2$. As $f(xy) = f(x)f(y)$ and $f(-x) = f(x)$ $f(x) = |x|^s$ for all $x \in \mathbf{Q}$. Finally, if $0 < s \leqslant 1$, we know that $x \mapsto |x|^s$ is an absolute value (*General Topology*, Chapter IX, § 3, no. 2); conversely, if s is such that $x \mapsto |x|^s$ is an absolute value on $\mathbf{Q}$, then $(1 + 1)^s \leqslant 1^s + 1^s$, that is $2^s \leqslant 2$, whence $s \leqslant 1$.

4. STRUCTURE OF FIELDS WITH A NON-ULTRAMETRIC ABSOLUTE VALUE

THEOREM 1 (Gelfand-Mazur). *Let* K *be an algebra over the field* $\mathbf{R}$ *with the two following properties:*

(1) K *is a (not necessarily commutative) field.*

(2) *There exists on* K *a norm* $x \mapsto \|x\|$ *compatible with the algebra structure on* K (*General Topology*, Chapter IX, § 3, no. 7, Definition 9).

Then the algebra K *is isomorphic to one of the algebras* $\mathbf{R}$, $\mathbf{C}$ *or* $\mathbf{H}$.

Recall (*loc. cit.*) that it may always be assumed that $\|xy\| \leqslant \|x\| . \|y\|$ for all x, y in K. We shall give K the topology (compatible with the algebra structure) defined by the norm.

(A) *First case:* K *is commutative and there exists* $j \in$ K *such that* $j^2 = -1$

Then there exists an isomorphism σ of the field $\mathbf{C}$ onto a subfield of K such that $\sigma(\xi + i\eta) = \xi.1 + \eta.j$ for ξ, η in $\mathbf{R}$. We shall prove by *reductio ad absurdum* that $\mathrm{K} = \sigma(\mathbf{C})$. Suppose then that there exists $x \in \mathrm{K} - \sigma(\mathbf{C})$; for all $z \in \mathbf{C}$, $x - \sigma(z)$ is therefore invertible in K; let us write $\mathrm{F}(z) = (x - \sigma(z))^{-1}$; as σ is continuous and the inverse is continuous on K (*General Topology*, Chapter IX, § 3, no. 7, Proposition 13 applied to the completion algebra of K), F is a continuous mapping of $\mathbf{C}$ to K. Moreover, we may write for $z \neq 0$

$$\mathrm{F}(z) = (\sigma(z))^{-1}(x(\sigma(z))^{-1} - 1)^{-1}.$$

But, as $(\sigma(z))^{-1} = \sigma(z^{-1})$ tends to 0 as z tends to infinity in $\mathbf{C}$, it is seen that $\mathrm{F}(z)$ tends to 0; in other words, $z \mapsto \|\mathrm{F}(z)\|$ is a continuous real-valued function, $\geqslant 0$ on $\mathbf{C}$, tending to 0 at the point at infinity and which can therefore be considered as a continuous function on the compact space $\check{\mathbf{C}}$ obtained by adjoining to $\mathbf{C}$ a point at infinity. The least upper bound α of $\|F\|$ on $\mathbf{C}$ is therefore finite and >0 and the set P of complex numbers z such that $\|F(z)\| = \alpha$ is closed and non-empty (*General Topology*, Chapter IV, § 6, no. 1, Theorem 1).

Let $z \in P$; let us write $y = x - \sigma(z)$ and let t be a complex number $\neq 0$ such that $\|\sigma(t)\| < \alpha^{-1}$, whence $\|\sigma(t).y^{-1}\| < 1$ by definition of α. The sequence of the $(\sigma(t)\,y^{-1})^n$ and that of the $n(\sigma(t)\,y^{-1})^n$ therefore tend to 0 in K as n tends to $+\infty$, for so do the corresponding sequences of norms in $\mathbf{R}$. On the other hand note that for every polynomial $H(T) = \prod_{k=1}^{p} (T - \sigma(c_k))$, where the c_k are distinct complex numbers, in the field K(T) of rational functions

$$\frac{H'(T)}{H(T)} = \sum_{k=1}^{p} \frac{1}{T - \sigma(c_k)}. \tag{5}$$

We apply this formula to the polynomial

$$H(T) = T^n - (\sigma(t))^n = \prod_{k=0}^{n-1} (T - \sigma(\omega_n^k t)),$$

where $\omega_n = \exp(2\pi i/n)$, and substitute for T the element $y \in K$, which is distinct from all the $\sigma(\omega_n^k t)$. It follows (in the commutative field K) that

$$\frac{ny^{n-1}}{y^n - (\sigma(t))^n} = \frac{1}{y - \sigma(t)} + \sum_{k=1}^{n-1} \frac{1}{y - \sigma(\omega_n^k t)}. \tag{6}$$

Taking account of the definitions of F and y, we obtain

$$\begin{aligned} (7)\quad & F(z + t) + \sum_{k=1}^{n-1} F(z + \omega_n^k t) - nF(z) \\ & = \frac{ny^{n-1}}{y^n - (\sigma(t))^n} - \frac{n}{y} = \frac{1}{y}\cdot\frac{n(\sigma(t)\,y^{-1})^n}{1 - (\sigma(t)\,y^{-1})^n}. \end{aligned}$$

But by virtue of the choice of t and the remarks made above, the last expression in (7) tends to 0 as n tends to $+\infty$; hence

$$\|F(z + t)\| = \lim_{n \to +\infty} \Big\|nF(z) - \sum_{k=1}^{n-1} F(z + \omega_n^k t)\Big\|. \tag{8}$$

Now, $\|F(z)\| = \alpha$ and $\|F(z + \omega_n^k t)\| \leqslant \alpha$ by definition of α, whence

$$\begin{aligned} & \Big\|nF(z) - \sum_{k=1}^{n-1} F(z + \omega_n^k t)\Big\| \geqslant \\ & \qquad n\|F(z)\| - \sum_{k=1}^{n-1} \|F(z + \omega_n^k t)\| \geqslant n\alpha - (n-1)\alpha = \alpha. \end{aligned}$$

Therefore by (8), letting n tend to $+\infty$, $\|F(z+t)\| \geqslant \alpha$ and by definition of α this implies

$$\|F(z+t)\| = \alpha,$$

in other words $z + t \in P$. This proves that the set P is *open* in **C**; as it is also closed and non-empty and **C** is connected, $P = \mathbf{C}$ and $\|F\|$ is therefore constant on **C**; as this function tends to 0 at the point at infinity, $\|F(z)\| = 0$ in **C** and in particular $\|F(0)\| = \|x^{-1}\| = 0$, which is absurd.

(B) *Second case; K is commutative and* -1 *is not the square of an element of* K

Let L be the commutative field obtained by adjoining to K a root j of $T^2 + 1$; L is a vector space over K admitting $(1, j)$ as a basis and L is obviously an algebra over **R**. Clearly the function $x + yj \to \|x\| + \|y\|$ is a norm on L compatible with its structure as a vector space over **R**; on the other hand, for $z = x + yj$, $z' = x' + y'j$ in L,

$$\begin{aligned} \|zz'\| &= \|xx' - yy'\| + \|xy' + x'y\| \leqslant \|xx'\| + \|yy'\| + \|xy'\| + \|x'y\| \\ &\leqslant \|x\|.\|x'\| + \|y\|.\|y'\| + \|x\|.\|y'\| + \|x'\|.\|y\| \\ &= (\|x\| + \|y\|)(\|x'\| + \|y'\|) = \|z\|.\|z'\|. \end{aligned}$$

The norm thus defined is consequently compatible with the **R**-algebra structure on L. By case (A) L is an **R**-algebra isomorphic to **C**; now the only sub-**R**-algebra of **C** distinct from **C** is **R** and hence K is isomorphic to **R**.

(C) *Third case:* K *is not commutative*

Let Z be the centre of K and x an element of K not in Z; the subfield $Z(x)$ of K is commutative and the norm induced by that on K is compatible with the **R**-algebra structure on $Z(x)$; as $Z \neq Z(x)$ and Z and $Z(x)$ are **R**-algebras isomorphic to **R** or **C** by virtue of (A) and (B), Z is necessarily isomorphic to **R** and $Z(x)$ to **C**. For all $x \in K$, $Z(x)$ is therefore of rank $\leqslant 2$ over Z. Now we have the following lemma:

Lemma 1. *Let* D *be a field with centre* L *such that, for all* $x \in D$, $L(x)$ *is an extension of* L *of degree* $\leqslant m$. *Then the rank of* D *over* L *is* $\leqslant m^2$.

We may obviously restrict our attention to the case where $D \neq L$. Then there exists in D a finite *separable* algebraic commutative extension E of L of degree >1 (*Algebra*, Chapter VIII, § 10, no. 3, Lemma 1); as $E = L(x)$ for some suitable x in E (*Algebra*, Chapter V, § 7, no. 7, Proposition 12 and Chapter VII, § 5, no. 7), by hypothesis $[E:L] \leqslant m$. Suppose the separable extension E is taken such that $[E:L]$ is finite and as great as possible and consider the *centralizer* $E' \supset E$ of E in D, which is a field of centre E such that

$$[D:E'] = [E:L] \leqslant m$$

(*Algebra*, Chapter VIII, § 10, no. 2, Theorem 2). If $E \neq E'$, there would exist in E' a finite separable algebraic extension F of E of degree >1 (*Algebra*, Chapter VIII, § 10, no. 3, Lemma 1); F would therefore be a finite separable algebraic extension of L (*Algebra*, Chapter V, § 7, no. 4, Proposition 7) of degree $>[E:L]$, contrary to the definition of E; therefore $E' = E$, whence $[D:L] = [D:E][E:L] \leqslant m^2$.

Applying this lemma to K with $m = 2$, it is seen that K is a non-commutative extension field of **R** of finite rank and hence isomorphic to the field of quaternions **H** (*Algebra*, Chapter VIII, § 11, no. 2, Theorem 2).

Remark (1) We shall give in the chapter devoted to normed algebras a shorter proof of the Gelfand-Mazur Theorem which is valid for every Hausdorff locally convex topological algebra K over **R** and whose principle is the following: it is reduced (as in cases (B) and (C)) to the case where K is a commutative algebra *over* **C**; if $x \in K - \mathbf{C}.1$, we consider as above the mapping $z \mapsto (x - z.1)^{-1}$ of **C** to K, which is continuous and differentiable on **C**. For every element x' of the *dual* K' of the locally convex space K, $z \mapsto \langle (x - z.1)^{-1}, x' \rangle$ is then a *bounded integral* function on **C** and therefore constant by Liouville's Theorem and we conclude as in part (A) of the proof of Theorem 1 that this necessarily implies $\langle (x - z.1)^{-1}, x' \rangle = 0$ for all $z \in \mathbf{C}$ and all $x' \in K'$; the Hahn-Banach Theorem shows that this conclusion is absurd, since $(x - z.1)^{-1} \neq 0$. Note that the argument in part (A) of the proof of Theorem 1 differs from the above only in appearance, for this argument is only a special case of that which serves to prove the maximum principle for analytic functions, the summation over the roots of unity and and passing to the limit being equivalent to calculating the integral $\int_\gamma \frac{F(z+t)\,dt}{t}$ along a circle of centre 0 and the use of Cauchy's formula being avoided here, thanks to the particular form of the function F.

THEOREM 2 (Ostrowski). *Let* K *be a (not necessarily commutative) field and* f *an element of* $\mathscr{V}(K)$ *which is not an ultrametric absolute value. Then there exist a unique real number* $s > 0$ *and an isomorphism* j *of* K *onto an everywhere dense subfield of one of the fields* **R**, **C** *or* **H** *such that* $f(x) = |j(x)|^s$ *for all* $x \in K$ (*). *For* f *to be an absolute value on* K, *it is necessary and sufficient that* $s \leqslant 1$.

By no. 2, Corollary to Proposition 3, K is of characteristic 0 and hence an algebra over **Q**; for all $x \in \mathbf{Q}$ we write $h(x) = f(x.1)$; clearly $h \in \mathscr{V}(\mathbf{Q})$ and therefore Proposition 4 of no. 3 may be applied; neither of cases (i) and (ii) of the statement of this proposition can hold, for this would imply $f(n.1) \leqslant 1$ for every integer $n > 0$ and f would be an ultrametric absolute value by virtue

(*) On **H** we write $|z|^2 = z.\bar{z} = \bar{z}.z$, $\bar{z}$ being the conjugate quaternion of z.

of no. 2, Proposition 3. Then there exists a real number $s > 0$ such that $h(x) = |x|^s$ for all $x \in \mathbf{Q}$, that is $f(x.1) = |x|^s$; we write $g = f^{1/s}$. Then $g \in \mathscr{V}(\mathrm{K})$ and $g(n.1) = n$ for every integer n; Proposition 2 of no. 1 therefore shows that g is an absolute value on K.

For $x \in \mathbf{Q}$ and $y \in \mathrm{K}$, $g(xy) = |x|g(y)$ and hence g is a *norm* on K compatible with its $\mathbf{Q}$-algebra structure (with the usual absolute value on $\mathbf{Q}$). The completion $\hat{\mathrm{K}}$ of K is therefore a normed algebra over $\hat{\mathbf{Q}} = \mathbf{R}$ (*General Topology*, Chapter IX, § 3, no. 7); let $\hat{g}$ be the norm on $\hat{\mathrm{K}}$ the continuous extension of g. As g is an absolute value on K, $\hat{\mathrm{K}}$ is a field and $\hat{g}$ an absolute value on $\hat{\mathrm{K}}$ (*General Topology*, Chapter IX, § 3, no. 3, Proposition 6). By Theorem 1 there exists an $\mathbf{R}$-algebra isomorphism $\hat{j}$ of $\hat{\mathrm{K}}$ onto one of the fields $\mathbf{R}$, $\mathbf{C}$ or $\mathbf{H}$ and $g'(x) = |\hat{j}(x)|$ is therefore an absolute value on $\hat{\mathrm{K}}$; as $\hat{\mathrm{K}}$ is finite-dimensional over R and g' and $\hat{g}$ coincide on the subfield $\mathbf{R}.1$ of $\hat{\mathrm{K}}$, $g' = \hat{g}$ by the following lemma:

Lemma 2. *Let* L *be a (not necessarily commutative) field and* K *a subfield of* L *such that* L *is a finite-dimensional left vector space over* K. *Let g be an absolute value on* L *and f its restriction to* K. *If* K *is complete and not discrete with respect to f,* L *is complete with respect to g; if further g' is another absolute value on* L *with the same restriction f to* K, *then $g' = g$.*

As the topology defined by g is Hausdorff and compatible with the left vector K-space structure on L, the first assertion follows from *Topological Vector Spaces*, Chapter I, § 2, no. 3, Theorem 2. Moreover the topologies on L defined by g and g' are identical (*loc. cit.*); there therefore exists a real number $s > 0$ such that $g' = g^s$ (*General Topology*, Chapter IX, § 3, no. 2, Proposition 5). Let x be an element of K such that $f(x) \neq 1$; the equation $g'(x) = g(x)$ proves that $s = 1$.

Returning to the proof of Theorem 2, it is seen that, if j denotes the restriction of $\hat{j}$ to K, j is an isomorphism of K onto an everywhere dense subfield of $\mathbf{R}$, $\mathbf{C}$ or $\mathbf{H}$ and $g(x) = |j(x)|$ for $x \in \mathrm{K}$, whence $f(x) = |j(x)|^s$.

Finally note that, if f is an absolute value on K, h is an absolute value on $\mathbf{Q}$ and $s \leqslant 1$ by no. 3, Proposition 4; conversely, if $s \leqslant 1$, $f = g^s$ is an absolute value on K since g is (*General Topology*, Chapter IX, § 3, no. 2); this proves the last assertion of the statement.

Remarks

(2) If K is a field and a normed algebra over $\mathbf{R}$, the norm is not necessarily an absolute value on K; for example, $\xi + i\eta \to |\xi| + |\eta|$ is a norm on $\mathbf{C}$ compatible with its $\mathbf{R}$-algebra structure.

(3) For a proof of case (C) of Theorem 1 not using the general results of *Algebra*, Chapter VIII, see Exercise 2.

7. THE APPROXIMATION THEOREM

1. THE INTERSECTION OF A FINITE NUMBER OF VALUATION RINGS

PROPOSITION 1. *Let* K *be a field,* $(A_i)_{1 \leqslant i \leqslant n}$ *a finite family of valuation rings of* K *and* $B = \bigcap_{i=1}^{n} A_i$. *We write* $\mathfrak{p}_i = B \cap \mathfrak{m}(A_i)$. *Then* $A_i = B_{\mathfrak{p}_i}$ *for all* i *and the field of fractions of* B *is* K.

Clearly $B_{\mathfrak{p}_i} \subset A_i$. To prove the converse inclusion we need the following lemma:

LEMMA 1. *Let* v_i $(1 \leqslant i \leqslant n)$ *be valuations on the field* K *and* $x \in K^*$. *Then there exists a polynomial* $f(X)$ *of the form*

$$(1)\quad f(X) = 1 + n_1X + \cdots + n_{k-1}X^{k-1} + X^k \qquad (k \geqslant 2,\ n_j \in \mathbf{Z} \text{ for } 1 \leqslant j \leqslant k-1)$$

such that $f(x) \neq 0$ *and the element* $z = f(x)^{-1}$ *enjoys the following properties for* $1 \leqslant i \leqslant n$:

$$\begin{aligned} v_i(z) &= 0 && \text{if } v_i(x) \geqslant 0 \\ v_i(z) + v_i(x) &> 0 && \text{if } v_i(x) < 0. \end{aligned}$$

Assuming this lemma for a moment, we show how it implies that $A_1 \subset B_{\mathfrak{p}_1}$. Let x be a non-zero element of A_1. We apply the lemma to x and valuations v_i associated with the A_i. Then $v_i(z) \geqslant 0$ and $v_i(zx) \geqslant 0$ for all i, hence $z \in B$ and $zx \in B$. As $v_1(x) \geqslant 0$, $v_1(z) = 0$ and hence $z \notin \mathfrak{p}_1$. Hence $x = xz/z \in B_{\mathfrak{p}_1}$. The field of fractions of B then contains A_1 and hence is K.

We now pass to the proof of the lemma. Let I be the set of indices i such that $v_i(x) \geqslant 0$. For all $i \in I$, let $\bar{x}_i$ denote the canonical image of x in $\kappa(A_i)$. For all $i \in I$ we construct a polynomial f_i as follows: if there exists a polynomial $g(X)$ of the form (1) such that $g(\bar{x}_i) = 0$ in $\kappa(A_i)$, we take f_i to be such a polynomial; otherwise we take $f_i = 1$. Then we write $f(X) = 1 + X^2 \prod_{i \in I} f_i(X)$. It is obviously a polynomial of the form (1). If $i \in I$, then $f(x) \in A_i$ and also $f(\bar{x}_i) \neq 0$ by construction; hence $f(x) \notin \mathfrak{m}(A_i)$, $v_i(f(x)) = 0$ and $v(z) = 0$. If $i \notin I$, then $v_i(x) < 0$, whence $v_i(f(x)) = kv_i(x)$ (§ 3, no. 1, Proposition 1) and

$$v_i(x) + v_i(z) = (1 - k)v_i(x) > 0$$

(since $k \geqslant 2$). Whence the lemma.

PROPOSITION 2. *With the hypotheses of Proposition* 1 *suppose further that* $A_i \not\subset A_j$ *for* $i \neq j$. *Then the* $\mathfrak{p}_i$ *are distinct maximal ideals of* B *and every maximal ideal of* B *is equal to one of the* $\mathfrak{p}_i$.

If $\mathfrak{p}_i \subset \mathfrak{p}_j$ for $i \neq j$, $A_i = B_{\mathfrak{p}_i} \supset B_{\mathfrak{p}_j} = A_j$. It is then sufficient to apply Chapter II, § 3, no. 5, Corollary to Proposition 17.

COROLLARY 1. *Suppose that* $A_i \not\subset A_j$ *for* $i \neq j$. *For every family of elements* $a_i \in A_i$ $(1 \leqslant i \leqslant n)$, *there exists* $x \in B$ *such that* $x \equiv a_i$ (mod. $\mathfrak{m}(A_i)$) *for* $1 \leqslant i \leqslant n$.

Since the $\mathfrak{p}_i$ are maximal ideals of B, $A_i/\mathfrak{m}(A_i) = B_{\mathfrak{p}_i}/\mathfrak{p}_i B_{\mathfrak{p}_i} = B/\mathfrak{p}_i$ and it may therefore be assumed that $a_i \in B$ for all i. The corollary then follows from the fact that the canonical mapping from B to $\prod_{i=1}^{n} (B/\mathfrak{p}_i)$ is surjective (Chapter II, § 1, no. 2, Proposition 5).

COROLLARY 2. *Suppose that* $A_i \not\subset A_j$ *for* $i \neq j$. *There exist elements* x_i $(1 \leqslant i \leqslant n)$ *of* K *such that* $v_i(x_i) = 0$ *and* $v_j(x_i) > 0$ *for* $i \neq j$.

For each index i apply Corollary 1 to the family (a_j) such that $a_i = 1$ and $a_j = 0$ for $j \neq i$.

COROLLARY 3. *Every valuation ring of* K *containing* B *contains one of the* A_i.

We may confine our attention to the case where $A_i \not\subset A_j$ for $i \neq j$. Let V be a valuation ring of K containing B. We write

$$\mathfrak{p} = \mathfrak{m}(V) \cap B.$$

There exists a maximal ideal $\mathfrak{p}_i$ of B containing $\mathfrak{p}$, whence

$$A_i = B_{\mathfrak{p}_i} \subset B_{\mathfrak{p}} \subset V.$$

2. INDEPENDENT VALUATIONS

DEFINITION 1. *Let* A *and* A′ *be two valuation rings of the same field* K. A *and* A′ *are called independent if* K *is the ring generated by* A *and* A′. *Two valuations on* K *are called independent if their rings are independent and dependent otherwise.*

An improper valuation on K is independent of every valuation on K. For two valuations *of height* 1 on K to be independent, it is necessary and sufficient that they be not equivalent (§ 4, no. 5, Proposition 6 (c)).

THEOREM 1 (Approximation Theorem for Valuations). *Let* v_i $(1 \leqslant i \leqslant n)$ *be valuations on a field* K *which are independent in pairs and* Γ_i *the order group of* v_i. *Let* $a_i \in K$ *and* $\alpha_i \in \Gamma_i$ $(1 \leqslant i \leqslant n)$. *Then there exists* $x \in K$ *such that* $v_i(x - a_i) \geqslant \alpha_i$ *for all* i.

If v_i is improper, then $\alpha_i = 0$ and the relation $v_i(x - a_i) \geqslant \alpha_i$ is true for all $x \in K$. We may therefore assume that the v_i are *not improper*.

Let A_i be the ring of v_i, $B = \bigcap_{i=1}^{n} A_i$ and $\mathfrak{p}_i = \mathfrak{m}(A_i) \cap B$. By Proposition 1 of no. 1, the a_i may be written $a_i = b_i/s$ $(b_i \in B,\ s \in B - \{0\})$; if we write $x = y/s$ and $\alpha'_i = \alpha_i + v_i(s)$, then $v_i(y - b_i) \geqslant \alpha'_i$. This shows that we may assume that $a_i \in B$ for all i; we may also assume that $\alpha_i > 0$ for all i. Let $\mathfrak{v}_i$ be the set of

$z \in K$ such that $v_i(z) \geqslant \alpha_i$; we write $\mathfrak{q}_i = \mathfrak{v}_i \cap B$. For $x \in B$, $v_i(x - a_i) \geqslant \alpha_i$ is equivalent to $x \equiv a_i\ (\mathfrak{q}_i)$. We therefore need to show that the canonical homomorphism $B \to \prod_{i=1}^{n} (B/\mathfrak{q}_i)$ is surjective, that is that $\mathfrak{q}_i + \mathfrak{q}_j = B$ for $i \neq j$ (Chapter II, § 1, no. 2, Proposition 5). As the maximal ideals of B are the $\mathfrak{p}_i$ (Proposition 2), it will suffice for this to show that $\mathfrak{q}_i \not\subset \mathfrak{p}_j$ for $i \neq j$.

Suppose that there exists i, j such that $\mathfrak{q}_i \subset \mathfrak{p}_j$ and $i \neq j$. We shall see shortly that the radical of $\mathfrak{q}_i$ is a *prime* ideal $\mathfrak{p}$ of B. Then $\mathfrak{p} \subset \mathfrak{p}_j$ and also $\mathfrak{p} \subset \mathfrak{p}_i$ since $\alpha_i > 0$ and hence $\mathfrak{q}_i \subset \mathfrak{p}_i$. Therefore $A_j = B_{\mathfrak{p}_j} \subset B_{\mathfrak{p}}$ (no. 1, Proposition 1) and similarly $A_i \subset B_{\mathfrak{p}}$. Now, as $\mathfrak{v}_i \neq (0)$ and $\mathfrak{v}_i = B_{\mathfrak{p}_i}\mathfrak{q}_i$ (Chapter II, § 2, no. 4, Proposition 10), $\mathfrak{q}_i \neq (0)$, whence $\mathfrak{p} \neq (0)$ and $B_{\mathfrak{p}} \neq K$. This contradicts the hypothesis that A_i and A_j are independent.

It remains to show that $\mathfrak{p}$ is prime. Now this follows from the following lemma:

Lemma 2. *Let A be a valuation ring and $\mathfrak{b}$ an ideal of A distinct from A. Then the radical $\mathfrak{r}$ of $\mathfrak{b}$ is a prime ideal.*

Suppose that $xy \in \mathfrak{r}$. Then there exists $n \geqslant 1$ such that $(xy)^n \in \mathfrak{b}$. Let v be a valuation associated with A. If, for example, $v(x) \geqslant v(y)$, then

$$v(x^{2n}) \geqslant v(x^n y^n),$$

whence $x^{2n} \in \mathfrak{b}$ and $x \in \mathfrak{r}$.

Corollary 1. *For every family of elements $\gamma_i \in \Gamma_i$ $(1 \leqslant i \leqslant n)$, there exists $x \in K$ such that $v_i(x) = \gamma_i$ $(1 \leqslant i \leqslant n)$.*

We may assume that $A_i \neq K$ for all i. Then, there exists for all i an $a_i \in K$ such that $v_i(a_i) = \gamma_i$ and an $\alpha_i \in \Gamma_i$ such that $\gamma_i < \alpha_i$. We apply Theorem 1 to these elements a_i: there exists $x \in K$ such that $v_i(x - a_i) > v_i(a_i)$; whence, as $x = a_i + (x - a_i)$, $v_i(x) = v_i(a_i) = \gamma_i$ (§ 3, no. 1, Proposition 1).

Corollary 2. *Let $\mathcal{T}_i$ be the topology defined on K by v_i; let K^n be given the topology the product of the $\mathcal{T}_i$. If the v_i are not improper, the diagonal of K^n is dense in K^n.*

Proposition 3. *Let v and v' be two non-improper valuations on the same field K. For v and v' to define the same topology on K, it is necessary and sufficient that they be dependent.*

Suppose that the topologies $\mathcal{T}_v$ and $\mathcal{T}_{v'}$, defined by v and v', are identical. Since $\mathcal{T}_v$ is Hausdorff, the diagonal of K^2 is closed and hence v and v' are dependent (Corollary 2 to Theorem 1).

Conversely, suppose that v and v' are dependent. Then their rings A and A′ are contained in the same ring A″ distinct from K and A″ is the ring of a valuation v'' (§ 4, no. 1, Proposition 1). It suffices to show that the topology $\mathcal{T}_{v''}$ is identical with $\mathcal{T}_v$. Let Γ and Γ'' be the order groups of v and v''. There exists an increasing homomorphism λ of Γ onto Γ'' such that $v'' = \lambda \circ v$ (§ 4, no. 3). If $\alpha'' \in \Gamma''$, let $\alpha \in \overset{-1}{\lambda}(\alpha'')$; the condition $v(x) \geqslant \alpha$ implies $v''(x) \geqslant \alpha''$. Let $\beta \in \Gamma$ and $\beta'' = \lambda(\beta)$; the condition $v(x) \leqslant \beta$ implies $v''(x) \leqslant \beta''$ and hence the condition $v''(x) > \beta''$ implies $v(x) > \beta$. As v and v'' are not improper, the inequalities in question define fundamental systems of neighbourhoods of 0 for $\mathscr{T}_v$ and $\mathscr{T}_{v''}$. Hence $\mathscr{T}_v = \mathscr{T}_{v''}$, which completes the proof.

Remarks

(1) Proposition 3 shows that the relation "v and v' are dependent" is an *equivalence relation.*

(2) Taking account of the relations between valuations of height 1 and ultrametric absolute values (§ 6, no. 2), Proposition 3 also follows, in the case of valuations of height 1, from the characterization of equivalent absolute values (*General Topology*, Chapter IX, § 3, no. 2, Proposition 5).

PROPOSITION 4. *Let $v_1, \ldots, v_n$ $(n \geqslant 2)$ be pairwise dependent valuations on the same field* K. *Then the rings* $A_1, \ldots, A_n$ *of* $v_1, \ldots, v_n$ *generate a subring of* K *distinct from* K.

For $n = 2$ Proposition 4 follows from Definition 1. Suppose it holds for $n - 1$ valuations. Then there exists a subring A of K distinct from K and containing $A_1, \ldots, A_{n-1}$; there also exists a subring $B \neq K$ containing A_{n-1} and A_n. As A and B contain A_{n-1}, they are comparable with respect to inclusion (§ 4, no. 1, Corollary to Proposition 1). The greater of these two therefore contains all the A_i.

3. THE CASE OF ABSOLUTE VALUES

THEOREM 2 (Approximation Theorem for Absolute Values). *Let f_i $(1 \leqslant i \leqslant n)$ be absolute values on the same field* K *which are not improper and no two of which are equivalent. Let a_i $(1 \leqslant i \leqslant n)$ be elements of* K *and* ε *a real number* >0. *Then there exists $x \in K$ such that $f_i(x - a_i) \leqslant \varepsilon$ for all i.*

Let K_i denote the field K with the topology defined by f_i. The result to be proved is equivalent to the following: in the product $P = K_1 \times \cdots \times K_n$, the closure $\overline{D}$ of the diagonal D is equal to P. This is obvious for $n = 1$. Suppose that this point has been established in the case of k absolute values for $k < n$.

We show first that there exists, for $2 \leqslant h \leqslant n$, an element x_h of K such that $f_1(x_h) < 1$, $f_2(x_h) > 1$ and $f_i(x_h) \neq 1$ for $3 \leqslant i \leqslant h$. We argue by induction on h. If $h = 2$, this follows from the fact that f_1 and f_2 are not equivalent. We therefore suppose that the existence of x_{h-1} has been shown and prove that of x_h. If $f_h(x_{h-1}) \neq 1$, we may take $x_h = x_{h-1}$; if $f_h(x_{h-1}) = 1$, we choose $z \in K^*$ such that $f_h(z) \neq 1$ and $x_h = z(x_{h-1})^s$ solves the problem for s sufficiently large. We have thus proved the existence of x_h.

As the integer q tends to infinity, $f_1(x_n^q)$ tends to 0, $f_2(x_n^q)$ tends to $+\infty$ and $f_i(x_n^q)$ tends to 0 or $+\infty$ for $i \geqslant 3$. Writing $y_q = x_n^q(1 + x_n^q)^{-1}$,

$$1 - y_q = (1 + x_n^q)^{-1};$$

hence the sequence (y_q) tends to 0 in K_1, to 1 in K_2 and to 0 or 1 in K_i for $i \geqslant 3$. By changing the numbering of the K_i, it may therefore be assumed that there exists an integer r $(1 \leqslant r < n)$ such that $\overline{D}$ contains the point $(e_1, \ldots, e_n)$ where $e_i = 1$ for $1 \leqslant i \leqslant r$ and $e_i = 0$ for $r + 1 \leqslant i \leqslant n$. Now, $\overline{D}$ is a vector sub-K-space of P. Hence $\overline{D}$ contains the diagonals D′ and D″ of

$$P' = K_1 \times \cdots \times K_r$$

and $P'' = K_{r+1} \times \cdots \times K_n$. By the induction hypothesis, $P' = \overline{D}'$ and $P'' = \overline{D}''$. Hence $\overline{D} = P$.

8. EXTENSIONS OF A VALUATION TO AN ALGEBRAIC EXTENSION

1. RAMIFICATION INDEX. RESIDUE CLASS DEGREE

Let K be a field, L an extension of K and A′ a valuation ring of L. As has been seen in § 1, no. 4, the ring $A = K \cap A'$ is a valuation ring of K and

$$\mathfrak{m}(A) = \mathfrak{m}(A') \cap K.$$

If v' is a valuation associated with A′, the restriction v of v' to K is a valuation on K associated with A; the order group Γ_v of v is a subgroup of the order group $\Gamma_{v'}$ of v'.

DEFINITION 1. *The index* $(\Gamma_v : \Gamma_{v'})$ *is called the ramification index of* v' *over* v (or over K) *and denoted by* $e(v'/v)$ (or $e(A'/A)$, or sometimes $e(L/K)$).

This index is a natural number or $+\infty$. If v'_0 is a valuation *equivalent* to v', $e(v'/v)$ is also called the ramification index of v'_0 *over* v. If $e(v'/v) = 1$, v' is called *unramified* over v.

On the other hand the residue field $\kappa(A)$ of v is identified with a subfield of the residue field $\kappa(A')$ of v'.

DEFINITION 2. *The degree* $[\kappa(A') : \kappa(A)]$ *is called the residue class degree of* v' *over* v (or over K) *and denoted by* $f(v'/v)$ (or $f(A'/A)$, or sometimes $f(L/K)$).

This degree is a natural number or $+\infty$.

LEMMA 1. *Let* K, K′, K″ *be three fields such that* $K \subset K' \subset K''$, v'' *a valuation on* K″ *and* v *and* v' *its restriction to* K *and* K′. *Then there are the relations:*

(1) $$e(v''/v) = e(v''/v')e(v'/v), \qquad f(v''/v) = f(v''/v')f(v'/v).$$

This is obvious.

LEMMA 2. *Let* K *be a field,* L *a finite extension of* K *of degree* n, v' *a valuation on* L *and* v *its restriction to* K. *Then the inequality*

$$e(v'/v)f(v'/v) \leqslant n \tag{2}$$

holds; in particular $e(v'/v)$ *and* $f(v'/v)$ *are finite.*

Let us take natural numbers r and s respectively not greater than $e(v'/v)$ and $f(v'/v)$. It suffices to show that $rs \leqslant n$. In view of the definition of r, there exist elements x_i of L $(1 \leqslant i \leqslant r)$ such that $v'(x_i) \not\equiv v'(x_j) \pmod{\Gamma_v}$ for $i \neq j$. In view of the definition of s, there exist elements y_k $(1 \leqslant k \leqslant s)$ of the ring A′ of v' whose canonical images $\bar{y}_k$ in $\kappa(\mathrm{A}')$ are linearly independent over $\kappa(\mathrm{A})$; obviously $v'(y_k) = 0$ for all k. We shall show that the rs elements x_iy_k are linearly independent over K, which will certainly establish the inequality $rs \leqslant n$.

Suppose then that there exists a non-trivial linear relation of the form

$$\sum_{i,k} a_{ik}x_iy_k = 0 \quad (a_{ik} \in \mathrm{K}). \tag{3}$$

Let us choose the indices j, m so that

$$v'(a_{jm}x_jy_m) \leqslant v'(a_{ik}x_iy_k)$$

for every ordered pair (i, k); then $a_{jm} \neq 0$. If $i \neq j$, then $v'(a_{ik}x_iy_k) = v'(a_{jm}x_jy_m)$ is impossible for this would imply

$$v'(x_i) - v'(x_j) = v'(a_{jm}) - v'(a_{ik}) \in \Gamma_v,$$

contrary to the choice of the x_i. Multiplying (3) by $(a_{jm}x_j)^{-1}$, we obtain a relation of the form

$$\sum_k b_ky_k + z = 0, \quad \text{where} \quad b_k = \frac{a_{jk}x_j}{a_{jm}x_j} \in \mathrm{A}', \qquad z \in \mathrm{A}'$$

and $v'(b_k) \geqslant 0$, $v'(z) > 0$. Whence, in $\kappa(\mathrm{A}')$, a relation of the form $\sum_k \bar{b}_k\bar{y}_k = 0$. As $b_m = 1$, this contradicts the hypothesis made on y_k.

PROPOSITION 1. *Let* K *be a field,* L *an algebraic extension of* K, v' *a valuation on* L, v *its restriction to* K *and* A *and* A′ *the rings of* v *and* v'. *Then* $\Gamma_{v'}/\Gamma_v$ *is a torsion group and* $\kappa(\mathrm{A}')$ *is an algebraic extension of* $\kappa(\mathrm{A})$.

Let (L_α) be the family of finite sub-extensions of L; let us write $\Gamma_\alpha = v'(\mathrm{L}_\alpha^*)$. The group $\Gamma_{v'}$ is the union of the right directed family consisting of the Γ_α; as the groups Γ_α/Γ_v are finite (Lemma 2), $\Gamma_{v'}/\Gamma_v$ is a torsion group. The argument is similar to prove that $\kappa(\mathrm{A}')$ is an algebraic extension of $\kappa(\mathrm{A})$.

COROLLARY 1. *The height of v' is equal to that of v.*

This follows from Proposition 1 and the following lemma:

LEMMA 3. *Let* G' *be a totally ordered group,* G *a subgroup of* G' *and* $\mathfrak{S}'$ (resp. $\mathfrak{S}$) *the set of isolated subgroups of* G' (resp. G). *The mapping* $H' \mapsto H' \cap G$ *maps* $\mathfrak{S}'$ *onto* $\mathfrak{S}$. *This mapping is bijective if* G'/G *is a torsion group.*

Clearly $H' \in \mathfrak{S}'$ implies $H' \cap G \in \mathfrak{S}$. Now let $H \in \mathfrak{S}$; let H' denote the set of $x' \in G'$ such that there exists $h \in H$ satisfying $-h \leqslant x' \leqslant h$; it is immediately verified that H' is an isolated subgroup of G'; then $H' \cap G = H$ since H is isolated; hence the mapping $H' \mapsto H' \cap G$ is surjective. Suppose finally that G'/G is a torsion group; let H'_1 and H'_2 be two isolated subgroups of G' such that $H'_1 \cap G = H'_2 \cap G$; then, for example, $H'_1 \supset H'_2$ (cf. § 4, no. 4); then H'_1/H'_2 is a totally ordered group and is isomorphic to a quotient group of $H'_1/(H'_1 \cap G)$ which itself is identified with a subgroup of G'/G; hence H'_1/H'_2 is a torsion group and therefore reduces to 0.

COROLLARY 2. *For v' to be improper* (resp. *of height* 1), *it is necessary and sufficient that v be improper* (resp. *of height* 1).

COROLLARY 3. *Suppose that* L *is a finite extension of* K. *For v' to be discrete, it is necessary and sufficient that v be discrete.*

If v' is discrete, Γ_v is isomorphic to a non-zero subgroup of $\mathbf{Z}$ (Corollary 2) and hence to $\mathbf{Z}$. Conversely, if v is discrete, Γ_v is isomorphic to $\mathbf{Z}$ and $\Gamma_{v'}/\Gamma_v$ is a finite group (Lemma 2); hence $\Gamma_{v'}$ is a finitely generated commutative group of rank 1 and torsion-free; consequently it is isomorphic to $\mathbf{Z}$.

2. EXTENSION OF A VALUATION AND COMPLETION

DEFINITION 3. *Let* K *be a field, v a valuation on* K *and* L *an extension of* K. *A family* $(v'_\iota)_{\iota \in I}$ *of valuations on* L *which extend v and such that every valuation on* L *extending v is equivalent to a unique v'_ι is called a complete system of extensions of v to* L.

PROPOSITION 2. *Let* K *be a field, v a valuation on* K, $\hat{K}$ *the completion of* K *with respect to v, $\hat{v}$ the continuous extension of v to* $\hat{K}$ *and* L *a finite extension of* K *of degree n.*

(a) *Let v' be a valuation on* L *extending v; let* $\hat{L}_{v'}$ *denote the completion of* L *with respect to v' and $\hat{v}'$ the continuous extension of v' to* $\hat{L}_{v'}$; *identifying* $\hat{K}$ *with the closure of* K *in* $\hat{L}_{v'}$,

$$e(\hat{v}'/\hat{v}) = e(v'/v), \qquad f(\hat{v}'/\hat{v}) = f(v'/v), \tag{4}$$

$$[\hat{L}_{v'} : \hat{K}] \leqslant n, \tag{5}$$

$$e(v'/v)f(v'/v) \leqslant [\hat{L}_{v'} : \hat{K}]. \tag{6}$$

(b) *Every set of pairwise independent valuations on* L *extending a non-improper valuation v is finite. Let $v'_1, \ldots, v'_s$ denote pairwise independent valuations on* L *extending*

v such that every valuation on L extending v is dependent on one of the v'_i; let L_i be the field L with the topology defined by v'_i and $\hat{\mathrm{L}}_i$ its completion; we write $n_i = [\hat{\mathrm{L}}_i : \hat{\mathrm{K}}]$. Then the canonical mapping

$$\phi : \hat{\mathrm{K}} \otimes_{\mathrm{K}} \mathrm{L} \to \prod_{i=1}^{s} \hat{\mathrm{L}}_i$$

(extending by continuity the diagonal mapping $\mathrm{L} \to \prod_{i=1}^{s} \mathrm{L}_i$) is surjective, its kernel is the Jacobson radical of $\hat{\mathrm{K}} \otimes_{\mathrm{K}} \mathrm{L}$ and

$$\sum_{i=1}^{s} n_i \leqslant n. \tag{7}$$

Let us first prove (a). Suppose that v is not improper. As v and $\hat{v}$ (resp. v' and $\hat{v}'$) have the same order group and the same residue field (§ 5, no. 3, Proposition 5 (b) and (f)), (4) holds. We deduce (6) from it by means of Lemma 2. Finally the vector sub-$\hat{\mathrm{K}}$-space of $\hat{\mathrm{L}}_{v'}$ generated by L is closed (§ 5, no. 2, Corollary to Proposition 4) and everywhere dense and hence equal to $\hat{\mathrm{L}}_{v'}$; this shows (5).

We now pass to (b). We may still assume that v is not improper. Let $(v'_1, \ldots, v'_r)$ be any finite family of pairwise independent valuations on L extending v. The image of L in $\prod_{i=1}^{r} \mathrm{L}_i$ under the diagonal mapping is everywhere dense (§ 7, no. 2, Theorem 1) and $\prod_{i=1}^{r} \mathrm{L}_i$ is dense in $\prod_{i=1}^{r} \hat{\mathrm{L}}_i$. Hence the canonical image of $\hat{\mathrm{K}} \otimes_{\mathrm{K}} \mathrm{L}$ in $\prod_{i=1}^{r} \hat{\mathrm{L}}_i$ is everywhere dense. On the other hand this image is a vector sub-$\hat{\mathrm{K}}$-space of $\prod_{i=1}^{r} \hat{\mathrm{L}}_i$; as $\prod_{i=1}^{r} \hat{\mathrm{L}}_i$ is of finite dimension over $\hat{\mathrm{K}}$ by (5), the image of $\hat{\mathrm{K}} \otimes_{\mathrm{K}} \mathrm{L}$ is closed (§ 5, no. 2, Corollary to Proposition 4) and hence equal to $\prod_{i=1}^{r} \hat{\mathrm{L}}_i$. As the dimension of $\hat{\mathrm{K}} \otimes_{\mathrm{K}} \mathrm{L}$ over $\hat{\mathrm{K}}$ is n, $\sum_{i=1}^{r} n_i \leqslant n$. This shows in particular that the integer r is bounded above by n and shows the first assertion of (b).

We now take $(v'_1, \ldots, v'_s)$ as in the statement. The fact that

$$\phi : \hat{\mathrm{K}} \otimes_{\mathrm{K}} \mathrm{L} \to \prod_{i=1}^{s} \hat{\mathrm{L}}_i$$

is surjective and relation (7) have already been shown. It remains to verify that the kernel $\mathfrak{n}$ of ϕ is the Jacobson radical $\mathfrak{r}$ of $\hat{\mathrm{K}} \otimes_{\mathrm{K}} \mathrm{L}$. As $\prod_{i=1}^{s} \hat{\mathrm{L}}_i$ is semi-simple, $\mathfrak{r} \subset \mathfrak{n}$. On the other hand, for every maximal ideal $\mathfrak{m}$ of $\hat{\mathrm{K}} \otimes_{\mathrm{K}} \mathrm{L}$, the quotient field $\mathrm{L}(\mathfrak{m}) = (\hat{\mathrm{K}} \otimes_{\mathrm{K}} \mathrm{L})/\mathfrak{m}$ is a composite extension of $\hat{\mathrm{K}}$ and L over K (*Algebra,*

Chapter VIII, § 8, Proposition 1). There exists a valuation w on $L(\mathfrak{m})$ extending $\hat{v}$ (§ 3, no. 3, Proposition 5); the restriction v' of w to L extends v. As $[L(\mathfrak{m}):\hat{K}]$ is finite, $L(\mathfrak{m})$ is complete with respect to w (§ 5, no. 2, Proposition 4). Now the closure of L in $L(\mathfrak{m})$ is a field containing $\hat{K}$ and L and hence is equal to $L(\mathfrak{m})$. Consequently $L(\mathfrak{m})$ is identified with the completion $\hat{L}_{v'}$ and $\mathfrak{m}$ is the kernel of the canonical mapping of $\hat{K} \otimes_K L$ onto $\hat{L}_{v'}$. Now, by hypothesis, there exists an index i such that v' and v'_i are dependent; whence $L_{v'} = L_i$ (§ 7, no. 2, Proposition 3). Thus $\mathfrak{n} \subset \mathfrak{m}$, which proves that $\mathfrak{n} \subset \mathfrak{r}$ and completes the proof.

Corollary 1. *If* K *is complete with respect to v and v is not improper, two valuations on* L *extending v are dependent.*

This follows since $\hat{K} \otimes_K L = L$.

Corollary 2. *If $\hat{K}$ or* L *is separable over* K, *the canonical mapping*

$$\phi: \hat{K} \otimes_K L \to \prod_{i=1}^{s} \hat{L}_i$$

is an isomorphism.

The Jacobson radical of $\hat{K} \otimes_K L$ is then zero (*Algebra*, Chapter VIII, § 7, no. 3, Theorem 1).

Remark. Proposition 2 (b) shows that every composite extension of $\hat{K}$ and L over K (*Algebra*, Chapter VIII, § 8) is isomorphic to one of the completions $\hat{L}_i$ and that these are composite extensions no two of which are isomorphic.

3. THE RELATION $\sum_i e_i f_i \leqslant n$

Let K be a field, v a valuation on K and L a finite extension of K of degree n. Let $(v'_1, \ldots, v'_r)$ be valuations on L extending v *no two of which are equivalent*; if they are *independent* (which is always the case if v is of height 1), then

$$\sum_{i=1}^{r} e(v'_i/v) f(v'_i/v) \leqslant n$$

by Proposition 2 (formulae (6) and (7)). We shall see that this result is true in the general case. To be precise:

Theorem 1. *Let* K *be a field, v a valuation on* K *and* L *a finite extension of* K *of degree n. Then:*

(a) *Every complete system $(v'_i)_{i \in I}$ of extensions of v to* L *is finite.*

(b) $\sum_{i \in I} e(v'_i/v) f(v'_i/v) \leqslant n$ *and a fortiori* Card(I) $\leqslant n$.

(c) *No two of the rings of the v'_i are comparable with respect to inclusion.*

Since the theorem is trivial if v is improper, we shall assume that v is not improper. Let $(v'_1, \ldots, v'_s)$ be any finite family of valuations on L extending v, no two of which are equivalent. We shall first prove that $\sum_{i=1}^{s} e(v'_i/v)f(v'_i/v) \leqslant n$. This will prove (a) and (b).

We argue by induction on s and suppose therefore that the inequality has been established for the case of $0, 1, \dots, s-1$ valuations. We distinguish two cases.

(1) Suppose that there exist at least two independent valuations v'_i. Then there exists (§ 7, no. 2, *Remark* 1) a partition $[1, s] = I_1 \cup \dots \cup I_t$ of $[1, s]$ such that:

(i) for v'_i and v'_j to be dependent, it is necessary and sufficient that i and j belong to the same I_k;

(ii) $\text{Card}(I_k) < s$ for all k.

We choose in each I_k an index $i(k)$. Let $\hat{L}_{i(k)}$ denote the completion of L with respect to $v'_{i(k)}$ and $n(k) = [\hat{L}_{i(k)} : \hat{K}]$. For all $i \in I_k$, v'_i defines on L the same topology as $v'_{i(k)}$ (§ 7, no. 2, Proposition 3) and hence may be extended to a valuation $\hat{v}'_i$ on $\hat{L}_{i(k)}$ whose restriction to $\hat{K}$ is $\hat{v}$. Since no two of the v'_i for $i \in I_k$ are equivalent, the same is true of the $\hat{v}'_i$. The induction hypothesis applied to the ordered pair $(\hat{K}, \hat{L}_{i(k)})$ shows, by virtue of Proposition 2 (a), formula (4), that $\sum_{i \in I_k} e(v'_i/v)f(v'_i/v) \leqslant n(k)$. As $\sum_{k=1}^{t} n(k) \leqslant n$ (Proposition 2 (b), formula (7)), certainly $\sum_{i=1}^{s} e(v'_i/v)f(v'_i/v) \leqslant n$.

(2) We now pass to the case where any two of the v'_i are dependent. Let A'_i be the ring of v'_i $(1 \leqslant i \leqslant s)$; writing A for the ring of v, $A'_i \cap K = A$ for all i. Let B′ be the subring of L generated by $A'_1, \dots, A'_s$; we write $B = B' \cap K$; then $B \supset A$. Then B is the ring of a valuation w on K and B′ the ring of a valuation w' which is not improper and extends w (§ 7, no. 2, Proposition 4); the field $\kappa(B')$ is an extension of $\kappa(B)$ of degree $f(w'/w)$. Consider the canonical images $\bar{A}'_i$, $\bar{A}$ of A'_i and A in $\kappa(B')$; then $\bar{A}$ is the ring of a valuation $\bar{v}$ on $\kappa(B)$ and the $\bar{A}'_i$ are rings of valuations $\bar{v}'_i$ on $\kappa(B')$ extending $\bar{v}$. As the A'_i generate B′, the $\bar{A}'_i$ generate $\kappa(B')$ and hence the $\bar{v}'_i$ are not all dependent (§ 7, no. 2, Proposition 4). From the first part of the proof,

$$\sum_{i=1}^{s} e(\bar{v}'_i/\bar{v})f(\bar{v}'_i/\bar{v}) \leqslant [\kappa(B') : \kappa(B)] = f(w'/w)$$

and hence

$$\sum_{i=1}^{s} e(w'/w)e(\bar{v}'_i/\bar{v})f(\bar{v}'_i/\bar{v}) \leqslant e(w'/w)f(w'/w) \leqslant n \qquad \text{(no. 1, Lemma 1).}$$

The proof of (a) and (b) will therefore be completed if we prove that

$$(8) \qquad f(\bar{v}'_i/\bar{v}) = f(\bar{v}'_i/v), \qquad e(w'/w)e(\bar{v}'_i/\bar{v}) = e(v'_i/v).$$

For this, we note that v and $\bar{v}$ (resp. v'_i and $\bar{v}'_i$) have the same residue field

(§ 4, no. 1, Corollary to Proposition 2); this proves the first equation. For the second there is, by virtue of the *Remark* in § 4, no. 3, the following commutative diagram, where the rows are exact sequences and the vertical arrows represent canonical injections:

$$\begin{array}{ccccccccc} 0 & \to & \Gamma_{\bar{v}} & \to & \Gamma_v & \to & \Gamma_w & \to & 0 \\ & & \downarrow & & \downarrow & & \downarrow & & \\ 0 & \to & \Gamma_{\bar{v}'_i} & \to & \Gamma_{v'_i} & \to & \Gamma_{w'} & \to & 0 \end{array}$$

We deduce that there is an exact sequence

$$0 \to \Gamma_{\bar{v}'_i}/\Gamma_{\bar{v}} \to \Gamma_{v'_i}/\Gamma_v \to \Gamma_{w'}/\Gamma_w \to 0$$

by Chapter I, § 1, no. 4, Proposition 2, which proves the second formula (8).

To complete the proof of Theorem 1, it remains to prove (c). If the ring of v'_i contains that of v'_j, $\Gamma_{v'_i}$ is identified with a quotient group $\Gamma_{v'_j}/\mathrm{H}$, H being an isolated subgroup (§ 4, no. 3). As the composite canonical mapping

$$\Gamma_v \to \Gamma_{v'_j} \to \Gamma_{v'_j}/\mathrm{H} = \Gamma_{v'_i}$$

is injective, $\mathrm{H} \cap \Gamma_v = \{0\}$, whence $\mathrm{H} = \{0\}$ (Lemma 3, no. 1). Then v'_i and v'_j are equivalent, whence $i = j$.

Remark. The intersection C of the rings A'_i of the valuations v'_i $(i \in \mathrm{I})$ is the *integral closure* of A in L (§ 1, no. 3, Corollary 3 to Theorem 3); it follows moreover from (c) and § 7, no. 1, Propositions 1 and 2 that C is a *semi-local* ring, that its maximal ideals are the intersections $\mathfrak{m}_i = \mathrm{C} \cap \mathfrak{m}(\mathrm{A}'_i)$ and that $\mathrm{A}'_i = \mathrm{C}_{\mathfrak{m}_i}$ for all $i \in \mathrm{I}$.

4. INITIAL RAMIFICATION INDEX

DEFINITION 4. *Let* G *be a totally ordered commutative group and* H *a subgroup of* G *of finite index. The number of major subsets of* G *consisting of strictly positive elements and containing all the elements* >0 *of* H *is called the initial index of* H *in* G *and denoted by* $\varepsilon(\mathrm{G}, \mathrm{H})$.

This initial index is a natural number by virtue of the following proposition:

PROPOSITION 3. *Under the hypotheses of Definition* 4, *if the set of strictly positive elements of* G *has no least element, then* $\varepsilon(\mathrm{G}, \mathrm{H}) = 1$ *for all* H. *If there exists a least element* >0 *of* G *and* G′ *denotes the subgroup it generates, then*

$$\varepsilon(\mathrm{G}, \mathrm{H}) = (\mathrm{G}' : (\mathrm{G}' \cap \mathrm{H})).$$

In the first case, let x be an element >0 in G. The set of $y \in \mathrm{G}$ such that $0 < y < x$ is infinite and hence there exist two elements of this set which are distinct and congruent modulo H; their difference is an element z of H such that $0 < z < x$. Hence every major subset which contains all the strictly positive elements of H contains x and hence all the elements >0 of G.

In the second case, let x be the least element >0 of G and let n be the least integer >0 such that $nx \in \mathrm{H}$. Clearly $n = (\mathrm{G}' : (\mathrm{G}' \cap \mathrm{H}))$. On the other hand, writing $\mathrm{M}(y)$ for the set of $z \in \mathrm{G}$ such that $y \leqslant z$ $(y \in \mathrm{G})$, it is immediately seen that the major sets of Definition 4 are just $\mathrm{M}(x), \mathrm{M}(2x), \ldots, \mathrm{M}(nx)$.

COROLLARY. *The initial index* $\varepsilon(\mathrm{G}, \mathrm{H})$ *divides the index* $(\mathrm{G} : \mathrm{H})$ *and is equal to it if* G *is isomorphic to* **Z**.

In particular, $\varepsilon(\mathrm{G}, \mathrm{H}) \leqslant (\mathrm{G} : \mathrm{H})$.

DEFINITION 5. *Let* K *be a field,* L *a finite extension of* K, *w a valuation on* L, *v its restriction to* K *and* Γ_w *and* Γ_v *their order groups. The initial index of* Γ_v *in* Γ_w *is called the initial ramification index of w with respect to v (or with respect to* K) *and denoted by* $\varepsilon(w/v)$.

From the above corollary, $\varepsilon(w/v)$ divides $e(w/v)$ with equality in the case of a discrete valuation.

PROPOSITION 4. *Under the hypotheses of Definition* 5, *let* A *and* $\mathfrak{m}$ (resp. A′ *and* $\mathfrak{m}'$) *be the ring and ideal of the valuation v* (resp. w). *Then*

$$[\mathrm{A}'/\mathfrak{m}\mathrm{A}' : \mathrm{A}/\mathfrak{m}] = \varepsilon(w/v) f(w/v).$$

The ideals of A′ containing $\mathfrak{m}\mathrm{A}'$ and distinct from A′ correspond to the major subsets of Γ_w consisting of elements >0 and containing the elements >0 of Γ_v (§ 3, no. 5, Corollary to Proposition 7). They are therefore equal in number to $\varepsilon(w/v)$ and, as they form a totally ordered set under inclusion, this number is equal to the length of the quotient ring $\mathrm{A}'/\mathfrak{m}\mathrm{A}'$. Now a module of length 1 over A′ is a 1-dimensional vector space over $\mathrm{A}'/\mathfrak{m}'$ and hence a module of length $f(w/v)$ over A; hence, as $\mathrm{A}'/\mathfrak{m}\mathrm{A}'$ is of length $\varepsilon(w/v)$ over A′, it is of length $\varepsilon(w/v) f(w/v)$ over A, that is over $\mathrm{A}/\mathfrak{m}$.

5. THE RELATION $\sum_i e_i f_i = n$

PROPOSITION 5. *Let* K *be a field, v a valuation on* K, A *its ring,* $\mathfrak{m}$ *its ideal,* L *a finite extension of* K *of degree n,* B *the integral closure of* A *in* L *and* $(v'_i)_{1 \leqslant i \leqslant s}$ *a complete system of extensions of v to* L. *Then*

$$[\mathrm{B}/\mathfrak{m}\mathrm{B} : \mathrm{A}/\mathfrak{m}] = \sum_{i=1}^{s} \varepsilon(v'_i/v) f(v'_i/v).$$

Let A_i be the ring of v'_i; then $\mathrm{A}_i = \mathrm{B}_{\mathfrak{m}_i}$, where $\mathfrak{m}_i$ runs through the family of maximal ideals of B (no. 3, *Remark*). Let $\mathfrak{q}_i$ be the saturation of $\mathfrak{m}\mathrm{B}$ with respect to $\mathfrak{m}_i$ (Chapter II, § 2, no. 4). By Chapter V, Corollary 3 to Proposition 1, no. 1, § 2, the canonical homomorphism $\mathrm{B}/\mathfrak{m}\mathrm{B} \rightarrow \prod_{i=1}^{s} \mathrm{B}/\mathfrak{q}_i$ is an isomorphism and $\mathfrak{m}_i$

is the only maximal ideal of B containing $\mathfrak{q}_i$. Hence $B/\mathfrak{q}_i$ is canonically isomorphic to $(B/\mathfrak{q}_i)_{\mathfrak{m}_i}$ (Chapter II, § 3, no. 3, Proposition 8), that is to

$$B_{\mathfrak{m}_i}/\mathfrak{m}B_{\mathfrak{m}_i} = A_i/\mathfrak{m}A_i.$$

Therefore there is a canonical isomorphism $B/\mathfrak{m}B \to \prod_{i=1}^{s} A_i/\mathfrak{m}A_i$, whence the result by virtue of Proposition 4 of no. 4.

COROLLARY. *With the same hypotheses and notation,*

$$[B/\mathfrak{m}B : A/\mathfrak{m}] = \sum_{i=1}^{s} \varepsilon(v'_i/v) f(v'_i/v) \leqslant \sum_{i=1}^{s} e(v'_i/v) f(v'_i/v) \leqslant n.$$

We know that $\varepsilon(v'_i/v) \leqslant e(v'_i/v)$ (no. 4, Corollary to Proposition 3) and $\sum_{=1}^{s} e(v'_i/v) f(v'_i/v) \leqslant n$ (no. 3, Theorem 1).

THEOREM 2. *With the hypotheses and notation of Proposition* 5, *the following conditions are equivalent:*

(a) B *is a finitely generated* A*-module;*
(b) B *is a free* A*-module;*
(c) $[B/\mathfrak{m}B : A/\mathfrak{m}] = n$*;*
(d) $\sum_{i=1}^{n} e(v'_i/v) f(v'_i/v) = n$ *and* $\varepsilon(v'_i/v) = e(v'_i/v)$ *for all* i.

The equivalence of (a) and (b) follows from Lemma 1, § 3, no. 6. Clearly (b) implies (c) (*Algebra*, Chapter II, § 1, no. 5, formula (19)). The equivalence of (c) and (d) follows from the Corollary to Proposition 5. It remains to show that (c) implies (b).

In general, if M is an A-module, we shall denote by V(M) the vector space $M/\mathfrak{m}M$ over $A/\mathfrak{m}$. Hypothesis (c) means that $\dim(V(B)) = n$. Let $x_1, \ldots, x_n$ be elements of B whose canonical images in V(B) form a basis of V(B) and let $L \subset B$ be the sub-A-module which they generate. As L is torsion-free and finitely generated, it is free (§ 3, no. 6, Lemma 1). We shall see that $B = L$. Let $y \in B$; we write $M = L + Ay$; this is also a free A-module. The canonical injections $L \to M \to B$ give canonical homomorphisms $V(L) \to V(M) \to V(B)$ As the ranks of L and M are $\leqslant n$, so are the dimensions of V(L) and V(M). Now, by hypothesis, $V(L) \to V(B)$ is surjective and V(B) is n-dimensional hence V(L) and V(M) are n-dimensional and $V(L) \to V(M)$ is surjective. As M is finitely generated, $L \to M$ is surjective (Chapter II, § 3, no. 2, Corollary 1 to Proposition 4), whence $L = M$, $y \in L$ and $B = L$. Hence B is free.

Remark (1) If v is *discrete,* $\varepsilon(v'_i/v) = e(v'_i/v)$ (no. 4) and condition (d) reduces to $\sum_{i=1}^{s} e(v'_i/v) f(v'_i/v) = n$.

COROLLARY 1. *With the same hypotheses and notation, suppose further that v is discrete and* L *separable. Then*

$$\sum_{i=1}^{s} e(v'_i/v) f(v'_i/v) = n.$$

The integral closure B of A is then a free A-module of rank n, since A is a principal ideal domain (Chapter V, § 1, no. 6, Corollary 2 to Proposition 18).

COROLLARY 2. *Let* K *be a field, v a discrete valuation on* K *with respect to which* K *is complete and* L *a finite extension of* K *of degree n. Then v admits a unique (up to equivalence) extension v' to* L, *the ring* A' *of v' is a finitely generated free module over the ring* A *of v and $e(v'/v) f(v'/v) = n$.*

All the extensions of v to L are dependent (no. 2, Corollary to Proposition 2); since they are discrete (no. 1, Corollary 3 to Proposition 1), they are therefore equivalent (§ 4, no. 5, Proposition 6 (c)). This shows the uniqueness of v'. The integral closure of A in L is therefore A' (§ 1, no. 3, Corollary 3 to Theorem 3). As v is discrete, the topology induced on A by that on K is the $\mathfrak{m}$-adic topology (where $\mathfrak{m} = \mathfrak{m}(A)$); the ring A is complete, for it is closed in K. We conclude that, since $A'/\mathfrak{m}A'$ is a finite-dimensional vector $(A/\mathfrak{m})$-space (no. 4, Proposition 4), A' is a finitely generated A-module (Chapter III, § 2, no. 9, Corollary 3 to Proposition 12). It is therefore free and $e(v'/v) f(v'/v) = n$ by Theorem 2.

COROLLARY 3. *Suppose that v is of height 1 and that the equivalent conditions of Theorem 2 hold; if $\hat{L}_i$ is the completion of* L *with respect to v'_i, the degree $n_i = [\hat{L}_i : \hat{K}]$ is equal to $e(v'_i/v) f(v'_i/v)$ for all i and the canonical homomorphism*

$$\phi : \hat{K} \otimes_K L \to \prod_{i=1}^{s} \hat{L}_i$$

(no. 2, Proposition 2). *is bijective. For all $x \in L$, the characteristic polynomial $Pc_{L/K}(x; X)$ is equal to the product of the characteristic polynomials $Pc_{\hat{L}_i/\hat{K}}(x; X)$ $(1 \leqslant i \leqslant s)$; in particular,*

$$(9) \quad \begin{cases} Tr_{L/K}(x) = \sum_{i=1}^{s} Tr_{\hat{L}_i/\hat{K}}(x) \\ N_{L/K}(x) = \prod_{i=1}^{s} N_{\hat{L}_i/\hat{K}}(x) \\ v(N_{L/K}(x)) = \sum_{i=1}^{s} n_i v'_i(x). \end{cases}$$

(The last relation in (9) is meaningful, for we may obviously assume that the v'_i, which are of height 1 by Corollary 2 to Proposition 1 of no. 1, take, as does v, their values in a subgroup of **R**.)

As no two of the v'_i are equivalent and they are of height 1, they are independent and Proposition 2 of no. 2 shows therefore that $e(v'_i/v)f(v'_i/v) \leqslant n_i$ for all i and $\sum_{i=1}^{s} n_i \leqslant n$. The first assertion therefore follows from these inequalities and the relation $\sum_{i=1}^{s} e(v'_i/v)f(v'_i/v) = n$. Under the isomorphism ϕ the endomorphism $z \mapsto z(1 \otimes x)$ of $\hat{K} \otimes_K L$ (for $x \in L$) is transformed into the endomorphism of $\prod_{i=1}^{s} \hat{L}_i$ leaving invariant each of the factors and reducing on each factor to multiplication by x (L being canonically imbedded in its completion $\hat{L}_i$); whence the assertion relating to the characteristic polynomial of x and the first two formulae of (9). Finally, let E be a finite quasi-Galois extension of $\hat{K}$, containing $\hat{L}_i$; as $\hat{K}$ is complete and $\hat{v}$ of height 1, there exists only one valuation (up to equivalence) w on E extending $\hat{v}$ (no. 2, Corollary 1 to Proposition 2); then, for every $\hat{K}$-automorphism σ of E, $w(\sigma(x)) = v'_i(x)$. Therefore

$$\hat{v}(N_{\hat{L}_i/\hat{K}}(x)) = n_i v'_i(x)$$

(*Algebra*, Chapter VIII, § 12, no. 2, formula (15)), which proves the formula of (9).

Corollary 4. *Under the hypotheses of Corollary* 3, *if* L *is a separable extension of* K, *each of the* $\hat{L}_i$ *is a separable extension of* $\hat{K}$. *If further* L *is a Galois extension of* K *with Galois group* $\mathscr{G}$ *and* $\mathscr{G}_i$ *denotes the decomposition group of the ideal of* v'_i *in* B (Chapter V, § 2, no. 2, Definition 2), *then* $\hat{L}_i$ *is a Galois extension of* $\hat{K}$ *whose Galois group is isomorphic to* $\mathscr{G}_i$.

Clearly $\hat{L}_i = \hat{K}(L)$; hence, if L is separable over K, $\hat{L}_i$ is separable over $\hat{K}$ (*Algebra*, Chapter V, § 7, no. 6, Proposition 10). Suppose now that L is Galois. Every automorphism $\sigma \in \mathscr{G}_i$ is *continuous* on L with the topology defined by v'_i, the fact that no two of the ideals of the v'_i are comparable with respect to inclusion (§ 7, no. 2, Corollary 1 to Theorem 1) necessarily implying that $v'_i = v'_i \circ \sigma$ by definition of $\mathscr{G}_i$; hence σ may be extended by continuity to a $\hat{K}$-automorphism $\hat{\sigma}$ of $\hat{L}_i$. This proves that the number of $\hat{K}$-automorphisms of $\hat{L}_i$ is at least equal to $\text{Card}(\mathscr{G}_i)$. But as the valuations v'_i are pairwise conjugate under $\mathscr{G}$ (Chapter V, § 2, no. 3, Proposition 6), $s = (\mathscr{G} : \mathscr{G}_i)$, whence

$$\text{Card}(\mathscr{G}_i) = n/s \leqslant n.$$

and on the other hand $n = sn_i$ by Corollary 3; this proves that $\hat{L}_i$ is a Galois extension of $\hat{K}$ and that the extensions by continuity of the automorphisms $\sigma \in \mathscr{G}_i$ are the only $\hat{K}$-automorphisms of $\hat{L}_i$.

Remark (2). Part of the above results extends to the case of valuations on a *not necessarily commutative* field K (cf. § 3, no. 1). Let L be an extension field of K and let v' be a valuation on L, v its restriction to K and A′ and A the respective rings of the valuations v' and v; then there is defined a ramification index $e(v'/v)$ as in no. 1; on the other hand, $\kappa(A)$ is identified with a subfield of $\kappa(A')$ and the (left) *residue rank* of v' with respect to v is defined to be the number $f(v'/v)$ equal to the dimension of the left vector $\kappa(A)$-space $\kappa(A')$, if this dimension is finite, and $+\infty$ in the opposite case. Then, if L is a left vector K-space of *finite* dimension n, Lemma 2 of no. 1 and its proof go over unchanged. Moreover, if K is *complete* with respect to v, the assertions of Corollary 2 to Theorem 2 of no. 5 (other than the existence of v') are also valid (n denoting the dimension of L as a left vector K-space) with the following proof:

In the first place the topology defined by v' on L is Hausdorff and compatible with its left vector K-space structure and hence two extensions of v to L give the same topology on L (§ 5, no. 2, Proposition 4), which proves that these extensions are the same up to equivalence (§ 6, no. 2). We show next that, if $\mathfrak{m} = \mathfrak{m}(A)$, $A'/\mathfrak{m}A'$ is a left vector $(A/\mathfrak{m})$-space of dimension $e(v'/v)f(v'/v)$. Writing $e = e(v'/v)$, we may assume that $v(K^*) = \mathbf{Z}$ and $v'(L^*) = e^{-1}\mathbf{Z}$; let u' be an element of L such that $v'(u') = e^{-1}$ and u an element of K such that $v(u) = 1$; hence $u = zu'^e$, where $z \in L$ is such that $v'(z) = 0$. As $\mathfrak{m}$ is generated by u (as a left or right ideal of A), $\mathfrak{m}A' = u'^eA' = A'u'^e$ and it suffices to prove that, for $0 \leqslant k \leqslant e-1$, $A'u'^k/A'u'^{k+1}$ is a left vector $(A/\mathfrak{m})$-space of dimension $f(v'/v)$. But $t \mapsto tu'^k$ is an isomorphism of the left A-module A' onto the left A-module $A'u'^k$ mapping $A'u'$ to $A'u'^{k+1}$ and which therefore gives by taking quotients an $(A/\mathfrak{m})$-isomorphism of $A'/A'u'$ onto $A'u'^k/A'u'^{k+1}$, whence our assertion by definition of $f(v'/v)$, u' generating the maximal ideal of A'. The proof is completed as when K and L are commutative (the fact that a finitely generated torsion-free A-module is free being proved as in § 3, no. 6, Lemma 1).

6. VALUATION RINGS IN AN ALGEBRAIC EXTENSION

PROPOSITION 6. *Let* K *be a field,* v *a valuation on* K, A *its ring,* L *an algebraic extension of* K *and* A' *the integral closure of* A *in* L. *Let* $\mathfrak{V}$ *be the set of valuation rings on* L *which extend* v *and* $\mathfrak{M}'$ *the set of maximal ideals of* A'. *Then the mapping* $V \mapsto \mathfrak{m}(V) \cap A'$ *is a bijection of* $\mathfrak{V}$ *onto* $\mathfrak{M}'$ *and* $\mathfrak{m}' \mapsto A'_{\mathfrak{m}'}$ *is the inverse bijection.*

Every maximal ideal $\mathfrak{m}'$ of A' is such that $\mathfrak{m}' \cap A$ is the maximal ideal of A (Chapter V, § 2, no. 1, Proposition 1) and $A'_{\mathfrak{m}'}$ is dominated by a valuation ring V of L (which is therefore the ring of a valuation on L extending v) (§ 1, no. 2, Corollary to Theorem 2). The field L is the union of a directed family of sub-extensions K_α of L which are finite over K and it will suffice, in order to see that $V = A'_{\mathfrak{m}'}$, to prove that $V \cap K_\alpha = A'_{\mathfrak{m}'} \cap K_\alpha$ for all α. Now, if we write $A'_\alpha = A' \cap K_\alpha$, A'_α is the integral closure of A in K_α and hence is the intersection of the rings of the valuations on K_α which extend v and these rings $V_{i\alpha}$ are finite in number and are the local rings $(A'_\alpha)_{\mathfrak{m}'_{i\alpha}}$ of A'_α $(1 \leqslant i \leqslant n)$, where the $\mathfrak{m}'_{i\alpha}$ are the distinct maximal ideals of A'_α (no. 3, *Remark*); but $\mathfrak{m}' \cap A'_\alpha$ is one of the $\mathfrak{m}'_{i\alpha}$ and $V \cap K_\alpha$ is therefore equal to the corresponding local ring $(A'_\alpha)_{\mathfrak{m}_{i\alpha}} \subset A'_{\mathfrak{m}'}$, which completes the proof that $V = A'_{\mathfrak{m}'}$. Conversely, if $V \in \mathfrak{V}$, then $A' \subset V$ (§ 3, no. 3, Proposition 6) and, if $\mathfrak{m}' = \mathfrak{m}(V) \cap A'$, then $\mathfrak{m}' \cap A = \mathfrak{m}$, hence $\mathfrak{m}'$ is a maximal ideal of A′ (Chapter V, § 2, no. 1, Proposition 1) and the above argument shows that $V = A'_{\mathfrak{m}'}$.

PROPOSITION 7. *Let* K *be a field,* L *a quasi-Galois extension of* K *and* f *and* f' *places of* L *with values in the same field* F. *Suppose that the restrictions of* f *and* f' *to* K *coincide. Then there exists a* K*-automorphism* s *of* L *such that* $f' = f \circ s$.

Let A be the ring of the place of K the common restriction of f and f'. The rings of f and f' contain the integral closure A′ of A in L (§ 1, no. 3, Corollary 3 to Theorem 3) and hence (Chapter V, § 2, no. 3, Corollary 1 to Proposition 6) there exists a K-automorphism s of L such that the restrictions of f' and $f \circ s$ to A′ are equal; if $\mathfrak{m}'$ is the common kernel of these restrictions, $\mathfrak{m}' \cap A$ is the maximal ideal of A, hence $\mathfrak{m}'$ is a maximal ideal of A′ and the places f' and $f \circ s$ coincide on the ring $A'_{\mathfrak{m}'}$; but by Proposition 6 the only valuation ring of L dominating $A'_{\mathfrak{m}'}$ is the ring $A'_{\mathfrak{m}'}$ itself and hence the rings of the places f' and $f \circ s$ are the same.

COROLLARY 1. *Let* K *be a field,* v *a valuation on* K, L *a quasi-Galois extension of* K *and* v' *and* v'' *two extensions of* v *to* L. *Then there exists a* K*-automorphism* s *of* L *such that* v'' *is equivalent to* $v' \circ s$.

Let f' and f'' be the places of K associated with v' and v''; replacing them if need be by equivalent places, it may be assumed that they both take their values in the algebraic closure of the residue field of v (no. 1, Proposition 1). Then there exists a K-automorphism s of L such that $f'' = f' \circ s$ (Proposition 7); thus v'' is equivalent to $v' \circ s$ by virtue of the correspondence between places and valuations (§ 3, no. 3).

COROLLARY 2. *Let* K *be a field,* f *a place of* K (resp. v *a valuation on* K) *and* L *a radicial extension of* K. *Then all the extensions of* f (resp. v) *to* L *are equivalent.*

L is a quasi-Galois extension and its only automorphism is the identity. Corollary 2 therefore follows from Proposition 7 (resp. Corollary 1).

PROPOSITION 8. *Let* K *be a field,* v *a valuation on* K, L *a finite quasi-Galois extension of* K *of degree* n *and* $(v'_i)_{1 \leqslant i \leqslant g}$ *a complete system of extensions of* v *to* L. *Then* $e(v'_i/v)$ *and* $f(v'_i/v)$ *have values* e *and* f *independent of* i. *Then* $efg \leqslant n$. *If the integral closure in* L *of the ring* A *of* v *is a finitely generated* A*-module, then* $efg = n$.

This follows immediately from Theorems 1 (no. 3) and 2 (no. 5).

7. THE EXTENSION OF ABSOLUTE VALUES

PROPOSITION 9. *Let* K *be a field,* L *an algebraic extension of* K *and* f *an absolute value on* K. *Then* f *can be extended to an absolute value on* L.

Suppose first that there exists a valuation v on K with real values such that $f(x) = e^{-v(x)}$. There exists a valuation v' on L whose restriction to K is equivalent to v (§ 3, no. 3, Proposition 5). Then v' is of height 0 or 1 (no. 1, Corollary 2 to Proposition 1) and therefore may be assumed to have real values. The restriction of the mapping $x \mapsto e^{-v'(x)}$ to K is an absolute value equivalent to f and hence of the form f^s with $s > 0$ (*General Topology*, Chapter IX, § 3, no. 2, Proposition 5). We conclude that

$$x \mapsto e^{-v'(x)/s}$$

is an absolute value on L extending f.

Suppose now that f is not ultrametric. Then K is identified with a subfield of $\mathbf{C}$ such that $f(x) = |x|^s$ where $0 \leqslant s \leqslant 1$ (§ 6, no. 4, Theorem 2). As $\mathbf{C}$ is algebraically closed, L is identified with a subfield of $\mathbf{C}$ and the absolute value $x \mapsto |x|^s$ extends f.

PROPOSITION 10. *Let* K *be a field,* f *an absolute value on* K *such that* K *is complete and not discrete with respect to* f *and* L *an algebraic extension of* K. *Then* f *can be extended uniquely to an absolute value* f' *on* L *and, if* L *is of finite degree* n*, then*

$$f'(x) = (f(N_{L/K}(x)))^{1/n}$$

for all $x \in L$.

The existence of f' follows from Proposition 9 and its uniqueness (over every finite sub-extension of L and therefore over the whole of L) from Lemma 2 of § 6, no. 4. Let f' be the unique extension of f to the algebraic closure of K and suppose L is of finite degree n. We know that $N_{L/K}(x) = \prod_{i=1}^{n} x_i$, where each x_i is a conjugate of x over K (*Algebra*, Chapter VIII, § 12, no. 2, Proposition 4). In view of the uniqueness of f', $f'(x_i) = f'(x)$ for all i, whence the stated formula.

PROPOSITION 11. *Let* K *be a field,* f *a non-ultrametric absolute value on* K, $\hat{K}$ *the completion of* K *with respect to* f, $\hat{f}$ *the continuous extension of* f *to* $\hat{K}$ *and* L *a finite extension of* K *of degree* n.

(a) *Let* f' *be an absolute value on* L *extending* f*; let* $\hat{L}_{f'}$ *denote the completion of* L *with respect to* f' *and let* $\hat{K}$ *be identified with the closure of* K *in* $\hat{L}_{f'}$*; then* $[\hat{L}_{f'} : \hat{K}] \leqslant n$.

(b) *The absolute values on* L *extending* f *are finite in number. If they are denoted by* $f'_1, \ldots, f'_s$ *and the completion of* L *with respect to* f'_i *by* $\hat{L}_i$, *the canonical mapping* $\hat{K} \otimes_K L \to \prod_{i=1}^{n} \hat{L}_i$ *is an isomorphism and*

$$\sum_{i=1}^{s} [\hat{L}_i : \hat{K}] = n. \tag{10}$$

The proof is the same as that for the analogous assertions in Proposition 2 (no. 2). The references

§ 7, no. 2, Theorem 1; § 5, no. 2, Corollary to Proposition 4 should be replaced by the following

§ 7, no. 3, Theorem 2; *Topological Vector Spaces*, Chapter I, § 2, no. 3, Corollary 1 to Theorem 2.

Observe that two extensions of f to L which define the same topology are equal (*General Topology*, Chapter IX, § 3, no. 2, Proposition 5). Finally, as f is not ultrametric, K is of characteristic 0 and hence the Jacobson radical of $\hat{K} \otimes_K L$ is zero.

Remarks

(1) Proposition 11 (b) shows that every composite extension of $\hat{K}$ and L over K is isomorphic to one of the completions $\hat{L}_i$ and that these are composite extensions no two of which are isomorphic.

(2) We know that the completions $\hat{K}$ and $\hat{L}_i$ are isomorphic to **R** or **C** (§ 6, no. 4, Theorem 2). If $\hat{K}$ is isomorphic to **C**, so is $\hat{L}_i$ for all i and (10) shows that the number of extensions f_i' is *equal to n*. If $\hat{K}$ is isomorphic to **R** (for example if $K = \mathbf{Q}$), let r_1 (resp. r_2) denote the number of indices i such that $\hat{L}_i$ is isomorphic to **R** (resp. **C**); then (10) may be written:

$$r_1 + 2r_2 = n. \tag{11}$$

PROPOSITION 12. *Let* K *be a field, f an absolute value on* K, L *a quasi-Galois extension of* K *and f' and f'' two extensions of f to* L. *Then there exists a* K*-automorphism s of* L *such that $f'' = f' \circ s$.*

If f is ultrametric, Corollary 1 to Proposition 7 (no. 6) shows that there exists a K-automorphism s of L such that f'' and $f' \circ s$ are equivalent absolute values; then there exists a real number $a > 0$ such that $f''(x) = (f'(s(x)))^a$ for all $x \in L$. If f is not improper, take $x \in K^*$ such that $f(x) \neq 1$, which shows that $a = 1$. If f is improper, so are f' and f'' (Corollary 2 to Proposition 1, no. 1) and s may be taken to be the identity automorphism.

If f is not ultrametric, there exist **Q**-isomorphisms u', u'' of L onto subfields of **C** and real exponents $a' > 0$, $a'' > 0$ such that $f'(x) = |u'(x)|^{a'}$ and

$$f''(x) = |u''(x)|^{a''}$$

for all $x \in L$ (§ 6, no. 4, Theorem 2). Taking $x = 2$, it is seen that $a' = a''$. The restrictions of u' and u'' to K extend by continuity to isomorphisms u_1 and u_2 of $\hat{K}$ onto **R** (resp. **C**). Then $u_2 \circ \overset{-1}{u_1}$ is an automorphism of the *valued* field **R** (resp. **C**) and is therefore the identity (resp. the identity or the automorphism $c: \zeta \to \bar{\zeta}$). Replacing if need be u' by $c \circ u'$, it is seen that the restrictions of u' and u'' to K may be assumed to coincide. Identifying K with a subfield of **C** by means of this common restriction, u' and u'' are K-isomorphisms of L onto subfields of **C**. As L is a quasi-Galois extension of K, there exists a K-automorphism s of L such that $u'' = u' \circ s$; since $a' = a''$, we deduce immediately that $f'' = f' \circ s$.

Remark (3). If $\hat{K}$ is isomorphic to **R**, Proposition 12 shows that all the completions $\hat{L}_i$ of L (in the notation of Proposition 11) are isomorphic to one another. Thus, with the notation of *Remark* 2 above, either $r_1 = n$ and $r_2 = 0$, or $r_1 = 0$ and $2r_2 = n$.

9. APPLICATION: LOCALLY COMPACT FIELDS

1. THE MODULUS FUNCTION ON A LOCALLY COMPACT FIELD

Let K be a locally compact field (not necessarily commutative). Recall that the function mod (or mod_K) has been defined (*Integration*, Chapter VII, § 1, no. 10, Definition 6) on K as follows: $\mathrm{mod}_K(0) = 0$ and for $x \neq 0$ in K, the number $\mathrm{mod}_K(x)$ is the modulus of the automorphism $y \mapsto xy$ of the additive group of K.

PROPOSITION 1. *If* K *is a locally compact field, the function* mod_K *belongs to* $\mathscr{V}(K)$ (§ 6, no. 1). *Moreover:*

(i) If $s > 0$ *is such that* $(\mathrm{mod}_K)^s = g$ *is an absolute value, then* g *defines the topology on* K.

(ii) *If* K *is not discrete and* mod_K *is an ultrametric absolute value, there exists a normed discrete valuation* v *on* K *whose ring is compact and whose residue field is finite with* q *elements, so that* $\mathrm{mod}_K = q^{-v}$. *The topology on* K *is defined by* v.

This follows from § 6, no. 1, Proposition 1, § 5, no. 1, Proposition 2 and *Integration*, Chapter VII, § 1, no. 10, Propositions 12 and 13.

PROPOSITION 2. *Let* K, K′ *be two (not necessarily commutative) locally compact fields such that* K *is a topological subfield of* K′ *and* K *is not discrete. Then:*

(i) K′ *is a finite dimensional left* (resp. *right*) *vector space over* K.

(ii) *If* K *is contained in the centre of* K′, *then, for all* $x \in K'$.

$$\mathrm{mod}_{K'}(x) = \mathrm{mod}_K(N_{K'/K}(x)). \tag{1}$$

As K is a complete valued field which is not discrete, assertion (i) follows from *Topological Vector Spaces*, Chapter I, § 2, no. 4, Theorem 3; assertion (ii) is just *Integration*, Chapter VII, § 1, no. 11, Proposition 17.

COROLLARY 1. *Every locally compact field whose centre is not discrete is of finite rank over its centre.*

The centre Z of a locally compact field K is closed in K and therefore locally compact.

COROLLARY 2. *Let* K′ *be a locally compact field and* K *a closed subfield of* K′ (*neither necessarily commutative*). *If* K′ *is a left* (resp. *right*) *vector space of finite dimension* n *over* K, *then*

$$\mathrm{mod}_{K'}(x) = (\mathrm{mod}_K(x))^n \quad \textit{for all} \quad x \in K. \tag{2}$$

In general it is known that in a (left or right) vector space of finite dimension n over K, the homothety with ratio $x \in K$ has modulus equal to $(\text{mod}_K(x))^n$; it suffices to apply this to K′.

2. EXISTENCE OF REPRESENTATIVES

PROPOSITION 3. *Let* K *be a (not necessarily commutative) locally compact field which is not discrete and whose topology is defined by a discrete valuation* v; *let* A *be the ring and* $\mathfrak{m}$ *the ideal of* v *and let us write* $\text{Card}(A/\mathfrak{m}) = q = p^f$ (p *prime*). *Then, there exists a system of representatives* S *of* $A/\mathfrak{m}$ *in* A *and a uniformizer* u *for* v *such that* $0 \in S$, $S^* = S \cap K^*$ *is a cyclic subgroup of* K^* *and* $u^{-1}Su = S$. *Moreover, every element of* A *may be written uniquely in the form* $\sum_{i=0}^{\infty} s_i u^i$, *where* $s_i \in S$.

We shall use the following lemma:

LEMMA 1. *Let* x, y *be two permutable elements of* A *such that* $x - y \in \mathfrak{m}^j$ $(j \geqslant 1)$; *then* $x^{p^n} - y^{p^n} \in \mathfrak{m}^{j+n}$ *for every integer* $n \geqslant 0$.

By induction on n, it is reduced to proving the lemma for $n = 1$. Then

$$x^p - y^p = (x - y)(x^{p-1} + x^{p-2}y + \cdots + y^{p-1});$$

the second factor is a sum of p terms congruent to one another mod. $\mathfrak{m}$ and, as $A/\mathfrak{m}$ is of characteristic p, $p.1 \in \mathfrak{m}$ in A and hence

$$x^{p-1} + x^{p-2}y + \cdots + y^{p-1} \in \mathfrak{m};$$

whence $x^p - y^p \in \mathfrak{m}^{j+1}$.

We know that the multiplicative group $(A/\mathfrak{m})^*$ is a cyclic group with $q - 1$ elements (*Algebra*, Chapter V, § 11, no. 1, Theorem 1); let x be a representative in A of a generator of this group; then $x^q - x \in \mathfrak{m}$, whence, by Lemma 1, $x^{q^{n+1}} - x^{q^n} \in \mathfrak{m}^{1+fn}$, since x^q and x are permutable. This proves that $(x^{q^n})_{n \geqslant 0}$ is a Cauchy sequence in A; as A is compact and hence complete, this sequence has a limit s in A which obviously satisfies $s \equiv x$ (mod. $\mathfrak{m}$) and $s^q = s$. As $s \neq 0$, $s^{q-1} = 1$, more precisely s is a *primitive* $(q - 1)$-th *root of unity* in A. Clearly the set S consisting of 0 and the powers s^j $(0 \leqslant j \leqslant q - 2)$ is a *system of representatives* of the classes of A mod. $\mathfrak{m}$ and is *invariant* under multiplication in A.

Now let a be a uniformizer for v and consider the inner automorphism $y \mapsto a^{-1}ya$ of K; it maps A to itself, $\mathfrak{m}$ to itself and therefore, taking quotients, it defines an automorphism of the field $A/\mathfrak{m}$; it is known (*Algebra*, Chapter V, § 11, no. 4, Proposition 5) that such an automorphism is of the form $z \mapsto z^{p^r}$, where $0 \leqslant r \leqslant f - 1$. Then $a^{-1}s^j a \equiv s^{jp^r}$ (mod. $\mathfrak{m}$) for $0 \leqslant j \leqslant q - 2$; as $a \in \mathfrak{m}$ and $s \notin \mathfrak{m}$, this implies that $s^{-j}as^{jp^r} \equiv a$ (mod. $\mathfrak{m}^2$).

Let us write

$$u = \sum_{j=0}^{q-2} s^{-j} a s^{jp^r}.$$

Then $u \equiv (q-1)a \equiv -a \pmod{\mathfrak{m}^2}$ since $p.1 \in \mathfrak{m}$; we conclude that u is also a uniformizer for v; moreover

(3) $$s^{-1}us^{p^r} = u$$

whence we deduce that $u^{-1}\,\mathrm{S}u = \mathrm{S}$.

Finally, for all $x \in \mathrm{A}$ there exists a unique sequence (s_i) $(i \in \mathbf{N})$ such that $s_i \in \mathrm{S}$ for all i and $x \equiv \sum_{i=0}^{n} s_i u^i \pmod{\mathfrak{m}^{n+1}}$ for all $n \geqslant 0$: it is immediate by induction on n, every element t of $\mathfrak{m}^{n+1}$ satisfying a relation of the form $t \equiv t'u^{n+1} \pmod{\mathfrak{m}^{n+2}}$, where t' is a uniquely determined element of S. Then $x = \sum_{i=0}^{\infty} s_i u^i$ and the family (s_i) satisfying this relation and such that $s_i \in \mathrm{S}$ for all i is determined uniquely.

3. STRUCTURE OF LOCALLY COMPACT FIELDS

The completions $\mathbf{R}$ and $\mathbf{Q}_p$ of the field $\mathbf{Q}$ with respect to the non-improper absolute values on $\mathbf{Q}$ (p any prime) are locally compact. On the other hand, for every power $q = p^f$ of a prime number p, the field $\mathbf{F}_q((\mathrm{T}))$ of formal power series over the finite field $\mathbf{F}_q$ with the valuation defined in § 3, no. 4, *Example* 3 is *locally compact*: for the maximal ideal of the valuation ring $\mathbf{F}_q[[\mathrm{T}]]$ is generated by T; we know that this ring is complete with the (T)-adic topology (Chapter III, § 2, no. 6, Proposition 6) and, as the residue field $\mathbf{F}_q$ is finite, Proposition 2 of § 5, no. 1 proves our assertion. Conversely:

THEOREM 1. *Let* K *be a (not necessarily commutative) locally compact field which is not discrete.*

(i) *If* K *is of characteristic* 0 *and* mod_K *is not an ultrametric absolute value then* K *is isomorphic to one of the fields* $\mathbf{R}$, $\mathbf{C}$ *or* $\mathbf{H}$.

(ii) *If* K *is of characteristic* 0 *and* mod_K *is an ultrametric absolute value,* K *is an algebra of finite rank over a p-adic field* $\mathbf{Q}_p$.

(iii) *If* K *is of characteristic* $p \neq 0$, *it is isomorphic to a field with centre a field of formal power series* $\mathbf{F}_q((\mathrm{T}))$ *(where q is a power of p) and of finite rank over its centre.*

(i) It follows from Ostrowski's Theorem (§ 6, no. 4, Theorem 2) that K is a topological field isomorphic to an everywhere dense subfield of $\mathbf{R}$, $\mathbf{C}$ or $\mathbf{H}$ and, as K is complete, it is isomorphic to $\mathbf{R}$, $\mathbf{C}$ or $\mathbf{H}$.

(ii) Let A be the ring of the absolute value mod_K and $\mathfrak{m}$ its maximal ideal. We know that $\mathrm{A}/\mathfrak{m}$ is a finite field (§ 5, no. 1, Proposition 2) and hence the absolute value induced by mod_K on $\mathbf{Q}$ has a finite residue field, which is only possible if it is equivalent to a p-adic absolute value (§ 6, no. 3, Proposition 4); the closure of $\mathbf{Q}$ in K is therefore isomorphic to $\mathbf{Q}_p$ and is contained in the centre of K since the latter is closed in K; we conclude using Proposition 2 of no. 1.

(iii) The second assertion follows from the first and the Corollary to Proposition 2 of no. 2. To show the first assertion, note that mod_K is necessarily an ultrametric absolute value (§ 6, no. 2, Corollary to Proposition 3); in the notation of the proof of Proposition 3 of no. 2, the centre Z of K consists of the elements which commute with both s and u; but by virtue of (3),

$$u^{-q}su^q = s^{qp^r} = s,$$

so that $u^q \in Z$ and we conclude that Z is not discrete. As Z is locally compact, we may confine our attention to the case where K is *commutative*. The sub-$\mathbf{F}_p$-algebra $\mathbf{F}_p[s]$ in K is then a finite field since $s^{q-1} = 1$ and obviously $y^q = y$ for every element of this field, which is therefore identical with S and isomorphic to $\mathbf{F}_q$ since $S \subset \mathbf{F}_p[s]$ has q elements. Since the sum of two elements of S is in S, the mapping which maps each formal power series $\sum_{i=0}^{\infty} s_i T^i \in \mathbf{F}_q[[T]]$ to the element $\sum_{i=0}^{\infty} s_i u^i$ is a bijective homomorphism of the ring $\mathbf{F}_q[[T]]$ onto the ring A, whence immediately the conclusion.

Corollary 1. *Every locally compact field which is not discrete is of finite rank over its centre.*

Corollary 2. *Every locally compact field is connected or totally disconnected; if it is connected, it is isomorphic to* $\mathbf{R}$, $\mathbf{C}$ *or* $\mathbf{H}$.

If the topology on a field K is defined by an ultrametric absolute value, K is totally disconnected with this topology.

Remark. Let s be an integer >0; the subfield $\mathbf{F}_q((T^s)) = L$ of $K = \mathbf{F}_q((T))$ is closed in K and $e(K/L) = s$ and $f(K/L) = 1$. It is therefore seen that there are closed subfields L of K which are not discrete and such that $e(K/L)$ (and *a fortiori* the degree $[K:L]$) is arbitrarily large (contrary to what happens for locally compact fields of characteristic 0, where every locally compact subfield L of such a field K necessarily contains $\mathbf{R}$ or $\mathbf{Q}_p$ and therefore $[K:L]$ is bounded).

10. EXTENSIONS OF A VALUATION TO A TRANSCENDENTAL EXTENSION

1. THE CASE OF A MONOGENOUS TRANSCENDENTAL EXTENSION

Lemma 1. *Let* K *be a field,* v *a valuation on* K, Γ *its order group,* Γ' *a totally ordered group containing* Γ *and* ξ *an element of* Γ'. *There exists a unique valuation* w *on* K(X) *such that, for* $P = \sum_j a_j X^j$ $(a_j \in K)$, $w(P) = \inf_j(v(a_j) + j\xi)$.

By Proposition 4 of § 3, no. 2, it suffices to show that the formula

$$w\Big(\sum_j a_j X^j\Big) = \inf_j(v(a_j) + j\xi) \tag{1}$$

defines a valuation on the ring K[X]. As

$$v(a_j + b_j) + j\xi \geqslant \inf(v(a_j), v(b_j)) + j\xi = \inf(v(a_j) + j\xi, v(b_j) + j\xi),$$

it follows that

$$w(P + Q) \geqslant \inf(w(P), w(Q)) \tag{2}$$

for P, Q in K[X], equality holding if $w(P) \neq w(Q)$. We show that

$$w(PQ) = w(P) + w(Q) \tag{3}$$

for $P = \sum_j a_j X^j$ and $Q = \sum_j b_j X^j$. Let i (resp. k) be the least of the integers j such that $v(a_j) + j\xi$ (resp. $v(b_j) + j\xi$) attains its minimum; let α (resp. β) denote this minimum. For j, j' in **N**,

$$w(a_j b_{j'} X^{j+j'}) = v(a_j) + j\xi + v(b_{j'}) + j'\xi \geqslant \alpha + \beta,$$

whence $w(PQ) \geqslant \alpha + \beta$ by (2). Consider now the term cX^{i+k} of degree $i + k$ in PQ; then $c = \sum_{n \in \mathbf{Z}} a_{i+n} b_{k-n}$; by the choice of i and k, the element

$$w(a_{i+n} b_{k-n} X^{i+k}) = v(a_{i+n}) + (i + n)\xi + v(b_{k-n}) + (k - n)\xi$$

takes its minimum value $\alpha + \beta$ once and once only with $n = 0$; hence $w(cX^{i+k}) = \alpha + \beta$, whence, by (1),

$$w(PQ) = \alpha + \beta = w(P) + w(Q).$$

Proposition 1. *Let* K *be a field,* v *a valuation on* K, Γ *its order group,* Γ' *a totally ordered group containing* Γ *and* ξ *an element of* Γ' *such that the relations* $n\xi \in \Gamma$, $n \in \mathbf{Z}$ *imply* $n = 0$. *Then there exists a unique valuation* w *on* K(X) *with values in* Γ' *and extending* v *such that* $w(X) = \xi$. *The residue field of* w *is equal to that of* v *and its order group is the subgroup* $\Gamma + \mathbf{Z}\xi$ *of* Γ'.

We show first the uniqueness of w. Let $P = \sum_j a_j X^j$ be an element of K[X]. Then $w(a_j X^j) = v(a_j) + j\xi$, which shows that the monomials $a_j X^j$ such that $a_j \neq 0$ have distinct values for w. It follows that $w(P) = \inf_j(v(a_j) + j\xi)$, which shows both the uniqueness of w on K[X] (hence also on K(X)) and the fact that the order group of w is $\Gamma + \mathbf{Z}\xi$. It is further seen that, if $P \neq 0$, we may write $P = aX^n(1 + u)$, where $a \in K^*$, $n \in \mathbf{N}$, $u \in K(X)$ and $w(u) > 0$; every element $R \neq 0$ of $K(X)$ can therefore be written in the form

$$R = bX^n(1 + u'),$$

where $b \in K^*$, $n \in \mathbf{Z}$, $u' \in K(X)$ and $w(u') > 0$; then $w(R) = v(b) + n\xi$, hence $w(R) = 0$ if and only if $v(b) = 0$ and $n = 0$; thus, when $w(R) = 0$, R and b are congruent modulo the ideal of w, which shows that the residue field of w is equal to that of v.

Finally the existence of w follows from Lemma 1.

PROPOSITION 2. *Let* K *be a field, v a valuation on* K, Γ *its order group and k its residue field. There exists a unique valuation w on* K(X) *extending v such that $w(X) = 0$ and the image t of* X *in the residue field k' of w is transcendental over k. The order group of w is equal to that of v and its residue field is $k(t)$.*

To show the uniqueness of w, it will suffice for us to show that, if $P = \sum_j a_j X^j$ is a non-zero element of K[X], then

$$w(P) = \inf_j(v(a_j)).$$

We may divide P by an element of K^* and suppose that $v(a_j) \geqslant 0$ for all j and that one of the $v(a_j)$ is zero. As $w(X) = 0$, P then belongs to the ring of w; writing $\bar{a}_j$ for the canonical image of a_j in k, the canonical image of P in the residue field k' is $\sum_j \bar{a}_j t^j$; as t is transcendental over k and one of the $\bar{a}_j$ is non-zero, this image is non-zero, whence

$$w(P) = 0 = \inf_j(v(a_j)).$$

We now show the existence of w. The formula $w(P) = \inf_j(v(a_j))$ (for $P = \sum_j a_j X^j$) defines a valuation w on K(X), by virtue of Lemma 1, and w obviously has the same order group as v. Then $w(X) = 0$. We show that the canonical image t of X in the residue field k' of w is transcendental over k: if $\sum_j \bar{a}_j t^j = 0$, where $\bar{a}_j \in k$ for all j, then, denoting by a_j a representative of $\bar{a}_j$ in the ring of v, $w\left(\sum_j a_j X^j\right) > 0$; whence $v(a_j) > 0$ for all j, hence $\bar{a}_j = 0$ for all j. We show finally that $k' = k(t)$: every element R of K(X) may be written $R = c\left(\sum_j a_j X^j\right) \Big/ \left(\sum_j b_j X^j\right)$, where c, a_j, b_j are in K, $v(a_j) \geqslant 0$ and $v(b_j) \geqslant 0$ for all j, one of the $v(a_j)$ and one of the $v(b_j)$ being zero; then $w(R) \geqslant 0$ if and only if $v(c) \geqslant 0$; denoting by f the canonical homomorphism of the ring of w onto k',

$$f(R) = f(c)\left(\sum_j f(a_j)t^j\right) \Big/ \left(\sum_j f(b_j)t^j\right),$$

which proves our assertion.

Remark. It should not be thought that the two types of extensions of v to K(X) which we have just met are the only ones; there may exist a third type of extension, where Γ'/Γ is a torsion group and k' a (not necessarily finite) algebraic extension of k. This third type is not necessarily provided by the procedure described in Lemma 1 (cf. § 3, Exercise 1).

2. THE RATIONAL RANK OF A COMMUTATIVE GROUP

DEFINITION 1. *The rational rank of a commutative group* G *is the dimension of the vector* $\mathbf{Q}$*-space* $G \otimes_{\mathbf{Z}} \mathbf{Q}$.

This dimension may also be defined as the least upper bound (finite or infinite) of the cardinals r such that there exist r elements of G which are linearly independent over $\mathbf{Z}$ (*Algebra*, Chapter II, § 7, no. 10, Proposition 26). The rational rank of G is *zero* if and only if G is a torsion group. For a subgroup of an additive group $\mathbf{R}^n$, the notion of rational rank coincides with that defined in *General Topology*, Chapter VII, § 1.

In the rest of this paragraph, we shall denote by $r(G)$ the rational rank of the commutative group G. If G′ is a subgroup of G, then (since $\mathbf{Q}$ is a flat $\mathbf{Z}$-module) we have the additive equation

$$r(G) = r(G') + r(G/G'). \tag{4}$$

PROPOSITION 3. *Let* G *be a totally ordered commutative group and* H *a subgroup of* G. *If* $h(G)$ *and* $h(H)$ *denote the heights of* G *and* H (§ 4, no. 4), *then the inequality*

$$h(G) \leqslant h(H) + r(G/H) \tag{5}$$

holds.

In fact, let $G_0 \subset G_1 \subset \cdots \subset G_n$ be a strictly increasing sequence of isolated subgroups of G. It is necessary to establish the inequality

$$n \leqslant h(H) + r(G/H). \tag{6}$$

It is obvious for $n = 0$. Suppose $n \geqslant 1$ and let us argue by induction on n. Applying the induction hypothesis to the group G_{n-1} and its subgroup $H \cap G_{n-1}$, we obtain

$$n - 1 \leqslant h(H \cap G_{n-1}) + r(G_{n-1}/(H \cap G_{n-1})). \tag{7}$$

Then we distinguish two cases:

(a) $H \cap G_{n-1} = H$, in other words $H \subset G_{n-1}$. Inequality (7) may be written

$$n \leqslant h(H) + r(G_{n-1}/H) + 1. \tag{8}$$

Now G/G_{n-1} is a totally ordered group not reduced to 0; it is therefore not a torsion group and $r(G/G_{n-1}) \geqslant 1$. Whence by (4), $r(G/H) \geqslant r(G_{n-1}/H) + 1$. Substituting in (8), we certainly obtain the desired inequality (6).

(b) $H \cap G_{n-1} \neq H$. As $H \cap G_{n-1}$ is an isolated subgroup of H, we conclude that $h(H) \geqslant h(H \cap G_{n-1}) + 1$. On the other hand, obviously $r(G/H) \geqslant r(G_{n-1}/(H \cap G_{n-1}))$. Substituting in (7), we again obtain (6).

COROLLARY. *For every totally ordered commutative group* G, $h(G) \leqslant r(G)$.
We set $H = \{0\}$ in Proposition 3.

PROPOSITION 4. *Let* G *be a totally ordered commutative group. Suppose that* G *is finitely generated and that* $h(G) = r(G)$. *Then* G *is isomorphic to* $\mathbf{Z}^{r(G)}$ *ordered lexicographically.*

Let us write $r = r(G) = h(G)$. If $r = 0$, then $G = \{0\}$. If $r = 1$, the structure of finitely generated commutative groups shows that there is an isomorphism j of G onto $\mathbf{Z}$ (*Algebra*, Chapter VII, § 4, no. 6, Theorem 3). Now $\mathbf{Z}$ has only two total orderings compatible with its group structure, namely the usual ordering and its opposite. Hence j or $-j$ is an isomorphism of the ordered group G onto $\mathbf{Z}$ with the usual ordering.

Suppose now that $r \geqslant 2$ and let us argue by induction on r. Let H be an isolated subgroup of G of height $r - 1$. Then $r(H) + r(G/H) = r$ (formula (4)), $r(H) \geqslant h(H) = r - 1$ and $r(G/H) \geqslant h(G/H) = 1$ (Corollary to Proposition 3), whence $r(H) = r - 1$ and $r(G/H) = 1$. The induction hypothesis shows that H is isomorphic to $\mathbf{Z}^{r-1}$ ordered lexicographically and the case $r = 1$ shows that G/H is isomorphic to $\mathbf{Z}$. As $\mathbf{Z}$ is a free $\mathbf{Z}$-module, H is a *direct factor* of G (*Algebra*, Chapter II, § 1, no. 11, Proposition 21). The following lemma then shows that G is isomorphic (not canonically) to the lexicographical product $H \times (G/H)$, which completes the proof.

LEMMA 2. *Let* H *be an isolated subgroup of a totally ordered commutative group* G. *If* H *is a direct factor of* G, *the ordered group* G *is isomorphic to the group* $(G/H) \times H$ *ordered lexicographically.*

Let j be an isomorphism of $(G/H) \times H$ onto G such that $j(0, x) = x$ for all $x \in H$ and $j(y, x)$ belongs to the coset y of H. As $(G/H) \times H$ is totally ordered, it amounts to showing that j is *increasing* (*Set Theory*, Chapter III, § 1, no. 12, Proposition 11). Let (y, x) be an element $\geqslant 0$ of $(G/H) \times H$ ordered lexicographically. If $y > 0$, the coset of H containing $j(y, x)$ is an element > 0, whence $j(y, x) > 0$, for, otherwise, $y \leqslant 0$ (§ 4, no. 2, Proposition 3). If $y = 0$ and $x \geqslant 0$, then $j(y, x) = x \geqslant 0$. Hence j is certainly increasing.

3. THE CASE OF ANY TRANSCENDENTAL EXTENSION

In this no. we shall use the following notation: K is a field, K′ an extension of K, v a valuation on K, v' an extension of v to K′, Γ and k (resp. Γ' and k') the order group and residue field of v (resp. v'). We shall write:

$$d(K'/K) = \dim.\mathrm{al}_K K' = \text{transcendence degree of } K' \text{ over } K;$$
$$s(v'/v) = \dim.\mathrm{al}_k k' = \text{transcendence degree of } k' \text{ over } k;$$
$$r(v'/v) = r(\Gamma'/\Gamma) = \text{rational rank of } \Gamma'/\Gamma,$$

if the right-hand sides are finite; otherwise, we shall make the convention that $d(K'/K) = +\infty$ (resp. $s(v'/v) = +\infty$, $r(v'/v) = +\infty$).

THEOREM 1. *Let $x_1, \ldots, x_s$ be elements of the ring of v' whose canonical images $\bar{x}_i$ in k' are algebraically independent over k and $y_1, \ldots, y_r$ elements of K' such that the canonical images of the $v'(y_j)$ in Γ'/Γ are linearly independent over $\mathbf{Z}$. Then the $r + s$ elements $x_1, \ldots, x_s, y_1, \ldots, y_r$ of K' are algebraically independent over K; the restriction of v' to $K(x_1, \ldots, x_s, y_1, \ldots, y_r)$ admits $k(\bar{x}_1, \ldots, \bar{x}_s)$ as residue field and*

$$\Gamma + \mathbf{Z}v'(y_1) + \cdots + \mathbf{Z}v'(y_r)$$

as order group.

Our assertion is obvious if $r + s = 0$. We argue by induction on $r + s$. If $r' \leqslant r$, $s' \leqslant s$ and $r' + s' < r + s$, the induction hypothesis shows that the hypotheses of Theorem 1 hold if K is replaced by $K(x_1, \ldots, x_{s'}, y_1, \ldots, y_{r'})$ and the families $(x_1, \ldots, x_s)$, $(y_1, \ldots, y_r)$ by $(x_{s'+1}, \ldots, x_s)$, $(y_{r'+1}, \ldots, y_r)$. The problem is therefore reduced to one of the two following cases:

(a) There is an element x in the ring v' such that $\bar{x}$ is transcendental over k; then it is necessary to show that x is transcendental over K and that the restriction of v' to $K(x)$ admits $k(\bar{x})$ as residue field and Γ as order group.

(b) There is an element y in K' such that the relations $nv'(y) \in \Gamma$ and $n \in \mathbf{Z}$ imply $n = 0$; it is necessary to show that y is transcendental over K and that the restriction of v' to $K(y)$ admits k as residue field and $\Gamma + \mathbf{Z}v'(y)$ as order group.

Now Proposition 1 of § 8, no. 1 shows that x (resp. y) cannot be algebraic over K. The other assertions of (a) (resp. (b)) follow immediately by Proposition 2 (resp. Proposition 1) of no. 1.

COROLLARY 1. *The inequality*

$$s(v'/v) + r(v'/v) \leqslant d(K'/K) \tag{9}$$

holds.

Further, if K' is a finitely generated extension of K and equality holds in (9), then Γ'/Γ is a finitely generated $\mathbf{Z}$-module and k' is a finitely generated extension of k.

Let r and s be natural numbers such that $r \leqslant r(v'/v)$ and $s \leqslant s(v'/v)$; we show that $r + s \leqslant d(K'/K)$ and this will prove (9). By hypothesis there exist elements $x_1, \ldots, x_s, y_1, \ldots, y_r$ of K' which satisfy the hypotheses of Theorem 1.

They are therefore algebraically independent over K, which shows the inequality $r + s \leqslant d(K'/K)$.

If K' is a finitely generated extension of K, $d(K'/K)$ is finite, hence $s(v'/v)$ and $r(v'/v)$ are also finite; we denote them by s and r. There exist elements $x_1, \ldots, x_s$, $y_1, \ldots, y_r$ of K' which satisfy the hypotheses of Theorem 1. If $r + s = d(K'/K)$, these elements form a transcendence basis of K' over K and K' is therefore a finite algebraic extension of $K'' = K(x_1, \ldots, y_r)$. Let Γ'' and k'' be the order group and residue field of the restriction of v' to K''. By Theorem 1, Γ''/Γ is a finitely generated **Z**-module and k'' is a finitely generated pure extension of k. On the other hand, as K' is a finite algebraic extension of K'', Γ'/Γ is a finite group and k' is a finite algebraic extension of k'' (§ 8, no. 1, Lemma 2). This proves the corollary.

COROLLARY 2. *Let h and h' be the heights of v and v'. Then*

$$s(v'/v) + h' \leqslant d(K'/K) + h. \tag{10}$$

By Proposition 3, $h' \leqslant r(v'/v) + h$.

COROLLARY 3. *Suppose that K' is a finitely generated extension of K, that Γ is isomorphic to $\mathbf{Z}^h$ (ordered lexicographically) and there is equality in formula (10). Then Γ' is isomorphic to $\mathbf{Z}^{h'}$ (ordered lexicographically) and k' is a finitely generated extension of k.*

If there is equality in (10), there is equality in (9), whence the fact that k' is a finitely generated extension of k and that Γ' is a finitely generated **Z**-module. Further, comparing (9) and (10), it is seen that $h' - h = r(\Gamma'/\Gamma)$, whence $h' = r(\Gamma')$ and Proposition 4 (no. 2) then shows that Γ' is isomorphic to $\mathbf{Z}^{h'}$ ordered lexicographically.

COROLLARY 4. *Suppose that v is improper (in which case $k = K$). Then*

$$h(\Gamma') + d(k'/K) \leqslant r(\Gamma') + D(k'/K) \leqslant d(K'/K). \tag{11}$$

If, in particular, v' is of height 1, then

$$d(k'/K) \leqslant d(K'/K) - 1; \tag{12}$$

further, if K' is a finitely generated extension of K and there is equality in (12), then v' is a discrete valuation and k' is a finitely generated extension of K.

It is a series of special cases of Corollaries 1, 2, 3.

EXERCISES

§ 1

1. Let K be a field; for every subring A of K let L(A) denote the set of local rings $A_\mathfrak{p}$ of A where $\mathfrak{p}$ runs through the prime ideals of A (these local rings being identified with subrings of K).

(a) For a local subring M of K to dominate a ring $A_\mathfrak{p} \in L(A)$, it is necessary and sufficient that $A \subset M$; the local ring $A_\mathfrak{p}$ dominated by M is then unique and corresponds to the prime ideal $\mathfrak{p} = \mathfrak{m}(M) \cap A$.

(b) Let M, N be two local subrings of K and P the subring of K generated by $M \cup N$. Show that the following conditions are equivalent:

(α) There exists a prime ideal $\mathfrak{p}$ of P such that $\mathfrak{m}(M) = \mathfrak{p} \cap M$, $\mathfrak{m}(N) = \mathfrak{p} \cap N$.

(β) The ideal $\mathfrak{a}$ generated in P by $\mathfrak{m}(M) \cup \mathfrak{m}(N)$ is distinct from P.

(γ) There exists a local subring Q of K dominating both M and N.

If these conditions are satisfied, M and N are called *related*.

(c) Let A, B be two subrings of K and C the subring of K generated by $A \cup B$. Show that the following conditions are equivalent:

(α) For every local ring $Q \subset K$ containing A and B, $A_\mathfrak{p} = B_\mathfrak{q}$, where $\mathfrak{p} = \mathfrak{m}(Q) \cap A$, $\mathfrak{q} = \mathfrak{m}(Q) \cap B$.

(β) For every prime ideal $\mathfrak{r}$ of C, $A_\mathfrak{p} = B_\mathfrak{q}$, where $\mathfrak{p} = \mathfrak{r} \cap A$, $\mathfrak{q} = \mathfrak{r} \cap B$.

(γ) If $M \in L(A)$ and $N \in L(B)$ are related, they are identical.

(δ) $L(A) \cap L(B) = L(C)$.

(Use (a) and (b)).

2. For an integral domain A to be a valuation ring, it is necessary and sufficient that every ideal of A be irreducible (*Algebra*, Chapter II, § 2, Exercise 16).

3. Show that every valuation ring is coherent (Chapter I, § 2, Exercise 12).

4. Let K be a field and $\mathfrak{F}$ a set of valuation rings of K totally ordered by inclusion. Show that the intersection of the rings belonging to $\mathfrak{F}$ is a valuation ring of K.

5. Let K be a field, A a integrally closed subdomain of K with K as a field of fractions and (A_α) a family of valuation rings of K whose intersection is A. Then if L is an extension of K, the integral closure of A in L is the intersection of the valuation rings of L which dominate one of the A_α.

¶ 6. Let R be a ring and A a subring of R such that $S = R - A$ is non-empty and the product of two elements of S belongs to S.

(a) Let a, a' be two elements of A and s, s' two elements of S such that $sa \in A$ and $s'a' \in A$. Show that one of the two elements sa', $s'a$ belongs to A. Deduce that the set $\mathfrak{p}_1$ of elements $a \in A$ for which there exists $s \in S$ such that $sa \in A$ is a prime ideal of A.

(b) Show that the set $\mathfrak{p}_0$ of $a \in A$ such that $sa \in A$ for all $s \in S$ is a prime ideal of R and A; it is the greatest ideal of R contained in A and $\mathfrak{p}_0 \subset \mathfrak{p}_1$.

(c) The set $\mathfrak{n}$ of elements $x \in R$ such that there exists $s \in S$ for which $sx = 0$ is both an ideal of R and an ideal of A and $\mathfrak{n} \subset \mathfrak{p}_0$.

(d) The ring A is integrally closed in R.

(e) If R is a field, A is a valuation ring of R, $\mathfrak{p}_0 = (0)$ and $\mathfrak{p}_1$ is the maximal ideal of A.

¶ 7. Let R be a ring. An ordered pair $(A, \mathfrak{p})$ consisting of a subring A of R and a prime ideal $\mathfrak{p}$ of A is called *maximal* in R if there exists no ordered pair $(A', \mathfrak{p}')$ consisting of a subring A′ of R and a prime ideal $\mathfrak{p}'$ of A′ such that $A \subset A'$, $A' \neq A$ and $\mathfrak{p}' \cap A = \mathfrak{p}$.

(a) For every subring B of R and every prime ideal $\mathfrak{q}$ of B there exists an ordered pair $(A, \mathfrak{p})$, maximal in R, such that $A \supset B$ and $\mathfrak{q} = \mathfrak{p} \cap B$.

(b) For an ordered pair $(A, \mathfrak{p})$ to be maximal in R, it is necessary and sufficient that, for all $s \in R - A$, there exist a finite family of elements $c_i \in \mathfrak{p}$ $(1 \leqslant i \leqslant n)$ such that the element $b = c_1 s + c_2 s^2 + \cdots + c_n s^n$ belongs to $A - \mathfrak{p}$ (use Chapter II, § 2, no. 5, Corollary 3 to Proposition 11).

(c) Show that, if the ordered pair $(A, \mathfrak{p})$ is maximal in R, the product of two elements of $R - A$ belongs to $R - A$. (If s, s' are in $R - A$ and $ss' = a \in A$, consider the least integers n, m such that there exists $c_i \in \mathfrak{p}$ $(1 \leqslant i \leqslant n)$ for which $b = \sum_i c_i s^i \in A - \mathfrak{p}$ and $c'_j \in \mathfrak{p}$ $(1 \leqslant j \leqslant m)$ for which

$$b' = \sum_i c'_j s'^j \in A - \mathfrak{p};$$

supposing for example $n \geqslant m$, consider the product $bb' \in A - \mathfrak{p}$ and obtain a contradiction from the definition of the integer n.)

(d) Under the hypotheses of (c), we write $S = R - A$; show that the ideal $\mathfrak{p}_1$ of A consisting of the $a \in A$ such that there exists $s \in S$ for which $sa \in A$ is contained in $\mathfrak{p}$ (use (b)). *A fortiori* the ideals $\mathfrak{p}_0$ of A consisting of the $a \in A$ such that $sa \in A$ for all $s \in S$ is contained in $\mathfrak{p}$ (Exercise 6 (b)). We write $A' = A/\mathfrak{p}_0$, $\mathfrak{p}' = \mathfrak{p}/\mathfrak{p}_0$ and $R' = R/\mathfrak{p}_0$; the ordered pair $(A', \mathfrak{p}')$ is then maximal in the integral domain R' and, for all non-zero $a' \in A'$, there exists $s' \in S' = R' - A'$ such that $s'a' \in S'$.

(e) Let $(A, \mathfrak{p})$ be a maximal ordered pair in the integral domain R such that, for all non-zero $a \in A$, there exists $s \in S = R - A$ such that $sa \in S$. We write $S_0 = S \cup \{1\}$ and denote by R_0 the ring of fractions $S_0^{-1}R$; the canonical homomorphism $R \to R_0$ is injective and therefore identifies R with a subring of R_0. Show that R_0 is a field identified with the field of fractions of R. Let $(B, \mathfrak{q})$ be a maximal ordered pair in R_0 such that $B \supset A$ and $\mathfrak{q} \cap A = \mathfrak{p}$. Show that $B \cap R = A$; in other words, A is the intersection of R and a valuation ring of R_0 and $\mathfrak{p}$ is the intersection of A and the maximal ideal of this valuation ring.

(f) Let K_0 be a field, A_0 the polynomial ring $K_0[X]$, B the ring of formal power series $A_0[[Y]]$ and R the ring of fractions B_Y consisting of the formal power series $Y^{-h}f(Y)$, where $f \in B$ and $h \geqslant 0$. Let P be the prime ideal YB of formal power series without constant term, $A = K_0 + \mathfrak{p}$, which is a subring of B. Show that A is the intersection of R and a valuation ring C of the field of fractions K of R and $\mathfrak{p}$ the intersection of A and the maximal ideal of C; moreover, the field of fractions of A is equal to that of R and every $a \neq 0$ in A is of the form $s^{-1}s'$, where s, s' are in $S = R - A$. However, the ordered pair $(A, \mathfrak{p})$ is not maximal in R.

¶ 8. Given a ring A, a polynomial $P(X_1, \ldots, X_n)$ with coefficients in A is called *dominated* if it is of the form $X^\alpha + \sum_\beta c_\beta X^\beta$, where $\beta < \alpha$ in the product ordered set $\mathbf{N}^n$ for all β such that $c_\beta \neq 0$. In a ring R, a subring A is called *paravaluative* if, for every dominated polynomial $P \in A[X_1, \ldots, X_n]$ (any n), $P(s_1, \ldots, s_n) \neq 0$ for all elements $s_i \in R - A$ $(1 \leqslant i \leqslant n)$.

(a) Show that, if A is paravaluative in R, the product of two elements of $S = R - A$ belongs to S.

(b) Let A be a subring of a ring R such that the product of two elements of $S = R - A$ belongs to S; let $\mathfrak{p}_0$ be the prime ideal of A and R consisting of the $a \in A$ such that $sa \in A$ for all $s \in S$ (Exercise 6 (b)). For A to be paravaluative in R, it is necessary and sufficient that $A/\mathfrak{p}_0$ be paravaluative in $R/\mathfrak{p}_0$.

(c) Let R be a ring and $h: R \to R'$ a ring homomorphism. If A' is a paravaluative subring of R', $A = \overset{-1}{h}(A')$ is paravaluative in R. Deduce that, if $(A, \mathfrak{p})$ is a maximal ordered pair in R (Exercise 7), A is paravaluative in R (use (b) and Exercise 7 (d) and (e)).

(d) Let R' be a ring and R a subring of R'. If A is a paravaluative subring of R, show that there exists a paravaluative subring A' of R' such that $A' \cap R = A$. (If $S = R - A$ and $S_0 = S \cup \{1\}$, consider the ring $T' = S_0^{-1}R'$ and the canonical homomorphism $h: R' \to T'$. Let B be the subring of T' generated by $h(A) \cup (h(S))^{-1}$ and $\mathfrak{b}$ the ideal of B generated by $(h(S))^{-1}$. Show that $\mathfrak{b} \neq \mathrm{B}$; consider a maximal ordered pair $(\mathrm{A}'', \mathfrak{p}'')$ in T' such that $\mathrm{B} \subset \mathrm{A}''$ and $\mathfrak{b} \subset \mathfrak{p}''$ and take $\mathrm{A}' = \overset{-1}{h}(\mathrm{A}'')$.)

Conclude from this that, if R is an integral domain, the paravaluative subrings of R are the intersections of R with the *valuation rings* of the field of fractions of R.

¶ 9. (a) Let R be a ring, A a subring of R, x an element of R and S the set of x^k $(k \geqslant 0)$ in R. Show that, for x to be integral over A, it is necessary and sufficient that $x/1 \in \mathrm{A}[1/x]$ in the ring $\mathrm{S}^{-1}\mathrm{R}$. If x is not integral over A, show that there exists a maximal ideal $\mathfrak{m}$ of $\mathrm{A}[1/x]$ containing $1/x$ and that the inverse image of $\mathfrak{m}$ under the canonical homomorphism $\mathrm{A} \to \mathrm{A}[1/x]$ is a maximal ideal of A.

(b) Show that the integral closure A′ of A in R is the intersection of the paravaluative subrings of R (Exercise 8) which contain A. (To see that A′ is contained in each of these rings, use Exercise 8 (a) and Exercise 6 (d); to see that, if $x \in \mathrm{R}$ is not integral over A, there exists a subring B of R which is paravaluative in R and such that $x \notin \mathrm{B}$, use (a), Exercise 7 (a) and Exercise 8 (c), arguing as in Theorem 3 of no. 3.)

§ 2

1. Let Ω be an extension of a field K, E and F two extensions of K contained in Ω and linearly disjoint over K and L an extension of K contained in E. Let f be a place of E with values in L which reduces to the identity automorphism on K. Show that there exists a unique place of $\mathrm{K}(\mathrm{E} \cup \mathrm{F})$ with values in $\mathrm{K}(\mathrm{L} \cup \mathrm{F})$, extending f and reducing to the identity automorphism on F. (If A is the ring of f, note that f can be extended uniquely to a homomorphism g from F[A] to $\mathrm{K}(\mathrm{L} \cup \mathrm{F})$ reducing to the identity on F and that, if $\mathfrak{m}$ is the ideal of f, every element $z \in \mathrm{F}[\mathrm{A}]$ such that $g(z) = 0$ can be written as xu, where $x \in \mathfrak{m}$, $u \in \mathrm{F}[\mathrm{A}]$ and $g(u) \neq 0$. Conclude by noting that $\mathrm{K}(\mathrm{E} \cup \mathrm{F})$ is the field of fractions of F[A].)

2. Let K, K′ be two extensions of a field k and f and f' places respectively of K and K′ with values in the same algebraically closed field L. Suppose that the restrictions of f and f' to k coincide. Show that there exist a composite extension (F, i, i') of K and K′ (*Algebra*, Chapter VIII, § 8) and a place g of F with values in L such that $f(x) = g(i(x))$ for $x \in \mathring{\mathrm{K}}$ and $f'(x) = g(i'(x'))$ for $x' \in \mathring{\mathrm{K}}'$. (Let V, V′ be the rings of f and f', A their common intersection with k and h: $\mathrm{V} \otimes_{\mathrm{A}} \mathrm{V}' \to \mathrm{L}$ the homomorphism such that $h(a \otimes a') = f(a)f'(a')$ for $a \in \mathrm{V}$, $a' \in \mathrm{V}'$. Using the fact that V and V′ are flat A-modules (§ 3, no. 5, Lemma 1), prove that, if $a \neq 0$, $a' \neq 0$, $a \otimes a' = (a \otimes 1)(1 \otimes a')$ is not a divisor of zero; if S is the multiplicative subset of $\mathrm{B} = \mathrm{V} \otimes_{\mathrm{A}} \mathrm{V}'$ consisting of these elements,

$S^{-1}B$ contains subfields K_1, K'_1 respectively isomorphic to K and K′ and is the ring generated by $K_1 \cup K'_1$. Show that, if $\mathfrak{q}$ is the kernel of h, there exists a prime ideal $\mathfrak{p}$ in B such that $\mathfrak{p} \subset \mathfrak{q}$ and $\mathfrak{p} \cap S = \varnothing$ and show that F may be taken as the field of fractions of $S^{-1}B/S^{-1}\mathfrak{p}$.)

3. (a) Let K be a field and f a place of K with values in an orderable field L (*Algebra*, Chapter VI, § 2, Exercise 8); show that K is orderable. (Note that, for a finite family of elements $(a_i)_{1 \leqslant i \leqslant n}$ of K, there is always an index j such that $f(a_i/a_j) \neq \infty$ for all i.)

If $(x_i)_{1 \leqslant i \leqslant n}$ is a sequence of elements of K such that $f\left(\sum_{i=1}^{n} x_i^2\right) \neq \infty$, show that $f(x_i) \neq \infty$ for all i.

(b) Let K be an ordered field and G a subgroup of K; show that the ring F(G) of elements of K which are not infinitely great with respect to G (*loc. cit.*, Exercise 11) is a valuation ring of K and the ideal I(G) of elements of F(G) infinitesimally small with respect to G is the maximal ideal of this ring; hence there is on K a place with values in $k(G) = F(G)/I(G)$, which reduces to the identity on G; this is called the *canonical place* of K with respect to G.

(c) Show that, if there exists no extension of G contained in K, distinct from G and comparable with G, $k(G)$ is algebraic over G (note that, if $t \in K$ does not belong to G, the field $G(t)$ contains an element $u \neq 0$ infinitesimally small with respect to G and that $G(t)$ is algebraic over $G(u)$).

¶ 4. An ordered field K is called *Euclidean* if, for all $x \geqslant 0$ in K, there exists $y \in K$ such that $x = y^2$.

(a) Let K be a Euclidean ordered field and G a subfield of K such that there exists no extension of G contained in K, distinct from G and comparable with G. Show that, if f is a place of K with values in an ordered field L, which is an algebraic extension of G, and reducing to the identity on G, then f is equivalent to the canonical place of K with respect to G. (f may be considered to take its values in a maximal ordered field N which is an algebraic extension of L. Observe that G is Euclidean and that, if $x \in G$, $y \in K$ and $x < y$, then $f(y) = \infty$ or $f(y) > x$. If A and $\mathfrak{p}$ are the ring and ideal of the place f, show successively that $\mathfrak{p} \subset I(G)$, $F(G) \subset A$, $I(G) \subset \mathfrak{p}$ and $A \subset F(G)$, noting that N is comparable with G.)

(b) Let K be a Euclidean ordered field, f a place of K with values in a maximal ordered field L and A and $\mathfrak{p}$ the ring and ideal of f. Let G be a maximal subfield among the subfields E of K such that $E \cap \mathfrak{p} = 0$. Show that $G \subset A$ and that G is algebraically closed in K and therefore Euclidean. Prove that there exists no extension of G contained in K, distinct from G and comparable with G. Moreover, the subfield of L generated by $f(A)$ is algebraic over $f(G)$ and f is therefore equivalent to the canonical place of K with respect to G.

(c) Let K be a maximal ordered field, L a maximal ordered subfield of K, distinct from K and such that K is comparable with L (*Algebra*, Chapter VI, § 2, Exercise 15). Show that there exists no place f of K with values in L reducing to the identity automorphism on L.

¶ 5. Let K be a field and f a place of K with values in a maximal ordered field L.

(a) For all $\alpha \in K$ show that there exists an extension of f, either to $K(\sqrt{\alpha})$ or to $K(\sqrt{-\alpha})$, which is a place with values in L. (Consider two cases according to whether there exists $a \in K$ such that $f(a^2\alpha)$ is neither 0 nor ∞ or $f(x^2\alpha)$ is equal to 0 or ∞ for all $x \in K$.)

(b) Deduce from (a) that there exist an extension E of K which is a maximal ordered field and an extension of f to E which is a place of E with values in L. (Reduce it to the case where K is Euclidean and use Exercise 4 (a).)

6. Let A be an integrally closed domain, K its field of fractions, $\mathfrak{p}$ a prime ideal of A and k the field of fractions of $A/\mathfrak{p}$. Show that, if $A_\mathfrak{p}$ is not a valuation ring, there exists a valuation ring V of K which dominates $A_\mathfrak{p}$ and whose residue field is a transcendental extension of k (use Chapter V, § 2, Exercise 14 (b)).

§ 3

¶ 1. (a) Let L be a field and T an indeterminate; in the field of formal power series L((T)), consider the series

$$s = c_0 + c_1T^{1!} + c_2T^{2!} + \cdots + c_nT^{n!} + \cdots$$

where the $c_i \in L$ are all $\neq 0$. Show that s is transcendental over the subfield L(T) of L((T)). (Argue by *reductio ad absurdum* by supposing that there exists a relation $a_0(T) + a_1(T)s + \cdots + a_q(T)s^q = 0$ where $a_i(T)$ are polynomials in L(T) such that $a_q \neq 0$; construct the equation satisfied by

$$s' = s - c_0 - c_1T^{1!} - \cdots - c_{n-1}T^{(n-1)!}$$

and show that for n sufficiently large its constant term is a polynomial of degree $< n!$, whence our contradiction.)

(b) Let k be a field, X, Y, Z, T four indeterminates, $K = k(X, Y, Z)$ and E an algebraic closure of $k(X)$; in E, for all n, let x_n denote an element such that $x_n^n = X$. In the field of formal power series E((T)), consider the element

$$s = x_1T + x_2T^{2!} + \cdots + x_nT^{n!} + \cdots$$

The elements s and T are algebraically independent over E by (a). Consider the homomorphism f: $K \to E((T))$ such that $f(X) = X, f(Y) = T, f(Z) = s$. If v is the discrete valuation on E((T)) defined by the order of formal power series (no. 3, *Example* 3), $v \circ f$ is a discrete valuation on K; show that the residue field of $v \circ f$ is the subfield $k(x_1, x_2, \ldots, x_n, \ldots)$ of E and is therefore infinite over k.

¶ 2. Let Γ be a totally ordered group and k a field.

(a) Let A, B be two subsets of Γ *well ordered* by the induced ordering; show that the set $A + B$ is well ordered and that, for all $\gamma \in A + B$, the set of ordered pairs $(\alpha, \beta) \in A \times B$ such that $\alpha + \beta = \gamma$ is finite. (To prove the first assertion, argue by *reductio ad absurdum* by showing that otherwise it would be possible to define a strictly decreasing sequence (γ_n) of elements of $A + B$ such that $\gamma_n = \alpha_n + \beta_n$, where $\alpha_n \in A$, $\beta_n \in B$, the sequence (α_n) is strictly increasing and the sequence (β_n) is strictly decreasing; conclude using *Set Theory*, Chapter III, § 4, Exercise 3.)

(b) Let $S(\Gamma, k)$ be the set of families $x = (x_\alpha)_{\alpha \in \Gamma}$ such that $x_\alpha \in k$ for all $\alpha \in \Gamma$ and the set of α for which $x_\alpha \neq 0$ is a *well ordered* subset of Γ. Show that $S(\Gamma, k)$ is an additive subgroup of k^Γ; moreover, if $x = (x_\alpha)$, $y = (y_\alpha)$ are two elements of $S(\Gamma, k)$, the set C of $\alpha + \beta$ for the ordered pairs (α, β) such that $x_\alpha \neq 0$ and $y_\beta \neq 0$ is well ordered by (a) and, for all $\gamma \in C$, the element $z_\gamma = \sum_{\alpha+\beta=\gamma} x_\alpha y_\beta$ is defined in k; if we write $z_\alpha = 0$ for $\alpha \notin C$, the element $z = (z_\alpha) \in k^\Gamma$ belongs to $S(\Gamma, k)$; it is denoted by xy. Show that with this law of composition and addition on the product group k^Γ the set $S(\Gamma, k)$ is a *field*; moreover, for all $x = (x_\alpha) \neq 0$ of $S(\Gamma, k)$, let λ be the least of the $\alpha \in \Gamma$ such that $x_\alpha \neq 0$; if we write $v(x) = \lambda$ (and $v(0) = +\infty$), show that v is a valuation on $S(\Gamma, k)$ with k as residue field and Γ as value group. v is called the *canonical valuation* on the field $S(\Gamma, k)$. The field k is canonically identified with a subfield of $S(\Gamma, k)$. For all $\alpha \in \Gamma$, let u_α denote the element $(x_\lambda)_{\lambda \in \Gamma}$ such that $x_\alpha = 1$, $x_\lambda = 0$ for $\lambda \neq \alpha$; for all non-zero $z \in S(\Gamma, k)$, there is then a unique ordered pair $(t, \alpha) \in k \times \Gamma$ such that $v(z) = \alpha$ and $v(z - tu_\alpha) > \alpha$. tu_α is called the *dominant term* of z.

¶ 3. Let K be a field, w a valuation on K, Γ the value group of w and k the residue field of w. We form the corresponding field $S(\Gamma, k)$ and preserve the notation of Exercise 2. For all $t \in k$, let t^0 be an element of the class t in the valuation ring of w; for all $\alpha \in \Gamma$, let u_α^0 be an element of K such that $w(u_\alpha^0) = \alpha$.

(a) Let M be a subset of $S(\Gamma, k)$ containing the elements tu_α for every ordered pair $(t, \alpha) \in k \times \Gamma$; suppose that there exists a bijection $x \mapsto x^0$ of M onto a subset M^0 of K and that the following conditions are fulfilled: (1) $(tu_\alpha)^0 = t^0 u_\alpha^0$ for every ordered pair $(t, \alpha) \in k \times \Gamma$; (2) if $x = (x_\alpha)$ belongs to M and C_x is the well-ordered subset of Γ consisting of the α such that $x_\alpha \neq 0$, then for every segment D of C_x the element $(x_\lambda)_{\lambda \in D}$ belongs to M; (3) if x, y are two elements of M and tu_α is the dominant term of $x - y$, then $w(x^0 - y^0 - t^0 u_\alpha^0) > w(x^0 - y^0)$. If z' is an element of K not belonging to M^0, show that there exists an element $z \notin M$ in $S(\Gamma, k)$ such that the mapping which coincides with $x \mapsto x^0$ on M and maps z' to z also satisfies the above conditions. (Note that, if $w(z') = \alpha$, there exists a t^0 such that

$$w(z' - t^0 u^0_\alpha) > \alpha;$$

write $z_\alpha = tu_\alpha$, then show by transfinite induction that the z_β of index $\beta > \alpha$ may be determined so that $z = (z_\lambda)_{\lambda \in \Gamma}$ solves the problem.)

(b) Deduce from (a) that $\text{Card}(K) \leqslant \text{Card}\, S(\Gamma, k)$.

4. Let A be a local integral domain, K its field of fractions, $\mathfrak{m}$ its maximal ideal and v a valuation on K whose ring B dominates A. Suppose that, either $\mathfrak{m}$ is a finitely generated ideal, or v is a discrete valuation; then there exists in $\mathfrak{m}$ an element x such that $v(x) = \inf_{y \in \mathfrak{m}} v(y)$. Let A′ be the subring of K generated by A and the elements yx^{-1}, where y runs through $\mathfrak{m}$; let A_1 be the local ring of A′ relative to the prime ideal $\mathfrak{p} \cap A'$, where $\mathfrak{p}$ is the ideal of the valuation v. Show that the ring A_1 does not depend on the choice of the element x satisfying the above conditions (A_1 is called "the prime quadratic transform of A in the direction of v"); if A is Noetherian, so is A_1.

¶ 5. Let k be a field and $K = k(X, Y)$ the field of rational functions in two indeterminates over k. We write $x_1 = X$, $y_1 = Y$ and for $n \geqslant 1$ define inductively the elements $x_{n+1} = y_n$ and $y_{n+1} = x_n y_n^{-1}$ of K.

(a) Write $A_n = k[x_n, y_n]$; the sequence (A_n) is increasing and, if

$$\mathfrak{m}_n = A_n x_n + A_n y_n,$$

$\mathfrak{m}_n$ is a maximal ideal of A_n and $\mathfrak{m}_n = A_n \cap \mathfrak{m}_{n+1}$; in the ring $A = \bigcap_n A_n$, $\mathfrak{m} = \bigcap_n \mathfrak{m}_n$ is a maximal ideal.

(b) Let v be the valuation on K defined by $v(X) = \rho = (1 + \sqrt{5})/2$, $v(Y) = 1$; let B and $\mathfrak{p}$ be the ring and ideal of v; show that $\mathfrak{m} = \mathfrak{p} \cap A$ (note that $\rho^2 - \rho - 1 = 0$ and use this fact to calculate $v(y_n)$). Show that for a monomial $X^i Y^j$ (i, j positive or negative integers) to belong to A, it is necessary and sufficient that $i\left(\dfrac{1 + \sqrt{5}}{2}\right) + j > 0$; for every monomial $z = X^i Y^j$, therefore, either $z \in A$ or $1/z \in A$.

(c) Show that $B = A_\mathfrak{m}$ (write an element $t \in B$ in the form of a quotient of two polynomials in X and Y and show in each of them the monomials at which v takes the least value).

(d) Show that, for all n, $(A_{n+1})_{\mathfrak{m}_{n+1}}$ is the prime quadratic transform of $(A_n)_{\mathfrak{m}_n}$ in the direction of v (Exercise 4).

6. (a) Let A be a valuation ring. Extend to finitely generated A-modules the results of *Algebra*, Chapter VII, § 4, nos. 1 to 4 (theory of invariant factors).

(b) Let K be a field, v a valuation on K, A the ring of v and Γ the order group of v. Let $U = (\alpha_{ij})$ be a square matrix of order n with elements in A and of determinant 0; show that there exists $\lambda \in \Gamma$ such that, for every square matrix $U' = (\alpha'_{ij})$ of order n with elements in A and satisfying the conditions $v(\alpha'_{ij} - \alpha_{ij}) > \lambda$ for every ordered pair (i, j), the invariant factors of U' are the same as those of U.

(c) Under the hypotheses of (b), generalize the following results of *Algebra*, Chapter IX, § 5, no. 1, Theorem and Exercise 1 and § 6, Exercise 10 (for the last, suppose that K is not of characteristic 2 and that $v(2) = 0$).

7. Show that the integral domain B defined in Chapter II, § 3, Exercise 2 (b) is a local ring whose maximal ideal is principal but is not a valuation ring.

8. Show that, if a valuation ring A is strongly Laskerian (Chapter IV, § 2, Exercise 28), A is a field or a discrete valuation ring (use Exercise 29 of Chapter IV, § 2).

§ 4

1. (a) Let G be a totally ordered group and M a major set in G_+ not containing 0. Show that there exists a *greatest* isolated subgroup H of G not meeting M.

(b) Let A be a valuation ring and v a valuation associated with A. For an ideal $\mathfrak{p} \neq 0$ of A to be prime, it is necessary and sufficient that $\mathfrak{p}$ correspond to a major set M in the totally ordered value group G of v such that, if H is the greatest isolated subgroup of G not meeting M, M is the complement of H_+ in G_+.

(c) With the notation of (b), for an ideal $\mathfrak{q}$ of A to be $\mathfrak{p}$-primary, it is necessary and sufficient that it satisfy one of the following conditions: either $H = \{0\}$ (and $\mathfrak{p}$ is then maximal) or $\mathfrak{q} = \mathfrak{p}$.

(d) For an ideal $\mathfrak{a}$ of A which is distinct from 0 and A to satisfy $\mathfrak{a}^2 = \mathfrak{a}$, it is necessary and sufficient that $\mathfrak{a}$ correspond to a major set M such that, if H is the greatest isolated subgroup of G not meeting M, the image M′ of M in the totally ordered group $G' = G/H$ has no least element; $\mathfrak{a}$ is then a prime ideal.

2. Let G be a totally ordered group.

(a) For every element $a > 0$ of G, let $H(a)$ be the smallest isolated subgroup containing a, $H^-(a)$ the greatest isolated subgroup not containing a; show that $H(a)/H^-(a)$ is isomorphic to a subgroup of $\mathbf{R}$.

(b) Conversely, if H is an isolated subgroup of G such that there exists a greatest element (with respect to inclusion) H′ in the set of isolated subgroups of G contained in H and $\neq$H, then $H = H(a)$ and $H' = H^-(a)$ for all $a \in H_+ \cap \complement H'_+$. Such an isolated subgroup H is called *principal* and H′ is called its *predecessor*.

(c) If H_1, H_2 are two distinct isolated subgroups of G such that $H_1 \subset H_2$, show that there exist two isolated subgroups H, H′ of G such that $H_1 \subset H' \subset H \subset H_2$ and H/H′ is isomorphic to a non-zero subgroup of **R**.

(d) Let Σ be the set (totally ordered with respect to inclusion) of isolated subgroups of G and let $\Theta \subset \Sigma$ be the set of principal isolated subgroups. Show that Σ is isomorphic to the *completion* (*Set Theory*, Chapter III, § 1, Exercise 15) of Θ.

3. (a) Let Θ be a totally ordered set and for each $s \in \Theta$ let A_s be a subgroup of **R** not reduced to 0. Let

$$\Gamma(\Theta, (A_s)_{s \in \Theta}) = G$$

be the subgroup of the product $\prod_{s \in \Theta} A_s$ consisting of the functions f whose support is a well-ordered subset of Θ. We define on G an ordering compatible with its group structure taking the set G_+ of elements $\geqslant 0$ to be the set consisting of the function 0 and the functions $f \neq 0$ such that $f(\theta) > 0$ for the least element θ of the support of f. Show that G is a totally ordered group and that Θ is canonically isomorphic to the set of principal isolated subgroups of G ordered by the relation $\supset$; moreover, if $H(a)$ corresponds to $s \in \Theta$ under this isomorphism, $H(a)/H^-(a)$ is isomorphic to A_s.

(b) Consider the subgroup G′ of the totally ordered group $\mathbf{Q} \times \mathbf{Q}$ (with the lexicographical ordering) generated by the elements (p_n^{-1}, np_n^{-1}), where (p_n) is the strictly increasing sequence of prime numbers. Show that the only isolated subgroup H′ of G′ distinct from 0 and G′ is the group $\{0\} \times \mathbf{Z}$; but G′ is not isomorphic to a subgroup of the product $H' \times (G'/H')$ ordered by the lexicographical ordering.

4. Let A be a valuation ring which is not a field.

(a) Show that for A to be a discrete valuation ring, it is necessary and sufficient that every prime ideal of A be principal.

(b) Show that, if A is the ring of a valuation of height one, the field of fractions of A is a finitely generated A-algebra.

* 5. Let K be a field and $\mathfrak{M}$ the set of subrings of K which are not fields. Show that a maximal element of $\mathfrak{M}$ is a valuation ring of height 1 of its field of fractions L and that K is an algebraic extension of L. The following conditions are equivalent:

(α) $\mathfrak{M}$ is not empty.

(β) $\mathfrak{M}$ admits a maximal element.

(γ) There exists a valuation of height 1 on K

(δ) K is not an algebraic extension of a finite prime field (cf. § 8, no. 1). *

¶ 6. (a) Let S be a *hyperstonian* compact space with no isolated point (*Integration*, Chapter V, § 5, Exercise 14). Let $\mathscr{C}_0(S)$ be the vector space of real-valued functions, finite or otherwise, defined on S, continuous on S and such that the sets $\overset{-1}{f}(-\infty)$ and $\overset{-1}{f}(+\infty)$ are nowhere dense in S (*Integration*, Chapter II, § 1, Exercise 13 (g)). Show that, for every measure $\mu > 0$ on S, there exist functions f in $\mathscr{C}_0(S)$ whose upper integral is infinite (note that for every nowhere dense closed subset N of S there exists a function f in $\mathscr{C}_0(S)$ which is equal to $+\infty$ on N; show that attention may be restricted to the case where the measure μ is normal and define f suitably as the upper bound of a sequence of functions in $\mathscr{C}_0(S)$).

(b) Let G be the ordered subgroup of $\mathscr{C}_0(S)$ consisting of the functions in $\mathscr{C}_0(S)$ whose values in $\mathbf{Z}$ or $\pm\infty$. Show that G is a complete lattice and is not isomorphic (as an ordered group) to any subgroup of a product group $\mathbf{R}^{I}$.

(c) Let K be a field and $A_0 = K[[X_s]]_{s \in S}$ the ring of formal power series with coefficients in K in a family of indeterminates (X_s) with the space S as indexing set (*Algebra*, Chapter IV, § 5, Exercise 1). Let $\mathfrak{b}$ be the set of formal power series $P \in A_0$ with the following property: there exists a *nowhere dense* set $N \subset S$ such that every monomial of P whose coefficient is $\neq 0$ contains an X_s for which $s \in N$. Show that $\mathfrak{b}$ is an ideal of A_0 and that the ring $A = A_0/\mathfrak{b}$ is an integral domain and completely integrally closed (to prove the latter point, apply Chapter V, § 1, no. 4, Proposition 14, arguing as in Chapter V, § 1, Exercise 10; use also the fact that in S every meagre set is nowhere dense).

(d) Let f be an element $\geqslant 0$ of the group G and let N be the nowhere dense set of points where $f(s) = \pm\infty$. Let p_f be the element of A the image of the formal power series $\prod_{s \notin N} (1 + X_s)^{f(s)}$; the mapping $f \mapsto p_f$ is an isomorphism of the additive monoid G_+ onto a multiplicative submonoid of A. Deduce that A cannot be the intersection of a family of valuation rings of height 1 of its field of fractions (show that this would imply that G is isomorphic to a subgroup of a product $\mathbf{R}^{I}$). (Cf. Exercises 8 and 9.)

¶ 7. A local integral domain A is said to be of *dimension* 1 if A is not a field and there is no prime ideal of A distinct from (0) and the maximal ideal $\mathfrak{m}$ of A. For every ideal $\mathfrak{a}$ of A, let $A : \mathfrak{a}$ denote the sub-A-module of the field of fractions of A which is the transporter of $\mathfrak{a}$ in A (which always contains A). In the following A is assumed to be of dimension 1.

(a) Show that, if A is strongly Laskerian (Chapter IV, § 2, Exercise 28), then $A : \mathfrak{m} \neq A$. (In the opposite case, note that $A : \mathfrak{m}^r = A$ for every integer $r > 0$; on the other hand, every ideal $\neq 0$ of A contained in $\mathfrak{m}$ is $\mathfrak{m}$-primary and so in particular is every principal ideal $Ax \subset \mathfrak{m}$ (where $x \neq 0$); note finally that $\mathfrak{m}^h \neq \mathfrak{m}^k$ for $h \neq k$ (Chapter IV, § 2, Exercise 29 (d)).)

(b) Show that for a strongly Laskerian local integral domain A of dimension 1 the following properties are equivalent:

(α) A is completely integrally closed.

(β) The maximal ideal $\mathfrak{m}$ is principal.

(γ) Every $\mathfrak{m}$-primary ideal is of the form $\mathfrak{m}^k$.

(δ) The ideal $\mathfrak{m}$ is invertible (Chapter II, § 5, no. 6), which is equivalent to $\mathfrak{m}.(A:\mathfrak{m}) = A$.

(ε) A is a discrete valuation ring.

(To show that (α) implies (δ), use (a) and observe that if $\mathfrak{m}.(A:\mathfrak{m})^k = \mathfrak{m}$ for all $k > 0$, A is not completely integrally closed (cf. Chapter VII, § 1, Exercise 4). To see that (δ) implies (γ), note that for every ideal $\mathfrak{a} \neq 0$ contained in $\mathfrak{m}$ we may write $\mathfrak{a} = \mathfrak{m}\mathfrak{a}_1$, where $\mathfrak{a}_1 \neq 0$ and use Chapter IV, § 2, Exercise 29 (d). Finally, to see that (γ) implies (β), observe that $\mathfrak{m} \neq \mathfrak{m}^2$ and that (γ) implies that $\mathfrak{m}/\mathfrak{m}^2$ is a 1-dimensional vector $(A/\mathfrak{m})$-space.)

(c) Show, in the notation of Chapter III, § 3, Exercise 15 (b), that the local ring $A_\mathfrak{m}$ is a Noetherian integral domain of dimension 1 but is not integrally closed.

(d) Let K be an algebraically closed field and A the subring of the field K(X, Y) of rational functions in two indeterminates over K consisting of the elements $f \in K(X, Y)$ such that 0 is substitutable for X in f and $f(0, Y)$ belongs to K. Show that A is a strongly Laskerian integrally closed local integral domain of dimension 1 but that it is not completely integrally closed. (If B is the polynomial ring K[X, Y], $\mathfrak{p}$ the prime ideal BX of B, note that A is the subring of $B_\mathfrak{p}$ equal to $K + \mathfrak{p}B_\mathfrak{p}$.) (*)

¶ 8. Let A be a local integral domain of dimension 1 (Exercise 7) and K its field of fractions. Show that for A to be the intersection of valuation rings (of K) of height 1, it is necessary and sufficient that A be completely integrally closed. Prove successively that:

(a) A subring of K containing A and the inverse of an element $\neq 0$ of the maximal ideal $\mathfrak{m}$ of A is equal to K (use the fact that every ideal of A distinct from 0 and A is $\mathfrak{m}$-primary).

(b) If $B \supset A$ is a subring of K distinct from K, there exists a valuation ring V of height 1 such that $B \subset V$ (use (a)).

(c) Let $z \in K - A$; show that $z\mathfrak{m} \subset \mathfrak{m}$ is impossible arguing as in Exercise 7 (b). Deduce that there exists a valuation ring V of height 1 such that $A \subset V$ and $z \notin V$. (If $z\mathfrak{m} \subset A$, use (a); otherwise, there is an $x \in \mathfrak{m}$ such that

$$xy = z \in K - A;$$

observe that $z \notin A[z^{-1}]$ and use (b) to show the existence of V.)

¶ 9. Let A be a Laskerian completely integrally closed domain (Chapter IV, § 2, Exercise 23). Suppose further, that, for all $x \neq 0$ in A, the prime ideals $\mathfrak{p}_i$

(*) There exist examples of local integral domains of dimension 1 which are *completely integrally closed* but are not valuation rings (cf. P. Ribenboim, Sur une note de Nagata relative à un problème de Krull, *Math. Zeitsch.*, v. LXIV (1956), pp. 159–168).

weakly associated with A/Ax (Chapter IV, § 1, Exercise 17) are all of *height* 1, that is to say that for all i, $\mathfrak{p}_i$ contains no prime ideal other than itself and 0 (cf. Chapter VII, § 1, no. 6). Show that under these conditions A is an intersection of valuation rings of height 1. (Show first that A is the intersection of all the rings $A_\mathfrak{p}$, where $\mathfrak{p}$ runs the set of prime ideals of height 1, using Exercise 17 (i) of Chapter IV, § 1. Then prove that these local rings $A_\mathfrak{p}$ are completely integrally closed, using condition (LA_I) of Chapter IV, § 2, Exercise 23. Finally apply Exercise 8.) The above hypotheses are in particular satisfied when A is a Laskerian completely integrally closed domain in which every ideal admits a single primary decomposition (Chapter IV, § 2, Exercise 26).

¶ 10. Let A be a ring in which the set of principal ideals is totally ordered by inclusion.

(a) Show that A is a local ring, in which the nilradical $\mathfrak{N}$ is a prime ideal. Show that, either $\mathfrak{N}^2 = \mathfrak{N}$, or $\mathfrak{N}$ is nilpotent; the ring $A/\mathfrak{N}$ is a valuation ring.

(b) If $\mathscr{L}$ is the totally ordered set of principal ideals of A, show that the set of ideals of A is totally ordered by inclusion and is isomorphic to the *completion* of the set $\mathscr{L}$ (*Set Theory*, Chapter III, § 1, Exercise 15).

(c) Suppose that $\mathfrak{N}^2 = 0$; then $\mathfrak{N}$ may be considered as a module over the valuation ring $V = A/\mathfrak{N}$; let K be the field of fractions of V and Γ the order group of a valuation v on K corresponding to V; show that there exists in Γ two major sets $M \subset M'$ such that $\mathfrak{N}$ is a V-module isomorphic to $\mathfrak{a}(M')/\mathfrak{a}(M)$ (notation of § 3, no. 4).

(d) Conversely, given a valuation ring V with field of fractions K and order group Γ and two major sets $M \subset M'$ in Γ, let $Q = \mathfrak{a}(M')/\mathfrak{a}(M)$ and define on the product additive group $A = V \times Q$ a multiplication by setting

$$(z, t)(z', t') = (zz', zt' + z't);$$

show that in A the set of principal ideals is totally ordered by inclusion; the set $\mathfrak{N}$ of ordered pairs $(0, t)$ where $t \in Q$ is the nilradical of A, $\mathfrak{N}^2 = 0$ and $A/\mathfrak{N}$ is isomorphic to V.

(e) For every prime number p, the ring $A = \mathbf{Z}/p^2\mathbf{Z}$ is such that the set of principal ideals of A is totally ordered by inclusion and the nilradical $\mathfrak{N}$ of A is such that $\mathfrak{N}^2 = 0$; but show that the ring A is not isomorphic to the ring constructed (starting with $V = A/\mathfrak{N}$ and $Q = \mathfrak{N}$) by the method in (d) (note that A contains no subring isomorphic to $\mathbf{Z}/p\mathbf{Z}$).

(f) Show that for every prime number p, the algebra A of the group U_p (*Algebra*, Chapter VII, § 2, Exercise 3) with respect to the prime field $\mathbf{F}_p$ is a ring in which the principal ideals form a set which is totally ordered by inclusion and the nilradical $\mathfrak{N}$ satisfies $\mathfrak{N}^2 = \mathfrak{N}$, but A is not isomorphic to the quotient of a valuation ring by an ideal.

11. Let K be a field with a valuation v of height 1.

(a) Let $P(X) = a_0X^n + a_1X^{n-1} + \cdots + a_n$ be a polynomial in K[X] of degree $n > 1$ and such that $a_n \neq 0$; show that there exists a strictly increasing sequence $(i_k)_{0 \leqslant k \leqslant r}$ of integers in the interval $[0, n]$ such that: (1) $i_0 = 0$, $i_r = n$; (2) $v(a_{i_k})$ is finite for $0 \leqslant k \leqslant r$; (3) for every index j such that $0 \leqslant j \leqslant n$ distinct from the i_k, such that $v(a_j)$ is finite, the point

$$(j, v(a_j)) \in \mathbf{R}^2$$

lies above the line passing through the points $(i_k, v(a_{i_k}))$ and $(i_{k+1}, v(a_{i_{k+1}}))$ and strictly above that line if $j < i_k$ or $j > i_{k+1}$. The union of the segments joining the points $(i_k, v(a_{i_k}))$ and $(i_{k+1}, v(a_{i_{k+1}}))$ is called the *Newton polygon* of P, the above segments are called the *sides* and the points $(i_k, v(a_{i_k}))$ the *vertices* of the polygon.

(b) Suppose that all the zeros of P belong to K. Show that for the valuations of all the zeros of P to be the same, it is necessary and sufficient that $r = 1$ (in other words, that the Newton polygon reduce to *a single side*). (To show that the condition is sufficient, consider the Newton polygon of a product P_1P_2 where all the zeros of P_1 are invertible in the ring of v, whilst all those of P_2 belong to the ideal of v.)

(c) Suppose that all the zeros of P belong to K; form the Newton polygon of P and write.

$$\rho_k = i_{k+1} - i_k, \qquad \sigma_k = (v(a_{i_{k+1}}) - v(a_{i_k}))/\rho_k.$$

Show that, for $0 \leqslant k \leqslant r - 1$, P admits exactly ρ_k zeros (counted with their orders of multiplicity) whose valuations are all equal to σ_k (use (b) and argue by induction on r).

(d) Generalize to the case of any valuation v (embed the order group Γ of v in the vector $\mathbf{Q}$-space $\Gamma_{(\mathbf{Q})}$, which has a natural totally ordered group structure).

§ 5

¶ 1. Let A be an integral domain, K its field of fractions and $\mathcal{T}$ a linear topology on A (Chapter III, § 2, Exercise 21).

(a) For the neighbourhoods of 0 under $\mathcal{T}$ to constitute a fundamental system of neighbourhoods of 0 for some topology $\mathcal{T}_K$ compatible with the ring structure on K, it is necessary and sufficient that $\mathcal{T}$ be the topology $\mathcal{T}_u(A)$ (Chapter III, § 2, Exercise 24); then A is a bounded subset with respect to $\mathcal{T}_K$ and $\mathcal{T}_K$ is a locally bounded Hausdorff topology (*General Topology*, Chapter III, § 6, Exercises 12 and 20 (e)). For K to be complete (resp. linearly compact. resp. strictly linearly compact (Chapter III, § 2, Exercise 21)) with $\mathcal{T}_K$, it is necessary and sufficient that A be so with $\mathcal{T}$.

(b) For the topology $\mathcal{T}_K$ (where $\mathcal{T} = \mathcal{T}_u(A)$) to be compatible with the field structure on K, it is necessary and sufficient that the Jacobson radical $\mathfrak{R}(A)$ of A be $\neq 0$.

¶ 2. Let K be a (commutative) field and $\mathscr{T}$ a non-discrete Hausdorff topology on K compatible with the ring structure on K. For the topology $\mathscr{T}$ to be defined by a valuation on K or an absolute value on K, it is necessary and sufficient that $\mathscr{T}$ be locally retrobounded (*General Topology*, Chapter III, § 6, Exercise 22. If there exist in K topologically nilpotent elements, use Exercise 22 (d) of *General Topology*, Chapter III, § 6 and Exercise 13 of *General Topology*, Chapter IX, § 3. In the opposite case, use Exercise 22 (f) of *General Topology*, Chapter III, § 6).

¶ 3. Let A be a Noetherian integral domain, K its field of fractions, $\mathscr{T}_u$ the topology $\mathscr{T}_u(A)$ and $\mathscr{T}_K$ the corresponding topology on K (Exercise 1).

(a) If A is a Zariski ring with Jacobson radical $\mathfrak{r} \neq 0$ and A is complete with the $\mathfrak{r}$-adic topology, show that A is complete with the topology $\mathscr{T}_u$ (*General Topology*, Chapter III, § 3, no. 5, Corollary 2 to Proposition 9).

(b) Suppose that $\mathscr{T}_K$ is not discrete and is defined by a valuation v on K. Show that v is a discrete valuation and that A is an open subring of the ring V of v; obtain the converse. (Using the hypothesis, which implies $A \neq K$, and Proposition 1 of § 4, no. 1, we may assume that $A \subset V$. Using the fact that A is open under $\mathscr{T}_K$, show that V is a finitely generated A-module and conclude using § 3, no. 5, Proposition 9. For the converse, observe that A is a local ring of which the maximal ideal $\mathfrak{m}$ and (0) are the only prime ideals and in which every ideal $\mathfrak{a} \neq 0$ is therefore $\mathfrak{m}$-primary). The integral closure of A is then V. Give an example where $A \neq V$ (take V to be a ring of formal power series $k[[T]]$, where k is a field).

(c) Deduce an example of a non-discrete complete topological field whose topology is not locally retrobounded (use (b) and Exercise 2).

4. Let v be a valuation on a field K, A the ring of v and $\mathfrak{m}$ its maximal ideal.

(a) For the topology defined by v on A to be identical with the $\mathfrak{m}$-adic topology, it is necessary and sufficient that A be a field or a discrete valuation ring.

(b) For the topology defined by v to make A into a strictly linearly compact ring (Chapter III, § 2, Exercise 21), it is necessary and sufficient that A be a field or a discrete valuation ring (use (a) and Chapter III, § 2, Exercise 22 (a)).

5. (a) Let K be a field, v a valuation on K and Γ the order group of v. For K to be linearly compact (Chapter III, § 2, Exercise 15) with the topology $\mathscr{T}_v$, it is necessary and sufficient that, for every *well-ordered* subset B of Γ and every family $(a_\beta)_{\beta \in B}$ of elements of K such that, for $\lambda < \mu < \nu$,

$$v(a_\lambda - a_\mu) < v(a_\mu - a_\nu),$$

there exists an element $a \in K$ such that $v(a - a_\lambda) = v(a_\lambda - a_\mu)$ for every ordered pair of indices λ, μ such that $\mu > \lambda$. (Use Exercise 4 of *Set Theory*,

Chapter III, § 2). If this is so, the ring A of the valuation v is also linearly compact with the discrete topology.

(b) Show that the field $S(\Gamma, k) = K$ defined in § 3, Exercise 2 is linearly compact with $\mathscr{T}_v$. Take Γ to be the totally ordered group $\mathbf{Q}$ of rational numbers and consider the subring K_0 of K consisting of the $x = (x_\alpha)$ such that the set of $\alpha \in \mathbf{Q}$ for which $x_\alpha \neq 0$ is finite or is the set of points of a strictly increasing sequence tending to $+\infty$. Show that K_0 is a field which is complete with respect to the valuation v_0 induced by the canonical valuation v on K and that v_0 has the same value group and residue field as v but that K_0 is not linearly compact (cf. § 10, Exercise 2).

¶ 6. (a) Let A be a valuation ring, M an A-module and M′ a submodule of M. Show that for M′ to be a pure submodule of M (Chapter I, § 2, Exercise 24), it is necessary and sufficient that, for all $\alpha \in A$, $M' \cap (\alpha M) = \alpha M'$.

(b) Let M′ be a pure submodule of M such that, if M′ is given the topology for which the $\alpha M'$ (where $\alpha \in A$, $\alpha \neq 0$) form a fundamental system of neighbourhoods of 0, M′ is linearly compact. Show that for every element $x \in M$ there exists $y' \in M'$ such that $x' = x + y'$ has the following property: for all $\alpha \in A$ such that $x + M' \in \alpha(M/M')$, $x' \in \alpha M$. (For each of the $\alpha \in A$ such that $x + M' \in \alpha(M/M')$, consider the subset S_α of M′ consisting of the y'_α such that $x + y'_\alpha \in \alpha M$.)

(c) Suppose that A is a linearly compact valuation ring; let K be the field of fractions of A. Show that, if M is a torsion-free A-module such that $M_{(K)}$ admits a countable basis, M is the direct sum of a countable family of A-modules of rank 1. (Consider M as the union of an increasing sequence (M'_i) of pure submodules such that M'_i is of rank i and for each i apply (b) to M'_i and its submodule M'_{i-1}.

(d) With the same hypotheses on A and M as in (c), let N be a pure submodule of M of finite rank. Show that N is a direct factor of M (observe that every A-module which is the direct sum of a finite number of A-modules of rank 1 is linearly compact and use (b)).

(e) With the same hypotheses on M and M′ as in (b), show that for all $x \in M$ there exists $y' \in M'$ such that the annihilator of $x' = x + y'$ is equal to the annihilator of the element $x + M' \in M/M'$. (Let $\mathfrak{a}$ be the annihilator of $x + M'$; for all $\alpha \in \mathfrak{a}$, let T_α be the subset of M′ consisting of the y' such that $\alpha(x + y') = 0$; show that the intersection of the T_α is non-empty.)

(f) Suppose that A is a valuation ring such that for every ideal $\mathfrak{a} \neq 0$ the A-module $A/\mathfrak{a}$ is linearly compact (with the discrete topology). Show that every finitely generated torsion A-module M is the direct sum of a finite number of monogenous A-modules. (If $(z_i)_{1 \leqslant i \leqslant n}$ is a system of generators of M, argue by induction on n, considering one of the z_i whose annihilator is the least and noting that the submodule of M which it generates is pure.)

¶ 7. Let K be a field and v a discrete valuation on K such that K is *complete* with $\mathscr{T}_v$: let A and $\mathfrak{p}$ denote the ring and ideal of v and $U = A - \mathfrak{p}$ the set of invertible elements of A. Let u be an isomorphism of K onto a subfield of K.

(a) Show that $u(\mathfrak{p}) \subset A$ and $u(U) \subset U$. (To prove the first point, observe that for all $z \in \mathfrak{p}$ the equation $x^n = 1 + z$ admits a solution in A for all $n > 0$ prime to the characteristic of the residue field of v, using Hensel's Lemma; if $v(u(z)) < 0$, deduce that the integer $v(u(z))$ would be divisible by every integer $n > 0$ prime to the characteristic of the residue field of v. Then deduce the second assertion from the first.)

(b) Deduce from (a) that, either $u(K^*) \subset U$, or u is continuous (consider the image under u of a uniformizer of v).

(c) Give an example of an algebraically closed field Ω such that, taking $K = \Omega((X))$, K is isomorphic to a subfield of K contained in $U \cup \{0\}$ (cf. *Algebra*, Chapter V, § 5, Exercise 13).

¶ 8. (a) Let K be a field, v a valuation on K of height 1 and A its ring. Let H be a compact subset of K (with the topology $\mathscr{T}_v$) and $a \neq 0$ a point of K. Show that there exists a polynomial $f \in K[X]$ without constant term and such that $f(a) = 1, f(H) \subset A$. (Prove that f may be taken to be a polynomial of the form

$$1 - (1 - a^{-1}X)(1 - c_1^{-1}X)^{n(1)} \ldots (1 - c_r^{-1}X)^{n(r)}$$

where the c_i are suitably chosen elements of H such that $v(c_i) < v(a)$ and the $n(i)$ are sufficiently large integers > 0.)

(b) Let X be a totally disconnected compact space; let the ring $\mathscr{C}(X; K)$ of continuous mappings from X to K be given the uniform convergence topology. Let B be a subring of $\mathscr{C}(X; K)$ containing the constants and separating the points of X; show that B is dense in $\mathscr{C}(X; K)$ (Use (a) and *General Topology*, Chapter X, § 4, Exercise 21 (b).)

¶ 9. Let A be a discrete valuation ring, π a uniformizer of A and K the field of fractions of A; suppose that A is *complete* with the π-adic topology. The injective A-modules (*Algebra*, Chapter II, § 2, Exercise 11) are identical with the divisible A-modules (*Algebra*, Chapter VII, § 2, Exercise 3) and are direct sums of A-modules isomorphic either to K or to K/A (*Algebra*, Chapter VII, § 2, Exercise 3). Moreover, every monogenous A-module is isomorphic to a submodule of K/A. The A-module $\mathrm{Hom}_A(M, K/A)$ is called the (*algebraic*) *toric dual* of an A-module M and denoted by M^*; the canonical mapping c_M: $M \to M^{**}$ is injective (*Algebra*, Chapter II, § 2, Exercise 13 (b)). For every submodule N of M the submodule N^0 of M^* consisting of the u such that $u(x) = 0$ is called the *orthogonal* of N in M^*; the dual of M/N is canonically identified with N^0 and the dual of N with M^*/N^0. The toric dual of a direct limit $\varinjlim M_\alpha$ is canonically isomorphic to the inverse limit $\varprojlim M_\alpha^*$.

(a) We know that the A-modules of rank one (*Algebra*, Chapter VII, § 4,

Exercise 22) are isomorphic to a module of one of the form A, K, K/A or $A/\pi^h A$. Show that the algebraic toric duals of K and $A/\pi^h A$ are respectively isomorphic to K and $A/\pi^h A$, that the toric dual of A is isomorphic to K/A and that that of K/A is isomorphic to A (use the knowledge about the sub-A-modules of K and the fact that A is complete (with respect to the dual of K/A)). Deduce that, for every A-module M of finite rank, M^* is a module of the same rank and that the canonical homomorphism c_M is bijective.

(b) Let M be an A-module and N a submodule of M of finite rank. Show that the orthogonal of N^0 in M^{**} is identified (by c_M) with N (use (a)).

(c) Show that an A-module M which is Noetherian (resp. Artinian) is of finite rank (embed M in its injective envelope (*Algebra*, Chapter II, § 2, Exercise 18)); then M^* is Artinian (resp. Noetherian).

(d) Let $\sigma(M^*, M)$ denote the topology (on M^*) of pointwise convergence on M (K/A being given the discrete topology). Show that, if N is a submodule of M, $\sigma(N^0, M/N)$ is induced on N^0 by $\sigma(M^*, M)$ and that $\sigma(M^*/N^0, N)$ is the quotient by N^0 of the topology $\sigma(M^*, M)$. (For the second point, note that, if P is a submodule of M such that $N \subset P$, the dual of P/N is identified with N^0/P^0). If $M = \varinjlim M_\alpha$, the topology $\sigma(M^*, M)$ is the inverse limit of the topologies $\sigma(M_\alpha^*, M_\alpha)$.

(e) The topologies $\sigma(K, K)$ and $\sigma(A, K/A)$ are the π-adic topologies; the topologies $\sigma(A/\pi^h A, A/\pi^h A)$ and $\sigma(K/A, A)$ are the discrete topologies. Deduce that for every A-module M, the module M^* with $\sigma(M^*, M)$ is *linearly compact* (Chapter III, § 2, Exercise 15; consider M as a quotient module of a free A-module).

(f) Let M, N be two A-modules; for every homomorphism $u: M \to N$, let ${}^t u$ denote the homomorphism $\mathrm{Hom}(u, 1_{K/A})$ from N^* to M^* such that

$$({}^t u(w))(x) = w(u(x))$$

for all $x \in M$ and all $w \in N^*$; show that ${}^t u$ is continuous for the topologies $\sigma(N^*, N)$ and $\sigma(M^*, M)$. If u is the endomorphism $x \mapsto \pi x$ of M, ${}^t u$ is the endomorphism $w \mapsto \pi w$ of M^*. For every submodule P of M, $(u(P))^0 = {}^t u^{-1}(P^0)$.

¶ 10. The hypotheses and notation are the same as in Exercise 9.

(a) If M is a discrete *topological* A-module, show that M is a torsion module. If further M is linearly compact, show that M is Artinian (if N is the kernel of the endomorphism $x \mapsto \pi x$, observe that N is linearly compact and discrete and can be considered as a vector $(A/\pi A)$-space; then use Exercise 20 (d) of Chapter II, § 2).

(b) Deduce from (a) that every linearly compact topological A-module is strictly linearly compact (Chapter II, § 2, Exercise 19).

(c) The *topological toric dual* of a topological A-module M is the submodule M′ of the algebraic toric dual M^* consisting of the continuous homomorphisms from M to K/A (the latter being given the discrete topology). If the topology on M is linear (Chapter II, § 2, Exercise 14) and N is a closed submodule of M, show that the topological toric dual of M/N is identified with $N^0 \cap M'$ and that of N with $M'/(N^0 \cap M')$ (to determine the dual of N, note that, for every continuous homomorphism u from N to K/A, there exists an open submodule U of M such that $u(x) = 0$ in $N \cap U$, then use Exercise 9).

(d) For a topological A-module M of finite rank, the topological toric dual M′ is equal to the algebraic toric dual M* (reduce it the case of modules of rank one).

(e) Let M be a Hausdorff topological A-module whose topology is linear; show that for all $x \neq 0$ in M, there exists $u \in M'$ such that $u(x) \neq 0$ (note that there exists an open submodule U of M such that $x \notin U$); in other words, the canonical mapping $M \to (M')^*$ is injective. Deduce that, if N is a closed submodule of M, $N^0 \cap M'$ is dense in N^0 with the topology $\sigma(M^*, M)$ (use (d)), and consequently $N = M \cap (N^0 \cap M')^0$ in M^{**}.

Show that M is dense in $(M')^*$ with the topology $\sigma((M')^*, M')$.

(f) Let M be a linearly compact topological A-module; show that the canonical injection $M \to (M')^*$ is a *bijection* and that the topology on M (identified with $(M')^*$) is equal to $\sigma(M, M')$. (Observe that, if U is an open submodule of M, M/U is Artinian by (a) and therefore U^0 is Noetherian (Exercise 9 (c)) and deduce the identities of the topologies considered on M; complete with the aid of (e).)

(g) Let M, N be two topological A-modules whose topologies are linear; for every *continuous* homomorphism $u: M \to N$, ${}^t u(N') \subset M'$ and ${}^t u$ is also used (by an abuse of notation) to denote the linear mapping from N′ to M′ with the same graph as ${}^t u$. If M and N are Hausdorff, show that the restriction to M of ${}^t({}^t u)$ coincides with u (M and N being respectively considered as canonically embedded in $(M')^*$ and (N'); moreover, for every submodule Q of N,

$$(\overset{-1}{u}(Q))^0 \cap M' = {}^t u(Q^0)$$

(use (e)).

(h) Let M be a linearly compact A-module; deduce from (g) and Exercise 9 (f) that for M to be torsion-free, it is necessary and sufficient that M′ be divisible and that for M to be divisible, it is necessary and sufficient that M′ be torsion-free.

§ 6

¶ 1. Every element z of the p-adic field $\mathbf{Q}_p$ (p a prime number) may be written uniquely as $p^h \sum_{k=0}^{\infty} c_k p^k$, where $h \in \mathbf{Z}$, $c_0 \neq 0$, $0 \leqslant c_k < p$ for all $k \geqslant 0$ ("p-adic expansion" of z).

(a) Show that, in order that $z \in \mathbf{Q}$, it is necessary and sufficient that there exist two integers $m \geqslant 0$, $n \geqslant 1$ such that $c_{k+n} = c_k$ for all $k \geqslant m$. (Observe that, if $z = a/b \in \mathbf{Q}$ where b is not divisible by p, then

$$a - b\sum_{k=0}^{n} c_k p^k = a_{n+1}p^{n+1}$$

where a_{n+1} is an integer and the sequence $|a_k|$ (ordinary absolute value) is bounded.)

(b) Suppose that the increasing sequence (k_n) of integers k such that $c_k \neq 0$ is such that $\limsup_{n\to\infty}(k_{n+1}/k_n) = +\infty$. Show that z is transcendental over $\mathbf{Q}$. (For $x \in \mathbf{Q}$, let $|x|$ and $|x|_p$ respectively denote the usual absolute value and a p-adic absolute value. Suppose that $\mathrm{P} \in \mathbf{Z}[\mathrm{X}]$ is an irreducible polynomial such that $\mathrm{P}(z) = 0$; if $z_n = p^n \sum_{k=0}^{\infty} c_k p^k$, show that $|\mathrm{P}(z_n)/(z - z_n)|_p$ tends to a limit $\neq 0$ as n tends to $+\infty$, using Taylor's formula; then obtain a contradiction from the hypothesis considering the usual absolute values $|\mathrm{P}(z_n)|$.)

2. Let D be a non-commutative field of characteristic $\neq 2$ and Z its centre. Suppose that every commutative subfield K of D containing Z is of degree $\leqslant 2$ over Z.

(a) Show without using Lemma 1 that $[\mathrm{D}:\mathrm{Z}] = 4$. (Let $a \in \mathrm{D} - \mathrm{Z}$ be such that $a^2 \in \mathrm{Z}$ and let $\sigma(x) = axa^{-1}$ for all $x \in \mathrm{D}$; note that D decomposes as a direct sum of two vector subspaces over Z, D_+ and D_- such that $\sigma(x) = x$ on D_+, $\sigma(x) = -x$ on D_-; note also that D_+ is a subfield of D and D_- is a 1-dimensional vector space over D_+; finally, show that D_+ cannot be distinct from $\mathrm{Z}(a)$.)

(b) Show that D is a quaternion algebra over Z. (Form a basis of D over Z with the aid of (a).)

§ 7

1. Let $(\phi_\iota)_{\iota\in\mathrm{I}}$ be a family of places of a field K taking their values in a finite number of fields; suppose further that for some $x \in \mathrm{K}$, the set of $\phi_\iota(x)$ is finite. Show that there exists a polynomial $f(\mathrm{X})$ of the form (1) of no. 1 such that $f(x) \neq 0$ and that the element $z = f(x)^{-1}$ enjoys the following properties: (i) $\phi_\iota(x) = \infty$ implies $\phi_\iota(xz) = 0$ and $\phi_\iota(z) = 0$; (ii) $\phi_\iota(x) \neq \infty$ implies $\phi_\iota(xz) \neq \infty$ and $\phi_\iota(z) \neq 0$. (Same proof as for Lemma 1).

2. With the hypotheses and notation of Proposition 1 of no. 1, let $\mathfrak{q}$ be a prime ideal of B; show that $\mathrm{B}_\mathfrak{q}$ is a valuation ring.

3. Let A_i $(1 \leqslant i \leqslant n)$ be pairwise independent valuation rings of a field K and let $\mathrm{A} = \bigcap_i \mathrm{A}_i$. For all i, let $\mathfrak{a}_i$ be an ideal $\neq 0$ of A_i and write $\mathfrak{a} = \bigcap_i \mathfrak{a}_i$;

show that, for all i, $\mathfrak{a}_i = A_i\mathfrak{a}$. Conversely, if $\mathfrak{b}$ is an ideal $\neq 0$ of A and $\mathfrak{b}_i = A_i\mathfrak{b}$, then $\mathfrak{b} = \bigcap_i \mathfrak{b}_i$ (use Theorem 1 of no. 2).

§ 8

1. Let K be a field and $A = K[[X, Y, Z]]$ the ring of formal power series in three indeterminates over K; let v' (resp. v'') be the valuation on A with values in the group $\mathbf{Z} \times \mathbf{Z}$ ordered lexicographically, such that $v'(X) = (1, 0)$, $v'(Y) = (0, 1)$, v' is improper in $K[[Z]]$ (resp. $v''(X) = (1, 0)$, v'' is improper in $K[[Y]]$, $v''(Z) = (0, 1)$). Let σ be the automorphism of A leaving invariant K and X and such that $\sigma(Y) = Z$, $\sigma(Z) = Y$; if B is the subring of A consisting of the elements invariant under σ and E (resp. F) the field of fractions of A (resp. B), the valuations v' and v'' (extended canonically to E) have the same restriction v to F, F is *complete* with the topology defined by v and E is a quadratic extension of F; the two valuations v', v'' on E are dependent but not equivalent.

2. Let K_0 be the field obtained by adjoining to the 2-adic field $\mathbf{Q}_2$ the roots of all the polynomials $X^{2^n} - 2$; let v be the unique valuation on K_0 extending the 2-adic valuation and let K be the completion of K_0 with respect to v. Show that the polynomial $X^2 - 3$ is irreducible in K[X]; let K′ be the field of roots of this polynomial and let v' be the extension of v to K′; then

$$n = [K' : K] = 2, \qquad e(v'/v) = f(v'/v) = 1.$$

3. Let k be the field of rational functions $\mathbf{F}_p(X_n)_{n \in \mathbf{N}}$ in an infinity of indeterminates over the prime field $\mathbf{F}_p$ and let $K = k(U, V)$ be the field of rational functions in two indeterminates over k.

(a) Show that the element $P(U) = \sum_{n=0}^{\infty} X_n^p U^{np}$ of the field of formal power series $k((U))$ is not algebraic over the field $k(U)$ of rational functions. The mapping $F(U, V) \mapsto F(U, P(U))$ of $k[U, V]$ to $k((U))$ can be extended to an isomorphism of K onto a subfield of $k((U))$; the restriction to this subfield of the valuation on $k((U))$ equal to the order of the formal power series (§ 3, no. 3, *Example* 3) is a discrete valuation v on K, whose residue field is k.

(b) Let K′ be the algebraic extension $K(V^{1/p})$ of K so that $[K' : K] = p$; if v' is the unique valuation on K′ extending v, show that $e(v'/v) = f(v'/v) = 1$. The ring of the valuation v' is therefore not a finitely generated module over the ring of the valuation v.

4. Let k be a field, $K = k(X, Y)$ the field of rational functions in two indeterminates over k and v the valuation on K with values in the group $\mathbf{Z} \times \mathbf{Z}$ ordered lexicographically, such that $v(X) = (0, 1)$, $v(Y) = (1, 0)$. Let K′ be the field $K(\sqrt{X})$; show that v has a single extension v' to K′ and that $e(v'/v) = 2$

$f(v'/v) = 1$, but the ring of the valuation v' is not a finitely generated module over the ring of the valuation v.

5. Let K be a field, A a valuation ring of K and L a finite algebraic extension of K. Let $A' \supset A$ be another valuation ring of K; let A'_i ($1 \leqslant i \leqslant m$) be the valuation rings of L such that $A'_i \cap K = A'$ and let k'_i be their respective residue fields. Let k be the residue field of A' and A" the valuation ring of K the canonical image of A; finally let A'_{ij} ($1 \leqslant j \leqslant n_i$) be the valuation rings of k'_i such that $A''_{ij} \cap k = A''$ and A_{ij} the inverse images in A'_i of the A''_{ij} ($1 \leqslant i \leqslant m$, $1 \leqslant j \leqslant n_i$).

(a) Show that the A_{ij} are valuation rings of L which are distinct and such that $A_{ij} \cap K = A$ and every valuation ring B of L such that $B \cap K = A$ is equal to one of the A_{ij}.

(b) Show that

$$e(A_{ij}/A) = e(A'_i/A')e(A''_{ij}/A'') \quad \text{and} \quad f(A_{ij}/A) = f(A''_{ij}/A'').$$

¶ 6. Let K be a field, v a valuation on K and A the ring of v.

(a) Suppose that A is Henselian (Chapter III, § 4, Exercise 3) in which case we also say, by an abuse of language, that K is *Henselian for v*. Then, for every algebraic extension L of K and every valuation v' on L extending v, the ring A' of v' is Henselian (Chapter III, § 4, Exercise 4).

(b) If K is complete with v and v is of height 1, A is Henselian. Give an example where K is complete with v and v is of height 2, but A is not Henselian (cf. Exercise 1).

(c) If K is *linearly compact* with v, show that K is Henselian. (In the notation condition (H) of Chapter III, § 4, Exercise 3, let $\mathscr{L}$ denote the set of prime ideals of A (totally) ordered by inclusion and $\mathfrak{L}$ the set of ordered pairs $(\mathscr{B}, \phi)$, where $\mathscr{B}$ is a well ordered subset (with respect to the relation $\supset$) of $\mathscr{L}$ and $\phi: \mathfrak{p} \mapsto (Q_\mathfrak{p}, Q'_\mathfrak{p})$ a mapping from $\mathscr{B}$ to the set $A[X] \times A[X]$, with the following properties: (1) $Q_\mathfrak{p}$ and $Q'_\mathfrak{p}$ are monic of respective degrees $\deg(\overline{Q})$ and $\deg(\overline{Q}')$; (2) if $\mathfrak{p} \supset \mathfrak{q}$ are two elements of $\mathscr{B}$, the coefficients of the polynomials $Q_\mathfrak{p} - Q_\mathfrak{q}$ and $Q'_\mathfrak{p} - Q'_\mathfrak{q}$ belong to $\mathfrak{p}$; (3) the coefficients of $P - Q_\mathfrak{p}Q'_\mathfrak{p}$ belong to $\mathfrak{p}$; (4) $\bar{f}(Q_\mathfrak{p}) = \overline{Q}$, $\bar{f}(Q'_\mathfrak{p}) = \overline{Q}'$. Define on $\mathscr{L}$ an ordering by setting $(\mathscr{B}, \phi) \leqslant (\mathscr{B}', \phi')$ when $\mathscr{B} \subset \mathscr{B}'$ and ϕ' is an extension of ϕ to $\mathscr{B}'$. Prove that $\mathscr{L}$ admits a maximal element $(\mathscr{B}_0, \phi_0)$ and that $\mathscr{B}_0$ has a last element equal to (0): obtain a contradiction from the opposite by considering two cases, according to whether or not $\mathscr{B}_0$ has a last element; in the former case, if $\mathfrak{p}$ is this last element, consider an element $c \in A$ such that $v(c)$ is the least value taken by v on the set of coefficients of $P - Q_\mathfrak{p}Q'_\mathfrak{p}$ and the least prime ideal $\mathfrak{p}'$ of A containing c; if $\mathfrak{p}' = \mathfrak{p}$, use Exercise 2 (a) of § 4 and the fact that a quotient of a linearly compact module by a closed submodule is linearly compact and hence complete. If on the contrary $\mathscr{B}_0$ has no last element, use directly the hypothesis that A is linearly compact.)

(d) In the notation of Exercise 5, show that for A to be Henselian, it is necessary and sufficient that A′ and A″ be so. (Note that, if $P \in A'[X]$, there exists $a \in A$ such that $a^n P(X/a) \in A[X]$ (where n is the degree of P); to verify axiom (H) of Chapter III, § 4, Exercise 3 for A′, attention may be confined to the polynomials of A[X]; use the fact that $A' = A_{\mathfrak{p}}$, where $\mathfrak{p}$ is a prime ideal of A contained in the maximal ideal $\mathfrak{m}$, and note that, if two polynomials in $(A/\mathfrak{p})[X]$ are strongly relatively prime, so are their images in $(A/\mathfrak{m})[X]$.)

(e) Suppose that A is Henselian; then, for every algebraic extension L of K, there exists (up to equivalence) only one valuation on L extending v. (If $P \in A[X]$ is a monic irreducible polynomial, x_i $(1 \leqslant i \leqslant m)$ the distinct roots of P in its splitting field N, show that, for every valuation v' on N extending v, the $v'(x_i)$ are all equal for $1 \leqslant i \leqslant m$.) Conversely, if the valuation ring A has this property, it is Henselian (observe that, in the integral closure A′ of A in L, there can only exist one maximal ideal lying above $\mathfrak{m}$ (Chapter V, § 1, Exercise 13) and conclude that, if $P \in A[X]$ is a monic irreducible polynomial its image in $(A/\mathfrak{m})[X]$ cannot be a product of two relatively prime polynomials).

7. Let K be a field, v a valuation on K, A the ring of v and $\mathfrak{m}$ the ideal of v. Suppose that A is Henselian; let L be a finite extension of K of degree n; let v' be the unique valuation on L extending v (Exercise 6 (e)), A′ its ring and $\mathfrak{m}'$ its ideal. Let x be any element of A′; show that the degree over $A/\mathfrak{m}$ of the class $\bar{x}$ of x in $A'/\mathfrak{m}'$ is a divisor of n and that the order of the class of $v'(x)$ in $\Gamma_{v'}/\Gamma_v$ is a divisor of n (consider the minimal polynomial of x over K).

¶ 8. Let K′ be a field, B the ring of a valuation v on K′, $\mathscr{G}$ a finite group of automorphisms of K′, K the subfield of K′ consisting of the elements invariant under $\mathscr{G}$ and $A = K \cap B$, which is a valuation ring of K, corresponding to the valuation $w = v|K$.

(a) The valuations on K′ which extend w are the $v \circ \sigma$, where $\sigma \in \mathscr{G}$ (no. 6, Corollary 1 to Proposition 6) and the integral closure A′ of A in K′ is the intersection of the $\sigma.B$, where σ runs through $\mathscr{G}$. If $\mathfrak{p}(B)$ is the intersection of A′ and the maximal ideal $\mathfrak{m}(B)$, then $\sigma.\mathfrak{p}(B) = \mathfrak{p}(\sigma.B)$. Show that the decomposition group $\mathscr{G}^Z(\mathfrak{p}(B))$ (Chapter V, § 2, no. 2) is the subgroup of $\mathscr{G}$ consisting of the σ such that $\sigma.B = B$; write $\mathscr{G}^Z = \mathscr{G}^Z(\mathfrak{p}(B))$ and denote by K^Z the subfield of K′ consisting of the elements invariant under $\mathscr{G}^Z$ and by B^Z the ring of the valuation induced on K^Z by v, which is equal to $B \cap K^Z$; recall that the residue fields of B^Z and A are the same and that the maximal ideal $\mathfrak{m}(B^Z)$ is generated by $\mathfrak{m}(A)B^Z$ (Chapter V, § 2, no. 2, Proposition 4).

(b) Let v^Z denote the restriction of v to K^Z and Γ and Γ^Z the order groups of w and v^Z. Show that $\Gamma = \Gamma^Z$. (Let $\mathscr{G}^D \supset \mathscr{G}^Z$ be the subgroup of $\mathscr{G}$ consisting of the $\sigma \in \mathscr{G}$ such that $\sigma.B$ is dependent on B (§ 7, no. 2). Argue by induction on the order of $\mathscr{G}$. If $(\mathscr{G} : \mathscr{G}^D) > 1$, let K″ be the field of invariants of $\mathscr{G}^D$ and $A'' = K'' \cap B$; show that the order group of $v|K''$ is equal to Γ: for this, use the Approximation Theorem (§ 7, no. 2, Theorem 1) in order to prove that, for all

$x \in K''$, there exists $y \in K''$ such that $v(y) = v(x)$ and $v(y_i) = 0$ for the conjugates y_i of y with respect to K, *distinct from* y; then $\mathscr{G}$ may be replaced by $\mathscr{G}^D$ and the induction hypothesis may be applied. If $\mathscr{G}^D = \mathscr{G}$, let B_0 be the valuation ring of K′ generated by the $\sigma.B$, where $\sigma \in \mathscr{G}$: let $\overline{K}' = \kappa(B_0)$ be its residue field, $\pi: B_0 \to \overline{K}'$ the canonical homomorphism, $\overline{B} = \pi(B)$ a valuation ring of $\overline{K}'$ and $\bar{v}$ the corresponding valuation on $\overline{K}'$ such that $v = \bar{v} \circ \pi$ in B_0, so that the order group $\overline{\Delta}$ of $\bar{v}$ is a subgroup of the order group Δ of v; if $\Delta_0 = \Delta/\overline{\Delta}$ and $\varpi: \Delta \to \Delta_0$ is the canonical homomorphism, $v_0 = \varpi \circ v$ is a valuation on K′ corresponding to B_0 and invariant under $\mathscr{G}$. By taking quotients, $\mathscr{G}$ defines a group of automorphisms $\overline{\mathscr{G}}$ of $\overline{K}'$; let $\mathscr{N}$ be the kernel of the canonical homomorphism $\mathscr{G} \to \overline{\mathscr{G}}$; show that $\mathscr{N} \subset \mathscr{G}^Z$ and deduce that, if $\mathscr{N} \neq \{e\}$, $\mathscr{G}$ may be replaced by $\mathscr{G}/\mathscr{N}$ and the induction hypothesis may be applied. Suppose finally that $\mathscr{N} = \{e\}$; let $A_0 = K \cap B_0$ and w_0 be the restriction of v_0 to K; show that the order group of w_0 is Δ_0 and that $\pi(A_0) = \overline{K}$ is the field of invariants of $\overline{\mathscr{G}}$ (cf. no. 1, Lemma 2); if $\bar{w}$ is the restriction to $\overline{K}$ of $\bar{v}$ and $\overline{\Gamma}$ its order group, Δ_0 is canonically isomorphic to $\Gamma/\overline{\Gamma}$. On the other hand, the group $\overline{\mathscr{G}}^Z(\mathfrak{p}(\overline{B}))$ is the canonical image of $\mathscr{G}^Z(\mathfrak{p}(B))$ in $\overline{\mathscr{G}}$; observe that $\overline{\mathscr{G}}^D \neq \overline{\mathscr{G}}$ and that the induction hypothesis may therefore be applied to prove that $\overline{\Gamma}$ is equal to the order group $\overline{\Gamma}^Z$ of the restriction of $\bar{v}$ to $\overline{K}^Z$.)

(c) Let $\mathscr{G}^T = \mathscr{G}^T(\mathfrak{p}(B))$ be the inertia group of $\mathfrak{p}(B)$ (Chapter V, § 2, no. 2) and let K^T denote the subfield of K′ consisting of the invariants of $\mathscr{G}^T$ and B^T the ring of the valuation induced on K^T by v, which is equal to $B \cap K^T$; recall that, if k and k' are the residue fields of A and B, the residue field k^T of B^T is the greatest separable extension of k contained in the quasi-Galois extension k' of k and that $\mathscr{G}^Z/\mathscr{G}^T$ is canonically isomorphic to the group of k-automorphisms of k' (or k^T) (Chapter V, § 2, Theorem 2 and Proposition 5). Let v^T be the restriction of v to K^T; show that the order group of v^T is also equal to Γ (apply Lemma 2 of no. 1 to the extension K^T of K^Z).

9. Let K be a field, L a finite algebraic extension of K, v' a valuation on L, A′ its ring, v the restriction of v' to K and $A = A' \cap K$ its ring.

(a) Suppose that A is Henselian (Exercise 6); show that the product $e(v'/v)f(v'/v)$ divides $n = [L:K]$ and that the quotient $n/e(v'/v)f(v'/v)$ is a power of the characteristic exponent of the residue field k of v. (Reduce it to the case where L is a Galois extension of K; use Exercises 6 (e) and 7 and Chapter V, § 2, Theorem 2 and Proposition 5.)

(b) Suppose further that n is not divisible by the characteristic exponent of the residue field k of v and that $f(v'/v) = 1$. Show that, for every integer m equal to the order of an element γ of the group $\Gamma_{v'}/\Gamma_v$ there exists $x \in L$ such that the class of $v'(x)$ mod. Γ_v is equal to γ and $x^m \in K$.

¶ 10. Let w be the 2-adic valuation on the 2-adic field $\mathbf{Q}_2$; on the field of rational functions $\mathbf{Q}_2(X)$ consider the discrete valuation v extending w and such that

$$v(a_0 + a_1X + \cdots + a_nX^n) = \inf_i(w(a_i))$$

(§ 10, no. 1, Lemma 1); let K be the completion of $\mathbf{Q}_2(X)$ with respect to this valuation and let its valuation also be denoted by v. Let L be the splitting field of the polynomial $(Y^2 - X)^2 - 2$ in K[Y] and let v' be the unique valuation on L extending v. Show that $[L:K] = 8$, $e(v'/v) = 4$, $f(v'/v) = 2$; if k (resp. k') is the residue field of v (resp. v'), k' is a radicial extension of k of degree 2; show that there exists no sub-extension E of L such that $[E:K] = 2$ for which the residue field of the restriction of v' to E is equal to k'. (Of necessity $E = K(\sqrt{\alpha})$, where $\alpha \in K$ would not be a square in K but such that $\alpha \equiv X$ (mod. 2); express $\sqrt{\alpha}$ with the aid of a suitable basis of L over $K(\sqrt{2})$ and observe that there exists in $K(\sqrt{2})$ no element whose square is congruent to X mod. 2.)

¶ 11. Let K be a field, L a finite Galois extension of K and $\mathcal{G}$ its Galois group. Let v be a valuation on L with residue field k and order group Γ; suppose that v is invariant under $\mathcal{G}$ and that the restriction $v|K$ has the same residue field k as v, so that $\mathcal{G} = \mathcal{G}^Z = \mathcal{G}^T$ (notation of Exercise 8).

(a) Let $x \in L^*$, $\sigma \in \mathcal{G}$. Show that $x^{-1}\sigma(x)$ is a unit for v, that its image $\varepsilon_\sigma(x)$ in k^* depends only on the valuation $v(x)$ and the class of σ modulo the commutator group $\mathcal{G}'$ of $\mathcal{G}$ and that there is thus defined a **Z**-bilinear mapping (also denoted by ε) $(\mathcal{G}/\mathcal{G}') \times \Gamma \rightarrow k^*$, which is equal to 1 on $(\mathcal{G}/\mathcal{G}') \times v(K^*)$.

(b) For all $\sigma \in \mathcal{G}$, let $\phi(\sigma)$ be the homomorphism $\Gamma/v(K^*) \rightarrow k^*$ mapping $\varepsilon_\sigma(x)$ to $v(x)$ for all $x \in L^*$; thus there is defined a homomorphism

$$\phi: \mathcal{G} \rightarrow \operatorname{Hom}_{\mathbf{Z}}(\Gamma/v(K^*), k^*).$$

Show that, if p is the characteristic exponent of k, the kernel $\mathcal{N}$ of ϕ has order a power of p. (In the contrary case, there would exist a $\sigma \neq e$ in $\mathcal{N}$ and a prime number $q \neq p$ such that $\sigma^q = e$; for some integer x in $L - K$, write $\sigma(x) = x + y$, where $v(y) > v(x)$, and calculate $\sigma^q(x)$ to obtain a contradiction.) Deduce that $\mathcal{G}$ is a *solvable* group.

(c) Suppose that $n = [L:K]$ is prime to p; show that ϕ is bijective, that $e(L/K) = n$ and that, if ν is the *lcm* of the orders of the elements of $\mathcal{G}$, k^* contains the ν-th roots of unity. (Use (b) and Lemma 2 of no. 1.)

¶ 12. Let K be a field, v a valuation on K, A the ring of the valuation v and Γ its order group.

(a) Let $f(X) = a_0X^n + a_1X^{n-1} + \cdots + a_n$ be a polynomial in A[X] such that $v(a_0) = 0$, all of whose roots belong to K; let x_i $(1 \leqslant i \leqslant r)$ be these distinct roots and let k_i be the order of multiplicity of x_i $(1 \leqslant i \leqslant r)$. Let

$$g(X) = b_0X^n + b_1X^{n-1} + \cdots + b_n = b_0(X - y_1)\ldots(X - y_n)$$

another polynomial in A[X] such that $v(b_0) = 0$ and the y_h belong to K. Show that there exists $\lambda_0 \in \Gamma$ such that $v(x_i - x_j) < \lambda_0$ for every ordered pair of distinct indices i, j and, for all $\lambda \geqslant \lambda_0$, there exists $\mu \geqslant \lambda$ such that the relations $v(a_i - b_i) \geqslant \mu$ for all i imply that, for $1 \leqslant i \leqslant r$, there are exactly k_i indices h such that $v(x_i - y_i) \geqslant \lambda$. (First evaluate $v(f(y_h))$ in two ways to show that necessarily $v(x_i - y_h) \geqslant \lambda$ for at least one i; then evaluate

$$v\left(\frac{f'(z)}{f(z)} - \frac{g'(z)}{g(z)}\right)$$

at suitable points $z \in K$.) If further $b_i = a_i$ except for $i = n - 1$, show that, so long as λ_0 is sufficiently large, the y_h are all distinct (evaluate $f'(y_h)/f(y_h)$ assuming that y_h is not a simple root of g).

(b) Suppose henceforth that K is the algebraic closure of a subfield K_0 such that $A_0 = A \cap K_0$ is *Henselian*. Suppose that the polynomial f belongs to $A_0[X]$ and is *irreducible and separable* over K_0. Let $z \in K$ be such that

$$v(z - v_i) > v(z - x_j)$$

for all $j \neq i$; then show that $K_0(x_i) \subset K_0(z)$. (Show that $z - x_i$ is equal to all its conjugates with respect to $K_0(x_i)$.)

(c) Suppose that $f \in A_0[X]$ is *separable* and let $f = a_0 \prod_{i=1}^{r} f_i$ be the decomposition of f into monic irreducible polynomials in $K_0[X]$ (which belong in fact to $A_0[X]$, cf. Chapter V, § 1, no. 3, Proposition 11). Show (in the notation of (a)) that, if μ is taken sufficiently large, the decomposition of g into monic irreducible factors in $K_0[X]$ can be written as $b_0 \prod_{i=1}^{r} g_i$, where, for all i, g_i is of the same degree as f_i (a decomposition said to be "of the same type" as that of f) and further the splitting fields of f_i and g_i are the same for all i. (Show first that g is separable; consider then a finite Galois extension of K_0 containing the roots of f and g and consider the way in which the Galois group of this extension permutes these roots; finally use (b).)

(d) Give an example where f is irreducible but not separable and where, for all $\mu \geqslant \lambda_0$, there exists a polynomial

$$g(X) = b_0X^n + b_1X^{n-1} + \cdots + b_n \in A_0[X]$$

where $v(a_i - b_i) \geqslant \mu$ for all i, which is not irreducible (take K_0 such that $K \neq K_0$ and K is a radicial extension of K_0).

¶ 13. (a) Let K be a field and v a non-improper valuation on K. Let E be a set filtered by an ultrafilter $\mathfrak{U}$ and $\xi \mapsto x_\xi$, $\xi \mapsto y_\xi$ two mappings of E to K with the topology $\mathcal{T}_v$. Show that, if the mapping $\xi \mapsto x_\xi y_\xi$ converges to 0 in K with respect to $\mathfrak{U}$, so does one of the mappings $\xi \mapsto x_\xi$, $\xi \mapsto y_\xi$. (Observe that, if

$\xi \mapsto x_\xi$ has no cluster point with respect to $\mathfrak{U}$, the mapping $z \mapsto x_\xi^{-1}$ is bounded on a subset of $\mathfrak{U}$.)

(b) Let f be a non-constant polynomial in K[X]. Show that, if $\xi \mapsto f(x_\xi)$ converges to 0 in K with respect to $\mathfrak{U}$, $\xi \mapsto x_\xi$ converges in $\hat{K}$ (decompose f into factors in an algebraic extension of K and use (a)).

(c) Suppose that K is algebraically closed in $\hat{K}$. Then the mapping $x \mapsto f(x)$ of K to itself is closed (use (b)). Deduce that, if g is a rational function in K(X) and B is a bounded closed subset of K, then $g(B - P)$ (where P is the set of poles of g in B) is closed in K.

(d) Suppose that the algebraic closure of K is a radicial extension of K. Then show that $\hat{K}$ is algebraically closed. (Apply (b) to an algebraic closure of $\hat{K}$ and a polynomial $f \in \hat{K}[X]$; use also Exercise 12 (a).)

¶ 14. (a) For a field K to be Henselian with a valuation v, it is necessary and sufficient that K be the greatest separable algebraic extension of K contained in $\hat{K}$ and that $\hat{K}$ be Henselian. (To see that the condition is necessary, use Exercise 12 (b); to show that it is sufficient, observe that, to verify condition (H) of Chapter III, § 4, Exercise 3, we may confine our attention to the case where P is separable over K, noting that, if Q is an irreducible factor of P in $\hat{K}[X]$ such that Q^{p^e} belongs to K[X], then Q is a *g.c.d.* of P and Q^{p^e} in $\hat{K}[X]$.) Consider the case where v is of height 1 (cf. Exercise 6 (b)).

(b) Deduce from (a) that there exists in the p-adic field $\mathbf{Q}_p$ a countable subfield which is Henselian with the p-adic valuation.

(c) Give an example of a field K which is Henselian with a discrete valuation which is not complete and such that $\hat{K}$ is a finite radicial extension of K (cf. Chapter V, § 1, Exercise 20).

¶ 15. (a) Let K be a field, v_1, v_2 two independent valuations (§ 7, no. 2) on K and K_1, K_2 the completions of K with respect to v_1 and v_2 respectively. Let L_1, L_2 be two fields such that $K \subset L_1 \subset K_1$, $K \subset L_2 \subset K_2$, which are Henselian (with the extensions by continuity of v_1 and v_2 respectively). Let $g_1 \in L_1[X]$, $g_2 \in L_2[X]$ be two separable polynomials with the same degree n. Show that there exists a polynomial $h \in K[X]$ which, in $L_i[X]$, has the same type of decomposition (Exercise 12 (c)) as g_i $(i = 1, 2)$. (Reduce it to the case where g_1 and g_2 are in K[X] and are monic; consider the polynomial

$$a^n g_1(X/a) + b^n g_2(X/a) - X^n,$$

where $v_1(a) \leqslant 0$, $v_2(a) > 0$ is arbitrarily large and $v_1(b) > 0$ is arbitrarily large.)

(b) Deduce from (a) that, if K is Henselian with v_1, the algebraic closure of L_2 is radicial over L_2 and K_2 is algebraically closed (take g_2 to be irreducible and $g_1 \in K[X]$ to be the product of n distinct factors of prime degree).

(c) Deduce from (b) that, if K is Henselian with v_1 and v_2, the algebraic closure of K is radicial over K.

(d) Show that, if K is Henselian with a discrete valuation v, the algebraic closure of K cannot be radicial over K and therefore K cannot be Henselian with any valuation independent of v (use *Algebra*, Chapter V, § 11, Exercise 12).

16. Let K be a Henselian field with a valuation v of height 1. If L is an *infinite* separable algebraic extension of K, show that L cannot be complete with the valuation extending v (and also denoted by v). (Form a sequence (x_p) of elements of L such that the degree n_p of x_p with respect to K tends to $+\infty$ and $v(x_{p+1} - x_p)$ is strictly greater than the valuations of the differences between x_p and its conjugates with respect to K; show that the sequence (x_p) is a Cauchy sequence which cannot converge in E, using Exercise 12 (b).)

¶ 17. Let K be a Henselian field with a valuation v which is not algebraically closed, and L a subfield of K such that [K: L] is finite. Show that under these conditions L is Henselian with the restriction of v. (In the contrary case, there would exist two extensions v_1, v_2 of $v|L$ to an algebraic closure Ω of K such that the restriction of v_1 and v_2 to a finite algebraic extension L′ of L would not be equivalent. Deduce that there would exist on a suitable finite quasi-Galois extension K′ of L containing K, two valuations v'_1, v'_2 which are not equivalent and for which K′ would be Henselian; using Exercise 15 (c), show that, if $E = L^{p^{-\infty}}$ (p being the characteristic exponent of Ω), $E \neq \Omega$ but Ω would be a finite extension of E; complete the argument with the aid of *Algebra*, Chapter VI, § 2, Exercise 31.)

¶ 18. Let K be a Henselian field with a valuation v, k the residue field of v and suppose that every finite algebraic extension of k is *cyclic* over k (which is for example the case when k is finite). Let A be the ring of v and $f \in A[X]$ a monic polynomial such that, if $f(X) = (X - \alpha_1) \dots (X - \alpha_n)$, where the α_i belong to an algebraic closure of K, the element $D = \prod_{i<j} (\alpha_i - \alpha_j)^2$ of A ("discriminant" of f) is such that $v(D) = 0$. Let $\bar{f}_j$ $(1 \leqslant j \leqslant s)$ be the irreducible factors of the canonical image $\bar{f}$ of f in $k[X]$ and let r_j be the degree of $\bar{f}_j$. Show that the Galois over K of the splitting field of f, considered as a group of permutations of the α_i is generated by a permutation σ which decomposes into cycles of respective lengths $r_1, r_2, \dots, r_s$ (observe that there exists only a single extension of k of given degree, up to a k-isomorphism). Deduce that, for D to be a *square* in A, it is necessary and sufficient that $n - s$ be *even* ("*Stickelberger's Theorem*"; examine under what condition D is invariant under σ).

19. Let K be a Henselian field with a valuation v. Let E be a finite-dimensional vector space over K and Q a non-degenerate quadratic form on E such that the relation $Q(x) = 0$ implies $x = 0$. Let

$$\Phi(x, y) = Q(x + y) - Q(x) - Q(y)$$

be the associated bilinear form. Show that $2v(\Phi(x, y)) \geqslant v(Q(x)) + v(Q(y))$; deduce that

$$v(Q(x + y)) \geqslant \inf(v(Q(x)), v(Q(y))).$$

¶ 20. Let K be a field of characteristic $\neq 2$ which is Henselian with a valuation v; let $\xi \mapsto \bar{\xi}$ be an involutive automorphism of K such that $v(\bar{\xi}) = v(\xi)$ and let Φ be a non-degenerate Hermitian form on a finite-dimensional vector space E over K.

(a) Let $U = (\alpha_{ij})$ be the matrix of Φ with respect to a basis (e_i) of E; show that there exists an element λ of the order group of v such that, if Φ' is another Hermitian form on E whose matrix $U' = (\alpha'_{ij})$ with respect to (e_i) satisfies the conditions $v(\alpha_{ij} - \alpha'_{ij}) > \lambda$ for every ordered pair (i, j), then Φ and Φ' are equivalent. (Argue as in *Algebra*, Chapter IX, § 6, Exercise 6, using Exercise 14 (a) above.)

(b) Suppose that $\xi \mapsto \bar{\xi}$ is the identity (hence Φ is a symmetric form) and that the valuation v is discrete, normed and such that $v(2) = 0$; let π be a uniformizer for v. Show that Φ is characterized, up to equivalence, by its index ν and by two symmetric bilinear forms Ψ_1, Ψ_2 *of index* 0 on vector spaces k^r and k^s respectively, where k is the residue field of v and $r + s = n - 2$. (With the aid of a Witt decomposition, reduce it to the case where Φ is of index 0; with the aid of Exercise 19, show that, for all $i \geqslant 0$, the set M_i of $x \in E$ such that $v(\Phi(x, x)) \geqslant i$ is a module over the ring A of the valuation v. Show that, if $x \in M_i$, $y \in M_{i+1}$, then $v(\Phi(x, y)) \geqslant i + 1$ and, by taking quotients, symmetric bilinear forms can therefore be derived from Φ on the vector k-spaces M_0/M_1 and M_1/M_2. Use Exercise 6 (c) of § 3 and the fact that the equation $\xi^2 = \alpha$ has a solution in A for $\alpha \equiv 1 \pmod{\pi}$.) Consider the case where k is finite (cf. *Algebra*, Chapter IX, § 6, Exercise 4).

21. Let K be a field, v a discrete valuation on K, A the ring of v, π a uniformizer of v and k the residue field of v.

(a) Let P, R be two polynomials in A[X], P being monic, and suppose that: (1) $\deg(R) < h.\deg(P)$, where $h \geqslant 1$ is an integer; (2) the canonical image $\bar{P}$ of P in k[X] is irreducible. Show then that, if the polynomial $Q = P^h + \pi R$ is reducible in $\hat{K}$[X], necessarily $h > 1$ and $\bar{P}$ divides the canonical image $\bar{R}$ of R in k[X]. Deduce that, for a given monic irreducible polynomial $r(X) \in k[X]$ and a given integer $h \geqslant 1$, there exists a separable irreducible polynomial $Q \in A[X]$ such that its canonical image $\bar{Q}$ is equal to r^h.

(b) Let $k' = k(\alpha)$ be an algebraic extension of k of degree m and h an integer $\geqslant 1$. Show that there exists an algebraic extension L of K of degree hm such that there is only a single valuation v' (up to equivalence) on L extending v, that $e(v'/v) = h$, $f(v'/v) = m$ and that the residue field of v' is isomorphic to k' (use (a)).

22. (a) If there exists a discrete valuation on a field K, show that the algebraic closure of K is infinite over K.

(b) Let K be a finitely generated extension of a field K_0. Show that, if K is not algebraic over K_0, there exists a discrete valuation v on K such that $v(x) = 0$ on K_0.

§ 9

1. (a) Let K be a field (not necessarily commutative) and $\mathcal{T}$ a locally compact topology on K compatible with the *ring* structure on K; show that $\mathcal{T}$ is compatible with the *field* structure on K. (Use the Theorem of R. Ellis (*General Topology*, Chapter X, § 3, Exercise 25) or argue directly reproducing the proofs of *Integration*, Chapter VII, § 1, no. 10 and those of Proposition 1 of this paragraph.)

(b) Give an example of a locally compact topology on a commutative field K, which is compatible with the additive group structure on K, but not with its ring structure. (Take K to be the field of fractions of a compact integral domain A (A being for example a ring of formal power series $k[[X, Y]]$, where k is finite) and take as fundamental system of neighbourhoods of 0 in K the neighbourhoods of 0 in A.)

¶ 2. (a) In a space $\mathbf{R}^n$ ($n \geqslant 2$), let U be a non-empty open set with non-empty complement; show that the frontier of U contains a non-empty perfect set (cf. *General Topology*, Chapter I, § 9, Exercise 17).

(b) Deduce from (a) that, if A is an everywhere dense subset of $\mathbf{R}^n$ which meets every perfect set in $\mathbf{R}^n$, A is convex.

(c) Show that there exists in $\mathbf{C}$ an everywhere dense subfield K which is *connected* and *locally connected* and is a *pure transcendental* extension of $\mathbf{Q}$ (apply *General Topology*, Chapter IX, § 5, Exercise 18 (b) and (c), constructing K by transfinite induction, using (b) and the method described in *Set Theory*, Chapter III, § 6, Exercise 24). Deduce that there exists a subfield $K' \supset K$ of $\mathbf{C}$ which is (algebraically) isomorphic to $\mathbf{R}$, connected and locally connected.

(d) Show, using (c), that there exists on $\mathbf{C}$ the topology of a connected and locally connected space, which is compatible with the field structure and for which the completion of $\mathbf{C}$ is an algebra over $\mathbf{C}$ which is the direct composition of two fields isomorphic to $\mathbf{C}$.

3. Let K be a totally disconnected non-discrete locally compact commutative field; let A be the ring of the absolute value mod_K on K and let U be the group of units of A. For every integer $n > 0$, let ${}_nU$ denote the group of n-th roots of unity in K and U^n the subgroup of U consisting of the n-th powers of the elements of U. Show that, if n is prime to the characteristic of K, U^n is an open subgroup of U and that

$$\mathrm{Card}(U/U^n) = \mathrm{mod}_K(n) . \mathrm{Card}({}_nU)$$

(use Exercise 14 of *Integration*, Chapter VII, § 2 and *Commutative Algebra*, Chapter III, § 4, no. 6, Corollary 1 to Theorem 2 to show that, if $\mathfrak{m}$ is the maximal ideal of A, the image under $x \mapsto x^n$ of $1 + \mathfrak{m}^k$ is $1 + n.\mathfrak{m}^k$ for k sufficiently large).

4. (a) Let K be a commutative field and v a valuation on K such that K is Henselian with v (§ 8, Exercise 6). Suppose further that K and the residue field k of v are of characteristic 0. Show that there exists a subfield K_0 of the ring A of v such that the canonical mapping $A \to k$, restricted to K_0, is an isomorphism of K_0 onto k. (Let H be a subfield of A such that the image of H under the canonical mapping is an isomorphism onto a subfield E of k; show that, if $E \neq k$, there exists $\alpha \notin H$ in A such that the subfield $H(\alpha)$ of K is contained in A and is canonically isomorphic to $E(\bar{\alpha})$, where $\bar{\alpha}$ is the class of α in k; distinguish two cases according to whether $\bar{\alpha}$ is algebraic or transcendental over E.)

(b) Suppose further that v is a discrete valuation and that K is complete with respect to v; deduce from (a) that K is isomorphic to the field of formal power series $k((T))$.

5. (a) Suppose that K is a commutative field, v a valuation of height 1 on K such that K is complete with respect to v and A the ring of v; suppose further that the residue field k of v is perfect and of characteristic $p > 0$. For every element $\xi \in k^*$ and every integer n, let x_n be an element of the class $\xi^{p^{-n}}$ in A; show that the sequence $(x_n^{p^n})$ is a Cauchy sequence in A whose limit is independent of the choice of the x_n in the classes $\xi^{p^{-n}}$. If this limit is denoted by $\phi(\xi)$, show that ϕ is the unique isomorphism u of the multiplicative group k^* to the multiplicative group K^* such that, for all $\xi \in k^*$, $u(\xi)$ is an element of A belonging to the class ξ.

(b) If K is also of characteristic p, show that ϕ, extended to k by $\phi(0) = 0$, is an isomorphism of the field k onto a subfield of K. Deduce a new proof of Theorem 1 (iii) of no. 3.

(c) Suppose that k is *finite*. Show that, if r is prime to p, the group $(K^*)^r$ of r-th powers of elements of K^* is of finite index in K^* (use Hensel's Lemma). Show that, if further v is discrete and K of characteristic 0, the same result is valid without restriction on r (observe that every element of $1 + p^2A$ is a p-th power).

§ 10

1. Let K be a field and P the prime subfield of K; the *absolute dimension* of K is the number $\dim.\mathrm{al}_P K$ if P is of characteristic $p > 0$ and the number $\dim.\mathrm{al}_P K + 1$ if P is of characteristic 0. Let v be a valuation on K, h its height, r its rational rank and k its residue field.

(a) Suppose that the absolute dimension n of K is finite. Then, if s is the absolute dimension k, $r + s \leqslant n$.

(b) Suppose further that K is a finitely generated extension of P. Then, if $r + s = n$, k is a finitely generated extension of its prime field and the order group of v is isomorphic to $\mathbf{Z}^r$; if $h + s = n$, k is a finitely generated extension of its prime field and the order group of v is isomorphic to $\mathbf{Z}^r$ ordered lexicographically; finally, if $s = n - 1$, v is a discrete valuation and k is a finitely generated extension of its prime field.

¶ 2. Let K be a field, v a valuation on K, L an extension of K and v' a valuation on L extending v; L is called an *immediate extension* (with respect to v') if $e(v'/v) = f(v'/v) = 1$. The completion $\hat{K}$ of K is an immediate extension of K.

(a) For L to be an immediate extension of K, it is necessary and sufficient that, for all $x \in L - K$, there exist $y \in K$ such that $v'(x - y) > v'(x)$.

(b) For every field K, show that there exists a *maximal* immediate extension L of K, that is, not admitting an immediate extension distinct from itself (use § 3, Exercise 3 (b)).

(c) Show that, if K is linearly compact with the topology defined by v, it admits no immediate extension other than itself (use (a), noting that, in the notation of (a), the set of $v'(x - y)$ for $y \in K$ has no greatest element in the value group of v').

(d) Suppose that K is not linearly compact; let B be a well-ordered subset of the order group of v and $(a_\beta)_{\beta \in B}$ a family of elements of K such that, for $\lambda < \mu < \nu$,

$$v(a_\lambda - a_\mu) < v(a_\mu - a_\nu),$$

but that there exists no $x \in K$ such that $v(x - a_\lambda) = v(a_\lambda - a_\mu)$ for every ordered pair (λ, μ) such that $\lambda < \mu$. For every extension E of K and every extension w of v to E, let $D_E(\lambda)$ denote the set of $z \in E$ such that $v(z - a_\lambda) \geqslant \gamma_\lambda$, where γ_λ is the common value of the $v(a_\lambda - a_\mu)$ for $\lambda < \mu$; the hypothesis is therefore that $\bigcap_{\beta \in B} D_K(\beta) = \varnothing$ (§ 5, Exercise 5 (a)). Let Ω be an algebraic closure of K and let v_0 be an extension of v to Ω; show that, if P is a polynomial in K[X], for there to exist $\lambda \in B$ such that $v(P(a_\mu)) = v(P(a_\lambda))$ for all $\lambda \leqslant \mu$, it is necessary and sufficient that no zero of P in Ω belong to $\bigcap_{\beta \in B} D_\Omega(\beta)$. If P has this property and E and w have the same meaning as above, show that, for all $z \in \bigcap_{\beta \in B} D_E(\beta)$, $w(P(z) - P(a_\mu)) > w(P(z)) = v(P(a_\mu))$ so long as μ is large enough (decompose P(X) into factors in Ω[X]). Deduce that one of the following situations holds:

(1) Either $\bigcap_{\beta \in B} D_\Omega(\beta) \neq \varnothing$ and, if θ is an element of this intersection whose degree over K is the least possible, $K(\theta)$ is an immediate extension of K distinct from K.

(2) Or $\bigcap_{\beta \in B} D_{\Omega}(\beta) = \varnothing$; then there is a valuation v' on $K(X)$ extending v such that $v(P(X)) = v(P(a_\mu))$ for μ large enough and $K(X)$, with v', is an immediate extension of K (use the criterion in (a)).

(e) Deduce from (c) and (d) that for K to be linearly compact, it is necessary and sufficient that K have no immediate extension other than itself.

CHAPTER VII

Divisors

All the rings considered in this chapter are assumed to be commutative and to possess a unit element. All the ring homomorphisms are assumed to map unit element to unit element. Every subring of a ring A is assumed to contain the unit element of A.

1. KRULL DOMAINS

1. DIVISORIAL IDEALS OF AN INTEGRAL DOMAIN

DEFINITION 1. *Let A be an integral domain and K its field of fractions. Every sub-A-module* $\mathfrak{a}$ *of K such that there exists an element* $d \neq 0$ *in A for which* $d\mathfrak{a} \subset A$ *is called a fractional ideal of A (or of K, by an abuse of language).*

Every *finitely generated* sub-A-module $\mathfrak{a}$ of K is a fractional ideal: for if $(a_i)_{1 \leqslant i \leqslant n}$ is a system of generators of $\mathfrak{a}$, we may write $a_i = b_i/d_i$, where $b_i \in A$, $d_i \in A$ and $d_i \neq 0$; if $d = d_1 \ldots d_n$, clearly $d\mathfrak{a} \subset A$. In particular the *monogenous* sub-A-modules of K are fractional ideals (recall that they have been called *fractional principal ideals* in *Algebra*, Chapter VI, § 1, no. 5). If A is *Noetherian*, every fractional ideal is a *finitely generated* A-module. Every sub-A-module of a fractional ideal of A is a fractional ideal. Every ideal of A is a fractional ideal; to avoid confusion, these will also be called the *integral* ideals of A.

We denote by I(A) the set of *non-zero* fractional ideals of A. Given two elements $\mathfrak{a}$, $\mathfrak{b}$ of I(A), we shall write $\mathfrak{a} \prec \mathfrak{b}$ (or $\mathfrak{b} \succ \mathfrak{a}$) for the relation "every fractional principal ideal containing $\mathfrak{a}$ also contains $\mathfrak{b}$"; clearly this relation is a *preordering* on I(A). Let R denote the associated equivalence relation "$\mathfrak{a} \prec \mathfrak{b}$ and $\mathfrak{b} \prec \mathfrak{a}$" (*Set Theory*, Chapter III, § 1, no. 2) and D(A) the quotient set I(A)/R; we shall say that the elements of D(A) are the *divisors* of A and, for every fractional ideal $\mathfrak{a} \in$ I(A), we shall denote by div $\mathfrak{a}$ (or $\text{div}_A\, \mathfrak{a}$) the canonical image of $\mathfrak{a}$ in D(A) and we shall say that div $\mathfrak{a}$ is the *divisor of* $\mathfrak{a}$; if $\mathfrak{a} = Ax$ is a fractional principal ideal, we write div(x) instead of div(Ax) and div(x) is called the *divisor of* x; the elements of D(A) of the form div(x) are called *principal divisors*. By taking the quotient, the preordering $\prec$ on I(A) defines on D(A) an *ordering* which we shall denote by $\leqslant$.

For all $\mathfrak{a} \in \mathrm{I}(A)$ there exists by hypothesis some $d \neq 0$ in A such that $\mathfrak{a} \subset Ad^{-1}$; the intersection $\tilde{\mathfrak{a}}$ of the fractional principal ideals containing $\mathfrak{a}$ is therefore an element of I(A). Clearly the relation $\mathfrak{a} \prec \mathfrak{b}$ is equivalent to the relation $\tilde{\mathfrak{a}} \supset \tilde{\mathfrak{b}}$; the relation $\mathfrak{a} \supset \mathfrak{b}$ therefore implies $\mathfrak{a} \prec \mathfrak{b}$. For two elements $\mathfrak{a}$, $\mathfrak{b}$ of I(A) to be equivalent modulo R, it is necessary and sufficient that $\tilde{\mathfrak{a}} = \tilde{\mathfrak{b}}$.

DEFINITION 2. *Every element* $\mathfrak{a}$ *of* I(A) *such that* $\mathfrak{a} = \tilde{\mathfrak{a}}$ *is called a divisorial fractional ideal of* A.

In other words a divisorial ideal is just a non-zero intersection of a non-empty family of fractional principal ideals. Every non-zero intersection of divisorial ideals is a divisorial ideal. If $\mathfrak{a}$ is divisorial, so is $\mathfrak{a}x$ for all $x \in K^*$, the mapping $\mathfrak{b} \mapsto \mathfrak{b}x$ being a bijection of the set of fractional principal ideals onto itself. For all $\mathfrak{a} \in \mathrm{I}(A)$, $\tilde{\mathfrak{a}}$ is the least divisorial ideal containing $\mathfrak{a}$ and is equivalent to $\mathfrak{a}$ modulo R. Moreover, if $\mathfrak{b}$ is a divisorial ideal equivalent to $\mathfrak{a}$ modulo R, then $\tilde{\mathfrak{a}} = \tilde{\mathfrak{b}} = \mathfrak{b}$. Hence $\tilde{\mathfrak{a}}$ is the unique divisorial ideal $\mathfrak{b}$ such that div $\mathfrak{a}$ = div $\mathfrak{b}$ (in other words, the restriction of the mapping $\mathfrak{a} \mapsto$ div $\mathfrak{a}$ to the set of *divisorial* ideals is *injective*).

Let $\mathfrak{a}$ and $\mathfrak{b}$ be two fractional ideals of K. Recall (Chapter I, § 2, no. 10) that $\mathfrak{b} : \mathfrak{a}$ denotes the set of $x \in K$ such that $x\mathfrak{a} \subset \mathfrak{b}$; this is obviously an A-module; if $\mathfrak{b} \in \mathrm{I}(A)$ and $\mathfrak{a} \in \mathrm{I}(A)$, then $\mathfrak{b} : \mathfrak{a} \in \mathrm{I}(A)$; for if d is a non-zero element of A such that $d\mathfrak{b} \subset A$ and $d\mathfrak{a} \subset A$ and a is a non-zero element of $A \cap \mathfrak{a}$, then $da(\mathfrak{b} : \mathfrak{a}) \subset A$; on the other hand, if $b \neq 0$ belongs to $\mathfrak{b}$, then $bd\mathfrak{a} \subset \mathfrak{b}$, hence $bd \in \mathfrak{b} : \mathfrak{a}$ and $\mathfrak{b} : \mathfrak{a} \neq 0$.

The definition of $\mathfrak{b} : \mathfrak{a}$ can also be written:

$$\mathfrak{b} : \mathfrak{a} = \bigcap_{x \in \mathfrak{a},\, x \neq 0} \mathfrak{b}x^{-1}. \tag{1}$$

PROPOSITION 1. (a) *If* $\mathfrak{b}$ *is a divisorial ideal and* $\mathfrak{a} \in \mathrm{I}(A)$, $\mathfrak{b} : \mathfrak{a}$ *is divisorial.*

(b) *Let* $\mathfrak{a}$, $\mathfrak{b}$ *be in* I(A). *In order that* div $\mathfrak{a}$ = div $\mathfrak{b}$, *it is necessary and sufficient that* $A : \mathfrak{a} = A : \mathfrak{b}$.

(c) *For all* $\mathfrak{a} \in \mathrm{I}(A)$, $\tilde{\mathfrak{a}} = A : (A : \mathfrak{a})$.

Assertion (a) follows immediately from equation (1) since, if $\mathfrak{b}$ is divisorial, so is $\mathfrak{b}x^{-1}$ for all $x \neq 0$.

To show (b), let P($\mathfrak{a}$) denote the set of fractional principal ideals containing $\mathfrak{a}$; the relation $Ax \in P(\mathfrak{a})$ is equivalent to $x^{-1}\mathfrak{a} \subset A$ and hence to $x^{-1} \in A : \mathfrak{a}$. As the relation div $\mathfrak{a}$ = div $\mathfrak{b}$ is by definition equivalent to $P(\mathfrak{a}) = P(\mathfrak{b})$, it is also equivalent to $A : \mathfrak{a} = A : \mathfrak{b}$.

Finally, as $\mathfrak{a}(A: \mathfrak{a}) \subset A$, $\mathfrak{a} \subset A: (A: \mathfrak{a})$. Replacing $\mathfrak{a}$ by $A: \mathfrak{a}$ in this formula, it is seen that $A: \mathfrak{a} \subset A: (A: (A: \mathfrak{a}))$; on the other hand, the relation $\mathfrak{a} \subset A: (A: \mathfrak{a})$ implies

$$A: \mathfrak{a} \supset A: (A: (A: \mathfrak{a})).$$

Therefore $A: \mathfrak{a} = A:(A: (A: \mathfrak{a}))$ and it follows from (b) that $\operatorname{div} \mathfrak{a} = \operatorname{div}(A: (A: \mathfrak{a}))$ As $A: (A: \mathfrak{a})$ is divisorial by (a), certainly $\tilde{\mathfrak{a}} = A: (A: \mathfrak{a})$, which proves (c).

Remark. In the course of the above proof it has been proved that $A: \mathfrak{a} = A: (A: (A: \mathfrak{a}))$ for every ideal $\mathfrak{a} \in I(A)$, which is a special case of *Set Theory*, Chapter III, § 1, no. 5, Proposition 2.

PROPOSITION 2. (i) *In* D(A) *every non-empty set bounded above admits a least upper bound. More precisely, if* $(\mathfrak{a}_\iota)$ *is a non-empty family of elements of* I(A) *which is bounded above, then*

$$\sup_\iota(\operatorname{div} \mathfrak{a}_\iota) = \operatorname{div}\Big(\bigcap_\iota \tilde{\mathfrak{a}}_\iota\Big).$$

(ii) *In* D(A) *every non-empty set bounded below admits a greatest lower bound. More precisely, if* $(\mathfrak{a}_\iota)$ *is a non-empty family of elements of* I(A) *which is bounded below, then*

$$\inf_\iota(\operatorname{div} \mathfrak{a}_\iota) = \operatorname{div}\Big(\sum_\iota \mathfrak{a}_\iota\Big).$$

(iii) *The set* D(A) *is a lattice.*

Let $(\mathfrak{a}_\iota)$ be a non-empty family of elements of I(A) which is bounded above. To say that a divisorial ideal $\mathfrak{b}$ bounds this family above amounts to saying that it is contained in all the $\tilde{\mathfrak{a}}_\iota$, that is that $\mathfrak{b}$ is contained in $\bigcap_\iota \tilde{\mathfrak{a}}_\iota$. Hence $\bigcap_\iota \tilde{\mathfrak{a}}_\iota \neq (0)$ and $\bigcap_\iota \tilde{\mathfrak{a}}_\iota$ is therefore a divisorial ideal, which shows (i).

Now let $(\tilde{\mathfrak{a}}_\iota)$ be a non-empty family of elements of I(A) which is bounded below. To say that a divisorial ideal $\mathfrak{b}$ bounds this family below means that it contains all the $\tilde{\mathfrak{a}}_\iota$, that is (since $\mathfrak{b}$ is divisorial) that it contains all the $\mathfrak{a}_\iota$, or also that $\mathfrak{b} \supset \sum_\iota \mathfrak{a}_\iota$. This proves (ii).

Finally, to prove (iii) it is sufficient by (i) and (ii) to prove that, if $\mathfrak{a}$, $\mathfrak{b}$ are in I(A), the set $\{\mathfrak{a}, \mathfrak{b}\}$ is bounded both above and below in I(A); now it is bounded above by $\mathfrak{a} \cap \mathfrak{b}$ (which is distinct from (0)). It is bounded below by $\mathfrak{a} + \mathfrak{b}$, for $\mathfrak{a} + \mathfrak{b} \in I(A)$: if d and d' are non-zero elements of A such that $d\mathfrak{a} \subset A$ and $d'\mathfrak{b} \subset A$, then $dd'(\mathfrak{a} + \mathfrak{b}) \subset A$.

COROLLARY. *If x, y and $x + y$ are in* K*, *then* $\operatorname{div}(x + y) \geqslant \inf(\operatorname{div}(x), \operatorname{div}(y))$.

$A(x + y) \subset Ax + Ay$ and hence $\operatorname{div}(x + y) \geqslant \operatorname{div}(Ax + Ay)$.

2. THE MONOID STRUCTURE ON D(A)

PROPOSITION 3. *Let* $\mathfrak{a}$, $\mathfrak{a}'$, $\mathfrak{b}$, $\mathfrak{b}'$ *be elements of* I(A). *The relations* $\mathfrak{a} \succ \mathfrak{a}'$ *and* $\mathfrak{b} \succ \mathfrak{b}'$ *imply* $\mathfrak{a}\mathfrak{b} \succ \mathfrak{a}'\mathfrak{b}'$.

We may restrict our attention to the case where $\mathfrak{b} = \mathfrak{b}'$. Then let Ax be a fractional principal ideal containing $\mathfrak{a}'\mathfrak{b}$; for every non-zero element y of $\mathfrak{b}$, $Ax \supset \mathfrak{a}'y$ and hence $Axy^{-1} \supset \mathfrak{a}'$, whence $Axy^{-1} \supset \mathfrak{a}$ and $Ax \supset \mathfrak{a}y$. Varying y, it is seen that $Ax \supset \mathfrak{a}\mathfrak{b}$, whence $\mathfrak{a}\mathfrak{b} \succ \mathfrak{a}'\mathfrak{b}$.

It follows from Proposition 3 that multiplication on I(A) defines, by passing to the quotient, a law of composition on D(A) which is obviously associative and commutative. It is written additively so that we may write:

$$\operatorname{div}(\mathfrak{a}\mathfrak{b}) = \operatorname{div} \mathfrak{a} + \operatorname{div} \mathfrak{b}, \tag{2}$$

for $\mathfrak{a}$, $\mathfrak{b}$ in I(A). Clearly div(1) is an identity element for this addition; this element is denoted by 0. Proposition 3 proves further that the order structure on D(A) is *compatible* with this addition (*Algebra*, Chapter VI, § 1, no. 1) and, more precisely (no. 1, Proposition 2 (ii)):

$$\begin{aligned}\inf(\operatorname{div} \mathfrak{a} + \operatorname{div} \mathfrak{b}, \operatorname{div} \mathfrak{a} + \operatorname{div} \mathfrak{c}) &= \inf(\operatorname{div}(\mathfrak{a}\mathfrak{b}), \operatorname{div}(\mathfrak{a}\mathfrak{c})) = \operatorname{div}(\mathfrak{a}\mathfrak{b} + \mathfrak{a}\mathfrak{c}) \\ &= \operatorname{div}(\mathfrak{a}(\mathfrak{b} + \mathfrak{c})) = \operatorname{div} \mathfrak{a} + \operatorname{div}(\mathfrak{b} + \mathfrak{c}) = \operatorname{div} \mathfrak{a} + \inf(\operatorname{div} \mathfrak{b}, \operatorname{div} \mathfrak{c}).\end{aligned}$$

For a fractional ideal $\mathfrak{a} \neq 0$ to be such that $\operatorname{div} \mathfrak{a} \geqslant 0$ in D(A), it is necessary and sufficient that $\mathfrak{a} \subset A$ (in other words, that $\mathfrak{a}$ be an *integral* ideal of A).

For two elements x, y of K^*, the relation $\operatorname{div}(x) = \operatorname{div}(y)$ is equivalent to $Ax = Ay$; the set of principal divisors of A with the order relation and the monoid law induced by that on D(A) is an *ordered group* canonically isomorphic to the multiplicative group of fractional principal ideals ordered by the opposite order relation to inclusion (*Algebra*, Chapter VI, § 1, no. 5). The relation S between two elements P, Q of D(A):

$$\text{“there exists } x \in K^* \text{ such that } P = Q + \operatorname{div}(x)\text{”}$$

is therefore an equivalence relation since the relation $P = Q + \operatorname{div}(x)$ is equivalent to $Q = P + \operatorname{div}(x^{-1})$; if P and Q are congruent modulo S, they are called *equivalent divisors* of A. Clearly moreover the relation S is compatible with the law of the monoid D(A) and the latter therefore defines, by taking quotients, a monoid structure on D(A)/S; this monoid is called the *divisor class monoid of* A.

PROPOSITION 4. *Let* $\mathfrak{a}$, $\mathfrak{b}$ *be two divisorial fractional ideals of* A. *The following properties are equivalent:*

(a) $\operatorname{div} \mathfrak{a}$ *and* $\operatorname{div} \mathfrak{b}$ *are equivalent divisors;*

(b) *there exists* $x \in K^*$ *such that* $\mathfrak{b} = x\mathfrak{a}$.

If div $\mathfrak{b}$ = div $\mathfrak{a}$ + div(x) for some $x \in K^*$, then div $\mathfrak{b}$ = div($x\mathfrak{a}$) and, as $\mathfrak{b}$ and $x\mathfrak{a}$ are divisorial, $\mathfrak{b} = x\mathfrak{a}$, which proves the proposition.

Let $\mathfrak{a}$ be an *invertible* fractional ideal (Chapter II, § 5, no. 6); then $\mathfrak{a} = A:(A:\mathfrak{a})$ (*loc. cit.*, Proposition 10) and hence $\mathfrak{a}$ is *divisorial* (no. 1, Proposition 1). The group J(A) of invertible fractional ideals is therefore identified with a subgroup of the monoid D(A) and the canonical image of J(A) in D(A)/S with the group of classes of *projective* A-modules *of rank* 1 (Chapter II, § 5, no. 7, Corollary 2 to Proposition 12 and *Remark* 1).

THEOREM 1. *Let* A *be an integral domain. For the monoid* D(A) *of divisors of* A *to be a group, it is necessary and sufficient that* A *be completely integrally closed.*

Suppose that D(A) is a group. Let $x \in K$; suppose that A[x] is contained in a finitely generated sub-A-module of K. Then we have seen (no. 1) that $\mathfrak{a} = A[x]$ is an element of I(A). Then $x\mathfrak{a} \subset \mathfrak{a}$ and hence div(x) + div $\mathfrak{a} \geqslant$ div $\mathfrak{a}$. Since D(A) is an ordered group, we conclude that div(x) $\geqslant 0$, whence $x \in A$. Thus A is completely integrally closed (Chapter V, § 1, no. 4, Definition 5).

Conversely, suppose that A is completely integrally closed. Let $\mathfrak{a}$ be a divisorial ideal. We shall show that div $\mathfrak{a}$ + div($A:\mathfrak{a}$) = 0, which will prove that D(A) is a group. As $\mathfrak{a}(A:\mathfrak{a}) \subset A$, it suffices (no. 1) to verify that every fractional principal ideal Ax^{-1} which contains $\mathfrak{a}(A:\mathfrak{a})$ also contains A. Now, for $y \in K^*$, the relation $Ay \supset \mathfrak{a}$ implies $y^{-1} \in A:\mathfrak{a}$, whence $y^{-1}\mathfrak{a} \subset \mathfrak{a}(A:\mathfrak{a}) \subset Ax^{-1}$ and hence $x\mathfrak{a} \subset Ay$. As $\mathfrak{a}$ is divisorial, we deduce that $x\mathfrak{a} \subset \mathfrak{a}$, whence $x^n\mathfrak{a} \subset \mathfrak{a}$ for all $n \in \mathbf{N}$. There exist elements x_0, x_1 of K^* such that $Ax_0 \subset \mathfrak{a} \subset Ax_1$; therefore $x^n x_0 \in Ax_1$, whence $x^n \in Ax_1x_0^{-1}$. As A is completely integrally closed, $x \in A$, that is $Ax^{-1} \supset A$, which completes the proof.

Note that, if A is completely integrally closed (and even Noetherian), a divisorial ideal of A is not necessarily invertible, in other words, in general J(A) $\neq$ D(A) (Exercise 2 and § 3, no. 2, Proposition 1).

COROLLARY. *Let* A *be a completely integrally closed domain and* $\mathfrak{a}$ *a divisorial fractional ideal of* A. *Then, for every fractional ideal* $\mathfrak{b} \neq 0$ *of* A, div($\mathfrak{a}:\mathfrak{b}$) = div $\mathfrak{a}$ − div $\mathfrak{b}$.

By virtue of formula (1) of no. 1:

$$\operatorname{div}(\mathfrak{a}:\mathfrak{b}) = \operatorname{div}\Big(\bigcap_{y \in \mathfrak{b},\, y \neq 0} y^{-1}\mathfrak{a}\Big) = \sup_{y \in \mathfrak{b},\, y \neq 0} \operatorname{div}(y^{-1}\mathfrak{a})$$

taking account of Proposition 2 and the fact that the fractional ideals $y^{-1}\mathfrak{a}$ are divisorial. But since D(A) is an ordered group (*Algebra*, Chapter VI, § 1, no. 8):

$$\sup_{y \in \mathfrak{b},\, y \neq 0} \operatorname{div}(y^{-1}\mathfrak{a}) = \sup_{y \in \mathfrak{b},\, y \neq 0} (\operatorname{div}\mathfrak{a} - \operatorname{div}(y))$$
$$= \operatorname{div}\mathfrak{a} - \inf_{y \in \mathfrak{b},\, y \neq 0} \operatorname{div}(y) = \operatorname{div}\mathfrak{a} - \operatorname{div}\mathfrak{b}.$$

3. KRULL DOMAINS

DEFINITION 3. *An integral domain* A *is called a Krull domain if there exists a family* $(v_\iota)_{\iota \in I}$ *of valuations on the field of fractions* K *of* A *with the following properties:*

(AK_I) *the valuations* v_ι *are discrete;*

(AK_{II}) *the intersection of the rings of the* v_ι *is* A*;*

(AK_{III}) *for all* $x \in K^*$, *the set of indices* $\iota \in I$ *such that* $v_\iota(x) \neq 0$ *is finite.*

It obviously suffices to verify condition (AK_{III}) for the elements x of A — (0).

Examples

(1) Every discrete valuation ring is a Krull domain.

(2) More generally, every *principal ideal domain* A is a Krull domain. For let $(p_\iota)_{\iota \in I}$ be a representative system of extremal elements of A and let v_ι be the valuation on the field of fractions of A defined by p_ι (Chapter VI, § 3, no. 3, *Example* 4). It is immediately seen that the family $(v_\iota)_{\iota \in I}$ satisfies properties (AK_I), (AK_{II}) and (AK_{III}).

(3) Let F be a field and $(R_i)_{1 \leqslant i \leqslant n}$ a *finite* family of subrings of F which are Krull domains. Then their *intersection* $S = \bigcap_{j=1}^{n} R_j$ is a Krull domain. For $1 \leqslant j \leqslant n$ let $(v_{j\iota})_{\iota \in I_j}$ be a family of valuations on the field of fractions of R_j satisfying (AK_I), (AK_{II}), (AK_{III}) (where A is replaced by R_j). Let $w_{j\iota}$ denote the restriction of $v_{j\iota}$ to the field of fractions of S. Then the family $(v_{j\iota})_{1 \leqslant j \leqslant n, \iota \in I_j}$ obviously satisfies (AK_{II}) (where A is replaced by S) and also (AK_{III}) since the set of indices j is finite. The valuations $w_{j\iota}$ are either discrete or improper. By retaining only those which are discrete, a family is obtained which obviously satisfies (AK_I), (AK_{II}) and (AK_{III}) (where A is replaced by S). Hence S is certainly a Krull domain.

(4) In particular, if A is a Krull domain and K′ a subfield of the field of fractions K of A, K′ ∩ A is a Krull domain.

THEOREM 2. *Let* A *be an integral domain. For* A *to be a Krull domain, it is necessary and sufficient that the two following conditions be satisfied:*

(a) A *is completely integrally closed;*

(b) *every non-empty family of divisorial integral ideals of* A *admits a maximal element (with respect to the relation* $\subset$).

Moreover, if P(A) *is the set of extremal elements of* D(A), P(A) *is then a basis of the* **Z**-*module* D(A) *and the positive elements of* D(A) *are the linear combinations of the elements of* P(A) *with coefficients* $\geqslant 0$.

Let A be a Krull domain. It is completely integrally closed (Chapter VI, § 4, no. 5, Corollary to Proposition 9). Let $(v_\iota)_{\iota \in I}$ be a family of valuations on the field of fractions K of A satisfying (AK_I), (AK_{II}) and (AK_{III}). The v_ι may be assumed to be normed (Chapter VI, § 3, no. 6, Definition 3). For all $\mathfrak{a} \in I(A)$, we shall write:

$$(3) \qquad v_\iota(\mathfrak{a}) = \sup_{\mathfrak{a} \subset Ax} (v_\iota(x));$$

then $v_\iota(\mathfrak{a}) \in \mathbf{Z}$, for, if a is a non-zero element of $\mathfrak{a}$, the relation $Ax \supset Aa$ implies that $v_\iota(x) \leqslant v_\iota(a)$ (by (AK_{II})), which shows that the family of $v_\iota(x)$ $(\mathfrak{a} \subset Ax)$ is bounded above. We establish the following properties:

(1) *Let $\mathfrak{a}$ be a divisorial fractional ideal; in order that $y \in \mathfrak{a}$, it is necessary and sufficient that $v_\iota(y) \geqslant v_\iota(\mathfrak{a})$ for all $\iota \in I$.*

As $\mathfrak{a}$ is divisorial, the relation $y \in \mathfrak{a}$ is equivalent to the relation "$\mathfrak{a} \subset Ax$ implies $y \in Ax$". Now, by (AK_{II}), the relation $y \in Ax$ is equivalent to "$v_\iota(y) \geqslant v_\iota(x)$ for all $\iota \in I$". Whence (1).

(2) *Let $\mathfrak{a}$ and $\mathfrak{b}$ be two divisorial fractional ideals of A; in order that $\mathfrak{a} \subset \mathfrak{b}$, it is necessary and sufficient that $v_\iota(\mathfrak{a}) \geqslant v_\iota(\mathfrak{b})$ for all $\iota \in I$.*

This follows immediately from property (1).

(3) *If $x \in K^*$, then $v_\iota(Ax) = v_\iota(x)$.*

If $Ay \supset Ax$, then $v_\iota(y) \leqslant v_\iota(x)$ by (AK_{II}) and the minimum value of $v_\iota(y)$ is taken at $y = x$.

(4) *For all $\mathfrak{a} \in I(A)$, the indices $\iota \in I$ such that $v_\iota(\mathfrak{a}) \neq 0$ are finite in number.*

There exist x, y in K^* such that $Ax \subset \mathfrak{a} \subset Ay$. By properties (2) and (3), $v_\iota(x) \geqslant v_\iota(\mathfrak{a}) \geqslant v_\iota(y)$ for all $\iota \in I$. It then suffices to apply (AK_{III}).

We have therefore shown the following lemma:

LEMMA 1. *If A is a Krull domain and $(v_\iota)_{\iota \in I}$ is a family of normed valuations on K satisfying (AK_I), (AK_{II}) and (AK_{III}), the mapping $\mathfrak{a} \mapsto (v_\iota(\mathfrak{a}))_{\iota \in I}$ is a decreasing injective mapping of the set of divisorial integer ideals of A (ordered by $\subset$) to the set of positive elements of the ordered group the direct sum $\mathbf{Z}^{(I)}$.*

This being so, every non-empty set of positive elements of $\mathbf{Z}^{(I)}$ has a minimal element (*Algebra*, Chapter VI, § 1, no. 13, Theorem 2). Hence A certainly satisfies property (b) of the statement.

Conversely, let A be an integral domain satisfying properties (a) and (b) of the statement. Since A is completely integrally closed, D(A) is an ordered group (no. 2, Theorem 1). This group is a lattice (no. 1, Proposition 2). By condition (b) of the statement, every non-empty family of positive elements of D(A) has a minimal element. Let P(A) be the set of extremal elements of D(A). Then (*Algebra*, Chapter VI, § 1, no. 13, Theorem 2) P(A) is a basis of the $\mathbf{Z}$-module D(A) and the positive elements of D(A) are the linear combinations with positive integer coefficients of the elements of P(A).

Thus, for $x \in K^*$, rational integers $v_P(x)$ are defined (for $P \in P(A)$) by writing:

$$\text{div}(x) = \sum_{P \in P(A)} v_P(x) \, . \, P. \tag{4}$$

We also write $v_P(0) = +\infty$.

From the relations

$$\text{div}(xy) = \text{div}(x) + \text{div}(y)$$

and

$$\text{div}(x + y) \geqslant \inf(\text{div}(x), \text{div}(y)),$$

for x, y and $x + y$ in K^*, we deduce that the v_P are *discrete valuations* on K. In order that $x \in A$, it is necessary and sufficient that $\text{div}(x) \geqslant 0$, that is that $v_P(x) \geqslant 0$ for all $P \in P(A)$. Thus the v_P satisfy conditions (AK_I) and (AK_{II}) and obviously also (AK_{III}).

Corollary. *For a Noetherian ring to be a Krull domain, it is necessary and sufficient that it be an integrally closed domain.*

An integrally closed Noetherian domain is completely integrally closed (Chapter V, § 1, no. 4).

There are non-Noetherian Krull domains, for example the polynominal ring $K[X_n]_{n \in \mathbf{N}}$ over a field K in an infinity of indeterminates (cf. Exercise 8).

4. ESSENTIAL VALUATIONS OF A KRULL DOMAIN

Let A be a Krull domain and K its field of fractions. The valuations defined by formula (4) of no. 3 (for $x \in K^*$) are called the *essential valuations* of K (or A).

We have remarked in the course of the proof of Theorem 2 that the valuations v_P satisfy properties (AK_I), (AK_{II}) and (AK_{III}) of Definition 3. Moreover, these discrete valuations v_P are *normed*: for every extremal divisor $P \in P(A)$, $P < 2P$ and hence, if $\mathfrak{a}$ and $\mathfrak{b}$ are the divisorial ideals corresponding to P and 2P, then $\mathfrak{a} \supset \mathfrak{b}$ and $\mathfrak{a} \neq \mathfrak{b}$; for $x \in \mathfrak{a} - \mathfrak{b}$, $\text{div}(x) \geqslant P$ and $\text{div}(x) \not\geqslant 2P$, whence $v_P(x) = 1$, which proves our assertion.

Proposition 5. *Let* A *be a Krull domain,* K *its field of fractions and* $(v_P)_{P \in P(A)}$ *the family of its essential valuations. Let* $(n_P)_{P \in P(A)}$ *be a family of rational integers which are zero except for a finite number of indices. Then the set of* $x \in K$ *such that* $v_P(x) \geqslant n_P$ *for all* $P \in P(A)$ *is the divisorial ideal* $\mathfrak{a}$ *of* A *such that* $\text{div}\, \mathfrak{a} = \sum_{P \in P(A)} n_P \, . \, P$.

Let $x \in K^*$. In order that $x \in \mathfrak{a}$, it is necessary and sufficient that $Ax \subset \mathfrak{a}$, hence that $\text{div}(x) \geqslant \text{div}\, \mathfrak{a}$ and hence, by (4), that $v_P(x) \geqslant n_P$ for all $P \in P(A)$.

PROPOSITION 6. *Let* A *be a Krull domain,* K *its field of fractions,* $(v_\iota)_{\iota \in I}$ *a family of valuations on* K *with the properties of Definition* 3 *and* A_ι *the ring of* v_ι. *Let* S *be a multiplicative subset of* A *not containing* 0 *and* J *the set of indices* $\iota \in I$ *such that* v_ι *is zero on* S. *Then* $S^{-1}A = \bigcap_{\iota \in J} A_\iota$; *in particular* $S^{-1}A$ *is a Krull domain.*

We write $B = \bigcap_{\iota \in J} A_\iota$. Then $S^{-1} \subset B$ and $A \subset B$ and hence $S^{-1}A \subset B$. Conversely, let $x \in B$. Let J' denote the finite set of indices ι such that $v_\iota(x) < 0$. If $\iota \in J'$, then $x \notin A_\iota$, hence $\iota \notin J$ and hence there exists $s_\iota \in S$ such that $v_\iota(s_\iota) > 0$. Let $n(\iota)$ be an integer >0 such that $v_\iota(s_\iota^{n(\iota)}x) \geqslant 0$; we write $s = \prod_{\iota \in J'} s_\iota^{n(\iota)}$. Then $v_\iota(sx) \geqslant 0$ for all $\iota \in I$ and hence $sx \in A$ and $x \in S^{-1}A$. Thus $B = S^{-1}A$.

COROLLARY 1. *Let* P *be an extremal divisor of* A *and* $\mathfrak{p}$ *the corresponding divisorial ideal. Then* $\mathfrak{p}$ *is prime, the ring of* v_P *is* $A_\mathfrak{p}$ *and the residue field of* v_P *is identified with the field of fractions of* $A/\mathfrak{p}$.

Let $S = A - \mathfrak{p}$. By Proposition 5, v_P is zero on S and >0 on $\mathfrak{p}$. Hence $\mathfrak{p}$ is the intersection of A and the ideal of v_P and is therefore prime. On the other hand, for every extremal divisor $Q \neq P$, $Q \ngeq P$ and hence the divisorial ideal $\mathfrak{q}$ corresponding to Q is not contained in $\mathfrak{p}$; thus $\mathfrak{q} \cap S \neq \varnothing$ and hence, by Proposition 5, v_Q is not zero on S. This being so, the corollary follows from Proposition 6 and Chapter II, § 3, no. 1, Proposition 3.

COROLLARY 2. *Let* A *be a Krull domain,* K *its field of fractions and* $(v_\iota)_{\iota \in I}$ *a family of valuations with the properties of Definition* 3. *Then every essential valuation of* A *is equivalent to one of the* v_ι.

Let P be an extremal divisor of A and $\mathfrak{p}$ the corresponding divisorial ideal. By Corollary 1, Proposition 5, Lemma 1 and assertion (1) in the proof of Theorem 2, no. 3, there exists $\iota \in I$ such that the ring A_ι of v_ι contains the ring $A_\mathfrak{p}$ of v_P. As v_ι and v_P are of height 1, they are therefore equivalent (Chapter VI, § 4, no. 5, Proposition 6).

PROPOSITION 7. *Let* A *be a Krull domain,* $(v_P)_{P \in P(A)}$ *the family of its essential valuations and* $\mathfrak{a} \in I(A)$. *Then the coefficient of* P *in* div $\mathfrak{a}$ *is* $\inf_{y \in \mathfrak{a}}(v_P(y))$. *If* $\mathfrak{p}$ *is the divisorial prime ideal corresponding to the extremal divisor* P, *then* $\mathfrak{a}A_\mathfrak{p} = \tilde{\mathfrak{a}}A_\mathfrak{p}$.

As $\mathfrak{a} = \sum_{y \in \mathfrak{a}} Ax$, Proposition 2 (b) (no. 1) shows that $\text{div}(\mathfrak{a}) = \inf_{x \in \mathfrak{a}}(\text{div}(Ax))$, whence our first assertion. The second follows immediately, since div $\tilde{\mathfrak{a}}$ = div $\mathfrak{a}$ and $A_\mathfrak{p}$ is the ring of the discrete valuation v_P.

PROPOSITION 8. *Let* A *be an integrally closed Noetherian domain.*

(a) *Let* P *be an extremal divisor of* A *and* $\mathfrak{p}$ *the corresponding divisorial prime ideal;*

for $n \in \mathbf{N}$, *let* $\mathfrak{p}^{(n)} = \mathfrak{p}^n A_\mathfrak{p} \cap A$; *then* $\mathfrak{p}^{(n)}$ *is the set of* $x \in A$ *such that* $v_P(x) \geqslant n$ *and is a* $\mathfrak{p}$-*primary ideal.*

(b) *Let* $\mathfrak{a}$ *be a divisorial integral ideal,* $n_1P_1 + \cdots + n_rP_r$ *the divisor of* $\mathfrak{a}$ (*the* P_i *being distinct extremal divisors*) *and* $\mathfrak{p}_i$ *the divisorial prime ideal corresponding to* P_i. *Then* $\mathfrak{a} = \bigcap_{i=1}^{r} \mathfrak{p}_i^{(n_i)}$ *is the unique reduced primary decomposition of* $\mathfrak{a}$ *and the* $\mathfrak{p}_i$ *are not immersed.*

By Corollary 1 to Proposition 6, the relation $x \in \mathfrak{p}^n A_\mathfrak{p} = (\mathfrak{p}A_\mathfrak{p})^n$ is equivalent to $v_P(x) \geqslant n$; on the other hand, as $A_\mathfrak{p}$ is a discrete valuation ring, $(\mathfrak{p}A_\mathfrak{p})^n$ is $(\mathfrak{p}A_\mathfrak{p})$-primary (Chapter IV, § 2, no. 1, *Example* 4) and hence $\mathfrak{p}^{(n)}$ is $\mathfrak{p}$-primary (Chapter IV, § 2, no. 1, Proposition 3); this shows (a). Proposition 5 certainly shows that $\mathfrak{a} = \bigcap_{i=1}^{r} \mathfrak{p}_i^{(n_i)}$. As $\mathfrak{p}_i \not\subset \mathfrak{p}_j$ for $i \neq j$ this primary decomposition is reduced: For if $\mathfrak{p}_i^{(n_i)} \supset \bigcap_{j \neq i} \mathfrak{p}_j^{(n_j)} \supset \prod_{i \neq j} \mathfrak{p}_j^{(n_j)}$, $\mathfrak{p}_{ji}$ would contain one of the $\mathfrak{p}_j$ for $j \neq i$ (Chapter II, § 1, no. 1, Proposition 1). The uniqueness follows from Chapter IV, § 2, no. 3, Proposition 5.

5. APPROXIMATION FOR ESSENTIAL VALUATIONS

As the essential valuations of a Krull domain are discrete and normed, no two of them are equivalent and hence they are independent (Chapter VI, § 7, no. 2). Corollary 2 to the approximation theorem (*loc. cit.*, Theorem 1) may therefore be applied to them: given some $n_i \in \mathbf{Z}$ and some essential valuations v_i finite in number and distinct, there exists $x \in K$ such that $v_i(x) = n_i$ for all i. But here there is a more precise result:

PROPOSITION 9. *Let* $v_1, \ldots, v_r$ *be distinct essential valuations of a Krull domain* A *and* $n_1, \ldots, n_r$ *rational integers. There exists an element* x *of the field of fractions* K *of* A *such* $v_i(x) = n_i$ *for* $1 \leqslant i \leqslant r$ *and* $v(x) \geqslant 0$ *for every essential valuation* v *of* A *distinct from* $v_1, \ldots, v_r$.

Let $\mathfrak{p}_1, \ldots, \mathfrak{p}_r$ be the divisorial ideals of A corresponding to the valuations $v_1, \ldots, v_r$. There exists $y \in K$ such that $v_i(y) = n_i$ for $1 \leqslant i \leqslant r$ (Chapter VI, § 7, no. 2, Corollary 1 to Theorem 1). The essential valuations $w_1, \ldots, w_s$ of A distinct from the v_i for which the integer $w_j(y) = -m_j$ is < 0 are finite in number; let $\mathfrak{q}_1, \ldots, \mathfrak{q}_s$ be the corresponding ideals. There exists no inclusion relation between $\mathfrak{p}_1, \ldots, \mathfrak{p}_r, \mathfrak{q}_1, \ldots, \mathfrak{q}_s$ since these ideals correspond to extremal divisors and these ideals are prime (Corollary 1 to Proposition 6). Hence the integral ideal $\mathfrak{a} = \mathfrak{q}_1^{m_1} \ldots \mathfrak{q}_s^{m_s}$ is contained in none of the $\mathfrak{p}_i$ (Chapter II, § 1, no. 1, Proposition 1) and is therefore not contained in their union (*loc. cit.*, Proposition 2). Therefore there exists $z \in \mathfrak{a}$ such that $z \notin p_i$ for $1 \leqslant i \leqslant r$; then $v_1(z) = \cdots = v_r(z) = 0$ and $w_j(z) \geqslant m_j$ for $1 \leqslant j \leqslant s$; hence the element $x = yz$ solves the problem.

COROLLARY 1. *Let A be a Krull domain, K its field of fractions and $\mathfrak{a}$, $\mathfrak{b}$ and $\mathfrak{c}$ three divisorial fractional ideals of A such that $\mathfrak{a} \subset \mathfrak{b}$. There exists $x \in K$ such that $\mathfrak{a} = \mathfrak{b} \cap x\mathfrak{c}$.*

Let $(v_\iota)_{\iota \in I}$ be the family of essential valuations of A and let (m_ι) (resp. (n_ι), (p_ι)) be the family of rational integers (zero except for a finite number of indices) such that $\mathfrak{a}$ (resp. $\mathfrak{b}$, $\mathfrak{c}$) is the set of $x \in K$ for which $v(x_\iota) \geqslant m_\iota$ (resp. n_ι, p_ι) for all $\iota \in I$ (Proposition 5, no. 4). The set J of $\iota \in I$ such that $m_\iota > n_\iota$ is finite. As $p_\iota = m_\iota = 0$ except for a finite number of indices, Proposition 9 shows that there exists $x \in K^*$ such that $v_\iota(x^{-1}) + m_\iota = p_\iota$ for $\iota \in J$ and

$$v_\iota(x^{-1}) + m_\iota \geqslant p_\iota$$

for $\iota \in I - J$. Then, for all $\iota \in I$, $m_\iota = \sup(n_\iota, v_\iota(x) + p_\iota)$. Whence $\mathfrak{a} = \mathfrak{b} \cap x\mathfrak{c}$.

COROLLARY 2. *Let A be a Krull domain. For a fractional ideal $\mathfrak{a}$ of A to be divisorial, it is necessary and sufficient that it be the intersection of two fractional principal ideals.*

The sufficiency is obvious (no. 1, Definition 2). The necessity follows from Corollary 1: take $\mathfrak{b}$ and $\mathfrak{c}$ to be principal and such that $\mathfrak{b} \supset \mathfrak{a}$.

6. PRIME IDEALS OF HEIGHT 1 IN A KRULL DOMAIN

DEFINITION 4. *Let A be an integral domain. A prime ideal $\mathfrak{p}$ of A is said to be of height 1 if it is minimal among the non-zero prime ideals of A.*

We shall also say that the ideal (0) is *of height* 0; a prime ideal *of height* $\leqslant 1$ is therefore by definition equal to (0) or of height 1.

We shall define below, in a general way, the height of a prime ideal.

THEOREM 3. *Let A be a Krull domain and $\mathfrak{p}$ an integral ideal of A. For $\mathfrak{p}$ to be the divisorial ideal corresponding to an extremal divisor, it is necessary and sufficient that $\mathfrak{p}$ be a prime ideal of height 1.*

If $\mathfrak{p}$ is the divisorial ideal corresponding to an extremal divisor, we know (no. 4, Corollary 1 to Proposition 6) that $\mathfrak{p}$ is prime and that $A_\mathfrak{p}$ is a discrete valuation ring; as $A_\mathfrak{p}$ has no prime ideals other than (0) and $\mathfrak{p}A_\mathfrak{p}$, (0) and $\mathfrak{p}$ are the only prime ideals of A contained in $\mathfrak{p}$ (Chapter II, § 3, no. 1, Proposition 3); hence $\mathfrak{p}$ is of height 1. Conversely, we shall show first that every prime ideal $\mathfrak{p} \neq (0)$ of A contains a divisorial prime ideal $\mathfrak{q}$ corresponding to an extremal divisor: for, as $A_\mathfrak{p} \neq K$, $A_\mathfrak{p}$ is the intersection of a non-empty family (A_ι) of essential valuation rings (no. 4, Proposition 6); each A_ι is of the form $A_{\mathfrak{q}_\iota}$ (no. 4, Corollary 1 to Proposition 6) and from $A_\mathfrak{p} \subset A_{\mathfrak{q}_\iota}$ we deduce that $\mathfrak{q}_\iota \subset \mathfrak{p}$. Thus, if $\mathfrak{p}$ is of height 1, then $\mathfrak{p} = \mathfrak{q}$, which shows that $\mathfrak{p}$ is the divisorial ideal corresponding to an extremal divisor.

COROLLARY 1. *In a Krull domain every non-zero prime ideal* $\mathfrak{m}$ *contains a prime ideal of height* 1. *If* $\mathfrak{m}$ *is not of height* 1, *then* $\operatorname{div} \mathfrak{m} = 0$ *and* $A : \mathfrak{m} = A$.

The first assertion has already been seen in the course of the proof of Theorem 3. If $\mathfrak{m}$ is not of height 1 and $\mathfrak{p}$ is a prime ideal of height 1 contained in $\mathfrak{m}$, then $\mathfrak{p} \subset \tilde{\mathfrak{m}}$ and $\mathfrak{p} \neq \tilde{\mathfrak{m}}$; as $\operatorname{div} \mathfrak{p}$ is extremal, necessarily $\operatorname{div} \mathfrak{m} = \operatorname{div} \tilde{\mathfrak{m}} = 0$; hence $\operatorname{div}(A : \mathfrak{m}) = 0$ and, as $A : \mathfrak{m}$ is divisorial (no. 1, Proposition 1), $A : \mathfrak{m} = A$.

COROLLARY 2. *Let* A *be a Krull domain,* K *its field of fractions,* v *a valuation on* K *which is positive on* A *and* $\mathfrak{p}$ *the set of* $x \in A$ *such that* $v(x) > 0$. *If the prime ideal* $\mathfrak{p}$ *is of height* 1, v *is equivalent to an essential valuation of* A.

Let B be the ring of v and $\mathfrak{m}$ its ideal. Then $\mathfrak{m} \cap A = \mathfrak{p}$ and hence $A_{\mathfrak{p}} \subset B$. Now $A_{\mathfrak{p}}$ is a discrete valuation ring (Theorem 3 and Corollary 1 to Proposition 6). As $\mathfrak{p} \neq (0)$, $B \neq K$ and hence $B = A_{\mathfrak{p}}$ (Chapter VI, § 4, no. 5, Proposition 6).

THEOREM 4. *Let* A *be an integral domain and* M *the set of its prime ideals of height* 1. *For* A *to be a Krull domain, it is necessary and sufficient that the following properties are satisfied:*

(i) *For all* $\mathfrak{p} \in M$, $A_{\mathfrak{p}}$ *is a discrete valuation ring.*

(ii) A *is the intersection of the* $A_{\mathfrak{p}}$ *for* $\mathfrak{p} \in M$.

(iii) *For all* $x \neq 0$ *in* A, *there exists only a finite number of ideals* $\mathfrak{p} \in M$ *such that* $x \in \mathfrak{p}$.

Moreover, the valuations corresponding to the $A_{\mathfrak{p}}$ *for* $\mathfrak{p} \in M$ *are the essential valuations of* A.

The conditions are trivially sufficient. Their necessity follows immediately from Theorem 3 of no. 4, Corollary 1 to Proposition 6 and the fact that the essential valuations of A satisfy the conditions of Definition 3 of no. 3.

PROPOSITION 10. *Let* A *be an integrally closed Noetherian domain and* $\mathfrak{a}$ *an integral ideal of* A. *The following conditions are equivalent:*

(a) $\mathfrak{a}$ *is divisorial;*

(b) *the prime ideals associated with* $A/\mathfrak{a}$ *are of height* 1.

Recall that, if $\mathfrak{a} = \bigcap_{i=1}^{n} \mathfrak{q}_i$ is a reduced primary decomposition of $\mathfrak{a}$ and $\mathfrak{p}_i$ denotes the prime ideal corresponding to $\mathfrak{q}_i$, the prime ideals associated with $A/\mathfrak{a}$ are just the $\mathfrak{p}_i$ (Chapter IV, § 2, no. 3, Proposition 4). The fact that (a) implies (b) then follows from Proposition 8 of no. 4. Conversely, if, in the above notation, the $\mathfrak{p}_i$ are of height 1, $A_{\mathfrak{p}_i}$ is a discrete valuation ring (Theorem 4); now, $\mathfrak{q}_i = \mathfrak{q}_i A_{\mathfrak{p}_i} \cap A$ (Chapter IV, § 2, no. 1, Proposition 3); denoting by v_i the essential valuation corresponding to $\mathfrak{p}_i$, there therefore exists an integer n_i such that $\mathfrak{q}_i$ is the set of $x \in A$ such that $v_i(x) \geqslant n_i$; this shows that the $\mathfrak{q}_i$ are divisorial (no. 4, Proposition 5), hence also is $\mathfrak{a}$.

7. APPLICATION: NEW CHARACTERIZATIONS OF DISCRETE VALUATION RINGS

PROPOSITION 11. *Let* A *be a local Krull domain (in particular an integrally closed local Noetherian domain) and* $\mathfrak{m}$ *its maximal ideal. The following conditions are equivalent:*

(a) A *is discrete valuation ring;*

(b) $\mathfrak{m}$ *is invertible;*

(c) $A : \mathfrak{m} \neq A$;

(d) $\mathfrak{m}$ *is divisorial;*

(e) $\mathfrak{m}$ *is the only non-zero prime ideal of* A.

As every non-zero ideal of a discrete valuation ring is principal (Chapter VI, § 3, no. 6, Proposition 9), it is invertible and hence (a) implies (b). If $\mathfrak{m}$ is invertible, its inverse is $A : \mathfrak{m}$ (Chapter II, § 5, no. 6, Proposition 10) and hence $A : \mathfrak{m} \neq A$; hence (b) implies (c). If $A : \mathfrak{m} \neq A$, then $A : (A : \mathfrak{m}) \neq A$; now $\mathfrak{m} \subset A : (A : \mathfrak{m})$; hence $\mathfrak{m} = A : (A : \mathfrak{m})$ since $\mathfrak{m}$ is maximal, so that $\mathfrak{m}$ is divisorial (no. 1, Proposition 1 (c)); thus (c) implies (d). The fact that (d) implies (e) follows from Theorem 3 of no. 6, Finally, if $\mathfrak{m}$ is the only non-zero prime ideal of A, it is of height 1 and hence $A_{\mathfrak{m}}$ is a discrete valuation ring (no. 6, Theorem 4); as A is local, $A_{\mathfrak{m}} = A$, which shows that (e) implies (a).

8. THE INTEGRAL CLOSURE OF A KRULL DOMAIN IN A FINITE EXTENSION OF ITS FIELD OF FRACTIONS

PROPOSITION 12. *Let* A *be a Krull domain,* K *its field of fractions,* K′ *a finite extension of* K *and* A′ *the integral closure of* A *in* K′. *Then* A′ *is a Krull domain. The essential valuations of* A′ *are the normed discrete valuations on* K′ *which are equivalent to the extensions of the essential valuations of* A.

Let $(v_\iota)_{\iota \in I}$ be the family of extensions to K′ of the essential valuations of A. Since the degree $n = [K' : K]$ is finite, the v_ι are discrete valuations on K′ (Chapter VI, § 8, no. 1, Corollary 3 to Proposition 1). Let B_ι be the ring of v_ι; then $A' \subset \bigcap_{\iota \in I} B_\iota$ (Chapter VI, § 1, no. 3, Theorem 3). Conversely, every element x of $\bigcap_{\iota \in I} B_\iota$ is integral over each of the essential valuation rings of A (Chapter VI, § 1, no. 3, Corollary 3 to Theorem 3); hence the coefficients of the minimal polynomial of x over K belong to A (Chapter V, § 1, no. 3, Corollary to Proposition 11), so that $x \in A'$; thus $A' = \bigcap_{\iota \in I} B_\iota$. Now let x be a non-zero element of A′; it satisfies an equation of the form $x^s + a_{s-1}x^{s-1} + \cdots + a_0 = 0$ where $a_i \in A$ and $a_0 \neq 0$; if $v_\iota(x) > 0$, then $v_\iota(a_0) > 0$; now the essential valuations v of A such that $v(a_0) > 0$ are finite in number and the valuations on K′ extending a given valuation on K are also finite in number (Chapter VI, § 8, no. 3, Theorem 1); hence $v_\iota(x) = 0$ except for a finite number of indices $\iota \in I$. Thus it has been proved that A′ is a Krull domain (no. 3, Definition 3).

It remains to show that the v_ι are equivalent to essential valuations of A' (no. 4, Corollary 2 to Proposition 6), that is (no. 6, Corollary 2 to Theorem 3) that the prime ideal $\mathfrak{p}_\iota$, consisting of the $x \in A'$ such that $v_\iota(x) > 0$, is of height 1. If this were not so, there would exist a prime ideal $\mathfrak{q}$ of A' such that $(0) \subset \mathfrak{q} \subset \mathfrak{p}_\iota$ distinct from (0) and $\mathfrak{p}_\iota$; then $(0) \subset \mathfrak{q} \cap A \subset \mathfrak{p}_\iota \cap A$ and $\mathfrak{q} \cap A$ would be distinct from (0) and $\mathfrak{p}_\iota \cap A$ (Chapter V, § 2, no. 1, Corollary 1 to Proposition 1); the prime ideal $\mathfrak{p}_\iota \cap A$ would therefore not be of height 1, which contradicts the fact that it corresponds to an essential valuation of A.

COROLLARY. *Let* $\mathfrak{p}$ (resp. $\mathfrak{p}'$) *be a prime ideal of* A (resp. A') *of height* 1 *and* v (resp. v') *the essential valuation of* A (resp. A') *corresponding to it. For* $\mathfrak{p}'$ *to lie above* $\mathfrak{p}$, *it is necessary and sufficient that the restriction of* v' *to* K *be equivalent to* v.

The valuation v' is equivalent to the extension of an essential valuation w of A (Proposition 12). Let $\mathfrak{q} = \mathfrak{p}' \cap A$, which is a prime ideal of A of height 1. For the restriction of v' to K to be equivalent to v, it is necessary and sufficient that $w = v$ and hence that $\mathfrak{q} = \mathfrak{p}$.

9. POLYNOMIAL RINGS OVER A KRULL DOMAIN

PROPOSITION 13. *Let* A *be a Krull domain and* $X_1, X_2, \ldots, X_n$ *indeterminates. The ring* $A[X_1, \ldots, X_n]$ *is a Krull domain.*

Arguing by induction on n, it is sufficient to show that, if X is an indeterminate, A[X] is a Krull domain. Let K be the field of fractions of A. The field of fractions of A[X] is K(X). Let I be the set of monic polynomials in K[X] which are irreducible over K; for all $f \in I$, let v_f be the valuation on K(X) defined by f (Chapter VI, § 3, no. 3, *Example* 4). On the other hand, for every essential valuation w of A, let $\bar{w}$ be the extension of w to K(X) defined by

$$\bar{w}\Big(\sum_j a_j X^j\Big) = \inf_j(w(a_j))$$

for $\sum_j a_j X^j \in K[X]$ (Chapter VI, § 10, no. 1, Lemma 1). Clearly the v_f and the $\bar{w}$ are discrete and normed and, for all $u \in K[X]$, $v_f(u) = 0$ (resp. $\bar{w}(u) = 0$) except for a finite number of valuations v_f (resp. $\bar{w}$).

To show the proposition, it therefore suffices to show that A[X] is the intersection of the rings of the valuations v_f and $\bar{w}$. Now the intersection of the rings of the valuations v_f is K[X]. On the other hand, for $\sum_j a_j X^j \in K[X]$, the relation $\bar{w}\Big(\sum_j a_j X^j\Big) \geqslant 0$ is equivalent to "$w(a_j) \geqslant 0$ for all j"; hence the relation "$\bar{w}\Big(\sum_j a_j X^j\Big) \geqslant 0$ for every valuation $\bar{w}$" is equivalent to "$w(a_j) \geqslant 0$ for all j and every essential valuation w of A." This proves our assertion.

Remark. The valuations v_f and $\bar{w}$ introduced in the proof of Proposition 13 are the *essential valuations* of A[X]. It will be sufficient for us to show that, if V is the set of valuations v_f (f irreducible) and $\bar{w}$ (w essential), then, for all $v' \in V$, there exists an element $g \in K(X)$ which is *not in* A[X] and such that $v''(g) \geqslant 0$ for all the valuations $v'' \in V$ distinct from v'; this will prove that $V - \{v'\}$ does not satisfy (AK_{II}) and the conclusion will follow therefore from no. 4, Corollary 2 to Proposition 6. Suppose first that v' is of the form $\bar{w}$: then we may take g to be an element $b \in K$ such that $w(b) < 0$, $w'(b) \geqslant 0$ for the essential valuations w' of A distinct from w, for then $v_f(b) = 0$ for every irreducible monic polynomial f in K[X]; the existence of an element b satisfying the above conditions follows from no. 5, Proposition 9. Suppose secondly that v' is of the form v_f for an irreducible monic polynomial $f \in K[X]$ of degree m; then we may take $g = a/f$ where $a \in A$. For $v_h(g) \geqslant 0$ for every irreducible monic polynomial $h \neq f$ in K[X]; it remains to choose $a \in A$ such that, for every essential valuation w of A, $w(a)$ is at least equal to the greatest lower bound of the elements $w(c_i)$, where the c_i are the coefficients of f ($1 \leqslant i \leqslant m$); now the existence of such an $a \in A$ follows from (AK_{III}) and no. 5, Proposition 9.

We may also say (no. 6, Theorem 4) that the *prime ideals* of A[X] *of height* 1 are:

(1) *the prime ideals of the form* $\mathfrak{p}A[X]$, *where* $\mathfrak{p}$ *is a prime ideal of* A *of height* 1;

(2) *the prime ideals of the form* $\mathfrak{m} \cap A[X]$, *where* $\mathfrak{m}$ *is a* (*necessarily principal*) *prime ideal of* K[X].

The latter are characterized by the fact that their intersection with A *is reduced to* 0.

10. DIVISOR CLASSES IN KRULL DOMAINS

Let A be a Krull domain. Recall that the group D(A) of divisors of A is the free commutative group generated by the set P(A) of its extremal elements (no. 3, Theorem 2) and that P(A) *is identified* with the set of prime ideals of A of height 1 (no. 6); for $\mathfrak{p} \in P(A)$ we shall denote by $v_{\mathfrak{p}}$ the *normed* essential valuation corresponding to $\mathfrak{p}$ (no. 4); recall that the ring of $v_{\mathfrak{p}}$ is $A_{\mathfrak{p}}$ (no. 4, Corollary 1 to Proposition 6). We shall denote by F(A) the subgroup of D(A) consisting of the principal divisors and by $C(A) = D(A)/F(A)$ the *divisor class group* of A (no. 2).

PROPOSITION 14. *Let* A *be a Krull domain and* B *a Krull domain containing* A. *Suppose that the following condition holds:*

(PDE) *For every prime ideal* $\mathfrak{P}$ *of* B *of height* 1, *the prime ideal* $\mathfrak{P} \cap A$ *is zero or of height* 1.

For $\mathfrak{p} \in P(A)$ *the* $\mathfrak{P} \in P(B)$ *such that* $\mathfrak{P} \cap A = \mathfrak{p}$ *are finite in number; we write*

$$i(\mathfrak{p}) = \sum_{\mathfrak{P} \in P(B),\ \mathfrak{P} \cap A = \mathfrak{p}} e(\mathfrak{P}/\mathfrak{p})\mathfrak{P},$$

where $e(\mathfrak{P}/\mathfrak{p})$ *denotes the ramification index of* $v_{\mathfrak{P}}$ *over* $v_{\mathfrak{p}}$ (Chapter VI, § 8, no. 1). *Then i defines, by linearity, an increasing homomorphism of* D(A) *to* D(B), *which enjoys the following properties:*

(a) *for every non-zero element x of the field of fractions of* A,

$$i(\operatorname{div}_A(x)) = \operatorname{div}_B(x);$$

(b) *for all* D, D′ *in* D(A),

$$i(\sup(D, D')) = \sup(i(D), i(D')).$$

Let $\mathfrak{p} \in P(A)$; consider a non-zero element a of $\mathfrak{p}$; the $\mathfrak{P} \in P(B)$ which contain a are finite in number (no. 6, Theorem 4); *a fortiori* the $\mathfrak{P} \in P(B)$ such that $\mathfrak{P} \cap A = \mathfrak{p}$ are finite in number.

We now show (a). By additivity, it may be assumed that $x \in A^* = A - \{0\}$. By definition, $\operatorname{div}_B(x) = \sum_{\mathfrak{P} \in P(B)} v_{\mathfrak{P}}(x).\mathfrak{P}$. For all $\mathfrak{P} \in P(B)$ such that $v_{\mathfrak{P}}(x) > 0$, $\mathfrak{P} \cap A$ is non-zero (for $x \in \mathfrak{P}$) and is therefore of height 1 by (PDE); setting $\mathfrak{p} = \mathfrak{P} \cap A$, by definition of the ramification index, $v_{\mathfrak{P}}(x) = e(\mathfrak{P}/\mathfrak{p})v_{\mathfrak{p}}(x)$ (since $v_{\mathfrak{p}}$ and $v_{\mathfrak{P}}$ are normed). As $\operatorname{div}_A(x) = \sum_{\mathfrak{p} \in P(A)} v_{\mathfrak{p}}(x).\mathfrak{p}$, and $i(\mathfrak{q}) = 0$ for all $\mathfrak{q} \in P(A)$ which is not of the form $\mathfrak{Q} \cap A$ where $\mathfrak{Q} \in P(B)$, we deduce (a).

To prove (b) we write

$$D = \sum_{\mathfrak{p} \in P(A)} n(\mathfrak{p}).\mathfrak{p} \quad \text{and} \quad D' = \sum_{\mathfrak{p} \in P(A)} n'(\mathfrak{p}).\mathfrak{p};$$

the coefficient of $\mathfrak{p}$ in sup(D, D′) is $\sup(n(\mathfrak{p}), n'(\mathfrak{p}))$. Let $\mathfrak{P}$ be an element of P(B). If $\mathfrak{P} \cap A = (0)$, the coefficients of $\mathfrak{P}$ in $i(D)$ and $i(D')$ and hence also in $\sup(i(D), i(D'))$ are zero; therefore the coefficient of $\mathfrak{P}$ in $i(\sup(D, D'))$ is zero. If $\mathfrak{P} \cap A \neq (0)$, it is a prime ideal $\mathfrak{p}$ of height 1 (by (PDE)); writing $e = e(\mathfrak{P}/\mathfrak{p})$, the coefficients of $\mathfrak{P}$ in $i(D)$, $i(D')$ and $i(\sup(D, D'))$ are respectively $en(\mathfrak{p})$, $en'(\mathfrak{p})$ and $e.\sup(n(\mathfrak{p}), n'(\mathfrak{p}))$; that of $\sup(i(D), i(D))$ is

$$\sup(e.n(\mathfrak{p}), e.n'(\mathfrak{p})) = e.\sup(n(\mathfrak{p}), n'(\mathfrak{p})).$$

This proves (b).

Under the hypotheses of Proposition 14, it follows from (a) that i defines, by taking quotients, a *homomorphism* $\bar{i}$, called *canonical*, of C(A) to C(B), which we shall also sometimes write as i, by an abuse of notation.

The condition (PDE) is fulfilled in the two following cases:

(1) B is *integral* over A; in this case, *for the prime ideal* $\mathfrak{P}$ *of* B *to be of height* 1, *it is necessary and sufficient that* $\mathfrak{p} = \mathfrak{P} \cap A$ *be of height* 1. For (0) is the only prime ideal of B lying above the ideal (0) of A (Chapter V, § 2, no. 1, Corollary 1 to Proposition 1); if $\mathfrak{P}$ is of height 1, then $\mathfrak{p} \neq 0$; if $\mathfrak{p}$ were not of height 1, there would exist a prime ideal $\mathfrak{p}'$ of A distinct from (0) and $\mathfrak{p}$ and such that

$0 \subset \mathfrak{p}' \subset \mathfrak{p}$; but then, as B is an integral domain and A is an integrally closed domain, there would be a prime ideal $\mathfrak{P}'$ of B such that $\mathfrak{P}' \cap A = \mathfrak{p}'$ and $\mathfrak{P}' \subset \mathfrak{P}$ (Chapter V, § 2, no. 4, Theorem 3), contrary to the hypothesis. Conversely, if $\mathfrak{p}$ is of height 1, there can exist no prime ideal $\mathfrak{P}'$ of B distinct from 0 and $\mathfrak{P}$ and such that $0 \subset \mathfrak{P}' \subset \mathfrak{P}$, otherwise $0 \subset \mathfrak{P}' \cap A \subset \mathfrak{p}$ and $\mathfrak{P}' \cap A$ would be distinct from 0 and $\mathfrak{p}$ by virtue of Chapter V, § 2, no. 1, Corollary 1 to Proposition 1.

(2) B is a *flat* A-module. More precisely:

PROPOSITION 15. *Let* A *and* B *be Krull domains such that* B *contains* A *and is a flat* A-*module. Then:*

(a) *condition* (PDE) *of Proposition* 14 *is fulfilled;*

(b) *for every divisorial ideal* $\mathfrak{a}$ *of* A, $B\mathfrak{a}$ *is the divisorial ideal of* B *which corresponds to the divisor* $i(\operatorname{div}_A(\mathfrak{a}))$.

To show (a), suppose that there exists a prime ideal $\mathfrak{P}$ of B of height 1 such that $\mathfrak{P} \cap A$ is neither 0 nor of height 1. Take an element $x \neq 0$ in $\mathfrak{P} \cap A$. The ideals $\mathfrak{p}_i$ of A of height 1 which contain x are finite in number and none contains $\mathfrak{P} \cap A$; there therefore exists an element y of $\mathfrak{P} \cap A$ such that $y \notin \mathfrak{p}_i$ for all i (Chapter II, § 2, no. 1, Proposition 2). Thus $\operatorname{div}_A(x)$ and $\operatorname{div}_A(y)$ are relatively prime elements of the ordered group P(A), so that $\sup(\operatorname{div}_A(x), \operatorname{div}_A(y)) = \operatorname{div}_A(x) + \operatorname{div}_A(y) = \operatorname{div}_A(xy)$; as the ideals $Ax \cap Ay$ and Axy are divisorial, we deduce that $Ax \cap Ay = Axy$. Since B is a flat A-module, $Bx \cap By = Bxy$ (Chapter I, § 2, no. 6, Proposition 6). This implies that $\sup(v_{\mathfrak{P}}(x), v_{\mathfrak{P}}(y)) = v_{\mathfrak{P}}(xy) = v_{\mathfrak{P}}(x) + v_{\mathfrak{P}}(y)$, which contradicts the inequalities $v_{\mathfrak{P}}(x) > 0$, $v_{\mathfrak{P}}(y) > 0$ (which hold since x and y are in $\mathfrak{P}$). Thus (a) has been proved by *reductio ad absurdum.*

We now show (b). If a is a divisorial ideal of A, it is the intersection of two fractional principal ideals (no. 5, Corollary 2 to Proposition 9), say

$$a = d^{-1}(Aa \cap Ab)$$

where a, b, d are in A^*; as B is flat over A, $Ba = d^{-1}(Ba \cap Bb)$ (Chapter I, § 2, no. 6, Proposition 6), which shows that $B\mathfrak{a}$ is divisorial. This shows also that $\operatorname{div}_B(B\mathfrak{a}) = \sup(\operatorname{div}_B(a), \operatorname{div}_B(b)) - \operatorname{div}_B(d)$; using Proposition 14 (a) and (b), it is seen that

$$\begin{aligned}\operatorname{div}_B(B\mathfrak{a}) &= \sup(i(\operatorname{div}_A(a)), i(\operatorname{div}_A(b))) - i(\operatorname{div}_A(d))\\ &= i(\sup(\operatorname{div}_A(a), \operatorname{div}_A(b))) - i(\operatorname{div}_A(d))\\ &= i(\operatorname{div}_A(Aa \cap Ab)) - i(\operatorname{div}_A(d))\\ &= i(\operatorname{div}_A(d^{-1}(Aa \cap Ab))) = i(\operatorname{div}_A(\mathfrak{a})).\end{aligned}$$

COROLLARY. *Let* A *be a local Krull domain and* B *a discrete valuation ring such that* B *dominates* A *and is a flat* A-*module. Then* A *is a field or a discrete valuation ring.*

Let $\mathfrak{M}$ be the maximal ideal of B. By (PDE), $\mathfrak{M} \cap A$ is zero or of height 1. As it is, by hypothesis the maximal ideal of A, our assertion follows from Proposition 11 of no. 7.

Remark. In the former of the two above cases, the mapping $i: D(A) \to D(B)$ is *injective*: as the elements of P(B) form a basis of D(B) and two distinct ideals of P(A) cannot be the traces on A of the same ideal of P(B), it amounts to verifying that $i(\mathfrak{p}) \neq 0$ for all $\mathfrak{p} \in P(A)$; now, this follows from Chapter V, § 2, no. 1, Theorem 1. It is similarly seen that, if B is a *faithfully flat* A-module, i is injective (Chapter II, § 2, no. 5, Corollary 4 to Proposition 11).

In what follows, we propose to study the canonical homomorphism $\bar{i}$ from C(A) to C(B) for certain ordered pairs of Krull domains A, B.

PROPOSITION 16. *Let A be a Zariski ring such that its completion $\hat{A}$ is a Krull domain. Then A is a Krull domain and the canonical homomorphism $\bar{i}$ from* C(A) *to* C($\hat{A}$) (*which is defined since $\hat{A}$ is a flat* A-*module;* cf. Chapter III, § 3, no. 4, Theorem 3) *is injective.*

As $\hat{A}$ is an integral domain and $A \subset \hat{A}$, A is an integral domain. Let L be the field of fractions of $\hat{A}$ and $K \subset L$ that of A. As $A = \hat{A} \cap K$ (Chapter III, § 3, no. 5, Corollary 4 to Proposition 9), A is a Krull domain (no. 3, *Example* 4). The fact that $\bar{i}: C(A) \to C(\hat{A})$ is injective follows from Proposition 15 (b) and the fact that, if $\mathfrak{b}\hat{A}$ is principal, $\mathfrak{b}$ is principal (Chapter III, § 3, no. 5, Corollary 3 to Proposition 9).

Now let A be a Krull domain and S a *multiplicative* subset of A not containing 0. The group D(A) (resp. $D(S^{-1}A)$ is the free commutative group with basis the set of $\operatorname{div}(\mathfrak{p})$ (resp. $\operatorname{div}(S^{-1}\mathfrak{p})$), where $\mathfrak{p}$ runs through the set of prime ideals of A of height 1 (resp. the set of prime ideals of A of height 1 such that $\mathfrak{p} \cap S = \varnothing$) (no. 4, Proposition 6) and, if $\mathfrak{p} \cap S = \varnothing$, then $i(\operatorname{div}(\mathfrak{p})) = \operatorname{div}(S^{-1}\mathfrak{p})$. Thus $D(S^{-1}A)$ is identified with the *direct factor* of D(A) generated by the elements $\operatorname{div}(\mathfrak{p})$ such that $\mathfrak{p} \cap S = \varnothing$ and admits as complement the free commutative subgroup of D(A) with basis the set of $\operatorname{div}(\mathfrak{p})$ such that $\mathfrak{p} \cap S \neq \varnothing$; we shall denote this complement by G. As $i: D(A) \to D(S^{-1}A)$ is surjective, so is $\bar{i}$: $C(A) \to C(S^{-1}A)$; and:

$$G/(G \cap F(A)) = (G + F(A))/F(A) = \operatorname{Ker}(\bar{i}); \tag{5}$$

for if an element of $D(S^{-1}A)$ is equal to $\operatorname{div}_{S^{-1}A}(x/s)$, where $x \in A$ and $s \in S$, it is the image under i of the principal divisor $\operatorname{div}_A(x)$ (Proposition 14).

Suppose now that S is generated by a family of elements $(p_\iota)_{\iota \in I}$ of A such that the principal ideals Ap_ι are all *prime*. Then, if $\mathfrak{p}$ is a prime ideal of A of height 1 such that $\mathfrak{p} \cap S \neq \varnothing$, $\mathfrak{p}$ contains a product of powers of the p_ι and therefore one of the p_ι, say p_α; as Ap_α is non-zero and prime and $\mathfrak{p}$ is of height 1, it follows that $\mathfrak{p} = Ap_\alpha$. In the above notation, therefore $G \subset F(A)$ and (5) shows that the kernel of $\bar{i}$ is zero. We have therefore shown the following result:

PROPOSITION 17. *Let* A *be a Krull domain and* S *a multiplicative subset of* A *not containing* 0. *Then the canonical homomorphism* $\bar{i}$ *from* C(A) *to* $C(S^{-1}A)$ *is surjective. If further* S *is generated by a family of elements* p_ι *such that the principal ideals* Ap_ι *are all prime, then* $\bar{i}$ *is bijective.*

As a second application of formula (5), consider the following situation: let R be a Krull domain; take A to be the polynomial ring $A = R[X]$ (no. 9, Proposition 13) and S to be the set $R - (0)$ of non-zero constant polynomials of A. The prime ideals $\mathfrak{p}$ of A of height 1 such that $\mathfrak{p} \cap S \neq \varnothing$ are those of the form $\mathfrak{p}_0 A$, where $\mathfrak{p}_0$ is a prime ideal of R of height 1 (no. 9, *Remark*). Hence, in the notation introduced above, G is identified with D(R) by identifying $\operatorname{div}_A(\mathfrak{p}_0 A)$ with $\operatorname{div}_R(\mathfrak{p}_0)$. On the other hand $G \cap F(A)$ is identified with F(R): for if an ideal $\mathfrak{a}_0$ of R generates a principal ideal $\mathfrak{a}_0 A = f(X)A$ in $A = R[X]$, then $f(0) \in \mathfrak{a}_0 A$ since $\mathfrak{a}_0 A$ is a graded ideal of the ring A (graded by the usual degree of polynomials) and hence $f(0) \in \mathfrak{a}_0$; further, for $a \in \mathfrak{a}_0$, $a = f(X)g(X)$ where $g(X) \in R$, whence, comparing terms of degree 0, $a = f(0)g(0)$; it follows that $\mathfrak{a}_0$ is the principal ideal of R generated by $f(0)$. Finally, denoting by K the field of fractions of R, $S^{-1}A$ is identified with the polynomials ring K[X], which is a principal ideal domain; hence $C(S^{-1}A) = (0)$. Thus, by (5), $C(A) = \operatorname{Ker}(\bar{i})$ is identified with C(R) and we have proved the following result:

PROPOSITION 18. *Let* R *be a Krull domain and* A *the polynomial ring* R[X]. *The canonical homomorphism of* C(R) *to* C(R[X]) *is bijective.*

2. DEDEKIND DOMAINS

1. DEFINITION OF DEDEKIND DOMAINS

Let A be an integral domain. Clearly the following conditions are equivalent:

(a) no two of the non-zero prime ideals of A are comparable with respect to inclusion;

(b) the non-zero prime ideals of A are maximal;

(c) the non-zero prime ideals of A are of height 1.

Definition 1. *A Krull domain all of whose non-zero prime ideals are maximal is called a Dedekind domain.*

Examples of Dedekind domains

(1) Every principal ideal domain is a Dedekind domain.

(2) Let K be a finite extension of **Q** and A the integral closure of **Z** in K. The ring A is a Krull domain (§ 1, no. 8, Proposition 12). Let $\mathfrak{p}$ be a non-zero prime ideal of A. Then $\mathfrak{p} \cap \mathbf{Z}$ is non-zero (Chapter V, § 2, no. 1, Corollary to Proposition 1) and hence is a maximal ideal of **Z**; hence $\mathfrak{p}$ is a maximal ideal of A (*loc. cit.*, Proposition 1). Therefore, A is a Dedekind domain. In general, A is not a principal ideal domain (*Algebra*, Chapter VII, § 1, Exercise 12).

(3) * Let V be an affine algebraic variety and A the ring of functions regular on V. Suppose that A is not a field (i.e. that V is not reduced to a point). For A to be a Dedekind domain, it is necessary and sufficient that V be an irreducible curve with no singular point: for to say that A is an integral domain amounts to saying that V is irreducible; to say that every non-zero prime ideal of A is maximal amounts to saying that A is a curve; finally, as A is Noetherian, to say that it is a Krull domain amounts to saying that it is integrally closed, that is that V is a normal curve, or also that it has no singular point. *

(4) A ring of fractions $S^{-1}A$ of a Dedekind domain A is a Dedekind domain if $0 \notin S$. For $S^{-1}A$ is a Krull domain (§ 1, no. 4, Proposition 6) and every non-zero prime ideal of $S^{-1}A$ is maximal by Chapter II, § 2, no. 5, Proposition 11.

2. CHARACTERIZATIONS OF DEDEKIND DOMAINS

Theorem 1. *Let* A *be an integral domain and* K *its field of fractions. The following conditions are equivalent:*

(a) A *is a Dedekind domain;*

(b) A *is a Krull domain and every non-improper valuation on* K *which is positive on* A *is equivalent to an essential valuation of* A;

(c) A *is a Krull domain and every fractional ideal* $\mathfrak{a} \neq (0)$ *of* A *is divisorial;*

(d) *every fractional ideal* $\mathfrak{a} \neq (0)$ *of* A *is invertible;*

(e) A *is a Noetherian integrally closed domain and every non-zero prime ideal of* A *is maximal;*

(f) A *is Noetherian and, for every maximal ideal* $\mathfrak{m}$ *of* A, $A_{\mathfrak{m}}$ *is either a field or a discrete valuation ring;*

(g) A *is Noetherian and, for every maximal ideal* $\mathfrak{m}$ *of* A, $A_{\mathfrak{m}}$ *is a principal ideal domain.*

We show first the equivalence of (a) and (b). Corollary 2 to Theorem 3, § 1, no. 6, shows immediately that (a) implies (b). Conversely, (b) implies (a), since, for every prime ideal $\mathfrak{p}$ of A, there exists a valuation ring of K which dominates $A_\mathfrak{p}$ (Chapter VI, § 1, no. 2, Corollary to Theorem 2).

The remainder of the proof is carried out by proving the following implications:

$$(a) \Rightarrow (c) \Rightarrow (d) \Rightarrow (e) \Rightarrow (f) \Rightarrow (g) \Rightarrow (a).$$

If A is a Dedekind domain and $\mathfrak{b}$ is a non-zero fractional ideal, then $\mathfrak{b}A_\mathfrak{p} = \tilde{\mathfrak{b}}A_\mathfrak{p}$ for every maximal ideal $\mathfrak{p}$ (§ 1, no. 4, Proposition 7) and hence $\mathfrak{b} = \tilde{\mathfrak{b}}$ (Chapter II, § 3, no. 3, Corollary 3 to Theorem 1); thus (a) implies (c).

We now show that (c) implies (d). If (c) holds, the mapping $\mathfrak{a} \mapsto \operatorname{div} \mathfrak{a}$ is a bijection of I(A) onto D(A) (cf. § 1, no. 1); as it is a homomorphism (§ 1, no. 2) and D(A) is a group, every element of I(A) is invertible.

We show that (d) implies (e). If (d) holds, every integral ideal $\neq (0)$ of A is finitely generated (Chapter II, § 5, no. 6, Theorem 4) and hence A is Noetherian; as I(A) is a group, D(A) is a group and A is therefore completely integrally closed (§ 1, no. 2, Theorem 1). Finally, if $\mathfrak{p}$ is a non-zero prime ideal of A and $\mathfrak{m}$ is a maximal ideal of A containing $\mathfrak{p}$, the ring $A_\mathfrak{m}$ is a principal ideal domain (Chapter II, § 5, no. 6, Theorem 4); as $\mathfrak{p}A_\mathfrak{m}$ is prime and non-zero, necessarily $\mathfrak{p}A_\mathfrak{m} = \mathfrak{m}A_\mathfrak{m}$ (a principal ideal domain being a Dedekind domain) whence $\mathfrak{p} = \mathfrak{m}$ (Chapter II, § 2, no. 5, Proposition 11) and $\mathfrak{p}$ is maximal.

We now show that (e) implies (f). If $\mathfrak{m}$ is a maximal ideal of A and (e) holds, $A_\mathfrak{m}$ is an integrally closed Noetherian domain and its maximal ideal $\mathfrak{m}A_\mathfrak{m}$ is, either (0), or the only non-zero prime ideal of $A_\mathfrak{m}$; hence $A_\mathfrak{m}$ is a field or a discrete valuation ring by Proposition 11 of § 1, no. 7.

The fact that (f) implies (g) is obvious.

We show finally that (g) implies (a). As A is the intersection of the $A_\mathfrak{m}$, where $\mathfrak{m}$ runs through the set of maximal ideals (Chapter II, § 3, no. 3, Corollary 4 to Theorem 1), (g) implies that A is integrally closed and Noetherian and hence that A is a Krull domain (§ 1, no. 3, Corollary to Theorem 2). On the other hand, it can be shown that every non-zero prime ideal of A is maximal as in the proof that (d) $\Rightarrow$ (e).

PROPOSITION 1. *A semi-local Dedekind domain is a principal ideal domain.*

Let A be a semi-local Dedekind domain, K its field of fractions, $\mathfrak{p}_1, \ldots, \mathfrak{p}_n$ its maximal ideals and $v_1, \ldots, v_n$ the corresponding essential valuations; these are the only essential valuations of A. Let $\mathfrak{a}$ be a non-zero integral ideal of A. Since it is divisorial, there exists (§ 1, no. 4, Proposition 5) integers $q_1, \ldots, q_n$ such that $\mathfrak{a}$ is the set of $x \in K$ such that $v_i(x) \geqslant q_i$ for $1 \leqslant i \leqslant n$. Let x_0 be an element of K such that $v_i(x_0) = q_i$ for $1 \leqslant i \leqslant n$ (Chapter VI, § 7, no. 2, Corollary 1 to Theorem 1). Then $\mathfrak{a}$ is the set of $x \in K$ such that $v_i(xx_0^{-1}) \geqslant 0$ for $1 \leqslant i \leqslant n$. Thus $\mathfrak{a} = Ax_0$.

If A is a Dedekind domain, it has been seen, in the proof of Theorem 1, that the group D(A) of divisors of A is identified with the *group* I(A) *of fractional ideals* $\mathfrak{a} \neq (0)$ (as A is Noetherian, every non-zero fractional ideal is finitely generated). The *divisor class* group C(A) of A (§ 1, no. 2) is then identified with the group of *classes of ideals* $\neq 0$ of A (defined in Chapter II, § 5, no. 7).

3. DECOMPOSITION OF IDEALS INTO PRODUCTS OF PRIME IDEALS

Let A be a Dedekind domain, I(A) the ordered multiplicative group of non-zero fractional ideals of A and D(A) the group of divisors of A. The isomorphism $\mathfrak{a} \mapsto \operatorname{div} \mathfrak{a}$ of I(A) onto D(A) maps the extremal divisors to the non-zero prime ideals of A (§ 1, no. 6, Theorem 3) and hence the multiplicative group I(A) admits as basis the set of non-zero prime ideals of A (§ 1, no. 3, Theorem 2). In other words, *every non-zero fractional ideal* $\mathfrak{a}$ *of* A *admits a unique decomposition of the form*:

$$\mathfrak{a} = \prod \mathfrak{p}^{n(\mathfrak{p})} \tag{1}$$

where the product extends to the non-zero prime ideals of A, the exponents $n(\mathfrak{p})$ being zero except for a finite number of them. Further $\mathfrak{a}$ is integral if and only if the $n(\mathfrak{p})$ are all positive. The relation (1) is called the *decomposition of* $\mathfrak{a}$ *into prime factors*. In particular, if $\mathfrak{a}$ is a principal ideal Ax, then, for all $\mathfrak{p}$, $n(\mathfrak{p}) = v_{\mathfrak{p}}(x)$, where $v_{\mathfrak{p}}$ denotes the essential valuation corresponding to $\mathfrak{p}$; this follows from formula (4) of § 1, no. 3. Let

$$\mathfrak{a} = \prod \mathfrak{p}^{m(\mathfrak{p})}, \qquad \mathfrak{b} = \prod \mathfrak{p}^{n(\mathfrak{p})}$$

be two non-zero fractional ideals of A. Then

$$\mathfrak{a}\mathfrak{b} = \prod \mathfrak{p}^{m(\mathfrak{p})+n(\mathfrak{p})} \tag{2}$$

$$\mathfrak{a} : \mathfrak{b} = \mathfrak{a}\mathfrak{b}^{-1} = \prod \mathfrak{p}^{m(\mathfrak{p})-n(\mathfrak{p})} \tag{3}$$

$$\mathfrak{a} + \mathfrak{b} = \prod \mathfrak{p}^{\inf(m(\mathfrak{p}),n(\mathfrak{p}))} \tag{4}$$

$$\mathfrak{a} \cap \mathfrak{b} = \prod \mathfrak{p}^{\sup(m(\mathfrak{p}),n(\mathfrak{p}))} \tag{5}$$

Relation (2) is obvious; relation (3) follows from it, the equation $\mathfrak{a} : \mathfrak{b} = \mathfrak{a}\mathfrak{b}^{-1}$ following from the equation

$$\operatorname{div}(\mathfrak{a} : \mathfrak{b}) = \operatorname{div} \mathfrak{a} - \operatorname{div} \mathfrak{b}$$

(§ 1, no. 2, Corollary to Theorem 1); formulae (4) and (5) follow from Proposition 2, § 1, no. 1.

These results apply in particular to the integral closure of **Z** in a finite extension of **Q**.

If A is a principal ideal domain, the above results again give those of *Algebra*, Chapter VII, § 1, no. 3.

4. THE APPROXIMATION THEOREM FOR DEDEKIND DOMAINS

In Dedekind domains there is an "approximation theorem" which strengthens both Theorem 1 of Chapter VI, § 7, no. 2 and Proposition 9 of § 1, no. 5:

PROPOSITION 2. *Let* A *be a Dedekind domain,* K *its field of fractions and* P *the set of non-zero prime ideals of* A*; for* $\mathfrak{p} \in P$ *let* $v_{\mathfrak{p}}$ *denote the corresponding essential valuation of* A. *Let* $\mathfrak{p}_1, \ldots, \mathfrak{p}_q$ *be distinct elements of* P *and* $n_1, \ldots, n_q$ *rational integers and* $x_1, \ldots, x_q$ *elements of* K. *Then there exists* $x \in K$ *such that* $v_{\mathfrak{p}_i}(x - x_i) \geqslant n_i$ *for* $1 \leqslant i \leqslant q$ *and* $v_{\mathfrak{p}}(x) \geqslant 0$ *for all* $\mathfrak{p} \in P$ *distinct from the* $\mathfrak{p}_i$.

Replacing if need be the n_i by greater integers, they may be assumed all to be positive. We examine first the case where the x_i are in A; it obviously amounts to finding an $x \in A$ satisfying the congruences

$$x \equiv x_i \pmod{\mathfrak{p}_i^{n_i}}$$

and the existence of x then follows from Chapter II, § 1, no. 2, Proposition 5.

We pass now to the general case. We may write $x_i = s^{-1}y_i$ where s, y_i are in A; writing $x = s^{-1}y$, it amounts to finding a $y \in A$ such that, on the one hand, $v_{\mathfrak{p}_i}(y - y_i) \geqslant n_i + v_{\mathfrak{p}_i}(s)$ and, on the other, $v_{\mathfrak{p}}(y) \geqslant v_{\mathfrak{p}}(s)$ for all $\mathfrak{p} \in P$ distinct from the $\mathfrak{p}_i$; as $v_{\mathfrak{p}}(s) = 0$ except for a finite number of indices $\mathfrak{p}$, it is thus reduced to the above case; whence the proposition.

Proposition 2 may be interpreted as a *density* theorem. To be precise, for all $\mathfrak{p} \in P$, let $\hat{K}_{\mathfrak{p}}$ (resp. $\hat{A}_{\mathfrak{p}}$) be the completion of K (resp. A) with respect to the discrete valuation $v_{\mathfrak{p}}$ and consider the product $\prod_{\mathfrak{p} \in P} \hat{K}_{\mathfrak{p}}$; an element $\boldsymbol{x} = (x_{\mathfrak{p}})$ of this product is called a *restricted adèle* of A if $x_{\mathfrak{p}} \in \hat{A}_{\mathfrak{p}}$ for all $\mathfrak{p} \in P$ with the exception of a finite number of them. Clearly the set $\boldsymbol{A}$ of restricted adèles is a *subring* of $\prod_{\mathfrak{p} \in P} \hat{K}_{\mathfrak{p}}$, which contains the product ring $\boldsymbol{A}_0 = \prod_{\mathfrak{p} \in P} \hat{A}_{\mathfrak{p}}$. Consider on $\boldsymbol{A}_0$ the product topology, with respect to which $\boldsymbol{A}_0$ is *complete*; there is on $\boldsymbol{A}$ a unique topology $\mathcal{T}$ which is compatible with its additive group structure and for which the neighbourhoods of 0 *in* $\boldsymbol{A}_0$ form a fundamental system $\mathfrak{S}$ of neighbourhoods of 0. The topology $\mathcal{T}$ is compatible with the *ring* structure on $\boldsymbol{A}$; for clearly axiom (AV_{II}) of *General Topology*, Chapter III, § 6, no. 3, holds, the topology induced by $\mathcal{T}$ on $\boldsymbol{A}_0$ being compatible with the ring structure on $\boldsymbol{A}_0$. On the other hand, for all $\boldsymbol{x} \in \boldsymbol{A}$ there exists a finite subset J of P such that, if we write $J' = P - J$, $\boldsymbol{K}_J \leqslant \prod_{\mathfrak{p} \in J} \hat{K}_{\mathfrak{p}}$, $\boldsymbol{A}_{J'} = \prod_{\mathfrak{p} \in J'} \hat{A}_{\mathfrak{p}}$, then

$\boldsymbol{x} \in K_J \times A_{J'}$ and, as $\hat{A}_\mathfrak{p}$ is open in $\hat{K}_\mathfrak{p}$ for all $\mathfrak{p}$, $\mathfrak{S}$ is a fundamental system of neighbourhoods of 0 for the product topology on $K_J \times A_{J'}$; since the latter is compatible with the ring structure on this product, axiom (AV_I) of *General Topology*, Chapter III, *loc. cit.* is also seen to hold, which proves our assertion. Clearly A_0 is an *open* subring of A and hence A is also a *complete* ring (*General Topology*, Chapter III, § 3, no. 3, Proposition 4).

For all $x \in K$, let $\Delta(x)$ be the element $(x_\mathfrak{p}) \in \prod_{\mathfrak{p} \in P} \hat{K}_\mathfrak{p}$ such that $x_\mathfrak{p} = x$ for all $\mathfrak{p} \in P$; as $x_\mathfrak{p} \in \hat{A}_\mathfrak{p}$ except for a finite number of values of $\mathfrak{p}$, $\Delta(x) \in A$; hence we have thus defined a homomorphism $\Delta: K \to A$ which is *injective* if $P \neq \varnothing$ (that is if A is not a field); the elements of $\Delta(K)$ are called *principal restricted adèles* and clearly $\Delta(A) \subset A_0$.

PROPOSITION 3. *The ring* A_0 (resp. A) *is identified with the completion of* A (resp. K) *with respect to the ring topology for which a fundamental system of neighbourhoods of* 0 *consists of all the integral ideals* $\neq (0)$ *of* A.

It is immediate that the topology considered on A (or K) is Hausdorff. Taking account of no. 3, the assertion concerning A_0 follows from Chapter III, § 2, no. 13, Proposition 17. This shows therefore that $\Delta(A)$ is dense in A_0; to see similarly that $\Delta(K)$ is dense in A, note that for all $\boldsymbol{x} = (x_\mathfrak{p}) \in A$ there is only a finite number of $\mathfrak{p} \in P$ such that $v_\mathfrak{p}(x_\mathfrak{p}) < 0$; by § 1, no. 5, Proposition 9 there is therefore an $s \in K$ such that $sx_\mathfrak{p} \in \hat{A}_\mathfrak{p}$ for all $\mathfrak{p} \in P$, in other words $\Delta(s)\boldsymbol{x} \in A_0$ and, as multiplication by $\Delta(s)$ is a homomorphism from A to itself, it suffices to apply the fact that $\Delta(A)$ is dense in A_0 to deduce that $\Delta(K)$ is dense in A.

We could of course also prove that $\Delta(K)$ is dense in A by using Proposition 2.

Consider now the multiplicative group $\mathbf{SL}(n, A)$ consisting of the matrices $U \in \mathbf{M}_n(A)$ such that $\det(U) = 1$; if $\mathbf{M}_n(A) = A^{n^2}$ is given the product topology, it induces on $\mathbf{SL}(n, A)$ a topology *compatible with the group structure* on $\mathbf{SL}(n, A)$. It suffices to verify that the mapping $U \mapsto U^{-1}$ is continuous on $\mathbf{SL}(n, A)$; but as U is unimodular, it is known (*Algebra*, Chapter III, § 6, no. 5, formula (17)) that the elements of U^{-1} are *minors* of U and hence polynomials in the elements of U, which proves our assertion. If K is identified with a subring of A by means of Δ, the group $\mathbf{SL}(n, K)$ is a subgroup of $\mathbf{SL}(n, A)$.

PROPOSITION 4. *The group* $\mathbf{SL}(n, K)$ *is dense in* $\mathbf{SL}(n, A)$.

Let G be the closure of $\mathbf{SL}(n, K)$ in $\mathbf{SL}(n, A)$; as K is dense in A (Proposition 3), G contains all the matrices of the form $I + \boldsymbol{a}.E_{ij}$ for $i \neq j$ and $\boldsymbol{a} \in A$. For all $\mathfrak{p} \in P$ and all $\lambda \in \hat{K}_\mathfrak{p}$, let $\lambda(\mathfrak{p})$ be the restricted adèle $\boldsymbol{x} = (x_\mathfrak{q})_{\mathfrak{q} \in P}$ such that $x_\mathfrak{p} = \lambda$ and $x_\mathfrak{q} = 0$ for $\mathfrak{q} \neq \mathfrak{p}$; the above shows that G contains the matrices $I + \lambda(\mathfrak{p})E_{ij}$ for $i \neq j$. But we know that the matrices of the form $I + \lambda E_{ij}$ for $\lambda \in \hat{K}_\mathfrak{p}$ generate the group $\mathbf{SL}(n, \hat{K}_\mathfrak{p})$ (*Algebra*, Chapter III). For every matrix $U \in \mathbf{SL}(n, A)$ let $U_{\mathfrak{p}}$ denote the canonical image of U in $\mathbf{SL}(n, \hat{K}_{\mathfrak{p}})$; it is therefore seen that, for all $\mathfrak{p} \in P$, G contains the matrices $U \in \mathbf{SL}(n, A)$ such that $U_{\mathfrak{q}} = I$ for all $\mathfrak{q} \neq \mathfrak{p}$. Since G is a group, it also contains all the matrices $U \in \mathbf{SL}(n, A)$ such that $U_{\mathfrak{p}} = I$ except for a *finite* number of $\mathfrak{p} \in P$; now, the definition of the topology on A shows immediately that the set of these matrices is dense in $\mathbf{SL}(n, A)$.

5. THE KRULL-AKIZUKI THEOREM

LEMMA 1. *Let* A *be a Noetherian domain in which every non-zero prime ideal is maximal and* M *a finitely generated torsion* A*-module. Then the length* $\operatorname{long}_A(M)$ *of* M *is finite.*

As M is a torsion module, every prime ideal associated with M is $\neq (0)$ and therefore maximal. The lemma then follows from Chapter IV, § 2, no. 5, Proposition 7.

LEMMA 2. *Let* A *be a ring,* T *an* A*-module and* (T_ι) *a right directed family of submodules of* T *with union* T. *Then* $\operatorname{long}_A(T) = \sup_\iota(\operatorname{long}_A(T_\iota))$.

$\operatorname{long}_A(T_\iota) \leqslant \operatorname{long}_A(T)$ for all ι. The lemma is obvious if no integer exceeds the $\operatorname{long}_A(T_\iota)$, both sides then being infinite. Otherwise, let ι_0 be an index for which $\operatorname{long}_A(T_\iota)$ takes its greatest value; then $T_{\iota_0} = T$ since the family (T_ι) is directed; whence our assertion in this case.

Remark. This proof does not assume that A is commutative.

LEMMA 3. *Let* A *be a Noetherian domain such that every non-zero prime ideal of* A *is maximal,* M *a torsion-free* A*-module of finite rank* r *and a non-zero element of* A. *Then* A/Aa *is an* A*-module of finite length and:*

$$\operatorname{long}_A(M/aM) \leqslant r.\operatorname{long}_A(A/Aa). \tag{6}$$

Lemma 1 shows that $\operatorname{long}_A(A/Aa)$ is finite. We show (6) first in the case where M is *finitely generated.* As M is torsion-free and of rank r, there exists a submodule L of M which is isomorphic to A^r and such that $Q = M/L$ is a finitely generated torsion A-module and hence of finite length (Lemma 1). For every integer $n \geqslant 1$, the kernel of the canonical surjection $M/a^nM \to Q/a^nQ$ is equal to $(L + a^nM)/a^nM$ and isomorphic to $L/(a^nM \cap L)$; as

$$a^nL \subset a^nM \cap L,$$

therefore

$$\operatorname{long}_A(M/a^nM) \leqslant \operatorname{long}_A(L/a^nL) + \operatorname{long}_A(Q/a^nQ) \leqslant \operatorname{long}_A(L/a^nL) + \operatorname{long}_A(Q). \tag{7}$$

Now, since M is torsion-free, multiplication by a defines an isomorphism of

M/aM onto aA/a^2M; similarly for L; whence, by induction on n, the formulae:

$$(8)\qquad \begin{aligned} \operatorname{long}_A(M/a^nM) &= n.\operatorname{long}_A(M/aM).\\ \operatorname{long}_A(L/a^nL) &= n.\operatorname{long}_A(L/aL).\end{aligned}$$

Taking account of (7) we deduce:

$$(9)\qquad \operatorname{long}_A(M/aM) \leqslant \operatorname{long}_A(L/aL) + n^{-1}\operatorname{long}_A(Q)$$

for all $n > 0$; as L is isomorphic to A^r, $\operatorname{long}_A(L/aL) = r \operatorname{long}_A(A/Aa)$; whence (6) by letting n tend to infinity in (9).

We now pass to the general case. Let (M_ι) be the family of finitely generated submodules of M. The module $T = M/aM$ is the union of the submodules $T_\iota = (M_\iota + aM)/aM = M_\iota/(M_\iota \cap aM)$. Now, T_ι is isomorphic to a quotient of M_ι/aM_ι and hence

$$\operatorname{long}_A(T_\iota) \leqslant r \operatorname{long}_A(A/Aa)$$

by what we have just proved. Whence

$$\operatorname{long}_A(T) \leqslant r \operatorname{long}_A(A/Aa)$$

by Lemma 2.

PROPOSITION 5 (Krull-Akizuki). *Let* A *be a Noetherian domain each of whose non-zero prime ideals is maximal,* K *its field of fractions,* L *a finite extension of* K *and* B *a subring of* L *containing* A. *Then* B *is Noetherian and every non-zero prime ideal of* B *is maximal. Moreover, for every ideal* $\mathfrak{b} \neq (0)$ *of* B, $B/\mathfrak{b}$ *is a finitely generated* A*-module.*

Let $\mathfrak{b}$ be a non-zero ideal of B. We shall show that $B/\mathfrak{b}$ is an A-module of finite length (hence, *a fortiori*, a B-module of finite length) and that $\mathfrak{b}$ is a finitely generated B-module.

A non-zero element y of $\mathfrak{b}$ satisfies an equation of the form:

$$a_r y^r + a_{r-1}y^{r-1} + \cdots + a_0 = 0 \qquad (a_i \in A,\ a_0 \neq 0).$$

This equation shows that $a_0 \in By \subset \mathfrak{b}$. Applying Lemma 3 to M = B, it is seen that B/a_0B is an A-module of finite length; so is $B/\mathfrak{b}$ which is a quotient module of it. Further the B-module $\mathfrak{b}$ contains, as a submodule, a_0B which is finitely generated; as $\mathfrak{b}/a_0B$ is of finite length (as a submodule of B/a_0B) and hence finitely generated, $\mathfrak{b}$ is certainly a finitely generated B-module.

The above shows first that B is Noetherian. On the other hand, if $\mathfrak{p}$ is a non-zero prime ideal of B, the ring $B/\mathfrak{p}$ is an integral domain and of finite length and hence is a field (*Algebra*, Chapter VIII, § 6, no. 4, Proposition 9), so that $\mathfrak{p}$ is maximal.

COROLLARY 1. *For every prime ideal* $\mathfrak{p}$ *of* A, *the set of prime ideals of* B *lying above* $\mathfrak{p}$ *is finite.*

Suppose first that $\mathfrak{p} = (0)$; then the only prime ideal $\mathfrak{q}$ of B such that $\mathfrak{q} \cap A = (0)$ is (0); otherwise, writing $S = A - \{0\}$, $S^{-1}\mathfrak{q}$ would be a non-zero prime ideal of $S^{-1}B$ (Chapter II, § 2, no. 5, Proposition 11) and $S^{-1}B$ is just the field of fractions of B, for it is a subring of L containing K (*Algebra*, Chapter V, § 3, no. 2, Proposition 3); whence an absurd conclusion. If now $\mathfrak{p} \neq (0)$, it follows from Proposition 5 that $B/\mathfrak{p}B$ is a finite-dimensional vector space over the field $A/\mathfrak{p}$, hence an *Artinian* ring and therefore has only a finite number of prime ideals (Chapter IV, § 2, no. 5, Proposition 9), which proves that there is only a finite number of prime ideals of B containing $\mathfrak{p}$.

COROLLARY 2. *The integral closure of* A *in* L *is a Dedekind domain.*

This integral closure is an integrally closed Noetherian domain all of whose non-zero prime ideals are maximal; it suffices therefore to apply Theorem 1 of no. 2.

In particular:

COROLLARY 3. *The integral closure of a Dedekind domain in a finite extension of its field of fractions is a Dedekind domain.*

PROPOSITION 6. *Let* A *be a Dedekind domain,* K *its field of fractions,* L *a finite extension of* K *and* B *the integral closure of* A *in* L. *Let* $\mathfrak{p}$ *be a non-zero prime ideal of* A, v *the corresponding essential valuation of* K *and*

$$B\mathfrak{p} = \prod_i \mathfrak{p}_i^{e(i)}$$

the decomposition of the ideal $B\mathfrak{p}$ *as a product of prime ideals. Then:*

(a) *the prime ideals of* B *lying above* $\mathfrak{p}$ *are the* $\mathfrak{p}_i$ *such that* $e(i) > 0$;

(b) *the valuations* v_i *on* L *corresponding to these ideals* $\mathfrak{p}_i$ *are, up to equivalence, the valuations on* L *extending* v;

(c) $[B/\mathfrak{p}_i : A/\mathfrak{p}] = f(v_i/v)$;

(d) $e_i = e(v_i/v)$ (cf. Chapter VI, § 8, no. 1, Definitions 1 and 2).

(a) To say that a prime ideal $\mathfrak{q}$ of B lies above $\mathfrak{p}$ amounts to saying that $\mathfrak{q} \supset \mathfrak{p}$, hence that $\mathfrak{q} \supset B\mathfrak{p}$ and that $\mathfrak{q}$ contains one of the $\mathfrak{p}_i$ such that $e(i) > 0$ (Chapter II, § 1, no. 1, Proposition 1).

(b) This follows, taking account of (a), from § 1, no. 8, Corollary to Proposition 12.

(c) The residue field of v is identified with $A/\mathfrak{p}$ and that of v_i with $B/\mathfrak{p}_i$ (§ 1, no. 4, Corollary 1 to Proposition 6).

(d) Let a (resp. a_i) be a uniformizer for v (resp. v_i). Then

$$aB_{\mathfrak{p}_i} = aA_{\mathfrak{p}}B_{\mathfrak{p}_i} = \mathfrak{p}A_{\mathfrak{p}}B_{\mathfrak{p}_i} = \mathfrak{p}B.B_{\mathfrak{p}_i} = \Big(\prod_j \mathfrak{p}_j^{e(j)}\Big)B_{\mathfrak{p}_i} = \prod_j (\mathfrak{p}_jB_{\mathfrak{p}_i})^{e(j)}$$
$$= (\mathfrak{p}_iB_{\mathfrak{p}_i})^{e(i)} = a_i^{e(i)}B_{\mathfrak{p}_i}$$

since $\mathfrak{p}_jB_{\mathfrak{p}_i} = B_{\mathfrak{p}_i}$ for $j \neq i$; whence (d), since $e(v_i/v) = v_i(a)$.

3. FACTORIAL DOMAINS

1. DEFINITION OF FACTORIAL DOMAINS

Definition 1. *A Krull domain all of whose divisorial ideals are principal is called a factorial* (or *unique factorization*) *domain.*

In other words, the group of divisor classes (§ 1, no. 2) is *reduced to* 0.

Examples

(1) Every principal ideal domain is factorial (and, recall, is a Dedekind domain). Conversely, every factorial Dedekind domain is a principal ideal domain by § 2, no. 2, Theorem 1 (c).

(2) In particular, if K is a field, the rings K[X] and K[[X]] are factorial domains (see Theorem 2 and Proposition 8 below for generalizations).

(3) * The local ring of a simple point of an algebraic variety is a factorial domain. The ring of germs of functions analytic at the origin of $\mathbf{C}^n$ is a factorial domain. *

2. CHARACTERIZATIONS OF FACTORIAL DOMAINS

Given a ring A, we need to consider the following condition:

(M) *Every non-empty family of integral principal ideals of* A *has a maximal element.*

Theorem 1. *Let* A *be an integral domain. The following conditions are equivalent:*

(a) A *is factorial;*

(b) *the ordered group of non-zero fractional principal ideals of* A *is a direct sum of groups isomorphic to* $\mathbf{Z}$ *(ordered by the product order);*

(c) *condition* (M) *is satisfied and the intersection of two principal ideals of* A *is a principal ideal;*

(d) *condition* (M) *is satisfied and, for every extremal element* p *of* A, *the ideal* Ap *is prime;*

(e) A *is a Krull domain and every prime ideal of height* 1 *is principal.*

We shall denote by K the field of fractions of A and by $\mathscr{P}^*$ (or $\mathscr{P}^*(A)$) the ordered group of non-zero fractional principal ideals of A. The proof will be carried out by proving the following implications:

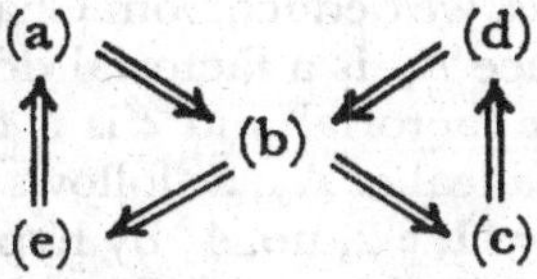

We show that (a) implies (b); if A is factorial, $\mathscr{P}^*$ is isomorphic to the group of divisors of A and hence to a direct sum of groups $\mathbf{Z}$ (§ 1, no. 3, Theorem 2).

Note now that the relation "the intersection of two integral principal ideals of A is a principal ideal" means that every ordered pair of elements of A admits a *lcm*, that is that $\mathscr{P}^*$ is a lattice-ordered group (*Algebra*, Chapter VI, § 1, no. 9, Proposition 8). The fact that (b) implies (c) (and even is equivalent to it) therefore follows from *Algebra*, Chapter VI, § 1, no. 13, Theorem 2. The fact that (c) implies (d) follows from *Algebra*, Chapter VI, § 1, no. 13, Proposition 14 (DIV).

The fact that (d) implies (b) follows from *Algebra*, Chapter VI, § 1, no. 13, Theorem 2 applied to the group $\mathscr{P}^*$.

We show that (b) implies (e). If (b) holds, there is an isomorphism of $\mathscr{P}^*$ onto $\mathbf{Z}^{(I)}$; let $(v_\iota(x))_{\iota \in I}$ denote the element of $\mathbf{Z}^{(I)}$ corresponding to the ideal Ax ($x \in K^*$). It is seen immediately that each v_ι is a discrete valuation on K, that A is the intersection of the rings of the v_ι and that, for $x \in K^*$, $v_\iota(x) = 0$ except for a finite number of indices ι; hence A is a Krull domain. On the other hand, let $\mathfrak{q}$ be a prime ideal of A of height 1; it contains a non-zero element a which is necessarily not invertible and hence also (by definition of a prime ideal) one of the extremal elements p of A; as Ap is prime and non-zero, $\mathfrak{q} = Ap$, which proves that $\mathfrak{q}$ is principal.

Finally we show that (e) implies (a). Let $\mathfrak{a}$ be a divisorial ideal of A. There exist prime ideals $\mathfrak{p}_i$ of A of height 1 such that $\operatorname{div} \mathfrak{a} = \sum_i n_i \operatorname{div} \mathfrak{p}_i$ where $n_i \in \mathbf{Z}$. If (e) holds, $\mathfrak{p}_i$ is of the form Ap_i, whence $\operatorname{div} \mathfrak{a} = \operatorname{div}\left(\prod_i Ap_i^{n_i}\right)$ and hence $\mathfrak{a} = \prod_i Ap_i^{n_i}$ since $\mathfrak{a}$ is divisorial.

Proposition 1. *Let* A *be a Krull domain. If every divisorial ideal of* A *is invertible, then, for every maximal ideal* $\mathfrak{m}$ *of* A, $A_\mathfrak{m}$ *is factorial. The converse is true if it is also assumed that every divisorial ideal of* A *is finitely generated* (in particular if A is *Noetherian*).

Suppose that every divisorial ideal of A is invertible; as $A_\mathfrak{m}$ is a Krull domain (§ 1, no. 4, Proposition 6), every divisorial ideal $\mathfrak{a}$ of $A_\mathfrak{m}$ is the intersection of two principal fractional ideals (§ 1, no. 5, Corollary 2 to Proposition 9); hence $\mathfrak{a} = \mathfrak{b}A_\mathfrak{m}$, where $\mathfrak{b}$ is a divisorial ideal of A (Chapter II, § 2, no. 4); as $\mathfrak{b}$ is invertible by hypothesis, we deduce from Chapter II, § 5, no. 6, Theorem 4 that $\mathfrak{a}$ is principal and hence $A_\mathfrak{m}$ is a factorial domain (no. 1, Definition 1). Conversely, if all the $A_\mathfrak{m}$ are factorial and $\mathfrak{c}$ is a finitely generated divisorial ideal of A, $\mathfrak{c}A_\mathfrak{m}$ is a divisorial ideal of $A_\mathfrak{m}$, as follows from § 1, no. 5, Corollary 2 to Proposition 9 and Chapter II, § 2, no. 4; by hypothesis $\mathfrak{c}A_\mathfrak{m}$ is principal and hence it follows from Chapter II, § 5, no. 6, Theorem 4 that $\mathfrak{c}$ is invertible.

3. DECOMPOSITION INTO EXTREMAL ELEMENTS

Let A be an integral domain, K its field of fractions and U the multiplicative group of invertible elements of A. Recall (*Algebra*, Chapter VI, § 1, no. 5) that there is a canonical isomorphism of K^*/U onto the group $\mathscr{P}^*$ of non-zero fractional principal ideals of A. Condition (b) of Theorem 1 may then be translated as follows:

PROPOSITION 2. *Let* A *be an integral domain. For* A *to be factorial, it is necessary and sufficient that there exist a subset* P *of* A *such that every* $a \in A - \{0\}$ *may be written uniquely in the form* $a = u \prod_{p \in P} p^{n(p)}$, *where* $u \in U$ *and the* $n(p)$ *are positive integers which are zero except for a finite number of them.*

If P satisfies this condition, clearly all its elements are *extremal* and every extremal element of A is associated with a unique element of P. Recall that P is then called a *representative system of extremal elements* of A (*Algebra*, Chapter VII, § 1, no. 3, Definition 2).

Suppose always that A is factorial. It has been seen (no. 2, Theorem 1) that the group $\mathscr{P}^*$ is a lattice. We may therefore apply the results of *Algebra*, Chapter VI, § 1, nos. 9 and 13. In particular, every element of K^* may be written in an essentially unique way in the form of an *irreducible fraction*. Any two elements a, b of K^* have a g.c.d. and a l.c.m; if $a = u \prod_{p \in P} p^{n(p)}$ and

$$b = u' \prod_{p \in P} p^{m(p)}$$

are decompositions of a and b as products of extremal elements, then:

$$\text{g.c.d.}(a, b) = w \prod_{p \in P} p^{\inf(m(p), n(p))} \tag{1}$$

$$\text{l.c.m.}(a, b) = w' \prod_{p \in P} p^{\sup(m(p), n(p))} \tag{2}$$

where w, w' are in U. We recover, in particular, the results of *Algebra*, Chapter VII, § 1, no. 3.

For all $p \in P$, the mapping $a \mapsto n(p)$ is a discrete valuation v_p on K whose ring is obviously A_{Ap}. It follows from Theorem 1(e) that the v_p are just the essential valuations of A and that the ideals Ap ($p \in P$) are just the prime ideals of A of height 1.

4. RINGS OF FRACTIONS OF A FACTORIAL DOMAIN

PROPOSITION 3. *Let* A *be a Krull domain and* S *a multiplicative subset of* A *not containing* 0.

(i) *If* A *is factorial,* $S^{-1}A$ *is factorial.*

(ii) *If* S *is generated by a family of elements* p_ι *such that the principal ideals* Ap_ι *are prime and* $S^{-1}A$ *is factorial, then* A *is factorial.*

This follows immediately from Definition 1 of no. 1 and § 1, no. 10, Proposition 17.

5. POLYNOMIAL RINGS OVER A FACTORIAL DOMAIN

Let A be a factorial domain, K its field of fractions and f a non-zero element of K[X]; an element c of K^* will be called a *content* of f if it is a g.c.d. of the coefficients of f. Let v be a valuation on K which is essential for A and $\bar{v}$ its canonical extension to K[X] (defined by $\bar{v}\left(\sum_i a_i X^i\right) = \inf v(a_i)$; cf. Chapter VI, § 10, no. 1, Proposition 2); then $\bar{v}(f) = v(c)$.

LEMMA 1 (Gauss). *Let* f, f' *be non-zero elements of* K[X] *and* c, c' *contents of* f, f'. *Then* cc' *is a content of* ff'.

Let d be a content of ff'. For every valuation v on K which is essential for A, let $\bar{v}$ denote its canonical extension to K[X]. Then

$$v(d) = \bar{v}(ff') = \bar{v}(f) + \bar{v}(f') = v(c) + v(c') = v(cc').$$

Hence $cc'd^{-1}$ is an invertible element of A.

THEOREM 2. *Let* A *be a factorial domain,* K *its field of fractions,* (p_ι) *a representative system of extremal elements of* A *and* (P_λ) *a representative system of irreducible polynomials of* K[X], *each* P_λ *having* 1 *as a content. Then:*

(i) A[X] *is a factorial domain;*

(ii) *the set of* p_ι *and* P_λ *is a representative system of extremal elements of* A[X].

Let f be a non-zero element of A[X]. In the ring K[X] f can be decomposed uniquely in the form:

$$f = a \prod_\lambda P_\lambda^{n(\lambda)} \qquad (a \in K^*, n(\lambda) \geqslant 0).$$

Lemma 1 proves that a is a content of f. Hence $a \in A$. As A is factorial, a can be decomposed uniquely in the form:

$$a = u \prod_{\iota} p_{\iota}^{m(\iota)} \qquad (u \text{ invertible in A, } m(\iota) \geqslant 0).$$

Whence the existence and uniqueness of the decomposition:

$$f = u \prod_{\iota} p_{\iota}^{m(\iota)} \prod_{\lambda} P_{\lambda}^{n(\lambda)}.$$

Note that this theorem proves that every element of A admits the *same* decomposition into extremal elements in A and A[X]. The g.c.d. of a family of elements of A is therefore the same in A and in A[X].

We may also use Proposition 18 of § 1, no. 10 to show that A[X] is a factorial domain if and only if A is a factorial domain.

COROLLARY. *If* A *is a factorial domain, the domain* $A[X_1, \ldots, X_n]$ *is factorial.*

Argue by induction on n.

This corollary may be extended to the case of an infinite family of indeterminates (cf. Exercise 2).

6. FACTORIAL DOMAINS AND ZARISKI RINGS

PROPOSITION 4. *Let* A *be a Zariski ring and* $\hat{A}$ *its completion. If* $\hat{A}$ *is a factorial domain,* A *is a factorial domain.*

This follows from no. 1, Definition 1 and § 1, no. 10, Proposition 16.

COROLLARY. *If the completion of a Noetherian local ring* A *is a factorial domain,* A *is a factorial domain.*

7. PRELIMINARIES ON AUTOMORPHISMS OF RINGS OF FORMAL POWER SERIES

LEMMA 2. *Let* $f(X_1, X_2, \ldots, X_n)$ *be a formal power series* $\neq 0$ *with coefficients in a ring* E. *There exist integers* $u(i) \geqslant 1$ $(1 \leqslant i \leqslant n-1)$ *such that*

$$f(T^{u(1)}, \ldots, T^{u(n-1)}, T) \neq 0.$$

Suppose that integers $u(i) \geqslant 1$ $(1 \leqslant i \leqslant k-1)$ are determined such that $f(X_n^{u(1)}, \ldots, X_n^{u(k-1)}, X_k, \ldots, X_n) \neq 0$. We shall determine an integer $u(k) \geqslant 1$ such that

$$f(X_n^{u(1)}, \ldots, X_n^{u(k-1)}, X_n^{u(k)}, X_{k+1}, \ldots, X_n) \neq 0.$$

The lemma will then be proved by induction.

Observe that the series $f(X_n^{u(1)}, \ldots, X_n^{u(k-1)}, X_k, \ldots, X_n)$ can be considered as a series in X_k and X_n with coefficients in $E[[X_{k+1}, \ldots, X_{n-1}]]$. Thus we see that it suffices to establish the lemma for $n = 2$.

Therefore let

$$f = \sum_{i,j} e_{ij} X^i Y^j \in E[[X, Y]]$$

where $f \neq 0$. Let $G \subset \mathbf{N} \times \mathbf{N}$ be the non-empty set of ordered pairs (i, j) such that $e_{ij} \neq 0$. Let $\mathbf{N} \times \mathbf{N}$ be given the lexicographical ordering. Let (c, d) be the least element of G. Choose an integer $p > d$. In the expansion of

$$f(T^p, T) = \sum_{(i,j) \in G} e_{ij} T^{ip+j}$$

we look for the terms of degree $cp + d$. If $ip + j = cp + d$, $i \geqslant c + 1$ is impossible, for this would give

$$ip + j \geqslant (c + 1)p + j \geqslant (c + 1)p > cp + d;$$

nor is $i < c$ possible, for (c, d) is the least element of G; therefore $i = c$ and then $j = d$. The term of degree $cp + d$ in $f(T^p, T)$ is therefore $e_{cd} T^{cp+d}$. Since $e_{cd} \neq 0$, $f(T^p, T) \neq 0$. Whence the lemma.

In the ring $E[[X_1, \ldots, X_n]]$, let $\mathfrak{a}$ be the ideal of formal power series without constant term. If $w_1, \ldots, w_n$ are elements of $\mathfrak{a}$, recall that the mapping $f(X_1, \ldots, X_n) \mapsto f(w_1, \ldots, w_n)$ is the unique endomorphism s of the ring $E[[X_1, \ldots, X_n]]$ such that $s(X_i) = w_i$ for $1 \leqslant i \leqslant n$ (Chapter III, § 4, no. 5, Proposition 6).

We take $w_1 = X_1 + X_n^{u(1)}, \ldots, w_{n-1} = X_{n-1} + X_n^{u(n-1)}$, $w_n = X_n$, where the $u(i)$ are integers $\geqslant 1$. Let s' be the endomorphism of $E[[X_1, \ldots, X_n]]$ which maps X_1 to $X_1 - X_n^{u(1)}, \ldots, X_{n-1}$ to $X_{n-1} - X_n^{u(n-1)}$ and X_n to X_n. Then $s'(s(X_i)) = X_i$ for $1 \leqslant i \leqslant n$ and hence $s' \circ s$ is the identity automorphism; similarly for $s \circ s'$. Hence s is an *automorphism*.

Lemma 3. *Let f be a non-zero element of* $E[[X_1, \ldots, X_n]]$. *There exist integers $u(i) \geqslant 1$ $(1 \leqslant i \leqslant n - 1)$ such that the automorphism s of* E *defined by*

$$s(X_i) = X_i + X_n^{u(i)} \quad (1 \leqslant i \leqslant n - 1)$$

and $s(X_n) = X_n$ maps f to an element g such that $g(0, \ldots, 0, X_n) \neq 0$.

$g(0, \ldots, 0, X_n) = f(X_n^{u(1)}, \ldots, X_n^{u(n-1)}, X_n)$. Lemma 3 is therefore a consequence of Lemma 2.

8. THE PREPARATION THEOREM

In this no. A will denote a *local* ring, $\mathfrak{m}$ its maximal ideal and $k = A/\mathfrak{m}$ its residue field. Suppose that A is *Hausdorff and complete* with the $\mathfrak{m}$-adic topology. Let $B = A[[X]]$; it is local ring whose maximal ideal $\mathfrak{N}$ is generated by $\mathfrak{m}$ and X; with the $\mathfrak{N}$-adic topology, B is Hausdorff and complete (Chapter III, § 2, no. 6, Proposition 6).

For every formal power series

$$f = \sum_{i=0}^{\infty} a_i X^i \in B,$$

we write

$$\bar{f} = \sum_{i=0}^{\infty} \bar{a}_i X^i \in k[[X]],$$

where $\bar{a}_i$ denotes the canonical image of a_i in k. The series $\bar{f}$ will be called the *reduced series* of f; if $\bar{f} \neq 0$, the order of $\bar{f}$ (that is the least integer s such that $a_s \notin \mathfrak{m}$) will be called the *reduced order* of f.

PROPOSITION 5. *Let $f \in B$ be a series whose reduced series is non-zero. Let s denote its reduced order and* M *the sub-A-module of* B *with basis* $\{1, X, \ldots, X^{s-1}\}$. *Then* B *is the direct sum of* M *and* fB *and* f *is not a divisor of zero in* B.

(a) We show that $fB \cap M = (0)$. Suppose that there is a relation:

$$(3) \qquad \left(\sum_{i=0}^{\infty} b_i X^i\right) f = r_0 + r_1 X + \cdots + r_{s-1} X^{s-1} \qquad (b_i \in A, r_j \in A).$$

We show that the b_i (and hence the r_j) are all zero, which will prove in particular that f is not a divisor of zero in B. Since A is Hausdorff, it suffices to show that $b_i \in \mathfrak{m}^n$ for all $i \geqslant 0$ and all $n \geqslant 0$. It is obvious for $n = 0$. We shall argue by double induction: we shall assume that $b_i \in \mathfrak{m}^{n-1}$ for all i and $b_i \in \mathfrak{m}^n$ for $i < k$ and show that this implies that $b_k \in \mathfrak{m}^n$. For this, we write $f = \sum_{i=1}^{\infty} a_i X^i$ and compare the coefficients of X^{s+k} in (3); then:

$$(4) \qquad (b_0 a_{s+k} + \cdots + b_{k-1} a_{s+1}) + b_k a_s + (b_{k+1} a_{s-1} + \cdots + b_{k+s} a_0) = 0.$$

The terms in the first bracket belong to $\mathfrak{m}^n$ since the $b_i \in \mathfrak{m}^n$ for $i < k$; similarly for those in the second, since the $b_i \in \mathfrak{m}^{n-1}$ for all i and the $a_i \in \mathfrak{m}$ for $i \leqslant s - 1$. Hence $b_k a_s \in \mathfrak{m}^n$ and, as a_s is an invertible element of A, $b_k \in \mathfrak{m}^n$.

(b) We show that $fB + M = B$. We write

$$g = a_s + a_{s+1} X + a_{s+2} X^2 + \cdots;$$

it is an invertible element of B. Then

$$f - X^s g = a_0 + a_1 X + \cdots + a_{s-1} X^{s-1};$$

if therefore we write $fg^{-1} - X^s = (f - X^s g) g^{-1} = -h$, the coefficients of h belong to $\mathfrak{m}$. Then let r be an element of B. By induction on n we define a sequence $(q^{(n)})$ of elements of B: we take $q^{(0)}$ to be the unique series satisfying:

(5) $$r \equiv X^s q^{(0)} \qquad (\text{mod. } M);$$

writing $h = \sum_{i=0}^{\infty} h_i X^i$ and $q^{(n)} = \sum_{i=0}^{\infty} q_i^{(n)} X^i$, the $q_i^{(n)}$ are defined by:

(6) $$q_i^{(n)} = \sum_{j=0}^{i+s} h_j q_{i+s-j}^{(n-1)}.$$

It follows immediately from (6) that:

(7) $$X^s q^{(n)} \equiv h q^{(n-1)} \qquad (\text{mod. } M).$$

As $h_j \in \mathfrak{m}$ for all j, it also follows from (6), by induction on n, that $q_i^{(n)} \in \mathfrak{m}^n$ for all i and all n. As A is complete, it follows that the series

$$q^{(0)} + q^{(1)} + \cdots + q^{(n)} + \cdots$$

converges to an element q of B. By (5) and (7),

(8) $$X^s(q^{(0)} + q^{(1)} + \cdots + q^{(n)}) \equiv r + h(q^{(0)} + \cdots + q^{(n-1)}) \qquad (\text{mod. } M).$$

As M is closed, at the limit (8) gives $r \equiv (X^s - h)q$ (mod. M), that is

$$r \in fg^{-1}q + M \subset fB + M.$$

We may also use the results of Chapter III, § 2 to show the relation $B = fB + M$ (cf. Exercise 12). The method followed here has the advantage of being applicable to convergent series.

COROLLARY. *With the hypotheses and notation of Proposition* 5, *suppose that* $s \geqslant 1$, *so that* $f \in B\mathfrak{m} + BX$. *Then the A-homomorphism h of* $B' = A[[T]]$ *to* $B = A[[X]]$ *such that* $h(T) = f$ (Chapter III, § 2, no. 9, Proposition 11 (a)) *defines on B a free B'-module structure admitting* $\{1, X, \ldots, X^{s-1}\}$ *as basis. In particular h is injective.*

Let the B'-module B be given the (T)-adic filtration, which consists of the f^nB for $n \geqslant 0$ (Chapter III, § 2, no. 1). Then B/fB is a free module over the ring $A = B'/TB'$ and the images of the X^i $(0 \leqslant i \leqslant s-1)$ in this A-module form a basis of it (Proposition 5); as moreover f is not a divisor of zero in B (Proposition 5), Bf^n/Bf^{n+1} is also a free (B'/TB')-module of rank s, so that condition (GR) of Chapter III, § 2, no. 8 is satisfied (replacing A by B' and M by B). On the other hand, since B' is Hausdorff and complete with respect to the (T)-adic filtration and gr(B) is a finitely generated gr(B')-module by the above, it is seen first (Chapter III, § 2, no. 9, Corollary 1 to Proposition 12) that B is a finitely generated B' module. The first assertion of the corollary then follows from Chapter III, § 2, no. 9, Proposition 13. The second follows immediately from it.

DEFINITION 2. *A polynomial* $F \in A[X]$ *is called distinguished if it is of the form*

$$F = X^s + a_{s-1}X^{s-1} + \cdots + a_0,$$

where $a_i \in \mathfrak{m}$ *for* $0 \leqslant i \leqslant s-1$.

Note that the product of two distinguished polynomials is a distinguished polynomial.

PROPOSITION 6 (Preparation Theorem). *Let* $f \in B$ *be a series whose reduced series is not zero and* s *its reduced order. Then there exists a unique ordered pair* (u, F) *such that* u *is an invertible element of* B, F *a distinguished polynomial of degree* s *and* $f = uF$.

We write $F = X^s + G$, where $G = g_0 + \cdots g_{s-1}X^{s-1}$ ($g_i \in A$). The relation $f = uF$ is equivalent to $F = u^{-1}f$, that is to $X^s = u^{-1}f - G$. Hence Proposition 5 shows the uniqueness of G and u^{-1} and therefore of F and u. It also shows that there exist $v \in B$ and a polynomial $G = g_0 + \cdots + g_{s-1}X^{s-1}$ ($g_i \in A$) such that $X^s = vf - G$; it remains to show that v is invertible in B and that $g_i \in \mathfrak{m}$ for all i. Now, writing $\bar{g}_i$ for the canonical image of g_i in k and $\bar{f}$, $\bar{v}$ for the reduced series of f, g,

$$X^s + \bar{g}_0 + \bar{g}_1X + \cdots + \bar{g}_{s-1}X^{s-1} = \bar{f}\bar{v};$$

since $\bar{f}$ is of order s, $\bar{g}_i = 0$ for all i and $\bar{v}$ is of order 0, hence v is invertible.

PROPOSITION 7. *Let* F *be a distinguished polynomial and* g, h *two formal power series of* B *such that* $F = gh$. *Then there exists an invertible element* u of B *such that* ug *and* $u^{-1}h$ *are distinguished polynomials and* $F = (ug)(u^{-1}h)$.

In fact, the reduced series of g and h are $\neq 0$; hence, by Proposition 6, there exist invertible elements u, v of B such that ug and vh are distinguished polynomials. Then $uvF = (ug)(vh)$ is a distinguished polynomial and uv is invertible. Passing to the reduced series, it is seen immediately that F and uvF have the same reduced order, that is the same degree. The uniqueness assertion in Proposition 6 therefore shows that $F = uvF$, whence $uv = 1$.

COROLLARY. *Suppose further that* A *is an integral domain and* F *a distinguished polynomial of degree* s. *For* F *to be extremal in* A[X], *it is necessary and sufficient that it be extremal in* B = A[[X]].

Suppose that F is not extremal in A[X], so that $F = f_1f_2$, where f_1 and f_2 are non-invertible elements of A[X]; the product of the dominant coefficients of f_1 and f_2 being equal to 1, these coefficients are invertible in A and the hypothesis implies that f_1 and f_2 are of degrees >0 and $<s$; as the reduced polynomials $\bar{f}_1, \bar{f}_2$ satisfy $\bar{f}_1\bar{f}_2 = X^s$, neither $\bar{f}_1$ nor $\bar{f}_2$ can be invertible in k[[X]], for, if $\bar{f}_1$ were invertible, $\bar{f}_2$ would be of order s, which is absurd. *A fortiori*, neither f_1 nor f_2 is invertible in B and F is not extremal in B.

Conversely, if F is not extremal in A[[X]], then F $= gh$, where neither g nor h is invertible in B; their reduced orders are therefore $\geqslant 1$; then the distinguished polynomials ug and $u^{-1}h$ of Proposition 7 are not constant, which shows that F is not extremal in A[X].

9. FACTORIALITY OF RINGS OF FORMAL POWER SERIES

PROPOSITION 8. *Let C be a ring which is either a field or a discrete valuation ring. Then the domain of formal power series* $C[[X_1, \dots, X_n]]$ *is factorial.*

Let $\mathfrak{p}$ be the maximal ideal of C and π a generator of $\mathfrak{p}$ (if C is a field, then $\pi = 0$). Let C be given the $\mathfrak{p}$-adic topology, which is Hausdorff. As C is a Noetherian local ring, $B = C[[X_1, \dots, X_n]]$ is a Noetherian local ring and its completion is $\hat{C}[[X_1, \dots, X_n]]$ (Chapter III, § 2, no. 6, Proposition 6). By the Corollary to Proposition 4 (no. 6), it suffices to prove that $\hat{C}[[X_1, \dots, X_n]]$ is factorial. Now, if C is a field, then $\hat{C} = C$; if C is a discrete valuation ring, the same is true of $\hat{C}$ (Chapter VI, § 5, no. 3, Proposition 5). We shall therefore assume in the remainder of the proof that C is *complete*.

Arguing by induction starting with the trivial case $n = 0$, we shall assume that it has been proved that $A = C[[X_1, \dots, X_{n-1}]]$ is factorial. We shall identify B with $A[[X_n]]$ and denote by $\mathfrak{m}$ the maximal ideal of A (generated by $\pi, X_1, \dots, X_{n-1}$). We shall prove that every non-zero element g of B is, in an essentially unique way, a product of extremal elements.

Let K be the field $C/C\pi$; as $B/B\pi$ is identified with $K[[X_1, \dots, X_n]]$, the ideal $B\pi$ is prime and π is extremal. If $\pi \neq 0$, $B_{B\pi}$ is therefore the ring of a normed discrete valuation w (Chapter VI, § 3, no. 6, Proposition 9); every non-zero element g of B may therefore be written as $g = \pi^{w(g)}f$, where $f \in B$ and f is not a multiple of π. It will therefore suffice to show that f is an essentially unique product of extremal elements. Now the canonical image of f in $K[[X_1, \dots, X_n]]$ is not zero; Lemma 3 (no. 7) therefore shows that there exists an automorphism of B which maps f to an element f' such that the coefficients of $f'(0, \dots, 0, X_n)$ are not all in $C\pi$; this means that the coefficients of the series f', considered as a formal power series in X_n, are not all in $\mathfrak{m}$. It will suffice to prove our assertion for f'.

In what follows, all the elements of B will be considered as formal power series in X_n with coefficients in A. By Proposition 6 of no. 8 (applicable since C and therefore A are separable and complete and the reduced series of f' is $\neq 0$), f' is associated, in B, with a unique distinguished polynomial F. By Proposition 7 of no. 8, every series which divides f' (or, what amounts to the same, which divides F) is associated with a distinguished polynomial which divides F and every decomposition of f' is, to within invertible factors, of the form $f' = uF_1 \dots F_q$, where u is invertible and the F_i are extremal distinguished polynomials (in B) such that $F = F_1 \dots F_q$. By the Corollary to Proposition 7 of no. 8, the F_i are also extremal in $A[X_n]$. Now, as A is factorial by the induction hypothesis, so is $A[X_n]$ (Theorem 2, no. 5); hence, since they are monic, the F_i are uniquely determined by F (up to a permutation). This shows the uniqueness of the decomposition $f' = uF_1 \ldots F_q$; its existence follows from the fact that B is Noetherian, which completes the proof.

Remarks

(1) There exist factorial rings A such that the ring A[[X]] is not factorial (Exercise 8). However, if A is a principal ideal domain, $A[[X_1, \ldots, X_n]]$ is factorial (Exercise 9).

(2) * We shall see later, by homological methods, that every regular local ring is factorial (cf. § 4, no. 7, Corollary 3 to Proposition 16). This will give another proof, conceptually simpler, of Proposition 8. *

4. MODULES OVER INTEGRALLY CLOSED NOETHERIAN DOMAINS

Throughout this paragraph, A will be a *commutative integral domain* with field of fractions K. Starting with no. 2, A will be further assumed to be *Noetherian and integrally closed* (and hence a Krull domain (§ 1, no. 3, Corollary to Theorem 2)); then P(A), D(A) and C(A) will respectively denote the set of prime ideals of A of height 1 (§ 1, no. 6), the divisor group of A (§ 1, no. 3) and the divisor class group of A (§ 1, no. 10), these latter being written *additively*.

The general method of studying finitely generated modules over an integrally closed Noetherian domain A consists of "*localizing*" the modules with respect to all the prime ideals $\mathfrak{p} \in P(A)$ *of height* 1 in A; as $A_\mathfrak{p}$ is then a *discrete valuation ring* (§ 1, no. 6, Theorem 4), the structure of finitely generated $A_\mathfrak{p}$-modules is well known (*Algebra*, Chapter VII, § 4) and therefore gives information about the structure of finitely generated A-modules. In the particular case where A is a Dedekind domain, we can arrive at as complete a theory as when A is a principal ideal domain (no. 10).

1. LATTICES

DEFINITION 1. *Let* V *be a finite-dimensional vector space over the field* K. A *lattice of* V *with respect to* A (or simply a *lattice of* V) *is defined to be any sub-*A*-module* M *of* V *satisfying the following condition:*

*There exist two free sub-*A*-modules* L_1, L_2 *of* V *such that* $L_1 \subset M \subset L_2$ *and* $rg_A(L_1) = rg_K(V)$.

Examples

(1) If we take V = K, the lattices of K are just the *fractional ideals* ≠ (0) of K (§ 1, no. 1, Definition 1).

(2) If $rg_K(V) = n$, every *free* sub-A-module L of V has a basis containing at most n elements, every subset of V which is free over A being free over K; for L to be a lattice of V, it is necessary and sufficient that L have a basis of n elements (in other words, that $rg_A(L) = n$).

(3) If A is a *principal ideal domain*, every lattice M of V is a finitely generated A-module (since A is Noetherian) which is torsion-free and hence a *free* A-module (*Algebra*, Chapter VII, § 4, no. 3, Corollary 2 to Theorem 2).

PROPOSITION 1. *For a sub-A-module* M *of* V *to be a lattice of* V, *it is necessary and sufficient that* KM = V *and that* M *be contained in a finitely generated sub-A-module of* V.

The conditions are obviously necessary, for a free sub-A-module of V with the same rank as V generates V. Conversely, if KM = V, M contains a basis $(a_i)_{1 \leqslant i \leqslant n}$ of V over K and hence it contains the free sub-A-module L_1 generated by the a_i; on the other hand, if $M \subset M_1$, where M_1 is a sub-A-module of V generated by a finite number of elements b_j and $(e_i)_{1 \leqslant i \leqslant n}$ is a basis of V over K, there exists an element $s \neq 0$ of A such that each of the b_j is a linear combination of the $s^{-1}e_i$ with coefficients *in* A; if L_2 is the free sub-A-modules of V generated by the $s^{-1}e_i$, then $M \subset L_2$.

COROLLARY. *Suppose that* A *is Noetherian; for a sub-A-module* M *of* V *to be a lattice of* V, *it is necessary and sufficient that* KM = V *and* M *be finitely generated.*

Remark (1). Recall that for every sub-A-module M of V the canonical mapping $M \otimes_A K \to V$ is injective and has image KM (*Algebra*, Chapter II, § 7, no. 10, Proposition 26); to say that KM = V means therefore that this mapping is *bijective.*

PROPOSITION 2. *Let* M *be a lattice of* V *and* M_1 *a sub-A-module of* V. *If there exist two elements* x, y *of* K^* *such that* $xM \subset M_1 \subset yM$, M_1 *is a lattice of* V; *conversely, if* M_1 *is a lattice of* V, *there exist two non-zero elements* a, b *of* A *such that* $aM \subset M_1 \subset b^{-1}M$.

If L_1, L_2 are two free lattices of V such that $L_1 \subset M \subset L_2$, the relations $xM \subset M_1 \subset yM$ imply $xL_1 \subset M_1 \subset yL_2$ and xL_1 and yL_2 are free lattices; conversely, if M_1 is a lattice and $(e_i)_{1 \leqslant i \leqslant n}$ a basis of L_2 over A, the relation $KM_1 = V$ implies the existence of $x = a/s \in K^*$ (where a and s are non-zero elements of A) such that $xe_i \in M_1$ for all i, whence $xM \subset xL_2 \subset M_1$ and *a fortiori* $aM \subset M_1$; exchanging the roles of M and M_1 it can be similarly shown that there exists $b \neq 0$ in A such that $bM_1 \subset M$.

PROPOSITION 3. (i) *If* M_1 *and* M_2 *are lattices of* V, *so are* $M_1 \cap M_2$ *and* $M_1 + M_2$.

(ii) *If* W *is a vector subspace of* V *and* M *is a lattice of* V, $M \cap W$ *is a lattice of* W.

(iii) *Let* V, $V_1, \ldots, V_k$ *be vector spaces of finite rank over* K *and let*

$$f\colon V_1 \times \cdots \times V_k \to V$$

be a multilinear mapping whose image generates V. *If* M_i *is a lattice of* V_i *for* $1 \leqslant i \leqslant k$, *the sub-*A*-module of* V *generated by* $f(M_1 \times \cdots \times M_k)$ *is a lattice of* V.

(iv) *Let* V *and* W *be two vector spaces of finite rank over* K, M *a lattice of* V *and* N *a lattice of* W. *The sub-*A*-module* N: M *of* $\mathrm{Hom}_K(V, W)$, *consisting of the* K*-linear mappings* f *such that* $f(M) \subset N$, *is a lattice of* $\mathrm{Hom}_K(V, W)$.

(i) By virtue of Proposition 2, there exist non-zero a and b in A such that $aM_1 \subset M_2 \subset b^{-1}M_1$; we conclude that $M_1 \cap M_2$ and $M_1 + M_2$ lie between aM_1 and $b^{-1}M_1$ and are therefore lattices by virtue of Proposition 2.

(ii) Let S be a complement of W in V, L_W a free lattice of W and L_S a free lattice of S, so that $L = L_W \oplus L_S$ is a free lattice of V. Then there exist x, y in K^* such that $xL \subset M \subset yL$. We deduce that $xL_W \subset M \cap W \subset yL_W$, which shows that $M \cap W$ is a lattice of W (Proposition 2).

(iii) As $KM_i = V_i$, clearly by linearity $f(M_1 \times \cdots \times M_k)$ generates the vector K-space V; on the other hand, for all i, there exists a finitely generated sub-A-module N_i of V_i such that $M_i \subset N_i$; the sub-A-module N of V generated by $f(N_1 \times \cdots \times N_k)$ is finitely generated and contains M and hence M is a lattice of V (Proposition 1).

(iv) Let P (resp. Q) be a free lattice of V (resp. W) containing M (resp. contained in N); obviously $N: M \supset Q: P$. Now it is immediate that $Q: P$ is isomorphic to $\mathrm{Hom}_A(P, Q)$, hence is a free A-module of rank $(\mathrm{rg}_A P)(\mathrm{rg}_A Q)$ (*Algebra*, Chapter II, § 1, no. 6, Corollary 1 to Proposition 6) and therefore a lattice of $\mathrm{Hom}_K(V, W)$. Similarly, if P′ (resp. Q′) is a free lattice of V (resp. W) contained in M (resp. containing N), then $Q': P' \supset N: M$ and $Q': P'$ is a lattice of $\mathrm{Hom}_K(V, W)$; whence the conclusion.

Remarks

(2) Proposition 3 (i) shows that the set R(V) of lattices of V is *lattice-ordered* with respect to inclusion; moreover, if M is a fixed lattice of V, the xM, where x runs through K^*, form a subset of R(V) which is both *coinitial* and *cofinal* (*Set Theory*, Chapter III, § 1, no. 7).

(3) In the notation of Proposition 3 (iv), the canonical mapping

$$N: M \to \mathrm{Hom}_A(M, N),$$

which maps every K-linear mapping $f \in N: M$ to the A-linear mapping from M to N which has the same graph as $f|M$, is *bijective*; for every A-linear mapping g: $M \to N$ can be imbedded in a K-linear mapping

$$g \otimes 1 : M \otimes_A K \to N \otimes_A K$$

and it has been seen that $M \otimes_A K$ and $N \otimes_A K$ are respectively identified with V and W.

In particular, if we take $W = K$, $N = A$, $\mathrm{Hom}_K(V, W)$ is just the *dual vector* K-*space* V^* of V and A: M is identified with the *dual* A-*module* M^* of M; we shall henceforth make this identification and we shall say that M^* is the *dual lattice* of M: it is therefore *the set of* $x^* \in V^*$ *such that* $\langle x, x^* \rangle \in A$ *for all* $x \in M$.

COROLLARY. *Let* U, V, W *be three vector spaces of finite rank over* K *and* $f: U \times V \to W$ *a left non-degenerate* K-*bilinear mapping* (*Algebra*, Chapter IX, § 1, no. 1, Definition 3). *If* M *is a lattice of* V *and* N *a lattice of* W, *the set* $N : {}_f M$ *of* $x \in U$ *such that* $f(x, y) \in N$ *for all* $y \in M$ *is a lattice of* U.

Let $s_f: U \to \mathrm{Hom}_K(V, W)$ be the K-linear mapping left associated with f (*Algebra*, Chapter IX, *loc. cit.*) such that $s_f(x)$ is the linear mapping $y \mapsto f(x, y)$; recall that to say that f is left non-degenerate means that s_f is *injective*. By Proposition 3 (iv), N: M is a lattice of $\mathrm{Hom}_K(V, W)$; as $N : {}_f M = s_f^{-1}(N : M)$ and s_f is injective, the corollary follows from Proposition 3 (ii).

Examples

(4) Let S be a (not necessarily associative) K-algebra of finite rank with a unit element; then the bilinear mapping $(x, y) \mapsto xy$ of $S \times S$ to S is (left and right) non-degenerate. If M and N are lattices of S with respect to A, so are M.N (Proposition 3 (iii)) and the set of $x \in S$ such that $xM \subset N$ (Corollary to Proposition 3). Note that there exists a *sub-A-algebra* of S containing the unit element of S which is a *lattice* of S; for consider a basis $(e_i)_{1 \leqslant i \leqslant n}$ of S such that e_1 is the unit element of S and let $e_i e_j = \sum_k c_{ijk} e_k$ be the multiplication table of S ($1 \leqslant i \leqslant n$, $1 \leqslant j \leqslant n$), so that $c_{1jk} = \delta_{jk}$, $c_{i1k} = \delta_{ik}$ (Kronecker symbols). Let $s \in A$ be a non-zero and such that $c'_{ijk} = s . c_{ijk} \in A$ for all triplets of indices (i, j, k); if we write $e'_i = s^{-1} e_i$ for $i \geqslant 2$, then

$$e'_i e'_j = s c'_{ij1} e_1 + \sum_{k \geqslant 2} c'_{ijk} e'_k$$

for $i \geqslant 2$ and $j \geqslant 2$; the lattice of S with basis e_1 and the e'_i ($2 \leqslant i \leqslant n$) is a sub-A-algebra of S with unit element e_1.

(5) Let V be a finite-dimensional vector space over K and f a non-degenerate bilinear form on V. If M is a lattice of V, it follows from the Corollary to Proposition 3 that the set M_f^* of $x \in V$ such that $f(x, y) \in A$ for all $y \in M$ is also a lattice of V; if $s_f: V \to V^*$ is the linear mapping left associated with f (which is bijective), $s_f(M_f^*)$ is just the dual lattice M^* of M.

PROPOSITION 4. *Let* B *be an integral domain,* A *a subring of* B *and* K *and* L *the respective fields of fractions of* A *and* B. *Let* V *be a finite-dimensional vector space over* K.

(i) *For every lattice* M *of* V *with respect to* A, *the image* BM *of* $M_{(B)} = M \otimes_A B$ *in* $V_{(L)} = V \otimes_K L$ *is a lattice of* $V_{(L)}$ *with respect to* B.

(ii) *Suppose further that* B *is a flat* A-*module. Then the canonical mapping* $M_{(B)} \to BM$ *is bijective. If further* B *is faithfully flat, the mapping which maps every lattice* M *of* V *with respect to* A *to the lattice* BM *of* $V_{(L)}$ *with respect to* B *is injective.*

(i) As KM = V, clearly $L.(BM) = V_{(L)}$; on the other hand M is contained in a finitely generated sub-A-module M_1 of V and hence BM is contained in BM_1 which is a finitely generated B-module; whence assertion (i) (Proposition 1).

(ii) $V_{(L)} = V \otimes_K L = V \otimes_A L$ (Chapter II, § 2, no. 7, Proposition 18) and, as L is a flat B-module, it is also a flat A-module (Chapter I, § 2, no. 7, Corollary 3 to Proposition 8). As B is a flat A-module, the canonical mapping $M \otimes_A B \to V \otimes_A B$ is injective; on the other hand, since V is a free K-module and K a flat A-module, V is a flat A-module (Chapter I, § 2, no. 7, Corollary 3 to Proposition 8) and hence the canonical mapping $V \otimes_A B \to V \otimes_A L$ is injective, which establishes the first assertion. To see also that the relation $BM_1 = BM_2$ implies $M_1 = M_2$ for two lattices M_1, M_2 of V with respect to A when B is a faithfully flat A-module, note first that $BM_1 \cap BM_2 = B(M_1 \cap M_2)$ (Chapter I, § 2, no. 6, Proposition 6); we may therefore confine our attention to the case where $M_1 \subset M_2$ and our assertion then follows from Chapter I, § 3, no. 1, Proposition 3 applied to the canonical injection $M_1 \to M_2$.

COROLLARY. *Suppose that* A *is a discrete valuation ring. Let* $\hat{A}$ *be its completion and let* $\hat{K}$ *be the field of fractions of* $\hat{A}$ (Chapter VI, § 5, no. 3). *The mapping* ϕ, *which maps every lattice* M *of* V *to the lattice* $\hat{A}M$ *of* $\hat{V} = V \otimes_K \hat{K}$ *with respect to* $\hat{A}$, *is bijective and its inverse maps every lattice* M′ *of* $\hat{V}$ *with respect to* $\hat{A}$ *to its intersection* $M' \cap V$ (V *being canonically identified with a vector sub*-K-*space of* $\hat{V}$).

If L is a free lattice of V, the lattices *a*L (for $a \in A$, $a \neq 0$) form a fundamental system of neighbourhoods of 0 for a topology $\mathscr{T}$ on V (compatible with its A-module structure), which (when a basis of L over A is taken) is identified with the *product topology* on K^n; by virtue of Proposition 2, a fundamental system of neighbourhoods of 0 for $\mathscr{T}$ also consists of *all the lattices* of V with respect to A; clearly $\hat{V}$ is the *completion* of V with respect to $\mathscr{T}$. Moreover, if $\mathfrak{m}$ is the maximal ideal of A, the topology $\mathscr{T}$ induces on every lattice M of V with respect to A the $\mathfrak{m}$-adic topology since M is a finitely generated A-module (Chapter III, § 3, no. 2, Theorem 2) and $\hat{A}M$ is the completion of M with respect to this topology (Chapter III, § 2, no. 12, Proposition 16); moreover, as M is open (and therefore closed) in V, $\hat{A}M \cap V = M$, which proves again the fact that ϕ is *injective* (which follows directly from Proposition 4, (ii), since $\hat{A}$ is a faithfully flat A-module). Finally, if M′ is a lattice of $\hat{V}$ with respect to $\hat{A}$, $M = M' \cap V$ is a lattice of V with respect to A, for every element of $\hat{A}$ is the product of an element of A and an invertible element of $\hat{A}$ and hence it follows from Proposition

2 that there exist a, b in $A - \{0\}$ such that $a\hat{A}L \subset M' \subset b\hat{A}L$, whence $aL \subset M' \cap V \subset bL$. Moreover M' is open in V and, as V is dense in $\hat{V}$, M' is the completion of $M' \cap V = M$; this proves that ϕ is *surjective*, whence the corollary.

Example (6) Let S be a multiplicative subset of A not containing 0; we apply Proposition 4 to $B = S^{-1}A$; then $L = K$, $BM = S^{-1}M$; hence $S^{-1}M$ is a lattice of V with respect to $S^{-1}A$. Moreover:

Proposition 5. *Let* V, W *vector spaces of finite rank over* K, M *a lattice of* V *and* N *a lattice of* W. *If* M *is finitely generated, then (in the notation of Proposition 3):*

$$S^{-1}(N: M) = S^{-1}N: S^{-1}M \tag{1}$$

in $\mathrm{Hom}_K(V, W)$.

Clearly the left hand side of (1) is contained in the right hand side. Conversely, let $f \in S^{-1}N: S^{-1}M$ and let $(x_i)_{1 \leqslant i \leqslant n}$ be a system of generators of M. There exists $s \in S$ such that $f(x_i) \in s^{-1}N$ for all i and hence $sf \in N: M$, which proves the proposition.

2. DUALITY; REFLEXIVE MODULES

Recall that from now on the domain A is assumed to be *Noetherian and integrally closed* and that P(A) (or simply P) denotes the set of prime ideals of A *of height* 1. Every lattice with respect to A is a *finitely generated* A-module (no. 1, Corollary to Proposition 1).

Let V be a vector space of finite rank over K, V* its dual and V** its bidual; we shall identify V and V** by means of the canonical mapping c_V (*Algebra*, Chapter II, § 7, no. 5, Theorem 6). Let M be a lattice of V; recall that the dual A-module M* of M is canonically identified with the *dual lattice* of M, the set of $x^* \in V^*$ such that $\langle x, x^* \rangle \in A$ for all $x \in M$; the *bidual* A-module M** of M is therefore a *lattice of* V which contains M. Moreover $M^{***} = M^*$, for the relation $M \subset M^{**}$ implies $(M^{**})^* \subset M^*$ and on the other hand $M^* \subset (M^*)^{**}$ by the above (cf. *Set Theory*, Chapter III, § 1, no. 5, Proposition 2).

If $\mathfrak{p}$ is a prime ideal, Proposition 5 applied with N = A gives the relation $(M^*)_\mathfrak{p} = (M_\mathfrak{p})^*$, which justifies the notation $M_\mathfrak{p}^*$ for both terms.

Theorem 1. *If* M *is a lattice of* V, *then* $M^* = \bigcap_{\mathfrak{p} \in P} M_\mathfrak{p}^*$.

Clearly M* is contained in each of the $M_\mathfrak{p}^*$. Conversely, suppose that $x^* \in \bigcap_{\mathfrak{p} \in P} M_\mathfrak{p}^*$; if $x \in M$, then $\langle x, x^* \rangle \in \bigcap_{\mathfrak{p} \in P} A_\mathfrak{p}$ and, as $A = \bigcap_{\mathfrak{p} \in P} A_\mathfrak{p}$ (§ 1, no. 6, Theorem 4), $x^* \in M^*$.

COROLLARY. $M^{**} = \bigcap_{\mathfrak{p} \in P} M_{\mathfrak{p}}$.

Theorem 1 applied to M^* shows that $M^{**} = \bigcap_{\mathfrak{p} \in P} M_{\mathfrak{p}}^{**}$. But as $A_{\mathfrak{p}}$ is a principal ideal domain (§ 1, no. 6, Theorem 4), $M_{\mathfrak{p}}$ is a finitely generated free $A_{\mathfrak{p}}$-module and hence $M_{\mathfrak{p}}^{**}$ is canonically identified with $M_{\mathfrak{p}}$ (*Algebra*, Chapter II, § 2, no. 7, Proposition 14), whence the corollary.

For any lattice M with respect to A, the canonical mapping $c_M : M \to M^{**}$ (*Algebra*, Chapter II, § 2, no. 7) identifies an element $x \in M$ with itself, for x is the unique element y of $V = V^{**}$ such that $\langle x, x^* \rangle = \langle y, x^* \rangle$ for all $x^* \in M^*$, since M^* generates V^*. We shall say that M is *reflexive* if $M^{**} = M$ (*loc. cit.*). As we have above $M^* = (M^*)^{**}$, it is seen that the *dual* of any lattice M is always *reflexive*.

Remark (1) Let M be a finitely generated A-module; it is immediate that the dual M^* of M, identified with a sub-A-module of $\operatorname{Hom}_A(M, K)$ is a *lattice* of the vector K-space $\operatorname{Hom}_A(M, K)$; in particular, every finitely generated *reflexive* A-module is isomorphic to a lattice of a suitable vector K-space.

THEOREM 2. *If* M *is a lattice of* V, *the following conditions are equivalent:*

(a) M *is reflexive.*

(b) $M = \bigcap_{\mathfrak{p} \in P} M_{\mathfrak{p}}$.

(c) $\operatorname{Ass}(V/M) \subset P$.

The equivalence of (a) and (b) follows from the Corollary to Theorem 1. If (b) holds, V/M is canonically identified with a sub-A-module of the product $\prod_{\mathfrak{p} \in P} (V/M_{\mathfrak{p}})$; but in fact, it is contained in the *direct sum* $\bigoplus_{\mathfrak{p} \in P} (V/M_{\mathfrak{p}})$: for if $L \subset M$ is a free lattice and $(e_i)_{1 \leq i \leq n}$ a basis of L, each of the coordinates x_i of a point $x \in V$ with respect to (e_i) belongs to $A_{\mathfrak{p}}$ except for a finite number of values of $\mathfrak{p} \in P$. The relation $V/M \subset \bigoplus_{\mathfrak{p} \in P} (V/M_{\mathfrak{p}})$ then implies:

$$\operatorname{Ass}(V/M) \subset \bigcap_{\mathfrak{p} \in P} \operatorname{Ass}(V/M_{\mathfrak{p}}).$$

As $V/M_{\mathfrak{p}}$ is an $A_{\mathfrak{p}}$-module, an element of $A - \mathfrak{p}$ cannot annihilate an element $\neq 0$ of $V/M_{\mathfrak{p}}$, since the elements of $A - \mathfrak{p}$ are invertible in $A_{\mathfrak{p}}$; the elements of $\operatorname{Ass}(V/M_{\mathfrak{p}})$ are therefore contained in $\mathfrak{p}$ and are $\neq 0$, since $V/M_{\mathfrak{p}}$ is a torsion $A_{\mathfrak{p}}$-module; as $\mathfrak{p}$ is of height 1, necessarily $\operatorname{Ass}(V/M_{\mathfrak{p}}) = \{\mathfrak{p}\}$ if $V/M_{\mathfrak{p}} \neq \{0\}$ and $\operatorname{Ass}(V/M_{\mathfrak{p}}) = \varnothing$ if $V/M_{\mathfrak{p}} = \{0\}$; hence $\operatorname{Ass}(V/M) \subset P$.

Finally, if condition (c) holds, then

$$\operatorname{Ass}(M^{**}/M) \subset \operatorname{Ass}(V/M) \subset P.$$

On the other hand, if $\mathfrak{p} \in P$, then it has been seen in the proof of the Corollary to Theorem 1 that $M_{\mathfrak{p}}^{**} = M_{\mathfrak{p}}$, whence $\mathfrak{p} \notin \operatorname{Ass}(M^{**}/M)$ (Chapter IV, § 1, no. 3, Corollary 1 to Proposition 7). We conclude that $\mathrm{Ass}(M^{**}/M) = \varnothing$, whence $M^{**} = M$ (Chapter IV, § 1, no. 1, Corollary 1 to Proposition 2).

Corollary. *Let* M, N *be two lattices of* V *with respect to* A *such that* N *is reflexive. In order that* $M \subset N$, *it is necessary and sufficient that, for all* $\mathfrak{p} \in P$, $M_{\mathfrak{p}} \subset N_{\mathfrak{p}}$.

The condition is obviously necessary and, if it is fulfilled, then

$$\bigcap_{\mathfrak{p}\in P} M_{\mathfrak{p}} \subset \bigcap_{\mathfrak{p}\in P} N_{\mathfrak{p}} = N.$$

As $M \subset M^{**} = \bigcap_{\mathfrak{p}\in P} M_{\mathfrak{p}}$, certainly $M \subset N$.

Examples

(1) Every *free* lattice is reflexive.

(2) Take V = K. For a fractional ideal $\mathfrak{a}$ of K to be a reflexive lattice, it is necessary and sufficient that it be a *divisorial ideal* by virtue of criterion (b) of Theorem 2 and § 1, no. 4, Propositions 5 and 7.

(3) Let M be a lattice with respect to A; if S is a multiplicative subset of A not containing 0, Proposition 5 of no. 1 shows that $S^{-1}(M^*) = (S^{-1}M)^*$; if M is reflexive, $S^{-1}M$ is therefore a reflexive lattice with respect to $S^{-1}A$.

Proposition 6. (i) *If* M_1 *and* M_2 *are reflexive lattices of* V, *so is* $M_1 \cap M_2$.

(ii) *If* W *is a vector subspace of* V *and* M *is a reflexive lattice of* V, $M \cap W$ *is a reflexive lattice of* W.

(iii) *Let* V, W *be two vector spaces of finite rank over* K *and* M (resp. N) *a lattice of* V (resp. W). *If* N *is reflexive, the lattice* N: M *of* $\mathrm{Hom}_K(V, W)$ (no. 1, Proposition 3) *is reflexive.*

(i) $(M_1 \cap M_2)_{\mathfrak{p}} = (M_1)_{\mathfrak{p}} \cap (M_2)_{\mathfrak{p}}$ for all $\mathfrak{p} \in P$ (Chapter II, § 2, no. 4, Theorem 1). If $M_1 = \bigcap_{\mathfrak{p}\in P} (M_1)_{\mathfrak{p}}$ and $M_2 = \bigcap_{\mathfrak{p}\in P} (M_2)_{\mathfrak{p}}$, then

$$M_1 \cap M_2 = \bigcap_{\mathfrak{p}\in P} (M_1 \cap M_2)_{\mathfrak{p}}$$

whence the conclusion by virtue of Theorem 2.

(ii) Similarly $(M \cap W)_{\mathfrak{p}} = M_{\mathfrak{p}} \cap W_{\mathfrak{p}} = M_{\mathfrak{p}} \cap W$, whence

$$M \cap W = \bigcap_{\mathfrak{p}\in P} (M \cap W)_{\mathfrak{p}},$$

which proves (ii).

(iii) As M is finitely generated, it follows from no. 1, Proposition 5 that $(N: M)_{\mathfrak{p}} = N_{\mathfrak{p}}M_{\mathfrak{p}}$; moreover, the relation $N = \bigcap_{\mathfrak{p}\in P} N_{\mathfrak{p}}$ implies:

$$N: M = \bigcap_{\mathfrak{p}\in P} (N_{\mathfrak{p}}: M_{\mathfrak{p}}).$$

For if $f \in \bigcap_{\mathfrak{p} \in \mathrm{P}} (\mathrm{N}_\mathfrak{p} : \mathrm{M}_\mathfrak{p})$ and $x \in \mathrm{M}$, then $f(x) \in \bigcap_{\mathfrak{p} \in \mathrm{P}} \mathrm{N}_\mathfrak{p} = \mathrm{N}$, whence $f \in \mathrm{N} : \mathrm{M}$; this shows that N: M is reflexive.

Remarks

(2) If M_1 and M_2 are reflexive lattices of V, the lattice $\mathrm{M}_1 + \mathrm{M}_2$ is not necessarily reflexive (cf. § 1, Exercise 2).

(3) If M is a finitely generated A-module and T its torsion submodule, the dual M^* of M is the same as the dual of M/T, since, for every linear form f on M, the image $f(\mathrm{T})$ is a torsion submodule of A and hence zero. As M/T is isomorphic to a lattice of a vector space over K, it is seen that the dual of *every* finitely generated A-module is *reflexive*.

Proposition 7. *Let* $0 \to \mathrm{M} \to \mathrm{N} \to \mathrm{Q} \to 0$ *be an exact sequence of* A-*modules. Suppose that* N *is finitely generated and torsion-free.*

(i) *If* M *is reflexive, then* $\mathrm{Ass}(\mathrm{Q}) \subset \mathrm{P} \cup \{\{0\}\}$ (*in other words, every ideal associated with* Q, *is, either* (0), *or of height* 1).

(ii) *Conversely, if* N *is reflexive and* $\mathrm{Ass}(\mathrm{Q}) \subset \mathrm{P} \cup \{\{0\}\}$, *then* M *is reflexive.*

As A is Noetherian, M is also finitely generated; if we write $\mathrm{V} = \mathrm{M}_{(\mathrm{K})}$, $\mathrm{W} = \mathrm{N}_{(\mathrm{K})}$, M (resp. N) is canonically identified with a lattice of V (resp. W) (no. 1, Proposition 1). Consider the two exact sequences:

$$0 \to \mathrm{V/M} \to \mathrm{W/M} \to \mathrm{W/V} \to 0$$

$$0 \to \mathrm{Q} \to \mathrm{W/M} \to \mathrm{W/N} \to 0.$$

(i) We deduce (Chapter IV, § 1, no. 1, Proposition 3) that:

$$\mathrm{Ass}(\mathrm{Q}) \subset \mathrm{ASS}(\mathrm{W/M}) \subset \mathrm{ASS}(\mathrm{V/M}) \cup \mathrm{ASS}(\mathrm{W/V}).$$

If M is reflexive, then $\mathrm{Ass}(\mathrm{V/M}) \subset \mathrm{P}$ (Theorem 2); on the other hand, clearly $\mathrm{Ass}(\mathrm{W/V})$ is, either empty, or reduced to $\{0\}$; whence (i).

(ii) Similarly:

$$\mathrm{Ass}(\mathrm{V/M}) \subset \mathrm{Ass}(\mathrm{W/M}) \subset \mathrm{Ass}(\mathrm{Q}) \cup \mathrm{Ass}(\mathrm{W/N}).$$

The hypotheses therefore imply that $\mathrm{Ass}(\mathrm{V/M}) \subset \mathrm{P} \cup \{\{0\}\}$. But V/M is a torsion A-module and hence $\{0\} \notin \mathrm{Ass}(\mathrm{V/M})$; Theorem 2 then shows that M is reflexive.

Proposition 8. *Let* R *and* S *be two commutative rings,* $\rho : \mathrm{R} \to \mathrm{S}$ *a ring homomorphism and* M *a finitely generated* R-*module. Suppose that* R *is Noetherian and that* S *is a flat* R-*module. Then, if* M *is reflexive, so is the* S-*module* $\mathrm{M}_{(\mathrm{S})} = \mathrm{M} \otimes_\mathrm{R} \mathrm{S}$.

We know (Chapter I, § 2, no. 10, Proposition 11) that there exists a canonical isomorphism $\omega_\mathrm{M} : (\mathrm{M}^*)_{(\mathrm{S})} \to (\mathrm{M}_{(\mathrm{S})})^*$, such that

$$\langle x \otimes 1, \omega_\mathrm{M}(x^* \otimes 1) \rangle = \rho(\langle x, x^* \rangle)$$

for $x \in M$, $x^* \in M^*$. As M is a quotient of a finitely generated free R-module L, M^* is isomorphic to a sub-R-module of the dual L^* and L^* is free and finitely generated; since R is Noetherian, M^* is therefore also a finitely generated R-module, whence an isomorphism $\omega_{M^*}: (M^{**})_{(S)} \to ((M^*)_{(S)})^*$ such that

$$\langle x^* \otimes 1, \omega_{M^*}(x^{**} \otimes 1)\rangle = \rho(\langle x^*, x^{**}\rangle)$$

for $x^* \in M$ and $x^{**} \in M^{**}$. On the other hand, there is an isomorphism ${}^t\omega_M: (M_{(S)})^{**} \to ((M^*)_{(S)})^*$, whence by composition a canonical isomorphism:

$$\phi = ({}^t\omega_M^{-1}) \circ (\omega_{M^*}): (M^{**})_{(S)} \to (M_{(S)})^{**}$$

such that, in the above notation:

$$(1) \qquad \langle \omega_M(x^* \otimes 1), \phi(x^{**} \otimes 1)\rangle = \rho(\langle x^*, x^{**}\rangle).$$

We consider now the canonical homomorphism $c_M: M \to M^{**}$ and show that the composite homomorphism:

$$(2) \qquad \psi: M_{(S)} \xrightarrow[c_M \otimes 1]{} (M^{**})_{(S)} \longrightarrow (M_{(S)})^{**}$$

is just the canonical homomorphism $c_{M_{(S)}}$. This follows immediately from (1) which gives the relations:

$$\langle \omega_M(x^* \otimes 1), \psi(x \otimes 1)\rangle = \rho(\langle x^*, c_M(x)\rangle)$$
$$= \rho(\langle x, x^*\rangle) = \langle x \otimes 1, \omega_M(x^* \otimes 1)\rangle$$

and from the fact that the elements $\omega_M(x^* \otimes 1)$ generate $(M_{(S)})^*$. This being so, the hypothesis that M is reflexive means that c_M is bijective, hence so is $c_M \otimes 1$ and therefore $\psi = c_{M_{(S)}}$ is bijective, which shows the proposition.

3. LOCAL CONSTRUCTION OF REFLEXIVE MODULES

We keep the notation and hypotheses of no. 2. We shall say that a property holds "*for almost all* $\mathfrak{p} \in P$" if the set of $\mathfrak{p} \in P$ for which it is not true is *finite*.

THEOREM 3. *Let* V *be a vector space of finite rank over* K *and* M *a lattice of* V *with respect to* A.

(i) *Let* N *be a lattice of* V *with respect to* A; *then, for every prime ideal* $\mathfrak{p}$ *of* A, $N_\mathfrak{p}$ *is a lattice of* V *with respect to* $A_\mathfrak{p}$ *and, for almost all* $\mathfrak{p} \in P$, $N_\mathfrak{p} = M_\mathfrak{p}$.

(ii) *Conversely, suppose given for all* $\mathfrak{p} \in P$ *a lattice* $N(\mathfrak{p})$ *of* V *with respect to* $A_\mathfrak{p}$ *such that* $N(\mathfrak{p}) = M_\mathfrak{p}$ *for almost all* $\mathfrak{p} \in P$. *Then* $N = \bigcap_{\mathfrak{p} \in P} N(\mathfrak{p})$ *is a reflexive lattice of* A *with respect to* A *and it is the only reflexive lattice* N′ *of* V *with respect to* A *such that* $N'_\mathfrak{p} = N(\mathfrak{p})$ *for all* $\mathfrak{p} \in P$.

(i) The first assertion follows from no. 1, Proposition 4. Moreover, there exist x, y in K^* such that $xN \subset M \subset yN$ (no. 1, Proposition 2); we know that, for almost all $\mathfrak{p} \in P$, $v_\mathfrak{p}(x) = v_\mathfrak{p}(y) = 0$ (§ 1, no. 6, Theorem 4), which shows that x and y are invertible in $A_\mathfrak{p}$ and hence $M_\mathfrak{p} = N_\mathfrak{p}$.

(ii) We may replace M by $x^{-1}M$ where $x \neq 0$ in A and assume that $N(\mathfrak{p}) \subset M_\mathfrak{p}$ for all $\mathfrak{p} \in P$. Let $\mathfrak{p}_1, \ldots, \mathfrak{p}_h$ be the elements of P such that $N(\mathfrak{p}) = M_\mathfrak{p}$ for $\mathfrak{p}$ distinct from the $\mathfrak{p}_i$ $(1 \leqslant i \leqslant h)$; we write:

$$Q = M \cap N(\mathfrak{p}_1) \cap \cdots \cap N(\mathfrak{p}_h).$$

As each of the $N(\mathfrak{p}_i)$ contain a free lattice with respect to $A_{\mathfrak{p}_i}$, it contains *a fortiori* a lattice of V with respect to A, hence Q contains a lattice of V with respect to A (no. 1, Proposition 3) and, as Q is contained in M, Q is a lattice with respect to A. To prove that $Q_\mathfrak{p} = N(\mathfrak{p})$ for all $\mathfrak{p} \in P$, we shall use the following lemma:

Lemma 1. *Let* $\mathfrak{p}$ *and* $\mathfrak{p}'$ *be two prime ideals of* A *such that* (0) *is the only prime ideal of* A *contained in* $\mathfrak{p} \cap \mathfrak{p}'$. *Then, for every sub-A-module* E *of* V, $(E_\mathfrak{p})_{\mathfrak{p}'} = K.E$.

Let S be the multiplicative subset $(A - \mathfrak{p})(A - \mathfrak{p}')$ of A; by Chapter II, § 2, no. 3, Proposition 7, $(E_\mathfrak{p})_{\mathfrak{p}'} = S^{-1}E$. Further, $A \subset S^{-1}A \subset K$; the prime ideals of $S^{-1}A$ correspond to the prime ideals $\mathfrak{q}$ of A such that $\mathfrak{q} \cap S = \varnothing$ (Chapter II, § 2, no. 5, Proposition 11) and by hypothesis (0) is the only prime ideal of A not meeting S; hence $S^{-1}A = K$ and $S^{-1}E = K.E$.

We now return to the proof of (ii). If $\mathfrak{p} \in P$ is distinct from the $\mathfrak{p}_i$ $(1 \leqslant i \leqslant h)$, Lemma 1 applied to $N(\mathfrak{p}_i)$ gives $(N(\mathfrak{p}_i))_\mathfrak{p} = ((N(\mathfrak{p}_i))_{\mathfrak{p}_i})_\mathfrak{p} = K.N(\mathfrak{p}_i) = V$, since the $\mathfrak{p}_i$ and $\mathfrak{p}$ are of height 1. Then

$$Q_\mathfrak{p} = M_\mathfrak{p} \cap (N(\mathfrak{p}_1))_\mathfrak{p} \cap \cdots \cap (N(\mathfrak{p}_h))_\mathfrak{p} = M_\mathfrak{p} = N(\mathfrak{p})$$

(Chapter II, § 2, no. 4). On the other hand, if $\mathfrak{p}$ is equal to $\mathfrak{p}_i$ $(1 \leqslant i \leqslant h)$, then $(N(\mathfrak{p}_i))_{\mathfrak{p}_j} = V$ for $i \neq j$ by the argument as above and $(N(\mathfrak{p}_i))_{\mathfrak{p}_i} = N(\mathfrak{p}_i)$, whence

$$Q_{\mathfrak{p}_i} = M_{\mathfrak{p}_i} \cap N(\mathfrak{p}_i) = N(\mathfrak{p}_i).$$

We have therefore proved that $Q_\mathfrak{p} = N(\mathfrak{p})$ for all $\mathfrak{p} \in P$. Then

$$N = Q^{**} = \bigcap_{\mathfrak{p} \in P} Q_\mathfrak{p}$$

is reflexive and satisfies the relations $N_\mathfrak{p} = Q_\mathfrak{p} = N(\mathfrak{p})$ for all $\mathfrak{p} \in P$; the uniqueness property follows immediately from Theorem 2 of no. 2.

Remark. Let L be a free lattice of V with respect to A. Since $A_\mathfrak{p}$ is a principal ideal domain for $\mathfrak{p} \in P$, $N(\mathfrak{p})$ is a free $A_\mathfrak{p}$-module of the same rank as L and there exists $u(\mathfrak{p}) \in \mathbf{GL}(V)$ such that $u(\mathfrak{p})(L_\mathfrak{p}) = N_\mathfrak{p}$; this condition moreover determines $u(\mathfrak{p})$ to within right multiplication by an element of $\mathbf{GL}(L_\mathfrak{p})$. The condition $N(\mathfrak{p}) = L_\mathfrak{p}$ for almost all $\mathfrak{p} \in P$ means that necessarily $u(\mathfrak{p}) \in \mathbf{GL}(L_\mathfrak{p})$

for almost all $\mathfrak{p} \in P$. The families $(u(\mathfrak{p}))_{\mathfrak{p} \in P}$ satisfying the latter property form a multiplicative group $\mathbf{GL}_a(V)$ containing as subgroup the product $\prod_{\mathfrak{p} \in P} \mathbf{GL}(L_\mathfrak{p})$. Theorem 3 then shows that *the set of reflexive lattices of* V *is in canonical one-to-one correspondence with the homogeneous space* $\mathbf{GL}_a(V)/\prod_{\mathfrak{p} \in P} \mathbf{GL}(L_\mathfrak{p})$. If a basis $(e_i)_{1 \leqslant i \leqslant n}$ of L over A is chosen, $\mathbf{GL}(V)$ (resp. $\mathbf{GL}(L_\mathfrak{p})$) is identified with the group of invertible matrices $\mathbf{GL}(n, K)$ (resp. $\mathbf{GL}(n, A_\mathfrak{p})$) and the group $\mathbf{GL}_a(V)$ with the group of systems of matrices of order n, $(U(\mathfrak{p}))_{\mathfrak{p} \in P}$, such that $U(\mathfrak{p}) \in \mathbf{GL}(n, K)$ for all $\mathfrak{p} \in P$ and $U(\mathfrak{p}) \in \mathbf{GL}(n, A_\mathfrak{p})$ for almost all $\mathfrak{p} \in P$. If A is a Dedekind domain, the group $\mathbf{GL}_a(V)$ is also identified with the group $\mathbf{GL}(n, \boldsymbol{A})$, where $\boldsymbol{A}$ is the ring of restricted adèles (§ 2, no. 4).

4. PSEUDO-ISOMORPHISMS

We preserve the notation and hypotheses of nos. 2 and 3.

PROPOSITION 9. *Let* M *be a finitely generated* A*-module. The following conditions are equivalent:*

(a) $M_\mathfrak{p} = 0$ *for every prime ideal* $\mathfrak{p}$ *of height* $\leqslant 1$.

(b) *The annihilator* $\mathfrak{a}$ *of* M *is an ideal* $\neq (0)$ *and* $A : \mathfrak{a} = A$ ($A : \mathfrak{a}$ *denoting, as in* § 1, no. 1, *the set of* $x \in K$ *such that* $x\mathfrak{a} \in A$).

We know (Chapter II, § 2, no. 2, Corollary 2 to Proposition 4) that the condition $M_\mathfrak{p} = 0$ is equivalent to $\mathfrak{a} \not\subset \mathfrak{p}$ and hence to $\mathfrak{a}A_\mathfrak{p} = A_\mathfrak{p}$ (Chapter II, § 2, no. 5, *Remark*); on the other hand, for every integral ideal $\mathfrak{b} \neq 0$ of A, the relation "$\mathfrak{b}A_\mathfrak{p} = A_\mathfrak{p}$ for all $\mathfrak{p} \in P$" is equivalent to $\operatorname{div} \mathfrak{b} = \operatorname{div} A = 0$ in $D(A)$ (§ 1, no. 4, Proposition 7), or also to $\operatorname{div}(A : \mathfrak{b}) = 0$ and, as $A : \mathfrak{b}$ is divisorial (§ 1, no. 1, Proposition 1), this relation is also equivalent to $A : \mathfrak{b} = A$. The proposition then follows by noting that to say that $\mathfrak{a} \not\subset \mathfrak{p}$ for $\mathfrak{p} = (0)$ means that $\mathfrak{a} \neq (0)$.

Remark (1) The equivalent conditions of Proposition 9 mean also that Ass(M) contains no prime ideal of height $\leqslant 1$. * They may also be interpreted by saying that Supp(M) is of *codimension* $\geqslant 2$ in Spec(A). *

DEFINITION 2. *An* A*-module* M *is called pseudo-zero if it is finitely generated and it satisfies the equivalent conditions of Proposition* 9.

This definition and Proposition 9 show that a pseudo-zero A-module is a *torsion A-module*; the converse is false.

Examples

(1) If A is a Dedekind domain, every prime ideal of A is of height $\leqslant 1$; to say that M is pseudo-zero means then that $\operatorname{Supp}(M) = \varnothing$ and hence that $M = 0$ (Chapter II, § 4, no. 4).

(2) Let k be a field and $A = k[X, Y]$ the polynomial ring over k in two indeterminates; if $\mathfrak{m}$ is the maximal ideal $AX + AY$ of A, the A-module $A/\mathfrak{m}$ is pseudo-zero; for its annihilator $\mathfrak{m}$ is not of height $\leqslant 1$ since it contains the principal prime ideals AX and AY and is distinct from them; therefore $A : \mathfrak{m} = A$ (§ 1, no. 6, Corollary 1 to Theorem 3).

DEFINITION 3. *Let* M *and* N *be two* A*-modules and* $f: M \to N$ *a homomorphism.* f *is called pseudo-injective* (resp. *pseudo-surjective, pseudo-zero*) *if* $\mathrm{Ker}(f)$ (resp. $\mathrm{Coker}(f)$, $\mathrm{Im}(f)$) *is pseudo-zero;* f *is called pseudo-bijective if it is both pseudo-injective and pseudo-surjective.*

A pseudo-bijective homomorphism is also called a *pseudo-isomorphism.*

Suppose that M and N are finitely generated; then, for $f: M \to N$ to be pseudo-injective (resp. pseudo-surjective, pseudo-zero), it is necessary and sufficient that, for all $\mathfrak{p} \in P \cup \{\{0\}\}$, $f_\mathfrak{p}: M_\mathfrak{p} \to N_\mathfrak{p}$ be injective (resp. surjective, zero); this follows from the flatness of the A-module $A_\mathfrak{p}$ (cf. Chapter I, § 2, no. 3, *Remark* 2).

Example (3) Let M be a torsion-free finitely generated A-module; then the canonical mapping $c_M: M \to M^{**}$ of M to its bidual is a *pseudo-isomorphism.* For M is identified with a lattice of $V = M \otimes_A K$ (no. 1, Proposition 1); we have seen that $M_\mathfrak{p} = M_\mathfrak{p}^{**}$ for all $\mathfrak{p} \in P$ (no. 2, Example 2) and, for $\mathfrak{p} = 0$, $M_\mathfrak{p}$ and $M_\mathfrak{p}^{**}$ are both equal to V.

THEOREM 4. *Let* E *be a finitely generated* A*-module,* T *the torsion submodule of* E *and* $M = E/T$. *There exists a pseudo-isomorphism*

$$f: E \to T \times M.$$

We shall first show two lemmas.

LEMMA 2. *Let* $(\mathfrak{p}_i)_{1 \leqslant i \leqslant k}$ *be a non-empty finite family of prime ideals of* A *of height* 1 *and* let $S = \bigcap_i (A - \mathfrak{p}_i)$; *then the ring* $S^{-1}A$ *is a principal ideal domain.*

$S^{-1}A$ is a semi-local ring whose maximal ideals are the $\mathfrak{m}_i = \mathfrak{p}_i S^{-1}A$ for $1 \leqslant i \leqslant k$, the local ring $(S^{-1}A)_{\mathfrak{m}_i}$ being isomorphic to $A_{\mathfrak{p}_i}$ (Chapter II, § 3, no. 5, Proposition 17) and hence a discrete valuation ring. The ring $S^{-1}A$ is therefore a Dedekind domain (§ 2, no. 2, Theorem 1 (f) and, as it is semi-local it is a principal ideal domain (§ 2, no. 2, Proposition 1).

LEMMA 3. *There exists a homomorphism* $g: E \to T$ *whose restriction to* T *is both a homothety and a pseudo-isomorphism.*

Let $\mathfrak{a}$ be the annihilator of T; as T is a finitely generated torsion A-module, $\mathfrak{a} \neq 0$. Let $\mathfrak{p}_i$ $(1 \leqslant i \leqslant k)$ be the prime ideals of height 1 containing $\mathfrak{a}$ (which are finite in number (§ 1, no. 6, Theorem 4)); if this number is 0, T is pseudo-zero (Proposition 9(a)) and we may take $g = 0$. Otherwise, let $S = \bigcap_i (A - \mathfrak{p}_i)$; by Lemma 2, $S^{-1}A$ is a principal ideal domain and hence $S^{-1}M$, which is a torsion-free finitely generated $S^{-1}A$-module, is *free* (*Algebra*, Chapter VII, § 4, no. 3, Corollary 2 to Theorem 2) and, as $S^{-1}M = (S^{-1}E)/(S^{-1}T)$, $S^{-1}T$ is a direct factor of $S^{-1}E$ (*Algebra*, Chapter II, § 1, no. 11, Proposition 21). Now,

$$\mathrm{Hom}_{S^{-1}A}(S^{-1}E, S^{-1}T) = S^{-1}\,\mathrm{Hom}_A(E, T)$$

(Chapter II, § 2, no. 7, Proposition 19); hence there exist $s_0 \in S$ and $g_0 \in \mathrm{Hom}_A(E, T)$ such that $s_0^{-1}g_0$ is a projector of $S^{-1}E$ onto $S^{-1}T$. If $h_0 \in \mathrm{Hom}_A(T, T)$ denotes the restriction of g_0 to T, there therefore exists $s_1 \in S$ such that $s_1h_0(x) = s_1s_0x$ for all $x \in T$; writing $s = s_1s_0$, $g = s_1g_0$, $h = s_1h_0$, h is therefore the homothety of ratio s on T and is the restriction of g to T. It remains to verify that h is a pseudo-isomorphism. Now, if $\mathfrak{p} = 0$ or if $\mathfrak{p} \in P$ is distinct from the $\mathfrak{p}_i$ $(1 \leqslant i \leqslant k)$, $T_\mathfrak{p} = 0$ (Chapter II, § 4, no. 4, Proposition 17) and $h_\mathfrak{p}: T_\mathfrak{p} \to T_\mathfrak{p}$ is an isomorphism; if on the contrary $\mathfrak{p}$ is equal to one of the $\mathfrak{p}_i$ $(1 \leqslant i \leqslant k)$, s is invertible in $A_{\mathfrak{p}_i}$ and $h_{\mathfrak{p}_i}$, the homothety of ratio s on $T_{\mathfrak{p}_i}$, is also an isomorphism, which completes the proof of Lemma 3.

We now prove Theorem 4. Let $g: E \to T$ be a homomorphism satisfying the properties of Lemma 3; let h be the restriction of g to T and let π be the canonical projection of E onto M. We show that the homomorphism $f = (g, \pi)$: $E \to T \times M$ solves the problem. There is the commutative diagram:

$$\begin{array}{ccccccccc}
0 & \longrightarrow & T & \longrightarrow & E & \longrightarrow & M & \longrightarrow & 0 \\
 & & \downarrow{\scriptstyle h} & & \downarrow{\scriptstyle f} & & \downarrow{\scriptstyle 1_M} & & \\
0 & \longrightarrow & T & \longrightarrow & T \times M & \longrightarrow & M & \longrightarrow & 0
\end{array}$$

where the rows are exact. The snake diagram (Chapter I, § 1, no. 4, Proposition 2) gives the exact sequence:

$$0 \to \mathrm{Ker}(h) \to \mathrm{Ker}(f) \to 0 \to \mathrm{Coker}(h) \to \mathrm{Coker}(f) \to 0$$

and hence $\mathrm{Ker}(f)$ is isomorphic to $\mathrm{Ker}(h)$ and $\mathrm{Coker}(f)$ to $\mathrm{Coker}(h)$. As h is a pseudo-isomorphism, so is f.

We can say that "to within a pseudo-isomorphism" Theorem 4 reduces the study of finitely generated A-modules to that of torsion-free modules on the one hand and to that of torsion modules on the other. Moreover, we have seen above (*Example* 3) that a torsion-free module is pseudo-isomorphic to its bidual and hence to a *reflexive* module. As for torsion modules, there is the following result, which determines them to within a pseudo-isomorphism:

THEOREM 5. *Let* T *be a finitely generated torsion* A*-module. There exist two finite families* $(n_\iota)_{\iota \in I}$ *and* $(\mathfrak{p}_\iota)_{\iota \in I}$, *where the* n_ι *are integers* $\geqslant 1$ *and the* $\mathfrak{p}_\iota$ *are prime ideals of* A *of height* 1 *such that, if we write* $T' = \bigoplus_{\iota \in I} A/\mathfrak{p}_\iota^{n_\iota}$, *there exists a pseudo-isomorphism of* T *to* T'. *Moreover, the families* $(n_\iota)_{\iota \in I}$ *and* $(\mathfrak{p}_\iota)_{\iota \in I}$ *with this property are unique to within a bijection of the indexing set and the* $\mathfrak{p}_\iota$ *contain the annihilator of* T.

Uniqueness: If $f\colon T \to T'$ is a pseudo-isomorphism and $\mathfrak{p} \in P$, $f_\mathfrak{p}\colon T_\mathfrak{p} \to T'_\mathfrak{p}$ is an isomorphism. Now, $T'_\mathfrak{p}$ is the direct sum of the $A_\mathfrak{p}/\mathfrak{p}^{n_\iota}A_\mathfrak{p}$, the sum being over the indices ι such that $\mathfrak{p}_\iota = \mathfrak{p}$; the $\mathfrak{p}^{n_\iota}A_\mathfrak{p}$ are therefore the *elementary divisors* of the torsion $A_\mathfrak{p}$-module $T_\mathfrak{p}$ (*Algebra*, Chapter VII, § 4, no. 7); their uniqueness has been proved in *Algebra*, Chapter VII, § 4, no. 7, Proposition 7.

Existence: We may confine our attention to the case where $T \neq 0$. Let $\mathfrak{a}$ be the annihilator (non-zero and distinct from A) of T, $\mathfrak{p}_i$ $(1 \leqslant i \leqslant k)$ the prime ideals of A of height 1 containing $\mathfrak{a}$ (which are finite in number (§ 1, no. 6, Theorem 4)) and $S = \bigcap_i (A - \mathfrak{p}_i)$. The semi-local ring $A' = S^{-1}A$ is a principal ideal domain (Lemma 2) and has maximal ideals the $\mathfrak{m}_i = \mathfrak{p}_iA'$; as $S^{-1}T$ is a finitely generated torsion A'-module, it is isomorphic to a finite direct sum $\bigoplus_{j \in I} A'/\mathfrak{m}_{\phi(j)}^{n_j}$, where ϕ is a mapping of a finite set I to $[1, k]$ (*Algebra*, Chapter VII, § 4, no. 7, Proposition 7); as $A'/\mathfrak{m}_{\phi(j)}^{n_j}$ is isomorphic to $S^{-1}(A/\mathfrak{p}_{\phi(j)}^{n_j})$ (Chapter II, § 2, no. 4), we have obtained a torsion A-module T' of the desired type and an isomorphism f_0 of $S^{-1}T$ onto $S^{-1}T'$. As $\operatorname{Hom}_{S^{-1}A}(S^{-1}T, S^{-1}T')$ is equal to $S^{-1} \operatorname{Hom}_A(T, T')$ (Chapter II, § 2, no. 7, Proposition 19), there exist $s \in S$ and a homomorphism $f\colon T \to T'$ such that $f_0 = s^{-1}f$. It remains to show that f is a pseudo-isomorphism: now, if $\mathfrak{p} = 0$ or if $\mathfrak{p} \in P$ is distinct from the $\mathfrak{p}_i$, $T_\mathfrak{p} = T'_\mathfrak{p} = 0$ (Chapter II, § 4, no. 4, Proposition 17); if on the contrary $\mathfrak{p}$ is one of the $\mathfrak{p}_i$ $(1 \leqslant i \leqslant k)$, s is invertible in $A_{\mathfrak{p}_i}$ and, as $f_{\mathfrak{p}_i} = s(f_0)_{\mathfrak{p}_i}$ and $(f_0)_{\mathfrak{p}_i}$ is an isomorphism of $T_{\mathfrak{p}_i} = (S^{-1}T)_{\mathfrak{m}_i}$ onto $T'_{\mathfrak{p}_i} = (S^{-1}T))_{\mathfrak{m}_i}$, so is $f_{\mathfrak{p}_i}$.

Remark (2) In the statement of Theorem 5, the modules $A/\mathfrak{p}_\iota^{n_\iota}$ may be replaced by $A/\mathfrak{p}_\iota^{(n_\iota)}$ (§ 1, no. 4, Proposition 8). For all $\mathfrak{p} \in P$, the canonical mapping $g\colon A/\mathfrak{p}^n \to A/\mathfrak{p}^{(n)} = A/(A \cap \mathfrak{p}^nA_\mathfrak{p})$ is a pseudo-isomorphism, as, for all $\mathfrak{q} \in P$ distinct from $\mathfrak{p}$, $A_\mathfrak{q}/\mathfrak{p}^nA_\mathfrak{q} = A_\mathfrak{q}/\mathfrak{p}^{(n)}A_\mathfrak{q} = 0$ and $A_\mathfrak{p}/\mathfrak{p}^nA_\mathfrak{p} = A_\mathfrak{p}/\mathfrak{p}^{(n)}A_\mathfrak{p}$.

* Given an exact sequence of A-modules, $E \to F \to G$, if E and G are pseudo-zero, so is F, as follows from Definition 2 and Chapter II, § 2, no. 4, Theorem 1. In the language of categories, we may then say that, in the category $\mathscr{C}$ of A-modules, the sub-category $\mathscr{C}'$ of pseudo-zero modules is *full* and we may then define the *quotient* category $\mathscr{C}/\mathscr{C}'$: the objects in this category are also A-modules but the set of morphisms from E to F (for E, F in $\mathscr{C}/\mathscr{C}'$) is the direct limit of the set

of commutative groups $\mathrm{Hom}_A(E', F')$, where E′ (resp. F′) runs through the set of submodules of E (resp. the set of quotient modules F/F″ of F) such that E/E′ (resp. F″) is pseudo-zero. Of course, for every ordered pair of A-modules E, F, there is a canonical homomorphism $\mathrm{Hom}_{\mathscr{C}}(E, F) \to \mathrm{Hom}_{\mathscr{C}/\mathscr{C}'}(E, F)$. To say that a homomorphism $u \in \mathrm{Hom}_A(E, F)$ is pseudo-zero (resp. pseudo-injective, pseudo-surjective, pseudo-bijective) means that its canonical image in $\mathrm{Hom}_{\mathscr{C}/\mathscr{C}'}(E, F)$ is zero (resp. a monomorphism, an epimorphism, an isomorphism). *

5. DIVISORS ATTACHED TO TORSION MODULES

We keep the same notation and hypotheses as in nos. 2, 3 and 4. Recall that D(A) (or simply D) denotes the *divisor group* of A, written additively: we know (§ 1, no. 3, Theorem 2) that D is the free **Z**-module generated by the elements of P.

Let T be a finitely generated torsion A-module. For all $\mathfrak{p} \in P$, $T_\mathfrak{p}$ is a finitely generated torsion $A_\mathfrak{p}$-module and hence a module of *finite length* (Chapter IV, § 2, no. 5, Corollary 2 to Proposition 7); we shall denote this length by $l_\mathfrak{p}(T)$. Now $T_\mathfrak{p} = 0$ for all $\mathfrak{p}$ not containing the annihilator of T and hence for almost all $\mathfrak{p}$ (§ 1, no. 6, Theorem 4), which justifies the following definition:

Definition 4. *If* T *is a finitely generated torsion* A*-module, the divisor:*

$$\chi(T) = \sum_{\mathfrak{p}\in P} l_\mathfrak{p}(T) . \mathfrak{p}.$$

is called the content of T.

Proposition 10. (i) *Let* $0 \to T_1 \to T_2 \to T_3 \to 0$ *be an exact sequence of finitely generated torsion* A*-modules. Then*

$$\chi(T_2) = \chi(T_1) + \chi(T_3).$$

(ii) *If there exists a pseudo-isomorphism* $f: T_1 \to T_2$, *then* $\chi(T_1) = \chi(T_2)$.

(iii) *In order that* $\chi(T) = 0$, *it is necessary and sufficient that* T *be pseudo-zero.*

In view of Definition 4, it suffices to consider for each $\mathfrak{p} \in P$ the values of $l_\mathfrak{p}$ for the torsion modules considered. Property (i) then follows from Chapter II, § 2, no. 4, Theorem 1 and the additivity of lengths in an exact sequence (*Algebra* Chapter II, § 1, no. 10, Proposition 16) and properties (ii) and (iii) follow immediately from the definitions in no. 4.

Corollary. *Let* $0 \to T_n \to T_{n-1} \to \cdots \to T_0 \to 0$ *be an exact sequence of finitely generated torsion* A*-modules. Then* $\sum_{i=0}^{n} (-1)^i \chi(T_i) = 0$.

In view of Chapter II, § 2, no. 4, Theorem 1, this follows again from the analogous property of the $l_\mathfrak{p}$ (*Algebra*, Chapter II, § 1, no. 10, Corollary 3 to Proposition 16).

Recall (Chapter II, § 5, no. 4) that we may speak of *the set* $\boldsymbol{F}(A)$ *of classes of finitely generated* A-*modules* with respect to the relation of isomorphism; for every finitely generated A-module M, let cl(M) denote the corresponding element of $\boldsymbol{F}(A)$; we shall denote by $\boldsymbol{T}(A)$ the subset of $\boldsymbol{F}(A)$ consisting of the classes of finitely generated torsion A-modules. Clearly χ defines a mapping of $\boldsymbol{T}(A)$ to D(A), also denoted by χ, such that $\chi(\mathrm{cl}(T)) = \chi(T)$.

PROPOSITION 11. *Let* G *be a commutative group written additively and* $\phi: \boldsymbol{T}(A) \to G$ *a mapping; for every finitely generated torsion* A-*module* T, *we also write, by an abuse of language,* $\phi(T) = \phi(\mathrm{cl}(T))$. *Suppose that the following conditions are satisfied:*

(1) *If* $0 \to T_1 \to T_2 \to T_3 \to 0$ *is an exact sequence of finitely generated torsion* A-*modules, then* $\phi(T_2) = \phi(T_1) + \phi(T_3)$.

(2) *If* T *is pseudo-zero, then* $\phi(T) = 0$.

Then there exists a unique homomorphism $\theta: D(A) \to G$ *such that* $\phi = \theta \circ \chi$.

As $\chi(A/\mathfrak{p}) = \mathfrak{p}$ for all $\mathfrak{p}$, necessarily $\theta(\mathfrak{p}) = \phi(A/\mathfrak{p})$ for all $\mathfrak{p} \in P$, which proves the uniqueness of θ, since the elements of P form a basis of D(A). Conversely, let θ be the homomorphism from D(A) to G such that $\theta(\mathfrak{p}) = \phi(A/\mathfrak{p})$ for all $\mathfrak{p} \in P$ and let us show that it solves the problem. For this, we write

$$\psi(T) = \phi(T) - \theta(\chi(T))$$

for every finitely generated torsion A-module T; clearly conditions (1) and (2) are also satisfied if ϕ is replaced by ψ. On the other hand, $\psi(A/\mathfrak{p}) = 0$ if $\mathfrak{p} \in P$; if $\mathfrak{p}$ is a prime ideal $\neq 0$ and not in P, the annihilator of $A/\mathfrak{p}$ is contained in no ideal of P, hence (no. 4, Theorem 5) $A/\mathfrak{p}$ is pseudo-zero and therefore $\psi(A/\mathfrak{p}) = 0$. This being so, every finitely generated torsion A-module T admits a decomposition series whose factors are isomorphic to A-modules of the form $A/\mathfrak{p}$, where $\mathfrak{p} \in \mathrm{Supp}(T)$ (Chapter IV, § 1, no. 4, Theorems 1 and 2), and hence $\mathfrak{p} \neq 0$ since T is a torsion module. By induction on the length of this decomposition series, we deduce (in view of property (1) for ψ) that $\psi(T) = 0$.

* We may, as in no. 4, consider the quotient category $\mathcal{T}/\mathcal{T}'$ of the category $\mathcal{T}$ of finitely generated torsion A-modules by the full sub-category $\mathcal{T}'$ of pseudo-zero finitely generated torsion A-modules. In the language of Abelian categories, Proposition 11 then expresses the fact that the *Grothendieck group* of the Abelian category $\mathcal{T}/\mathcal{T}'$ is canonically isomorphic to D(A). *

PROPOSITION 12. *If* $\mathfrak{a}$ *is an ideal* $\neq 0$ *of* A,

$$\chi(A/\mathfrak{a}) = \chi((A : \mathfrak{a})/A) = \operatorname{div} \mathfrak{a}.$$

Let $\mathfrak{p} \in P$. Then $\mathfrak{a}A_\mathfrak{p} = \mathfrak{p}^{n_\mathfrak{p}}A_\mathfrak{p}$ where $n_\mathfrak{p} \geqslant 0$, since $A_\mathfrak{p}$ is a discrete valuation ring. As $(A/\mathfrak{a})_\mathfrak{p} = A_\mathfrak{p}/\mathfrak{a}A_\mathfrak{p}$, $l_\mathfrak{p}(A/\mathfrak{a}) = n_\mathfrak{p}$, whence $\chi(A/\mathfrak{a}) = \sum_{\mathfrak{p} \in P} n_\mathfrak{p}\mathfrak{p} = \operatorname{div} \mathfrak{a}$ (§ 1, no. 4, Proposition 7).

On the other hand, $(A: \mathfrak{a})_{\mathfrak{p}} = A_{\mathfrak{p}}: \mathfrak{a}A_{\mathfrak{p}} = \mathfrak{p}^{-n_{\mathfrak{p}}}A_{\mathfrak{p}}$, hence $l_{\mathfrak{p}}((A: \mathfrak{a})/A) = n_{\mathfrak{p}}$ and we conclude in the same way.

6. RELATIVE INVARIANT OF TWO LATTICES

We keep the notation and hypotheses of nos. 2 to 5. Let V be a vector space of finite rank n over K and M a lattice of V with respect to A. Let W be the exterior power $\bigwedge^n V$, which is a vector space of *rank* 1 over K and let M_W denote the lattice of W generated by the image of M^n under the canonical mapping $V^n \to \bigwedge^n V$ (no. 1, Proposition 3 (iii); note that M_W is not necessarily isomorphic to $\bigwedge^n M$ (*Algebra*, Chapter III, § 5, Exercise 9)). If e is a basis of W over K, we may write $M_W = \mathfrak{a}.e$, where $\mathfrak{a}$ is a fractional ideal $\neq 0$ of A.

Let M′ be another lattice of V and let us write $M'_W = \mathfrak{a}'.e$, where $\mathfrak{a}'$ is a fractional ideal $\neq 0$ of A; the divisor $\operatorname{div}(\mathfrak{a}) - \operatorname{div}(\mathfrak{a}')$ *does not depend on the choice of basis e of* W, $\mathfrak{a}$ and $\mathfrak{a}'$ being multiplied by the same element of K^* when the basis is changed; we shall write $\chi(M, M') = \operatorname{div}(\mathfrak{a}) - \operatorname{div}(\mathfrak{a}')$ and say that this divisor is the *relative invariant of* M′ *with respect to* M. Clearly, if M, M′, M″ are three lattices of V, then:

$$\chi(M, M') + \chi(M', M'') + \chi(M'', M) = 0 \tag{3}$$

$$\chi(M, M') + \chi(M', M) = 0. \tag{4}$$

For all $\mathfrak{p} \in P$, it follows immediately from the definitions that $(M_W)_{\mathfrak{p}} = (M_{\mathfrak{p}})_W$; moreover, since $M_{\mathfrak{p}}$ is then a *free* $A_{\mathfrak{p}}$-module since $A_{\mathfrak{p}}$ is a principal ideal domain, a basis of $M_{\mathfrak{p}}$ over $A_{\mathfrak{p}}$ is a basis of V over K, hence $(M_{\mathfrak{p}})_W = \bigwedge^n (M_{\mathfrak{p}})$ (Chapter II, § 2, no. 8) and the fractional ideal $\mathfrak{a}_{\mathfrak{p}} = \mathfrak{a}A_{\mathfrak{p}}$ is a principal ideal domain. If we set $\mathfrak{a}_{\mathfrak{p}} = \mathfrak{p}^{n_{\mathfrak{p}}}A_{\mathfrak{p}}$, $\mathfrak{a}' = \mathfrak{p}^{n'_{\mathfrak{p}}}A_{\mathfrak{p}}$, then:

$$\chi(M, M') = \sum_{\mathfrak{p} \in P} (n_{\mathfrak{p}} - n'_{\mathfrak{p}}).\mathfrak{p},$$

which may also be written as:

$$\chi(M, M') = \sum_{\mathfrak{p} \in P} \chi(M_{\mathfrak{p}}, M'_{\mathfrak{p}}) \tag{5}$$

identifying $D(A_{\mathfrak{p}})$ with the sub-**Z**-module of D(A) generated by $\mathfrak{p}$.

PROPOSITION 13. *Let* M *be a lattice of* V *and* u *a* K*-automorphism of* V. *Then:*

$$-\chi(M, u(M)) = \operatorname{div}(\det(u)). \tag{6}$$

For all $\mathfrak{p} \in P$, then $\bigwedge^n (u(M)_{\mathfrak{p}}) = \bigwedge^n (u(M_{\mathfrak{p}}))$; if $(e_i)_{1 \leqslant i \leqslant n}$ is a basis of $M_{\mathfrak{p}}$, then

$$\bigwedge^n (M_{\mathfrak{p}}) = A_{\mathfrak{p}}.e_1 \wedge e_2 \wedge \cdots \wedge e_n,$$

and $\bigwedge^{n}(u(M_{\mathfrak{p}})) = A_{\mathfrak{p}}.\det(u)e_1 \wedge e_2 \wedge \cdots \wedge e_n$, whence the proposition by virtue of formula (5).

PROPOSITION 14. *If* M, M′ *are two lattices of* V *such that* $M' \subset M$, M/M' *is a finitely generated torsion* A*-module and:*

$$(7) \qquad \chi(M, M') = -\chi(M/M').$$

Clearly $M/M' \subset V/M'$ is a finitely generated torsion module; on the other hand, for all $\mathfrak{p} \in P$, we know (*Algebra*, Chapter VII, § 4, no. 2, Theorem 1) that there exist bases $(e_i)_{1 \leqslant i \leqslant n}$ of $M_{\mathfrak{p}}$ and $(e'_i)_{1 \leqslant i \leqslant n}$ of $M'_{\mathfrak{p}}$ such that $e'_i = \pi^{\nu_i} e_i$ for $1 \leqslant i \leqslant n$ and integers $\nu_i \geqslant 0$, π being a uniformizer of $A_{\mathfrak{p}}$. Therefore (in the notation introduced above) $n'_{\mathfrak{p}} - n_{\mathfrak{p}} = \sum_{i=1}^{n} \nu_i$; and on the other hand, $(M/M')_{\mathfrak{p}} = M_{\mathfrak{p}}/M'_{\mathfrak{p}}$ is isomorphic to the torsion $A_{\mathfrak{p}}$-module $\bigoplus_{i=1}^{n} A_{\mathfrak{p}}/\mathfrak{p}^{\nu_i}A_{\mathfrak{p}}$ and hence its length is $\sum_{i=1}^{n} \nu_i$, which proves the proposition, in view of (5) and Definition 4 of no. 5.

COROLLARY. *Let* L_1, L_2 *be two free* A*-modules of the same rank n and let* $f: L_1 \rightarrow L_2$ *be a homomorphism. Let U be the matrix of f with respect to bases of* L_1 *and* L_2. *For* Coker(f) *to be a torsion* A*-module, it is necessary and sufficient that* $\det(U) \neq 0$ *and then:*

$$(8) \qquad \chi(\mathrm{Coker}(f)) = \mathrm{div}(\det(U)).$$

L_1 and L_2 can be considered as lattices in $V_1 = L_1 \otimes_A K$ and $V_2 = L_2 \otimes_A K$ respectively, f extending to a K-homomorphism $f_{(K)}: V_1 \rightarrow V_2$. Then

$$(\mathrm{Coker}(f))_{(K)} = \mathrm{Coker}(f_{(K)})$$

and to say that Coker(f) is a torsion A-module means that $\mathrm{Coker}(f_{(K)}) = 0$; now, it amounts to the same to say that $f_{(K)}$ is surjective or that $\det(U) \neq 0$, whence the first assertion. On the other hand, we may write $f(L_1) = u(L_2)$, where u is an endomorphism of L_2 of determinant $\det(U)$; as

$$\mathrm{Coker}(f) = L_2/u(L_2),$$

formula (8) follows from (7) and (6).

Example. If $A = \mathbf{Z}$, the divisor group of A is identified with the multiplicative group $\mathbf{Q}^*_+$ of rational numbers >0. For every finite commutative group T, $\chi(T)$ is the *order* of T; the above corollary shows that the order of the group Coker(f) is equal to the *absolute value* of $\det(U)$ (cf. *Algebra*, Chapter VII, § 4, no. 7, Corollary 3 to Theorem 3).

7. DIVISOR CLASSES ATTACHED TO FINITELY GENERATED MODULES

We keep the notation and hypotheses of nos. 2 to 6. Recall that C(A) (or simply C) denotes the *divisor class group* of A, the quotient of D(A) by the subgroup of principal divisors. For every divisor $d \in D$, we shall denote by $c(d)$ its class in C.

PROPOSITION 15. *Let* M *be a finitely generated* A*-module. There exists a free submodule* L *of* M *such that* M/L *is a torsion module and the element* $c(\chi(M/L))$ *of* C *does not depend on the free submodule* L *chosen.*

We write $S = A - \{0\}$ and let $V = S^{-1}M = M \otimes_A K$; if n is the rank of V over K, there exist n elements e_i $(1 \leqslant i \leqslant n)$ of M whose canonical images in V form a basis of V; these elements are obviously linearly independent in M and hence generate a free submodule L of M such that $S^{-1}(M/L) = S^{-1}M/S^{-1}L = 0$, so that M/L is a torsion module.

Now let L_1 be another free submodule of M of rank n. Since $S^{-1}L = S^{-1}L_1$, there exists $s \in S$ such that $sL_1 \subset L$; we may therefore limit ourselves to proving that, if $L_1 \subset L_2$ are two free submodules of M of rank n, then

$$c(\chi(M/L_1)) = c(\chi(M/L_2)).$$

Now, $\chi(M/L_1) = \chi(M/L_2) + \chi(L_2/L_1)$ and it follows from no. 6, Corollary to Proposition 14 that $\chi(L_2/L_1)$ is a principal divisor and therefore

$$c(\chi(L_2/L_1)) = 0.$$

The element $c(\chi(M/L))$ will be denoted by $-c(M)$ in what follows; we shall say that $c(M)$ is the *divisor class attached to* M.

PROPOSITION 16. (i) *Let* $0 \rightarrow M_1 \xrightarrow{f} M_2 \xrightarrow{g} M_3 \rightarrow 0$ *be an exact sequence of finitely generated* A*-modules. Then*

$$c(M_2) = c(M_1) + c(M_3).$$

(ii) *If there exists a pseudo-isomorphism from* M_1 *to* M_2, *then* $c(M_1) = c(M_2)$.
(iii) *If* T *is a torsion module, then* $c(T) = -c(\chi(T))$.
(iv) *If* $\mathfrak{a} \neq 0$ *is a fractional ideal of* K, *then*

$$c(\mathfrak{a}) = c(\operatorname{div}(\mathfrak{a})).$$

(v) *If* L *is a free* A*-module, then* $c(L) = 0$.

To prove (i), consider a free sub-module L_1 (resp. L_3) of M_1 (resp. M_3) such that M_1/L_1 (resp. M_3/L_3) is a torsion module. Since L_3 is free and g is surjective, there exists in $\overset{-1}{g}(L_3)$ a free complement L_{23} of Ker(g) which is isomorphic to L_3 (*Algebra*, Chapter II, § 1, no. 11, Proposition 21); but Ker(g) $= f(M_1)$

contains $f(L_1) = L_{12}$ which is free since f is injective. The sum $L_2 = L_{12} + L_{23}$ is direct and L_2 is therefore a *free* submodule of M_2. There is moreover the commutative diagram:

$$\begin{array}{ccccccccc} 0 & \longrightarrow & L_1 & \longrightarrow & L_2 & \longrightarrow & L_3 & \longrightarrow & 0 \\ & & \downarrow & & \downarrow & & \downarrow & & \\ 0 & \longrightarrow & M_1 & \xrightarrow[f]{} & M_2 & \xrightarrow[g]{} & M_3 & \longrightarrow & 0 \end{array}$$

where the rows are exact and the vertical arrows are injections. We therefore obtain from the snake diagram (Chapter I, § 1, no. 4, Proposition 2) the exact sequence:

$$0 \to M_1/L_1 \to M_2/L_2 \to M_3/L_3 \to 0.$$

As M_1/L_1 and M_3/L_3 are torsion modules, this exact sequence shows first that so is M_2/L_2 and then, by virtue of Proposition 10 of no. 5, that:

$$\chi(M_2/L_2) = \chi(M_1/L_1) + \chi(M_3/L_3)$$

which proves (i).

Assertions (iii) and (v) are obvious from the definition. We prove (ii). Therefore let $f\colon M_1 \to M_2$ be a pseudo-isomorphism and let L_1 be a free submodule of M_1 such that M_1/L_1 is a torsion module. We set $L_2 = f(L_1)$; as $\mathrm{Ker}(f)$ is pseudo-zero, it is a torsion module, hence $\mathrm{Ker}(f) \cap L_1 = 0$ and therefore L_2 is free. Let $\bar{f}\colon M_1/L_1 \to M_2/L_2$ be the homomorphism derived from f by taking quotients; $\mathrm{Ker}(\bar{f})$ is isomorphic to $\mathrm{Ker}(f)$ and $\mathrm{Coker}(\bar{f})$ to $\mathrm{Coker}(f)$ and hence $\bar{f}$ is a pseudo-isomorphism; moreover

$$\mathrm{Coker}(\bar{f}) = M_2/f(M_1)$$

is a torsion module and so is $f(M_1)/L_2 = \bar{f}(M_1/L_1)$, hence M_2/L_2 is a torsion module and it follows from no. 5, Proposition 10 (ii) that

$$\chi(M_1/L_1) = \chi(M_2/L_2).$$

Finally it remains to prove (iv). Let $x \in K^*$ be such that $\mathfrak{a} \subset xA$. By considering the exact sequence $0 \to \mathfrak{a} \to xA \to xA/\mathfrak{a} \to 0$, we obtain

$$c(\mathfrak{a}) = c(xA) - c(xA/\mathfrak{a}) = -c(xA/\mathfrak{a})$$

by (i) and (v). But $xA/\mathfrak{a}$ is isomorphic to $A/x^{-1}\mathfrak{a}$, whence, by virtue of (iii),

$$c(xA/\mathfrak{a}) = -c(\chi(A/x^{-1}\mathfrak{a})) = -c(\mathrm{div}(x^{-1}\mathfrak{a})) = -c(\mathrm{div}(\mathfrak{a}))$$

(no. 5, Proposition 12). This completes the proof.

When M is a *lattice* of V with respect to A, $\chi(M/L) = -\chi(M, L)$ (no. 6, Proposition 14); let $(e_i)_{1 \leqslant i \leqslant n}$ be a basis of L, $e = e_1 \wedge e_2 \wedge \cdots \wedge e_n$ and

$M_W = \mathfrak{a}.e$ (notation of no. 6); then $\chi(M, L) = \text{div}(\mathfrak{a})$, whence $c(M) = c(\text{div}(\mathfrak{a}))$, which generalizes Proposition 16 (v).

COROLLARY 1. *Let* $0 \to M_n \xrightarrow{u} M_{n-1} \to \cdots \to M_0 \to 0$ *be an exact sequence of finitely generated A-modules. Then*

$$\sum_{i=0}^{n} (-1)^i c(M_i) = 0.$$

We argue by induction on n, the case $n = 2$ being Proposition 16 (i). If $M'_{n-1} = \text{Coker}(u)$, there are the two exact sequences:

$$0 \to M_n \to M_{n-1} \to M'_{n-1} \to 0$$

$$0 \to M'_{n-1} \to M_{n-2} \to \cdots \to M_0 \to 0.$$

The first shows that M'_{n-1} is finitely generated and the induction hypothesis gives

$$(-1)^{n-1} c(M'_{n-1}) + \sum_{i=0}^{n-2} (-1)^i c(M_i) = 0$$

and

$$c(M'_{n-1}) = c(M_{n-1}) - c(M_n),$$

whence the corollary.

An exact sequence

$$0 \to L_n \to L_{n-1} \to \cdots \to L_0 \to E \to 0$$

where the L_i $(0 \leqslant i \leqslant n)$ are *finitely generated free* A-modules, is called a *finite free resolution* of the A-module E.

COROLLARY 2. *If a divisorial fractional ideal* $\mathfrak{a} \neq 0$ *of* A *admits a finite free resolution, it is principal.*

In fact we apply Corollary 1 to a finite free resolution of $\mathfrak{a}$:

$$0 \to L_n \to L_{n-1} \to \cdots \to L_0 \to \mathfrak{a} \to 0.$$

By virtue of Proposition 16 (v), $c(\mathfrak{a}) = 0$ and hence, by virtue of Proposition 16 (iv), $\text{div}(\mathfrak{a})$ is principal; as $\mathfrak{a}$ is assumed to be divisorial, it is principal (§ 1, no. 1).

COROLLARY 3. *If every divisorial ideal* $\neq 0$ *of* A *admits a finite free resolution,* A *is factorial.*

This is an immediate consequence of Corollary 2 and § 3, no. 1, Definition 1.

* We shall see later that a *regular* local ring satisfies the hypothesis of Corollary 3 and therefore is *factorial.* *

If M is a finitely generated A-module, we shall denote its *rank* by $r(M)$ (recall that it is the rank over K of $M_{(K)} = M \otimes_A K$); if $0 \to M_1 \to M_2 \to M_3 \to 0$ is an exact sequence of finitely generated A-modules, the sequence

$$0 \to (M_1)_{(K)} \to (M_2)_{(K)} \to (M_3)_{(K)} \to 0$$

is also exact and hence $r(M_2) = r(M_1) + r(M_3)$. We write

$$\gamma(M) = (r(M), c(M)) \in \mathbf{Z} \times C(A);$$

γ therefore satisfies property (i) of Proposition 16 and, if M is pseudo-zero, $\gamma(M) = 0$ (since M is a torsion module). There exists a unique mapping from $\boldsymbol{F}(A)$ to $\mathbf{Z} \times C(A)$, also denoted by γ, such that $\gamma(M) = \gamma(\text{cl}(M))$ for every finitely generated A-module M. We shall see that the above properties essentially *characterize* γ:

PROPOSITION 17. *Let G be a commutative group written additively and ϕ a mapping from the set $\boldsymbol{F}(A)$ of classes of finitely generated A-modules to G; for every finitely generated A-module M we also write, by an abuse of language, $\phi(M) = \phi(\text{cl}(M))$. Suppose the following conditions are satisfied:*

(1) *If $0 \to M_1 \to M_2 \to M_3 \to 0$ is an exact sequence of finitely generated A-modules, then $\phi(M_2) = \phi(M_1) + \phi(M_3)$.*

(2) *If T is pseudo-zero, then $\phi(T) = 0$.*

Then there exists a unique homomorphism $\theta: \mathbf{Z} \times C \to G$ such that $\phi = \theta \circ \gamma$.

By virtue of Proposition 16 (iv), every element of $\mathbf{Z} \times C$ is of the form $(r(M), c(M))$ for some suitable finitely generated A-module M; whence the uniqueness of θ. We apply Proposition 11 of no. 5 to the restriction of $-\phi$ to $\boldsymbol{T}(A)$: then there exists a homomorphism $\theta_0: D \to G$ such that

$$-\phi(T) = \theta_0(\chi(T))$$

for every finitely generated torsion A-module T. Let x be a non-zero element of A; applying property (1) to the exact sequence:

$$0 \longrightarrow A \xrightarrow{h_x} A \longrightarrow A/xA \longrightarrow 0$$

where h_x is multiplication by x, we obtain $\phi(A/xA) = 0$, whence $\theta_0(\text{div}(x)) = 0$. Taking quotients, θ_0 therefore defines a homomorphism $\theta_1: C \to G$ and $\phi(T) = \theta_1(c(T))$ for every torsion A-module T. We show now that the homomorphism θ defined by $\theta(n, z) = n.\phi(A) + \theta_1(z)$ solves the problem. For this, we write $\phi'(M) = \phi(M) - \theta(\gamma(M))$ for every finitely generated A-module M; clearly condition (1) is still satisfied if ϕ is replaced by ϕ'. Moreover, $\phi'(M) = 0$ when M is a torsion module or a free module; but as for every finitely generated A-module M, there exists a free sub-module L of M such that M/L is a torsion module (Proposition 15), property (1) shows that $\phi'(M) = 0$ for every finitely generated A-module M.

* In the language of Abelian categories, Proposition 17 shows that $\mathbf{Z} \times C(A)$ is canonically isomorphic to the *Grothendieck group* of the quotient category $\mathscr{F}/\mathscr{F}'$, where $\mathscr{F}$ is the category of finitely generated A-modules and $\mathscr{F}'$ the full sub-category of $\mathscr{F}$ consisting of the pseudo-zero modules. *

8. PROPERTIES RELATIVE TO FINITE EXTENSIONS OF THE RING OF SCALARS

In this no., A and B denote *two integrally closed Noetherian domains such that* $A \subset B$ *and* B *is a finitely generated* A*-module* and K and L the fields of fractions of A and B respectively. We shall write $\operatorname{div}_A, \chi_A, c_A, \gamma_A, r_A$ instead of div, χ, c, γ, r respectively where A-modules are concerned and use analogous notation for B-modules.

We know (§ 1, no. 10) that for a prime ideal $\mathfrak{P}$ of B to be of height 1, it is necessary and sufficient that $\mathfrak{p} = \mathfrak{P} \cap A$ be of height 1; moreover (*loc. cit.*, Proposition 14) for $\mathfrak{p} \in P(A)$, there is only a finite number of prime ideals $\mathfrak{P} \in P(B)$ lying above $\mathfrak{p}$. To abbreviate, we shall denote by $\mathfrak{P}|\mathfrak{p}$ the relation "$\mathfrak{P}$ lies above $\mathfrak{p}$" (that is $\mathfrak{p} = \mathfrak{P} \cap A$); we shall then denote by $e_{\mathfrak{P}/\mathfrak{p}}$ or $e(\mathfrak{P}/\mathfrak{p})$ the ramification index $e(v_{\mathfrak{P}}/v_{\mathfrak{p}})$ of the valuation $v_{\mathfrak{P}}$ over the valuation $v_{\mathfrak{p}}$ (Chapter VI, § 8, no. 1) and by $f_{\mathfrak{P}/\mathfrak{p}}$ or $f(\mathfrak{P}/\mathfrak{p})$ the residue class degree $f(v_{\mathfrak{P}}/v_{\mathfrak{p}})$ (*loc. cit.*); recall that the discrete valuations $v_{\mathfrak{p}}$ and $v_{\mathfrak{P}}$ are *normed* and that $f_{\mathfrak{P}/\mathfrak{p}}$ is the degree of the field of fractions of $B/\mathfrak{P}$ over the field of fractions of $A/\mathfrak{p}$. We set $n = r_A(B)$, where B is considered as an A-module; hence by definition $n = [L : K]$ and, for all $\mathfrak{p} \in P(A)$, n is also the rank of the free $A_{\mathfrak{p}}$-module B for all $\mathfrak{P}|\mathfrak{p}$. Then it follows from Chapter VI, § 8, no. 5, Theorem 2 that for all $\mathfrak{p} \in P(A)$:

$$\sum_{\mathfrak{P}/\mathfrak{p}} e_{\mathfrak{P}/\mathfrak{p}} f_{\mathfrak{P}/\mathfrak{p}} = n. \tag{9}$$

This being so, as D(A) and D(B) are free **Z**-modules, we define an increasing homomorphism of ordered groups $N: D(B) \to D(A)$ (also denoted by $N_{B/A}$), by the condition:

$$N(\mathfrak{P}) = f_{\mathfrak{P}/\mathfrak{p}} . \mathfrak{p} \quad \text{for} \quad \mathfrak{P} \in P(B), \quad \text{where} \quad \mathfrak{p} = \mathfrak{P} \cap A. \tag{10}$$

On the other hand (§ 1, no. 10, Proposition 14) we have defined an increasing homomorphism of ordered groups $i: D(A) \to D(B)$ (also denoted by $i_{B/A}$), by the condition:

$$i(\mathfrak{p}) = \sum_{\mathfrak{P}/\mathfrak{p}} e_{\mathfrak{P}/\mathfrak{p}} . \mathfrak{P} \quad \text{for} \quad \mathfrak{p} \in P(A). \tag{11}$$

Clearly for every family (d_ι) (resp. (d'_ι)) of divisors of A (resp. B):

$$N(\sup(d'_\iota)) = \sup(N(d'_\iota)), \qquad N(\inf(d'_\iota)) = \inf(N(d'_\iota)) \tag{12}$$

$$i(\sup(d_\iota)) = \sup(i(d_\iota)), \qquad i(\inf(d_\iota)) = \inf(i(d_\iota)). \tag{13}$$

Formula (9) shows that:

$$N \circ i = n . 1_{D(A)}. \tag{14}$$

For all $a \in A$ (§ 1, no. 10, Proposition 14):

$$i(\mathrm{div}_A(a)) = \mathrm{div}_B(a). \tag{15}$$

We deduce (by means of (13)) that, for every fractional ideal $\mathfrak{a}$ of A, also:

$$i(\mathrm{div}_A(\mathfrak{a})) = \mathrm{div}_B(\mathfrak{a}B). \tag{16}$$

For every element $b \in B$, we know (Chapter V, § 1, no. 3, Corollary to Proposition 11) that $N_{L/K}(b) \in A$; moreover (Chapter VI, § 8, no. 5, formula (9)):

$$v_{\mathfrak{p}}(N_{L/K}(b)) = \sum_{\mathfrak{P}/\mathfrak{p}} f_{\mathfrak{P}/\mathfrak{p}} v_{\mathfrak{P}}(b) \tag{17}$$

whence:

$$N(\mathrm{div}_B(b)) = \mathrm{div}_A(N_{L/K}(b)). \tag{18}$$

Formulae (15) and (18) show that, by taking quotients, the homomorphisms N and i define homomorphisms which will also be denoted, by an abuse of language, by:

$$N: C(B) \to C(A), \qquad i: C(A) \to C(B).$$

Note that the homomorphism $i: C(A) \to C(B)$ is not in general injective (§ 3, Exercise 7).

Recall that for every B-module R, $R_{[A]}$ denotes the A-module obtained from R by restricting the scalars to A (*Algebra*, Chapter II, § 1, no. 13).

PROPOSITION 18. (i) *For* R *to be a pseudo-zero* B*-module, it is necessary and sufficient that the* A*-module* $R_{[A]}$ *be pseudo-zero.*

(ii) *For* R *to be finitely generated torsion* B*-module, it is necessary and sufficient that* $R_{[A]}$ *be a finitely generated torsion* A*-module and then:*

$$\chi_A(R_{[A]}) = N(\chi_B(R)). \tag{19}$$

(iii) *For* R *to be finitely generated* B*-module, it is necessary and sufficient that* $R_{[A]}$ *be a finitely generated* A*-module and then:*

$$c_A(R_{[A]}) = N(c_B(R)) + r_B(R)c_A(B) \tag{20}$$

$$r_A(R_{[A]}) = n . r_B(R) \qquad \text{(recall that } n = r_A(B)). \tag{21}$$

As B is a finitely generated A-module, for R to be a finitely generated B-module, it is necessary and sufficient that $R_{[A]}$ be a finitely generated A-module. Moreover, if $\mathfrak{b}$ is the annihilator of R, $\mathfrak{b} \cap A = \mathfrak{a}$ is the annihilator of $R_{[A]}$; as B is integral over A, there is no ideal other than 0 lying above the ideal 0 of A (Chapter V, § 2, no. 1, Corollary 1 to Proposition 1) and hence it amounts to the same to say that $\mathfrak{a} \neq 0$ or that $\mathfrak{b} \neq 0$.

(i) By virtue of this last remark, we may confine our attention to the case where R is a torsion B-module. If $\mathfrak{b}$ is contained in a prime ideal $\mathfrak{P} \in P(B)$, $\mathfrak{a}$ is contained in $\mathfrak{P} \cap A = \mathfrak{p}$, which is of height 1. Conversely, if $\mathfrak{a}$ is contained in a prime ideal $\mathfrak{p} \in P(A)$, there exists a prime ideal $\mathfrak{P}$ of B which contains $\mathfrak{b}$ and lies above $\mathfrak{p}$ (Chapter V, § 2, no. 1, Corollary 2 to Theorem 1). Assertion (i) follows from these remarks and no. 4, Definition 2.

(ii) For every finitely generated torsion B-module R, we write

$$\phi(R) = \chi_A(R_{[A]});$$

clearly (for finitely generated torsion B-modules) ϕ satisfies conditions (1) and (2) of Proposition 11 of no. 5 (taking account of (i)). There therefore exists a homomorphism $\theta: D(B) \to D(A)$ such that $\phi(R) = \theta(\chi_B(R))$ for every finitely generated torsion B-module R. The homomorphism θ is determined by its value for every B-module of the form $B/\mathfrak{P}$ where $\mathfrak{P} \in P(B)$, since $\chi_B(B/\mathfrak{P}) = \mathfrak{P}$. Now, for every prime ideal $\mathfrak{q} \neq \mathfrak{p} = \mathfrak{P} \cap A$ in P(A), $\mathfrak{p} \not\subset \mathfrak{q}$ and hence $(B/\mathfrak{P})_\mathfrak{q} = 0$. On the other hand, if we set $S = A - \mathfrak{p}$, $\mathfrak{P}.S^{-1}B$ is a maximal ideal of $S^{-1}B$ and $(B/\mathfrak{P})_\mathfrak{p} = S^{-1}B/\mathfrak{P}.S^{-1}B$ is isomorphic to the field of fractions of $B/\mathfrak{P}$ (Chapter II, § 2, no. 5, Proposition 11), that is to the residue field of $v_\mathfrak{P}$; its length as an $A_\mathfrak{p}$-module is therefore $f_{\mathfrak{P}/\mathfrak{p}}$; which proves that $\theta = N$ (no. 5, Definition 4).

(iii) If T is the torsion submodule of R, $T_{[A]}$ is the torsion submodule of $R_{[A]}$ and $(R/T)_{[A]} = R_{[A]}/T_{[A]}$; to prove (21) we may therefore confine our attention to the case where R is torsion-free. Then R is identified with a sub-B-module of $R_{(L)}$ and contains a basis $(e_i)_{1 \leqslant i \leqslant m}$ of $R_{(L)}$ over L. If $(b_j)_{1 \leqslant j \leqslant n}$ is a basis of L over K consisting of elements of B, the $b_j e_i$ form a basis of $R_{(L)}$ over K consisting of elements of R, whence (21). On the other hand let M be a free sub-B-module of R such that R/M is a torsion B-module; as $M_{[A]}$ is a direct sum of $r_B(R)$ A-modules isomorphic to B, by Proposition 16 (i)

$$c_A(M_{[A]}) = r_B(R).c_A(B).$$

Moreover, $c_A((R/M)_{[A]}) = -c_A(N(\chi_B(R/M)))$ by virtue of (19); but by definition of the homomorphism $N: C(B) \to C(A)$, $c_A(N(d)) = N(c_B(d))$ for all $d \in D(B)$ and, as $c_B(\chi(R/M)) = -c_B(R)$ by definition, finally

$$c_A((R/M)_{[A]}) = N(c_B(R));$$

then it suffices to apply Proposition 16 (i) to obtain (20).

PROPOSITION 19. *Let* R *be a finitely generated* B*-module. For* R *to be reflexive, it is necessary and sufficient that* $R_{[A]}$ *be a reflexive* A*-module.*

We have remarked in the proof of Proposition 18 that for R to be a torsion-free B-module, it is necessary and sufficient that $R_{[A]}$ be a torsion-free A-module. We may therefore assume that R is a lattice of $W = R \otimes_B L$ with respect to B. We shall use the following lemma:

LEMMA 4. *Let* W *be a vector space of finite rank over* L *and let* R *be a lattice of* W *with respect to* B. *Then, for all* $\mathfrak{p} \in P(A)$, $(R_{[A]})_{\mathfrak{p}} = \bigcap_{\mathfrak{P}/\mathfrak{p}} R_{\mathfrak{P}}$.

If $S = A - \mathfrak{p}$, the prime ideals of the ring $S^{-1}B$ are generated by the prime ideals of B not meeting S, in other words the ideals $\mathfrak{P}_i$ $(1 \leqslant i \leqslant m)$ lying above $\mathfrak{p}$ and the ideal (0); this shows that $S^{-1}B$ is a semi-local ring whose maximal ideals are the $\mathfrak{m}_i = \mathfrak{P}_i(S^{-1}B)$ for $1 \leqslant i \leqslant m$; moreover the local ring $(S^{-1}B)_{\mathfrak{P}_i}$ is isomorphic to $B_{\mathfrak{m}_i}$ (Chapter II, § 2, no. 5, Proposition 11) and hence is a discrete valuation ring. The ring $S^{-1}B$ is therefore a Dedekind domain (§ 2, no. 2, Theorem 1 (f)) and, as it is semi-local, it is a principal ideal domain (§ 2, no. 2, Proposition 1). This being so, $(R_{[A]})_{\mathfrak{p}}$ is equal to $S^{-1}R$ considered as an $A_{\mathfrak{p}}$-module; by the above, $S^{-1}R$ is a *free* lattice of W with respect to $S^{-1}B$ and Theorem 2 of no. 2 may therefore be applied to it, giving $S^{-1}R = \bigcap_i (S^{-1}R)_{\mathfrak{m}_i}$: but $(S^{-1}R)_{\mathfrak{m}_i} = R_{\mathfrak{P}_i}$, which proves the lemma.

Returning to the proof of Proposition 19, by Lemma 4,

$$\bigcap_{\mathfrak{P} \in P(B)} R_{\mathfrak{P}} = \bigcap_{\mathfrak{p} \in P(A)} (R_{[A]})_{\mathfrak{p}}$$

and the conclusion follows from no. 2, Theorem 2.

COROLLARY. *The ring* B *is a reflexive* A*-module.*

PROPOSITION 20. (i) *For a finitely generated* A*-module* M *to be pseudo-zero, it is necessary and sufficient that* $M \otimes_A B$ *be a pseudo-zero* B*-module.*

(ii) *If* M *is a finitely generated torsion* A*-module,* $M \otimes_A B$ *is a finitely generated* B*-module and:*

$$\chi_B(M \otimes_A B) = i(\chi_A(M)). \tag{22}$$

(iii) *If* M *is a finitely generated* A*-module,* $M \otimes_A B$ *is a finitely generated* B*-module and:*

$$c_B(M \otimes_A B) = i(c_A(M)) \tag{23}$$

$$r_B(M \otimes_A B) = r_A(M). \tag{24}$$

(i) Let $\mathfrak{P}$ be a prime ideal of B, $\mathfrak{p} = \mathfrak{P} \cap A$; then $(M \otimes_A B)_{\mathfrak{P}} = M \otimes_A B_{\mathfrak{P}}$ (Chapter II, § 2, no. 7, Proposition 18) and on the other hand

$$M \otimes_A B_{\mathfrak{P}} = (M \otimes_A A_{\mathfrak{p}}) \otimes_{A_{\mathfrak{p}}} B_{\mathfrak{P}} = M_{\mathfrak{p}} \otimes_{A_{\mathfrak{p}}} B_{\mathfrak{P}};$$

the relation $M_{\mathfrak{p}} = 0$ is therefore equivalent to $(M \otimes_A B)_{\mathfrak{P}} = 0$ (Chapter II, § 4, no. 4, Lemma 4). It suffices to apply this remark to the ideal $\mathfrak{P} = (0)$ and the ideals $\mathfrak{P} \in P(B)$ to prove (i), taking account of no. 4, Definition 2.

To prove (ii), we shall use the following lemma:

LEMMA 5. *Let* M_1, M_2 *be two finitely generated* A-*modules, and* $f: M_1 \to M_2$ *an injective homomorphism. Then the kernel of* $f \otimes 1_B: M_1 \otimes_A B \to M_2 \otimes_A B$ *is pseudo-zero.*

Let $\mathfrak{p}$ be a prime ideal of A of height $\leqslant 1$. Then $(M_i \otimes_A B)_{\mathfrak{p}} = (M_i)_{\mathfrak{p}} \otimes_{A_{\mathfrak{p}}} B_{\mathfrak{p}}$ $(i = 1, 2)$ (Chapter II, § 2, no. 7, Proposition 18) and $(f \otimes 1_B)_{\mathfrak{p}} = f_{\mathfrak{p}} \otimes 1_{B_{\mathfrak{p}}}$; the hypothesis that f is injective implies that so is $f_{\mathfrak{p}}$ (Chapter II, § 2, no. 4, Theorem 1); on the other hand, in view of the choice of $\mathfrak{p}$, $A_{\mathfrak{P}}$ is a principal ideal domain and $B_{\mathfrak{p}}$ a finitely generated torsion-free $A_{\mathfrak{p}}$-module and hence *free*; we conclude that $f_{\mathfrak{p}} \otimes 1_{B_{\mathfrak{p}}}$ is itself injective. If $I = \operatorname{Ker}(f \otimes 1)$, then $I_{\mathfrak{p}} = \operatorname{Ker}((f \otimes 1)_{\mathfrak{p}})$ (Chapter II, § 2, no. 4, Theorem 1); therefore $I_{\mathfrak{p}} = 0$, whence *a fortiori* $I_{\mathfrak{P}} = (I_{\mathfrak{p}})_{\mathfrak{P}} = 0$ for $\mathfrak{P}|\mathfrak{p}$, which proves the lemma (no. 4, Definition 2).

We return now to the proof of (ii). For every finitely generated torsion A-module M, write $\phi(M) = \chi_B(M \otimes_A B)$; it follows from (i) that, if M is pseudo-zero, $\phi(M) = 0$. On the other hand, consider an exact sequence of finitely generated torsion A-modules:

$$0 \to M_1 \to M_2 \to M_3 \to 0.$$

It follows from Lemma 4 that there is an exact sequence of B-modules:

$$0 \to I \to M_1 \otimes_A B \to M_2 \otimes_A B \to M_3 \otimes_A B \to 0$$

where I is pseudo-zero. Using no. 5, Corollary to Proposition 10, we therefore obtain $\phi(M_2) = \phi(M_1) + \phi(M_3)$. We therefore conclude from Proposition 11 of no. 5 that there exists a homomorphism $\theta: D(A) \to D(B)$ such that $\phi(M) = \theta(\chi_A(M))$ for every finitely generated torsion A-module M. To prove that $\theta = i$, it suffices to show that $\phi(A/\mathfrak{p}) = i(\mathfrak{p})$ for all $\mathfrak{p} \in P(A)$; now $(A/\mathfrak{p}) \otimes_A B = B/\mathfrak{p}B$ and, for all $\mathfrak{P} \in P(B)$, $(B/\mathfrak{p}B)_{\mathfrak{P}} = B_{\mathfrak{P}}/\mathfrak{p}B_{\mathfrak{P}}$; the last module is 0 if $\mathfrak{P}$ does not lie above $\mathfrak{p}$; if on the contrary $\mathfrak{P}|\mathfrak{p}$, $B_{\mathfrak{P}}/\mathfrak{p}B_{\mathfrak{P}}$ is a $B_{\mathfrak{P}}$-module of length $e(\mathfrak{P}/\mathfrak{p})$ by definition of the ramification index (Chapter VI, § 8, no. 1); therefore $\chi_B(B/\mathfrak{p}B) = \sum_{\mathfrak{P}/\mathfrak{p}} e_{\mathfrak{P}/\mathfrak{p}} . \mathfrak{P} = i(\mathfrak{p})$, which completes the proof of (ii).

Formula (24) is immediate, for

$$(M \otimes_A B) \otimes_B L = M \otimes_A L = (M \otimes_A K) \otimes_K L$$

and the rank of $(M \otimes_A K) \otimes_K L$ over L is equal to the rank of $M \otimes_A K$ over K. To show (23), consider a free submodule H of M such that $Q = M/H$ is a torsion A-module. Applying Lemma 4 as above, we obtain an exact sequence of B-modules:

$$0 \to I \to H \otimes_A B \to M \otimes_A B \to Q \otimes_A B \to 0$$

where I is pseudo-zero. It therefore follows from no. 7, Proposition 16 (ii) and (v) and Corollary 1 to Proposition 16 that

$$c_B(M \otimes_A B) = c_B(Q \otimes_A B) = -c_B(\chi_B(Q \otimes_A B)) = -c_B(i(\chi_A(Q)))$$

by virtue of (ii); but by definition of the homomorphism $i\colon C(A) \to C(B)$, $c_B(i(\chi_A(Q))) = i(c_A(\chi_A(Q))) = -i(c_A(M))$, which completes the proof of (23).

Remarks
(1) If M is a reflexive A-module, $M \otimes_A B$ is not necessarily reflexive (exercise 6). However it is so when B is a *flat* A-module (no. 2, Proposition 8).
(2) Let C be a third integrally closed Noetherian domain such that $B \subset C$ and C is a finitely generated B-module (and hence also a finitely generated A-module). Then we have the transitivity equations:

$$N_{C/A} = N_{B/A} \circ N_{C/B}, \tag{25}$$

$$i_{C/A} = i_{C/B} \circ i_{B/A} \tag{26}$$

which follow immediately from the transitivity equations for the ramification indices and the residue class degrees (Chapter VI, § 8, no. 1, Lemma 1).

9. A REDUCTION THEOREM

We preserve the notation and hypotheses of nos. 2 to 7.

LEMMA 6. *Let* R *be a commutative ring and* $\mathfrak{p}_i$ $(1 \leqslant i \leqslant n)$ *distinct prime ideals of* R.

(i) *For* $1 \leqslant i \leqslant n$, *let* H_i *be a subset of* $R/\mathfrak{p}_i$ *satisfying the following condition: there exists no element* $\alpha_i \in R/\mathfrak{p}_i$ *such that* $\alpha_i + H_i$ *contains an ideal* $\neq 0$ *of* $R/\mathfrak{p}_i$. *Then there exists* $a \in R$ *such that, for* $1 \leqslant i \leqslant n$, *the canonical image of* a *in* $R/\mathfrak{p}_i$ *does not belong to* H_i.

(ii) *If* $\operatorname{Card}(H_i) < \operatorname{Card}(R/\mathfrak{p}_i)$, *the* H_i *satisfy the condition in* (i).

(i) We argue by induction on n, the case $n = 0$ being trivial. Therefore let $n \geqslant 1$. We may make a permutation on the indices i and may assume that $\mathfrak{p}_1$ is minimal among the $\mathfrak{p}_i$ and therefore, for $2 \leqslant i \leqslant n$, there exists $c_i \in \mathfrak{p}_i$ such that $c_i \notin \mathfrak{p}_1$. By the induction hypothesis there exists $b \in R$ such that the canonical image of b in $R/\mathfrak{p}_i$ does not belong to H_i for $2 \leqslant i \leqslant n$. For all $x \in R$ we write

$a_x = b + xc_2c_3 \ldots c_n$; as $c_i \in \mathfrak{p}_i$, obviously $a_x \equiv b \pmod{\mathfrak{p}_i}$ for $2 \leqslant i \leqslant n$. It therefore suffices to prove that there exists $x \in R$ such that the canonical image of a_x in $R/\mathfrak{p}_1$ does not belong to H_1. Now, the set of canonical images of the a_x in $R/\mathfrak{p}_1$, where x runs through R, is just $\beta + \mathfrak{c}$, where β is the canonical image of b and $\mathfrak{c}$ the ideal of $R/\mathfrak{p}_1$ generated by the canonical image of $c_2c_3 \ldots c_n$; by virtue of the choice of the c_i, $\mathfrak{c} \neq 0$ since $R/\mathfrak{p}_1$ is an integral domain and the hypothesis on H_1 implies the existence of an x which solves the problem.

(ii) As $R/\mathfrak{p}_i$ is an integral domain, every ideal $\neq 0$ of $R/\mathfrak{p}_i$ has cardinal equal to that of $R/\mathfrak{p}_i$ and the same is true of any translation of an ideal by an element of $R/\mathfrak{p}_i$, whence the conclusion.

THEOREM 6. *Let* M *be a torsion-free finitely generated* A*-module. There exists a free submodule* L *of* M *such that* M/L *is isomorphic to an ideal of* A.

We shall denote by n the *rank* of M (rank of $V = M \otimes_A K$ over K) and consider M as a lattice of V with respect to A. Then, for all $\mathfrak{p} \in P(A)$, $M_\mathfrak{p}$ is a lattice of V with respect to $A_\mathfrak{p}$ (no. 1, *Example* 6) and, as $A_\mathfrak{p}$ is a principal ideal domain, $M_\mathfrak{p}$ is a *free* $A_\mathfrak{p}$-module of rank n. We shall write:

$$M(\mathfrak{p}) = M_\mathfrak{p}/\mathfrak{p}M_\mathfrak{p}.$$

We shall denote by $k(\mathfrak{p})$ the field of fractions of $A/\mathfrak{p}$ (isomorphic to the residue field of $A_\mathfrak{p}$); $M(\mathfrak{p}) = M \otimes_A k(\mathfrak{p})$ is therefore a vector space of rank n over $k(\mathfrak{p})$. For all $x \in M$, we shall denote by $x(\mathfrak{p})$ the canonical image of x in $M(\mathfrak{p})$.

LEMMA 7. *Let* x_i $(1 \leqslant i \leqslant m)$ *be linearly independent elements of* M (*over* A *or* K, *which amounts to the same*) *and let* L *be the sub-*A*-module of* M *generated by the* x_i. *Then, for almost all* $\mathfrak{p} \in P$, *the* $x_i(\mathfrak{p}) \in M(\mathfrak{p})$ *are linearly independent over* $k(\mathfrak{p})$; *for them to be linearly independent over* $k(\mathfrak{p})$ *for all* $\mathfrak{p} \in P$, *it is necessary and sufficient that* M/L *be torsion-free.*

Let $x_{m+1}, \ldots, x_n$ be elements of M which, with $x_1, \ldots, x_m$, form a basis of V and let N be the free sub-A-module of M generated by the x_i $(1 \leqslant i \leqslant n)$. It follows from no. 3, Theorem 3 that $N_\mathfrak{p} = M_\mathfrak{p}$ for almost all $\mathfrak{p} \in P$; as $x_1(\mathfrak{p}), \ldots, x_n(\mathfrak{p})$ form a basis of $N(\mathfrak{p})$ over $k(\mathfrak{p})$, this establishes the first assertion.

If M/L is torsion-free, so is $(M/L)_\mathfrak{p} = M_\mathfrak{p}/L_\mathfrak{p}$ for all $\mathfrak{p} \in P$ (no. 1, *Example* 6) and, as $A_\mathfrak{p}$ is a principal ideal domain, $M_\mathfrak{p}/L_\mathfrak{p}$ is free. Therefore $M_\mathfrak{p}$ is the direct sum of $L_\mathfrak{p}$ and a free $A_\mathfrak{p}$-module E of rank $n - m$; hence $M(\mathfrak{p})$ is the direct sum of $L(\mathfrak{p})$ and the vector $k(\mathfrak{p})$-space $E/\mathfrak{p}E$ of rank $n - m$; therefore $L(\mathfrak{p})$ is of rank m and, as it is generated by the $x_i(\mathfrak{p})$ $(1 \leqslant i \leqslant m)$, the latter are linearly independent.

Conversely, suppose that the $x_i(\mathfrak{p})$ $(1 \leqslant i \leqslant m)$ are linearly independent over $k(\mathfrak{p})$ for all $\mathfrak{p} \in P$. Then $L_\mathfrak{p}$ is a direct factor of $M_\mathfrak{p}$ for all $\mathfrak{p}$ (Chapter II, § 3, no. 2, Corollary 1 to Proposition 5) and therefore $M_\mathfrak{p}/L_\mathfrak{p} = (M/L)_\mathfrak{p}$ is torsion-free for all $\mathfrak{p} \in \mathrm{P}$. We conclude that $\mathrm{P} \cap \mathrm{Ass}(\mathrm{M}/\mathrm{L}) = \varnothing$ by Chapter IV, § 1, no. 2, Corollary to Proposition 5. But as L is reflexive, it follows from no. 2, Proposition 7 (i) that the only prime ideal which can belong to $\mathrm{Ass}(\mathrm{M}/\mathrm{L})$ is the ideal (0); hence M/L is torsion-free.

LEMMA 8. *Suppose that the rank n of* M *is* $\geqslant 2$; *then there exists an element* $x \neq 0$ *of* M *such that* $\mathrm{M}/\mathrm{A}x$ *is torsion-free.*

Let $y \neq 0$ be an element of M. By Lemma 7, the set Y of $\mathfrak{p} \in \mathrm{P}$ such that $y(\mathfrak{p}) = 0$ is finite. If $\mathrm{Y} = \varnothing$, it follows from Lemma 7, applied to the sequence (x_i) consisting of the single element y, that $\mathrm{M}/\mathrm{A}y$ is torsion-free. Suppose therefore that $\mathrm{Y} \neq \varnothing$ and write $\mathrm{S} = \bigcap_{\mathfrak{p} \in \mathrm{Y}} (\mathrm{A} - \mathfrak{p})$; we know (no. 4, Lemma 2) that $\mathrm{S}^{-1}\mathrm{A}$ is a semi-local principal ideal domain whose maximal ideals are the $\mathfrak{p}\mathrm{S}^{-1}\mathrm{A}$, where $\mathfrak{p} \in \mathrm{Y}$, the corresponding local rings being the $\mathrm{A}_\mathfrak{p}$. Therefore

$$\mathrm{S}^{-1}\mathrm{A}/\mathfrak{p}\mathrm{S}^{-1}\mathrm{A} = k(\mathfrak{p}),$$

whence

$$\mathrm{S}^{-1}\mathrm{M}/\mathfrak{p}\mathrm{S}^{-1}\mathrm{M} = (\mathrm{M}/\mathfrak{p}\mathrm{M}) \otimes_\mathrm{A} \mathrm{S}^{-1}\mathrm{A} = \mathrm{M} \otimes_\mathrm{A} ((\mathrm{A}/\mathfrak{p}) \otimes_\mathrm{A} \mathrm{S}^{-1}\mathrm{A}) = \mathrm{M} \otimes_\mathrm{A} k(\mathfrak{p}) = \mathrm{M}(\mathfrak{p})$$

for all $\mathfrak{p} \in \mathrm{Y}$. By Chapter II, § 1, no. 2, Proposition 6, there exists an element $z/s \in \mathrm{S}^{-1}\mathrm{M}$ ($z \in \mathrm{M}$, $s \in \mathrm{S}$) whose canonical images in the $\mathrm{M}(\mathfrak{p})$ for $\mathfrak{p} \in \mathrm{Y}$ are all $\neq 0$. By definition of S therefore $z(\mathfrak{p}) \neq 0$ for all $\mathfrak{p} \in \mathrm{Y}$. We may further assume that y and z are *linearly independent* over K. For in the opposite case consider an element $t \in \mathrm{M}$ which is linearly independent of y (such exists since $n \geqslant 2$); on the other hand take an element $a \neq 0$ belonging to $\bigcap_{\mathfrak{p} \in \mathrm{Y}} \mathfrak{p}$ (which is not reduced to 0 since A is an integral domain) and write $z' = z + at$: clearly y and z' are linearly independent over K and $z'(\mathfrak{p}) = z(\mathfrak{p}) \neq 0$ for all $\mathfrak{p} \in \mathrm{Y}$.

Supposing therefore that y and z are linearly independent over K, let Z be the set of $\mathfrak{p} \in \mathrm{P} - \mathrm{Y}$ such that $y(\mathfrak{p})$ and $z(\mathfrak{p})$ are linearly independent over $k(\mathfrak{p})$; it follows from Lemma 7 that this set is *finite*. For all $\mathfrak{p} \in \mathrm{Z}$, we may therefore write $z(\mathfrak{p}) = \lambda(\mathfrak{p})y(\mathfrak{p})$ where $\lambda(\mathfrak{p}) \in k(\mathfrak{p})$. Now, $\mathrm{Card}(\mathrm{A}/\mathfrak{p}) \geqslant 2$ for all $\mathfrak{p} \in \mathrm{P}$; it therefore follows from Lemma 6 that there exists $b \in \mathrm{A}$ such that, for all $\mathfrak{p} \in \mathrm{Z}$, the canonical image of b in $\mathrm{A}/\mathfrak{p}$ is distinct from $\lambda(\mathfrak{p})$. We now show that the element $x = z - by$ solves the problem; it suffices (by virtue of Lemma 7 applied with $m = 1$) to verify that $x(\mathfrak{p}) \neq 0$ for *all* $\mathfrak{p} \in \mathrm{P}$. Now:

— if $\mathfrak{p} \in \mathrm{Y}$, then $x(\mathfrak{p}) \neq 0$ by construction;
— if $\mathfrak{p} \in \mathrm{Z}$, then $x(\mathfrak{p}) = \mu_\mathfrak{p} y(\mathfrak{p})$ where $\mu_\mathfrak{p} \neq 0$ by virtue of the choice of b and hence $x(\mathfrak{p}) \neq 0$ since $y(\mathfrak{p}) \neq 0$;
— if $\mathfrak{p} \in \mathrm{P} - (\mathrm{Y} \cup \mathrm{Z})$, $y(\mathfrak{p})$ and $z(\mathfrak{p})$ are linearly independent and hence $x(\mathfrak{p}) \neq 0$.

Having established these lemmas, we pass to the proof of Theorem 6. We argue by induction on n, the case $n \leqslant 1$ being trivial since M itself is then isomorphic to an ideal of A. Suppose therefore that $n \geqslant 2$; by virtue of Lemma 8 there exists a free submodule L_0 of M of rank 1 such that M/L_0 is torsion-free; M/L_0 is therefore of rank $n - 1$. By the induction hypothesis there exists a free submodule L_1 of M/L_0 such that $(M/L_0)/L_1$ is isomorphic to an ideal of A. Let L be the inverse image of L_1 in M; L/L_0 is isomorphic to L_1 and, since L_1 is free, L is isomorphic to $L_0 \oplus L_1$ (*Algebra*, Chapter II, § 1, no. 11, Proposition 21) and hence free; as M/L is isomorphic to $(M/L_0)/L_1$, the theorem has been shown.

Remark. If M is reflexive, the same is not necessarily true of M/L (Exercise 9).

10. MODULES OVER DEDEKIND DOMAINS

We now assume that A is a *Dedekind domain*; then we know that the ideals $\mathfrak{p} \in P$ are *maximal* and that they are the only prime ideals $\neq 0$ of A (§ 2, no. 1); the group D(A) is identified with the group I(A) of fractional ideals $\neq 0$ of A.

PROPOSITION 21. *Let* A *be a Dedekind domain. Every pseudo-zero* A*-module is zero. Every pseudo-injective* (resp. *pseudo-surjective, pseudo-bijective, pseudo-zero*) A*-module homomorphism is injective* (resp. *surjective, bijective, zero*).

The first assertion has already been shown (no. 4, *Example* 1); the others follow from it immediately.

PROPOSITION 22. *Let* A *be a Dedekind domain and* M *a finitely generated* A*-module. The following properties are equivalent:*

(a) M *is torsion-free.*
(b) M *is reflexive.*
(c) M *is projective.*

We already know (with no hypothesis on the integral domain A) that (b) implies (a) (no. 2, *Remark* 1) and that (c) implies (b) (*Algebra*, Chapter II, § 2, no. 7, Corollary 4 to Proposition 14). If M is torsion-free, it is identified with a lattice of $V = M \otimes_A K$ with respect to A; $M_\mathfrak{p}$ is therefore a free $A_\mathfrak{p}$-module for every *maximal* ideal $\mathfrak{p} \in P$, since $A_\mathfrak{p}$ is a principal ideal domain. The conclusion then follows from Chapter II, § 5, no. 2, Theorem 1 (b).

COROLLARY. *Let* M *be a finitely generated* A*-module and let* T *be its torsion submodule. Then* T *is a direct factor of* M.

As M/T is torsion-free and finitely generated, it is projective by Proposition 22 and the corollary therefore follows from *Algebra*, Chapter II, § 2, no. 2, Proposition 4.

PROPOSITION 23. *Let* A *be a Dedekind domain and* T *a finitely generated torsion* A-*module. There exist two finite families* $(n_\iota)_{\iota \in I}$ *and* $(\mathfrak{p}_\iota)_{\iota \in I}$, *where the* n_ι *are integers* $\geqslant 1$ *and the* $\mathfrak{p}_\iota$ *elements of* P, *such that* T *is isomorphic to the direct sum* $\bigoplus_{\iota \in I} (A/\mathfrak{p}_\iota^{n_\iota})$. *Further, the families* $(n_\iota)_{\iota \in I}$ *and* $(\mathfrak{p}_\iota)_{\iota \in I}$ *are unique to within a bijection of the indexing set.*

This follows from no. 4, Theorem 5, taking account of the fact that a pseudo-isomorphism is here an isomorphism.

PROPOSITION 24. *Let* A *be a Dedekind domain and* M *a finitely generated torsion-free* A-*module of rank* $n \geqslant 1$. *Then there exists an ideal* $\mathfrak{b} \neq 0$ *of* A *such that* M *is isomorphic to the direct sum of the modules* A^{n-1} *and* $\mathfrak{b}$. *Moreover, the class of the ideal* $\mathfrak{b}$ *is determined uniquely by this condition.*

Theorem 6 of no. 9 shows that there exists a free submodule L of M such that M/L is isomorphic to an ideal $\mathfrak{a}$ of A. If $\mathfrak{a} = 0$, we take $\mathfrak{b} = A$. Otherwise, $\mathfrak{a}$ is of rank 1, hence $L = A^{n-1}$ and $\mathfrak{a}$ is a *projective* module (Proposition 22); M is therefore isomorphic to the direct sum of L and $\mathfrak{a}$ (*Algebra*, Chapter II, § 2, no. 2, Proposition 4), which proves the first part of the proposition. Moreover, it follows from no. 7, Proposition 16 (i), (iv) and (v) that $c(M) = c(\mathfrak{b})$ whence the uniqueness of the class of $\mathfrak{b}$.

Remarks

(1) Propositions 23 and 24 and the Corollary to Proposition 22 determine completely the structure of finitely generated A-modules. Proposition 24 shows that a torsion-free finitely generated A-module is determined up to isomorphism by its *rank* and the *divisor class* which is attached to it.

(2) It can be shown that over a Dedekind domain a projective module which is not finitely generated is necessarily *free* (Exercise 21) and that every submodule of a projective module is projective (Exercise 20).

EXERCISES

§ 1

1. Let K be a field and A the subring of the ring of formal power series K[[T]] consisting of the series in which the coefficient of T is 0; A is a Noetherian local integral domain which is not integrally closed.

(a) Show that every principal ideal of A, which is non-zero and distinct from A, contains a unique element of the form $T^n + \lambda T^{n+1}$ where $\lambda \in K$, $n \geqslant 2$; moreover, this ideal contains all the powers T^m where $m \geqslant n + 2$. Deduce that the non-principal ideals of A are the ideals $AT^n + AT^{n+1}$ where $n \geqslant 2$.

(b) Show that every fractional ideal of A is divisorial (prove that every non-principal ideal is the intersection of two principal ideals). Describe the structure of the monoid D(A). The only prime ideals of A are the maximal ideal $\mathfrak{m} = AT^2 + AT^3$ and (0).

2. (a) In the factorial polynomial domain K[X, Y] in two indeterminates over a field K, give an example of two principal ideals $\mathfrak{a}$, $\mathfrak{b}$ such that $\mathfrak{a} + \mathfrak{b}$ is not divisorial.

(b) Let K be the quadratic extension of the field $\mathbf{Q}(X)$ of rational functions in one indeterminate over $\mathbf{Q}$, obtained by adjoining to $\mathbf{Q}(X)$ a root y of the polynomial $Y^2 - 2X \in \mathbf{Q}(X)[Y]$. Let A be the subring of K generated by $\mathbf{Z}$, X and y; A is an integrally closed Noetherian domain. Show that $\mathfrak{p} = AX + Ay$ is a prime ideal of height 1 but that $\mathfrak{p}^2$ is not divisorial (show that $X^{-1}\mathfrak{p}^2$ is a prime ideal strictly containing $\mathfrak{p}$). Moreover $\mathfrak{p}.(A : \mathfrak{p}) \neq A$.

(c) Let A be an integrally closed Noetherian local integral domain, whose maximal ideal $\mathfrak{m}$ is not of height $\leqslant 1$; let $\mathfrak{p}$ be a prime ideal of A of height 1; for every integer $i > 0$ let $\mathfrak{a}_i = \mathfrak{p} + \mathfrak{m}^i$; show that $\operatorname{div}\left(\bigcap_i \mathfrak{a}_i\right)$ is distinct from $\sup(\operatorname{div}(\mathfrak{a}_i))$.

3. Let A be a Noetherian integral domain and $\mathfrak{p}$ a prime ideal of height 1.

Show that $A : \mathfrak{p} \neq A$ (in the field of fractions of A). (If $c \neq 0$ is an element of $\mathfrak{p}$, note that $\mathfrak{p} \in \mathrm{Ass}(A/Ac)$ and deduce that $Ac : \mathfrak{p} \neq Ac$ by virtue of Chapter IV, § 2, Exercise 30). Is the result still valid for a non-Noetherian integral domain (consider a non-discrete valuation ring of height 1)?

4. Let A be a completely integrally closed domain and $\mathfrak{m}$ a maximal ideal of A. Show that, either $\mathfrak{m}$ is invertible, or $\mathfrak{m}$ is not divisorial and $A : \mathfrak{m} = A$. (Note that, if $\mathfrak{m}$ is not invertible, necessarily $\mathfrak{m}.(A : \mathfrak{m})^k = \mathfrak{m}$ for every integer $k > 0$.)

5. Let A be an integral domain, K its field of fractions and U the subgroup of K^* consisting of the invertible elements of A; consider the group $K^*/U = G$ as ordered by the relation derived by passing to the quotient from the relation $x^{-1}y \in A$ in K^* (*Algebra*, Chapter VI, § 1, no. 5).

(a) For a fractional ideal $\mathfrak{a} \neq 0$ of K to be divisorial, it is necessary and sufficient that the image of $\mathfrak{a} - \{0\}$ in G be a *major* set in G (*Algebra*, Chapter VI, § 1, Exercise 30); for every fractional ideal $\mathfrak{a} \neq 0$ of K, the image of $\mathfrak{a} - \{0\}$ is the least major set containing the image of $\mathfrak{a} - \{0\}$. If A is a Krull domain, G is an ordered group isomorphic to a directed subgroup of an ordered group of the form $\mathbf{Z}^{(I)}$.

(b) Show that for a homomorphism w of G to a totally ordered group Γ to be such that the restriction to $A - \{0\}$ of the composite mapping $K^* \to K^*/U \xrightarrow{w} \Gamma$ is a valuation, it is necessary and sufficient that w be an *increasing* homomorphism.

(c) Let H be the directed subgroup of the product ordered group $\mathbf{Z} \times \mathbf{Z}$ consisting of the ordered pairs (s_1, s_2) such that $s_1 + s_2 \equiv 0 \pmod{2}$. Show that there exists no integral domain A such that K^*/U is an ordered group isomorphic to H (deduce from (b) that such a domain would necessarily be a Krull domain, but Proposition 9 of no. 5 would not hold for this ring (cf. Exercise 22)).

6. Let A be an integral domain.

(a) For a divisorial ideal $\mathfrak{a}$ of A to be such that $\mathrm{div}(\mathfrak{a})$ is invertible in $D(A)$, it is necessary and sufficient that $\mathfrak{a} : \mathfrak{a} = A$ (cf. Exercise 5 and *Algebra*, Chapter VI, § 1, Exercise 30). In particular, in order that $\mathfrak{a} : \mathfrak{a} = A$ for every divisorial ideal $\mathfrak{a}$ of A, it is necessary and sufficient that A be completely integrally closed.

(b) In order that $\mathfrak{a} : \mathfrak{a} = A$ for every finitely generated ideal $\mathfrak{a} \neq 0$ of A, it is necessary and sufficient that A be integrally closed.

7. Let K be a field and $(v_\iota)_{\iota \in I}$ a family of discrete valuations on K satisfying conditions (AK_I) and (AK_{II}) of no. 3 and such that, for every integer r, every family $(n_h)_{1 \leqslant h \leqslant r}$ of rational integers and every family $(\iota_h)_{1 \leqslant h \leqslant r}$ of distinct elements of I, there exists $x \in K$ such that $v_{\iota_h}(x) = n_h$ for $1 \leqslant h \leqslant r$ and

$v_\iota(x) \geqslant 0$ for ι distinct from the ι_h. Show that the intersection of the valuation rings of the v_ι is a Krull domain A whose field of fractions is K and that $(v_\iota)_{\iota \in I}$ is the family of essential valuations of A. (Show first that K is the field of fractions of A and hence that A is a Krull domain; then prove that for all $\iota \in I$ the prime ideal of A consisting of those x such that $v_\iota(x) > 0$ is necessarily of height 1, arguing by *reductio ad absurdum* and using Corollary 2 to Proposition 6 of no. 4.)

8. Let A be a Krull domain; show that, for every set I, the polynomial ring $A[X_\iota]_{\iota \in I}$ is a Krull domain. (Observe that, if B is a Krull domain, the valuations induced on B by the essential valuations of $B[X_1, \ldots, X_n]$ are the essential valuations of B.)

¶ 9. (a) Let B be a discrete valuation ring, $A = B[[X]]$ the ring of formal power series in one indeterminate over B and $C = A_X$ the ring of fractions $S^{-1}A$, where S is the multiplicative set of X^h ($h \geqslant 0$). Let v be a normed valuation on B; for every element $f = \sum_{n \geqslant h} b_n X^n \neq 0$ of C with $b_h \neq 0$ in B (h positive or negative), we write $s(f) = v(b_h)$. Show that s is a *Euclidean stathm* on C (*Algebra*, Chapter VII, § 1, Exercise 7), such that $s(fg) = s(f) + s(g)$ for non-zero f, g in C. (If $s(f) = p \geqslant s(g) = q$ and h is the least of the degrees of the terms $\neq 0$ in f, show that there exists $u \in C$ such that $f - ug = X^{h+1}f_1$ where $f_1 \in C$ and, by arguing by *reductio ad absurdum* deduce the existence of a process of "Euclidean division" in C.) If $s(f) = 0$, f is invertible in C.

* Show that A is not a Dedekind domain. *

(b) Deduce from (a) that, if B is a Krull domain, $A = B[[X]]$ is also a Krull domain. (If K is the field of fractions of B, note that A is the intersection of $K[[X]]$ and a family (C_ι) of principal ideal domains with the same field of fractions as A and that every element of A is invertible in almost all the C_ι.)

10. Let A be a Krull domain, K its field of fractions, L a field containing K and (B_α) a right directed family of Krull domains containing A and contained in L such that the field of fractions L_α of B_α is a finite algebraic extension of K and B_α is the integral closure of A in L_α. For every essential valuation v of A and all α, let $e_\alpha(v)$ be the sum of the ramification indices of the valuations on B_α which extend v. Show that for the union of the B_α to be a Krull domain, it is necessary and sufficient that for every essential valuation v of A, the set of $e_\alpha(v)$ be bounded.

¶ 11. A divisor d in an integral domain A is called *finitely generated* if it is of the form $\operatorname{div}(\mathfrak{a})$, where $\mathfrak{a}$ is a finitely generated fractional ideal (which is not necessarily divisorial; cf. (b)).

(a) Show that, if A is a Krull domain, every divisor of D(A) is of the form div($\mathfrak{a}$), where $\mathfrak{a} = Ax + Ay$, x, y being elements $\neq 0$ of the field of fractions of A (use Corollary 2 to Proposition 9 of no. 5).

(b) Let K be a field, $A = K[X_n]_{n \in \mathbf{N}}$ the ring of polynomials over K in a denumerable set of indeterminates, which is a Krull ring (Exercise 8). Let A′ be the subring of A generated by the unit element and the monomials $X_i X_j$ for all pairs (i, j) (equivalently, A′ is the set of polynomials all whose terms have even degree). If L and L′ are the fields of fractions of A and A′, show that $A' = A \cap L'$, hence A′ is a Krull ring. Let $\mathfrak{p}$ be the ideal in A′ generated by the products $X_0 X_i$ for all $i \geqslant 0$; show that $\mathfrak{p}$ is a prime ideal of height 1 (hence *divisorial*) but that *it is not finitely generated.*

12. (a) Let A be an integrally closed domain, $f(X) = \sum_i a_i X^i$ and $g(X) = \sum_j b_j X^j$ be two polynomials in A[X]; show that, if $c \in A$ is such that all the coefficients of $f(X)g(X)$ belong to Ac, then all the products $a_i b_j$ belong to Ac (reduce it to the case where A is a valuation ring).

(b) If A is the ring defined in Exercise 1, give an example of two polynomials f, g of degree 1 in A[X] and an element $c \in A$ for which the conclusion of (a) does not hold (take $c = T^4 + T^5$).

(c) Under the hypotheses of (a), let $\mathfrak{a}$, $\mathfrak{b}$, $\mathfrak{c}$ be the ideals of A generated respectively by the coefficients of f, g and fg: show that $\operatorname{div}(\mathfrak{c}) = \operatorname{div}(\mathfrak{a}) + \operatorname{div}(\mathfrak{b})$. Give an example where $\tilde{\mathfrak{c}} \neq \widetilde{\mathfrak{a}\mathfrak{b}}$. Show that, if $\mathfrak{c}$ is principal, $\mathfrak{c} = \mathfrak{a}\mathfrak{b}$ and $\mathfrak{a}$ and $\mathfrak{b}$ are invertible; if further A is a local ring, $\mathfrak{a}$ and $\mathfrak{b}$ are principal.

13. Let A be an integral domain and $(p_\iota)_{\iota \in I}$ a family of non-zero elements such that the principal ideals Ap_ι are prime; let S be the multiplicative subset, generated by the p_ι.

(a) Show that $A = S^{-1}A \cap \left(\bigcap_\iota A_{Ap_\iota}\right)$.

(b) Suppose that every non-empty family of *principal* ideals of A has a maximal element. Show that each of the rings A_{Ap_ι} is a discrete valuation ring (cf. Chapter VI, § 3, no. 5, Proposition 9). Deduce that, if $S^{-1}A$ is integrally closed (resp. completely integrally closed), A is integrally closed (resp. completely integrally closed). If $S^{-1}A$ is a Krull domain, show that A is a Krull domain.

14. Let A be a Krull domain and S a *saturated* multiplicative subset of A not containing 0 (Chapter II, § 2, Exercise 1). Show that, if the canonical homomorphism $\bar{i}: C(A) \to C(S^{-1}A)$ (no. 10, Proposition 17) is bijective, S is gene-

rated by a family (p_ι) of elements such that each Ap_ι is a prime ideal (note that every divisorial prime ideal which meets S is principal).

¶ 15. (a) Let A be an integral domain and a, b two elements $\neq 0$ of A such that $Aa \cap Ab = Aab$. Show that in the polynomial ring A[X] the principal ideal generated by $aX + b$ is prime (prove that every polynomial $f(X) \in A[X]$ such that $f(-b/a) = 0$ is a multiple of $aX + b$, arguing induction on the degree of f).

(b) Let A be a Krull domain and a, b two elements of A such that Aa and $Aa + Ab$ are prime and distinct. Show that $A[X]/(aX + b)$ is a Krull domain and that $C(A[X]/(aX + b))$ is isomorphic to C(A). (Note, with the aid of (a), that $A[X]/(aX + b)$ is isomorphic to $A[b/a] = B$; show that Ba is prime in B. Observe that $A[a^{-1}] = B[a^{-1}]$ is a Krull domain and use Exercise 12.)

(c) Let A be an integrally closed Noetherian domain and a, b two elements of the Jacobson radical $\mathfrak{r}$ of A. Suppose that the ideal $Aa + Ab$ is a *non-divisorial* prime ideal; show that Aa and Ab are prime ideals. (Prove first that $Aa \cap Ab = Aab$ considering the ideals of Ass(A/Aa). Show secondly that the relations $xy \in Ab$ and $y \notin Aa + Ab$ imply $x \in Ab + Aa^h$ for all h, by induction on h; deduce that then $x \in Ab$. Finally, deduce that, if $xy \in Ab$, $x \notin Ab$, $y \notin Ab$, then necessarily $x \in Ab + Aa^h$ and $y \in Ab + Aa^h$ for all h, arguing by induction on h; deduce a contradiction.)

16. Let $A = \bigoplus_{n \in \mathbf{Z}} A_n$ be a graded Krull domain. Let $D_g(A)$ (resp. $F_g(A)$) denote the group generated by the divisors of the *graded* integral divisorial ideals (resp. by the divisors div(a), where a is *homogeneous* in A); write $C_g(A) = D_g(A)/F_g(A)$. Let S be the multiplicative subset of homogeneous elements $\neq 0$ of A.

(a) Show that every prime ideal of A of height 1 which meets S is graded (use Chapter III, § 1, no. 4, Proposition 5).

(b) Let $B = S^{-1}A$, which is a graded ring (Chapter II, § 2, no. 9); write $B = \bigoplus_{n \in \mathbf{Z}} B_n$; show that B_0 is a field and that, if $A \neq A_0$, B is isomorphic to the ring $B_0[X, X^{-1}]$ of rational functions $P(X)/X^k$, where $P(X) \in B_0[X]$.

(c) Show that $C_g(A)$ and C(A) are canonically isomorphic (use (a), (b) and equation (5) of no. 10).

17. Let $A = \bigoplus_{n \in \mathbf{Z}} A_n$ be a graded Krull domain and $p \in A_1$ be such that Ap is prime and $\neq 0$.

(a) Show that, if B is the ring $A_{(p)}$ defined in Chapter III, § 1, Exercise 1 (a), B is a Krull domain and the groups C(A) and C(B) are canonically isomorphic. (Show first that C(B) is isomorphic to $C(B[p, p^{-1}])$ and observe that $B[p, p^{-1}] = A[p^{-1}]$.)

(b) Under the hypotheses of Exercise 15 (b), show that the groups C(A) and C(A[X, Y]/(*a*X + *b*Y)) are isomorphic.

¶ 18. Let $A = \bigoplus_{n \in \mathbf{Z}} A_n$ be a graded Krull domain with positive degrees, such that A_0 is a field and let $\mathfrak{m} = \bigoplus_{n \geqslant 1} A_n$, a maximal ideal of A. Let S be the multiplicative set of homogeneous elements $\neq 0$ of A, so that the ring $S^{-1}A$ is a principal ideal domain (Exercise 16 (b)).

(a) Let $\mathfrak{p}$ be a prime ideal of A of height 1 which meets $A - \mathfrak{m}$; then $\mathfrak{p}$ is not graded (otherwise $\mathfrak{p} = A$) and the ideal $S^{-1}\mathfrak{p}$ of $S^{-1}A$ is principal, generated by an element of the form $x = 1 + x_1 + \cdots + x_n$ where $x_j \in (S^{-1}A)_j$. Writing an element of $\mathfrak{p}$ of the form $1 + a_1 + \cdots + a_q$ $(a_i \in A_i)$ as a multiple of x in $S^{-1}A$ and using Chapter V, § 1, no. 3, Proposition 11, show that $x \in \mathfrak{p}$; writing finally every element of $\mathfrak{p}$ as a multiple of x in $S^{-1}A$, show that $\mathfrak{p} = Ax$.

(b) Deduce from (a) that C(A) and $C(A_{\mathfrak{m}})$ are canonically isomorphic (use Proposition 17 of no. 10).

19. Let A be a local Krull domain and $\mathfrak{m}$ its maximal ideal; if $A' = (A[X])_{\mathfrak{m}A[X]}$, show that C(A) and C(A') are canonically isomorphic. (Apply the criterion of Proposition 17 of no. 10, using Exercise 12 (c).)

¶ 20. An integral domain A is called *Bezoutian* (or a *Bezout domain*) if every *finitely generated* ideal of A is principal. Every Bezoutian Noetherian domain is a principal ideal domain. Every valuation ring is a Bezoutian domain (and therefore a Bezoutian domain is not therefore necessarily Noetherian nor completely integrally closed).

(a) Show that every Bezoutian domain is integrally closed (cf. Exercise 6). If an integral domain is the union of a right directed family of Bezoutian subdomains, it is Bezoutian. If A is a Bezoutian domain, so is $S^{-1}A$ for every multiplicative subset S of A such that $0 \notin S$.

(b) Let v_i $(1 \leqslant i \leqslant n)$ be independent valuations on a field K and A_i the ring of the valuation v_i. Show that the intersection A of the A_i is a Bezoutian domain. (If an ideal $\mathfrak{a}$ of A is finitely generated, the set of $v_i(x)$ for $x \in \mathfrak{a}$ admits a least element α_i in the value group of v_i; for all i, let $x_i \in \mathfrak{a}$ be such that $v_i(x_i) = \alpha_i$. Using the Approximation Theorem (Chapter VI, § 7, no. 2, Theorem 1), show that there are elements $a_i \in A$ such that $x = \sum_{i=1}^{n} a_i x_i \in \mathfrak{a}$ satisfies the relations $v_i(x) = \alpha_i$ for $1 \leqslant i \leqslant n$.) If the v_i are discrete valuations, A is a principal ideal domain.

(c) Let K be an algebraically closed field of characteristic $\neq 2$ and let A be the subring of an algebraic closure of K(X) generated by K and two sequences of elements (x_n), $(1/x_n)$ $(1 \leqslant n < +\infty)$, where $x_1 = X$ and $x_{n-1} = x_n^2$. Show that A is Bezoutian (use (a)). If $\mathfrak{p}$ is a prime ideal $\neq 0$ of A, show that $\mathfrak{p}$ is generated by a sequence of elements of the form $x_n - a_n$, where $a_n \in K$, $a_n \neq 0$ and $a_n^2 = a_{n-1}$ (consider for all n the intersection $\mathfrak{p} \cap K[x_n, 1/x_n]$). Show that $\mathfrak{p}$ is not finitely generated.

21. An integral domain A is called *pseudo-Bezoutian* (resp. *pseudo-principal*) if, in the notation of Exercise 5, the group K*/U is lattice-ordered (completely lattice ordered). Every Bezoutian * (resp. factorial) * domain is pseudo-Bezoutian * (resp. pseudo-principal) *. Every pseudo-principal domain is pseudo-Bezoutian. A valuation ring whose order group is **R** is pseudo-principal but not a principal ideal domain. Every pseudo-Bezoutian (resp. pseudo-principal) domain is integrally closed (resp. completely integrally closed) (use Exercise 6). * Every Noetherian pseudo-Bezoutian domain is factorial. * Give an example of an integrally closed Noetherian (and therefore Krull) domain which is not pseudo-Bezoutian (cf. Exercise 2). If A is pseudo-Bezoutian, so is $S^{-1}A$ for every multiplicative subset S of A such that $0 \notin S$ (use Chapter II, § 2, Exercise 1).

¶ 22. (a) Let Γ be a lattice-ordered additive group and A the algebra of Γ over a field k; A is an integral domain (*Algebra*, Chapter II, § 11, no. 4, Proposition 8). Every element $x \neq 0$ in A may be written uniquely as $x = \sum_{i=1}^{n} \alpha_i e^{\nu_i}$, where the ν_i are distinct elements of Γ, the e^{ν_i} the corresponding elements of the canonical basis of A over k and the α_i elements $\neq 0$ of k. Write $\phi(x) = \inf(\nu_1, \ldots, \nu_n)$ in Γ; show that $\phi(x + y) \geqslant \inf(\phi(x), \phi(y))$ if x, y and $x + y$ are $\neq 0$ in A and $\phi(xy) = \phi(x) + \phi(y)$ if $xy \neq 0$ in A. (To show the second assertion, establish first the following lemma: given a finite family (ξ_j) of elements of Γ, in order that $\inf_i(\xi_j) = 0$, it is necessary and sufficient that for all $\eta > 0$ in Γ there exist an index j and an element $\zeta \in \Gamma$ such that $0 < \zeta \leqslant \eta$ and $\inf(\xi_j, \zeta) = 0$. Apply this lemma reducing it to the case where $\phi(x) = \phi(y) = 0$.)

(b) Deduce from (a) that, if K is the field of fractions of A, ϕ can be extended to a homomorphism from K* to Γ (also denoted by ϕ) such that

$$\phi(x + y) \geqslant \inf(\phi(x), \phi(y))$$

if x, y and $x + y$ are $\neq 0$ in K. Deduce that, if B is the set of $x \in K$ such that $x = 0$ or $\phi(x) \geqslant 0$, B is a domain whose field of fractions is K and such that, if U is the group of invertible elements of B, K*/U is isomorphic to Γ (and in particular B is a pseudo-Bezoutian domain). Derive examples of a pseudo-Bezoutian domain which is not completely integrally closed, a completely integrally closed pseudo-Bezoutian domain which is not pseudo-principal (cf.

Algebra, Chapter VI, § 1, Exercise 31) * and a pseudo-principal domain which is not factorial. *

* (c) Derive from (b) an example of an integral domain which is the union of a right directed family of factorial subdomains and is completely integrally closed but not pseudo-Bezoutian. (Let θ be an irrational number in [0, 1] and for every integer j let q_j be the greatest integer such that $q_j/2^j < \theta$. For all j define on the product group $\mathbf{Q} \times \mathbf{Q}$ a lattice-ordered group structure by taking the set $(G_j)_+$ of elements $\geqslant 0$ of this group to be the ordered pairs (ξ, η) such that $\xi \geqslant 0$ and $0 \leqslant \eta \leqslant (q_j 2^{-j})\xi$). *

23. (a) Let A be a pseudo-Bezoutian domain (Exercise 21) and K its field of fractions; for every polynomial $f \in A[X]$ a g.c.d. of the coefficients of f (determined to within an invertible element of A) is called a *content* of f. Let f, g be two polynomials of A[X]; show that for f to divide g in A[X], it is necessary and sufficient that f divide g in K[X] and that a content of f divide a content of g (use Exercise 12) (cf. Exercise 30 (c)).

(b) Deduce from (a) that, if A is a pseudo-Bezoutian (resp. pseudo-principal) domain, then so is A[X]. Moreover, if (a_ι) is a finite (resp. any) family of elements $\neq 0$ of A and d a g.c.d. of this family in A, d is also a g.c.d. of this family in A[X].

(c) Deduce from (b) that, if A is pseudo-Bezoutian (resp. pseudo-principal) then $A[X_\lambda]_{\lambda \in L}$ is pseudo-Bezoutian (resp. pseudo-principal) for every family $(X_\lambda)_{\lambda \in L}$ of indeterminates.

24. Let K be an algebraically closed field and A = K[X, Y] the polynomial ring in two indeterminates over K, which is a Krull domain * (and even a factorial domain). * For every ordered pair $(\alpha, \beta) \in K^2$ let $w_{\alpha,\beta}$ be the discrete valuation on A such that, for every polynomial $f \neq 0$, $w_{\alpha,\beta}(f)$ is the least degree of the monomials $\neq 0$ in $f(X + \alpha, Y + \beta)$. Show that A is the intersection of the rings of the valuations $w_{\alpha,\beta}$, none of which is an essential valuation of A.

25. An integrally closed domain A is said to be of *finite character* if there exists a family $(v_\iota)_{\iota \in I}$ of valuations on the field of fractions of A satisfying properties (AK_{II}) and (AK_{III}).

(a) Show that, if A is integrally closed and of finite character, for every element $x \neq 0$ in A, there can only be a finite number of extremal elements of the ordered group of principal divisors which are $\leqslant \operatorname{div}(x)$. (Let $J \subset I$ be the finite set of indices such that $v_\iota(x) > 0$. If m is the number of elements in J, show that there cannot be $m + 1$ elements y_j $(1 \leqslant j \leqslant m + 1)$ dividing x and such that the elements $\operatorname{div}(y_j)$ are extremal, observing that, for all j, $\operatorname{div}(y_j)$ must be prime to $\sum_{k \neq j} \operatorname{div}(y_k)$.)

* (b) Show that the ring of integral functions of one complex variable (Chapter V, § 1, Exercise 12) is a pseudo-principal domain (Exercise 21) but not a domain of finite character (use (a)). *

¶ 26. Let A be an integral domain and K its field of fractions. A valuation v on K is called *essential* for A if the ring of v is a local ring $A_{\mathfrak{p}}$ at a prime ideal $\mathfrak{p}$ of A (the intersection of A and the ideal of v).

(a) Let v be an essential valuation for A of height 1 and let $\mathfrak{p}$ be the prime ideal A consisting of those x such that $v(x) > 0$. Show that $\mathfrak{p}$ is of height 1. If $x \in K$ does not belong to $A_{\mathfrak{p}}$, then $A \cap x^{-1}A \subset \mathfrak{p}$. If w is a valuation on K whose ring B contains A and whose ideal $\mathfrak{q}$ satisfies $\mathfrak{q} \cap A \subset \mathfrak{p}$, show that w is equivalent to v.

(b) An integrally closed domain A is said to be *of finite character and real type* if there exists a family $(v_\iota)_{\iota \in I}$ of valuations on K *of height* 1 satisfying properties (AK_{II}) and (AK_{III}). Show that under these conditions, if $z \in K$ does not belong to A and $\mathfrak{p}$ is a prime ideal of A such that $A \cap z^{-1}A \subset \mathfrak{p}$, there exists $\iota \in I$ such that $v_\iota(z) < 0$ and that the prime ideal $\mathfrak{q}_\iota$ of $x \in A$ such that $v_\iota(x) > 0$ is contained in $\mathfrak{p}$. (Argue by *reductio ad absurdum* by considering the finite number of v_{ι_k} such that $v_{\iota_k}(z) < 0$; prove the existence of an $a \in A$ such that $a \notin \mathfrak{p}$ and $v_{\iota_k}(a) > 0$ for all k; deduce that $a^n z \notin A$ for every integer $n > 0$ and show that this implies a contradiction.) Conclude from this that every essential valuation for A of height 1 is equivalent to one of the v_ι (use (a)). Every finite intersection of subdomains of K which are of finite character and real type is also a domain of finite character and real type.

(c) Suppose that the hypotheses of (b) are satisfied and moreover that *all* the v_ι are essential. Show that, for every prime ideal $\mathfrak{p}$ of A of height 1, $A_{\mathfrak{p}}$ is the ring of one of the valuations v_ι (use (b) taking $z^{-1} \in \mathfrak{p}$). For all $x \in A$, the principal ideal Ax then admits a unique reduced primary decomposition (Chapter IV, § 2, Exercise 20), the prime ideals corresponding to this decomposition being the prime ideals of height 1 containing x.

(d) Suppose that the hypotheses of (b) are satisfied. Let S be a multiplicative subset of A not containing 0 and $J \subset I$ the set of $\iota \in I$ such that $v_\iota(x) = 0$ on S. Show that the family $(v_\iota)_{\iota \in I-J}$ satisfies properties (AK_{II}) and (AK_{III}) for the ring $S^{-1}A$; this family consists of essential valuations for $S^{-1}A$ if $(v_\iota)_{\iota \in I}$ consists of essential valuations for A.

(e) Generalize Proposition 9 of no. 5 to the case where the hypotheses of (c) are satisfied.

(f) Suppose the hypotheses of (b) are satisfied. Let K′ be a finite extension of K and A′ the integral closure of A in K′. Show that the valuations (no two of which are equivalent) on K′ which extend the v_ι satisfy properties (AK_{II}) and (AK_{III}) for A′ and the latter is therefore a domain of finite character and real type; if the v_ι are all essential for A, so are their extensions for A′ (argue as in no. 8, Proposition 12 and use also Chapter VI, § 8, no. 3, *Remark*).

(g) If A is a domain of finite character and real type, so is A[X]; if the conditions of (c) are fulfilled, determine the essential valuations for A[X] (argue as in no. 9). Generalize Exercise 8 similarly.

27. In Exercise 22 take Γ to be a direct sum of a family $(\Gamma_\iota)_{\iota \in I}$, where the Γ_ι are subgroups of the additive group **R**. Show that the ring B defined in Exercise 22 (b) is a domain of finite character and real type and that it is the intersection of a family of essential valuation rings for B; moreover, every prime ideal $\neq 0$ of B is maximal.

¶ 28. An integrally closed domain A is said to be *of finite character and rational type* if there exists a family $(v_\iota)_{\iota \in I}$ of valuations on its field of fractions K satisfying properties (AK_{II}) and (AK_{III}) and whose value groups are *subgroups of the additive group* **Q**. Show that the family of essential valuations for A of height 1 satisfy (AK_{II}) and (AK_{III}). (For all $\iota \in I$, let $\mathfrak{p}_\iota$ be the prime ideal of A consisting of the x such that $v_\iota(x) > 0$ and let V_ι be the ring of the valuation v_ι. Show that, if there are two indices α, β such that $\mathfrak{p}_\beta \subset \mathfrak{p}_\alpha$, then A is the intersection of the V_ι of index $\iota \neq \alpha$. For this, argue by *reductio ad absurdum* by showing that otherwise there would be an element $x \in K^*$, an element $y \in \mathfrak{p}_\beta$ and two positive integers r, s such that $v_\beta(x^r y^s) > 0$, $v_\alpha(x^r y^s) = 0$ and $v_\iota(x^r y^s) \geqslant 0$ for all $\iota \in I$, contrary to the hypothesis.)

29. Let A be an integrally closed domain and K its field of fractions.

(a) Let L be an algebraic extension of K and B the integral closure of A in L. Show that for every fractional ideal $\mathfrak{a}$ of A, if we write $\mathfrak{b} = \mathfrak{a}B$, $\tilde{\mathfrak{b}} \cap A = \tilde{\mathfrak{a}}$. (Reduce it to the case where $A \subset \tilde{\mathfrak{a}}$, in other words $A : \mathfrak{a} \subset A$ and prove that $B : \mathfrak{b} \subset B$; for this, let $x \in B : \mathfrak{b}$ and let c_i $(1 \leqslant i \leqslant n)$ be the coefficients of its minimal polynomial over K; note that for all $y \in \mathfrak{a}$ the elements $c_i y^i$ $(1 \leqslant i \leqslant n)$ belong to A and deduce that the c_i belong to A.)

(b) Let C be a polynomial ring $A[X_\lambda]_{\lambda \in L}$ in any family of indeterminates. Show that for every fractional ideal $\mathfrak{a}$ of A, if we write $\mathfrak{c} = \mathfrak{a}C$, $\tilde{\mathfrak{c}} \cap A = \tilde{\mathfrak{a}}$ (same method).

¶ 30. An integral domain is called *regularly integrally closed* if, in the monoid D(A), every finitely generated divisor (Exercise 11) is a *regular* element. Every completely integrally closed domain is regularly integrally closed; every pseudo-Bezoutial domain (Exercise 21) is regularly integrally closed.

(a) If A is regularly integrally closed and d, d', d'' are three finitely generated divisors, the relation $d + d'' \leqslant d' + d''$ implies $d \leqslant d'$.

(b) Deduce from (a) that, for an integral domain A to be regularly integrally closed, it is necessary and sufficient that, for every fractional ideal $\mathfrak{a}$ of A such that $\operatorname{div}(\mathfrak{a})$ is a finitely generated divisor, $\mathfrak{a} : \mathfrak{a} = A$; in particular A is integrally closed (Exercise 6 (b)). (Use the fact that $A : (\mathfrak{b}\mathfrak{c}) = (A : \mathfrak{b}) : \mathfrak{c}$ for two fractional ideals $\mathfrak{b}$, $\mathfrak{c}$ of A, taking $\mathfrak{b} = \mathfrak{a}$, $\mathfrak{c} = A : \mathfrak{a}$.) Moreover, $\operatorname{div}(\mathfrak{a})$ is then invertible in D(A).

(c) Let A be a regularly integrally closed domain and K its field of fractions; the monoid $D_f(A)$ of finitely generated divisors of A generates in D(A) a lattice-ordered group $G_f(A)$. For every polynomial $p \in K[X]$, let $d(p)$ denote the divisor of the fractional ideal of A generated by the coefficients of p; for every rational function $r = p/q$ of K(X), where p, q are in K[X] and $q \neq 0$, $d(p) - d(q)$ is well defined on $G_f(A)$ (independently of the expression of r as a quotient of two polynomials); it is denoted by $d(r)$ (cf. Exercise 12 (c)). If we then write $\gamma(r) = ((r), d(r))$, where (r) is the fractional principal ideal of K[X] generated by r, γ is an isomorphism of the ordered group $\mathscr{P}^*(A[X])$ (notation of § 3, no. 2) onto a subgroup of the product ordered group

$$\mathscr{P}^*(K[X]) \times G_f(A) \subset \mathscr{P}^*(K[X]) \times D(A).$$

Deduce that A[X] is a regularly integrally closed domain and that $D_f(A[X])$ is isomorphic to $\mathscr{P}^*(K[X]) \times D_f(A)$ (use (a) and (b) and Exercise 29 (b)).

(d) Let B be an integrally closed domain, K its field of fractions and A the subring of the polynomial ring K[X, Y] consisting of the polynomials whose constant term belongs to B. Show that, if $B \neq K$, A is integrally closed but not regularly integrally closed (show that the divisor inf(div(X), div(Y)) corresponds to a divisorial ideal $\mathfrak{a}$ such that $\mathfrak{a} : \mathfrak{a} \neq A$).

¶ 31. (a) Let A be a regularly integrally closed domain (Exercise 30) and K its field of fractions; for every polynomial $P \in K[X]$, let $c(P)$ denote the divisor div($\mathfrak{a}$), where $\mathfrak{a}$ is the fractional ideal of K generated by the coefficients of P. Let B denote the subring of K(X) consisting of the rational functions P/Q such that $c(P) \geqslant c(Q)$ (show that this condition depends only on the element P/Q, using Exercise 30 (a) and 12 (c)). Then $B \cap K = A$.

(b) Show that B is a Bezoutian domain (Exercise 20) (observe that, if P, Q are two polynomials of A[X], $P + X^m Q$ divides P and Q in B if m is sufficiently large).

(c) Show that, if A is a Krull domain, B is a principal ideal domain (consider an increasing sequence of finitely generated fractional principal ideals of B). Give an example where A is completely integrally closed but B is not a principal ideal domain (cf. Exercise 22).

(d) Deduce from (c) an example of a Noetherian domain B for which there exists a subfield K of its field of fractions such that $B \cap K$ is not Noetherian.

32. Let A be an integral domain $(a_i)_{1 \leqslant i \leqslant n}$ a finite family of elements of A, $\mathfrak{a}$ the ideal $\sum_i Aa_i$ and R the submodule of A^n generated by the element $(a_1, \ldots, a_n)$. Show that, for the torsion submodule of the A-module $M = A^n/R$ to be a direct factor of M, it is necessary and sufficient that

$$\mathfrak{a}(A : \mathfrak{a}) + (A : (A : \mathfrak{a})) = A.$$

§ 2

1. (a) Show that, if $\mathfrak{a}$ is an ideal $\neq 0$ in a Dedekind domain A, the ring $A/\mathfrak{a}$ is a principal ideal ring (*Algebra*, Chapter VII, § 1, Exercise 5) (note that $A/\mathfrak{a}$ is semi-local and argue as in Proposition 1 of no. 2). Deduce again that every fractional ideal of A is generated by at most two elements (cf. Exercise 11 (a)).

(b) In the polynomial ring $A = k[X, Y]$ over a field k, let $\mathfrak{m}$ be the ideal $AX + AY$. Show that for every integer n the minimum number of generators of the ideal $\mathfrak{m}^n$ is $n + 1$.

2. (a) Let $\mathfrak{a}$, $\mathfrak{b}$ be two integral ideals of a Dedekind domain A; show that there exists an integral ideal $\mathfrak{c}$ which is prime to $\mathfrak{b}$ and such that $\mathfrak{ac}$ is principal (use Proposition 9 of § 1, no. 5).

(b) Let $\mathfrak{a}$, $\mathfrak{b}$ be two integral ideals of A; show that there exists $x \neq 0$ in the field of fractions of A such that $x\mathfrak{a}$ is an integral ideal of A which is prime to $\mathfrak{b}$ (same method). Deduce that the module $\mathfrak{a}/\mathfrak{ab}$ is isomorphic to $A/\mathfrak{b}$.

3. Let A be a Dedekind domain, K its field of fractions, P the set of prime ideals $\neq 0$ of A and for all $\mathfrak{p} \in P$ let $v_\mathfrak{p}$ be the corresponding essential valuation on K. For every sub-A-module M of K and all $\mathfrak{p} \in P$ we write $v_\mathfrak{p}(M) = \inf v_\mathfrak{p}(x)$ (taken in $\overline{\mathbf{R}}$); if $M \neq 0$, $v_\mathfrak{p}(M) < +\infty$ for all $\mathfrak{p} \in P$ and $v_\mathfrak{p}(M) \leqslant 0$ for almost all $\mathfrak{p} \in P$. If M, N are two sub-A-modules of K, show that the relation $M \subset N$ is equivalent to $v_\mathfrak{p}(N) \leqslant v_\mathfrak{p}(M)$ for all $\mathfrak{p} \in P$ (use Proposition 9 of § 1, no. 5). Conversely, for every family $(\nu_\mathfrak{p})_{\mathfrak{p} \in P}$ of elements equal to an integer or $-\infty$ and such that $\nu_\mathfrak{p} \leqslant 0$ for almost all $\mathfrak{p} \in P$, there exists a unique sub-A-module M of K such that $v_\mathfrak{p}(M) = \nu_\mathfrak{p}$ for all $\mathfrak{p} \in P$.

4. Let A be a Dedekind domain and K its field of fractions. If L is a subfield of K such that A is integral over $A \cap L$, then $A \cap L$ is a Dedekind domain (show first that $A \cap L$ is a Krull domain and then that every prime ideal of $A \cap L$ is maximal, using Chapter V, § 2, no. 1, Theorem 1).

5. (a) Let k be a field, K the field $k(X, Y)$ of rational functions in two indeterminates over k and A the polynomial ring $K[Z]$ in one indeterminate, which is a principal ideal domain. Let L be the subfield $k(Z, X + YZ)$ of $k(X, Y, Z)$; show that $A \cap L$ is not a Dedekind domain (prove that there are prime ideals which are not maximal and $\neq 0$ in this ring).

(b) Let A be a Dedekind domain which is not a principal ideal domain and for which the ideal class group C(A) is finite * (the ring of integers of a finite algebraic extension of $\mathbf{Q}$ has the latter property). * Let $(\mathfrak{a}_j)_{1 \leqslant j \leqslant r}$ be a representative system of the group I(A) of fractional ideals $\neq 0$ modulo the subgroup of fractional principal ideals, consisting of integral ideals of A and let $\mathfrak{p}_i$ $(1 \leqslant i \leqslant s)$ be the prime ideals dividing at least one of the $\mathfrak{a}_j$. Let S (resp. T) denote the multiplicative set of elements $x \in A$ such that $v_{\mathfrak{p}_i}(x) = 0$ for all i (resp. consisting of 1 and the $x \in A$ such that $v_{\mathfrak{q}}(x) = 0$ for all the prime ideals $\mathfrak{q}$ distinct from the $\mathfrak{p}_i$ and $v_{\mathfrak{p}_i}(x) \geqslant 1$ for all i; the hypothesis implies that T is not reduced to 1). Show that $S^{-1}A$ and $T^{-1}A$ are principal ideal domains and that $A = (S^{-1}A) \cap (T^{-1}A)$.

6. Show that in a Dedekind domain the notions of primary ideal, irreducible ideal, primal ideal (Chapter IV, § 2, Exercise 33), quasi-prime ideal (Chapter IV, § 2, Exercise 34) and power of a prime ideal are identical.

7. Let A be a Noetherian domain. Show that the following properties are equivalent:

(α) A is a Dedekind domain.

(β) For every maximal ideal $\mathfrak{m}$ of A, there exists no ideal $\mathfrak{a}$ distinct from $\mathfrak{m}$ and $\mathfrak{m}^2$ such that $\mathfrak{m}^2 \subset \mathfrak{a} \subset \mathfrak{m}$.

(γ) For every maximal ideal $\mathfrak{m}$ of A, the set of primary ideals for $\mathfrak{m}$ is totally ordered by inclusion.

(δ) For every maximal ideal $\mathfrak{m}$ of A, every primary ideal for $\mathfrak{m}$ is a product of prime ideals.

(Prove that each of the properties (γ) and (δ) implies (β); then show that (β) implies that $A_{\mathfrak{m}}$ is a field or a discrete valuation ring (cf. Chapter VI, § 3, no. 5, Proposition 9).)

Give an example of a local integral domain satisfying properties (β), (γ) and (δ), which is not a valuation ring (cf. Chapter VI, § 3, Exercise 7).

¶ 8. Let A be an integral domain in which every prime ideal $\neq 0$ is invertible. Show that A is a Dedekind domain. (Prove first that every prime ideal $\mathfrak{p}' \neq 0$ of A is maximal, noting that, if $\mathfrak{p}$ is a prime ideal such that $\mathfrak{p} \neq \mathfrak{p}'$ and $\mathfrak{p} \supset \mathfrak{p}'$, then $\mathfrak{p}' : \mathfrak{p} = \mathfrak{p}'$ (Chapter II, § 1, Exercises 8 (b)) and on the other hand that $\mathfrak{p}' : \mathfrak{p} = \mathfrak{p}'\mathfrak{p}^{-1}$, whence a contradiction is deduced. Deduce that A is Noetherian (Chapter II, § 1, Exercise 6 (b)) and finally apply Chapter II, § 5, no. 6, Theorem 4 to show that $A_{\mathfrak{m}}$ is a field or a discrete valuation ring for every maximal ideal $\mathfrak{m}$ of A.)

¶ 9. Let A be an integral domain.

(a) Let $(\mathfrak{p}_i)_{1 \leqslant i \leqslant m}$, $(\mathfrak{p}'_j)_{1 \leqslant j \leqslant n}$ be two finite families of *invertible* prime ideals such that $\mathfrak{p}_1\mathfrak{p}_2 \ldots \mathfrak{p}_m = \mathfrak{p}'_1\mathfrak{p}'_2 \ldots \mathfrak{p}'_n$. Show that $m = n$ and that there exists a permutation π of $[1, n]$ such that $\mathfrak{p}'_i = \mathfrak{p}_{\pi(i)}$ for all i such that $1 \leqslant i \leqslant n$.

(b) Let $\mathfrak{p}$ be a prime ideal of A and a an element of $A - \mathfrak{p}$; if $\mathfrak{p} \subset Aa + \mathfrak{p}^2$, show that $\mathfrak{p} = \mathfrak{p}(Aa + \mathfrak{p})$; deduce that, if $\mathfrak{p}$ is invertible, then $Aa + \mathfrak{p} = A$.

(c) Show that, if every ideal of A is a product (not necessarily *a priori* unique) of prime ideals of A, A is a Dedekind domain. (Show first that every invertible prime ideal $\mathfrak{p}$ is maximal, considering an element $a \in A - \mathfrak{p}$ and decompositions of $Aa + \mathfrak{p}$ and $Aa^2 + \mathfrak{p}$ as products of prime ideals and applying (a) in the ring $A/\mathfrak{p}$ to show that $Aa^2 + \mathfrak{p} = (Aa + \mathfrak{p})^2$; then use (b). Prove then that every prime ideal $\mathfrak{p} \neq 0$ is invertible, by decomposing as a product of prime ideals a principal ideal Ab where $b \in \mathfrak{p}$, observing that the factors of this product are invertible and, by applying the above, show that $\mathfrak{p}$ is necessarily equal to one of these factors. Conclude with the aid of Exercise 8.)

¶ 10. Let A be a ring in which every ideal is a product of prime ideals.

(a) Show that, for every prime ideal $\mathfrak{p}$ of A, $A/\mathfrak{p}$ is a Dedekind domain (Exercise 9).

(b) Show that the set of minimal prime ideals $\mathfrak{p}_\iota$ of A is finite (decompose (0) as a product of prime ideals).

(c) Suppose that $A/\mathfrak{p}_\iota$ is not a field. Show that, if $y \in \mathfrak{p}_\iota$, then, for *all* $x \in A - \mathfrak{p}_\iota$, there exists $z \in \mathfrak{p}_\iota$ such that $y = zx$. (Consider in A the decompositions as products of prime ideals of Ax and $Ax + Ay$ and the corresponding decompositions in $A/\mathfrak{p}_\iota$.) Deduce that, for all non-zero $y \in \mathfrak{p}_\iota$, $Ay = \mathfrak{p}_\iota$ (consider $\mathfrak{p}_\iota/Ay$, decomposing Ay as a product of prime ideals); show finally that $\mathfrak{p}_\iota = \mathfrak{p}_\iota^2$ and therefore that there exists in $\mathfrak{p}_\iota$ an idempotent e_ι such that $\mathfrak{p}_\iota = Ae_\iota$ (cf. Chapter II, § 4, Exercise 15). Then A is the direct composition of the ring Ae_ι and the ring $A(1 - e_\iota)$, isomorphic to the Dedekind domain $A/\mathfrak{p}_\iota$.

(d) Suppose now that all the $\mathfrak{p}_\iota$ are maximal, so that A is a direct composition of rings of the form $A/\mathfrak{p}_\iota^{\gamma_\iota}$ (Chapter II, § 1, no. 2, Proposition 5) and we may therefore confine our attention to the case where A is primary or where the unique prime ideal $\mathfrak{p}$ of A is nilpotent. Show that in this case the hypothesis implies that A is a principal ideal ring (*Algebra*, Chapter VII, § 1, Exercises 5 and 6).

(e) Conclude from (c) and (d) that A is the product of a finite number of Dedekind domains and a principal ideal ring.

11. Let K be a field and $(v_\iota)_{\iota \in I}$ a family of discrete valuations on K satisfying conditions (AK_I) and (AK_{III}) of § 1, no. 3 and such that, for every integer r, every family $(n_h)_{1 \leqslant h \leqslant r}$ of integers $\geqslant 0$, every family $(\iota_h)_{1 \leqslant h \leqslant r}$ of distinct elements of I and every family $(a_h)_{1 \leqslant h \leqslant r}$ of elements of K, there exists $x \in K$ such that $v_{\iota_h}(x - a_h) \geqslant n_h$ for $1 \leqslant h \leqslant r$ and $v_\iota(x) = 0$ for ι distinct from the ι_h. Show that the intersection A of the valuation rings of the v_ι is a Dedekind domain whose field of fractions is K and that $(v_\iota)_{\iota \in I}$ is the family of essential valuations of A. (Use Exercise 7 of § 1; then prove that two distinct prime ideals of height 1 are relatively prime and deduce that every prime ideal $\neq 0$ of A is maximal.)

¶ 12. Let A be an integral domain and K its field of fractions. Show that the following properties are equivalent:

(α) For every prime ideal $\mathfrak{p}$ of A, the local ring $A_\mathfrak{p}$ is a valuation ring.

(β) For every maximal ideal $\mathfrak{m}$ of A, the local ring $A_\mathfrak{m}$ is a valuation ring.

(γ) Every finitely generated ideal $\neq 0$ in A is invertible (and therefore the finitely generated fractional ideals $\neq 0$ form a *group*).

(δ) Every torsion-free finitely generated A-module is projective.

(ε) For all $x \in K$, the fractional ideal $A + Ax$ is invertible.

(ζ) For all $x \neq 0$ in K, there exists $y \in K$ such that $y \equiv 0 \pmod{1}$, $y \equiv 0 \pmod{x}$ and $y \equiv 1 \pmod{1 - x}$.

(η) If $(\mathfrak{a}_i)$ is a finite family of ideals of A, for the system of congruences $x \equiv c_i \pmod{\mathfrak{a}_i}$ to have a solution it is necessary and sufficient that $c_i \equiv c_j \pmod{(\mathfrak{a}_i + \mathfrak{a}_j)}$ for every pair of indices i, j ("Chinese remainder theorem").

(θ) If $\mathfrak{a}, \mathfrak{b}, \mathfrak{c}$ are three ideals of A, then $\mathfrak{a} \cap (\mathfrak{b} + \mathfrak{c}) = \mathfrak{a} \cap \mathfrak{b} + \mathfrak{a} \cap \mathfrak{c}$.

(ι) If $\mathfrak{a}, \mathfrak{b}, \mathfrak{c}$ are three ideals of A, then $\mathfrak{a} + (\mathfrak{b} \cap \mathfrak{c}) = (\mathfrak{a} + \mathfrak{b}) \cap (\mathfrak{a} + \mathfrak{c})$.

($\varkappa$) A is integrally closed and every finitely generated ideal of A is divisorial.

(λ) A is integrally closed and for $z \neq 0$ in K there exist x, y in A such that $z = x + yz^2$.

(μ) A is integrally closed and, if $\mathfrak{a}, \mathfrak{b}, \mathfrak{c}$ are three finitely generated fractional ideals $\neq 0$, the relation $\mathfrak{a}\mathfrak{b} = \mathfrak{a}\mathfrak{c}$ implies $\mathfrak{b} = \mathfrak{c}$.

(ν) For every finitely generated A-module M, the torsion submodule of M is a direct factor of M.

(σ) Every ring B such that $A \subset B \subset K$ is an integrally closed domain.

Then A is called a *Prüferian* domain (or a *Prüfer domain*).

(To prove the equivalence of (α), (β), (γ) and (δ), use Chapter II, § 5, no. 2, Theorem 1 and no. 6, Theorem 4. To prove that (ε) implies (γ), use the identity $(\mathfrak{a} + \mathfrak{b})(\mathfrak{b} + \mathfrak{c})(\mathfrak{c} + \mathfrak{a}) = (\mathfrak{a} + \mathfrak{b} + \mathfrak{c})(\mathfrak{a}\mathfrak{b} + \mathfrak{b}\mathfrak{c} + \mathfrak{c}\mathfrak{a})$ between three fractional ideals in an integral domain. To prove that (ζ) implies (η), argue by induction on the number of $\mathfrak{a}_i$. The equivalence of (η), (θ) and (ι) has been shown in *Algebra*, Chapter VI, § 1, Exercise 25. Prove that (λ) implies (ε) by noting that (λ) implies the relation

$$x(A + Az) \subset z(A + Az),$$

using Exercise 6 (b) of § 1 and noting that

$$(A + Az)(Ay + Ax/z) = A.$$

To prove that (μ) implies (λ) noting that always

$$z(A + Az) \subset (A + Az^2)(A + Az).$$

Prove that ($\varkappa$) implies (λ) noting that every fractional principal ideal At which contained A and Az^2 also contains Az. To prove that (ν) implies (γ), consider, for a finitely generated integral ideal $\mathfrak{a} \neq 0$ of A, a non-zero element $c \in \mathfrak{a}(A : \mathfrak{a})$ and apply Exercise 32 of § 1 to the ideal $\mathfrak{b} = c\mathfrak{a}$. To prove that ($\sigma$) implies ($\beta$), reduce it to the case where A is a local ring, consider an element $z \in K - A$ and show that $z^{-1} \in A$, noting that the hypothesis implies that $z \in A[z^2]$.)

¶ 13. (a) In a Prüferian domain A, let $\mathfrak{a}$, $\mathfrak{b}$ be two finitely generated ideals; show that, if $\mathfrak{b} \not\subset \mathfrak{a}$, there exists a finitely generated ideal $\mathfrak{c}$ such that $\mathfrak{a} \subset \mathfrak{c}$, $\mathfrak{a} \neq \mathfrak{c}$ and $\mathfrak{b}\mathfrak{c} \subset \mathfrak{a}$ (consider the ideal $\mathfrak{a} + \mathfrak{b}$).

(b) Deduce from (a) that, if $\mathfrak{a}$ is a finitely generated ideal in a Prüferian domain A, then in the ring $A/\mathfrak{a}$ every element which is not a divisor of zero is invertible. In particular, a prime ideal of A cannot be finitely generated if it is maximal.

(c) Show that in a Prüferian domain A two prime ideals $\mathfrak{p}$, $\mathfrak{p}'$ neither of which is contained in the other are relatively prime (consider, for every maximal ideal $\mathfrak{m}$ of A, the ideal $\mathfrak{p}A_{\mathfrak{m}} + \mathfrak{p}'A_{\mathfrak{m}}$).

(d) In a Prüferian domain A let $\mathfrak{a}$ be a finitely generated ideal. For $\mathfrak{a}$ to be primal (Chapter IV, § 2, Exercise 33), it is necessary and sufficient that $\mathfrak{a}$ be contained only in a single maximal ideal of A (use (b)); $\mathfrak{a}$ is then irreducible and quasi-prime (Chapter IV, § 2, Exercise 34), so that, for finitely generated ideals, these three notions coincide. For $\mathfrak{a}$ to be primary, it is necessary and sufficient that it be contained only in a single prime ideal (necessarily maximal) $\mathfrak{m}$; for $\mathfrak{a}$ to be strongly primary (Chapter IV, § 2, Exercise 27), it is necessary and sufficient also that $\mathfrak{m}A_{\mathfrak{m}}$ be principal.

14. Let A be a Prüferian domain. Show that for an A-module M to be flat, it is necessary and sufficient that it be torsion-free (use Exercise 12 (δ)). Deduce that A is a coherent ring (Chapter I, § 2, Exercise 12).

15. (a) If A is a Prüferian domain, $A/\mathfrak{p}$ is Prüferian for every prime ideal $\mathfrak{p}$ of A.

(b) If A is Prüferian, so is $S^{-1}A$ for every multiplicative subset S of A not containing 0.

(c) Let K be a field and (A_λ) a non-empty right directed family of subrings of K. Show that, if the A_λ are Prüferian domains, so is their union A (use Exercise 12 (ζ)).

16. Let A be a Prüferian domain, K its field of fractions and L a (finite or otherwise) algebraic extension of K; show that the integral closure A′ of A in L is a Prüferian domain. (Let $x \in L$ and let $a_0X^n + a_1X^{n-1} + \cdots + a_n$ be a polynomial in A[X] of which x is a root; if we write

$$a_0X^n + a_1X^{n-1} + \cdots + a_n = (X - x)(b_0X^{n-1} + \cdots + b_{n-1}),$$

show that the b_i and the b_ix belong to A′, using § 1, Exercise 12 (a). If $\mathfrak{a} = \sum_{i=0}^{n} Aa_i$, $\mathfrak{b} = \sum_{j=0}^{n-1} Bb_j$, show then that the ideal $\mathfrak{b}.B\mathfrak{a}^{-1}$ is the inverse of $B + Bx$.)

¶ 17. Let A be an integral domain.

(a) For A to be Bezoutian (§ 1, Exercise 20), it is necessary and sufficient that it be Prüferian and pseudo-Bezoutian (§ 1, Exercise 21).

(b) For A to be a Dedekind domain, it is necessary and sufficient that it be a Prüferian Krull domain (show that every divisorial ideal of A is finitely generated, using Exercise 12 ($\varkappa$) and § 1, Exercise 11 (a); deduce that every prime ideal $\neq 0$ of A is maximal (Exercise 13 (b)), then that every prime ideal of A is finitely generated and conclude with the aid of Chapter II, § 1, Exercise 6).

(c) For A to be a Dedekind domain, it is necessary and sufficient that it be Prüferian and strongly Laskerian (Chapter IV, § 2, Exercise 28). (Observe that, if A is Prüferian and strongly Laskerian, every maximal ideal $\mathfrak{m}$ of A is such that $A_{\mathfrak{m}}$ is a discrete valuation ring (Chapter VI, § 3, Exercise 8), then that for all non-zero $x \in A$ there is a product of a finite number of maximal ideals of A contained in the ideal Ax and conclude that A is a Krull domain.)

18. Let A be a ring, the union of a right directed family of subrings A_α which are Dedekind domains. Suppose that, for every prime ideal $\mathfrak{p}$ of A_α, there exists $\beta \geqslant \alpha$ such that there are at least two distinct prime ideals of A_β lying above $\mathfrak{p}$.

(a) Show that the ring of all algebraic integers (the integral closure of $\mathbf{Z}$ in $\mathbf{C}$) satisfies the above conditions (cf. Chapter V, § 2, Exercise 6).

(b) Let v be a non-improper valuation on A, $z \in A$ such that $v(z) > 0$ and let $\mathfrak{a}$ be the ideal of A consisting of the x satisfying $v(x) \geqslant v(z)$. Show that $A : \mathfrak{a} = A$ and hence that $\mathfrak{a}$ is not invertible, although for every maximal ideal $\mathfrak{m}$ of A, $\mathfrak{a}A_{\mathfrak{m}}$ is a principal ideal in the valuation ring $A_{\mathfrak{m}}$ (cf. Chapter II, § 5, no. 6, Theorem 4).

(c) Deduce from (a) and (b) an example of a Prüferian domain which is completely integrally closed but not a Krull domain (cf. Exercises 15 (c) and 17 (b) and Chapter V, § 1, Exercise 14).

¶ 19. An integral domain A is called *pseudo-Prüferian* if the set of finitely generated divisors of D(A) (§ 1, Exercise 11) is a *group*. A pseudo-Prüferian domain is regularly integrally closed (§ 1, Exercise 29). A pseudo-Bezoutian domain (§ 1, Exercise 21) is pseudo-Prüferian; a Prüferian domain is pseudo-Prüferian; a Krull domain is pseudo-Prüferian.

(a) Show by examples that the converses of the last three assertions are not necessarily true.

(b) Let θ be an irrational number > 0 and Γ the group $\mathbf{R}^2$ ordered by taking the set of ordered pairs (α, β) such that $\alpha \geqslant 0$ and $\beta \geqslant \theta\alpha$ as the set of positive elements; the ordered group Γ is a complete lattice. Let B be the pseudo-principal domain derived from Γ by the procedure of § 1, Exercise 22 and let A be the subring the intersection of B and the algebra over k of the subgroup $\mathbf{Q}^2$ of $\mathbf{R}^2$. Show that A is a completely integrally closed domain but is not pseudo-Prüferian (note that, if K is the field of fractions of A and U the group of invertible elements of A, the ordered group K^*/U is isomorphic to the group $\Gamma \cap \mathbf{Q}^2$, ordered by the ordering induced by that on Γ).

(c) If the domain A is pseudo-Prüferian, show that the polynomial domain A[X] is pseudo-Prüferian (cf. § 1, Exercise 30 (c)).

(d) Let A be a pseudo-Prüferian domain and K its field of fractions. Show that the integral closure B of A in an algebraic extension L of K is a pseudo-Prüferian domain (method of Exercise 16, using § 1, Exercise 29 (a)).

* (e) Deduce from Exercise 30 (d) of § 1 an example of an increasing sequence (A_n) of factorial domains with the same field of fractions whose union is not a regularly integrally closed domain (nor *a fortiori* a pseudo-Prüferian domain) (in the example quoted, take B to be a discrete valuation ring). *

20. Let A be a Dedekind domain, K its field of fractions, L a finite algebraic extension of K and B the integral closure of A in L, which is a Dedekind domain. Let $\mathfrak{f}$ be an ideal of B. For there to exist a ring C such that $A \subset C \subset B$ and such that $\mathfrak{f}$ is the conductor of B in A (Chapter V, § 1, no. 5), it is necessary and sufficient that, for every prime ideal $\mathfrak{p}'$ of B containing $\mathfrak{f}$ such that the field $B/\mathfrak{p}'$ is isomorphic to $A/(\mathfrak{p}' \cap A)$ (prime ideals of *residue degree* 1), the intersections of $\mathfrak{f}$ and the transporter $\mathfrak{f} : \mathfrak{p}'$ of $\mathfrak{p}'$ in $\mathfrak{f}$ with A be equal. (Note that, if there exists such a ring C, the conductor of B in the ring $C_0 = A + \mathfrak{f}$ is also equal to $\mathfrak{f}$; the existence of C is therefore equivalent to the fact that $A + \mathfrak{f}$ contains no ideal of B distinct from $\mathfrak{f}$ and containing $\mathfrak{f}$. To prove that the latter condition is equivalent to that of the statement, reduce it to the case where A is a discrete valuation ring.)

¶ 21. (a) Let A be an integral domain, f, g two polynomials in A[X] and $h = fg$. Let $\mathfrak{a}$, $\mathfrak{b}$, $\mathfrak{c}$ denote the ideals of A generated respectively by the coefficients of f, g, h. Show that, if $\deg(g) = n$, then $\mathfrak{a}^{n+1}\mathfrak{b} = \mathfrak{a}^n\mathfrak{c}$. (Let $f(X) = \sum_{i=1}^{m} a_i X^i$, $g(X) = \sum_{j=1}^{n} b_j X^j$; for every increasing sequence

$$\sigma = (i_k)_{1 \leqslant k \leqslant n+1}$$

of integers $\leqslant m$ and all j, let

$$u_{\sigma,j} = a_{i_1} a_{i_2} \ldots a_{i_j+1} b_j a_{i_j+2} \ldots a_{i_n+1};$$

take on the set of $u_{\sigma,j}$ a total ordering such that, for $j < j'$, $u_{\sigma,j} < u_{\tau,j'}$ for all σ, τ, and, for all j, $u_{\sigma,j} \leqslant u_{\tau,j}$ if and only if $\sigma \leqslant \tau$ in the lexicographical ordering on $[0, m]^{n+1}$. Then argue by induction on this totally ordered set.)

(b) Deduce from (a) that, if A is Prüferian (Exercise 12), then $\mathfrak{c} = \mathfrak{a}\mathfrak{b}$.

(c) Taking A to be the polynomial ring $\mathbf{Z}[Y]$, give an example of two polynomials f, g of the first degree in A[X] such that $\mathfrak{c} \neq \mathfrak{a}\mathfrak{b}$.

¶ 22. (a) Let A be a Noetherian domain each of whose prime ideals $\neq 0$ is maximal, so that every ideal $\mathfrak{a} \neq 0$ may be written uniquely as a product $\prod_i \mathfrak{q}_i$ of primary ideals relative to the distinct prime ideals $\mathfrak{p}_i$ containing $\mathfrak{a}$.

Let $\mathfrak{a}$ be an *invertible* ideal in A and let $\mathfrak{b}$ be any ideal $\neq 0$ in A; show that there exists an ideal $\mathfrak{c}$ such that $\mathfrak{b} + \mathfrak{c} = A$ and $\mathfrak{a}\mathfrak{c}$ is principal (Observe that, if $(\mathfrak{p}_i)_{1 \leqslant i \leqslant n}$ is a finite family of distinct maximal ideals of A and we write $\mathfrak{r}_i = \prod_{j \neq i} \mathfrak{p}_j$, then $\mathfrak{a} = \sum_i \mathfrak{a}\mathfrak{r}_i$ and $\bigcap_i \mathfrak{a}\mathfrak{r}_i = \mathfrak{a}\mathfrak{p}_1\mathfrak{p}_2 \ldots \mathfrak{p}_n$; using Chapter II, § 1, no. 2, Proposition 6, deduce the existence of an element $x \in \mathfrak{a}$ such that $x^{-1}\mathfrak{a} + \mathfrak{b} = A$.) Deduce that $\mathfrak{a}$ is generated by two elements (cf. Exercise 1).

(b) Let K be a field and A the subring of the ring of formal power series $K[[T]]$ consisting of the series $a_0 + T^n P(T)$, where $a_0 \in K$, $P(T) \in K[[T]]$, for a given integer n. Show that A is a Noetherian local integral domain with a single prime ideal $\mathfrak{m} \neq 0$, but that the smallest cardinal of a system of generators of $\mathfrak{m}$ is n.

§ 3

1. For an integral domain to be factorial, it is necessary and sufficient that there exist a mapping $x \mapsto s(x)$ of $A - \{0\}$ to $\mathbf{N}$ such that $s(xy) = s(x) + s(y)$, the relation $s(x) = 0$ implies that x is invertible in A and finally, for any two elements x, y of $A - \{0\}$, neither of which divides the other, there exist elements a, b, z, t of $A - \{0\}$ such that $ax + by = zt$, $s(z) < s(x)$, t being prime to x and to y. (If this condition is fulfilled, show first that every non-zero element of A is a product of extremal elements and then that, for every extremal element p, Ap is a prime ideal; conclude with the aid of Theorem 1 (d)).

2. (a) Let A be an integral domain, the union of a right directed family (A_λ) of subrings. Suppose that each of the A_λ is factorial and that, if $\lambda \leqslant \mu$, every extremal element of A_λ is extremal in A_μ. Show that A is a factorial domain whose set of extremal elements is the union of the sets of extremal elements of each of the A_λ (cf. § 2, Exercise 19 (e)).

(b) Deduce from (a) that for every factorial domain A the polynomial domain $A[X_\lambda]_{\lambda \in L}$ in any family of indeterminates is factorial.

(c) Let A be a factorial domain such that every domain of formal power series $A[[X_1, \ldots, X_n]]$ in a finite number of indeterminates is factorial. Show that every domain of formal power series $A[[X_\lambda]]_{\lambda \in L}$ in any family of indeterminates (*Algebra*, Chapter IV, § 5, Exercise 1) is factorial (same method).

3. (a) Let $A = \bigoplus_{n \geqslant 0} A_n$ be a graded algebra with positive degrees over a field k; suppose that $A_0 = k$ and that A is a Krull domain. Show that for A to be factorial, it is necessary and sufficient that every graded prime ideal $\mathfrak{p}$ of A of height 1 be of the form $\mathfrak{p} = Aa$, where a is a homogeneous element (use Exercise 16 of § 1). Show that every homogeneous element $\neq 0$ of A is a product of homogeneous extremal elements.

(b) Let A be a graded k-algebra satisfying the conditions of (a). Let k' be an extension of k and suppose that $A \otimes_k k'$ is a factorial domain; show that A is factorial. (If a graded ideal $\mathfrak{a}$ of A is such that $\mathfrak{a} \otimes_k k'$ is principal in $A \otimes_k k'$, show that $\mathfrak{a}$ is principal; then use (a).)

4. (a) Show that the ring $A = \mathbf{Q}[X, Y]/\mathfrak{p}$, where $\mathfrak{p}$ is the principal ideal generated by $X^2 + Y^2 - 1$ in $\mathbf{Q}[X, Y]$, is a non-factorial Krull domain (if x is the image of X in A, show that x is an extremal element of A but that Ax is not a maximal ideal of A).

(b) Show that the ring $A \otimes_{\mathbf{Q}} \mathbf{Q}(i)$ is factorial (prove that this ring is isomorphic to the quotient of $\mathbf{Q}(i)$ [X, Y] by the principal ideal $(XY - 1)$; compare with Exercise 3 (b)).

¶ 5. (a) Let k be a Noetherian factorial domain and B the polynomial ring $k[X_1, \ldots, X_n]$ where $n \geqslant 3$; let g_i $(0 \leqslant i \leqslant r)$ be elements of $k[X_3, \ldots, X_n]$ where g_0 is extremal. Writing $g = X_1X_2 - \sum_{i=0}^{r} g_i X_1^i$, consider the quotient ring $A = B/gB$. Show that A is factorial. (Let S be the multiplicative subset of A generated by 1 and the image of X_1 in A; apply Proposition 3 of no. 4 to $S^{-1}A$.)

(b) Let k be a field of characteristic $\neq 2$ and F a homogeneous polynomial of the second degree in $k[X_1, \ldots, X_n]$ where $n \geqslant 5$, such that the corresponding polynomial function on k^n is a non-degenerate quadratic form. Show that the domain $k[X_1, \ldots, X_n]/(F)$ is factorial. (Prove first that a homogeneous polynomial G of the second degree in $k[X_1, \ldots, X_n]$, such that the corresponding polynomial function is a non-degenerate quadratic form, is extremal for $n \geqslant 3$. Then prove the proposition when k is algebraically closed, using (a); finally pass to the general case with the aid of Exercise 3 (b).)

(c) If $F = X_1X_2 - X_3X_4$, show that the domain $k[X_1, X_2, X_3, X_4]/(F)$ is not factorial. (Show that the images of the X_i in this ring are extremal elements.)

¶ 6. (a) Let K be an algebraically closed field of characteristic $\neq 2$. Determine the graded prime ideals of the ring

$$K[X, Y, Z]/(X^2 + Y^2 + Z^2).$$

(b) Let k be an ordered field, a, b, c elements >0 of k and A the ring $k[X, Y, Z]/(aX^2 + bY^2 + cZ^2)$. Show that A is a factorial domain. (Reduce it to the case where k is a maximal ordered field, using Exercise 3 (b)); then prove, using (a), that every graded prime ideal of A of height 1 is principal and apply Exercise 3 (a).)

(c) Let k be an ordered field, a, b elements >0 of k and B the ring

$$k[X, Y]/(aX^2 + bY^2 + 1).$$

Show that B is a factorial domain (use (b) and Exercise 17 (a) of § 1.)

(d) Show that the domain $C = \mathbf{Q}[X, Y]/(X^2 + 2Y^2 + 1)$ is factorial, but that the domain $C \otimes_{\mathbf{Q}} \mathbf{Q}(i)$ (where $i^2 = -1$) is not factorial.

¶ 7. Let K be a Noetherian factorial domain and F an extremal element of the polynomial ring $K[X_1, \ldots, X_n]$; suppose that, when a weight $q(i) > 0$ $(1 \leqslant i \leqslant n)$ is attributed to each X_i, F is isobaric and of weight $q > 0$. Let A be the domain generated, in an algebraic closure Ω of the field of fractions of $K(X_1, \ldots, X_n)$, by K, the X_i $(1 \leqslant i \leqslant n)$ and a root z of the polynomial $Z^c - F$, where c is an integer prime to q. Show that A is factorial in the two following cases:

(1) $c \equiv 1 \pmod{q}$;

(2) every finitely generated projective K-module is free (which holds for example when K is a field or a principal ideal domain or a local ring). (In the first case, consider the ring of fractions $A[1/z]$; show that it is a factorial domain and apply Proposition 3 of no. 4. In the second case, consider an integer d such that $cd \equiv 1 \pmod{q}$ and let $z' \in \Omega$ such that $z = z'^d$; the domain $B = A[z']$ is factorial by virtue of the first case and is a free A-module. Consider B as a graded ring taking z' to be of weight q, each of the X_i of weight $cdq(i)$ and reduce it to proving that for two homogeneous elements u, v of A the ideal $Au \cap Av$ is principal, using Exercise 16 of § 1; consider finally the ideal $Bu \cap Bv$ in B and use Chapter I, § 3, no. 6, Proposition 12.) In particular, if K is a field and a, b, c are three integers >0 which are relatively prime in pairs, the domain

$$A = K[X, Y, Z]/(Z^a - X^b - Y^c)$$

is factorial.

¶ 8. Let A be an integral domain and x, y, z three non-zero elements of A, where x is extremal and $Ax \cap Ay = Axy$. Let S be the multiplicative set of the x^n $(n \geqslant 0)$ and let $B = S^{-1}A$; consider the domains of formal power series $A[[T]]$ and $B[[T]]$.

(a) Let i, j, k be three integers $\geqslant 0$ such that $ijk - ij - jk - ki \geqslant 0$ and $z^i \in Ax^j + Ay^k$. Consider in $A[[T]]$ the element $v = xy - z^{i-1}T$. Show that there exists an integer $t > 0$ and a series

$$v' = y^t x^{-1} + b_1 x^{-2} T + \cdots + b_{n-1} x^{-n} T^{n-1} + \cdots$$

in $B[[T]]$ such that $vv' \in A[[T]]$ (determine the b_n inductively, proceeding in $\mathbf{N}$ by intervals of length ij; in the interior of each interval, take

$$b_{n+1} = b_n z^{i-1} x^{-1};$$

at the end of each interval, use the inequality $ijk + ij - jk - ki \geqslant 0$).

(b) Suppose that $z^{i-1} \notin Ax + Ay$. Show that there exists in the ring $A[[T]]$ no formal power series of constant term y^k (k an integer >0) and which is an element associated with v in B[[T]] (calculate the coefficient of T in the product of v and an invertible element of B[[T]]).

(c) Deduce from (a), (b) and Exercise 7 an example of a factorial domain A such that the domain of formal power series A[[T]] is not factorial. (In the above notation and assuming that the conditions of (a) and (b) on x, y, z, i, j, k are fulfilled, show that vv' cannot be a product of extremal factors u_h $(1 \leqslant h \leqslant r)$ in A[[T]]; observe that the u_h are formal power series whose constant terms are powers of y. Then consider the ring $C = S'^{-1}A[[T]]$, where S' consists of the series whose constant term is in S; B[[T]] is the completion of the Zariski ring C; show that $v' \in C$ and that v and the u_h are extremal in C; obtain finally a contradiction with (b).)

(d) Deduce from (c) an example of a Noetherian factorial local integral domain A whose completion $\hat{A}$ is a non-factorial Krull domain.

¶ 9. (a) Let A be a Noetherian domain such that, for every maximal ideal $\mathfrak{m}$ of A, $A_{\mathfrak{m}}$ is a discrete valuation ring. If B is the ring of formal power series $A[[X_1, \ldots, X_n]]$, show that, for every maximal ideal $\mathfrak{n}$ of B, the domain $B_{\mathfrak{n}}$ is factorial (consider its completion, using Proposition 8 of no. 9). Deduce that every divisorial ideal of B is a projective B-module.

(b) Let C be a Noetherian ring such that every finitely generated projective C-module is free; show that the ring of formal power series C[[X]] has the same property (cf. Chapter II, § 3, no. 2, Proposition 5).

(c) Deduce from (a) and (b) that, if A is a principal ideal domain, the domain of formal power series $A[[X_1, \ldots, X_n]]$ is factorial.

10. (a) A Prüferian (§ 2, Exercise 12) factorial domain is a principal ideal domain.

(b) A pseudo-Bezoutian (§ 1, Exercise 21) Krull domain is factorial.

11. Let K be a field and A the polynomial domain K[X, Y], which is factorial; if L is the field $K(X^2, Y/X) \subset K(X, Y)$, show that the domain $A \cap L$ is not factorial.

12. Show Proposition 5 of no. 8 using Chapter III, § 2, no. 8, Corollary 3 to Theorem 1.

13. Extend the Corollary to Proposition 7 of no. 8 to the case where the complete Hausdorff local ring A is not an integral domain (use *Algebra*, Chapter VIII, § 6, Exercise 6 (b)).

14. Let A be a complete Noetherian local ring whose residue field is of characteristic $p > 0$. In the ring of formal power series A[[T]] consider the elements $\omega_n = (1 - T)^{p^n}$ and $\gamma_n = 1 - \omega_n$ for every integer $n > 0$. Show that γ_n is, to within a sign, a distinguished polynomial (no. 8); deduce that $A_n = A[[T]]/(\gamma_n)$ is identified with the algebra over A of the group

$G_n = \mathbf{Z}/p^n\mathbf{Z}$. Show that the intersection of the principal ideals (γ_n) is reduced to 0; deduce that A[[T]] is identified with the inverse limit $\varprojlim A_n$.

¶ 15. Let K be a field which is complete with respect to the valuation v, A the ring of the valuation, k its residue field and P a polynomial in $A[X_1, \ldots, X_n]$ of total degree d with the following property: there exists an algebraic extension K' of K such that in $K'[X_1, \ldots, X_n]$ P is a product of polynomials of total degree 1. Suppose further that there exist two polynomials Q, R in $A[X_1, \ldots, X_n]$ such that Q is of total degree s and contains a monomial aX_1^s where $\phi(a) \neq 0$ (ϕ denoting the canonical homomorphism $A \to k$), R is of degree $\leqslant d - s$ and $\overline{P} = \overline{Q}.\overline{R}$ (notation of Chapter III, § 4). Show that there then exist in $A[X_1, \ldots, X_n]$ two polynomials Q_0, R_0 of respective degrees s, $d - s$, such that $P = Q_0R_0$, $\overline{Q} = \overline{Q}_0$, $\overline{R} = \overline{R}_0$ and Q_0 contains a monomial $a_0X_1^s$ where $\phi(a) = \phi(a_0)$. (Consider P, Q, R as polynomials in X_1 with coefficients in the ring B, the completion of $A[X_2, \ldots, X_n]$ with respect to the valuation obtained by extending v by the method of Chapter VI, § 10, no. 1, Proposition 2; then apply Hensel's Lemma; finally use the initial hypothesis on P.)

16. Let B be a discrete valuation ring whose residue field k is finite and is not a prime field; let k_0 be the prime subfield of k and let A be the subring of B consisting of the elements whose classes in the residue field belong to k_0. Let π be a uniformizer of B and $(\theta_i)_{1 \leqslant i \leqslant m}$ a system of invertible elements of B such that the classes $\bar{\theta}_i$ mod. π of the θ_i form a system of representatives of k^* mod. k_0^*. Show that the elements $p_i = \theta_i\pi$ and the element π are extremal in A and that every element in A is a product of an invertible element and powers of the p_i and π, although A is not integrally closed.

17. (a) Let A be an integral domain; show that in the ring $A[X_{ij}]$, where (X_{ij}) is a family of n^2 indeterminates $(1 \leqslant i \leqslant n, 1 \leqslant j \leqslant n)$, the element $\det(X_{ij})$ is extremal. (Reduce it to the case where A is a field; observe that the factors of $\det(X_{ij})$ would necessarily be homogeneous polynomials and argue by induction on n.)

(b) Let K be an infinite field and F a polynomial in $K[Y_1, \ldots, Y_m]$, which is also written as F(Y); for every square matrix $s = (\alpha_{ij})$ of order m with elements in K let $F(s.Y)$ denote the polynomial F, where the element $\sum_{j=1}^{m} \alpha_{ij}Y_j$ has been substituted for each Y_i. Show that, if F is extremal, so is $F(s.Y)$ for every invertible matrix s. If there exists an integer $k \geqslant 0$ such that

$$F(s.Y) = (\det(s))^k F(Y)$$

for every invertible matrix s, F is necessarily homogeneous in each of the Y_j;

moreover, if $F = GH$ where G and H are two polynomials in $K[Y_1, \ldots, Y_m]$, there exist two integers p and q such that $p + q = k$ and

$$G(s.Y) = (\det(s))^p G(Y), \qquad H(s.Y) = (\det(s))^q H(Y)$$

for every invertible matrix s (use (a)).

(c) Let A be a ring which is not reduced to 0; consider the polynomial ring $A[X_{ij}]$, where the X_{ij} are $n(n+1)/2$ (resp. $2n(2n-1)/2$) indeterminates with $1 \leqslant i \leqslant j \leqslant n$ (resp. $1 \leqslant i < j \leqslant 2n$); let $U = (\xi_{ij})$ (resp. $V = (\eta_{ij})$) be the square matrix of order n (resp. $2n$) over $A[X_{ij}]$ such that $\xi_{ij} = X_{ij}$ for $1 \leqslant i \leqslant j \leqslant n$ and $\xi_{ij} = X_{ji}$ for $i > j$ (resp. $\eta_{ii} = 0$ for $1 \leqslant i \leqslant 2n$, $\eta_{ij} = X_{ij}$ for $1 \leqslant i < j \leqslant 2n$, $\eta_{ij} = -X_{ji}$ for $i > j$). Show that $\det(U)$ (resp. $\mathrm{Pf}(V)$) is an extremal element in $A[X_{ij}]$ (argue as in (b), considering $\det(s.U.{}^ts)$ and $\mathrm{Pf}(s.V.{}^ts)$).

18. Let K be a field and $f = g/h$ an element of the field of rational functions $K(U, V)$ in two indeterminates, where g and h are two relatively prime polynomials of $K[U, V]$. Show that in the field of rational functions

$$K(X_1, Y_1, \ldots, X_n, Y_n)$$

in $2n$ indeterminates the determinant $\det(f(X_i, Y_j))$ is equal to:

$$\Big(\prod_{i,j} h(X_i, Y_j)\Big)^{-1} V(X_1, \ldots, X_n) V(Y_1, \ldots, Y_n) F(X_1, Y_1, \ldots, X_n, Y_n)$$

where F is a polynomial in $K[X_1, Y_1, \ldots, X_n, Y_n]$ and $V(X_1, \ldots, X_n)$ is the Vandermonde determinant (*Algebra*, Chapter III, § 6, no. 4). Consider the particular case where $f = 1/(U + V)$ ("*Cauchy's identity*").

19. If U is a square matrix of order n, Δ its determinant and Δ_p the determinant of the p-th exterior power of U (*Algebra*, Chapter III, § 6, no. 3), show that:

$$\Delta_p = \Delta^{\binom{n-1}{p-1}}$$

(use Exercise 11 of *Algebra*, Chapter III, § 6, and Exercise 17 (a) above).

20. Let A be a factorial domain and $f = \sum_{k=0}^{n} a_k X^k$ a polynomial in $A[X]$: suppose that there exists an extremal element p in A such that:

(1) there exists an index $k \leqslant n$ such that a_k is not divisible by p but a_i is divisible by p for $i < k$;

(2) a_0 is divisible by p but not by p^2.

Show that under these conditions one of the irreducible factors of f in $A[X]$ is of degree $\geqslant k$ (argue in $(A/pA)[X]$). Consider the particular case where $k = n$ ("*Eisenstein's irreducibility criterion*").

21. Show that in $\mathbf{Z}[X]$ the following polynomials are irreducible:
$X^n - a$, where one of the prime factors of a has exponent 1;
$X^{2^k} + 1$, (replace X by X + 1);
$X^4 + 3X^3 + 3X^2 - 5$;
$5X^4 - 6X^3 - aX^2 - 4X + 2$.
(Use Exercise 20.)

¶ 22. (a) Let k be an ordered field and A the quotient of the polynomial ring $k[X, Y, Z]$ by the principal ideal $(X^2 + Y^2 + Z^2 - 1)$; let x, y, z denote the canonical images of X, Y, Z in A. Show that A is a factorial domain (consider the ring of fractions A_{z-1} (notation of Chapter II, § 5, no. 1), using no. 4, Proposition 3).

(b) Let $(e_i)_{1 \leqslant i \leqslant 3}$ be the canonical basis of A^3 and let M be the quotient of A^3 by the monogenous sub-A-module N generated by $xe_1 + ye_2 + ze_3$; show that M is a projective A-module (form a complementary submodule of N in A^3).

* (c) Show that, if $k = \mathbf{R}$, the A-module M is not free (identify M with a submodule of the module of continuous sections of the fibre bundle of tangent vectors to the unit sphere and use the fact that there exists no continuous field of tangent vectors $\neq 0$ at every point of the sphere). *

¶ 23. Let B be a Krull domain, E its field of fractions, Δ a derivation of E such that $\Delta(B) \subset B$, K the subfield of E the kernel of Δ and A the Krull domain $B \cap K$; suppose that K is of characteristic $p > 0$; then $E^p \subset K$ and $B^p \subset A$, so that B is the integral closure of A in E and the canonical homomorphism $\bar{i}: C(A) \to C(B)$ is defined (§ 1, no. 10). Let U denote the group of invertible elements of B.

(a) If $b \in E$ is such that $\mathrm{div}_B(b)$ is the canonical image of a divisor in D(A), show that $\Delta b/b \in B$ (note that for every prime ideal $\mathfrak{P}$ of B of height 1 there exists $b' \in K$ such that $v_{\mathfrak{P}}(b) = v_{\mathfrak{P}}(b')$ and observe that $B_{\mathfrak{P}}$ is invariant under Δ). Let L be the additive subgroup of B consisting of the $\Delta b/b$ ("logarithmic derivatives") which belong to B (for $b \in E$ or $b \in B$, which amounts to the same, and $b \neq 0$) and let L′ be the subgroup of L consisting of the $\Delta u/u$, where $u \in U$. Show that, for every divisor $d \in D(A)$ such that the image of d is a principal divisor in D(B), the class mod. L′ of $\Delta b/b$ for all b such that $i(d) = \mathrm{div}_B(b)$ depends only on the class of d in C(A) and derive a canonical injective homomorphism ϕ of $\mathrm{Ker}(\bar{i})$ to L/L′.

(b) Show that, if $\Delta(B)$ is contained in no prime ideal of B of height 1 and $[E:K] = p$, ϕ is *bijective*. (Reduce it to showing that, if $b \in E$ is such that $\Delta b/b \in B$ and $\mathfrak{P}$ is a prime ideal of B of height 1 such that $v_{\mathfrak{P}}(b)$ is not a multiple of p, then $e(\mathfrak{P}/\mathfrak{p}) = 1$, where $\mathfrak{p} = \mathfrak{P} \cap A$; for this, deduce from the hypothesis that, if t is a uniformizer of $B_{\mathfrak{P}}$, then $\Delta t/t \in B_{\mathfrak{P}}$, so that $\mathfrak{P}B_{\mathfrak{P}}$ is invariant under Δ and that Δ therefore defines by taking quotients a derivation $\bar{\Delta}$ of the residue field $k = B_{\mathfrak{P}}/\mathfrak{P}B_{\mathfrak{P}}$; show that $\bar{\Delta} \neq 0$ and deduce that $f(\mathfrak{P}/\mathfrak{p}) = p$.)

(c) Let k be a field of characteristic 2, B the polynomial ring $k[X, Y, Z]$ and Δ the derivation of $E = k(X, Y, Z)$ such that $\Delta(X) = Y^4$, $\Delta(Y) = X^2$, $\Delta(Z) = XYZ$; the field K the kernel of Δ is such that $[E : K] = 4$. Then $\Delta(Z)/Z \in B$, but show that $\operatorname{div}_B(Z)$ is not the canonical image of a divisor in D(A). (Argue by *reductio ad absurdum* supposing that $e(\mathfrak{P}/\mathfrak{p}) = 1$ for $\mathfrak{P} = BZ$, $\mathfrak{p} = \mathfrak{P} \cap A$; there would then be in A a uniformizer of $B_{\mathfrak{P}}$, necessarily of the form $a(X, Y, Z)Z$ where $b = a(X, Y, 0) \neq 0$; deduce that $\Delta b/b = -XY$ and obtain a contradiction by calculating $\Delta(-XY)$.)

24. (a) Let E be a field of characteristic 2, Δ a derivation of E and K the subfield of E the kernel of Δ; suppose that $[E : K] = 2$. Show that $\Delta^2 = a\Delta$ where $a \in K$ and that for an element $t \in E$ to be of the form $\Delta x/x$, it is necessary and sufficient that $\Delta t = at + t^2$.

(b) Let B be a factorial local integral domain of characteristic 2, $\mathfrak{m}$ its maximal ideal, E its field of fractions and Δ a derivation of E; suppose that the subfield K of E, the kernel of Δ, satisfies $[E : K] = 2$; moreover, suppose that there exist two elements x, y of $\mathfrak{m}$ such that Δx and Δy generate the ideal $\mathfrak{q}$ of B generated by $\Delta(B)$. Show then that, if $t = \Delta z/z$ belongs to $\mathfrak{q}$, there exists an invertible element u of B such that $t = \Delta u/u$ (write $t = r\Delta x + s\Delta y$, where r, s are in B, and use (a)).

¶ 25. Let k be a field of characteristic 2, B the ring of formal power series $k[[X, Y]]$, E its field of fractions and Δ the k-derivation of E defined by $\Delta(X) = Y^{2j}$, $\Delta(Y) = X^{2i}$ (i, j integers $\geqslant 0$); the subring A of B consisting of the $x \in B$ such that $\Delta x = 0$ is the ring of formal power series in $k[[X, Y, Z]]$ where X^2 is substituted for X, Y^2 for Y and $X^{2i+1} + Y^{2j+1}$ for Z.

(a) Show that the group C(A) contains a vector space over k of dimension $N(i, j)$ equal to the number of ordered pairs of integers (a, b) such that $0 \leqslant a < i$, $0 \leqslant b < j$ and $(2j + 1)a + (2i + 1)b \geqslant 2ij$. (Use Exercise 23 (b) and Exercise 24 (a), note that the elements of L are the formal power series $F \in B$ such that $\Delta F = E^2$; attributing to X the weight $2j + 1$ and to Y the weight $2i + 1$, decompose F as an infinite sum of isobaric polynomials; if L_q is the subgroup of L consisting of the $F \in L$ whose isobaric components are of weight $\geqslant q$, L/L' is isomorphic to the direct sum of the groups C_q/C_{q+1}, where $C_q = L_q/(L' \cap L_q)$; calculate these groups for $q \geqslant 4ij$, using Exercise 24 (b).)

(b) Show that the ideal AX^2 of A is prime; if $A' = A[X^{-2}]$, deduce that C(A') and C(A) are isomorphic (no. 4, Proposition 3). Show that A' is a Dedekind domain (consider the ring $B' = B[X^{-2}]$, which is integral over A' and a principal ideal domain, and use Chapter V, § 2, no. 4, Theorem 3). Deduce an example of a Dedekind domain whose ideal class group is infinite.

26. Let A be a factorial domain and K its field of fractions. For elements $_i$ $(1 \leqslant i \leqslant r)$ of $B = A[X_1, \ldots, X_n]$ to be such that the ideal $\sum_{i=1}^{r} Bf_i$ is equal

to B, it is necessary and sufficient that there exist polynomials v_i $(1 \leqslant i \leqslant r)$ in $K[X_1, \ldots, X_n]$ such that $\sum_{i=1}^{r} v_i f_i = 1$ and which satisfy further the following condition: if we write $v_i = w_i/d$, where $d \in A$, the polynomials w_i belong to $A[X_1, \ldots, X_n]$ and the *g.c.d.* of the set of coefficients of all the w_i $(1 \leqslant i \leqslant r)$ is equal to 1, then, for every extremal element p of A dividing d, the ideal generated by the classes of the f_i in the ring $(A/Ap)[X_1, \ldots, X_n]$ is the whole of this ring.

¶ 27. (a) Let K be a field and B the ring generated, in an algebraic closure of the field of rational functions K(U, V, X, Y) in 4 indeterminates, by the polynomial ring K[U, V, X, Y] and a root z of the polynomial

$$F = Z^7 - U^6X^2 - V^4Y^3.$$

Show that B is a factorial domain (cf. Exercise 7).

(b) Let $\mathfrak{p}$ be the (prime) ideal generated by X, Y, U, V and z in A and write $C = A_{\mathfrak{p}}[[T]]$; show that C is a Noetherian local integral domain, which is not factorial, but whose associated graded domain gr(C) is factorial (use Exercise 8).

§ 4

1. Let A be an integrally closed Noetherian domain, and V a vector space of finite rank over the field of fractions of A. Show that, if (M_λ) is any family of reflexive lattices of V all containing the same lattice N, the lattice $M = \bigcap_\lambda M_\lambda$ is reflexive (consider the dual lattices M_λ^*).

* 2. Let A be an integrally closed Noetherian domain and E, F two finitely generated A-modules. Suppose that E is torsion-free, that $\mathrm{Hom}_A(E, F)$ is reflexive and that $\mathrm{Ext}^1_A(E, F) = 0$. Show then that F is reflexive. (Prove first that F is torsion-free; if $T = F^{**}/c_F(F)$, calculate $\mathrm{Ass}(\mathrm{Hom}_A(E, T))$ in two ways using the exact sequence:

$$0 \to \mathrm{Hom}(E, F) \to \mathrm{Hom}(E, F^{**}) \to \mathrm{Hom}(E, T) \to 0$$

and Chapter IV, § 1, no. 4, Proposition 10.) *

3. Let A be an integrally closed Noetherian domain, K its field of fractions, M a reflexive lattice of a vector space of finite rank over K and L a *free* lattice of V containing M. Show that there exists a free lattice L_1 of V such that $M = L \cap L_1$. (Consider the finite set I of prime ideals $\mathfrak{p}$ of A of height 1 such that $L_\mathfrak{p} \neq M_\mathfrak{p}$ and the principal ideal domain $S^{-1}A$, where $S = \bigcap_{\mathfrak{p} \in I} (A - \mathfrak{p})$; show that there exist a free lattice L_0 of V such that $M_\mathfrak{p} = (L_0)_\mathfrak{p}$ for all $\mathfrak{p} \in I$ and an $s \in S$ such that $M \subset s^{-1}L_0 \subset L_1$.)

4. Let k be a field and A the polynomial ring $k[X, Y]$.

(a) Let (e_1, e_2) be the canonical basis of the A-module A^2 and let E be the sub-A-module of A^2 generated by $(X - Y)e_1$, $e_1 + Xe_2$ and $e_1 + Ye_2$; let F be the monogenous submodule of E generated by $(X - Y)^2e_1$. Show that the A-module $M = E/F$ is not the direct sum of its torsion submodule and a torsion-free module.

(b) Show that the torsion A-module A/AXY is not a direct sum of monogenous submodules of the form $A/\mathfrak{P}_i^{n_i}$, where the $\mathfrak{P}_i$ are prime ideals of height 1.

5. Let A be an integrally closed Noetherian domain and M a finitely generated A-module. Show that, if $\mathfrak{a}$, $\mathfrak{b}$ are two ideals of A, the A-module

$$((\mathfrak{a}M) \cap (\mathfrak{b}M))/(\mathfrak{a} \cap \mathfrak{b})M$$

is pseudo-zero; give an example where it is not zero.

6. Let k be a field, B the polynomial ring $k[X, Y]$ and A the submodule $k[X^2, XY, Y^2]$ of B.

(a) Show that A is an integrally closed Noetherian domain and that B is a finitely generated A-module.

(b) Show that the ideal $\mathfrak{p} = AX^2 + AXY$ of A is a prime ideal of height 1 (and hence divisorial) but that the B-module $\mathfrak{p} \otimes_A B$ is a torsion module $\neq 0$ and hence not reflexive; the ideal $\mathfrak{p}B$ of B is not divisorial, the canonical mapping $\mathfrak{p} \otimes_A B \to \mathfrak{p}B$ is not injective and $\mathfrak{p}$ is not a flat A-module.

7. Let A be an integrally closed Noetherian domain, E a finitely generated torsion-free A-module and E^* its dual.

(a) Show that the canonical homomorphism $E^* \otimes_A E \to \mathrm{End}_A(E)$ is a pseudo-isomorphism.

(b) Deduce from (a) that for E to be a projective A-module, it is necessary and sufficient that $E^* \otimes_A E$ be a reflexive A-module. (Note that, if $E^* \otimes_A E$ is reflexive, the canonical homomorphism $E^* \otimes_A E \to \mathrm{End}_A(E)$ is bijective.)

¶ * 8. Let A be an integrally closed Noetherian domain and M_1, M_2 two finitely generated A-modules.

(a) Show that the A-modules $\mathrm{Tor}_i^A(M_1, M_2)$, $\mathrm{Ext}_A^i(M_1, M_2)$ are pseudo-zero for $i \geqslant 2$ (reduce it to the case where A is a principal ideal domain).

(b) If M_1 is torsion-free, show that $\mathrm{Tor}_1^A(M_1, M_2)$ and $\mathrm{Ext}_A^1(M_1, M_2)$ are pseudo-zero (same method).

(c) If M_1 is a torsion A-module, show that (in the notation of no. 5):

$$\chi(M_1 \otimes_A M_2) - \chi(\mathrm{Tor}_1^A(M_1, M_2)) = r(M_2)\chi(M_1)$$

$$\chi(\mathrm{Hom}_A(M_1, M_2)) - \chi(\mathrm{Ext}_A^1(M_1, M_2)) = -r(M_2)\chi(M_1)$$

$$\chi(\mathrm{Hom}_A(M_2, M_1)) - \chi(\mathrm{Ext}_A^1(M_2, M_1)) = r(M_2)\chi(M_1).$$

(d) Show that for every ordered pair of finitely generated A-modules M_1, M_2:

$$c(M_1 \otimes_A M_2) - c(\mathrm{Tor}_1^A(M_1, M_2)) = r(M_1)c(M_2) + r(M_2)c(M_1)$$

$$c(\mathrm{Hom}_A(M_1, M_2)) - c(\mathrm{Ext}_A^1(M_1, M_2)) = r(M_1)c(M_2) - r(M_2)c(M_1). *$$

9. Let k be a field and A the polynomial ring $k[X, Y]$. On the A-module $M = A^2$ consider the linear form f such that $f(e_1) = X$, $f(e_2) = Y$ ((e_1, e_2) being the canonical basis). Show that the kernel L of f is a monogenous free A-module, but that the quotient M/L (which is isomorphic to an ideal of A) is not reflexive.

¶ 10. Let A be a commutative ring. For every submodule R of a finitely generated free A-module $L = A^n$, let $\mathfrak{c}_1(R)$ denote the ideal generated by the $\langle x, x^* \rangle$, where x runs through R and x^* runs through the dual L^*; write $\mathfrak{c}_k(R) = \mathfrak{c}_1\left(\mathrm{Im}\left(\bigwedge^k R\right)\right)$, $\mathrm{Im}\left(\bigwedge^k R\right)$ being the canonical image of the k-th exterior power of R in $\bigwedge^k L$. If $R_1 \subset R_2$ are two submodules of L, then $\mathfrak{c}_k(R_1) \subset \mathfrak{c}_k(R_2)$ for all k.

(a) Let M be a finitely generated A-module; M is isomorphic to a quotient module L/R, where $L = A^n$ for a suitable n. Show that the sequence of ideals $(\mathfrak{a}_k)_{k \geqslant 0}$ such that $\mathfrak{a}_k = \mathfrak{c}_{n-k}(R)$ for $k \leqslant n$ and $\mathfrak{a}_k = A$ for $k > n$ is independent of the expression of M in the form L/R. (Consider first the case where L/R and L/R′ are isomorphic; then note that A^n/R is isomorphic to $A^{n+h}/(R \times A^h)$ for all $h > 0$.) We write $\mathfrak{d}_k(M) = \mathfrak{a}_k$ for all $k \geqslant 0$ and say that these are the *determinantal ideals* associated with M. Then $\mathfrak{d}_k(M) \subset \mathfrak{d}_{k+1}(M)$ for $k \geqslant 0$.

(b) If A is a principal ideal domain and $M = L/R$ where L is free and finitely generated, show that the ideals $\mathfrak{c}_{k+1}(R)(\mathfrak{c}_k(R))^{-1}$ are the *invariant factors* of R in L.

(c) Let $r = \gamma(M)$ be the least of the cardinals of the systems of generators of M and let r_0 be the least integer h such that $\mathfrak{d}_h(M) = A$. Show that $r_0 \leqslant r$ and give an example where $r_0 < r$ (take A to be a Dedekind domain).

(d) If $\mathfrak{a}$ is the annihilator of M, show that $\mathfrak{d}_0(M) \subset \mathfrak{a}$ and $\mathfrak{a}^{r-k} \subset \mathfrak{d}_k(M)$ for $k \leqslant r = \gamma(M)$.

(e) Suppose that M is a direct sum of submodules M_i ($1 \leqslant i \leqslant h$). Show that

$$\mathfrak{d}_k(M) = \sum \mathfrak{d}_{k_1}(M_1)\ldots\mathfrak{d}_{k_h}(M_h)$$

where the sum is over the finite sequences $(k_i)_{1 \leqslant i \leqslant h}$ such that $\sum_{i=1}^{h} k_i = k$.

(f) If N is a finitely generated submodule of M, then $\mathfrak{d}_k(M/N) \subset \mathfrak{d}_k(M)$ for all $k \geqslant 0$. Show that:

$$\sum_{j=0}^{k} \mathfrak{d}_j(N)\mathfrak{d}_{k-j}(M/N) \subset \mathfrak{d}_k(M).$$

(g) Let K be a field, A the polynomial ring K[X, Y, Z], $\mathfrak{m}$ the maximal ideal AX + AY + AZ and $\mathfrak{q}$ the ideal generated by X, Y^2, YZ and Z^2. Consider the A-module $M = A/\mathfrak{q}$ and its submodule $N = \mathfrak{m}/\mathfrak{q}$. Show that $\mathfrak{d}_0(M) = \mathfrak{q}$, $\mathfrak{d}_1(M) = A$, $\mathfrak{d}_0(N) = \mathfrak{m}^2$ and $\mathfrak{d}_1(N) = \mathfrak{m}$.

11. Let A be a Krull domain and M, N two finitely generated lattices in a vector space V of finite rank n over the field of fractions of A. For all $c \neq 0$ in A such that $cN \subset M$, consider the determinantal ideals $\mathfrak{d}_k(M/cN)$ (Exercise 10) and write:

$$d_k(N, M) = \operatorname{div}(\mathfrak{d}_k(M/cN)) - (n - k)\operatorname{div}(c).$$

(a) Show that $d_k(N, M)$ does not depend on the choice of c such that $cN \subset M$; $d_k(N, M)$ is called the *determinantal divisor* of index k of N with respect to M.

(b) $d_k(N, M) \geqslant d_{k+1}(N, M)$ and $d_n(N, M) = 0$. Show that, if we write

$$e_k(N, M) = d_{n-k}(N, M) - d_{n-k+1}(N, M),$$

then

$$e_k(N, M) \leqslant e_{k+1}(N, M)$$

for $1 \leqslant k \leqslant n$; the divisors $e_k(N, M)$ are called the *invariant factors* of N with respect to M (reduce it to the case where A is a principal ideal domain).

(c) Show that $d_0(N, M) = \chi(M, N)$ (no. 5).

(d) If M is a lattice in V but N is a lattice in a subspace W of V of rank $q < n$, the invariant factors of N with respect to $M \cap W$ are called the *invariant factors* of N with respect to M. Show how Exercises 8 to 10 and 14 to 16 of *Algebra*, Chapter VII, § 4 may be extended (assuming if need be in certain cases that A is an integrally closed Noetherian domain).

12. Let A be an integrally closed Noetherian domain and M, N two torsion-free finitely generated A-modules of the same rank r such that $N \subset M$; let $j: N \to M$ be the canonical injection.

(a) Show that, for all k, $\bigwedge^k j: \bigwedge^k N \to \bigwedge^k M$ is pseudo-injective, that $\operatorname{Coker}\left(\bigwedge^k j\right)$ is a torsion A-module and that:

$$\chi\left(\operatorname{Coker}\left(\bigwedge^k j\right)\right) = \binom{r-1}{k-1}\chi(\operatorname{Coker} j).$$

(b) Using (a), show that, if M is a torsion-free finitely generated A-module, the torsion submodule of $\bigwedge^k M$ is pseudo-zero for all k and

$$c\left(\bigwedge^k M\right) = \binom{r-1}{k-1}c(M),$$

where M is of rank r; in particular, $c\left(\bigwedge^r M\right) = c(M)$.

13. Let k be a field and A the polynomial ring $k[X, Y]$.

(a) Let $\mathfrak{m}$ be the maximal ideal AX + AY of A; show that there exists a pseudo-isomorphism of $\mathfrak{m}$ to A, but that there exists no pseudo-isomorphism of A to $\mathfrak{m}$.

(b) Let $\mathfrak{m}'$ be the maximal ideal $A(X - 1) + AY$ of A; show that $c(\mathfrak{m}) = c(\mathfrak{m}') = 0$, but that there exists no pseudo-isomorphism of $\mathfrak{m}$ to $\mathfrak{m}'$ nor of $\mathfrak{m}'$ to $\mathfrak{m}$.

(c) Let $\mathfrak{p} = AX$, $\mathfrak{q} = AY$, which are prime ideals of height 1. In the A-module $L = A^2$, consider the submodules $M = \mathfrak{p}e_1 \oplus \mathfrak{q}e_2$, $N = Ae_1 \oplus \mathfrak{p}\mathfrak{q}e_2$ (e_1, e_2 being the vectors of the canonical basis of L). Show that M and N are isomorphic and that there exist a pseudo-isomorphism from L/M to L/N and a pseudo-isomorphism from L/N to L/M but that there exists no pseudo-isomorphism from L to itself mapping M to N or N to M (observe that a pseudo-isomorphism from L to itself is necessarily an automorphism of L, with the aid of Proposition 10).

¶ 14. Let A be a Prüferian domain (§ 2, Exercise 12) and M, N two finitely generated lattices in a vector space V of finite rank n over the field of fractions of A. For all $c \neq 0$ in A such that $cN \subset M$, consider the determinantal ideals $\mathfrak{d}_k(M/cN)$ (Exercise 10) and write:

$$\mathfrak{d}_k(N, M) = c^{k-n}\mathfrak{d}_k(M/cN).$$

(a) Show that $\mathfrak{d}_k(N, M)$ does not depend on the choice of c such that $cN \subset M$; these are called the *determinantal ideals* of N with respect to M.

(b) $\mathfrak{d}_k(N, M) \subset \mathfrak{d}_{k+1}(N, M)$ and $\mathfrak{d}_n(N, M) = A$. Show that, if we write $\mathfrak{e}_k(N, M) = \mathfrak{d}_{n-k}(N, M)(\mathfrak{d}_{n-k+1}(N, M))^{-1}$, the integral ideals $\mathfrak{e}_k(N, M)$ are such that $\mathfrak{e}_k(N, M) \supset \mathfrak{e}_{k+1}(N, M)$ for $1 \leqslant k \leqslant n$; the finitely generated ideals $\mathfrak{e}_k(N, M)$ are called the *invariant factors* of N with respect to M (consider the $A_\mathfrak{m}$-modules $M_\mathfrak{m}$ and $N_\mathfrak{m}$ for every maximal ideal $\mathfrak{m}$ of A).

(c) If M is a lattice of V but N is a lattice in a subspace W of V of rank $q < n$, the invariant factors of N with respect to $M \cap W$ are called the *invariant factors* of N with respect to M. Show how Exercises 8 to 10 and 14 to 16 of *Algebra*, Chapter VII, § 4 may be extended.

15. Let A be the polynomial ring $\mathbf{Z}[X, Y, T, U, V, W]$, L the free A-module A^4, $(e_i)_{1 \leqslant i \leqslant 4}$ its canonical basis and M the submodule of L generated by the four vectors Xe_1, Ye_2, $Te_3 + Ue_4$, $Ve_3 + We_4$. Show that

$$\mathfrak{d}_1(L/M)\mathfrak{d}_3(L/M) \not\subset (\mathfrak{d}_2(L/M))^2$$

(compare with Exercise 14 (b)).

¶ 16. (a) Let A be a Prüferian domain and M a finitely generated lattice in a vector space V of finite rank n over the field of fractions K of A. Show that there exists a basis $(e_i)_{1 \leqslant i \leqslant n}$ of V and n fractional ideals $\mathfrak{a}_i$ $(1 \leqslant i \leqslant n)$ such that M is equal to the direct sum of the $\mathfrak{a}_i e_i$. (Argue by induction on n: if $(u_i)_{1 \leqslant i \leqslant n}$ is a basis of V, consider the finitely generated fractional ideal $\mathfrak{b}_1$ generated by the u_1-coordinates of the elements of M; by considering the A-module $\mathfrak{b}_1^{-1}$M, reduce it the case where $\mathfrak{b}_1 = A$.)

Deduce that, if W is a subspace of V, $M \cap W$ is a direct factor of M.

(b) Suppose that M is a submodule of A^n equal to the direct sum of the $\mathfrak{a}_i e_i$, where the $\mathfrak{a}_i$ are finitely generated (integral) ideals of A and $(e_i)_{1 \leqslant i \leqslant n}$ the canonical basis of A^n. Show that there exists a basis $(u_i)_{1 \leqslant i \leqslant n}$ of A^n and ideals $\mathfrak{b}_i$ of A such that M is the direct sum of the $\mathfrak{b}_i u_i$ and $\mathfrak{b}_i$ *divides* $\mathfrak{b}_{i+1}$ for $1 \leqslant i \leqslant n - 1$. (Reduce it to the case where $n = 2$ and $\mathfrak{a}_1 + \mathfrak{a}_2 = A$.)

17. Let k be a field, A the polynomial ring $k[X, Y]$, L the free A-module A^2 and (e_1, e_2) the canonical basis of L. Let M be the submodule of L generated by the two vectors $(X + 1)e_1 + Ye_2$, $Ye_1 + Xe_2$. Show that there exists no pseudo-isomorphism from L to itself, whose restriction to M is a pseudo-isomorphism from M to the submodule N of the form $\mathfrak{a}u + \mathfrak{b}v$ where (u, v) is a basis of the A-module L and $\mathfrak{a}$, $\mathfrak{b}$ two ideals of A, or whose restriction to N is a pseudo-isomorphism from N to M. (By considering the determinantal ideals (Exercise 10) and noting that M is reflexive, attention may be confined to the case where N would also be reflexive and then necessarily $\mathfrak{a} = A$, $\mathfrak{b} = AP$, where

$$P(X, Y) = X(X + 1) - Y^2$$

is an extremal element of A; show finally that, for every basis (u, v) of L, $M \cap Au$ can neither contain Au nor be contained in APu.)

¶ 18. Let A be a Dedekind domain, K its field of fractions and M, N two lattices in a vector space V of finite rank n over K. Let $\mathfrak{e}_k = \mathfrak{e}_k(N, M)$ be the invariant factors of N with respect to M (Exercise 14). Show that there exists a basis $(u_i)_{1 \leqslant i \leqslant n}$ of V such that M is equal to a direct sum $\bigoplus_i \mathfrak{a}_i u_i$, where the $\mathfrak{a}_i$ are fractional ideals, and N is equal to the direct sum $\bigoplus_i \mathfrak{e}_i \mathfrak{a}_i u_i$. (Use the theory of finitely generated modules over a principal ideal domain (*Algebra*, Chapter VII, § 4, no. 2, Theorem 1) and the approximation theorem in the unimodular group $\mathbf{SL}(n, A)$ (§ 2, no. 4).)

Extend this result to the case where A is the union of a right directed family of subrings which are Dedekind domains (for example the integral closure of a Dedekind domain in the algebraic closure of its field of fractions).

19. Let A be a Dedekind domain and $\mathfrak{a}$, $\mathfrak{b}$ fractional ideals of A. Show that the A-modules $A \oplus \mathfrak{a}\mathfrak{b}$ and $\mathfrak{a} \oplus \mathfrak{b}$ are isomorphic.

20. Let A be a Dedekind domain, P a projective A-module and N a submodule of P. Show that N is projective and a direct sum of modules isomorphic to ideals of A. (Attention may be confined to the case where P is free. Then

proceed as in Exercise 16, using a transfinite induction argument based on that of *Algebra*, Chapter VII, § 3, Theorem 1.)

21. Let P be a projective module over a Dedekind domain A. Show that, if P is not finitely generated, it is free. (By virtue of Exercise 20, we may write P as an infinite direct sum $\bigoplus_{\lambda} (\mathfrak{b}_\lambda \oplus \mathfrak{c}_\lambda)$, where, for each λ, $\mathfrak{b}_\lambda$ and $\mathfrak{c}_\lambda$ are ideals of A. Using Exercise 19, $P = L \oplus Q$, where L is isomorphic to $A^{(I)}$ (I infinite) and $Q = \bigoplus_{\alpha \in I} \mathfrak{a}_\alpha$, where the $\mathfrak{a}_\alpha$ are ideals of A. Then apply Exercise 3 of *Algebra*, Chapter II, § 2.)

¶ 22. Let A be a local ring, $\mathfrak{m}$ its maximal ideal and E a finitely generated A-module.

(a) If $r = \gamma(E)$ is the least of the cardinals of a system of generators of E, r is also the least integer h such that the determinantal ideal $\mathfrak{d}_h(E)$ is equal to A and also the least of the integers k such that $\bigwedge^{k+1} E = 0$ (note that r is the rank of $E/\mathfrak{m}E$ over $A/\mathfrak{m}$).

(b) If we write $\mathfrak{e}(E) = \mathfrak{d}_{r-1}(E)$, show that $\bigwedge^{r} E$ is isomorphic to $A/\mathfrak{e}(E)$. For an ideal $\mathfrak{a}$ of A to contain $\mathfrak{e}(E)$, it is necessary and sufficient that $E/\mathfrak{a}E$ be a free $(A/\mathfrak{a})$-module (reduce it to the case where $\mathfrak{a} = 0$).

(c) If $\mathfrak{e}(E)$ is a principal ideal $A\alpha$, show that E contains a direct factor isomorphic to $A/A\alpha$ (write E as a quotient of A^r by a submodule R and note that there is in R an element of the form αy, where y is an element of a basis of A^r).

(d) Let $\mathfrak{d}'_h(E)$ be the annihilator of $\bigwedge^{h+1} E$ for $0 \leqslant h \leqslant r - 1$. Show that the following conditions are equivalent:

(α) The ideals $\mathfrak{d}_h(E)$ $(0 \leqslant h \leqslant r - 1)$ are principal.

(β) The ideals $\mathfrak{d}'_h(E)$ $(0 \leqslant h \leqslant r - 1)$ are principal.

(γ) E is the direct sum of r modules isomorphic to $A/A\lambda_h$, where λ_{h+1} divides λ_h for $0 \leqslant h \leqslant r - 1$.

Moreover, if these conditions hold, then $\mathfrak{d}'_h(E) = A\lambda_h$ for $0 \leqslant h \leqslant r - 1$ and $\mathfrak{d}_h(E) = \lambda_h \lambda_{h+1} \ldots \lambda_{r-1} A$ for $0 \leqslant h \leqslant r - 1$. (Proceed by induction on r, using (c).)

(e) Suppose further that A is an integral domain. Show that, if $\mathfrak{p}$ is a prime ideal of A such that $E/\mathfrak{p}E$ is a free $(A/\mathfrak{p})$-module and $E_\mathfrak{p}$ a free $A_\mathfrak{p}$-module, then E is a free A-module (using (b), show that $\gamma(E_\mathfrak{p}) = \gamma(E)$ and that $\mathfrak{e}(E_\mathfrak{p}) = \mathfrak{e}(E)$).

23. Let A be a ring, E a finitely generated A-module, r the least of the integers k such that $\bigwedge^{k+1} E = 0$ and $\mathfrak{e}(E)$ the annihilator of $\bigwedge^{r} E$. Show that, for every ideal $\mathfrak{a} \supset \mathfrak{e}(E)$ in A, the $(A/\mathfrak{a})$-module $E/\mathfrak{a}E$ is flat (reduce it to the case where A is a local ring and use Exercise 22 (b)). Deduce that, if $\mathfrak{r}$ is the Jacobson radical of A and if, for every maximal ideal $\mathfrak{m}$ of A, the rank of the vector $(A/\mathfrak{m})$-space $E/\mathfrak{m}E$ is the same, then $E/\mathfrak{r}E$ is a flat $(A/\mathfrak{r})$-module.

24. Let A be an integrally closed Noetherian domain. For a lattice M (with respect to A) to be reflexive, it is necessary and sufficient that it satisfy the following condition: for every ordered pair (a, b) of elements of A such that $a \neq 0$ and the homothety of ratio b on A/aA is injective, the homothety of ratio b on M/aM is then injective. (To see that the condition is necessary, consider two elements x, y of M such that $ax + by = 0$ and show that, for all $\mathfrak{p} \in P(A)$, $y \in aM_{\mathfrak{p}}$. To show that the condition is sufficient, use criterion (c) of Theorem 2 and observe (using Proposition 8 of § 1, no. 4) that the hypothesis may be written as

$$\mathrm{Ass}(M/aM) = \mathrm{Ass}(A/aA)$$

for all $a \neq 0$ in A and that for every module E such that $M \subset E \subset V$ there exists $a \neq 0$ in A such that $aM \subset aE \subset M$.)

25. Let $A = \bigoplus_{n \geqslant 0} A_n$ be a Noetherian graded ring with positive degrees. Show that, if A is a factorial domain, then A_0 is a factorial domain and the A_n are reflexive. (To show that A_0 is a factorial domain, use criterion (c) of § 3, no. 2, Theorem 1; to show that the A_n are reflexive, use Exercise 24.)

¶ 26. Let A be a Noetherian ring and M a finitely generated A-module. For the symmetric algebra S(M) to be a factorial domain, it is necessary and sufficient that A be a factorial domain and that the $S^n(M)$ be reflexive A-modules. (To see that the condition is sufficient, observe first that, if $T = A - \{0\}$, S(M) is identified with a subring of $T^{-1}S(M)$; then show that every extremal element p of A is extremal in S(M), by reducing it to proving that, if p divides a product xy of two homogeneous elements in S(M), it divides one of them; finally, apply Proposition 3 of § 3, no. 4.)

HISTORICAL NOTE

(Numbers in brackets refer to the bibliography at the end of this Note.)

"Abstract" commutative algebra is a recent creation but its development can only be understood as a function of that of the theory of algebraic numbers and algebraic geometry, which gave birth to it.

It has been conjectured without too much improbability that the famous "proof" Fermat claimed to possess of the impossibility of the equation $x^p + y^p = z^p$ for p an odd prime and x, y, z integers $\neq 0$ depended on the decomposition

$$(x + y)(x + \zeta y) \dots (x + \zeta^{p-1}y) = z^p$$

in the ring $\mathbf{Z}[\zeta]$ (where $\zeta \neq 1$ is a p-th root of unity) and on a divisibility argument in this ring, assuming it to be a *principal ideal domain.* In any case an analogous argument is found outlined by Lagrange ([2], vol. II, p. 531); it is by arguments of this type, with certain variations (notably changes of variable aimed at lowering the degree of the equation) that Euler ([1], vol. I, p. 488) (*) and Gauss ([3], vol. II, p. 387) show Fermat's Theorem for $p = 3$, Gauss (*loc. cit.*) and Dirichlet ([4], vol. I, p. 42) for $p = 5$ and Dirichlet the impossibility of the equation $x^{14} + y^{14} = z^{14}$ ([4], vol. I, p. 190). Finally, in his first research on the theory of numbers, Kummer believed he had obtained in this way a general proof and it was no doubt this mistake (which Dirichlet pointed out to him) that led him to his study of the arithmetic of cyclotomic fields from which he was eventually to succeed in deducing a correct version of his proof for prime numbers $p < 100$ [7d].

On the other hand, the celebrated memoir of Gauss of 1831 on biquadratic

(*) In the proof, Euler argues as if $\mathbf{Z}[\sqrt{-3}]$ were a principal ideal domain, which is not the case; however, his argument can be rendered correct by considering the conductor of $\mathbf{Z}[\rho]$ (ρ a cube root of unity) on $\mathbf{Z}[\sqrt{-3}]$ (cf. Sommer, *Introduction à la théorie des nombres algébriques* (trans. A. Lévy), Paris (Hermann), 1911, p. 190).

residues, whose results are derived from a detailed study of divisibility in the ring $\mathbf{Z}[i]$ of "Gaussian integers" ([3], vol. II, p. 109), showed clearly the interest that the extension of divisibility to algebraic numbers could hold out for the classical problems of the theory of numbers (*); therefore it is not surprising that between 1830 and 1850 this theory was the subject of numerous works by German mathematicians, first Jacobi, Dirichlet and Eisenstein, and then, a little later, Kummer and his pupil and friend Kronecker. We shall not speak here of the theory of units, which is too specialized a branch of the theory of numbers, where progress was very rapid, Eisenstein obtaining the structure of the group of units for cubic fields and Kronecker for cyclotomic fields, just before Dirichlet in 1846 ([4], vol. I, p. 640) proved the general theorem, at which Hermite had almost arrived independently ([8], vol. I, p. 159). The question (central to the whole theory) of decomposition into prime factors appeared much more difficult. Since Lagrange had given examples of numbers of the form $x^2 + Dy^2$ (x, y, D integers) with divisors which are not of the form $m^2 + Dn^2$ ([2], vol. II, p. 465), it was effectively known that the ring $\mathbf{Z}[\sqrt{-D}]$ could not in general be expected to be a principal ideal domain and Euler's temerity was followed by considerable circumspection; when Dirichlet, for example, proves that the relation $p^2 - 5q^2 = r^5$ (p, q, r integers) is equivalent to

$$p + q\sqrt{5} = (x + y\sqrt{5})^5$$

for integers x, y, he restricts himself to pointing out that "*there are analogous theorems for many other prime numbers* [than 5]" ([4], vol. I, p. 31). In Gauss's memoir of 1831 and the work of Eisenstein on cubic residues [6a], it is certainly true that there are advanced studies on arithmetic in the principal ideal domains $\mathbf{Z}[i]$ and $\mathbf{Z}[\rho]$ ($\rho = (-1 + i\sqrt{3})/2$, a cube root of unity) in perfect analogy with the theory of rational integers and in these examples at least the close connection between arithmetic in quadratic fields and the theory of binary quadratic forms developed by Gauss was very apparent; but the general case lacked a "dictionary" which would have allowed quadratic fields to be treated by a simple translation from Gauss's theory (†).

In fact, it is not for quadratic fields but for cyclotomic fields (and for reasons which will only appear clearly much later (cf. p. 585)) that the problem was

(*) The research of Gauss on division on the lemniscate and elliptic functions related to this curve, not published during his lifetime, but dating from about 1800, must have led him from this time to consider the arithmetical properties of the ring $\mathbf{Z}[i]$, division by numbers in this ring playing an important role in the theory; see what Jacobi say on this subject ([5], vol. VI, p. 275) and also the calculations related to these questions found in Gauss's papers ([3], vol. II, p. 411; see also [3], vol. X_2, p. 33 *et seq.*).

(†) The reader will find an exact discription of this correspondence between quadratic forms and quadratic fields in Sommer, *loc. cit.*, pp. 205–229.

first solved. From 1837 onwards, Kummer, originally an analyst, turns to the arithmetic of cyclotomic fields which was to occupy him almost exclusively for 25 years. Like his predecessors, he studies divisibility in the ring $\mathbf{Z}[\zeta]$, where ζ is a p-th root of unity $\neq 1$ (p an odd prime); he quickly sees that here also rings are encountered which are not principal ideal domains, blocking all progress in the extension of the laws of arithmetic [7a] and it is only in 1845, after 8 years' efforts, that the light dawns, thanks to his definition of "ideal numbers" ([7c] and [7d]).

What Kummer does amounts exactly, in modern language, to defining the *valuations* on the field $\mathbf{Q}(\zeta)$: they are in one-to-one correspondence with his "ideal prime numbers", the "exponent" with which such a factor appears in the "decomposition" of a number $x \in \mathbf{Z}[\zeta]$ is just the value at x of the corresponding valuation. As the conjugates of x also belong to $\mathbf{Z}[\zeta]$ and their product $N(x)$ (the "norm" of x (*)) is a rational integer, the "ideal prime factors" to be defined must also be "factors" of the rational prime numbers and in order to define them it was sufficient just to say what were the "ideal prime divisors" of a prime number $q \in \mathbf{Z}$. For $q = p$ Kummer had already effectively proved [7a] that the principal ideal $(1 - \zeta)$ was prime and that its $(p - 1)$-th power was the principal ideal (p); this case therefore raised no new problem. For $q \neq p$ the idea which seems to have guided Kummer is to replace the cyclotomic equation $\Phi_p(z) = 0$ by the congruence $\Phi_p(u) \equiv 0 \pmod{q}$, in other words to decompose the cyclotomic polynomial $\Phi_p(X)$ *over the field* $\mathbf{F}_q$ and to associate with each irreducible factor of this polynomial an "ideal prime factor" A simple case (explicitly mentioned in the Note [7b] where Kummer announces his results without proof) is that where $q \equiv 1 \pmod{p}$; if $q = mp + 1$ and $\gamma \in \mathbf{F}_q$ is a primitive $(q - 1)$-th root of 1, then, in $\mathbf{F}_q[X]$,

$$\Phi_p(X) = \prod_{k=1}^{p-1} (X - \gamma^{km})$$

(*) The notion of norm of an algebraic number goes back to Lagrange: if α_i $(1 \leqslant i \leqslant n)$ are the roots of a polynomial of degree n, he even considers the "norm form" $N(x_0, x_1, \ldots, x_{n-1}) = \prod_{i=1}^{n} (x_0 + \alpha_i x_1 + \cdots + \alpha_i^{n-1} x_{n-1})$ in the variables x_i which was no doubt suggested to him by his research on the solution of equations and "Lagrange resolvents" ([2], vol. VII, p. 170). It is to be noted that it is the multiplicative property of the norm which leads Lagrange to his identity on binary quadratic forms, whence Gauss was able to obtain the "composition" of these forms ([2], vol. II, p. 522). On the other hand, when the theory of algebraic numbers comes into being about 1830, it is very often in the form of the solution of equations $N(x_0, \ldots, x_{n-1}) = \lambda$ (in particular with $\lambda = 1$ in research on units) or the study of "norm forms" (also called "decomposable forms") that the problems are presented; and even in recent works, the properties of these particular Diophantine equations are used fruitfully, notably in the theory of p-adic numbers (Skolem, Chabauty).

since $\gamma^{pm} = 1$. Then associating with each factor $X - \gamma^{km}$ an "ideal prime factor" $\mathfrak{q}_k$ of q, Kummer says that an element $x \in \mathbf{Z}[\zeta]$, of which P is the minimal polynomial over $\mathbf{Q}$, is *divisible by* $\mathfrak{q}_k$ if in $\mathbf{F}_q$ $P(\gamma^{km}) = 0$; to sum up, in modern language, he write the quotient ring $\mathbf{Z}[\zeta]/q\mathbf{Z}[\zeta]$ as a direct composition of fields isomorphic to $\mathbf{F}_q$. For $q \not\equiv 1 \pmod{p}$, the irreducible factors of $\Phi_p(X)$ in $\mathbf{F}_q[X]$ are no longer of first degree and it would therefore be necessary to substitute for X in P(X) "Galois imaginary" roots of the factors of Φ_p in $\mathbf{F}_q[X]$. Kummer avoids this difficulty by passing, as we would say today, to the *decomposition field* K of q; if f is the least integer such that $q^f \equiv 1 \pmod{p}$ and $p - 1 = ef$, K is just the subfield of $\mathbf{Q}(\zeta)$ consisting of the invariants of the subgroup of order f of the Galois group (cyclic of order $p - 1$) of $\mathbf{Q}(\zeta)$ over $\mathbf{Q}$; in other words it is the unique subfield of $\mathbf{Q}(\zeta)$ which is of degree e over $\mathbf{Q}$; it had been well known since Gauss's *Disquisitiones*, being generated by the "periods"

$$\eta_k = \zeta_k + \zeta_{k+f} + \zeta_{k+2f} + \cdots + \zeta_{k+(e-1)f}$$

($0 \leqslant k \leqslant e - 1$, $\zeta_\nu = \zeta^{g^\nu}$ where g is a primitive root of the congruence $z^{p-1} \equiv 1 \pmod{p}$), which form a normal basis for it. If R(X) is the minimal polynomial (monic and with rational integer coefficients) of any of these "periods" η, Kummer, starting from Gauss's formulae, proves that, over the field $\mathbf{F}_q$, R(X) also decomposes into distinct factors of the first degree $X - u_j$ ($1 \leqslant j \leqslant e$) and it is with each of the u_j that he now associates an "ideal prime factor" $\mathfrak{q}_j$. To define "divisibility by $\mathfrak{q}_j$", Kummer writes every $x \in \mathbf{Z}[\zeta]$ in the form $x = \sum_{k=0}^{f-1} \zeta^k y_k$, where each $y_k \in K$ may itself be written uniquely as a polynomial of degree $\leqslant e - 1$ in η with rational integer coefficients; he says that x is divisible by $\mathfrak{q}_j$ if and only if, when u_j is substituted for η in each of the y_k, the elements of $\mathbf{F}_q$ obtained are *all* zero. But it was also necessary to define the "exponent" of $\mathfrak{q}_j$ in x. For this, Kummer introduces what we would now call a *uniformizer* for $\mathfrak{q}_j$, that is an element $\rho_j \in K$ such that $N(\rho_j) \equiv 0 \pmod{q}$, $N(\rho_j) \not\equiv 0 \pmod{q^2}$ and finally such that ρ_j is divisible by $\mathfrak{q}_j$ (in the sense defined above) but by *none other* of the ideal factors $\neq \mathfrak{q}_j$ of q. The existence of such a ρ_j had effectively been proved by Kronecker in his dissertation the previous year ([9a], p. 23); then writing $\rho'_j = N(\rho_j)/\rho_j$, Kummer says that the exponent of $\mathfrak{q}_j$ in x is equal to h if $x\rho_j'^h \equiv 0 \pmod{q^h}$ but $x\rho_j'^{h+1} \not\equiv 0 \pmod{q^{h+1}}$; he begins of course by proving that the relation $x\rho'_j \equiv 0 \pmod{q}$ is equivalent to the fact that x is divisible by $\mathfrak{q}_j$ (in the above sense). Once these definitions were made, the extension to $\mathbf{Z}[\zeta]$ of the usual laws of divisibility for "ideal numbers" no longer offered serious difficulty; and from his first memoir [7c] Kummer could even, using Dirichlet's "box method", show that the "classes" and "ideal factors" were *finite* in number (*).

(*) He does no more than reproduce an argument of Kronecker in his disser-

We shall not pursue the history of Kummer's later works on cyclotomic fields, concerning the determination of the class number and the application to the proof of Fermat's theorem in certain cases. Let us just mention the way in which, in 1859, he extends his method to obtain (at least partially) the "ideal prime numbers" in a "Kummerian field" $\mathbf{Q}(\zeta, \mu)$, where μ is a root of the irreducible polynomial $P(X) = X^p - \alpha$, where $\alpha \in \mathbf{Z}[\zeta]$ [7e]. It is interesting that Kummer envisages the problem precisely by considering $\mathbf{Q}(\zeta,\mu)$ as a cyclic extension *of the field* $\mathbf{Q}(\zeta)$ taken as "base field" (†): he starts with an "ideal prime number" $\mathfrak{q}$ of $\mathbf{Z}[\zeta]$ which he assumes divides neither p nor α and this time he examines (in modern terms) the polynomial $\bar{P}(X) = X^p - \bar{\alpha}$ in the *residue field* k of the valuation on $\mathbf{Q}(\zeta)$ corresponding to $\mathfrak{q}$ ($\bar{\alpha}$ being the canonical image of α in k). As $\mathbf{Q}(\zeta)$ is the field of the p-th roots of unity, $\bar{P}$ is, either irreducible over k, or the product of factors of the first degree. In the first case, Kummer says that $\mathfrak{q}$ remains prime in $\mathbf{Z}[\zeta, \mu]$; in the second, he introduces elements w_i ($1 \leqslant i \leqslant p$) of $\mathbf{Z}[\zeta]$ whose images in k are the roots of $\bar{P}$ and he associates with each index i an ideal prime factor $\mathfrak{r}_i$ of $\mathfrak{q}$; then writing $W_i(X) = \prod_{j \neq i} (X - w_j)$, he says that, for a polynomial f with coefficients in $\mathbf{Z}[\zeta]$, $f(\mu)$ contains the factor $\mathfrak{r}_i$ m times if

$$f(w_i)W_i^m(w_i) \equiv 0 \qquad (\text{mod. } \mathfrak{q}^m)$$

but

$$f(w_i)W_i^{m+1}(w_i) \not\equiv 0 \qquad (\text{mod. } \mathfrak{q}^{m+1}).$$

To sum up, he obtains in this way the valuations on $\mathbf{Q}(\zeta, \mu)$ which are *unramified* over $\mathbf{Q}$, which are sufficient for the applications he has in view.

* * *

Kummer had had the chance to meet, in studying particular fields to which his research on Fermat's Theorem had led at first, a number of fortuitous circumstances which made their study much more accessible. The extension to the

---

tation, relating to the classes of solutions of equations of the form

$$N(x_0, x_1, \ldots, x_{n-1}) = a$$

([9a], p. 25). On the other hand, Kummer makes several allusions to results obtained by Dirichlet on equations of this type (for any algebraic number field); but these results have neither been published nor found among Dirichlet's papers.

(†) In his memoir on quadratic forms with coefficients in the ring of Gaussian integers ([4], vol. I, pp. 533–618) Dirichlet had in various places been led to consider the relative norm of the field $Q(\sqrt{D}, i)$ over its quadratic subfield $\mathbf{Q}(\sqrt{D})$. Similarly, Eisenstein, studying the 8-th roots of unity, considers the field they generate as a quadratic extension of $\mathbf{Q}(i)$ and uses the norm relative to this subfield ([6b], p. 253). But the work of Kummer is the first example of profound arithmetical study of a "relative field".

general case of Kummer's results presented considerable difficulties and was to take years of effort.

With Kronecker and Dedekind, who play the principal roles there, the history of the theory of algebraic numbers, during the 40 years following Kummer's discovery, is not dissimilar (but happily without the same acrimony) to that of the rivalry of Newton and Leibniz 180 years earlier concerning the invention of Infinitesimal Calculus. Pupil and later colleague of Kummer in Berlin, Kronecker (whose thesis, as we have seen, had served as an essential point in Kummer's theory) was greatly interested in "ideal numbers" with the aim of applying them to his own research; and we admire his astonishing penetration when we see him, as early as 1853 ([9b], p. 10) announce the general theorem on the structure of Abelian extensions of **Q** and, what is perhaps still more remarkable, create, in the years which follow, the theory of complex multiplication and discover the first germ of class field theory ([9c] and [9d]). A letter from Kronecker to Dirichlet in 1857 ([9], vol. 5, pp. 418–421) shows that he already possessed at that time a generalization of Kummer's theory, which moreover Kummer himself confirms in one of his own works ([7e], p. 57) and Kronecker will make many an allusion to this theory in his memoirs between 1860 and 1880 (*).

But although at that time none of the mathematicians of the German school of the Theory of Numbers was unaware of the existence of these works of Kronecker, the latter seems only to have communicated the principles of his methods to a restricted circle of friends and pupils and when he finally decided to publish them in his memoir of 1881 on the discriminant [9e] and above all in his great "Festschrift" of 1882 [9f], Dedekind could not refrain from expressing his surprise ([10], vol. III, p. 427), having imagined the processes were completely different, from the echoes he had heard ([10], vol. III, p. 287). Kronecker moreover was far from possessing to the same degree Dedekind's remarkable gifts of exposition and clarity and it is therefore not surprising that it is chiefly the methods of the latter, already published in 1871, which have formed the framework of the theory of algebraic numbers; however interesting it may be, Kronecker's method of the "adjunction of indeterminates", where the Theory of Numbers is concerned, is scarcely more in our eyes than a variant of Dedekind's (cf. Chapter VII, § 1, Exercise 31) and it is chiefly in another direction, oriented towards Algebraic Geometry, that Kronecker's ideas acquire all their importance for the history of Commutative Algebra, as we shall see later.

For reasons which could only clearly be seen much later, a first preliminary to any attempt at a general theory was of course the clarification of the notion of algebraic integer. This is obtained about 1845–50, although it is difficult

(*) On the evolution of his ideas on this subject, see the very interesting introduction to his memoir of 1881 on the discriminant ([9e], p. 195).

enough to date its appearance precisely; it seems probable that it is the idea of a system stable under addition and multiplication (or, more precisely, what we now call a **Z**-algebra of finite rank) which, more or less consciously, led to the general definition of algebraic integers: in fact this definition is inevitably hit upon when a **Z**-algebra of the form $\mathbf{Z}[\theta]$ is restricted to being of finite rank, by analogy with the ring $\mathbf{Z}[\zeta]$ generated by a root of unity, which was always at the centre of arithmeticians' preoccupations at this time. At any rate, when, independently, Dirichlet ([4], vol. I, p. 640), Hermite ([8], vol. I, pp. 115 and 146) and Eisenstein ([6c], p. 236) introduce the notion of algebraic integer, they do not appear to consider that they are dealing with a new concept nor to judge that it will be useful to make a detailed study thereof; only Eisenstein shows effectively (*loc. cit.*) that the sum and product of two algebraic integers are algebraic integers, without moreover claiming that this result is original.

A much more subtle point was the determination of the rings in which a generalization of Kummer's theory could be expected. The latter, in his first note [7b], does not hesitate to affirm that he can regain by his method Gauss's theory of binary quadratic forms by considering the rings $\mathbf{Z}[\sqrt{D}]$ (D an integer); he never developed this idea, but it certainly seems that neither he nor any one else before Dedekind perceived that unique composition into "ideal" prime factors is impossible in the ring $\mathbf{Z}[\sqrt{D}]$ when $D \equiv 1 \pmod{4}$ (although the example of the cube roots of unity showed that the ring $\mathbf{Z}[\rho]$ considered from the time of Gauss is distinct from $\mathbf{Z}[\sqrt{-3}]$) (*). Before Dedekind and Kronecker, the only rings studied are always of the type $\mathbf{Z}[\theta]$ or sometimes certain particular rings of the type $\mathbf{Z}[\theta, \theta']$ (†). As far as Kronecker is concerned, it is possible that the idea of considering *all* the integers of an algebraic extension was first suggested to him by the study of the field of algebraic functions, where this ring arises naturally as the set of functions which are "finite at infinite distance"; in any case he insists in his memoir of 1881 on the discriminant (written and announced at the Academy of Berlin as early as 1862) on this characterization of the "integers" in these fields [9e]. Dedekind gives no indication as to the origin of his own ideas on this point, but in his very first publication on number fields in 1871 the ring of all integers of such a field plays a capital role in his theory; it is also Dedekind who clarifies the relation between such a ring and its subrings with the same field of fractions, by the introduction of the notion of *conductor* [10c].

(*) Although Kronecker must have been led to study the arithmetic of the rings $\mathbf{Z}[\sqrt{-D}]$ $(D > 0)$ by his work on complex multiplication, he published nothing on this subject and the characterization of the integers of any quadratic field $\mathbf{Q}(\sqrt{D})$ is given explicitly for the first time by Dedekind in 1871 ([10c], pp. 105–106).

(†) We have seen earlier the example of the ring $\mathbf{Z}[\zeta, \mu]$ introduced by Kummer [7e]. Earlier, Eisenstein had been led to envisage a subring generated by two elements of the ring of integers in the field of the 21st roots of unity [6b].

But here was not the only difficulty. To generalize the ideas of Kummer, it was necessary first to get rid of passing via the decomposition field, which naturally could have no analogue in the case of a non-Abelian field. This detour seems moreover at first sight very surprising and artificial, for, starting with the irreducible polynomial $\Phi_p(X)$ of $\mathbf{Z}[X]$, one may wonder why Kummer does not push his ideas to their logical conclusion and what prevents him from using the theory of "Galois imaginary" numbers which were well known at the time. The obstacle comes more clearly to light in an unfortunate attempt at generalization made as early as 1865 by Selling, a pupil of Dedekind: given an irreducible polynomial $P \in \mathbf{Z}[X]$, Selling decomposes the corresponding polynomial $\bar{P}(X)$ into irreducible factors in $\mathbf{F}_q[X]$; the roots of this polynomial therefore belong to a finite extension $\mathbf{F}_r$ of $\mathbf{F}_q$; but Selling, in order to define in Kummer's way the exponent of an "ideal prime factor" of q in an integer of the splitting field of $P(X)$, does not hesitate to speak, *in the field* $\mathbf{F}_r$, of congruences modulo a *power of* q ([11], p. 26); and a little later when he tries to approach the question of ramification, he "adjoins" to $\mathbf{F}_r$ "imaginary roots" of an equation of the form $x^h = q$ ([11], p. 34). Clearly these bold steps (which would be justified were the finite field $\mathbf{F}_q$ replaced by the q-adic field) could at that time only end in nonsense. Fortunately, Dedekind in 1857 [10a], under the name of "theory of higher congruences", took up again in another form the theory of finite fields (*): he interprets the elements of the latter as "residues" of the polynomials of $\mathbf{Z}[X]$ with respect to a "double modulus" consisting of the linear combinations with coefficients in $\mathbf{Z}[X]$ of a prime number p and an irreducible monic polynomial $P \in \mathbf{Z}[X]$ (which is no doubt for him, as for Kronecker, the origin of the general idea of *module* at which they were to arrive independently a little later). According to his own testimony ([10d], p. 218) it seems that Dedekind had begun by attacking the problem of the "ideal factors" of p in a field $\mathbf{Q}(\xi)$, where $P \in \mathbf{Z}[X]$ is the minimal polynomial of ξ, as follows (certainly at least in the "unramified" case, that is when the polynomial $\bar{P}$ in $\mathbf{F}_p[X]$ corresponding to P has no multiple root): he writes, in $\mathbf{Z}[X]$,

$$P = P_1 P_2 \ldots P_h + p.G$$

where the $\bar{P}_i$ are irreducible and distinct in $\mathbf{F}_p[X]$; it may be assumed that G is not divisible (in $\mathbf{Z}[X]$) by any of the P_i and for all i he writes $W_i = \prod_{j \neq i} P_j$; then, if $f \in \mathbf{Z}[X]$, it will be said that $f(\xi)$ contains the "ideal factor" $\mathfrak{p}_i$ of p

(*) It is known that certain results of this theory, published first by Galois, had been obtained (in the language of congruences) by Gauss about 1800; after the death of Gauss, Dedekind was charged with the publication of part of his works and had rediscovered in particular in the papers left by Gauss the memoir on finite fields ([3], vol. II, pp. 212–240).

corresponding to P_i k times if

$$fW_i^k \equiv 0 \qquad (\text{modd. } p^k, \mathrm{P})$$

and

$$fW_i^{k+1} \not\equiv 0 \qquad (\text{modd. } p^{k+1}, \mathrm{P}).$$

The relationship with the method followed by Kummer for "Kummerian fields" is here manifest and Kummer's original definition for cyclotomic fields can, in this way, easily be recovered (see for example the work of Zolotareff [14] who, at first independently of Dedekind, developed these ideas a little later).

However, neither Dedekind nor Kronecker who appears also to have made analogous attempts, could progress further in this direction, both of them halted by the difficulties presented by ramification ([10d], p. 218 and [9f], p. 325) (*). If the ring of integers A of the number field K under consideration admits a basis (over **Z**) consisting of the powers of the same integer θ, it is not difficult to generalize the above method for ramified prime numbers in $\mathbf{Z}[\theta]$ (as Zolotareff indicates (*loc. cit.*)). But there are fields K where no basis of this type exists in the ring A; and Dedekind even finished by discovering that there are cases where certain prime numbers p (the "extraordinary factors of the discriminant" of the field K) are such that, *for all* $\theta \in A$, applying the above method of the minimal polynomial of θ over **Q** leads to attributing to p multiple ideal factors when in fact p is unramified in A (†); he admits that he was held up for a long time by this unforeseen difficulty, before managing to surmount it by the creation from scratch of the theory of modules and ideals, in a masterly exposition (and already in a wholly modern style, in contrast with the discursive style of his contemporaries) in what is without doubt his masterpiece, the famous "11th supplement" to Dirichlet's book on the Theory of Numbers [10f]. This work saw three successive versions, but already in the first (published as the "10th supplement" to the second edition of Dirichlet's book in 1871 [4 *bis*]) the essentials of the method are present and almost at one stroke the theory of algebraic numbers passes from sketches and earlier gropings to a fully mature discipline already possessing its essential tools: from the beginning, the ring of all integers of a number field is placed at the centre of the theory; Dedekind proves the existence of a basis of this ring over **Z** and

(*) Zolotareff circumvents the difficulty by a refinement of his method which appears of little more than anecdotal interest [14].

(†) Kronecker claims to have come across the same phenomenon in a subfield of the field of the 13th roots of unity which he does not describe more precisely ([9f], p. 384). The example of an extraordinary factor of the discriminant given by Dedekind is treated in detail in Hasse, *Zahlentheorie* (Berlin, Akad. Verlag, 1949), p. 333; a little later, Hasse gives an example of a field K where there is no extraordinary factor of the discriminant, but where there exists no $\theta \in A$ such that $A = \mathbf{Z}[\theta]$ (*loc. cit.*, p. 335).

deduces from it the definition of the discriminant of the field as the square of the determinant formed by the elements of a basis of the ring of integers and their conjugates; however he only gives in the 11th supplement the characterization of ramified prime numbers (as prime factors of the discriminant) for quadratic fields ([10f], p. 202), whereas he was in possession of the general theorem from 1871 onwards (*). The central result of the work is the existence and uniqueness theorem for the decomposition of ideals into prime factors, for which Dedekind starts by developing an elementary theory of "modules"; in fact, in the 11th supplement, he reserves this name for sub-**Z**-modules of a number field, but the conception he forms of them and the results he shows are already expounded in a way which is immediately applicable to general modules (†); amongst other things must be noted, as early as 1871, the introduction of the notion of "transporter" which plays an important role (as well as the "ascending chain condition") in the first proof of the unique factorization theorem. In the two following editions, Dedekind was also to give two other proofs of this theorem which he justly considered as the cornerstone of his theory. It should be noted here that it is in the third proof that fractional ideals are made use of (already introduced as early as 1859 by Kummer for cyclotomic fields) and the fact that they form a *group* is established; we shall return later to the second proof (p. 594).

All these results (except for the language) were no doubt already known to Kronecker about 1860 as particular cases of his more general conceptions of which we speak later (whereas Dedekind recognizes that he only surmounted the last difficulties of his theory in 1869–70 ([10e], p. 351)) (‡); as far as number fields are concerned, it must in particular be underlined that, already at this time, Kronecker knew that the whole theory is applicable without essential change starting with a "base field" k which is itself a number field (other than **Q**), a point of view to which the theory of complex multiplication led naturally; he had thus recognized, for certain fields k, the existence of algebraic extensions $K \neq k$ *unramified over* k ([9f], p. 269), a fact which cannot hold for $k = \mathbf{Q}$ (as follows from minorations of Hermite and Minkowski for the discriminant). Dedekind was never to develop this last point of view (although he indicates its possibility in his memoir of 1882 on the different) and the first systematic exposition of "relative field" theory is due to Hilbert [16d].

(*) He only gives the proof of this theorem in his memoir of 1882 on the different [10e].

(†) In his memoir of 1882 on algebraic curves (jointly with H. Weber) [10 *bis*], he uses the theory of modules over the ring **C**[X] in the same way.

(‡) Kronecker had however not succeeded in obtaining by his methods the complete characterization of ramified ideals in the case of number fields. On the other hand he does have this characterization for fields of algebraic functions of one variable and proves moreover that in this case there is no "extraordinary factor" of the discriminant [9e].

Finally, in 1882 [10e], Dedekind completes the theory by introducing the *different*, which gives him a new definition of the discriminant and allows him to define precisely the exponents of the ideal prime factors in the decomposition of the latter. It is also about this time that he becomes interested in the particular features presented by Galois extensions, introducing the notions of decomposition group and inertia group (in his memoir [10g] which was only published in 1894) and even (in papers not published during his lifetime ([10], vol. II, pp. 410–411)) a sketch of ramification groups, which Hilbert (independently of Dedekind) was to develop a little later ([16c] and [16d]).

Thus, about 1895, the theory of algebraic numbers had completed the first stage of its development; the tools forged during this formation period will allow it almost immediately to enter the next stage, general class field theory (or, what amounts to the same, the theory of Abelian extensions of number fields) which carries on to our own day and which we shall not describe here. From the point of view of Commutative Algebra, it may be said that at the same time the history of Dedekind domains is practically completed, setting aside their axiomatic characterization, and also the structure of finitely generated modules over these domains (which, in the case of number fields, will only be substantially elucidated by Steinitz in 1912 [20b]) (*).

* * *

The later progress in Commutative Algebra arises chiefly from quite different problems, issuing from Algebraic Geometry (which will moreover influence the Theory of Numbers directly even before the "abstract" developments of the present period).

We shall not concern ourselves here with the detailed history of Algebraic Geometry which, until the death of Riemann, scarcely touches our subject. Let it suffice to recall that it was mainly concerned with the study of algebraic curves in the complex projective plane, usually approached by projective geometric methods (with or without the use of coordinates). There was a parallel development, with Abel, Jacobi, Weierstrass and Riemann, of the theory of "algebraic functions" of one complex variable and their integrals; mathematicians were obviously conscious of the connection between this theory and the geometry of plane algebraic curves and even on occasions were known to "apply Analysis to Geometry"; but the methods used for the study of algebraic functions were chiefly of a "transcendental" nature, even before Riemann (†); this character is still further accentuated in the work of the latter,

(*) A start to the study of modules over a ring of algebraic numbers had already been made by Dedekind [10h].

(†) It must be noted however that Weierstrass, in his research on Abelian functions (which goes back to 1857 but was only expounded in his lectures about 1865 and only published in his complete Works ([17], vol. IV)), gives, in contrast with the introduction of "Riemann surfaces" and arbitrary analytic functions defined on such a surface. Almost immediately after the death of Riemann, Roch and above all Clebsch recognized the possibility of obtaining from the profound results obtained by Riemann's transcendental methods numerous striking applications to the projective geometry of curves, which was of course to incite contemporary geometers to give purely "geometric" proofs of these results; this programme, incompletely followed by Clebsch and Gordan, was completed by Brill and M. Noether several years later [13], with the aid of the study of systems of variable points on a given curve and auxiliary curves (the "adjoints") passing through such systems of points. But even for his contemporaries, Riemann's transcendental methods (and notably his use of topological notions and of "Dirichlet's principle") appeared to rest on uncertain foundations; and although Brill and Noether are rather more careful than most contemporary "synthetic" geometers (see below p. 593), their geometric-analytic methods are not safe from all reproach. It is essentially to give the theory of plane algebraic curves a solid basis that Dedekind and Weber published in 1882 their great memoir on this subject [10 *bis*]: "*The research published below*", they say, "*is intended to lay the foundations of the theory of algebraic functions of one variable, one of Riemann's principal creations, in a way which is at the same time simple, rigorous and entirely general. In earlier research on this subject, in general restrictive hypotheses have been made on the singularities of the functions considered and the would-be exceptional cases are, either mentioned in passing as limiting cases, or entirely neglected. Similarly, certain fundamental theorems on continuity or analyticity are accepted, whose "evidence" depends on geometric intuitions of a varied nature*" ([10 *bis*], p. 181) (*).

---

to Riemann, a purely algebraic definition of the genus of a curve, as the least integer p such that there are rational functions on the curve with poles at $p + 1$ given arbitrary points. It is interesting to point out that, seeking to obtain elements which serve as functions with only one pole on the curve, Weierstrass, before finally using for this purpose transcendental functions, had, according to Kronecker ([9e], p. 197), urged the latter to extend to algebraic functions of one variable the results he had at that time just obtained for number fields ("ideal prime factors" effectively playing the role desired by Weierstrass).

(*) It is well known that, in spite of the efforts of Dedekind, Weber and Kronecker, the laxness in the conception of what constituted a correct proof, already visible in the German school of Algebraic Geometry of the years 1870–1880, was only to be aggravated more and more in the work of French and above all Italian geometers of the next two generations, who, following the German geometers and developing their methods, attack the theory of algebraic surfaces: a "scandal" often denounced (chiefly since 1920) by algebraists, but to a certain extent justified by the brilliant successes achieved by these "non-rigorous" methods, contrasting with the fact that, until about 1940, the orthodox successors of Dedekind had shown themselves incapable of formulating with sufficient flexibility and power the algebraic notions which would have allowed correct proofs to be given of these results.

The essential idea of their work is to model the theory of algebraic functions of one variable on the theory of algebraic numbers as Dedekind had just developed it; to do this, they must first look at it from an "affine" point of view (in contrast with their contemporaries, who invariably considered algebraic curves as imbedded in complex projective space): they therefore start with a finite algebraic extension K of the field $\mathbf{C}(X)$ of rational functions and the ring A of "integral algebraic functions" in K, i.e. the elements of this field which are integral over the polynomial ring $\mathbf{C}[X]$; their fundamental result, which they obtain without using any topological consideration (*), is that A is a Dedekind domain, to which may be applied *mutatis mutandis* (and even, as Dedekind and Weber remark, without yet clearly seeing the reason ([10], vol. I, p. 268), in a simpler way) all the results of the "11th supplement". Having done this, they prove that their theorems are in fact birationally invariant (in other words, depend only on the field K) and in particular do not depend on the choice of the "line at infinity" made at the beginning. What is no doubt still more interesting for us, is that, in order to define the points of the "Riemann surface" corresponding to K (and in particular the "points at infinity", which cannot correspond to ideals of A), they are led to introduce the notion of *place* of the field K: They find themselves in the same situation as Gelfand will find himself in 1940 when founding the theory of normed algebras, knowing a set K of elements which are not given to start with as functions and yet one wants to consider as such; and, to obtain the defining set of hypothetical functions, they have for the first time the idea (which Gelfand followed and which has become commonplace through being used at every turn in modern mathematics) of associating with a point x a set E and with a set $\mathscr{F}$ of mappings of E to a set G the mapping $f \mapsto f(x)$ of $\mathscr{F}$ to G, in other words of considering, in the expression $f(x)$, *f as variable and x as fixed*, in contrast with the classical tradition. Finally, they have no difficulty, starting from the notion of place, in defining "positive divisors" ("Polygon" in their terminology) which include the ideals of A as particular cases and correspond to the "systems of points" of Brill and Noether; but, although they write principal divisors and divisors of differentials as "quotients" of positive divisors, they do not give the general definition of *divisors* and it is only in 1902 that Hensel and Landsberg introduce, by analogy with fractional ideals, this notion which will always embarrass the champions of purely "geometric" methods (obliged in spite of themselves to define them with the name "virtual systems", but uneasy at not being able to give them a "concrete" interpretation).

The same year 1882 also sees appear Kronecker's great memoir awaited for more than 20 years [9f]. Much more ambitious than the work of Dedekind-

(*) They underline that, thanks to this fact, all their results would remain valid if the field $\mathbf{C}$ were replaced by the field of all algebraic numbers ([10], vol. I, p. 240).

Weber, it is also unfortunately much more vague and obscure. Its central theme is (in modern language) the study of the ideals of a finite integral algebra over one of the polynomial rings $\mathbf{C}[X_1, \ldots, X_n]$ or $\mathbf{Z}[X_1, \ldots, X_n]$; Kronecker limits himself *a priori* to those ideals which are finitely generated (the fact that they all are was only to be proved (for the ideals of $\mathbf{C}[X_1, \ldots, X_n]$)) some years later by Hilbert in the course of his work on invariants [16a]). As far as $\mathbf{C}[X_1, \ldots, X_n]$ or $\mathbf{Z}[X_1, \ldots, X_n]$ is concerned, this naturally led to associating with each ideal of one of these rings the "algebraic variety" consisting of the zeros common to all the elements of the ideal; and the study of geometry in 2 and 3 dimensions during the 19th century was to lead intuitively to the idea that every variety is a union of a finite number of "irreducible" varieties whose "dimensions" are not necessarily all the same. It seems that the proof of this fact is the aim Kronecker sets himself, although he nowhere says so explicitly and no definition of "irreducible variety" can be found in his memoir, nor of "dimension". In fact, he limits himself to indicating summarily how a general elimination method (*) gives, starting with a system of generators of the ideal considered, a finite number of algebraic varieties for each of which, in a suitable coordinate system, a certain number of coordinates are arbitrary and the others are "algebraic functions" of them (†). But if it is indeed the decomposition into irreducible varieties at which Kronecker is aiming, it must be recognized that he only arrives there in the elementary case of a *principal* ideal, where he proves effectively, extending a classical lemma of Gauss on $\mathbf{Z}[X]$ ([3], vol. I, p. 34), that the domains $\mathbf{C}[X_1, \ldots, X_n]$ and $\mathbf{Z}[X_1, \ldots, X_n]$ are factorial; and, in the general case, it is questionable whether Kronecker was in possession of the notion of prime ideal (what he calls "Primmodulsystem" is an ideal which is *indecomposable as a product* of two others ([9f], p. 336); this is all the more astonishing as the definition already given by Dedekind in 1871 was perfectly general).

It must however be said that Kronecker's elimination method, suitably applied, certainly leads to the decomposition of an algebraic variety into its irreducible components: this is clearly established by E. Lasker at the beginning of his great memoir of 1905 on polynomial ideals [19]; he defines correctly the

(*) By a linear change of coordinates, it may be assumed that the generators F_i $(1 \leqslant i \leqslant r)$ of the ideal are polynomials where the term of highest degree in X_1 is of the form $c_i X_1^{m_i}$, where c_i is a constant $\neq 0$. It may also be assumed that the F_i have no common factor. Consider then for $2r$ indeterminates u_i, v_i $(1 \leqslant i \leqslant r)$ the polynomials $\sum_{i=1}^{r} u_i F_i$ and $\sum_{i=1}^{r} v_i F_i$ as *polynomials in* X_1; form their Sylvester resultant, which is a polynomial in the u_i and v_i with coefficients in $\mathbf{C}[X_2, \ldots, X_n]$ (resp. $\mathbf{Z}[X_2, \ldots, X_n]$); by annihilating these coefficients, a system of equations is obtained whose solutions $(x_2, \ldots, x_n)$ are precisely the projections of the solutions $(x_1, \ldots, x_n)$ of the system of equations $F_i(x_1, x_2, \ldots, x_n) = 0$ $(1 \leqslant i \leqslant r)$. The application of the method may then be continued by induction on n.

(†) It is this number of arbitrary coordinates that he calls the *dimension* ("Stufe").

notion of irreducible variety (in $\mathbf{C}^n$) as an algebraic variety V such that a product of two polynomials can only be zero on the whole of the variety V if one of them is and he also gives a definition which is independent of the choice of axes. In the interesting historical considerations he inserts in this work, Lasker shows that he is interested, not only in the purely algebraic tendencies of Kronecker and Dedekind, but also in the problems raised by the geometric methods of the school of Clebsch and M. Noether and notably in the famous theorem proved by the latter in 1873 [12]. He is essentially concerned, as we would say today, with determining the ideal $\mathfrak{a}$ of polynomials of $\mathbf{C}[X_1, \ldots, X_n]$ which are zero at the points of a given set M of $\mathbf{C}^n$; usually M was the "algebraic variety" of zeros common to polynomials f_i finite in number and for a long time it seems that it was accepted (of course without justification) that, at least for $n = 2$ or $n = 3$, the ideal $\mathfrak{a}$ was simply generated by the f_i (*). M. Noether had shown that even for $n = 2$ and for two polynomials f_1, f_2 this is generally false and he had given sufficient conditions for $\mathfrak{a}$ to be generated by f_1 and f_2. Ten years later, Netto proves that, with no hypothesis on f_1 and f_2, a *power* of $\mathfrak{a}$ is always contained in the ideal generated by f_1 and f_2 [15], a theorem which Hilbert generalized in 1893 in his celebrated Nullstellensatz [16b]. No doubt inspired by this result, Lasker, in his memoir, introduces the general notion of *primary* ideal (†) in the rings $\mathbf{C}[X_1, \ldots, X_n]$ and $\mathbf{Z}[X_1, \ldots, X_n]$ (after having given for these rings the definition of prime ideal, by transcribing Dedekind's definition) and shows (‡) the existence of a primary decomposition

(*) See the remarks of M. Noether at the beginning of his memoir [13]. It is interesting to note on this subject that, according to Lasker, Cayley, about 1860, had conjectured that for every twisted algebraic curve in $\mathbf{C}^3$ there were a finite number of polynomials generating the ideal of polynomials of $\mathbf{C}[X, Y, Z]$ which are zero on the curve (in other words, a particular case of Hilbert's finiteness theorem [16a]).

(†) Examples of primary ideals which are not powers of prime ideals had been encountered by Dedekind in "orders", i.e. rings of algebraic numbers with a given number field as field of fractions ([10], vol. III, p. 306). Kronecker also gives as an example of an ideal "indecomposable" as a product of two other non-trivial ideals, the ideal of $\mathbf{Z}[X]$ generated by p^2 and $X^2 + p$, where p is a prime number (an ideal which is primary for the prime ideal generated by X and p ([9f], p. 341)).

(‡) Lasker proceeds by induction on the maximal dimension h of the irreducible components of the variety V of zeros of the ideal $\mathfrak{a}$ under consideration. In modern terms, he considers first the prime ideals $\mathfrak{p}_i$ $(1 \leqslant i \leqslant r)$ containing $\mathfrak{a}$, which correspond to the irreducible components of maximal dimension h of V. With each $\mathfrak{p}_i$ he associates the saturation $\mathfrak{q}_i$ of $\mathfrak{a}$ with respect to $\mathfrak{p}_i$ (cf. Chapter IV, § 2, no. 3, Proposition 5); he then considers the transporter $\mathfrak{b}_i = \mathfrak{a} : \mathfrak{q}_i$ of $\mathfrak{q}_i$ in $\mathfrak{a}$, takes in $\sum_i \mathfrak{b}_i$ an element c belonging to none of the $\mathfrak{p}_i$ and shows on the one hand that $\mathfrak{a}$ is the intersection of the $\mathfrak{q}_i$ and $\mathfrak{a} + (c) = \mathfrak{a}'$ and on the other that the variety V' of for every ideal in these rings (*). He does not seem to be concerned with questions of *uniqueness* in this decomposition; it is Macaulay who, a little later [21] introduces the distinction between "immersed" and "non-immersed" primary ideals and shows that the latter are determined uniquely, but not the former. It should finally be noted that Lasker also extends his results to the ring of *convergent power series* in a neighbourhood of a point, by using Weierstrass's "preparation theorem". This part of his memoir is no doubt the first place this ring had been considered from a purely algebraic point of view and the methods which Lasker develops on this occasion were strongly to influence Krull when in 1938 he created the general theory of local rings (cf. [29d], p. 204 and *passim*).

* * *

The movement of ideas which will give birth to modern Commutative Algebra begins to take shape about 1910. If the general notion of field was reached by the beginning of the 20th century, in contrast the first work where the general notion of ring is defined is probably that of Fraenkel in 1914 [23]. At this time, there were already as examples of rings, not only the integral domains of the Theory of Numbers and Algebraic Geometry, but also rings of power series (formal and convergent) and finally algebras (commutative or not) over a base field. However, for the theory of fields as well as that of rings, the catalyst role seems to have been played by Hensel's theory of *p-adic numbers*, which Fraenkel and also Steinitz [20a] mention specially as the starting point of their research.

Hensel's first publication on this subject goes back to 1897; he there starts from the analogy shown by Dedekind and Weber between the points of a Riemann surface of an algebraic function field K and the prime ideals of a number field k; he proposes to carry over to the Theory of Numbers "Puiseux expansions" (classical from the middle of the 19th century) which, in a neighbourhood of any point of the Riemann surface of K, allow every element $x \in K$ to be expressed in the form of a convergent series of powers of the "uniformizer" at the point considered (a series with only a finite number of terms

---

zeros of $\mathfrak{a}'$ has only irreducible components of dimension $\leqslant h-1$, which allows him to conclude by induction.

(*) It is interesting to note that Dedekind's second proof of the unique decomposition theorem proceeds by first establishing the existence of a unique reduced primary decomposition; and in a passage not published in the 11th supplement, Dedekind observes explicitly that this part of the proof is valid not only for the ring A of all integers of a number field K, but also for all the "orders" of K ([10], vol. III, p. 303). It is only then, after showing explicitly that A is "completely integrally closed" (to within terminology) that he proves, using this fact, that the primary ideals of the above decomposition are in fact powers of prime ideals ([10], vol. III, p. 307).

with negative exponent). Hensel's shows similarly that, if $\mathfrak{p}$ is a prime ideal of k lying above a prime number p, a "p-adic series" may be associated with every $x \in k$, of the form $\sum_i \alpha_i p^i$ $\left(\text{or } \sum_i \alpha_i p^{i/e} \text{ when } \mathfrak{p} \text{ is ramified over } p\right)$, the α_i being taken in a given representative system of the field of residues of the ideal $\mathfrak{p}$; but his great originality lies in having had the idea of considering such "expansions" even when they correspond to *no element* of k, by analogy with the expansions in integral series of transcendental functions on a Riemann surface [18a].

Throughout the rest of his career, Hensel devotes himself to polishing and perfecting little by little his new calculus; and if his manner seems to us hesitant or ponderous, it must not be forgotten that at the beginning at least he had at his disposal none of the topological or algebraic tools of modern mathematics which would have facilitated his task. In his first publications he moreover scarcely speaks of topological notions and on the whole for him the ring of $\mathfrak{p}$-adic integers ($\mathfrak{p}$ a prime ideal in the ring of integers A of a number field k) is, in modern terms, the inverse limit of the rings $A/\mathfrak{p}^n$ for n increasing indefinitely, in a purely algebraic sense; and to establish the properties of this ring and its field of fractions, it is necessary at each step to use more or less painfully *ad hoc* arguments (for example to prove that the $\mathfrak{p}$-adic integers form an integral domain). The idea of introducing topological notions into a $\mathfrak{p}$-adic field does not appear in Hensel's works before 1905 [18d]; and it is only in 1907, after having published the book where he reexpounds the theory of algebraic numbers according to his ideas [18f]), that he arrives at the definition and essential properties of $\mathfrak{p}$-adic absolute values [18e], starting with which he will be able to develop, modeling it on Cauchy's theory, a new "$\mathfrak{p}$-adic analysis" which he will be able to apply fruitfully in the Theory of Numbers (notably by using the $\mathfrak{p}$-adic exponential and logarithm) and whose importance has been growing ever since.

Hensel had well seen, from the beginning, the simplifications his theory brought to classical expositions, by allowing the problems to be "localized" and the work to be carried out in a field where not only are the divisibility properties trivial, but also, thanks to the fundamental lemma which he discovered as early as 1902 [18c], the study of polynomials whose "reduced" polynomials mod p have no multiple roots is reduced to the study of polynomials over a finite field. He had given as early as 1897 [18b] striking examples of these simplifications, notably on questions related to the discriminant (in particular, a short proof of the criterion he had given a few years earlier for the existence of "extraordinary divisors"). But for a long time it seems that the p-adic numbers inspired considerable distrust in contemporary mathematicians; a current attitude no doubt towards ideas that are "too abstract", but which was also justified in part by the rather excessive enthusiasm of their author (so frequent in mathematics among zealots of new theories). Not content to apply his theory fruitfully to algebraic numbers, Hensel, impressed as all his contemporaries were, by the proofs of the transcendence of e and π and perhaps misled by the adjective "transcendental" applied both to numbers and to functions, had come to think that there existed a connection between his p-adic numbers and transcendental real numbers and he had thought for a moment that he had obtained a simple proof of the transcendence of e and even of e^e ([18d], p. 556) (*).

Soon after 1910, the situation changes, with the rising of the next generation, influenced by the ideas of Fréchet and F. Riesz on topology and by those of Steinitz on algebra, and from the start devoted to "abstraction"; it will know how to assimilate and put in their true place Hensel's works. As early as 1913, Kürschak [22] gives a general definition of the notion of absolute value, recognizes the importance of ultrametric absolute values (of which the p-adic absolute value was an example), proves (by modelling the proof on the case of real numbers) the existence of the completion of a field with respect to an absolute value and above all shows generally the possibility of extending an absolute value to any algebraic extension of the given field. But he had not seen that the ultrametric character of an absolute value was already revealed in the prime field; this point was established by Ostrowski, to whom also is due the determination of all the absolute values on the field $\mathbf{Q}$ and the fundamental theorem characterizing fields with a non-ultrametric absolute value as subfields of $\mathbf{C}$ [24]. In the years from 1920 to 1935, the theory will be completed by a more detailed study of absolute values which are not necessarily discrete, including amongst others the examination of various circumstances which arise in passing to an algebraic or transcendental extension (Ostrowski, Deuring, F. K. Schmidt); on the other hand, in 1931, Krull introduces and studies the general notion of valuation [29b] which will be greatly used in the years that follow by Zariski and his school of Algebraic Geometry (†). We must also mention here, although it lies outside our scope, the deeper studies on the structure of complete valued fields and complete local rings, which date from the same period (Hasse-Schmidt, Witt, Teichmüller, I. Cohen).

* * *

The work of Fraenkel mentioned above (p. 594) only treated a very special

(*) This research at all cost of a narrow parallelism between p-adic series and Taylor series also leads Hensel to pose himself strange problems: he proves for example that every p-adic integer may be written in the form of a series $\sum_{k=0}^{\infty} a_k p^k$ where the a_k are rational numbers chosen so that the series converges not only in $\mathbf{Q}_p$, but also in $\mathbf{R}$ (perhaps by analogy with Taylor series which converge at several places at once?) ([16e] and [16f]).

(†) An example of a valuation of height 2 had already been introduced incidentally by H. Jung in 1925 [27].

type of ring (Artinian with only a single prime ideal, which is moreover assumed to be principal). With the exception of Steinitz's work on fields [20a], the first important works on the study of general commutative rings are E. Noether's two great memoirs on ideal theory: that of 1921 [25a], devoted to primary decomposition, which takes up again in all generality and completes on many points the results of Lasker and Macaulay; and that of 1927 characterizing Dedekind domains axiomatically [25b]. Just as Steinitz had shown for fields, it is seen in these memoirs how a small number of abstract ideas, such as the notion of irreducible ideal, the chain conditions and the idea of an integrally closed domain (the last two, as we have seen, already brought to light by Dedekind) can by themselves lead to general results which seemed inextricably bound up with results of pure computation in the cases where they had previously been known.

With these memoirs of E. Noether, joined to the slightly later works of Artin-van der Waerden on divisorial ideals [31] and Krull relating these ideals to essential valuations [29b], the long study of the decomposition of ideals started a century earlier (*) is thus complete, at the same time as modern Commutative Algebra is being inaugurated.

The innumerable later research works on Commutative Algebra are grouped most easily according to several important directions of development:

(A) *Local rings and topologies*

Although the germ was contained in all the earlier works on the Theory of Numbers and Algebraic Geometry, the general idea of localization came to light very slowly. The general notion of ring of fractions is only defined in 1926 by H. Grell, a pupil of E. Noether, and only for integral domains [28]; its extension to more general rings will only be given in 1944 by C. Chevalley for Noetherian rings and in 1948 by Uzkov in the general case. Until about 1940, Krull and his school are practically alone in using in general arguments the consideration of the local rings $A_\mathfrak{p}$ of an integral domain A; these rings will only begin to appear explicitly in Algebraic Geometry with the works of Chevalley and Zariski starting in 1940 (†).

The general study of local rings themselves only begins in 1938 with Krull's

(*) Following the definition of divisorial ideals, a considerable number of research works (Prüfer, Krull, Lorenzen, etc.) were undertaken on ideals which are *invariant* under other operations $\mathfrak{a} \mapsto \mathfrak{a}'$ satisfying axiomatic conditions analogous to the properties of the operation $\mathfrak{a} \mapsto A : (A : \mathfrak{a})$ which gives birth to divisorial ideals; the results obtained in this way have as yet found no application in Algebraic Geometry nor in the Theory of Numbers.

(†) In the works of Hensel and his pupils on the Theory of Numbers, the local rings $A_\mathfrak{p}$ are systematically neglected to the benefit of their completions, no doubt because of the possibility of applying Hensel's lemma to the latter.

great memoir [29d]. The most important results of this work concern dimension theory and regular rings, of which we shall not speak here; but here for the first time appears the completion of any arbitrary Noetherian local ring and also a still imperfect form of the graded ring associated with a local ring (*); the latter will only be defined about 1948 by P. Samuel [36] and independently in research on Algebraic Topology by Leray and H. Cartan. Krull, in the above mentioned work, hardly uses topological language; but already in 1928 [29a], he had proved that, in a Noetherian ring A, the intersection of the powers of an ideal $\mathfrak{a}$ is the set of $x \in A$ such that $x(1 - a) = 0$ for some $a \in \mathfrak{a}$; it is easily deduced from this that, for every ideal $\mathfrak{m}$ of A, the $\mathfrak{m}$-adic topology on A induces on an ideal $\mathfrak{a}$ the $\mathfrak{m}$-adic topology on $\mathfrak{a}$; in his memoir of 1938, Krull completes this result by proving that in a Noetherian local ring every ideal is closed. These theorems were soon afterwards extended by Chevalley to Noetherian semi-local rings and then by Zariski to the rings which bear his name [33b]; to Chevalley also goes back the introduction of "linear compactness" in topological rings, as also the determination of the structure of complete semi-local rings [32b].

(B) *Passage from the local to the global*

Since Weierstrass, an analytic function of one variable (and in particular an algebraic function) has habitually been associated with the set of its "expansions" at all the points of the Riemann surface where it is defined. In the introduction to his book on the Theory of Numbers ([18f], p. V), Hensel similarly associates with each element of an algebraic number field k the set of elements corresponding to it in the *completions* of k with respect to *all* the absolute values on k (†). It may be said that it is this point of view which, in modern Commutative Algebra, has replaced the decomposition formula of an ideal as a product of prime ideals (extending in a certain sense Kummer's initial point of view). Hensel's remark amounts implicitly to embedding k in the *product* of all its completions; this is what Chevalley does explicitly in 1936 with his theory of

---

(*) If $\mathfrak{m}$ is the maximal ideal of the Noetherian local ring A under consideration and $(\alpha_i)_{1 \leqslant i \leqslant r}$ a minimal system of generators of $\mathfrak{m}$, Krull defines for $x \neq 0$ in A the "initial forms" of x as follows: if j is the greatest integer such that $x \in \mathfrak{m}^j$, the initial forms of x are all the homogeneous polynomials of degree j, $P(X_1, \ldots, X_r)$ with coefficients in the residue field $k = A/\mathfrak{m}$, such that $x \equiv P(\alpha_1, \ldots, \alpha_r) \pmod{\mathfrak{m}^{j+1}}$. With each ideal $\mathfrak{a}$ of A he associates the graded ideal of $k[X_1, \ldots, X_r]$ generated by the initial forms of all the elements of $\mathfrak{a}$ ("Leitideal"); these two notions for him take the place of the associated graded ring.

(†) Hensel takes, as non-ultrametric absolute values on a field K of degree n over $\mathbf{Q}$, the functions $x \mapsto |x^{(i)}|$ (where the $x^{(i)}$ for $1 \leqslant i \leqslant n$ are the conjugates of x) currently used since Dirichlet; Ostrowski was to show a little later that these functions are essentially the only non-ultrametric absolute values on K.

"idèles" [32a], which perfects earlier analogous ideas of Prüfer and von Neumann (the latter confining themselves to embedding k in the product of its $\mathfrak{p}$-adic completions) (*). Although this is somewhat outside our scope, it is important to mention here that, thanks to an appropriate topology on the group of idèles, all the techniques of locally compact groups (including Haar measure) can thus be very effectively applied to the Theory of Numbers.

In a more general context, Krull's theorem [29b] characterizing an integrally closed domain as an intersection of valuation rings (which amounts also to embedding the domain under consideration in a product of valuation rings) often facilitates the study of these rings, although the method is only really tractable for essential valuations of Krull domains. Moreover Krull frequently exhibits [29e] (quite elementary) examples of the "passage from the local to the global" method consisting of showing a property of an integral domain A by verifying it for the "localized" rings $A_{\mathfrak{p}}$ of A at all its prime ideals (†); more recently, Serre perceived that this method works for arbitrary commutative rings A, that it is applicable also to A-modules and that it is even sufficient often to "localize" at the maximal ideals of A (Chapter II, § 3, Theorem 1): a point of view which is closely connected with ideas about "spectra" and sheaves defined over these spectra (see below, p. 602).

(C) *Integers and integral closure*

We have seen that the notion of algebraic integer, first introduced for number fields, had already been extended by Kronecker and Dedekind to algebraic function fields, although in this case it might appear rather artificial (not corresponding to a projective notion). E. Noether's memoir of 1927, followed by the work of Krull starting in 1931, showed the interest that these notions present for more general rings (‡). Krull in particular is responsible for the

(*) Because of this remark by Hensel, the non-ultrametric absolute values on a number field K have habitually been called (by an abuse of language) the "places at infinity" of K, by analogy with the process by which Dedekind and Weber define the "points at infinity" of the Riemann surface of an affine curve (cf. p. 591).

(†) In speaking of the "passage from the local to the global", there is often an allusion to much more difficult questions, connected with class field theory, and the best known examples of which are those treated in Hasse's memoirs ([26a] and [26b]) on quadratic forms over an algebraic number field k; he shows there among other things that for an equation $f(x_1, \ldots, x_n) = a$ to have a solution in k^n (f a quadratic form, $a \in k$), it is necessary and sufficient that it have a solution in each of the completions of k. According to Hasse, the idea of this type of theorem had been suggested to him by his master Hensel [26c]. The extension of this "principle of Hasse" to groups other than the orthogonal group is one of the objectives of the modern theory of "adelizations" of algebraic groups.

(‡) Krull and E. Noether limit themselves to integral domains, but the extension of their methods to the general case is not difficult; the most interesting memoir theorems on the lifting of prime ideals to integral algebras [29c], as also for extending the theory of decomposition and inertia groups of Dedekind-Hilbert [29b]. As for E. Noether, we owe to her the general formulation of the normalization lemma (*) (from which follows amongst other things Hilbert's Nullstellensatz) as also the first general criterion (transcribing the classical arguments of Kronecker and Dedekind) for the integral closure of an integral domain to be *finite* over that domain.

Finally, it should be pointed out here that one of the reasons for the present importance of the notion of integrally closed domain is due to Zariski's studies on algebraic varieties; he discovered that "normal" varieties (that is those whose local rings are integrally closed domains) are distinguished by particularly pleasant properties, notably the fact that they have no "singularity of codimension 1"; and it has then been seen that analogous phenomena are true for "analytic spaces". Therefore "normalization" (that is the operation which, for the local rings of a variety, consists of taking their integral closures) has become a powerful weapon in the arsenal of modern Algebraic Geometry.

(D) *The study of modules and the influence of Homological Algebra*

One of the striking characteristics of the work of E. Noether and W. Krull in Algebra is the tendency to "linearization", extending the analogous development given to field theory by Dedekind and Steinitz; in other words, ideals are considered above all as *modules* and so all the constructions of Linear Algebra (quotient, product and more recently tensor product and formation of homomorphism modules) are brought to bear on them, producing in general modules which are no longer ideals. It is thus quickly seen that in many questions (for commutative or non-commutative rings), interest should not be confined to the study of ideals of a ring A, but on the contrary the theorems should be stated in general for A-modules (sometimes subjected to certain finiteness conditions).

The intervention of Homological Algebra has strongly reinforced the above tendency, since this branch of Algebra is essentially concerned with questions of a *linear* nature. We shall not retrace its history here; but it is interesting to point out that several fundamental notions of Homological Algebra (such as that of projective module and that of Tor functor) came into being on the occasion of a close study of the behaviour of modules over a Dedekind domain relative to the tensor product, a study undertaken by H. Cartan in 1948.

---

on this subject is that where I. Cohen and Seidenberg extend Krull's lifting theorems, indicating exactly the limits of their validity [35]. It should be mentioned that E. Noether had explicitly mentioned the possibility of such generalizations in her memoir of 1927 ([25b], p. 30).

(*) A particular case had already been asserted by Hilbert in 1893 ([16b], p. 316).

Conversely, it could be foreseen that the new classes of modules introduced naturally by Homological Algebra as "universal annihilators" of the Ext functors (projective modules and injective modules) and the Tor functors (flat modules) would throw new light on Commutative Algebra. It happens that chiefly projective modules and still more flat modules have shown themselves useful: the importance of the latter arises above all from the remark, made first by Serre [38b], that localization and completion introduce flat modules naturally, thus "explaining" in a much more satisfactory way the properties of these operations already known and rendering them much easier to use. It should moreover be mentioned (as we shall see in later chapters) that the applications of Homological Algebra are far from being limited to this and that it is playing a more and more important role in Algebraic Geometry.

(E) *The notion of spectrum*

The most recent in date of the new notions of Commutative Algebra has a complex history. Hilbert's spectral theorem introduced ordered sets of orthogonal projectors of a Hilbert space, forming a "Boolean algebra" (or rather a *Boolean lattice*) (*), in one-to-one correspondence with a Boolean lattice of classes of measurable subsets (for a suitable measure) of $\mathbf{R}$. No doubt his earlier work on operators on Hilbert spaces, about 1935, led M. H. Stone to study Boolean lattices generally and notably to look for "representations" of them by subsets of a set (or classes of subsets with respect to a certain equivalence relation). He observes that a Boolean lattice becomes a *commutative ring* (moreover of a very special type), if multiplication is defined on it by $xy = \inf(x, y)$ and addition by $x + y = \sup(\inf(x, y'), \inf(x', y))$. In the particular case where the Boolean lattice in question is the set $\mathfrak{P}(X)$ of all subsets of a *finite* set X, it is immediately seen that the elements of X are in a natural one-to-one correspondence with the *maximal ideals* of the corresponding "Boolean" ring; and Stone obtains precisely his general representation theorem for a Boolean lattice by similarly considering the set of maximal ideals of the corresponding ring and associating with each element of the Boolean lattice the set of maximal ideals which contains it [30a].

On the other hand, the set of both open and closed subsets of a topological space was a well-known classical example of a Boolean lattice. In a second paper [30b], Stone showed that in fact *every* Boolean lattice is also isomorphic to a Boolean lattice of this nature. For this it was of course necessary to define a *topology* on the set of maximal ideals of a "Boolean" ring; which was very

(*) A *Boolean lattice* is a lattice-ordered set E, with a least element α and a greatest element ω, where each of the laws sup and inf is *distributive* with respect to the other and, for all $a \in E$, there exists a unique $a' \in E$ such that $\inf(a, a') = \alpha$ and $\sup(a, a') = \omega$ (cf. *Set Theory*, Chapter III, § 1, Exercise 17).

simply accomplished by taking as closed sets for each ideal $\mathfrak{a}$ the set of maximal ideals containing $\mathfrak{a}$.

We shall not speak here of the influence of these ideas on Functional Analysis, where they played an important role in the birth of the theory of normed algebras developed by I. Gelfand and his school. But in 1945, Jacobson observes [34] that the process of defining a topology, invented by Stone, can in fact be applied to *any* ring A (commutative or not) provided the set of ideals taken is not the set of maximal ideals but the set of two-sided "primitive" ideals (i.e. the two-sided ideals $\mathfrak{b}$ such that $A/\mathfrak{b}$ is a primitive ring); for a commutative ring, these of course turn out to be the maximal ideals. On his part, Zariski, in 1944 [33a], uses an analogous method to define a topology on the set of *places* of an algebraic function field. However, these topologies remained for the majority of algebraists mere curiosities, by reason of the fact that they are not usually Hausdorff and a quite understandable repugnance was felt about working on such unusual objects. This distrust was only overcome when A. Weil showed, in 1952, that every algebraic variety can be given a natural topology of the above type and that this topology allows the definition, in perfect analogy with the case of differentiable or analytic manifolds, of the notion of *fibre bundle* [37]; soon afterwards, Serre had the idea of extending to these varieties thus topologized the theory of *coherent sheaves*, thanks to which the topology renders in the case of "abstract" varieties the same services as the usual topology when the base field is $\mathbf{C}$, notably as far as applying the methods of Algebraic Topology is concerned ([38a] and [38b]).

From then on it was natural to use this geometric language throughout Commutative Algebra. It was quickly seen that considering maximal ideals is usually insufficient to obtain useful assertions (*) and that the adequate notion is that of the set of *prime* ideals of the ring topologized in the same manner. With the introduction of the notion of spectrum, there now exists a dictionary allowing every theorem of Commutative Algebra to be expressed in a geometric language very close to that of the Algebraic Geometry of the Weil-Zariski period; which has moreover immediately brought about a considerable enlargement of the scope of the latter, so that Commutative Algebra is scarcely more than the most elementary part of it [39].

(*) The inconvenience of limiting attention to the "maximal spectrum" arises from the fact that, if $\phi: A \to B$ is a ring homomorphism and $\mathfrak{n}$ a maximal ideal of B, $\overset{-1}{\phi}(\mathfrak{n})$ is not necessarily a maximal ideal of A, whereas for every prime ideal $\mathfrak{p}$ of B, $\overset{-1}{\phi}(\mathfrak{p})$ is a prime ideal of A. Hence in general a mapping of the set of maximal ideals of B to the set of maximal ideals of A cannot naturally be associated with ϕ.

BIBLIOGRAPHY

1. L. EULER, Vollständige Anleitung zur Algebra (= *Opera Omnia* (1), vol. I, Leipzig-Berlin (Teubner), 1911).
2. J. L. LAGRANGE, *Oeuvres*, 14 volumes, Paris (Gauthier-Villars), 1867–1892.
3. C. F. GAUSS, *Werke*, 12 volumes, Göttingen, 1870–1927.
4. P. G. LEJEUNE-DIRICHLET, *Werke*, 2 volumes, Berlin (Reimer), 1889–1897.

4 (*bis*). P. G. LEJEUNE-DIRICHLET, Vorlesungen über Zahlentheorie, 2te Aufl., Braunschweig (Vieweg), 1871.

5. C. G. J. JACOBI, *Gesammelte Werke*, 7 volumes, Berlin (Reimer), 1881–1891.
6. G. EISENSTEIN: (a) Beweis der Reciprocitätsgesetze für die cubischen Reste in der Theorie der aus dritten Wurzeln der Einheit zusammengesetzen Zahlen, *Crelle's Journal*, **27** (1844), pp. 289–310; (b) Zur Theorie der quadratischen Zerfällung der Primzahlen $8n + 3$, $7n + 2$ und $7n + 4$, *Crelle's Journal*, **37** (1848), pp. 97–126; (c) Über einige allgemeine Eigenschaften der Gleichung von welcher die Teilung der ganzen Lemniscate abhängt, nebst Anwendungen derselben auf die Zahlentheorie, *Crelle's Journal*, **39** (1850), pp. 160–179 and 224–287.
7. E. KUMMER: (a) Sur les nombres complexes qui sont formés avec les nombres entiers réels et les racines de l'unité, *J. de Math.*, (1), **12** (1847), pp. 185–212; (b) Zur Theorie der complexen Zahlen, *Crelle's Journal*, **35** (1847), pp. 319–326; (c) Ueber die Zerlegung der aus Wurzeln der Einheit gebildeten complexen Zahlen in Primfactoren, *Crelle's Journal*, **35** (1847), pp. 327–367; (d) Mémoire sur les nombres complexes composés de racines de l'unité et des nombres entiers, *J. de Math.*, (1), **16** (1851), pp. 377–498; (e) Über die allgemeinen Reciprocitätsgesetze unter den Resten und Nichtresten der Potenzen deren Grad eine Primzahl ist (*Abh. der Kön. Akad. der Wiss. zu Berlin* (1859), Math. Abhandl., pp. 19–159).
8. C. HERMITE, *Oeuvres*, 4 Volumes, Paris (Gauthier-Villars), 1905–1917.
9. L. KRONECKER, *Werke*, 5 volumes, Leipzig (Teubner), 1895–1930: (a) De unitatibus complexis, vol. I, pp. 5–71 (= *Inaug. Diss.*, Berolini, 1845); (b) Über die algebraisch auflösbaren Gleichungen I, vol. IV, pp. 1–11 (= *Monatsber. der Kön. Preuss. Akad. der Wiss.*, 1853, pp. 365–374); (c) Über die elliptischen Functionen für welche complexe Multiplication stattfindet vol. IV, pp. 177–183 (= *Monatsber. der Kön. Preuss. Akad. der Wiss.*, 1857, pp. 455–460); (d) Über die complexe Multiplication der elliptischen Functionen, vol. IV, pp. 207–217 (= *Monatsber. der Kön. Preuss. Akad. der Wiss.*, 1862, pp. 363–372); (e) Über die Discriminante algebraischer Functionen einer Variabeln, vol. II, pp. 193–236 (= *Crelle's Journal*, **91** (1881), pp. 301–334); (f) Grundzüge einer arithmetischen Theorie der

algebraischen Grössen, vol. II, pp. 237–387 (=*Crelle's Journal*, **92** (1882), pp. 1–122).

10. R. Dedekind, *Gesammelte mathematische Werke*, 3 volumes, Braunschweig (Vieweg), 1932: (a) Abriss einer Theorie der höheren Kongruenzen in bezug auf einen reellen Primzahl-Modulus, vol. I, pp. 40–66 (=*Crelle's Journal*, **54** (1857), pp. 1–26; (b) Sur la Théorie des Nombres entiers algébriques, vol. III, pp. 262–296 (=*Bull. Sci. Math.*, (1), **11** (1876), pp. 278–288 and (2), **1** (1877), pp. 17–41, 69–92, 144–164, 207–248); (c) Über die Anzahl der Ideal-Klassen in den verschiedenen Ordnungen eines endlichen Körpers, vol. I, pp. 105–157 (=*Festschrift der Technischen Hochschule in Braunschweig zur Säkularfeier des Geburtstages von C. F. Gauss*, Braunschweig, 1877, pp. 1–55); (d) Über den Zusammenhang zwischen der Theorie der Ideals und der Theorie der höheren Kongruenzen, vol. I, pp. 202–230 (=*Abh. Kön. Ges. Wiss. zu Göttingen*, **23** (1878), pp. 1–23); (e) Über die Discriminanten endlicher Körper, vol. I, pp. 351–396 (=*Abh. Kön. Ges. Wiss. zu Göttingen*, **29** (1882), pp. 1–56); (f) Über die Theorie der ganzen algebraischen Zahlen, vol. III, pp. 1–222 (=Supplement XI von Dirichlets *Vorlesungen über Zahlentheorie*, 4 Aufl. (1894), pp. 434–657); (g) Zur Theorie der Ideale, vol. II, pp. 43–48 (=*Nachr. Göttingen*, 1894, pp. 272–277); (h) Über eine Erweiterung des Symbols $(\mathfrak{a}, \mathfrak{b})$ in der Theorie der Moduln, vol. II, pp. 59–85 (=*Nachr. Göttingen*, 1895, pp. 183–208).

10 (*bis*). R. Dedekind-H. Weber, Theorie der algebraischen Funktionen einer Veränderlichen, *Crelle's Journal*, **92** (1882), pp. 181–290 (=R. Dedekind, *Ges. Math. Werke*, vol. I, pp. 238–349).

11. E. Selling, Ueber die idealen Primfactoren der complexen Zahlen, welche aus den Wurzeln einer beliebigen irreductiblen Gleichung rational gebildet sind, *Zeitschr. für Math. und Phys.*, **10** (1865), pp. 17–47.

12. M. Noether, Über einen Satz aus der Theorie der algebraischen Funktionen, *Math. Ann.*, **6** (1873), pp. 351–359.

13. A. Brill-M. Noether, Ueber algebraischen Funktionen, *Math. Ann.*, **7** (1874), pp. 269–310.

14. G. Zolotareff, Sur la théorie des nombres complexes, *J. de Math.* (3), **6** (1880), pp. 51–84 and 129–166.

15. E. Netto, Zur Theorie der Elimination, *Acta Math.*, **7** (1885), pp. 101–104.

16. D. Hilbert: (a) Über die Theorie der algebraischen Formen, *Math. Ann.*, **36** (1890), pp. 473–534; (b) Über die vollen Invariantensysteme, *Math. Ann.*, **42** (1893), pp. 313–373; (c) Grundzüge einer Theorie des Galoischen Zahlkörpers, *Gött. Nachr.*, (1894), pp. 224–236; (d) Zahlbericht, *Jahresber. der D. M. V.*, **4** (1897), pp. 175–546 (translated into French by A. Lévy and Th. Got under the title "*Théorie des corps de nombres algébriques*", Paris (Hermann), 1913).

17. K. Weierstrass, *Mathematische Werke*, 7 volumes, Berlin (Mayer und Müller), 1894–1927.

18. K. HENSEL: (a) Über eine neue Begründung der Theorie der algebraischen Zahlen, *Jahresber. der D. M. V.*, **6** (1899), pp. 83–88; (b) Ueber die Fundamentalgleichung und die ausserwesentlichen Diskriminantentheiler eines algebraischen Körpers, *Gött. Nachr.*, (1897), pp. 254–260; (c) Neue Grundlagen der Arithmetik, *Crelle's Journal*, **127** (1902), pp. 51–84; (d) Über die arithmetische Eigenschaften der algebraischen und transzendenten Zahlen, *Jahresber. der D. M. V.*, **14** (1905), pp. 545–558; (e) Ueber die arithmetischen Eigenschaften der Zahlen, *Jahresber. der D. M. V.*, **16** (1907), pp. 299–319, 388–393, 474–496; (f) *Theorie der algebraischen Zahlen*, Leipzig (Teubner), 1908.
19. E. LASKER, Zur Theorie der Moduln und Ideale, *Math. Ann.*, **60** (1905), pp. 20–116.
20. E. STEINITZ: (a) Algebraische Theorie der Körper, *Crelle's Journal*, **137** (1910), pp. 167–308; (b) Rechteckige Systeme und Moduln in algebraischen Zahlkörpern, *Math. Ann.* **71** (1912), pp. 328–354 and **72** (1912), pp. 297–345.
21. F. S. MACAULAY, On the resolution of a given modular system into primary systems including some properties of Hilbert numbers, *Math. Ann.*, **74** (1913), pp. 66–121.
22. J. KÜRSCHAK, Über Limesbildung und allgemeine Körpertheorie, *Crelle's Journal*, **142** (1913), pp. 211–253.
23. A. FRAENKEL, Über die Teiler der Null und die Zerlegung von Ringen, *Crelle's Journal*, **145** (1914), pp. 139–176.
24. A. OSTROWSKI, Über einige Lösungen der Funktionalgleichung $\phi(x)\phi(y) = \phi(x.y)$, *Acta Math.*, **41** (1917), pp. 271–284.
25. E. NOETHER: (a) Idealtheorie in Ringbereichen, *Math. Ann.*, **83** (1921), pp. 24–66; (b) Abstrakter Aufbau der Idealtheorie in algebraischen Zahl- und Funktionenkörpern, *Math. Ann.*, **96** (1927), pp. 26–61.
26. H. HASSE: (a) Ueber die Darstellbarkeit von Zahlen durch quadratischen Formen im Körper der rationalen Zahlen, *Crelle's Journal*, **152** (1923), pp. 129–148; (b) Ueber die Äquivalenz quadratischer Formen im Körper der rationalen Zahlen, *Crelle's Journal*, **152** (1923), pp. 205–224; (c) Kurt Hensels entscheidender Anstoss zur Entdeckung des Lokal-Global-Prinzips, *Crelle's Journal*, **209** (1960), pp. 3–4.
27. H. JUNG, *Algebraischen Flächen*, Hannover (Helwing), 1925.
28. H. GRELL, Beziehung zwischen den Idealen verschiedener Ringe, *Math. Ann.*, **97** (1927), pp. 490–523.
29. W. KRULL: (a) Primidealketten in allgemeine Ringbereichen, *Sitz. Ber. Heidelberg Akad. Wiss.*, 1928; (b) Allgemeine Bewertungstheorie, *Crelle's Journal*, **167** (1931), pp. 160–196; (c) Beiträge zur Arithmetik kommutativer Integritätsbereiche, III, *Math. Zeitschr.*, **42** (1937), pp. 745–766; (d) Dimensionstheorie in Stellenringen, *Crelle's Journal*, **179** (1938), pp. 204–226; (e) *Idealtheorie*, Berlin (Springer), 1935.

30. M. H. Stone: (a) The theory of representations for Boolean algebras, *Trans. Amer. Math. Soc.*, **40** (1936), pp. 37–111; (b) Applications of the theory of Boolean rings to general topology, *Trans. Amer. Math. Soc.*, **41** (1937), pp. 375–481.

31. B. L. van der Waerden, *Moderne Algebra*, vol. II, Berlin (Springer), 1931.

32. C. Chevalley: (a) Généralisation de la théorie du corps de classes pour les extensions infinies, *J. de Math.*, (9), **15** (1936), pp. 359–371; (b) On the theory of local rings, *Ann. of Math.*, **44** (1943), pp. 690–708.

33. O. Zariski: (a) The compactness of the Riemann manifold of an abstract field of algebraic functions, *Bull. Amer. Math. Soc.*, **50** (1944), pp. 683–691; (b) Generalized semi-local rings, *Summa Bras. Math.*, **1** (1946), pp. 169–195.

34. N. Jacobson, A topology for the set of primitive ideals in an arbitrary ring, *Proc. Nat. Acad. Sci. U.S.A.*, **31** (1945), pp. 333–338.

35. I. Cohen-A.Seidenberg, Prime ideals and integral independence, *Bull. Amer. Math. Soc.*, **52** (1946), pp. 252–261.

36. P. Samuel, La notion de multiplicité en Algèbre et en Géométrie algébrique, *J. de Math.*, (9), **30** (1951), pp. 159–274.

37. A. Weil, *Fibre-spaces in Algebraic Geometry* (Notes by A. Wallace), Chicago Univ., 1952.

38. J. P. Serre: (a) Faisceaux algébriques cohérents, *Ann. of Math.*, **61** (1955), pp. 197–278; (b) Géométrie algébrique et géométrie analytique, *Ann. Inst. Fourier*, **6** (1956), pp. 1–42.

39. A. Grothendieck, *Éléments de géométrie algébrique*, Publ. math. Inst. Htes Et. Scient., 1960.

INDEX OF NOTATION

The reference numbers indicate the chapter, section and sub-section (or exercise) in that order.

1_E (E a set), U . V, UV (U, V additive subgroups), $\mathfrak{a}^0$ ($\mathfrak{a}$ an ideal): Preliminary conventions of Chapter I
E:F: I.2.10
$A[S^{-1}]$, a/s (A a ring, S a subset of A, $a \in A$, s a product of elements of S): II.2.1 i_A^S: II.2.1
$S^{-1}A$, $A_\mathfrak{p}$ (S a multiplicative subset, $\mathfrak{p}$ a prime ideal): II.2.1
$M[S^{-1}]$, m/s, i_M^S (M an A-module, S a subset of A, $m \in M$, s a product of elements of S): II.2.2
$S^{-1}M$, $M_\mathfrak{p}$ (M an A-module, S a multiplicative subset of A, $\mathfrak{p}$ a prime ideal of A): II.2.2
$S^{-1}u$, $u_\mathfrak{p}$ (u an A-module homomorphism): II.2.2
$\mathfrak{r}(\mathfrak{a})$ ($\mathfrak{a}$ an ideal): II.2.6
V(M), V(f) (M a subset of the ring A, $f \in A$): II.4.3
Spec(A): II.4.3
X_f ($f \in A$, X = Spec(A)): II.4.3
$\mathfrak{J}(Y)$ (Y a subset of Spec(A)): II.4.3
${}^a h$ (h a ring homomorphism): II.4.3
Supp(M) (M an A-module): II.4.4
A_f, M_f, u_f (A a ring, M an A-module, u an A-homomorphism, $f \in A$): II.5.1
$rg_\mathfrak{p}(P)$ (P a projective module): II.5.3
rg(P) (P a projective module): II.5.3
P(A), cl(M) (A a ring, M a projective A-module of rank 1): II.5.4
$\mathfrak{C}$, $\mathfrak{C}(A)$: II.5.7
det(u), χ_u (u an endomorphism of a projective module of rank n): II.5. Ex. 9
$A^{(d)}$, $M^{(d,k)}$, $M^{(d)}$ (A a graded ring, M a graded A-module): III.1.3
$A_{(\mathfrak{p})}$, $M_{(\mathfrak{p})}$ (A a graded ring, $\mathfrak{p}$ a graded prime ideal of A, M a graded A-module): III.1.4

$\mathrm{gr}_n(\mathrm{G})$, $\mathrm{gr}(\mathrm{G})$ (G a filtered group): III.2.3
$\mathrm{gr}(h)$ (h a homomorphism compatible with the filtrations): III.2.4
$\mathbf{Z}_n$ (n an integer >1): III.2.12
$\hat{\mathbf{Z}}$: III.2.13
$\mathrm{A}\{\mathrm{X}_1, \ldots, \mathrm{X}_p\}$ (A a linearly topologized ring): III.4.2
$f(b_1, \ldots, b_p)$ (f a restricted formal power series): III.4.2
$\mathbf{f} \circ \mathbf{g}$, $M_\mathbf{f}$, $M_\mathbf{f}(\mathbf{X})$, $\mathrm{J}_\mathbf{f}$, $\mathrm{J}_\mathbf{f}(\mathbf{X})$, $\mathbf{X}$, $\mathbf{1}_n$ ($\mathbf{f}$, $\mathbf{g}$ systems of formal power series, $\mathbf{g}$ without constant term): III.4.4
$\mathbf{f}(\mathbf{x})$ ($\mathbf{f}$ a system of formal power series, $\mathbf{x}$ a system of topological nilpotent elements): III.4.5
$\mathfrak{m}^{\times n}$ ($\mathfrak{m}$ an ideal): III.4.5
$\mathrm{Ass}_\mathrm{A}(\mathrm{M})$, $\mathrm{Ass}(\mathrm{M})$ (M an A-module): IV.1.1
$\mathrm{Ass}_f(\mathrm{M})$: IV.1. Ex. 17
$\mathrm{A}^\mathscr{G}$ (A an algebra, $\mathscr{G}$ a group operating on A): V.1.9
$\mathscr{G}^\mathrm{Z}(\mathfrak{p}')$, $\mathscr{G}^\mathrm{Z}$, $\mathrm{A}^\mathrm{Z}(\mathfrak{p}')$, A^Z ($\mathscr{G}$ a group operating on a ring A′, $\mathfrak{p}'$ a prime ideal of A′): V.2.2
$\mathscr{G}^\mathrm{T}(\mathfrak{p}')$, $\mathscr{G}^\mathrm{T}$, $\mathrm{A}^\mathrm{T}(\mathfrak{p}')$, A^T ($\mathscr{G}$ a group operating on a ring A′, $\mathfrak{p}'$ a prime ideal of A′): V.2.2
$\mathrm{K}^\mathrm{Z}(\mathfrak{p}')$, K^Z, $\mathrm{K}^\mathrm{T}(\mathfrak{p}')$, K^T (K the field of fractions of an integrally closed domain A, $\mathfrak{p}'$ a prime ideal of the integral closure of A in a quasi-Galois extension of K): V.2.3
$\mathrm{Y}^\mathbf{p}$ (where $\mathbf{p} = (p_1, \ldots, p_m)$, the p_i being integers $\geqslant 0$): V.3.1
$\mathfrak{m}(\mathrm{A})$, $\kappa(\mathrm{A})$, $\mathrm{U}(\mathrm{A})$ (A a local ring): VI
$\tilde{\mathrm{K}}$, ∞: VI.2.1
$+\infty$: VI.3.1
Γ_A, v_A: VI.3.2
$\mathfrak{a}(\mathrm{M})$ (M a major set): VI.3.5
$h(\mathrm{G})$ (G a totally ordered group): VI.4.4
$\mathscr{T}_v$ (v a valuation): VI.5.2
$e(v'/v)$, $e(\mathrm{A}'/\mathrm{A})$, $e(\mathrm{L}/\mathrm{K})$: VI.8.1
$f(v'/v)$, $f(\mathrm{A}'/\mathrm{A})$, $f(\mathrm{L}/\mathrm{K})$: VI.8.1
$\varepsilon(\mathrm{G}, \mathrm{H})$ (G a totally ordered group, H a subgroup of G of finite index): VI.8.4
$\varepsilon(v'/v)$ (v a valuation, v' an extension of v): VI.8.4
$\mathrm{mod}(x)$, $\mathrm{mod}_\mathrm{K}(x)$ (K a non-discrete locally compact field, $x \in \mathrm{K}$): VI.9.1
$r(\mathrm{G})$ (rational rank of a commutative group): VI.10.2
$d(\mathrm{K}'/\mathrm{K})$, $s(v'/v)$, $r(v'/v)$ (v a valuation on K, v' an extension of v to a transcendental extension K′ of K): VI.10.3
$\mathrm{I}(\mathrm{A})$, $\mathrm{D}(\mathrm{A})$ (A an integral domain): VII.1.1
$\mathfrak{a} \prec \mathfrak{b}$, $\mathrm{div}(\mathfrak{a})$, $\mathrm{div}(x)$ ($\mathfrak{a}$, $\mathfrak{b}$ fractional ideals, x an element of the field of fractions): VII.1.1
$\tilde{\mathfrak{a}}$ ($\mathfrak{a}$ a fractional ideal): VII.1.1
$d_1 \leqslant d_2$ (d_1, d_2 divisors): VII.1.1

$\mathfrak{b} : \mathfrak{a}$ ($\mathfrak{a}$, $\mathfrak{b}$ fractional ideals): VII.1.1
J(A) (A an integral domain): VII.1.2
P(A) (A a Krull domain): VII.1.3
$\mathfrak{p}^{(n)}$ ($\mathfrak{p}$ a divisorial prime ideal): VII.1.4
$v_{\mathfrak{p}}$ ($\mathfrak{p}$ a prime ideal of height 1 in a Krull domain): VII.1.10
F(A), C(A) (A a Krull domain): VII.1.10
$e(\mathfrak{P}/\mathfrak{p})$ ($\mathfrak{p} \in$ P(A), $\mathfrak{P} \in$ P(B), A and B Krull domains, A $\subset$ B, $\mathfrak{P} \cap$ A $= \mathfrak{p}$): VII.1.10
i (homomorphism from D(A) to D(B), or of C(A) to C(B)): VII.1.10
$\bar{i}$ (homomorphism from C(A) to C(B)): VII.1.10
A, A_0, Δ(K) (rings of restricted adèles): VII.2.4
$\mathfrak{P}^*$, $\mathfrak{P}^*$(A) (A an integral domain): VII.3.2
M* (dual lattice of a lattice M): VII.4.2
$l_{\mathfrak{p}}$(T), χ(T) (T a torsion A-module, $\mathfrak{p}$ a prime ideal of height 1): VII.4.5
F(A), T(A), cl(M): VII.4.5
χ(M, M′) (M, M′ lattices): VII.4.6
$c(d)$ (d a divisor): VII.4.7
c(M), r(M), γ(M) (M a lattice): VII.4.7
$\mathfrak{P}|\mathfrak{p}$, $e_{\mathfrak{P}/\mathfrak{p}}$, $f_{\mathfrak{P}/\mathfrak{p}}$, $f(\mathfrak{P}/\mathfrak{p})$ (A $\subset$ B Krull domains such that B is a finite A-algebra $\mathfrak{p} \in$ P(A), $\mathfrak{P} \in$ P(B), $\mathfrak{P} \cap$ A $= \mathfrak{p}$): VII.4.8
$N_{B/A}$, N, $i_{B/A}$: VII.4.8

INDEX OF TERMINOLOGY

The reference numbers indicate the chapter, section and sub-section (or exercise) in that order.

Adèle, restricted, principal restricted adèle: VII.2.4
$\mathfrak{m}$-adic filtration: III.2.1
— topology: III.2.5
n-adic integers: III.2.12
Algebra, Azumaya: II.1.7
— finitely generated: III.1.1
— integral, finite, over a ring: V.1.1
Algebraic closure of a field in an algebra: V.1.2
Algebraically closed field in an algebra: V.1.2
— dependent, independent, elements: III.1.1
— free, related, family: III.1.1
Almost all $\mathfrak{p} \in P(A)$ (property valid for): VII.4.3
— nilpotent endomorphism: IV.1.4
Approximation theorem for absolute values: VI.7.3
— theorem for valuations: VI.7.2
Artin-Rees Lemma: III.3.1
Associated (filtered module) with a graded module: III.2.1
— (filtered ring) with a graded ring: III.2.1
— (filtration) with a graduation: III.2.1
— (graded homomorphism) with a homomorphism compatible with filtrations: III.2.4
— (graded module) with a filtered module: III.2.3
— (graded ring) with a filtered ring: III.2.3
— (mapping) with a ring homomorphism: II.4.3
— (prime ideal) with a module: IV.1.1
— ring, place, valuation: VI.3.3

Canonical decomposition of a place: VI.2.3
— factorization of a valuation: VI.3.2
— homomorphism of the decomposition group of a prime ideal $\mathfrak{p}'$ of A′ into the automorphism group of $A'/\mathfrak{p}'$: V.2.2
Class, divisor, attached to a finitely generated module: VII.4.7
Classes, divisor (monoid of): VII.1.2
Closure, algebraic, of a field in an algebra: V.1.2
— integral, of an integral domain: V.1.2
— integral, of a ring in an algebra: V.1.2
Commutative diagram: I.1.2
Compatible (filtration) with a ring structure, module structure: III.2.1
— (homomorphism) with filtrations: III.2.4
Complete system of extensions of a valuation: VI.8.2
Completely integrally closed domain: V.1.4
Component, irreducible (of a topological space): II.4.1
Conditions, Hensel's: III.4.5
Conductor of a submodule: V.1.5
Content of a polynomial over a pseudo-Bezoutian domain: VII.1.Ex.23
— of a torsion module: VII.4.5
Criterion, Eisenstein's irreducibility: VII.3.Ex.20

Decomposition, canonical, of a place: VI.2.3
— complete, of a prime ideal: V.2.2
— field of a prime ideal: V.2.2
— group, ring, of a prime ideal: V.2.2
— of an ideal in a Dedekind domain into prime factors: VII.2.3
— primary: IV.2.2 and Ex.20
— reduced primary: IV.2.3 and Ex.20
Decreasing filtration: III.2.1
Dedekind domain: VII.2.1
Defined (topology) by a filtration: III.2.5
Defining ideal: III.3.2
Degree, residue class, of one valuation over another: VI.8.1
Dependence, integral (equation of): V.1.1
Derived (module filtration) from a ring filtration: III.2.1
Diagram, commutative: I.1.2
— snake: I.1.4
Discrete filtration: III.2.1
— valuation: VI.3.6
Distinguished polynomial: VII.3.8
Divisor, principal divisor: VII.1.1
— determinantal: VII.4.Ex.11
— finitely generated: VII.1.Ex.11

Divisorial fractional ideal: VII.1.1
Divisors, equivalent: VII.1.2
Domain, Bezoutian (or Bezout): VII.1.Ex.20
— completely integrally closed: V.1.4
— Dedekind: VII.2.1
— factorial: VII.3.1
— integrally closed: V.1.2
— integrally closed, of finite character: VII.1.Ex.25, 26 and 28
— integrally Noetherian: V.3.Ex.6
— Krull: VII.1.3
— local integral, of dimension 1: VI.4.Ex.7
— Prüferian (or Prüfer): VII.2.Ex.12
— pseudo-Bezoutian: VII.1.Ex.21
— pseudo-principal: VII.1.Ex.21
— pseudo-Prüferian: VII.2.Ex.19
— regularly integrally closed: VII.1.Ex.30
Dominating (local ring) a local ring: VI.1.1
Dual, algebraic toric, of a module: VI.5.Ex.9
— lattice: VII.4.2
— topological toric: VI.5.Ex.10

Element, topologically nilpotent: III.4.3
Elements, algebraically dependent, independent: III.1.1
— strongly relatively prime: III.4.1
Endomorphism, almost nilpotent: IV.1.4
Equivalent divisors: VII.1.2
— valuations: VI.3.2
Essential graded ideal: III.1.4
— valuations: VII.1.4
Euclidean ordered field: VI.2.Ex.4
Exhaustive filtration: III.2.1
Extension, quasi-Galois: V.2.2

Factor, invariant: VII.4.Ex.11 and 14
Factorial domain: VII.3.1
Factorization, canonical, of a valuation: VI.3.2
Faithfully flat module: I.3.1
Family, algebraically free, related: III.1.1
— formally free: III.2.9
Field, algebraically closed in an algebra: V.1.2
— decomposition: V.2.3
— projective: VI.2.1
— residue, of a local ring: II.3.1

Field, residue, of a place: VI.2.3
— residue, of a valuation: VI.3.2
— value, of a place: VI.2.2
Filtered group, ring, module: III.2.1
Filtration, $\mathfrak{m}$-adic: III.2.1
— associated with a graduation: III.2.1
— compatible with a ring structure, module structure: III.2.1
— discrete: III.2.5
— $\mathfrak{m}$-good: III.3.1
— increasing, decreasing, separated, exhaustive: III.2.1
— induced, product, quotient: III.2.1
— module, derived from a ring filtration: III.2.1
— trivial: III.2
Finite algebra over a ring: V.1.1
— (place) at an element: VI.2.2
Finitely generated algebra: III.1.1
Finitely presented: I.2.8
Flat for M, M-flat (module): I.2.2
— module: I.2.3
Formally free family: III.2.9
Fractional ideal: VII.1.1
Function, order: III.2.2

Gaussian integer: V.1.1
Gauss's lemma: VII.3.5
Gelfand-Mazur Theorem: VI.6.4.
Generated by a subset (multiplicative subset): II.2.1
Generators, formal system of: III.2.9
$\mathfrak{m}$-good filtration: III.3.1
Group, decomposition: V.2.2
— filtered: III.2.1
— inertia: V.2.2
— of classes of invertible modules: II.5.7
— of operators, locally finite: V.1.9
— operating on a ring: V.1.9
— order, of a valuation: VI.3.2
— ordered, of height n, of height $+\infty$: VI.4.4

Height $\leqslant 1$ (prime ideal of): VII.1.6
— of an ordered group, of a valuation: VI.4.4
Henselian ring: III.4.Ex.3
Hensel's conditions: III.4.5
— Theorem: III.4.3

Homomorphism, canonical, of the decomposition group of a prime ideal $\mathfrak{p}'$
— from A' to the automorphism group of A'/$\mathfrak{p}'$: V.2.2
— compatible with filtrations: III.2.4
— graded, associated with a homomorphism compatible with filtrations: III.2.4
— local: II.3.1
— pseudo-injective, pseudo-surjective, pseudo-zero, pseudo-bijective: VII.4.4

Ideal, determinantal: VII.4.Ex.10 and 14
— essential graded: III.1.4
— immersed prime: IV.2.3
— integral, fractional ideal: VII.1.1
— invertible fractional: II.5.7
— lying above an ideal: V.2.1
— minimal prime: II.2.6
— of a place: VI.2.3
— of a valuation: VI.3.2
— prime: II.1.1
— prime, associated with a module: IV.1.1
— prime, decomposing completely: V.2.2
— prime, of height $\leqslant 1$: VII.1.6
— primary, $\mathfrak{p}$-primary: IV.2.1 and Ex.20
— unramified: V.2.Ex.18 and 19
Ideally Hausdorff module: III.5.1
Ideals, relatively prime: II.1.2
Identity, Cauchy's: VII.3.Ex.18
Immersed prime ideal: IV.2.3
Improper valuation: VI.3.1
Increasing filtration: III.2.1
Independent valuation rings: VI.7.2
— valuations: VI.7.2
Index, initial, of a subgroup of an ordered group, initial ramification index of a valuation: VI.8.4
— ramification: VI.8.1
Induced filtration: III.2.1
Induction, Noetherian (principle of): II.4.2
Inertia field: V.2.3
— ring, group: V.2.2
Initial ramification index: VI.8.4
Integer, algebraic: V.1.1
— Gaussian: V.1.1
— over a ring: V.1.1

Integers, *n*-adic: III.2.12
Integral algebra over a ring: V.1.1
— closure: V.1.2
— ideal: VII.1.1
Integrally closed domain: V.1.2
— closed in an algebra (ring): V.1.2
Invariant, relative, of one lattice with respect to another: VII.4.6
Invertible fractional ideal: II.5.7
— submodule: II.5.6
Irreducible component: II.4.1
— set: II.4.1
— space: II.4.1
Isolated subgroup: VI.4.2

Jacobson ring: V.3.4

Krull-Akizuki Theorem: VII.3.5
Krull domain: VII.1.3
Krull's Theorem: III.3.2

Laskerian module: IV.2.Ex.23
Lattice: VII.4.1
— dual: VII.4.2
— reflexive: VII.4.2
Lemma, Artin-Rees: III.3.1
— Gauss's: VII.3.5
— normalization: V.3.1
Linearly topologized ring: III.4.2
Local homomorphism: II.3.1
— ring: II.3.1
Locally finite group of operators: V.1.9
Lying above (ideal) an ideal: V.2.1

Major set: VI.3.5
Mapping, continuous, associated with a ring homomorphism: II.4.3
Minimal polynomial: V.1.3
— prime ideal: II.2.6
Module, coherent: I.2.Ex.11
— faithfully flat: I.3.1
— filtered: III.2.1
— filtered, associated with a graded module: III.2.1
— finitely presented: I.2.8
— flat: I.2.3
— flat for M, M-flat: I.2.2

Module graded, associated with a filtered module: III.2.3
— ideally Hausdorff: III.5.1
— of fractions defined by a subset of a ring: II.2.2
— projective, of rank n: II.5.3
— pseudo-coherent: I.2.Ex.11
— pseudo-zero: VII.4.4
Monoid, divisor class: VII.1.2
Morphism for laws of composition not everywhere defined: VI.2.1
Multiplicative subset: II.2.1

Nilradical of a ring: II.2.6
Noetherian space: II.4.2
Non-degenerate submodule: II.5.5
Normalization lemma: V.3.1
Normed discrete valuation: VI.3.6
Nullstellensatz: V.3.3

Order group of a valuation: VI.3.2
— of an element for a valuation: VI.3.2
— reduced, of a formal power series: VII.3.8
Ordered pair of rings with the linear extension property: I.3.7
Ostrowski's theorem: VI.6.4

Place, finite at x: VI.2.2
— of a field: VI.2.2
— trivial: VI.2.2
Point, generic, of an irreducible space: II.4.Ex.2
Polygon, Newton: VI.4.Ex.11
Polynomial, distinguished: VII.3.8
— minimal: V.1.3
Preparation theorem: VII.3.8
Presentation of a module, — finite: I.2.8
n-presentation: I.2.Ex.6
Presented, finitely (module): I.2.8
Primary decomposition: IV.2.2 and Ex.20
— $\mathfrak{p}$-primary, ideal, submodule: IV.2.1 and Ex.20
Prime ideal: II.1.1
— spectrum: II.4.3
Principal divisor: VII.1.1
— restricted adèle: VII.2.4
Principle of Noetherian induction: II.4.2
Product filtration: III.2.1
Projective field: VI.2.1

Pseudo-injective, pseudo-surjective, pseudo-zero, pseudo-bijective (homomorphism): VII.4.4
Pseudo-isomorphism: VII.4.4
Pseudo-zero module: VII.4.4

Quasi-Galois extension: V.2.2
Quotient filtration: III.2.1

Radical of an ideal: II.2.6
Rank at 𝔭 of a projective module: II.5.3
— of a projective module: II.5.3
— rational, of a commutative group: VI.10.2
— residue: VI.8.5
Rational rank of a commutative group: VI.10.2
Reduced order: VII.3.8
— primary decomposition: IV.2.3
— ring: II.2.6
— series: VII.3.8
Reflexive lattice: VII.4.2
Related local rings: VI.1.Ex.1
Relatively prime ideals: II.1.2
Representative system of extremal elements: VII.3.3
Residue class degree of a valuation: VI.8.1
— field: II.3.1
— rank of a valuation: VI.8.5
Resolution, finite free, of a module: VII.4.7
Restricted adèle: VII.2.4
— formal power series: III.4.2
Ring, absolutely flat: I.2.Ex.17
— coherent (left, right): I.2.Ex.17
— decomposition: V.2.2
— filtered: III.2.1
— filtered, associated with a graded ring: III.2.1
— graded, associated with a filtered ring: III.2.3
— inertia: V.2.2
— integrally closed in an algebra: V.1.2
— Jacobson: V.3.4
— linearly topologized: III.4.2
— local: II.3.1
— local, dominating a local ring: VI.1.1
— local, of A at 𝔭, of 𝔭 (𝔭 a prime ideal): II.3.1
— of a place: VI.2.3
— of a valuation: VI.3.2

Ring of fractions defined by a subset of a ring: II.2.1
— reduced: II.2.6
— semi-local: II.3.5
— total, of fractions: II.2.1
— unramified: V.2.Ex.19
— valuation, valuation ring of a field: VI.1.1
— Zariski: III.3.3
Rings, independent valuation: VI.7.2

Saturation of a submodule with respect to a multiplicative subset (with respect to a prime ideal): II.2.4
Semi-local ring: II.3.5
Series, reduced: VII.3.8
— restricted formal power: III.4.2
Set, irreducible: II.4.1
— major, in a totally ordered group: VI.3.5
Space, irreducible: II.4.1
— Noetherian: II.4.2
Special topology: II.4.3
Spectrum, prime, of a ring: II.4.3
Strongly Laskerian module: IV.2.Ex.28
— primary submodule: IV.2.Ex.27
— relatively prime elements: III.4.1
Subgroup, isolated, of an ordered group: VI.4.2
Submodule, invertible: II.5.6
— non-degenerate: II.5.5
— primary, $\mathfrak{p}$-primary: IV.2.1
Subset, multiplicative, of a ring: II.2.1
— multiplicative, generated by a subset: II.2.1
— saturated multiplicative: II.2.Ex.1
Support of a module: II.4.4
System, complete, of extensions of a valuation: VI.8.2
— formal, of generators: III.2.9
— representative, of extremal elements: VII.3.3

Theorem, approximation, for absolute values: VI.7.3
— approximation, for valuations: VI.7.2
— Gelfand-Mazur: VI.6.4
— Hensel's: III.4.3
— Hilbert's zeros (Nullstellensatz): V.3.3
— Krull's: III.3.1
— Krull-Akizuki: VII.3.5
— Ostrowski's: VI.6.3

Theorem, preparation: VII.3.8
— Stickelberger's: VI.8.Ex.18
— Zariski's Principal: V.3.Ex.7
Topologically nilpotent element: III.4.3
Topology defined by a filtration: III.2.5
— spectral: II.4.3
— Zariski: II.4.3
Transporter: I.2.10
Trivial filtration: III.2.1
— place: VI.2.2

Ultrametric absolute value: VI.6.1
Uniformizer for a discrete valuation: VI.3.6
Unramified valuation: VI.8.1

Valuation, valuation of an element *x*: VI.3.1 and 2
— discrete, normed discrete valuation: VI.3.6
— essential: VII.1.4 and Ex.26
— improper: VI.3.1
— ring: VI.1.1
— unramified: VI.8.1
Valuations, equivalent: VI.3.2
— independent: VI.7.2
Value field of a place: VI.2.2
— ultrametric absolute: VI.6.1

Weakly associated (prime ideal) with a module: IV.1.Ex.17

Zariski, ring: III.3.3
— topology: II.4.3

TABLE OF IMPLICATIONS

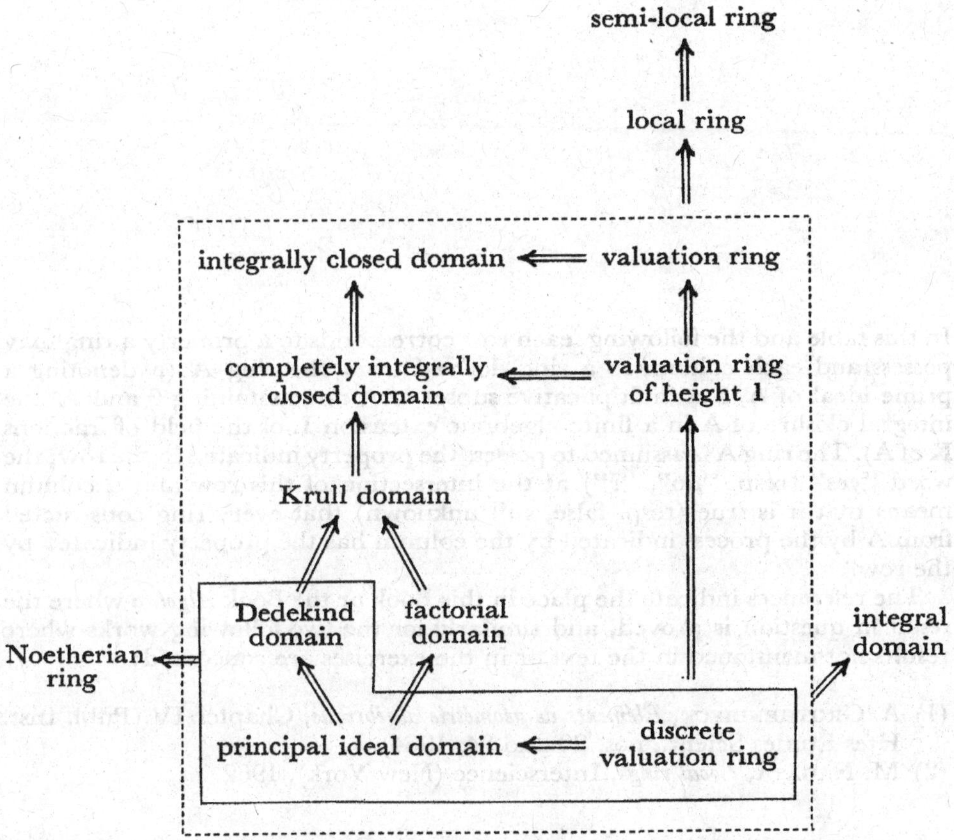

In the case of *Noetherian* rings this table reduces to the following:

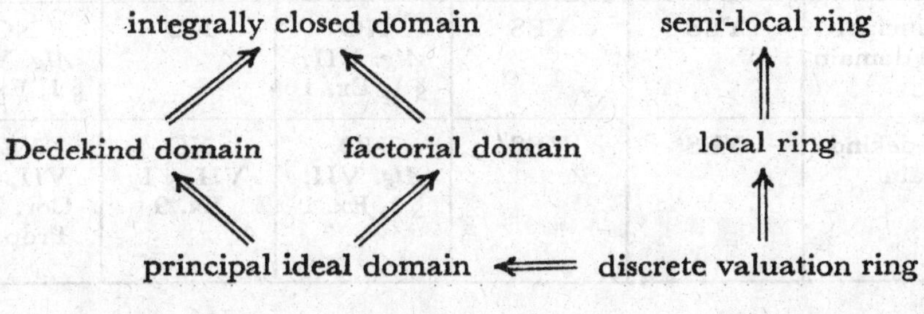

TABLE OF INVARIANCES — I

In this table and the following, each row corresponds to a property a ring may possess and each column to a ring derived from the ring A ($\mathfrak{p}$ denoting a prime ideal of A, S a multiplicative subset of A not containing 0 and A′ the integral closure of A in a finite algebraic extension L of the field of fractions K of A). The ring A is assumed to possess the property indicated in the row; the word "yes" (resp. "no", "?") at the intersection of this row and a column means that it is true (resp. false, still unknown) that every ring constructed from A by the process indicated by the column has the property indicated by the row.

The references indicate the place in this Book or the Book *Algebra* where the result in question is proved, and similarly for the two following works where results not mentioned in the text or in the exercises are concerned:

(1) A. Grothendieck, *Éléments de géométrie algébrique*, Chapter IV (Publ. Inst. Htes Études Scient., nos. 20 and 24, 1964).
(2) M. Nagata, *Local rings*, Interscience (New York), 1962.

	$A/\mathfrak{p}$	$S^{-1}A$	A[X]	A[[X]]	A′
A a principal ideal domain	YES	YES	NO *Alg.* VII, § 1, Ex. 1	NO	NO *Alg.* VII, § 1, Ex. 12
A a Dedekind domain	YES	YES	NO *Alg.* VII, § 1, Ex. 1	NO VII, § 1, Ex. 9	YES VII, § 2, Cor. 2 to Prop. 9)

	A/$\mathfrak{p}$	$S^{-1}A$	A[X]	A[[X]]	A′
A a factorial domain	NO V, § 1, Ex. 9	YES VII, § 3, Prop. 3	YES VII, § 3, Th. 2	NO VII, § 3, Ex. 9	NO *Alg.* VII, § 1, Ex. 12
A a Noetherian integrally closed domain	NO V, § 1, Ex. 9	YES V, § 1, Cor. 1 to Prop. 16	YES V, § 1, Cor. 1 to Prop. 13	YES V, § 1, Prop. 14	? (YES if L is separable; V, § 1, Cor. 1 to Prop. 18)
A a field or a discrete valuation ring	YES VI, § 3, no. 6	YES VI, § 3, no. 6	NO *Alg.* VII, § 1, Ex. 1	NO VII, § 1, Ex. 9	NO V, § 2, Ex. 6
A a field or a valuation ring of height 1	YES VI, § 4	YES VI, § 4, Prop. 1	NO *Alg.* VII, § 1, Ex. 1	NO VII, § 1, Ex. 9	NO V, § 2, Ex. 6
A a valuation ring	YES VI, § 1, Th. 1	YES VI, § 1, Th. 1	NO *Alg.* VII, § 1, Ex. 1	NO VII, § 1, Ex. 9	NO V, § 2, Ex. 6
A a complete valuation ring	YES VI, § 5, Prop. 1	YES VI, § 7, Prop. 3	NO *Alg.* VII, § 1, Ex. 1	NO VII, § 1, Ex. 9	NO V, § 2, Ex. 6
A a Krull domain	NO V, § 1, Ex. 9	YES VII, § 1, Prop. 6	YES VII, § 1, Prop. 13	YES VII, § 1, Ex. 9	YES VII, § 1, Prop. 12

TABLE OF INVARIANCES — II

In this table $\mathfrak{a}$ denotes an ideal of A distinct from A, S a multiplicative subset of A and A′ the integral closure of A which is assumed to be an integral domain.

	$A/\mathfrak{a}$	$S^{-1}A$	A[X]	A[[X]]	A′
A local	YES	NO II, § 2, Prop. 11	NO	YES *Alg.* IV, § 5, Prop. 4	NO V, § 2, Ex. 20
A local and complete	? YES if A is Noetherian (III, § 3, Prop. 6)	NO II, § 2, Prop. 11	NO	YES III, § 2, Prop. 6	? YES if A is Noetherian (1)
A semi-local	YES	NO IV, § 2, Ex. 23(c)	NO	YES *Alg.* IV, § 5, Prop. 4	? YES if A is Noetherian V, § 2, Ex. 7
A semi-local and complete	? YES if A is Noetherian (III, § 3, Prop. 6)	NO IV, § 2, Ex. 23(c)	NO	YES III, § 2, Prop. 6	
A Noetherian	YES *Alg.* VIII, § 2, Prop. 6	YES II, § 2, Cor. 2 to Prop. 10	YES III, § 2, Cor. 1 to Th. 2	YES III, § 2, Cor. 6 to Th. 2	NO V, § 1, Ex. 21 YES if A is locally complete (1) A′ is always a Krull domain (2)

INVARIANCES UNDER COMPLETION

(a) Let A be a ring and $\mathfrak{m}$ an ideal of A distinct from A. Let A be given the $\mathfrak{m}$-adic topology and let $\hat{A}$ denote its Hausdorff completion.

	$\hat{A}$
A Hausdorff	YES
A Noetherian	YES (III, § 3, Proposition 8)
A local	YES (III, § 2, Proposition 19)
A semi-local	YES (III, § 2, Corollary to Proposition 19)
A a Zariski ring	YES (III, § 3, Proposition 8)

(b) Suppose now that A is local and Noetherian and that $\mathfrak{m}$ is its maximal ideal.

	$\hat{A}$
A an integral domain	NO (III, § 3, Exercise 15 (b))
A an integrally closed domain	NO (2) YES for excellent rings (1)
A a discrete valuation ring	YES (VI, § 5, Proposition 5)
A reduced	NO (2) YES for excellent rings (1)